ENCYCLOPEDIA OF PHYSICS

EDITED BY

S. FLÜGGE

VOLUME XLII

NUCLEAR REACTIONS III

WITH 276 FIGURES

SPRINGER-VERLAG
BERLIN HEIDELBERG GMBH
1957

HANDBUCH DER PHYSIK

HERAUSGEGEBEN VON

S. FLÜGGE

BAND XLII

KERNREAKTIONEN III

MIT 276 FIGUREN

SPRINGER-VERLAG
BERLIN HEIDELBERG GMBH
1957

ISBN 978-3-642-45880-4 ISBN 978-3-642-45878-1 (eBook)
DOI 10.1007/978-3-642-45878-1

Ursprünglich erschienen bei Springer-Verlag OHG. Berlin · Göttingen · Heidelberg 1957
Softcover reprint of the hardcover 1st edition 1957

Contents.

Nuclear Isomerism.

By

D. E. ALBURGER.

With 101 Figures.

A. Introduction.

Nuclear isomers may be defined as atoms whose nuclei exist in excited states for measurable lengths of time. All of the known isomers as well as other low-lying energy levels ordinarily found in radioactive decay processes are bound states in the sense that they are stable against the emission of protons or neutrons. While it is to be expected that any bound excited state has a finite and definable lifetime it is clear that the measurement criterion for labeling a state as isomeric depends on whether or not we have the experimental techniques at our disposal for determining its lifetime.

In the present article we shall use the term "isomer" to denote a state other than a ground state whose half-life *has been measured* directly or indirectly regardless of the characteristics of the radiation emitted. We shall be concerned mainly with nuclear energy levels which occur in radioactive decays and on which there is experimental information about the lifetime. Additional data will be presented on the lifetimes of levels measured indirectly by means of several recently developed methods including COULOMB excitation.

Previous reviews of nuclear isomerism have been given by SEGRÈ and HELMHOLZ [*Se 49*][1], GOLDHABER and HILL [*Go 52*], GOLDHABER and SUNYAR [*Si 55*] (Chap. 16), and KORSUNSKII [*Ko 54a*].

B. History of nuclear isomerism.

1. Discovery and early experiments. Nuclear isomerism had its experimental beginnings in 1921 when HAHN [*Ha 21*] discovered that ${}_{90}Th^{234}$ (known at that time as UX_1) decayed by beta-ray emission not only to the well-known daughter ${}_{91}Pa^{234}$, (then known as UX_2, half-life 1.14 min) but that a beta-ray activity of 6.7 hour half-life also followed the Th^{234} decay. It was concluded that the second substance, called UZ, must have had the same atomic number and the same atomic weight as UX_2 or in present-day terminology the two were both isotopic and isobaric (see Fig. 93). No explanation for this curious behavior was proposed. The possibility that nuclear isomers might exist had already been suggested in 1917 by SODDY [*So 17*] who imagined that nuclei of given mass and charge might be able to occur in different forms, these states being distinguishable from each other by radioactive or other properties.

It was not until a decade had passed and artificial radioactivity had been discovered in 1934 by JOLIOT and CURIE that another case of isomerism was uncovered. SZILARD and CHALMERS [*Sz 35*] found suggestions of isomerism in indium but the first well-established example was that of bromine as reported by

[1] References are listed at the end of this article, p. 94 seq.

AMALDI *et al.* [*Am 35*] and by KURCHATOV *et al.* [*Ku 35*]. There are two stable bromine isotopes, Br^{79} and Br^{81}, yet when bromine was irradiated with slow neutrons three activities were produced having half-lives of 18 min, 4.4 and 34 hr. Assuming that a third as yet undetected stable isotope was not present and that simple neutron capture was the only nuclear reaction taking place it was then necessary to assign at least two of the half-lives to either Br^{80} or to Br^{82}. Later it was shown by BOTHE and GENTNER [*Bo 37*] and by SNELL [*Sn 37*] that the 18 min and 4.4 hr activities are associated with Br^{80} (see Fig. 23) and that the 34 hr activity can be assigned to Br^{82} (see Fig. 25).

2. VON WEIZSÄCKER'S hypothesis and subsequent developments. The first proposed explanation of isomerism was put forward in 1936 by VON WEIZSÄCKER [*Vo 36*]. He suggested that isomers were atoms whose nuclei were in excited states which differed by several units of nuclear angular momentum, or spin, from any lower energy level including the ground state. If such an isomeric state happened to be reached through heavy particle excitation, beta decay or gamma-ray transitions from higher energy states, the nucleus would have to make the relatively difficult transition involving large spin change to a lower energy level or else decay by some other path such as alpha-particle or beta-ray emission if energetically possible. VON WEIZSÄCKER showed that while gamma-ray transitions requiring only one or two units of spin change could be expected to take place with half-lives of the order of 10^{-13} sec, those of larger spin change might result in longer and measurable half-lives. Each additional unit of spin change could, according to the theory, increase the half-life of a transition of given energy by a factor of 10^6.

VON WEIZSÄCKER'S hypothesis seemed to be very reasonable and in fact it can still be taken as a simplified explanation of long-lived isomeric transitions. It furnished considerable incentives to experimenters to discover more cases of isomerism and to establish the characteristics of the states. While research was proceeding along these lines it became apparent that the half-life and energy values of an isomeric transition might not be sufficient information to establish the multipole order of the radiation. Calculation of the lifetime[1] of a particular transition requires assumptions about the structure of the nucleus and the radiating mechanism. Early attempts were made to calculate lifetimes using such models as a radiating alpha-particle or proton moving in the average field of the other nucleons, the liquid drop model where the radiation arises from vibrations of the drop, or a radiating mechanism associated with a rotating charged drop. These models were mostly speculative because of insufficient experimental evidence.

Meanwhile the theoretical physicists became interested in trying to find other ways of identifying the multipole orders of isomeric transitions which did not require specific assumptions about nuclear structure, as in the lifetime calculations. In particular the internal conversion process, which will be discussed later, was found to be sensitive to the nature of the radiation field and might give the additional information desired. With the development of beta-ray spectrometers the measurement of conversion coefficients in the various atomic shells became feasible and the theorists concentrated on calculating conversion coefficients and conversion ratios as guides to the experimentalists. Notable in this respect were the K-shell conversion coefficients calculated by HEBB and UHLENBECK [*He 38*] and by DANCOFF and MORRISON [*Da 39*] and the K/L ratio curves

[1] The lifetime or mean life τ of a radioactive substance is longer than the half-life $T_{\frac{1}{2}}$ by a factor of 1/0.693.

computed by HEBB and NELSON [*He 40*]. In the latter work the electric multipole conversion coefficients in the K and L shells were calculated for $Z=35$. To make the results applicable to a range of atomic numbers Z and of transition energies E, the ratios of K to L conversion for each multipole were plotted against Z^2/E. Using the curves for a transition in a nucleus of Z other than 35 involved the assumption that the amplitudes and shapes of the curves remained essentially the same for neighboring Z's. Departures of 10 to 20% for the range $25 < Z < 50$ were expected.

Prior to World War II little progress was made in grouping and classifying isomers on the basis of their conversion coefficients and conversion ratios not only because relatively few of these quantities had been measured with sufficient accuracy but also because it was not certain whether or not the theoretical calculations were reliable in view of the approximations used. We now know that these early theoretical results were in error by enough to lead to erroneous multipole assignments, had accurately measured conversion data been available.

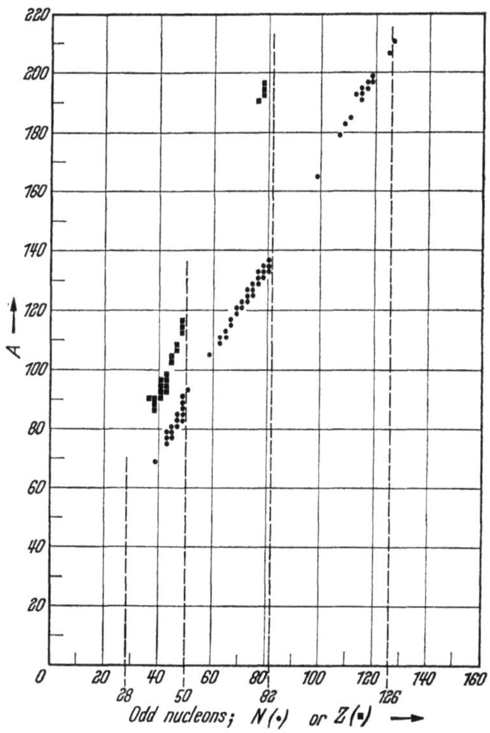

Fig. 1. Distribution of long-lived isomers of odd mass number A, plotted against the number of odd nucleons (N or Z). Based on a figure first given by GOLDHABER and HILL [*Go 52*] (Courtesy Dr. M. GOLDHABER).

3. The nuclear shell model[1]. With the development of the nuclear reactor during World War II and the subsequent construction of many reactor laboratories throughout the world after the war a tremendous new surge of interest began in problems of nuclear structure. The years 1945 to 1950 saw the introduction of new instruments and techniques and the accumulation of large amounts of experimental information on nuclear decay schemes and isomeric transitions. Not too much progress was made in their interpretation during this period although certain attempts at classifying isomers were undertaken, such as that of AXEL and DANCOFF [*Ax 49*].

Around 1950 a significant advance in nuclear physics was made when the shell model was introduced independently by MAYER [*Ma 51*] and by HAXEL, JENSEN and SUESS [*Ha 50*]. (For recent reviews see [*Si 55*], Chap. 15, [*Ma 55f*] and [*Pr 54c*].) The shell model came about partly through the data of beta decay and isomeric transitions. It was noticed that when total beta disintegration energy was plotted against neutron or proton number the curves exhibited smooth portions separated from each other by discontinuities at certain unique numbers of protons or neutrons. Isomers were found in regions of the periodic table which appeared to be bounded at these same proton or neutron numbers. This is illustrated in Fig. 1 which shows a plot of the distribution of long-lived isomers of odd mass number. The shell model was devised to account for such

[1] A detailed report on the shell model is given by LANE and ELLIOTT in Vol. XXXIX of this Encyclopedia.

departures from systematic behavior. It treated nuclei as though each nucleon were moving in the average potential produced by all of the other nucleons. Strong spin-orbit coupling was assumed, the strength of the interaction being adjusted so as to lead to a certain sequence of available states. These levels correspond roughly to the quantum states of a harmonic oscillator. Nuclei can be imagined as being formed by first adding one particle at a time to the lowest state until the PAULI exclusion principle forbids the addition of any more; the next level is then filled, and so on, until the total number of particles corresponding to the particular nucleus under consideration is reached. Neutrons and protons fill their own sets of states independently. The complete filling of a given level is referred to as the closure of a sub-shell. However, the so-called "magic" numbers, while they occur at closures of sub-shells, do not for the most part happen at points where all possible particles belonging to given major oscillator quantum numbers have been added to the nucleus. The magic numbers are somewhat artificial in that they refer to certain sub-shell closures where the step in energy to the position of the next sub-shell level is considerably larger than average and where simultaneously there are unusually large differences between the spins of adjacent levels just before the closure of the shell at the magic number. Although the spin-orbit coupling is adjusted so as to fit the experimental data it is significant that a single coupling strength is sufficient to account for the occurrence of all of the shells observed, i.e. at 2, 8, 20, 50, 82 and 126 total protons or neutrons. Table 1 gives the order of level spins and parities predicted by the shell model.

Table 1. *Spins and parities in shells, and oscillator levels according to the shell model*[1].

Shell	I	II	III	IV		V	VI	
Occupation numbers	1−2	3−8	9−20	21−28	29−50	51−82	83−126	127→
j	$\frac{1}{2}$	$\frac{3}{2}$ $\frac{1}{2}$	$\frac{5}{2}$ $\frac{3}{2}$ $\frac{1}{2}$	$\frac{7}{2}$	$\frac{5}{2}$ $\frac{3}{2}$ $\frac{1}{2}$ $\frac{9}{2}$	$\frac{7}{2}$ $\frac{5}{2}$ $\frac{3}{2}$ $\frac{1}{2}$ $\frac{11}{2}$	$\frac{9}{2}$ $\frac{7}{2}$ $\frac{5}{2}$ $\frac{3}{2}$ $\frac{1}{2}$ $\frac{13}{2}$	$\frac{11}{2}$ $\frac{9}{2}$ etc.
parity	+	− −	+ + +	−	− − − +	+ + + + −	− − − − − +	+ + ...
Oscillator level $\hbar\omega$	0	1	2	3		4	5	6

To account for isomerism we may imagine that we have constructed in the manner prescribed above a nucleus such as ${}_{38}Sr^{87}_{49}$. The 38 protons and 48 of the neutrons form an even-even core having a net spin of zero; according to the shell model, as may be seen from Table 1, the 49th neutron has available to it, as a level of lowest energy, either a $p_{\frac{1}{2}}$ or $g_{\frac{9}{2}}$ state. (The spectroscopic notation s, p, d, f, g, h, i etc. indicates orbital momenta of 0, 1, 2, 3, 4, 5, 6 etc. respectively and the subscript gives the total angular momentum resulting from the vector addition of the orbital angular momentum with the spin $+\frac{1}{2}$ or $-\frac{1}{2}$ of the odd particle.) Experimentally it is found that for Sr^{87}, as well as for a number of neighboring nuclei, the $g_{\frac{9}{2}}$ level is filled by the odd neutron in the ground state. If now we excite Sr^{87} presumably only the loosely bound odd neutron is involved in the excitation; the shell model predicts that the state of next highest energy is a $p_{\frac{1}{2}}$ level. Assuming that we can somehow place the odd neutron in this state, as is accomplished in the decay of Y^{87} (see Fig. 30), the only path to the ground state is a transition in which there are four units of spin change and a

[1] Taken from J. H. D. JENSEN's article in [*Si 55*], Chap. 15.

parity change. As we shall see later this results in a magnetic 2^4-pole radiation, or $M4$, in which the 0.39 Mev transition has a half-life of 2.8 hours.

Parallel with the development of the shell model more theoretical work was being done on internal conversion and on the radiating mechanism of nuclei. Relative L_{III}/L_I-shell magnetic conversion coefficients had been calculated by TRALLI and LOWEN [*Tr 49*]. However, probably the most important advance came with the relativistic calculations by ROSE and co-workers [*Ro 51*] of conversion in the K-electron shell[1] for all magnetic and electric radiations up to and including a multipole order of 5, covering a wide range of transition energies and atomic numbers up to $Z=96$. At the same time the problem of estimating the radiation probabilities of nuclear transitions was under study by WEISSKOPF [*Bl 52*]. Using an independent-particle model, WEISSKOPF formulated relations of lifetime as a function of energy for various electric and magnetic multipole orders. These calculations will be discussed in more detail later.

4. The GOLDHABER and SUNYAR empirical classification. While beta decay systematics, neutron binding energies and the ground state spins and magnetic moments of nuclei all gave convincing evidence in support of the shell model there were a number of inconsistencies between the data of isomeric transitions and the level assignments predicted by the model. In 1951 the whole general problem was attacked with a fresh approach by GOLDHABER and SUNYAR [*Go 51*]. They summarized the difficulties as follows:

(a) Ten cases of isomers with apparently no parity change had been found in the $g_{\frac{9}{2}}$ shell and three similar cases were found in the $h_{\frac{11}{2}}$ shell. Parity change was predicted by the shell model as may be seen from Table 1.

(b) No $E5$ (electric 2^5-pole) crossover transitions had been found in the $h_{\frac{11}{2}}$ shell.

(c) The interpretation of experimental K/L internal conversion ratios using the early theoretical calculations had indicated that $E5$ and $M4$ transitions of similar energy had half-lives of the same order.

(d) No isomers corresponding to a spin change of 3 had been observed.

(e) No isomers had been observed with half-lives in the millisecond region.

GOLDHABER and SUNYAR took the view that the older calculations, particularly those of the K/L ratios, were probably not reliable. They therefore decided to use a semi-empirical approach, basing firm assignments of spins and parities only on the new and well-founded K-shell conversion coefficients. Their conclusions, arrived at from a study of a large amount of existing data, were as follows:

(a) Most of the experimental results could be brought into agreement with the shell model predictions provided that one of the shell model rules not be adherred to strictly, namely that j-j coupling of an odd number of particles in the $g_{9/2}$ shell should lead to a $g_{\frac{9}{2}}$ state as the configuration of lowest energy. This exception seemed reasonable since the rule could only be expected to hold theoretically for nuclear forces of zero range.

(b) The early theoretical K/L ratio curves were not reliable. To replace them, sets of empirical K/L ratio curves were constructed using experimental results and the more certain multipole order assignments mentioned above.

[1] Recent experimental evidence [A. WAPSTRA and G. NIJGH, Nuclear Phys. **1**, 245 (1956)] indicates that the calculations of ROSE *et al.* which neglect the finite nuclear size, give $M1$ K-conversion coefficients which are too high by $\sim 35\%$ for $Z \sim 80$. SLIV *et al.* [L. A. SLIV, Ž. eksp. teor. Fiz. **21**, 77 (1951); L. SLIV and M. LISTENGARTEN, Ž. eksp. teor. Fiz. **22**, 29 (1952)] have predicted nuclear size effects of this magnitude for $M1$ radiation and smaller corrections for other multipoles; i.e. α_{E2} in the same region is predicted to be lower by $\sim 3\%$.

(c) The lifetimes of $M4$ transitions were found to follow with remarkable uniformity an empirical law containing the energy and mass number (see Sect. 43).

(d) For $E3$, $E4$, and $E5$ transitions the multipole order was shown to have been overestimated previously by one unit.

(e) Electric transitions, except for some of the $E2$ transitions, were found to have lifetimes longer than expected on the basis of the WEISSKOPF formula and were comparable in magnitude to the half-lives of magnetic transitions of the same spin change.

(f) The only mixed transitions that occur are $M1+E2$.

(g) For even-even nuclei the first excited state has a spin of 2 and even parity in most cases.

5. Further experimental refinements and theoretical conversion calculations. The success of GOLDHABER and SUNYAR's classification of isomers was demonstrated by the large number of papers published during the years immediately following, the results of which added new and confirming information. Meanwhile further refinements, both experimental and theoretical, were being worked out to assist in making multipole order assignments. Using a high resolution magnetic beta-ray spectrometer GRAVES, LANGER and MOFFAT [*Gr 52*] made considerably more accurate determinations than in previous work of the $K/(L+M)$ ratios for a number of transitions known to be $M4$. Their curve of $K/(L+M)$ versus Z^2/E differed only slightly from the $M4$ curve of GOLDHABER and SUNYAR. On the other hand it had been noticed by SWAN and HILL [*Sw 53c*] and by ALBURGER [*Al 53*] that the K/L ratios of transitions in some of the heavy elements showed substantial departures ($\sim 30\%$) from the empirical curves which had been based largely on data taken from the middle of the periodic table. Their work suggested that the plotting of K/L ratios versus the quantity Z^2/E was not a satisfactory procedure for a very wide range of Z.

Since the accurate theoretical K-shell conversion coefficients were already available it became of first importance for similar calculations to be carried out on the L-shell which would then allow the K/L ratios to be computed. GELLMAN, GRIFFITH and STANLEY [*Ge 52*] had made such computations for three Z values 49, 84 and 92, three energy values and three multipoles, $E1$, $E2$ and $M1$. The need for extending the calculations to higher multipole order and wider ranges of Z and of E was apparent. A program of L-shell conversion calculations was therefore carried out by ROSE and co-workers ([*Si 55*], Chap. 14 and Appendix 4). All three sub-shells L_{I}, L_{II} and L_{III} were calculated for all electric and magnetic multipole orders up to and including 5. Approximately the same ranges of energy and of Z were covered as in the case of the K-shell conversion coefficients. Although the complete table for the L-shell has become available only recently, preliminary values were distributed privately during the course of the work. They were of considerable assistance to research in progress at that time. It was found that not only were the early theoretical K/L ratio curves of HEBB and NELSON considerably in error, but that there was no universal set of curves of K/L versus Z^2/E for all Z.

6. The collective model of BOHR and MOTTELSON[1]. The observation by GOLDHABER and SUNYAR that certain $E2$ transitions proceed with a greater probability than predicted by the WEISSKOPF formula represented a deficiency in the single-particle representation. In 1953 the related process of COULOMB excitation (see Sect. 37) was first observed experimentally. To explain these effects as well as

[1] For details see the article of S. MOSZKOWSKI in vol. XXXIX of this Encyclopedia.

several other nuclear phenomena BOHR and MOTTELSON introduced the collective model of the nucleus [*Bo 53d*]. (See also [*Si 55*], Chapt. 17.)

They pictured the nucleus, particularly in regions between closed shells, as an elliptically-shaped body which is incompressible but whose surface can be deformed by wave motions involving the collective action of many nucleons. Such surface wave patterns are quantized energy states which can be excited by the electric field of passing charged particles and between which transitions can occur giving rise to the emission of gamma radiation. The energies E_J of these so-called "rotational" states are given by

$$E_J = \frac{1}{2} I \omega^2 = \frac{\hbar^2}{2I} [J(J+1) - J_0(J_0+1)] \tag{6.1}$$

where J_0 is the spin of the ground state, J the spin of the excited state and I is the moment of inertia of the deformed nucleus. For odd-A nuclei the spin sequence is J_0, J_0+1, J_0+2 etc., while for even-even nuclei the spins of the rotational set follow the sequence $0^+, 2^+, 4^+, 6^+ \ldots$.

Between closed shells the nucleus is highly deformed giving rise to large quadrupole moments and therefore large moments of inertia. The first rotational excited state is then of the order of 100 keV above the ground state whereas near the closed shells this level may be at $\sim$1 Mev. The model has had surprising success in predicting the ratios of the energies of the excited states. As an example the nucleus Ta^{181} has a ground-state spin of $\frac{7}{2}$ and exhibits gamma rays of 137 and 166 kev under COULOMB excitation corresponding to states at 137 and 303 kev. The energy ratio of these states is 2.21 ± 0.02 in close agreement with the ratio $\frac{20}{9}$ found from Eq. (6.1) by giving J_0 the value $\frac{7}{2}$ and J the values $\frac{9}{2}$ and $\frac{11}{2}$. Another success of the model is that it can explain the speed-up of $E2$ transitions as being a consequence of collective effects.

In addition to the usual angular momentum quantum number, rotational states are characterized by a quantum number K which represents the projection of the total angular momentum on the axis of symmetry of the deformed nucleus. All of the members of a given rotational set have the same K-value, i.e. for even-even nuclei the rotational set to which the ground state belongs has $K=0$. Transitions among the members of a given set do not involve a change in K. Other rotational sets can exist for which K has a different value and when K is required to change, as in transitions from a member of one set to a member of another set, an additional inhibition in the transition rate can be expected. These K-forbidden transitions are discussed in Sect. 13.

C. Radiations from isomeric nuclei.

7. Multipole order. A nuclear energy level may be characterized by its energy E above the ground state, by the total angular momentum or spin of the nucleus I in units of $\hbar$, and by its parity π. Of the three properties, parity is the least obvious physically. It refers to the quantum mechanical wave function ψ describing the nuclear state. If a coordinate system is established with respect to a particular nucleus the nuclear state can be described in principle by a function $\psi(x, y, z)$. Actually we do not yet know the correct ψ function for any nuclear state, but we shall assume that such a description is possible. We then take the particular arrangement of nucleons at a given instant and imagine that each nucleon is transported in such a way that its spin direction is preserved and that its coordinates have the same values but have changed sign. When all of the nucleons have been transposed, their relative positions and internuclear distances remain unchanged and we have constructed a nucleus similar in all respect to the original one except that it has become a mirror image. Evidently the energy and spin of the nucleus will be the same so that the state can still be described, in magnitude at least, by the function ψ. However, in the process described, ψ may have undergone a change of sign. If $\psi(x, y, z) = \psi(-x, -y, -z)$ the parity of the state is said to be "even" or plus. On the other hand, if $\psi(x, y, z) = -\psi(-x, -y, -z)$ the parity is "odd" or minus. This explanation should be taken only as a mathematical concept and it does not imply that nuclei can spontaneously undergo transitions to their mirror images.

In an isomeric transition the nucleus changes from its initial state E_i, I_i and π_i to a final state E_f, I_f and π_f of lower energy having a spin and parity either of which may or may not have changed. The angular momentum taken away by the photon or a conversion electron is not necessarily equal to the numerical spin change of the nucleus $I_i - I_f$ because of the possibility that the final spin vector may have a different direction in space than the initial spin vector. For example, suppose that the initial spin is 2 and directed upwards and that final spin is 1. If the latter is also directed upwards then $\Delta I = 1$, but if I_f were directed at a different angle the spin taken away could be as great as $\Delta I = 3$ even though the numerical nuclear spin change is only 1. Quantum mechanics of course demands that ΔI be an integer in units of $\hbar$ and conservation of angular momentum requires that

$$|I_i - I_f| \leq \Delta I \leq |I_i + I_f|. \tag{7.1}$$

Consider another example where the initial and final spins are both $\frac{1}{2}$. In this case the directions of I_i and I_f must be opposite if a photon is emitted because of the requirement that photons, being of a transverse nature, must carry away at least one unit of angular momentum.

The spin and parity carried away by a photon completely determine the multipole character of the radiation. Fields corresponding to angular momenta of 1, 2, 3, 4 etc. are referred to as being dipole, quadrupole, octapole, 2^4-pole, etc. respectively and whether or not they are electric or magnetic radiations is governed by the parity change if any. Table 2 summarizes the multipole characters of electromagnetic transitions. Parity change or no parity change are indicated by "yes" and "no" respectively. By remembering the parity change conditions for the two dipoles and that they alternate regularly with ΔI, the parity change for any multipole can be recalled quickly.

Table 2. *Spin and parity changes for various multipoles.*

ΔI	1	2	3	4	5
Electric	$E1$ yes	$E2$ no	$E3$ yes	$E4$ no	$E5$ yes
Magnetic	$M1$ no	$M2$ yes	$M3$ no	$M4$ yes	$M5$ no

Suppose that we know the initial and final spins and parities and we ask whether it is possible to predict what multipole radiations will result. As an example choose $I_i - I_f = 2$, no parity change. The number of possible multipoles which could occur depends on the magnitudes of I_i and I_f. Thus if $I_i = 2$ and $I_f = 0$ the radiation must necessarily be pure $E2$; however, if $I_i = 3$ and $I_f = 1$ then we could have $E2$, $M3$ or $E4$ radiations according to Eq. (7.1) and Table 2. In the most general case we cannot unambiguously state what type of radiation will appear without a knowledge of limiting factors or other selection rules. Qualitatively we know from the considerations of VON WEIZSÄCKER that the higher the multipole order the longer the associated half-life. We should thus expect that in the above example where $E2$, $M3$ and $E4$ all conform with the selection rules the resulting radiation would be mostly that of lowest multipole order, i.e. $E2$. Experimentally this is found to happen except in some cases of $M1$ or $E1$ transitions where the emission probability for $E2$ or $M2$ radiation respectively may be comparable.

$E1+M2$ or $M1+E2$ are known as mixtures. (A few examples have been found of weak mixtures among higher multipoles.) Classically the transition of a particular nucleus would give rise to only a single multipole and the term "mixture" would apply to an assemblage of nuclei some of which, for example, may emit $M1$ radiation and others $E2$ radiation, depending on whether or not the nuclear spin vector has undergone a change of direction. The ratio of the numbers of the two types of events, i.e. $M2/E1$ or $E2/M1$ is known as the mixing ratio α^2, where α is the ratio of the radiation amplitudes. This number sometimes may be obtained experimentally from internal conversion or angular correlation measurements. According to quantum mechanics, however, we cannot distinguish between those nuclei of a mixture whose spin orientations are presumed to change and those which do not. The classical picture is no longer tenable, and we must take into account the possible interference between the two multipole fields. This is particularly important in angular correlation measurements where the interference term may have a very large effect. Fortunately, it has been proved from theorems of detailed balance and time reversal that the two mixing fields must be either in phase or 180° out of phase. If it were not for this circumstance the calculations of the angular correlations for mixed radiations would be difficult and comparison with experiment would often be inconclusive.

A special class of multipole order is the $E0$ or electric monopole. Theoretically an $E0$ component can be present in transitions between any two states of the same spin and parity but it is only the $0^+ \to 0^+$ transitions which have been observed thus far. Transitions of this type occur in the even-even nuclei O^{16} (6.03 Mev), Ca^{40} (3.46 Mev), Ge^{72} (0.68 Mev), Zr^{90} (1.75 Mev, see Fig. 32) and Po^{214} (RaC′, 1.41 Mev). In a zero-zero transition the emission of gamma radiation is completely forbidden because the requirement that a quantum must take away at least one unit of angular momentum cannot be fulfilled. Such transitions therefore must go by internal conversion or by nuclear pair emission. The theory of zero-zero transitions was first worked out by DRELL and ROSE [*Dr 52*] and has been reviewed more recently by CHURCH and WENESER [*Ch 55*].

8. Internal conversion. The electromagnetic field generated by a nucleus when it makes a transition between two states almost always results either in the emisson of a single photon or in the emission of a conversion electron. The latter process, known as internal conversion, should not be thought of in a real sense as a double event consisting of the emission of a photon by the nucleus followed by the interaction of the photon with an atomic electron. According to quantum mechanics the wave function of an atomic electron has a finite value within the bounds of the nucleus and the electron is therefore capable of participating in a nuclear transition and of taking away the transition energy and spin and parity changes. Since a given orbital electron has a unique binding energy the conversion process of emission of the electron into an unbound state results in a conversion electron whose energy is sharp and has a value equal to the difference between the transition energy and the electron binding energy.

Most of the experimental information and all of the theoretical calculations thus far have to do with the K, L and M atomic electron shells. The K shell consists of two $1s_{\frac{1}{2}}$ electrons of principal quantum number $n=1$, orbital angular momentum $l=0$ (designated by the spectroscopic notation s) and total angular momentum $j=\frac{1}{2}$. Since j can have $2j+1$ orientations, there are two possible K electrons; they are the closest of all to the nucleus and therefore they are the most tightly bound. The second or L shell has a principal quantum number of 2 and the orbital angular momentum can then be 0 or 1. When $l=0$ we have two possible $2s_{\frac{1}{2}}$ electrons. This pair comprises the L_{I} sub-shell which is the most strongly bound of all L electrons. For $l=1$, or the p electrons, there are two possible ways in which the orbit and spin can combine vectorially; firstly j can be $\frac{1}{2}$ corresponding to two $2p_{\frac{1}{2}}$ electrons in the L_{II} sub-shell. Finally when $j=\frac{3}{2}$ the spin can assume four possible orientations and hence there are four $2p_{\frac{3}{2}}$ electrons in the L_{III} sub-shell. L_{III} electrons are the most weakly bound of the three. It should further be pointed out that the binding energies of the various electrons of a given sub-shell are all the same. Altogether there are 8 electrons in the L shell. Similar considerations apply to the M shell where it can be shown that there are five sub-shells and a total of 18 possible electrons.

Only those electrons in the first few atomic electron shells are internally converted to an appreciable extent as may be seen from the fact that the portion of an electronic wave function within the nucleus in general decreases with increasing orbit radius.

9. Internal conversion coefficients. Among the most important data on isomeric transitions are their conversion coefficients. The K-conversion coefficient α_K (not the same α as used in defining the mixing ratio α^2) is defined by

$$\alpha_K = \frac{N_K}{N_\gamma} \tag{9.1}$$

or as the ratio of the number of K-shell internal conversion electrons corresponding to the isomeric transition emitted from a source per unit time to the number of gamma rays per unit time due to the same transition. In essence, it is a measure of the relative probabilities that a given atom will emit a K-conversion electron or a gamma ray. It is a common error to think of the K-conversion coefficient as the ratio of K electrons to total transitions. Such a definition, while more logical in some ways, would obviously never lead to a conversion coefficient greater than 1. Actually according to definition (9.1) α_K can have any value from ~ 0 to $\sim\infty$ corresponding to cases from extremely weak to almost complete conversion.

Conversion coefficients for the other shells or sub-shells are similarly defined, as for example α_L is the ratio of the total number of L-conversion electrons to gamma rays, $\alpha_{L_{\mathrm{I}}}$ is the ratio of L_{I} electrons to gamma rays, etc. If N_{Tot} is the total number of isomeric transitions per unit time then

$$\left.\begin{aligned} N_{\mathrm{Tot}} &= N_\gamma + N_K + N_{L_{\mathrm{I}}+L_{\mathrm{II}}+L_{\mathrm{III}}} + N_M + \cdots \\ &= N_\gamma (1 + \alpha_K + \alpha_L + \alpha_M + \cdots). \end{aligned}\right\} \tag{9.2}$$

Accurate theoretical calculations of the K and L-shell conversion coefficients[1] have been made by ROSE and co-workers ([*Ro 51*], [*Si 55*] Chap. 14 and Appendix 4). The variables which determine the value of the coefficient for a given shell are the transition energy E, which is designated by the variable k when given in units of mc^2 (when $k=1$ the transition energy is 0.511 Mev), the multipole order, the magnetic or electric nature of the radiation and the atomic number Z of the nucleus. The qualitative features of K conversion are as follows:

[1] See Ref. 1, p. 5.

(a) For a given Z and multiple order α_K decreases with energy in a roughly logarithmic fashion. An increase in energy by a factor of two reduces the conversion coefficient by a factor of between 5 and 10.

(b) At a fixed energy and Z, α_K increases with multipole order. In going from an electric (or magnetic) radiation to the electric (or magnetic) one of the next highest multipole the value of α_K increases by a factor of between 2 and 4.

(c) If energy and multipole order are held fixed α_K increases with Z. As an example a 0.5 Mev $E2$ transition has a conversion coefficient varying from 7.3×10^{-3} at $Z=54$ to 3.1×10^{-2} at $Z=92$, an increase by a factor of 4.

Since we almost always know the transition energy and the atomic number of the nucleus before attempting a comparison between experimental numbers and the theoretical calculations we are concerned mainly with the second of the above features in finding the multipole order. Fig. 2 shows a typical set of K-shell conversion coefficients, those for $Z=64$. We see that, while the coefficients for the various electric transitions α_1, α_2 etc. are well separated from each other and the same applies to the various magnetic transitions β_1, β_2 etc., the two sets of curves are intermixed so that ambiguities can arise. Without taking any information other than Z and E into account it is frequently impossible to make a definite assignment of multipole order on the basis of α_K alone even if the accuracy of the measurement is high.

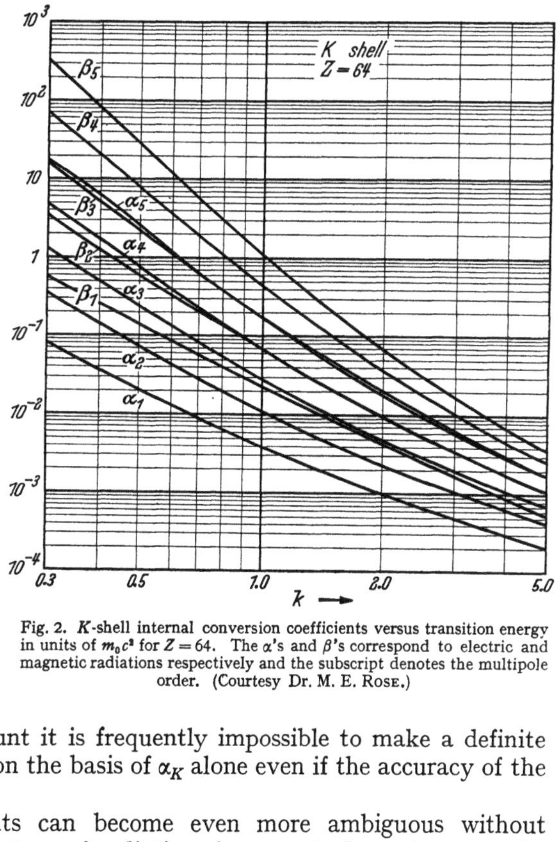

Fig. 2. K-shell internal conversion coefficients versus transition energy in units of m_0c^2 for $Z=64$. The α's and β's correspond to electric and magnetic radiations respectively and the subscript denotes the multipole order. (Courtesy Dr. M. E. Rose.)

Multipole order assignments can become even more ambiguous without supporting information if a mixture of radiations is present. In such a case the observed coefficient is the linear combination of the coefficients for the two components weighted according to the mixing ratio. For example if a mixture consists of 80% $M1$ and 20% $E2$ radiation and if the corresponding K-conversion coefficients are 1×10^{-2} and 5×10^{-2} respectively, the resulting α_K will be 1.8×10^{-2}. This value could fall close to that of a third multipole; if pure radiation rather than a mixture were assumed to be present a misinterpretation could result. On the other hand if one knows that a mixture is present and what the nature of the two components must be, the observed conversion coefficient can lead to a determination of the mixing ratio. This is possible only when the individual conversion coefficients differ appreciably from each other and when the value of α^2 is neither so small nor so large that the effect of the weaker component is concealed by the experimental error in the determination of α_K. No account need be taken of interference between radiations in considerations of the conversion coefficients of mixtures.

The majority of conversion coefficient measurements are made on K-shell electrons although there are occasionally cases where K-conversion is energetically impossible or where the K-conversion electron has too low an energy to be measured conveniently. A determination of α_L may be feasible and this can lead to an identification of multipole order.

10. Internal conversion ratios. The measurement of internal conversion ratios is often simpler than finding α_K or α_L alone. N_γ cancels out in the ratio of any two conversion coefficients and hence it is not necessary to know the gamma-ray intensity or some other reference establishing the number of isomeric transitions. Using the calculations of ROSE *et al.* on the K and L-shell conversion coefficients ([*Si 55*], Chap. 14) the desired ratio for given E and Z values may be computed and then compared with the experimental results. Fig. 3 shows the $K/L_{\mathrm{I+II+III}}$ ratio curves for $Z=85$. A few of the qualitative features of the behavior of K/L ratios are helpful to remember.

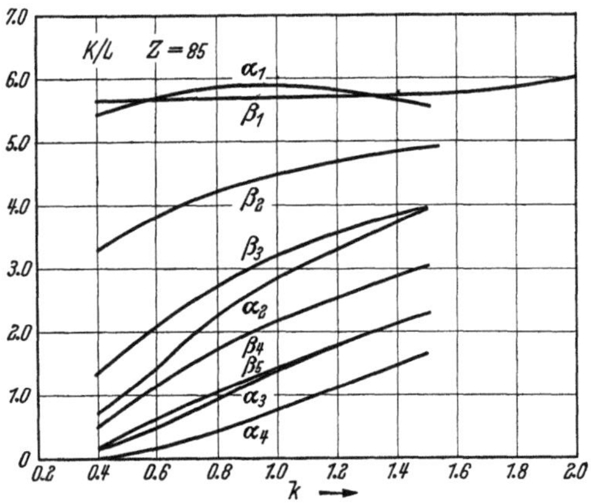

Fig. 3. $K/L_{\mathrm{I+II+III}}$ internal conversion ratios versus transition energy in units of m_0c^2 for $Z=85$. The α's and β's correspond to electric and magnetic radiations respectively and the subscript denotes the multipole order. (Courtesy Dr. M. E. ROSE.)

(a) At a given E and Z the K/L ratio decreases with increasing multipole order.

(b) For a given Z and multipole order the K/L ratio generally increases with increasing energy except possibly for dipole radiations.

(c) For a fixed Z the K/L ratio has a maximum possible value to which the K/L ratios for all multipole orders approach as the energy is increased. This maximum value is ~ 12 at $Z=25$ and ~ 6 at $Z=85$.

The identification of multipole orders from K/L ratios alone can be ambiguous even for pure radiation as may be seen from Fig. 3. When a mixture is present the observed K/L ratio will be the weighted average of the ratios for the two components. If the characteristics of the components are known the K/L ratio can be used to derive the mixing ratio unless the mixings either too small or too large to be detected within the accuracy of the measurement.

It is sometimes more convenient to use only the L sub-shell conversion coefficient ratios for purposes of multipole order assignments. The general trends of their behavior are less clearly defined than for the K/L ratios, but from the computations one sees that L_{I} conversion predominantes over L_{II} and L_{III} conversion in transitions of low multipole order and becomes less important at higher multipole orders. Other details may be found in the article by ROSE ([*Si 55*], Chap. 14).

It is clear from the foregoing discussions of conversion coefficients and their ratios that an individual measurement of one particular quantity may not lead to an unambiguous assignment; however, depending on the case, a combination of two or more measurements may be sufficient i.e., determinations of both α_K and $K/L_{\mathrm{I+II+III}}$ for example, or a measurement of one conversion datum together with an angular correlation or lifetime determination.

11. Directional angular correlation[1]. A method sometimes used to assist in determining the multipole orders of isomeric transitions is the study of angular correlations between successive radiations. Because of the experimental difficulties and more limited application, angular correlation has not occupied as important a position in studying nuclear isomers as the measurement of conversion coefficients, conversion ratios and lifetimes. It is limited to cases where at least one other radiation is in prompt or nearly prompt cascade with the isomeric transition. Usually the prompt cascade radiation consists of a gamma ray following the isomeric transition.

The theory of angular correlation for two successive gamma-ray transitions γ_1 and γ_2 considers three states, the initial one of spin I_i, the intermediate state of spin I_0 and the final state of spin I_f. Emission of either gamma ray alone shows no preferred direction in space because of the random orientation of nuclei in the source. However, if the members of a given cascade are observed by two detectors in coincidence, the emission of the first gamma ray serves to define a direction in space relative to which there is a probability distribution for the direction of the intermediate spin I_0. Since this distribution is in general not isotropic the directional probability distribution for the emission of γ_2 can be anisotropic with respect to γ_1. The second transition takes place between one of the magnetic sub-states of I_0 and one of the sub-states of I_f. For each such transition there will exist a unique directional distribution; the overall distribution for an assemblage of nuclei will be the average of the single distributions weighted according to the relative populations of the substates of I_0. These populations are determined by the first transition and hence by I_i and I_0.

The early theoretical work on angular correlation was done by HAMILTON [*Ha 40*]. More recently, BIEDENHARN and ROSE [*Bi 53*] have carried out calculations from which the directional correlation between γ_1 and γ_2 can be derived for almost any situation apt to be met in practice. The directional correlation $W(\vartheta)$ of two successive gamma rays can be expressed as a series of LEGENDRE polynomials $P(\cos\vartheta)$ as follows:

$$W(\vartheta) = 1 + A_2 P_2(\cos\vartheta) + A_4 P_4(\cos\vartheta) + \cdots \tag{11.1}$$

where

$$P_2(\vartheta) = \tfrac{1}{2}(3\cos^2\vartheta - 1), \tag{11.2}$$

$$P_4(\vartheta) = \tfrac{1}{8}(35\cos^4\vartheta - 30\cos^2\vartheta + 3), \tag{11.3}$$

and A_2, A_4 etc. are the correlation coefficients.

There are no known correlations with terms higher than $P_4(\cos\vartheta)$. Combining Eqs. (11.2) and (11.3) with Eq. (11.1) it follows that

$$\left.\begin{aligned} W(\vartheta) &= \left(1 - \frac{A_2}{2} + \frac{3A_4}{8}\right) + \left(\frac{3A_2}{2} - \frac{30A_4}{8}\right)\cos^2\vartheta + \left(\frac{35A_4}{8}\right)\cos^4\vartheta \\ &= \text{const} \times (1 + a_2\cos^2\vartheta + a_4\cos^4\vartheta). \end{aligned}\right\} \tag{11.4}$$

Eq. (11.4) gives the coincidence intensity as a function of the angle formed by lines from the source to the centers of the two detectors.

Two important features of an angular correlation curve are its shape and its anisotropy. Anisotropy is defined by the expression

$$\text{Anisotropy} = \frac{W(\vartheta)_{180^\circ} - W(\vartheta)_{90^\circ}}{W(\vartheta)_{90^\circ}} = a_2 + a_4. \tag{11.5}$$

[1] Details are given in the article by DEVONS and GOLDFARB.

It is the fraction by which the correlation curve rises (positive anisotropy) or decreases (negative anisotropy) in going from 90 to 180°. The shape of the curve is determined by the magnitudes and signs of A_2 and A_4. It is these coefficients which are predicted by the theory when the spins I_i, I_0 and I_f are given and when the multipole orders and mixing ratios α^2, if any, of both γ_1 and γ_2 are specified. From Eq. (11.4) several other features of angular correlation curves are apparent.

(a) All correlation curves are symmetric about 90°. Thus to completely determine a correlation it is sufficient to measure it between 90 and 180°; experimentally this is the most convenient range of angles. Data may be taken on either side of 180° since by symmetry the corresponding angles are equivalent.

(b) The derivative $dW(\vartheta)/d\vartheta$ contains both sines and cosines in each term and hence the slope of any correlation curve is zero at both 90 and 180°.

A basic requirement of angular correlation theory is that the orientation of the nuclear spin I_0 in the intermediate state must remain essentially constant during the lifetime of that state. If the lifetime is extremely short no change in direction can occur. But when the lifetime is longer it sometimes can happen that electromagnetic fields are present tending to make the nucleus precess by an appreciable amount during its lifetime. A large amount of precession results in a "loss-of-memory" effect which is a term used to indicate that the intermediate state spin direction is no longer well-defined when referred to the direction of the first gamma ray emitted. Precession can therefore "wash out" an otherwise anisotropic angular correlation making it isotropic, or more nearly so. Magnetic fields result from crystal structure effects, vacancies in electron shells or external application by laboratory means. From the point of view of isomeric transitions considerable effort has been spent in ensuring that observed correlations have not been influenced by such effects. On the other hand if one knows all of the necessary information about a decay scheme the angular correlation of a radioactive substance in various chemical states or crystal structures, or when external magnetic fields are applied can be used as a tool in solid state studies.

The analysis of angular correlation data, which will be discussed in more detail in Sect. 28, amounts to determining the values of the correlation coefficients and comparing them with the theoretical predictions for various assumptions as to the spins of the three states and the multipole orders of the gamma rays. Just as in the cases of the internal conversion coefficients and conversion ratios discussed previously, the correspondence between angular correlation experiment and theory is not always unique. However, when combined with the results of conversion or lifetime measurements an angular correlation determination can be crucial in obtaining the multipole orders of pure or mixed radiations.

12. Radiative half-life of single-particle transitions. The most striking feature of an isomer is its half-life and, as first suggested by VON WEIZSÄCKER [*Vo 36*], the half-life is intimately connected with nuclear spin change. While other means of identifying multipole radiations, such as from their conversion coefficients and conversion ratios, or from angular correlations, are based on well-understood electromagnetic calculations the radiative half-life results directly from the effects of nuclear structure. Because our detailed knowledge about the nucleus is at present very meager theorists have constructed various models of the nucleus with which experimental results are compared. One which has had considerable success is the independent-particle model which forms the basis for nuclear shell theory. According to this model each nucleon moves independently in a common potential produced by all of the other nucleons. The probability for the emission of gamma radiation is calculated on the assumption that one of the protons in the nucleus makes a transition between two states having specified quantum numbers and differing in angular momentum by l units. The results obtained by WEISSKOPF, which are discussed in detail elsewhere [*Bl 52*], may be expressed by two equations. For electric radiation of multipole order l the radiative transition probability λ (where $\lambda = 1/\tau$ and τ is the mean lifetime of the state) is given by

$$\lambda_E(l) \approx \frac{4.4\,(l+1)}{l\,[(2l+1)!!]^2} \left(\frac{3}{l+3}\right)^2 \left(\frac{E_\gamma}{197}\right)^{2l+1} \times R^{2l} \times 10^{21}\ \mathrm{sec}^{-1} \tag{12.1}$$

and for magnetic radiation of multipole order l

$$\lambda_M(l) \approx \frac{1.9\,(l+1)}{l\,[(2l+1)!!]^2} \left(\frac{3}{l+3}\right)^2 \left(\frac{E_\gamma}{197}\right)^{2l+1} \times R^{2l-2} \times 10^{21}\ \mathrm{sec}^{-1} \tag{12.2}$$

with the symbol

$$(2l+1)!! = 1 \cdot 3 \cdot 5 \ldots (2l+1).$$

In both equations E_γ is expressed in Mev and R is the nuclear radius in units of 10^{-13} cm. For a nucleus of mass number A, the generally accepted radius R (in 10^{-13} cm) is given by[1]

$$R = 1.45\, A^{\frac{1}{3}}. \tag{12.3}$$

WEISSKOPF has emphasized that Eqs. (12.1) and (12.2) can be expected to be only very rough estimates which might differ from observation by factors of the order of 10^3. He has therefore omitted a statistical weight factor S representing the effect of the number of magnetic substates of the initial and final states in the transition. As an example of the statistical weight effect we may consider a transition between states of spins $\frac{9}{2}$ and $\frac{1}{2}$ (which would be $M4$ if the parity changes). If the $\frac{9}{2}$ spin state has the higher energy a nucleus in any one of its magnetic sub-states has $2I+1$ or two possible paths to make a transition to the $I=\frac{1}{2}$ state. Conversely when the upper state has a spin of $\frac{1}{2}$ there are ten possible paths to the $\frac{9}{2}$ spin state. Hence in the latter situation the transition probability could be expected to be 5 times as great as in the former. It is actually found for $M4$ transitions that when the statistical weight correction is made the transition matrix elements are remarkably uniform (see Fig. 100).

The WEISSKOPF estimates of transition probabilities based on the independent-particle model have been reduced to a more usable form by MOSZKOWSKI ([*Si 55*], Chap. 13). Table 3 summarizes the radiation probabilities where the statistical weight factor S has been included.

Table 3. *Single proton transition probabilities. E_γ is in Mev and A is the nuclear mass number.*

Electric multipole	λ in sec^{-1}	Magnetic multipole	λ in sec^{-1}
$E1$	$1.5 \times 10^{14}\ A^{\frac{2}{3}}\ E_\gamma^3\ S$	$M1$	$2.8 \times 10^{13}\ E_\gamma^3\ S$
$E2$	$1.6 \times 10^{8}\ A^{\frac{4}{3}}\ E_\gamma^5\ S$	$M2$	$1.2 \times 10^{8}\ A^{\frac{2}{3}}\ E_\gamma^5\ S$
$E3$	$1.1 \times 10^{2}\ A^{2}\ E_\gamma^7\ S$	$M3$	$1.8 \times 10^{2}\ A^{\frac{4}{3}}\ E_\gamma^7\ S$
$E4$	$5.0 \times 10^{-5}\ A^{\frac{8}{3}}\ E_\gamma^9\ S$	$M4$	$1.5 \times 10^{-4}\ A^{2}\ E_\gamma^9\ S$
$E5$	$1.6 \times 10^{-11}\ A^{\frac{10}{3}}\ E_\gamma^{11}\ S$	$M5$	$7.5 \times 10^{-11}\ A^{\frac{8}{3}}\ E_\gamma^{11}\ S$

[1] Cf. the much smaller radii resulting from more recent experiments: D. L. HILL in Vol. XXXIX of this Encyclopedia.

To give a better feeling for the dependence of the transition probability on energy and multipole order, Fig. 4 shows the curves of λ versus E_γ for a nucleus of mass 100. We point out a few of the features which can be deduced by inspection.

(a) The magnetic and electric curves of the same multipole order have the same slope corresponding to E^{2l+1}.

(b) The separation between the electric (or magnetic) curves is such that the transition probability decreases (or lifetime increases) by a factor of $\sim 10^5$ to $\sim 10^8$ for each increase in multipole order by one unit.

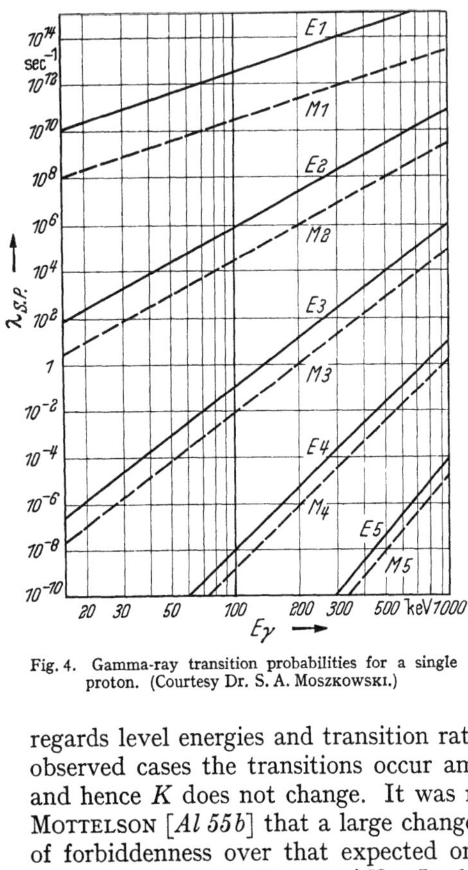

Fig. 4. Gamma-ray transition probabilities for a single proton. (Courtesy Dr. S. A. MOSZKOWSKI.)

(c) For a given multipole order a magnetic transition has a smaller probability than the electric by a factor of 10 to 100.

Although the deviations from the WEISSKOPF formulae are expected to be as great as factors of 10^3 the dependence of lifetime on multipole order is a factor of between 10^5 and 10^8 per unit change as pointed out above. Hence, in spite of our very rough knowledge of nuclear structure, we can expect that lifetime measurements may lead to unambiguous identifications of multipole orders in many cases. In Sect. 43 we shall show how the observed radiative lifetimes correspond to the single-particle estimates.

13. "K" isomerism and "core" isomerism. As discussed in Sect. 6, the collective model of BOHR and MOTTELSON has had considerable success in describing the low-lying states of nuclei in regions between closed shells both as regards level energies and transition rates. For almost all of the experimentally observed cases the transitions occur among the members of one rotational set, and hence K does not change. It was recognized by ALAGA, ALDER, BOHR and MOTTELSON [*Al 55b*] that a large change in K could cause an additional amount of forbiddenness over that expected on the basis of spin change alone. They introduced the quantity $\nu = \Delta K - L$ where L is the usual multipole order of the radiation.

An example of the K-forbiddenness effect on gamma-ray transitions has been established recently by SCHARFF-GOLDHABER, MCKEOWN and MIHELICH [*Sc 56a*] in the nucleus Hf^{180} (see Fig. 74). States belonging to the rotational set $K=0$ are observed having spins of 0^+ (ground), 2^+, 4^+, 6^+ and 8^+ which are fed in the decay of a 5.5 hour isomeric state. It was determined from gamma-ray conversion and intensity data that the isomeric state decays $\sim 90\%$ via a 57 kev $E1$ transition to the 8^+ state and $\sim 10\%$ via a 501 kev $E3$ transition to the 6^+ level. A spin and parity of 9^- for the isomer are thus indicated. As compared with the WEISSKOPF formula the 57 kev transition is too slow by a factor $\sim 10^{15}$ and the 501 kev transition is too slow by a factor of $\sim 10^9$.

If one assumes that the 9^- level belongs to a rotational set for which $K=9$, then $\Delta K=9$ for both of these transitions. The K-forbiddenness factor ν is thus 8 for the 57 kev transition and 6 for the 501 kev transition. This could explain the high forbiddenness was well as the difference of $\sim 10^6$ in their amounts of forbiddenness. Theoretical calculations of K-forbidden transition rates have not yet been made.

Another unusual type of isomerism has been discovered in the nucleus Mo^{93} (see Fig. 34). GOLDHABER has proposed [*Go 53a*] that this case can be explained if the odd neutron remains in a $g_{\frac{7}{2}}$ state throughout the three-step isomeric transition while the even-even core goes from $8^+ \to 4^+ \to 2^+ \to 0^+$. The spin of the 7 hour state may thus be as high as $\frac{23}{2}$. This so-called "core" isomerism is associated with complex states which are not yet well understood.

14. Transition half-life. The half-life actually observed for an isomeric state will always be shorter than the radiative half-life discussed in Sect. 12 since the latter does not take into account the presence of electrons surrounding the nucleus. Internal conversion can be regarded as an independent process which is in competition with a radiative transition. By itself it has a characteristic decay period depending on such factors as Z, E and multipole order. In principle one could attempt to calculate the lifetime expected for internal conversion in the various electron shells but this would be difficult. In practice internal conversion coefficients, which are independent of assumptions about nuclear structure, give directly the ratios of internal conversion transition probabilities to radiative transition probabilities.

The relationship between radiative and transition half-lives may be derived from Eq. (9.2) which may be written as

$$N_{\text{Tot}} = N_\gamma (1 + \alpha_T) \tag{14.1}$$

where α_T is the total internal conversion coefficient. Since the numbers of events N are proportional to the relative transition probabilities λ and $\lambda = 1/\tau$ we may write

$$\lambda_{\text{Tot}} = \lambda_\gamma (1 + \alpha_T) \tag{14.2}$$

or

$$\tau_\gamma = \tau_{\text{Tot}} (1 + \alpha_T) \tag{14.3}$$

$$T_{\frac{1}{2}\gamma} = T_{\frac{1}{2}\text{Tot}} (1 + \alpha_T). \tag{14.4}$$

When $\alpha_T \ll 1$ the observed half-life will be very nearly the same as the radiative half-life, but if α is large internal conversion will have a strong influence. Eq. (14.4) shows that the observed half-life, $T_{\frac{1}{2}\text{Tot}}$, can be very much shorter than the radiative half-live. In general we are most interested in τ_γ because of its relationship to nuclear structure; we can compute this from Eq. (14.3) provided the lifetime of the state is known and we have measured or deduced the total conversion coefficient.

For many years it was thought that the lifetime of a radioactive substance could not be changed by any physical means. However, Eq. (14.4) shows that if we were able to remove all of the electrons surrounding an isomeric nucleus the total half-life of the nuclear state would necessarily lengthen. It has been pointed out earlier that internal conversion occurs mainly in the K, L and M shells. The physical removal of these strongly bound electrons for an appreciable length of time might be expected to be nearly impossible. Since the internal conversion process depends on the magnitude of the electronic wave function within the nucleus one might at best to be able to distort the wave function and thereby bring about a change in lifetime.

Such an experiment has been carried out successfully by BAINBRIDGE, GOLDHABER and WILSON [*Ba 53*]. The isomer Tc^{99m} (*m* indicates the isomeric or metastable state) decays by a 2-kev $E3$ transition of 6 hour half-life, this transition being followed by one of 140 kev (see Fig. 39). The 2-kev transition is almost totally converted and since its energy is below the binding energies of both the K and L shells, conversion takes place mainly in the M and N shells. As we have seen from Eq. (14.4) the lifetime in this case is governed mainly by the internal conversion process. Furthermore the decay is unique in that the conversion of electrons at relatively large distances from the nucleus determines the decay constant. Chemical combinations involve only the outermost electrons of an atom, but at the same time the wave functions of inner electrons can be expected to be slightly sensitive to outside influences. If Tc is combined chemically with various elements differences in the shapes of the M and N-shell electronic wave functions should arise depending on the structure of each particular type of molecule.

In the experiment of BAINBRIDGE *et al.* the decay rate of Tc^{99m} was observed by detecting the 140 kev gamma ray (which is in equilibrium with the 2-kev transition and displays the same half-life) and by comparing the rates when the activity as present as Tc metal, $KTcO_4$, Tc_2S_7 or $TcReS_7$. Using a highly sensitive balanced double ion chamber they showed that the half-life varied according to the chemical form by as much as 0.3%. The experiment was a fundamental one in that it proved for the first time that internal conversion is a competitive process and that it does not follow gamma-ray emission. The magnitude of the effect was found to be reasonable from the point of view of solid state physics. Furthermore it was proposed that the changes in the decay constants of isomers, if observable, could be used in understanding the structure of molecules.

For historical completeness it should be mentioned that the experiment described above was not the first time that a detectable change had been made in the decay rate of a radioactive substance. From analogous arguments SEGRÈ [*Se 47*] had suggested that an electron capture rate should be influenced by changes in the molecular arrangement. The electron capture decay of Be^7 was studied by LEININGER, SEGRÈ and WIEGAND [*Le 51*] who found a slightly slower decay rate for Be^7F_2 as compared with Be^7O. Later KRAUSHAAR, WILSON and BAINBRIDGE [*Kr 53*] repeated the experiment using the double ion chamber technique. They confirmed the results of LEININGER *et al.* and for the same two molecules they found a difference in the decay constant of 0.06%.

15. Partial half-lives. Isomeric states sometimes have available two or more paths for decay, such as two nuclear transitions or beta-ray emission in competition with a nuclear transition. Each path has its own characteristic partial half-life T_1, T_2 etc. The observed half-life T of the state is given by

$$\frac{1}{T} = \frac{1}{T_1} + \frac{1}{T_2} + \cdots. \tag{15.1}$$

Evidently when one of the partial half-life values is much shorter than all of the others, it will be the main one to determine the observed half-life of the state. Partial half-lives can never be measured directly and the decay rates of each of a number of competing radiations will be exactly equal to each other and to the overall half-life of the parent state.

Partial half-life values may be inferred from the relative intensities of the competing radiations together with the observed half-life of the state. If we take a short time interval dt following the formation of N nuclei in the isomeric state then $dN_1 = N\lambda_1 dt$, $dN_2 = N\lambda_2 dt$, etc. will be the numbers of competing transitions occurring during dt, where λ_1, λ_2 etc. are the partial decay constants. The ratios $dN_1/dN_2/\ldots$ for this interval will thus be equal to $\lambda_1/\lambda_2/\ldots$. Since this holds true for any interval the integrated total numbers of transitions will also be proportional to $\lambda_1/\lambda_2/\ldots$. Inasmuch as λ and T are related by the equation

$$\lambda T = -\ln \tfrac{1}{2} = 0.693 \tag{15.2}$$

it follows that

$$N_1/N_2/\ldots = \frac{1}{T_1}\Big/\frac{1}{T_2}\Big/\ldots \tag{15.3}$$

D. Experimental methods.

I. The measurement of internal conversion coefficients.

As discussed in Sect. 9 the conversion coefficient for a particular shell, say the K shell, is the ratio N_K/N_γ where N_K and N_γ are the numbers of K electrons and of uncoverted gamma rays per unit time respectively. The determination of this ratio may be carried out in a number of ways.

16. Direct measurements of N_e and N_γ. For an activity of sufficiently long half-life, total conversion coefficients can sometimes be found by direct measurements of N_e and N_γ using simple detectors of known solid angle and efficiency. With present-day techniques, a scintillation spectrometer, consisting of a NaI(Tl) crystal mounted on a photomultiplier tube together with the usual circuit components, is capable of 5 to 10% accuracy in the determination of absolute gamma-ray strengths ([*Si 55*], Chap. 5). The procedure is to place the source on the axis of a cylindrical NaI crystal of standard size (1½ inches in diameter by 1 inch thick is most common) and at an accurately measured distance from the crystal face. Allowance must be made for the distance between the face of the crystal and the surface of its container. The total net counting rate from the photomultiplier resulting from the radiation but exclusive of noise and possible x-ray counts is then measured by pulse height analysis. Using the tables of McGowan [*Mc 54*] which give the results of calculations of the total efficiency times solid angle at a number of gamma-ray energies and source-to-crystal distances one can find the absolute gamma-ray emission rate in a case where the spectrum consists of only one component. For a complex spectrum it may be more convenient to determine only the number of counts in the full-energy-loss peak corresponding to the desired gamma ray. Curves of the ratio of the full-energy-loss peak intensity to the total intensity as a function of gamma-ray energy have been determined empirically for several crystal sizes by Heath, Bell and Davis (see [*Si 55*], Chap. 5, Fig. 6). Using these curves, which are accurate to better than 10%, one may deduce what the total counting rate would be corresponding to a given gamma ray of the spectrum simply by dividing the full-energy peak intensity by the appropriate peak-to-total ratio. One may then find the absolute gamma-ray emission rate as before from the tables of McGowan.

The determination of the number of conversion electrons may not be as easy. The simplest case would be that in which a single isomeric transition occurs and it is neither preceded nor followed by beta-ray emission. A measurement of the total number of conversion electrons may then be feasible using an end-window Geiger counter and a collimating system of accurately known geometry. Of course one has to make the necessary corrections for the detection of unconverted gamma rays and x-rays following internal conversion. Such a counter is not capable of resolving the K, L, etc. conversion electrons. In order to derive the individual conversion coefficients an auxiliary measurement of the relative conversion intensities has to be made by means of a magnetic spectrometer. When beta rays are present, or when the decay scheme has other conversion lines of comparable intensity, this method of finding the total conversion intensity of a particular transition may be difficult or impossible.

17. Conversion intensities based on the transmission of a beta-ray spectrometer. Several ways of determining conversion coefficients make use of magnetic field beta-ray spectrometers for finding the intensities of the internal conversion electrons. The momentum resolution of these devices is usually sufficient to detect conversion lines in the presence of beta rays and to resolve the conversion lines of some of the shells from each other. One may calculate directly the total emission rate of a given conversion line from a source by measuring the maximum number of such electrons which can pass through the spectrometer and dividing this number by the solid angle of the instrument expressed as a fraction of a sphere. A separate measurement of the gamma-ray strength with a NaI scintillation spectrometer will then allow the conversion coefficient to be computed.

The main cause of error in this type of measurement usually lies in the spectrometer transmission. For a source of negligible dimensions the transmission is determined (and properly defined) as the ratio of the net counts per unit time when the magnetic field of the instrument is set at a value corresponding to the peak of the conversion line, to the total number of these electrons emitted per unit time from the source. Generally transmission is a geometrical factor depending on the particular settings of the spectrometer baffles and it is independent of the magnetic field strength in a variable field instrument. Under the above conditions a characteristic "window" curve will result when counting rate is plotted versus field strength in passing over a conversion line. Its full width at half maximum height $\Delta H \varrho_{\text{spec}}$ expressed in percentage of the momentum setting $H\varrho$ is referred to as the resolution, i.e. $(\Delta H \varrho_{\text{spec}}/H\varrho)\times 100$. In a measurement where the source area and/or the source thickness cannot be neglected the observed line width will be greater than that of the "window" curve and the peak counting rate will be smaller than it would be for a point source of the same strength. Contributions to line width due to the source area $\Delta H \varrho_{\text{area}}$ and source thickness $\Delta H \varrho_{\text{thick}}$ compound with the spectrometer window width resulting in a line width

$$\Delta H \varrho = \sqrt{(\Delta H \varrho_{\text{spec}})^2 + (\Delta H \varrho_{\text{area}})^2 + (\Delta H \varrho_{\text{thick}})^2}\,. \tag{17.1}$$

When the spectrometer transmission and window width $\Delta H \varrho_{\text{spec}}$ are known accurately the peak counting rate can be used to find a conversion line intensity even if source area and thickness effects are present, provided these effects do not too strongly distort the shape of the line. The areas under the observed conversion line and under a "window" curve of the same height at the same momentum setting are determined. The ratio of these areas times the peak height of the measured curve gives the peak height which would result if the source were a point. An actual determination of the areas under the curves may be carried out using standard methods. To a first approximation the ratio of the measured line width $\Delta H\varrho$ to the window width $\Delta H \varrho_{\text{spec}}$ is equal to the ratio of areas, and this may be a sufficiently accurate factor for correcting the peak counting rate.

An alternative approach is to define an *effective* transmission according to the peak counting rate for a source of fixed dimensions. No corrections for a source under test are necessary if it has accurately the same dimensions as the standard source with respect to which the effective transmission has been defined. One must be careful to have the distribution of activity over the source area the same in both the unknown and calibration sources.

The value of high resolution in beta-ray spectroscopy is obvious as regards the accurate measurement of internal conversion line momenta and the resolution of complex line structure. When conversion lines are superimposed on a continuous beta-ray distribution there is an additional advantage of high resolution in that the peak heights of the lines are larger relative to the underlying beta-ray continuum than at lower resolution. To illustrate this point suppose that a resolution of 1% is being used and that a particular conversion line rises only 10% above a beta-ray continuum. Let the transmission of the spectrometer be 1% and assume that there are no source area or thickness contributions at any resolution setting of the instrument. Now suppose that the resolution could be improved to 0.2% at which setting the transmission is 0.1%. The number of conversion electrons detected will have dropped in proportion to the transmission i.e. to $\frac{1}{10}$ of the initial yield. On the other hand the line width is now $\frac{1}{5}$ of its former value so that the number of beta rays detected at a given setting of the field will be reduced by the product of the line width by the transmission, or to $\frac{1}{50}$ of its initial value. Thus at the expense of a factor of 10 in conversion line yield the ratio of the conversion line intensity to the beta-ray background intensity has improved by a factor of 5. The conversion electron yield now amounts to 50% of the beta-ray background intensity.

18. Comparison of conversion and photo lines with a standard in a beta-ray spectrometer. This method is neither absolute nor does it depend on any knowledge about the decay scheme. It consists of four steps. The number of internal conversion electrons, say N_{K1}, from the unknown source is measured in a beta-ray spectrometer. An absorber is then arranged to remove the internal conversion electrons and a photoelectric converter foil is placed close to the source giving a photoelectron yield $N_{\mathrm{P.E.1}}$. The same two measurements are then carried out on a calibration source of the same dimensions and having a transition with a known conversion coefficient, giving yields N_{K2} and $N_{\mathrm{P.E.2}}$. The yields of photoelectrons must be corrected using an empirical formula for the photoelectric cross-section $\sigma_{\mathrm{P.E.}}$ as a function of energy. We then have

$$\alpha_{K1} \sim \frac{N_{K1}}{N_{\mathrm{P.E.1}}}\,\sigma_{\mathrm{P.E.1}}; \qquad \alpha_{K2} \sim \frac{N_{K2}}{N_{\mathrm{P.E.2}}}\,\sigma_{\mathrm{P.E.2}} \tag{18.1}$$

and combining these relations it follows that

$$\alpha_{K1} = \alpha_{K2}\,\frac{N_{K1}}{N_{K2}} \times \frac{N_{\mathrm{P.E.2}}}{N_{\mathrm{P.E.1}}} \times \frac{\sigma_{\mathrm{P.E.1}}}{\sigma_{\mathrm{P.E.2}}}\,. \tag{18.2}$$

The method essentially uses a photoelectric conversion line as a measure of the relative intensities of unconverted gamma rays leaving the source. It works well when the transition energies do not differ by more than 30 to 50%. For too large an energy difference the empirical photoelectric cross-section function may not be valid, and even if it were, the change in the angular distribution of photoelectrons with energy might make the yields of $N_{\mathrm{P.E.1}}$ and $N_{\mathrm{P.E.2}}$ vary according to the source-converter geometry.

19. Comparison within a given source. When the internal conversion coefficient of one transition in a source has been measured, others in the same decay scheme may be determined by comparison with the known transition. The spectrum of internal conversion lines must be established by means of a beta-ray spectrometer; the relative intensities of uncoverted gamma rays are then found with a NaI scintillation spectrometer or by the photoelectric converter method described above. Inserting these intensity ratios into Eq. (18.2) (where $N_{\mathrm{P.E.}}/\sigma_{\mathrm{P.E.}}$ may be replaced by N_γ) leads to a value for the unknown conversion coefficient.

20. Conversion following known beta-ray branching. The number of internal conversion electrons may sometimes be referred to the number of beta rays either preceding or following the isomeric transition. Suppose in the simplest case that each such transition were preceded by a beta ray. A spectrometer measurement of the area under the beta-ray spectrum, excluding the conversion lines and AUGER electron lines associated with the conversion process, gives the total number of disintegrations N_β times the solid angle of the spectrometer. The total number of conversion electrons N_e times solid angle may be found in the same experiment from the total area under all of the conversion lines. Area measurements must be made using a plot of the number of counts per unit momentum interval versus momentum. If a spectrometer has a constant momentum resolution the momentum interval accepted is proportional to the momentum setting. Hence the proper plot for area measurements is obtained by dividing the yield at each point by the momentum setting $H\varrho$. Evidently

$$N_\beta = N_\gamma + N_e \tag{20.1}$$

and the total conversion coefficient is

$$\alpha_T = \frac{N_e}{N_\gamma} = \frac{N_e}{N_\beta - N_e}. \tag{20.2}$$

The separate conversion coefficients α_K, α_L, etc. may be determined using Eq. (20.2) and the relative conversion line intensities. Suppose that $K/(L+M+\cdots) = 3$, then

$$\alpha_{(L+M+\ldots)} = \tfrac{1}{3}\,\alpha_K, \tag{20.3}$$

$$\alpha_T = \alpha_K + \alpha_{(L+M+\ldots)} = \alpha_K + \tfrac{1}{3}\,\alpha_K = \tfrac{4}{3}\,\alpha_K \tag{20.4}$$

or

$$\alpha_K = \frac{3}{4}\,\frac{N_e}{N_\beta - N_e} = \frac{N_K}{N_\beta - \frac{4}{3}N_K} \tag{20.5}$$

in this example.

In a more complex spectrum where several gamma rays and beta-ray groups of known branching are present the various branching ratios may be used to reduce the problem to one of determining the number of times the isomeric state is formed, the fraction of these which decay through the isomeric transition and the relative numbers of conversion electrons.

21. Coefficients based on conversion electron comparison measurements alone. A measurement of relative internal conversion electron intensities together with a knowledge of the conversion coefficient of a single transition may lead to the conversion coefficient of another transition in the scheme. Supporting information is usually necessary such as previous knowledge about the relative gamma-ray intensities or a coincidence experiment which has shown for instance a one-to-one correspondence in the intensities of the two transitions. From the total conversion electron intensity N_{e1} of the known transition we may use $\alpha_T = N_e/N_\gamma$ and $N = N_e + N_\gamma$ to derive the total number of these transitions as

$$N_1 = N_{e1}\left(1 + \frac{1}{\alpha_{T1}}\right). \tag{21.1}$$

If we know that the number N_2 of transitions of unknown coefficient is equal to $R N_1$ the total conversion coefficient of the unknown transition may be derived readily as

$$\alpha_{T2} = \frac{N_{e2}}{R N_{e1}\left(1 + \frac{1}{\alpha_{T1}}\right) - N_{e2}}. \tag{21.2}$$

22. Coefficients based on gamma-ray comparison measurements alone. The equations in the preceding section may be used to find an unknown total electron conversion coefficient α_{T2} in terms of a known one α_{T1} together with the gamma-ray intensities and the factor R where $N_2 = R N_1$. The conversion coefficient is given by

$$\alpha_{T2} = \frac{R N_{\gamma 1}(1 + \alpha_{T1}) - N_{\gamma 2}}{N_{\gamma 2}}. \tag{22.1}$$

With a calibrated scintillation spectrometer the relative intensities of γ_1 and γ_2 may be determined. However, one may see from Eq. (22.1) in the simple case when $R = 1$ that either α_{T1} or α_{T2} must be > 0.1 in order for the value of the numerator to be outside of the errors of $\sim 10\%$ in the determinations of $N_{\gamma 1}$ and $N_{\gamma 2}$. This scheme therefore works well only when at least one of the transitions is highly converted. There is of course the requirement that neither transition is so highly converted that its gamma rays cannot be detected at all.

23. Conversion coefficients from x-ray measurements. When a decay scheme is simple one can find the K-conversion coefficient by measuring the numbers of K x-rays and of unconverted gamma rays with a scintillation spectrometer, provided there are no other x-rays such as those resulting from electron capture or from gamma rays converted to an extent comparable to the one under study. Each K-conversion event produces a vacancy in the K shell and the filling of these vacancies by outer electrons results in the emission of either K x-rays or AUGER electrons. The AUGER effect may be thought of as an internal conversion of an x-ray transition. K x-ray transitions can produce AUGER electrons from the L, M or outer shells. The ratio of the number of K x-rays to the total number of K-shell vacancies is known as the K fluorescent yield, a number which varies from ~ 0.15 at $Z = 20$ to ~ 0.95 at $Z = 80$ (see [*Si 55*], Chap. 22 (1), Fig. 2). L fluorescent yield is similarly defined and for lead it has a value of about 0.35. In order to find the number of K-shell vacancies one must divide the observed K x-ray yield, corrected for the detector efficiency at that energy, by the fluorescent yield constant.

A further correction must be made in using a NaI scintillation detector if the energy of the x-ray exceeds the K-shell binding energy of iodine. Photoelectric conversion of the incident x-ray in the iodine K shell will occur, and this will result in iodine K x-rays produced in the scintillator. Some of these secondary x-rays may be able to leave the crystal resulting in the so-called "escape" peak which has an energy equal to the difference between the energy of the incident x-ray and the iodine x-ray. A proper value is obtained for the incident x-ray intensity only if the escape peak intensity is added to the usual full-energy-loss peak.

As an example of this type of measurement suppose that a 1 Mev transition takes place in lead and that the associated unconverted gamma rays and x-rays are detected by an $1\frac{1}{2}$ inch diameter by 1 inch high NaI crystal. A lead K x-ray peak will appear at ~ 75 kev together with a resolved escape peak having an energy of $75 - 37 \approx 38$ kev and an intensity relative to the 75 kev peak of $\sim 15\%$

for this size of crystal, depending on the geometry. The areas under both peaks may be measured or one may find only the full-energy-loss peak intensity N_x and correct for escape using factors which have been calculated for various energies and source-crystal geometries [*Ax 53*]. In this example the number of K-shell vacancies (and therefore K-conversion electrons) from the source is given by

$$N_K = \frac{1.15\,N_x}{0.95\,\Omega_x\,\varepsilon_x}, \qquad (23.1)$$

where Ω_x is the effective solid angle of the detector for x-rays and ε_x is the x-ray detection efficiency (generally almost unity). The unconverted gamma-ray intensity is found in the usual manner from total area or full-energy-loss peak area. In taking the ratio of N_K/N_γ it may not be accurate to assume that the effective solid angle of the detector is the same for both the x-rays and the gamma rays. Generally the solid angle for x-ray detection is larger than for gamma-ray detection because the x-rays are absorbed in the surface layers of the NaI crystal nearest the source while the gamma rays are absorbed more uniformly throughout the volume of the crystal, depending on their energy.

24. Coefficients from coincidence measurements. Under certain conditions the unconverted gamma rays and the associated x-rays of an isomeric transition can both be measured in coincidence with a prompt cascade gamma ray in the decay scheme. Two NaI scintillation detectors may be used. When the proper corrections for fluorescent yield, escape peak and variations in solid angle for x-ray and gamma-ray detection have been made, the K-conversion coefficient is found from the ratio of K-shell vacancies, derived from the x-ray yield, to gamma-ray intensity, just as in the preceding section. X-rays following electron capture to the isomeric state may cause confusion with the internal conversion x-rays unless the lifetime of the state is longer than the resolving time of the coincidence circuit. Corrections may have to be made for the coincidence detection of the isomeric gamma ray in the reference gamma-ray detector and the x-rays associated with the prompt gamma ray in the other detector.

II. Measurement of internal conversion ratios.

25. General considerations. To find the relative intensities of the conversion electrons of a given transition requires the use of a device which is capable of resolving those lines of interest. Generally one must use a beta-ray spectrometer for this purpose although in exceptional cases an organic crystal scintillation detector will resolve the K and $(L+M)$ lines. Most beta-ray spectrometers employ magnetic fields. The transmission and momentum resolution for given baffle settings are constants for such spectrometers and the electron momentum focussed is directly proportional to the magnetic field strength.

Table 4. *Atomic electron binding energies in* kev.

Shell	$Z=28$	$Z=82$
K	8.34	88.0
L_I	1.02	15.86
L_{II}	0.88	15.20
L_{III}	0.86	13.03
M_I	0.11	3.85
...	...	...

The momentum separation of the various conversion lines depends on Z and on the transition energy. We may illustrate with a few examples. For the lead nucleus, $Z=82$, the atomic electron binding energies are given in the third column of Table 4. If a 1-Mev transition is converted in lead, the momentum of a particular conversion electron is found by subtracting its binding energy from the transition energy and then looking up the corresponding momentum in a table of electron energy versus electron momentum $H\varrho$ ([*Si 55*], appendix 8). The K electron evidently has the lowest energy and momentum and the K and L lines in this case are separated by about 5% in momentum; the L and M lines are separated by about 1%. A spectrometer of 1% resolution will just resolve the K, L and $M+N+\cdots$ lines. The L_{I} and L_{II} lines are only 0.05% apart and there are few if any spectrometers capable of resolving them.

The percentage separation in momentum of conversion lines for a given Z will always be larger for lower energy transitions because the fixed binding energy differences will be greater in proportion to the transition energy. Thus for a 100 kev transition in lead the L_{I} and L_{II} lines are 0.4% apart, or 8 times as great, percentage-wise, as the corresponding line separation at 1 Mev.

Lighter nuclei have lower atomic electron binding energies and therefore smaller momentum separations between conversion lines as may be seen from the second column of Table 4. The K and L electrons of a 1 Mev transition at $Z=28$ will be separated by 0.5% in momentum or only $\frac{1}{10}$ as much as for the same transition energy in lead.

26. Conversion ratios in variable field spectrometers. Internal conversion spectra are measured in variable field instruments by plotting the spectrometer yield versus the magnetic field setting. A minimum of 8 to 10 points is necessary to define the amplitude and shape of each conversion line and, of course, due regard must be taken of counting statistics. Extraction of the desired conversion ratios from the data will depend on the particular situation. Peak height after subtraction of the appropriate background often may serve as a measure of conversion intensity since it will be governed by the constant transmission of the instrument. This will apply if the half-widths of the various lines are equal in percentage of momentum. Thus if the three L-conversion lines of a given transition are all present and the spectrometer line width is large compared with their separation the L conversion line width may not be noticeably greater than the K line width; the L peak will then be the same as if all of the L-conversion electrons had identical energies and the height will be equal to the sum of the three lines. When the resolution is improved the L line will become broader than the K line and the height of the L peak will no longer be a measure of total L conversion. Similar considerations apply when the $M+N\ldots$ electrons are unresolved from the L line. For extremely high resolution the natural line width for K conversion in heavy elements is greater than for L conversion. This results in K lines which are broader than the individual L lines. Differences in line width may also occur because of source thickness, i.e. the K line, being of lower energy, may be more seriously affected by scattering in the source than the L electron lines.

In all cases where variations in line width are present one must either find the areas under the curves or make other suitable corrections. To a first approximation the product of peak height by the line half-width in percent is proportional to the intensity. Areas may be found in two equivalent ways. If the spectrum is plotted as spectrometer yield versus momentum the area under each peak is measured with, for example, a planimeter and divided by the momentum value at the center of the line. Alternatively one may divide the spectrometer yield at each point by the corresponding momentum and thus make a plot of $N/H\varrho$ versus $H\varrho$. The measured areas will then be proportional to the conversion intensities.

Source thickness sometimes results in relatively large low energy tails on the conversion lines. Measurements of line area must either include the tail or cut off at a fixed number of half-widths below the peak position.

As described in Sect. 10 the most useful conversion ratios for assigning multipole orders are the K/L and $L_{\mathrm{I}}/L_{\mathrm{II}}/L_{\mathrm{III}}$ ratios. Using beta-ray spectrometers of moderate resolution the M, N etc. electrons are frequently unresolved from the L line and for this reason values of $K/(L+M)$ are often quoted in the literature. For purposes of comparison with theory one would like to estimate the value of K/L when only $K/(L+M)$ is known. A convenient rule of thumb is that L/M ratios are usually about equal to K/L ratios. Suppose for example that the $K/(L+M)$ ratio is 3.0, then from the relationship

$$\frac{K}{L+M} = 3.0 = \frac{K}{L + \frac{L}{3}} = \frac{3}{4}\,\frac{K}{L} \tag{26.1}$$

it follows that the K/L ratio is ~ 4.0. An error of 50% in the assumed L/M ratio will result in an error of less than 25% in the K/L ratio deduced.

27. Conversion ratios using fixed field spectrometers. Fixed magnetic field beta-ray spectrometers with photographic recording are used frequently for the measurement of internal conversion spectra. They generally have better resolution than variable field instruments and hence they are valuable for very complex line spectra. The photographic plate is usually placed inside the vacuum chamber so that there are no windows for the electrons to pass through. This feature together with good shielding by the magnet against stray fields makes this type of instrument particularly useful for the detection of very low energy conversion lines.

The relative intensities of lines on a photographic plate may be measured with a densitometer. This works well for lines close enough to each other so that the variation of plate sensitivity with electron energy is small, whereas for wide energy separations the relative intensities are subject to large errors. The measured K/L ratios of low energy transitions in heavy elements may thus be unreliable whereas the $L_{\mathrm{I}}/L_{\mathrm{II}}/L_{\mathrm{III}}$ ratios would probably be quite accurate. Photometer data are analyzed as regards line width and peak height using considerations similar to those applying to spectra measured by variable field beta-ray spectrometers.

III. Angular correlation.

28. Experimental methods and analysis. The experimental arrangement for studying the angular correlation between two gamma rays usually consists of one fixed NaI scintillation detector, the source, and a second NaI detector movable in a plane about an axis passing through the source. If an electron-gamma correlation is to be measured one of the NaI crystals is replaced by an organic scintillator, such as anthracene, covered with a thin aluminum foil. The following precautions are generally taken.

(a) The apparatus should be capable of accurate and reproducible angular settings. At 180° the center lines of the detectors should coincide and pass through the center of the source.

(b) The solid angle subtended by each detector should be kept to a minimum consistent with reasonable coincidence yield. A total acceptance angle of 15° for each detector in a given plane is usually considered to be good geometry.

(c) Shielding of each detector from radiation scattered by the other detector is often necessary particularly in cases of low energy gamma rays. This may be accomplished by conical lead absorbers. Graded absorbers are sometimes used in front of the detectors to avoid complications from source x-rays. Such absorbers may consist of lead sandwiched by copper and then by aluminum. Copper will absorb the fluorescence x-rays produced in the lead and aluminum will absorb the copper fluorescence x-rays. The energy of the aluminum fluorescence x-rays is too low to be detected ordinarily. Difficulties sometimes arise if positrons are emitted by the source inasmuch as the 0.5 Mev annihilation gamma rays are always emitted at 180° with respect to each other. When one of the gamma rays whose correlation is under study has an energy near 0.5 Mev the 180° point may be unusable.

(d) It is important to minimize the source size and especially its radial extent. The source length along the axis of rotation of the movable detector is not so serious whereas the radial extent should be kept small in order to minimize effects of gamma-ray scattering.

(e) As described in Sect. 11 one must attempt to make the chemical composition of the source such that the true correlation is observed in the experiment. With liquid sources the effects of crystal structure often present in solids will be avoided.

(f) Pulse height analysis of each detector output is desirable for singling out the corresponding gamma ray.

(g) Source strength should be chosen in order to give a maximum yield consistent with a reasonable real-to-chance coincidence ratio. As in any coincidence experiment this ratio varies inversely with the source strength. The chance rate, either calculated from the single rates and the known coincidence resolving time or actually measured, must of course be subtracted before plotting the data.

(h) In order to minimize the effects of circuit instabilities the order of angles at which data are taken should be randomly chosen from a preselected set of angles and the data should be normalized properly. If the source strength is constant and if the counting times at all points are the same, it is sufficient to normalize by dividing the yield at each point by the product of the total number of single counts of the individual detectors in their respective coincidence circuit channels. Such normalization corrects for drifts of each channel with respect to the corresponding pulse height distribution and also corrects for slight variations of distance between the source and the movable detector. Automatic apparatus is sometimes used for angular correlation measurements and this case a uniform sequence of angles is usually necessary.

The simplest first analysis of angular correlation data is to sketch a smooth curve through the experimental points and to normalize so that the curve (though not necessarily the 90° point) has a value of unity at 90°. A rough and uncorrected value of the anisotropy, if any, can be derived at once. When the geometry is reasonably good the correct anisotropy and the values of the correlation coefficients may be derived. Rose [*Ro 53*] has developed a mathematical procedure for making a least squares fit of the general correlation function Eq. (11.1) to the data. Values of the correlation coefficients and their probable errors may be derived. These uncorrected coefficients must then be corrected for the solid angles of the detectors and for source size. The complete detector correction requires that the dimensions be specified and that one know the efficiency per unit volume as a function of distance from the center of the detector. Rose has worked out these corrections for cylindrical NaI crystals. An approximation is to replace the detector by an imaginary circle or square at an appropriate distance from the source. Corrections for such geometries have been calculated by Feingold and Frankel [*Fe 55*]. Finally it is necessary to correct for source size if this is appreciable. The calculations of Feingold and Frankel correct exactly for the length of the source along the axis of rotation of the movable detector.

Once the proper correlation coefficients have been obtained by making suitable corrections to the experimental data they may be compared with the theoretical coefficients as discussed in Sect. 11.

IV. The measurement of isomeric half-lives.

29. Direct measurement. When an isomeric state is formed by heavy particle irradiation its half-life sometimes can be measured directly by ordinary counting techniques. Frequently, however, other activities may be present or the state is either preceded or followed by other radiations. Depending on the particular decay scheme the isomeric half-life may then have to be found by the analysis of complex decay curves or by using detectors which respond only to the radiations associated with the isomeric transition. For example, the rate of decay of the isomer's conversion lines in a beta-ray spectrometer can be used provided the decay is not determined by a radioactive parent of comparable or longer half-life. Such direct measurements which involve the removal of the sample from an accelerator or a reactor followed by the runing of a decay curve can be made for half-lives as short as a few seconds. Speed of transfer to the detector is essential for very short half-life measurements and the counting rate decay must be measured with automatic recording equipment.

30. Pulsed activation. If the activity does not have to be removed from the accelerator, for instance in an arrangement where a detector can be placed next to a cyclotron target, even shorter half-lives can be determined. The technique would then be to use a pulsed beam or to interrupt a steady beam with a fast shutter and immediately begin to record the counting rate. Half-lives of $\sim 10^{-2}$ sec have been measured in this way by ALVAREZ [*Al 50*] using a pulsed linear proton accelerator.

AXEL and co-workers ([*Ax 55*] and [*Ve 55*]) have extended the pulsed activation technique to half-lives of $\sim 10^{-5}$ sec. The irradiation pulse of a betatron induces the activity and at the same moment triggers a circuit which has an electronic gate movable in time with respect to the trigger pulse. A plot of the number of pulses in the gate recorded for different amounts of delay between the irradiation time and the opening of the gate will give the half-life of the activity.

31. Rapid chemical separation. The remaining techniques which will be discussed depend upon the presence of a relatively long-lived radioactive parent feeding the isomeric state. In such cases the isomer will be in equilibrium with the parent and the radiations normally will display the parent's lifetime. However, if the isomer can be separated chemically from its parent no new atoms of the isomeric activity can be formed in the separated sample, and its decay will be characteristic of the isomeric half-life. Half-lives down to about 0.5 sec have been measured in this way.

32. The delayed coincidence method. Other measurements of lifetime when the isomer is in equilibrium with a parent depend upon the parent activity not only for furnishing a continuous supply of isomeric nuclei but the radiations leading to the isomeric state serve as a time zero reference for determining the half-life. The delayed coincidence method can be used to measure half-lives between 10^{-4} and 10^{-10} sec. It consists of placing two detectors close to the radioactive source, one detector being adjusted to respond to the radiation leading to the isomeric state and the other being set to respond to the radiation from either the isomeric transition itself or to another radiation occurring promptly with the isomeric transition. The pulse outputs of the two detectors are fed to a coincidence circuit whose time resolution is adjusted according to the particular experiment. Since a lifetime is present the two radiations will not be in coincidence although, when the circuit is adjusted to detect prompt coincidences, some of the events will take place within the circuit resolving time. The number of counts when no artificial delays are introduced will be larger than for any delay setting since in any decay the probability is greatest that the delayed radiation will follow the originating radiation immediately. Many of the associated pairs of pulses will not be counted under these conditions however, because the delayed radiation may occur later than the circuit resolving time. In order to count this type of event as a coincidence, pulses from the parent radiation

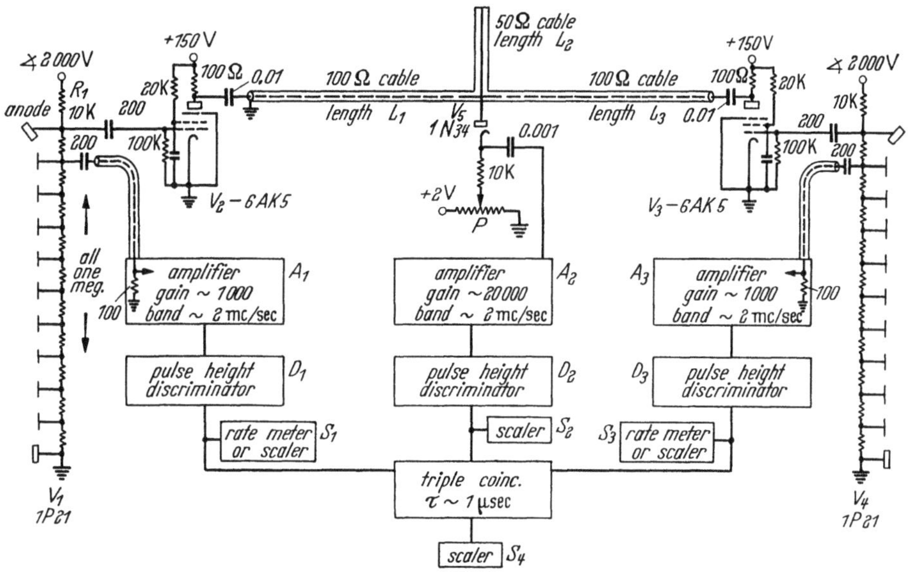

Fig. 5. Bell, Graham and Petch circuit for delayed coincidence measurement of short half-lives. (Courtesy Dr. R. E. Bell.)

must be delayed before being inserted into the coincidence circuit. A half-life value can thus be determined from the distribution of delayed coincidence intensity versus time delay. Delaying the parent pulse is usually accomplished by allowing it to pass through a delay line or through various lengths of coaxial cable before entering the coincidence circuit.

The achievement of short time resolution has been the subject of intensive development in recent years. Improvements have been made to the point where the response time of scintillation crystals (trans-stilbene and anthracene) is the limiting factor in time resolution. With present-day techniques the best resolving time is about 10^{-9} sec. The problem is then to find ways of measuring as short half-lives as possible within this resolving time limitation.

Until very recently the ordinary delayed coincidence technique has been the only available coincidence method for determining the shortest half-lives in radioactive substances, i.e. half-lives down to $\sim 10^{-10}$ sec. A typical circuit for this purpose, as designed by Bell, Graham and Petch [*Be 52a*], is shown schematically in Fig. 5. In each half of the circuit pulses from the anode of a 1P21 photomultiplier tube are fed through a capacitor to the grid of a normally conducting 6AK5 pentode amplifier tube. The photo tube output pulses are negative in polarity and their amplitudes are adjusted to be large in order to cut off the 6AK5 before the anode pulse has reached more than a small fraction of its final amplitude. The outputs of the two 6AK5 tubes are connected through 100 Ω coaxial cables to a common terminal. A length of 50 Ω coaxial cable short circuited at its far end and a biased diode (tube V5) are also connected to the common terminal. When one of the 6AK5 tubes is cut off a positive polarity step-like voltage wave is propagated down the 100 Ω connecting coaxial cable and then down the 50 Ω cable to the shorted end where it is reflected as a negative polarity step. That portion of the wave proceeding to the plate of the opposite 6AK5 will not be reflected appreciably owing to proper impedance matching by the use of a 100 Ω plate resistor. At the common terminal there is thus produced a rectangular positive pulse whose width in time is determined by how long it has taken the pulse step to travel from the terminal to the end of the shorted line and back. Suppose now that two radiations actuating the photomultipliers are in true coincidence. Each half of the circuit will independently produce a short positive pulse at the terminal and the net effect, when the pulses overlap in time, will be a pulse having an amplitude of up to twice as great as the amplitude of one alone. Only those pulses of amplitude greater than single pulses alone will be detected in the output if the bias on the diode is adjusted properly. Evidently the function of the 6AK5 tube together with the shorted line is to act as an equalizer, presenting to the diode pulses of uniform shape associated in time with the initial rise of the photomultiplier output pulse. It is necessary to impose restrictions as to what amplitudes of photomultiplier pulses will be detected in the final output by connecting each photomultiplier tube output through an amplifier A to a pulse height analyzer D which, together with the diode output, operates a triple coincidence output circuit of relatively slow resolving time. (In Fig. 5 the positive pulse from the last dynode is used for purposes of pulse height analysis.) The analyzers select a narrow range of pulse heights and insure that all of the final output pulses will be associated with the desired portions of the spectrum. Such discrimination eliminates the coincidence time jitter which would otherwise result if pulses with a wide range of amplitudes (and therefore rates of rise) were accepted.

The time resolution of a coincidence circuit is defined as the difference in time that two pulses may occur and still be detected as a coincidence. At first glance the resolution of the circuit in Fig. 5 would appear to be governed only by the length of the shorted line $L2$; to a large extent this is true. However, for very short pulses the diode will not perform as a peak-reading voltmeter; the counting of the output pulse will depend rather on a certain minimum amount of charge being passed through the diode and the charge in turn will depend on (a) the exactness of pulse overlap and (b) the diode bias. By increasing the bias a more nearly exact overlap is demanded in order to result in a large enough output pulse to be counted and this will improve the resolving time. At a fixed bias setting the resolving time can be varied by changing the length of the shorted line.

The entire process of detecting a coincidence event in this circuit is completed even before the photomultiplier output pulses have reached their full values. At a relatively much later time, i.e. after ~ 1 μsec, the trailing edges of the phototube pulse outputs bring the 6AK5 tubes back into the conducting region. The associated pulses at the terminal are of negative polarity and could not be detected by the biased diode in any case.

In order for prompt coincidences to be detected, the lengths of coaxial cable between the common terminal and the plates of the 6AK5 amplifier tubes must be of equal length providing that there are no unbalanced time delays in the system. Delayed coincidences on the other hand may be detected by arbitrarily changing the length of the cable (L_1 or L_3) in that half of the circuit which is actuated by the parent radiation. In the usual experiment a complete delayed coincidence distribution is obtained by varying the lengths of L_1 and L_3 in turn and measuring the coincidence rate for each length. Cable length may be converted to time when plotting the data by using the appropriate cable propagation velocity. For 100 Ω cable the delay is $\sim 4\times 10^{-11}$ sec per cm of length, or one millimicrosecond for each 10 inches of length.

Typical delayed coincidence data taken by SUNYAR [*Su 55*] using a similar circuit are shown in Fig. 6. The presence of a half-life in the source is revealed by comparing the observed delayed coincidence distribution with a so-called "prompt" distribution as obtained using a source in which the delay between the two detected radiations is known to be immeasurably short. In general there are two effects introduced by a lifetime. One is the occurence of a wing on one side of the curve representing the detection of the delayed coincidence events. When the logarithm of the yield is plotted against time delay the wing of the curve approaches a straight line whose slope leads directly to a half-life value. The second effect, which is useful when the half-life is so short that a wing is not clearly evident, is that the mean center of the distribution for a delayed source is displaced along the time axis from the center of the curve for a prompt source by an amount equal to the mean lifetime. A centroid type of analysis for determining the amount of displacement can be carried out which makes use of all of the points of both curves [*Ba 50*], [*Ne 50*], [*Ba 55e*].

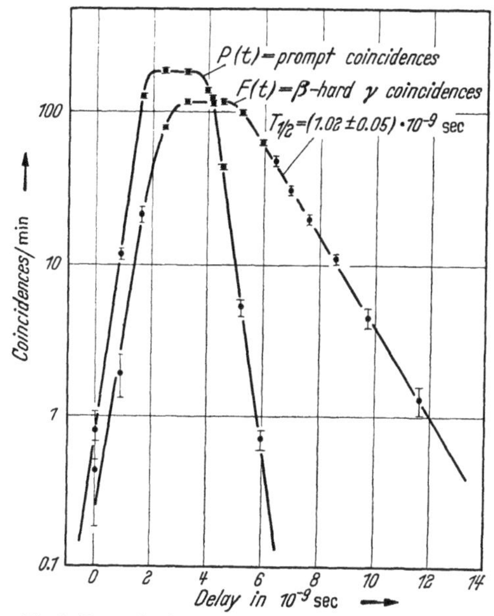

Fig. 6. Delayed coincidence distribution using a source of Ta^{182}. (Ref. [*Su 54b*], Fig. 1.) (Courtesy Dr. A. W. SUNYAR.)

33. The SUNYAR pulse height analysis method. Very recently a new technique has been developed by SUNYAR [*Su 55a*] which appears to be a significant advance in the determination of very short half-lives. A circuit of the type shown in Fig. 5 is used in such a way that the pulse height in the final output is proportional to the amount of overlap of the two pulses presented to the diode. This requires slowing down the response of the diode so that it cannot operate as a peak-reading voltmeter. Also an arbitrary delay is introduced between the common terminal and the 6AK5 tube associated with the delayed radiation so as to reduce the amount of overlap of the pulses. The delay not only makes the variation in the size of the output pulse more sensitive to the relative time positions of the two 6AK5 pulses, but the pulse amplitude will then depend uniquely on the actual time sequence of the two events. If one measures the pulse height spectrum from the diode for a prompt source, restricted of course by a slow triple coincidence circuit which allows only those pulses to be observed which correspond to the desired photo tube output amplitudes, one observes a distribution of intensity versus pulse height having a broad maximum. When a source is used which has a measurable half-life the amount of pulse overlap at the diode will be reduced on the average and the pulse height curve intensity maximum will shift to a lower pulse height value.

The amount by which the pulse height maximum shifts for a given delay may be determined in several ways. One may measure the distributions using a prompt source under the above conditions and then after changing the length of coaxial cable between the common terminal and one of the 6AK5 tubes. The shift in pulse height can then be calibrated as a function of time delay. Another scheme is to measure the pulse height distribution for a prompt source, move one of the detectors to a different distance from the source and remeasure the pulse height distribution. The shift in pulse height will then be governed by the change in distance between the source and the detector and by the velocity of light. SUNYAR has been able to detect the time of flight of a nuclear gamma ray over a path length of 1 cm corresponding to a time difference of 3×10^{-11} sec. The mean life of the 0.411 Mev state in Au^{198} has been measured as $\sim 3.5 \times 10^{-11}$ sec using this technique. The results of several other such measurements are included in Table 5 (p. 83).

SUNYAR'S method has reduced the measurable lower limit of half-life by a factor of ~ 5 below the capabilities of the ordinary delayed coincidence technique. This has largely been the result of several distinct advantages which are:

(a) Except for the first calibration method mentioned above no changes in the lengths of cables are necessary; thus the circuit can be left undisturbed and only the test and prompt sources need be interchanged.

(b) By the use of a multi-channel analyser all of the delayed coincidence events are recorded during a given counting period. This greatly reduces the time necessary to find the distribution of pulse heights with the desired accuracy, as compared with ordinary delayed coincidence where one delay point at a time is measured. Uncertainties due to drifts in the circuit are therefore expected to be much less serious.

34. The time-channel analyzer. In this device the pulses from the parent radiation detector trigger a circuit having an electronic gate which opens at a given time after the initiating pulse. During the interval the gate is open delayed radiations will be counted. By maintaining the width of the gate constant but sliding its position with respect to the trigger pulse a delayed radiation intensity distribution may be obtained. Circuits have also been designed which contain a number of adjacent gates, each with its own count register, for the more rapid accumulation of data. An instrument of this type has been built by VON DARDEL [*Vo 51*]. The range of half-lives which can be measured with the time-channel analyzer, i.e. $\sim 10^{-3}$ to $\sim 10^{-6}$ sec, is limited at short half-lives by the speed with which it is possible to open and close electronic gate circuits and at long half-lives by intensity and background considerations.

Unlike the delayed coincidence technique the time channel analyzer cannot measure half-lives which are well within the resolving time. In this method the resolving time is related to the pulse width and it is a measure of how closely the sliding gate can be adjusted toward the initiating pulse, detect delayed pulses and yet not detect truly prompt coincident pairs of pulses.

Another distinguishing feature of the time-channel analyzer as well as the method described in the next section is that a single detector can be used if it is sensitive to both the parent and the delayed radiations. The output of the detector is fed to the trigger circuit and also to the gated amplifier. If a pulse into the trigger circuit happens to result from the parent radiation there is a possibility that the next pulse from the detector will be associated with the delayed radiation and that it will be counted as such in the corresponding time channel. The latter pulse will have no effect on the trigger circuit which is insensitive until the time cycle is completed. When the parent radiation has not been detected and the delayed radiation pulse enters the trigger circuit the normal cycle of operation will occur but there will, of course, be no associated delayed pulses and hence no counts possible other than random events.

35. Triggered oscilloscope method. Half-lives in the range of 10^{-3} to 10^{-6} sec may also be measured with a triggered oscilloscope or synchroscope. When two detectors are used the one sensitive to the parent radiation triggers the horizontal sweep circuit. The delayed radiation detector is connected to the vertical deflection plates resulting in pulses at various distances from the origin of the sweep trace. Measurement of the intensity distribution of pulses along the sweep trace leads to the usual exponential decay curve after the random contribution has been subtracted.

A single detector may also be used, as in the measurement of the 145 μsec isomeric state in Pb^{206} by ALBURGER and PRYCE [*Al 53a*], see Fig. 91). In their experiment a source of Bi^{206}, the parent activity, was placed close to a single NaI crystal detector the pulse output of which was fed through an amplifier to both the sweep trigger and the vertical deflection circuits of a synchroscope. A slotted mask placed in front of the oscilloscope tube allowed a fixed time interval of the sweep to be observed by an optical pulse detecting system. The latter consisted of a lens focusing an image of the oscilloscope screen onto a photomultiplier which recorded one count for each pulse appearing in the slot of the mask. By placing the slot at various distances from the trace origin the distribution of delayed pulses was obtained.

36. General remark on delayed coincidence and time-channel methods. In an actual experimental measurement of a half-life using the delayed coincidence, time-channel or triggered oscilloscope methods it is desirable to select the resolving time or channel width so as to obtain the optimum data, if only a single channel point is recorded at a time. Consider for example, a half-life measurement of a 100 μsec isomeric state using a time-channel analyzer with a single sliding channel. Adjusting the channel width to only 5 or 10 μsec would mean that a relatively small number of counts would be recorded in a given time at each delay setting. On the other hand while a channel width of 500 μsec would still lead to the correct exponential decay curve it would be undesirable from the point of view of chance background, i.e. if the channel is set to open at 100 μsec and to close at 600 μsec, most of the real counts would appear in the early part of the channel whereas the random counts occur uniformly. Hence in the 300 to 600 μsec portion of the channel most of the counts could be randoms.

Generally the conditions for obtaining the best compromise between counting rate and the real-to-chance ratio is when the resolving time or the channel width is about the same or perhaps slightly less than the value of the half-life being measured. For very short half-lives it is not possible to have the optimum conditions and, as described in Sect. 33, the half-life being measured may be as little as $\frac{1}{100}$ of the circuit resolving time.

37. COULOMB excitation. Although the possibility of COULOMB excitation of the nucleus was proposed as early as 1938 by WEISSKOPF [*We 38*] it was not until 1953 that the effect was first observed by HUUS and ZUPANČIČ [*Hu 53c*] and by

McCLELLAND and GOODMAN [*Mc 53c*]. This process may be thought of as the interaction between the electric field of a nucleus and that of a closely passing bombarding particle. If the target nucleus has a quadrupole moment, implying that its shape is not spherically symmetrical, the incident particle can induce oscillations in the nuclear surface through the electric interaction (see Sect. 6).

In the present discussion we will be concerned mainly with the electric quadrupole or $E2$ excitation of the first one or two rotational states above the ground state since the COULOMB excitation of higher levels is not usually observed. The radiative half-life of a rotational excited state may be derived from the cross-section for its excitation by COULOMB excitation. As shown by HUUS and ZUPANČIČ [*Hu 53c*] and by ALDER and WINTHER [*Al 54b*] the cross-section is given by

$$\sigma = \frac{2\pi^2}{25} \frac{m^2 v_f^2}{Z_2^2 \hbar^2} \times g_2(\zeta) \times B(E2) \times 10^{-48}, \tag{37.1}$$

where $g_2(\zeta)$ is a numerically tabulated function of the adiabatic parameter

$$\zeta = \frac{Z_1 Z_2 e^2}{\hbar}\left(\frac{1}{v_f} - \frac{1}{v_i}\right), \tag{37.2}$$

m is the mass of the bombarding particle and v_i and v_f its velocities before and after the interaction, and Z_1 and Z_2 are the charges of the bombarding and target nuclei. $B(E2)$ is proportional to the reduced upward electric transition probability and it is the quantity which we seek for the purposes of half-life determination. It is commonly expressed in units of 10^{-48} cm^4 and differs from the $B(E2)$ in references [*Hu 53c*] and [*Al 54b*] by the factor $10^{-48} e^2$.

Following the calculations of HEYDENBURG and TEMMER [*He 55c*] if we use the approximation $\Delta E/E \ll 1$, which holds for low-lying levels and bombarding energies of a few Mev, then Eq. (37.2) reduces to

$$\zeta = \frac{Z_1 Z_2 e^2}{\hbar v} \frac{\Delta E}{E}. \tag{37.3}$$

As an example of the calculations we consider alpha-particle bombardment in which case Eq. (37.1) may be written

$$\sigma(E) \approx \frac{15.1\, E}{Z_2^2} g_2(\zeta)\, B(E2) \tag{37.4}$$

where values of the various constants have been inserted, $\sigma(E)$ is expressed in barns (1 barn $= 10^{-24}$ cm^2) and E is expressed in Mev (1 Mev $= 1.59 \times 10^{-6}$ erg).

In the usual experiment a thick target is bombarded with protons or alpha particles and the yield Y' of gamma rays per bombarding particle following COULOMB excitation is measured. Scintillation counters of known solid angle and efficiency with respect to the target gamma rays are generally employed and corrections may have to be made for absorption, back-scattering, x-ray contributions, and possible angular distribution effects of the emitted gamma rays. Accurate measurements of the bombarding beam current and energy are of course necessary. Inasmuch as the bombarding particle continually slows down during its passage in the target and the cross-section is energy dependent it is necessary to integrate the cross-section along the particle's path. If one makes the assumption that the stopping power dE/dx varies as $E^{-\frac{1}{2}}$ and that the stopping power per atom varies as $Z_2^{\frac{1}{2}}$ then the yield Y of excited states per incident particle is given by

$$Y = N\sigma(E_0)\left[\frac{E_0\, y(\zeta)}{(dE/dx)_0}\right], \tag{37.5}$$

where

$$y(\zeta_0) = \frac{2}{3}\zeta_0^{\frac{5}{3}}\frac{1}{g_2(\zeta)}\int_{\zeta_0}^{\infty}\frac{g_2(\zeta)}{\zeta^{\frac{8}{3}}}\,d\zeta, \qquad (37.6)$$

the subscript zero referring to the incident particle and N being the number of atoms per cm^3.

Once a state has been excited by COULOMB interaction it may make a radiative transition to the ground state, undergo internal conversion, or decay via an intermediate state if one exists. In the absence of the last possibility the yield Y' of gamma rays will be smaller than the yield of excited states Y by a factor of $1/(1+\alpha_T)$ where α_T is the total internal conversion coefficient. In order to make the conversion correction an experimental determination of this quantity may be used, if available. Otherwise the theoretical conversion coefficients can be applied in certain cases such as pure $E2$ transitions between 2^+ and 0^+ states. The total conversion coefficient is then the sum of α_K, $\alpha_{L_{\rm I}}$, $\alpha_{L_{\rm II}}$ and $\alpha_{L_{\rm III}}$, as found from the tables [*Si 55*] plus a small correction for $M+N$ conversion. For cases of mixed transitions such as $E2+M1$ it is possible to derive the total conversion coefficient only if the mixing ratio is known, unless a measurement of α_T has been made. The correction for decay via a branch to an intermediate level may be estimated from the intensity of the corresponding gamma ray to that of the ground state gamma ray. However, the total branching intensity must also include a factor for the total internal conversion of the transition to the intermediate level.

Assuming that the various corrections outlined above can be made, one may arrive at a proper value Y of excited nuclei per incident particle. Substituting the expression for the cross-section Eq. (37.4) into the yield function Eq. (37.5) and solving for $B(E2)$ we have for alpha-particle bombardment

$$B(E2)_{\rm up} = \frac{A\,Z_2^2\,(dE/dx)_0}{9.06\,E_0^2\,g_2(\zeta_0)\,y(\zeta_0)} \times Y \times 10^{-48}\ {\rm cm}^4 \qquad (37.7)$$

where A is the atomic weight of the target, $(dE/dx)_0$ is the stopping power of the target for alpha particles in Mev-cm^2/g and E_0 is the incident energy in Mev. In the literature on COULOMB excitation a value for the quantity $\varepsilon B(E2)$ is often given. It is the number obtained by inserting the thick target yield of gamma rays per incident particle Y' into Eq. (37.7) and hence it does not include the internal conversion or branching corrections. ε is always smaller than unity and its exact value will be known only if reliable conversion and branching correction factors have been determined.

The transition probability for a photon of multipole order l has already been given in Eqs. (12.1) and (12.2). One may also express the transition probability by the equation

$$\lambda(\gamma) = \frac{8\pi(l+1)}{l\,[(2l+1)!!]^2}\,\frac{1}{\hbar}\left(\frac{\Delta E}{\hbar c}\right)^{2l+1} B(\gamma) \qquad (37.8)$$

with $(2l+1)!! = 1\cdot 3\cdot 5\ldots(2l+1)$, where $B(\gamma)$, the reduced transition probability discussed above, depends on the matrix elements of the multipole operator between the initial and final states. For $E2$ transitions this may be written [*Bo 53d*]

$$\lambda(E2)_{\rm up} = \frac{4\pi}{75\hbar}\left(\frac{\Delta E}{\hbar c}\right)^5 e^2\,B(E2) \times 10^{-48}\ {\rm sec}^{-1} \qquad (37.9)$$

where ΔE is the excitation energy in ergs and $B(E2)$ is expressed in units of 10^{-48} cm^4. Expressing (ΔE) in Mev and inserting the values of the constants into Eq. (37.9) we have

$$\lambda(E2)_{\rm up} = 1.16 \times 10^{13} \times (\Delta E_{\rm Mev})^5 \times B(E2). \qquad (37.10)$$

The next stage in deriving the radiative half-life is to take into account the statistical weight factor. As explained previously in Sect. 12, the relative probabilities for the upward and downward transitions will depend on the numbers of magnetic substates of the two levels. Thus if the ground and first excited states are 0^+ and 2^+ respectively the downward transition probability will be only $\frac{1}{5}$ of the upward transition probability.

The $E2$ partial mean radiative lifetime is given by the relation

$$\tau_\gamma(E2) = \frac{1}{\lambda(E2)_{\text{down}}}. \tag{37.11}$$

The value of $\tau_\gamma(E2)$ is of most interest for comparison with theoretical calculations. However, if one wishes to derive the total mean life τ or the total half-life $T_{\frac{1}{2}}$ of the state several corrections may have to be made. In case the ground state transition has an $M1$ admixture, this will shorten the radiative lifetime as derived from the inverse of the upward electric transition probability, since the $M1$ component may appear in the decay but cannot be involved in the $E2$ excitation of the level. Further corrections must be made for internal conversion and for branching via an intermediate level. One may note that in going from thick target gamma-ray yield Y' to the total lifetime of the state both of the latter correction factors have to be applied twice.

Summarizing the above discussion, one first determines experimentally the yield Y of excited states per incident alpha particle and substitutes this into Eq. (37.7) in order to determine $B(E2)$. The upward $E2$ transition probability follows from Eq. (37.10) and the downward $E2$ transition probability results by applying the statistical weight factor. Similar considerations apply for proton excitation except for different values of the parameters.

As an example[1] of the calculations suppose that a 1 Mev 2^+ state is excited by COULOMB excitation from a 0^+ ground state in a nucleus of $Z=30$ and that the resulting $\varepsilon B(E2)=0.11$. For this case internal conversion is negligible and ε is close to unity. Inserting the values of ΔE and $B(E2)$ into Eq. (37.10) it follows that

$$\lambda(E2)_{\text{up}} = 1.28\times 10^{12}\ \text{sec}^{-1} \tag{37.12}$$

and including the statistical weight factor

$$\lambda(E2)_{\text{down}} = 0.2\times 1.28\times 10^{12}\ \text{sec}^{-1} = 2.6\times 10^{11}\ \text{sec}^{-1}, \tag{37.13}$$

$$\tau_\gamma(E2) = 3.9\times 10^{-12}\ \text{sec}. \tag{37.14}$$

38. The nuclear recoil method. The half-lives of states excited in nuclear reactions may sometimes be determined by the nuclear recoil method as used by THIRION and TELEGDI [*Th 53*]. This consists of measuring the distance that an excited recoil nucleus moves before it emits a gamma ray. In the experiment of THIRION and TELEGDI an accelerator beam produces a reaction in a very thin target such that excited nuclei can detach from the target and recoil forward *in vacuo* with a reasonably well-defined velocity. If the lifetime is long enough the excited nuclei will decay at a measurable distance from the target depending on the recoil velocity and on the half-life of the state. A gamma-ray collimator and detector are arranged so as to examine at 90° to the beam a limited region dx in front of the target. The measurement consists of moving either the collimator and counter together or the target along a line parallel with the beam and noting the variation of counting rate. From the shape of the observed distribution and other parameters of the experiment the half-life for gamma-ray

[1] See Zn^{64} in Table 5.

emission can be deduced. The method is useful for measuring half-lives of $\lesssim 10^{-10}$ sec depending on the mass of the target nucleus and on the incident particle mass and velocity. Thus in the case of the $O^{16}(d, p)O^{17}$ reaction a state at 0.87 Mev is excited in O^{17} which has a lifetime of 2.5×10^{-10} sec. At $E_d = 1.3$ Mev THIRION and TELEGDI found that almost all of the O^{17} nuclei decayed within a distance of 1 mm from the target. A higher bombarding energy would evidently increase the dispersion and hence would allow shorter half-lives to be measured.

39. DOPPLER shift and DOPPLER broadening. If in the above experiment one were to assume that all of the product nuclei recoiled in the forward direction and that accurate measurements of the energies of the gamma rays were made by observing them in the forward direction and then at 90° to the beam, the DOPPLER effect would cause the gamma rays at 0° to appear somewhat higher in energy than at 90°. Suppose, however, that a thick target is used and that the lifetime is longer than the slowingdown time of the recoil nuclei. In such a case the gamma-ray energy observed in any direction would be the same. From the slowing-down time, which can be estimated, one can place a lower limit on the half-life of a state if there is no shift in energy between 0 and 90°. On the other hand if a shift in energy is observed which is just the amount expected for a nucleus having its initial recoil velocity then the lifetime would be shorter than the slowing-down time, thus leading to an upper limit for the half-life. For an intermediate energy shift it would be possible to estimate the actual value of a half-life. The experimental situation is not as simple as outlined here since the recoil nuclei will have an angular distribution with respect to the beam direction. Such effects tend to wash out the unique dependence of energy on angle.

In the limiting case where the directional distribution of recoil nuclei is isotropic the mean energy of gamma rays observed in all directions is the same and is equal to the transition energy. However the recoil effect causes a continuous spectrum, or DOPPLER broadening, about the mean energy. Such a case is that of the $B^{10}(n, \alpha)Li^7$ reaction studied by ELLIOTT and BELL [*El 48*], [*El 49*]. The slow neutron which induces this reaction imparts a negligible velocity to the B^{11} compound nucleus, but the emission of the α particle leading to a 0.478 Mev excited state in Li^7 is energetic and can occur in any direction. (The energy release or Q of the reaction is 2.8 Mev.) ELLIOTT and BELL measured the DOPPLER broadening of these gamma rays in sources of various chemical compositions by comparing the width of the K-shell photoline from a converter in a lens spectrometer with that of a 0.47 Mev monoenergetic gamma ray in the decay of Ir^{192}. They showed that the amount of broadening varied according to the target material and they were able to derive a lifetime of 7.5×10^{-14} sec for this state based on estimates of the slowing-down time of the recoil nuclei in the different samples.

40. Nuclear resonance fluorescence. The natural energy spread or width Γ of a nuclear energy level is related to the lifetime τ for the emission of radiation by the well-known HEISENBERG uncertainty relationship

$$\Gamma \cdot \tau \sim \hbar. \tag{40.1}$$

Thus an experimental determination of Γ can lead directly to a lifetime value for the state. One of the processes by which gamma rays can be scattered by an atom is that of resonant scattering by the nucleus. If gamma radiation is incident on a nucleus and has an energy of just the right amount necessary to supply the nucleus with energy corresponding to a particular level, the gamma radiation can be strongly absorbed resonantly. Decay of the state would occur by re-emission of a gamma ray of very nearly the same energy. According to electromagnetic theory [*He 36*] the cross-section for resonant absorption is given by

$$\sigma = g \cdot \frac{\lambda^2}{8\pi} \frac{\Gamma^2}{(E - E_0)^2 + \frac{\Gamma^2}{4}}, \tag{40.2}$$

where E_0 is the energy of the level, E the energy supplied by the incident gamma ray, λ the wavelength of the incident gamma ray and g is a statistical weight factor. Evidently σ is maximum when $E = E_0$, but if $E - E_0$ is large compared to Γ then σ will be very small since the energy supplied is on the wing of the resonance curve far from the peak value.

Resonant scattering is difficult to observe because of the recoil of the nucleus in the gamma-ray emission and absorption processes. As an example consider the decay of Au^{198} in which beta-ray emission leads to a 0.411 Mev state in Hg^{198} (see Fig. 85). Neglect for the moment the recoil occurring in beta-ray emission and assume that the Hg^{198} nucleus in the 0.411 Mev state is at rest. Emission of the 0.411 Mev gamma ray is itself accompanied by recoil of the mercury nucleus. Because of conservation of energy and momentum the actual gamma-ray energy will be slightly less than that of the excited state by an amount $E_0^2/2Mc^2$ where M is the mass of the nucleus. If now the emitted gamma ray is incident on a scatterer consisting of Hg^{198} nuclei in their ground states a part of its energy amounting to $E_0^2/2Mc^2$ must be taken for energy and momentum conservation if it is to be absorbed. Thus because of recoil both in emission and in absorption, the energy supplied to the scattering nucleus is lower than the resonant energy by a total amount E_0^2/Mc^2. For the particular case of Au^{198} this amounts to 1 ev. The width of the level Γ on the other hand is known to be only 2×10^{-5} ev.

There are several ways of increasing the resonant scattering cross-section to the point where observations can be realized experimentally. The first positive result was obtained with a method first suggested by MOON [*Mo 50*]. This consistent of making up the 1 ev recoil energy loss discussed above by means of the DOPPLER effect. If the source is moved toward the scatterer with a sufficiently high velocity the emitted gamma ray can be made to have an energy with respect to the scatterer which is somewhat above the level energy, or more nearly suitable for resonant absorption. MOON and co-workers [*Mo 53a*], [*Da 53*] accomplished this by placing a source of Au^{198} on the tips of a high speed rotor. Tip velocities as high as 1 km/sec were achieved. Using a mercury scatterer and a detector shielded from direct radiation they were able to observe a maximum in the scattered yield as a function of the rotor velocity. In deriving the cross-section at the resonance peak the DOPPLER spread in gamma-ray energies due the thermal motion causes the yield to be less than if no thermal effects were present and this must be taken into account.

One can make use of thermal DOPPLER broadening itself to observe nuclear resonant scattering as was first accomplished by MALMFORS ([*Ma 53f*], *Si 55*], Chap. 18). In this procedure the source is stationary but it is heated to very high temperature. As the temperature is raised the DOPPLER distribution becomes broader and there is a greater possibility that some of the gamma rays will have the correct energy for resonant scattering. MALMFORS showed that heating a Au^{198} source to its melting point (1063° C) should increase the resonant yield by a factor of 200 over the yield at room temperature. He was then able to observe resonant scattering in the presence of other elastic scattering processes by comparison experiments using mercury and lead scatterers. Similar research has also been done by METZGER and TODD [*Me 54*] who found a half-life value for the Hg^{198} state in agreement with the result of MOON *et al.* described above.

Prior to the work of MOON and of MALMFORS another method had been suggested by KUHN [*Ku 29*] and by POLLARD and ALBURGER [*Po 48*] for observing resonant scattering. Emission of a beta ray preceding the gamma ray, which was neglected above, will supply recoil energy to the nucleus. If the beta ray is directed away from the scatterer the motion of the recoil nucleus toward the scatterer will shift the emitted gamma-ray energy to a higher value provided that appreciable slowing down of the nucleus does not take place between the beta and gamma-ray emission events. Although the resonant condition is satisfied for only one particular narrow range of recoil velocities there will at least be a greater probability of resonant scattering than if there were no beta-ray recoil effect. No resonance yield was observed by POLLARD and ALBURGER apparently because the slowing-down time was shorter than the lifetime of the state and hence the beta-ray recoil was ineffective.

As suggested by MOON and ILAKOVAC [*Mo 54e*], [*Il 54*] the slowing-down time of the excited recoil nucleus following beta-ray emission can be increased with a liquid source. Using Zn^{63} they were able to observe a positive resonant scattering effect from copper whereas the effect with solid sources could not be established with certainty. METZGER [*Me 55*] proposed that gaseous radioactive sources would be even more suitable for increasing the slowing-down time and he has been able to observe resonant scattering and to deduce the lifetimes of the nuclear states in several cases. (For a more complete discussion see [*Me 56a*].)

E. Present knowledge of nuclear isomers.

The data of nuclear isomers were compiled and reviewed in 1952 by GOLDHABER and HILL [*Go 52*] (see also [*Ko 54a*]). In addition to their critical analysis of a large amount of well-established information they pointed out numerous problems such as uncertain mass assignments and doubtful information on the levels in many of the isotopes. During the intervening years some of the outstanding problems have been solved and many new isomeric states have been discovered. In this section data on nuclear isomers are presented based on information available up to about the end of 1955[1]. Isomers which are of tentative or doubtful assignment have been omitted.

41. Isomers of half-life greater than 10^{-6} sec. As a result of the development of new methods of determining short half-lives, particularly that of COULOMB excitation, many short-lived states have been established. Some of these are not reached in radioactive decay processes. For a consistent presentation of the data, decay schemes are given only in cases of isomers having half-lives greater than an arbitrarily chosen figure of 10^{-6} sec. They are presented in Figs. 6a through 94[2]. Information on those levels in the figures for which the measured half-lives are $< 10^{-6}$ sec may be found in Table 5 (p. 83).

The decay schemes in general include not only the disintegrations leading to or from isomeric nuclei but those of neighboring isobars since the data on these may contribute to information about the isomeric levels. Other details of the figures are as follows.

(a) Ground states and metastable states (except for states with measured half-lives of $< 10^{-6}$ sec) are indicated by heavy horizontal lines. If it is not known which of two activities is the ground state, both are drawn at the same vertical level. Both, in the cases of isomeric states where the separation is not known and

[1] Many 1956 references have been added during the course of publication of this volume.
[2] Figs. 6a—94 are reproduced on p. 40—83.

in electron capture where the disintegration energy is unknown, the level position may be only very approximate.

(b) Level assignments in parentheses are considered tentative.

(c) Where orbital angular momenta are given the nuclear shell model with strong spin-orbit coupling has been used.

(d) Ground states of even-even nuclei have been assumed to be 0^+.

(e) The symbol ε is used to indicate orbital electron capture.

Because of space limitations no discussion of the experimental evidence is given. Level assignments are generally those found in the most recent and complete papers listed in the references. Only those articles published since 1952 are included; earlier papers may be found from the references listed by GOLDHABER and HILL [*Go 52*] or in the Table of isotopes published by HOLLANDER, PERLMAN and SEABORG [*Ho 53c*]. The Nuclear Science Abstracts[1] annual cumulations were found to be very helpful in preparing the figures and references.

42. Isomers of half-life less than 10^{-6} sec. Isomers having half-lives of less than 10^{-6} sec are listed in Table 5 on p. 83—89. In the third column the half-life of the state $T_{\frac{1}{2}}$, the mean life of the state τ or the radiative mean life τ_γ are given alternatively in accord with the references. The fourth column contains the experimental $B(E2)_{up}$ as found from COULOMB excitation (see Sect. 37). In some cases a COULOMB excited level is included even though the corresponding lifetime

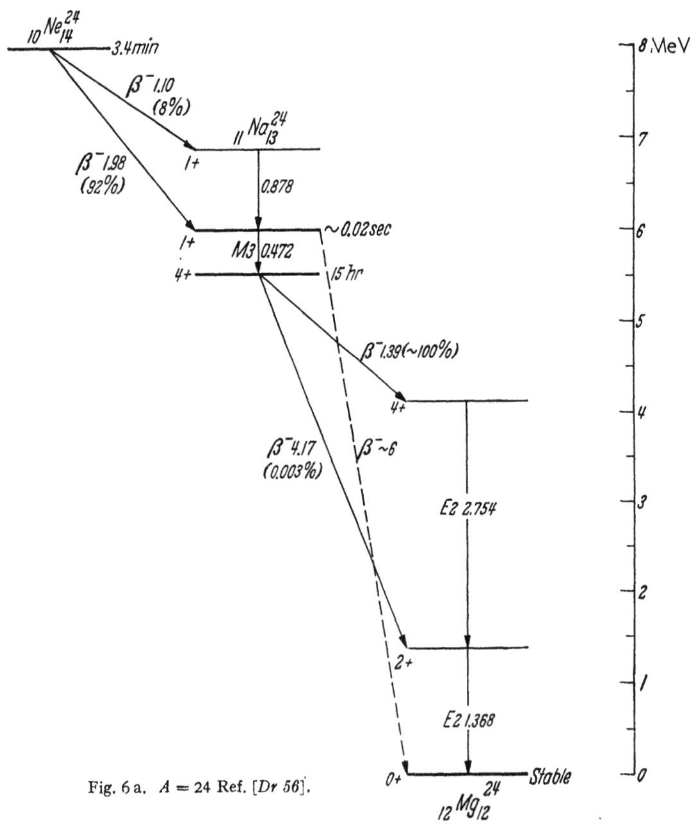

Fig. 6a. $A = 24$ Ref. [*Dr 56*].

[1] U. S. Atomic Energy Commission, Technical Information Service, Oak Ridge, Tennessee (Compiled by WAY, KING, KUNDU, MCGINNIS and VAN LIESHOUT. Nuclear Data Group, National Research Council, Washington, D. C.)

information has not been reported. The lifetime may be estimated using Eq. (37.10) after making corrections for statistical weight and conversion. Information on the spins and parities of the isomeric or upper state and the lower (usually ground) state is given in the fifth column.

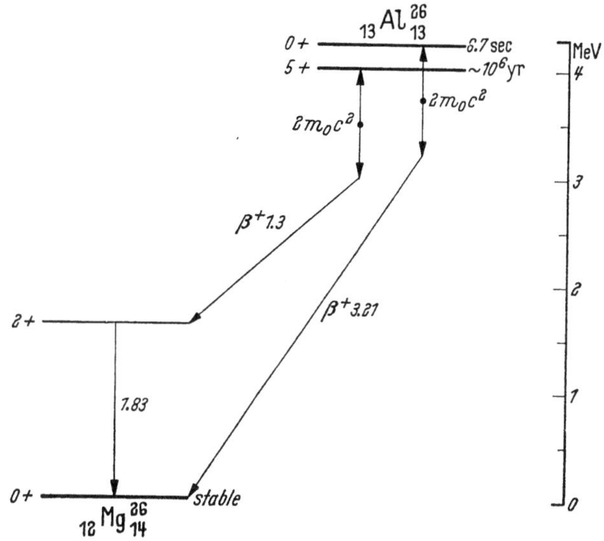

Fig. 7. $A = 26$. Refs. Al^{26}: [*Br 54, El 55, Gr 54a, Ha 54, Ha 55, Ka 55, Ki 53, Kl 54, Mo 53, Mo 54, Si 54, St 53, Sw 50*].

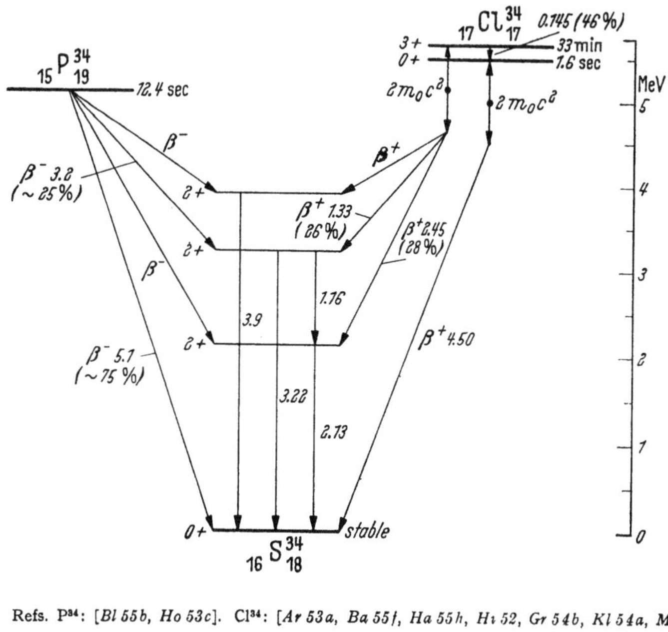

Fig. 8. $A = 34$. Refs. P^{34}: [*Bl 55b, Ho 53c*]. Cl^{34}: [*Ar 53a, Ba 55f, Ha 55h, Hi 52, Gr 54b, Kl 54a, Ma 54j, St 53c*].

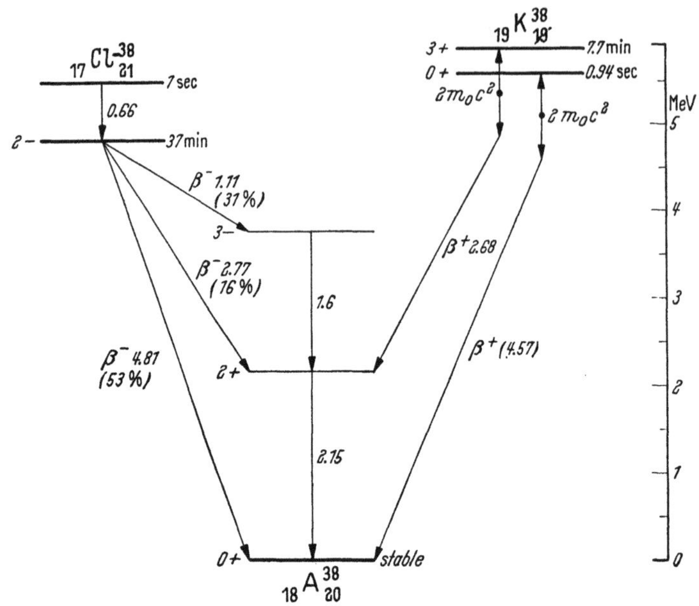

Fig. 9. $A = 38$. Refs. Cl^{38}: [*Ho 53c, Kr 54a, Ma 55g, Sc 54*]. A^{38}: [*Fl 53*]. K^{38}: [*Gr 54b, Kl 54a, St 53*].

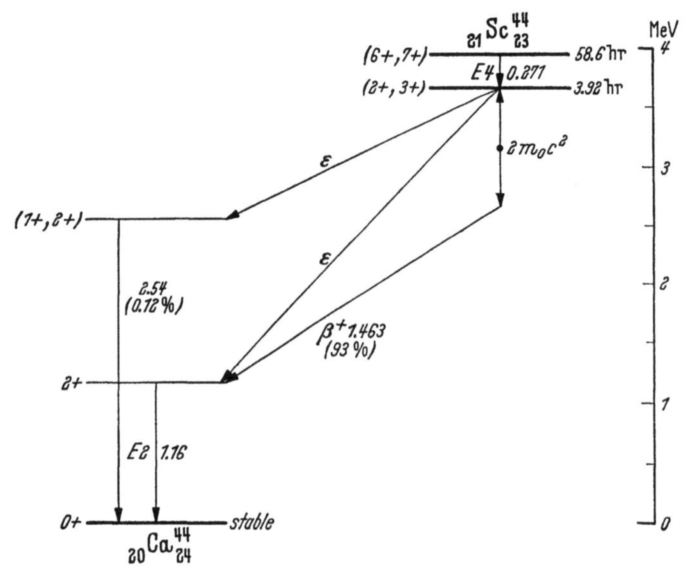

Fig. 10. $A = 44$. Refs. Sc^{44}: [*An 54, Bl 55c, La 54*].

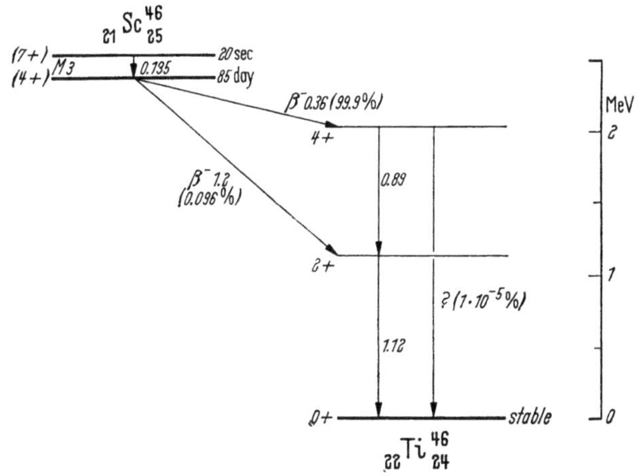

Fig. 11. $A = 46$. Refs. Sc^{46}: [*Az 55a, Ha 55d, Ke 54, Ke 54a, Ko 53, Li 53, Mu 54, St 54, Wh 53, Wh 53a, Yo 53*].

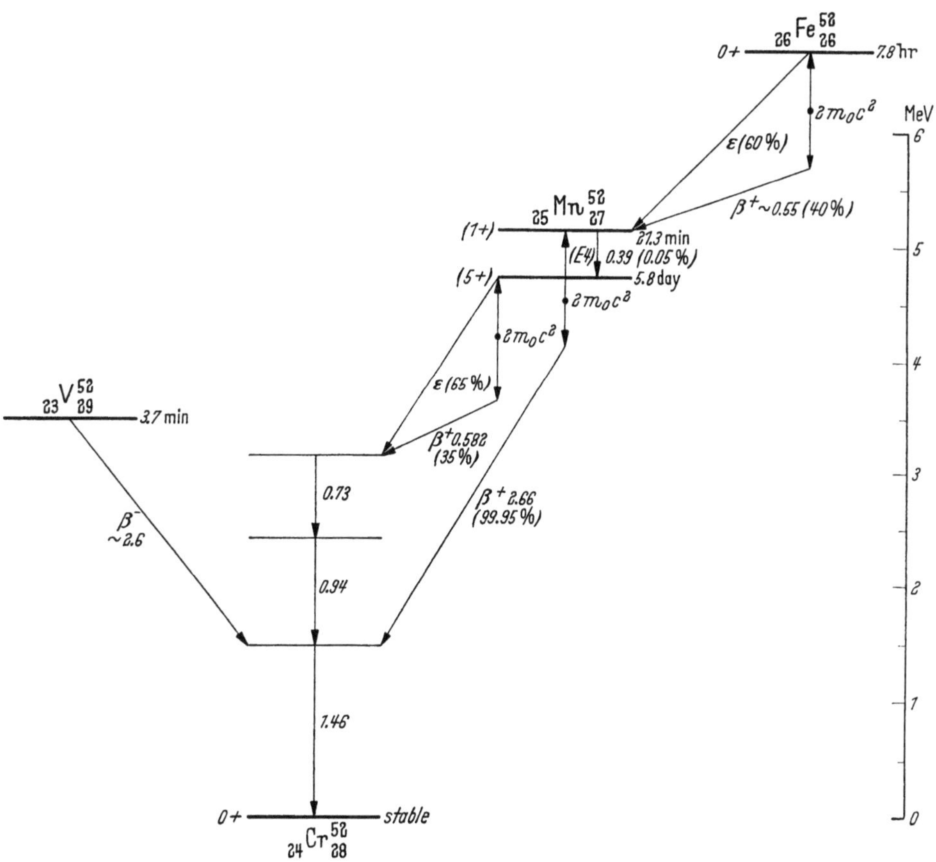

Fig. 12. $A = 52$. Refs. V^{52}: [*Ko 54, Le 54, Sa 53, We 54, Wi 53*]. Mn^{52}: [*Ba 55, Bu 54, Ke 54a*]

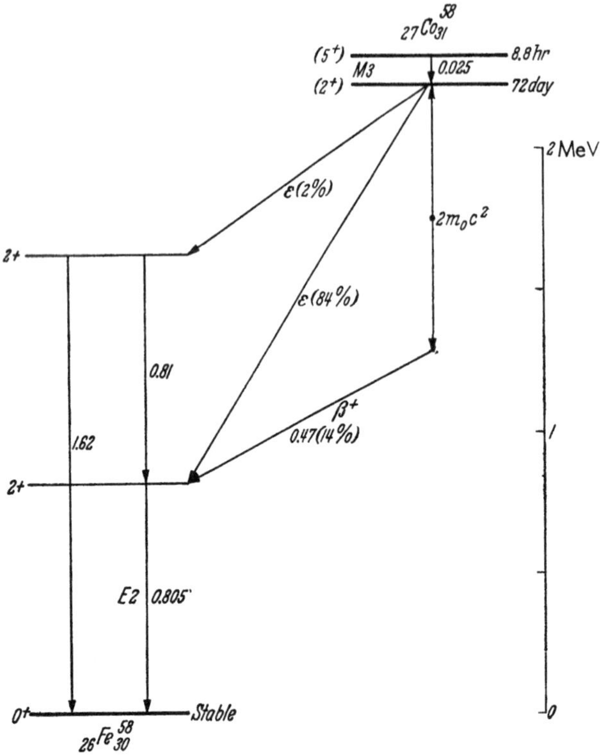

Fig. 13. $A = 58$. Refs. Co^{58}: [*Bi 52, Ch 52, Co 55b, Da 52, Fr 56, Ho 52, Ro 56, Ro 56a Wh 55*].

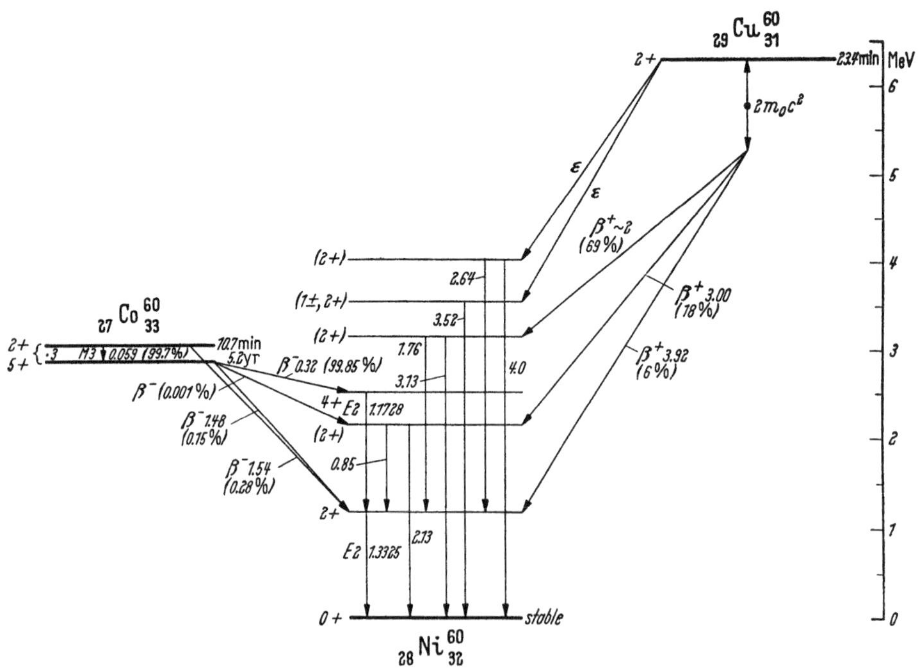

Fig. 14. $A = 60$. Refs. Co^{60}: [*Ba 53b, Ba 55a, Bi 52, Bo 54, Ch 53, Co 55a, En 53, Go 53, Ka 53, Ke 54, Ke 54a, Kl 52, Kl 53, Kr 53a, La 53, Le 54a, Li 53a, Lo 53, Po 55a, Wi 54, Wh 55a, Wo 55a, Yo 53*]. Cu^{60}: [*Nu 54*].

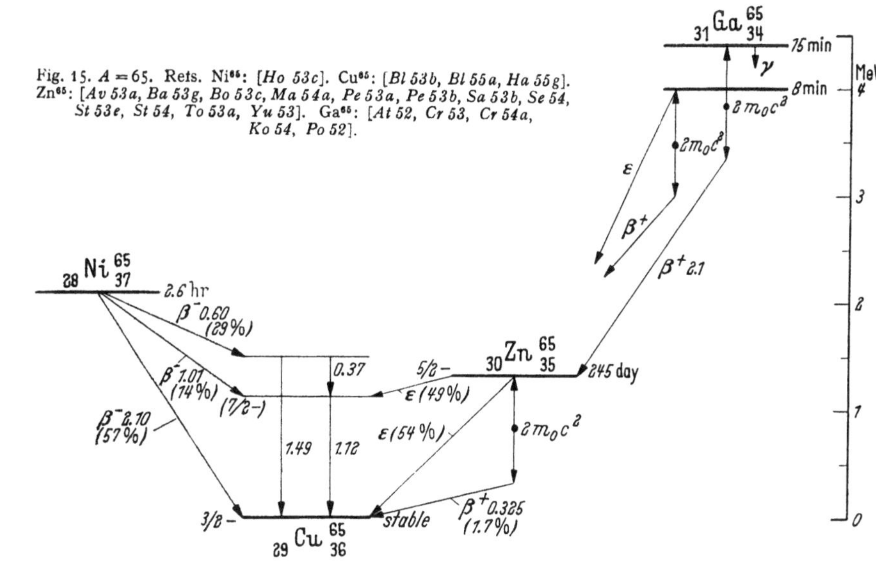

Fig. 15. $A = 65$. Refs. Ni^{65}: [*Ho 53c*]. Cu^{65}: [*Bl 53b, Bl 55a, Ha 55g*]. Zn^{65}: [*Av 53a, Ba 53g, Bo 53c, Ma 54a, Pe 53a, Pe 53b, Sa 53b, Se 54, St 53e, St 54, To 53a, Yu 53*]. Ga^{65}: [*At 52, Cr 53, Cr 54a, Ko 54, Po 52*].

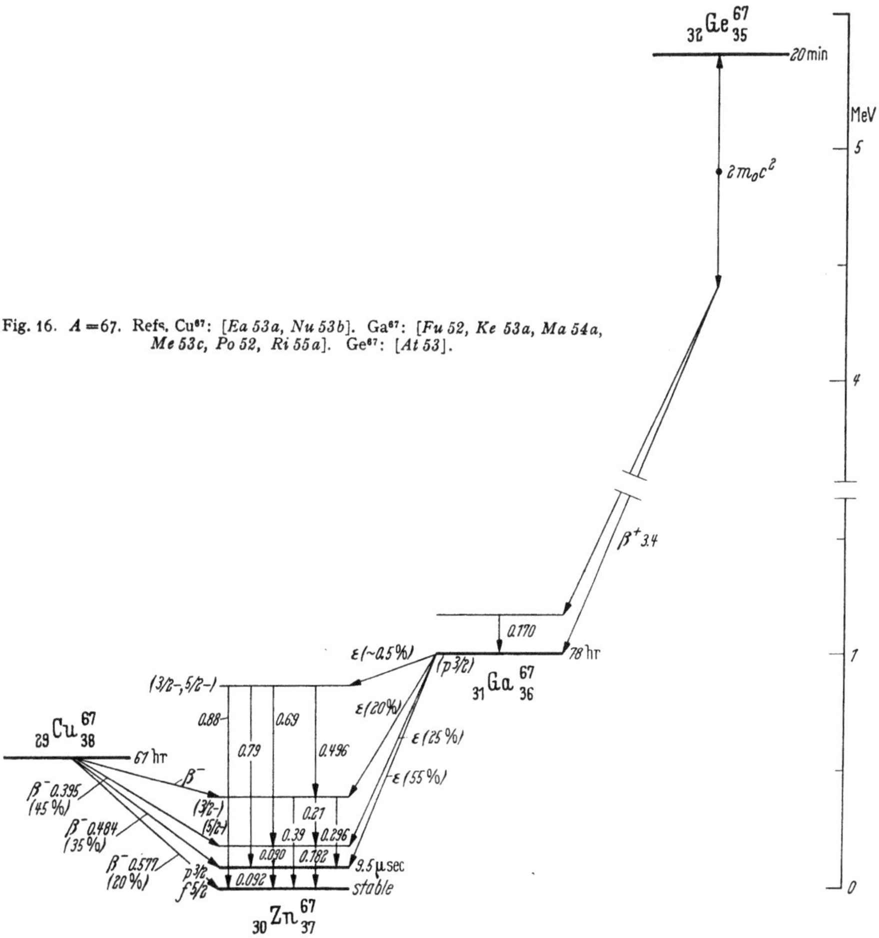

Fig. 16. $A = 67$. Refs. Cu^{67}: [*Ea 53a, Nu 53b*]. Ga^{67}: [*Fu 52, Ke 53a, Ma 54a, Me 53c, Po 52, Ri 55a*]. Ge^{67}: [*At 53*].

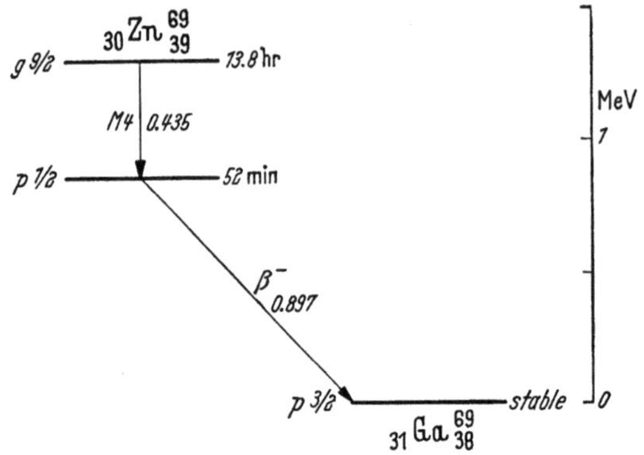

Fig. 17. $A = 69$. Refs. Zn^{69}: [*Do 53, Dr 55, Du 53, Le 55*]. Ga^{69}: [*Da 54, De 53*].

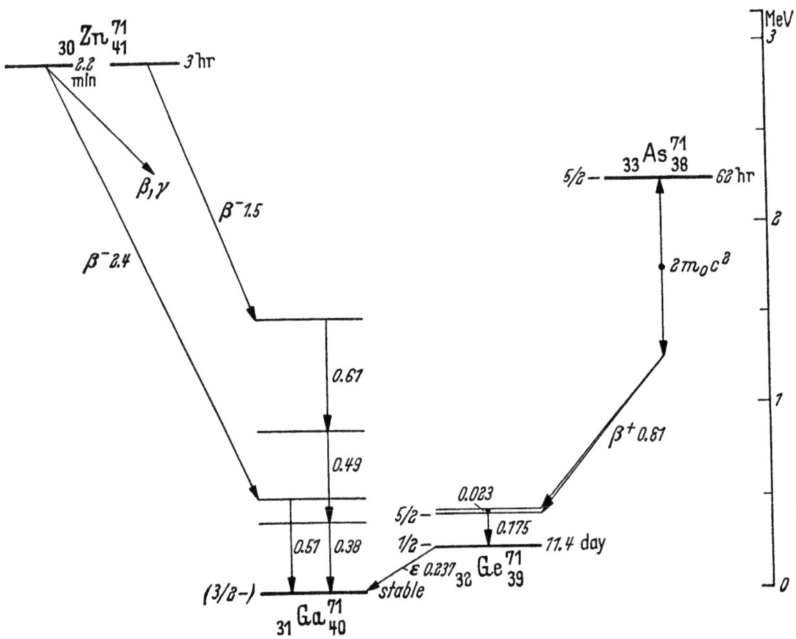

Fig 18. $A = 71$. Refs. Zn^{71}: [*Le 55*]. Ga^{71}: [*Da 54*]. Ge^{71}: [*Bi 55b, La 54g, Sa 53a, Sa 54b*]. As^{71}: [*At 53a, Gr 55a, St 53c, Th 54c*].

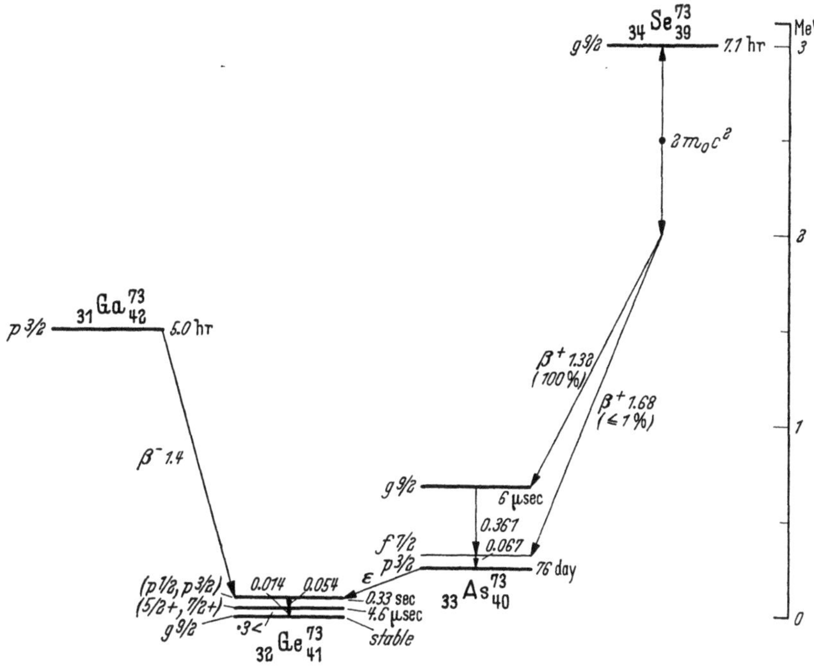

Fig. 19. $A = 73$. Refs. Ga^{73}: [*Ho 53c*]. Ge^{73}: [*Ak 54, Je 53a*]. As^{73}: [*Ba 53h, St 53f, We 53b*]. Se^{73}: [*Ha 55c, Ho 53d*].

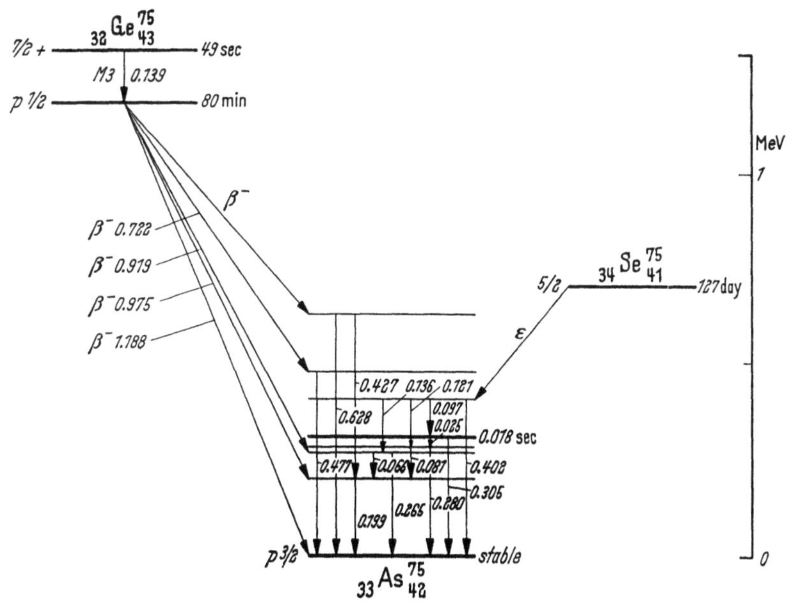

Fig. 20. $A = 75$. Refs. Ge^{75}: [*Ba 53d, Bu 54a, Sc 55, Sm 52*]. As^{75}: [*Je 53b, Sc 56, Ti 53*]. Se^{75}: [*Aa 55, Je 53b Lu 55, Sc [55, Sc 56*].

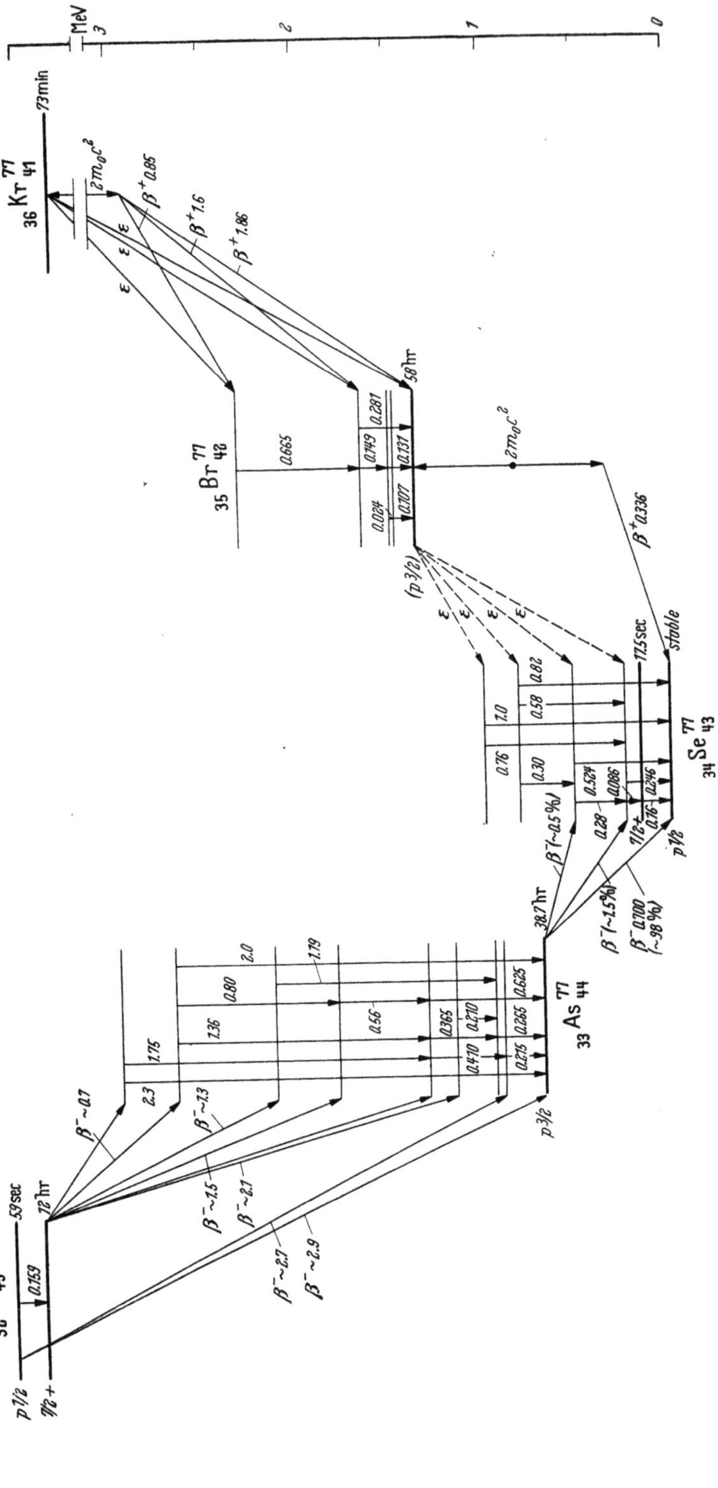

Fig. 21. $A = 77$. Refs. Ge^{77}: [Bu 54a, Sa 53a]. As^{77}: [Bi 55b, Bu 53, Er 54, La 55a, Ra 53, Re 53, Sa 53a, Sc 55a, Su 53]. Se^{77}: [Da 54a, Or 52, Se 54, Wa 53]. Br^{77}: [Bu 53, Se 54]. Kr^{77}: [Th 55a].

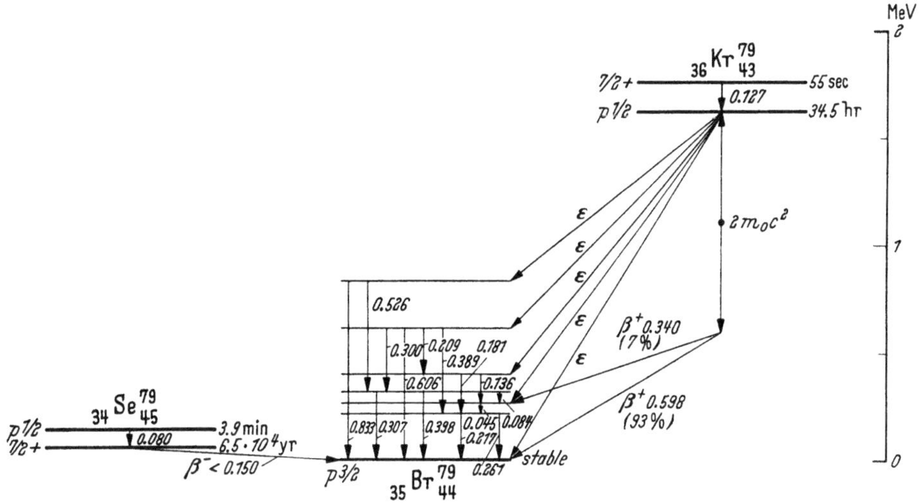

Fig. 22. $A = 79$. Refs. Se^{79}: [*Bi 54, Cu 53, Dr 55, Ha 53, Yt 54*]. Br^{79}: [*Ki 54, Ko 53a, Ma 53*]. Kr^{79}: [*Be 52, La 54a, La 55e, Ra 55a, Th 54e, Th 55a*].

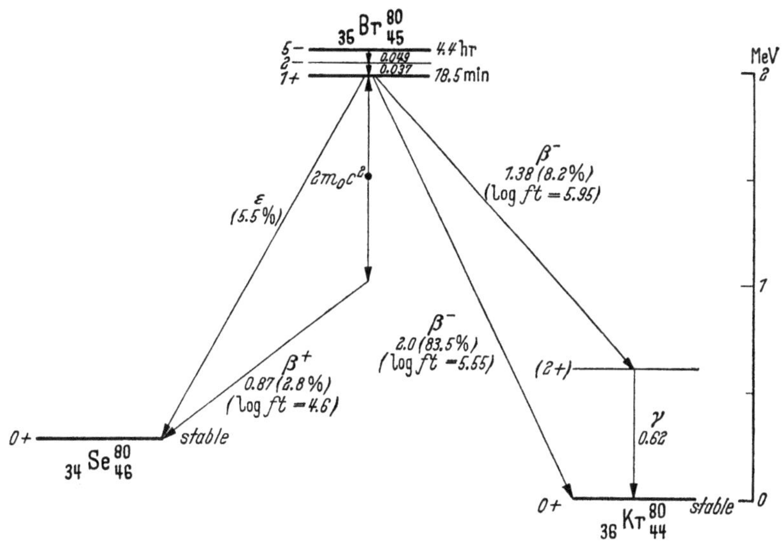

Fig. 23. $A = 80$. Refs. Br^{80}: [*La 54b, La 54c, Li 54, Sc 53, Sc 54, Sh 55*].

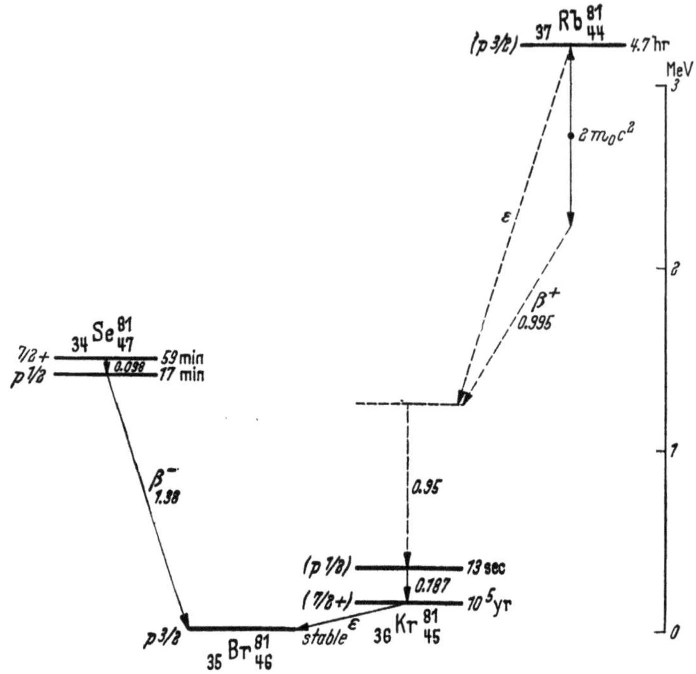

Fig. 24. $A = 81$. Refs. Se^{81}: [*Dr 54, Dr 55, Yt 54*]. Br^{81}: [*Ki 54*]. Rb^{81}: [*Ho 54*].

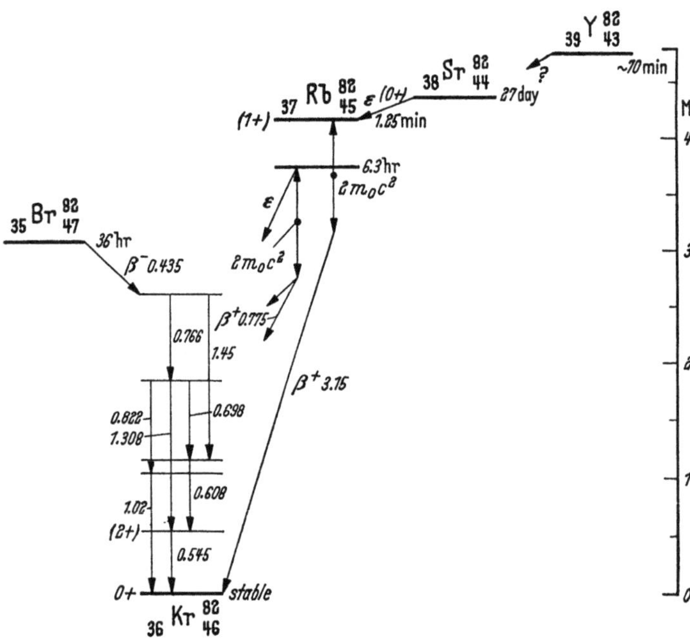

Fig. 25. $A = 82$. Refs. Br^{82}: [*Ba 53d, Be 56, Dz 52, Ho 53c, Lu 54b*]. Rb^{82}: [*Be 56, Hu 52b, Kr 53d, Li 53f*]. Sr^{82}: [*Kr 53d, Li 53f*]. Y^{82}: [*Ca 52*].

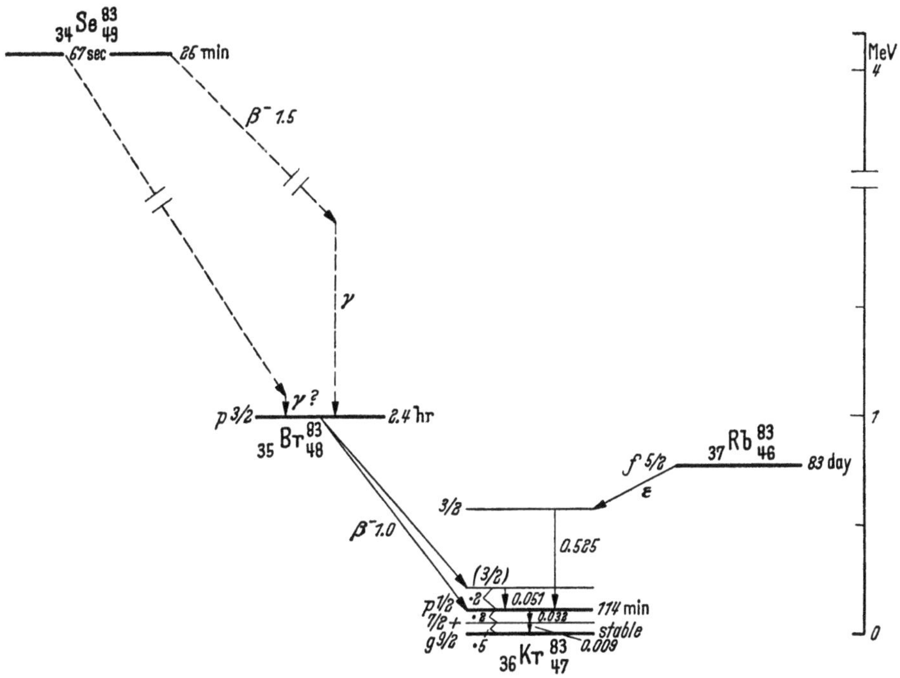

Fig. 26. $A = 83$. Refs Br^{83}: [*Sw 53*]. Kr^{83}: [*Br 54a, De 52*]. Rb^{83}: [*Ca 52a, Pe 55*].

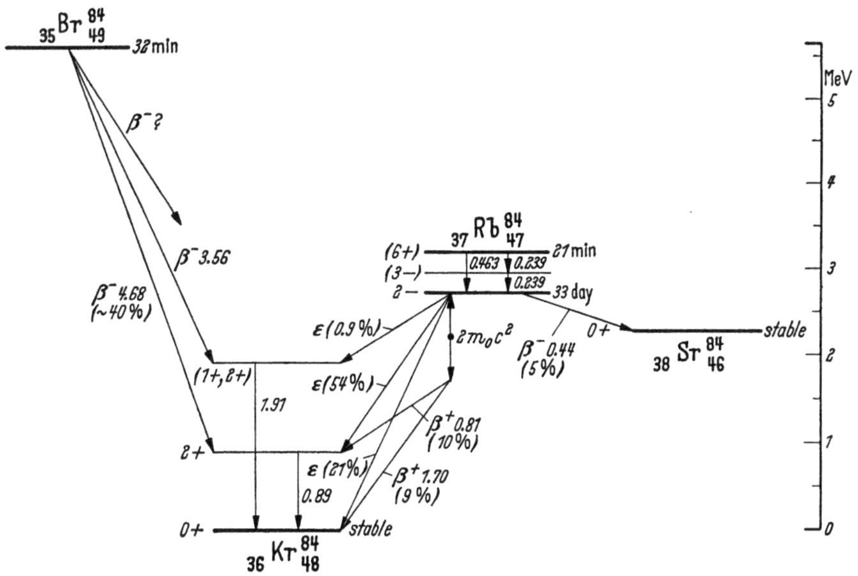

Fig. 27. $A = 84$. Refs. Br^{84}: [*Du 51, La 52*]. Rb^{84}: [*Ca 53a, Hu 52b, We 55*].

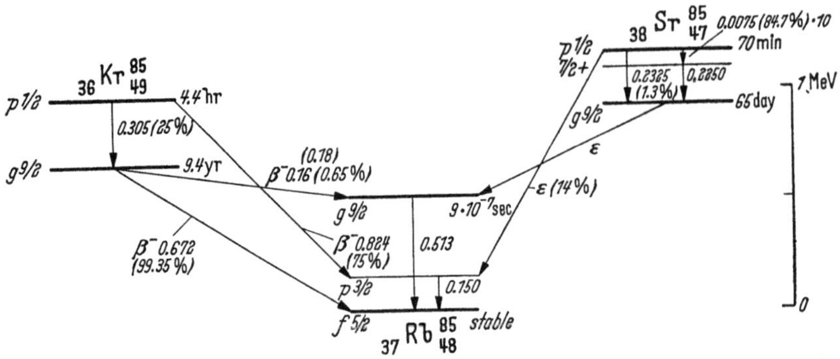

Fig. 28. $A = 85$. Refs. Kr^{85}: [*Be 52, Be 54, Bl 55d, Ra 55, Th 55a, Tu 53, Wa 53a*]. Rb^{85}: [*Kr 55, Se 55*].

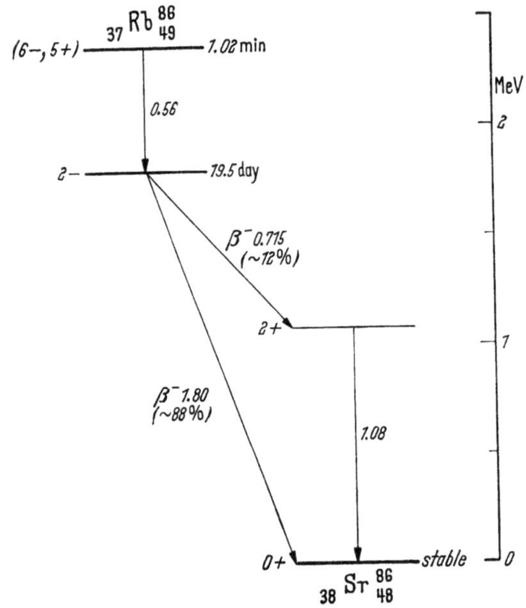

Fig. 29. $A = 86$.
Refs. Rb^{86}: [*Be 53b, Ca 54a, Em 55, Ha 53a, La 55b, Ly 54, Ma 54d, Ni 55, Po 54a, Sc 53d*].

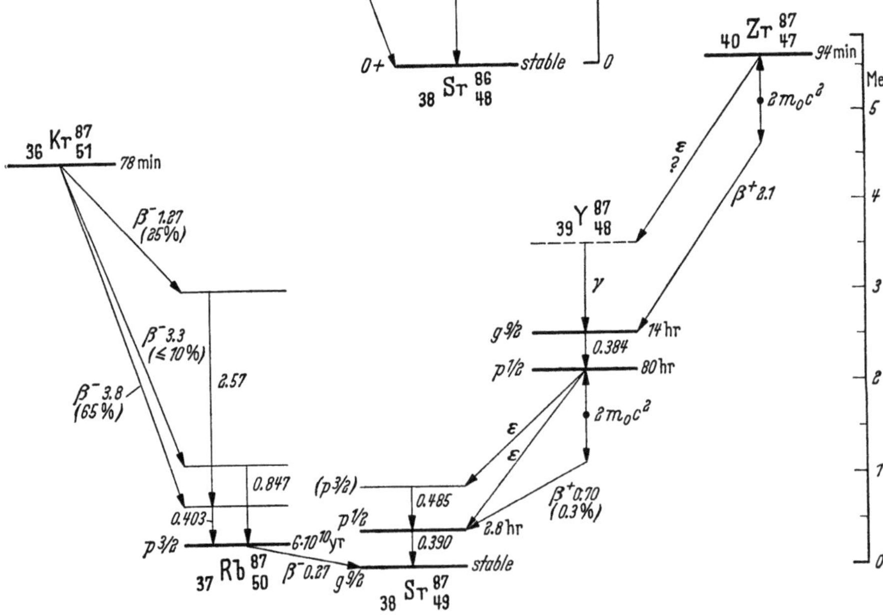

Fig. 30. $A = 87$. Refs. Kr^{87}: [*Th 55a*]. Rb^{87}: [*Ba 52a, Fl 54, Ge 54, Kr 55, Le 52, Ma 54, Tr 54, We 56*]. Sr^{87}: [*An 54a, An 55, Gr 52, Je 53*]. Y^{87}: [*Gr 52*].

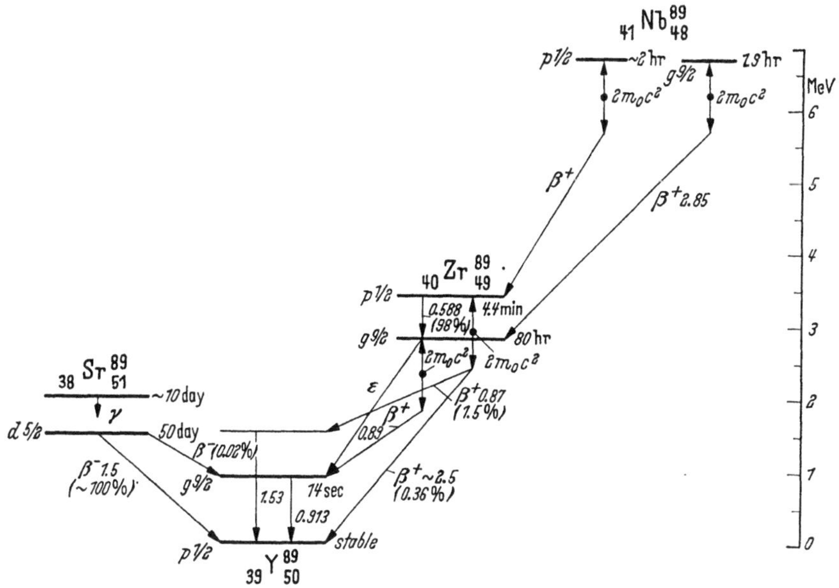

Fig. 31 $A = 89$. Refs. Sr^{89}: [*He 54, He 55, Ly 55*]. Y^{89}: [*Br 54b*]. Zr^{89}: [*Ax 54, Ka 53a, Sh 53*]. Nb^{89}: [*Di 54b, Ma 55*].

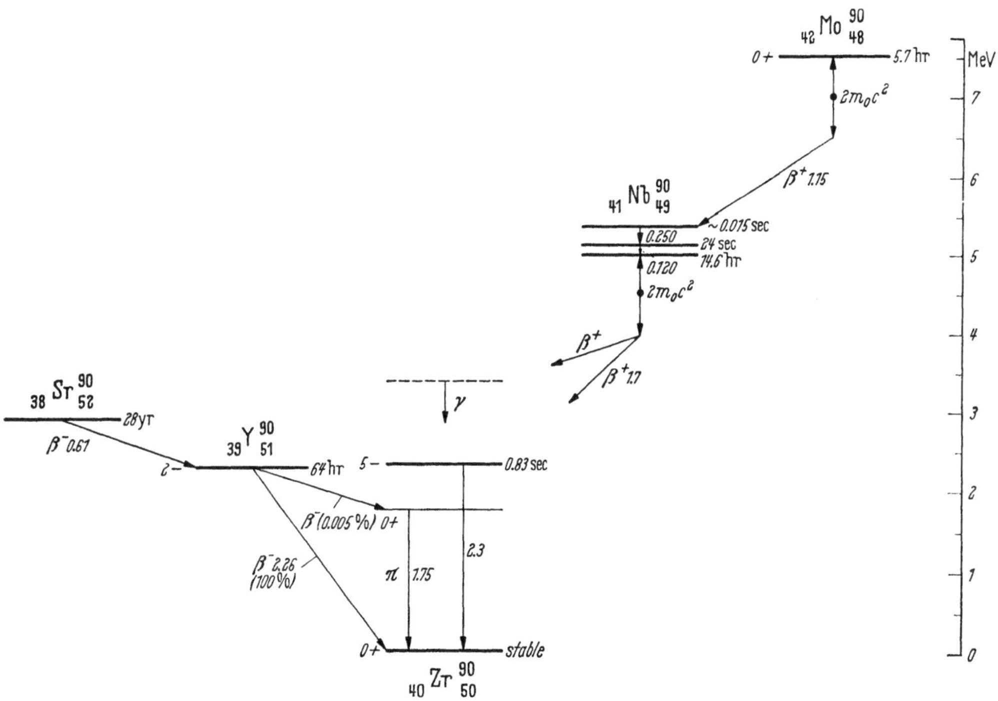

Fig. 32. $A = 90$. Refs. Sr^{90}: [*Ho 53c, Re 55a, Wi 55*]. Y^{90}: [*Ch 54a, Jo 55, Ma 53g, Ma 55h, Sa 55, Sc 55b, Vo 55*]. Zr^{90}: [*Ca 55a, Fo 55b*]. Nb^{90}: [*Di 53, Ho 54b, Ma 55i*]. Mo^{90}: [*Di 53, Ma 55i*].

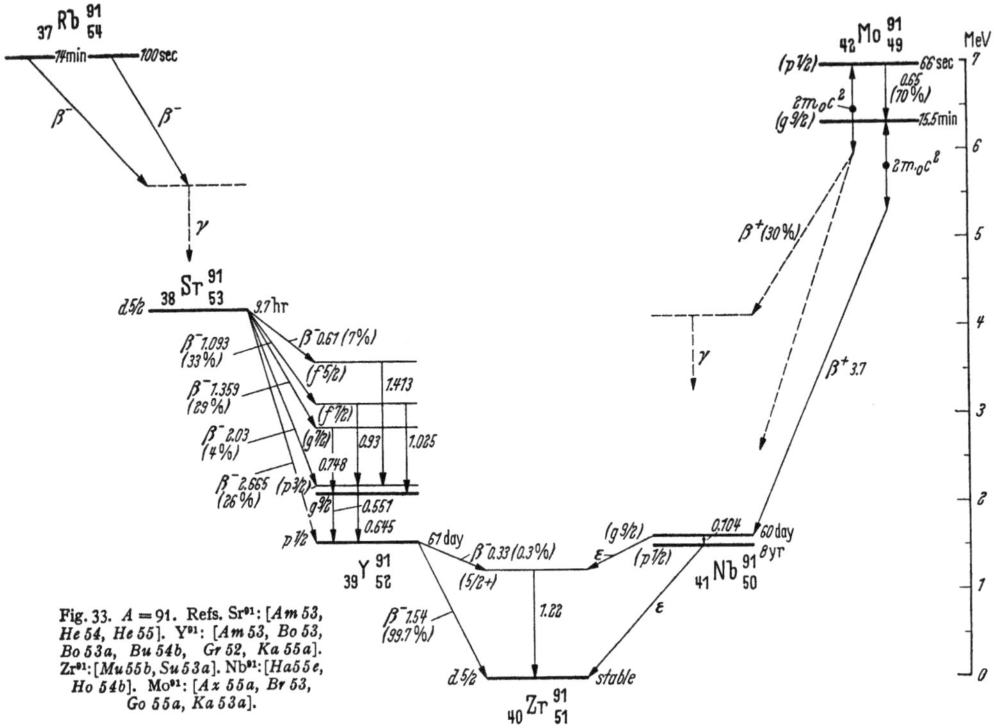

Fig. 33. $A = 91$. Refs. Sr^{91}: [*Am 53, He 54, He 55*]. Y^{91}: [*Am 53, Bo 53, Bo 53a, Bu 54b, Gr 52, Ka 55a*]. Zr^{91}: [*Mu 55b, Su 53a*]. Nb^{91}: [*Ha 55e, Ho 54b*]. Mo^{91}: [*Ax 55a, Br 53, Go 55a, Ka 53a*].

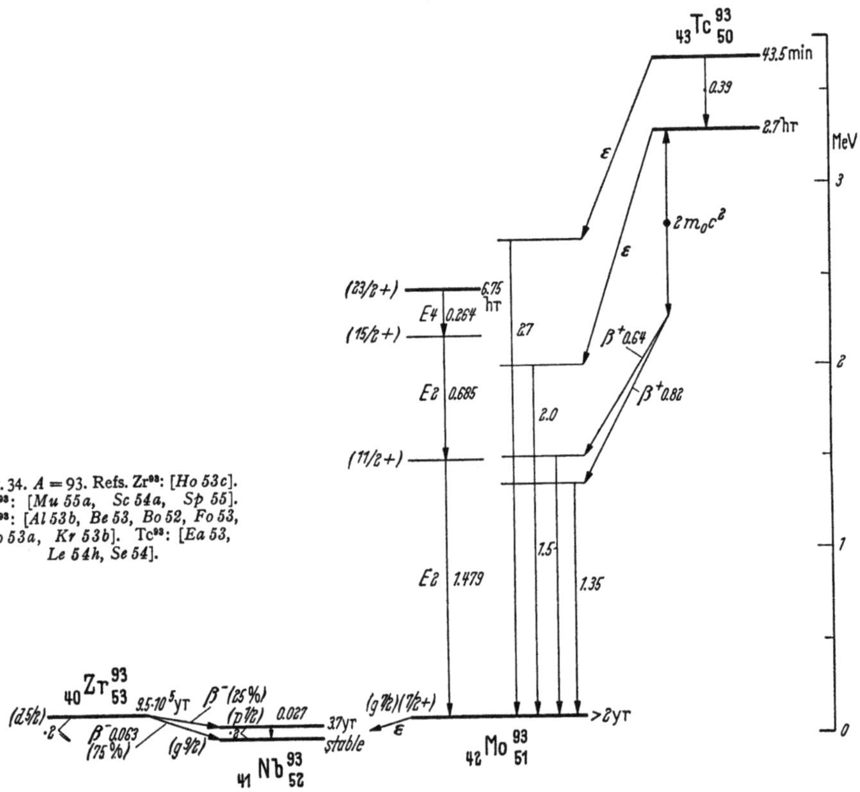

Fig. 34. $A = 93$. Refs. Zr^{93}: [*Ho 53c*]. Nb^{93}: [*Mu 55a, Sc 54a, Sp 55*]. Mo^{93}: [*Al 53b, Be 53, Bo 52, Fo 53, Go 53a, Kr 53b*]. Tc^{93}: [*Ea 53, Le 54h, Se 54*].

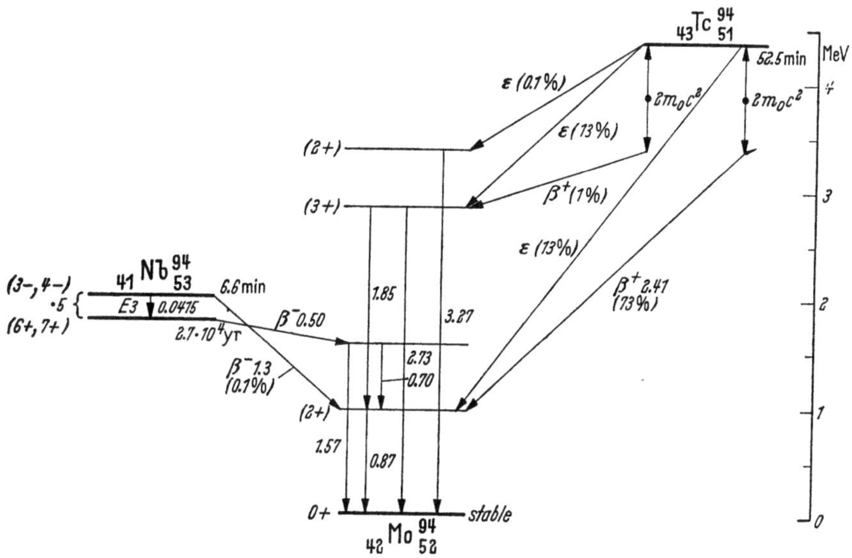

Fig. 35. $A = 94$. Refs. Nb^{94}: [*Do 53a, Ro 54, Ro 55a*].

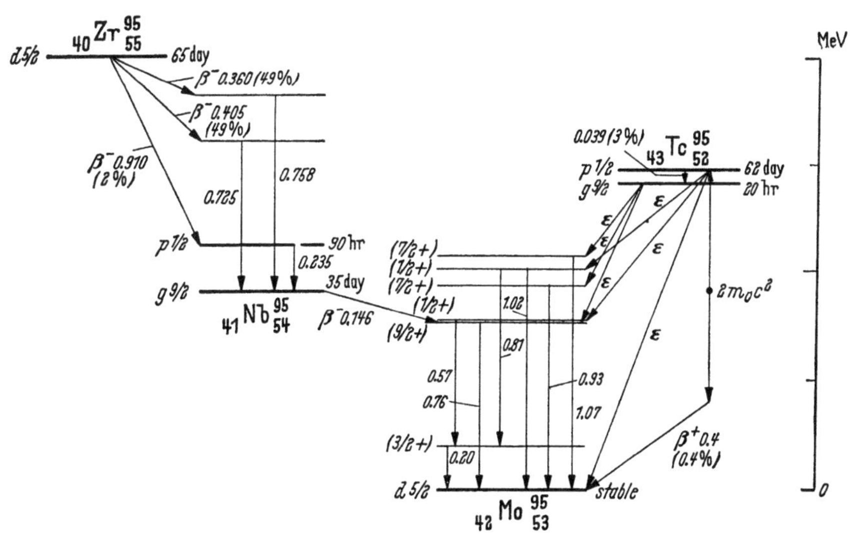

Fig. 36. $A = 95$. Refs. Zr^{95}: [*Co 53, Ma 53a, Mi 54, Ro 55, Sl 53*]. Nb^{95}: [*Co 53, Do 53, Dr 55, Ho 54b, Ma 53a, Ma 53b Mi 54, Sh 52, Sl 53, St 54*]. Mo^{95}: [*Mu 55b, Ow 56 Wo 54*].

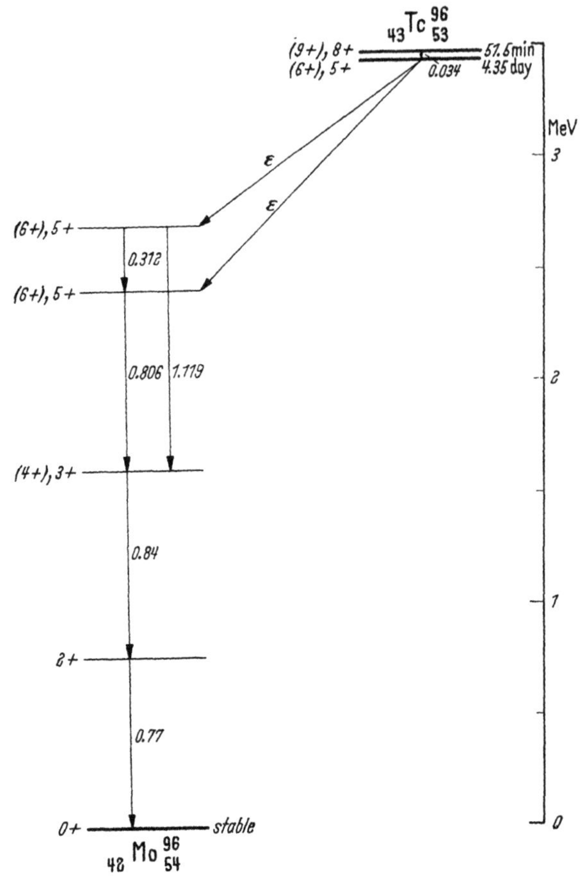

Fig. 37. $A = 96$. Refs. Tc^{96}: [*Ea 53*].

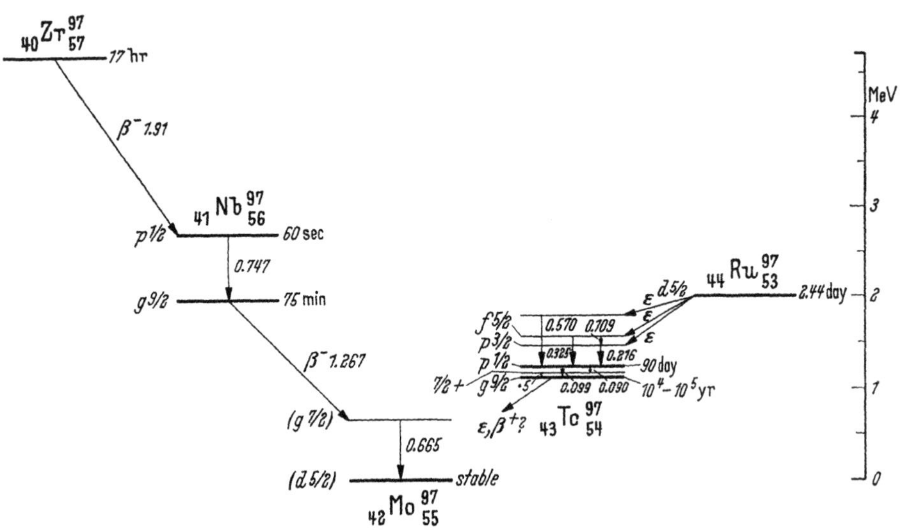

Fig. 38. $A = 97$. Refs. Zr^{97}: [*Sa 55*]. Nb^{97}: [*Me 53, Sa 55*]. Mo^{97}: [*Mu 55b, Ow 56, Wo 54*]. Tc^{97}: [*Bo 54*]. Ru^{97}: [*Co 55c*].

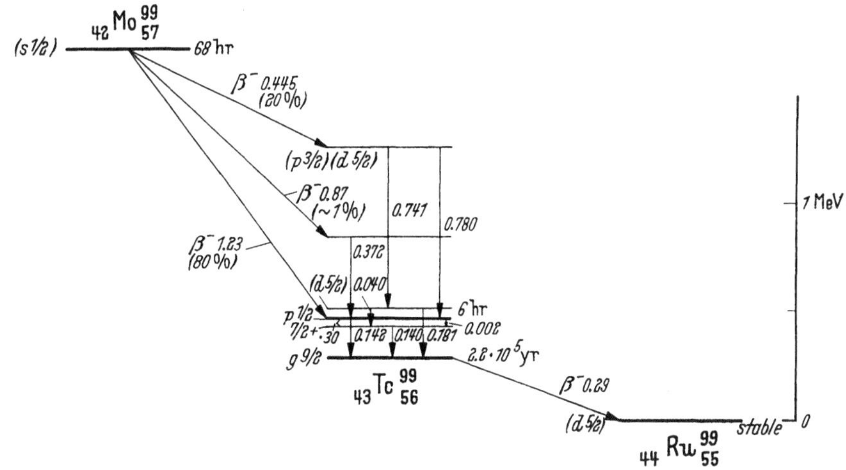

Fig. 39. $A = 99$. Refs. Mo^{99}: [*Ca 54*, *Le 54b*, *Le 55b*, *Va 54*]. Tc^{99}: [*Ba 53*, *Ke 53*]. Ru^{99}: [*Gr 52a*, *Mu 53*].

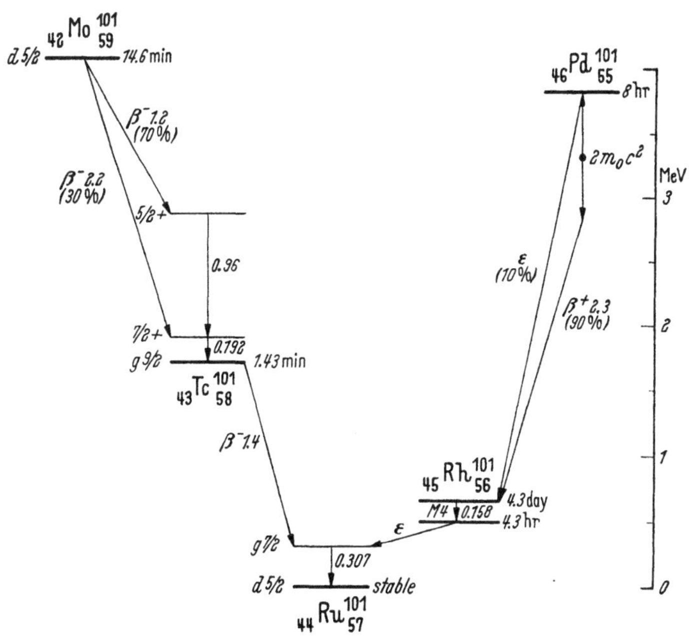

Fig. 40 $A = 101$. Refs. Mo^{101}: [*Wi 54b*]. Tc^{101}: [*Wi 54b*]. Ru^{101}: [*Gr 52a*, *Mu 53*]. Rh^{101}: [*Fa 55*, *Fu 52*]. Pd^{101}: [*Ho 53c*].

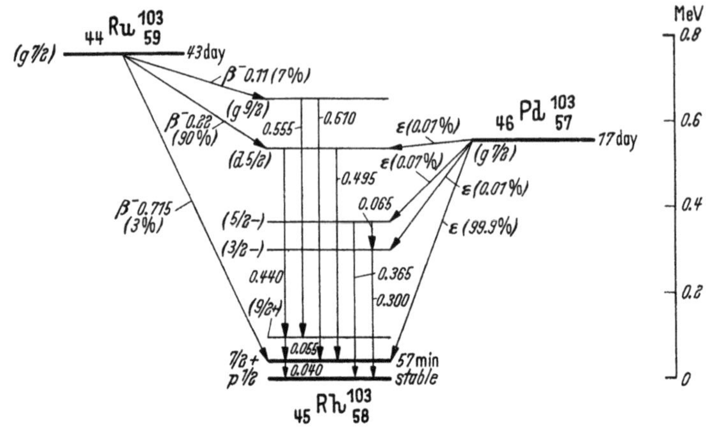

Fig. 41. $A = 103$. Refs. Ru^{103}: [*En 53, Fo 55a, Ra 54, Sa 55a*]. Rh^{103}: [*Av 55, So 55*]. Pd^{103}: [*Av 53, Av 55, Ri 54, Sa 55a*].

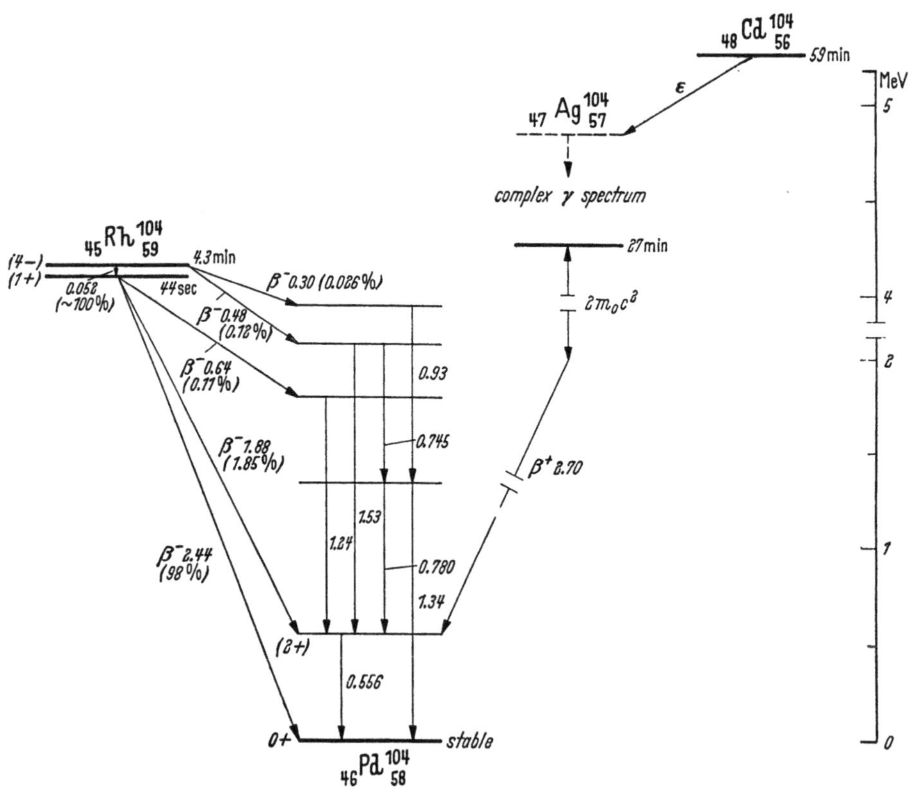

Fig. 42. $A = 104$. Refs. Rh^{104}: [*Bu 55, Ge 53, Jo 53*].

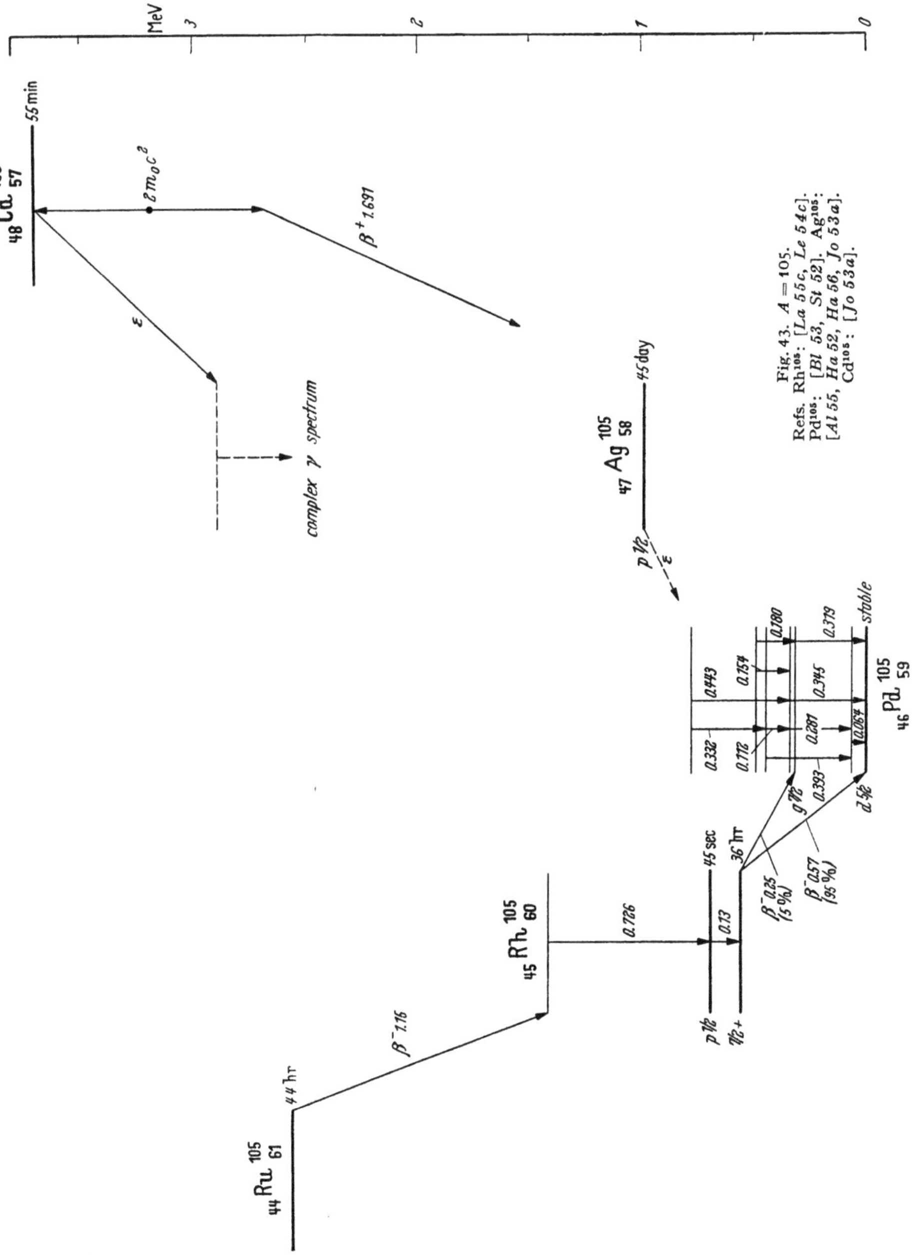

Fig. 43. $A = 105$.
Refs. Rh^{105}: [*La* 55*c*, *Le* 54*c*]. Pd^{105}: [*Bl* 53, *St* 52]. Ag^{105}: [*Al* 55, *Ha* 52, *Ha* 56, *Jo* 53*a*]. Cd^{105}: [*Jo* 53*a*].

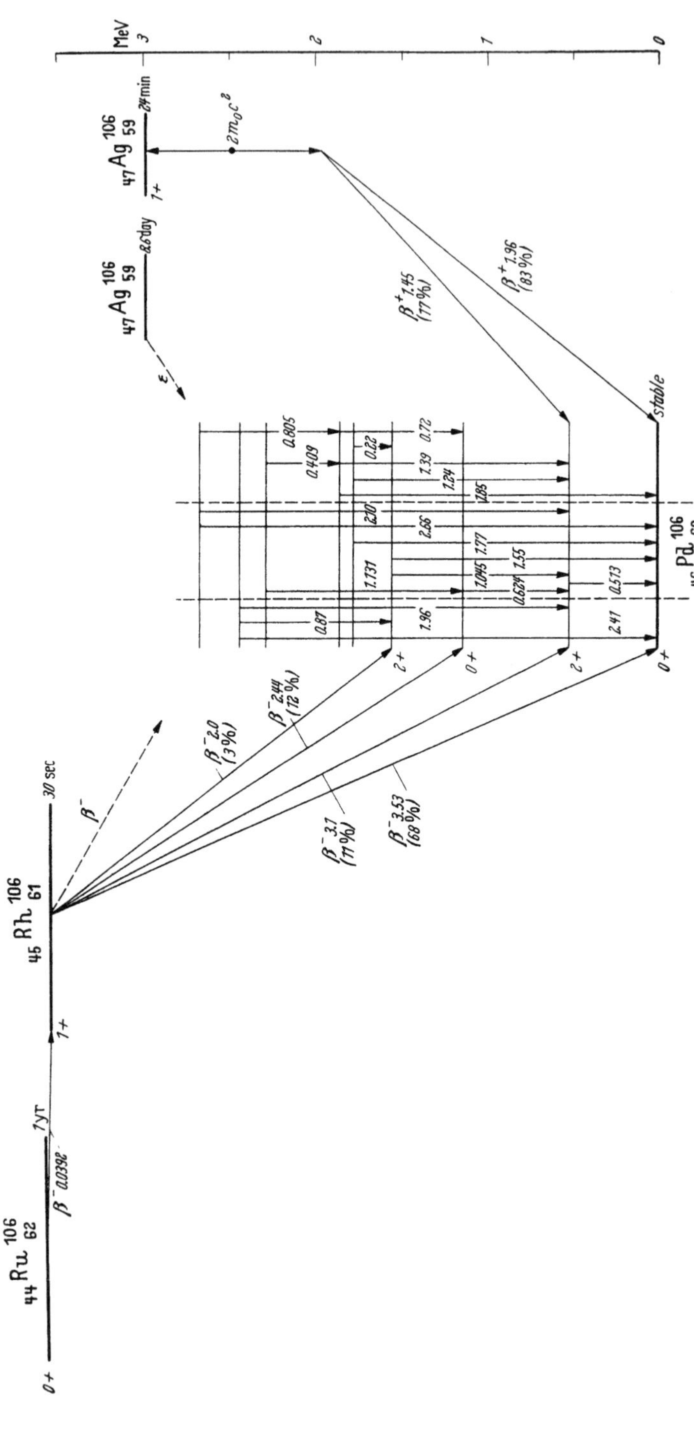

FIG. 44. $A = 106$. Transitions between dashed lines are common to both Rh^{106} and 8.6 day Ag^{106}. Refs. Rh^{106}: [*Al 52, Al 55, Ba 55b, Ka 53b, Kl 53a, Kr 53a, Lu 55, Wi 54a*]. Ag^{106}: [*Al 55, Be 53a, Ha 52*].

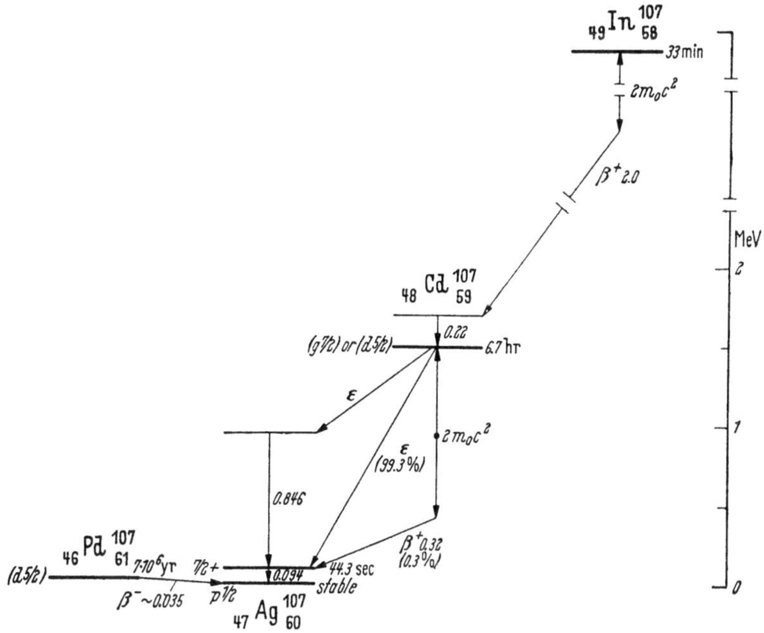

Fig. 45. $A = 107$. Refs. Ag^{107}: [*Br 53a, Br 54b, Jo 53a, So 54, We 53*]. Cd^{107}: [*Ma 54a*]. In^{107}: [*Ca 55*].

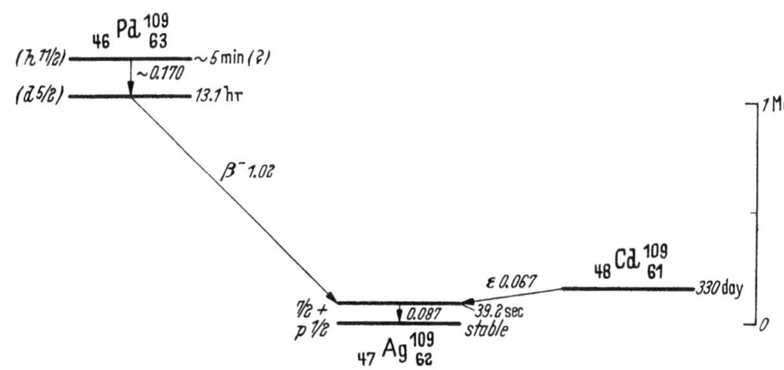

Fig. 46. $A = 109$. Refs. Pd^{109}: [*Me 53a, Mo 54a*]. Ag^{109}: [*Av 53, Br 53a, Br 54b, Jo 53a, Mo 54a, Nu 53, So 54, Wa 54, We 53*]. Cd^{109}: [*Be 54a, De 53a*].

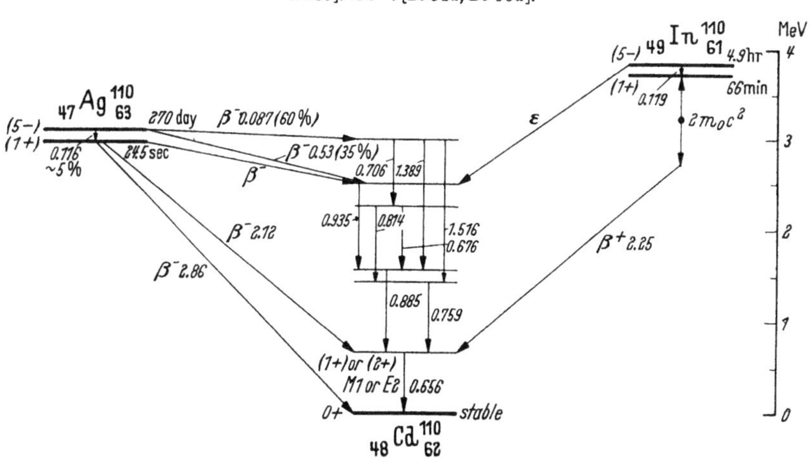

Fig. 47. $A = 110$. Refs. Ag^{110}: [*Az 54, Bo 54b, Ja 54, Jo 52, Ro 55*]. In^{110}: [*Bl 53a*].

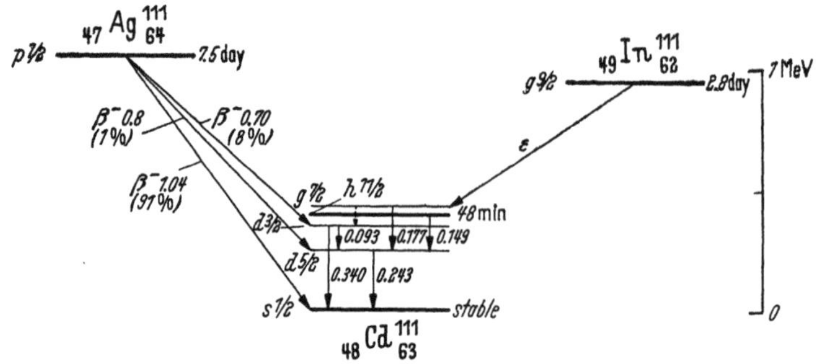

Fig. 48. $A = 111$. Refs. Ag^{111}: [*Br 53b, Le 54d, Wo 55*]. Cd^{111}: [*Al 53c, Al 54a, He 55e, Kr 53c, Le 55c, Ri 55b, St 55*]. In^{111}: [*Gi 54, Gr 52, No 55, St 53a, St 55a*].

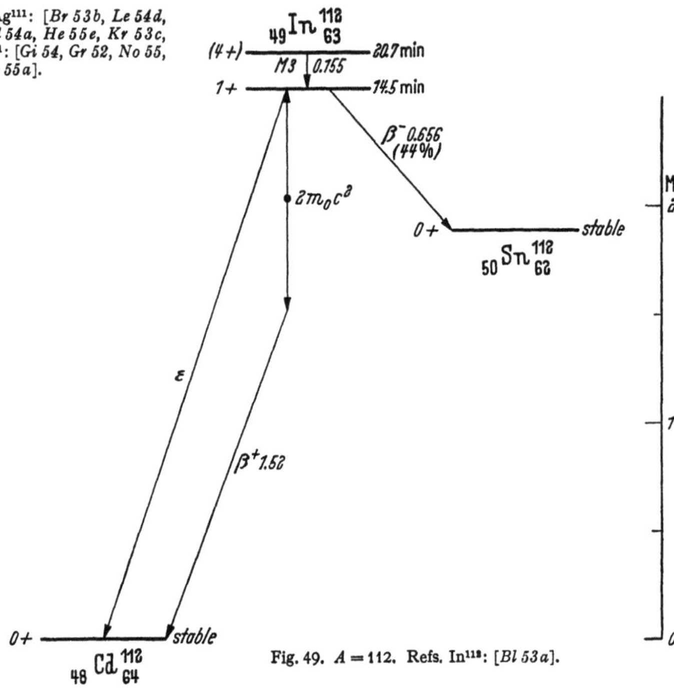

Fig. 49. $A = 112$. Refs. In^{112}: [*Bl 53a*].

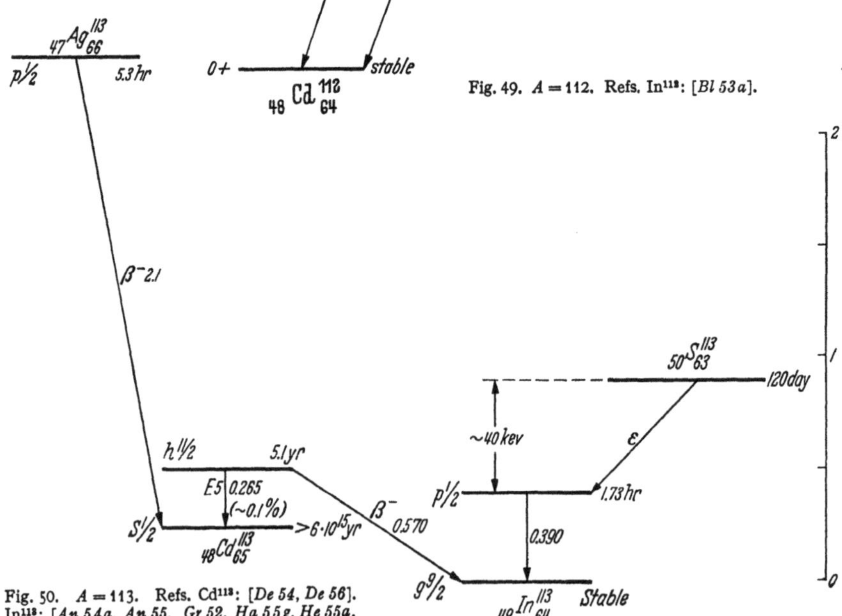

Fig. 50. $A = 113$. Refs. Cd^{113}: [*De 54, De 56*]. In^{113}: [*An 54a, An 55, Gr 52, Ha 55g, He 55a, Ti 53*]. Sn^{113}: [*Br 53c, De 53b*].

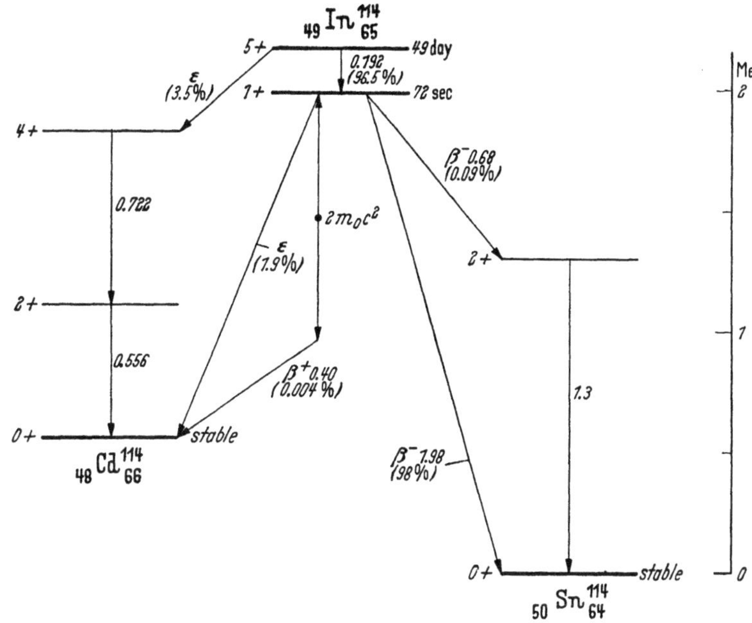

Fig. 51. $A = 114$. Refs. In^{114}: [*Al 53d, Br 55b, Da 55, Gr 52, Gr 55c, Jo 54, Jo 56a, Lu 54, Ma 54a, Nu 53a, St 52a, St 55e, St 56*].

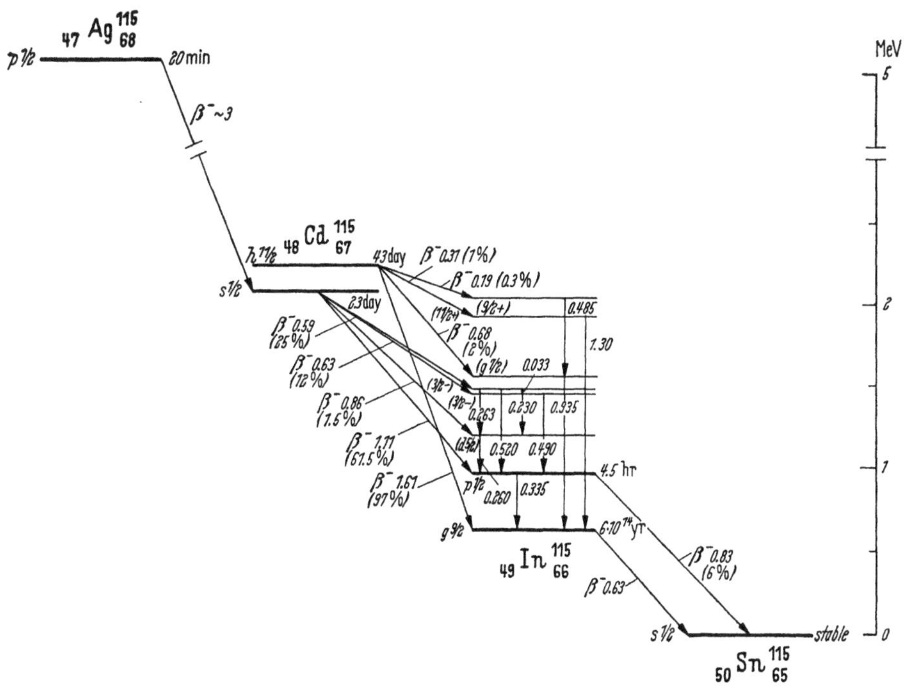

Fig. 52. $A = 115$. Refs. Cd^{115}: [*Va 55*]. In^{115}: [*An 54a, An 55, Gr 52, Ti 53*].

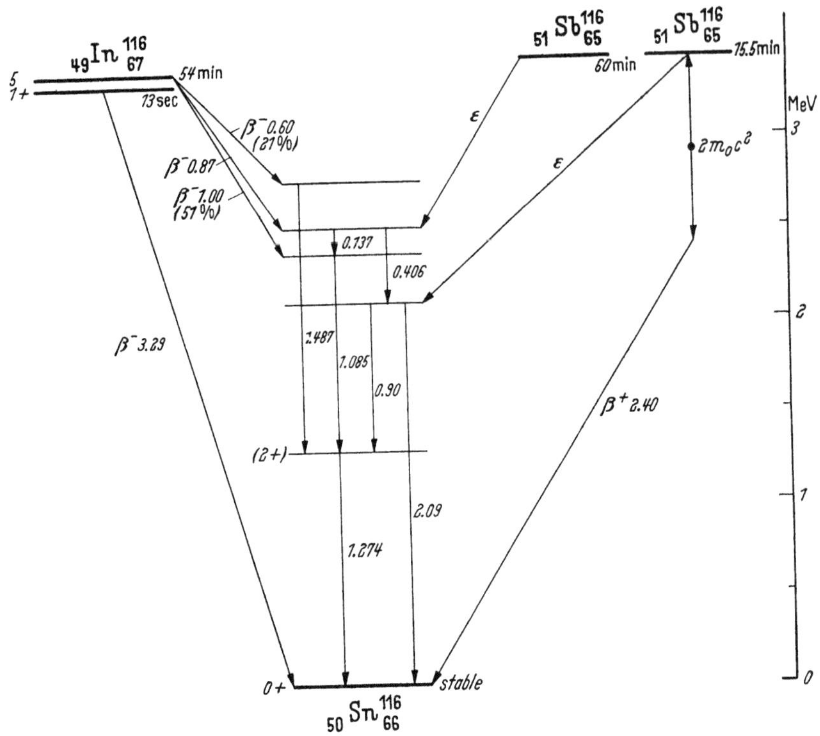

Fig. 53. $A = 116$. Refs. In^{116}: [*Bo 54b, Do 52, Go 55b, Lo 53, Wi 53a*]. Sb^{116}: [*At 54, Bl 50, St 53b*].

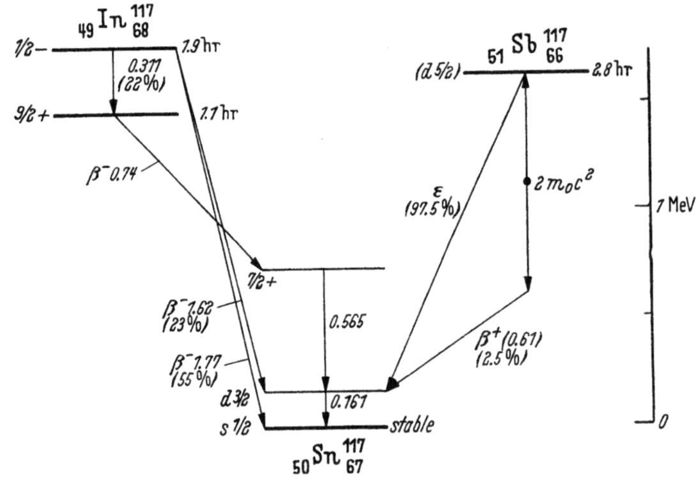

Fig. 54. $A = 117$. Refs. In^{117}: [*Co 53a, Le 54e, Mc 55*]. Sn^{117}: [*Go 54*]. Sb^{117}: [*Mc 55*].

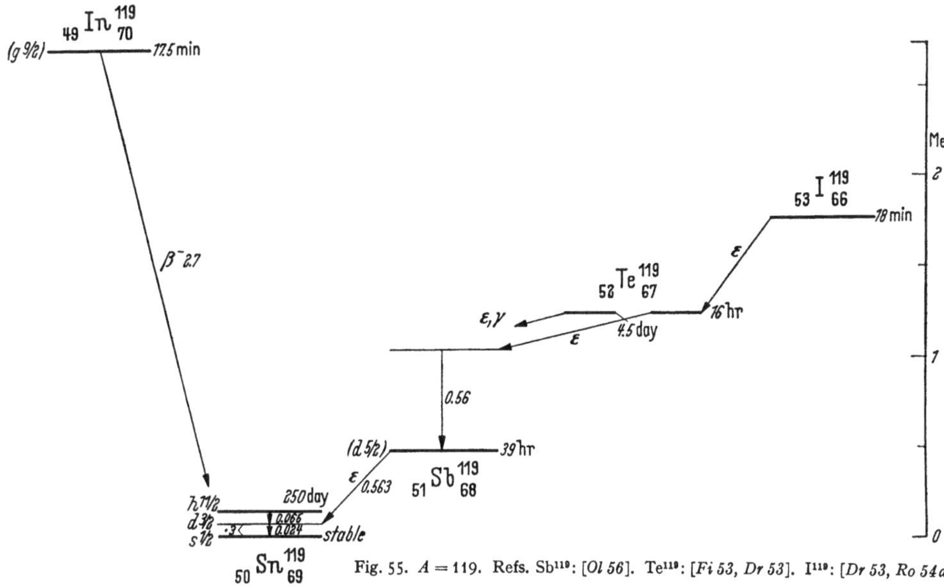

Fig. 55. $A = 119$. Refs. Sb[119]: [*Ol 56*]. Te[119]: [*Fi 53, Dr 53*]. I[119]: [*Dr 53, Ro 54a*].

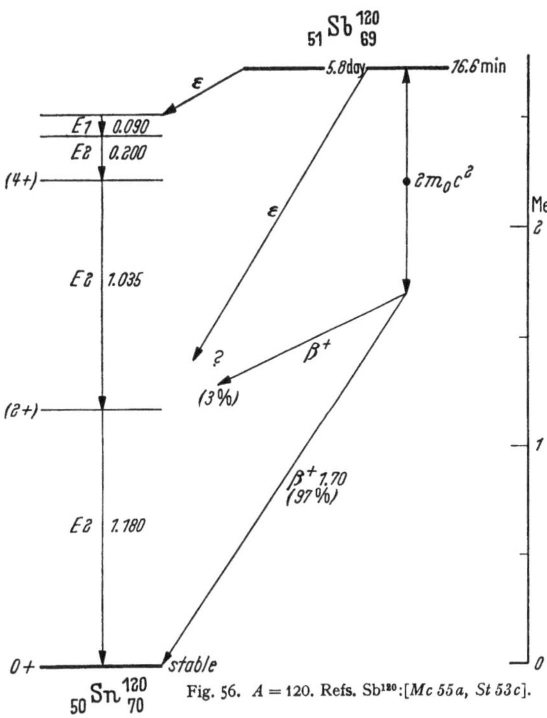

Fig. 56. $A = 120$. Refs. Sb[120]: [*Mc 55a, St 53c*].

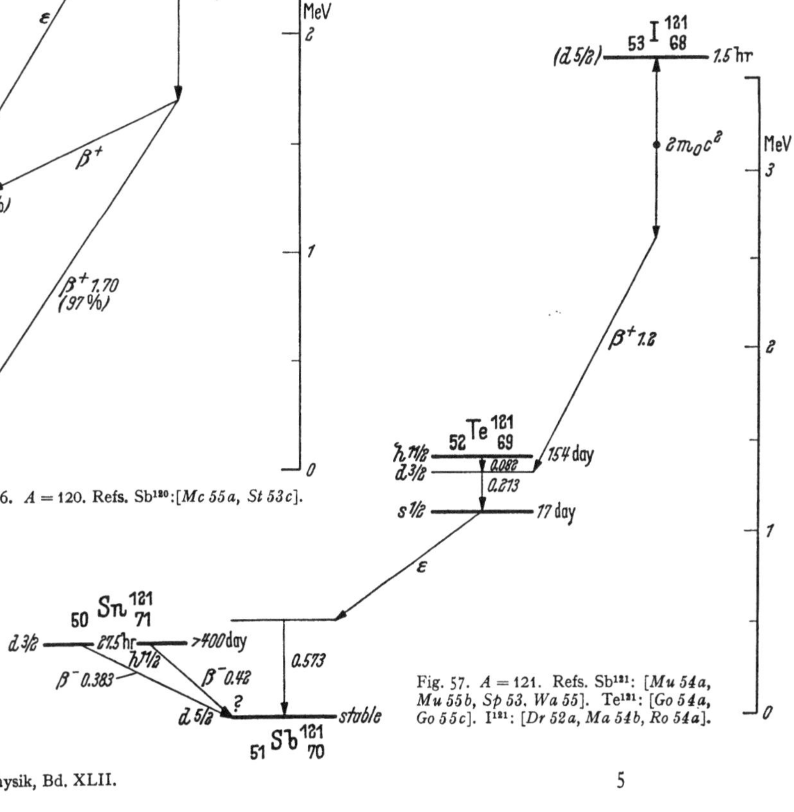

Fig. 57. $A = 121$. Refs. Sb[121]: [*Mu 54a, Mu 55b, Sp 53, Wa 55*]. Te[121]: [*Go 54a, Go 55c*]. I[121]: [*Dr 52a, Ma 54b, Ro 54a*].

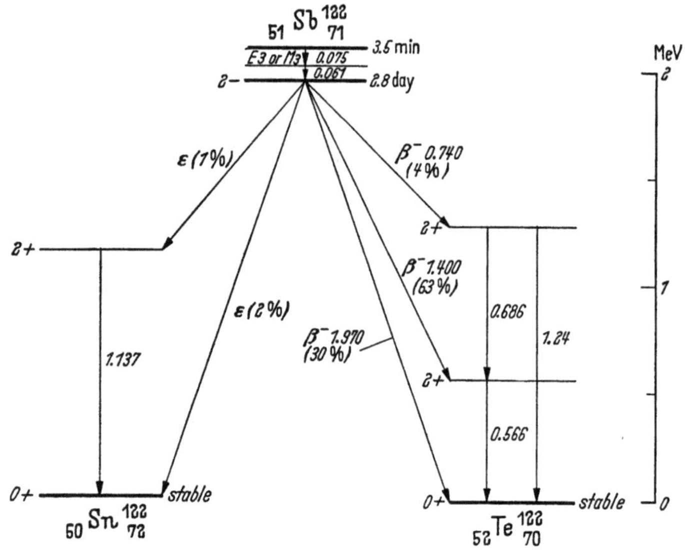

Fig. 58. $A = 122$. Refs. Sb[122]: [*Co 54, Fa 55a, Gl 55, Le 55d, Mo 54b, Ro 55*].

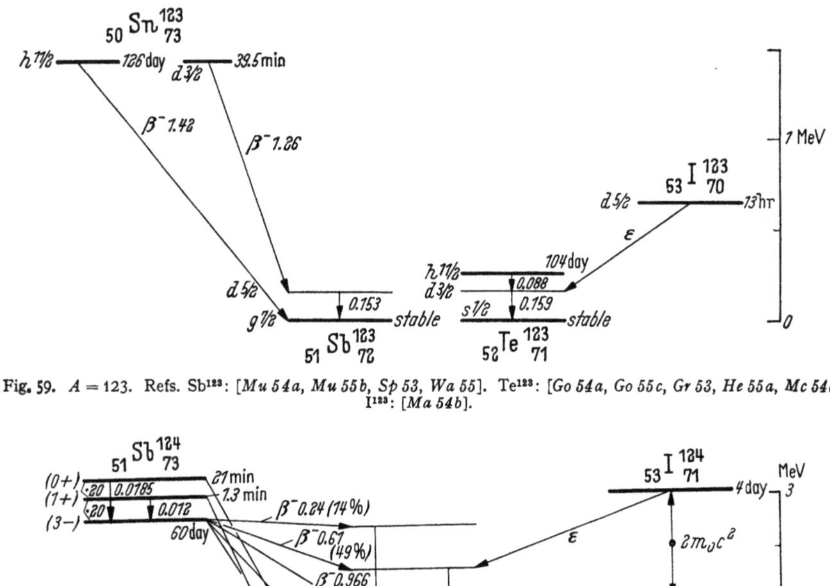

Fig. 59. $A = 123$. Refs. Sb[123]: [*Mu 54a, Mu 55b, Sp 53, Wa 55*]. Te[123]: [*Go 54a, Go 55c, Gr 53, He 55a, Mc 54a*]. I[123]: [*Ma 54b*].

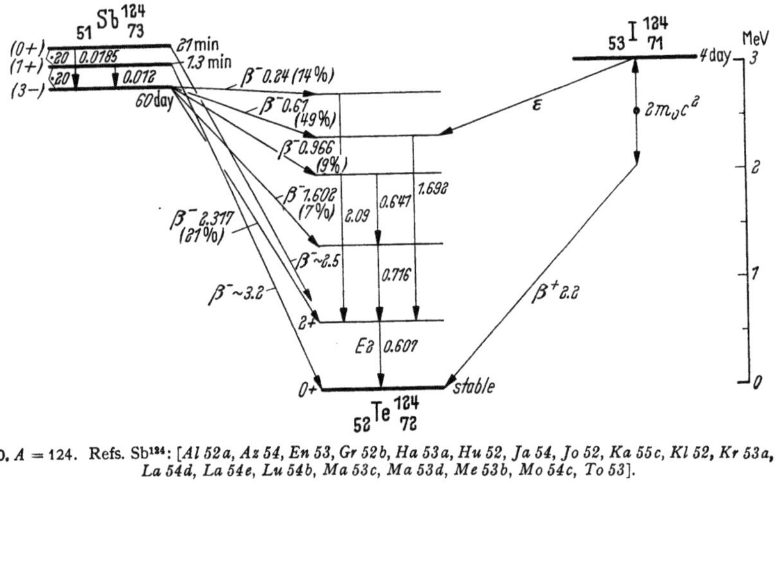

Fig. 60. $A = 124$. Refs. Sb[124]: [*Al 52a, Az 54, En 53, Gr 52b, Ha 53a, Hu 52, Ja 54, Jo 52, Ka 55c, Kl 52, Kr 53a, La 53a, La 54d, La 54e, Lu 54b, Ma 53c, Ma 53d, Me 53b, Mo 54c, To 53*].

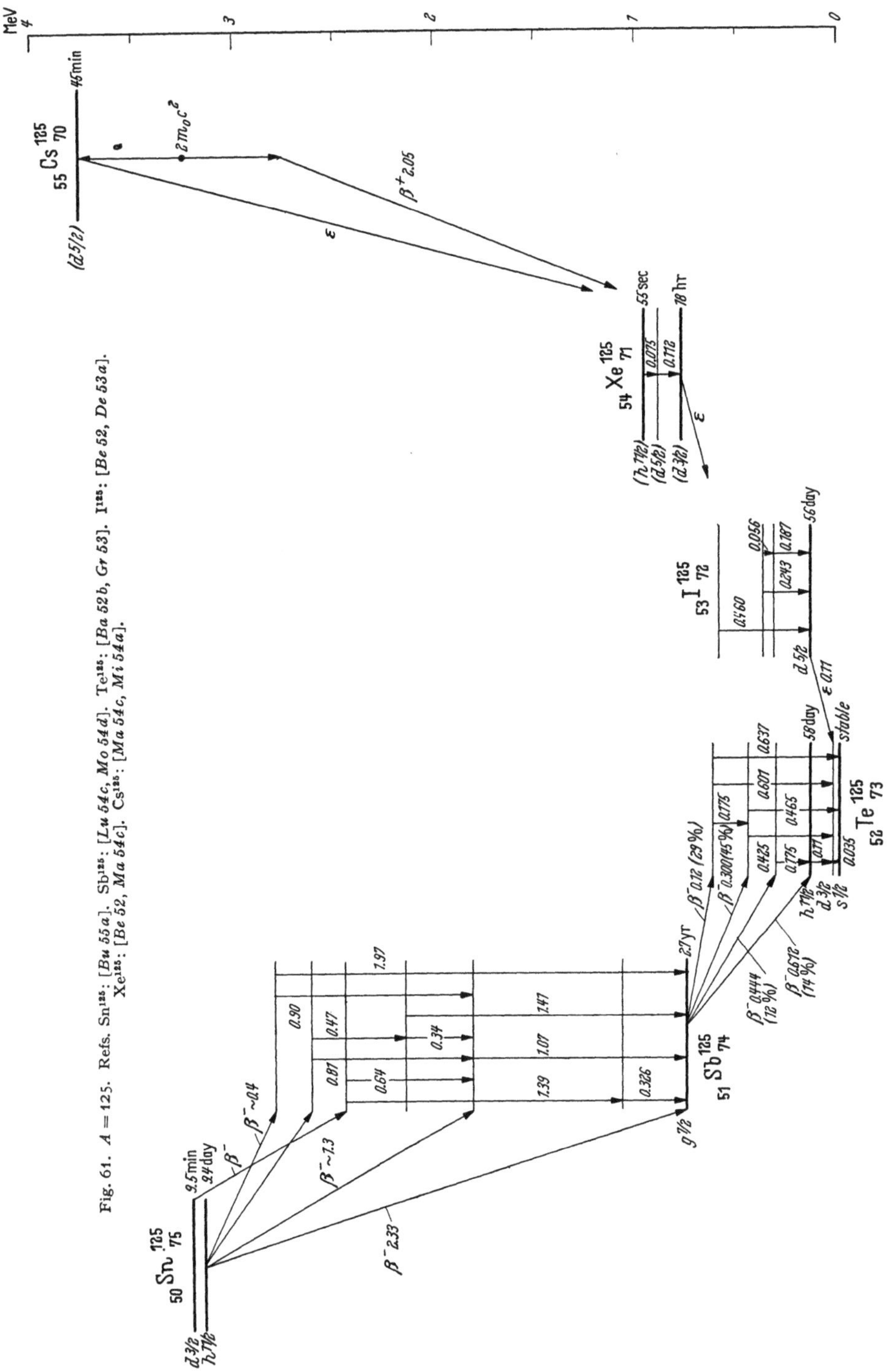

Fig. 61. $A = 125$. Refs. Sn^{125}: [*Bu 55a*]. Sb^{125}: [*Lu 54c, Mo 54d*]. Te^{125}: [*Ba 52b, Gr 53*]. I^{125}: [*Be 52, De 53a*]. Xe^{125}: [*Be 52, Ma 54c*]. Cs^{125}: [*Ma 54c, Mi 54a*].

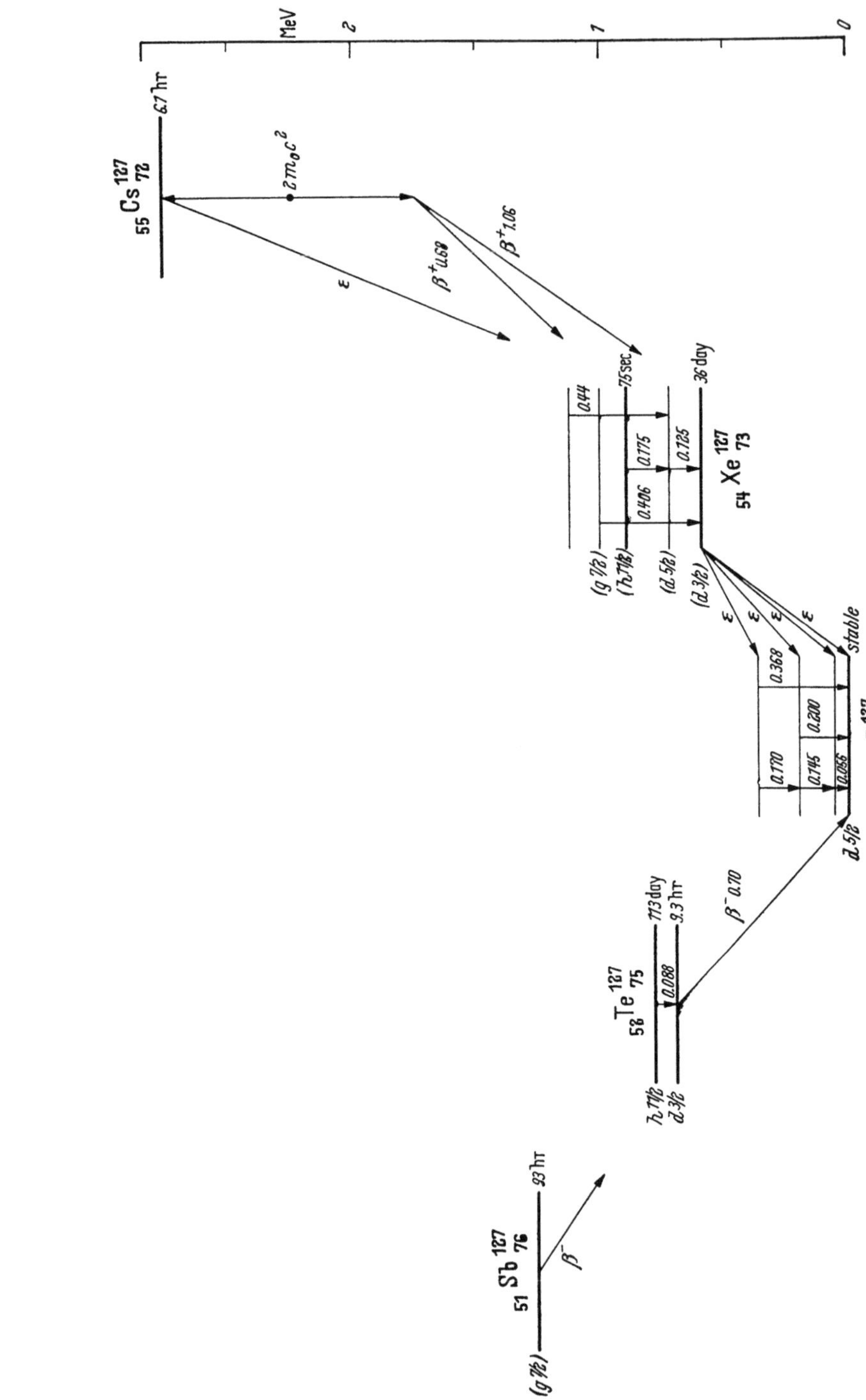

Fig. 62. $A = 127$. Refs. Te^{127}: [*Da 55a*]. I^{127}: [*Ja 53, Ja 53a, Ka 52, Mu 54b, Mu 55b*]. Xe^{127}: [*Ba 54, Be 52, Ma 54c, Ma 55a*]. Cs^{127}: [*Ma 54c, Mi 54a, Wa 53b*].

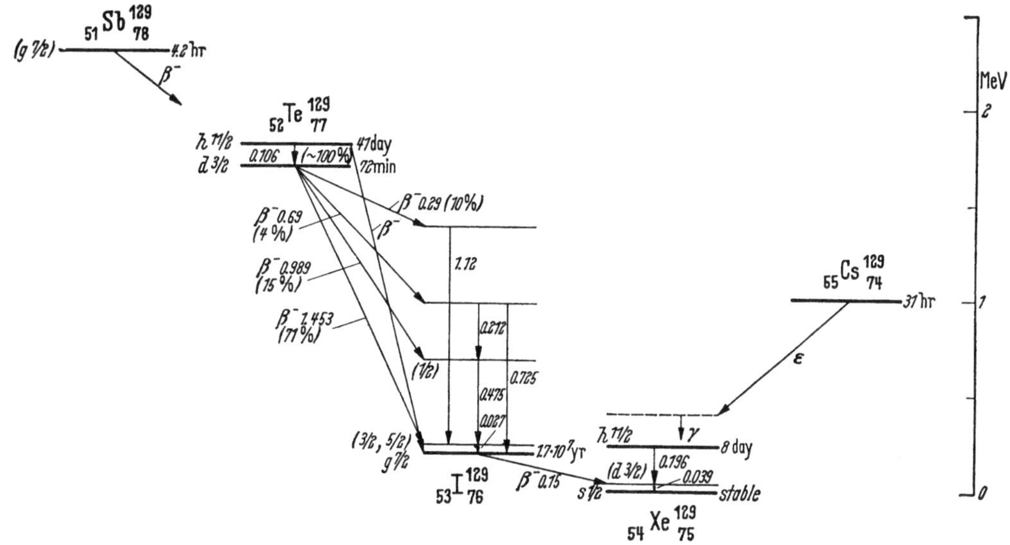

Fig. 63. $A = 129$. Refs. Te^{129}: [*Da 55a, Gr 55d, Ma 55j, St 55f*]. I^{129}: [*De 54a, Li 53b*]. Xe^{129}: [*Be 54, Br 54c, Th 54*]. Cs^{129}: [*Ne 55, Ro 55b, Wa 53b*].

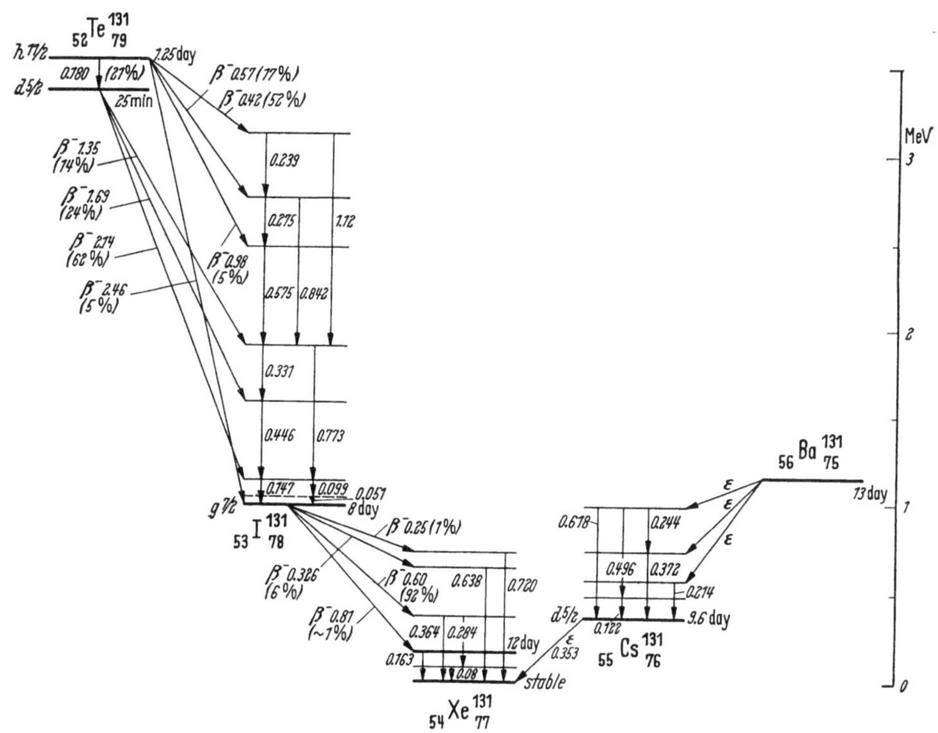

Fig. 64. $A = 131$. Refs. Te^{131}: [*Ge 52a, He 55b, Ma 55j*]. I^{131}: [*Al 52b, Ba 53c, Ha 52a, Ho 53, Li 53c Lo 53, Ni 54, Sa 54, Sc 53a, Se 53, Wo 52*]. Xe^{131}: [*Be 52, Be 54, Bo 52a, Br 54c, Sn 56*]. Cs^{131}: [*Be 53b, Sa 54a*]. Ba^{131}: [*Be 56a, Co 53b, El 52, Lu 55, Pa 53*].

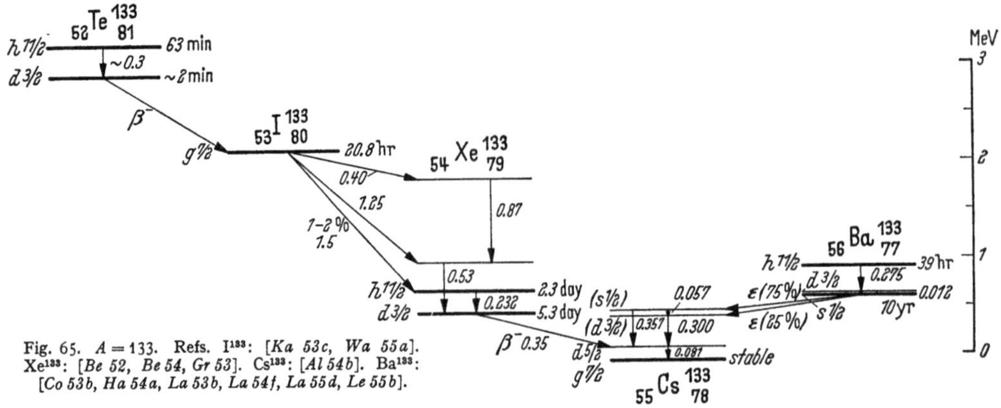

Fig. 65. $A = 133$. Refs. I^{133}: [*Ka 53c, Wa 55a*]. Xe^{133}: [*Be 52, Be 54, Gr 53*]. Cs^{133}: [*Al 54b*]. Ba^{133}: [*Co 53b, Ha 54a, La 53b, La 54f, La 55d, Le 55b*].

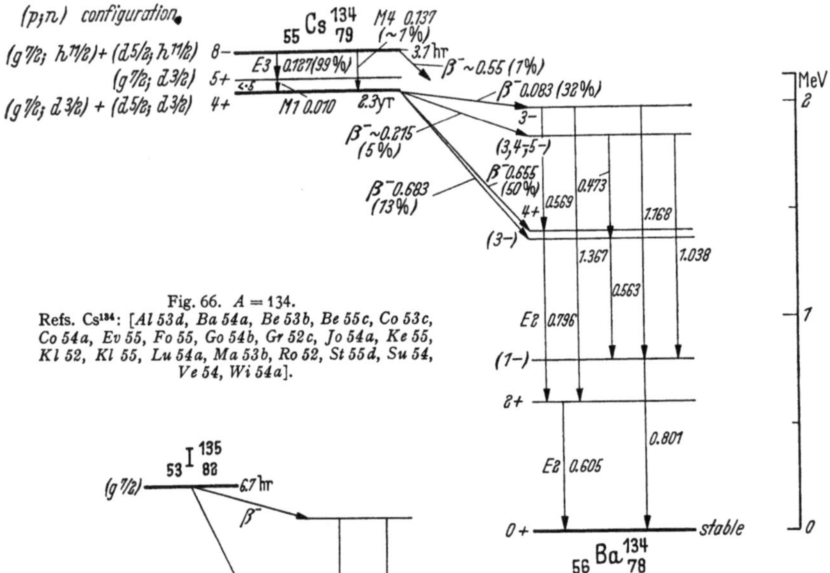

Fig. 66. $A = 134$.
Refs. Cs^{134}: [*Al 53d, Ba 54a, Be 53b, Be 55c, Co 53c, Co 54a, Ev 55, Fo 55, Go 54b, Gr 52c, Jo 54a, Ke 55, Kl 52, Kl 55, Lu 54a, Ma 53b, Ro 52, St 55d, Su 54, Ve 54, Wi 54a*].

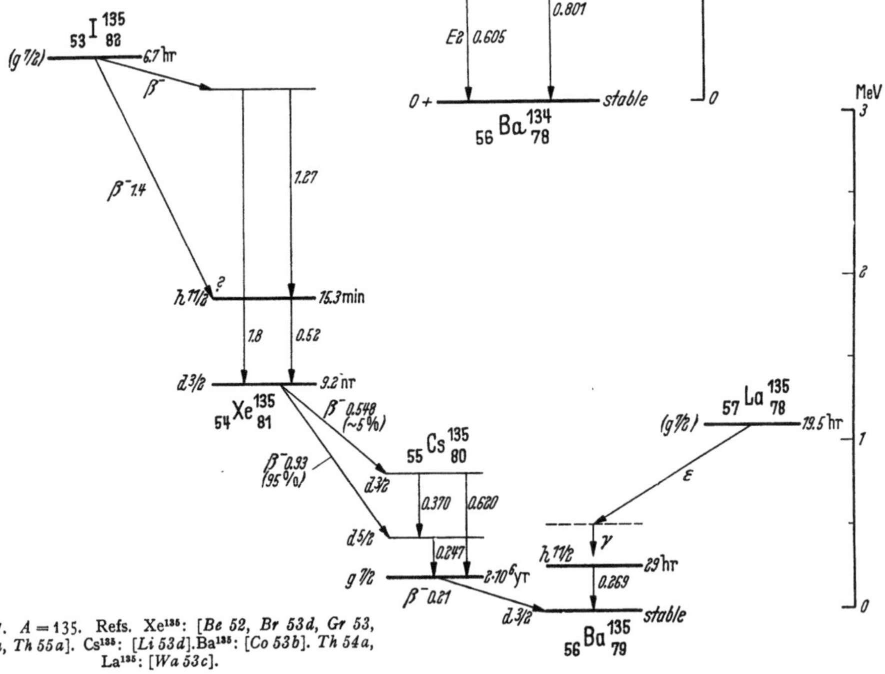

Fig. 67. $A = 135$. Refs. Xe^{135}: [*Be 52, Br 53d, Gr 53, Th 54a, Th 55a*]. Cs^{135}: [*Li 53d*]. Ba^{135}: [*Co 53b*]. *Th 54a*, La^{135}: [*Wa 53c*].

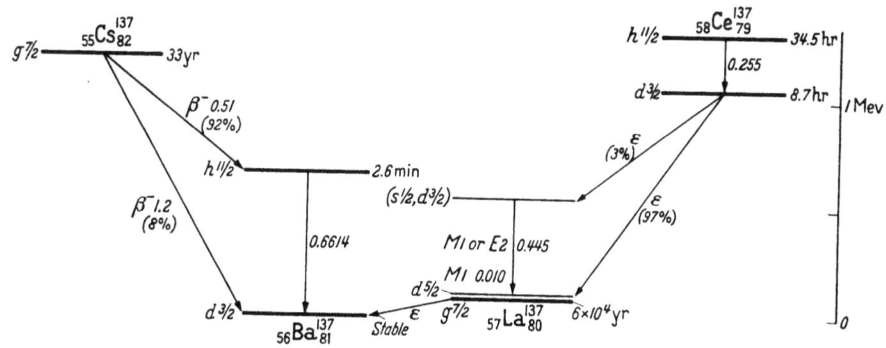

Fig. 68. $A = 137$. Refs. Cs^{137} and Ba^{137}: [*Az 54a, Bh 54, Br 53c, Br 55c, Do 53, Dr 55, Gr 52, Li 53e, Ma 53b, Ma 54d, Mu 52, Ol 54, Ve 54, Wa 54a, Wi 53b, Wi 55a*]. La^{137}: [*Br 56*]. Ce^{137}: [*Br 55a, Br 56*].

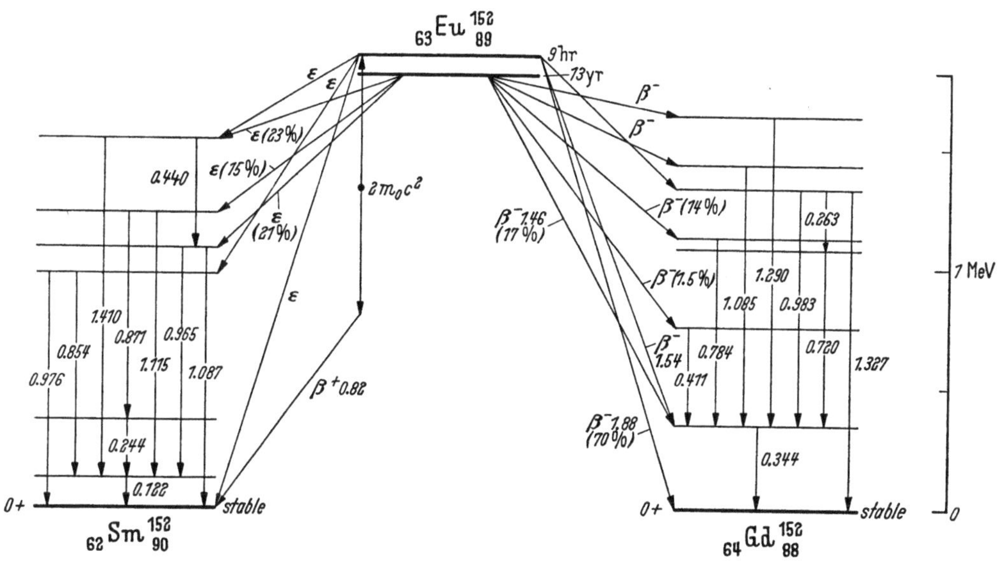

Fig. 69. $A = 152$. Refs. Eu^{152}: [*Ch 54, Dz 55a, Gr 55b, Ka 53d, Le 54f, Lo 53, Sl 54, Su 55*].

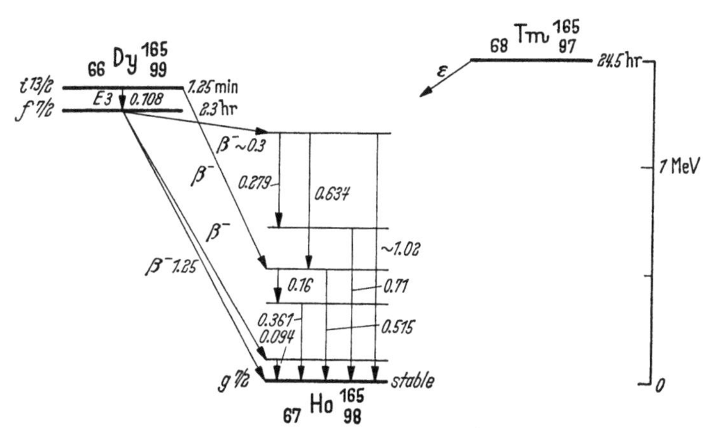

Fig. 70. $A = 165$. Refs. Dy^{165}: [*Jo 53b, Ma 54f, We 54*]. Ho^{165}: [*Po 54*]. Tm^{165}: [*Ha 53b, Mi 54a*].

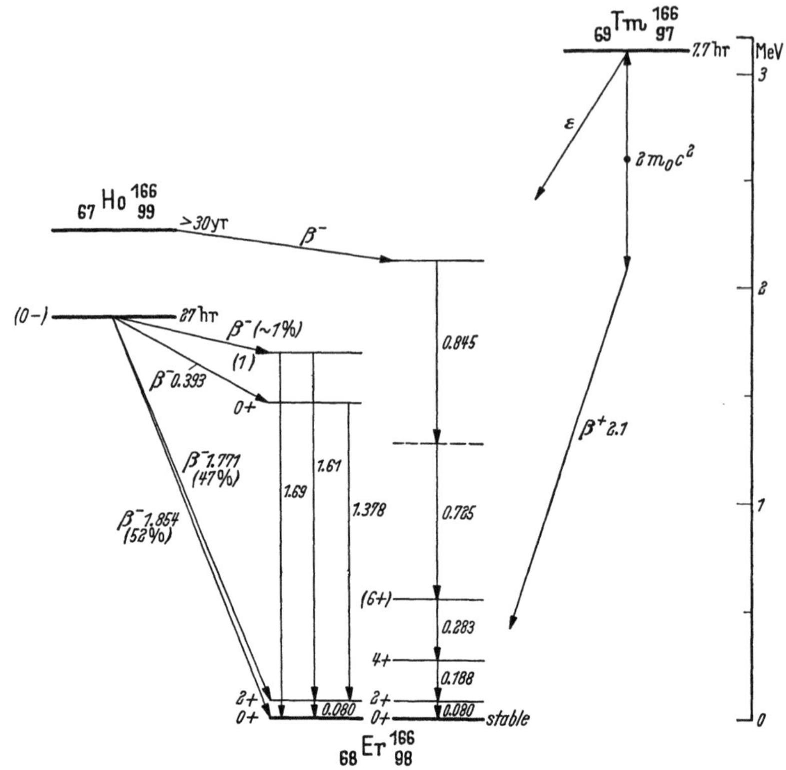

Fig. 71. $A = 166$. Refs. Ho^{166}: [*Bu 52, Fr 55, Gr 55, Mi 55, Su 54a*]. Tm^{166}: [*Mi 54a, Ne 55*].

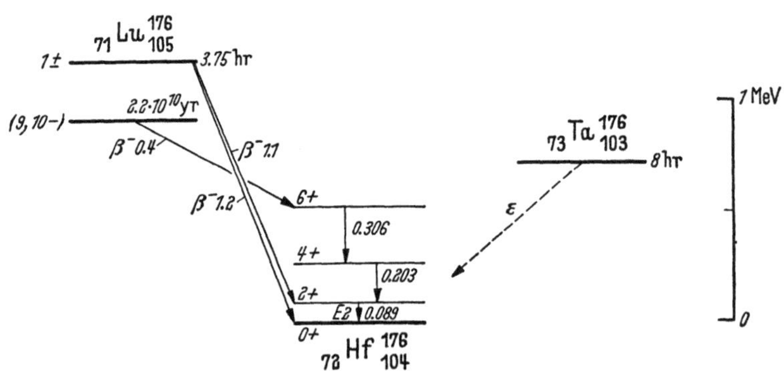

Fig. 72. $A = 176$. Refs. Lu^{176}: [*Ar 53, Ar 54, Di 54, Di 55a*].

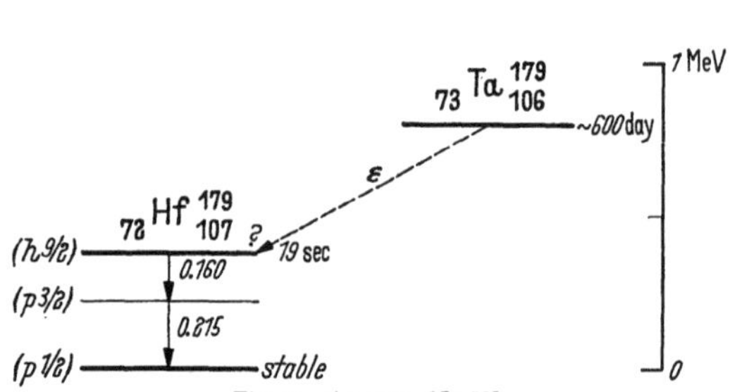

Fig. 73. $A = 179$. [*Go 52*].

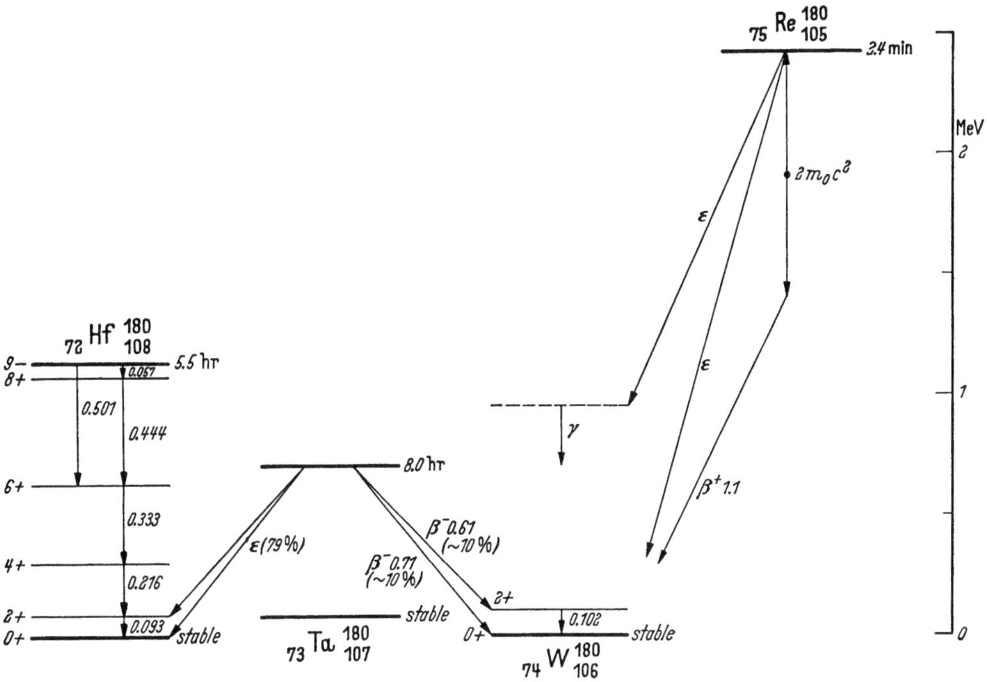

Fig. 74. $A = 180$. Refs. Hf^{180}: [*Mi 54g, Sc 56a*]. Ta^{180}: [*Eb 55, Su 55, Wh 55b*]. W^{180}: [*So 55a*]. Re^{180}: [*Fi 55*].

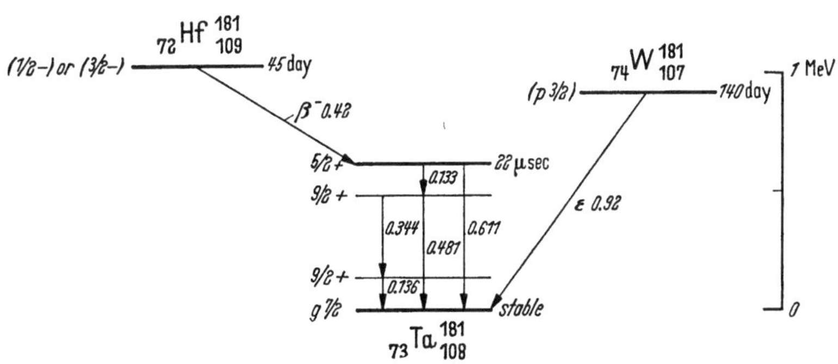

Fig. 75. $A = 181$. Refs. Hf^{181}: [*Ba 53d, Ba 53e, De 52a, En 53, Mc 54a, Mc 54b, Mu 53a, Sw 53a*]. Ta^{181}: [*Br 52, Br 53e, Gi 55, Go 55, He 55d, He 55f, Ka 55b*]. W^{181}: [*Bi 55, Co 53d, Kr 55b*].

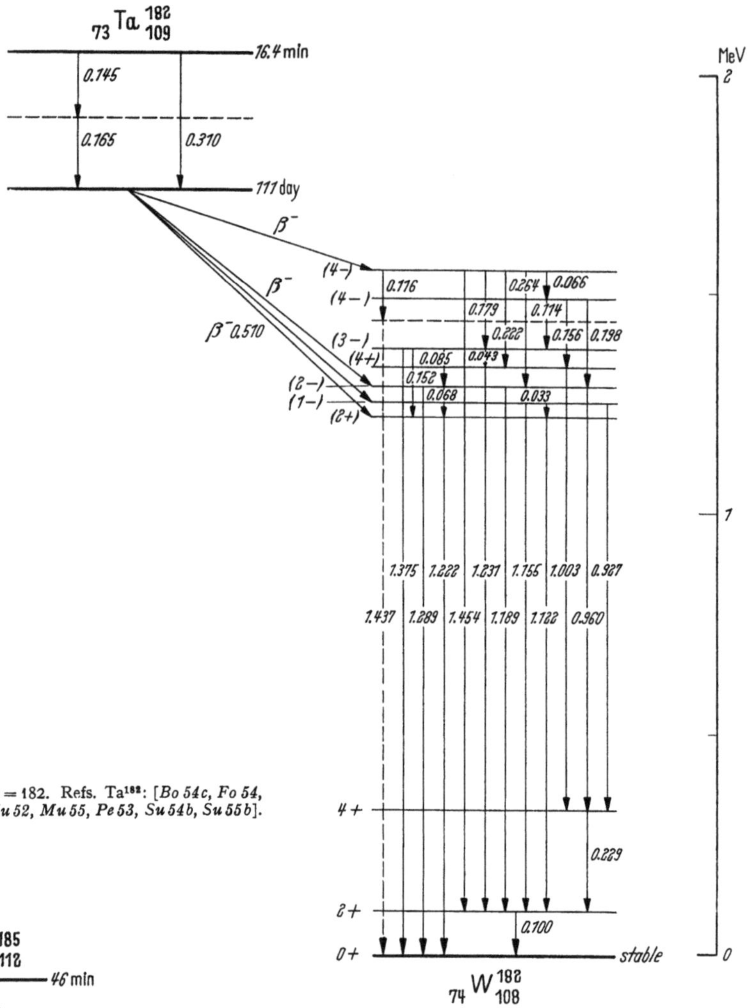

Fig. 76. $A = 182$. Refs. Ta^{182}: [*Bo 54c, Fo 54, Mi 54b, Mu 52, Mu 55, Pe 53, Su 54b, Su 55b*].

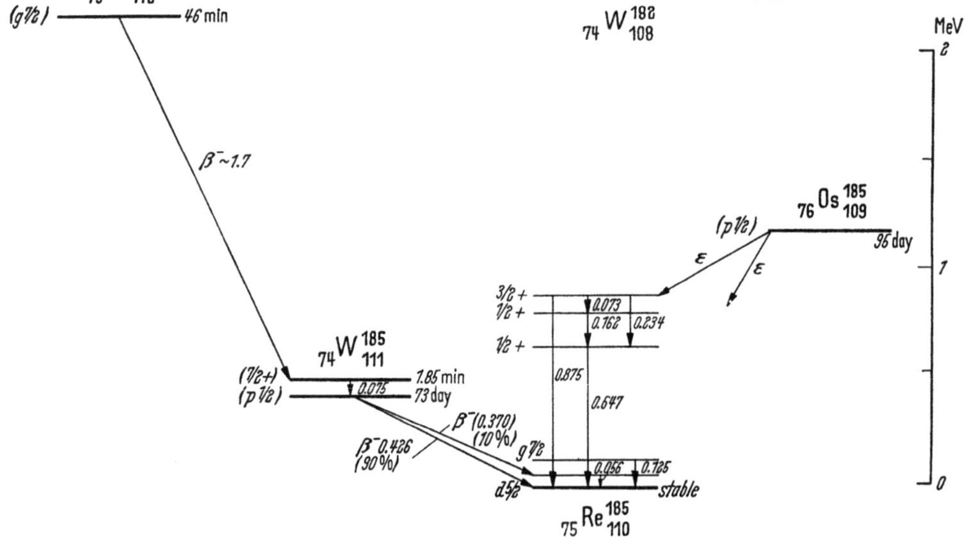

Fig. 77. $A = 185$. Refs. W^{185}: [*Bi 55a, Kr 55b, La 53c, Mi 54f, Po 55b*]. Re^{185}: [*Ja 54a*]. Os^{185}: [*Ho 53c, Na 56, Pr 55*].

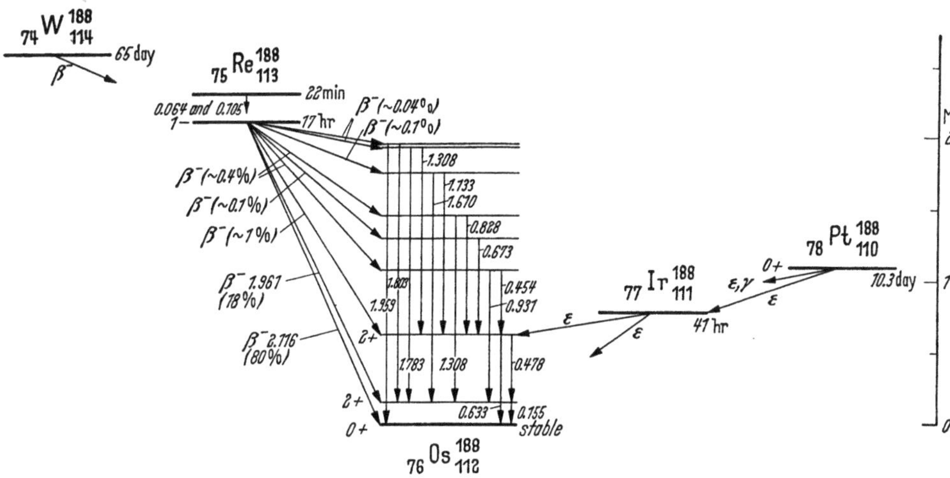

Fig. 78. $A = 188$. Refs. Re^{188}: [*Dz 54, Fl 53a, Gu 54, Jo 56, Mc 53b, Mi 53a, Po 56, Su 55*]. Ir^{188}: [*Na 54a, Sm 55*]. Pt^{188}: [*Na 54a, Sm 55*].

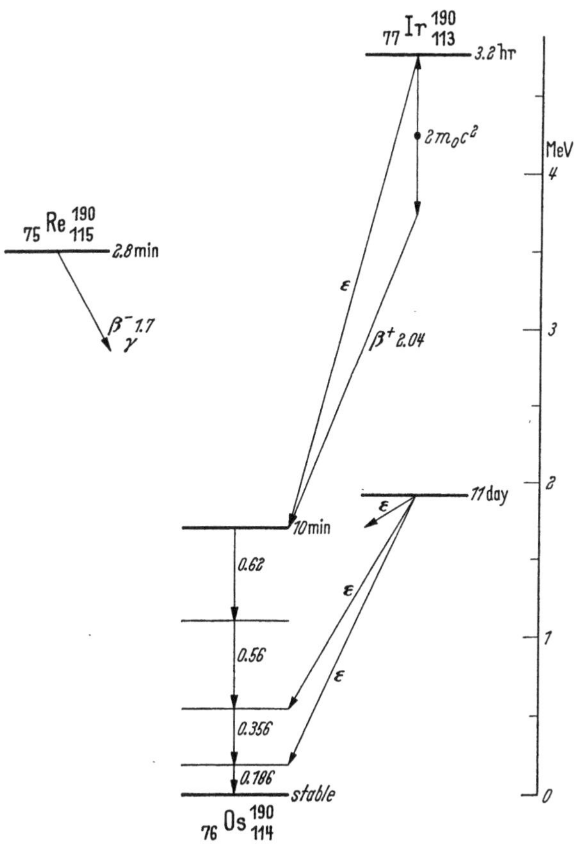

Fig. 79. $A = 190$. Refs. Re^{190}: [*At 55a*]. Os^{190}: [*At 53b, At 55*]. It^{190}: [*At 55*].

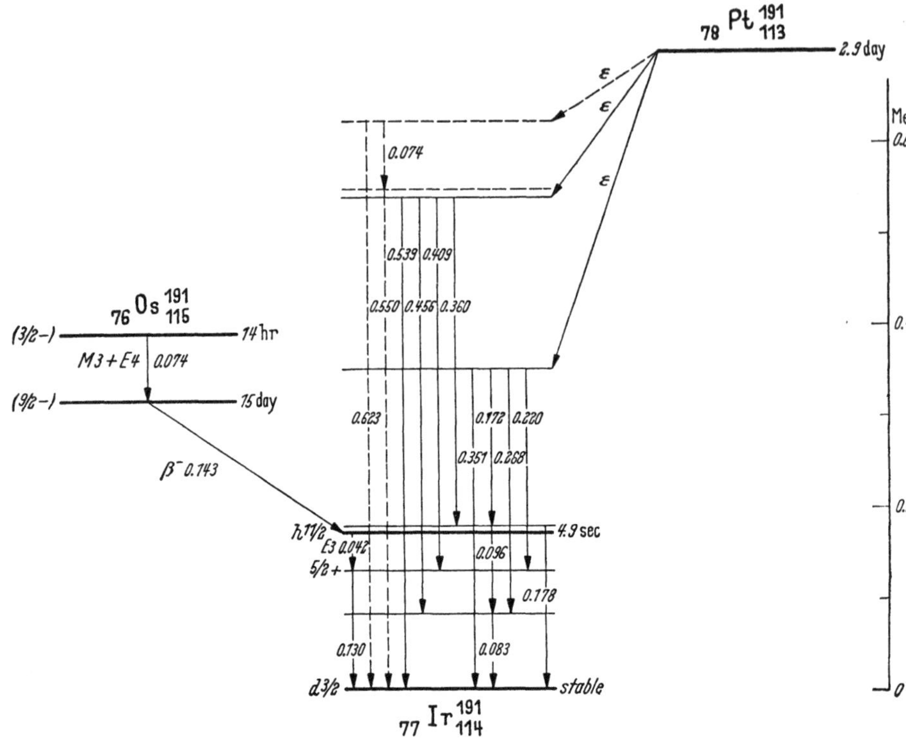

Fig. 80. $A = 191$. [Refs. Os^{191}: [*Hi 53, Mc 54a, Mi 55a*]. Ir^{191}: [*Bu 54c, Fi 55, Mi 54c, Na 54, Vo 53*]. Pt^{191}: [*Co 54b, Gi 54a, Sm 55, Su 55, Sw 53b, To 54, To 55c*].

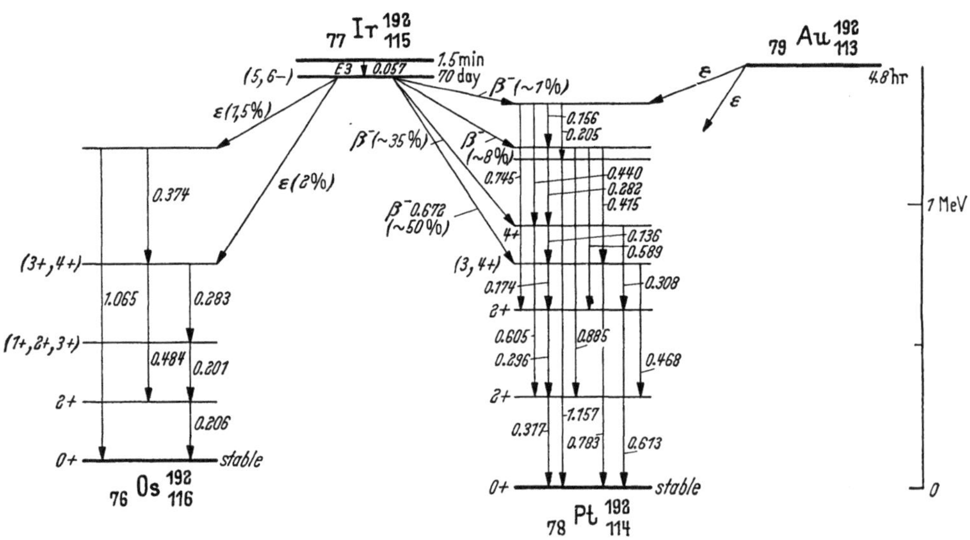

Fig. 81. $A = 192$. Refs. Os^{192}: [*Fr 52*]. Ir^{192}: [*Ba 52c, Ba 55g, Gl 55a, Gr 55f, Ja 54, Jo 54b, Lu 54a, Mi 54d, Mu 52, Pr 54, Ta 55, We 53a, We 54*]. Au^{192}: [*Ew 53, Gi 54a*].

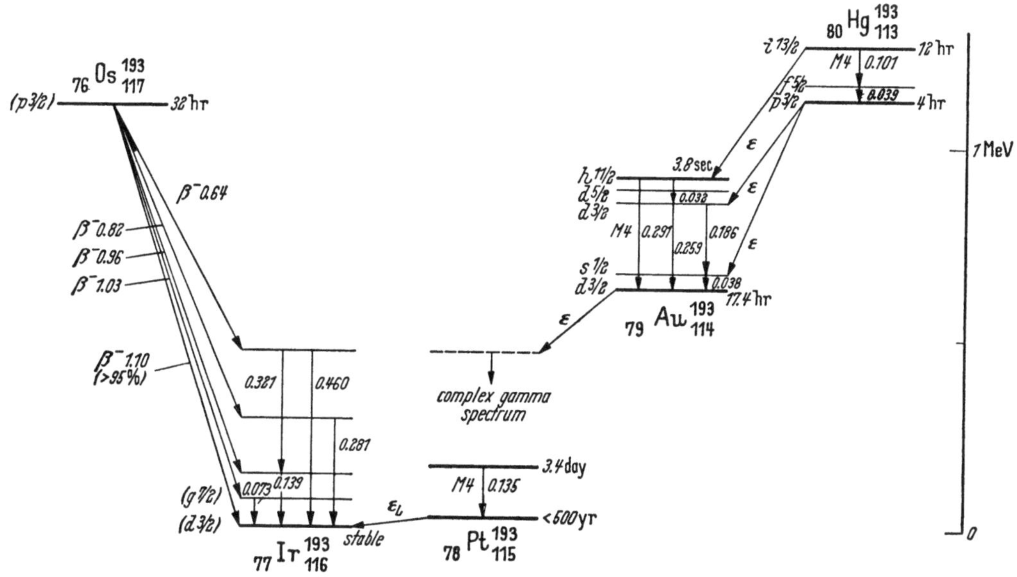

Fig. 82. $A = 193$. Refs. Os^{193}: [*Co 53c, De 54b, Na 56, Sa 53a, Sw 52*]. Ir^{193}: [*Vo 53*]. Pt^{193}: [*Br 55, Co 54b, Na 56a, Sw 53*]. Au^{193}: [*Br 55, Ew 53, Fi 55, Gi 54a*]. Hg^{193}: [*Br 55, Br 55d, Gi 54a*].

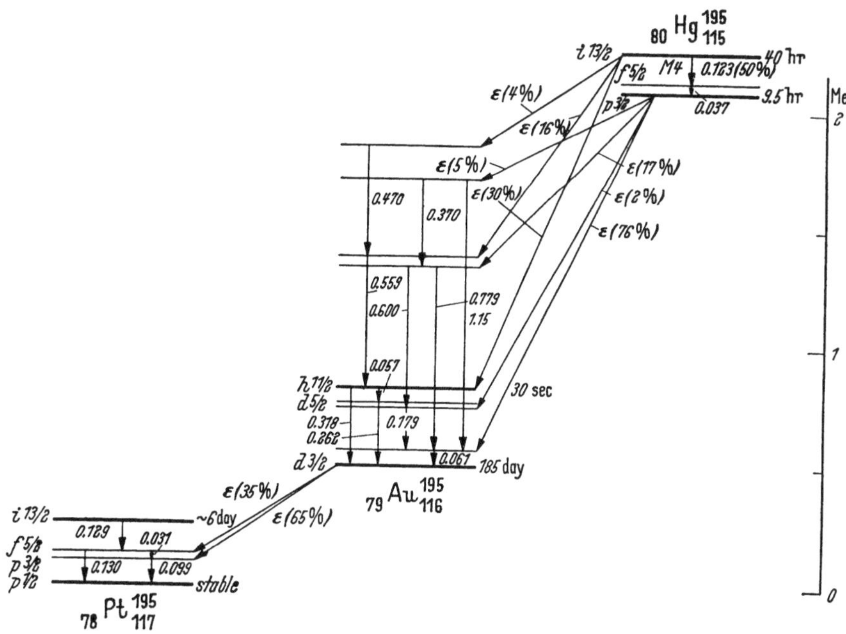

Fig. 83. $A = 195$. Refs. Pt^{195}: [*Co 54b*]. Au^{195}: [*Bi 54a, Br 54d, Fi 55, Gi 54a, Hu 52a, Hu 53*]. Hg^{195}: [*Br 54d, Br 54e, Br 55d, Gi 54a, Hu 52a, Hu 53, Jo 55b*].

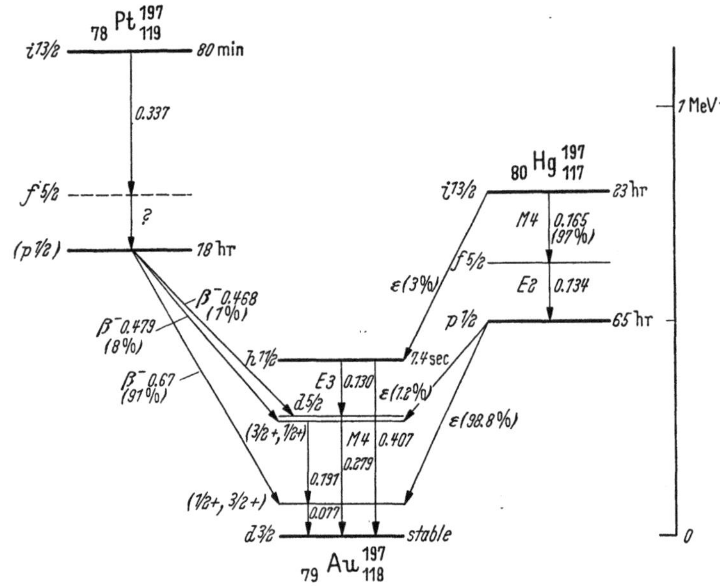

Fig. 84. $A = 197$. Refs. Pt^{197}: [*Po 55, Su 55*]. Au^{197}: [*Fi 55, Hu 53, Mi 53, Po 54, Vo 53*]. Hg^{197}: [*Bi 54b, Br 54d, Br 54g, Co 55, Gi 55, Hu 53, Jo 55b, Mi 53*].

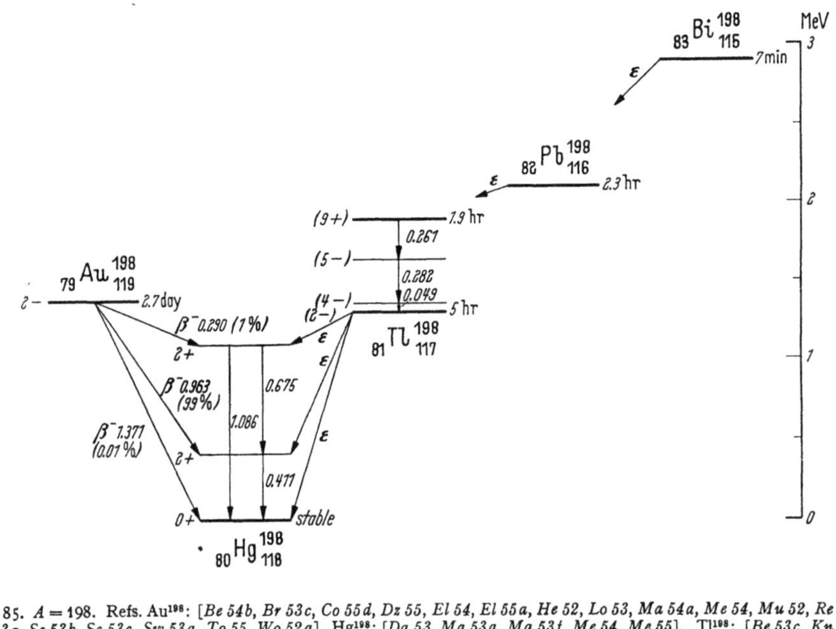

Fig. 85. $A = 198$. Refs. Au^{198}: [*Be 54b, Br 53c, Co 55d, Dz 55, El 54, El 55a, He 52, Lo 53, Ma 54a, Me 54, Mu 52, Re 55, Sc 53a, Sc 53b, Sc 53c, Sw 53a, To 55, Wo 52a*]. Hg^{198}: [*Da 53, Ma 53a, Ma 53f, Me 54, Me 55*]. Tl^{198}: [*Be 53c, Kn 55, Mi 54a, Pa 54*]. Pb^{198}: [*An 55a, Ho 53c*]. Bi^{198}: [*Ho 53c*].

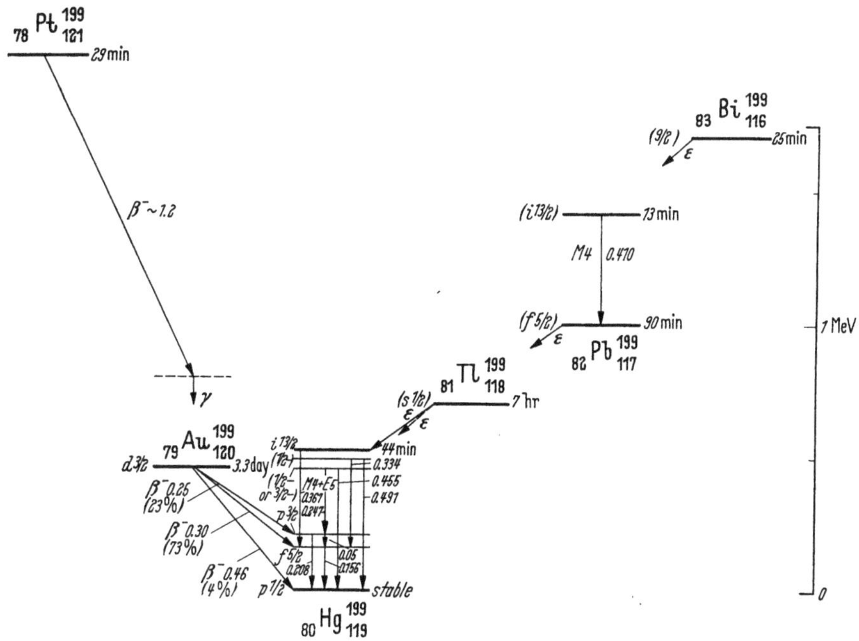

Fig. 86. $A = 199$. Refs. Pt^{199}: [*Le 54i*]. Au^{199}: [*Be 55d, Gr 55e, Ha 55f, Re 55*]. Hg^{199}: [*Bo 55, Me 54a, Sw 53a*]. Tl^{199}: [*Be 53c, Mi 54a*]. Pb^{199}: [*An 55a, St 56a*]. Bi^{199}: [*Ho 53c*].

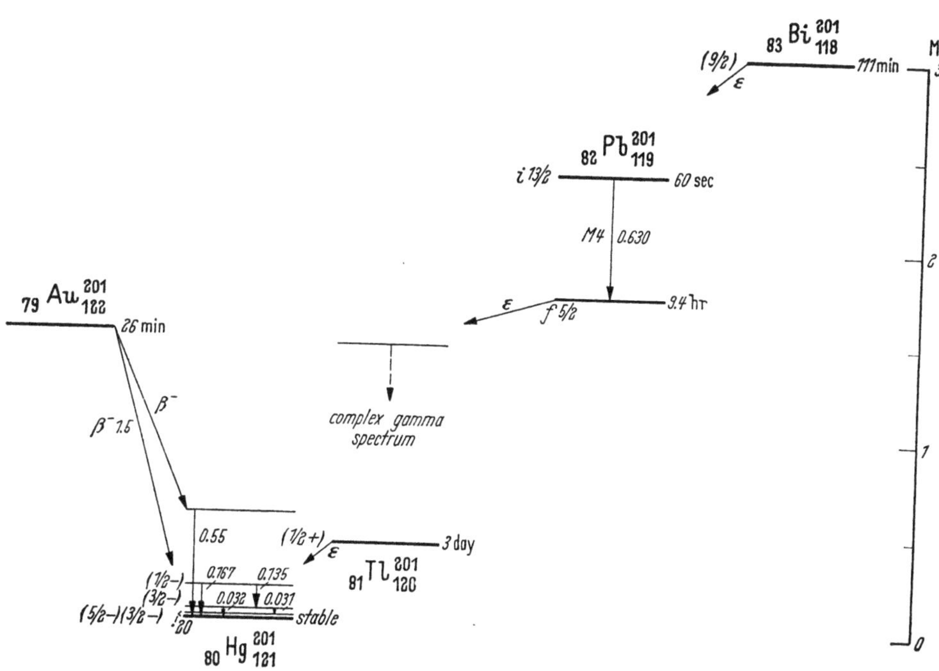

Fig. 87. $A = 201$. Refs. Au^{201}: [*Bu 52a*]. Hg^{201}: [*Be 53c, De 54c, Mu 55a*]. Tl^{201}: [*Be 53c, Be 55*]. Pb^{201}: [*Be 55, Fi 55, St 56a, Wa 54c*]. Bi^{201}: [*Ho 53c, St 56a*].

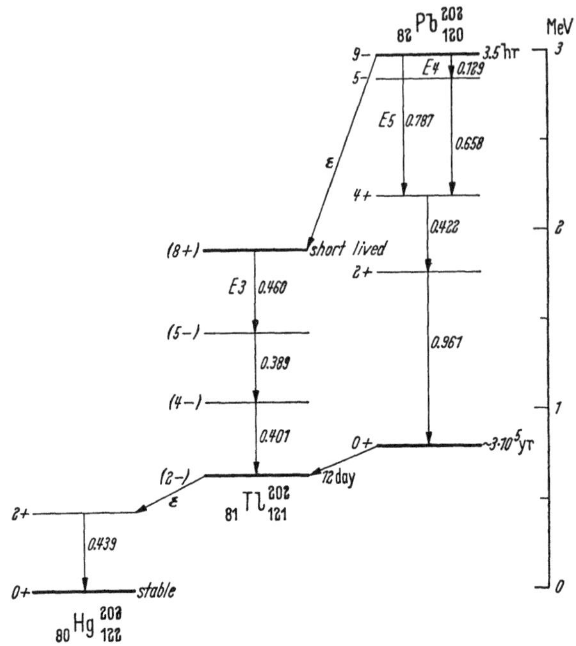

Fig. 88. $A = 202$. Refs. Tl^{202}: [*Be 53c, Hu 54, Me 55b*]. Pb^{202}: [*Be 55, Be 55b, Ho 52a, Hu 54, Ma 54g, Ma 54i*].

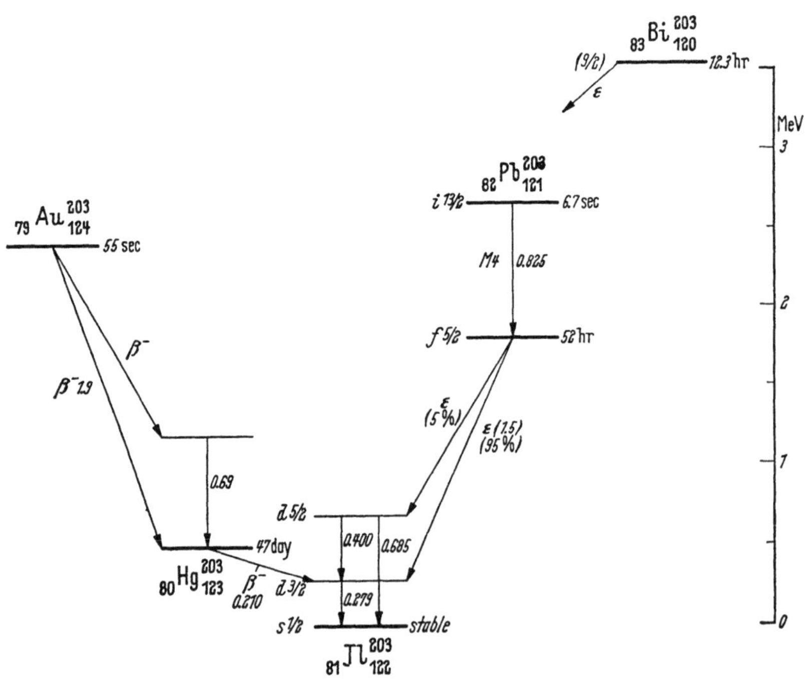

Fig. 89. $A = 203$. Refs. Au^{203}: [*Bu 52a*]. Hg^{203}: [*Az 55a, Be 53c, De 55b, Do 55, En 53, Ma 55d, Sw 53a, Th 54b, Wa 54b*]. Tl^{203}: [*Gu 53, Me 54b*]. Pb^{203}: [*Be 53b, Fi 55, Pr 54b, St 56a, Va 54a, Wa 54c*]. Bi^{203}: [*Ho 53c, St 56a*].

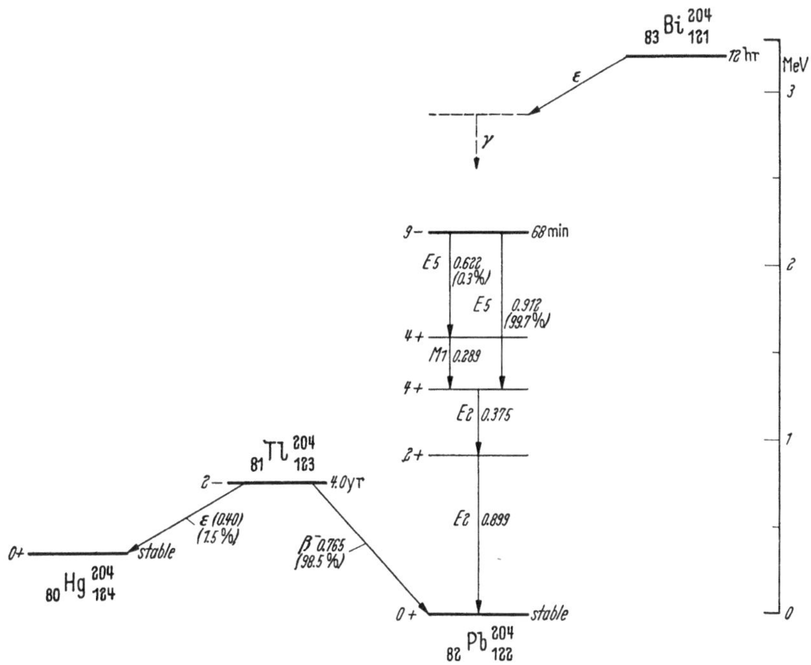

Fig. 90. $A = 204$. Refs. Tl[204]: [*Ch 55a, De 52b, Fe 54, Ha 53c, Ho 54a, Ja 54, Jo 55a, Kn 54, Lo 53, To 55a, Yu 54*]. Pb[204]: [*Be 55, Fr 53, Fr 54, He 56, Hu 56, Kr 54, Kr 55a, Ma 54g*].

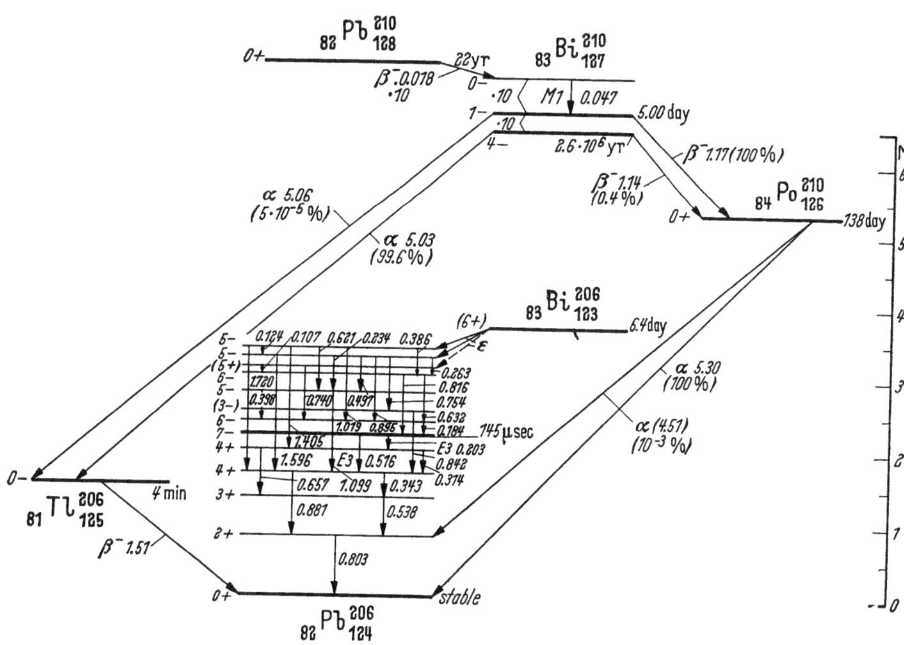

Fig. 91. $A = 206$ and $A = 210$. Refs. Tl[206]: [*Mi 54a, Ne 54, Sa 53*]. Bi[206]: [*Al 53a*]. Pb[210]: [*Ba 53f, Da 53a, Da 54b, Ew 52, Hu 53a, Ja 53b, Ko 53b, Le 55e, To 55b, Wu 53*]. Bi[210]: [*Bo 53a, Fr 53b, Hu 53b, Li 55, Lo 53, Ma 53e, Mi 55b, No 53, Pl 54, Pn 53, Sm 54*]. Po[210]: [*Ad 54, Al 53a, Cu 53a, Du 55, Ei 54, Gi 53, Ha 55a, Ri 55, Ro 55c*].

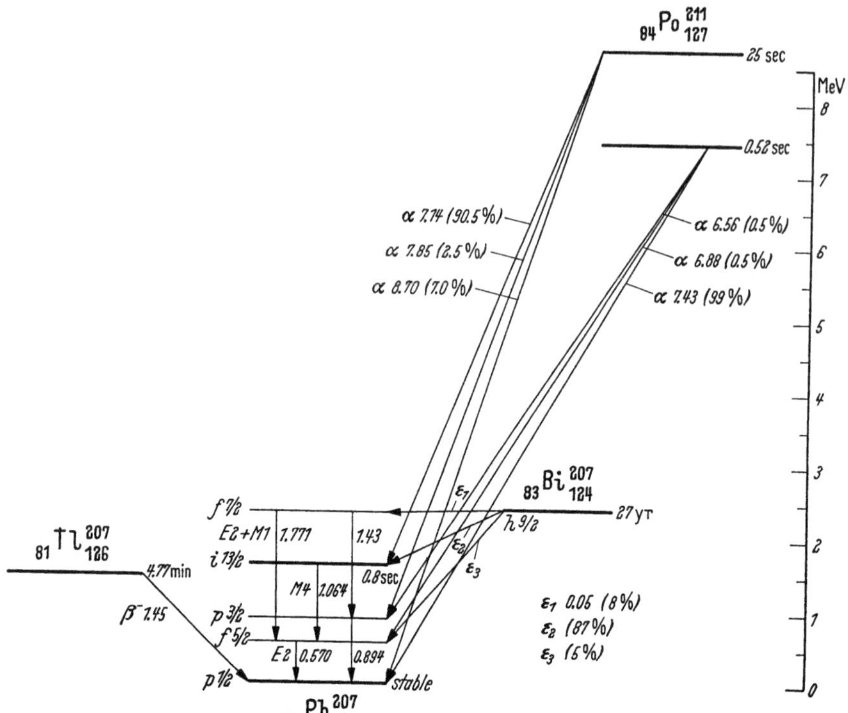

Fig. 92. $A = 207$ and $A = 211$. Refs. Tl^{207}: [*Sa 53*]. Pb^{207}: [*Be 55a, Ca 53, Fr 53a, Ho 52a, Mc 53, Re 54*]. Bi^{207}: [*Al 53, Al 55a, Ch 55, Ha 55b, La 55, Ma 54h, Mc 53a, Pr 54a, Wa 54b, Ya 55, Ya 55a*]. Po^{211}: [*Fr 53a, Ho 53a, Je 54, Mi 54e, Sp 54*].

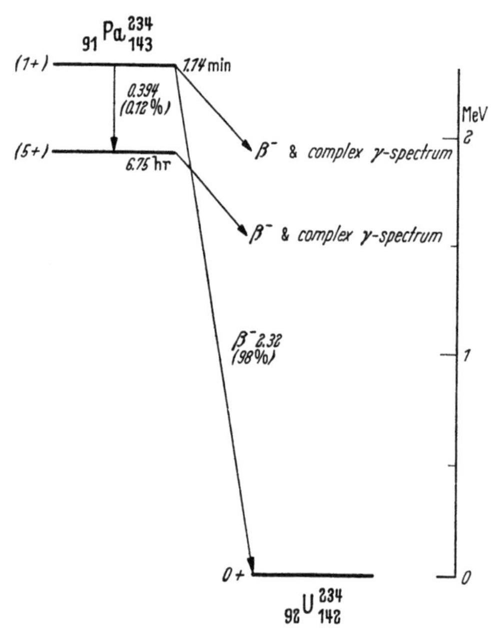

Fig. 93. $A = 234$. Refs. Pa^{234}: [*Bo 53b, Cr 54, De 53c, De 55c, Ho 53b, Ho 55, Jo 54c, St 53d, Zi 54*]. U^{234}: [*En 33a, Fl 52, Va 53*].

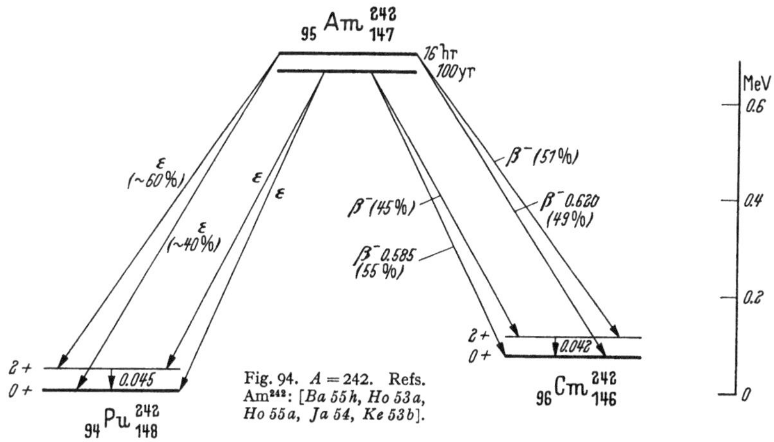

Fig. 94. $A = 242$. Refs. Am^{242}: *[Ba 55h, Ho 53a, Ho 55a, Ja 54, Ke 53b]*.

Table 5. *Experimental data on isomers of half-life less than 10^{-6} sec. The reduced transition probability $B(E2)$ for* Coulomb *excitation is expressed in units of 10^{-48} cm^4.*

Abbreviations include gnd. for ground state, Coul. exc. for Coulomb excitation, Del. coin. for delayed coincidence, Res. fluor. for resonance fluorescence, and Coin. pulse ht for coincidence pulse height.

Nucleus	E_{trans} (Mev)	$T_{1/2}$, τ or τ_γ in sec [1]	$B(E2)_{up}$	$I_{upper} \to I_{lower}$	Mother isotope or reaction	Method	References
$_3Li^7_4$	0.479	$\tau = 7.5 \times 10^{-14}$		$1/2^- \to 3/2^-$ gnd.	$B^{10}(n,\alpha)Li^7$	Doppler effect	[El 49]
$_5B^{10}_5$	0.72	$\tau = 7 \times 10^{-10}$		$1^+ \to 3^+$ gnd.	$Be^9(d,n)B^{10}$	Recoil	[Th 53]
		$\tau = 9.5 \times 10^{-10}$				Recoil	[Se 55a]
$_7N^{14}_7$	2.31	$\tau = \sim 2 \times 10^{-13}$		$0^+ \to 1^+$ gnd.	$N^{14}(p,p')N^{14}$	Doppler effect	[Th 55]
$_8O^{16}_8$	6.06	$T_{1/2} = 5 \times 10^{-11}$		$0^+ \to 0^+$ gnd.	$F^{19}(p,\alpha)O^{16}$	Recoil	[Aj 55]
	6.13	$T_{1/2} = (0.5\text{–}1) \times 10^{-12}$		$3^- \to 0^+$ gnd.	$F^{19}(p,\alpha)O^{16}$	Recoil	[Aj 55]
$_8O^{17}_9$	0.88	$\tau = 2.5 \times 10^{-10}$		$1/2^+ \to 5/2^+$ gnd.	$O^{16}(d,p)O^{17}$	Recoil	[Th 53]
$_9F^{19}_{10}$	0.109	$\tau = 1.0 \times 10^{-9}$		$1/2^- \to 1/2^+$ gnd.	$F^{19}(p,p')F^{19}$	Recoil	[Th 54d]
	0.197	$\tau = 8 \times 10^{-8}$	$0.009 \times \frac{1}{\varepsilon}$	$5/2^+ \to 1/2^+$ gnd.	$F^{19}(p,p')F^{19}$	Recoil	[Th 54d], [Ba 55c]
$_{11}Na^{23}_{12}$	0.446		$0.041 \times \frac{1}{\varepsilon}$			Coul. exc.	[Te 55]
$_{13}Al^{28}_{15}$	0.032	$\tau = 2.7 \times 10^{-9}$		$(2^+) \to (3^+)$ gnd.	$Al^{27}(d,p)Al^{28}$	Recoil	[Se 55a]
$_{19}K^{41}_{22}$	1.3	$T_{1/2} = 6.6 \times 10^{-9}$		$f_{7/2} \to d_{3/2}$ gnd.	A^{41}	Del. coin.	[En 53], [Go 52]
$_{22}Ti^{46}_{24}$	0.89	$\tau_\gamma = 13 \times 10^{-12}$	$0.056 \times \frac{1}{\varepsilon}$	$2^+ \to 0^+$ gnd.		Coul. exc.	[He 55c]
$_{22}Ti^{47}_{25}$	0.160	$\tau_\gamma = 2.6 \times 10^{-8}$	$0.04 \times \frac{1}{\varepsilon}$	$7/2^- \to 5/2^-$ gnd.		Coul. exc.	[Te 54], [He 55c]
$_{22}Ti^{48}_{26}$	0.99	$\tau_\gamma = 14 \times 10^{-12}$	$0.031 \times \frac{1}{\varepsilon}$	$2^+ \to 0^+$ gnd.		Coul. exc.	[He 55c]
$_{23}Va^{50}_{27}$	0.225	$\tau_\gamma = 1.3 \times 10^{-8}$	$0.011 \times \frac{1}{\varepsilon}$	$(7^+) \to 6^+$ gnd.		Coul. exc.	[He 55c]
$_{23}Va^{51}_{28}$	0.325	$\tau_\gamma = 3.3 \times 10^{-9}$	$0.0055 \times \frac{1}{\varepsilon}$	$5/2^- \to 7/2$ gnd.		Coul. exc.	[Ma 55e], [He 55c]
		$\tau_\gamma = 2.8 \times 10^{-10}$			Ti^{51}	Coin. pulse ht	[De 55d]
$_{24}Cr^{53}_{29}$	0.155		0.015	$? \to 7/2$ gnd.		Coul. exc.	[Ma 55e]
$_{25}Mn^{55}_{30}$	0.131		0.087	$? \to 5/2$ gnd.		Coul. exc.	[Ma 55e], [Te 54]

[1]) Values of τ_γ derived from the Coulomb excitation $B(E2)$ refer to the partial mean $E2$ radiative lifetime only while in all other cases τ_γ represents the total mean radiative lifetime. See for examble the two valus for the 0.325 Mev transition in V^{51} which decays with a strong $M1$ component.

Table 5. (Continued.)

Nucleus	E_{trans} (Mev)	$T_{\frac{1}{2}}$, τ or τ_γ in sec	$B(E2)_{up}$	$I_{upper} \to I_{lower}$	Mother isotope or reaction	Method	References
${}_{26}Fe^{56}_{30}$	0.85	$\tau_\gamma = 8.7\times10^{-12}$ $\tau_\gamma \gtrsim 8\times10^{-12}$	$0.10\times\frac{1}{\varepsilon}$	$2^+\to0^+$ gnd.		Coul. exc. Res. fluor.	[*He 55c*] [*Il 54*]
${}_{26}Fe^{57}_{31}$	0.014	$T_{\frac{1}{2}} = 1.1\times10^{-7}$		${}^3/_2^-\to{}^1/_2^-$ gnd.	Co^{57}	Del. coin.	[*Go 52*]
	0.137	$\tau_\gamma = 8.9\times10^{-8}$	$0.057\times\frac{1}{\varepsilon}$	${}^5/_2^-\to{}^1/_2^-$ gnd.		Coul. exc.	[*Le 55a*], [*Te 54*], [*He 55c*]
${}_{28}Ni^{60}_{32}$	1.33	$\tau_\gamma = 1.1\times10^{-12}$		$2^+\to0^+$ gnd.		Res. fluor.	[*Me 56*]
${}_{29}Cu^{63}_{34}$	0.69	$\tau_\gamma = (1.3-5.2)\times10^{-10}$	$0.010\times\frac{1}{\varepsilon}$	$?\to{}^3/_2$ gnd.		Coul. exc.	[*He 55c*]
	0.99	$\tau_\gamma = (0.8-3.0)\times10^{-11}$ $\tau_\gamma = 6\times10^{-13}$	$0.029\times\frac{1}{\varepsilon}$	$?\to{}^3/_2$ gnd.		Coul. exc. Res. fluor.	[*He 55c*] [*Il 54*], [*Mo 54e*]
${}_{29}Cu^{65}_{36}$	0.815	$\tau_\gamma = (0.7-2.6)\times10^{-10}$	$0.0087\times\frac{1}{\varepsilon}$	$?\to{}^3/_2$ gnd.		Coul. exc.	[*He 55c*]
	1.15	$\tau_\gamma = (0.4-1.5)\times10^{-11}$	$0.027\times\frac{1}{\varepsilon}$	$?\to{}^3/_2$ gnd.		Coul. exc.	[*He 55c*]
${}_{30}Zn^{64}_{34}$	1.00	$\tau_\gamma = 3.7\times10^{-12}$	$0.11\times\frac{1}{\varepsilon}$	$2^+\to0^+$ gnd.		Coul. exc.	[*He 55c*]
${}_{30}Zn^{66}_{36}$	1.04	$\tau_\gamma = 4.0\times10^{-12}$	$0.087\times\frac{1}{\varepsilon}$	$2^+\to0^+$ gnd.		Coul. exc.	[*He 55c*]
${}_{32}Ge^{70}_{38}$	1.02	$\tau_\gamma = 3.9\times10^{-12}$	$0.098\times\frac{1}{\varepsilon}$	$2^+\to0^+$ gnd.		Coul. exc.	[*He 55c*]
${}_{32}Ge^{71}_{39}$	0.175	$T_{\frac{1}{2}} = 7\times10^{-8}$		${}^5/_2^-\to{}^1/_2^-$ gnd.	As^{71}	Del. coin.	[*Gr 55a*]
${}_{32}Ge^{72}_{40}$	0.68 0.83	$T_{\frac{1}{2}} = 3\times10^{-7}$ $\tau_\gamma = 6.2\times10^{-12}$ $\tau_\gamma = 4.6\times10^{-12}$	$0.16\times\frac{1}{\varepsilon}$	$0^+\to0^+$ gnd. $2^+\to0^+$ gnd.	Ga^{72}; As^{72}	Del. coin. Coul. exc. Res. fluor.	[*Go 52*] [*He 55c*] [*Me 56a*]
${}_{32}Ge^{74}_{42}$	0.593	$\tau_\gamma = 2.3\times10^{-11}$ $\tau_\gamma = 1.9\times10^{-11}$	$0.25\times\frac{1}{\varepsilon}$	$2^+\to0^+$ gnd.		Coul. exc. Res. fluor.	[*He 55c*] [*Me 55a*], [*Me 56a*]
${}_{32}Ge^{76}_{44}$	0.566	$\tau_\gamma = 2.9\times10^{-11}$	$0.23\times\frac{1}{\varepsilon}$	$2^+\to0^+$ gnd.		Coul. exc.	[*He 55c*]
${}_{33}As^{75}_{42}$	0.200		$0.025\times\frac{1}{\varepsilon}$	$?\to{}^3/_2^-$ gnd.		Coul. exc.	[*He 55c*]
	0.283		$0.071\times\frac{1}{\varepsilon}$	$?\to{}^3/_2^-$ gnd.		Coul. exc.	[*He 55c*]
	0.574		$0.072\times\frac{1}{\varepsilon}$	$?\to{}^3/_2^-$ gnd.		Coul. exc.	[*He 55c*]
	0.814		$0.066\times\frac{1}{\varepsilon}$	$?\to{}^3/_2^-$ gnd.		Coul. exc.	[*He 55c*]
${}_{34}Se^{74}_{40}$	0.635	$\tau_\gamma = 1.9\times10^{-11}$	$0.21\times\frac{1}{\varepsilon}$	$2^+\to0^+$ gnd.		Coul. exc.	[*He 55c*]
${}_{34}Se^{76}_{42}$	0.567	$\tau_\gamma = 1.7\times10^{-11}$	$0.43\times\frac{1}{\varepsilon}$	$2^+\to0^+$ gnd.		Coul. exc.	[*He 55c*]
${}_{34}Se^{77}_{43}$	0.244	$\tau_\gamma = 1.5\times10^{-9}$	$0.12\times\frac{1}{\varepsilon}$	$({}^3/_2^-)\to{}^1/_2^-$ gnd.		Coul. exc.	[*He 55c*]
	0.457	$\tau_\gamma = 4.9\times10^{-11}$	$0.25\times\frac{1}{\varepsilon}$	$({}^5/_2^-)\to{}^1/_2^-$ gnd.		Coul. exc.	[*He 55c*]
${}_{34}Se^{78}_{44}$	0.615	$\tau_\gamma = 1.3\times10^{-11}$	$0.36\times\frac{1}{\varepsilon}$	$2^+\to0^+$ gnd.		Coul. exc.	[*He 55c*]
${}_{34}Se^{80}_{46}$	0.654	$\tau_\gamma = 1.5\times10^{-11}$	$0.23\times\frac{1}{\varepsilon}$	$2^+\to0^+$ gnd.		Coul. exc.	[*He 55c*]
${}_{34}Se^{82}_{48}$	0.880	$\tau_\gamma = 1.3\times10^{-11}$	$0.056\times\frac{1}{\varepsilon}$	$2^+\to0^+$ gnd.		Coul. exc.	[*He 55c*]

Table 5. (Continued).

Nucleus	E_{trans} (Mev)	$T_{\frac{1}{2}}$, τ or τ_γ in sec	$B(E2)_{up}$	$I_{upper} \to I_{lower}$	Mother isotope or reaction	Method	References
${}_{41}Nb^{97}_{56}$	0.665	$\tau > 1.5\times10^{-12}$		$(g_{\frac{7}{2}}) \to (d_{\frac{5}{2}})$ gnd.		Res. fluor.	[*Me 53*]
${}_{42}Mo^{94}_{52}$	0.871	$\tau_\gamma = 2.8\times10^{-12}$	$0.29\times\frac{1}{\varepsilon}$	$2^+ \to 0^+$ gnd.		Coul. exc.	[*He 55c*]
${}_{42}Mo^{95}_{53}$	0.204	$\tau_\gamma = 2.2\times10^{-9}$	$0.070\times\frac{1}{\varepsilon}$	$({}^3/_2{}^+) \to {}^5/_2{}^+$ gnd.		Coul. exc.	[*He 55c*], [*Ma 55e*]
${}_{42}Mo^{96}_{54}$	0.778	$\tau_\gamma = 4.5\times10^{-12}$	$0.31\times\frac{1}{\varepsilon}$	$2^+ \to 0^+$ gnd.		Coul. exc.	[*He 55c*]
${}_{42}Mo^{98}_{56}$	0.786	$\tau_\gamma = 5.0\times10^{-12}$	$0.27\times\frac{1}{\varepsilon}$	$2^+ \to 0^+$ gnd.		Coul. exc.	[*He 55c*], [*St 55b*]
${}_{42}Mo^{100}_{58}$	0.528	$\tau_\gamma = 1.5\times10^{-11}$	$0.66\times\frac{1}{\varepsilon}$	$2^+ \to 0^+$ gnd.		Coul. exc.	[*He 55c*], [*St 55b*]
${}_{43}Tc^{99}_{56}$	0.181	$T_{\frac{1}{2}} = 3.5\times10^{-9}$		$(d_{\frac{5}{2}}) \to g_{\frac{9}{2}}$ gnd.	Mo^{99}	Del. coin.	[*Le 55b*]
${}_{44}Ru^{99}_{55}$	0.090	$\tau_\gamma = (0.9\text{–}1.8)\times10^{-7}$	$0.054\times\frac{1}{\varepsilon}$	$({}^1/_2{}^+, {}^3/_2{}^+) \to {}^5/_2{}^+$ gnd.		Coul. exc.	[*He 55c*]
${}_{44}Ru^{100}_{56}$	0.540	$\tau_\gamma = 3.0\times10^{-11}$	$0.30\times\frac{1}{\varepsilon}$	$2^+ \to 0^+$ gnd.		Coul. exc.	[*He 55c*]
${}_{44}Ru^{101}_{57}$	0.127	$\tau_\gamma = 6.8\times10^{-8}$	$0.061\times\frac{1}{\varepsilon}$	$({}^9/_2{}^+) \to {}^5/_2{}^+$ gnd.		Coul. exc.	[*He 55c*]
	0.307	$\tau_\gamma = 1.1\times10^{-9}$	$0.036\times\frac{1}{\varepsilon}$	$({}^7/_2{}^+) \to {}^5/_2{}^+$ gnd.		Coul. exc.	[*He 55c*]
	0.522	$\tau_\gamma = (2\text{–}9)\times10^{-11}$	$0.041\times\frac{1}{\varepsilon}$	$? \to {}^5/_2{}^+$ gnd.		Coul. exc.	[*He 55c*]
${}_{44}Ru^{102}_{58}$	0.473	$\tau_\gamma = 2.6\times10^{-11}$	$0.63\times\frac{1}{\varepsilon}$	$2^+ \to 0^+$ gnd.		Coul. exc.	[*He 55c*]
${}_{44}Ru^{104}_{60}$	0.362	$\tau_\gamma = 6.9\times10^{-11}$	$1.04\times\frac{1}{\varepsilon}$	$2^+ \to 0^+$ gnd.		Coul. exc.	[*He 55c*]
${}_{45}Rh^{103}_{57}$	0.295	$\tau_\gamma = 2.9\times10^{-10}$	0.25	${}^3/_2{}^- \to p_{\frac{1}{2}}$ gnd.		Coul. exc.	[*St 55b*], [*He 55c*]
	0.365	$T_{\frac{1}{2}} = 6.39\times10^{-11}$	0.36	${}^5/_2{}^- \to p_{\frac{1}{2}}$ gnd.		Coul. exc.	[*St 55b*]
${}_{46}Pd^{104}_{58}$	0.550	$\tau_\gamma = 1.7\times10^{-11}$	$0.46\times\frac{1}{\varepsilon}$	$2^+ \to 0^+$ gnd.		Coul. exc.	[*He 55c*], [*St 55b*]
${}_{46}Pd^{105}_{59}$	0.266		$0.04\times\frac{1}{\varepsilon}$	$? \to {}^5/_2$ gnd.		Coul. exc.	[*He 55c*], [*Ma 55e*]
	0.433		$0.17\times\frac{1}{\varepsilon}$	$? \to {}^5/_2$ gnd.		Coul. exc.	[*He 55c*], [*Ma 55e*]
${}_{46}Pd^{106}_{60}$	0.513	$\tau_\gamma = 2.0\times10^{-11}$	$0.59\times\frac{1}{\varepsilon}$	$2^+ \to 0^+$ gnd.		Coul. exc.	[*He 55c*], [*St 55b*], [*Ma 55e*]
${}_{46}Pd^{108}_{62}$	0.424	$\tau_\gamma = 4.0\times10^{-11}$	$0.78\times\frac{1}{\varepsilon}$	$2^+ \to 0^+$ gnd.		Coul. exc.	[*He 55c*], [*St 55b*], [*Ma 55e*]
${}_{46}Pd^{110}_{64}$	0.370	$\tau_\gamma = 5.7\times10^{-11}$	$1.04\times\frac{1}{\varepsilon}$	$2^+ \to 0^+$ gnd.		Coul. exc.	[*He 55c*], [*St 55b*], [*Ma 55e*]
${}_{47}Ag^{107}_{60}$	0.319	$\tau_\gamma = 3.3\times10^{-10}$	0.16	${}^3/_2 \to p_{\frac{1}{2}}$ gnd.		Coul. exc.	[*Fa 55b*], [*He 55c*]
	0.419	$\tau_\gamma = 9.7\times10^{-11}$	0.23	${}^5/_2 \to p_{\frac{1}{2}}$ gnd.		Coul. exc.	[*Fa 55b*], [*He 55c*]
${}_{47}Ag^{107,109}_{60,62}$	0.325	$T_{\frac{1}{2}} = 4.71\times10^{-12}$ or 7.63×10^{-12}	0.23	${}^3/_2 \to p_{\frac{1}{2}}$ gnd.		Coul. exc.	[*St 55b*], [*Ma 55e*]
	0.427	$T_{\frac{1}{2}} = 2.99\times10^{-11}$	0.36	${}^5/_2 \to p_{\frac{1}{2}}$ gnd.		Coul. exc.	[*St 55b*]
${}_{47}Ag^{109}_{62}$	0.306	$\tau_\gamma = 3.6\times10^{-10}$	0.18	${}^3/_2 \to p_{\frac{1}{2}}$ gnd.		Coul. exc.	[*Fa 55b*], [*He 55c*]
	0.412	$\tau_\gamma = 1.0\times10^{-10}$	0.31	${}^5/_2 \to p_{\frac{1}{2}}$ gnd.		Coul. exc.	[*Fa 55b*], [*He 55c*]

Table 5. (Continued.)

Nucleus	E_{trans} (Mev)	$T_{\frac{1}{2}}$, τ or τ_γ in sec	$B(E2)_{up}$	$I_{upper} \to I_{lower}$	Mother isotope or reaction	Method	References
${}_{48}Cd^{110}_{62}$	0.654	$\tau_\gamma = 8.3\times10^{-12}$	$0.41\times\frac{1}{\varepsilon}$	$2^+\to0^+$ gnd.		Coul. exc.	[*He 55c*]
${}_{48}Cd^{111}_{63}$	0.243	$T_{\frac{1}{2}} = 8\times10^{-8}$		$d_{\frac{5}{2}}\to s_{\frac{1}{2}}$ gnd.	Ag^{111}	Del. coin.	[*Go 52*]
	0.340	$\tau_\gamma = 1.9\times10^{-10}$	$0.19\times\frac{1}{\varepsilon}$	$d_{\frac{3}{2}}\to s_{\frac{1}{2}}$ gnd.		Coul. exc.	[*He 55c*], [*Ma 55e*]
${}_{48}Cd^{112}_{64}$	0.620	$\tau_\gamma = 9.7\times10^{-12}$	$0.46\times\frac{1}{\varepsilon}$	$2^+\to0^+$ gnd.		Coul. exc.	[*He 55c*], [*Ma 55e*]
${}_{48}Cd^{113}_{65}$	0.290	$\tau_\gamma = 7\times10^{-10}$	$0.46\times\frac{1}{\varepsilon}$	$^3/_2\to{}^1/_2$ gnd.		Coul. exc.	[*He 55c*], [*Ma 55e*]
${}_{48}Cd^{114}_{66}$	0.550	$\tau_\gamma = 1.5\times10^{-11}$	$0.55\times\frac{1}{\varepsilon}$	$2^+\to0^+$ gnd.		Coul. exc.	[*He 55c*], [*Ma 55e*]
${}_{48}Cd^{116}_{68}$	0.508	$\tau_\gamma = 2.0\times10^{-11}$	$0.62\times\frac{1}{\varepsilon}$	$2^+\to0^+$ gnd.		Coul. exc.	[*He 55c*]
${}_{49}In^{115}_{66}$	0.500		0.058	$?\to{}^9/_2$ gnd.		Coul. exc.	[*Ma 55e*]
${}_{50}Sn^{119}_{69}$	0.024	$T_{\frac{1}{2}} = 1.85\times10^{-8}$		$d_{\frac{3}{2}}\to s_{\frac{1}{2}}$ gnd.	Sb^{119}	Del. coin.	[*Ol 56*]
${}_{52}Te^{120}_{68}$	0.56	$\tau_\gamma = 1.4\times10^{-11}$	$0.55\times\frac{1}{\varepsilon}$	$2^+\to0^+$ gnd.		Coul. exc.	[*He 55c*]
${}_{52}Te^{122}_{70}$	0.57	$\tau_\gamma = 1.4\times10^{-11}$	$0.47\times\frac{1}{\varepsilon}$	$2^+\to0^+$ gnd.		Coul. exc.	[*He 55c*]
${}_{52}Te^{123}_{71}$	0.159	$T_{\frac{1}{2}} = 1.9\times10^{-10}$		$d_{\frac{3}{2}}\to s_{\frac{1}{2}}$ gnd.	I^{123}	Del. coin.	[*Gr 53*]
			0.018			Coul. exc.	[*Fa 55b*]
${}_{52}Te^{124}_{72}$	0.608	$\tau_\gamma = 1.3\times10^{-11}$	$0.39\times\frac{1}{\varepsilon}$	$2^+\to0^+$ gnd.		Coul. exc.	[*He 55c*]
${}_{52}Te^{125}_{13}$	0.035	$T_{\frac{1}{2}} = 1.6\times10^{-9}$		$d_{\frac{3}{2}}\to s_{\frac{1}{2}}$ gnd.	Te^{125m}	Del. coin.	[*Gr 53*]
	0.466		0.44	$?\to s_{\frac{1}{2}}$ gnd.		Coul. exc.	[*Fa 55b*]
	0.668		0.26	$?\to s_{\frac{1}{2}}$ gnd.		Coul. exc.	[*Fa 55b*]
${}_{52}Te^{126}_{74}$	0.662	$\tau_\gamma = 1.0\times10^{-11}$	$0.32\times\frac{1}{\varepsilon}$	$2^+\to0^+$ gnd.		Coul. exc.	[*He 55c*]
${}_{52}Te^{128}_{76}$	0.75	$\tau_\gamma = 6.2\times10^{-12}$	$0.28\times\frac{1}{\varepsilon}$	$2^+\to0^+$ gnd.		Coul. exc.	[*He 55c*]
${}_{52}Te^{130}_{73}$	0.85	$\tau_\gamma = 3.5\times10^{-12}$	$0.26\times\frac{1}{\varepsilon}$	$2^+\to0^+$ gnd.		Coul. exc.	[*He 55c*]
${}_{53}I^{127}_{74}$	0.212			$?\to{}^5/_2$ gnd.		Coul. exc.	[*Ma 55e*]
${}_{54}Xe^{131}_{77}$	0.080	$T_{\frac{1}{2}} = 5\times10^{-10}$		$s_{\frac{1}{2}}\to d_{\frac{3}{2}}$ gnd.	I^{131}	Del. coin.	[*Go 52*]
	0.364	$\tau = \sim4\times10^{-11}$		$d_{\frac{5}{2}}\to d_{\frac{3}{2}}$ gnd.	I^{131}	Coin. pulse ht	[*Su 55a*]
${}_{55}Cs^{133}_{78}$	0.081	$T_{\frac{1}{2}} = 6\times10^{-9}$		$d_{\frac{5}{2}}\to g_{\frac{7}{2}}$ gnd.	Xe^{133}, Ba^{133}	Del. coin.	[*Gr 53*], [*Le 55b*]
${}_{55}Cs^{135}_{80}$	0.25	$T_{\frac{1}{2}} = 2.8\times10^{-10}$		$d_{\frac{5}{2}}\to g_{\frac{7}{2}}$ gnd.	Xe^{135}	Del. coin.	[*Gr 53*]
${}_{57}La^{139}_{82}$	0.163	$T_{\frac{1}{2}} = 1.5\times10^{-9}$			Ba^{139}	Del. coin.	[*Ge 55*]
${}_{59}Pr^{141}_{82}$	0.145	$\tau_\gamma = 3.5\times10^{-9}$		$^7/_2{}^+\to{}^5/_2{}^+$ gnd.		Res. fluor.	[*Me 55a*]
		$T_{\frac{1}{2}} = 1.0\times10^{-9}$			Ce^{141}	Del. coin.	[*De 55a*], [*Su 55a*]
${}_{60}Nd^{145}_{85}$	0.070		$\sim0.03\times\frac{1}{\varepsilon}$	$?\to{}^7/_2$ gnd.		Coul. exc.	[*He 55c*]
${}_{60}Nd^{146}_{86}$	0.455		$0.84\times\frac{1}{\varepsilon}$	$2^+\to0^+$ gnd.		Coul. exc.	[*He 55c*]
${}_{60}Nd^{148}_{88}$	0.300	$\tau = 2.6\times10^{-10}$	$1.5\times\frac{1}{\varepsilon}$	$2^+\to0^+$ gnd.		Coul. exc.	[*He 55c*], [*Si 55a*]
${}_{60}Nd^{150}_{90}$	0.128	$\tau = 7\times10^{-9}$	$1.24\times\frac{1}{\varepsilon}$	$2^+\to0^+$ gnd.		Coul. exc.	[*He 55c*], [*Si 55a*]
${}_{61}Pm^{147}_{86}$	0.092	$T_{\frac{1}{2}} = 2.44\times10^{-9}$			Nd^{147}	Del. coin.	[*Gr 53*]
${}_{62}Sm^{148}_{86}$	0.56		$2.06\times\frac{1}{\varepsilon}$	$2^+\to0^+$ gnd.		Coul. exc.	[*He 55c*], [*Ma 55c*]

Table 5. (Continued).

Nucleus	E_{trans} (Mev)	$T_{\frac{1}{2}}$, τ or τ_γ in sec	$B(E2)_{up}$	$I_{upper} \to I_{lower}$	Mother isotope or reaction	Method	References
${}_{62}Sm^{150}_{88}$	0.337		$2.32 \times \frac{1}{\varepsilon}$	$2^+ \to 0^+$ gnd.		Coul. exc.	[*He 55c*], [*Ma 55c*]
${}_{62}Sm^{152}_{90}$	0.122	$T_{\frac{1}{2}} = 1.4 \times 10^{-9}$		$2^+ \to 0^+$ gnd.	Eu^{152}	Del. coin.	[*Su 55*]
			$1.36 \times \frac{1}{\varepsilon}$			Coul. exc.	[*He 55c*], [*Ma 55c*]
${}_{62}Sm^{154}_{92}$	0.082		$0.48 \times \frac{1}{\varepsilon}$	$2^+ \to 0^+$ gnd.		Coul. exc.	[*He 55c*], [*Ma 55c*]
${}_{63}Eu^{151}_{88}$	0.300		$0.64 \times \frac{1}{\varepsilon}$	$? \to {}^5/_2$ gnd.		Coul. exc.	[*He 55c*], [*Ma 55c*]
${}_{63}Eu^{153}_{90}$	0.070	$T_{\frac{1}{2}} = 1.4 \times 10^{-10}$		${}^1/_2{}^+ \to {}^3/_2{}^+$	Sm^{153}	Del. coin.	[*Gr 54*]
	0.103	$T_{\frac{1}{2}} = 4 \times 10^{-9}$		${}^3/_2{}^+ \to {}^5/_2{}^+$ gnd.	Sm^{153}	Del. coin.	[*Gr 54*], [*Mc 54a*], [*Su 55a*]
	0.082		$0.36 \times \frac{1}{\varepsilon}$	${}^7/_2 \to {}^5/_2{}^+$ gnd.		Coul. exc.	[*He 55c*], [*Ma 55c*]
	0.187		$0.49 \times \frac{1}{\varepsilon}$	${}^9/_2 \to {}^5/_2{}^+$ gnd.		Coul. exc.	[*He 55c*], [*Ma 55c*]
${}_{64}Gd^{154}_{90}$	0.123	$T_{\frac{1}{2}} = 1.2 \times 10^{-9}$		$2^+ \to 0^+$ gnd.	Eu^{154}	Del. coin.	[*Su 55*]
			$1.9 \times \frac{1}{\varepsilon}$			Coul. exc.	[*He 55c*], [*Ma 55c*]
${}_{64}Gd^{155}_{91}$	0.145		$0.085 \times \frac{1}{\varepsilon}$	$? \to {}^7/_2$ gnd.		Coul. exc.	[*He 55c*], [*Ma 55c*]
${}_{64}Gd^{156}_{92}$	0.089		$0.88 \times \frac{1}{\varepsilon}$	$2^+ \to 0^+$ gnd.		Coul. exc.	[*He 55c*], [*Ma 55c*]
${}_{64}Gd^{157}_{93}$	0.131		$0.61 \times \frac{1}{\varepsilon}$	$? \to {}^7/_2$ gnd.		Coul. exc.	[*He 55c*], [*Ma 55c*]
${}_{64}Gd^{158}_{94}$	0.079		$0.63 \times \frac{1}{\varepsilon}$	$2^+ \to 0^+$ gnd.		Coul. exc.	[*He 55c*], [*Ma 55c*]
${}_{64}Gd^{160}_{96}$	0.076		$0.50 \times \frac{1}{\varepsilon}$	$2^+ \to 0^+$ gnd.		Coul. exc.	[*He 55c*], [*Ma 55c*]
${}_{65}Tb^{159}_{94}$	0.136		$0.041 \times \frac{1}{\varepsilon}$			Coul. exc.	[*He 55c*], [*Ma 55c*]
${}_{66}Dy^{160}_{94}$	0.085	$T_{\frac{1}{2}} = 1.8 \times 10^{-9}$		$2^+ \to 0^+$ gnd.		Del. coin.	[*Mc 52a*]
${}_{66}Dy^{161-164}_{95-98}$	0.076		$0.23 \times \frac{1}{\varepsilon}$			Coul. exc.	[*He 55c*]
${}_{66}Dy^{161,163}_{95,97}$	0.166		$0.29 \times \frac{1}{\varepsilon}$	$? \to {}^7/_2$ gnd.		Coul. exc.	[*He 55c*]
${}_{67}Ho^{165}_{98}$	0.094		$0.54 \times \frac{1}{\varepsilon}$	$? \to {}^7/_2$ gnd.		Coul. exc.	[*He 55c*]
	0.206		$0.036 \times \frac{1}{\varepsilon}$	$? \to {}^7/_2$ gnd.		Coul. exc.	[*He 55c*]
${}_{68}Er^{162-170}_{94-102}$	0.079		$0.48 \times \frac{1}{\varepsilon}$			Coul. exc.	[*He 55c*]
${}_{68}Er^{164}_{96}$	0.091	$T_{\frac{1}{2}} = 1.4 \times 10^{-9}$		$2^+ \to 0^+$ gnd.	Ho^{164}	Del. coin.	[*Br 54f*]
${}_{68}Er^{166}_{98}$	0.080	$T_{\frac{1}{2}} = 1.8 \times 10^{-9}$		$2^+ \to 0^+$ gnd.	Ho^{166}	Del. coin.	[*Gr 55*]
	0.184	$T_{\frac{1}{2}} = \sim 8 \times 10^{-11}$		$4^+ \to 2^+$ first	Ho^{166}	Del. coin.	[*Mi 55*]
${}_{68}Er^{(167)}_{(99)}$	0.172		$0.081 \times \frac{1}{\varepsilon}$	$? \to {}^7/_2$ gnd.		Coul. exc.	[*He 55c*]
${}_{69}Tm^{169}_{100}$	0.109		$1.12 \times \frac{1}{\varepsilon}$	$? \to {}^1/_2$ gnd.		Coul. exc.	[*He 55c*]
	cascade)						
	0.308	$T_{\frac{1}{2}} = 6.4 \times 10^{-7}$		${}^7/_2{}^+ \to {}^3/_2{}^+$ 1st	Y^{169}	Del. coin.	[*Mi 56*]
	0.063	$T_{\frac{1}{2}} = 4.5 \times 10^{-8}$		${}^9/_2{}^- \to {}^7/_2{}^+$ 4th	Y^{169}	Del. coin.	[*Mi 56*]
	0.094	$T_{\frac{1}{2}} = 4.0 \times 10^{-7}$		${}^{11}/_2{}^- \to {}^9/_2{}^-$ 5th	Y^{169}	Del. coin.	[*Mi 56*]

Table 5. (Continued.)

Nucleus	E_{trans} (Mev)	$T_{\frac12}$, τ or τ_γ in sec	$B(E2)_{up}$	$I_{upper}\to I_{lower}$	Mother isotope or reaction	Method	References
$_{70}Yb^{170}_{100}$	0.084	$T_{\frac12}=1.57\times10^{-9}$		$2^+\to0^+$ gnd.	Tm^{170}	Del. coin.	[*Gr 52d*]
$_{70}Yb^{170-176}_{100-106}$	0.078		$0.28\times\frac{1}{\varepsilon}$			Coul. exc.	[*He 55c*]
$_{70}Yb^{(171)}_{(101)}$	0.110		$0.20\times\frac{1}{\varepsilon}$	$?\to{}^1/_2$ gnd.		Coul. exc.	[*He 55c*]
$_{70}Yb^{(173)}_{(103)}$	0.180		$0.086\times\frac{1}{\varepsilon}$	$?\to{}^5/_2$ gnd.		Coul. exc.	[*He 55c*]
$_{71}Lu^{175}_{104}$	0.114		$0.72\times\frac{1}{\varepsilon}$	$?\to{}^7/_2$ gnd.		Coul. exc.	[*He 55c*]
	0.250		$0.20\times\frac{1}{\varepsilon}$	$?\to{}^7/_2$ gnd.		Coul. exc.	[*He 55c*]
$_{71}Lu^{(176)}_{(105)}$	0.180		$(1.14)\times\frac{1}{\varepsilon}$			Coul. exc.	[*He 55c*]
$_{71}Lu^{177}_{106}$	0.150	$T_{\frac12}=1.3\times10^{-7}$			Yb^{177}	Del. coin.	[*Go 52*]
$_{72}Hf^{176}_{104}$	0.089	$T_{\frac12}=1.35\times10^{-9}$		$2^+\to0^+$ gnd.	Lu^{176}	Del. coin.	[*Mc 52*]
			$0.56\times\frac{1}{\varepsilon}$			Coul. exc.	[*St 55b*], [*He 55c*]
$_{72}Hf^{177}_{105}$	0.112		$0.77\times\frac{1}{\varepsilon}$			Coul. exc.	[*He 55c*]
	0.250		$0.55\times\frac{1}{\varepsilon}$			Coul. exc.	[*He 55c*]
$_{72}Hf^{178}_{106}$	0.090		$0.85\times\frac{1}{\varepsilon}$	$2^+\to0^+$ gnd.		Coul. exc.	[*He 55c*]
$_{72}Hf^{179}_{107}$	0.119		$0.67\times\frac{1}{\varepsilon}$			Coul. exc.	[*He 55c*]
$_{72}Hf^{180}_{108}$	0.093	$T_{\frac12}=1.4\times10^{-9}$		$2^+\to0^+$ gnd.	Ta^{180}	Del. coin.	[*Su 55*]
			$0.78\times\frac{1}{\varepsilon}$			Coul. exc.	[*He 55*]
$_{73}Ta^{181}_{108}$	0.137	$T_{\frac12}=3.78\times10^{-11}$	$0.70\times\frac{1}{\varepsilon}$	${}^9/_2{}^+\to g_{\frac72}$ gnd.		Coul. exc.	[*St 55b*], [*He 55c*], [*Be 55e*]
	0.303	$T_{\frac12}=1.13\times10^{-11}$	$0.15\times\frac{1}{\varepsilon}$	$?\to g_{\frac72}$ gnd.		Coul. exc.	[*St 55b*], [*He 55c*]
	0.481	$T_{\frac12}=1.04\times10^{-8}$		${}^9/_2{}^+\to g_{\frac72}$ gnd.	Hf^{181}	Del. coin.	[*En 53*], [*De 52a*], [*Di 55*]
$_{74}W^{182}_{108}$	0.100	$T_{\frac12}=1.27\times10^{-9}$		$2^+\to0^+$ gnd.	Ta^{182}	Del. coin.	[*Su 55*], [*Su 54b*]
$_{74}W^{182,184,186}_{108,110,112}$	0.112	$T_{\frac12}=7.88\times10^{-10}$	5.44	$2^+\to0^+$ gnd.		Coul. exc.	[*St 55b*]
$_{75}Re^{187}_{112}$	0.072	$T_{\frac12}=5.5\times10^{-7}$		2nd→ 1st	W^{187}	Del. coin.	[*Su 53b*]
$_{76}Os^{186}_{110}$	0.137	$T_{\frac12}=1.8\times10^{-9}$		$2^+\to0^+$ gnd.	Re^{186}	Del. coin.	[*Mc 53b*], [*Mc 51*]
$_{76}Os^{188}_{112}$	0.155	$T_{\frac12}=1.7\times10^{-9}$		$2^+\to0^+$ gnd.	Re^{188}	Del. coin.	[*Mc 53b*]
		$T_{\frac12}=6.5\times10^{-10}$				Del. coin.	[*Su 55*]
$_{77}Ir^{191}_{114}$	0.082	$T_{\frac12}=3.9\times10^{-9}$			Pt^{191}	Del. coin.	[*Su 55*]
$_{77}Ir^{193}_{116}$	0.073	$T_{\frac12}=6.2\times10^{-9}$		${}^7/_2{}^+\to{}^3/_2{}^+$ gnd.	Os^{193}	Del. coin.	[*De 54b*]
$_{78}Pt^{194}_{116}$	0.330	$T_{\frac12}=3.85\times10^{-11}$	1.4	$2^+\to0^+$ gnd.		Coul. exc.	[*St 55b*]
$_{78}Pt^{195}_{117}$	0.029			${}^3/_2{}^-\to p_{\frac12}$ gnd.		Coul. exc.	[*Be 55e*]
	0.126		0.5	$f_{\frac52}\to p_{\frac12}$ gnd.		Coul. exc.	[*Be 55e*]
	0.210		0.5	${}^3/_2{}^-\to p_{\frac12}$ gnd.		Coul. exc.	[*Be 55e*]
	0.240			${}^5/_2{}^-\to p_{\frac12}$ gnd.		Coul. exc.	[*Be 55e*]
$_{78}Pt^{196}_{118}$	0.358	$T_{\frac12}=3.49\times10^{-11}$	1.4	$2^+\to0^+$ gnd.		Coul. exc.	[*St 55b*]
$_{78}Pt^{198}_{120}$	0.403	$T_{\frac12}=1.88\times10^{-11}$	1.3	$2^+\to0^+$ gnd.		Coul. exc.	[*St 55b*]

Table 5. (Continued).

Nucleus	E_{trans} (Mev)	$T_{\frac{1}{2}}$, τ or τ_γ in sec	$B(E2)_{up}$	$I_{upper} \to I_{lower}$	Mother isotope or reaction	Method	References
$_{79}Au^{197}_{118}$	0.077	$T_{\frac{1}{2}} = 1.90 \times 10^{-10}$		$^3/_2{}^+ \to d_{\frac{3}{2}}$ gnd.	Pt^{197}	Del. coin.	[*Su 55*]
			0.18			Coul. exc.	[*Be 55e*]
	0.268		0.13	$(^3/_2{}^+, {}^1/_2{}^+ \to d_{\frac{3}{2}}$ gnd.		Coul. exc.	[*Be 55e*]
	0.277	$T_{\frac{1}{2}} = 4.11 \times 10^{-11}$	0.31	$d_{\frac{5}{2}} \to d_{\frac{3}{2}}$ gnd.		Coul. exc.	[*St 55b*], [*Be 55e*]
	0.550	$T_{\frac{1}{2}} = 4.43 \times 10^{-12}$	0.42	$^7/_2 \to d_{\frac{3}{2}}$ gnd.		Coul. exc.	[*St 55b*]
$_{80}Hg^{197}_{117}$	0.133	$T_{\frac{1}{2}} = 8 \times 10^{-9}$		$f_{\frac{5}{2}} \to p_{\frac{1}{2}}$ gnd.	$Hg^{197\,m}$	Del. coin.	[*Co 55*], [*Go 52*]
$_{80}Hg^{198}_{118}$	0.411	$T_{\frac{1}{2}} = 2.2 \times 10^{-11}$		$2^+ \to 0^+$ gnd.		Res. fluor.	[*Da 53*], [*Mo 53a*], [*Ma 53d*]
		$\tau = 3.15 \times 10^{-11}$				Res. fluor.	[*Me 54*]
		$\tau = \sim 3.5 \times 10^{-11}$			Au^{198}	Coin. pulse ht	[*Su 55a*]
$_{80}Hg^{199}_{119}$	0.159	$T_{\frac{1}{2}} = 2.4 \times 10^{-9}$		$f_{\frac{5}{2}} \to p_{\frac{1}{2}}$ gnd.	Au^{199}	Del. coin.	[*Go 52*]
	0.209	$\tau = 3.1 \times 10^{-10}$		$p_{\frac{3}{2}} \to p_{\frac{1}{2}}$ gnd.		Res. fluor.	[*Me 54a*]
		$T_{\frac{1}{2}} = 7 \times 10^{-11}$			Au^{199}	Del. coin.	[*Gr 55e*]
$_{80}Hg^{202}_{122}$	0.439	$\tau = 2.4 \times 10^{-11}$				Res. fluor.	[*Me 55b*]
$_{81}Tl^{203}_{122}$	0.280	$T_{\frac{1}{2}} = 3.25 \times 10^{-10}$	0.13			Coul. exc.	[*St 55b*], [*Ba 55d*]
		$\tau_\gamma = \sim 10^{-9}$				Res. fluor.	[*Me 54b*]
		$T_{\frac{1}{2}} = 1.2 \times 10^{-10}$			Hg^{203}	Del. coin.	[*De 55b*]
		$T_{\frac{1}{2}} = 3 \times 10^{-10}$			Hg^{203}	Del. coin.	[*Az 55*]
	0.410	$T_{\frac{1}{2}} = 1.12 \times 10^{-12}$	0.14			Coul. exc.	[*St 55b*]
$_{81}Tl^{205}_{124}$	0.205	$T_{\frac{1}{2}} = 2.48 \times 10^{-10}$	0.14			Coul. exc.	[*St 55b*], [*Ba 55d*]
$_{82}Pb^{204}_{122}$	0.374	$T_{\frac{1}{2}} = 2.6 \times 10^{-7}$		$4^+ \to 2^+$ first	$Pb^{204\,m}$	Del. coin.	[*Kr 54*]
$_{82}Pb^{206}_{124}$	0.803	$T_{\frac{1}{2}} = 7.7 \times 10^{-12}$	0.16	$2^+ \to 0^+$ gnd.		Coul. exc.	[*St 55b*]
$_{82}Pb^{207}_{125}$	0.570	$T_{\frac{1}{2}} = 1.0 \times 10^{-10}$	0.028	$f_{\frac{5}{2}} \to p_{\frac{1}{2}}$ gnd.		Coul. exc.	[*St 55b*]
		$\tau = 1.2 \times 10^{-10}$			Bi^{207}	Coin. pulse ht	[*Su 55a*]
$_{82}Pb^{208}_{126}$	0.583	$T_{\frac{1}{2}} = 2.4 \times 10^{-10}$		$5^- \to 3^-$ first	Tl^{208}	Del. coin.	[*El 54a*]
$_{82}Pb^{209}_{127}$	0.120	$T_{\frac{1}{2}} = 3.1 \times 10^{-9}$		$p_{\frac{1}{2}}$ 3rd $\to d_{\frac{5}{2}}$ 2nd	Tl^{209}	Del. coin.	[*St 56b*]
$_{82}Pb^{212}_{130}$	0.239	$T_{\frac{1}{2}} = (2\text{–}20) \times 10^{-14}$				x-ray diffraction	[*Kn 54a*]
$_{84}Po^{210}_{126}$	0.047	$T_{\frac{1}{2}} = 1.5 \times 10^{-9}$			At^{210}	Del. coin.	[*Su 55*]
$_{89}Ac^{227}_{138}$	0.027	$T_{\frac{1}{2}} = 4.2 \times 10^{-8}$			Pa^{231}	Del. coin.	[*Te 53*]
$_{90}Th^{232}_{142}$	0.76		0.13			Coul. exc.	[*St 55b*]
$_{91}Pa^{231}_{140}$	0.026 and 0.085	$T_{\frac{1}{2}} = 4.1 \times 10^{-8}$			Th^{231}	Del. coin.	[*St 55c*]
$_{94}Pu^{239}_{145}$	0.105 and 0.061	$T_{\frac{1}{2}} = 1.93 \times 10^{-7}$			Np^{239}	Del. coin.	[*En 55*]

F. Comparative lifetimes.

43. Discussion of the systematics. While the problem of identifying the multipole orders of isomeric transitions and of measuring their half-lives are still receiving considerable experimental attention, the procedures are much more straightforward and reliable than they were a few years ago. Theoretical and experimental conversion coefficients and ratios, combined in many cases with lifetime or angular correlation measurements and the shell model predictions have meant that almost positive identifications of many nuclear level configurations have been possible.

It is natural that with our increased confidence in assignments of multipole orders for isomeric transitions the emphasis should shift in the direction of investigating the details of nuclear structure. We are concerned here with the mechanism responsible for the transition of a nucleus from an isomeric state to a level of lower energy. The effect of nuclear structure, aside from determining the arrangements of nucleons or configurations of the two levels, should be exhibited most strongly by the transition rate. Thus if the transition can be represented best by the jump of a single particle, say a proton, then the WEISSKOPF formula for predicting the transition probability and therefore the lifetime of the state should be applicable. On the other hand, if the transition involves two nucleons or the collective motion of many nucleons, a transition rate would result which is considerably different from that of a single particle. Another possibility is that the transition is predominantly that of a single particle but that the remaining core of nucleons cooperates in some way so as to make the transition faster than that expected of a single particle alone.

One might hope to gain information about nuclear structure by studying the systematic behavior, if any, of transition rates with mass number A, atomic number Z or neutron number N. In doing so it is obvious that all isomers of a given multipole order should be grouped together. It is further desirable to reduce all transitions for a given multipole order to an equal footing by removing any factors which may affect their lifetimes but which have little to do with the specific nuclear mechanism. Two such factors are the transition energy E and the mass number A. For purposes of comparison a nuclear model must be used in order to apply the proper powers of E and A. GOLDHABER and SUNYAR ([*Si 55*], Chap. 16) have used the single-particle model according to which the transition probabilities follow the relationships given in Table 3. Since $\tau_\gamma = 1/\lambda$ the transition radiative lifetimes τ_γ should be proportional to the negative powers of E and A as given in Table 3. Hence if we multiply the observed value of τ_γ by the corresponding positive powers of E and A the effects of energy and mass number will be removed. The quantities

$$\tau_\gamma A^{\frac{2\Delta I}{3}} E_\gamma^{2\Delta I+1} \tag{43.1}$$

for electric transitions and

$$\tau_\gamma A^{\frac{2\Delta I-2}{3}} E_\gamma^{2\Delta I+1} \tag{43.2}$$

for magnetic transitions are known as comparative lifetimes.

Data on comparative lifetimes have been compiled and analysed in considerable detail by GOLDHABER and SUNYAR ([*Si 55*], Chap. 16). Their results are summarized in Figs. 95 through 100 where the $\log_{10}$ of the comparative lifetime has been plotted against neutron number. In every case the theoretical single proton lifetime corresponding to unit matrix element is given as a horizontal line. Thus, if a point is above the single-particle line, the lifetime is longer and the transition probability is less than predicted by the WEISSKOPF formula. As may be seen by inspection of the figures the lifetimes of almost all but the $E2$ transitions are longer than the single-particle value. $E2$ transitions are unique in that they vary smoothly as a function of neutron number and are a factor of 10 to 100 faster than the single-particle rate in regions between the closed shells. GOLDHABER and SUNYAR have interpreted this feature as evidence of collective effects corresponding to the rotational states of BOHR and MOTTELSON [*Bo 53d*] (see Sect. 6).

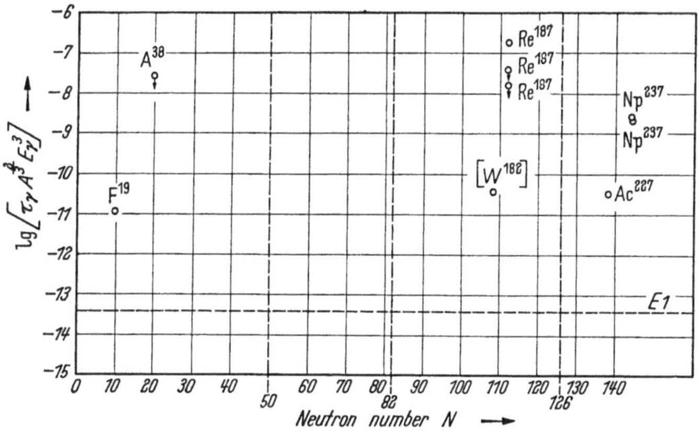

Fig. 95. Comparative lifetimes for low energy $E1$ transitions, plotted against neutron number N. The theoretical line corresponds to matrix elements equal to unity for single proton transitions. (Courtesy Drs. M. GOLDHABER and A. W. SUNYAR.)

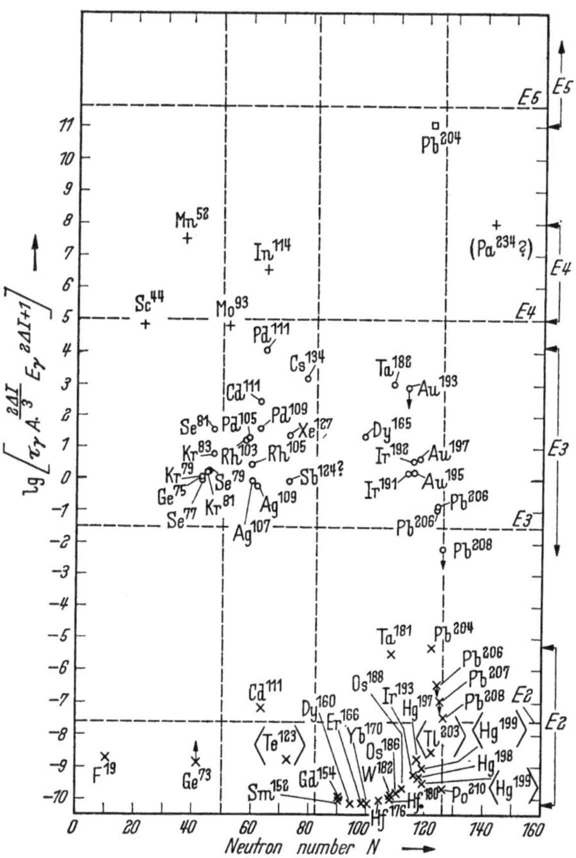

Fig. 96. Comparative lifetimes for $E2$, $E3$, $E4$ and $E5$ transitions, plotted against neutron number N. The theoretical lines, corresponding to matrix elements equal to unity for single particle proton transitions, are also shown. Points marked ⟨⟩ are based on mixed transitions. (Courtesy of Drs. M. GOLDHABER and A. W. SUNYAR.)

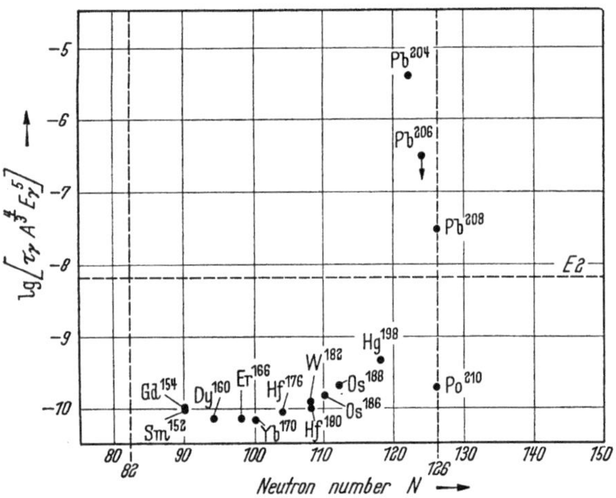

Fig. 97.

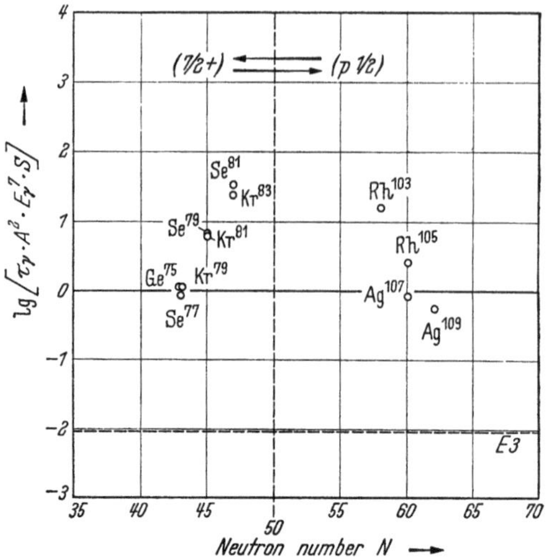

Fig. 98.

Fig. 97. Comparative lifetimes for $E2$ transitions in even-even nuclei, plotted against neutron number N. The theoretical line, corresponding to matrix elements equal to unity for single particle proton transitions, is also shown. (Courtesy of Drs. M. GOLDHABER and A. W. SUNYAR.)

Fig. 98. Comparative lifetimes of $^7/_2{}^+ \xleftrightarrow{E3} p_{\frac{1}{2}}$ transitions, plotted against neutron number N. The statistical weight correction was made as though we were dealing with single particle transitions. (Courtesy of Drs. M. GOLDHABER and A. W. SUNYAR.)

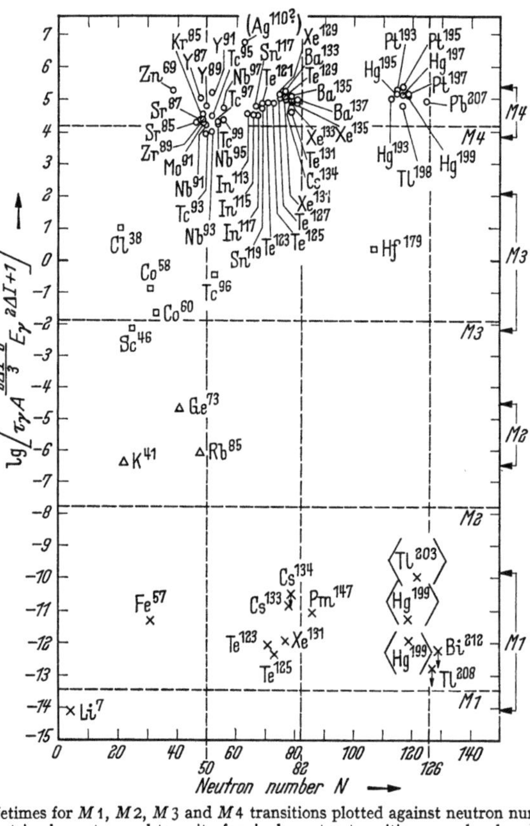

Fig. 99. Comparative lifetimes for $M1$, $M2$, $M3$ and $M4$ transitions plotted against neutron number N. The theoretical lines corresponding to matrix elements equal to unity for single proton transitions are also shown. Points marked ⟨⟩ are based on mixed transitions. (Courtesy of Drs. M. GOLDHABER and A. W. SUNYAR.)

In the case of magnetic transitions the matrix elements for $M4$ transitions show a remarkable uniformity whereas other magnetic transition matrix elements scatter appreciably.

Such studies have indicated that collective effects predominate in certain regions of the periodic table and for certain multipole orders whereas for other

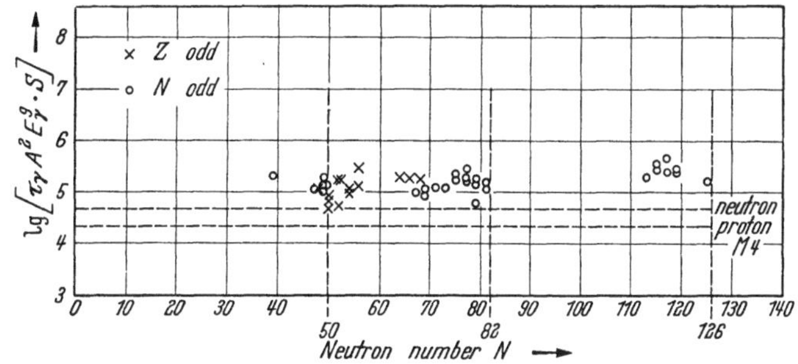

Fig. 100. Comparative lifetimes for $M4$ transitions ocurring in odd A nuclei. Statistical weight corrections have been applied. (Courtesy of Drs. M. GOLDHABER and A. W. SUNYAR.)

regions the single-particle representation is more satisfactory. While theory can make partial explanations of the data on transitions rates there is still no universal quantum-mechanical description of the nucleus as regards isomeric transitions.

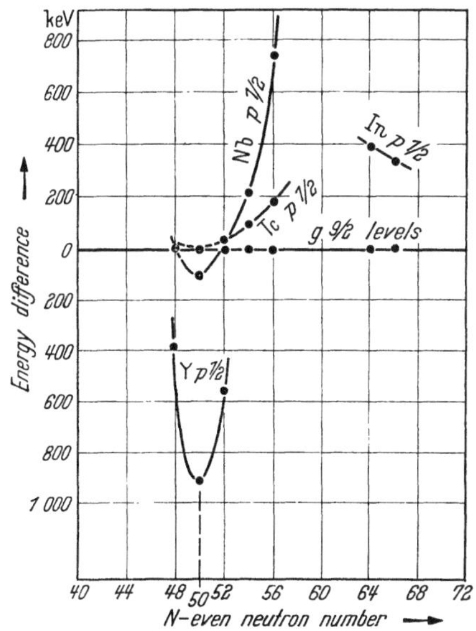

Fig. 101. Energy difference between $2p_{\frac{1}{2}}$ levels and $1g_{\frac{9}{2}}$ levels for odd-proton isomers, plotted as a function of N. (Courtesy of Drs. M. GOLDHABER and R. D. HILL.)

The explanation of the energy spacings of nuclear states also remains as a challenge for a more detailed theory. It has already been pointed out (Sect. 6) that the spacings of rotational levels are well accounted for by the collective model. Those isomeric levels which are long-lived and whose spin and parity assignments are consistent with the shell model have not been explained with similar accuracy as regards energy position, although certain regularities in the spacings exist, as was first discovered by HILL [*Hi 50*]. Regularities have been found both in odd-A nuclei [*Go 52*] and in even-even nuclei [*Sc 53e*], [*Pr 53*]. An example of the behavior of level spacings is shown in Fig. 101 where the energy separation between $p_{\frac{1}{2}}$ and $g_{\frac{9}{2}}$ levels in odd-proton isomers is plotted as a function of neutron number. Such information on the systematic variation of level spacing is of value not only for theoretical analysis but for predicting the occurence of unknown isomers. Failure to observe such a predicted isomeric transition sometimes happens. An interesting case is the recent search by STOCKENDAL *et al.* [*St 56a*] for an $M4$ isomeric transition in Pb^{205}. The nuclei Pb^{197}, Pb^{199}, Pb^{201}, Pb^{203} and Pb^{207} all exhibit $M4$ isomeric transitions between $i_{\frac{13}{2}}$ and $f_{\frac{5}{2}}$ states whose energies vary in a smooth fashion. The non-appearance of a similar transition in Pb^{205} may result from the presence of other interfering levels to which competing transitions take place.

Acknowledgements.

The author is indebted to Drs. M. GOLDHABER, G. SCHARFF-GOLDHABER and A. W. SUNYAR for helpful comments on the text and to Drs M. GOLDHABER and R. D. HILL for the use of some of the figures from Ref. [*Go 52*]. Thanks are due to many colleagues who furnished information on isomeric decay schemes prior to publication.

References.

[*Aa 54*] AAMODT, FLETCHER, SILVEY and TOWNES: Phys. Rev. **94**, 789(A) (1954).
[*Aa 55*] AAMODT, L. C., and P. C. FLETCHER: Phys. Rev. **98**, 1224 (1955).
[*Ad 54*] ADER, M.: J. Phys. Radium **15**, 60 (1954).
[*Aj 55*] AJZENBERG, F., and T. LAURITSEN: Rev. Mod. Phys. **27**, 77 (1955).
[*Ak 54*] AKSENOV, S. I., and K. V. VLADIMIRSKII: Dokl. Akad. Nauk SSSR. **96**, 37 (1954).
[*Al 50*] ALVAREZ, L. W.: Phys. Rev. **80**, 519 (1950).
[*Al 52*] ALBURGER, D. E.: Phys. Rev. **88**, 339 (1952).
[*Al 52a*] ALBURGER, D. E.: Phys. Rev. **88**, 1257 (1952).
[*Al 52b*] ALMQVIST, S., and S. A. E. JOHANSSON: Nature, Lond. **170**, 583 (1952).
[*Al 53*] ALBURGER, D. E.: Phys. Rev. **92**, 1257 (1953).
[*Al 53a*] ALBURGER, D. E., and M. H. L. PRYCE: Phys. Rev. **92**, 514 (1953); **95**, 1482 (1954).
[*Al 53b*] ALBURGER, D. E., and S. THULIN: Phys. Rev. **89**, 1146 (1953).
[*Al 53c*] ALBERS-SCHÖNBERG, ALDER, BRAUN, HEER and NOVEY: Phys. Rev. **91**, 1287 (1953).
[*Al 53d*] ALKHAZOV, LEMBERG and GRINBERG: Izv. Akad. Nauk SSSR., Ser. Fiz. **17**, 487 (1953).
[*Al 54*] ALTHOFF, K. H., and H. KRÜGER: Naturwiss. **41**, 368 (1954).
[*Al 54a*] ALBERS-SCHÖNBERG, HEER, NOVEY and SCHERRER: Helv. phys. Acta **27**, 547, 637 (1954).
[*Al 54b*] ALDER, K., and A. WINTHER: Phys. Rev. **96**, 237 (1954).
[*Al 55*] ALBURGER, D. E., and B. J. TOPPEL: Phys. Rev. **100**, 1357 (1955).
[*Al 55a*] ALBURGER, D. E., and A. W. SUNYAR: Phys. Rev. **99**, 695 (1955).
[*Al 55b*] ALAGA, ALDER, BOHR and MOTTELSON: Kgl. danske Vid. Selsk., mat.-fys. Medd. **29**, Nr. 9 (1955).
[*Am 35*] AMALDI, D'AGOSTINO, FERMI, PONTECORVO and SEGRÈ: Ric. sci. **61**, 581 (1935).
[*Am 53*] AMES, BUNKER, LANGER and SORENSON: Phys. Rev. **91**, 68 (1953).
[*An 54*] ANDERSON, G.: Phil. Mag. **45**, 621 (1954).
[*An 54a*] ANTONOVA, I. A., and I. V. ESTULIN: Izv. Akad. Nauk SSSR., Ser. Fiz. **18**, 79 (1954).
[*An 55*] ANTONOVA, I., and U. ESTULIN: J. Phys. Radium **16**, 534 (1955).
[*An 55a*] ANDERSSON, ARBMAN, BERGSTRÖM and WAPSTRA: Phil. Mag. **46**, 70 (1955).
[*Ar 53*] ARNOLD, J. R., and T. SUGIHARA: Phys. Rev. **90**, 332 (1953).
[*Ar 53a*] ARBER, W., and P. STÄHELIN: Helv. phys. Acta **26**, 433, 584 (A), 691 (1953). — Phys. Rev. **92**, 1076 (1953).
[*Ar 54*] ARNOLD, J. R.: Phys. Rev. **93**, 743 (1954).
[*At 52*] ATEN, DE WIJS and BOELHOUWER: Physica, Haag **18**, 1032 (1952).
[*At 53*] ATEN, DE VRIES-HAMERLING and LINDNER: Physica, Haag **19**, 1046 (1953).
[*At 53a*] ATTERLING, H., and S. THULIN: Nature, Lond. **171**, 927 (1953).
[*At 53b*] ATEN, A. H. W., and G. D. DE FEYFER: Physica, Haag **19**, 1143 (1953).
[*At 54*] ATEN, MANASSEN and DE FEYFER: Physica, Haag **20**, 665 (1954).
[*At 55*] ATEN, DE FEYFER, STERK and WAPSTRA: Physica, Haag **21**, 740 (1955).
[*At 55a*] ATEN, A. H. W., and G. D. DE FEYFER: Physica, Haag **21**, 543 (1955).

[Av 53] AVIGNON, P.: J. Phys. Radium **14**, 636 (1953).
[Av 53a] AVIGNON, P.: C. R. Acad. Sci., Paris **237**, 157 (1953).
[Av 55] AVIGNON, MICHALOWICZ and BOUCHEZ: J. Phys. Radium **16**, 404 (1955).
[Ax 49] AXEL, P., and S. M. DANCOFF: Phys. Rev. **76**, 892 (1949).
[Ax 53] AXEL, P.: Brookhaven National Laboratory Report BNL 271 ($T-44$), 1953. Unpublished.
[Ax 54] AXEL, P., and J. D. FOX: Phys. Rev. **95**, 613(A) (1954).
[Ax 55] AXEL, P.: Private communication.
[Ax 55a] AXEL, FOX and PARKER: Phys. Rev. **97**, 975 (1955).
[Az 54] AZUMA, T.: Phys. Rev. **94**, 638 (1954). — J. Phys. Soc. Japan **10**, 167 (1955).
[Az 54a] AZUMA, T.: J. Phys. Soc. Japan **9**, 1 (1954).
[Az 55] AZUMA, R. E., and G. M. LEWIS: Phil. Mag. **46**, 1034 (1955).
[Az 55a] AZUMA, R. E.: Phil. Mag. **46**, 1031 (1955).

[Ba 50] BAY, Z.: Phys. Rev. **77**, 419 (1950).
[Ba 52] BASHILOV, ANTON'EVA and DZHELEPOV: Izv. Akad. Nauk SSSR., Ser. Fiz. **16**, 264 (1952).
[Ba 52a] BÄHNISCH, HUSTER and WALCHER: Naturwiss. **39**, 379 (1952).
[Ba 52b] BALFOUR, J. G.: Quoted by S. C. CURRAN, Physica, Haag **18**, 1161 (1952).
[Ba 53] BAINBRIDGE, GOLDHABER and WILSON: Phys. Rev. **84**, 1260 (1951); **90**, 430 (1953).
[Ba 53a] BAKER, BLEANEY, BOWERS, SHAW and TRENAM: Proc. Phys. Soc. Lond. A **66**, 305 (1953).
[Ba 53b] BARTHOLOMEW, BROWN, HOWELL, SHOREY and YAFFE: Canad. J. Phys. **31**, 714 (1953).
[Ba 53c] BARTHOLOMEW, BROWN, HAWKINGS, MERRITT and YAFFE: Canad. J. Phys. **31**, 120 (1953).
[Ba 53d] R. BALLINI: Ann. Physik **8**, 441 (1953).
[Ba 53e] BASHILOV, ANTON'EVA, DZHELEPOV and DOLGENTSEVA: Izv. Akad. Nauk SSSR., Ser. Fiz. **17**, 437 (1953).
[Ba 53f] BASHILOV, DZHELEPOV and CHERVENSKAYA: Izv. Akad. Nauk SSSR., Ser. Fiz. **17**, 428 (1953).
[Ba 53g] BASHILOV, ANTON'EVA, BRODER and DZHELEPOV: Izv. Akad. Nauk SSSR., Ser. Fiz. **17**, 468 (1953).
[Ba 53h] BARLOUTAUD, BALLINI and SARTORI: C. R. Acad. Sci., Paris **237**, 886 (1953).
[Ba 54] BALESTRINI, S.: Phys. Rev. **95**, 1502 (1954).
[Ba 54a] BASHILOV, ANTON'EVA, BLINOV and DZHELEPOV: Izv. Akad. Nauk SSSR., Ser. Fiz. **18**, 43 (1954).
[Ba 55] BACKOFEN, E. W., and R. H. HERBER: Phys. Rev. **97**, 743 (1955).
[Ba 55a] BAY, HENRI, and MCLERNON: Phys. Rev. **97**, 561 (1955).
[Ba 55b] BARÓ, SEELMANN-EGGEBERT and ZABALA: Z. Naturforsch. **10**a, 80 (1955). — Publ. com. nacl. energia atomica Argentina, Ser. quim. **1**, 29 (1955).
[Ba 55c] BARNES, C. A.: Phys. Rev. **97**, 1226 (1955).
[Ba 55d] BARLOUTAUD, GRJEBINE and RIOU: C. R. Acad. Sci., Paris **240**, 1207 (1955).
[Ba 55e] BAY, HENRI and KANNER: Phys. Rev. **100**, 1197 (1955).
[Ba 55f] BASILE, SCHUHL and SEBAOUN: C. R. Acad. Sci., Paris **241**, 387 (1955).
[Ba 55g] BAGGERLY, MARMIER, BOEHM and DUMOND: Phys. Rev. **100**, 1364 (1955).
[Ba 55h] BARANOV, S. A., and K. N. SHLYAGIN: Proc. Acad. Sci. USSR. on Peaceful Uses of Atomic Energy, July 1—5, 1955.
[Be 52] BERGSTRÖM, I.: Ark. Fysik **5**, 191 (1952).
[Be 52a] BELL, GRAHAM and PETCH: Canad. J. Phys. **30**, 35 (1952).
[Be 53] BERNAS, R., and S. BEYDON: C. R. Acad. Sci., Paris **236**, 194 (1953).
[Be 53a] BENDEL, SHORE, BROWN and BECKER: Phys. Rev. **90**, 888 (1953).
[Be 53b] BELLAMY, E. H., and K. F. SMITH: Phil. Mag. **44**, 33 (1953).
[Be 53c] BERGSTRÖM, HILL and DE PASQUALI: Phys. Rev. **92**, 918 (1953).
[Be 54] BERGSTRÖM, THULIN, WAPSTRA and ÅSTRÖM: Ark. Fysik **7**, 255 (1954).
[Be 54a] BERTOLINI, BISI, LAZZARINI and ZAPPA: Nuovo Cim. **11**, 539 (1954).
[Be 54b] BELL, R. E., and L. YAFFE: Canad. J. Phys. **32**, 416 (1954).
[Be 55] BERGQVIST, BERGSTRÖM, HERRLANDER, HULTBERG, SLÄTIS, SOKOLOWSKI, WAPSTRA and WIEDLING: Phil. Mag. **46**, 65 (1955).
[Be 55a] BENDEL, TOMS and TOBIN: Phys. Rev. **99**, 672(A) (1955).
[Be 55b] BERGSTRÖM, I., and A. H. WAPSTRA: Phil. Mag. **46**, 61 (1955).
[Be 55c] BERTOLINI, BETTONI and LAZZARINI: Nuovo Cim. **1**, 746 (1955).
[Be 55d] BELL, GRAHAM and YAFFE: Canad. J. Phys. **33**, 457 (1955).
[Be 55e] BERNSTEIN, E. M., and H. W. LEWIS: Phys. Rev. **100**, 1345 (1955).
[Be 56] BENCZER, N., and C. S. WU: Bull. Amer. Phys. Soc. **1**, 41 (1956).
[Be 56a] BEGGS, ROBINSON and FINK: Phys. Rev. **101**, 149 (1956).

[*Bh 54*] BHATTACHERJEE, WALDMAN and MILLER: Phys. Rev. **95**, 404 (1954).
[*Bi 52*] BISHOP, DANIELS, GOLDSCHMIDT, HALBAN, KURTI and ROBINSON: Phys. Rev. **88**, 1432 (1952).
[*Bi 53*] BIEDENHARN, L. C., and M. E. ROSE: Rev. Mod. Phys. **25**, 729 (1953).
[*Bi 54*] BIRD, G. R., and C. H. TOWNES: Phys. Rev. **94**, 1203 (1954).
[*Bi 54a*] BISI, A., and L. ZAPPA: Nuovo Cim. **12**, 539 (1954).
[*Bi 54b*] BITTER, DAVIS, RICHTER and YOUNG: Phys. Rev. **96**, 1531 (1954).
[*Bi 55*] BISI, TERRANI and ZAPPA: Nuovo Cim. **1**, 651 (1955).
[*Bi 55a*] BISI, TERRANI and ZAPPA: Nuovo Cim. **1**, 291 (1955).
[*Bi 55b*] BISI, GERMAGNOLI, ZAPPA and ZIMMER: Nuovo Cim. **2**, 290 (1955).
[*Bl 50*] BLASER, BOEHM and MARMIER: Helv. phys. Acta **23**, 623 (1950).
[*Bl 52*] BLATT, J. M., and V. F. WIESSKOPF: Theoretical Nuclear Physics. John Wiley & Sons 1952.
[*Bl 53*] BLAISE, J., and H. CHANTREL: J. Phys. Radium **14**, 135 (1953).
[*Bl 53a*] BLEULER, BLUE, CHOWDARY, JOHNSON and TENDAM: Phys. Rev. **90**, 464 (1953).
[*Bl 53b*] BLEANEY, BOWERS and TRENAM: Proc. Phys. Soc. Lond. A **66**, 410 (1953).
[*Bl 55*] BLEANEY, B., and W. LOW: Proc. Phys. Soc. Lond. A **68**, 55 (1955).
[*Bl 55a*] BLEANEY, BOWERS and PRYCE: Proc. Roy. Soc. Lond., Ser. A **228**, 166 (1955).
[*Bl 55b*] BLEULER, E., and H. MORINAGA: Phys. Rev. **99**, 658(A) (1955); Phys. Rev. **103**, 1423 (1956)
[*Bl 55c*] BLUE, J. W., and E. BLEULER: Phys. Rev. **100**, 1324 (1955).
[*Bl 55d*] BLADES, A. T., and H. G. THODE: Z. Naturforsch. **10**a, 838 (1955).
[*Bo 37*] BOTHE, W., and W. GENTNER: Naturwiss. **25**, 284 (1937). — Z. Physik **112**, 45 (1939).
[*Bo 52*] BOYD, G. E., and R. A. CHARPIE: Phys. Rev. **88**, 681 (1952).
[*Bo 52a*] BOHR, KOCH and RASMUSSEN: Ark. Fysik **4**, 455 (1952).
[*Bo 53*] BOLEY, F. I., and D. S. DUNAVAN: Phys. Rev. **90**, 158 (1953).
[*Bo 53a*] BOLGIANO, MADANSKY and RASETTI: Phys. Rev. **89**, 679 (1953).
[*Bo 53b*] BOUISSIERES, MARTY and TEILLAC: C. R. Acad. Sci., Paris **237**, 324 (1953).
[*Bo 53c*] BOUCHEZ, HUBERT, PERRIN and SAKAI: J. Phys. Radium **14**, 29(A), 273 (1953). — C. R. Acad. Sci., Paris **236**, 1249 (1953).
[*Bo 53d*] BOHR, A., and B. R. MOTTELSON: Kgl. danske Vid. Sels., mat.-fys. Medd. **27**, Nr. 16 (1953).
[*Bo 54*] BOLLA, TERRANI and ZAPPA: Nuovo Cim. **12**, 875 (1954).
[*Bo 54a*] BOYD, G.: Phys. Rev. **95**, 113 (1954).
[*Bo 54b*] BOLEY, F. I.: Phys. Rev. **94**, 1078 (1954).
[*Bo 54c*] BOEHM, MARMIER and DUMOND: Phys. Rev. **95**, 864 (1954).
[*Bo 55*] BOLOTIN, H. H., and R. G. WILKINSON: Phys. Rev. **99**, 671(A) (1955).
[*Br 52*] BROWN, R. M., and D. H. TOMBOULIAN: Phys. Rev. **88**, 1158 (1952).
[*Br 53*] BROLLEY, J. E.: Phys. Rev. **89**, 877 (1953); **88**, 618 (1952).
[*Br 53a*] BRUNNER, HUBER, JOLY and MAEDER: Helv. phys. Acta **26**, 588 A (1953).
[*Br 53b*] BRABANT, COCHRAN and CASWELL: Phys. Rev. **91**, 210(A) (1953).
[*Br 53c*] BROYLES, THOMAS and HAYNES: Phys. Rev. **89**, 715 (1953).
[*Br 53d*] BROWN, F., and L. YAFFE: Canad. J. Chem. **31**, 242 (1953).
[*Br 53e*] BROWN, H. N., and R. A. BECKER: Phys. Rev. **90**, 328 (1953).
[*Br 54*] BROWNE, C. P.: Phys. Rev. **95**, 860 (1954).
[*Br 54a*] BRUN, OESER, STAUB and TELSCHOW: Helv. phys. Acta **27**, 173(A) (1954).
[*Br 54b*] BRUN, OESER, STAUB and TELSCHOW: Phys. Rev. **93**, 172 (1954).
[*Br 54c*] BRUN, OESER, STAUB and TELSCHOW: Phys. Rev. **93**, 904 (1954).
[*Br 54d*] BRADEN, WYLY and PATRONIS: Phys. Rev. **95**, 758 (1954).
[*Br 54e*] BRUNNER, HALTER, HUBER, JOLY and MAEDER: Helv. phys. Acta **27**, 512 (A), 572 (1954).
[*Br 54f*] BROWN, H. N., and R. A. BECKER: Phys. Rev. **96**, 1372 (1954).
[*Br 54g*] BREITENBERGER, E.: Nature, Lond. **173**, 737 (1954).
[*Br 55*] BRUNNER, GUHL, HALTER and HUBER: Helv. phys. Acta **28**, 85 (1955).
[*Br 55a*] BROSI, A. R., and B. H. KETELLE: Phys. Rev. **100**, 169 (1955).
[*Br 55b*] BRAZOS, J. N., and R. M. STEFFEN: Phys. Rev. **99**, 1645(A) (1955); **102**, 753 (1956).
[*Br 55c*] BROWN, HALL and WALTER: J. Inorg. a. Nucl. Chem. **1**, 241 (1955).
[*Br 55d*] BRUNNER, GUHL, HALTER and LEISI: Helv. phys. Acta **28**, 475 (1955).
[*Br 56*] BROSI, A. R., and B. H. KETELLE: Phys. Rev. **103**, 917 (1956).
[*Bu 52*] BUTEMENT, F. D. S.: Proc. Phys. Soc. Lond. A **65**, 254 (1952).
[*Bu 52a*] BUTEMENT, F. D. S., and R. SHILLITO: Proc. Phys. Soc. Lond. A **65**, 945 (1952).
[*Bu 53*] BUNKER, PRESTWOOD and STARNER: Phys. Rev. **91**, 1021 (1953).
[*Bu 54*] BURGUS, COWAN, HADLEY, HESS, SHULL, STEVENSON and YORK: Phys. Rev. **95**, 750 (1954).

[*Bu 54a*] BURSON, JORDAN and LE BLANC: Phys. Rev. **96**, 1555 (1954).
[*Bu 54b*] BUNKER, MIZE and STARNER: Phys. Rev. **94**, 1694 (1954).
[*Bu 54c*] BUTEMENT, F. D. S., and A. J. POË: Phil. Mag. **45**, 31, 1090 (1954).
[*Bu 55*] BUNKER, MIZE and STARNER: Phys. Rev. **99**, 659(A) (1955).
[*Bu 55a*] BURSON, LE BLANC and MARTIN: Phys. Rev. **99**, 660(A) (1955).
[*Bu 55b*] BURKHARDT, WINHOLD and DUPREE: Phys. Rev. **100**, 199 (1955).

[*Ca 52*] CARETTO, A. A., and E. I. WIIG: J. Amer. Chem. Soc. **74**, 5235 (1952).
[*Ca 52a*] CASTNER, S. V., and D. H. TEMPLETON: Phys. Rev. **88**, 1126 (1952).
[*Ca 53*] CAMPELL, E. C.: Oak Ridge National Laboratory Report ORNL-1620, 1953. Unpublished.
[*Ca 53a*] CAIRD, R. S., and A. C. G. MITCHELL: Phys. Rev. **89**, 573 (1953).
[*Ca 54*] CAPPELLER, U., and R. KLINGELHÖFER: Z. Physik **139**, 402 (1954).
[*Ca 54a*] CAIRD, R. S., and A. C. G. MITCHELL: Phys. Rev. **94**, 412 (1954).
[*Ca 55*] CASSETT, W. A., and W. W. MEINKE: Phys. Rev. **100**, 1372 (1955).
[*Ca 55a*] CAMPBELL, PEELLE, MAIENSCHEIN and STELSON: Phys. Rev. **98**, 1172(A) (1955).
[*Ch 52*] CHENG, DICK and KURBATOV: Phys. Rev. **88**, 887 (1952).
[*Ch 53*] CHATTERJEE, S., and A. K. SAHA: Z. Physik **135**, 141 (1953).
[*Ch 54*] CHURCH, E. L., and M. GOLDHABER: Phys. Rev. **95**, 626(A) (1954).
[*Ch 54a*] CHETHAM-STRODE, A., and E. M. KINDERMANN: Phys. Rev. **93**, 1029 (1954).
[*Ch 55*] CHURCH, E. L., and J. WENESER: Phys. Rev. **100**, 943 (1955). **103**, 1035 (1956).
[*Ch 55a*] CHENG, RIDOLFO, POOL and KUNDU: Phys. Rev. **98**, 231(A) (1955).
[*Cr 53*] CRASEMANN, B.: Phys. Rev. **90**, 995 (1953).
[*Cr 54*] CROSS, W. G., and T. A. EASTWOOD: Phys. Rev. **95**, 628(A) (1954).
[*Cr 54a*] CRASEMANN, B.: Phys. Rev. **93**, 1034 (1954).
[*Co 52*] CORK, LE BLANC, NESTER and STUMPF: Phys. Rev. **88**, 685 (1952).
[*Co 53*] CORK, LE BLANC, MARTIN, NESTER and BRICE: Phys. Rev. **90**, 579 (1953).
[*Co 53a*] CORYELL, LÉVÊQUE and RICHTER: Phys. Rev. **89**, 903(A) (1953).
[*Co 53b*] CORK, LE BLANC, NESTER and BRICE: Phys. Rev. **91**, 76 (1953).
[*Co 53c*] CORK, LE BLANC, NESTER, MARTIN and BRICE: Phys. Rev. **90**, 444 (1953).
[*Co 53d*] CORK, NESTER, LE BLANC and BRICE: Phys. Rev. **92**, 119 (1953).
[*Co 53e*] CORK, BRICE, NESTER, LE BLANC and MARTIN: Phys. Rev. **89**, 1291 (1953).
[*Co 54*] CORK, BRICE, HICKMAN and SCHMID: Phys. Rev. **93**, 1059 (1954).
[*Co 54a*] COHEN, V. W., and D. A. GILBERT: Phys. Rev. **95**, 569 (1954); **97**, 243 (1955).
[*Co 54b*] CORK, BRICE, SCHMID, HICKMAN and NINE: Phys. Rev. **94**, 1218 (1954).
[*Co 55*] COBURN, H. H., and S. FRANKEL: Phys. Rev. **99**, 671(A) (1955).
[*Co 55a*] COLOMBO, ROSSI and SCOTTI: Nuovo Cim. **1**, 522 (1955).
[*Co 55b*] CORK, BRICE and SCHMID: Phys. Rev. **99**, 703 (1955).
[*Co 55c*] CORK, BRICE, SCHMID and HELMER: Phys. Rev. **100**, 188 (1955).
[*Co 55d*] CONNORS, MILLER and WALDMAN: Phys. Rev. **100**, 1237(A) (1955).
[*Cu 53*] CUNINGHAME, J. G.: Phil. Mag. **44**, 900 (1953).
[*Cu 53a*] CURTIS, M. L.: Phys. Rev. **92**, 1489 (1952).

[*Da 39*] DANCOFF, S. M., and P. MORRISON: Phys. Rev. **55**, 122 (1939).
[*Da 52*] DANIELS, GRACE, HALBAN, KURTI and ROBINSON: Phil. Mag. **43**, 1297 (1952).
[*Da 53*] DAVEY, W. G., and P. B. MOON: Proc. Phys. Soc. Lond. A **66**, 956 (1953).
[*Da 53a*] DAMON, P. E., and R. R. EDWARDS: Phys. Rev. **90**, 280 (1953).
[*Da 54*] DALY, R. T., and J. H. HOLLOWAY: Phys. Rev. **96**, 539 (1954).
[*Da 54a*] DAVIS, S. P.: Phys. Rev. **93**, 159 (1954).
[*Da 54b*] DAMON, P. E., and R. R. EDWARDS: Phys. Rev. **95**, 1698 (1954).
[*Da 55*] DAUBIN, S. C., and D. R. HAMILTON: Phys. Rev. **99**, 683 (1955).
[*Da 55a*] DAY, EAKINS and VOIGT: Phys. Rev. **100**, 796 (1955).
[*De 52*] DE VRIES, VERINGA and CLAY: Prov. Kon. Ned. Akad. Wetensch., Ser. B **55**, 303 (1952).
[*De 52a*] DE WAARD, H.: Physica, Haag **18**, 1151 (1952).
[*De 52b*] DER MATEOSIAN, E., and A. SMITH: Phys. Rev. **88**, 1186 (1952).
[*De 53*] DEHMELT, H. G.: Phys. Rev. **92**, 1240 (1953).
[*De 53a*] DER MATEOSIAN, E.: Phys. Rev. **92**, 938 (1953).
[*De 53b*] DESCHAMPS, Y., and P. AVIGNON: C. R. Acad. Sci., Paris **236**, 478 (1953).
[*De 53c*] DE HAAN, SIZOO and KRAMER: Physica, Haag **19**, 1201 (1953).
[*De 54*] DER MATEOSIAN, E.: Phys. Rev. **95**, 646(A) (1954).
[*De 54a*] DER MATEOSIAN, E., and C. S. WU: Phys. Rev. **95**, 458 (1954).
[*De 54b*] DE WAARD, H.: Physica, Haag **20**, 41 (1954).
[*De 54c*] DEHMELT, ROBINSON and GORDY: Phys. Rev. **93**, 480 (1954).
[*De 55*] DEVONS, MANNING and BUNBURY: Proc. Phys. Soc. Lond. A **68**, 18 (1955).

[De 55a] DE WAARD, H., and T. R. GERHOLM: Physica, Haag **21**, 599 (1955).
[De 55b] DE WAARD, H.: Phys. Rev. **99**, 1045 (1955).
[De 55c] DE HAAN, SIZOO and KRAMER: Physica, Haag **21**, 803 (1955).
[De 55d] DEUTSCH, M., and A. W. SUNYAR: Unpublished.
[De 56] DER MATEOSIAN, E., and M. GOLDHABER: Bull. Amer. Phys. Soc. **1**, 207 (1956).
[Di 53] DIAMOND, R. M.: Phys. Rev. **89**, 1149 (1953).
[Di 54] DIXON, MCNAIR and CURRAN: Phil. Mag. **45**, 683 (1954).
[Di 54a] DIXON, D., and A. MCNAIR: Phil. Mag. **45**, 1099 (1954).
[Di 54b] DIAMOND, R. M.: Phys. Rev. **95**, 410 (1954).
[Di 55] DICK, FOUCHER, PERRIN and VARTAPETIAN: C. R. Acad. Sci., Paris **240**, 1335 (1955).
[Di 55a] DIXON, MCNAIR and CURRAN: J. Phys. Radium **16**, 538 (1955).
[Do 52] DOWNES, PRICE, SHER and WALSH: Brookhaven National Laboratory Report BNL-216 (T-33), 1952. Unpublished.
[Do 53] DOLLSHNYUK, DRABKIN, ORLOV and RUSINOV: Dokl. Akad. Nauk SSSR. **92**, 1141 (1953).
[Do 53a] DOUGLAS, MEWHERTER and SCHUMAN: Phys. Rev. **92**, 369 (1953).
[Do 55] DOERNER, R. K., and A. H. WEBER: Phys. Rev. **99**, 672(A) (1955).
[Dr 52] DRELL, S. D., and M. E. ROSE: Progr. Theor. Phys., Japan **7**, 125 (1952). — DRELL, S. D.: Oak Ridge National Laboratory Report ORNL 792, 1950. Unpublished.
[Dr 52a] DROPESKY, B., and E. O. WIIG: Phys. Rev. **88**, 683 (1952).
[Dr 53] DROPESKY, B.: Thesis, Univ. of Rochester, 1953. Unpublished.
[Dr 54] DRABKIN, G. M., and L. I. RUSINOV: Dokl. Akad. Nauk SSSR. **97**, 417 (1954).
[Dr 55] DRABKIN, ORLOV and RUSINOV: Izv. Akad. Nauk SSSR., Ser. Fiz. **19**, 324 (1955).
[Dr 56] DROPESKY, B. J., and A. W. SCHARDT: Phys. Rev. **102**, 426 (1956).
[Du 51] DUFFIELD, R. B., and L. M. LANGER: Phys. Rev. **81**, 203 (1951).
[Du 53] DUFFIELD, R. B., and L. M. LANGER: Rev. **89**, 854 (1953).
[Du 55] DUQUESNE, M.: C. R. Acad. Sci., Paris **241**, 36 (1955).
[Dz 52] DZHELEPOV, B., and A. SILANT'EV: Dokl. Akad. Nauk SSSR. **85**, 533 (1952).
[Dz 54] DZHELEPOV, NOVOSIL'TSEVA TISHKIN: Izv. Akad. Nauk SSSR., Ser. Fiz. **18**, 76 (1954).
[Dz 55] DZHELEPOV, ZHUKOVSKII, PRIKHODTSEVA and KHOL'NOV: Izv. Akad. Nauk SSSR., Ser. Fiz. **19**, 271 (1955).
[Dz 55a] DZELEPOV, ZHUKOVSKIĬ and NEDOVESOV: Izv Akad. Nauk SSSR., Ser. Fiz. **19**, 296 (1955).

[Ea 53] EASTERDAY, H. T., and H. A. MEDICUS: Phys. Rev. **89**, 752 (1953).
[Ea 53a] EASTERDAY, H. T.: Phys. Rev. **91**, 653 (1953).
[Eb 55] EBERHARDT, GEISS and LANG: Z. Naturforsch. **10**a, 796 (1955).
[Ei 53] EICHHOLZ, G. G.: Phys. Rev. **89**, 525 (1953).
[Ei 54] EICHELBERGER, JORDON, ORR and PARKS: Phys. Rev. **96**, 719 (1954).
[El 48] ELLIOTT, L. G., and R. E. BELL: Phys. Rev. **74**, 1869 (1948).
[El 49] ELLIOTT, L. G., and R. E. BELL: Phys. Rev. **76**, 168(A) (1949).
[El 52] ELLIOTT, CHEN, HASKINS and KURBATOV: Phys. Rev. **88**, 263 (1952).
[El 54] ELLIOTT, PRESTON and WOLFSON: Canad. J. Phys. **32**, 153 (1954).
[El 54a] ELLIOTT, GRAHAM, WALKER and WOLFSON: Phys. Rev. **93**, 356 (1954).
[El 55] ELBEK, MADSEN and NATHAN: Phil. Mag. **46**, 663 (1955).
[El 55a] ELLIOTT, PRESTON and WOLFSON: Canad. J. Phys. **33**, 607 (1955).
[Em 55] EMERY, BRADLEY and VEALL: Nature, Lond. **175**, 34 (1955). — Proc. Phys. Soc. Lond. A **68**, 346 (1955).
[En 53] ENGELDER, T. C.: Phys. Rev. **90**, 259 (1953).
[En 53a] ENGELKEMEIR, D., and L. B. MAGNUSSON: Phys. Rev. **94**, 1395 (1954).
[En 55] ENGELKEMEIR, D., and L. B. MAGNUSSON: Phys. Rev. **99**, 135 (1955).
[Er 54] ERDÖS, JORDON, SCHMOUKER and STOLL: Helv. phys. Acta **27**, 187 (A) (1954).
[Ev 55] EVERETT, A. E., and M. J. GLAUBMAN: Phys. Rev. **100**, 955(A) (1955).
[Ew 52] EWAN, G. T., and M. A. S. ROSS: Nature, Lond. **170**, 760 (1952).
[Ew 53] EWAN, G. T., and A. L. THOMPSON: Trans. Roy. Soc. Canada **47**, 126(A) (1953).

[Fa 55] FARMER, D. J.: Phys. Rev. **99**, 659(A) (1955).
[Fa 55a] FARRELLY, KOERTS, BENCZER, VAN LIESHOUT and WU: Phys. Rev. **99**, 1440 (1955).
[Fa 55b] FAGG, WOLICKI, BONDELID, DUNNING and SNYDER: Phys. Rev. **100**, 1299 (1955).
[Fe 54] FEUVRAIS, L., and T. YUASA: C. R. Acad. Sci., Paris **239**, 1627 (1954).
[Fe 55] FEINGOLD, A. M., and S. FRANKEL: Phys. Rev. **97**, 1025 (1955).
[Fi 53] FINK, R. W.: Thesis, Univ. of Rochester 1953. Unpublished.
[Fi 55] FISCHER, V. K.: Phys. Rev. **99**, 764 (1955).
[Fl 52] FLEMING, GHIORSO and CUNNINGHAM: Phys. Rev. **88**, 642 (1952).

[*Fl 53*] Fleming, W. H., and H. G. Thode: Phys. Rev. **90**, 857 (1953).
[*Fl 53a*] Flammersfeld, A.: Z. Naturforsch. **8**a 217 (1953).
[*Fl 54*] Flinta, J., and S. Eklund: Ark. Fysik **7**, 401 (1954).
[*Fo 53*] Forsthoff, Goeckermann and Naumann: Phys. Rev. **90**, 1004 (1953).
[*Fo 54*] Fowler, Kruse, Keshishian, Klotz and Mellor: Phys. Rev. **94**, 1082 (1954).
[*Fo 55*] Forster, H. H., and J. S. Wiggins: Phys. Rev. **99**, 660(A) (1955).
[*Fo 55a*] Forster, H. H., and A. Rosen: Nuovo Cim. **1**, 972 (1955). — Phys. Rev. **98**, 1172(A) (1955).
[*Fo 55b*] Ford, K. W.: Phys. Rev. **98**, 1516 (1955).
[*Fr 52*] Fremlin, J. H., and M. C. Walters: Proc. Phys. Soc. Lond. A **65**, 911 (1952).
[*Fr 53*] Frauenfelder, Lawson, Jentschke and de Pasquali: Phys. Rev. **92**, 1241 (1953).
[*Fr 53a*] Friedlander, Wilson, Ghiorso and Perlman: Phys. Rev. **91**, 498(A) (1953).
[*Fr 53b*] Fred, Tomkins and Barnes: Phys. Rev. **92**, 1324 (1953).
[*Fr 54*] Frauenfelder, Lawson and Jentschke: Phys. Rev. **93**, 1126 (1954).
[*Fr 55*] Fraser, J. S., and J. C. D. Milton: Phys. Rev. **98**, 1173(A) (1955).
[*Fr 56*] Frauenfelder, Levine, Rossi and Singer: Phys. Rev. **103**, 352 (1956).
[*Fu 52*] Fultz, Nash, Woodward and Pool: Phys. Rev. **88**, 170(A) (1952).

[*Ga 53*] Gauthe, B., and J. M. Blum: C. R. Acad. Sci., Paris **236**, 1255 (1953).
[*Ge 52*] Gellman, Griffith and Stanley: Phys. Rev. **85**, 944 (1952).
[*Ge 52a*] Geiger, K.: Z. Naturforsch. **7**a, 579 (1952).
[*Ge 53*] Germagnoli, Malvicini and Zappa: Nuovo Cim. **10**, 1388 (1953).
[*Ge 54*] Geese-Bähnisch, I., and E. Huster: Naturwiss. **41**, 495 (1954).
[*Ge 55*] Gerholm, T. R., and H. de Waard: Physica, Haag **21**, 601 (1955).
[*Gi 53*] Ginnings, Ball and Vier: J. Res. Nat. Bur. Stand. **50**, 75 (1953). — J. Franklin Inst. **255**, 241 (1953).
[*Gi 54*] Gimmi, Heer and Scherrer: Z. Physik **138**, 394 (1954). — Helvet. phys. Acta **27**, 180(A) (1954).
[*Gi 54a*] Gillon, Gopalakrishnan, de Shalit and Mihelich: Phys. Rev. **93**, 124 (1954).
[*Gi 55*] Gimmi, Heer and Scherrer: Helvet. phys. Acta **28**, 470 (1955).
[*Gl 55*] Glaubman, M. J.: Phys. Rev. **98**, 645 (1955).
[*Gl 55a*] Glazunov, Dzhelepov and Khol'nov: Izv. Akad. Nauk SSSR., Ser. Fiz. **19**, 294 (1955).
[*Go 51*] Goldhaber, M., and A. W. Sunyar: Phys. Rev. **83**, 906 (1951).
[*Go 52*] Goldhaber, M., and R. W. Hill: Rev. Mod. Phys. **24**, 179 (1952).
[*Go 53*] Gorodetzky, Knipper, Armbruster and Gallman: J. Phys. Radium **14**, 550 (1953).
[*Go 53a*] Goldhaber, M.: Phys. Rev. **89**, 1146 (1953).
[*Go 54*] Golden, R. K., and S. Frankel: Phys. Rev. **95**, 613(A) (1954).
[*Go 54a*] Goldberg, N., and S. Frankel: Phys. Rev. **93**, 1425 (1954).
[*Go 54b*] Goodman, L. S., and S. Wexler: Phys. Rev. **95**, 570 (1954).
[*Go 55*] Goodlove, T. F., and J. G. Carver: Phys. Rev. **99**, 1634(A) (1955).
[*Go 55a*] Gove, Smith and Becker: Phys. Rev. **100**, 1236(A) (1955).
[*Go 55b*] Goodman, L. S., and S. Wexler: Phys. Rev. **100**, 1796(A) (1955).
[*Go 55c*] Goldberg, N., and S. Frankel: Phys. Rev. **100**, 1350 (1955).
[*Gr 52*] Graves, Langer and Moffat: Phys. Rev. **88**, 344 (1952).
[*Gr 52a*] Griffiths, J. H. E., and J. Owen: Proc. Phys. Soc. Lond. A **65**, 951 (1952).
[*Gr 52b*] Gormov, K. et al: Dokl. Akad. Nauk SSSR. **86**, 255 (1952).
[*Gr 52c*] Gromov, K., and B. Dzhelepov: Dokl. Akad. Nauk SSSR. **85**, 299 (1952).
[*Gr 52d*] Graham, Wolfson and Bell: Canad. J. Phys. **30**, 459 (1952).
[*Gr 53*] Graham, R. L., and R. E. Bell: Canad. J. Phys. **31**, 377 (1953).
[*Gr 54*] Graham, R. L., and J. Walker: Phys. Rev. **94**, 794(A) (1954).
[*Gr 54a*] Green, Harris and Cooper: Phys. Rev. **96**, 817(A) (1954).
[*Gr 54b*] Green, D., and J. R. Richardson: Phys. Rev. **96**, 858(A) (1954).
[*Gr 55*] Graham, Wolfson and Clark: Phys. Rev. **98**, 1173(A) (1955).
[*Gr 55a*] Graves, W. E., and A. C. G. Mitchell: Phys. Rev. **97**, 1033 (1955).
[*Gr 55b*] Grodzins, L.: Bull. Amer. Phys. Soc. **1**, 163, 164 (1956).
[*Gr 55c*] Grodzins, L., and H. T. Motz: Phys. Rev. **100**, 1236(A) (1955); **102**, 761 (1956).
[*Gr 55d*] Graves, W. E., and A. C. G. Mitchell: Phys. Rev. **101**, 701 (1956).
[*Gr 55e*] Graham, Bell, Yaffe and Geiger: Phys. Rev. **99**, 1646(A) (1955).
[*Gr 55f*] Grard, Danguy and Franeau: J. Phys. Radium **16**, 839 (1955).
[*Gu 53*] Gutowsky, H. S., and B. R. Garvey: Phys. Rev. **91**, 81 (1953).
[*Gu 54*] Guss, Killian and Porter: Phys. Rev. **95**, 627(A) (1954).

[*Ha 21*] HAHN, O.: Chem. Ber. **54**, 1131 (1921).
[*Ha 40*] HAMILTON, D. R.: Phys. Rev. **58**, 122 (1940).
[*Ha 50*] HAXEL, JENSEN and SUESS: Z. Physik **128**, 301 (1950).
[*Ha 52*] HAYWARD, R. W.: Phys. Rev. **85**, 760(A) (1952).
[*Ha 52a*] HASKINS, J. R., and J. D. KURBATOV: Phys. Rev. **88**, 884 (1952).
[*Ha 53*] HARDY, SILVEY, TOWNES, BURKE, STRANDBERG, PARKER and COHEN: Phys. Rev. **85**, 494 (1952); **92**, 1532 (1953).
[*Ha 53a*] HAMILTON, LEMONICK and PIPKIN: Phys. Rev. **92**, 1191 (1953).
[*Ha 53b*] HANDLEY, T. H., and E. L. OLSON: Phys. Rev. **92**, 1260 (1953).
[*Ha 53c*] HARBOTTLE, G.: Phys. Rev. **91**, 1234 (1953).
[*Ha 54*] HASLAM, ROBERTS and ROBB: Canad. J. Phys. **32**, 361 (1954).
[*Ha 54a*] HAYWARD, HOPPES and ERNST: Phys. Rev. **93**, 916(A) (1954).
[*Ha 55*] HANDLEY, T. H., and W. S. LYON: Phys. Rev. **99**, 755 (1955).
[*Ha 55a*] HAYWARD, HOPPES and MANN: J. Res. Nat. Bur. Stand. **54**, 47 (1955).
[*Ha 55b*] HARBOTTLE, G.: Private communication.
[*Ha 55c*] HAYWARD, R. W., and D. D. HOPPES: Phys. Rev. **98**, 1172(A) (1955); **101**, 93 (1956).
[*Ha 55d*] HAYASHI, KAWAMURA and AOKI: J. Phys. Soc. Japan **10**, 334 (1955).
[*Ha 55e*] HAYWARD, HOPPES and ERNST: Phys. Rev. **98**, 231(A) (1955).
[*Ha 55f*] HAYNES, S. K., and W. T. ACHOR: J. Phys. Radium **16**, 635 (1955).
[*Ha 55g*] HARRISON, CRAWFORD and HOPKINS: Phys. Rev. **100**, 841 (1955).
[*Ha 55h*] HANDLER, H. E., and J. R. RICHARDSON: Phys. Rev. **98**, 281 (A) (1955).
[*Ha 56*] HAYWARD, R. W., and D. D. HOPPES: Bull. Amer. Phys. Soc. **1**, 42 (1956).
[*He 36*] HEITLER, W.: The Quantum Theory of Radiation. Oxford University Press 1936.
[*He 38*] HEBB, M. H., and G. E. UHLENBECK: Physica, Haag **5**, 605 (1938).
[*He 40*] HEBB, M. H., and E. NELSON: Phys. Rev. **58**, 486 (1940).
[*He 52*] HEDGRAN, A., and D. A. LIND: Ark. Fysik **5**, 177 (1952).
[*He 54*] HERRMANN, G.: Z. Elektrochem. **58**, 626 (1954).
[*He 55*] HERRMANN, G., and F. STRASSMANN: Z. Naturforsch. **10**a, 146 (1955).
[*He 55a*] HEINTZE, J.: Z. Naturforsch. **10**a, 77 (1955).
[*He 55b*] HEBB, E.: Phys. Rev. **97**, 987 (1955).
[*He 55c*] HEYDENBURG, N. P., and G. M. TEMMER: Phys. Rev. **100**, 150 (1955); **99**, 1609 (1955); **104**, 967, (1956); **104**, 981 (1956).
[*He 55d*] HEER, RÜETSCHI, GIMMI and KÜNDIG: Helv. phys. Acta **28**, 336 (1955).
[*He 55e*] HEER, E.: J. Phys. Radium **16**, 600 (1955).
[*He 55f*] HEER, RÜETSCHI and SCHERRER: Z. Naturforsch. **10**a, 834 (1955).
[*He 56*] HERRLANDER, STOCKENDAL, MCDONNELL and BERGSTRÖM: Nuclear Phys. (to be published).
[*Hi 50*] HILL, R. W.: Phys. Rev. **79**, 1021 (1950).
[*Hi 52*] HINTZ, N. M., and N. F. RAMSEY: Phys. Rev. **88**, 19 (1952).
[*Hi 53*] HILL, R. D., and J. W. MIHELICH: Phys. Rev. **89**, 323 (1953).
[*Hi 54*] HINTENBERGER, HERR and VOSHAGE: Phys. Rev. **95**, 1690 (1954).
[*Ho 52*] HOFFMAN, D. C., and D. S. MARTIN: J. Phys. Chem. **56**, 1097 (1952).
[*Ho 52a*] HOPKINS, N. J.: Phys. Rev. **88**, 680 (1952).
[*Ho 53*] HOYT, H. C., and J. W. M. DUMOND: Phys. Rev. **91**, 1027 (1953).
[*Ho 53a*] HOFF, R. W.: Universtiy of California Radiation Laboratory Report, UCRL-2325, 1953. Unpublished.
[*Ho 53b*] HOK, O. P., and G. J. SIZOO: Physica, Haag **19**, 1205 (1953).
[*Ho 53c*] HOLLANDER, PERLMAN and SEABORG: Rev. Mod. Phys. **25**, 469 (1953).
[*Ho 53d*] HOOGE, F. N., and A. H. W. ATEN: Physica, Haag **19**, 1047 (1953).
[*Ho 54*] HOBSON, HUBBS, NIERENBERG and SILSBEE: Phys. Rev. **96**, 1450 (1954).
[*Ho 54a*] HORROCKS, D. L., and A. F. VOIGT: Phys. Rev. **95**, 1205 (1954).
[*Ho 54b*] HOK, KRAMER and MEIJER: Physica, Haag **20**, 1200 (1954).
[*Ho 55*] HOK, O. P.: Phys. Rev. **99**, 1613 (1955).
[*Ho 55a*] HOFF, JAFFE, PASSELL, STEPHENS, HULET and THOMPSON: Phys. Rev. **100**, 1403 (1955).
[*Hu 52*] HUTCHINSON, D. R., and M. L. WIEDENBECK: Phys. Rev. **88**, 699 (1952).
[*Hu 52a*] HUBER, JOLY, SCHERRER and VERSTER: Helv. phys. Acta **25**, 621 (1952).
[*Hu 52b*] HUDDLESTON, C. M., and A. C. G. MITCHELL: Phys. Rev. **88**, 1350 (1952).
[*Hu 53*] HUBER, HALTER, JOLY, MAEDER and BRUNNER: Helv. phys. Acta **26**, 591 (A) (1953).
[*Hu 53a*] HUSTER, E.: Naturwiss. **40**, 197 (1953). — Z. Physik **136**, 303 (1953). — Phys. Rev. **92**, 1076 (1953).
[*Hu 53b*] HUGHES, D. J., and H. PALEVSKY: Phys. Rev. **92**, 1206 (1953).
[*Hu 53c*] HUUS, T., and C. ZUPANČIČ: Kgl. danske Vid. Selsk., mat.-fys. Medd. **28**, Nr. 1 (1953).

[*Hu 54*] HUIZENGA, J. R., and C. M. STEVENS: Phys. Rev. **96**, 548 (1954).
[*Hu 56*] HUIZENGA, KROHN and RABOY: Bull. Amer. Phys. Soc. **1**, 43 (1956). — Phys. Rev. **102**, 1063 (1956).

[*Il 54*] ILAKOVAC, K.: Proc. Phys. Soc. Lond., Ser. A **67**, 601 (1954).

[*Ja 53*] JACCARINO, KING and STROKE: Quoted by LIVINGSTON et al., Phys. Rev. **92**, 1271 (1953).
[*Ja 53a*] JACCARINO, KING, SATTEN and STROKE: Phys. Rev. **94**, 1798 (1953).
[*Ja 53b*] JAFFE, A. A., and S. G. COHEN: Phys. Rev. **89**, 454 (1953).
[*Ja 54*] JAFFE, H.: University of California Radiation Laboratory report UCRL-2537, 1954. Unpublished.
[*Ja 54a*] JAVAN, A., and A. ENGELBRECHT: Phys. Rev. **96**, 649 (1954).
[*Je 53*] JEFFRIES, C. D., and P. B. SOGO: Phys. Rev. **91**, 1286 (1953).
[*Je 53a*] JEFFRIES, C. D.: Phys. Rev. **92**, 1262 (1953).
[*Je 53b*] JENSEN, LASLETT, MARTIN, HUGHES and PRATT: Phys. Rev. **90**, 557 (1953).
[*Je 54*] JENTSCHKE, JUVELAND and KINSEY: Phys. Rev. **96**, 231 (1954).
[*Jo 52*] JOHANSSON, S. A. E., and S. ALMQVIST: Ark. Fysik **5**, 427 (1952). — Nature, Lond. **170**, 583 (1952).
[*Jo 53*] JORDAN, CORK and BURSON: Phys. Rev. **90**, 862 (1953).
[*Jo 53a*] JOHNSON, F. A.: Canad. J. Phys. **31**, 1136 (1953).
[*Jo 53b*] JORDAN, CORK and BURSON: Phys. Rev. **92**, 1218 (1953).
[*Jo 54*] JOHNS, MCMULLEN, DONNELLY and NABLO: Canad. J. Phys. **32**, 35 (1954).
[*Jo 54a*] JOSHI, M. C., and B. V. THOSAR: Phys. Rev. **96**, 1022 (1954).
[*Jo 54b*] JOHNS, M. W., and S. V. NABLO: Phys. Rev. **96**, 1599 (1954).
[*Jo 54c*] JOHANSSON, S. A. E.: Phys. Rev. **96**, 1075 (1954).
[*Jo 55*] JOHNSON, JOHNSON and LANGER: Phys. Rev. **98**, 1517 (1955).
[*Jo 55a*] JOSHI, M. C.: Current Sci. **24**, 227 (1955).
[*Jo 55b*] JOLY, BRUNNER, HALTER and HUBER: Helv. phys. Acta **28**, 403 (1955).
[*Jo 56*] JOHNS, MCMULLEN, WILLIAMS and NABLO: Canad. J. Phys. **34**, 169 (1956).
[*Jo 56a*] JOHNS, WILLIAMS and BRODIE: Canad. J. Phys. **34**, 147 (1956).

[*Ka 52*] KAMEI, T.: J. Phys Soc. Japan **7** 649 (1952).
[*Ka 53*] KASTNER, J., and G. N. WHYTE: Phys. Rev. **91**, 332 (1953).
[*Ka 35a*] KATZ, BAKER and MONTALBETTI: Canad. J. Phys. **31**, 250 (1953).
[*Ka 53b*] KAHN, B., and W. S. LYON: Phys. Rev. **92**, 902 (1953).
[*Ka 53c*] KATCOFF, S., and W. RUBINSON: Phys. Rev. **91**, 1458 (1953).
[*Ka 53d*] KASTNER, J.: Canad. J. Phys. **31**, 169 (1953).
[*Ka 55*] KAVANAGH, MILLS and SHERR: Phys. Rev. **97**, 248 (1955).
[*Ka 55a*] KAHN, B., and W. S. LYON: Phys. Rev. **98**, 58 (1955).
[*Ka 55b*] KAMEI, T.: Phys. Rev. **99**, 789 (1955).
[*Ka 55c*] KATO, Y., and M. MORITA: Progr. Theor. Phys., Japan **13**, 276 (1955).
[*Ke 53*] KESSLER, K. G., and R. E. TREES: Phys. Rev. **92**, 303 (1953).
[*Ke 53a*] KETELLE, BROSI and PORTER: Phys. Rev. **90**, 567 (1953).
[*Ke 53b*] KEENAN, PENNEMAN and MCINTEER: J. Chem. Phys. **21**, 1802 (1953). — Phys. Rev. **87**, 204 (A) (1952).
[*Ke 54*] KEISTER, G. L., and F. H. SCHMIDT: Phys. Rev. **93**, 140 (1954).
[*Ke 54a*] KEISTER, G. L.: Phys. Rev. **96**, 855 (A) (1954).
[*Ke 55*] KEISTER, LEE and SCHMIDT: Phys. Rev. **97**, 451 (1955).
[*Ki 53*] KING, R. W., and D. C. PEASLEE: Phys. Rev. **90**, 1001 (1953).
[*Ki 54*] KING, J. G., and V. JACCARINO: Phys. Rev. **94**, 1610 (1954).
[*Kl 52*] KLOEPPER, LENNOX and WIEDENBECK: Phys. Rev. **88**, 695 (1952).
[*Kl 53*] KLEMA, E. D., and F. K. MCGOWAN: Phys. Rev. **91**, 616 (1953).
[*Kl 53a*] KLEMA, E. D., and F. K. MCGOWAN: Phys. Rev. **92**, 1469 (1953).
[*Kl 54*] KLUYVER, VAN DER LEUN and ENDT: Phys. Rev. **94**, 1795 (1954).
[*Kl 54a*] KLINE, R. M., and D. J. ZAFFARANO: Phys. Rev. **96**, 1620 (1954).
[*Kl 55*] KLEMA, E. D.: Phys. Rev. **100**, 66 (1955).
[*Kn 54*] KNIGHT, BRAID and RICHARDSON: Proc. Phys. Soc. Lond. A **67**, 881 (1954).
[*Kn 54a*] KNOWLES, J. W.: Phys. Rev. **94**, 795 (A) (1954).
[*Kn 55*] KNIGHT, J. D., and E. W. BAKER: Phys. Rev. **100**, 1334 (1955).
[*Ko 53*] KOICKI, BALLINI and CHAMINADA: C. R. Soc. Acad. Sci., Paris **236**, 1155 (1953).
[*Ko 53a*] KOJIMA, TSUKADA, OGAWA and SHIMAUCHI: J. Chem. Phys. **21**, 1415 (1953).
[*Ko 53b*] KOBAYASKI, Y., and G. MIYAMOTO: J. Phys. Soc. Japan **8**, 135, 273 (1953).
[*Ko 54*] KOESTER, L.: Z. Naturforsch. **9**a, 104 (1954).
[*Ko 54a*] KORSUNSKIĬ, M. I.: Isomerism of Atomic Nuclei, Moscow; Gostekhizdat 1954.
[*Kr 53*] KRAUSHAAR, WILSON and BAINBRIDGE: Phys. Rev. **90**, 610 (1953).

[*Kr 53a*] KRAUSHAAR, J. J., and M. GOLDHABER: Phys. Rev. **89**, 1081 (1953).
[*Kr 53b*] KRAUSHAAR, J. J.: Phys. Rev. **92**, 318 (1953).
[*Kr 53c*] KRAUSHAAR, J. J., and R. V. POUND: Phys. Rev. **92**, 523 (1953).
[*Kr 53d*] KRUGER, P., and N. SUGARMAN: Phys. Rev. **90**, 158 (1953).
[*Kr 54*] KROHN, V. E., and S. RABOY: Phys. Rev. **95**, 1354 (1954).
[*Kr 54a*] KRAUSHAAR, MIHELICH and SUNYAR: Phys. Rev. **95**, 456 (1954).
[*Kr 55*] KRÜGER, H., and U. MEYER-BERKHOUT: Naturwiss. **42**, 94 (1955).
[*Kr 55a*] KROHN, V. E., and S. RABOY: Phys. Rev. **97**, 1017 (1955).
[*Kr 55b*] KREGER, MCISAAC, MACKIN and LAI: Phys. Rev. **100**, 955(A) (1955).
[*Ku 29*] KUHN, W.: Phil. Mag. **8**, 625 (1929).
[*Ku 35*] KURCHATOV, KURCHATOV, MYSSOWSKI and ROUSSINOV: C. R. Acad. Sci., Paris **200**, 1201 (1935).

[*La 52*] LANGER, L. M., and R. B. DUFFIELD: Quoted in Ref. [*Hu 52b*].
[*La 53*] LAWSON, FRAUENFELDER and JENTSCHKE: Phys. Rev. **91**, 649 (1953).
[*La 53a*] LANGER, LAZAR and MOFFAT: Phys. Rev. **91**, 338 (1953).
[*La 53b*] LANGEVIN, M.: C. R. Acad. Sci., Paris **236**, 689 (1953).
[*La 53c*] LAZAR, MOFFAT and LANGER: Phys. Rev. **91**, 498(A) (1953).
[*La 54*] LANGEVIN, M., and N. MARTY: J. Phys. Radium **15**, 127 (1954).
[*La 54a*] LANGEVIN, M., and P. RADVANYL: C. R. Acad. Sci., Paris **238**, 77, 232 (1954).
[*La 54b*] LABERRIQUE-FROLOW, LANGEVIN and BERNAS: C. R. Acad. Sci., Paris **238**, 677 (1954); **236**, 1246 (1953).
[*La 54c*] LABERRIQUE-FROLOW, J., and N. MARTY: J. Phys. Radium **15**, 584 (1954).
[*La 54d*] LAZAR, N. H.: Phys. Rev. **95**, 292 (1954).
[*La 54e*] LANGER, L. M., and J. W. STARNER: Phys. Rev. **93**, 253 (1954).
[*La 54f*] LANGEVIN, M.: C. R. Acad. Sci., Paris **238**, 1310 (1954).
[*La 54g*] LANGEVIN, M.: C. R. Acad. Sci., Paris **238**, 2518 (1954); **239**, 1625 (1954).
[*La 55*] LAZAR, N. H., and E. D. KLEMMA: Phys. Rev. **98**, 710 (1955).
[*La 55a*] LANGEVIN, M.: J. Phys. Radium **16**, 238 (1955).
[*La 55b*] LABERRIQUE-FROLOW, J., and M. LEDERER: J. Phys. Radium **16**, 346 (1955).
[*La 55c*] LABERRIQUE-FROLOW, J.: C. R. Acad. Sci., Paris **240**, 287 (1955).
[*La 55d*] LANGEVIN, M.: C. R. Acad. Sci., Paris **240**, 289 (1955).
[*La 55e*] LANGEVIN, M.: J. Phys. Radium **16**, 516 (1955).
[*Le 51*] LEININGER, SEGRÈ and WIEGAND: Phys. Rev. **81**, 280 (1951); **76**, 897 (1949).
[*Le 52*] LEWIS, G. M.: Phil. Mag. **43**, 1070 (1952).
[*Le 54*] LE BLANC, CORK, BURSON and JORDAN: Phys. Rev. **93**, 1124 (1954).
[*Le 54a*] LEMMER, H. R., and M. A. GRACE: Proc. Phys. Soc. Lond. A **67**, 1051 (1954).
[*Le 54b*] LEVI, C., and L. PAPINEAU: C. R. Acad. Sci., Paris **239**, 1782 (1954).
[*Le 54c*] LEVI, C., and L. PAPINEAU: C. R. Acad. Sci., Paris **238**, 1407 (1954).
[*Le 54d*] LEMONICK, A., and F. M. PIPKIN: Phys. Rev. **95**, 1356 (1954).
[*Le 54e*] LE BLANC, CORK and BURSON: Phys. Rev. **93**, 916(A) (1954).
[*Le 54f*] LEE, M. R., and R. KATZ: Phys. Rev. **93**, 155 (1954).
[*Le 54g*] LEVY, H. B., and I. PERLMAN: Phys. Rev. **94**, 152 (1954).
[*Le 54h*] LEVI, C., and L. PAPINEAU: C. R. Acad. Sci., Paris **238**, 2313 (1954).
[*Le 54i*] LE BLANC, CORK and BURSON: Phys. Rev. **95**, 627(A) (1954).
[*Le 55*] LE BLANC, CORK and BURSON: Phys. Rev. **97**, 750 (1955).
[*Le 55a*] LEMMER, SEGAERT and GRACE: Proc. Phys. Soc. Lond. A **68**, 701 (1955).
[*Le 55b*] LEHMANN, P., and J. MILLER: C. R. Acad. Sci., Paris **240**, 1525 (1955).
[*Le 55c*] LEHMANN, P., and J. MILLER: C. R. Acad. Sci., Paris **240**, 298 (1955).
[*Le 55d*] LE BLANC, CORK and BURSON: Phys. Rev. **98**, 39 (1955).
[*Le 55e*] LEWIS, G. M.: Proc. Phys. Lond. Soc. A **68**, 735 (1955).
[*Li 53*] LINDQVIST, T.: Ark. Fysik **6**, 123 (1953).
[*Li 53a*] LINDSTRÖM, HEDGRAN and ALBURGER: Phys. Rev. **89**, 1303 (1953).
[*Li 53b*] LIVINGSTON, R., and H. ZELDES: Phys. Rev. **90**, 609 (1953).
[*Li 53c*] LIVINGSTON, BENJAMIN, COX and GORDY: Phys. Rev. **92**, 1271 (1953).
[*Li 53d*] LIDOFSKY, ALPEROVITCH and WU: Phys. Rev. **90**, 387(A) (1953).
[*Li 53e*] LINDSTRÖM, SIEGBAHN and WAPSTRA: Proc. Phys. Soc. Lond. B **66**, 54 (1953).
[*Li 53f*] LITZ, RING and BALKWELL: Phys. Rev. **92**, 288 (1953).
[*Li 54*] LIDOFSKY, GOLD and WU: Phys. Rev. **94**, 780(A) (1954).
[*Li 54a*] LINDEN, K., and N. STARFELDT: Ark. Fysik **7**, 109 (1954).
[*Li 55*] LIDOFSKY, BENCZER, MACKLIN and WU: Phys. Rev. **98**, 1186(A) (1955).
[*Lo 53*] LOCKETT, E. E., and R. H. THOMAS: Nucleonics **11**, No. 3, 14 (1953).
[*Lu 54*] LU, D. C., and W. H. KELLY: Phys. Rev. **95**, 121 (1954).
[*Lu 54a*] LU, D. C., and M. L. WIEDENBECK: Phys. Rev. **94**, 501 (1954).
[*Lu 54b*] LU, KELLY and WIEDENBECK: Phys. Rev. **95**, 1533 (1954).

[*Lu 54c*] Lu, Kelly and Wiedenbeck: Phys. Rev. **95**, 121 (1954).
[*Lu 55*] Lu, Kelly and Wiedenbeck: Phys. Rev. **97**, 139 (1955).
[*Ly 54*] Lyon, W. S., and J. E. Strain: Phys. Rev. **95**, 1500 (1954).
[*Ly 55*] Lyon, W. S., and R. R. Rickard: Phys. Rev. **100**, 112 (1955).

[*Ma 51*] Mayer, M. G.: Phys. Rev. **78**, 16, 22 (1951).
[*Ma 53*] Manring, Brown and Williams: Phys. Rev. **90**, 348(A) (1953).
[*Ma 53a*] Mandeville, Shapiro, Mendenhall, Zucker and Conklin: Phys. Rev. **89**, 559 (1953).
[*Ma 53b*] Maerker, R. E., and R. D. Birkhoff: Phys. Rev. **89**, 1159 (1953).
[*Ma 53c*] Maerker, R. E., and R. D. Birkhoff: Phys. Rev. **88**, 699 (1953).
[*Ma 53d*] Mann, K. C., and R. M. Pearce: Proc. Roy. Soc. Canada **47**, 130(A) (1953).
[*Ma 53e*] Martin, D. G. E., and G. Parry: Phil. Mag. **44**, 344 (1953).
[*Ma 53f*] Malmfors, K. G.: Ark. Fysik **6**, 49 (1953).
[*Ma 53g*] Makiej, B.: Acta phys. polon. **12**, 32 (1953).
[*Ma 54*] MacGregor, M. H., and M. L. Wiedenbeck: Phys. Rev. **94**, 138 (1954).
[*Ma 54a*] Maeder, Müller and Wintersteiger: Helv. phys. Acta **27**, 3 (1954).
[*Ma 54b*] Mathur, H. B., and E. K. Hyde: Phys. Rev. **96**, 126 (1954).
[*Ma 54c*] Mathur, H. B., and E. K. Hyde: Phys. Rev. **95**, 708 (1954).
[*Ma 54d*] Macq, P.: Ann. Soc. sci. Bruxelles **67**, 309 (1954).
[*Ma 54e*] Marty, N.: C. R. Acad. Sci., Paris **238**, 2516 (1954). — J. Phys. Radium **15**, 605 (1954).
[*Ma 54f*] Mayquez, E. R.: An. Real Soc. españ. Fis. y Quim., Ser. A **50**, 95 (1954).
[*Ma 54g*] Maeder, Wapstra, Nijgh and Ornstein: Physica, Haag **20**, 521 (1954).
[*Ma 54h*] Marmier, P., and F. Boehm: Private communication.
[*Ma 54i*] Maeder, Wapstra, Nijgh and Ornstein: Phys. Rev. **93**, 1433 (1954).
[*Ma 54j*] Mayer-Kuckuk, Th.: Z. Naturforsch. **9**a, 338 (1954).
[*Ma 55*] Mathur, Hyde, Levine and Kofstad: Phys. Rev. **97**, 117 (1955).
[*Ma 55a*] Mathur, H. B.: Phys. Rev. **97**, 707 (1955).
[*Ma 55b*] Marmier, P., and F. Boehm: Phys. Rev. **97**, 103 (1955).
[*Ma 55c*] Mark, H., and G. T. Paulissen: Phys. Rev. **100**, 813 (1955).
[*Ma 55d*] Marty, N.: C. R. Acad. Sci., Paris **240**, 291 (1955).
[*Ma 55e*] Mark, McClelland and Goodman: Phys. Rev. **98**, 1245 (1955).
[*Ma 55f*] Mayer, M. G., and J. D. H. Jensen: Elementary Theory of Nuclear Shell Structure. John Wiley & Sons 1955.
[*Ma 55g*] Macq, P.: Bull. Cl. Sci. Acad. roy. Belg. **41**, 467 (1955).
[*Ma 55h*] Marathe, E. V.: J. Sci. Industr. Res., India B **14**, 354 (1955).
[*Ma 55i*] Mathur, H. B., and E. K. Hyde: Phys. Rev. **98**, 79 (1955).
[*Ma 55j*] Mallmann, Aten, Bes and de McMillan: Phys. Rev. **99**, 7 (1955).
[*Mc 51*] McGowan, F. K.: Phys. Rev. **81**, 1066 (1951).
[*Mc 52*] McGowan, F. K.: Phys. Rev. **87**, 542 (1952).
[*Mc 52a*] McGowan, F. K.: Phys. Rev. **85**, 142 (1952).
[*Mc 53*] McGowan, F. K.: Phys. Rev. **92**, 524 (1953).
[*Mc 53a*] McGowan, F. K., and E. C. Campbell: Phys. Rev. **92**, 523 (1953).
[*Mc 53b*] McMullen, C. C., and M. W. Johns: Phys. Rev. **91**, 418 (1953).
[*Mc 53c*] McClelland, C., and C. Goodman: Phys. Rev. **91**, 760 (1953).
[*Mc 54*] McGowan, F. K.: Oak Ridge National Laboratory, tables available on request. See Phys. Rev. **93**, 163 (1954).
[*Mc 54a*] McGowan, F. K.: Phys. Rev. **93**, 163 (1954).
[*Mc 54b*] McGowan, F. K.: Phys. Rev. **93**, 471 (1954).
[*Mc 55*] McGinnis, C. L.: Phys. Rev. **97**, 93 (1955).
[*Mc 55a*] McGinnis, C. L.: Phys. Rev. **98**, 1172(A) (1955).
[*Me 53*] Metzger, F. R., and W. B. Todd: Phys. Rev. **91**, 1286 (1953).
[*Me 53a*] Meinke, W. W.: Phys. Rev. **90**, 410 (1953).
[*Me 53b*] Metzger, F. R.: Phys. Rev. **90**, 328 (1953).
[*Me 53c*] Meyerhof, Mann and West: Phys. Rev. **92**, 758 (1953).
[*Me 54*] Metzger, F. R., and W. B. Todd: Phys. Rev. **95**, 853 (1954).
[*Me 54a*] Metzger, F. R., and W. B. Todd: Phys. Rev. **94**, 794(A) (1954).
[*Me 54b*] Metzger, F. R., and W. B. Todd: Phys. Rev. **95**, 627(A) (1954).
[*Me 55*] Metzger, F. R.: Phys. Rev. **97**, 1258; **98**, 1186(A) (1955).
[*Me 55a*] Metzger, F. R.: Phys. Rev. **99**, 613(A) (1955).
[*Me 55b*] Metzger, F. R.: Phys. Rev. **98**, 200 (1955).
[*Me 56*] Metzger, F. R.: Phys. Rev. **103**, 983 (1956).
[*Me 56a*] Metzger, F. R.: Phys. Rev. **101**, 286 (1956).
[*Mi 53*] Mihelich, J. W., and A. de-Shalit: Phys. Rev. **91**, 78 (1953).

[*Mi 53a*] MIHELICH, J. W.: Phys. Rev. **89**, 907 (A) (1953).
[*Mi 54*] MITTELMAN, P. S.: Phys. Rev. **94**, 99 (1954).
[*Mi 54a*] MICHEL, M. C., and D. H. TEMPLETON: Phys. Rev. **93**, 1422 (1954).
[*Mi 54b*] MIHELICH, J. W.: Phys. Rev. **95**, 626(A) (1954).
[*Mi 54c*] MIHELICH, MCKEOWN and GOLDHABER: Phys. Rev. **96**, 1450 (1954).
[*Mi 54d*] MIZE, STARNER and BUNKER: Phys. Rev. **96**, 444 (1954).
[*Mi 54e*] MIHELICH, SCHARDT and SEGRÈ: Phys. Rev. **95**, 1508 (1954).
[*Mi 54f*] MIJATOVIC, A. M.: Bull. Inst. Nucl. Sci., Boris Kidrich **4**, 75 (1954).
[*Mi 54g*] MIHELICH, SCHARFF-GOLDHABER and MCKEOWN: Phys. Rev. **94**, 794(A) (1954).
[*Mi 55*] MILTON, FRASER and MILTON: Phys. Rev. **98**, 1173(A) (1955).
[*Mi 55a*] MIHELICH, J. W., and M. GOLDHABER: Phys. Rev. **98**, 1185(A) (1955).
[*Mi 55b*] MICHALOWICA, A., and R. BOUCHEZ: J. Phys. Radium **16**, 578 (1955).
[*Mi 56*] MIHELICH, WARD and JACOB: Phys. Rev. **103**, 1285 (1956).
[*Mo 50*] MOON, P. B.: Proc. Phys. Soc. Lond. **63**, 1189 (1950); **64**, 76 (1951).
[*Mo 53*] MONTALBETTI, KATZ and GOLDENBERG: Phys. Rev. **91**, 659 (1953).
[*Mo 53a*] MOON, P. B., and A. STORRUSTE: Proc. Phys. Soc. Lond A **66**, 585 (1953).
[*Mo 54*] MOSZKOWSKI, S. A., and D. C. PEASLEE: Phys. Rev. **93**, 455 (1954).
[*Mo 54a*] MOREAU, J.: J. Phys. Radium **15**, 380 (1954).
[*Mo 54b*] MOREAU, J.: C. R. Acad. Sci., Paris **239**, 1130 (1954).
[*Mo 54c*] MOREAU, J.: C. R. Acad. Sci., Paris **239**, 820 (1954).
[*Mo 54d*] MOREAU, J.: Ark. Fysik **7**, 391 (1954).
[*Mo 54e*] MOON, P. B., and K. ILAKOVAC: Phys. Rev. **93**, 254 (1954).
[*Mu 52*] MULLER, HOYT, KLEIN and DUMOND: Phys. Rev. **88**, 775 (1952).
[*Mu 53*] MURAKAWA, K.: J. Phys. Soc. Japan **8**, 535 (1953).
[*Mu 53a*] MURDOCH, H. S.: Proc. Phys. Soc. Lond. A **66** 944 (1953).
[*Mu 54*] MURDOCH, H. S., and A. J. WEBB: Proc. Phys. Soc. Lond. A **67**, 286 (1954).
[*Mu 54a*] MURAKAWA, K.: Phys. Rev. **93**, 1232 (1954).
[*Mu 54b*] MURAKAWA, K., and S. SUWA: Z. Physik **137**, 575 (1954).
[*Mu 55*] MURRAY, BOEHM, MARMIER and DUMOND: Phys. Rev. **97**, 1007 (1955).
[*Mu 55a*] MURAKAWA, K.: Phys. Rev. **98**, 1285 (1955).
[*Mu 55b*] MURAKAWA, K.: Phys. Rev. **100**, 1369 (1955).

[*Na 54*] NAUMANN, R. A., and J. B. GEBHART: Phys. Rev. **96**, 1452 (1954).
[*Na 54a*] NAUMANN, R. A.: Phys. Rev. **96**, 90 (1954).
[*Na 56*] NABLO, S. V., and M. W. JOHNS: Bull. Amer. Phys. **1**, 42 (1956).
[*Na 56a*] NAUMANN, R. A.: Bull. Amer. Phys. Soc. **1**, 42 (1956).
[*Ne 50*] NEWTON, T. D.: Phys. Rev. **78**, 490 (1950).
[*Ne 54*] NEWSON, H. W., and R. H. ROHRER: Phys. Rev. **94**, 654 (1954).
[*Ne 55*] NERVIK, W. E., and G. T. SEABORG: Phys. Rev. **97**, 1092 (1955).
[*Ni 54*] NIJGH, ORNSTEIN and GROBBEN: Physica, Haag **20**, 243 (1954).
[*Ni 55*] NIDAY, J. B.: Phys. Rev. **98**, 42 (1955).
[*No 53*] NOVAKOW, E. T.: Bull. Sci. Cons. Acad. Yugoslavia **1**, 11 (1953).
[*No 55*] NOBLE, L. M., and R. M. STEFFEN: Phys. Rev. **100**, 955(A) (1955).
[*Nu 53*] NUSSBAUM, WAPSTRA, VERSTER, CERFONTAIN and CERFONTAIN: Physica, Haag **19**, 385 (1953).
[*Nu 53a*] NUSSBAUM, R. H., and R. VAN LIESHOUT: Physica, Haag **19**, 451 (1953).
[*Nu 53b*] NUSSBAUM, WAPSTRA and VERSTER: Physika, Haag **19**, 131 (1953).
[*Nu 54*] NUSSBAUM, VAN LIESHOUT, WAPSTRA, VERSTER, TEN HAUF, NIJGH and ORNSTEIN: Physica, Haag **20**, 555 (1954). — Phys. Rev. **93**, 255 (1954).
[*Nu 55*] NUSSBAUM, R. H.: Thesis, University of Amsterdam 1955.

[*Ol 54*] OLSEN, J. L., and G. D. O'KELLEY: Phys. Rev. **93**, 1125 (1954).
[*Ol 56*] OLSEN, MANN and LINDNER: Bull. Amer. Phys. Soc. **1**, 41 (1956).
[*Or 52*] ORRING, J.: Ark. Fysik **4**, 469 (1952).
[*Ow 56*] OWEN, J., and I. M. WARD: Phys. Rev. **102**, 591 (1956).

[*Pa 53*] PAYNE, W., and M. GOODRICH: Phys. Rev. **91**, 497(A) (1953).
[*Pa 54*] PASSELL, MICHEL and BERGSTRÖM: Phys. Rev. **95**, 999 (1954).
[*Pe 53*] PEARCE, R. M., and K. C. MANN: Canad. J. Phys. **31**, 592 (1953).
[*Pe 53a*] PERKINS, J. F., and S. K. HAYNES: Phys. Rev. **92**, 687 (1953).
[*Pe 53b*] PERRIN, N.: J. Phys. Radium **14**, 273 (1953).
[*Pe 55*] PERLMAN, M. L., and J. P. WELKER: Phys. Rev. **100**, 81 (1955).
[*Pl 54*] PLASSMANN, E. A., and L. M. LANGER: Phys. Rev. **96**, 1593 (1954).
[*Pn 53*] PNIEWSKI, J.: Acta phys. polon. **11**, 215 (1953). — Nature, Lond. **171**, 694 (1953).

[*Po 48*] POLLARD, E. C., and D. E. ALBURGER: Phys. Rev. **74**, 926 (1948).
[*Po 52*] POOL, M. L.: Physica, Haag **18**, 1304 (1952).
[*Po 54*] PORSCHEN, W., and W. RIEZLER: Z. Naturforsch. **9**a, 701 (1954).
[*Po 54a*] POHM, LEWIS, TALBOY and JENSEN: Phys. Rev. **95**, 1523 (1954).
[*Po 55*] POTNIS, MANDEVILLE and BURLEW: Phys. Rev. **99**, 671 (A) (1955).
[*Po 55a*] POPPEMA, STEENLAND, BUEN and GORTER: Physica, Haag **21**, 233 (1955).
[*Po 55b*] POË, A. J.: Phil. Mag. **46**, 611 (1955).
[*Po 56*] POTNIS, DUBEY and MANDEVILLE: Bull. Amer. Phys. Soc. **1**, 42 (1956).
[*Pr 53*] PREISWERK, P., and P. STÄHELIN: Nuovo Cim. **10**, 1219 (1953).
[*Pr 54*] PRINGLE, TURCHINETZ and TAYLOR: Phys. Rev. **95**, 115 (1954).
[*Pr 54a*] PRESCOTT, J. R.: Proc. Phys. Soc. Lond. A **67**, 540 (1954).
[*Pr 54b*] PRESCOTT, J. R.: Proc. Phys. Soc. Lond. A **67**, 254 (1954).
[*Pr 54c*] PRYCE, M. H. L.: Rep. Progr. Phys. **17**, 1 (1954).
[*Pr 55*] PRUETT, C. H., and R. G. WILKINSON: Phys. Rev. **100**, 1237 (A) (1955).

[*Ra 53*] RASETTI, F., and E. C. BOOTH: Phys. Rev. **91**, 1192 (1953).
[*Ra 54*] RAAD, MIDDLEKOOP, VAN NOOYEN and ENDT: Physica, Haag **20**, 1278 (1954).
[*Ra 55*] RASMUSSEN, E., and V. MIDDELBOE: Z. Physik **144**, 160 (1955).
[*Ra 55a*] RADVANYI, P.: J. Phys. Radium **16**, 509 (1955). — Ann. Physik **10**, 584 (1955).
[*Re 53*] REYNOLDS, LEDDICOTTE and MAHLMAN: Phys. Rev. **91**, 333 (1953).
[*Re 54*] REID, J. M., and K. G. MCNEILL: Phil. Mag. **45**, 957 (1954). — Proc. Phys. Soc. Lond. A **66**, 1179 (1953).
[*Re 55*] REYNOLDS, CHRISTENSEN, HAMILTON, LEMONICK, PIPKEN and STROKE: Phys. Rev. **99**, 613 (A) (1955).
[*Re 55a*] REED, G. W.: Phys. Rev. **98**, 1327 (1955).
[*Ri 54*] RIETJENS, VAN DEN BOLD and ENDT: Physica, Haag **20**, 107 (1954).
[*Ri 55*] RIOU, M.: J. Phys. Radium **16**, 583 (1955).
[*Ri 55a*] RIETJENS, L. H. TH., and H. J. VAN DEN BOLD: Physica, Haag **21**, 701 (1955).
[*Ri 55b*] RIETJENS, VAN DEN BOLD and HEYLIGERS: Physica, Haag **21**, 899 (1955).
[*Ro 51*] ROSE, GOERTZEL, SPINRAD, HARR and STRONG: Phys. Rev. **83**, 79 (1951).
[*Ro 52*] ROBINSON, B. L., and L. MADANSKY: Phys. Rev. **88**, 1065 (1952).
[*Ro 53*] ROSE, M. E.: Phys. Rev. **91**, 610 (1953).
[*Ro 54*] ROLLIER, M. A., and E. SAELAND: Phys. Rev. **94**, 1079 (1954).
[*Ro 54a*] ROSSI, JONES, HOLLANDER and HAMILTON: Phys. Rev. **93**, 256 (1954).
[*Ro 54b*] ROLLIER, M. A.: Gazz. chim. ital. **84**, 658 (1954).
[*Ro 55*] ROHR, R. C., and R. D. BIRKHOFF: Phys. Rev. **98**, 1266 (1955).
[*Ro 55a*] ROLLIER, SAELAND, MORPURGO u. CAGLIERIS: Acta chem. scand., Copenh. **9**, 57 (1955).
[*Ro 55b*] ROBINSON, B. L., and R. W. FINK: Phys. Rev. **98**, 231 (A) (1955).
[*Ro 55c*] ROJO, HAKEEM and GOODRICH: Phys. Rev. **99**, 1629 (A) (1955).
[*Ro 56*] ROBINSON, B. L., and R. W. FINK: Bull. Amer. Phys. Soc. **1**, 40 (1956).
[*Ro 56a*] ROSSI, FRAUENFELDER, LEVINE and SINGER: Bull. Amer. Phys. Soc. **1**, 163 (1956).

[*Sa 53*] SARGENT, YAFFE and GRAY: Canad. J. Phys. **31**, 235 (1953).
[*Sa 53a*] SARAF, VARMA and MANDEVILLE: Phys. Rev. **91**, 1216 (1953).
[*Sa 53b*] SAKAI, M., and P. HUBERT: C. R. Acad. Sci., Paris **236**, 1249 (1953).
[*Sa 54*] SAKAI, MURRAY and KURBATOV: Phys. Rev. **96**, 826 (A) (1954).
[*Sa 54a*] SARAF, B.: Phys. Rev. **94**, 642 (1954).
[*Sa 54b*] SARAF, B.: Phys. Rev. **95**, 97 (1954).
[*Sa 55*] SARAF, VARMA and MANDEVILLE: Phys. Rev. **98**, 1206 (A) (1955).
[*Sa 55a*] SARAF, B.: Phys. Rev. **97**, 715 (1955).
[*Sc 53*] SCHARFF-GOLDHABER, G., and M. MCKEOWN: Phys. Rev. **92**, 356 (1953).
[*Sc 53a*] SCHIFF, D., and F. R. METZGER: Phys. Rev. **90**, 849 (1953).
[*Sc 53b*] SCHRADER, NELSON and JACOBS: Phys. Rev. **90**, 159 (1953).
[*Sc 53c*] SCHRADER, C. D.: Phys. Rev. **92**, 928 (1953).
[*Sc 53d*] SCHWARTZ, PERLMAN and BERNSTEIN: Phys. Rev. **91**, 883 (1953).
[*Sc 53e*] SCHARFF-GOLDHABER, G.: Phys. Rev. **90**, 587 (1953).
[*Sc 54*] SCHARFF-GOLDHABER, G., and M. MCKEOWN: Phys. Rev. **95**, 613 (A) (1954).
[*Sc 54a*] SCHUMAN, R. P.: Phys. Rev. **96**, 121 (1954).
[*Sc 55*] SCHARDT, A. W., and J. P. WELKER: Phys. Rev. **99**, 810 (1955).
[*Sc 55a*] SCHMOUKER, ERDÖS, JORDAN et STOLL: J. Phys. Radium **16**, 169 (1955).
[*Sc 55b*] SCHMELING, P.: Z. Naturforsch. **10**a, 79 (1955).
[*Sc 56*] SCHARDT, A. W.: Bull. Amer. Phys. Soc. **1**, 85 (1956).
[*Sc 56a*] SCHARFF-GOLDHABER, MCKEOWN and MIHELICH: Bull. Amer. Phys. Soc. **1**, 206 (1956).

[Se 47] SEGRÈ, E.: Phys. 71, 274 (1947).
[Se 49] SEGRÈ, E., and A. C. HELMHOLZ: Rev. Mod. Phys. **21**, 271 (1949).
[Se 53] SELIGER, CAVALLO and CULPEPPER: Phys. Rev. **90**, 443 (1953).
[Se 54] SEHR, R.: Z. Physik **137**, 523 (1954).
[Se 55] SENITZKY, RABI and PERL: Phys. Rev. **98**, 1537(A) (1955).
[Se 55a] SEVERIENS, J. C., and S. S. HANNA: Phys. Rev. **100**, 1254 (A) (1955).
[Sh 52] SHPINEL, V. S.: Ž. eksp. teor. Fiz. **22**, 255 (1952).
[Sh 53] SHORE, BENDEL, BROWN and BECKER: Phys. Rev. **91**, 1203 (1953).
[Sh 55] SHAKHBAZYAN, B. A., and L. I. RUSINOV: Izv. Akad. Nauk SSSR., Ser. Fiz. **19**, 308 (1955).
[Si 54] SIMANTON, RIGHTMIRE, LONG and KOHMAN: Phys. Rev. **96**, 1711 (1954).
[Si 55] SIEGBAHN, K.: Beta- and Gamma-Ray Spectroscopy. North Holland Publishing Co. 1955.
[Si 55a] SIMMONS, VAN PATTER, FAMULARO and STUART: Phys. Rev. **97**, 89 (1955).
[Sl 53] SLÄTIS, H., and L. ZAPPA: Ark. Fysik **6**, 81 (1953).
[Sl 54] SLATTERY, LU and WIEDENBECK: Phys. Rev. **96**, 465 (1954).
[Sm 52] SMITH, CAIRD and MITCHELL: Phys. Rev. **88**, 150 (1952).
[Sm 54] SMITH, K. F.: Quoted by E. A. PLASSMANN and L. M. LANGER, Phys. Rev. **96**, 1593 (1954).
[Sm 55] SMITH, W. G., and J. M. HOLLANDER: Phys. Rev. **98**, 1258 (1955).
[Sn 37] SNELL, A. H.: Phys. Rev. **52**, 1007 (1937).
[Sn 56] SNELL, A. H., and F. PLEASONTON: Bull. Amer. Phys. Soc. **1**, 42 (1956).
[So 17] SODDY, F.: Nature, Lond. **99**, 414, 433 (1917).
[So 54] SOGO, P. B., and C. D. JEFFRIES: Phys. Rev. **93**, 174 (1954).
[So 55] SOGO, P. B., and C. D. JEFFRIES: Phys. Rev. **98**, 1316 (1955).
[So 55a] SOFTKY, S. D.: Phys. Rev. **98**, 736 (1955).
[Sp 53] SPRAGUE, G., and D. H. TOMBOULIAN: Phys. Rev. **92**, 105 (1952).
[Sp 54] SPIESS, F. N.: Phys. Rev. **94**, 1292 (1954).
[Sp 55] SPECK, R. D., and F. A. JENKINS: Phys. Rev. **98**, 282(A) (1955).
[St 52] STEUDEL, A.: Z. Physik **132**, 429 (1952).
[St 52a] STEFFEN, R. M., and W. ZOBEL: Phys. Rev. **88**, 170(A) (1952).
[St 53] STÄHELIN, P.: Helv. phys. Acta **26**, 691 (1953). — Phys. Rev. **92**, 1076 (1953).
[St 53a] STEFFEN, R. M.: Phys. Rev. **89**, 903(A) (1953).
[St 53b] STÄHELIN, P., and P. PREISWERK: Nuovo Cim. **10**, 1219 (1953).
[St 53c] STOKER, P. H., and O. P. HOK: Physica, Haag **19**, 279 (1953).
[St 53d] STOKER, HEERSCHAP and HOK: Physica, Haag **19**, 433 (1953).
[St 53e] STURCKEN, E. F., and A. H. WEBER: Phys. Rev. **91**, 484(A) (1953).
[St 54] STURCKEN, O'FRIEL and WEBER: Phys. Rev. **93**, 1053 (1954).
[St 55] STEFFEN, R. M., and W. ZOBEL: Phys. Rev. **97**, 1188 (1955); **103**, 126 (1956)
[St 55a] STEFFEN, R. M.: Phys. Rev. **99**, 659(A) (1955).
[St 55b] STELSON, P. H., and F. ;. MCGOWAN: Phys. Rev. **99**, 112 (1955).
[St 55c] STROMINGER, D., and J. O. RASMUSSEN: Phys. Rev. **100**, 844 (1955).
[St 55d] STEWART, SCHARENBERG and WIEDENBECK: Phys. Rev. **99**, 691 (1955).
[St 55e] STEFFEN, R. M., and J. N. BRAZOS: Phys. Rev. **99**, 1646(A) (1955).
[St 55f] STRIBEL, TH.: Z. Naturforsch. **10**a, 797 (1955).
[St 56] STEFFEN, R. M., and L. M. NOBLE: Bull. Amer. Phys. Soc. **1**, 42 (1956).
[St 56a] STOCKENDAL, MCDONELL, SCHMORAK and BERGSTRÖM: To be published.
[St 56b] STROMINGER, STEPHENS and RASMUSSEN: Phys. Rev. **103**, 748 (1956).
[Su 53] SUGARMAN, N.: Phys. Rev. **89**, 570 (1953).
[Su 53a] SUWA, S.: J. Phys. Soc. Japan **8**, 734 (1953).
[Su 53b] SUNYAR, A. W.: Phys. Rev. **90**, 387(A) (1953).
[Su 54] SUNYAR, MIHELICH and GOLDHABER: Phys. Rev. **95**, 570 (1954).
[Su 54a] SUNYAR, A. W.: Phys. Rev. **93**, 1345 (1954).
[Su 54b] SUNYAR, A. W.: Phys. Rev. **93**, 1122 (1954).
[Su 54c] SUTTLE, A. D., and W. F. LIBBY: Phys. Rev. **95**, 866 (1954).
[Su 55] SUNYAR, A. W.: Phys. Rev. **98**, 653 (1955).
[Su 55a] SUNYAR, A. W.: Private communication.
[Su 55b] SUNYAR, A. W., and P. AXEL: Private communication.
[Sw 50] SWANN, MANDEVILLE and WHITEHEAD: Phys. Rev. **79**, 598 (1950).
[Sw 52] SWAN, J. B., and R. D. HILL: Phys. Rev. **88**, 831 (1952).
[Sw 53] SWINBANK, P., and J. WALKER: Proc. Phys. Soc. Lond. A **66**, 1093 (1953).
[Sw 53a] SWAN, J. B., and R. D. HILL: Phys. Rev. **91**, 424 (1953).
[Sw 53b] SWAN, PORTNOY and HILL: Phys. Rev. **90**, 257 (1953).
[Sw 53c] SWANN, J. B., and R. D. HILL: Austral. J. Phys. **6**, 371 (1953).
[Sz 35] SZILARD, L., and T. A. CHALMERS: Nature, Lond. **135**, 98 (1935).

[*Ta 55*] TAYLOR, H. W., and R. W. PRINGLE: Phys. Rev. **99**, 1345 (1955).
[*Te 53*] TEILLAC, RIOU and DESNEIGES: C. R. Acad. Sci., Paris **237**, 41 (1953).
[*Te 54*] TEMMER, G. M., and N. P. HEYDENBURG: Phys. Rev. **96**, 426 (1954).
[*Te 55*] TEMMER, G. M., and N. P. HEYDENBURG: Phys. Rev. **98**, 1198(A) (1955).
[*Th 53*] THIRION, J., and V. L. TELEGDI: Phys. Rev. **92**, 1253 (1953).
[*Th 54*] THULIN, S.: Ark. Fysik **7**, 269 (1954).
[*Th 54a*] THULIN, S.: Phys. Rev. **94**, 734 (1954).
[*Th 54b*] THULIN, S., and K. NYBÖ: Ark. Fysik **7**, 289 (1954).
[*Th 54c*] THULIN, MOREAU and ATTERLING: Ark. Fysik **8**, 219 (1954).
[*Th 54d*] THIRION, BARNES and LAURITSEN: Phys. Rev. **94**, 1076 (1954).
[*Th 54e*] THULIN, MOREAU and ATTERLING: Ark. Fysik **8**, 229 (1954).
[*Th 55*] THIRION, J., and R. BARLOUTAUD: C. R. Acad. Sci., Paris **240**, 2136 (1955).
[*Th 55a*] THULIN, S.: Ark. Fysik **9**, 137 (1955).
[*Ti 53*] TING, Y., and D. WILLIAMS: Phys. Rev. **89**, 595 (1953).
[*To 53*] TOMLINSON, RIDGWAY and GOPALAKRISHNAN: Phys. Rev. **91**, 484(A) (1953).
[*To 53a*] TOBAILEM, J.: J. Phys. Radium **14**, 553 (1953).
[*To 54*] TOMLINSON, NAUMANN and MIHELICH: Phys. Rev. **94**, 794(A) (1954).
[*To 55*] TOBAILEM, J.: J. Phys. Radium **16**, 48 (1955).
[*To 55a*] TOBAILEM, J., and J. ROBERT: J. Phys. Radium **16**, 340 (1955).
[*To 55b*] TOBAILEM, J.: Phys. Radium **16**, 235 (1955).
[*To 55c*] TOMLINSON, E. P., and R. A. NAUMANN: Phys. Rev. **100**, 955(A) (1955).
[*Tr 39*] TRALLI, N., and I. S. LOWEN: Phys. Rev. **76**, 1541 (1949).
[*Tr 54*] TRISCHKA, J. W., and R. BRAUNSTEIN: Phys. Rev. **96**, 968 (1954).
[*Tu 53*] TURNER, J. F.: Atomic Energy Research Establishment report AERE N/R 1254, 1953. Unpubished.

[*Va 53*] VALLADAS, G.: C. r. Acad. Sci., Paris **237**, 1673 (1953).
[*Va 54*] VARMA, J., and C. E. MANDEVILLE: Phys. Rev. **94**, 91 (1954).
[*Va 54a*] VARMA, J.: Phys. Rev. **94**, 1688 (1954). — J. Franklin Inst. **257**, 247 (1954).
[*Va 55*] VARMA, J., and C. E. MANDEVILLE: Phys. Rev. **97**, 977 (1955).
[*Ve 54*] VERHAEGHE, J., and J. DEMUYNCK: C. R. Acad. Sci., Paris **239**, 1374 (1954).
[*Ve 55*] VEGORS, S. H., and P. AXEL: Phys. Rev. **100**, 1238(A) (1955); **101**, 1067 (1956).
[*Vo 36*] WEIZSÄCKER, C. F. v.: Naturwiss. **24**, 813 (1936).
[*Vo 51*] DARDEL, G. v.: Ark. Fysik **2**, 337 (1951).
[*Vo 53*] SIEMENS, W. v.: Ann. Physik **13**, 136 (1953).
[*Vo 55*] VOLCHOK, H. L., and J. L. KULP: Phys. Rev. **97**, 102 (1955).

[*Wa 53*] WALCHLI, H. E.: Phys. Rev. **90**, 331 (1953)
[*Wa 53a*] WANLESS, R. K., and H. G. THODE: Canad. J. Phys. **31**, 517 (1953).
[*Wa 53b*] WAPSTRA, VERSTER and BOELHOUWER: Physica, Haag **19**, 138 (1953).
[*Wa 53c*] WAPSTRA, A. H.: Physica, Haag **19**, 671 (1953).
[*Wa 54*] WAPSTRA, A. H.: Ark. Fysik **7**, 265 (1954).
[*Wa 54a*] WAPSTRA, A. H.: Ark. Fysik **7**, 275 (1954).
[*Wa 54b*] WAPSTRA, A. H.: Ark. Fysik **7**, 279 (1954).
[*Wa 54c*] WAPSTRA, MAEDER, NIJGH and ORNSTEIN: Physica, Haag **20**, 169 (1954).
[*Wa 55*] WANG, T. C.: Phys. Rev. **99**, 566 (1955).
[*Wa 55a*] WAHL, A. C.: Phys. Rev. **99**, 730 (1955).
[*We 38*] WEISSKOPF, V. F.: Phys. Rev. **53**, 1018 (1938).
[*We 53*] WESSEL, G., and H. LEW: Phys. Rev. **91**, 476(A) (1953).
[*We 53a*] WEBER, G., and A. FLAMMERSFELD: Z. Naturforsch. **8**a, 580 (1953).
[*We 53b*] WELKER, SCHARDT, FRIEDLANDER and HOWLAND: Phys. Rev. **92**, 401 (1953).
[*We 54*] WEBER, G.: Z. Naturforsch. **9**a, 115 (1954).
[*We 55*] WELKER, J. P., and M. L. PERLMAN: Phys. Rev. **100**, 74 (1955).
[*We 56*] WETHERILL, ALDRICH, TILTON and DAVIS: Bull. Amer. Phys. Soc. **1**, 31 (1956).
[*Wh 53*] WHALEN, PORTER and COOK: Phys. Rev. **89**, 902(A) (1953).
[*Wh 53a*] WHITTLE, C. E., and F. T. PORTER: Phys. Rev. **90**, 498 (1953).
[*Wh 55*] WHEATLEY, GRIFFING and HILL: Phys. Rev. **99**, 334 (1955).
[*Wh 55a*] WHEATLEY, HUISKAMP, DIDDENS, STEENLAND and TOLHOEK: Physica, Haag **21**, 841 (1955).
[*Wh 55b*] WHITE, COLLINS and ROURKE: Phys. Rev. **97**, 566 (1955).
[*Wi 53*] WIEDLING, T.: Phys. Rev. **91**, 767 (1953).
[*Wi 53a*] WILHELMI, BRUNSZ and DABROWSKI: Bull. Acad. Polon. Sci. **1**, 105 (1953).
[*Wi 53b*] WILES, SMITH, HORSLEY and THODE: Canad. J. Phys. **31**, 419 (1935).
[*Wi 53c*] WIEDLING, T.: Ark. Fysik **6**, 39 (1953).

[*Wi 54*] WIEDLING, T.: Ark. Fysik **7**, 69 (1954).
[*Wi 54a*] WINTERSTEIGER, V. Z.: Bull. Inst. Nucl. Sci., Boris Kidrich **4**, 79 (1954).
[*Wi 54b*] WILES, D. R.: Phys. Rev. **93**, 181 (1954).
[*Wi 55*] WILES, D. M., and R. H. TOMLINSON: Canad. J. Phys. **33**, 133 (1955).
[*Wi 55a*] WILES, D. M., and R. H. TOMLINSON: Phys. Rev. **99**, 188 (1955).
[*Wo 52*] WOLFSON, J. L.: Canad. J. Phys. **30**, 715 (1952).
[*Wo 52a*] WOLFSON, J. L., and L. G. ELLIOTT: Proc. Phys. Soc. Canada **46**, 142(A) (1952).
[*Wo 54*] WOODWARD, E. C.: Phys. Rev. **93**, 954(A) (1954).
[*Wo 55*] WOODGATE, G. K., and R. W. HELLWARTH: Nature, Lond. **176**, 395 (1955).
[*Wo 55a*] WOLFSON, J. L.: Canad. J. Phys. **33**, 886 (1955).
[*Wu 53*] WU, BOEHM and NAGEL: Phys. Rev. **91**, 319 (1953).

[*Ya 55*] YAVIN, A. I., and F. H. SCHMIDT: Phys. Rev. **100**, 171 (1955).
[*Ya 55a*] YAMPOL'SKII, LEIPUNSKII, GEN and TIKHOMIROV: Izv. Akad. Nauk SSSR., Ser. Fiz. **19**, 338 (1955).
[*Yo 53*] YOSHIZAWA, Y.: J. Phys. Soc. Japan **8**, 435 (1953).
[*Yt 54*] YTHIER, C., and G. HERRMANN: Z. Elektrochem. **58**, 630 (1954).
[*Yu 53*] YUASA, T.: C. R. Acad. Sci., Paris **237**, 1077 (1953).
[*Yu 54*] YUASA, LABERRIGUE-FROLOW and FEUVRAIS: C. R. Acad. Sci., Paris **238**, 1500 (1954). — J. Phys. Radium **16**, 39, 165 (1955).

[*Zi 54*] ZIJP, TOM and SIZOO: Physica, Haag **20**, 727 (1954).

Alpha Radioactivity.

By

I. PERLMAN and J. O. RASMUSSEN.

With 33 Figures.

Introduction.

The study of alpha-radioactivity has generated many of the fundamental concepts of atomic and nuclear structure. Important discoveries came from both the effort to understand the mechanism of the alpha-emission process and from the observation of events produced by the high velocity particles. In this latter category we should recall that it was from the analysis of large-angle scattering of alpha-particles that RUTHERFORD [1] conceived of the atomic nucleus as the center of mass and positive charge. He also made the fundamental deduction that the subatomic particles bearing the charge within the nucleus must exhibit strong short range attractive force, otherwise the nucleus could not exist. As further examples, the discoveries of nuclear transmutation by RUTHERFORD[1] and of artificial radioactivity by I. CURIE and F. JOLIOT[2] came about from the irradiation of light elements with alpha particles.

In this review we shall deal principally with those phenomena concerned more directly with the *mechanism* of alpha-emission. Nevertheless, some derivative topics will be mentioned or partially developed because they currently receive much support from the study of alpha-radioactivity. Among these topics is the identification and classification of nuclear levels in the heavy element region. Much information on this subject is obtained from the study of alpha-spectra and furthermore the explanation of the degree of population of the various available levels is one of the central objectives in the current development of alpha emission theory.

It will become clear that current directions of interest in alpha-radioactivity had to await the events starting about 1945 in which mucn new experimental information became available. Up to that time fewer than thirty alpha-emitters had been reported and almost all of these lay in the natural radioactive series. At present there are about 160 species known and many of these have been studied intensively. Before pursuing the discussion of these a brief historical sketch will be presented in the following paragraphs.

As early as 1900 Mme. CURIE[3], after an experiment in which she showed that alpha rays differed from RÖNTGEN rays in that they became degraded in energy in traversing matter, suggested that alpha rays were heavy projected particles. Further work by STRUTT (1901)[4] and CROOKES (1902)[5] on the ionization produced in gases served to reinforce this conclusion. Measurements of the deflection of

[1] E. RUTHERFORD: Phil. Mag. **37**, 581 (1919).
[2] I. CURIE and F. JOLIOT: C. r. Acad. Sci., Paris **198**, 254, 259 (1934).
[3] M. CURIE: C. r. Acad. Sci., Paris **130**, 76 (1900).
[4] R. J. STRUTT: Phil. Trans. Roy. Soc. Lond., Ser. A **196**, 507 (1901).
[5] W. CROOKES: Proc. Roy. Soc. Lond., Ser. A **69**, 413 (1902).

alpha-particles in magnetic and electrostatic fields by RUTHERFORD (1903)[1] demonstrated clearly that these rays were indeed heavy charged particles and also afforded a measure of their velocities. Further deflection experiments (particularly those of MACKENZIE[2]) showed that a mixture of alpha-emitters produced a mixture of alpha-groups of different velocity each of which was homogeneous. Almost three decades were to pass before ROSENBLUM[3] demonstrated that these homogeneous alpha groups often had "fine structure".

Along with the early deflection work another type of measurement came into use for characterizing alpha-groups. Mme. CURIE[4] had demonstrated that alpha-particles from a *thin* polonium source had a definite range and an extension of this work by BRAGG[5] firmly brought in the concept of *range* as a distinguishing feature of each alpha-emitter.

The relation between the range (or velocity) of the alpha-group and the half-life for emission was noted at an early date by RUTHERFORD[6]. However, it remained for GEIGER and NUTTALL[7] to examine this relationship systematically and to show that the logarithm of the decay constant changed linearly with the logarithm of the range. It was even possible to predict from the range the half-life of ionium which had not been measured and to deduce that there was an extremely short-lived alpha activity associated with RaC. This was later shown to be RaC′ (Po^{214}). Among the other significant deductions made was that alpha-groups of low energy would require a very long time for emission, and it became clear in this way why all the energies observed did not vary continuously to include low values.

In 1911, RUTHERFORD (see [*1*]) produced the concept of the atomic nucleus as the center of mass and positive charge in the atom. Soon thereafter, considerable speculation arose on the structure of the nucleus and the relation of the alpha emission process to it. Among the early interesting theories of the alpha-emission process was that of LINDEMANN[8] who was able to obtain an expression which followed the GEIGER-NUTTALL relation. By 1919, RUTHERFORD[9] had succeeded in transmuting light elements by irradiating with alpha-particles and the concepts of the potential barrier and short range attractive forces in the nucleus were brought into sharper focus. With regard to heavy atoms such as uranium the puzzle yet remained as to how the alpha-particle could leave the nucleus even though particles of even higher energy could not get into the nucleus. The development of wave mechanics was the necessary prelude to the explanation.

In 1928, GAMOW [*2*] and CONDON and GURNEY [*3*] independently showed that the wave nature of matter permits the alpha-particle to penetrate the region of potential energy higher than its kinetic energy. Excellent quantitative agreement was obtained for the dependence of the decay constant upon the decay energy, and the principal feature of the GEIGER-NUTTALL relation was now understood as was the anomaly of how an alpha particle could leave a heavy nucleus but not be free to enter.

Despite the remarkable success of the theory it was soon obvious that important details were not yet explained. In the beginning it was noted that an alpha

---

[1] E. RUTHERFORD: Phys. Z. **4**, 235 (1903). — Phil. Mag. **5**, 177 (1903).
[2] D. R. MACKENZIE: Phil. Mag. **10**, 538 (1905).
[3] S. ROSENBLUM: J. Phys. **1**, 438 (1930).
[4] M. CURIE: C. r. Acad. Sci., Paris **130**, 76 (1900).
[5] W. H. BRAGG: Phil. Mag. **8**, 719 (1904); **10**, 600 (1905); **11**, 617 (1906). — W. H. BRAGG and R. KLEEMAN: Phil. Mag. **8**, 726 (1904); **10**, 318 (1905).
[6] E. RUTHERFORD: Phil. Mag. **13**, 110 (1907).
[7] H. GEIGER and J. M. NUTTALL: Phil. Mag. **22**, 613 (1911); **23**, 439 (1912).
[8] F. A. LINDEMANN: Phil. Mag. **30**, 560 (1915).
[9] E. RUTHERFORD: Phil. Mag. **37**, 581 (1919).

emitter such as AcX did not lie on the appropriate GEIGER-NUTTALL line and the new theory did not help in this regard. The situation took on an added degree of interest and complexity when ROSENBLUM[1] showed by magnetic analysis that the alpha particles from a particular substance had "fine structure". The demands on alpha decay theory now took on the added dimension of explaining the relative competition between several alpha groups of a particular emitter. Ensuing work showed that the energy dependence was not the only factor involved, indeed cases began to appear in which the alpha-group of highest intensity did not have the highest energy. This problem takes us to the present day. Although a coherent quantitative theory to explain all alpha-spectra is still incomplete a number of the factors influencing alpha-emission rates are recognized. Much of the material in this article is concerned with just these problems.

Part A of this article has to do with the alpha-energies and nuclear levels and emphasizes separately the total decay energies of the alpha-emitters (Sects. 1 to 7), the spectrum of alpha-groups for each species (Sects. 8 to 11), and patterns of energy levels in the heavy element region (Sects. 12 to 15). The total disintegration energy is now a property which can be predicted with fair accuracy. It is, of course, simply related to mass differences between parent and product, and it is now clear that these mass differences for the most part vary in regular fashion.

The next two major parts (B and C) are concerned with alpha-decay rates. A division is made between the even-even emitters and the other types in order to develop the theory first for the simplest cases. The first group in Sects. 16 to 18 in Part B have to do with the semi-empirical correlations of decay rates. These have been of great practical value as a guide to experimental work and point out most clearly what is demanded of the theory. Following this are sections devoted to the theory: Sects. 19 to 27 are on the one-body classical theory, Sects. 28 to 33 are on the effects of non-central fields and Sects. 34 to 37 consider multi-body aspects.

Part C takes up those aspects of alpha-emission which are different for nuclei having unpaired nucleons. The first several Sects. (40 to 42) treat the semi-empirical correlations and parallel those for the even-even type (16 to 18). The second group (Sects. 43 to 48) treats the directions which the theoretical interpretations are taking.

Finally, Part D discusses the energy level diagram of several cases selected to illustrate different types encountered. Information obtained from the alpha-emission is in this way correlated with that obtained from other sources.

Unfortunately, a number of interesting topics related to the alpha-emission process are omitted in this review. One of these, the experimental techniques employed in studying nuclear spectra, has become too elaborate to be dealt with in limited space. Some of the details will be found elsewhere[2]. Others, such as the interaction of alpha particles with matter, could be eliminated because detailed accounts can be found elsewhere[3].

A. Alpha energies and nuclear states.

I. Alpha disintegration energies.

1. Conditions for alpha instability. We are concerned here with denoting the specific conditions which must apply in order that particular nuclear species can be alpha emitters, and further, to see in general how these conditions are

[1] S. ROSENBLUM: J. Phys. **1**, 438 (1930).
[2] Cf. especially Vol. XLV of this Encyclopedia.
[3] Viz. in Vol. XXXIV.

met in different regions of the system of nuclei. From straightforward energy considerations it is seen that any nuclear species will be alpha-unstable if the sum of the separation energies of two neutrons and two protons is less than the binding energy of the alpha particle, 28.3 Mev.

As discussed by KOHMAN[1] and others [4], the slope of the packing fraction curve gives a rough indication of where alpha instability may be expected, but many of the important features are not revealed because the packing fraction curve only reflects the gross structure of nucleon-binding energies. The closed shells have a dramatic effect upon alpha energies and appear to be dividing lines between regions of alpha stability and alpha instability. Actually, the sudden

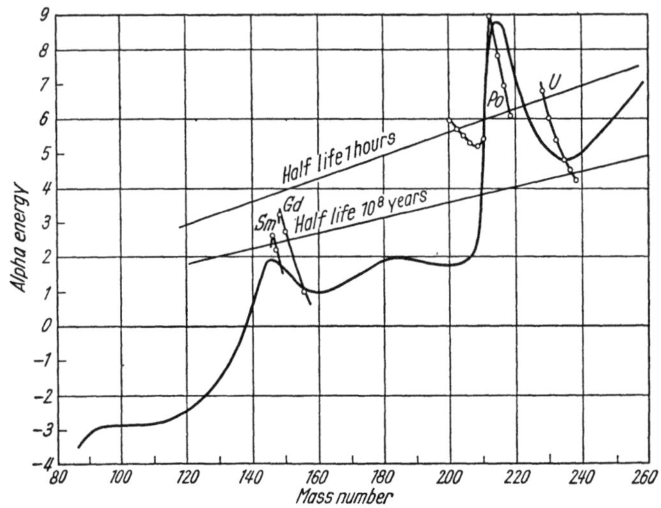

Fig. 1. Alpha-energy profile. The continuous curve shows alpha energies as a function of mass number for nuclides lying along the bottom of the energy surface valley. The segments connecting isotopes of uranium, polonium, gadolinium and samarium indicate the effects of change in neutron number for these elements. The half-life guide lines denote the alpha energies which would be required to provide these half lives (for nuclides lying on the main curve). All species which are beta-stable and which lie below the "10^8 years" curve are sufficiently long lived to have persisted since the creation of the elements. [This figure follows the form published by T. P. KOHMAN, Phys. Rev. 76, 448 (1949)].

appearance of alpha radioactivity in crossing a closed shell means only that the lifetimes have suddenly become short enough to permit detection of the instability.

From existing atomic mass data it is possible to determine (with varying degrees of accuracy) alpha energies in regions where the actual decay process is too slow to permit observation and also where it is energetically impossible. Such regions can then be made continuous with those in which accurate direct measurements are possible and in this way to display a profile of alpha energies throughout the system of nuclei. Fig. 1 contains such a plot of alpha energy as a function of mass number (heavy continuous line). This line attempts to show alpha energies for nuclei which lie along the "line of stability" of the energy surface; that is, it pertains to the most beta-stable isobar for each mass number.

Alpha energies are by no means simply a function of the mass number any more than a packing fraction curve reflects the masses of all isobars. It is more illuminating to consider an "alpha-energy surface" in analogy to a mass or energy surface. It has been found convenient to correlate alpha energies by relating

[1] T. P. KOHMAN: Phys. Rev. 76, 448 (1949).

the isotopes of each element separately[1-5] and on the alpha-energy surface this would correspond to contours of constant atomic number. Several such curves derived from experimental energy measurements have been entered in Fig. 1. That for uranium shows monotonic variation of alpha energy with mass number while the polonium curve exhibits a sharp inversion. These properties will be returned to presently.

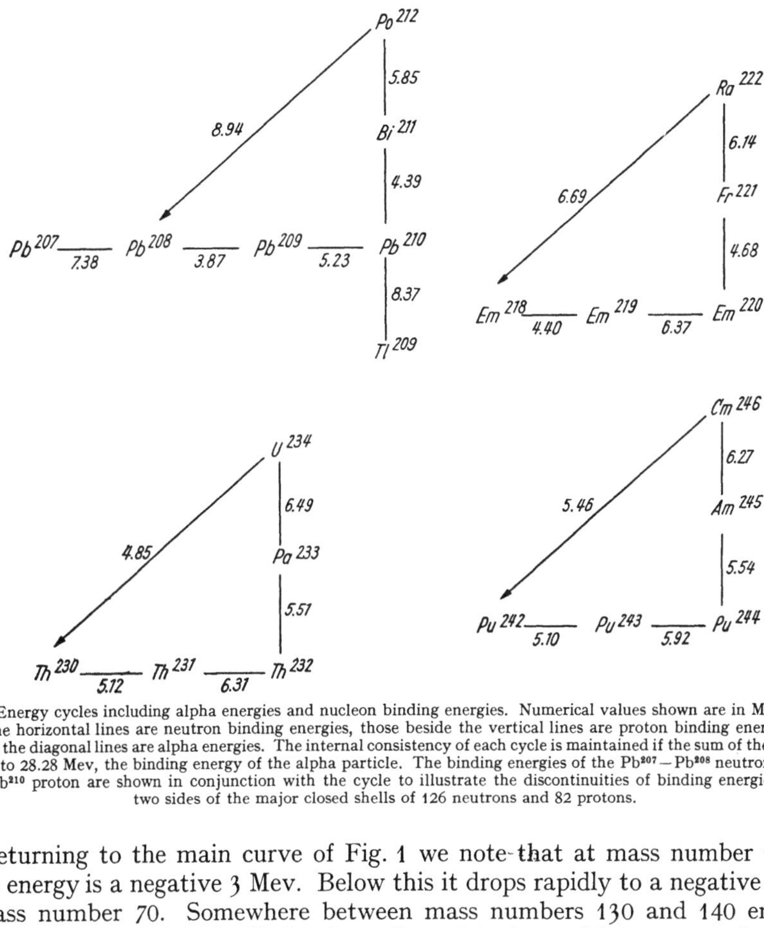

Fig. 2. Energy cycles including alpha energies and nucleon binding energies. Numerical values shown are in Mev; those below the horizontal lines are neutron binding energies, those beside the vertical lines are proton binding energies, and those on the diagonal lines are alpha energies. The internal consistency of each cycle is maintained if the sum of the energies is equal to 28.28 Mev, the binding energy of the alpha particle. The binding energies of the $Pb^{207}-Pb^{208}$ neutron and the $Tl^{209}-Pb^{210}$ proton are shown in conjunction with the cycle to illustrate the discontinuities of binding energies on the two sides of the major closed shells of 126 neutrons and 82 protons.

Returning to the main curve of Fig. 1 we note that at mass number 90 the alpha energy is a negative 3 Mev. Below this it drops rapidly to a negative 8 Mev at mass number 70. Somewhere between mass numbers 130 and 140 energies become positive and it is likely that all nuclei above this mass number (and within the band of beta-stability) are unstable toward alpha emission.

The prominent peaks in the curve above mass number 140 and again above 210 are consequences of major closed shells. Other irregularities, no doubt, occur, but mass data are not sufficiently refined to permit placement of any more detail in the curve. The effect of closed shells on alpha energies expressed in terms of the more familiar behavior of neutron and proton binding energies is illustrated by the sequence of energy cycles of Fig. 2. Selection was made of the alpha

[1] G. FOURNIER: C. r. Acad. Sci., Paris **184**, 878 (1927). — K. FAJANS: Radioelements and Isotopes, Chap. I. New York: McGraw-Hill Book Company, Inc. 1931.

[2] J. SCHINTLMEISTER: Wien. Chem. Ztg. **46**, 106 (1943).

[3] A. BERTHELOT: J. Phys. Radium VIII **3**, 17 (1942).

[4] B. KARLIK: Acta Phys. Austriaca **2**, 182 (1948).

[5] I. PERLMAN, A. GHIORSO and G. T. SEABORG: Phys. Rev. **74**, 1730 (1948); **77**, 26 (1950).

emitters Po^{212}, Ra^{222}, U^{234}, and Cm^{246} which represent points at the respective mass numbers on the curve of Fig. 1. It is seen that the alpha energy of Po^{212} is determined by the binding energies of two neutrons beyond Pb^{208} and two protons beyond Pb^{210}. These are the 127th and 128th neutrons and the 83rd and 84th protons all of which have low binding energies because they are just past closed shells. The sum of these four binding energies when subtracted from 28.3 Mev (the binding energy of the alpha particle) gives the large alpha energy, 8.9 Mev. As the closed shells are left behind, the nucleon-binding energies increase rather sharply as shown for the cycles pertinent to the decay of Ra^{222} and of U^{234}. The alpha energies consequently decrease. Superimposed upon the factors causing an increase in nucleon-binding energies beyond a closed shell are those which produce a gradual decrease as one progresses to higher and higher atomic numbers. The minimum in alpha energy which results from these opposite effects occurs at U^{234} in this illustration, and the energies then increase again. This increase in alpha energy may be seen by comparing the energy cycles for Cm^{246} and U^{234} in Fig. 2, and the increase is illustrated by the final upturn of the main curve in Fig. 1.

The guide lines in Fig. 1 labeled with half-lives of 1 hr and 10^8 yr show the approximate alpha energies necessary to produce the indicated half-lives for alpha decay. The sudden appearance of natural alpha radioactivity above mass number 210 is readily seen, as is the fact that a beta-stable uranium isotope (U^{238}) has a half-life longer than 10^9 yr, and therefore has persisted through geological time. Isotopes of other elements with sufficiently long alpha-decay lifetimes turn out to be beta unstable. (The exception is Th^{232}.) In the rare-earth region, on the contrary, almost all of the beta-stable species have alpha lifetimes too long for detection. The exception is Sm^{147} which is an alpha emitter found in nature and Sm^{146} which is beta-stable but has an alpha decay half-life which is too short to have persisted to the present day. A number of neutron-deficient isotopes of this region have been prepared artificially and these have *both* electron capture and alpha decay lifetimes too short to have persisted through geological time.

Further discussion of these and other alpha emitters will be found in Sects. 3 to 7.

2. The energy surface. We shall now make a more detailed examination of alpha energies in the heavy-element region. For this purpose it is convenient to refer to a "mass surface" or "energy surface" of the heavy-element region. One manner of exhibiting the array of relative masses is shown in Fig. 3a. Here the neutron number is plotted against the mass decrement (Δ) which is simply the difference between the mass (M) and the mass number (A)[1]. Contour lines are shown connecting points of constant mass number (A) and constant atomic number (Z). The "line of stability" which follows the bottom of the valley is also shown. The actual points in Fig. 3a would not lie on a smooth surface as shown because of the differences in mass depending upon whether the nucleons are paired or not. The data were normalized to make a smooth surface by subtracting a pairing energy term from the odd-odd species and adding a term to those with both even neutrons and protons. Since the alpha-decay process does not change the nuclear type, the same alpha-decay energy would be derived from mass differences whether the actual masses were used or normalized values used such as shown in Fig. 3a.

---

[1] The data upon which this plot is based were taken from Ref. [5].

In order to obtain an alpha energy from Fig. 3a the mass decrement of the alpha particle (3.87 milli-mass units) is subtracted from the difference between the mass decrement of the alpha emitter (Δ_i) and its decay products (Δ_f)

$$E_\alpha = \Delta_i - (\Delta_f + \Delta_\alpha). \tag{2.1}$$

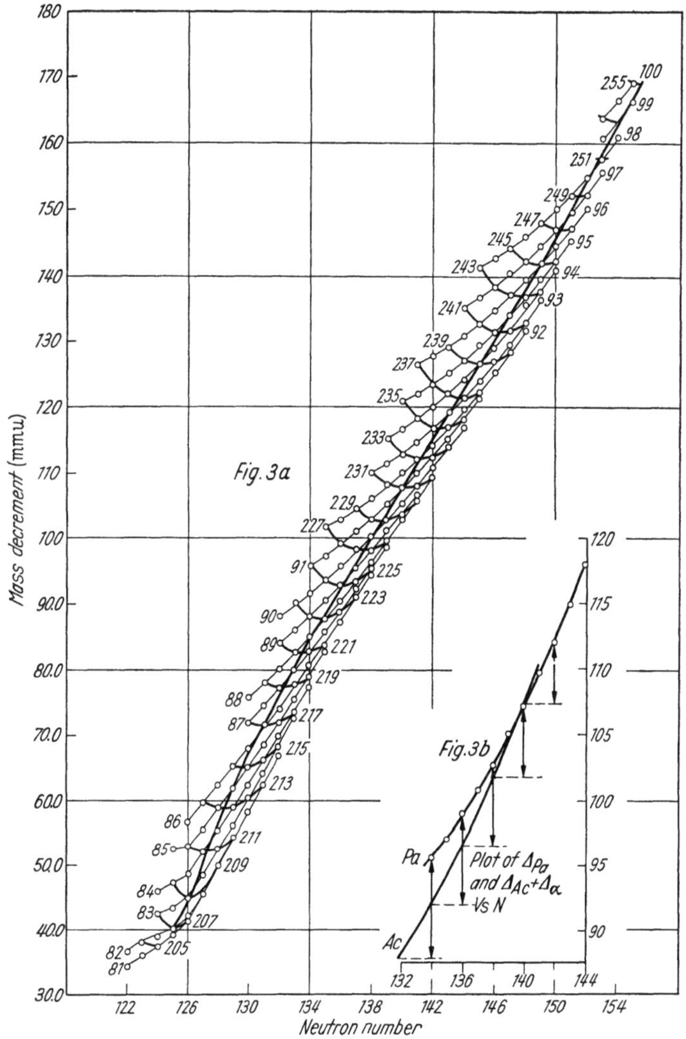

Fig. 3a and b. Energy surface expressed in terms of mass decrements. Fig. 3a shows mass decrement (M-A) in millimass units plotted against neutron number. Data were obtained from [5] and these include nucleon pairing energy terms used to normalize the different nuclear types. Contours are shown at constant A (for odd values) and at constant Z (all values). The heavy line running along the bottom of the valley is the "line of stability" and goes through points of greatest beta stability for each mass number. Fig. 3b illustrates alpha energy variation with neutron number for a series of protactinium isotopes. The mass decrements for protactinium isotopes (Δ_{Pa}) are taken from Fig. 3a and make up the curve labelled *Pa*. The curve labelled *Ac* consists of $\Delta_{Ac} + \Delta_\alpha$. The energy differences between points on the two curves which are related through alpha decay are simply the alpha-energies, indicated by lengths of arrows. These data illustrate the increase in alpha energy with decrease in neutron number.

The curves of Fig. 3b are used to illustrate this derivation of alpha energy. Here the curve labeled "Pa" containd mass decrements of a series of protactinium isotopes, the (Δ_i) of Eq. (2.1), while that labeled "Ac" consists of the mass decrements of actinium isotopes (Δ_f) to which have been added the mass decrement of the alpha particle (Δ_α). The alpha energies, shown by the vertical arrows, are simply the differences between points on the two curves related by alpha emission. It is seen that the alpha energy increases with decreasing neutron number. This can be shown to be an expected consequence of a "regular" energy surface as defined by any of the semi-empirical statistical treatments of nuclear masses.

Fig. 4 is an idealized sketch of such an energy surface and entered upon it are the uranium family and an artificially-produced chain collateral to the uranium

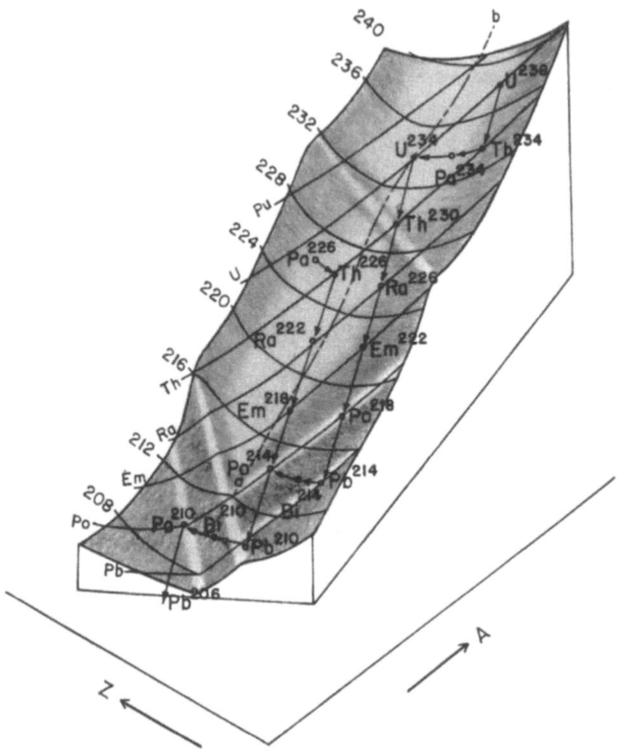

Fig. 4. Energy surface of heavy-element region; a schematic representation of relative nuclear stability. The uranium family and one of the collateral series are entered showing directions of alpha- and beta-decay relative to the surface. The line *ab* is the "line of stability" which passes through or near the most beta-stable isobar for each mass number. Note that alpha decay proceeds skew to the line of stability and a sequence of alpha emitters must eventually produce a β-unstable nucleus.

series. Since this is a normalized surface the beta-decay steps (along constant mass number) are not portrayed accurately. For example, it would appear that Em^{222} should decay to Ra^{222} by sliding to the center of the trough through two successive β^- transitions. However, the intermediate nucleus Fr^{222} really lies on another surface above that of the even-even nuclei and the decay of Em^{222} to Fr^{222} is energetically impossible by a small margin.

It should be noted that alpha decay cuts across the valley in the progression downhill. If a chain of alpha emitters is sufficiently long it will eventually progress high enough up the side of the wall to produce a β^--unstable nucleus which will then decay toward the line of stability. In the series shown, this happens at Th^{234} (UX_1), Pb^{214} (RaB) and Pb^{210} (RaD).

The marked change in the lower part of the energy surface displays the effects of the closed shells at 82 protons and 126 neutrons. Some of the consequences will be discussed presently.

3. Systematics of alpha energies. A convenient manner to display the trends in alpha energies is shown in Fig. 5 in which the decay energy is plotted against the mass number with points of constant atomic number joined[1]. Over a large area of this chart it is seen that for each element there is a nearlyperfect monotonic increase in alpha energy with successive decreases in mass number. This regularity is the reflection of a regular trough-like energy surface as already mentioned. Differences in slope or spacing of the lines in Fig. 5 can be interpreted in terms of departure from extreme regularity of the surface such as small changes in the curvature of the trough or of its slope. Individual irregularities must occur wherever a particular nucleon-binding energy does not follow the smooth trends.

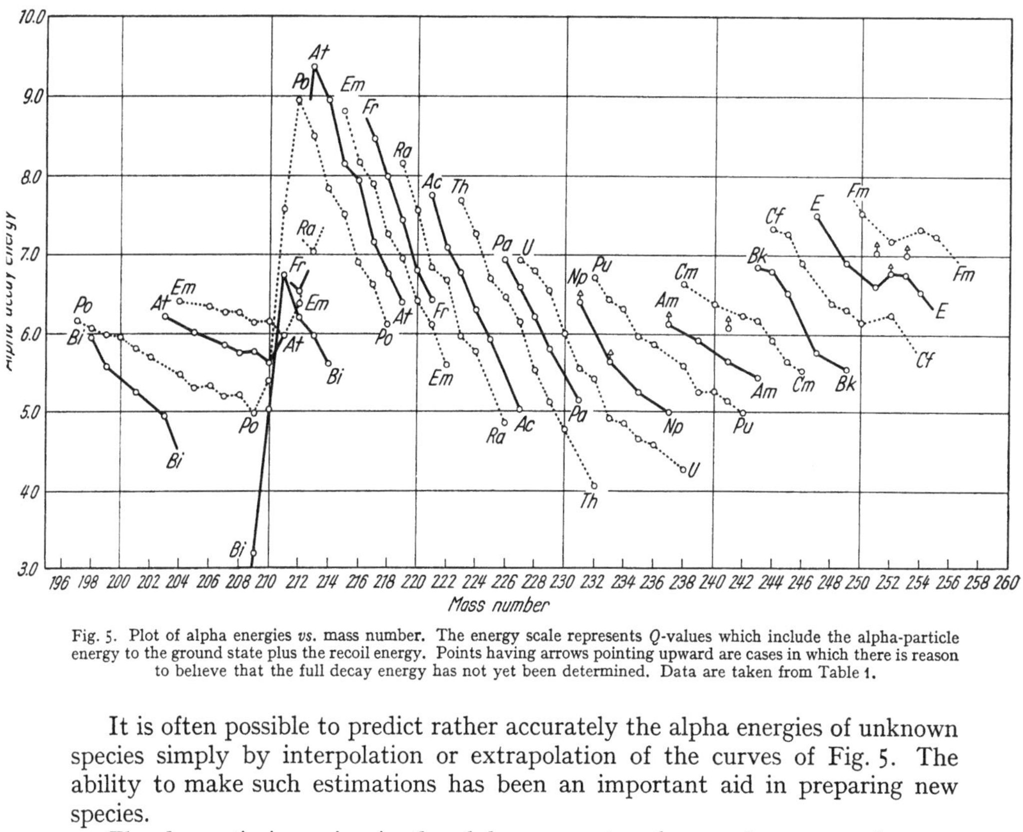

Fig. 5. Plot of alpha energies *vs.* mass number. The energy scale represents Q-values which include the alpha-particle energy to the ground state plus the recoil energy. Points having arrows pointing upward are cases in which there is reason to believe that the full decay energy has not yet been determined. Data are taken from Table 1.

It is often possible to predict rather accurately the alpha energies of unknown species simply by interpolation or extrapolation of the curves of Fig. 5. The ability to make such estimations has been an important aid in preparing new species.

The dramatic inversion in the alpha-energy trend around mass number 212 is a consequence of the major closed shells in this region[2] [*6*]. We can see what happens more precisely by following the curve for the polonium isotopes. From Po^{218} down to Po^{212} the curve follows the normal trend; then Po^{211} is seen to have considerably lower alpha energy than Po^{212}. This can readily be shown to be a reflection of the fact that the binding energy of the 126th neutron in lead (Pb^{207}—Pb^{208}) is greater than that of the 128th neutron in polonium (Po^{211}—Po^{212}). Similarly, since the binding energy of the 125th neutron in lead (Pb^{206}—Pb^{207}) is greater than that of the 127th neutron in polonium (Po^{210}—Po^{211}), the alpha energy of Po^{210} is lower than that of Po^{211}. After the neutron shell of 126 is well past and the neutron-binding energies change monotonically for both parents and daughters,

[1] For further discussion and references see [*6*].

[2] B. Karlik: Acta Phys. Austriaca **2**, 182 (1948).

the regular trend of increasing alpha energy with decreasing neutron number is resumed (see region between Po^{208} and lower isotopes).

The curve for bismuth is seen to parallel the polonium curve with a wide energy spacing between the two. This energy spacing is presumably a consequence of the break in binding energies at the 82-proton shell. The reappearance of alpha radioactivity in highly neutron-deficient isotopes of bismuth was the clue needed to establish the generality of this effect of crossing the region of 126 neutrons[1].

It will be noted that the alpha energies for the isotopes with 128 neutrons would be expected to become progressively greater for each higher proton number. Accordingly, half-lives will be very short in this region so that preparation and identification of such nuclides would be difficult. However, more neutron-deficient isotopes should be more stable just as Po^{210} is more stable than Po^{212}. Such a region has been found and includes isotopes of emanation, francium and even radium, showing that the effect of 126 neutrons extends at least this high[2-4]. Undoubtedly these points shown on Fig. 5 join with those of higher neutron number by going through sharp peaks higher than those shown for polonium and astatine.

The question arises as to whether or not there is evidence for other closed shells or subshells on the basis of alpha-decay data. A situation similar to that at 126 neutrons, but considerably more subdued, seems to occur at neutron number 152[5]. The evidence came intitially from the alpha-energy trend of californium isotopes as seen in Fig. 5 showing that the one with 154 neutrons has a higher energy than that with 152 neutrons. The data on einsteinium and fermium isotopes are not out of line with this concept and the lines are drawn in Fig. 5 according to expectations.

It will be noted in Fig. 5 that the energy increments from isotope to isotope along many of the curves are not very uniform. Aside from the marked inversion in trend in the region of lead and the probable small inversion in conjunction with 152 neutrons, it is seen that in some places the isotopes seem bunched and in others relatively spread out. It has been suggested that there are subshells at 92 protons and at 88 protons in explanation of some of these irregularities. Other inferences concerning alpha energies as related to possible subshells have been discussed by BRONIEWSKY[6].

4. Decay energies from energy-balance cycles. The ability to predict decay energies is of inestimable value in the preparation of new isotopes in the heavy-element region. Through judicious use of these decay energies it is possible to predict lifetimes and consequently to design the experiment accordingly. The method of prediction of decay energies of wide application consists of constructing a self-consistent system of energy-balance cycles from alpha- and beta-energies which are either measured or estimated by interpolation or extrapolation of curves such as those shown in Fig. 5.

To illustrate this method of correlating decay energies, a segment of the decay cycle representation of $4n+1$ type[7] nuclei is shown in Fig. 6. A few examples of the uses of these cycles will be mentioned. It is noted that by making use of

---

1 D. H. TEMPLETON and I. PERLMAN: Phys. Rev. **73**, 1211 (1948).
2 E. K. HYDE, A. GHIORSO and G. T. SEABORG: Phys. Rev. **77**, 765 (1950).
3 F. F. MOMYER jr. and E. K. HYDE: J. Inorg. Nucl. Chem. **1**, 274 (1955).
4 W. E. BURCHAM: Proc. Phys. Soc. Lond. A **67**, 555 (1954).
5 GHIORSO, THOMPSON, HIGGINS, HARVEY and SEABORG: Phys. Rev. **95**, 293 (1954).
6 A. BRONIEWSKI: Canad. J. Phys. **29**, 193 (1951).
7 The type "$4n+1$" means that all mass numbers are divisible by 4 with remainder 1. All nuclei connected by α- and β-decay processes are of the same type.

three measured-decay energies, the alpha energy of the 6.8 day beta-emitter U^{237} is calculated to be 4.25 Mev. This is almost identical with the decay energy of U^{238} which has a half-life 4.5×10^9 yr and one would expect the partial alpha half-life of U^{237} to be at least that long.

Consequently, the alpha-branching of U^{237} would be only of the order of 10^{-12} so that this mode of decay would be most difficult to observe. On the contrary, the alpha-branching of Pu^{241} was similarly estimated to be about 10^{-5} which was within reach for measurement. The alpha energy as subsequently measured is shown in Fig. 6.

Another use of these cycles has to do with predictions of beta-stability. If one considers the possibility that Cm^{245} is a β^--emitter and that its preparation would therefore also produce Bk^{245}, the idea should be rejected because it is seen that Bk^{245} is unstable with respect to Cm^{245} by about 0.7 Mev. The estimated

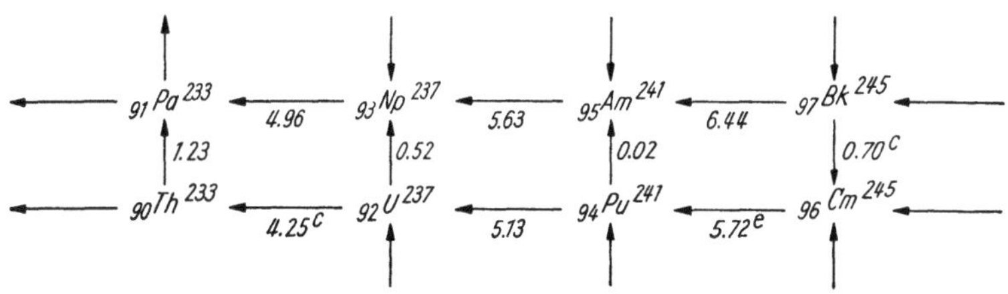

Fig. 6. Energy balance cycles. Segment of cycles for $4n+1$-type nuclei. Vertical arrows pointing downward indicate electron-capture decay; those pointing upward indicate β-emission. Horizontal arrows show alpha emission. The superscript "c" indicates that the decay energy was calculated from the other three members of the cycle; "e" means that the decay energy is estimated from some such chart as that shown in Fig. 5.

Note: The value shown for Cm^{245} alpha decay was estimated, but has since been measured as 5.62 Mev. This changes the calculated value for the electron capture of Bk^{245} to 0.80 Mev.

alpha energy of Cm^{245} which went into this calculation could not possibly be in error by an amount to reverse this conclusion and it was subsequently measured and shown to be close to the value listed.

An extension of these cycles to still higher elements gives a means of making predictions into a region where measurements have not been made, and these predictions serve as an important guide in designing the experiments. Other cycles can be devised to join different nuclear types through measured neutron-binding energies. With a single neutron-binding energy measurement joining two series, other neutron-binding energies can be calculated.

Energy balance cycles showing alpha and beta energies in the transuranium region will be found in the article by HYDE and SEABORG [7]. In the same article are also presented neutron- and proton-binding energies and isotopic masses. Cycles covering the entire heavy-element region will be found in [5].

5. Table of alpha energies. Table 1 in Sect. 11 consists of a listing of the alpha groups found in the heavy-element region. The Q-values for alpha decay are shown in Column 9 and these are the numerical data appearing in Fig. 5. The Q-value is, of course, the total disintegration energy and in most cases is obtained by adding the recoil energy of the nucleus to the energy for the alpha group leading to the ground state. In some cases the highest energy alpha group detected leads to an excited state in which case the de-excitation energy of this state must also be included. The determination of the total decay energy is uncertain in some instances and when this is the case appropriate notation is made.

6. Alpha emitters just below lead. The removal of neutrons from any element increases the potential toward alpha decay and this is the basis for the main trend in Fig. 5. Alpha-active isotopes of gold and mercury have been prepared by removing many neutrons from the stable isotopes[1]. In the case of gold, the stable isotope Au^{197} is estimated to have an alpha energy of only 1 to 2 Mev, while the isotope observed with an alpha energy of 5.1 Mev is believed to lie in the mass number range 183 to 187. As neutrons are removed, successive isotopes become more unstable toward orbital electron capture also, but since alpha- decay lifetimes are extremely sensitive to energy, this mode of decay should at some point become discernible.

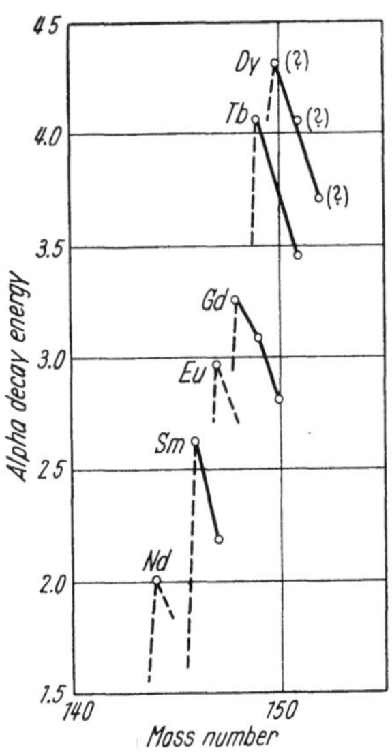

Fig. 7. Rare earth alpha emitters. Plot of alpha energy *vs.* mass number for alpha emitters in the rare earth region. This family of curves is to be compared with that for the heavy elements (Fig. 5) particularly in the similarity of the effect of the closed shell of 82 neutrons with that of 126 neutrons.

7. Rare earth alpha emitters. Among the rare-earth elements we pass through a region where stable or slightly neutron-deficient nuclides can decay by alpha emission to the closed shell of 82 neutrons. Such a nuclide with 84 neutrons is Sm^{146} which is beta-stable but missing in nature because of its relatively short alpha half life ($\sim 5\times 10^7$ yr)[2]. The alpha energies are summarized in Fig. 7 and although the curves are fragmentary as compared with those in the heavy-element region (Fig. 5), the basic structure as related to the 82-neutron shell is unmistakable. The point assigned to Nd^{144} is of special interest because Nd^{144} is a component of natural neodymium[3].

The other alpha emitter existing in nature is the well-known samarium isotope, Sm^{147}, which has three neutrons beyond the closed shell and would correspond to Po^{213} in the heavy element region. The other species shown in Fig. 7 are all electron-capture unstable and their energies relative to each other conform will with their positions relative to the 82-neutron shell[4,1].

II. Complex alpha spectra.

8. Types of alpha spectra. Just as for other decay processes, the appearance of multiple groups in the alpha-emission process may be considered to be the result of competition in populating available energy levels. It will be seen that the transition probability or partial half-life is influenced by a number of factors, and among these is the sharp dependence of lifetime with decay energy. Because of this dependence it is not to be expected that transitions to high-lying levels (say 1 Mev) would be readily observed. If all other factors are equal an alpha transition to the ground state would be about 10^6 times faster than one leading to an excited state at 1 Mev above ground.

The highest energy group is not always the most prominent despite the sensitive energy dependence mentioned above. In many cases there are selection processes which are strong enough to delay higher-energy groups and therefore make the lower-energy groups prominent. One of the most extreme cases noted

[1] J. O. RASMUSSEN, S. G. THOMPSON and A. GHIORSO: Phys. Rev. **89**, 33 (1953).
[2] D. C. DUNLAVEY and G. T. SEABORG: Phys. Rev. **92**, 206 (1953).
[3] E. C. WALDRON, V. A. SCHULTZ and T. P. KOHMAN: Phys. Rev. **93**, 254 (1954).
[4] J. O. RASMUSSEN: Thesis, University of California 1952 (UCRL 1473 rev.).

is that of the decay of Cm^{243} to Pu^{239} in which two groups differing in energy by 230 kev show an 80-fold greater abundance of the lower-energy group[1] (see Table 1). From energy dependence considerations alone one would expect a ratio of 15 in the opposite order which means that by some selective mechanism this higher-energy group is "hindered" by a factor of 1200 relative to the lower-energy group.

The observed alpha spectra do fall into certain patterns which can be correlated with nuclear type and systematic trends in available energy levels. The energy level patterns in the heavy-element region are discussed in Sects. 12 to 15 while the transition probabilities to these levels lie in the province of "kinetics of alpha decay" and are discussed in later sections. Here we shall examine some alpha spectra and tabulate information on alpha groups obtained from all heavy-element alpha emitters.

9. Spectra of even-even emitters. The portion of the alpha spectrum observable by alpha-particle spectroscopy[2] is fairly simple for even-even species. The four

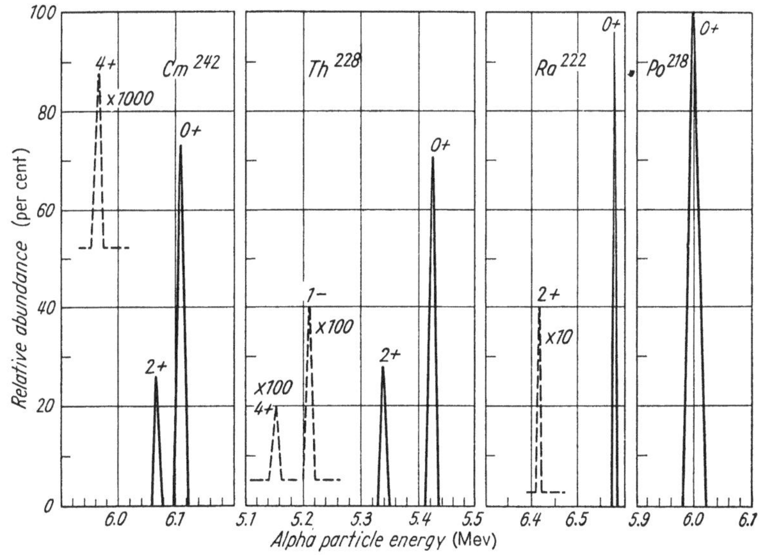

Fig. 8. Selected alpha spectra of even-even species. Representation of alpha spectra for Cm^{242}, Th^{228}, Ra^{222} and Po^{218}. Solid lines show abundances according to the ordinate scale; broken lines are on an expanded scale as indicated by each peak. The spin and parity designations refer to the states populated by the particular alpha groups. Note the differences in energy spacing between the 0^+ (ground state) and 2^+ (first excited state) in progressing from the highest element to the lower ones. The peaks shown in this figure are those which have been observed directly; others in lower intensity can be inferred from gamma-ray data.

spectra shown in Fig. 8 illustrate this point. Starting with a transuranium isotope, Cm^{242}, it is seen that the group of highest energy is most abundant and that there is a nearby group (~40 kev difference) also in good abundance. Somewhat farther away (~140 kev) is another group in extremely low intensity. As far as is known all even-even alpha emitters from plutonium to the highest elements exhibit a spectrum which is virtually identical with this in the energy spacing of these three groups. The symbols 0^+, 2^+, 4^+ indicate the spins and parities of the first three levels of Pu^{238} which are reached by the respective alpha groups. These levels have been recognized as members of a rotational

[1] F. Asaro, S. G. Thompson and I. Perlman: Phys. Rev. **92**, 694 (1953).

[2] It will be seen that gamma-ray measurements can indicate the positions of levels populated by alpha decay with a sensitivity which is often several orders of magnitude greater than the direct observation of alpha groups.

band (see Sect. 13). The existence of other alpha groups in the decay of Cm^{242} can be inferred from gamma-ray data (see Table 1), but more highly refined techniques would be required to observe them directly because of their low intensities. The alpha- and gamma-spectra associated with Cm^{242} decay are discussed in detail in Sect. 54.

Turning to the decay of Th^{228}, the spectrum is seen to be quite similar to that of Cm^{242} except that a state of spin 1 and odd parity has appeared between the 4^+ and 2^+ states. Such odd-parity states appear as low-lying levels only in a limited region and are discussed in Sects. 13 and 53. Another difference between this spectrum and that of Cm^{242} is the wider spacing between the 0^+, 2^+, 4^+ sequence. The trend noted is that of increase in rotational level spacing in progressing toward a closed shell configuration, in this case the approach to the region of 126 neutrons and 82 protons. The spacing increases to such an extent that in Em^{218} (decay of Ra^{222}) the 2^+ state is seen at 325 kev above the ground state and there are no intervening states. Consequently the spectrum is quite simple at least in terms of alpha groups which can readily be observed. Finally, we have in Fig. 8, the single-line spectrum for Po^{218}. The 2^+ state for Pb^{214}, like that for other lead isotopes with even neutron number, probably lies well above the ground state and is hence so slightly populated by Po^{218} alpha decay that only the ground state transition can be observed.

10. Spectra of odd-nucleon types. All of the low-lying energy levels[1] in even-even nuclei appear to be due to collective aspects of internal motion and it is, of course, the low-lying states which are most prominent in alpha spectra. The spectra leading to these states are fairly regular with the ground states most heavily populated. In nuclei with unpaired nucleons the low-lying states are, in general, more dense and consequently the alpha spectra are more complex. For these nuclei, states due to collective modes are superimposed upon each of a number of possible particle states and the selection rules which determine the relative alpha populations of the different configurations are not yet explicitly codified. Nevertheless certain types of spectra are now recognized and it is these which will be illustrated here. The meaning of some selected samples is discussed in detail in Sects. 55 and 56.

In Fig. 9 are shown four spectra of alpha emitters having an odd proton or an odd neutron. For both Cm^{243} and Am^{241} it is immediately seen that the alpha group in highest intensity is not the one of highest energy. In the case of Cm^{243} the most intense group populates an energy level of Pu^{239} which is 286 kev above the ground state. The transitions to the ground state and near-lying first excited state (indicated by vertical arrows in Fig. 9) have not yet been observed partly because of their low intensities and because they fall in the same position as the intense groups of Cm^{242} which were present with the Cm^{243}. The general observation to be made from this spectrum is that there is a "favored" alpha group analogous to the favored transition of an even-even nucleus but in the odd-nucleon case this group does not in general lead to the ground state. The low-intensity high-energy groups are interpreted as populating another rotational band based upon the ground state configuration. Transitions to this band are highly hindered. The interpretation of these spectra will be found in Sects. 55 and 56.

The spectrum of Am^{241} also shows a favored transition to an excited state (60 kev above ground) and the rotational band based upon this state. The transitions to the ground state and its first rotational member are highly hindered as evidenced by the low intensities of the alpha groups.

[1] By "low-lying" we refer to energies up to several hundred kilovolts which is still well below the energy required to unpair nucleons.

The spectrum of U^{233} is shown to illustrate that sometimes the favored transition does lead to the ground state in which case the spectrum to the low-lying levels can look much like that of an even-even nucleus.

In the region just below thorium the density of low-lying levels appears to be very great because the alpha spectra become quite complex. The 12 alpha groups of Th^{227} are shown in Fig. 9 and thus far they have not been interpreted. The appearance of several groups in almost equal intensity probably means that there are at least that many particle states in this region. It is unlikely that the population of different members of the same rotational band can proceed with equal probability.

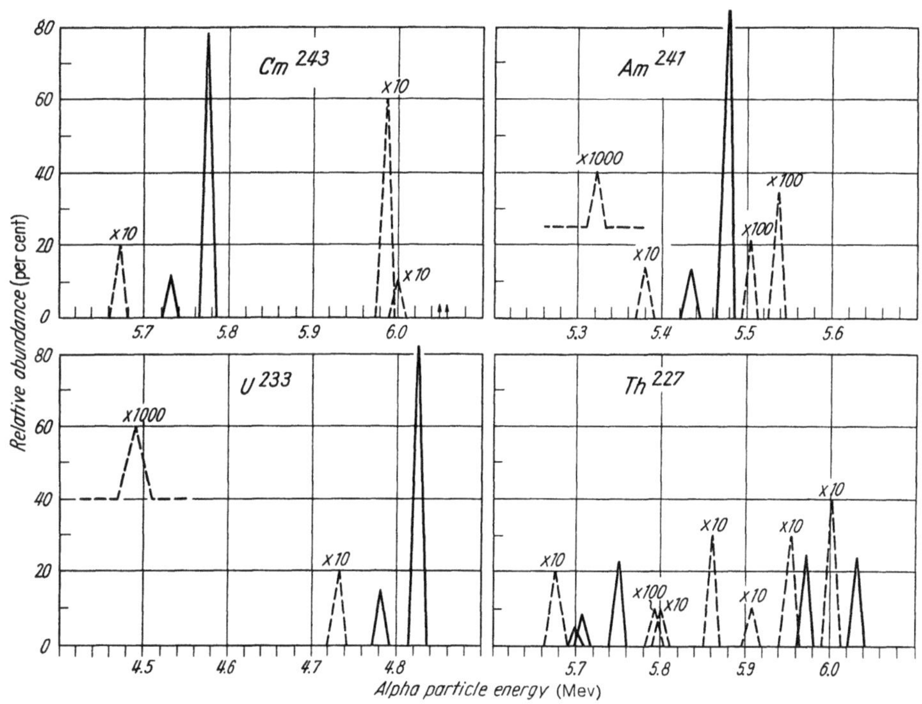

Fig. 9. Selected alpha spectra of odd-nucleon species. Representation of alpha spectra for Cm^{243}, Am^{241}, U^{233}, Th^{227}. Solid lines show abundances according to the ordinate scale; broken lines are on an expanded scale as indicated by each peak. Note that the ground-state transitions are not in general the most intense. The small arrows at about 6.05 Mev on the Cm^{243} spectrum indicate positions of groups leading to the ground state and first excited state. These have not been observed directly in the alpha spectrum.

11. Table of alpha spectra. Table 1 presents a list of alpha groups associated with all heavy-element alpha emitters. Included are those groups which could not be seen directly but which are inferred from gamma-ray data. The gamma-ray intensities are used to deduce the relative abundances of the alpha groups. These data are of value in developing alpha-decay theory since it is primarily the relative intensities of competing groups which must be explained. It will also be noted that "long-range alpha groups" are listed. These are the high-energy groups which arise from the decay of excited states of alpha emitters and represent favorable competition between alpha emission and de-excitation of the levels by gamma emission. In all known cases, the levels are initially populated by β^--decay processes. The energies for long-range alpha groups are, of course, greater than the Q-values which represent transitions from ground state to ground state.

Table 1. *Alpha spectra of the heavy elements.*

Alpha emitter	Alpha half-life	Identification	Alpha particle energy (Mev)	Method of meas.	Energy ref.	Excited state energy	Relative abund. (%)	Adopted Q (Mev)	Comments about Q
Fm^{255}	21.5 h	*B*	7.08	ion ch	1	0	~50.	7.19	ins evid
			~7.04	ion ch	1	~40	~50		
Fm^{254}	3.24 h	*A*	7.20	ion ch	1	0	83	7.32	
			7.16	α-γ	2	42	17		
			7.06	α-γ	2	139	0.4		
Fm^{253} > 10 d		*C*	6.85	ion ch	3			6.96	ins evid
Fm^{252}	~30 h	*B*	7.04	ion ch	4, 3			7.15	e-e
Fm^{251} ~7 h	~24 d	*B*	6.9	ion ch	4			7.0	ins evid
Fm^{250} 30 m	~2 h	*B*	7.4	ion ch	4			7.5	e-e
E^{254}	320 d	*A*	6.42	ion ch	1			6.52	ins evid
E^{253}	20.03 d	*A*	6.633	spect	5	0	90.2	6.740	
			6.592	spect	5	41.7	7.7		
			6.545	spect	5	90	1.8		
			6.493	spect	5	142	0.3		
			6.25	ion ch	6, 1	393 γ	0.1		
E^{252} ~140 d	~0.8 y	*B*	6.64	ion ch	4			6.75	ins evid
E^{251} 1.5 d	0.8 y	*B*	6.48	ion ch	4			6.58	ins evid
E^{249} ~2 h	63 d	*B*	6.76	ion ch	4			6.87	ins evid
E^{248} 25 m	~8 d	*B*	6.87	ion ch	7a			6.98	ins evid
E^{246} 7.3 m		*D*	7.3	ion ch	8			7.4	ins evid
Cf^{252}	2.2 y	*A*	6.112	spect	9	0	84.5	6.211	
			6.069	spect	9	43.4c	15.5		
			5.970	α-γ	9	143	0.2		
Cf^{250}	11 y	*A*	6.024	spect	9	0	83	6.122	
			5.980	spect	9	44	17		
Cf^{249}	5×10² y	*A*	6.19	ion ch	6	0	} ~3	6.29	
			6.14	ion ch	10	53 γ			
			5.94	ion ch	10	~250	} ~6		
			5.90	ion ch	10	~290			
			5.80	ion ch	6	394 γ	~90		
Cf^{248}	250 d	*B*	6.26	ion ch	12			6.36	
Cf^{246}	35.7 h	*A*	6.753	spect	13	0	78	6.865	
			6.711	spect	13	42.12c′	22		
			6.615	α-γ	14	140	0.16		
			6.467	α-γ	14	291	0.015		
Cf^{245} 44 m	2.2 h	*A*	7.11	ion ch	15			7.23	
Cf^{244}	~25 m	*A*	7.17	ion ch	15			7.29	e-e
Bk^{249} 280 d	8×10⁴ y	*A*				0		5.53	α-γ
			5.40	ion ch	16	~40 γ	94		
			~5.06	α-γ	17	~390	6		
Bk^{247}	~10⁴ y	*B*				0		5.85	α-γ
			5.67	ion ch	16	84.3c′	~40		
			5.51	ion ch	16	265 γ	~60		
			5.30	ion ch	16	~460	~5		
Bk^{245} 4.95 d	14 y	*B*	6.37	ion ch	16, 11	0	32	6.48	α-γ
			6.16	ion ch	16, 11	206 γ	41		
			5.90	ion ch	16, 11	480 γ	27		

Table 1. (Continued.)

Alpha emitter	Alpha half-life	Identification	Alpha particle energy (Mev)	Method of meas.	Energy ref.	Excited state energy	Relative abund. (%)	Adopted Q (Mev)	Comments about Q
Bk^{244} 4.35 h	8 y	*B*	6.67	ion ch	16			6.78	ins evid
Bk^{243} 4.5 h	0.5 y	*A*	6.72	ion ch		0	30	6.83	
			6.55	ion ch		192 γ	53		
			6.20	ion ch		550 γ	17		
Cm^{246}	3×10^3 y	*A*	5.4	ion ch	18			5.5	e-e
Cm^{245}	1.4×10^4 y	*A*						5.62	α-γ
			5.45	ion ch	19	~80	~15		
			5.36	ion ch	20	172 γ	~77		
			5.31	γ-γ	21	~230	~8		
Cm^{244}	19 y	*A*	5.798	spect		0	76.7	5.895	
			5.756	spect	22	42.88c	23.3		
			5.658	spect	22	141.8c′	0.016		
			5.511	γ	22	292	4×10^{-3}		
Cm^{243}	35 y	*A*						6.159	α-γ
			6.003	spect	21	57.31c′	1		
			5.985	spect	23	75.76c′	6		
			5.777	spect	23	285.7c′	78		
			5.732	spect	23	330.3c′	13		
			5.679	spect	21	384 (391.8c′?)	2		
Cm^{242}	162.5 d	*A*	6.110	spect	23	0	73.7	6.213	
			6.066	spect	23	44.11c	26.3		
			5.965	spect	23	146.0c	0.035		
			5.811	γ-γ	23	303.7c	6×10^{-3}		
			5.605	γ-γ	24	514	3×10^{-5}		
			5.515	γ-γ	24	605	1×10^{-4}		
			5.20	γ-γ	24	~930	3×10^{-5}		
			5.12	γ	24	~1010	4×10^{-6}		
Cm^{241} 35 d	10 y	*A*	5.95	ion ch	25			6.05	ins evid
Cm^{240}	26.8 d	*A*	6.27	ion ch	25			6.38	e-e
Cm^{238} 2.4 h		*B*	6.52	ion ch	26			6.63	e-e
Am^{243}	7.9×10^3 y	*A*	5.339	spect	27	0	0.17	5.428	
			5.308	spect	27	31	0.16		
			5.266	spect	27	74	86.9		
			5.224	spect	27	117	11.5		
			5.169	spect	27	173	1.3		
Am^{241}	461 y	*A*	5.535	spect	28	0	0.42	5.628	
			5.503	spect		33.20c	0.24		
			5.476	spect		59.57c	84.3		
			5.433	spect		103.0c	13.6		
			5.379	spect		158.6c	1.4		
			5.314	spect	29	225	0.015		
			5.27	γ	30	267.4c′	0.004		
Am^{239} 12 h	33 y	*A*				0		5.90	α-γ
			5.75	ion ch		48 γ			
Am^{237} ~1.3 h	~3 y	*B*	6.01	ion ch				6.11	ins evid
Pu^{242}	3.76×10^5 y	*A*	4.898	spect	31	0	74	4.980	
			4.854	spect	31	45	26		
Pu^{241} 13.2 y	3×10^5 y	*A*				0		5.121	γ
			4.893	spect	31	145 γ	75		
			4.848	spect	31	191	25		

Table 1. (Continued.)

Alpha emitter	Alpha half-life	Identification	Alpha particle energy (Mev)	Method of meas.	Energy ref.	Excited state energy	Relative abund. (%)	Adopted Q (Mev)	Comments about Q
Pu^{240}	6580 y	*A*	5.162	spect		0	75.5	5.249	
			5.118	spect		45	24.5		
			5.014	spect	31	151	0.1		
Pu^{239}	24360 y	*A*	5.150	spect		0	72.5	5.238	ins evid
			5.137	spect		13.5	16.8		
			5.099	spect		51.7c	10.7		
			4.98	γ-γ	33	172	0.02		
			4.78	γ-γ	33	379	0.006		
			4.73	γ-γ	33	~430	0.006		
Pu^{238}	89.6 y	*A*	5.495	spect	34	0	72	5.589	
			5.452	spect	34	43.50c	28		
			5.352	spect	34	143.31c	0.095		
			5.204	γ-γ	34	296.4 γ	0.004		
			5.004	γ-γ	35	499	7×10^{-6}		
Pu^{236}	2.7 y	*A*	5.763	spect	36	0	68.9	5.862	
			5.716	spect	36	47.5	30.9		
			5.610	spect	36	156	0.18		
			5.448	α-γ	36	321	0.002		
Pu^{235} 26 m	~2.5 y	*B*	5.85	ion ch				5.95	ins evid
Pu^{234} 9.0 h	~9 d	*A*	6.19	ion ch		0	75	6.30	
			6.14	conv emuls	37	47	25 γ		
Pu^{233} 20 m		*B*	6.30	ion ch	38			6.41	ins evid
Pu^{232} 36 m		*B*	6.58	ion c				6.70	e-e
Np^{237}	2.20×10^{6} y	*A*	4.872	ion ch	39	0	3.1	4.956	
			4.816	ion ch	39	57	3.5		
			4.787	ion ch	39	86	53		
			4.767	ion ch	39	107	29		
			4.713	ion ch	39	162	1.7		
			4.674	ion ch	39	201	3.3		
			4.644	ion ch	39	232	6.0		
			4.589	ion ch	39	288	0.5		
			4.52	ion ch	39	~360	0.02		
Np^{235} 410 d	3.2×10^{4} y	*A*	5.06	ion ch		0		5.15	α-γ
Np^{233} 35 m	~7 y	*A*	5.53	ion ch				5.63	ins evid
Np^{231} ~ 50 m		*A*	6.28	ion ch				6.39	ins evid
U^{238}	4.49×10^{9} y	*A*	4.182	ion ch	40	0	77	4.253	
			4.135	conv emuls	41	48	23		
			4.02	γ-α	42	~160	~ 0.1		
U^{236}	2.39×10^{7} y	*A*	4.499	ion ch		0	73	4.577	
			4.450	conv emuls		49 γ	27		
			4.339	γ	43	163	~ 0.5		
U^{235}	7.13×10^{8} y	*A*	4.58	ion ch		0	10	4.66	
			4.47 ?	ion ch		~110	~3		
			4.40	ion ch		186 γ	83		
			4.20	ion ch		~390	4		
U^{234}	2.48×10^{5} y	*A*	4.768	spect	29	0	72	4.851	
			4.717	spect	29	52.4	28		
			4.60	γ-α	44	170	0.3		

Table 1. (Continued.)

Alpha emitter	Alpha half-life	Identification	Alpha particle energy (Mev)	Method of meas.	Energy ref.	Excited state energy	Relative abund. (%)	Adopted Q (Mev)	Comments about Q
U^{233}	1.62×10^{5} y	*A*	4.816	spect	29	0	83.5	4.900	
			4.773	spect	29	42.8 γ	14.9		
			4.717	spect	29	98.9 γ	1.6		
			4.489	spect	29	333	0.03		
U^{232}	73.6 y	*A*	5.318	spect	45	0	68	5.411	
			5.261	spect	45	57.5 c′	32		
			5.132	spect	45	186.1 c′	0.32		
			4.998	γ	45	326	0.01		
U^{231} 4.3 d	21 y	*A*	5.45	ion ch				5.54	ins evid
U^{230}	20.8 d	*A*	5.884	spect	46	0	67.2	5.988	
			5.813	spect	46	72.13 c	32.1		
			5.662 c	spect	47; 46	226.4 c	0.4 γ		
			5.658 γ-α	spect	48; 46	230.4 c′	0.3 γ		
U^{229} 58 m	5 h	*A*	6.42	ion ch				6.53	ins evid
U^{228} 9.3 m	12 m	*A*	6.67	ion ch				6.79	e-e
U^{227} 1.3 m		*B*	6.8	ion ch				6.9	ins evid
Pa^{231}	3.43×10^{4} y	*A*	5.046	spect	49	0	10	5.135	
			5.017	spect	49	29	23		
			5.001	spect	49	46	24		
			4.971	spect	49	76	2.3		
			4.938	spect		110	22		
			4.921	spect	49	127	2.8		
			4.839	spect	49	211	1.4		
			4.722	spect	49	329	11		
			4.696	spect	49	356	1.4		
			4.666	spect	49	387	2.1		
Pa^{229} 1.5 d	1.6 y	*A*	5.69	ion ch				5.79	ins evid
Pa^{228} 22 h	~50 d	*A*	6.09	ion ch		0	75	6.20	ins evid
			5.85	ion ch		~240	25		
Pa^{227} 38.3 m	~45 m	*A*	6.46	ion ch				6.58	ins evid
Pa^{226} 1.8 m		*B*	6.81	ion ch				6.93	ins evid
Th^{232}	1.39×10^{10} y	*A*	3.994	ion ch	40	0	76	4.064	
			3.936	conv emuls	51	59	24		
Th^{230}	8.0×10^{4} y	*A*	4.682	spect		0	74	4.765	
			4.615	spect		67.62 c′	26		
			4.476	γ-α	52	210 γ	0.2		
			4.433	γ-γ	53	253	0.03		
			4.368	γ-γ	69	320	0.001		
			4.273	γ-γ	53	416	8×10^{-6}		
			4.245	γ-γ	53	445	8×10^{-6}		
Th^{229}	7340 y	*A*	5.02	ion ch		0	~10	5.11	ins evid
			4.94	ion ch		~80	~20		
			4.85	ion ch		~170	~70		
Th^{228}	1.90 y	*A*	5.421	spect	54	0	71	5.518	
			5.338	spect	54	84.47 c	28		
			5.208	spect	54	217	0.4		
			5.173	spect	54	253	0.2		
			5.137	spect	53	289	0.03		

Table 1. (Continued.)

Alpha emitter	Alpha half-life	Identification	Alpha particle energy (Mev)	Method of meas.	Energy ref.	Excited state energy	Relative abund. (%)	Adopted Q (Mev)	Comments about Q
Th^{227}	18.17 d	*A*	6.030	spect		0	24	6.138	
			6.001	spect	55	29.8c	4		
			5.970	spect	55	61.3c	25		
			5.952	spect	55	79	3		
			5.907	spect	55	125	1		
			5.859	spect	55	174	3		
			5.800	spect	55	234	1		
			5.792	spect	55	242	0.1		
			5.749	spect	55	286.2c	23		
			5.706	spect	55	330.0c	4		
			5.702	spect	55	334.5	6		
			5.693	spect	55	343.2	3		
			5.686	spect	55	350.6	1		
			5.661	spect	55	376	2		
Th^{226}	30.9 m	*A*	6.330	spect	46	0	79	6.444	
			6.220	spect	46	111.1c	19		
			6.095	spect	46	242 γ	1.7		
			6.029	spect	46	309 γ	0.6		
Th^{225} 8.0 m	9 m	*A*	6.57	ion ch				6.69	ins evid
Th^{224} short		*A*	7.13	ion ch				7.26	e-e
Th^{223} short		*B*	7.55	ion ch				7.69	ins evid
Ac^{227} 22.0 y	1.8×10^{3} y	*A*	4.942	spect				5.031	ins evid
Ac^{225}	10.0 d	*A*	5.818	spect	56	0	56	5.923	
			5.781	spect	56	36.6c	28		
			5.719	spect	56	99.6c	12		
			5.627	spect	56	194	4		
Ac^{224} 2.9 h	~29 h	*A*	6.17	ion ch				6.28	ins evid
Ac^{223}	2.2 m	*A*	6.64	ion ch				6.76	ins evid
Ac^{222} 5.5 s		*B*	6.96	ion ch				7.09	ins evid
Ac^{221} short		*E*	7.6	range emuls				7.7	ins evid
Ra^{226}	1622 y	*A*	4.777	spect		0	94.3	4.863	
			4.593	spect		187	5.7		
			4.34	γ-γ	57	450	0.01		
Ra^{224}	3.64 d	*A*	5.681	spect		0	95.1	5.784	
			5.445	spect		241c	4.9		
			5.15	γ-γ	57	540	0.009		
			5.04	γ-γ	57	650	0.012		
Ra^{223}	11.68 d	*A*	5.860	spect		0	weak	5.962	c, ins evid
			5.735	spect	31	122c	9		
			5.704	spect		154c	53		
			5.592	spect	31	270c	24		
			5.525	spect	31	336c	9		
			5.487 ?	spect		375	2		
			5.418	spect	31	444c	3		
Ra^{222}	38 s	*A*	6.551	spect	46	0	96	6.671	
			6.232	γ-α	48	324.6c	4		
			5.91	γ-γ	35	650	0.008		
			5.765	γ-γ	35	800	0.03		
			5.72	γ-γ	35	850	0.002		
Ra^{221}	30 s	*A*	6.71	ion ch				6.83	ins evid

Table 1. (Continued.)

Alpha emitter	Alpha half-life	Identification	Alpha particle energy (Mev)	Method of meas.	Energy ref.	Excited state energy	Relative abund. (%)	Adopted Q (Mev)	Comments about Q
Ra^{220} short		*A*	7.43	ion ch				7.57	e-e
Ra^{219} short		*B*	8.0	ion ch				8.1	ins evid
Ra^{213} 2.7 m		*B*	6.90	ion ch				7.03	ins evid
Fr^{221}	4.8 m	*A*	6.332	spect	56	0	84	6.449	
			6.116	spect	56	220	16		
Fr^{220} 27.5 s		*A*	6.69	ion ch				6.81	ins evid
Fr^{219}	0.02 s	*A*	7.30	ion ch				7.44	ins evid
Fr^{218} short		*B*	7.85	ion ch				8.00	ins evid
Fr^{217} short		*E*	8.3	range emuls				8.5	ins evid
Fr^{212} 19.3 m	44 m	*A*	6.411	spect	58	0	37	6.534	ins evid
			6.387	spect	58	24	39		
			6.342	spect	58	70	24		
Em^{222}	3.825 d	*A*	5.486	spect		0	99.9	5.587	
			4.985	γ	59	510	∼0.07		
Em^{221} 25 m	2.1 h	*A*	6.0	ion ch	60			6.1	ins evid
Em^{220}	54.5 s	*A*	6.282	spect		0	99.7	6.398	
			5.747	spect	61	545	0.3		
Em^{219}	3.92 s	*A*	6.807	spect	62	0	69	6.94	ins evid
			6.542	spect		270	15		
			6.417	spect		397	12		
			6.197	spect		622	4		
Em^{218}	0.019 s	*A*	7.127	spect	46	0	99.8	7.260	
			6.529	γ	46	609	0.2		
Em^{217}	$\sim 10^{-3}$ s	*A*	7.74	ion ch				7.89	ins evid
Em^{216} short		*A*	8.01	ion ch				8.16	e-e
Em^{215} short		*B*	8.6	ion ch				8.8	ins evid
Em^{212} 23 min		*A*	6.264	spect	58		∼100	6.384	
Em^{211} 16 h	62 h	*A*	5.847	spect	58	0	33.5	5.960	
			5.779	spect	58	68.8c	64.5		
			5.613	spect	58	239	2		
Em^{210} 2.7 h	2.8 h	*A*	6.037	spect	58		∼100	6.154	
Em^{209} 30 m	3 h	*B*	6.037	spect	58			6.154	ins evid
Em^{208} 23 m	∼2 h	*B*	6.141	spect	58		∼100	6.261	e-e
Em^{207} 11 m	4.5 h	*B*	6.14	ion ch	65			6.26	ins evid
Em^{206} 6.2 m	9 m	*B*	6.22	ion ch	66			6.34	e-e
Em^{204} 3 m		*D*	6.3	ion ch	66			6.4	e-e
At^{219}	0.9 m	*A*	6.27	ion ch				6.39	ins evid
At^{218}	∼2 s	*E*	6.63	ion ch				6.75	ins evid
At^{217}	0.018 s	*A*	7.05	spect	56		∼100	7.18	
At^{216}	$\sim 3\times 10^{-4}$ s	*A*	7.79	ion ch				7.94	ins evid

Table 1. (Continued.)

Alpha emitter	Alpha half-life	Identification	Alpha particle energy (Mev)	Method of meas.	Energy ref.	Excited state energy	Relative abund. (%)	Adopted Q (Mev)	Comments about Q
At^{215}	~10^{-4} s	*A*	8.00	ion ch				8.15	ins evid
At^{214} short		*B*	8.78	ion ch				8.95	ins evid
At^{213} short		*E*	9.2	range emuls				9.4	ins evid
At^{211} 7.5 h	18.3 h	*A*	5.862	spect			~100	5.975	
At^{210} 8.3 h	2.0×10^{2} d	*A*	5.519	spect		0	32	5.626	ins evid
			5.437	spect		83	31		
			5.355	spect		167	37		
At^{209} 5.5 h	~5 d	*B*	5.642	spect	67		~100	5.752	ins evid
At^{208} 1.7 h	14 d	*A*	5.65	ion ch		0	~97	5.76	ins evid
			5.52	α-γ	68	~130	~3		
At^{207} 1.8 h	~20 h	*B*	5.75	ion ch				5.86	ins evid
At^{205} 25 m		*B*	5.90	ion ch				6.02	ins evid
At^{203} 7 m		*D*	6.10	ion ch				6.22	ins evid
$At^{<203}$ 1.7 m		*D*	6.35	ion ch				6.48	ins evid
$At^{<202}$ 43 s		*D*	6.50	ion ch				6.63	ins evid
Po^{218}	3.05 m	*A*	5.998	spect			~100	6.110	
Po^{217} short		*A*	6.54	ion ch	60			6.66	ins evid
Po^{216}	0.158 s	*A*	6.775	spect	63		~100	6.90	
Po^{215}	1.83×10^{-3} s	*A*	7.38	spect	63		~100	7.52	
Po^{214}	1.637×10^{-4} s	*A*	7.680	spect	63		100	7.827	
		long range groups	8.277	spect	63		4×10^{-5}	8.435	
			8.938	spect	63		4×10^{-5}	9.108	
			9.065	spect	63		0.0022	9.238	
			9.313	spect	63		4×10^{-5}	9.490	
			9.489	spect	63		1.4×10^{-4}	9.669	
			9.657	spect	63		4×10^{-5}	9.841	
			9.779	spect	63		1.1×10^{-4}	9.965	
			9.905	spect	63		4×10^{-5}	10.093	
			10.074	spect	63		1.7×10^{-4}	10.265	
			10.146	spect	63		4×10^{-5}	10.339	
			10.326	spect	63		1.1×10^{-4}	10.523	
			10.506	spect	63		2×10^{-5}	10.706	
Po^{213}	4.2×10^{-6} s	*A*	8.35	spect	56		~100	8.51	ins evid
Po^{212}	3.04×10^{-7} s	*A*	8.780	spect	63		100	8.949	
		long range groups	9.492	spect	63		0.0035	9.675	
			10.422	spect	63		0.002	10.622	
			10.543	spect	63		0.018	10.746	
Po^{211}	0.52 s	*A*	7.44	spect	63	0	99.0	7.578	
			6.88	spect		569c	0.53		
			6.56	spect		~900	0.50		
Po^{211m}	25 s	*B*	8.70	ion ch	70	0	7	8.91	c′
			7.85	ion ch	70	~870	2.5		
			7.14	ion ch	70	1632c′	90.5		
Po^{210}	138.40 d	*A*	5.299	spect	71	0	100	5.402	
			4.510	γ-α		804*c*	0.0012γ		
Po^{209}	~100 y	*A*	4.877	spect		0	99.4	4.972	
			4.62	γ-α		260.5*c′*	0.6		
Po^{208}	2.83 y	*A*	5.108	spect			~100	5.208	

Table 1. (Continued.)

Alpha emitter	Alpha half-life	Identification	Alpha particle energy (Mev)	Method of meas.	Energy ref.	Excited state energy	Relative abund. (%)	Adopted Q (Mev)	Comments about Q
Po^{207} 5.7 h	~6 y	*B*	5.10	ion ch				5.20	ins evid
Po^{206} 9 d	180 d	*B*	5.218	spect			~100	5.321	
Po^{205} 1.5 h	~84 d	*B*	5.2	ion ch				5.3	ins evid
Po^{204} 3.8 h	~16 d	*B*	5.370	spect	72			5.477	e-e
Po^{202} 52 m	43 h	*B*	5.575	spect	72			5.689	e-e
Po^{201} 18 m		*B*	5.671	spect	72			5.786	ins evid
Po^{200} 11 m		*B*	5.770	spect	72			5.888	e-e
Po^{199} ~11 m		*B*	5.846	spect	72			5.966	ins evid
Po^{198} ~6 m		*D*	5.935	spect	72			6.057	e-e
Po^{197} ~4 m		*D*	6.040	spect	72			6.165	ins evid
Bi^{214} 19.7 m	~34 d	*A*	5.505	spect		0	45	5.610	ins evid
			5.444	spect		62	55		
			5.33 ?	spect		~180	(17)		
Bi^{213} 47 m	1.6 d	*A*	5.86	ion ch				5.97	ins evid
Bi^{212} 60.5 m	2.99 h	*A*	6.086	spect		0	27.2	6.203	
			6.047	spect		40	69.9		
			5.765	spect		327	1.7		
			5.622	spect		473	0.15		
			5.603	spect		492	1.1		
			5.481 ?	spect		617	0.016		
Bi^{211} 2.16 m	2.17 m	*A*	6.620	spect	63	0	82.6		
			6.273	spect	63	354	17.4		
Bi^{210}	2.6×10^{6} y	*A*	4.94	ion ch	73			5.03	
Bi^{209}	$\sim10^{17}$ y	*F*	3.0	range emuls	71			3.1	ins evid
Bi^{203} 12 h	2×10^{4} y	*B*	4.85	range emuls				4.95	ins evid
Bi^{201} 62 m	4 y	*B*	5.15	ion ch				5.25	ins evid
Bi^{199} ~25 m	~0.5 y	*B*	5.47	ion ch range mica				5.58	ins evid
Bi^{198} 7 m	10 d	*B*	5.83	ion ch				5.95	ins evid
$Bi^{<198}$ 1.7 m		*E*	6.2	ion ch				6.3	ins evid

Table 1 (Explanation of columns).

Column 1 indicates the alpha emitters with measured spectra. Those emitters in which alpha decay was deduced only by the chemical separation of the daughter are not included.

Column 2 shows the partial alpha half-life. If the partial alpha half-life is unknown or is measureably different from the overall half life, the latter is shown in column 1.

Column 3 indicates the degree of certainty that the half-life and the most prominent alpha group belong to the emitter listed in column 1. The meaning of the symbols is as follows:

A = Element certain, mass number certain.
B = Element certain, mass number probable.

C = Element probable, mass number probable.
D = Element certain, mass number uncertain or not well established.
E = Element probable and mass number not well-established or known.
F = Insufficient evidence.

Column 4 shows the measured alpha particle energies. Where there is some uncertainty about the existence of an alpha group, a question mark appears after the energy. Where there is considerable uncertainty about the existence of alpha groups, they are omitted from this table.

Column 5 indicates the method of measurement which best defines the energy and existence of the various alpha groups.

spect = Magnetic or electrostatic spectrograph.
ion ch = Ionization chamber coupled with some form of pulse-height analyzer.
range air (or mica) = Range determination in air (or mica).
range emuls = Range of alpha tracks in a photographic emulsion.

γ-α = Pulse height analyzed alpha spectrum observed in coincidence with gamma rays or electrons. The listed energy is the measured alpha energy or that deduced from the energy of the gamma ray (or electron), whichever is known better.

γ-γ = The alpha was deduced from gamma ray-gamma ray coincidences and a knowledge of the decay scheme.

conv emuls = Conversion electrons were observed in coincidence with alpha particles in a photographic emulsion. The listed energy is equal to the energy of the ground state alpha group minus the corrected gamma energy corresponding to the conversion electrons. A small correction in the gamma energy is necessary to compensate for the difference in recoil of the alpha groups populating the states spanned by the gamma ray.

α-γ = A gamma ray was observed in coincidence with alpha particles and the energy of the alpha group was deduced from a knowlege of the decay scheme.

γ or c = A gamma ray or conversion line was observed and the energy of the alpha group was deduced from a knowledge of the decay scheme.

Column 6 gives the reference for the energy determinations shown in columns 4 and 5. Where no references are given they may be found in the *Table of Isotopes*, Rev. Mod. Phys. **25**, 469 (1953) by HOLLANDER, PERLMAN and SEABORG.

Column 7 indicates the energies of the excited states corresponding to the various alpha groups. Where the excited state energy was deduced from measurements other than the type shown in column 4, the following symbols are used following the energy value in column 7:

γ = Deduced from a gamma ray measurement.
c = Deduced from a conversion line measurement.
γ' (or c') = Deduced from a gamma ray (or conversion line) measurement of a nuclide other than the alpha emitter which decays to the same residual nuclide as the alpha emitter.

The absence of a symbol in column 7 does not mean necessarily that the value was taken from the measurement referred to in column 5 and 6 but only that the measurement was the same *type*.

Column 8 shows the relative abundances of the various alpha groups. These values were determined from the same type of measurement as shown in columns 5 and 6 but are not necessarily the same measurement. One exception is a low energy group of Po^{210} where the abundance was deduced from a gamma abundance. The designation "$\sim$100" signifies low-energy alpha groups have been looked for without success with a high resolution instrument, and very low limits may be set on their abundances.

Column 9. In a large fraction of cases the "highest-energy group" of column 4 is either known to be the ground state transition or is assumed to be so in the absence of information regarding the complexity of the alpha spectrum. The Q-values, unless otherwise stated under "comments", were calculated by adding the recoil energy of the residual nucleus to the alpha particle energy listed in column 4. The recoil energy is $4E/(A-4)$, where E is the alpha particle energy and A is the mass number of the emitter.

Column 10. The comments in this column for the most part reinforce the decision on the decay energy.

ins evid = Insufficient evidence to know whether or not the alpha energy measured is that of the ground state.

e-e = No direct evidence, but since the nucleus is of the even-even type, it can be assumed that the measured energy is that of the ground state transition.

γ, c, c', α-γ = These designations indicate, respectively, that gamma rays, conversion electrons from alpha emitters, conversion electrons from beta emitters, or coincidences between alpha particles and gamma rays have been observed which show some doubt that the highest-energy alpha group is the ground-state transition or the most accurate measurement of it. Where the evidence is not sufficiently definite to deduce an alpha decay energy based on anything other than the highest-energy alpha group, this is reflected by the value in columns 4 and 9 differing only by the recoil energy. Where the evidence is sufficiently definite to deduce the decay energy, it will be found that columns 4 and 9 differ by more than the recoil energy. In Ra^{223} and Po^{211m} decay a more accurate disintegration energy was obtained by adding to the particle energy of the most prominent alpha group its nuclear recoil energy and the energy of the gamma ray thought to span the corresponding energy level and the ground state.

References for Table 1.

1. JONES, SCHUMAN, BUTLER, COWPER, EASTWOOD and JACKSON: Phys. Rev. **102**, 203 (1956).
2. ASARO, STEPHENS, THOMPSON and PERLMAN: Phys. Rev. **98**, 19 (1956).
3. FRIEDMAN, GINDLER, BARNES, SJOBLOM and FIELDS: Phys. Rev. **102**, 585 (1956).
4. HARVEY, CHETHAM-STRODE, GHIORSO, THOMPSON and CHOPPIN: Unpublished data (1955).
5. HUMMEL, ASARO, STEPHENS, THOMPSON and PERLMAN: Unpublished data (1955).
6. A. GHIORSO and B. G. HARVEY: Unpublished data (1954) mentioned in [7],
7. F. S. STEPHENS jr.: Ph. D. Thesis, University of California (U.C. Radiation Laboratory Unclassified Report UCRL-2970), June 1955.

7a. A. CHETHAM-STRODE and L. W. HOLM: To be published (1956).

8. GHIORSO, ROSSI, HARVEY and THOMPSON: Phys. Rev. **93**, 257 (1956).
9. ASARO, STEPHENS, HARVEY and PERLMAN: Phys. Rev. **100**, 137 (1955).
10. L. B. MAGNUSSON: Unpublished data mentioned in [*11*].
11. MAGNUSSON, FRIEDMAN, ENGELKEMEIR, FIELDS and WAGNER: Phys. Rev., to be published (1956).
12. HULET, THOMPSON and GHIORSO: Phys. Rev. **95**, 1703 (1954).
13. HUMMEL, STEPHENS, ASARO, CHETHAM-STRODE and PERLMAN: Phys. Rev. **98**, 22 (1955).
14. ASARO, STEPHENS, CHETHAM-STRODE and PERLMAN: Unpublished data (1955).
15. A. CHETHAM-STRODE jr., G. R. CHOPPIN and B. G. HARVEY: Phys. Rev. **102**, 747 (1956).
16. A. CHETHAM-STRODE jr.: Ph. D. Thesis, University of California (U. C. Radiation Laboratory Unclassified Report UCRL-3322), June 1956.
17. A. CHETHAM-STRODE jr., F. S. STEPHENS jr. and F. ASARO: Unpublished data (1955)
18. FIELDS, STUDIER, DIAMOND, MECH, INGHRAM, PYLE, STEVENS, FRIED, MANNING GHIORSO, THOMPSON, HIGGINS and SEABORG: Phys. Rev. **102**, 180 (1956).
19. Private communication from E. K. HULET, 1956.
20. E. K. HULET, S. G. THOMPSON and A. GHIORSO: Phys. Rev. **95**, 1703 (1954).
21. ASARO, THOMPSON and PERLMAN: Unpublished data (1954).
22. HUMMEL, ASARO, THOMPSON and PERLMAN: Unpublished data (1955).
23. F. ASARO, S. G. THOMPSON and I. PERLMAN: Phys. Rev. **92**, 694 (1953).
24. F. ASARO, B. G. HARVEY and I. PERLMAN: Unpublished data (1955/56).
25. R. A. GLASS: Ph. D. Thesis, University of California (U.C. Radiation Laboratory Unclassified Report UCRL-2560), April 1954.
26. G. H. HIGGINS: Ph. D. Thesis, University of California (U.C. Radiation Laboratory Unclassified Report UCRL-1796), June 1952.
27. STEPHENS, HUMMEL, ASARO and PERLMAN: Phys. Rev. **98**, 261 (1955) and unpublished data (1955).
28. F. ASARO and I. PERLMAN: Phys. Rev. **93**, 1423 (1954).
29. L. L. GOL'DIN, E. F. TRET'YAKOV and G. I. NOVIKOVA: Conf. Acad. Sci. USSR. on Peaceful Uses of Atomic Energy, Phys. Math. Sci., p. 226 (July 1955) UCRL translation 242.
30. JAFFE, PASSELL, BROWNE and PERLMAN: Phys. Rev. **97**, 142 (1955).
31. F. ASARO and I. PERLMAN: Unpublished data mentioned in [*32*]; also F. ASARO, Ph. D. Thesis, University of California (U.C. Radiation Laboratory Unclassified Report UCRL-2180), June 1952.
32. I. PERLMAN and F. ASARO: Ann. Rev. Nucl. Sci. **4**, 157 (1954).
33. F. ASARO and I. PERLMAN: Unpublished data (1956).
34. F. ASARO and I. PERLMAN: Phys. Rev. **94**, 381 (1954).
35. F. ASARO, F. S. STEPHENS jr. and I. PERLMAN: Unpublished data (1955).
36. J. P. HUMMEL, F. ASARO, G. H. HIGGINS and I. PERLMAN: Unpublished data (1956).
37. D. C. DUNLAVEY and G. T. SEABORG: Unpublished data listed in [*32*].
38. T. D. THOMAS and R. A. GLASS: Unpublished data (1956).
39. MAGNUSSON, ENGELKEMEIR, FREEDMAN, PORTER and WAGNER: Phys. Rev. **100**, 1237A (1955).

40. H. G. JACKSON and B. G. HARVEY: Private communication mentioned in [32].
41. B. ZAJAC: Phil. Mag. **43**, 264 (1952).
42. G. VALLADAS: Dr. Thesis, Commissariat a L'Energie Atomique Rapport C.E.A. No. 483 (1955).
43. J. R. GROVER and G. T. SEABORG: Private communication (1952), reported in University of California Radiation Laboratory Unclassified Report UCRL-2455, p. 22 (Jan. 1954).
44. G. VALLADAS: C. r. Acad. Sci., Paris **237**, 1673 (1953); see also [42].
45. F. ASARO and I. PERLMAN: Phys. Rev. **99**, 37 (1955).
46. F. ASARO and I. PERLMAN: Phys. Rev., in press (1956); see also [48].
46. F. ASARO and I. PERLMAN: Phys. Rev., in press (1956); see also [48].
47. W. G. SMITH, F. ASARO and J. M. HOLLANDER: Phys. Rev., in press (1956).
48. F. S. STEPHENS jr., F. ASARO and I. PERLMAN: Phys. Rev. **96**, 1568 (1954).
49. J. P. HUMMEL, F. ASARO and I. PERLMAN: Unpublished data (1955); see also [50] and [29].
50. J. P. HUMMEL, F. ASARO and I. PERLMAN: Phys. Rev. **98**, 261 A (1955).
51. S. W. PEAT and M. A. S. ROSS: Proc. Phys. Soc. Lond. A **68**, 923 (1955).
52. G. VALLADAS and R. BERNAS: C. r. Acad. Sci., Paris **91**, 2230 (1953).
53. F. S. STEPHENS jr., F. ASARO and I. PERLMAN: Unpublished data (1955).
54. F. ASARO, F. S. STEPHENS jr. and I. PERLMAN: Phys. Rev. **92**, 1495 (1953).
55. PILGER, HUMMEL, ASARO and PERLMAN: Unpublished data (1955). See University of California Radiation Laboratory Unclassified Report UCRL-3240, p. 72 (Dec. 1955).
56. STEPHENS, HUMMEL, ASARO and PERLMAN: Unpublished data (1956); see also [7].
57. F. S. STEPHENS jr., F. ASARO and I. PERLMAN: Unpublished data (1956).
58. F. F. MOMYER jr., F. ASARO and E. K. HYDE: J. Inorg. Nucl. Chem. **1**, 267 (1955).
59. L. MADANSKY and F. RASETTI: Phys. Rev. **102**, 464 (1956).
60. F. F. MOMYER jr. and E. K. HYDE: Phys. Rev. **101**, 136 (1956).
61. F. ASARO and I. PERLMAN: Unpublished data (1953) listed in [32].
62. G. H. BRIGGS [63] recalculated the energy of the main Em^{219} alpha group from the original data [64] using later values for the reference energies.
63. G. H. BRIGGS: Rev. Mod. Phys. **26**, 1 (1954). This is a compilation of the best values for the energies of natural alpha particles.
64. M. CURIE and S. ROSENBLUM: C. r. Acad. Sci., Paris **196**, 1598 (1933).
65. The alpha particle energy of Em^{207} was increased 20 kev above the value given in [66] in order to correspond to the best available value for Em^{208}.
66. F. F. MOMYER jr. and E. K. HYDE: J. Inorg. Nucl. Chem. **1**, 274 (1955).
67. R. W. HOFF, J. P. HUMMEL and F. ASARO: Unpublished data (1954); see also University of California Radiation Laboratory Unclassified Report UCRL-3157, p. 67 (Sept. 1955).
68. STONER, ASARO, STEPHENS and HYDE: University of California Radiation Laboratory Unclassified Report UCRL-2455, p. 14 (Jan. 1954).
69. PERLMAN, STEPHENS, HUMMEL, PILGER and ASARO: University of California Radiation Laboratory Unclassified Report UCRL-3068, p. 37 (July 1955).
70. W. JENTSCHKE, A. C. JUVELAND and G. H. KINSEY: Phys. Rev. **96**, 231 (1954).
71. F. ASARO and I. PERLMAN: Rev. Mod. Phys. **26**, 237 (1954). This is a compilation of alpha disintegration energies.
72. S. ROSENBLUM and H. TYRÉN: C. r. Acad. Sci., Paris **239**, 1205 (1954).
73. H. B. LEVY and I. PERLMAN: Phys. Rev. **94**, 152 (1954).

There are a considerable number of instances in which it is not certain that the highest energy group listed represents the ground-state transition, but the necessary information to determine total decay energy is lacking. Here, the Q-value is calculated from the highest energy group (usually the only group seen) and notation is made that there is insufficient evidence to consider this figure reliable. For the even-even alpha emitters in this category the assignment of a proper Q-value is a good deal more certain because it is known empirically that for these, the main alpha group invariably does lead to the ground state.

A statement should be made regarding the completeness of the information in Table 1. To illustrate differences in available data for different species we may compare the spectra of Cm^{242} and Cm^{240}. For Cm^{242}, eight alpha groups are listed; for Cm^{240}, only a single group is listed. From what is known about the regularities of alpha spectra, it is fairly certain that the alpha spectra of these two substances are similar. In the case of Cm^{242}, large amounts of the isotopically-pure substance can be made and consequently, the spectrum can be examined with high resolution and with great sensitivity. Close-lying groups may be resolved in instruments requiring intense sources and also very rare events may be sought, for example, in the range of one event in 10^5 to 10^7 disintegrations. As seen in Fig. 8 only three alpha groups have been observed directly; the other five groups of Table 1 are inferred from gamma-ray data which at present is obtained with greater sensitivity. In the case of Cm^{240}, it is much more difficult to prepare sizeable amounts and it is not possible to prepare it free of other curium isotopes. The only measurements made utilized an instrument of relatively low resolution and not even the expected intense group of energy some 45 kev lower than the ground-state transition has been resolved. As for gamma-ray analysis, the problem becomes quite difficult in view of the presence of other isotopes, particularly those which decay by orbital electron capture.

III. Nuclear states and nuclear models.

12. General remarks. Before one can hope to formulate any fundamental understanding of alpha decay kinetics, especially of odd nuclear types, it will be necessary to have considerable knowledge of the nuclear states involved. Thus, it is appropriate here to consider briefly the present state of this knowledge in the regions where alpha emission is prevalent.

One wishes first to know the energies of as many as possible of the low-lying nuclear states. Alpha-ray spectroscopy has played a key role in the establishment of the lowest levels (Sects. 8 to 11). Next, one desires to classify states according to spin and parity. Such information is gained from experimental determination of ground-state spins by various methods and determination of the multipolarity of gamma transitions. Additional information on the details of nuclear states is given by nuclear magnetic and electric moments, by actual rates of nuclear transitions, and by energy level patterns.

For discussion of alpha-decay properties and for correlation of level energies and other nuclear properties with nuclear models, it is most logical to segregate heavy alpha emitters into three more-or-less distinct classes according to neutron number: (a) a region comprising decays with parent and daughter having $N \leq 126$; (b) a region with $126 < N \leq 138$; (c) a region with $N > 138$. Alpha decays (i.e. bismuth isotopes and 127-neutron isotones) crossing the closed shells of 82 protons or 126 neutrons constitute special cases.

For nuclei near the Pb^{208} doubly-closed configuration the nuclear models holding greatest promise are of the spherical well form deriving from the original MAYER-JENSEN shell models[1,2] [*8*]. For nuclei more than one nucleon removed from Pb^{208} the modifications treating specific nucleon-nucleon interactions among the nucleons outside the closed shells are of greatest utility. (See the work of PRYCE [*9*], ALBURGER and PRYCE[3], and TRUE[4].)

For nuclei in the region with $N > 138$ it seems clear that a stabilized spheroidal deformation has set in, and the BOHR-MOTTELSON [*10*] "strong coupling" model has given many useful correlations of nuclear properties in this region.

---

[1] M. GOEPPERT-MAYER: Phys. Rev. **75**, 1969 (1949).
[2] HAXEL, JENSEN and SUESS: Phys. Rev. **75**, 1766 (1949).
[3] D. E. ALBURGER and M. H. L. PRYCE: Phys. Rev. **95**, 1482 (1954).
[4] W. W. TRUE: Phys. Rev. **101**, 1342 (1956).

On the low side of the border $N=138$ the GOLDHABER-WENESER[1] or WILETS-JEAN[2] models offer hope, at least for even-even nuclei. The GOLDHABER-WENESER model couples the motion of two or more individual nucleons with collective vibrational motion about a spherical equilibrium shape. The WILETS-JEAN model treats the collective oscillations with stable total deformation but with little or no shape stability. For odd-nucleon types in the regions where the above two models apply the theory is less developed. Calculations along the lines of the intermediate coupling treatment of BOHR and MOTTELSON [*10*] and of CHOUDHURY[3] might be of use in filling the gap.

13. Energies and other properties of nuclear states of even-even nuclei. The ground states of even-even nuclei are presumed all to have zero spin and even parity.

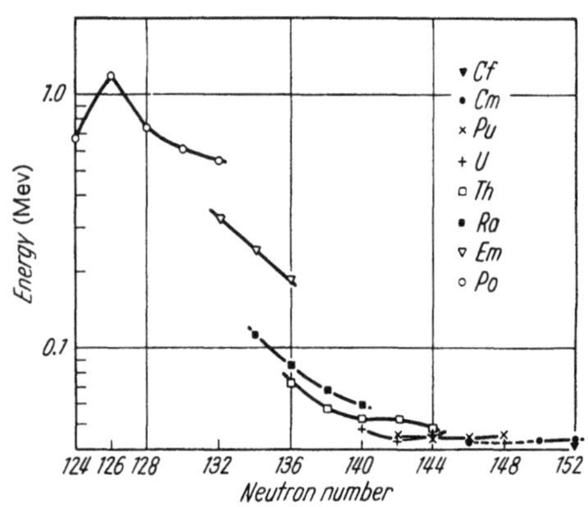

Fig. 10. Energies of first excited states of even-even nuclei in the heavy element region. Most of the data were obtained from alpha-spectra and these will be found in Table 1. All points shown are believed to represent 2^+ states. Note the progressive increase in energy in approaching the 126 neutron shell.

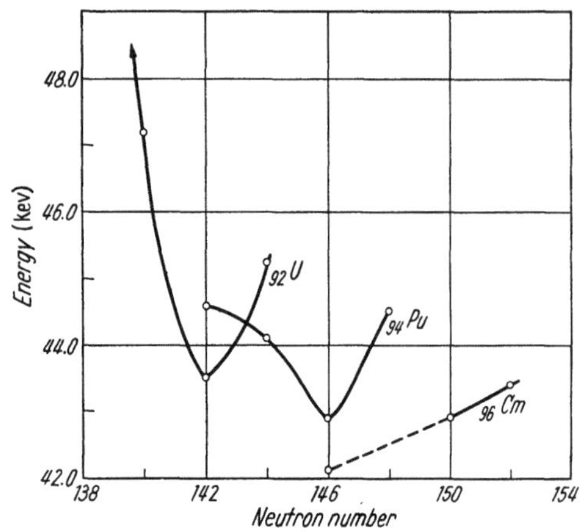

Fig. 11. Energies on an expanded scale for first excited states of even-even nuclei of atomic number ≥ 92. These energies are based on conversion electron spectroscopic measurements and are accurate to at least 0.1%. These data are from a compilation by J. M. HOLLANDER, Phys. Rev. 103, 1590 (1956).

The first excited states of the heavy even-even nuclei (except Pb^{208})[4] have, wherever determined, spin two and even parity. Their energies have rather a regular dependence on mass number, being very high near Pb^{208} and descending to a broad flat region. Figs. 10 and 11 show plots of these energies *vs.* neutron number.

The spectra of higher excited states of even-even nuclei become quite complicated in the regions near the closed shells. The discussion of these levels is not too essential in considering alpha decay since the higher excited levels of even-even nuclei in this region usually lie too high in energy to be detectably populated by alpha decay, although "long-range" alpha emission occurs from highly excited levels in Po^{212} and Po^{214} populated by beta decay of Bi^{212} and Bi^{214}. As the levels dip in energy further from the closed shell, a new regularity appears in the even spin-even parity excited states. That is, these energies approach a close agreement with a simple rotational energy formula.

$$E_I = \frac{\hbar^2}{2\mathfrak{J}} I(I+1) \tag{13.1}$$

[1] G. GOLDHABER and J. WENESER: Phys. Rev. **98**, 212 (1955).

[2] L. WILETS and M. JEAN: To be published. Preliminary report in C. r. Acad. Sci., Paris **241**, 1108 (1955).

[3] D. C. CHOUDHURY: Dan. Mat. fys. Medd. **28**, No. 4 (1954).

[4] Pb^{208} has a first excited state of spin 3, odd parity.

where I is the spin and $\mathfrak{J}$ is the moment of inertia. Population by alpha decay of rotational band levels as high as the 8^+ has been observed in Pu^{238} following alpha decay of Cm^{242} (see Table 1).

For a number of even-even nuclei in the heavy region, odd parity states of spin one have also been observed at moderate energies, lying lowest around 88 protons and 136 neutrons. The 1^- states characteristically decay[1] by $E1$ gamma transitions to the 0^+ and 2^+ states, and the relative gamma intensities indicate that the component of total angular momentum along the nuclear symmetry axis (K-quantum number) is 0.

The 1^-states are usually considered odd states of rotation belonging to the ground rotational band. In the usual form of the theory of spheroidally deformed nuclei, the nucleus is presumed to be symmetric with respect to rotation of π about a principal axis perpendicular to the cylindrical symmetry axis. Then for properly symmetrized strong coupling wave functions (cf. [*10*], Eq. (II.15)) the states of odd rotational angular momentum must vanish. If the symmetry of the nucleus with respect to the rotation of π is not perfect, states of odd rotational angular momentum may appear but displaced to higher energy with respect to the even members of their band. CHRISTY[2] has thus proposed that a pear-shaped deformation must set in for nuclei in the region of low-lying 1^- states.

It is beyond the scope of the present paper to speculate on details of the theory of the 1^- states. We do wish to point out the significance of the fairly rapid $E1$ transitions (see Sect. 53) since available evidence indicates the lifetimes of the 1^-states to be less than a millimicrosecond. If one grants the rotational nature of the 1-states, the occurrence of $E1$ transitions implies a separation between the center of mass and the center of charge in these deformed nuclei. It may well be that the two-fold rotational symmetry is broken down in these nuclei not so much by pear-shaped deformations as by opposite displacements of proton and neutron centers of mass along the cylindrical symmetry axis. The conditions for such displacements probably involve the availability of near-lying levels of opposite parity for both protons and neutrons near the top of the "FERMI sea". Such conditions are fulfilled in the region of Ra^{226} and Ra^{224}, since even-odd Ra^{225} and odd-even Ac^{227} exhibit low-energy $E1$ gamma transitions to ground.

14. Region of spheroidal nuclei. The appearance of rotational bands is associated by BOHR and MOTTELSON [*10*], [*11*] with onset of a stabilized spheroidal nuclear shape. An important quantity in the alpha decay treatments [*12*] in the rotational band region is the intrinsic quadrupole moment Q_0. The magnitude of Q_0 may be determined either from the lifetime of the first excited state or from the COULOMB excitation cross section for this state. The latter method has been applied by DIVATIA *et al.*[3] and from gamma-ray yield data with assumed total conversion coefficients of 340 for Th^{232} and 700 for U^{238} they calculate first excited state lifetimes of 7.8×10^{-10} and 4.4×10^{-10} sec, respectively. Their results may also be expressed in terms of intrinsic quadrupole moment values of $\pm7.1\times10^{-24}\,\mathrm{cm}^2$ and $\pm8.5\times10^{-24}\,\mathrm{cm}^2$, respectively. From other evidence we believe these intrinsic quadrupole moments to be positive. For both Am^{241} and Am^{243} ($I=\frac{5}{2}$) MANNING *et al.*[4] have determined spectroscopic quadrupole moments of $+4.9$ barns, corresponding to $Q_0=+14$ barns.

In the region of the heavy deformed nuclei many rotational bands have been identified, not only for even-even nuclei, but also for odd-mass nuclei. Most of these bands have been revealed by alpha spectroscopy [*4*] and some more recently by COULOMB excitation[5]. Heavy nuclei exhibiting such bands with three or more member levels definitely identified are the following: Th^{229}, U^{235} (2 bands), Np^{237}, Np^{239}, Pu^{239}, and Bk^{249}.

---

[1] STEPHENS, ASARO and PERLMAN: Phys. Rev. **100**, 1543 (1955).
[2] R. F. CHRISTY: Unpublished (1954).
[3] DIVATIA, DAVIS, MOFFAT and LIND: Phys. Rev. **100**, 1266 A (1955); and verbal report quoted in National Research Council Nuclear Data Cards 56-1-118 and 56-1-120 (1956).
[4] MANNING, FRED and TOMKINS: Phys. Rev. **102**, 1108 (1956).
[5] J. O. NEWTON: Nature, Lond. **175**, 1028 (1955).

The occurrence of such bands implies an approximate conservation of angular momentum along the nuclear symmetry axis. The quantum number measuring this angular momentum component, usually designated as K, has proved to be of great usefulness in classifying nuclear states and correlating gamma, beta, and alpha transition rates[1–3]. There is a selection rule in the model for deformed nuclei that the multipolarity L of a radiation must equal or exceed ΔK, the change in K between states. (The intrinsic nucleonic angular momentum along the symmetry axis is designated Ω and will be equal to K for the low states, which do not involve shape vibrational excitation.)

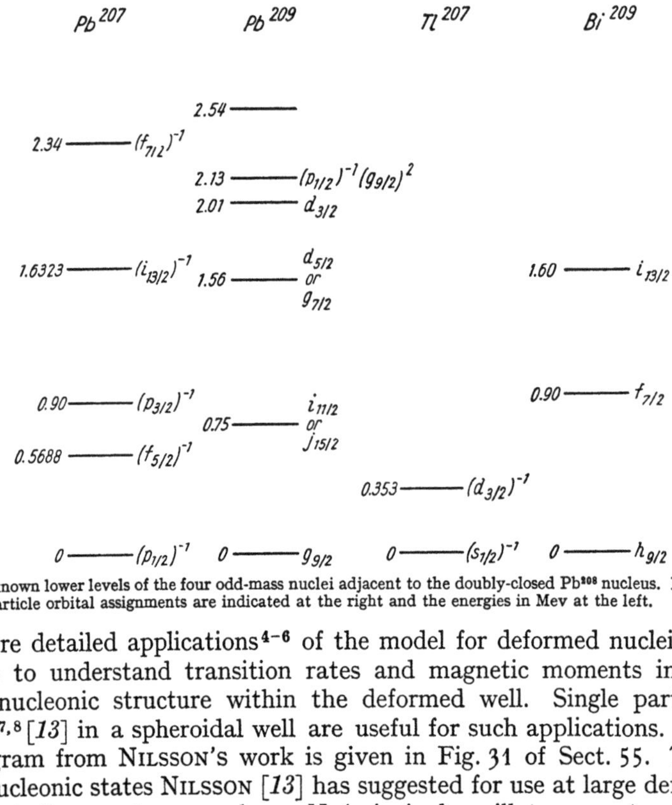

Fig. 12. The known lower levels of the four odd-mass nuclei adjacent to the doubly-closed Pb^{208} nucleus. Probable single-particle orbital assignments are indicated at the right and the energies in Mev at the left.

In more detailed applications[4–6] of the model for deformed nuclei, attempts are made to understand transition rates and magnetic moments in terms of intrinsic nucleonic structure within the deformed well. Single particle wave functions[7,8] [*13*] in a spheroidal well are useful for such applications. An energy level diagram from NILSSON's work is given in Fig. 31 of Sect. 55. To further classify nucleonic states NILSSON [*13*] has suggested for use at large deformations the asymptotic quantum numbers, N (principal oscillator quantum number), n_z (or μ_z) (z-axis oscillator quantum number), and Λ (nucleon orbital angular momentum component along the symmetry axis) appropriate to a three-dimensional anisotropic harmonic oscillator with no spin-orbit interaction. The violation of selection rules in these quantum numbers for beta and gamma transitions appears generally to have a retarding effect on transitions though not so severe as violation of K-selection rules[5,9].

[1] ALAGA, ALDER, BOHR and MOTTELSON: Dan. Mat. fys. Medd. **29**, No. 9 (1955).
[2] RASMUSSEN, STEPHENS, STROMINGER and ÅSTRÖM: Phys. Rev. **99**, 47 (1955).
[3] D. M. CHASE and L. WILETS: Phys. Rev. **101**, 1038 (1956).
[4] B. R. MOTTELSON and S. G. NILSSON: Z. Physik **144**, 217 (1955).
[5] D. STROMINGER: Ph. D. Thesis, University of California 1956 (Radiation Laboratory Report UCRL 3374).
[6] J. M. HOLLANDER, W. G. SMITH and J. O. RASMUSSEN: Phys. Rev. **102**, 1372 (1956).
[7] K. GOTTFRIED: Ph. D. Thesis, Massachusetts Institute of Technology 1955 (unpublished).
[8] M. RICH: Bull. Amer. Phys. Soc., Ser. II **1**, No. 5, F 10 (1956).
[9] G. ALAGA: Phys. Rev. **100**, 432 (1955).

Concerning possible collective vibrational excitation of deformed nuclei less is presently known. Probably some recently discovered states populated weakly in alpha decay of radium isotopes to $Em^{218,220,222}$ are the second excited 2^+ states predicted by GOLDHABER-WENESER or WILETS-JEAN models. Some states near 1 Mev are populated in Cm^{242} decay by mildly hindered groups; one may speculatively associate these states with beta- and gamma- vibrational states of spheroidal nuclei (see Cm^{242} treatment in Sect. 54).

15. Region of spherical nuclei. In the region near Pb^{208} the equilibrium nuclear shape is probably spherical, and treatments involving the coupling of nucleons and/or holes outside the closed configuration seem most promising[1] [9].

The simplest cases will be the nuclei immediately adjoining Pb^{208}. Their level systems with present probable shell model assignments are shown in Fig. 12, the assignments being somewhat different from PRYCE's [9] by virtue of more recent work.

We will have occasion to use some of these shell model level assignments in Sects. 49 to 52 where alpha decay across closed shells (Po^{211}, Bi^{210}, Bi^{211}, and Bi^{212}) is discussed in detail; these decays across the closed shell are of an especially highly hindered variety.

B. Kinetics of alpha decay (even-even type).

I. Correlations of decay rates of even-even nuclei.

16. Ground state transitions. As early as 1911, GEIGER and NUTALL found that a plot of the logarithms of decay constants for alpha emitters against the logarithms of the ranges resulted in a family of straight lines, one for each decay series. The aspect of these relations of greatest interest was that the slopes were about the same and demonstrated a sharp dependence of decay constant upon disintegration energy. After some two decades this relationship was interpreted by GAMOW [2] and by CONDON and GURNEY [3] in terms of the potential barrier penetration problem.

The theory for alpha decay related the decay constant to three other parameters: the decay energy, the nuclear charge (atomic number), and the nuclear radius. Only the radius is not subject to direct measurement, but since no model of the nucleus permits widely varying nuclear radii in a limited region there was no difficulty in testing the major premise of the theory. In the main, the wide range of decay constants could be explained quantitatively in terms of the decay energy and the nuclear charge by letting the radius assume a simple $A^{\frac{1}{3}}$ dependence. As the theory is now developing, it is becoming difficult to ascribe fundamental significance to the nuclear radius as used in this way. Discussion of this problem will be found in Sects. 19 to 27. Nevertheless, the concept is familiar and useful in explaining some of the observed trends and will be adhered to in this section.

The theory for the alpha-decay process has in recent years played an indirect but important practical role in the preparation of the many new alpha-emitting species. According to the regularities discussed in Sect. 3, it is possible to predict with fair accuracy the energy of any alpha-emitter. The theory then allows the calculation of the half-life in a manner summarized in Sect. 27. Such information, quite obviously, constitutes a powerful guide for the preparation of new species. However, the calculations are too laborious to be made repeatedly, and it has become the practice to display them graphically. Several such sets of curves

---
[1] W. W. TRUE: Phys. Rev. **101**, 1342 (1956).

have been published[1,2] [6] and one is reproduced in Fig. 13. It should be stated at the start that there is excellent agreement between theory and experiment only for the ground state to ground state transitions in the even-even nuclei.

It will be noted in Fig. 13 that nuclear radius does not appear. The manner by which it has been eliminated as an independent parameter is as follows. In essence, for each atomic number, the radius is not independent of the alpha-energy. This follows from the nearly monotonic change in alpha energy with mass number (see Sect. 3), and the radius in turn is taken to be a simple function of $A^{\frac{1}{3}}$. Therefore, to define a point on the curve for element Z, one simply takes the measured

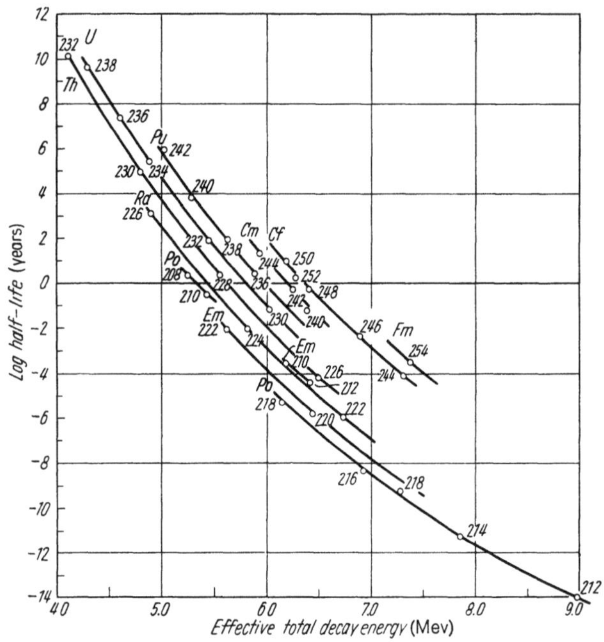

Fig. 13. Experimental values of half-life *vs.* alpha energy for even-even nuclei. The effective decay energy (Q_{eff}) is the alpha decay energy corrected for electron screening (see Sect. 25).

or estimated alpha energy for Z^A as the abscissa coordinate and calculates the half-life using A to define the radius. Since there will not generally be two values A for the particular decay energy the calculated half-life will be unique. The curves of Fig. 13 are calculated in the manner mentioned, and the points shown are experimental.

Aside from the utility of such curves for predicting the decay properties of unknown species, there has arisen another important function, namely, to serve as a baseline for discussing transitions which do not conform to the simple theory. Transitions to excited states of even-even nuclei and all transitions of odd nuclear types fall into this category.

For some purposes it is more convenient to present the information of Fig. 13 in another manner as illustrated by Fig. 14. Here the logarithm of the half-life is plotted against the reciprocal square root of the decay energy and the resulting family of straight lines is obtained[3,4]. The lines as shown were defined by least

[1] A. BERTHELOT: J. Phys. Radium VII **3**, 52 (1942).

[2] S. BISWAS: Indian J. Phys. **23**, 51 (1949).

[3] A. BOHR, P. O. FRÖMAN and B. R. MOTTELSON: Dan. Mat. fys. Medd. **29**, No. 10 (1955).

[4] C. J. GALLAGHER and J. O. RASMUSSEN: University of California Radiation Laboratory Report, UCRL 3176 (1955).

squares fitting using only those points for which accurate decay energies and half-lives are available. The relation of these lines to barrier penetration theory can be found from inspection of Eq. (20.7) which predicts a nearly straight line dependence of the functions plotted.

Each line in Fig. 14 representing a single element can be expressed analytically as follows:

$$\operatorname{Log} t_{\frac{1}{2}}\,(\mathrm{sec}) = \frac{A}{\sqrt{Q_{\mathrm{eff}}\,(\mathrm{Mev})}} + B. \tag{16.1}$$

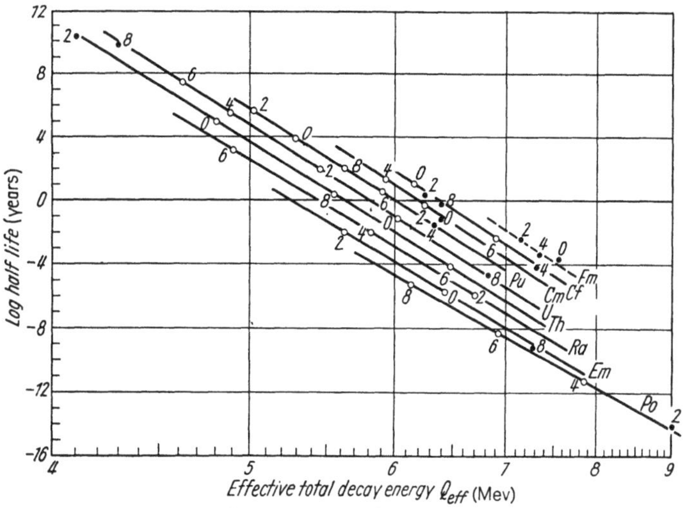

Fig. 14. Plot of logarithms of partial alpha half-lives for ground-state transitions *versus* the inverse square root of the effective total alpha-decay energy (Q_{eff} = alpha-particle energy + recoil energy + electron screening correction). The points are experimental, and the straight lines are based on a least squares analysis of the points where energies have been determined by magnetic spectrographs (excluding Em^{218} which exhibits an apparently anomalous decay rate). There is only one point for element fermium (Fm); hence, the slope of the fermium line was arrived at by extrapolation from the slopes of lower elements. Points used in the analysis are indicated by triangles, and points not used, by circles. Nuclei with 126 or fewer neutrons are not shown on this plot. Constants in the equations for these lines are given in Table 2. The last digit in the mass number of the alpha emitter is given beside each point.

Here Q_{eff} is the effective decay energy which consists of the alpha-particle energy plus the recoil energy and plus an additive correction ($\sim$40 kev) for the orbital electron screening effect (see Sect. 25). Table 2 lists the values of A and B for each element.

Table 2. *Semi-empirical constants*[1] *from correlation of ground state decay rates of even-even nuclei.*

Element	A	B	Remarks	Element	A	B	Remarks
100 Fm	156.38	− 53.3742	Extrapolated	90 Th	144.19	− 53.2644	
98 Cf	152.86	− 52.9506		88 Ra	139.17	− 52.1476	
96 Cm	152.44	− 53.6825		86 Em	137.46	− 52.4597	
94 Pu	146.23	− 52.0899		84 Po	129.35	− 49.9229	
92 U	147.49	− 53.6565					

A few alpha emitters for which reliable data are available were not used for determination of the lines of Fig. 14. These are the polonium alpha emitters Po^{210}, Po^{208}, and Po^{206}, which in common with all other even-even alpha emitters with neutron number equal to or less than 126 decay more slowly than the simple theory would predict. Reasons for this behavior are advanced in Sect. 39. Po^{212} and Em^{218} have also been excluded from the determination of the lines since they appear to behave somewhat abnormally.

[1] Constants to use with Eq. (16.1) or (40.1), where $t_{\frac{1}{2}\alpha}$ is in seconds and Q_{eff} in Mev.

17. Hindrance factor in alpha-decay. It has already been mentioned that only the ground state transitions of the even-even alpha-emitters obey the simple barrier penetration theory. Other transitions are slower in varying degree than the demands of the theory and are consequently said to be *hindered*. It is convenient to express this effect quantitatively in terms of a *hindrance factor (F)* which is defined as the factor by which the observed alpha half-life is greater than that calculated. For some purposes rather subtle effects may be of importance, and we shall want to define this factor somewhat differently. The essence of the redefinition is that the hindrance factors for all even-even ground state transitions are taken to be unity whether or not the points fall precisely on the calculated curves of Fig. 13 or the best semi-empirical curves of Fig. 14. The reason for this small change will appear in the following discussion of transitions to excited states.

18. Transitions to excited states. Corresponding to each ground state transition entered in Figs. 13 and 14, it would be possible to enter points for transitions leading to excited states. The alpha-particle energies can be obtained from

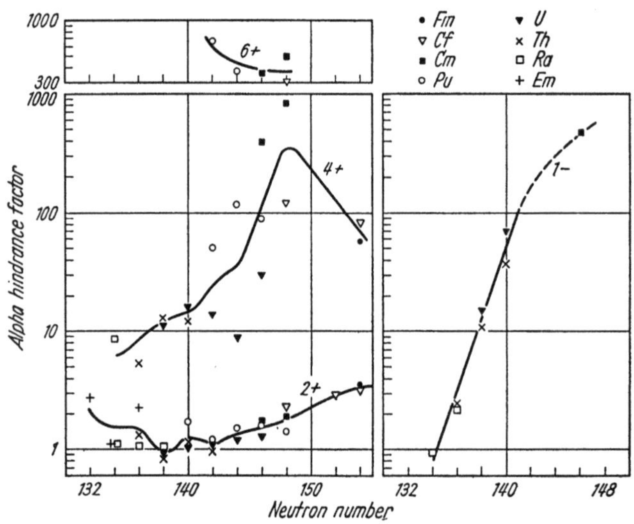

Fig. 15. Plots of hindrance factors for alpha decay of even-even nuclei to excited states. a) Even parity, even spin final states. b) Odd parity, spin one final states. Numerical values of the factors to these states are tabulated below:

Alpha emitter	Spin of state					Alpha emitter	Spin of state				
	2	4	6	8	1		2	4	6	8	1
Fm^{254}	3.5	57				U^{234}	1.1	14			
Cf^{252}	3.2	82				U^{232}	1.0	16			71
Cf^{250}	2.9					U^{230}	0.91	11			15
Cf^{246}	2.3	120	280			Th^{232}	0.96				
Cm^{244}	1.9	830	480			Th^{230}	1.1	12	(8200)		38
Cm^{242}	1.7	390	350	5100	480	Th^{228}	0.85	13			11
Pu^{242}	1.4					Th^{226}	1.3	5.3			2.4
Pu^{240}	1.6	89				Ra^{226}	0.99				
Pu^{238}	1.5	116	360	15000		Ra^{224}	1.0				2.3
Pu^{236}	1.2	50	640			Ra^{222}	1.1	8.7			0.97
Pu^{234}	1.7					Em^{222}	2.2				
U^{238}	1.3	30				Em^{220}	1.1				
U^{236}	1.2	8.8				Em^{218}	2.7				

Table 1, and the partial alpha half-life for each transition may be calculated from its abundance and the total alpha half-life. It would be noted that such transitions are almost invariably hindered, some to a marked degree.

The transitions which have been well studied in the even-even nuclei are those to members of the rotational band based upon the ground state. The energy levels populated bear the designations $0^+, 2^+, 4^+ \ldots$ where the 0^+ state is the ground state. Fig. 15 shows the calculated hindrance factors for the various types of transitions. The hindrance factors for the ground state transitions are taken to be unity. The hindrance factor to an excited state is then obtained by normalizing the ground state transition to the proper curve of Fig. 14 and shifting the measured partial alpha half-life of the transition to the excited state accordingly. This normalization has the effect of simply referring hindrance factors of excited states to the ground state rather than to use the theory as an absolute guide.

Let us recapitulate the function of the hindrance factor. As well as can be told, the ground state transition of almost any even-even alpha-emitter can be described adequately in terms of simple barrier penetration. This transition is therefore taken to be "allowed" or unhindered. The hindrance factor of any other transition then expresses a retardation of decay rate which must be explained by some extension of the basic theory.

One of the factors long recognized as affecting the decay rate is the spin change. This aspect and others of the classical theory are discussed in Sects. 19 to 27.

Other factors are now recognized which influence the alpha-emission lifetime and these are treated in succeeding sections. The curves of Fig. 14 will also be used (Sect. 40) to define hindrance factors for odd nucleon alpha-emitters.

II. Decay rate theory for purely central fields.

19. Introduction. In the preceding chapter it was shown that the rates of ground state alpha transitions of even-even nuclei with few exceptions are amenable to simple exponential correlations with decay energy and atomic number. This sensitive exponential energy dependence of alpha decay rates is markedly different from the energy dependence of gamma and beta transition rates, where power laws of the energy are generally more applicable. An early triumph of quantum mechanics was the proposal by GAMOW [*2*] and by CONDON and GURNEY [*3*] that alpha decay is essentially a coulombic barrier penetration process. With widely varying assumptions about the fundamental process of formation of alpha particles from nuclear matter the barrier rate formulations correlate the experimental rate data for even-even nuclei (rates varying over a factor of 10^{20}) quite well so long as one parameter, usually the coefficient of $A^{\frac{1}{3}}$ in an "effective nuclear radius" expression, is left free to be adjusted for the group of alpha emitters. The only significant breaks in the smooth correlations occur at 126 neutrons, those nuclei with 126 or less neutrons showing slower decay rates by factors of 5 to 20 than would be expected from correlations with nuclei of 128 or more neutrons.

In our first detailed applications of rate theory we will be concerned with the even-even nuclei, which possess zero angular momentum in their ground states.

After outlining some of the principal modern decay treatments and giving their formulas we will apply the experimental data in three ways, calculating (a) the effective nuclear radius for alpha decay as calculated by KAPLAN'S[1] approximation to the PRESTON-SEXL [*14*], [*15*] decay formula, (Table 8), (b) the normalized *S-wave* "surface probability" near the nucleus according to the

[1] I. KAPLAN: Phys. Rev. **81**, 962 (1951).

treatment of G. H. WINSLOW [16] (Table 7). (c) the reduced derivative width δ^2 according to the precise collision matrix formulation of R. G. THOMAS [17] (Table 7).

20. A simple semi-classical treatment. Let us consider first a very simple approximate treatment of alpha decay along the lines first given by VON LAUE[1]. In common with other one-body treatments the alpha daughter nucleus is simply assumed to give rise to a potential function for the alpha particle. As the potential is usually idealized, it is taken as purely coulombic outside the "effective nuclear radius", R, and is taken to be a constant value (less than the alpha decay energy) for distances less than R. An alpha particle is initially confined in a virtual state within the well, where it makes frequent collisions with the wall with a small probability of quantum-mechanical penetration of the negative energy or barrier region. This three-dimensional problem with spherically symmetric potential can readily be reduced to a one-dimensional radial problem by familiar methods (Sect. 22). For *S-wave* alpha emission the effective potential in the one-dimensional problem is as shown in Fig. 16.

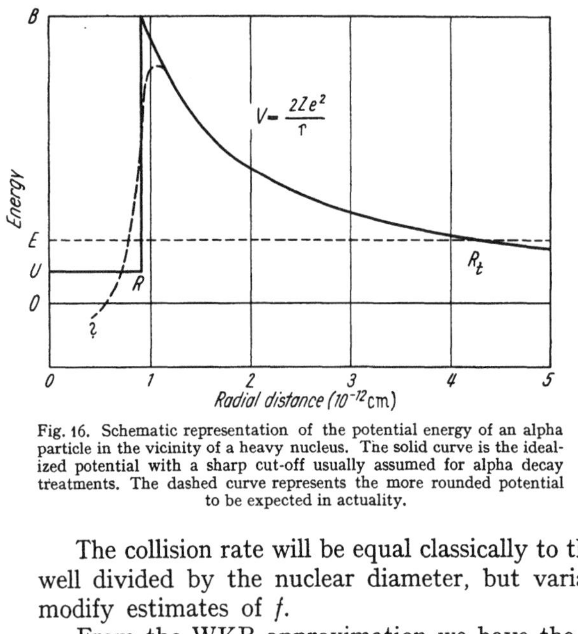

Fig. 16. Schematic representation of the potential energy of an alpha particle in the vicinity of a heavy nucleus. The solid curve is the idealized potential with a sharp cut-off usually assumed for alpha decay treatments. The dashed curve represents the more rounded potential to be expected in actuality.

The decay constant λ will be a simple product of the frequency factor, f, giving the collisions per second with the wall, and the quantum-mechanical penetration factor P.

$$\lambda = f P. \tag{20.1}$$

The collision rate will be equal classically to the alpha particle velocity in the well divided by the nuclear diameter, but variations in the model sometimes modify estimates of f.

From the WKB approximation we have the penetration factor below:

$$P \approx e^{-\frac{2}{\hbar}\int\limits_{R}^{\frac{2Ze^2}{E}} \sqrt{2M\left(\frac{2Ze^2}{r}-E\right)}\,dr} \tag{20.2}$$

where Z is the charge of the recoil nucleus; M, the reduced mass of the system $\left(\text{i.e., } M=\frac{M_\alpha M_r}{M_\alpha + M_r}\right)$; e, the elemental charge in electrostatic units. The integral is evaluated through the barrier.

The integral of (20.2) can be evaluated analytically, and using BETHE's [18] notation we write,

$$P \approx e^{-2g(Z,R)\gamma(x)} \tag{20.3}$$

where $x = E/B$, B, the barrier height being equal to $2Ze^2/R$. $g(Z, R)$ is a function independent of alpha decay energy and equal to

$$g = \frac{2e}{\hbar}\sqrt{MZR}. \tag{20.4}$$

[1] M. VON LAUE: Z. Physik **52**, 726 (1928).

The function $\gamma(x)$ is given by

$$\gamma(x) = x^{-\frac{1}{2}} \arccos(x^{\frac{1}{2}}) - (1-x)^{\frac{1}{2}}. \tag{20.5}$$

BETHE [18] has given a graph of γ vs. x. For more accurate calculations we have included on p. 157 a five-place numerical tabulation (Table 9) of γ with argument y, where $y=(1-x)^{\frac{1}{2}}$. WINSLOW and SIMPSON [19] have plotted vs. x a function $f_0(x)$ which is $x^{\frac{1}{2}}$ times BETHE's $\gamma(x)$.

Formula (20.1) with (20.3) fairly satisfactorily correlates the variation of even-even ground state alpha decay rates with energy and atomic number.

One might in a semi-classical fashion include an extra factor in the penetrability expression to take into account special reflections due to the potential discontinuity at R. That is,

$$P \approx \frac{(E-U)^{\frac{1}{2}}}{(B-E)^{\frac{1}{2}}} e^{-2g\gamma}. \tag{20.6}$$

In optical analogy the new factor gives the transmission across a discontinuity in refractive index. The factor appears in more refined quantum-mechanical treatments in the one-body model.

The nature of the function in (20.3) can be illustrated by a TAYLOR expansion about $x=0$. Thus,

$$\log P = \frac{\sqrt{2MB}\,R}{\hbar}\left[\frac{\pi B^{\frac{1}{2}}}{E^{\frac{1}{2}}} - 4 + \frac{E}{B} - \frac{3E^2}{4B^2} + \cdots\right]. \tag{20.7}$$

From the above we see that the main energy dependence of $\log P$ is an inverse square root dependence. Such is the mathematical plausibility argument why plots of log alpha half-life vs. $E^{-\frac{1}{2}}$ should be nearly linear for constant Z (cf. Sect. 16 and Fig. 14).

There is an extensive literature on alpha decay theory, and a variety of special treatments in various degrees of approximation have been published. The better one-body treatments do not bring in any large corrections to the semi-classical formula (20.1), with (20.6).

The infinitely steep walls of the idealized potential are not realistic, and there have been treatments on sloping wall and rounded potentials by several authors[1] [17], [20].

A slight increase of the barrier has recently been suggested by CORBEN[2] as arising from vacuum polarization effects.

21. Concerning applications to experimental data. It has been traditional in applying alpha decay formulas to experimental data to leave the channel radius or "effective nuclear radius for alpha decay" as a parameter to be calculated. These alpha decay radius values from one-body models are generally of the same order of magnitude but slightly smaller (see Table 8) than radii from cross-section experiments involving alpha particles as projectiles (see Table 4). Some representative results for channel radii from various alpha decay theories are given in Table 3. The values are strongly model-dependent, and there is still uncertainty on internal details giving rise to the frequency factor f. Until the fundamental process of alpha particle formation is better understood, such calculated "alpha decay radii" must be viewed with reservations.

With the present limited understanding of the intra-nuclear mechanics of the alpha decay process, it seems more logical, as WINSLOW [16] has argued, to

[1] M. L. CHAUDHURY: Phys. Rev. **88**, 137 (1952).
[2] H. C. CORBEN: Bull. Amer. Phys. Soc., Ser. II **1**, 181 (1956).

Table 3. *Effective nuclear radii from alpha decay rate theories.*

Model	Apparent radius at $A=232$ (10^{-13} cm)	Reference
One body model (PRESTON) (lowest virtual level, $f\sim 3\cdot 10^{20}\,\text{sec}^{-1}$)	9.4	I. KAPLAN: Phys. Rev. **81**, 962 (1951) [*14*].
One body model (BISWAS-PATRO) ($f=v/R\sim 2\cdot 10^{21}\,\text{sec}^{-1}$)	9.1	S. BISWAS and J. PATRO: Indian J. Phys. **22**, 540 (1948). I. PERLMAN and T. J. YPSILANTIS: Phys. Rev. **79**, 30 (1950)
Extreme many body model (BETHE) ($f\sim 10^{15}\,\text{sec}^{-1}$)	12.6	[*18*]

Table 4. *Effective nuclear radii from alpha particle cross section studies.*

Type of determination	Radius expression (in 10^{-13} cm)	Corresponding radius at $A=232$ (10^{-13} cm)	Reference
Alpha scattering (20–40 Mev)	$\sim(1.4–1.5)\,A^{\frac{1}{3}}+2.5$	11.1–11.7	D. L. HILL, Vol. XXXIX, this Encyclopedia.
Total inelastic alpha cross sections (240 Mev)	$1.84\,A^{\frac{1}{3}}+0.35$	11.7	MILLBURN, BIRNBAUM, CRANDALL and SCHECHTER: Phys. Rev. **95**, 1268 (1954).
Alpha-fission cross sections (15–40 Mev)	—	10.3	J. JUNGERMANN: Phys. Rev. **79**, 640 (1950).

summarize numerical calculations from experiment by some parameter not dependent on intra-nuclear assumptions. One such parameter is the squared normalized alpha wave amplitude near the nuclear surface[1]. For these calculations WINSLOW makes use of just that part of alpha decay theory which is undisputed and is common to all models. From a tabulation of these experimental wave amplitudes alone one may hope to draw certain conclusions, but the tabulation may also serve as a point of departure for any more detailed study of the internal mechanism of alpha decay. The squared alpha wave amplitudes when corrected for centrifugal barrier and non-central interaction effects we shall call reduced transition probabilities *(RTP)*, and they bear a close analogy to the reduced transition probabilities for gamma transitions or ft values for beta transitions, where one makes use of the general solutions to the problem without assumptions about the finer details of the intra-nuclear parts of the processes

22. COULOMB wave functions. Let us now develop further the theory of alpha decay in central fields and in particular introduce the standard notation and mathematical relations involving the COULOMB functions.

The general time-independent wave equation for an alpha particle in a repulsive nuclear COULOMB potential may be written in spherical polar coordinates (center of mass system) as follows:

$$H\psi = E\psi \tag{22.1}$$

$$-\frac{\hbar^2}{2M}\nabla_{\boldsymbol{r}}^2\psi(\boldsymbol{r}) + \frac{2Ze^2}{r}\psi(\boldsymbol{r}) = E\psi(\boldsymbol{r}) \tag{22.2}$$

where M is the reduced mass $\frac{M_\alpha M_r}{M_\alpha + M_r}$ of the system, Ze is the charge of the recoil nucleus (of mass M_r), E is the total energy of the system (usually includes a

[1] We shall call this quantity the "surface probability".

correction to the experimental alpha plus recoil energy to get the decay energy that a bare nucleus with no orbital electrons would exhibit), and ∇_r^2 is the Laplacian operator. Since the potential is purely central, i.e., a function of the radial distance alone, the solutions may be conveniently expressed as

$$\psi_{nLm}(\boldsymbol{r}) = r^{-1} R_n(r)\, Y_L^m(\vartheta, \varphi) \tag{22.3a}$$

where $Y_L^m(\vartheta, \varphi)$ are the normalized spherical harmonics.

With such a substitution the radial function must satisfy the equation

$$-\frac{\hbar^2}{2M}\frac{d^2 R_n}{dr^2} + \left[\frac{2Ze^2}{r} - E + \frac{\hbar^2 L(L+1)}{2Mr^2}\right] R_n = 0. \tag{22.3b}$$

This equation may be rearranged to a standard form involving the dimensionless parameters ϱ and η, where $\varrho = kr$ and $\eta = \frac{Z_\alpha Z e^2}{\hbar v}$ with k, the wave number, defined as $k = \frac{(2ME)^{\frac{1}{2}}}{\hbar}$

$$\frac{d^2 u}{d\varrho^2} + \left[1 - \frac{2\eta}{\varrho} - \frac{L(L+1)}{\varrho^2}\right] u = 0. \tag{22.4}$$

Two linearly independent solutions of this COULOMB equation (the confluent hyper-geometric equation) are usually defined, a solution $F_L(\eta, \varrho)$ regular at the origin and a solution $G_L(\eta, \varrho)$ irregular at the origin. Many of the properties and approximations to these functions have been recently summarized by FRÖBERG[1]. Tables of the functions for small values of ϱ and η have been published, the most extensive being those of the U.S. Bureau of Standards Group[2]. The tables do not extend nearly to the range of η values for the alpha-emitting nuclei ($\eta \sim 20$), but there are good approximation methods for calculating these functions. For alpha wave functions near the nuclear surface the most completely worked out approximation method appears to be the RICCATI I method, as treated by ABRAMOWITZ[3]. The equations for determination of F_0 and G_0 are given by FRÖBERG[1] in his Eqs. (9.1) to (9.3).

Asymptotically, as $\varrho \to \infty$

$$\left.\begin{aligned} F_L &\to \sin\left(\varrho - \eta \log 2\varrho - \frac{L\pi}{2} + \sigma_L\right), \\ G_L &\to \cos\left(\varrho - \eta \log 2\varrho - \frac{L\pi}{2} + \sigma_L\right) \end{aligned}\right\} \tag{22.5}$$

where

$$\sigma_L = \arg \Gamma(i\eta + L + 1).$$

In any rigorous alpha decay treatment we will usually be interested in the linear combination $G_L + iF_L$, which represents a pure outgoing wave. The Wronskian relation is

$$G_L F_L' - F_L G_L' \equiv 1 \tag{22.6}$$

where primes denote derivatives with respect to ϱ.

More widely used approximations for the COULOMB functions as applied to alpha decay theory are the JWKB approximation and the steepest descents approximation to the contour integral form for the COULOMB functions.

---

[1] C. E. FRÖBERG: Rev. Mod. Phys. **27**, 399 (1955).
[2] Tables of COULOMB Wave Functions, Vol. I, NBS, Appl. Math. Ser. **17** (Washington, D.C., 1952).
[3] M. ABRAMOWITZ: Quart. Appl. Math. **7**, 75 (1949).

The JWKB approximation yields the following expansion for the logarithmic derivatives

$$\left.\begin{matrix} G'_L/G_L \text{ upper sign} \\ F'_L/F_L \text{ lower sign} \end{matrix}\right\} = S_0 + S_1 + S_2 + \cdots$$

with $S_0 = \mp \varkappa$

$$\left.\begin{aligned} S_1 &= -\frac{\varkappa'}{2\varkappa}, \\ S_2 &= \pm \frac{3(\varkappa')^2}{8\varkappa^3} \mp \frac{\varkappa''}{4\varkappa^2}, \\ \varkappa &= +\left[\frac{2\eta}{\varrho} + \frac{L(L+1)}{\varrho^2} - 1\right]^{\frac{1}{2}}. \end{aligned}\right\} \tag{22.7}$$

where

The second order and higher terms have generally been neglected in alpha decay treatments. Thus, to first order, after applying JWKB turning point formulas[1] at the transition point ($\varkappa = 0$), we write in the barrier region,

$$\left.\begin{aligned} G_L &= \varkappa^{-\frac{1}{2}} e^{\int_p^\tau \varkappa \, d\xi}, \\ F_L &= \frac{1}{2} \varkappa^{-\frac{1}{2}} e^{-\int_p^\tau \varkappa \, d\xi} \end{aligned}\right\} \tag{22.8}$$

where τ is the value of ϱ for which $\varkappa = 0$.

As discussed earlier in connection with Eq. (20.2), the $L = 0$ integral has a simple analytical expression (20.3).

More generally for all L, after performing the integration, the result can be expressed, in agreement with THOMAS' [17] Eq. (28) and WINSLOW'S [21] Eq. (11), (see also BETHE'S [18] Eq. (630) to (632)) as follows:

$$\left.\begin{aligned} G_L(\eta, \varrho) &= \varkappa^{-\frac{1}{4}} (\eta^2 + l^2)^{-\frac{l}{2}} \left(l\varkappa + \frac{l^2}{\varrho} + \eta\right)^l \\ &\times \exp\left[\pi\eta - \eta \arccos \frac{\eta - \varrho}{\sqrt{\eta^2 + l^2}} - \varrho\varkappa\right] \end{aligned}\right\} \tag{22.9}$$

where $\varkappa$ is defined above (22.7) and the LANGER[2] modification substituting for l the value $L + \frac{1}{2}$ [and substituting $(L + \frac{1}{2})^2$ for $L(L+1)$ in Eq. (22.7) for $\varkappa$] is recommended. WINSLOW and SIMPSON [21] have shown the accuracy of the above approximation for the values of η and ϱ encountered in alpha decay theory to be within 0.1%. We shall not bother to reproduce the expression for F_L, since it is so much smaller than G_L that we generally only have need for G_L and the logarithmic derivative G'_L/G_L in practical alpha decay formulations.

The ratio $(G_L/G_0)^{-2}$ is of some interest, since it enters as the centrifugal barrier reduction of the penetrability. Values calculated by WINSLOW and SIMPSON [21] and by THOMAS [17] are given in Table 5. The THOMAS example corresponds to emission of a 5 Mev alpha particle by uranium from a radius $R = 9.6 \times 10^{-13}$ cm. PRESTON [14] and THOMAS [17] derive frequency factors that increase with L for the literal one-body model. With such a model the decay rate is predicted to increase for small L values. Here in Table 5 we are concerned only with the barrier penetrability, a monotonically decreasing function of L.

[1] L. I. SCHIFF: Quantum Mechanics. New York: McGraw-Hill 1949. Eq. (28.17) and (28.18).

[2] R. E. LANGER: Phys. Rev. 51, 669 (1937).

Table 5. *Values of centrifugal reduction factor* $(G_L/G_0)^{-2}$.

η	ϱ	$L=0$	1	2	3	4	5	6	References
20.0	10.0	1.000	0.844	0.601	0.363	0.1854	0.0806	0.0299	[21]
25.4	9.42	1.000	0.854	0.588	0.354	0.181	0.082	—	[17]

It can be shown to a rough approximation that the logarithm of the reduction factor is inversely proportional to the square roots of the atomic number and the nuclear radius but is not very dependent on decay energy. A rough expression is

$$\left(\frac{G_L}{G_0}\right)^{-2} \approx e^{-\frac{2(L+\frac{1}{2})^2}{\sqrt{2\eta\varrho}}} = e^{-\frac{\hbar(L+\frac{1}{2})^2}{\sqrt{MRZe^2}}}. \tag{22.10}$$

The values of Table 5 should be fairly representative for the heavy element alpha emitters. The corresponding values given by DEVANEY [22] and by and WEISSKOPF [23] in their text are in error, WINSLOW and SIMPSON [21] having pointed out that their factors are just the squares of what they should be.

23. Concerning time-dependent alpha decay treatments. While the majority of alpha decay theoretical treatments do not specifically introduce time-dependent wave functions, there are additional interesting insights to be gained by such considerations. RASETTI [24] has given an interesting one-body alpha decay treatment embodying time-dependent wave functions. WINSLOW and SIMPSON [25] have given a more detailed treatment along the same general lines as RASETTI. KEMBLE [26] has also treated alpha decay in this general fashion. At time zero an initial alpha particle wave function is constructed of finite spatial extension in a quasi-stationary state within the nuclear potential well. This initial state is expressed as a wave packet formed by the infinitely extending regular solutions in the given potential, the energy distribution being about the virtual level within the potential well. A time-dependent wave function results, and this solution consists in the region outside the nucleus of purely outgoing COULOMB waves decaying in time at any given spatial position according to the exponential decay law. The reader is referred to these original works for details.

24. Treatment in collision matrix formalism. A rigorous but simple formulation of alpha decay along the lines of the time-independent R-matrix theory of WIGNER and collaborators[1] for nuclear reactions has been made by R. G. THOMAS [17]. The decay constants of radioactive states in this formulation are $2/\hbar$ times the imaginary parts of the energies of the poles of the collision matrix. THOMAS first derives quite generally a formula giving the alpha decay constant in terms only of behavior of the internal alpha wave solution at the nuclear surface, as expressed by a parameter $\delta_{\lambda L}^2$ called the reduced width for the derivative of the radial alpha wave function (λ-th group with L angular momentum), is simply related to the reduced width $\gamma_{\lambda L}^2$ (commonly used in nuclear reaction applications) as follows:

$$\delta_{\lambda L} = \varrho\,\frac{F_L F_L' + G_L G_L'}{F_L^2 + G_L^2}\,\gamma_{\lambda L} \approx \varrho\,\frac{G_L'}{G_L}\,\gamma_{\lambda L}. \tag{24.1}$$

The following equations express the decay rate in his formalism:

$$\lambda_L = \frac{2\varrho\,\gamma_L^2}{\hbar(G_L^2+F_L^2)} \approx \frac{2\delta_L^2}{\varrho\,\hbar\,(G_L')^2} = \frac{2\delta_L^2}{\varrho\,\hbar\left(\frac{G_L'}{G_L}\right)^2 G_L^2}. \tag{24.2}$$

[1] E. P. WIGNER and L. EISENBUD: Phys. Rev. **72**, 29 (1947). — T. TEICHMAN and E. P. WIGNER: Phys. Rev. **87**, 123 (1952).

We shall consider what various alpha decay models yield for the value of δ^2. The formal definition of δ_λ, as given by THOMAS, is

$$\delta_\lambda = \gamma_L + \sqrt{\frac{R\hbar^2}{2M}} \int\limits_S \psi_c^* \operatorname{grad}_n X_\lambda \, dS. \tag{24.3}$$

The integral is carried over the nuclear surface at radius R; $X_\lambda = \frac{u(r)}{r} \; Y_L^m(\vartheta,\varphi)$ is the alpha wave function in the channel; φ_c is a product wave function including internal coordinates of the alpha particle, the residual nucleus, and a spherical harmonic if there is angular momentum associated with that alpha channel. γ_λ, the reduced width, will be generally much smaller than δ_λ and THOMAS neglects it for applications of the formula. Formally,

$$\gamma_\lambda = \sqrt{\frac{\hbar^2}{2MR}} \int\limits_S \psi_c^* X_\lambda \, dS. \tag{24.4}$$

Reducing (24.3) to a more directly applicable form we may write

$$\delta^2 \approx \frac{R\hbar^2}{2M} \left| \frac{du}{dr} \right|^2_{r=R} |\psi_c|^2. \tag{24.5}$$

The factor $|\psi_c|^2$ may be regarded as a formation probability factor and would be unity in a one-body treatment.

For the usual square-well one-body model THOMAS' expression is

$$\left. \begin{aligned} \delta_L^2 &= \frac{\hbar^2 z_L^2 \varrho^2 \left(\frac{G_L'}{G_L}\right)^2}{M R^2 \left[\varrho\left(\frac{G_L'}{G_L}\right) + L\right]\left[\varrho\left(\frac{G_L'}{G_L}\right) - L - 1\right] + z_L^2} \\ &\approx \frac{\hbar^2}{MR^2} z_L^2 \end{aligned} \right\} \tag{24.6}$$

where z_L is a zero (usually taken as the first) of the spherical BESSEL function $j_{L+\frac{1}{2}}(z)$. The first z_0 is simply π, and substitution of this value in the above equations for $L=0$ leads to results in agreement with other one-body treatments. That is $\delta_0^2 \approx 1.2$ Mev.

It may be of some interest to consider the reduced widths obtained from higher virtual levels than the lowest. (We discuss in Sect. 36 a recent estimate of 140 Mev for the kinetic energy of the alpha particle in the nucleus, whereas the lowest s-level is at about 0.5 Mev.) From Eq. (24.6) for $L=0$ we substitute a typical average value of -20 for $\varrho\, G'/G$ and have

$$\delta^2 \approx \frac{\hbar^2 z^2}{MR^2\left(1 + \frac{z^2}{400}\right)} \approx 0.12 \frac{z^2}{1 + \frac{z^2}{400}} \text{Mev}. \tag{24.7}$$

For $L=0$ the zeroes of the BESSEL function are given by $z = n\pi$, with n a positive integer. For the lowest solution $\delta^2 \approx 1.2$ Mev, increasing for higher solutions and approaching the limit $\delta^2 \to 48$ Mev.

The formula for the $L=0$ reduced width is easily modified for the WINSLOW surface well model [*21*], [*27*] by substituting ΔR, the width of the surface well, for R, the nuclear radius. Evidently higher reduced widths than in the ordinary model will be obtained in the one body form of the surface well model, since $\Delta R < R$.

δ^2 is quite simply connected with the frequency factor f used in the early part of this chapter. Indeed, the reduced derivative width is approximately PLANCK's constant h times the semi-classical frequency factor.

$$\delta^2 \approx 2\pi \hbar f. \tag{24.8}$$

δ^2 is 2π times BETHE's [18] G_α, the "α width in the absence of barrier" [cf. BETHE's Equation (593)].

25. Influence of the atomic electron cloud. The alpha decay rate treatments are all formulated in terms of the decay of isolated nuclei, stripped of electrons. For many years the influence of the electron cloud was quite ignored. Credit for focussing attention on this question goes to AMBROSINO and PIATIER[1], who in 1951 proposed that additive corrections to the observed alpha decay energy should be made before substitution into decay rate formulas derived for bare nuclei. Subsequent works of others[2] ([21], p. 14 and [17]) generally concur in the necessity for energy correction but support corrections of about half the size proposed by AMBROSINO and PIATIER.

To understand the nature of these corrections let us imagine we have an assemblage of bare nuclei undergoing alpha decay with a measurable rate and alpha particle energy. What, if any, change will there be in these two observables as electrons are added to form normal atoms?

First, the decay rate will hardly change at all upon addition of orbital electrons. One might predict theoretically a negligibly slight increase in decay rate. To the extent that some electronic charge lies within the classical turning radius for alpha decay (4×10^{-12} cm) the potential barrier for alpha decay will be slightly lowered. A rough estimate with relativistic K-electron wave functions indicated the effect is negligible.

Second, the average observed alpha particle energy will decrease by an energy of the order of 40 kev in the heavy element region. In the main, this energy simply represents the work done against coulombic attraction of electrons during removal of the positively charged alpha particle from the center of the diffuse electron cloud to infinity. This is the potential energy of the alpha particle in the atomic electron field of the parent atom, and from FOLDY's work[3] is well represented by the expression $65.3 Z_i^{\frac{7}{5}}$ ev[4] (for Cm, 39.4 kev). If the alpha moved out with a velocity large compared to the average velocity of the orbital electrons, this potential would represent the difference between observed decay energy (alpha particle plus recoil nucleus kinetic energy) of the bare nucleus and that of the atom. However, the alpha actually moves at a lower velocity than the more tightly bound electrons, and some adiabatic adjustment of the electron wave functions is to be expected during alpha emission. With complete adiabatic readjustment, according to the results of SERBER and SNYDER[5] is associated an energy of $91.4 Z_i^{\frac{2}{5}}$ (for Cm, -567 ev). By THOMAS' estimate for a 4.5 Mev alpha particle from thorium there should be about 87% of the full adiabatic correction realized on the average. The best value for the alpha decay energy difference between bare nucleus and atom would then be

$$\varDelta \mathrm{E} = 65.3 Z_i^{\frac{7}{5}} - 80 Z_i^{\frac{2}{5}} \,\mathrm{ev}. \tag{25.1}$$

[1] G. AMBROSINO and H. PIATIER: C. R. Acad. Sci., Paris **232**, 400 (1951).

[2] RASMUSSEN, THOMPSON and GHIORSO: University of California Radiation Laboratory Report 1473 (1951).

[3] L. L. FOLDY: Phys. Rev. **83**, 397 (1951).

[4] Since FOLDY's formula is based on non-relativistic HARTREE calculations of total electron binding energy, his formula may give slightly small answers for heavy elements.

[5] R. SERBER and H. S. SNYDER: Phys. Rev. **87**, 152 (1952).

The only important change necessary to use the conventional decay rate formulas is to add the above energy difference to the observed total alpha decay energy before substituting into the rate formulas.

The 13% of the adiabatic energy term (70 ev) not realized in the alpha decay energy correction represents the average electronic excitation of the recoil atom. In principle this average energy of excitation plus the 80 ev total energy of ionization of helium should be added to the experimental alpha decay energy for relative atomic mass calculations [5]. The average energy of excitation is usually to be associated with ionization of outer-shell electrons.

Occasionally the alpha emission process will lead to ejection of a tightly bound electron. The ejection of K and L electrons has received attention both theoretically and experimentally, and some of the results are summarized in Table 6.

Table 6. *Some theoretical and experimental K- and L-shell ionization probabilities for alpha decay of* Po^{210}.

Shell	Theoretical		Experimental		
	MIGDAL[1]	LEVINGER[2]	BARBER and HELM[3]	RUBINSON and BERNSTEIN[4]	ROY and GOES[5]
K	$2.6 \cdot 10^{-6}$	$1.0 \cdot 10^{-7}$	$1.4 \cdot 10^{-6}$	—	—
Total L	$0.24 \cdot 10^{-4}$	$0.50 \cdot 10^{-4}$	—	$8.8 \cdot 10^{-4}$	—
Total (all shells)	—	—	—	—	$2.7 \cdot 10^{-3}$

MIGDAL [28], who first treated the problem of ionization induced by alpha decay, carried out a time-dependent perturbation treatment which effectively makes a semi-classical electric multipole expansion of the field from the moving alpha particle, similar to the semi-classical treatments of the COULOMB excitation theory. MIGDAL used non-relativistic unscreened hydrogen-like electron wave functions, assumed uniform alpha particle velocity, and was able to derive analytical expressions for the ionization probability and the energy distribution of ejected electrons in the continuum.

MIGDAL considered only electric dipole terms and obtained a result of K ionizations per alpha disintegration of

$$W_k = \left(\frac{v}{v_k}\right)^2 \frac{2.2}{Z^2} \tag{25.2}$$

where v is the velocity of the alpha particle, v_k the average velocity of a K-electron, and Z the atomic number of the alpha emitter.

LEVINGER[2] in addition treated electric quadrupole terms, which he found to be of significance for the p electrons of the L-shell. His treatment differs most sharply from MIGDAL'S in his reduction of the dipole terms by a factor of about 25, due to special recoil considerations. SCHAEFFER[6] has treated the problem for K-electrons and estimates the dipole term as even slightly enhanced over MIGDAL'S value. Recently SCHWARTZ[7] has given theoretical arguments in support of MIGDAL'S formulation of the dipole interaction over LEVINGER'S.

SCHAEFFER'S treatment finds monopole interaction of some importance and for the constant alpha velocity treatment gets a total ionization probability

[1] A. MIGDAL: J. Phys. USSR. **4**, 449 (1941).
[2] J. S. LEVINGER: Phys. Rev. **90**, 11 (1953). — J. Phys. Radium **16**, 556 (1955).
[3] W. C. BARBER and R. H. HELM: Phys. Rev. **86**, 275 (1952).
[4] W. RUBINSON and W. BERNSTEIN: Phys. Rev. **82**, 334 (1951); **86**, 545 (1952).
[5] R. ROY and M. L. GOES: C. r. Acad. Sci., Paris **237**, 1515 (1953).
[6] G. W. SCHAEFFER: M. A. Thesis, University of Toronto, Canada, 1953 (unpublished).
[7] H. M. SCHWARTZ: Phys. Rev. **100**, 135 (1955).

about twice that of MIGDAL. He then corrects for the variation in alpha particle velocity, lowering the result by about a factor of two, numerically about equalling MIGDAL'S value. SCHAEFFER feels that a relativistic treatment might further lower the probability in closer accord with the experimental value of BARBER and HELM[1].

LEVINGER[2] has estimated that correction for electron screening effects should give an increase of about 40% in the K ionization.

We can generally conclude that the theory (except LEVINGER'S) and experiment are in agreement within a factor of two for K-shell ionization.

The discrepancy (see Table 6) regarding L-shell ionization is still unresolved. The theoretical values of MIGDAL and of LEVINGER are an order of magnitude lower than the experimental value of RUBINSON and BERNSTEIN[3]. LEVINGER feels that relativistic and screening corrections will not be very large for the L-electron problem and cannot explain the discrepancy.

The coincidence experiments of LAGASSE and DOYEN[4] are a neat confirmation of the MIGDAL theoretical energy distribution of ejected electrons.

26. Some numerical comparisons with decay rate data of even-even nuclei. In Table 7 are presented data and calculated values of two parameters measuring the intrinsic decay rate calculated from experimental decay energy and rate data for even-even nuclei. An unusual procedure is used here. The ground state alpha groups are not treated separately but the total half-lives for alpha decay were used in order to put all alpha emitters on an even basis for comparison. (Some cases probably have unresolved fine structure of appreciable intensity, Cf^{248}, Cm^{240}.) The alpha particle energy for the ground-state transition, corrected for electron screening, appears in Col. 3. The logarithm to base 10 of the square of the irregular COULOMB function G_0 evaluated at $R = 9.3 \times 10^{-13}$ cm for heavy elements and at 8.0×10^{-13} cm for rare earths is given in Col. 4. In Col. 5 is the logarithmic derivative of G_0^2 (R in units of 10^{-13}) at $R = 9.3$. Finally, the parameters measuring intrinsic decay rate are given. In Col. 6 are WINSLOW'S [*16*] surface probabilities; in Col. 7 are THOMAS' [*17*] reduced derivative widths δ^2.

In Table 8 are summarized ASARO'S[5] calculations of the effective nuclear radius using the alpha decay formulas of PRESTON [*14*] [Eqs. (4.3) and (4.4)] or using a slight modification due to KAPLAN[6] in which PRESTON'S Eq. (4.3) is supplanted by the equation

$$\mu^2 = \frac{0.52}{Q_\alpha\,(\text{Mev})},$$

μ^2 being the ratio of the alpha energy inside the well to that at infinity. Unlike the calculations of Table 7 these calculations use the partial alpha half life for the ground state group alone, and in the cases of Cf^{248} and Cm^{240}, where ion chamber measurements have not resolved the expected alpha fine structure, reasonable branching ratios were assumed. In the last columns are given the r_0 values from which the effective nuclear radius[7] of the alpha decay daughter (mass number A) can be calculated by the formula

$$R = r_0 A^{\frac{1}{3}} 10^{-13}\,\text{cm}.$$

[1] W. C. BARBER and R. H. HELM: Phys. Rev. **86**, 275 (1952).
[2] J. S. LEVINGER: Phys. Rev. **90**, 11 (1953). — J. Phys. Radium **16**, 556 (1955).
[3] W. RUBINSON and W. BERNSTEIN: Phys. Rev. **82**, 334 (1951); **86**, 545 (1952).
[4] A. LAGASSE and J. DOYEN: C. r. Acad. Sci., Paris **239**, 670 (1954).
[5] F. ASARO: Unpublished calculations (1955).
[6] I. KAPLAN: Phys. Rev. **81**, 962 (1951).
[7] For nuclear radii cf. D. L. HILL'S article in Vol. XXXIX of this Encyclopedia. Cf. also Sect. 36, this article.

Table 7. *Alpha surface probabilities and reduced derivative widths for even-even nuclei.* (Evaluated at $R=9.3$ for heavy region and $R=8.0$ for rare earths; R in units of 10^{-13} cm.)

Alpha Emitter	Log $t_{\frac{1}{2}}\alpha$ (sec)	E_α†	Log $G_0^2(R)$	$-\frac{d}{dR}$ (Log G_0^2)	Surface probability	δ^2 (Mev)
Fm^{254}	4.061	7.242	24.49	1.778	0.088	1.96
Cf^{252}	7.821	6.154	28.32	1.797	0.120	2.74
Cf^{250}	8.472	6.066	28.76	1.801	0.072	1.65
Cf^{248}	7.298	6.302	27.60	1.791	0.072	1.64
Cf^{246}	5.109	6.795	25.34	1.771	0.059	1.31
Cm^{244}	8.782	5.839	29.00	1.785	0.062	1.39
Cm^{242}	7.147	6.151	27.39	1.772	0.064	1.42
Cm^{240}	6.365	6.291	26.70	1.767	0.080	1.76
Pu^{242}	13.198	4.932	33.42	1.798	0.069	1.57
Pu^{240}	11.317	5.202	31.66	1.787	0.088	1.98
Pu^{238}	9.451	5.535	29.66	1.773	0.063	1.39
Pu^{236}	7.930	5.790	28.26	1.762	0.080	1.82
Pu^{234}	5.908	6.230	26.05	1.744	0.051	1.08
U^{238}	17.151	4.219	37.67	1.802	0.146	3.35
U^{236}	14.877	4.538	35.11	1.789	0.073	1.65
U^{234}	12.894	4.802	33.18	1.778	0.081	1.81
U^{232}	9.366	5.357	29.65	1.756	0.076	1.65
U^{230}	6.255	5.928	26.56	1.733	0.076	1.60
U^{228}	2.842	6.709	23.04	1.697	0.055	1.12
Th^{232}	17.642	4.031	38.14	1.786	0.143	3.20
Th^{230}	12.402	4.719	32.66	1.757	0.075	1.64
Th^{228}	7.778	5.458	28.06	1.725	0.074	1.55
Th^{226}	3.268	6.373	23.55	1.687	0.070	1.40
Ra^{226}	10.709	4.813	30.88	1.728	0.061	1.29
Ra^{224}	5.498	5.717	25.67	1.689	0.057	1.14
Ra^{222}	1.580	6.590	21.94	1.652	0.082	1.58
Em^{222}	5.520	5.521	25.69	1.671	0.057	1.12
Em^{220}	1.736	6.317	21.99	1.635	0.065	1.22
Em^{218}	−1.721	7.162	18.80	1.597	0.120	2.06
Em^{216}	—	8.045	16.07	1.555	—	—
Em^{214}	—	—	—	—	—	—
Em^{212}	3.140	6.297	22.05	1.636	0.0030	0.056
Em^{210}	3.987	6.071	23.02	1.646	0.0039	0.075
Em^{208}	3.799	6.173	22.56	1.641	0.0019	0.0366
Po^{218}	2.262	6.032	22.29	1.621	0.040	0.741
Po^{216}	−0.801	6.808	19.21	1.586	0.036	0.640
Po^{214}	−3.786	7.714	16.27	1.543	0.038	0.63
Po^{212}	−6.517	8.810	13.40	1.489	0.025	0.395
Po^{210}	7.078	5.332	25.64	1.652	0.0014	0.0273
Po^{208}	7.966	5.142	26.67	1.660	0.0020	0.0388
Po^{206}	6.891	5.252	26.05	1.655	0.0056	0.109
Po^{204}	6.136	5.404	25.23	1.648	0.0049	0.0939
Po^{202}	5.193	5.624	24.12	1.638	0.0032	0.0614
Po^{200}	—	5.874	22.93	1.627	—	—
Gd^{148}	9.65	3.18	30.13	1.636	0.153	2.78
Sm^{146}	15.2	2.57	34.88	1.633	0.027	0.49
Nd^{144}	22.7	1.92	42.49	1.631	0.040	0.72

† E_α is particle energy plus screening correction used for these calculations (Mev).

Table 8. *Calculated alpha decay "nuclear radii" from one-body theory.*

Alpha emitter	Partial alpha half-life	Abundance gd. state group (α_0)	Partial α_0 half-life	Alpha particle energy (Gd. state)	Effective decay energy (Q_{eff})	Radius formula coefficient (r_0)	
						PRESTON	KAPLAN
Fm^{254}	3.3 h	0.83	4.0 h	7.22	7.378	1.48_0	—
Cf^{252}	2.2 y	0.845	2.6 y	6.112	6.252	1.512	—
Cf^{250}	10.0 y	0.83	12.1 y	6.024	6.163	1.495	—
Cf^{248}	250 d	0.80†	313 d	6.26	6.404	—	1.497
Cf^{246}	35.7 h	0.78	45.8 h	6.753	6.906	1.492	1.495
Cm^{244}	19 y	0.767	25 y	5.798	5.935	1.498	—
Cm^{242}	162.5 d	0.737	220.5 d	6.110	6.253	1.502	1.504
Cm^{240}	26.8 d	0.70†	38.3 d	6.27	6.416	1.503	—
Pu^{242}	3.76×10^5 y	0.74	5.08×10^5 y	4.898	5.019	1.516	—
Pu^{240}	6580 y	0.755	8.71×10^3 y	5.162	5.289	1.520	1.522
Pu^{238}	89.6 y	0.72	1.24×10^2 y	5.495	5.628	1.509	—
Pu^{236}	2.7 y	0.689	3.9 y	5.763	5.901	—	1.51_3
U^{238}	4.49×10^9 y	0.77	5.84×10^9 y	4.182	4.290	—	1.548
U^{236}	2.39×10^7 y	0.73	3.28×10^7 y	4.499	4.613	—	1.523
U^{234}	2.48×10^5 y	0.74	3.35×10^5 y	4.763	4.883	—	1.532
U^{232}	73.6 y	0.68	108 y	5.318	5.448	1.527	—
U^{230}	20.8 d	0.679	30.6 d	5.884	6.026	1.532	—
Th^{232}	1.39×10^{10} y	0.76	1.83×10^{10} y	3.994	4.110	—	1.551
Th^{230}	8.0×10^4 y	0.763	1.05×10^5 y	4.682	4.801	1.534	—
Th^{228}	1.90 y	0.71	2.68 y	5.421	5.554	1.534	1.536
Th^{226}	30.9 m	0.79	39.1 m	6.330	6.480	1.542	—
Ra^{226}	1622 y	0.943	1720 y	4.777	4.898	—	1.545
Ra^{224}	3.64 d	0.948	3.84 d	5.681	5.819	—	1.546
Ra^{222}	38 s	0.95	40 s	6.551	6.706	1.545	—
Em^{222}	3.825 d	—	3.825 d	5.486	5.621	—	1.552
Em^{220}	54.5 s	0.997	54.7 s	6.282	6.432	—	1.560
Em^{218}	0.019 s	0.998	0.019 s	7.127	7.294	—	1.585
Em^{212}	23 m	—	23 m	6.262	6.417	—	1.445
Em^{210}	2.8 h	—	2.8 h	6.037	6.188	—	1.461
Po^{218}	3.05 m	—	3.05 m	5.998	6.143	—	1.543
Po^{216}	0.158 s	—	0.158 s	6.774	6.935	—	1.541
Po^{214}	1.637×10^{-4} s	—	1.637×10^{-4} s	7.680	7.859	1.537	1.545
Po^{212}	3.04×10^{-7} s	—	3.04×10^{-7} s	8.776	8.978	—	1.527
Po^{210}	138.4 d	—	138.4 d	5.299	5.435	—	1.422
Po^{208}	2.93 y	0.997	2.94 y	5.108	5.241	—	1.442
Po^{206}	180 d	—	180 d	5.218	5.354	—	1.460

A comparison between Tables 7 and 8 reveals, of course, the same fluctuations in all three parameters, "surface probability", δ^2, and r_0^2.

27. Summary of useful numerical formulas. Below are summarized a few formulas useful for numerical work on alpha decay rates. The nomenclature follows general practice. The numerical constants were calculated using 1955 revised values of COHEN, DUMOND, LAYTON, and ROLLETT[1]. A value of 4.00278 atomic mass units was used for the mass of the alpha particle, based on the mass

† Value assumed.

[1] COHEN, DUMOND, LAYTON and ROLLETT: Rev. Mod. Phys. **27**, 363 (1955).

of 4.003873 ± 0.000015 for the helium atom given by AJZENBERG and LAURITSEN[1].

Q_{eff}, the total alpha disintegration energy of a bare nucleus equals alpha particle energy (E) in the lab system plus recoil energy $\left(\frac{M_\alpha E}{M_r}\right)$ plus ΔE_{sc}, the orbital electron "screening" correction. Q_{eff} will be expressed in Mev in the following formulas.

$$\Delta E_{\text{sc}} = 65.3\,(Z+2)^{\frac{7}{5}} - 80\,(Z+2)^{\frac{2}{5}}\ \text{ev}, \tag{27.1}$$

where Z is the atomic number of the daughter nucleus. η, an argument of the standard COULOMB functions

$$\eta = \frac{2Ze^2}{\hbar v} = 0.630035_5 \frac{Z}{\sqrt{Q_{\text{eff}}}} \sqrt{\frac{A}{A+4}}, \tag{27.2}$$

where v is the velocity of separation of alpha particle and recoil nucleus for large separation distance and A is the atomic weight of the daughter. More precisely $\frac{A}{A+4}$ should be replaced by $\frac{M_r}{M_r+M_\alpha}$.

ϱ, an argument of the standard COULOMB functions

$$\varrho = k r.$$

k, the wave number for large separation distance r

$$k = \frac{M v}{\hbar} = 0.437\,56_6 \sqrt{\frac{A\,Q_{\text{eff}}}{A+4}} \cdot 10^{13}\ \text{cm}^{-1}. \tag{27.3}$$

x, the dimensionless ratio of total energy Q_{eff} to the potential energy $\frac{2Ze^2}{r}$

$$x = \frac{Q_{\text{eff}}\, r}{2Ze^2} = \frac{\varrho}{2\eta} = 0.347\,25_5 \frac{Q_{\text{eff}}\, r}{Z}, \tag{27.4}$$

where r is in units of 10^{-13} cm.

$$G_L(\varrho,\eta) = G(\varrho,\eta)\,\frac{G_L(\varrho,\eta)}{G(\varrho,\eta)} \text{ the irregular COULOMB function[2].}$$

$$\left.\begin{aligned} \operatorname{Log} G^2 &= \frac{\sqrt{8\varrho\eta}}{\log 10}\,\gamma(x) - \frac{1}{2}\operatorname{Log}\left(\frac{1-x}{x}\right) \\ &= 0.644\,96_1 \sqrt{\frac{ZrA}{A+4}}\,\gamma(x) - \frac{1}{2}\operatorname{Log}\left(\frac{1-x}{x}\right), \end{aligned}\right\} \tag{27.5}$$ [3]

$$\operatorname{Log}\left(\frac{G_L^2}{G^2}\right) = \frac{2}{\varrho \log 10}\left[\sqrt{x(1-x)} - \frac{x}{4\varrho(1-x)}\right]\left(L+\frac{1}{2}\right)^2 + \cdots. \tag{27.6}$$ [4]

The dimensionless logarithmic derivative is given as follows:

$$\frac{d \operatorname{Log} G_L}{d \operatorname{Log} r} = -\varrho\varkappa - \frac{\varrho}{2\varkappa}\frac{d\varkappa}{d\varrho}, \tag{27.7}$$

$$\varkappa = \left[\frac{2\eta}{\varrho} + \frac{(L+\frac{1}{2})^2}{\varrho^2} - 1\right]^{\frac{1}{2}}. \tag{27.8}$$

---

[1] F. AJZENBERG and T. LAURITSEN: Rev. Mod. Phys. **24**, 321 (1952).

[2] It is to be noted that in the LANGER modification of the WKB approximation employed G is not quite equal to G_0. At the nuclear radius for most heavy alpha emitters G_0 exceeds G by about 1%.

[3] The function γ and its first difference are tabulated to five decimal places in Table 9 with argument y, where $y = (1-x)^{\frac{1}{2}}$.

[4] This is essentially the logarithm of the centrifugal barrier reduction factor. The $(L+\frac{1}{2})^4$ and $(L+\frac{1}{2})^6$ terms of this expansion are given in WINSLOW's [21] Eq. (18). He has found for a representative case ($\eta = 20$, $\varrho = 10$) that in the above expansion of the logarithm the fourth power term is equal to 0.002 for $L = 3$, and the sixth power term is 0.001 for $L = 6$.

Table 9. *Table of* γ (y) $y=\sqrt{1-x}$ *where* $x=E/B$, $\gamma=x^{-\frac{1}{2}}\,\text{arc}\cos x^{\frac{1}{2}}-(1-x)^{\frac{1}{2}}$

y	$\gamma(y)$	δ_1	y	$\gamma(y)$	δ_1
0.750	0.53215	0.00350	0.811	0.80572	0.00578
0.751	0.53565	0.00352	0.812	0.81150	0.00584
0.752	0.53917	0.00355	0.813	0.81734	0.00588
0.753	0.54272	0.00357	0.814	0.82322	0.00594
0.754	0.54629	0.00361	0.815	0.82916	0.00600
0.755	0.54990	0.00363	0.816	0.83516	0.00605
0.756	0.55353	0.00366	0.817	0.84121	0.00611
0.757	0.55719	0.00368	0.818	0.84732	0.00617
0.758	0.56087	0.00372	0.819	0.85349	0.00623
0.759	0.56459	0.00374	0.820	0.85972	0.00629
0.760	0.56833	0.00377	0.821	0.86601	0.00635
0.761	0.57210	0.00380	0.822	0.87236	0.00640
0.762	0.57590	0.00383	0.823	0.87876	0.00647
0.763	0.57973	0.00386	0.824	0.88523	0.00654
0.764	0.58359	0.00389	0.825	0.89177	0.00660
0.765	0.58748	0.00392	0.826	0.89837	0.00666
0.766	0.59140	0.00396	0.827	0.90503	0.00673
0.767	0.59536	0.00398	0.828	0.91176	0.00679
0.768	0.59934	0.00401	0.829	0.91855	0.00687
0.769	0.60335	0.00405	0.830	0.92542	0.00693
0.770	0.60740	0.00408	0.831	0.93235	0.00700
0.771	0.61148	0.00411	0.832	0.93935	0.00707
0.772	0.61559	0.00414	0.833	0.94642	0.00715
0.773	0.61973	0.00418	0.834	0.95357	0.00722
0.774	0.62391	0.00421	0.835	0.96079	0.00729
0.775	0.62812	0.00424	0.836	0.96808	0.00737
0.776	0.63236	0.00428	0.837	0.97545	0.00744
0.777	0.63664	0.00431	0.838	0.98289	0.00753
0.778	0.64095	0.00435	0.839	0.99042	0.00760
0.779	0.64530	0.00438	0.840	0.99802	0.00768
0.780	0.64968	0.00442	0.841	1.00570	0.00776
0.781	0.65410	0.00446	0.842	1.01346	0.00785
0.782	0.65856	0.00449	0.843	1.02131	0.00793
0.783	0.66305	0.00453	0.844	1.02924	0.00801
0.784	0.66758	0.00457	0.845	1.03725	0.00810
0.785	0.67215	0.00460	0.846	1.04535	0.00819
0.786	0.67675	0.00465	0.847	1.05354	0.00828
0.787	0.68140	0.00468	0.848	1.06182	0.00838
0.788	0.68608	0.00472	0.849	1.07020	0.00846
0.789	0.69080	0.00476	0.850	1.07866	0.00856
0.790	0.69556	0.00480	0.851	1.08722	0.00865
0.791	0.70036	0.00485	0.852	1.09587	0.00876
0.792	0.70521	0.00488	0.853	1.10463	0.00885
0.793	0.71009	0.00493	0.854	1.11348	0.00895
0.794	0.71502	0.00497	0.855	1.12243	0.00905
0.795	0.71999	0.00501	0.856	1.13148	0.00916
0.796	0.72500	0.00506	0.857	1.14064	0.00927
0.797	0.73006	0.00510	0.858	1.14991	0.00938
0.798	0.73516	0.00514	0.859	1.15929	0.00948
0.799	0.74030	0.00519	0.860	1.16877	0.00959
0.800	0.74549	0.00524	0.861	1.17837	0.00971
0.801	0.75073	0.00528	0.862	1.18808	0.00983
0.802	0.75601	0.00533	0.863	1.19791	0.00994
0.803	0.76134	0.00537	0.864	1.20785	0.01007
0.804	0.76671	0.00543	0.865	1.21792	0.01019
0.805	0.77214	0.00547	0.866	1.22811	0.01031
0.806	0.77761	0.00552	0.867	1.23842	0.01045
0.807	0.78313	0.00557	0.868	1.24887	0.01057
0.808	0.78870	0.00563	0.869	1.25944	0.01070
0.809	0.79433	0.00567	0.870	1.27014	0.01084
0.810	0.80000	0.00572	0.871	1.28098	0.01098

Table 9. (Continued.)

y	$\gamma(y)$	δ_1	y	$\gamma(y)$	δ_1
0.872	1.29196	0.01112	0.910	1.84750	0.01959
0.873	1.30308	0.01127	0.911	1.86709	0.01994
0.874	1.31435	0.01140	0.912	1.88703	0.02029
0.875	1.32575	0.01156	0.913	1.90732	0.02067
0.876	1.33731	0.01171	0.914	1.92799	0.02104
0.877	1.34902	0.01187	0.915	1.94903	0.02145
0.878	1.36089	0.01202	0.916	1.97048	0.02185
0.879	1.37291	0.01219	0.917	1.99233	0.02226
0.880	1.38510	0.01235	0.918	2.01459	0.02269
0.881	1.39745	0.01252	0.919	2.03728	0.02314
0.882	1.40997	0.01269	0.920	2.06042	0.02359
0.883	1.42266	0.01287	0.921	2.08401	0.02407
0.884	1.43553	0.01305	0.922	2.10808	0.02455
0.885	1.44858	0.01323	0.923	2.13263	0.02505
0.886	1.46181	0.01342	0.924	2.15768	0.02558
0.887	1.47523	0.01362	0.925	2.18326	0.02611
0.888	1.48885	0.01381	0.926	2.20937	0.02667
0.889	1.50266	0.01401	0.927	2.23604	0.02724
0.890	1.51667	0.01421	0.928	2.26328	0.02784
0.891	1.53088	0.01443	0.929	2.29112	0.02846
0.892	1.54531	0.01464	0.930	2.31958	0.02911
0.893	1.55995	0.01487	0.931	2.34867	0.02976
0.894	1.57482	0.01508	0.932	2.37843	0.03044
0.895	1.58990	0.01532	0.933	2.40887	0.03116
0.896	1.60522	0.01556	0.934	2.44003	0.03190
0.897	1.62078	0.01580	0.935	2.47193	0.03266
0.898	1.63658	0.01604	0.936	2.50459	0.03347
0.899	1.65262	0.01631	0.937	2.53806	0.03430
0.900	1.66893	0.01656	0.938	2.57236	0.03517
0.901	1.68549	0.01683	0.939	2.60753	0.03606
0.902	1.70232	0.01711	0.940	2.64359	0.03701
0.903	1.71943	0.01739	0.941	2.68060	0.03799
0.904	1.73682	0.01767	0.942	2.71859	0.03901
0.905	1.75449	0.01798	0.943	2.75760	0.04008
0.906	1.77247	0.01828	0.944	2.79768	0.04120
0.907	1.79075	0.01859	0.945	2.83888	0.04237
0.908	1.80934	0.01892	0.946	2.88125	0.04359
0.909	1.82826	0.01924			

"Surf. prob." is our abbreviation for WINSLOW's *[16]* normalized squared alpha wave amplitude, evaluated arbitrarily at a radial distance of 9.3×10^{-13} cm for trans-lead alpha emitters and at 8.0×10^{-13} cm for rare earth nuclei. In WINSLOW's notation it is written $R\,|\Phi_{J,I}(R)|^2$.

$$\left.\begin{aligned}\text{surf. prob.} &= \frac{M R \log 2}{\hbar k\, t_{\frac{1}{2}\alpha}}\, G_0^2(R)\\ &= 0.998\cdot 10^{-22}\sqrt{\frac{A}{A+4}}\,\frac{R\,G_0^2(R)}{Q_{\text{eff}}^{\frac{1}{2}}\, t_{\frac{1}{2}\alpha}\,(\text{sec})}\,.\end{aligned}\right\}\qquad(27.9)$$

δ_L^2 is THOMAS' *[17]* reduced derivative width. (It is 2π times BETHE's *[18]* G_α, "alpha width in the absence of a potential barrier".) We have evaluated it at the same radii as for "surf. prob." above.

$$\left.\begin{aligned}\delta_L^2 &= \frac{\hbar \log 2}{2 k R\, t_{\frac{1}{2}\alpha}}\left(\frac{d\,\text{Log}\, G_L}{d\log r}\right)_{r=R}^2 G_L^2(R)\,,\\ \delta_L^2\,(\text{Mev}) &= 5.2114\cdot 10^{-22}\sqrt{\frac{A+4}{A\, Q_{\text{eff}}}}\,\frac{G_L^2(R)}{R\, t_{\frac{1}{2}\alpha}}\left(\frac{d\,\text{Log}\, G_L}{d\,\text{Log}\, r}\right)_{r=R}^2\,.\end{aligned}\right\}\qquad(27.10)$$

III. Decay rate theory including non-central fields.

28. Introduction. Whenever an alpha particle leaves the nucleus with the daughter nucleus in an excited state, we may consider semi-classically that the daughter nucleus has associated with it an electromagnetic radiation field or fields associated with the matrix elements of gamma transitions to other states. In calculation of internal conversion coefficients one calculates relative probabilities of the radiation field ejecting a bound electron into the continuum or of its creating a photon. In the same general manner, the radiation field (particularly the electric components) acts on the escaping alpha particle while it is in the vicinity of the nucleus, and there will be an opportunity for nuclear transitions to take place by inducing an energy increase and angular momentum change in the motion of the alpha particle. To be more general, we must consider that the escaping alpha particle may also lose energy and excite the daughter nucleus to higher excited states, again through the medium of the non-central electromagnetic transition fields. In a strict sense, then, it is not exact to consider only a single nuclear state of the daughter nucleus in solving the SCHRÖDINGER equation for alpha decay, as is done in Sects. 19 to 26.

These refinements to alpha decay theory have not received much attention until quite recently. M. A. PRESTON[1] [*29*] gave a general approximate treatment of the problem in 1949. Recent attention has been given to the special importance of coupling by large $E2$ matrix elements in even-even nuclei of the rotational band region[2,3].

29. Formulation relating gamma transition rates and non-central alpha decay interactions. One can make a general derivation of the alpha decay problem including non-central fields of electric multipole nature[4] and relate the interactions to reduced gamma transition probabilities.

Consider the total HAMILTONian for the alpha decay system where the alpha particle is at a distance beyond special short range nuclear forces. This may be written as

$$H = T + H_{\text{nuc}} + V, \tag{29.1}$$

where T is the kinetic energy operator for the alpha-recoil nucleus system excluding internal motion of the recoil nucleus, H_{nuc} is the HAMILTONian for internal structure of the recoil nucleus, and V represents the complete coulombic potential energy interaction.

$$\left.\begin{aligned} T &= -\frac{\hbar^2}{2M}\nabla_r^2 \\ V &= \sum_{p=1}^{A} \frac{2e \cdot e_p}{|\boldsymbol{r}-\boldsymbol{r}_p|}, \end{aligned}\right\} \tag{29.2}$$

where the summation is carried out over all nucleons in the recoil nucleus, and the position vectors $\boldsymbol{r}$ and $\boldsymbol{r}_p$ are taken with the center of mass of the recoil nucleus as origin; $\boldsymbol{r}$ gives the position of the alpha and $\boldsymbol{r}_p$ of the nucleon.

---

1 M. A. PRESTON: Phys. Rev. **82**, 515 (1951).

2 J. O. RASMUSSEN: University of California Radiation Laboratory Report 2431 (1953) (unpublished).

3 R. F. CHRISTY: Phys. Rev. **98**, 1205 A (1955).

4 Magnetic multipole fields will probably be of little importance in alpha decay, since the alpha particle has no intrinsic spin.

Further, by an expansion of $|\boldsymbol{r}-\boldsymbol{r}_p|^{-1}$ valid for $r > r_p$,

$$V = 2e \sum_p \sum_{\lambda=0}^{\infty} \frac{e_{\lambda p} r_p^{\lambda} P_\lambda(\cos\gamma)}{\lambda!\, r^{\lambda+1}} \tag{29.3}$$ [1]

where $P_\lambda(\cos\gamma)$ is a LEGENDRE polynomial and γ is the angle between $\boldsymbol{r}_p$ and $\boldsymbol{r}$.

The $\lambda = 0$ term of the summation is simply the ordinary central coulombic term

$$V_{\text{cent}} = \frac{2Ze^2}{r},$$

and it will be shown that the succeeding terms are quite generally related to electric transition probabilities and in special cases to such quantities as the intrinsic nuclear quadrupole moment.

The SCHRÖDINGER wave equation $H\psi = E\psi$ will in general not be separable when non-central interactions are included. It will be a partial differential equation of $3(A+4)$ independent variables. (A being the mass number of the recoil nucleus.) To simplify the problem and reduce it to a set of ordinary differential equations in r, one usually expands the total wave function in terms of an orthonormal set of functions in all variables except r, with the expansion coefficients functions of r. One can, for example, construct such a set using eigenstates of the recoil nucleus, thus utilizing the orthogonality properties of these eigenfunctions[2]. Different and more specialized expansions have been used where the spheroidal nuclear model is applicable[3] [*12*] and these will be taken up later in this paper.

In the present derivation we wish to deal only with total wave functions which conserve the angular momentum and parity of the parent alpha-emitting nucleus. Such functions can be constructed in familiar fashion by summations weighted by CLEBSCH-GORDAN coefficients (cf. BLATT and WEISSKOPF's [*23*] Appendix A). That is, the orthonormal functions for the expansion can be represented by

$$\mathfrak{Y}_{l I_f \tau}^{IM}(\vartheta, \varphi, x_i) = \sum_m (l I_f m\ M-m | l I_f I M) \cdot Y_{lm}(\vartheta, \varphi)\, \Phi_{I_f M-m \tau}(x_i), \tag{29.4}$$

where l is the angular momentum associated with the alpha particle motion, m is its component on a space-fixed axis, I and I_f are initial and final nuclear angular momenta, M and M_f their components in space. τ represents all other quantum numbers specifying the nuclear state.

The orthogonality relation of these functions is as follows:

$$\int \mathfrak{Y}_{l I_f \tau}^{IM*}\, \mathfrak{Y}_{l' I_f' \tau'}^{IM}\, d\omega\, dx_i = \delta_{ll'}\, \delta_{I_f I_f'}\, \delta_{\tau\tau'}. \tag{29.5}$$

Letting the total wave function be expressed as

$$\psi = \sum_{l' I_f' \tau'} r^{-1} u_{l' I_f' \tau'}(r)\, \mathfrak{Y}_{l' I_f' \tau'}^{IM}(\vartheta, \varphi, x_i) \tag{29.6}$$

---

[1] We define an $e_{\lambda p}$ in a slightly different manner from BOHR and MOTTELSON's [*10*] Eq. (VII.3). We let $e_{\lambda p}$ be the effective charge of a nucleon which includes the recoil correction. $e_{\lambda p} = \left[1 + (-)^\lambda \frac{Z}{A^\lambda}\right] e$ for protons, and $e_{\lambda p} = (-)^\lambda \frac{Ze}{A^\lambda}$ for neutrons except for $\lambda = 0$, where the recoil term vanishes. Strictly speaking, this special definition would apply to summation over one or a limited number of nucleons with the center of mass of the residual Z protons and $A-Z$ neutrons defining the origin of the coordinate system. If the summation were really carried over all nucleons with the nuclear center of mass as the origin, no special definition of $e_{\lambda p}$ need be made; it will be the unit charge e for a proton and zero for a neutron.

[2] M. A. PRESTON: Phys. Rev. **75**, 90 (1949).

[3] J. O. RASMUSSEN: University of California Radiation Laboratory Report 2431 (1953) (unpublished).

we substitute into the SCHRÖDINGER equation, $(H-E)\,\psi=0$ with the HAMILTONian of (29.1), multiply by $\mathfrak{Y}_{l I_f \tau}^{IM*}$ and integrate over all space for all variables except r. Hereafter we write $\mathfrak{Y}_{l I_f \tau}^{IM}$ as $|\tau\, l\, I_f;\, IM)$. There result the ordinary differential equations

$$\left.\begin{aligned}-\frac{\hbar^2}{2M}\frac{d^2u_{lI_f\tau}}{dr^2}+\left(\frac{\hbar^2}{2Mr^2}l(l+1)+E_{I_f\tau}^{\mathrm{nuc}}-E\right)u_{lI_f\tau}+\frac{2Ze^2}{r}u_{lI_f\tau}+\\+\sum_{l'I_f'\tau'}u_{l'I_f'\tau'}\left(\tau l I_f;IM\left|\sum_{\lambda=1}^{\infty}\frac{2e}{\lambda!\,r^{\lambda+1}}\sum_p r_p^\lambda e_{\lambda p}P_\lambda(\cos\gamma)\right|\tau' l' I_f';IM\right)=0.\end{aligned}\right\}\quad(29.7)$$

From RACAH's [30] formula (38) we can evaluate the desired matrix elements of the coulombic interaction between the alpha and the protons of the recoil nucleus as follows:

$$\left.\begin{aligned}&\frac{2e}{\lambda!\,r^{\lambda+1}}\left(\tau l I_f;IM\left|\sum_{p=1}^{A}e_{\lambda p}r_p^\lambda P_\lambda(\cos\gamma)\right|\tau' l' I_f';IM\right)\\&=\frac{2e}{\lambda!\,r^{\lambda+1}}(-)^{l+I_f'-I}(l\|\boldsymbol{C}_\alpha^{(\lambda)}\|l')\cdot W(l I_f l' I_f';I\lambda)\cdot\left(I_f\tau\left\|\sum_p e_{\lambda p}r_p^\lambda\boldsymbol{C}_p^{(\lambda)}\right\|I_f'\tau'\right)\end{aligned}\right\}\quad(29.8)$$

where we use RACAH's notation throughout.

By RACAH's [30] Eqs. (50) and (51)

$$(l\|\boldsymbol{C}_\alpha^{(\lambda)}\|l')=(-)^l\left[\frac{(2l+1)(2l'+1)}{2\lambda+1}\right]^{\frac{1}{2}}(l\,l'\,0\,0|l\,l'\,\lambda\,0)\quad(29.9)$$

we may establish the connection between $\sum\limits_p(I_f\tau\,\|r_p^\lambda\boldsymbol{C}_p^{(\lambda)}\|\,I_f'\tau')$ and the electric 2^λ-pole reduced transition probability connecting the states.

The electric 2^λ-pole transition probability from BOHR and MOTTELSON's [10] Eq. (VII.I) is as follows:

$$T(\lambda)=\frac{8\pi(\lambda+1)}{\lambda[(2\lambda+1)!!]^2}\,\frac{1}{\hbar}\left(\frac{\omega}{c}\right)^{2\lambda+1}B_e(\lambda)\quad(29.10)$$

where $B_e(\lambda)$, the reduced transition probability, is given by

$$B_e(\lambda)=\sum_{M_f\,\mu}\left|\left(I_fM_f\tau\left|\sum_{p=1}^{A}e_{\lambda p}r_p^\lambda Y_{\lambda\mu}(\vartheta_p,\varphi_p)\right|I_f'M_f'\tau'\right)\right|^2.\quad(29.11)$$

The transition operator transforms with rotation of coordinates like a single spherical harmonic of order λ, so we may use RACAH's equation to evaluate matrix elements of it. By RACAH's Eq. (29) the above is equal to

$$\left.\begin{aligned}B&=\sum_{M_f\mu}\left|(-)^{I_f-M_f}\left(\tau_fI_f\left\|\sum_p e_{\lambda p}r_p^\lambda\left(\frac{2\lambda+1}{4\pi}\right)^{\frac{1}{2}}C^{(\lambda)}\right\|\tau_f'I_f\right)\cdot V(I_fI_f'\lambda;-M_fM_f'\mu)\right|^{2}\\&=\left(\frac{2\lambda+1}{4\pi}\right)\left(\tau_fI_f\left\|\sum_p e_{\lambda p}r_p^\lambda C^{(\lambda)}\right\|\tau_f'I_f'\right)^2\cdot\sum_{M_f\mu}V(I_fI_f'\lambda;-M_fM_f'\mu)^2\end{aligned}\right\}\quad(29.12)$$

and by RACAH's Eq. (20a) the summation reduces to $(2I_f'+1)^{-1}$, so

$$\underset{I_f'-I_f}{B(\lambda)}=\left(\frac{2\lambda+1}{4\pi}\right)\frac{\left(\tau_fI_f\left\|\sum_p e_{\lambda p}r_p^\lambda C^{(\lambda)}\right\|\tau_f'I_f'\right)^2}{2I_f'+1}$$

Therefore

$$\left(\tau_fI_f\left\|\sum_p e_{\lambda p}r_p^\lambda C^{(\lambda)}\right\|\tau_f'I_f'\right)=\pm\left[\frac{4\pi(2I_f'+1)\,B_e(\lambda)}{2\lambda+1}\right]^{\frac{1}{2}}.\quad(29.13)$$

Thus, combining (29.8), (29.9), and (29.13) we may write for the coefficients of the coupling terms in (29.7)

$$\left.\begin{aligned}&\left(\tau l I_f; I M\left|\frac{2e}{\lambda!\, r^{\lambda+1}} \sum_p e_{\lambda p} r_p^\lambda P_\lambda(\cos\gamma_p)\right| \tau' l' I'_f; I M\right)\\&\quad = \pm(-)^{I'_f - I}\,\frac{2e\left[(2l+1)(2l'+1)\,4\pi(2I'_f+1)\,B_{I'_f\to I_f}(\lambda)\right]^{\frac{1}{2}}}{\lambda!\, r^{\lambda+1}(2\lambda+1)}\times\\&\quad\times(l l' 0 0|l l' \lambda 0)\, W(l I_f l' I'_f; I \lambda)\,.\end{aligned}\right\} \tag{29.14}$$

With this relationship the magnitude of non-central coupling terms in alpha decay may be calculated from a knowledge of the gamma transition probability or coulomb excitation cross section. There is an uncertainty in sign associated with this procedure, corresponding to the two phase possibilities for a gamma transition. In some cases information on relative phases of gamma radiation may be gained from gamma angular distribution experiments with mixed multipolarities In other cases, the sign may be determined from a theoretical model.

30. Numerical calculation of some $E1$ and $E2$ interaction matrix elements. In order to get an idea of the importance in actual alpha emitters of including coupling induced by electric multipole fields, let us consider numerical examples.

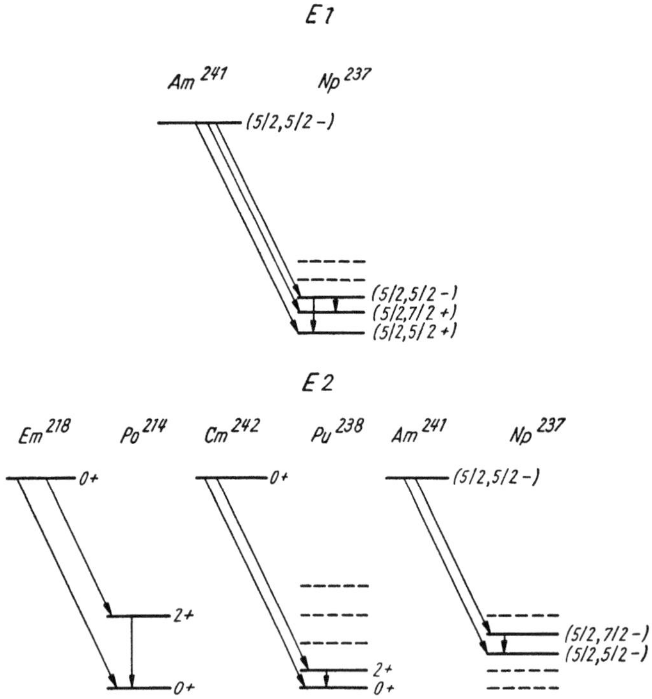

Fig. 17. Partial alpha-decay schemes with solid arrows designating the gamma transitions for which non-central matrix elements for the electric interaction with the alpha particle are given in Table 10. The state assignments are given at the right, the convention for the odd nuclei being (K, I, π).

Throughout the strong deformation region of heavy isotopes ($A > 224$) many low energy $E1$ transitions have been identified. About half of these decay from metastable states of measurable half life, meaning a large retardation ($\sim 10^3$–10^5) from the single proton transition lifetime formula. Throughout the whole region of alpha emitters many $E\,2$ transitions have been identified, and in the region $A > 224$ these transitions are usually of the collective rotational type, exhibiting transition probabilities much larger than single proton formulas estimate.

We shall consider briefly first $E\,1$ coupling, where the coupling effects beyond the range of specifically nuclear forces are expected to be relatively unimportant. A special section will be devoted to $E\,2$ coupling, the inclusion of which is of essential importance in any fundamental treatment of alpha decay in the region of the enhanced $E\,2$ rotational transitions.

In Fig. 17 are shown the decay schemes of alpha emitters to be considered in numerical calculation of coupling terms here. The gamma transitions and levels taken into account in the examples are shown as solid lines and others as dashed lines.

In Table 10 are calculated according to Eq. (29.14) values of the matrix elements coupling the alpha decay Eq. (29.7) for actual alpha emitters. By far the largest coupling terms are found for the enhanced $E\,2$ transitions in the rotational region.

Table 10. *Representative non-central coupling term magnitudes in alpha emission*

Multi-polarity	Alpha emitter	I	I_f	l	E kev	I'_f	l'	E' kev	B $I'_f \to I_f$ (cf. footnote 3)	Coupling energy at $r = 1 \cdot 10^{-12}$ cm Mev	Determined from
$E\,1$	Am^{241}	$5/2$	$5/2$	1	0	$5/2$	0	60	$1.7 \cdot 10^{-38}$ ev · cm³	$\pm 1.2 \cdot 10^{-4}$	Delay coincidence[1] lifetime of 60-kev state (6.3×10^{-8} sec)
		$5/2$	$5/2$	1	0	$5/2$	2	60	$1.7 \cdot 10^{-38}$ ev · cm³	$\pm 0.98 \cdot 10^{-4}$	
		$5/2$	$5/2$	3	0	$5/2$	2	60	$1.7 \cdot 10^{-38}$ ev · cm³	$\pm 0.90 \cdot 10^{-4}$	
		$5/2$	$7/2$	1	33	$5/2$	0	60	$1.4 \cdot 10^{-38}$ ev · cm³	$\pm 1.1 \cdot 10^{-4}$	
		$5/2$	$7/2$	1	33	$5/2$	2	60	$1.4 \cdot 10^{-38}$ ev · cm³	$\pm 0.28 \cdot 10^{-4}$	
$E\,2$	Cm^{242}	0	0	0	0	2	2	44	$2.3 \cdot 10^{-55}$ ev · cm⁵	$+0.58$	Q_0 value of 9×10^{-24} cm² estimated from coulomb excitation work on Th^{232} and U^{238} (see Sect. 14)
	Am^{241}	$5/2$	$5/2$	0	60	$5/2$	2	60	—		$Q_0 \approx +14 \times 10^{-24}$ cm² from Am^{241} spectroscopic determination[2] used
		$5/2$	$5/2$	0	60	$7/2$	2	103	$1.0 \cdot 10^{-54}$ ev · cm⁵	$+0.54$	
		$5/2$	$5/2$	2	60	$7/2$	2	103	(same)	-0.62	
										-0.40	
	Em^{218}	0	0	0	0	2	2	609	$1.1 \cdot 10^{-56}$ ev · cm⁵	± 0.13	Lifetime ($6 \cdot 10^{-12}$ sec) estimation from long range alpha particle data (see Table 27 of Ref. [*10*])

31. Approximate treatment of *E* 1 interaction in alpha decay of Am^{241}. The $E\,1$ coupling terms for Am^{241} seem very small, but we should like to carry through an approximate treatment to determine whether the $E\,1$ coupling can provide a simple explanation for the intensity of the highly hindered alpha group to the ground state. PRESTON[4] has suggested that some rare alpha groups might be explained by assuming the wave of the rare group to have zero amplitude at the nuclear surface. The wave would attain finite amplitude at larger distances by virtue of coupling terms to more abundant alpha groups. PRESTON's method[4]

[1] BELING, NEWTON and ROSE: Phys. Rev. **87**, 670 (1952).

[2] MANNING, FRED and TOMKINS: Phys. Rev. **102**, 1108 (1956).

[3] B is the reduced gamma transition probability in the notation of BOHR and MOTTELSON [*10*].

[4] M. A. PRESTON: Phys. Rev. **82**, 515 (1951).

for solving the coupled equations for small energy difference should be suited to Am^{241}, since the energy difference of the coupled alpha groups is quite small. Another method due to DANCOFF[1] we believe to be incorrect, in that the total rate of electric transitions from one alpha group to another is interpreted as the rate of emergence of the latter alpha group from the barrier. This method gives answers vastly over-estimating the importance of coupling.

Let us consider the *S-wave* to the 60-kev state as having unit amplitude at the nuclear surface and any waves to the ground state as having zero amplitude. Considering just these two states and only *S-* and *P-waves*, the wave equation of (29.7) reduces to two equations:

$$\left.\begin{aligned} &\frac{d^2 u_0}{d r^2} - \frac{2M}{\hbar^2}\left(\frac{2Ze^2}{r} + E_0 - E\right) u_0 \pm \frac{2M}{\hbar^2}\frac{\varepsilon}{r^2} u_1 = 0, \\ &\frac{d^2 u_1}{d r^2} - \left[\frac{2M}{\hbar^2}\left(\frac{2Ze^2}{r} + E_1 - E\right) + \frac{2}{r^2}\right] u_1 \pm \frac{2M}{\hbar^2}\frac{\varepsilon}{r^2} u_0 = 0 \end{aligned}\right\} \tag{30.1}$$

with $E_0 = 60$ kev and $E_1 = 0$.

The diagonal energy difference $(H_{11} - H_{00})$ is $\frac{\hbar^2}{M r^2} + E_1 - E_0$, which shows a different radial dependence from that of the off-diagonal energies. $\frac{\hbar^2}{M r^2}$ has the value of about 0.11 Mev near the nuclear surface ($r = 1 \times 10^{-12}$ cm) Thus, the diagonal energy difference will be positive out to about 1.4×10^{-12} cm and negative thereafter. Rather than attempting an exact solution of Eq. (30.1) we shall solve two simpler approximate problems for order of magnitude estimates of the coupling effects.

In the first we let the diagonal energy difference be zero for all r. Then by addition of Eq. (30.1) we obtain an uncoupled equation in $u_0 + u_1$, and by subtraction, an equation in $u_1 - u_0$. These equations differ only in the sign of an ε/r^2 diagonal energy term. The effect of such a term on the penetration factor for alpha decay has been derived in connection with centrifugal barrier effects. That is, the addition of an ε/r^2 term to the potential barrier causes an extra attenuation of an outgoing wave approximately given by the factor $\exp\left[\frac{-\varepsilon}{\hbar e}\sqrt{\frac{M}{ZR}}\right]$. By this method if we take the boundary conditions at the nuclear surface of $\left|\frac{u_1}{u_0}\right|^2_{r=R} = 0$, then we find $\left|\frac{u_1}{u_0}\right|^2_{r\to\infty} = 2.5 \times 10^{-8}$, representing a build-up many orders of magnitude less than the experimental relative intensity ratio $\left|\frac{u_1}{u_0}\right|^2 = 3.5 \times 10^{-3}$.

In the second approximate method we set $E_1 - E_0$ equal to zero and retain $\frac{\hbar^2}{M r^2}$ as the diagonal energy difference. In a similar manner we find the linear combinations which decouple the equations. With the same boundary conditions as before we get $\left|\frac{u_1}{u_0}\right|^2_{r\to\infty} = 1.0 \cdot 10^{-7}$, a factor of four larger than by the first treatment, but still far less than the experimental intensity ratio.

We conclude that alpha coupling by the $E1$ radiation field is quite negligible in the Am^{241} case. It is possible that there could be larger $E1$ coupling effects in the diffuse surface region or nuclear interior due to short range nuclear forces. However, this phase of the problem cannot be as precisely formulated as the coulomb interaction problem.

---

[1] S. M. DANCOFF: U.S. Atomic Energy Commission Unclassified Document 2853 (1950) (unpublished).

It might be argued that external coupling would be larger in cases where $E\,1$ transitions are not so retarded as in the Am^{241} case. These retarded $E\,1$ transitions are probably fairly typical, at least of odd-mass nuclei, in the heavy region. It seems likely that all the known low-energy $E\,1$ transitions of odd-A nuclei in the heavy region are much slower than given by the single-particle lifetime formula, by virtue principally of selection rules[1] in the asymptotic quantum numbers appropriate to large spheroidal deformation (see Sect. 56). We have no measured $E\,1$ lifetimes in the even-even nuclei, as only limits have been set at present. If $E\,1$ transition matrix elements of single-proton magnitude were taken in the Am^{241} problem above, the coupling could build up the relative wave amplitude $\left|\frac{u_1}{u_0}\right|^2$ from zero at the nuclear surface to order 10^{-2} at large distance.

32. Alpha decay of spheroidal nuclei; the $E\,2$ interaction. An important special case where the non-central coupling is especially large occurs for nuclei with large spheroidal deformation. These nuclei, odd-mass as well as even-even types, generally exhibit rotational band structure in their level schemes. Many of these rotational band levels are interconnected by especially large $E\,2$ matrix elements (cf. Table 10 for $E\,2$ interactions for Cm^{242} and Am^{241}). The formulation of the alpha decay equations may be made in the same general manner as in the preceding section. The necessary equations giving the reduced transition probabilities within a given rotational band are given by BOHR and MOTTELSON [*10*], Eq. (VII.18) and (VII.19).

These expressions together with Eq. (29.14) allow the calculation of the magnitude of rotational quadrupole coupling terms in odd or even nuclear types. A general expression[2] giving the sign of the term as well may be derived by using BOHR-MOTTELSON strong coupling wave functions [*10*], Eq. (II.15), to evaluate the invariant $(I_f \tau \| \sum_p e\, r_p^2 \boldsymbol{C}_p^{(2)} \| I_f' \tau')$. One makes use of the transformation [*10*], (Eq. VII.13), of the electric multipole operator from the space-fixed to the body-fixed coordinate system.

The general result for the quadrupole matrix element connecting states of a given rotational band (i.e., K and Ω unchanged) is

$$\left.\begin{aligned}
&\left(K\Omega\, l\, I_f;\, I\, M \left| \frac{2e}{r^3}\, \frac{Q_0\, e\, P_2(\cos\gamma)}{2} \right| K\Omega\, l'\, I_f';\, I\, M\right) \\
&\quad = (-)^{I_f + I_f' - K - I}\, \frac{Q_0\, e^2}{r^3}\, \frac{[(2I_f+1)(2I_f'+1)(2l+1)(2l'+1)]^{\frac{1}{2}}}{5} \times \\
&\quad \times (I_f I_f' K - K | I_f I_f' 2 0) \cdot (l l' 0 0 | l l' 2 0)\, W(l I_f l' I_f'; I\, 2),
\end{aligned}\right\} \qquad (32.1)$$

where γ is the angle between the alpha direction vector and the nuclear symmetry axis.

For decay of even-even nuclei ($I=0$) to the ground rotational band ($K=\Omega=0$) the above expression reduces[3] simply to

$$\frac{[(2l+1)(2l'+1)]^{\frac{1}{2}}}{5}\, (l l' 0 0 | l l' 2 0)^2,$$

[1] G. ALAGA: Phys. Rev. **100**, 432 (1955).

[2] J. O. RASMUSSEN and B. SEGALL: University of California Chemistry Division Quarterly Report (UCRL 2531) (April, 1954) (unpublished).

[3] J. O. RASMUSSEN: University of California Radiation Laboratory Report 2431 (1953) (unpublished).

which for $l = l'$ is equal to $\frac{l(l+1)}{(2l+3)(2l-1)}$ and for $l' = l+2$ is equal to $\frac{3(l+1)(l+2)}{2(2l+3)\sqrt{(2l+1)(2l+5)}}$ and which vanishes whenever l and l' differ by other than 0 or ± 2.

For treating alpha emission of odd mass nuclei in the region of validity of the BOHR-MOTTELSON spheroidal deformation model it is for some purposes better to use a different representation for the functions in terms of which expansion (29.6) is made. RASMUSSEN and SEGALL [12] have used an expansion in which not only the nucleon coordinates (of recoil nucleus) are referred to a rotating body-fixed coordinate system (as in BOHR-MOTTELSON strong deformation model wave functions), but also the coordinates of the alpha particle are referred to this body-fixed system. The cylindrical symmetry axis of the nuclear spheroid is taken as the polar axis of the body-fixed system. We write the new functions

$$\mathsf{Y}^{IM}_{K_f l m;\tau} = |\tau K_f l m; IM) = \chi_\tau(x'_p)\sqrt{\frac{2I+1}{8\pi^2}}\,\mathfrak{D}^I_{M\,K_f+m}(\Theta_i)\,Y_{lm}(\vartheta',\varphi') \tag{32.2}$$

where Θ_i are the three EULERian angles specifying the position of the body-fixed nuclear coordinate system in space, the primed coordinates are with reference to the body-fixed system (the x'_p being coordinates of the nucleons in the recoil nucleus, and the ϑ', φ' being the alpha angular coordinates).

$$\sqrt{\frac{2I+1}{8\pi^2}}\,\mathfrak{D}^I_{M\,K_f+m}(\Theta_i)$$

is the normalized symmetric-top rotational wave function, for angular momentum of I, component on the space-fixed z-axis of M, and component on the body-fixed z'-axis of K_f+m. The Y representation contrasts with the $\mathfrak{Y}$ representation used in (29.6) mainly in that m supplants I_f for the specification of the state of alpha particle motion. In general, a state in the Y representation is a linear combination of several different states in the $\mathfrak{Y}$ representation. The transformation Eq. (11) of Ref. [12] is

$$\mathsf{Y}^{IM}_{K_f l m;\tau} = \sum_{I_f}(-)^{m+I_f-I}(I\,l\,K_f+m\,-m\,|\,I\,l\,I_f K_f)\,\mathfrak{Y}^{IM}_{I_f K_f l;\tau}. \tag{32.3}$$

When the expansion

$$\Psi = \sum_{m' l'} r^{-1} w_{l'm'}(r)\,\mathsf{Y}^{IM}_{K_f l' m';\tau} \tag{32.4}$$

is substituted into the SCHRÖDINGER equation with HAMILTONian (29.1), multiplied by $\mathsf{Y}^{IM*}_{K_f l m;\tau}$ and integrated over all coordinates except r, one gets a set of coupled ordinary differential equations similar to (29.7).

$$\left.\begin{aligned}&\left[\frac{d^2}{dr^2}+\frac{2M}{\hbar^2}(E-E_{\tau K_f})-\frac{l(l+1)}{r^2}-\frac{4MZe^2}{\hbar^2 r}\right]w_{lm}(r)\\ -&\frac{2M}{\hbar^2}\sum_{l'm'}w_{l'm'}(lm|H_{\rm rot}|l'm')-\frac{2M}{\hbar^2}\sum_{l'm'}w_{l'm'}\left(lm\left|V-\frac{2Ze^2}{r}\right|l'm'\right)=0.\end{aligned}\right\} \tag{32.5}$$

$H_{\rm rot}$ is the rotational energy HAMILTONian for the recoil nucleus, and it gives rise to the characteristic rotational band structure of energy levels. From the eigenvalue equation

$$H_{\rm rot}\,\mathfrak{Y}^{IM}_{I_f K_f l;\tau} = \frac{\hbar^2}{2\mathfrak{J}}I_f(I_f+1)\,\mathfrak{Y}^{IM}_{I_f K_f l;\tau} \tag{32.6}$$

and the expansion (32.3) the required matrix elements of H_{rot} may be evaluated.

$$\left.\begin{aligned}(l\,m\,|\,H_{\text{rot}}\,|\,l'\,m') = \frac{\hbar^2}{2\mathfrak{J}}\,\delta_{l\,l'} \sum_{I_f} (-)^{m-m'}\, I_f(I_f+1) \times \\ \times (I\,l\,K_f + m - m\,|\,I\,l\,I_f\,K_f)\,(I\,l\,K_f + m' - m'\,|\,I\,l\,I_f\,K_f)\,.\end{aligned}\right\} \tag{32.7}$$

The last term in (32.5) involves matrix elements of the quadrupole interaction, and in this representation has a simpler form than in (32.1)

$$\left.\begin{aligned}&\left(\tau\,K_f\,l\,m;\,I\,M\left|\frac{2e}{r^3}\,\frac{Q_0\,e\,P_2(\cos\gamma)}{2}\right|\tau\,K_f\,l'\,m';\,I\,M\right) \\ &\qquad = \frac{Q_0\,e^2}{r^3}\int Y^*_{l\,m}(\vartheta,\varphi)\,P_2(\cos\vartheta)\,Y_{l'\,m'}(\vartheta,\varphi)\,d\omega\,.\end{aligned}\right\} \tag{32.8}$$

The integral is easily evaluated and contains no off-diagonal elements in m. The integral is equal to the $c^2(l'\,m', l\,m)$ tabulated by CONDON and SHORTLEY[1] but can be simply given in terms of CLEBSCH-GORDAN coefficients as shown by RACAH [*30*]

$$c^2(l'\,m', l\,m) = \delta_{m\,m'}\,\frac{[(2l+1)\,(2l'+1)]^{\frac{1}{2}}}{5}\,(l\,l'\,0\,0\,|\,l\,l'\,2\,0)\,(l\,l' - m\,m\,|\,l\,l'\,2\,0)\,. \tag{32.9}$$

An illustration of the usefulness of this representation is provided in the consideration of the favored alpha decay cases[2] [*31*] in odd mass number deformed nuclei. Favored alpha decay transitions are those in which the intrinsic wave function of the odd nucleon remains essentially unchanged. In such cases $\Delta K = \Delta\Omega = 0$. When the alpha particle is in the surface region of the nucleus we would expect nearly to conserve the component of total angular momentum of the system along the symmetry axis; hence, that near the nucleus all alpha components except those with $m=0$ will be negligibly small. The matrix elements of the electric potential have no off-diagonal elements in m. Hence, the only build-up of alpha components with $m \neq 0$ will result from the matrix elements of H_{rot}, the nuclear rotational energy. In the limiting case of $H_{\text{rot}} = 0$ (infinite nuclear moment of inertia) the $m \neq 0$ components in favored alpha decay would be identically zero. We could set down exactly the branching of a given l value between various nuclear rotational states I_f. The expansion (31.3) gives the desired branching. That is, a given l-group branches as the squares of CLEBSCH-GORDAN coefficients, $(I\,l\,K_f\,0\,|\,I\,l\,I_f\,K_f)^2$. The intensity relationship is quite similar to relationships for branching of gamma and beta transitions to various members of a rotational band. It is not obvious how good the relationship in alpha decay will be, since the matrix elements in H_{rot} are not zero. BOHR, FRÖMAN, and MOTTELSON [*31*] take H_{rot} into account approximately by reducing the alpha intensity to higher rotational members by the reduction in barrier penetrability appropriate to the energy differences of the final states. They have used their approximation to calculate expected intensities to various final states of odd nuclei based on the empirical relative alpha intensities for neighboring even-even nuclei. The comparison of their approximation with experimental intensities is given in Table 16.

33. Numerical results for alpha decay of spheroidal nuclei. Some numerical work on the alpha decay of deformed nuclei has been carried out.

---

[1] E. U. CONDON and G. H. SHORTLEY: Theory of Atomic Spectra, p. 175. London: Cambridge University Press 1935.

[2] J. O. RASMUSSEN: Ark. Fysik **7**, 185 (1953).

R. F. CHRISTY[1] has developed a useful analytical approximation with a new form of the WKB method. This approximation, which is most applicable in the limit of very large nuclear moments of inertia, treats the wave function from each point on the nuclear surface as penetrating outward radially while diffusing transversely in angle. A total solution corresponding to an arbitrary wave function at the nuclear surface can be constructed by superposition of the solutions from points on the surface.

L. DRESNER[2] also has made use of the WKB method to derive a formula for the alpha penetration problem. He, like CHRISTY, considers the limiting case of infinitely large nuclear moment of inertia.

Since CHRISTY'S equations have not been published, we are unable to make a detailed comparison between his and DRESNER'S WKB treatments. One significant difference seems to be that DRESNER has dropped from his Eq. (III.2) the term that would give rise to the centrifugal barrier reduction (or diffusion in angle), while CHRISTY has retained this feature. RASMUSSEN and SEGALL [*12*] have derived the alpha decay equations in prolate spheroidal coordinates and have done both outward numerical integrations for Cm^{242} with simple assumptions on nuclear surface boundary conditions and inward integrations for Cm^{242} and Th^{228} with experimental alpha intensities setting the boundary conditions.

The work of CHRISTY[1] and of RASMUSSEN and SEGALL [*12*] has shown that the boundary conditions making the alpha wave function constant over the spheroidal nuclear surface does not give for Cm^{242} the correct $l=4$ alpha group intensity, the experimental value being smaller than that calculated. Work of NOSOV[3] indicates that the assumption of constant wave function over the surface may be more consistent with experimental Cm^{242} alpha intensities if the nucleus is not prolate but oblate spheroidal. Other evidence weighs strongly for prolate deformation in the heavy region, so we are inclined to discount this interpretation. Furthermore the work of RASMUSSEN and SEGALL and of DRESNER shows that the one-body model, with the alpha in its lowest virtual level within the well, does not give a low enough $l=4$ group intensity, although the values are somewhat lower than for the uniform boundary condition.

CHRISTY has found that by decreasing the alpha wave on the nuclear surface near the poles or by introducing some fourth order deformation in the nuclear surface the relative alpha group intensities observed in Cm^{242} can be theoretically obtained.

RASMUSSEN and SEGALL have for Cm^{242} and Th^{228} with $l=0$, 2, and 4 groups taken the experimental alpha intensities to set the boundary conditions outside the barrier and have carried solutions of the wave equation in to the spheroidal nuclear surface. (Recourse is made to a WKB approximation method.) There is an ambiguity in the choice of relative phases, the experimental intensities giving only a measure of wave amplitudes. For the treatment including three groups there are in general four choices of phase leading to different physically acceptable solutions. Most choices lead to a narrow wave distribution at the equator on the nuclear surface, and we are inclined to believe one of the choices of phase other than these to be the true one. The most direct evidence in support of our choice of relative $l=0$, $l=2$ phase comes from the interpretation of the α-γ angular correlation in Am^{241}, in which the phase difference determines the sign of the interference term in the correlation function (see Sect. 56). Fig. 18 shows the

---

[1] R. F. CHRISTY: Phys. Rev. **98**, 1205 A (1955).

[2] L. DRESNER: 1955 (unpublished).

[3] V. G. NOSOV: Dokl. Akad. Nauk SSSR. **103**, 65 (1955). English translation, University of California Radiation Laboratory Translation 258.

alpha probability ($|\psi|^2$) as a function of position on the nuclear surface from these inward integrations for the most likely choice of relative phases.

These results are of special interest in that they demonstrate most clearly the inadequacy of a one-body model, and they are suggestive of a model in which the alpha wave function at the surface is a measure of the angular overlap of the wave function of the most lightly bound nucleons. The distribution seems more dependent on proton than on neutron number.

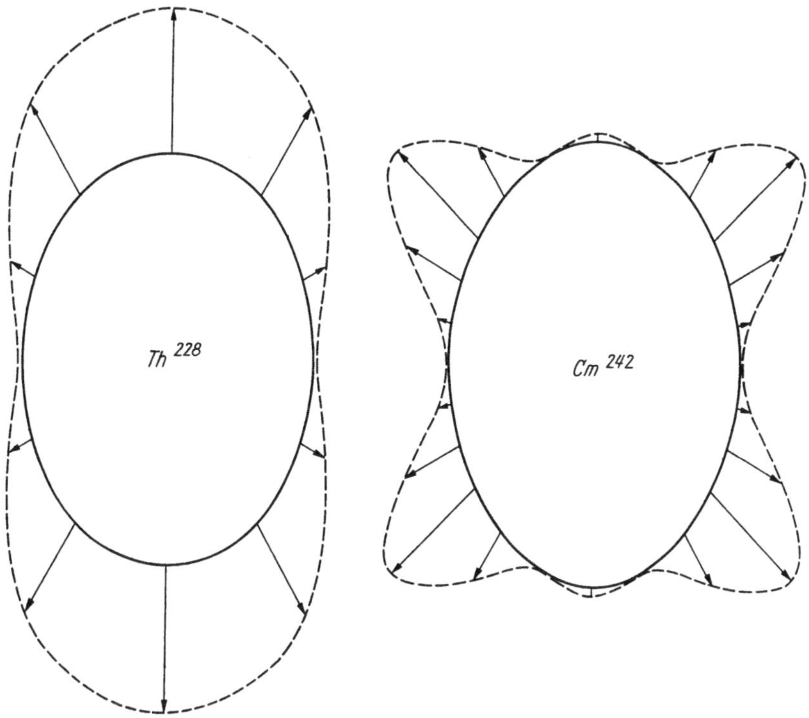

Fig. 18. Plots of the alpha-particle probability functions (wave function squared) on the spheroidal nuclear surfaces for alpha emitters Cm^{242} and Th^{228}, from calculations of RASMUSSEN and SEGALL [12]. (See Sects. 32 and 33 regarding relative phase assumptions involved in the calculations.)

The results of Fig. 18 are subject to some modification, since the deformations and intrinsic quadrupole moments assumed for the study appear to be about 50% too large compared to recent quadrupole moment determinations (Th^{232}, U^{238}, Am^{241}, Am^{243}) (see Sect. 14).

The coupled equations of alpha decay in spherical coordinates lend themselves to solution by high-speed digital computers, and computations in progress[1] may shed more light on the whole interesting problem of deducing alpha angular distributions near the nuclear surface.

On the basis of numerical work to date it is felt that the observed regular variations of $l=2$ and $l=4$ group hindrance (cf. Fig. 15) represent a regular shift of the area of highest alpha formation probability in the nuclear surface from near the poles (thorium alpha emitters) to zones midway between poles and equator (curium), and, perhaps, somewhat further toward the equator for higher elements. The development and shift of higher order deformations in the nuclear surface would provide an alternative interpretation, though.

[1] J. O. RASMUSSEN and E. HANSEN: Unpublished.

IV. Alpha decay as a multibody process. (The even-even nuclei.)

34. Introduction. There have been numerous approaches suggested for treating the internal mechanics of the alpha decay process in a more realistic manner than in the one-body model. There is a general feeling that for a given radius of cut-off of the coulombic potential barrier a realistic model would give smaller decay rates for alpha decay than the one-body model. BETHE *[18]* early advanced arguments that the alpha particle should probably not be thought of as existing within the nucleus a large fraction of the time.

The attempts at many-body modifications of alpha theory may be classified into three categories: (1) statistical model ("level spacing") approaches analogous to the statistical theory of nuclear reactions, (2) "shell model" (overlap integral) approaches based on the individual particle (nucleon) nuclear models, and (3) the "resonating group" approaches treating configuration mixing involving alpha particles explicitly in some configurations.

35. Statistical model. BETHE *[18]* suggested that the formation and emission of an alpha particle ($\sim$6 Mev) in the absence of barrier should occur about as rapidly as the emission of a neutron from a nucleus with slightly more excitation energy ($\sim$6 Mev) than needed to emit a neutron. Namely, he suggested a "decay constant for alpha emission in the absence of a barrier" of $\sim 10^{15}\ \text{sec}^{-1}$, (i.e. $\delta^2 \approx 6$ ev), whereas PRESTON'S lowest virtual level one-body treatment yields $\sim 2\times 10^{20}\ \text{sec}^{-1}$ ($\delta^2 \approx 1.2$ Mev). The alpha decay channel radii calculated on this theory seem a little too large to accord with our present ideas of nuclear sizes.

COHEN[1] suggested that BETHE'S approach should be modified, increasing BETHE'S fundamental "reduced decay constant" of $10^{15}\ \text{sec}^{-1}$ by the ratio of the "average level spacing" near ground to the average level spacing at an excitation of $\sim$6 Mev, thus giving a reduced decay constant $\sim 10^{19}\ \text{sec}^{-1}$ ($\delta^2 \approx$ 50 kev) which leads to more reasonable channel radii. The proportionality of neutron level widths to average level spacings had been noted experimentally and derived theoretically in the statistical theory of nuclear reactions of FESHBACH, PEASLEE, and WEISSKOPF[2]. The concept of an average level spacing at the ground state is a vague one. COHEN suggested taking $\sim$100 kev for this average from the energies of first excited states in even-even nuclei. On this point RASMUSSEN[3] *[32]* asserted that the spacings to first excited states were not relevant, since they are generally of a different spin from ground. He suggested that the spacing to the next 0^+ level would be more appropriate and proposed in the absence of experimental information that this energy might be of the order of the lowest order surface vibrational quantum energy, $\sim$1 Mev.

DEVANEY *[22]* suggested taking first excited state energies for the average level spacing in the absence of better information.

Various authors[3] *[22]*, *[23]* have given theoretical derivations basing alpha decay rate on an average level spacing. In terms of the one-body model the level spacing treatment reduces the decay rate by the ratio of the real nuclear level spacing to that in the one-body model ($\sim$1 Mev). One assumes the reciprocal of the logarithmic derivative of the alpha wave function at the nuclear surface $S_i(W)$ to be a function of the form $1/S_i(W) = \tan z(W)$ with virtual levels

[1] B. L. COHEN: Phys. Rev. **80**, 105 (1950).

[2] FESHBACH, PEASLEE and WEISSKOPF: Phys. Rev. **71**, 150 (1947).

[3] J. O. RASMUSSEN: Ph. D. Thesis, University of California 1952. University of California Radiation Laboratory Report UCRL-1473 rev.

at $z(W)=n\pi$, where $z(W)$ is some monotonic function of the total energy of the system. It can be shown [17] that δ^2, the reduced derivative width, is related to S_i and its derivative with respect to energy as follows:

$$\delta^2 = -\frac{S_i^2}{\partial S_i/\partial E} = \left[\frac{\partial(1/S_i)}{\partial E}\right]^{-1}. \tag{34.1}$$

The assumption is made[1] that $dz/dE=\pi/D^*$, where D^* is an average level spacing. Then at a resonance

$$\delta^2 = \frac{D^*}{\pi}. \tag{34.2}$$

THOMAS [17] has briefly discussed a modification of the level spacing approach also. He gives as a formula

$$\langle\delta_\lambda^2\rangle_{\text{ave}} \approx q_\alpha T_\alpha \left(\frac{D}{W}\right) \tag{34.3}$$

with D the actual level spacing, T_α the kinetic energy of the alpha particle within the nucleus, w the characteristic *single-particle level spacing* within the nucleus ($W=\pi\hbar^2 K/MR$ where $K\sim 10^{13}\,\text{cm}^{-1}$, characteristic of a nucleon of mass M within the nucleus). $q_\alpha \gtrsim 1$ is a probability factor for the existence of the alpha particle on the nuclear surface. THOMAS mentions nuclear reaction work as supporting values near unity for q_α.

36. Surface well model. Another modification of the level spacing approach was made by WINSLOW [21], [27] who proposed a "surface well model". The attractive nuclear potential for the alpha particle is presumed to exceed the coulombic repulsive potential when the alpha particle is very near the nuclear surface but not yet "within" the nuclear matter. The boundary conditions from the level spacing estimate are applied at the inner surface, and the alpha will move in a surface region where its kinetic energy is positive before it enters the coulombic barrier. WINSLOW derives the decay rate theory for the one-body model of a single alpha particle trapped in the surface well and obtains a higher decay probability for given barrier radius than the traditional model yields. This result is to be expected, since the confinement of the alpha particle within the small volume of the surface means a greater kinetic energy for the lowest state and classically a higher collision rate with the barrier. WINSLOW then introduces the many-body concept of an equilibrium between the nuclear configuration with the alpha particle in the surface well and the configuration with none in the well but with constituent nucleons moving individually within the nucleus. The ratio of the probability of the former configuration to the total of configurations is designated as the preformation factor, P_L, which should reduce the one-body decay rate.

$$\delta^2 \approx P_L \cdot T_\alpha. \tag{35.1}$$

The level spacing model of DEVANEY [22] is used by WINSLOW to estimate P_L as

$$P_L \sim \frac{D}{2\pi\hbar}\left[\frac{v}{2\Delta R} + \frac{D}{2\pi\hbar}\right]^{-1} \approx \frac{D\cdot\Delta R}{\pi\hbar v}$$

where ΔR is the width of the surface well. Taking from BLATT and WEISSKOPF'S Table 3.2 [23] a D for Po^{212} decay of 0.7 Mev, WINSLOW finds that an inner (nuclear matter) radius of $1.31\times 10^{-13} A^{\frac{1}{3}}$ cm and $\Delta R = 1.2\times 10^{-13}$ cm give agreement with the experimental rate. He points out that the slower decay rate of

[1] This assumption is not correct in the square well one-body case, where $z(W)$ varies as $W^{\frac{1}{2}}$.

Po^{210} could be accounted for by a decrease of ΔR to 0.6×10^{-13} cm, leaving the inner radius unchanged.

The surface well picture may prove valuable in the future development of a detailed model of the alpha formation process. The square surface well model, as WINSLOW points out, represents an extreme idealization. The necessity of adjustment of the potential in the surface well to give the experimentally observed alpha decay energy to the lowest virtual state seems artificial, though, and the procedure of making the inside potential infinite needs more theoretical study.

WINSLOW's surface well model shows a regular decrease in decay rate with increasing angular momentum instead of the increase to a maximum at $L=2$ as found in the traditional model [*14*].

37. Concerning the intra-nuclear alpha potential energy. The recent paper of TOLHOEK and BRUSSARD [*33*] offers fresh and promising lines of attack on the question of the relation between the nuclear density distribution and the attractive potential for alpha particles and also on the fundamental mechanism of formation of the alpha particle from constituent nucleons.

By a four-step energy cycle argument, assembling and breaking up an alpha particle inside and outside the nucleus, they estimate an intra-nuclear potential energy of ~ -134 Mev for alpha particles. This large attractive potential "tails off" beyond the nuclear matter radius to counteract the coulombic potential for some distance and gives a large effective channel radius for alpha decay. They assume an attractive potential between alpha and an element of nuclear matter of the form $-V_0 \exp(-r^2/\beta^2)$ with $\beta=1.6\times10^{-13}$ cm, thus yielding a potential from nuclear forces of

$$V(\boldsymbol{x}) = \frac{-V_0}{\varrho_0 \pi^{\frac{3}{2}} \beta^3} \int d\xi\, \varrho(\xi)\, e^{-\frac{(\boldsymbol{x}-\xi)^2}{\beta^2}} \qquad (36.1)$$

where $V_0=167$ Mev, $\varrho(\xi)$ is the density function for nuclear matter (for which they substitute a specific form), and $\boldsymbol{x}$ is the position vector of the alpha particle. The results of the high energy electron scattering work are used to estimate the density function, and a function with the surface fall-off thickness parameter $s=1.26\times10^{-13}$ cm (as calculated by RAVENHALL and YENNIE[1] for gold) was used.

It has been conventional to argue [*18*] that the alpha particle in the nuclear potential well, being a BOSE-EINSTEIN particle, must be in its lowest level. For a well of nuclear volume this lowest level involves about 0.5 Mev of kinetic energy, in strong contrast to the $\sim$140 Mev of the above-discussed estimate. DEVANEY [*22*] has also discussed the question of the intra-nuclear alpha particle kinetic energy. He points out that the alpha particle, as a conglomerate of four FERMI-DIRAC particles, can only be considered to obey BOSE-EINSTEIN statistics when the perturbations on its constituent particles are small compared to the binding of the particles in the conglomerate. This condition is not fulfilled in nuclear matter, and the four-particle must obey the exclusion principle to some extent because of its FERMI-DIRAC constituents. DEVANEY estimates a kinetic energy of 5 Mev if the FERMI-DIRAC statistics were obeyed by the constituents, and he then proposes an average figure of 3 Mev to be most nearly correct.

We cannot resolve the vast discrepancy between these estimates and feel the matter is still open to question. The TOLHOEK-BRUSSARD well depth of

[1] D. G. RAVENHALL and D. R. YENNIE: Phys. Rev. **96**, 239 (1954).

-134 Mev represents a large departure from the customary one-body model estimates. The question of the potential (and consequently the average alpha particle kinetic energy) seems a difficult one, and there are unanswered questions attached to the energy cycle estimate. First, the energy cycle argument [*33*] has as its weakest link the energy associated with the assembly of nucleons into an alpha particle within nuclear matter; in view of the saturation nature of nuclear forces and the extra kinetic energy of alpha particle internal motion the formation of the alpha cluster in nuclear matter is in all probability a rather endothermic process. The problem is related to the problem of treating residual nucleon-nucleon interactions in the individual particle model. These interactions are responsible for configuration mixing in the shell model, and perhaps eventually the studies of specific nucleon-nucleon interactions within the nuclear well can shed light on questions concerning alpha particle clusters in nuclear matter.

38. Nucleon overlap model. TOLHOEK and BRUSSARD [*33*] propose a "shell model" picture for the probability of formation of alpha particles. The probability of finding an alpha particle in a particular position is equated approximately to the probability of finding the four constituent nucleons within the distance of the alpha particle radius (taken as $\beta_\alpha \approx 1.6 \times 10^{-13}$ cm). Then the alpha-clustering probability P_α of four particular nucleons is just proportional to the square of their overlap integral. Making the rough assumption of nucleon wave functions constant throughout the nuclear volume they obtain

$$P_\alpha = 64\left(\frac{\beta_\alpha}{R}\right)^9. \tag{37.1}$$

The frequency factor for an alpha hitting the potential barrier is taken as

$$f = \frac{v_i}{2R} P_\alpha n_\alpha \tag{37.2}$$

where n_α is the number of ways the energetic alpha particle can be formed from nucleons in outer orbits. n_α is taken by them to be about 3. Applying the above frequency factor with a potential falling off as determined by Eq. (36.1) to alpha decay of Po^{214} they calculate a nuclear matter radius of 6.8×10^{-13} cm, which corresponds to a constant of r_0 in the $R = r_0 A^{\frac{1}{3}}$ formula of 1.13×10^{-13} cm, consistent with the charge radii from electron scattering work. For this Po^{214} solution the preformation factor P_α is equal to 1.4×10^{-4}. This preformation factor of 1.4×10^{-4} will reduce the conventional one-body reduced width ($\delta^2 \approx$ 1 Mev), but the 300-fold increase in kinetic energy acts in the opposite direction with a factor of about 40 (see Sect. 24 on one-body δ^2 estimates). The decreased reduced width is consistent with experimental data for the larger effective channel width derived by TOLHOEK and BRUSSARD from the exponential tailing of the strong nuclear attractive force.

TOLHOEK and BRUSSARD also present a so-called dynamical estimate of the alpha formation probability, based on collision cross-section and mean free path arguments. The estimate yielded similar numerical results to the overlap integral estimate and was cited as adding confidence to the overlap integral results.

The nucleon overlap approach to the problem of alpha particle formation, while needing much further study, seems realistic and offers hope for future theoretical work. The great successes of the individual nucleon models, contrasted with the limited utility of the alpha particle nuclear model, supports the idea that within the body of nuclear matter the state of motion is more easily described by representations involving independent nucleon wave functions than by representations involving clustered groups of nucleons. The tendency toward some clustering is expressible in the individual nucleon representation by configuration mixing. As we move from the center of a heavy nucleus into the surface region and beyond, we must expect an increasing tendency toward association or clustering in the alpha particle grouping. That is, the total probability of finding the energetic nucleons at a given distance falls off with distance, but the relative probability of their being associated in an alpha particle must increase with distance. This conclusion follows from the fact that for alpha emitters there is some small probability of finding alpha particles at any distance, though unassociated nucleons cannot be found at large distances. It would seem logical to set the channel radius near the transition between the inner region of independent nucleons and the outer region of associated nucleons. The matching of boundary conditions on the alpha wave function could be made at this radius. The channel radius is obviously arbitrary, but if set too far out, the individual nucleon representation will encounter difficulty in adequately representing the alpha cluster probability at the channel radius. On the other hand, if the channel radius is set too small, the treatment of the alpha particle as experiencing simply a real potential may fail close to the channel radius. The optimum channel radius R_c probably lies near the outside of the nuclear density distribution. If R_c lies in a region of positive kinetic energy of the alpha, then a form of WINSLOW's surface well picture may be applicable. If R_c lies in the negative kinetic energy or barrier region, more conventional treatments with COULOMB functions determining the boundary conditions would be applicable.

The theoretical analysis of relative intensities to rotational band members for alpha decay of spheroidal nuclei promises to add much to our understanding of the process of alpha formation. These studies appear to indicate that the mean free path of alpha particles in nuclear matter is quite small. The explanation in terms of the individual nucleon wave function overlap seems promising. For a more detailed discussion refer to Sects. 32 and 33.

WHEELER[1] has set up a formalism incorporating mixing of configurations specifically composed of groups or clusters, such as the alpha particle. Although the use of a one-body model frequency factor reduced by a "pre-formation" factor lies essentially in the resonating group category, no rigorous attempt has been made to formulate such an approach for alpha emitters with due attention to redundancy of coordinates and the supplementary conditions on them, and the partial observance of the exclusion principle (discussed by DEVANEY [*22*]).

39. The abnormal decay rates of alpha emitters with neutron number 126 or less. It is easy to recognize from the individual nucleon approach that the alpha formation probability would be lower for the even-even polonium isotopes with 126 or less neutrons. The $6h$ proton radial function is concentrated at the outermost part of the nucleus and overlaps well with the $7i$ or $6g$ neutron functions beyond 126 neutrons, but the $6h$ protons would overlap poorly with the $4p$ neutrons (most lightly bound) for neutron numbers of 126 and just below.

It also would seem reasonable that the cut-off radius for the barrier might be somewhat less for daughter nuclei with no nucleons outside the closed shells than for those possessing a few. The extra slowness of polonium alpha emitters has been discussed in terms of a shrinkage of the nuclear radius at the closed shell [*6*].

[1] J. A. WHEELER: Phys. Rev. **52**, 1083, 1107 (1937).

It may be possible to decide whether formation probability (overlap) or radius shrinkage is the more important effect by examining in detail the reduced transition probabilities (*RTP*) of some polonium and emanation alpha emitters.

Of even-even alpha emitters with daughters from two to six nucleons beyond the closed shells, three (Em^{218}, Em^{220}, Ra^{222}) can be taken as a basis of the normal alpha decay. Their mean *RTP*[1] is 0.086. Po^{212} decays to Pb^{208} and should reflect any shrinkage of the effective channel radius at the closed shell but should not suffer any reduction in alpha formation probability from the overlap integral arguments. Its *RTP* is 0.025, a factor of 3.4 smaller than the base comparison. (The heavier poloniums have a mean *RTP* a factor of 2.3 smaller than the base comparison.) The six even-even alpha emitters of polonium and emanation with 126, 124, and 122 neutrons exhibit a mean *RTP* of 0.0027, a factor of 32 smaller than the base comparison. Semi-quantitatively we may divide the retardation factor for these alpha emitters of $N \leq 126$ into a factor of three due to effective radius decrease and a factor of 10 due to lowered formation probability.

C. Kinetics of alpha decay of odd nucleon types.

I. Correlations of decay rates of odd nuclei.

40. Hindrance factors for odd nuclei. In contrast to the regular behavior of decay rates of even-even alpha emitters the odd nucleon types show great variation in their rates. Sometimes transition probabilities of alpha groups are comparable to those of neighboring even-even nuclei; in other cases the rates are much slower.

It is most in the region of regular decay rate behavior of even-even nuclei to try to compare the reduced decay rates of alpha groups of an odd nucleus to the ground state rates of its nearest even-even neighbors. The factor by which the alpha group in question decays slower than this "normal" rate is usually called the hindrance factor (which we designate by F) (see Sects. 17 and 42).

One might calculate effective nuclear radii for the even-even neighbors using some form of one-body rate theory and apply an average radius value to calculation of the "normal" decay rate for the odd emitter. A simpler and nearly equivalent method is to construct a family of normal half-life vs. energy curves (Fig. 14, Sect. 16) that closely represent the behavior of the even-even ground decay groups and then to read normal alpha half-lives for odd groups from these curves. For the odd-Z isotopes it is necessary to define lines between the neighboring even-Z lines, and these lines are usually chosen to give a mean decay rate between even-even neighbors of the same decay energy.

As discussed in Sect. 16, in the region above the closed shells the partial alpha half-lives of even-even alpha emitters of a given element may be well represented by expression (16.1):

$$\operatorname{Log} t_{\frac{1}{2}\alpha}(\mathrm{sec}) = A\, Q_{\mathrm{eff}}^{-\frac{1}{2}}(\mathrm{Mev}) + B.$$

The justification of this particular form by decay rate theory was discussed in Sect. 20. The constants A and B obtained from a least squares analysis were tabulated in Table 2.

---

[1] Em^{218} seems somewhat anomalously to decay too fast compared to its neighbors. Exclusion of Em^{218} from the comparison base would give an average *RTP* (between Em^{220} and Ra^{222}) of 0.073.

In Table 11 we supplement Table 2 with the interpolated constants for the odd-Z lines used as a basis for hindrance factor calculations.

Fig. 19 shows a plot of the "normal" decay rate lines with points for all the known ground state groups of even-odd nuclei. The distance of a point above its line is proportional to the logarithm of its hindrance factor, F. That is,

$$\operatorname{Log} F = \operatorname{Log} t_{\frac{1}{2}\alpha} - (A\,Q_{\mathrm{eff}}^{-\frac{1}{2}} + B)\,. \quad (40.1)$$

Table 11. *Interpolated semi-empirical constants*[1] *for hindrance factor calculations.*

Element	A	B	Remarks
99 E	155.04	−53.3141	Extrapolated
97 Bk	152.65	−53.3166	
95 Am	149.33	−52.8862	
93 Np	146.86	−52.8732	
91 Pa	145.84	−53.4604	
89 Ac	141.68	−52.7060	
87 Fr	138.31	−52.3037	
85 At	133.40	−51.1913	

Tables 12, 13, and 14 list F for all alpha groups of even-odd, odd-even, and odd-odd nuclei with 128 or more neutrons. (This excludes the region around 126 neutrons where decay rate properties of even-even nuclei change sharply and the hindrance factor concept is not well defined.)

The distribution of logarithms of hindrance factors for all the established alpha groups of odd nuclei beyond the closed shells ($Z \geq 84$, $N \geq 128$) are given as histograms in Fig. 20 (odd-evens), Fig. 21 (even-odds), and Fig. 22 (odd-odds). The bold line in each case divides the alpha emitters with $N \leq 138$ (above) from those with $N > 138$ (below), roughly the dividing line between the region of spherical nuclear shape and that of stable spheroidal shape.

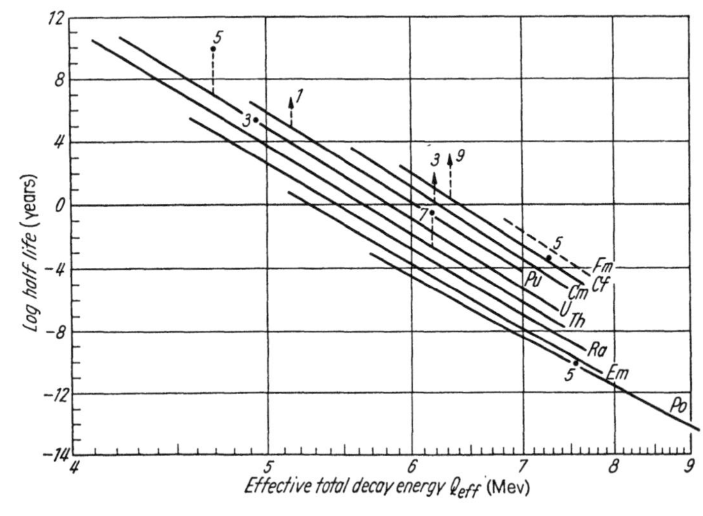

Fig. 19. Half-life *vs.* energy relation for ground-state transitions of even-odd nuclei using Log $t_{\frac{1}{2}}$ and $Q_{\mathrm{eff}}^{-\frac{1}{2}}$ coordinates. The lines are those of Fig. 14 pertaining to the even-even nuclei. The lengths of the dashed lines are the logarithms of the hindrance factors (F).

In the $N \leq 138$ region one notes in Figs. 20, 21, and 22 generally many alpha groups with low hindrance factors for all odd types including odd-odd. Exceptional are the alpha emitters Ra^{223}, Ac^{225}, and Th^{227}, which exhibit great complexity with many highly hindered groups.

[1] Constants for the even-Z elements are listed in Table 2. The constants are for use in Eq. (40.1), where $t_{\frac{1}{2}\alpha}$ is in seconds and Q_{eff} is in Mev.

Table 12. *Hindrance factors for even-odd nuclei.*

Alpha emitter	Alpha particle energy Mev	Hindrance factor F	Alpha emitter	Alpha particle energy Mev	Hindrance factor F
Fm^{255}	7.08	2.6	U^{231}	5.45	1.1
	7.04	1.8	U^{229}	6.42	2.3
Fm^{251}	6.9	5.2	Th^{229}	5.02	110.0
Cf^{249}	6.19	N.R.[1]		4.94	18.0
	6.04			4.85	1.3
	5.94	N.R.[1]	Th^{227}	6.030	120.0
	5.90			6.001	500.0
	5.80	2.9		5.970	58.0
Cf^{245}	7.11	1.4		5.952	380.0
Cm^{245}	5.45	42.0		5.907	700.0
	5.36	2.5		5.859	140.0
	5.31	5.5		5.800	200.0
Cm^{243}	6.003	1700.0		5.792	1850.0
	5.985	280.0		5.749	4.9
	5.777	1.4		5.706	7.6
	5.732	4.8		5.699	12.0
	5.679	16.0		5.661	19.0
Cm^{241}	5.95	12.0	Th^{225}	6.57	2.5
Pu^{241}	4.893	3.2	Ra^{223}	5.860	large
	4.848	5.1		5.735	59.0
Pu^{239}	5.150	2.9		5.704	6.8
	5.137	10.0		5.592	4.2
	5.099	9.3		5.525	4.9
	4.98	900.0		5.487	14.0
	4.78	140.0		5.418	3.9
	4.73	62.0	Ra^{221}	6.71	3.4
Pu^{235}	5.85	1.7	Em^{221}	6.0	7.7
U^{235}	4.58	950.0	Em^{219}	6.807	14.0
	4.47?	450.0		6.542	5.8
	4.40	4.6		6.417	2.2
	4.20	2.3		6.197	0.76
U^{233}	4.816	1.2	Em^{217}	7.74	4.5
	4.773	3.3	Po^{215}	7.38	1.2
	4.717	12.0	Po^{213}	8.35	2.0
	4.489	14.0			

Odd-odd nuclei of the $N > 138$ region are probably generally very highly hindered. E^{252} and E^{254} constitute two of the three known cases in the whole region. It may be significant that these alpha emitters lie beyond 152 neutrons, where a small discontinuity ("subshell effect") in neutron binding energies is observed. Also, it is probably significant that E^{253} exhibits essentially unhindered ("favored") decay to the ground state of Bk^{249}, a unique case for odd-even nuclei of the $N > 138$ region. One might be justified in treating the region of $N > 152$ as distinct from the other regions, although we have not done so.

It is evident that the hindrance factor as defined and calculated above is not necessarily a very fundamental measure of the alpha group probability at the nuclear surface. The centrifugal barrier effects associated with angular momentum of alpha decay are not taken into account, and neither are the effects of the large non-central electric quadrupole interactions in the spheroidal region.

[1] Alpha groups not resolved.

Table 13. *Hindrance factors for odd-even nuclei.*

Alpha emitter	Alpha particle energy Mev	Hindrance factor F	Alpha emitter	Alpha particle energy Mev	Hindrance factor F
E^{253}	6.633	1.2		4.787	3.5
	6.592	8.8		4.767	4.8
	6.545	24.0		4.713	34.0
	6.493	80.0		4.674	9.2
	6.25	17.0		4.644	3.1
E^{251}	6.48	3.0		4.589	15.0
E^{249}	6.76	5.8		4.52	120.0
Bk^{249}	5.40	6.4	Np^{235}	5.06	1.7
	5.06	0.70	Np^{233}	5.53	0.32
Bk^{247}	5.67	70.0	Pa^{231}	5.046	230.0
	5.51	5.9		5.017	68.0
	5.30	3.9		5.001	51.0
Bk^{245}	6.37	450.0		4.971	330.0
	6.16	34.0		4.938	21.0
	5.90	2.5		4.921	130.0
Bk^{243}	6.72	670.0		4.839	73.0
	6.55	68.0		4.722	1.5
	6.20	4.9		4.696	7.6
Am^{243}	5.339	1500.0		4.666	3.2
	5.308	1000.0	Pa^{229}	5.69	5.7
	5.266	1.1	Pa^{227}	6.46	1.5
	5.224	4.5	Ac^{227}	4.942	3.3
	5.169	18.0	Ac^{225}	5.818	7.1
Am^{241}	5.535	520.0		5.781	4.6
	5.503	600.0		5.719	10.5
	5.476	1.2		5.627	11.0
	5.433	4.2	Ac^{223}	6.64	3.1
	5.379	20.0	Fr^{221}	6.332	3.3
	5.314	750.0		6.116	1.9
	5.27	1500.0	Fr^{219}	7.30	1.2
Am^{239}	5.75	2.3	At^{219}	6.27	1.9
Am^{237}	6.01	4.6	At^{217}	7.05	0.60
Np^{237}	4.872	230.0	At^{215}	8.00	3.7
	4.816	140.0			

Centrifugal barrier effects on the wave function in the external region are well worked out and should be incorporated in more detailed comparison of alpha decay rates, where spin changes are known. The large electric quadrupole interactions in the region of deformed nuclei should be taken into account, but this is a problem of greater difficulty.

Table 14. *Hindrance factors for odd-odd nuclei.*

Alpha emitter	Alpha particle energy Mev	Hindrance factor (F)
E^{254}	6.42	1.7
E^{252}	6.64	16.0
Bk^{244}	6.67	1900.0
Pa^{228}	6.09	63.0
	5.85	13.0
Ac^{224}	6.17	22.0
Fr^{220}	6.69	7.8
At^{218}	6.63	2.0
At^{216}	7.79	2.6

For the odd nuclei not included in the hindrance factor Tables 12, 13, and 14 we have calculated by the prescription of WINSLOW [*16*] [our Eq. (27.9)] the alpha surface probabilities, assuming $L=0$ and, where the true angular momentum values may be inferred (see Sects. 49 to 52), the reduced transition probabilities (RTP) which

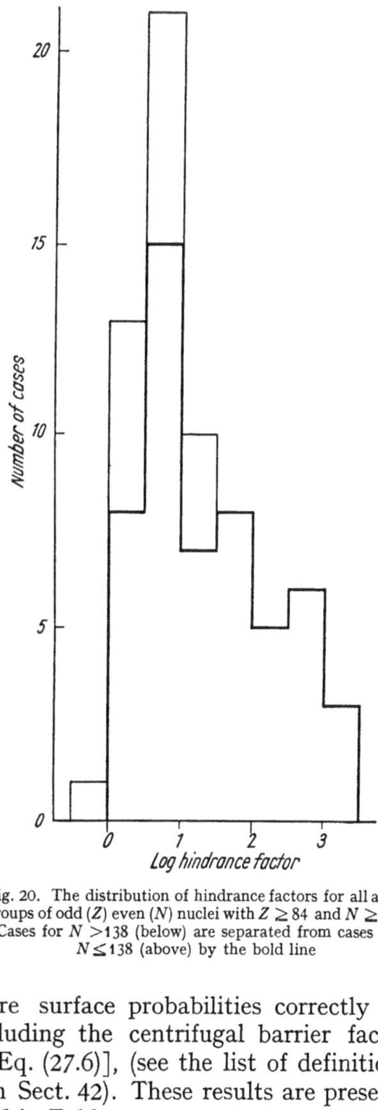

Fig. 20. The distribution of hindrance factors for all alpha groups of odd (Z) even (N) nuclei with $Z \geq 84$ and $N \geq 128$. Cases for $N > 138$ (below) are separated from cases for $N \leq 138$ (above) by the bold line

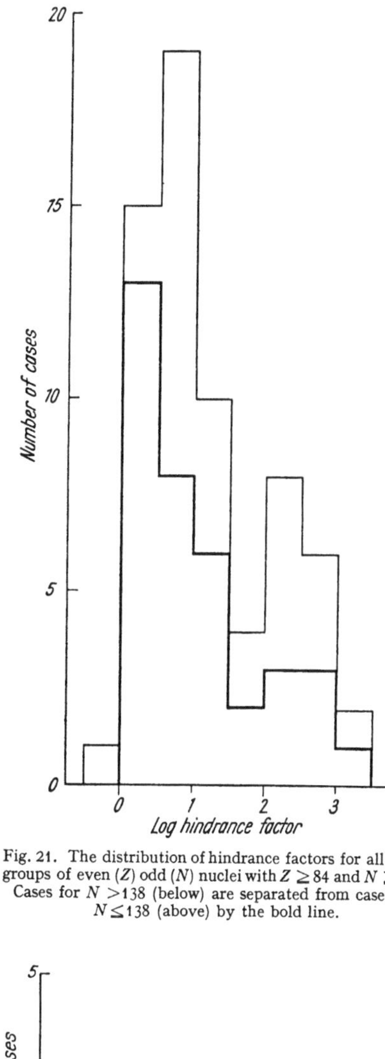

Fig. 21. The distribution of hindrance factors for all alpha groups of even (Z) odd (N) nuclei with $Z \geq 84$ and $N \geq 128$. Cases for $N > 138$ (below) are separated from cases for $N \leq 138$ (above) by the bold line.

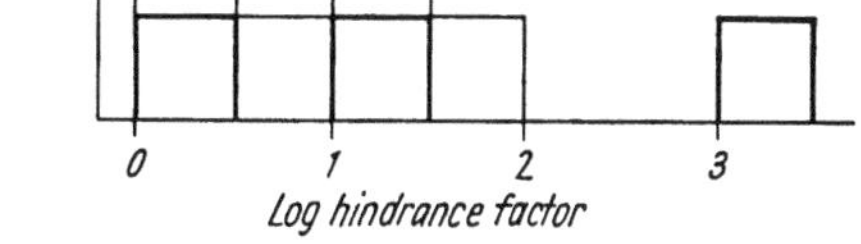

Fig. 22. The distribution of hindrance factors for all alpha groups of odd (Z) odd (N) nuclei with $Z \geq 84$ and $N \geq 128$. Cases for $N > 138$ (below) are separated from cases for $N \leq 138$ (above) by the bold line.

are surface probabilities correctly including the centrifugal barrier factor [Eq. (27.6)], (see the list of definitions in Sect. 42). These results are presented in Table 15.

41. Decays across the major closed shells. The order of magnitude of reduced hindrance factor (cf. Sect. 42) for decay across closed shells, compared to the even-even alpha emitters above the closed shells (Em^{218}, Em^{220}, Ra^{222}) (mean $RTP = 0.086$) may be summarized as follows:

α) Decays crossing one closed configuration. The odd and even mass bismuth isotopes with $N \geq 126$ have reduced hindrance factors of order 200. The Po^{211} $(^9/_2{}^+)$ groups to $\boldsymbol{p}$ configurations experience a similar reduced hindrance of about 100 and to $\boldsymbol{f}_{\frac{5}{2}}$ about 1000. We tentatively associate this retardation mainly

Table 15. *Surface probabilities and reduced transition probabilities for odd nuclei with $Z < 84$ or $N < 128$.*

Alpha emitter	Measured alpha particle energy Mev	Log $t_{\frac{1}{2}\alpha}$ sec	Neg. logarithm of surface probability base 10	Lowest alpha angular momentum[1]	Reduced transition probability[2] RTP
Fr^{212}	6.411	3.85	3.43	—	—
	6.387	3.83	3.31	—	—
	6.342	4.04	3.33	—	—
Em^{211}	5.847	5.82	3.38	(2)	$(7.1 \cdot 10^{-4})$
	5.779	5.54	2.77	—	—
	5.613	7.05	3.45	—	—
Em^{209}	6.037	4.03	2.39	—	—
Em^{207}	6.14	4.21	3.08	—	—
At^{211}	5.862	4.819	2.92	(0)	$(1.2 \cdot 10^{-3})$
At^{210}	5.519	7.73	4.17	—	—
	5.437	7.75	3.77	—	—
	5.355	7.67	3.26	—	—
At^{209}	5.42	5.64	2.73	(0)	$(1.9 \cdot 10^{-3})$
At^{208}	5.65	6.10	3.20	—	—
	5.52	7.56	4.00	—	—
At^{207}	5.75	4.86	2.45	(0)	$(3.5 \cdot 10^{-3})$
Po^{211}	7.44	−0.28	4.20	5	$7.2 \cdot 10^{-4}$
	6.88	1.99	4.63	3	$5.9 \cdot 10^{-5}$
	6.56	2.02	3.49	3	$9.5 \cdot 10^{-4}$
Po^{211m}	8.70	2.55	10.55	(10)	$(1.5 \cdot 10^{-7})$
	7.85	3.00	8.73	(8)	$(5.9 \cdot 10^{-7})$
	7.14	1.83	5.36	(3)	$(2.9 \cdot 10^{-5})$
Po^{209}	4.877	9.50	2.88	2	$2.3 \cdot 10^{-3}$
	4.62	11.90	3.64	—	—
Po^{207}	5.10	8.28	2.98	—	—
Po^{205}	5.2	6.86	2.13	—	—
Bi^{214}	5.505	6.82	4.14	—	—
	5.444	6.73	3.75	—	—
Bi^{213}	5.86	5.14	4.15	(5)	$(9.0 \cdot 10^{-4})$
Bi^{212}	6.086	4.60	4.60	5	$3.2 \cdot 10^{-4}$
	6.047	4.19	4.02	3	$2.7 \cdot 10^{-4}$
	5.765	5.80	4.38	—	—
	5.622	6.86	4.76	—	—
	5.603	5.99	3.80	—	—
	5.481 ?	7.83	5.04	—	—
Bi^{211}	6.620	2.198	4.31	5	$6.1 \cdot 10^{-4}$
	6.273	2.875	3.65	3	$5.8 \cdot 10^{-4}$
Bi^{210}	4.94	13.92	8.14	4	$4.2 \cdot 10^{-8}$
$Bi^{210m}_{(Ra\,E)}$	(4.96) ?	11.41	5.80	(0)	$(1.6 \cdot 10^{-6})$
Bi^{203}	4.85	11.80	5.57	—	—
Bi^{201}	5.15	8.10	3.61	—	—
Bi^{199}	5.47	7.20	4.40	—	—
Bi^{198}	5.83	5.96	4.88	—	—

with the reduced overlap integral where one constituent nucleon forming the alpha particle is concentrated well within the nuclear volume and its radial wave function overlaps poorly with the three high angular momentum nucleons beyond the shell. The smaller RTP to $f_{\frac{5}{2}}$ than to the p configurations in Po^{211} decay is the reverse of the simple expectation, though. The high spin Po^{211}

[1] Parentheses indicate uncertainty in assignment of angular momenta.

[2] Reduced transition probability is calculated for the lowest permitted alpha angular momentum, but this does not imply that mixed angular momenta will not be encountered in reality.

siomer has reduced hindrance factors (assuming $^{19}/_2{}^-$) of 5.7×10^5 to $p_{\frac{1}{2}}$, 1.5×10^5 to $p_{\frac{3}{2}}$, and 2.5×10^3 to $i_{\frac{13}{2}}$.

β) Decays crossing both closed configurations, Bi^{210}. The low spin isomer (RaE) has a reduced hindrance of 5.4×10^4 for $L=0$, while the high spin isomer has reduced hindrance of 2.0×10^6 with $L=4$ assumed.

γ) Decays for $Z\geqq 84$ and $N\leqq 126$. As discussed in Sect. 39 most even-even nuclei of the region below 126 neutrons have a reduced hindrance factor of about 30. Several of the odd alpha emitters exhibit more than one group. They range in hindrance from about that of their immediate even-even neighbors to as much as an additional factor of 50.

42. Summary of proposed terminology. *Hindrance factor* (F). Ratio of the experimental partial alpha half life to a "normal" value based on ground-state transitions of even-even neighbors. For excited state transitions in an even-even nucleus the basis of comparison is to be taken as its own ground-state transition rate. For all odd nuclei we have defined the "normal" rate from the semi-empirical correlation of even-even decay rates in Sect. 16. Formally, $\mathrm{Log}\,F = \mathrm{Log}\; t_{\frac{1}{2}\alpha} - (A_z\, Q_{\mathrm{eff}}^{-\frac{1}{2}} + B_z)$. Centrifugal barrier and non-central interaction effects are ignored in calculating hindrance factors.

Reduced hindrance factor. The same as hindrance factor except that centrifugal barrier effects (and non-central interaction effects where important) have been taken into account. The correction of F for centrifugal barrier effects consists simply in multiplying F by $(G_L/G_0)^{-2}$. See Table 5 and Eq. (27.6).

Surface probability. A quantum-mechanical probability expression for a given alpha decay group, proposed by WINSLOW [*16*] and denoted as $R\,|\Phi_{J,I}(R)|^2$. The outgoing COULOMB wave $\Phi_{J,I}(r)$ is normalized such that the total outgoing flux is proper for the given partial alpha half-life if there is a unit probability of finding the alpha in the nuclear region. The probability function $|\Phi_{J,I}(R)|^2$ is made dimensionless by multiplying by R. For the calculations of the present paper the function is evaluated at $R=9.3\times 10^{-13}$ cm for heavy elements and at $R=8.0\times 10^{-13}$ cm for rare earths. For the present paper the "surface probability" will be calculated without considering centrifugal barrier or non-central interaction effects, just as is the hindrance factor.

Reduced transition probability (RTP). This is the real "surface probability", correctly taking into account centrifugal barrier and non-central interaction effects.

Reduced derivative width (δ_L^2). A fundamental measure of the hypothetical "decay rate in the absence of a barrier". The decay rate constant without barrier, λ_0, is equal to δ_L^2 divided by PLANCK's constant h [see Eq. (27.1)].

II. Alpha decay as a multibody process (the odd nuclei).

43. General observations. It was noted early that odd nuclei tend to decay at a generally slower rate than neighboring even-even nuclei. From the known data on alpha groups of odd nuclei we see by referring to the histogram distributions (Figs. 20, 21, and 22 of Sect. 40) of hindrance factors that the most probable value of the hindrance factor is not much greater than unity, with the probability gradually falling off toward the higher hindrance factors. The only possibly significant separate groups in the distributions lie near hindrance factor 500.

The attempts to explain hindrance factors solely on the basis of the centrifugal barrier for higher angular momentum waves cannot possibly account for the hindrance factors near 1000. An angular momentum of 5, the maximum allowable for the ground state alpha group of Am^{241} (Initial, $^5/_2{}^-$; Final, $^5/_2{}^+$) has an extra centrifugal barrier reduction factor of only 12, compared to the actual hindrance factor of 520. Following such arguments as this, PERLMAN *et al.* [*6*] suggested that the large hindrance factors are to be associated mainly with abnormally small preformation factors for the assembly of an alpha particle from unpaired nucleons.

44. Favored alpha decay intensities. The development of the BOHR-MOTTELSON model for deformed nuclei and the accumulation of more precise data on alpha fine structure and nuclear excited states has made possible some really quantitative correlations of decay rates of a few selected alpha groups in odd nuclei. It was suggested[1] that the main three alpha groups of Am^{241} populate excited levels of a single rotational band, with the spacings suggesting the assignment of $I=\frac{5}{2}$ to the base state of the band. The spin of the Am^{241} ground state was known to be $\frac{5}{2}$, and the main alpha decay group to the 60-kev state exhibited no hindrance from normal even-even alpha decay rate behavior (hindrance factor about unity). Thus, it seemed logical to theorize that the odd proton wave functions in Am^{241} and the 60-kev state of Np^{237} were essentially the same. PERLMAN and ASARO [*4*] were soon to point out cases in other odd-mass nuclei where essentially unhindered alpha decay groups populated three members of a rotational band (Am^{243}, U^{233}, Cm^{243}). BOHR, FRÖMAN, and MOTTELSON [*31*] (referred to hereafter as BFM) directed further attention to these special decay groups and proposed the term "favored" for those alpha transitions in which the intrinsic wave function of the odd nucleon remains essentially unchanged. They, furthermore, derived a simple approximate relationship between the intensities of favored alpha decay to various rotational band members and the intensities of alpha decay groups in neighboring even-even nuclei (see Sect. 32). Their formula is exact only in the limit of infinite nuclear moment of inertia. For favored alpha decay from a nucleus with spin $I_0=K_0$ to a band in the daughter of like K, the $L=0$ alpha group can only go to the base level by conservation of angular momentum. The $L=2$ group can generally divide between more than one band member. The relative intensities of the $L=2$ branches by the BFM theory are proportional to $P(Z,E)\times(I_0 2 K_0 O|I_0 2 I_f K_0)^2$, where the P's are barrier penetration factors and the other factor is the square of a CLEBSCH-GORDAN coefficient. The $L=4$ wave generally populates more levels, and the relative branching is determined in a similar way. The relative intensities of $L=0$, $L=2$, and $L=4$ waves are estimated from the relative alpha intensities to rotational band members in neighboring even-even nuclei. For much of the heavy region the $L=4$ group is almost negligible. The relative transition probabilities are given by

$$\lambda \propto P_0(Z, E_f) \sum_L C_L (I_i L K O | I_i L I_f K)^2 \tag{44.1}$$

where C_L is the reciprocal of the hindrance factor for the alpha group of angular momentum L in neighboring even-even nuclei.

As a demonstration of the kind of agreement between this theory and experiment we reproduce in Table 16 the Table 4 of BFM, brought up to date with the most recent data and using revised estimates for the C_L's.

There is good semi-quantitative agreement, and most discrepancies may possibly be explainable when numerical work is performed without the assumption of infinite nuclear moment of inertia.

[1] J. O. RASMUSSEN: Ark. Fysik 7, 185 (1953).

Table 16. *Alpha group intensities in favored alpha decay.*

Parent nucleus	I_i	Observed relative intensities				Calculated relative intensities (BFM) [31]				Assumed constants	
		$I=K_f$	K_f+1	K_f+2	K_f+3	$I=K_f$	K_f+1	K_f+2	K_f+3	C_2	C_4
E^{253}	$(^{11}/_2)$					100	5.3	0.5	0	0.27	0[1]
	or	100	8.5	2.0	0.3						
	$(^7/_2)$					100	7.0	1.2	0.1	0.27	0.013
Cm^{243}	$(^5/_2)$	100	17	<2	—	100	12	1.9	—	0.56	—
Am^{243}	$^5/_2$	100	13	1.5	—	100	12	1.9	—	0.56	0.002
Am^{241}	$^5/_2$	100	16	1.2	0.02	100	13	2.2	0.02	0.59	0.005
Pu^{241}	$^5/_2$	100	33	—	—	100	13	—	—	0.59	—
Pu^{239}	$^1/_2$	100	23	15	—	100	20	9.5	—	0.63	—
U^{233}	$^5/_2$	100	18	2.4	—	100	16	2.9	—	0.87	0.067
Pa^{231}	$^3/_2$	100	19	—	—	100	15	—	—	1.0	—

Two cases were calculated for E^{253}. The level spacings support the $K=\frac{11}{2}$ assignment, but in view of possible level perturbations the $K=\frac{7}{2}$ possibility has also been calculated. (These two spins are the only high spins that appear reasonable for $Z=99$ by NILSSON's [*13*] spheroidal well calculations.) The intensity comparison rather supports the lower K.

The theory of intensities of favored alpha decay really avoids the basic problem of alpha formation, since it makes use of experimental intensity data from even-even nuclei. This theory neatly exploits the similarity between favored alpha decay and decay of even-even nuclei.

45. Hindered decay; selection rules; and approaches to the problem. Going beyond the favored decay groups to the hindered groups of odd nuclei we find a territory still uncharted by theory. Here we only hope to define the problem and indicate some of the possible paths that may bear future exploration. The empirical hindrance factor distributions of Sect. 40 offer few clues, as they show only slight tendency toward any grouping.

At the outset we may set down the strict selection rules for alpha decay, based on conservation of total angular momentum and parity.

$$|I_i - I_f| \leq L \leq I_i + I_f. \tag{45.1}$$

Only even values of L are allowed if the parity of initial and final nuclear states is the same and only odd values of L allowed if there is a parity change. If either I_i or I_f is zero, it is possible to have cases where alpha decay is strictly forbidden. For example, decay of an even-even nucleus (0^+) cannot populate odd spin states of even parity or even spin states of odd parity.

Alpha decay rate behavior stands in marked contrast to that of beta and gamma decay in that angular momentum change *per se* plays a relatively minor role. Alpha decay involving an angular momentum of eight units (Cm^{242}) has been detected in competition with $L=0$ emission. In contrast, the associated angular momentum is the dominant factor separating beta and gamma emission rates into groups. The relative insensitivity of alpha transition rates to angular momentum means that mixtures of L values may frequently be encountered in alpha decay of odd nuclei, a matter of special importance to angular correlation studies involving alpha radiation.

[1] Assumed zero because necessary CLEBSCH-GORDAN coefficients not readily calculable.

For theoretical interpretation of the alpha emission process for groups other than "favored" we can recognize three areas that need consideration and exploration: First, the influence of non-central interactions, both electromagnetic and short-range nuclear; second, configuration mixing in parent or daughter that may allow alpha emission from a paired nucleon structure; and third, the direct formation of alphas from constituent nucleons in unpaired configurations.

46. Influence of non-central interactions. The first area, non-central interactions, has been dealt with in Sects. 28 to 33, and we only take time briefly to recapitulate here.

The coupled radial wave equations in the region beyond the nucleus have been precisely formulated to include the effects of electric multipole radiation fields. Numerical application of the equations has been quite limited to date, but it seems clear that the fast rotational electric quadrupole fields play an important role in the relative alpha group intensities to states connected by these $E2$ matrix elements. As yet, only the $E2$ coupling within a given rotational band has been treated numerically, but relatively large $E2$ matrix elements may also connect states of different bands where there has been configuration mixing.

$E1$ matrix elements of the order of magnitude of the single proton estimate[1] or larger could play a significant role. However, many of the low-energy $E1$ transitions in the alpha emitter region are of measured half-life and retarded by more than 10^4 from the single particle lifetime. The numerical example of Am^{241} in Sect. 30 showed that the intensity of the hindered ground state group could not be accounted for by assuming the group to be of zero amplitude at the nuclear surface and considering the build-up due to $E1$ coupling to the favored group to the 60-kev state. There is theoretical reason to believe that the odd mass nuclei with $E1$ lifetimes too short to be measured may still be considerably retarded from the single proton formula[1] expectation. Hence, electric dipole coupling probably has negligible influence on alpha group intensities of odd-mass nuclei. (The lifetimes of $E1$ transitions from 1^- states of even-even nuclei are more uncertain, only upper limits being known.)

The possibility exists that short-range nuclear forces or surface polarizations may exert important coupling on the alpha particle in the nuclear surface region. The formulas of Sects. 28 to 33 could be adapted to study this coupling if some form and strength of the interaction were to be assumed.

47. Influence of configuration mixing. In the second area, we must consider configuration mixing in parent or daughter that may allow alpha emission from the even-even paired structure of the nucleus.

In the region of strongly deformed nuclei the ordinary BOHR-MOTTELSON wave function (Eq. (II.15) of Ref. [*10*]) for an odd-mass nucleus will consider all nucleons paired except the odd one, and the odd nucleon will move within the deformed well such that the projection (Ω) of its angular momentum about the cylindrical symmetry axis is conserved. Certain perturbations, as the rotation-particle-coupling *(RPC)*[1] and residual nucleon-nucleon interactions will mix the zero order wave functions somewhat. KERMAN's analysis[2] of the levels of W^{183} shows the effect of the rotation-particle-coupling in mixing of bands of the same parity but differing in Ω (and K) by one. Other interactions may connect still other pairs of intrinsic states.

---

[1] S. A. MOSZKOWSKI: Chap. XIII in Beta and Gamma Ray Spectroscopy; KAI SIEGBAHN, Editor; p. 391. Amsterdam: North-Holland Publishing Co. 1955.

[2] A. K. KERMAN: Dan. Mat. fys. Medd. **30**, No. 15 (1956).

Obviously, we can again, as for favored alpha decay, exploit empirical knowledge of the alpha decay rates of neighboring even-even nuclei to interpret decay of odd-mass alpha emitters wherever configuration mixing of states gives in parent and daughter some configurations with identical odd nucleon wave functions. Expression of these concepts might be treated in the formalism of the fractional parentage coefficients.

Knowing the general importance of the rotation-particle-coupling, we may predict that high hindrance factors in deformed nuclei should not usually occur where initial and final states in the alpha transition differ by one unit in Ω (and K) and S- or D-*wave* alpha emission is allowed by spin and parity selection rules. No systematic test of this hypothesis will be attempted here, but we know of no violations in the highly hindered ($\sim$500) groups, most of which involve a parity change and some of which (i.e. Cm^{243}) involve an Ω change of two units with no parity change.

Alpha decay groups involving a parity change may also be partly or wholly accountable by configuration mixing, especially in the region (near $A=224$) where the $I=1^-$, $K=0$ states in even-even nuclei lie lowest and where the hindrance factors for decay to them are relatively low. That is, one would consider the mixing in of configurations in which the odd nucleon is promoted to an orbital of opposite paritiy (from the principal configuration) and is coupled with an even-even core in an odd parity state of collective motion.

48. Direct formation from unpaired nucleons; the overlap model. The third area to consider for non-favored alpha decay groups includes the process of direct formation of alpha particles from constituent nucleons, at least two of which are not paired to each other in the individual nucleon configuration. The elucidation of the participation of this third mechanism would be of great fundamental interest.

The direct formation mechanism may be amenable to theoretical treatment by a Tolhoek-Brussard-type individual nucleon approach [*33*] in which the alpha formation probability inside the nucleus is proportional to the overlap of the four consituent nucleon wave functions. It seems obvious that the overlap will be a good deal less for orbitals not in a pair than for paired orbitals, since the configuration interaction whereby paired nucleons may increase their overlap would generally be absent for the nucleons not paired together. For alpha emitters decaying across a closed shell (the Bi isotopes and 127-neutron isotones) the radial overlap integral should be extremely poor, since it involves nucleons in different shells with greatly differing angular momentum (see Sect. 41).

D. Illustrative individual decay schemes.

I. Alpha decay across a closed shell.

49. Bismuth-211 and Bismuth-213. Bi^{211} (AcC) decays across the 82-proton shell according to the decay scheme of Fig. 23. With the spherical well shell-model assignments of Pryce [*9*] (refer to Sect. 15, Fig. 12) to ground-state alpha group must be pure $L=5$ and the lower-energy group may have $L=3$ or 5. When Winslow's [*16*] (see our Table 15) calculated surface probabilities at 9.3×10^{-13} cm are corrected for angular momenta 5 an 3, respectively, the reduced-transition probabilities are 6.1×10^{-4} for the ground-state transition and 5.8×10^{-4} for the other transition. It is interesting to note how nearly equal these values are though both are highly hindered in comparison with nearby even-even nuclei

(Em^{218}, Em^{220}, Ra^{222}: $RTP \approx 0.1$). The extraordinary slowness may be associated qualitatively with a lowered formation probability (poor overlap of constituent protons in parent nucleus).

The 353-kev gamma transition in Tl^{207} has been shown[1] to have a K-conversion coefficient $\alpha_k = 0.18 \pm 0.03$ and a K/L conversion ratio of 5.5. With the screened relativistic theoretical conversion coefficients of ROSE[2] the transition appeared to be an $M1-E2$ mixture, but with the recent refined theoretical coefficients of SLIV[3] the best interpretation would be pure $M1$. [Theoretical $\alpha_k(M1) = 0.20$.]

GORODETZKY *et al.*[4] have performed alpha-gamma angular correlation measurements with the 353-kev gamma transition and find an essentially isotropic dis-

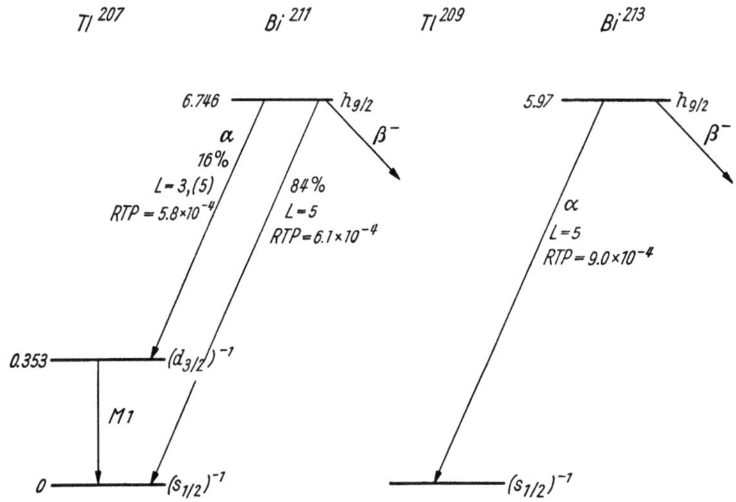

Fig. 23. Decay schemes (alpha branch) for Bi^{211} and Bi^{213} with probable shell model assignments. Allowed angular momentum values L are shown as well as reduced transition probabilities *(RTP)* (see Sect. 42 and Table 15) for the alpha transitions.

tribution. The lifetime of the state was measured[1] as less than, or equal to, 1.2×10^{-9} sec; with such a short lifetime attenuation effects due to extranuclear fields would not be expected to be too serious. From the lack of angular correlation it was suggested[4] that the intermediate state be assigned $s_{\frac{1}{2}}$ and the ground state of Tl^{207}, $d_{\frac{3}{2}}$, the reverse of PRYCE's [9] proposals and of our Fig. 23. We still prefer the $s_{\frac{1}{2}}$ ground-state assignment, mainly on grounds of the beta decay of Tl^{207} being "favored" first forbidden ($\log ft = 5.2$) by the analysis of KING and PEASLEE[5]. With the Fig. 23 spin assignments isotropy in the alpha-gamma correlation should occur with 87% $L=3$ and 13% $L=5$ alpha radiation, a reasonable admixture (or with 13% $L=3$ and 87% $L=5$).

Bi^{213} exhibits a single alpha group to the ground state in Tl^{209}. We assume angular momentum 5 for the alpha-decay group and calculate a reduced transition probability of 9.0×10^{-4}, quite comparable to those in Bi^{211}.

50. The polonium-211 isomers. Po^{211} (AcC′) decays with a 0.52-sec half-life across the 126-neutron shell to the well-studied Pb^{207}. A 25-sec alpha-emitting

[1] FALK-VAIRANT, TEILLAC and VICTOR: J. Phys. Radium **13**, 313 (1952).
[2] M. E. ROSE: Privately circulated tables and Appendix IV of Beta and Gamma Ray Spectroscopy; KAI SIEGBAHN, editor. Amsterdam: North-Holland Publishing Co. 1955.
[3] L. A. SLIV: Unpublished tables (1956).
[4] GORODETZKY, GALLMANN, KNIPPER and ARMBRUSTER: C. r. Acad. Sci., Paris **237**, 245 (1953).
[5] R. W. KING and D. C. PEASLEE: Phys. Rev. **94**, 1285 (1954).

isomer was discovered by SPIESS[1]. Further work of JENTSCHKE *et al.*[2] led to the decay scheme of Fig. 24. The spin of the ground state, AcC′, is expected to be $\frac{9}{2}$, involving mainly a configuration with the two protons beyond Pb^{208} as $(h_{\frac{9}{2}})^2_{J=0}$ and the odd neutron as $g_{\frac{9}{2}}$. Arguments[7] have been given for the spin of the isomer being at least $\frac{19}{2}$. The most reasonable configuration might involve $\left[(j_{\frac{15}{2}})\,(h_{\frac{9}{2}})^2_{J=2}\right]_{\frac{19}{2}}$ the two $h_{\frac{9}{2}}$ protons coupled to a $J=2$ state, with the odd neutron in $j_{\frac{15}{2}}$.

The possible alpha angular momenta for AcC′ are as shown in Fig. 24. We calculate, taking the lowest values of L in all cases, reduced transition probabilities of 7.2×10^{-4} to $p_{\frac{1}{2}}$ ground state, 5.9×10^{-5} to $f_{\frac{5}{2}}$, and 9.5×10^{-4} to $p_{\frac{3}{2}}$. The similarity in the values to the p states is interesting, as is their similarity to the Bi^{211} values. The greater retardation of the transition to $f_{\frac{5}{2}}$ is the reverse of expectation in terms of the naive radial wave-function overlap consideration (see Sect. 48).

If we take, speculatively, the assignment $^{19}/_{2}{}^{-}$ for the 1.30-Mev isomer, the allowed angular momenta are 10 to the $p_{\frac{1}{2}}$; 8 or 10 to the $p_{\frac{3}{2}}$; and 3, 5, 7, 9, 11, 13, or 15 to the $i_{\frac{13}{2}}$. The calculated reduced transition probabilities (lowest L) are as follows:

to $p_{\frac{1}{2}}$, 1.5×10^{-7},
to $p_{\frac{3}{2}}$, 5.9×10^{-7},
to $i_{\frac{13}{2}}$, 2.9×10^{-5}.

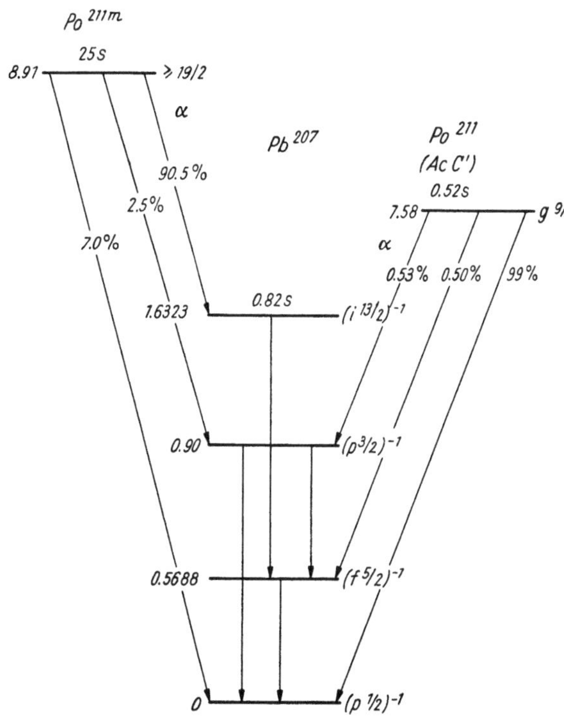

Fig. 24. Decay schemes for the Po^{211} isomers with probable shell model assignments.

The extra amount by which these transitions are slower than from the AcC′ isomer might be attributed to the unpaired proton configuration.

51. Bismuth-212. Bi^{212} (ThC) decays 33.7% by alpha emission and the remainder by beta minus decay. Six alpha groups are known, the two main groups populating ground state and a 39.85-kev excited state which decays to ground state by an $M1$ transition. The weaker alpha groups populate a group of four upper excited states.

The spin assignments 5^+ and 4^+ for ground state and first excited state were proposed theoretically by PRYCE [*9*] and are supported by angular-correlation studies and analysis by J. W. HORTON[3]. The spins of the upper four levels are somewhat uncertain and are omitted from our Fig. 25. Pryce has suggested that

[1] F. N. SPIESS: Phys. Rev. **94**, 1292 (1954).
[2] W. JENTSCHKE, A. C. JUVELAND and G. H. KINSEY: Phys. Rev. **96**, 231 (1954).
[3] J. W. HORTON: Phys. Rev. **101**, 717 (1956).

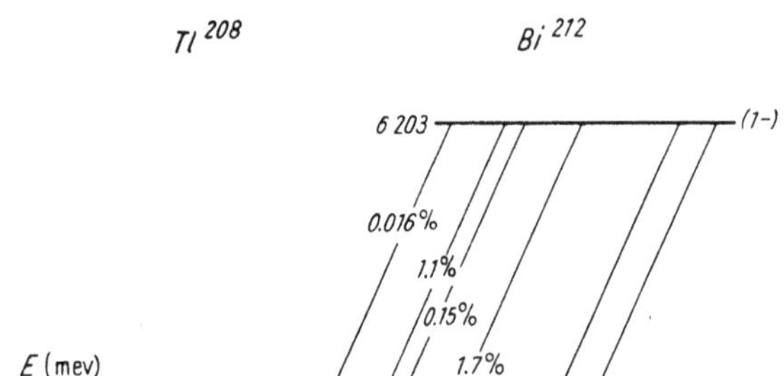

Fig. 25. Alpha branch decay scheme of Bi²¹² (ThC) with probable shell model configurations. The gamma transitions shown have been studied by conversion electron spectroscopy and are all of *M*1 multipolarity. The spins of the upper quartet of levels are uncertain and are not given here.

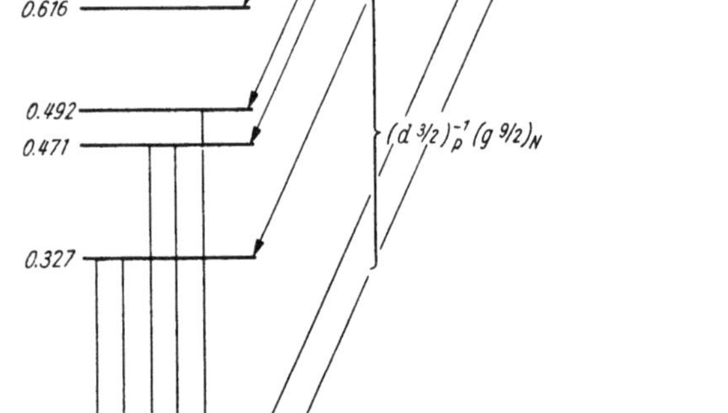

Fig. 26. Theoretical anisotropy for Bi²¹² alpha-gamma correlation with 40-kev *M*1 gamma. The anisotropy is plotted as a function of the mixing percentage of $L=3$ and $L=5$ alpha radiation. The upper and lower branches of the loop correspond to the two possible phase differences between the $L=3$ and $L=5$ alpha waves.

they constitute a quartet of levels with a $(d_{\frac{3}{2}})_P^{-1}(g_{\frac{9}{2}})_N$ configuration and spins of 3, 4, 5, and 6. All the gamma transitions drawn on Fig. 25 are of *M*1 multipolarity, the assignment for the energetic gamma rays being due to work by O. B. NIELSEN[1]. NIELSEN'S results are consistent with PRYCE'S spin predictions [9, cf. second reference listed] for the 327- and 471-kev levels but are inconsistent with the prediction of spin 6 for the 492-kev level.

The alpha group to the ground state exhibits a reduced transition probability of 3.2×10^{-4}, and that to the upper state, 2.7×10^{-4}. It is interesting that the values are quite comparable to one another and to those in Bi^{211} and Bi^{213}. Alpha-gamma angular correlation offers a test for the alpha angular momenta of the group to the 40-kev state. Several workers have performed alpha-gamma angular correlation experiments, and we quote the value obtained by HORTON[2], for the anisotropy $W(\pi)/W(\pi/2)-1=-0.229$. HORTON[2] has discussed many possible spin assignments. Here we consider only those of Fig. 25. Fig. 26 plots the theoretical anisotropy as a function of the mixing percentage between $L=3$ and 5. The calculation assumes pure *M*1 gamma radiation. The experimental anisotropy agrees with pure $L=3$ or nearly pure $L=5$. $L=3$ seems the better choice on decay rate grounds. Actually it seems rather surprising to find such pure radiation in this case, since the centrifugal barrier factor for $L=5$ is only 4.6 times the factor for $L=3$. Angular cor relation involving alpha radiation is

[1] O. B. NIELSEN: Dan. Mat. fys. Medd. **30**, No. 11 (1955).
[2] J. W. HORTON: Phys. Rev. **101**, 717 (1956).

especially susceptible to attenuation; therefore, it is conceivable that the unperturbed anisotropy slightly exceeds that measured, though the lifetime[1] is extremely short ($t_{\frac{1}{2}} < 7 \cdot 10^{-11}$ sec).

52. The bismuth-210 isomers. The long-lived isomer of Bi^{210} decays by alpha emission across both closed shells to Tl^{206} with an alpha-particle energy of 4.93 Mev and a half-life of $2.6 \cdot 10^6$ yr[2]. Its decay scheme is shown in Fig. 27. From LEVY and PERLMAN's[3] determination of β^- branching ($\log ft \sim 18.7$) and lack of isomeric transition the most likely assignment is (4^-) · Tl^{206} is $(1^-)\,(s_{\frac{1}{2}})_P^{-1}\,(p_{\frac{1}{2}})_N^{-1}$. An L-value of 4 is permitted, and we calculate a reduced-transition probability of $4.2 \cdot 10^{-8}$. This result must be taken as somewhat uncertain, with the possibility that the spin may be in error or that the decay may not proceed to the ground state of Tl^{206}.

The five-day beta-emitter Bi^{210m} (RaE) is probably (1^-) with mainly a mixture of configurations $(i_{\frac{13}{2}})_P\,(j_{\frac{15}{2}})_N$ and $(f_{\frac{7}{2}})_P\,(g_{\frac{9}{2}})_N$, and it has been observed to undergo alpha branching to the extent of $5 \cdot 10^{-5}\%$[4], or $1.7 \cdot 10^{-4}\%$[5]. L values and 0 of 2 should be allowed. Assuming $L = 0$ and taking the latter branching ratio value we calculate the RTP as $1.6 \cdot 10^{-6}$.

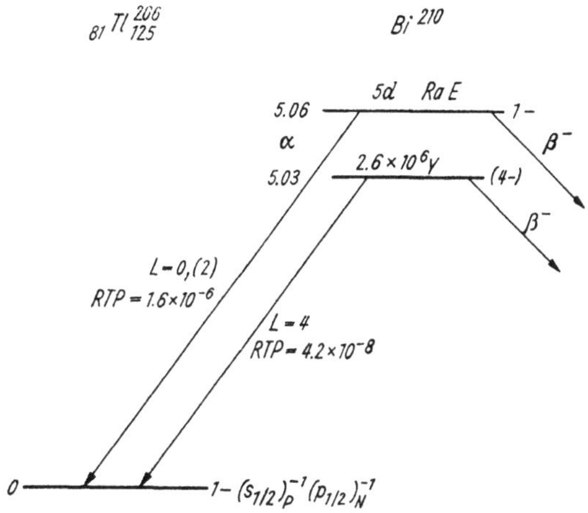

Fig. 27. Alpha branch decay schemes of the Bi^{210} isomers. Angular momentum values (L) and reduced transition probabilities (RTP) (see Sect. 42 and Table 15) are indicated on the figure.

II. Alpha decay in the region of spheroidal nuclei.

53. Thorium-230 and the levels of radium-226. The spectrum of levels exhibited by Ra^{226} (Fig. 28) is an interesting one, for in addition to the normal even-parity rotational sequence $A\,0$, $A\,2$, $A\,4$, and perhaps $A\,6$ at 416 kev, there appear at least one ($B\,1$) and probably two other ($B\,3$, $B\,5$) odd-parity states at surprisingly low energy for even-even nuclei. Th^{230} has been the subject of many careful α-γ and γ-γ angular correlation measurements, and this work constitutes an outstanding example of the wealth of information obtainable by the angular correlation method.

Alpha spectroscopy[6] has directly established the two lowest levels. The next three levels are firmly established by α-γ and γ-γ coincidence studies[7]. The evidence for the 416- and 445-kev levels is not entirely conclusive, consisting of

---

[1] R. L. GRAHAM and R. E. BELL: Canad. J. Phys. **31**, 377 (1953).
[2] D. J. HUGHES and H. PALEVSKY: Phys. Rev. **92**, 1206 (1953).
[3] H. B. LEVY and I. PERLMAN: Phys. Rev. **94**, 152 (1954).
[4] E. BRODA and N. FEATHER: Proc. Roy. Soc. Lond., Ser. A **190**, 20 (1947).
[5] FINK, WARREN, ROBINSON and EDWARDS: Bull. Amer. Phys. Soc., Ser. II **1**, 171 (1956).
[6] ROSENBLUM, VALADARES, BLANDIN-VIAL and BERNAS: C. r. Acad. Sci., Paris **238**, 1496 (1954). — HUMMEL, ASARO and PERLMAN: Unpublished data (1955).
[7] BOOTH, MADANSKY and RASETTI: Phys. Rev. **102**, 800 (1956). — G. VALLADAS and R. BERNAS: C. r. Acad. Sci., Paris **91**, 2230 (1953). — PERLMAN, ASARO, STEPHENS, HUMMEL and PILGER: University of California Radiation Laboratory Unclassified Report 2932 (Chemistry Division Quarterly Report, Dec., 1954 to Feb., 1955) p. 59.

the observation[1] of two weak gamma rays of 206 ± 5 kev and 237 ± 5 kev in coincidence with the 124-kev gamma ($A\,4$-$A\,2$).

The spin and parity assignments of the levels $A\,2$, $A\,4$, and $B\,1$ rest firmly on alpha-gamma angular correlation work (cf. especially VALLADAS, *et al.*[2], FALK-VAIRANT and PETIT[3], and STEPHENS[4] with additional references therein), internal conversion coefficient determinations, and gamma-gamma angular correlations.

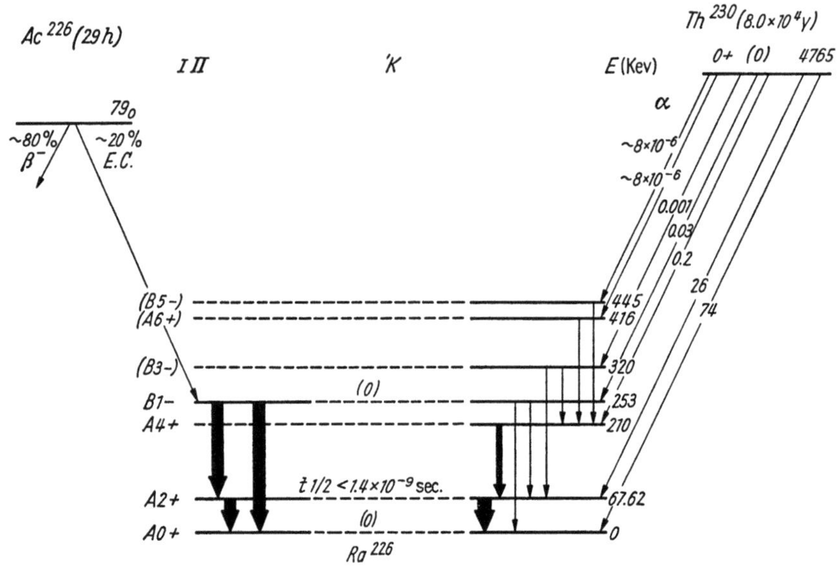

Fig. 28. Alpha-decay scheme of Th^{230} and electron-capture decay scheme of Ac^{226} to Ra^{226}. The vertical arrows representing the experimentally observed gamma transitions indicate qualitatively by their width the relative transition intensities. Spin and parity assignments are given on the left hand side of a level, those in parentheses being somewhat uncertain. The K-quantum number for a rotational band is shown in parentheses near the center of the base level of the band. The energies in kev are given at the right of the levels. Each known level is drawn in both decay schemes but is dashed in the decay scheme where it is not detectably populated.

Hindrance factors for Th^{230}.

Final state		Alpha intensity (%)	Hindrance factor	Centrifugal barrier factor	Final state		Alpha intensity (%)	Hindrance factor	Centrifugal barrier factor
Energy (kev)	Spin and parity				Energy (kev)	Spin and parity			
0	0^+	74	(1)	1	320	(3^-)	0.001	370	(2.8)
67.62	2^+	26	1.1	1.7	416	(6^+)	$\sim 8 \cdot 10^{-6}$	8200	(40)
210	4^+	0.2	12	5.4	445	(5^-)	$\sim 8 \cdot 10^{-6}$	4900	(14)
253	1^-	0.03	38	1.2					

Since a good rotational level sequence is exhibited by Ra^{226}, we may reason that there is a stable spheroidal nuclear shape and that the K-quantum number may be fairly good. Band A necessarily has $K = 0$, the spin of its lowest member. The K-quantum number applying to state $B\,1$ can be deduced from the relative reduced gamma transition probabilities for the two $E\,1$ gamma rays depopulating it. Study of the electron-capture decay of Ac^{226}, which mainly populates $B\,1$, has given a good relative intensity figure.

[1] STEPHENS, ASARO and PERLMAN: To be published (1956).
[2] VALLADAS, TEILLAC, FALK-VAIRANT and BENOIST: J. Phys. Radium **16**, 123 (1955).
[3] P. FALK-VAIRANT and G. Y. PETIT: C. r. Acad. Sci., Paris **240**, 296 (1955).
[4] F. S. STEPHENS: Ph. D. Thesis, University of California Radiation Laboratory Unclassified Report 2970 (1955), unpublished.

Experimentally, the relative reduced transition probabilities are

$$\frac{B_{B1\to A0}}{B_{B1\leftarrow A2}} = 1.32\left(\frac{185}{253}\right)^3 = 0.50.$$

Theoretically,

$$\frac{B_{1\to 0}}{B_{1\to 2}} = \frac{(1\,1\,K_i\;-K_i\,|\,1\,1\,00)^2}{(1\,1\,K_i\;-K_i\,|\,1\,1\,20)^2} = \begin{cases} 2.00 \text{ for } K_i = 1, \\ 0.50 \text{ for } K_i = 0. \end{cases}$$

The assignment $K_i = 0$ is clearly demanded, and the similar comparison for 1^- levels in Em^{218}, Ra^{222}, Ra^{224}, Ra^{226}, Th^{226}, Th^{228}, and Pu^{238} also favors $K=0$ (cf. Sect. 13). If the 1^- level were an ordinary nucleonic excited state involving two unpaired nucleons, the BOHR-MOTTELSON theory would require $K=1$. Hence the state $B1$ must be regarded as a state of collective excitation; more specifically $B1$ is to be regarded as a member of the ground rotational band (see Sect. 13 for discussion).

The state at 320 kev is thought to be 3^- and $K=0$, on the basis of its decay properties. It is probably to be considered as another odd member of the ground rotational band. Curiously, the effective energy constant $\hbar^2/2\mathfrak{J}$ as calculated from the spacing between $B1$ and $B3$ is only 60% of that from the spacing between $A0$ and $A2$.

The very tentative spin and parity assignments of Fig. 28 for the 416- and 445-kev states are based principally on expectations from the rotational energy formula, but their observed decay only to $A4$ is consistent with the assignments.

It is interesting to compare the levels with a rotational energy formula. In order to secure agreement with the even parity energy levels it is necessary to employ a three-parameter equation

$$E_I = A\,I(I+1) - B\,I^2(I+1)^2 + C\,I^3(I+1)^3 \tag{53.1}$$

in which case, with the energies 67.62, 210, and 416 kev for $A2$, $A4$, and $A6$ one obtains

$$A = 11.69\text{ kev}, \quad B = 0.075\text{ kev}, \quad \text{and } C = 0.0008\text{ kev}.$$

For the odd-parity levels

$$E_I = 240 + 6.75\,I(I+1) \tag{53.2}$$

gives a satisfactory fit within experimental uncertainty. The perturbation from the simple formula appears to be of the opposite sense from that of the even-parity levels (see discussion in work of A. K. KERMAN[1]), but the level energies are not known precisely enough to establish the point.

The alpha-gamma angular correlation involving the 68-kev $A2$-$A0$ gamma has been the object of careful study[2,3]. Theoretically, for the sequence $0 \xrightarrow[2]{\alpha} 2 \xrightarrow[2]{\gamma} 0$ one should obtain an angular correlation function of form $(\sin 2\vartheta)^2$. That is, the coincidence rate should be zero at $\vartheta = \pi$ or $\pi/2$. The experimental function has maxima and minima at the proper angles, but they are less pronounced, and the function does not get very near zero at the minima. Such *attenuation* of the correlation function is attributed to the interaction of external magnetic fields on the nuclear magnetic moment of the intermediate state or of electric field gradients on the nuclear electric quadrupole moment. One conventionally defines

[1] A. K. KERMAN: Dan. Mat. fys. Medd. **30**, No. 15 (1956).
[2] G. M. TEMMER and J. M. WYCKOFF: Phys. Rev. **92**, 913 (1953).
[3] VALLADAS, TEILLAC, FALK-VAIRANT and BENOIST: J. Phys. Radium **16**, 123 (1955).

attenuation factors G_2 and G_4 for the separate terms in the expansion of the correlation function in LEGENDRE functions. That is, for the above alpha-gamma sequence we have for no attenuation

$$W(\vartheta) = 1 + 0.714 P_2(\cos\vartheta) - 1.714 P_4(\cos\vartheta) \tag{53.3}$$

while in the actual case with attenuation

$$W(\vartheta) = 1 + 0.714 G_2 P_2(\cos\vartheta) - 1.714 G_4 P_4(\cos\vartheta) \tag{53.4}$$

VALLADAS *et al.*[1] find $G_2 = 0.47_3$ and $G_4 = 0.56_5$.

As discussed in the papers of TEMMER and WYCKOFF[2] and VALLADAS *et al.*[1], the attenuation here is to be attributed predominantly to electric rather than magnetic interactions, since mangetic interaction should give $G_2 > G_4$ in disagreement with experiment. By theories of ALDER[3] and ABRAGAM and POUND[4] the attenuation coefficients are given as functions of $\omega\tau$, where τ is the mean life of the intermediate state and $2\pi\omega$ is the smallest non-vanishing LARMOR precession frequency for the quadrupole interaction. The above attenuations are $(\omega\tau)_1 \sim 1$.

The attenuation coefficients for the $0 \xrightarrow[4]{\alpha} 4 \xrightarrow[2]{\gamma} 2$ sequence have been determined[5] as $G_2 = 0.74$ and $G_4 = 0.61$, consistent with electric interaction with $(\omega\tau)_2 \sim 0.13$.

Using ABRAGAM and POUND's relationship between ω and the quadrupole moment and BOHR and MOTTELSON's [*10*] relationships between lifetime, quadrupole moment, and intrinsic quadrupole moment, VALLADAS *et al.*[6] calculate an expected theoretical ratio

$$\frac{(\omega\tau)_2}{(\omega\tau)_1} = 0.11 \pm 0.04$$

in agreement with their experimental ratio of 0.13.

Alpha-gamma angular correlation has been done[5] with state $B1$ as intermediate, confirming the spin assignment of unity. The attenuation is small: $0.8 < G_2 < 1$. That is, assuming electric quadrupole interactions, $\omega\tau < 1$.

54. Curium-242 and the levels of plutonium-238. The energy levels of Pu^{238} have been investigated thoroughly both from the alpha decay[7] of Cm^{242} and the β^- decay[8] of Np^{238}. Some work on electron capture[9] of Am^{238} has also been done. The level pattern and alpha spectrum (Fig. 29) are expected to be representative for even-even nuclei in the region of pronounced spheroidal distortion.

The energy levels $A0$, 2, 4, 6, and 8 are believed to comprise a rotational band based upon the ground state. A part of this sequence of states is well established. The alpha groups to states $A0$, $A2$, and $A4$ have been observed

---

[1] G. M. TEMMER and J.M. WYCKOFF: Phys. Rev. **92**, 913 (1953).
[2] VALLADAS, TEILLAC. FALK-VAIRANT and BENOIST: J. Phys. Radium **161** 123 (1955).
[3] K. ALDER: Helv. phys. Acta **25**, 234 (1952).
[4] ABRAGAM and POUND: Phys. Rev. **92**, 943 (1953).
[5] P. FALK-VAIRANT and G. Y. PETIT: C. r. Acad. Sci., Paris **240**, 296 (1955).
[6] VALLADAS, TEILLAC, FALK-VAIRANT and BENOIST: J. Phys. Radium **16**, 123 (1955).
[7] ASARO, PERLMAN and THOMPSON: Phys. Rev. **92**, 694 (1953). — F. ASARO: Unpublished results (1956).
[8] FREEDMAN, JAFFEY and WAGNER: Phys. Rev. **79**, 410 (1950). — RASMUSSEN, SLÄTIS and PASSELL: Phys. Rev. **99**, 42 (1955). — RASMUSSEN, STEPHENS, STROMINGER and ÅSTRÖM: Phys. Rev. **99**, 47 (1955). — S. A. BARANOV and K. N. SHLYAGIN: Atomnia Energia **1**, 52 (1956).
[9] R. J. CARR: Ph. D. Thesis, University of California 1956 (Radiation Laboratory Report UCRL 3395).

directly, and the energy level spacing agrees with the $I(I+1)$ dependence expected of a rotational band. Conversion coefficients for the transitions $A2 \rightarrow A0$ and $A4 \rightarrow A2$ have shown them to be $E2$ and no crossover ($A4 \rightarrow A0$) could be found. From this information the state $A4$ is assigned uniquely 4^+ only on the basis of the $I(I+1)$ level spacing although the other data support this assign-

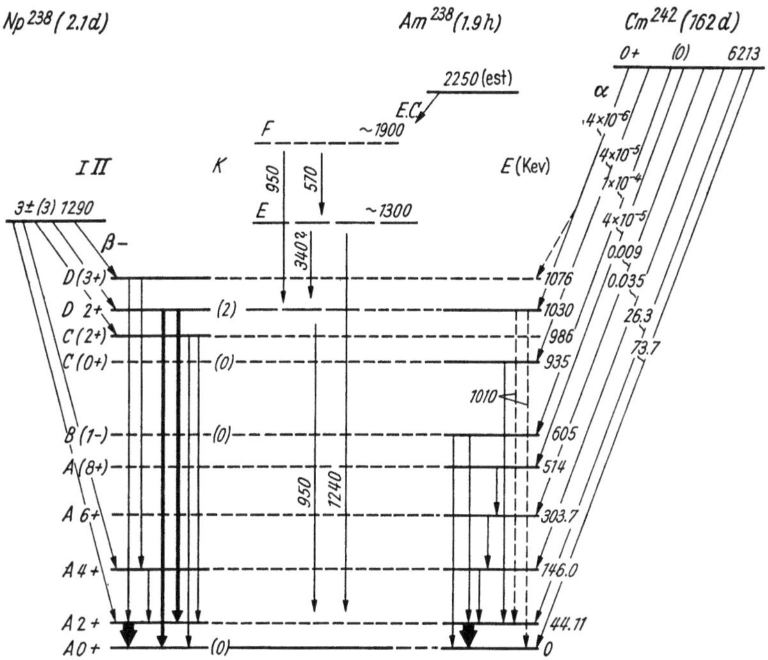

Fig. 29. Alpha-decay scheme of Cm^{242}, beta-decay scheme of Np^{238}, and preliminary electron-capture decay scheme of Am^{238} to the common daughter nucleus Pu^{238}. For comments on the conventions followed in drawing these decay schemes see the caption to Fig. 28. The placement of the level $C2$ and its association with the level $C0$ in a common rotational band is quite tentative.

Hindrance factors for Cm^{242}.

Final state		Alpha intensity (%)	Hindrance factor	Centrifugal barrier factor	Final state		Alpha intensity (%)	Hindrance factor	Centrifugal barrier factor
Energy (kev)	Spin and parity				Energy (kev)	Spin and parity			
0	0^+	73.7	(1.0)	1.0	514	(8^+)	3×10^{-6}	5000	(340)
44	2^+	26.3	1.7	1.6	605	(1^-)	1×10^{-4}	500	(1.2)
146	4^+	0.035	390	4.9	935	(0^+)	3×10^{-5}	20	(1)
304	(6^+)	0.006	350	(29)	~1030	(2^+)	4×10^{-6}	45	(1.6)

ment. However, the same sequence has been investigated in other even-even nuclei and alpha-gamma and gamma-gamma angular correlations show quite definitely that the third member is a 4^+ state (see Sect. 53).

The 44-kev $E2$ transition ($A2 \rightarrow A0$) has an L-shell conversion coefficient of several hundred, and since this state is by far the most heavily populated of the excited states, the most prominent electromagnetic radiation consists of L x-rays. The 102-kev transition ($A4 \rightarrow A2$) is found to be in coincidence with L x-rays as expected. The remainder of the cascading sequence $A8 \rightarrow A6 \rightarrow A4$ was characterized by showing that each gamma ray was in coincidence with the one below. The spins of these states ($A8$ and $A6$) were assigned on the basis of the cascading transitions and because the energy level spacings agree with the expectations for higher members of the rotational band. The low intensity of population of these states makes it difficult to obtain more detailed information, such as conversion coefficients, for these transitions. It will be noted later that the 6^+ and 8^+ states are not seen in the β^- decay of Np^{238}, and this fact conforms with a consistent set of level assignments for Np^{238} and the higher energy levels of Pu^{238}.

The energy level $B1$, assigned 1^-, was deduced from a pair of gamma-rays of 605 kev and about 560 kev. The higher-energy component was not in coincidence with any photons, and the lower-energy component was in coincidence with L x-rays proving that it leads to level $A2$. Such pairs of gamma rays have been seen in other heavy even-even nuclei[1] and have been proved to be $E1$ transitions arising from 1^- states. In all cases examined in which the 1^- assignment was established, the reduced transition probabilities of the competing $E1$ transitions were found in a ratio characteristic for the K-quantum number equal to zero for the 1^- state as well as for the rotational band based upon the ground state. This same relationship of the reduced transition probabilities was found in the present case. Beyond this reasoning by analogy there is no other supporting evidence for the $(0, 1^-)$ assignment of state $B1$ and it must, therefore, be considered provisional.

The best information on the higher levels comes from the study of Np^{238} beta decay. The best high-energy conversion electron data are those of BARANOV and SHLYAGIN[2], and they show five gamma transitions near 1 Mev: 1032, 1030, 988, 942 and 927 kev[3]. The K-lines of the two highest-energy transitions are barely identifiable as two separate lines[2], but coincidence measurements showed earlier[4] that there are indeed two gamma rays of about 1030 kev, the more intense component being not in coincidence with L x-rays. Hence this gamma leads to the ground state, definitely establishing a level at 1030 kev designated $D2$ in Fig. 29. The 988-kev gamma-ray is found in about equal intensity to the 1030-kev gamma-rays, and its energy agrees with the $D2$-$A2$ difference, but also with the $C2$-$A0$ difference, and BARANOV and SHLYAGIN[2] divide the line intensity between these transitions. The 942-kev gamma-ray is probably in coincidence with L x-rays, and hence does not go to the ground state. Whether this gamma goes to level $A4$ or to $A2$ was regarded as an unsettled question by RASMUSSEN *et al.*[4]. BARANOV and SHLYAGIN'S[2] work favors the latter alternative of directing γ_{942} to level $A2$, thus defining a level at 986 kev.

Level $A4$ is populated in about 3% of the beta transitions. Gamma-gamma coincidence work[3] suggested population of $A4$ by a hard gamma-ray of energy consistent with either the 942 or 927 kev gamma-rays. Electron-electron coincidence work[1] suggested population of $A4$ directly by a hard beta group (1133 kev), roughly to the extent of 2.8% of total disintegrations. There is evidently population of $A4$ by both beta- and gamma-rays, although the relative amounts must surely be regarded as somewhat uncertain.

Most of the gamma transitions must be of electric quadrupole nature, and the data are consistent with all being $E2$. Both of the recent studies[2,4] proposed sets of relative intensities for all transitions, but with the unresolved beta and

---

[1] STEPHENS, ASARO and PERLMAN: Phys. Rev. **96**, 1568 (1954). — F. S. STEPHENS jr. and I. PERLMAN: To be published.

[2] S. A. BARANOV and K. N. SHLYAGIN: Atomnia Energia **1**, 52 (1956).

[3] BARANOV and SHLYAGIN [5] reported the absence of the 927 kev transition seen by RASMUSSEN, SLÄTIS and PASSELL (s. footnote 8, p. 192), but in private communication to J. M. HOLLANDER, BARANOV reported later observing it. The K-line is certainly much weaker than reported by RASMUSSEN, SLÄTIS, and PASSELL.

[4] RASMUSSEN, STEPHENS, STROMINGER and ÅSTRÖM: Phys. Rev. **99**, 47 (1955).

gamma transitions these relative intensities must be based on balancing transition intensities to and from the lower levels. The proposed sets of intensities are presently to be taken with some reservation.

The levels $D2$ and $D3$ have been designated[1,2] as $(2, 2^+)$ and $(2, 3^+)$ states and could possibly have the predominant character of a so-called gamma vibration (shape vibration) band predicted theoretically by BOHR and MOTTELSON [*10*]. The spacing between $D2$ and $D3$ is roughly 46 kev, and this corresponds to the same or slightly lower rotational moment of inertia as the ground state. If the levels are of vibrational nature, there must, however, also be a sizeable admixture of an excited nucleonic structure, since the beta decay would necessarily be much slower if the levels were purely vibrational.

The assignments of level $D2$ to be a $(2, 2^+)$ state receives its strongest support from the relative gamma-ray intensities. For large spheroidal distortion, where K may be a fairly good quantum number, one can calculate relative transition rates to different members of a rotational band. (The reason why all transitions from state $D2$ are expected to be pure $E2$ will be dealt with presently.) The reduced transition probabilities should go simply as the squares of appropriate CLEBSCH-GORDAN coefficients involving I- and K-quantum numbers. In particular, this analysis often permits an unambiguous assignment of the K-quantum number. When applied to the present case, it turns out that only if $K=2$ for state $D2$ would one expect to find transitions $D2 \rightarrow A0$ and $D2 \rightarrow A2$ in the relative intensities observed and have the $D2 \rightarrow A4$ transition of unobservably low intensity.

One of the demands of these assignments is that the transition $D2 \rightarrow A2$, be $E2$. The reason why it does not exhibit measurable $M1$ admixture is because of the selection rule that the multipolarity must exceed or be equal to ΔK.

The most reasonable spin assignment for Np^{238} with the decay scheme of Fig. 29, would seem to be 3 with even or odd parity admissible.

The alpha group shown to populate the state $D2$ is assigned from rather fragmentary evidence. A definite, but very weak, gamma-ray of $\sim$1010 kev was observed in the Cm^{242} spectrum, and this could be a mixture of the 988- and 1030-kev transitions from level $D2$ seen in the beta decay of Np^{238}.

The level designated $C0$ is based on a weak $\sim$890-kev gamma transition observed in alpha decay of Cm^{242}. It was found to be in coincidence with L x-rays. Our tentative interpretation of this level, placed at 935 kev, is as a 0^+ level, perhaps to be associated with a BOHR-MOTTELSON [*10*] beta vibrational excitation.

It is very interesting to note that the 986-kev level from beta decay of Np^{238} is spaced above the level $C0$ by about 50 kev which is of the right magnitude to be a rotational excitation. In Fig. 29 we have associated these levels with a single rotational band, calling the 986-kev level $C2$. It should be emphasized that the interpretation of the states $C0$ and $C2$ is highly speculative and is advanced here mainly as a guide to further measurements.

Attention will be called here to the obvious need of relating the observed intensities of alpha population of the several levels with alpha-decay theory. The hindrance factor, which expresses the ratio of the partial alpha half-life of each transition to that of the ground state transition after removing the energy dependence, is given in the caption to Fig. 29. Also given is the factor taken from the treatment of WINSLOW [*21*] which expresses the retardation due to the centrifugal barrier resulting from alpha-particle angular momentum.

---

[1] RASMUSSEN, STEPHENS, STROMINGER and ÅSTRÖM: Phys. Rev. **99**, 47 (1955).

[2] The usual convention of A. BOHR and co-workers of listing quantum numbers in the order (K, I, π) is used here.

The assignment of the ground-state rotational band follows partly from the identification of cascading $E2$ transitions and partly from the level spacing. The BOHR-MOTTELSON rotational formula is of the form

$$E = A\,I(I+1) - B\,I^2(I+1)^2 \qquad (54.1)$$

where $B \ll A$ in the region of applicability of the model. The agreement between experiment and theory may be judged from Table 17.

Table 17. *Energies of ground rotational band members in* Pu^{238}

Member of band (I, π)	0^+	2^+	4^+	6^+	8^+
Measured energy	0	44.11	146.0	303.7	514[1]
Calculated energies based upon $E = A\,I(I+1)$; A from 44.11-kev 2^+ state		(44.11)	147.0	308.8	529.3
Calculated energies including $I^2(I+1)^2$ term; $A = 7.37$, $B = 0.0034$ from measured energies of 2^+ and 4^+ states		(44.11)	(146.0)	303.4	513.0

55. Curium-243 and the levels of plutonium-239. Our knowledge of this alpha emitter and the levels of its daughter nucleus Pu^{239} is considerable as a result of study of the three isotopes decaying to Pu^{239}; i.e., Np^{239}, Am^{239}, and Cm^{243}, as well as coulombic excitation of Pu^{239}. The essential results of these studies[2–5] are embodied in the decay scheme of Fig. 30.

The levels may be grouped into rotational bands, levels $A\text{-}\frac{1}{2}$, $A\text{-}\frac{3}{2}$, $A\text{-}\frac{5}{2}$, and $A\text{-}\frac{7}{2}$ comprising one band, and the measured[6] ground-state spin of $\frac{1}{2}$ identifies this band, with its irregular spacings, as an "anomalous" $K = \Omega = \frac{1}{2}$ band with energy level spacings given by

$$E_I = \frac{\hbar^2}{2\mathfrak{J}}\left[I\,(I+1) + (-)^{I+\frac{1}{2}}\,a\left(I+\frac{1}{2}\right)\right] \qquad (55.1)$$

with a, the decoupling parameter, depending on details of the intrinsic nucleonic structure. Levels $B\text{-}\frac{5}{2}$ and $B\text{-}\frac{7}{2}$ evidently belong to a common rotational band, and the occurrence of three $M1$ gamma transitions to levels of spins $\frac{3}{2}$, $\frac{5}{2}$, and $\frac{7}{2}$ fixes the spin and relative parity of level $B\text{-}\frac{5}{2}$. Level $C\text{-}\frac{5}{2}$ is the opposite parity from the lower levels and is presumed of spin $\frac{5}{2}$ on the basis of the transitions depopulating it. All the above levels presumably receive some alpha population, although experimental difficulties (interference from Cm^{242}) has hitherto prevented the actual observance of any alpha decay to the lowest two levels.

COULOMB excitation gives rise to observation of gamma transitions $A\text{-}\frac{5}{2} \rightarrow A\text{-}\frac{3}{2}$ and $A\text{-}\frac{5}{2} \rightarrow A\text{-}\frac{1}{2}$. Level $D\text{-}\frac{7}{2}$ is populated by decay of Np^{239} or Am^{239}, but its excitation has not been observed in alpha decay, perhaps on account of its high energy. Spins of $\frac{5}{2}$ or $\frac{7}{2}$ are consistent with experiment, but $\frac{7}{2}$ seems more likely from inspection of the NILSSON diagram [*13*], Fig. 31.

[1] This energy is based upon a very weak gamma ray whose energy is known only to ± 10 kev. The close agreement with the calculated energy is therefore fortuitous.

[2] (Np^{239}) J. M. HOLLANDER, W. G. SMITH and J. W. MIHELICH: Phys. Rev. **102**, 740 (1956). — S. A. BARANOV and K. N. SHLYAGIN: Atomnia Energia **1**, 52 (1956). — D. ENGELKEMEIR and L. B. MAGNUSSON: Phys. Rev. **99**, 135 (1955). — H. W. LEFEVRE, E. M. KINDERMAN and H. H. VAN TUYL: Phys. Rev. **100**, 1374 (1955). Other references in "Table of Isotopes" by HOLLANDER, PERLMAN and SEABORG, Rev. Mod. Phys. **25**, 469 (1953).

[3] (Am^{239}) W. G. SMITH, W. M. GIBSON and J. M. HOLLANDER: University of California Radiation Laboratory Report 3356 (1956) (to be published).

[4] (Cm^{243}) F. ASARO, S. G. THOMPSON and I. PERLMAN: Phys. Rev. **92**, 694 (1953).

[5] (coul. exc.) J. O. NEWTON: Nature, Lond. **175**, 1028 (1955) and private communication (Oct. 1955).

[6] M. VAN DEN BERG and P. F. A. KLINKENBERG: Physica, Haag **20**, 37, 461 (1954).

The rotational spacing factor $\hbar^2/2\mathfrak{J}$ is 6.25 kev for band A and 6.39 kev for band B. This is to be compared with 4.74 kev and 6.20 kev for bands A and B, respectively, in nearby Np^{237} and to 7.35 kev for Pu^{238}.

Level B-$\frac{5}{2}$ has a measured lifetime[1] of 1.1×10^{-9} sec representing a slow-down from the single-particle formula of about a factor of 10^4 for the main $M1$ transition B-$\frac{5}{2}\rightarrow A$-$\frac{3}{2}$ depopulating it. This slowness is attributed to the violation of the K selection rule, since $\Delta K=2$ exceeds the gamma multipolarity.

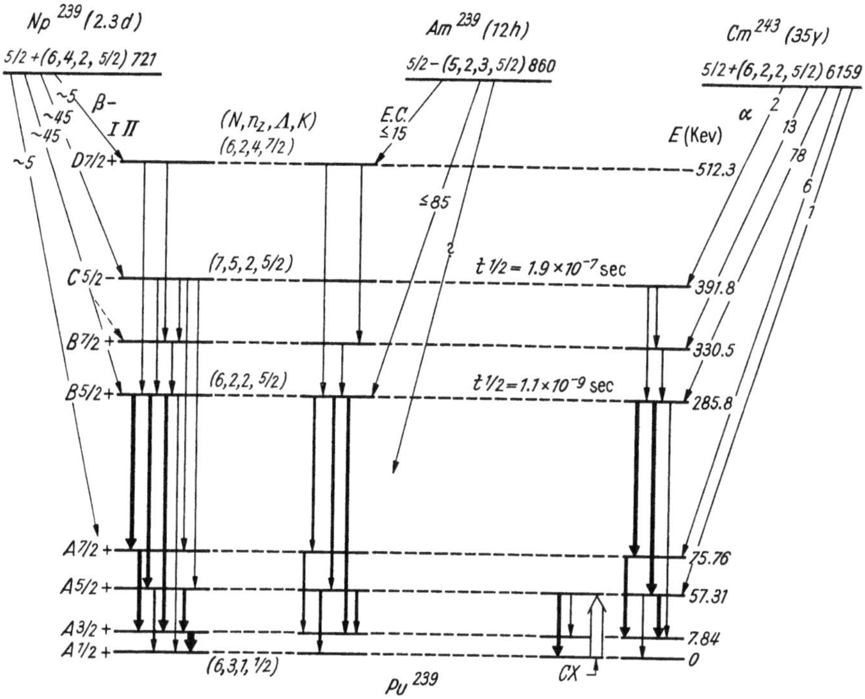

Fig. 30. Alpha decay scheme of Cm^{243}, beta decay scheme of Np^{239}, and electron capture decay of Am^{239} to the common daughter nucleus Pu^{239}. Also shown is the coulombic excitation scheme. The vertical arrows in each scheme are drawn only for experimentally observed transitions. Hindrance factors for the various alpha groups have been listed in Table 12. Proposed assignments of the asymptotic quantum numbers (N, n_z, Λ, K) are given as well as spins and parities.

Level C-$\frac{5}{2}$ has a measured lifetime[2] of $1.9\cdot10^{-7}$ sec, representing a retardation of $\sim2\times10^6$ on the single-proton lifetime estimate for $E1$ transition C-$\frac{5}{2}\rightarrow B$-$\frac{5}{2}$. The slowness is rather commonly observed for $E1$ transitions in the heavy region and may be associated[3,4] with a violation of selection rules in the asymptotic quantum numbers n_z, or Λ (defined in Sect. 14). With the assignments in Fig. 30 there is $\Delta n_z=3$, exceeding the multipolarity.

The relative reduced-transition probabilities of the $E2$ transition A-$\frac{5}{2}\rightarrow A$-$\frac{1}{2}$ and the $E2$ component of mixed $M1$-$E2$ transition A-$\frac{5}{2}\rightarrow A$-$\frac{3}{2}$ have been shown[5] to be proportional to the squares of the appropriate Clebsch-Gordan coefficients $(I_i\,2K0|I_i\,2I_f\,K)^2$, as they should be for K a fairly good quantum number.

[1] R. L. Graham and R. E. Bell: Phys. Rev. **83**, 222 A (1951).
[2] D. Engelkemeir and L. B. Magnusson: Phys. Rev. **99**, 135 (1955).
[3] J. O. Rasmussen and D. Strominger: Bull. Amer. Phys. Soc., Ser. II **1**, No. 4 Paper R4 (1956).
[4] D. Strominger: Ph. D. Thesis, University of California (1956) (Radiation Laboratory Report UCRL 3374).
[5] Hollander, Smith and Mihelich: Phys. Rev. **102**, 740 (1956).

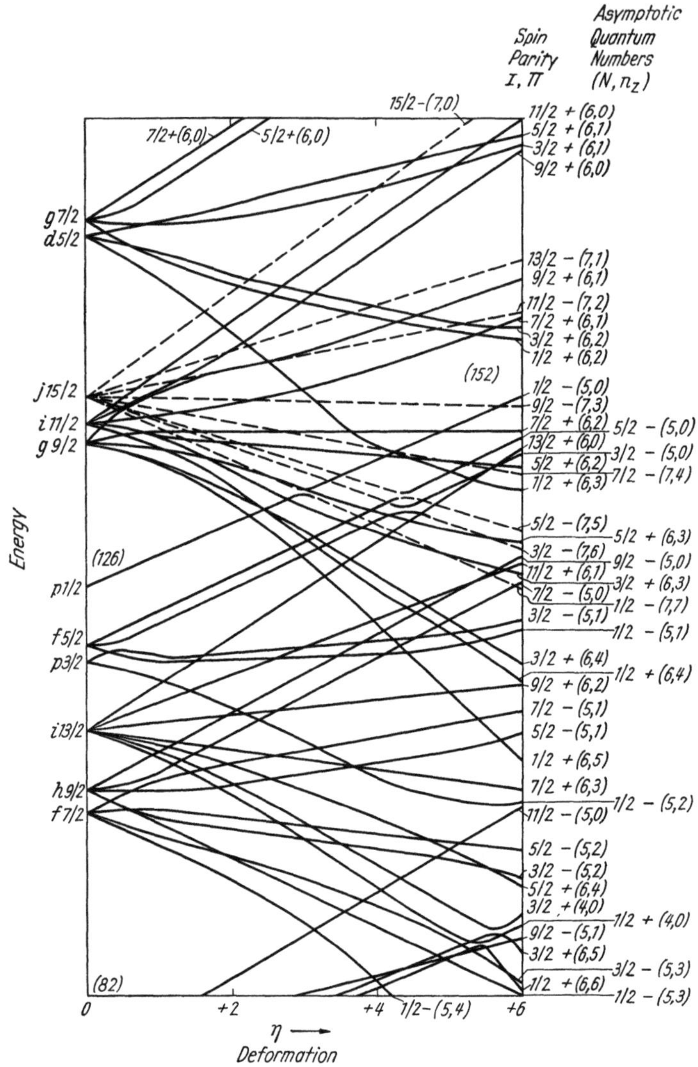

Fig. 31. Plot of eigenvalues for a nucleon in a prolate spheroidal three-dimensional harmonic oscillator potential with strong spin-orbit coupling according to the calculations of S. G. Nilsson [13]. As calculations of the 7th oscillator shell were not available, the $j_{\frac{15}{2}}$ levels (dashed lines) were estimated. The vertical positioning of the $j_{\frac{15}{2}}$ levels is rather arbitrary, in this diagram being somewhat lower than proposed by Nilsson [13]. The lower position gives a natural gap in levels at large deformation for 152 particles thus offering some *rationale* for the observed neutron "subshell" at 152.

The levels are labelled in the ordinary shell model convention according to l and j at the extreme left of Fig. 31, corresponding to a spherical well. At the right hand side, corresponding to large prolate deformation, the levels are labelled according to the Ω-quantum number (the spin of the base state of a nuclear rotational band except in some cases for $\Omega=\frac{1}{2}$) and the parity of the level. In parentheses are the asymptotic quantum numbers N and n_z most appropriate to the eigenfunction at the largest deformation ($\eta=6$) here plotted. The asymptotic quantum number Λ can be determined as follows: Λ can differ from Ω only by $\frac{1}{2}$ and takes the even or odd value according to whether $N-n_z$ is even or odd. (N is the principal oscillator quantum number, n_z is the symmetry-axis oscillator quantum number, Λ is the symmetry-axis component of orbital angular momentum, and Ω is the symmetry-axis component of total nucleonic angular momentum. For details see Nilsson [13].)

The ground state magnetic moment ($|\mu|=0.4$) and decoupling parameter ($a=1.418$) for the band have been compared with theoretical calculations[1]

[1] D. Strominger: Ph. D. Thesis, Universiy of California (1956) (Radiation Laboratory Report UCRL 3374).

using NILSSON'S spheroidal well nucleonic wave functions [13], and the assignment to the particular NILSSON state giving agreement was made.

Likewise, a consistent interpretation of the beta- and electron-capture ft values in terms of ALAGA'S selection rules[1] in N, n_z, and Λ can be obtained with the assignments of Fig. 30. It is necessary to assume assignment of $^5/_2{}^+$ to Np^{239}, the same as Np^{237}, but in contradiction to the spectroscopically measured spin[2] of $\frac{1}{2}$. The violation of selection rules in n_z and Λ for the allowed beta transitions from Np^{239} to bands B and D slows them to a comparable rate ($\log ft \sim 7$) with the unhindered first-forbidden transition to C ($\log ft \sim 6.5$). The most dramatic effect of the asymptotic quantum number selection rules results from the violation $\Delta N = 2$ for the allowed transition from Am^{239} to C-$\frac{5}{2}$; the transition is too weak to be observed and must have a $\log ft > 8$.

The main alpha group to level B-$\frac{5}{2}$ has a hindrance factor of only 1.4 and is presumed to be of the "favored" type in which the odd-neutron in state B-$\frac{5}{2}$ is in nearly the same state as in Cm^{243}. See Table 16 for comparison of favored alpha groups with the BOHR-FRÖMAN-MOTTELSON [31] theory.

The alpha transitions to the ground band fall in the category of highly-hindered transitions. It is significant that the $E2$ radiations between the ground band and the favored band are weak, indicating that the electric coupling of the hindered group with the favored is not large. Beyond this we are unable to say why the ground transitions are so hindered. The alpha transition to state C-$\frac{5}{2}$ is hindered by a factor of 16. The parity change restricts the transition to $L = 1$, 3, or 5.

56. Americium-241 and the levels of neptunium-237. Here again (Fig. 32) the three radioactivities decaying to Np^{237} have been carefully studied[3-5], and COULOMB excitation experiments[6] have been carried out.

Levels A-$\frac{5}{2}$, A-$\frac{7}{2}$, and A-$\frac{9}{2}$ form a regular rotational band as shown by COULOMB excitation. Levels B-$\frac{5}{2}$, B-$\frac{7}{2}$, and B-$\frac{9}{2}$ form a second rotational band of opposite parity to the first. The base spins of these bands and their K-values are $\frac{5}{2}$. These levels constitute the main levels receiving alpha decay, although alpha decay to level A-$\frac{9}{2}$ has not been seen and is probably almost unobservable.

The rotational spacing factor $\hbar^2/2\mathfrak{J}$ for band A is 4.74 kev, the smallest known for any nucleus, while $\hbar^2/2\mathfrak{J}$ for band B is 6.20 kev.

Level B-$\frac{5}{2}$ is a metastable state[7] of half-life 6.3×10^{-8} sec, decaying by $E1$ transitions B-$\frac{1}{2} \rightarrow A$-$\frac{5}{2}$ and B-$\frac{5}{2} \rightarrow A$-$\frac{7}{2}$. Like other $E1$ transitions of this region these violate a selection rule in n_z ($\Delta n_z = 2$). The gyromagnetic ratio for state

---

1 G. ALAGA: Phys. Rev. **100**, 432 (1955).

2 J. G. CONWAY and R. D. MCLAUGHLIN: Phys. Rev. **96**, 541 (1954).

3 (Am^{241}) JAFFE, PASSELL, BROWNE and PERLMAN: Phys. Rev. **97**, 142 (1955). — F. ASARO and I. PERLMAN: Phys. Rev. **93**, 1423 (1954). — P. P. DAY: Phys. Rev. **97**, 689 (1955). — MILSTED, ROSENBLUM and VALADARES: C. r. Acad. Sci., Paris **239**, 259, 700 (1954). — J. L. WOLFSON: Private communication (1954). — J. M. HOLLANDER, W. G. SMITH and J. O. RASMUSSEN: Phys. Rev. **102**, 1372 (1956). — GOL'DIN, TRET'YAKOV and NOVITSOVA: Conf. Acad. Sci. USSR on Peaceful Use of Atomic Energy, Phys. Math. Sci. p. 226 (July 1955). — J. F. TURNER: Phil. Mag. **46**, 687 (1955). — S. A. BARANOV and K. N. SHLYAGIN: Conf. Acad. Sci. USSR. on Peaceful Use of Atomic Energy, Phys. Math. Sci. p. 251 (July 1955).

4 (U^{237}) WAGNER, FREEDMAN, ENGELKEMEIR and HUIZENGA: Phys. Rev. **89**, 502 (1953). — L. MELANDER and H. SLÄTIS: Phys. Rev. **74**, 709 (1948). — Ark. Mat. Astronom. Fys., Ser. A **36**, No. 15 (1949). — S. A. BARANOV and K. N. SHLYAGIN: Conf. Acad. Sci. USSR. on Peaceful Use of Atomic Energy, Phys. Math. Sci. p. 251 (July 1955). — Zhur. Exp. Theor. Fys. **30**, 225 (1956). — HOLLANDER, CANAVAN and RASMUSSEN: To be published (1956).

5 (Pu^{237}) R. W. HOFF: Ph. D. Thesis, University of California 1954 (Radiation Laboratory Report UCRL 2325). — R. W. HOFF and J. L. OLSEN: Private communication (1956).

6 J. O. NEWTON: Nature, Lond. **175**, 1028 (1955) and private communication (Oct. 1955).

7 BELING, NEWTON and ROSE: Phys. Rev. **87**, 670 (1952).

B-$\frac{5}{2}$ has been measured[1] as $+0.8 \pm 0.2$ by the attenuation of the alpha-gamma angular correlation in an applied magnetic field.

The alpha decay to band A is of the highly-hindered type. Alpha decay to band B appears to be of the favored type. Here, it would seem, is an interesting opportunity to test the favored alpha-decay hypothesis that the odd-nucleon wave function remains essentially unchanged. The measured magnetic moment[2] of Am^{241} is $+1.4$ nuclear magnetons while that of state B-$\frac{5}{2}$ of Np^{237} by the attenuation measurement[2] is $+2.0 \pm 0.5$ nuclear magnetons. The moments are

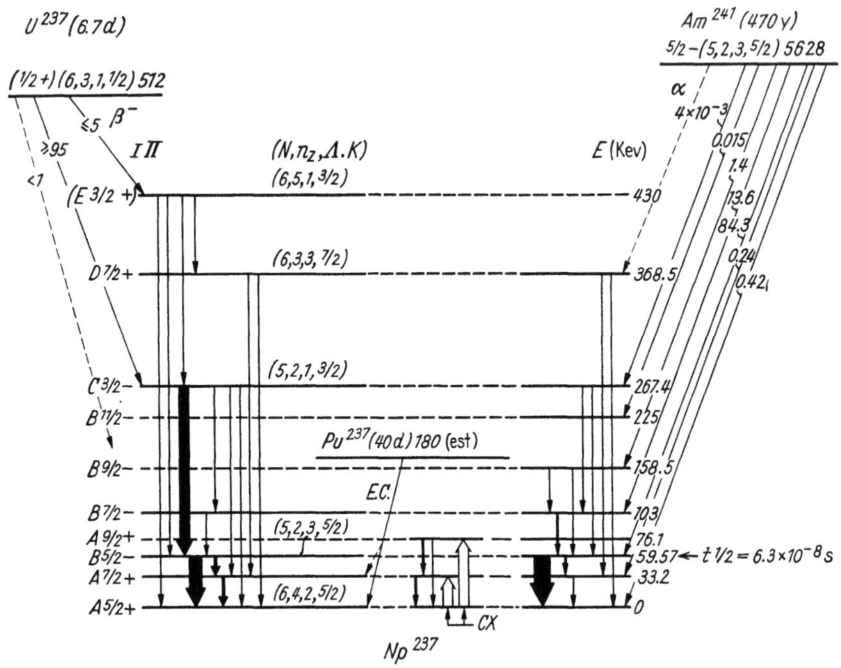

Fig. 32. Alpha decay scheme of Am^{241}, beta decay scheme of U^{237}, and electron-capture decay scheme of Pu^{237} to the common daughter nucleus Np^{237}. Also shown is the coulombic excitation scheme. The vertical arrows in each scheme again are drawn only for experimentally observed transitions. Proposed assignments of the asymptotic quantum numbers (N, n_z, Λ, K) are given as well as spins and parities. Hindrance for the various alpha groups have been listed in Table 13.

not the same but the check is probably close enough to be consistent with the favored decay hypothesis, particularly when it is remembered that magnetic moments are often extremely sensitive to details of the wave function.

The $M1$-$E2$ mixing ratios for cascade radiation in band B have been measured by conversion electron intensity work[3] and, coupled with the experimental magnetic moment for level B-$\frac{5}{2}$, provide a test in satisfactory agreement with the model of a single free proton in a spheroidal well. This agreement is in contrast to similar tests in the ground band which do not agree with the extreme single-particle picture. The unusually large magnetic moment[4] of $+6.0 \pm 2.5$ nuclear magnetons for Np^{237} is not calculable on the extreme single-particle picture, though the even-parity assignment of Fig. 32 gives a larger $(+3\,nm)$ moment than other possible states.

---

[1] KROHN, NOVEY and RABOY: Phys. Rev. **98**, 1187 (1955).
[2] T. E. MANNING, M. FRED and F. S. TOMKINS: Phys. Rev. **102**, 1108 (1956).
[3] HOLLANDER, SMITH and RASMUSSEN: Phys. Rev. **102**, 1372 (1956).
[4] BLEANEY, LLEWELLYN, PRYCE and HALL: Phil. Mag. **45**, 992 (1954).

The U^{237} main beta transition to state C-$\frac{3}{2}$ has $\log ft \sim 6.1$ and is classified with the state assignments of the figure as first forbidden ($\Delta I = 1$, yes), unhindered. Beta decay to A-$\frac{5}{2}$ should be second forbidden, and is not observed. Beta decay to B-$\frac{5}{2}$ would by spin and parity change be unique first forbidden ($\Delta I = 2$, yes), but it is hindered by the $\Delta \Lambda$ selection rule. This beta group has not been observed, and experimentally a limit has been set of $\log ft > 9$.

State D-$\frac{7}{2}$ cannot be significantly populated directly by beta decay as it has ($\Delta I = 3$) and would be second forbidden. The spin $\frac{5}{2}$ is also consistent with experiment, but $\frac{7}{2}$ seems more likely by inspection of the NILSSON diagram (Fig. 31). Experimentally somewhat less certain is the level designated tentatively E-$\frac{3}{2}$. For most of the gamma-rays depopulating this level only one conversion line has been observed. The beta branch to this level is quite weak. There are also a few weak conversion lines not accounted for by the present decay scheme.

Pu^{237} decays predominantly to the ground state (or possibly some to other members of the ground band). An upper limit¹ on decay giving the 60-kev gamma (B-$\frac{5}{2} \rightarrow A$-$\frac{5}{2}$) has been set as less than 2% of total K-capture. The L/K-capture ratio is 0.88, about normal for allowed or $\Delta I = 0, 1$, yes, first forbidden for the decay energy ~ 180 kev, estimated² from closed decay energy cycles.

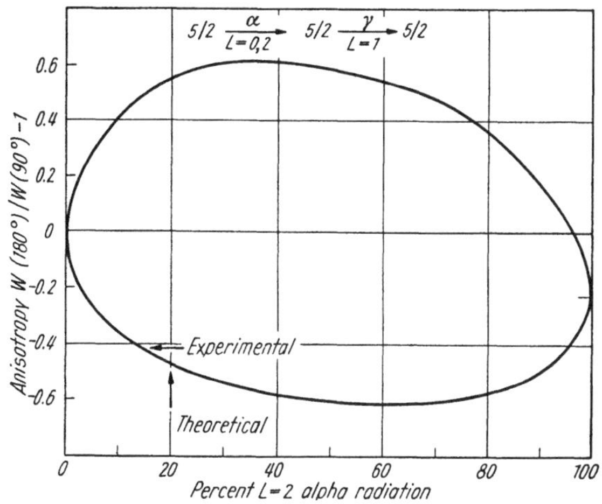

Fig. 33. Plot of the anisotropy of the Am^{241} alpha-gamma angular correlation involving the main alpha group and the 60-kev gamma ray. The anisotropy is given as a function of the percentage $L = 2$ alpha admixed with $L = 0$. The theoretical mixture, 20% $L = 2$, predicted by BOHR, FRÖMAN and MOTTELSON [*31*] is indicated on the plot. The recent experimental value of NOVEY, KROHN and RABOY [private communication (1956)] is indicated also.

The alpha decay hindrance factors are given in Table 13.

The decay to band B is of favored type, and one is referred to Table 16 for comparison of intensities with the BOHR-FRÖMAN-MOTTELSON [*31*] theory.

The BFM formulation can be used to predict that the main alpha group to level B-$\frac{5}{2}$ is about 80% $L = 0$ and 20% $L = 2$. The alpha-gamma angular correlation with gamma B-$\frac{5}{2} \rightarrow A$-$\frac{5}{2}$ will be sensitive to this admixture. Fig. 33 plots the anisotropy for various mixtures of $L = 0$ and 2, with the gamma radiation assumed pure dipole.

The experimental anisotropy, $\frac{W(\pi)}{W(\pi/2)} - 1 = -0.41$, obtained by NOVEY³, KROHN and RABOY with solution sources, is slightly lower than the theoretical.

The angular correlation is especially important on another score, as it determines the $L = 0$ and $L = 2$ waves to be nearly in phase instead of 180° out of

¹ Note added in proof: Recent work by KALKSTEIN, STEPHENS, CARETTO, and HOLLANDER has shown positively that Pu^{237} does decay to the extent of about 2% to the 60-kev state of Np^{237}. They can set an upper limit on electron capture to the 33-kev state of < 0.5%. Fig. 32 is subject, thus, to slight modificaton.

² R. A. GLASS: Private communication (1956).

³ NOVEY, KROHN and RABOY: Private communication (1956).

phase. This allows us to select the more likely cases for the inward integration of coupled alpha-decay equations for nearby even-even isotopes (see Sect. 32).

For Fig. 33, we have used for the two possible phase differences a formula given by FRAUENFELDER[1]. The formula is subject to modification where strong non-central electric interactions exist, as they do here. The greater the inequality in the mixing ratio, the greater the modification. Evaluation of the phase shift modification must await numerical studies of the appropriate alpha decay differential equations, but any modification would be expected to bring both branches of the loop of Fig. 33 closer to a straight line connecting the points for pure $L=0$ or $L=2$.

Acknowledgments.

We wish to express our thanks to all who have helped by corresponding with us on various matters pertaining to this work. We also wish to thank many of our colleagues at the University of California Radiation Laboratory for supplying us with data and discussing topics with us. Especially are we indebted to Dr. F. ASARO for his indispensable assistance in assembling and assessing the alpha decay data of Table 1 and to Mr. C. J. GALLAGHER for calculation of alpha decay hindrance factors.

General references.

[1] RUTHERFORD, E.: Radioactive Substances and their Radiations. London: Cambridge University Press 1913. — A work such as this, written during the early formative period of radioactivity, provides insight into the development rather than the results of one of the most brilliant chapters in the history of science.

[2] GAMOW, G.: Z. Physik **51**, 204 (1928). — The original paper propounding the wave mechanical explanation of the alpha decay process. The explanation was independently advanced by CONDON aud GURNEY [3].

[3] CONDON, E. U., and R. W. GURNEY: Nature, Lond. **122**, 439 (1928). — Phys. Rev. **33**, 127 (1929). — The original papers proposing, independently of GAMOW [2], the wave mechanical explanation of the alpha decay process.

[4] PERLMAN, I., and F. ASARO: Ann. Rev. Nucl. Sci. **4**, 157 (1954). — This review article on alpha radioactivity presents for the first time numerous results from high resolution alpha spectroscopic work. The paper touches on a large number of topics related to alpha emission, including the then new spectroscopic investigations revealing population of numerous nuclear rotational bands.

[5] GLASS, R. A., S. G. THOMPSON and G. T. SEABORG: J. Nucl. Inorg. Chem. **1**, 3 (1955). — This paper presents a comprehensive set of nuclear masses for the heavy region, calculated from alpha- and beta-decay energy balance cycles supplemented by a few experimental neutron binding energies. By estimating alpha decay energies the table is extrapolated to include masses of as-yet-undiscovered nuclei.

[6] PERLMAN, I., A. GHIORSO and G. T. SEABORG: Phys. Rev. **77**, 26 (1950). — This paper systematizes the trends of alpha decay energies and alpha decay rates, drawing on the large number of new artificial alpha emitters produced and studied for the first time in the preceding decade. Many of the principles developed in the present article were laid down in this 1950 paper.

[7] HYDE, E. K., and G. T. SEABORG: This Volume. Berlin: Springer 1957. — This article has a thorough modern review of the transuranium isotopes. Its presentation of chemical and nuclear properties makes it an indispensable reference for those interested in the transuranium region.

[8] FEENBERG, E.: Shell Theory of the Nucleus. Princeton: Princeton University Press 1955. — MAYER, M. G., and J. H. D. JENSEN: Elementary Theory of Nuclear Shell Structure. New York: John Wiley & Sons; London: Chapman & Hall 1955. — PRYCE, M. H. L.: Nuclear Shell Structure. Rep. Progr. Phys. **17**, 1 (1954). — These works thoroughly cover the theory and applications of the spin-orbit coupling spherical well nuclear shell models, which have affected nuclear science so profoundly since their inception in 1949.

---

[1] H. FRAUENFELDER: Chap. XIX(I), Beta and Gamma Ray Spectroscopy; KAI SIEGBAHN, Editor, Eq. (48). Amsterdam: North Holland Publishing Co. 1955.

[9] PRYCE, M. H. L.: Proc. Phys. Soc. Lond. A **65**, 773, 962 (1952). — In this paper the spherical well shell model is applied to the problem of the level systems of nuclei near Pb^{208}, which has closed neutron and proton configurations. This work is an indispensable reference for theoretical studies or correlations involving nuclei of this region.

[10] BOHR, A., and B. R. MOTTELSON: Dan Mat. fys. Medd. **27**, No. 16 (1953). — This work gives the original comprehensive application of the unified nuclear model, which introduces features of collective motion into the individual particle model. Most of the data constituting the strongest support of the model have been accumulated subsequent to this 1953 paper. The paper contains a wealth of useful formulas for application of the model.

[11] BOHR, A., and B. R. MOTTELSON: Dan Mat. fys. Medd. **30**, No. 1 (1955). — This paper is concerned largely with the rotational properties of nuclei in the regions of stabilized spheroidal deformation. A large number of data on nuclear rotational bands are presented, but the main emphasis is on the theoretical problem of explaining the empirical moments of inertia.

[12] RASMUSSEN, J. O., and B. SEGALL: Phys. Rev. **103**, 1298 (1956). — This paper develops the theory of alpha decay of nuclei with stable spheroidal deformation. The derived wave equations are applied numerically to the cases Th^{228} and Cm^{242} to give information regarding the most probable surface zones for formation of energetic alpha particles.

[13] NILSSON, S. G.: Dan. Mat. fys. Medd. **29**, No. 16 (1955). — This paper presents tables and graphs of eigenfunctions and eigenvalues for a nucleon in a three-dimensional harmonic oscillator potential with varying degrees of spheroidal distortion. Semi-empirical terms for spin-orbit interaction and an $\boldsymbol{l} \cdot \boldsymbol{l}$ interaction were included. These tables have many applications in correlating nuclear properties in the regions of spheroidal nuclear shapes.

[14] PRESTON, M. A.: Phys. Rev. **71**, 865 (1947). — A careful derivation (by GAMOW's complex eigenvalue method) of the decay rate probability of an alpha particle trapped in the lowest virtual level (for given angular momentum) of the one-body square potential well is given. The rather involved derivation of the COULOMB function from the contour integral form leads to essentially the same result as does the ordinary WKB method.

[15] SEXL, T.: Z. Physik **81**, 163 (1933). — This paper contains an interesting review of several early treatments of alpha decay rate theory in the five years following the original papers [2] and [3]. SEXL's own thorough treatment of the problem is along the lines described in the preceding reference [14] and seems indeed to form the basis for PRESTON's treatment.

[16] WINSLOW, G. H.: On the One-Body Model of Alpha Radioactivity, IV. Calculated Wave Amplitudes near the Nucleus. Argonne National Laboratory Report ANL-5381, Jan., 1955 (unpublished). — This work contains an extensive set of tables calculating squares of the radial alpha wave function near the nuclear surface from data for known alpha emitters. Trends of these calculated values are discussed. Tables and graphs for the calculation of centrifugal barrier effects are given.

[17] THOMAS, R. G.: A Formulation of the Theory of Alpha-particle Decay from Time-independent Equations. Progr. Theor. Phys. **12**, 253 (1954). — This work concisely presents alpha decay rate theory in a formalism closely related to the collision matrix theory often applied to nuclear reactions. Results are derived for the one-body model. Many-body modifications are also discussed. The paper contains a lucid discussion of electron screening effects and contains several references to modern literature on alpha decay rate theory.

[18] BETHE, H. A.: Rev. Mod. Phys. **9**, 69 (1937). For alpha decay and barrier penetration see pp. 161—171 and pp. 177—78. — This work on alpha decay is still a valuable reference work nearly two decades after its appearance. The penetration factor formulas and symbols are still widely used. Although the particular form of the many-body model BETHE proposed is not generally favored today, this treatment of the one-body model and discussion of the fundamental problems are lucid and timely even today.

[19] WINSLOW, G. H., and O. C. SIMPSON: On the One-body Model of Alpha Radioactivity, I. Survey of the Existing Results. Argonne National Laboratory Report ANL-4841, June, 1952 (unpublished). — This paper gives a good review and numerical comparisons of several earlier one-body alpha decay rate formulations. Graphs and tables serving as aids to computation of barrier penetrabilities are also presented.

[20] WINSLOW, G. H., and O. C. SIMPSON: On the One-body Model of Alpha Radioactivity, IIb. *S*-state Decay with the Limiting Square Well Model. Argonne National Laboratory Report ANL-4910, Jan., 1953 (unpublished). — A one-body treatment of alpha decay with sloping potential walls at the nuclear surface is given.

[21] WINSLOW, G. H., and O. C. SIMPSON: On the One-body Model of Alpha Radioactivity, III. Non-S-state Decay, Preformation Factors, Model Variations, General Discussion. Argonne National Laboratory Report ANL-5277, April, 1954 (unpublished). — This paper tests WKB approximations of the COULOMB functions. The authors also discuss the atomic electron screening correction. They present a "surface well" one-body model and propose many-body modifications. Some of the material of this paper is published in — [27].

[22] DEVANEY, J. J.: Phys. Rev. **91**, 587 (1953). — This paper develops an alpha decay rate theory of many-body type based on statistical considerations of average nuclear level spacings near ground.

[23] BLATT, J. M., and V. F. WEISSKOPF: Theoretical Nuclear Physics. New York: John Wiley & Sons, Inc. 1952. For alpha decay see pp. 568—578. — In the space devoted to alpha decay the statistical many-body model based on average level spacings is developed and applied to nuclear data for calculation of effective nuclear radii. Selection rules for alpha decay are presented.

[24] RASETTI, F.: Elements of Nuclear Physics, p. 100. New York: Prentice-Hall, Inc. 1936. — The one-body model of alpha decay is treated by construction of a wave packet of time-dependent wave functions, and the behavior with time is derived. This is a more general and fundamental approach than the usual treatments by the complex eigenvalue or other methods using time-independent wave functions, but the numerical results are essentially the same.

[25] WINSLOW, G. H., and O. C. SIMPSON: On the One-body Model of Alpha Radioactivity, IIa. S-state Decay with the Consistent Square Well Model. Argonne National Laboratory Report ANL-4901, Dec., 1952 (unpublished). — The method of RASETTI [24] is employed and extended to derive a one-body decay rate expression, and they demonstrate that the methods of BETHE [18] and PRESTON [14] lead to the same results as theirs when sufficiently accurate approximations are employed

[26] KEMBLE, E. C.: The Fundamental Principles of Quantum Mechanics, p. 192. New York: McGraw-Hill Co. 1937. — KEMBLE also gives an interesting treatment of the one-body alpha decay model using wave packets of time-dependent wave functions.

[27] WINSLOW, G. H.: Phys. Rev. **96**, 1032 (1954). — A "surface well" one-body model and its many-body modification are treated and discussed. Peripheral subjects such as the atomic electron screening correction are also discussed.

[28] MIGDAL, A.: J. Phys. USSR. **4**, 449 (1941). — Formulas for the probability of orbita electron ejection during alpha emission are derived. Expressions for the energy distribution of ejected electrons are also given.

[29] PRESTON, M. A.: Phys. Rev. **75**, 90 (1949). — The effects of non-central interactions between alpha particle and recoil nucleus are explored, and the general methods underlying later developments (e.g. [12]) are set forth.

[30] RACAH, G.: Phys. Rev. **62**, 438 (1942). — This paper contains a wealth of formulas useful for quantum-mechanical treatment of systems of interacting particles. The very general tensor algebra developed here offers powerful methods for attacking problems of non-central interactions in alpha decay, as well as a host of other problems including many-electron or many-nucleon systems.

[31] BOHR, A., P. O. FRÖMAN and B. R. MOTTELSON: Dan. Mat. fys. Medd. **29**, No. 10 (1955). — This paper is notable for its applications of the spheroidal well nuclear model to alpha decay and for its extension of the concept of "favored" alpha decay groups.

[32] RASMUSSEN, J. O., S. G. THOMPSON and A. GHIORSO: Phys. Rev. **89**, 33 (1953). — Data on a number of artificial alpha emitters in the rare earth region are presented, and correlations of decay rates with alpha decay theories are attempted.

[33] TOLHOEK, H. A., and P. J. BRUSSARD: Physica, Haag **21**, 449 (1955). — This paper presents fresh approaches to alpha decay rate theory. A very strong attractive nuclear potential for alpha particles is proposed. The diffuseness of the nuclear surface and the finite range of nuclear forces are included in the derivation of the alpha particle potential energy function. An overlap integral expression for the probability of formation of an alpha particle from constituent nucleons in nuclear matter is derived.

The Transuranium Elements.

By

Earl K. HYDE and GLENN T. SEABORG.

With 44 Figures.

A. General considerations. The place of the heaviest elements in the periodic system.

I. Historical views.

1. Speculations antedating discovery of transuranium elements. The place of the very heaviest elements in the periodic system is of great scientific interest and is of crucial importance in understanding the physical and chemical properties of these elements. Ideas on this question have varied considerably over the years. Periodic charts have commonly placed thorium, protactinium, and uranium in positions immediately below the elements hafnium, tantalum, and tungsten. The latter group of elements are members of a transition series in which the $5d$ electron shell is being filled. Hence this earlier placement of the three heavy elements corresponded to the assumption that they are members of a $6d$ transition series. Indeed thorium, protactinium, and, to a much smaller extent, uranium do show considerable resemblance in chemical properties to the $4d$ and $5d$ transition series elements. The electronic configuration beyond the radon core on this basis would be written thus: thorium ($6d^2 7s^2$), protactinium ($6d^3 7s^2$), and uranium ($6d^4 7s^2$).

After the appearance of N. BOHR's classic paper[1] on the quantized nuclear atom many authors suggested that a $5f$ transition series should start in the vicinity of uranium before completion of the $6d$ shell. Because the complexity of the theoretical problem made it impossible to calculate exact results and because experimental data were neither plentiful nor definitive there was a wide difference of opinion as to where the new series might be expected to start. Some of these early suggestions were the following: SUGIURA and UREY[2] predicted that the first entry of the electron into the $5f$ electron shell should occur at the element with atomic number 95; BOHR[3] chose element 94 as the starting point of the series; KARAPETOFF[4] and WU and GOUDSMIT[5] suggested element 93; MAYER[6] suggested elements protactinium or uranium; GOLDSCHMIDT[7] chose protactinium; SWINNE[8], MCLENNAN, MCFAY, and SMITH[9] and PERRIN[10] and VILLAR[11] suggested thorium.

[1] N. BOHR: Phil. Mag. **26**, 1, 476 (1913).
[2] Y. SUGIURA and H. C. UREY: Kgl. danske Vid. Selsk., mat.-fys. Medd. **7**, No. 13, 3 (1926).
[3] N. BOHR: Nature, Lond. **112**, 29 (1923).
[4] V. KARAPETOFF: J. Franklin Inst. **210**, 609 (1930).
[5] T. Y. WU and S. GOUDSMIT: Phys. Rev. **43**, 496 (1933).
[6] M. G. MAYER: Phys. Rev. **60**, 184 (1941).
[7] V. M. GOLDSCHMIDT. Travaux du Congrès Jubilaire Mendeléev II, 387 (1937).
[8] R. SWINNE: Z. Elektrochem. **31**, 417 (1925).
[9] J. C. MCLENNAN, A. B. MCFAY and H. G. SMITH: Proc. Roy. Soc. Lond. A **112**, 76 (1926).
[10] J. PERRIN: Grains de Matière et de Lumière, II—30. Paris: Hermann & Cie. 1935.
[11] G. E. VILLAR: J. Chem. Educat. **19**, 329 (1942). — Ann. Acad. brasil. Sci. **12**, 51 (1940).

2. The actinide element concept. The intensive study of the heaviest elements in the last few years including the synthesis and study of transuranium elements up to atomic number 101 has made it certain that, considering the series as a whole, the $5f$ series of elements begins at actinium in the same sense that the rare-earth series begins with lanthanum. SEABORG has stated the evidence leading to this view in several previous publications[1–3].

The two series are not strictly comparable, however, because the tripositive state which would be expected to be characteristic of the actinide series does not appear at all in thorium and protactinium and does not become the most stable oxidation state in aqueous solution until americium is reached. For the elements uranium through americium several oxidation states are well established. This is a very real difference from the lanthanide series. It has been interpreted by some as evidence that these lower elements are not members of an actinide series and that they must be members of a $6d$ transition series. Actually this evidence does not distinguish between the two series but emphasizes the important fact that the $7s$, $6d$, and $5f$ levels lie quite close together in energy in this region.

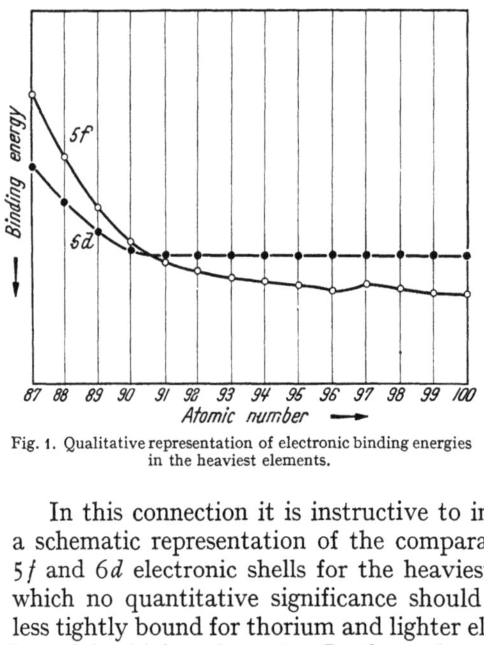

Fig. 1. Qualitative representation of electronic binding energies in the heaviest elements.

Additional evidence, particularly magnetic susceptibility data, has established reasonably well that $5f$ electrons are present in ionic compounds of uranium, neptunium, and plutonium. In the case of protactinium it is not settled whether the electronic configuration beyond the radon core contains $5f$ electrons and in the case of thorium it seems probable that no $5f$ electrons are to be expected.

In this connection it is instructive to introduce a figure by SEABORG giving a schematic representation of the comparative electronic binding energy of the $5f$ and $6d$ electronic shells for the heaviest elements. This diagram (Fig. 1), to which no quantitative significance should be attached, shows the $5f$ electrons less tightly bound for thorium and lighter elements and progressively more tightly bound for higher elements. In the region near the crossing point the $5f$ and $6d$ shells lie so close together that the energy necessary for the shift from one shell to the other may lie within the range of chemical binding energies. This has been pointed out by SEABORG, by STARKE[4] and by others. Thus, for example, the number of $5f$ electrons found for the gaseous atom may not necessarily be the same as that for the same element in the metallic state, and, in turn, it is possible that neither of these structures would correspond directly to the chemical compounds in which hydration and lattice energies play an important role.

[1] G. T. SEABORG: Paper (21.1) of "The Transuranium Elements", National Nuclear Energy Series, Division IV, Vol. 14B, McGraw-Hill Book Co., Inc., New York 1949: Hereafter in this text this book will be referred to as N.N.E.S. Vol. 14B "The Transuranium Elements".

[2] G. T. SEABORG: Chap. 17 of "The Actinide Elements", National Nuclear Energy Series, Division IV, Vol. 14A, McGraw-Hill Book Co., Inc., New York, 1954: Hereafter in this text this book will be referred to as N.N.E.S. Vol. 14A "The Actinide Elements".

[3] G. T. SEABORG: Nucleonics **5** (5), 16–36 (1949).

[4] K. STARKE: Naturwiss. **34**, 69 (1947).

SEABORG has prepared a list of suggested electronic configurations for the heaviest elements which reflect these considerations of the relative stability of the $5f$ and $6d$ electrons and which correspond to spectroscopic data where known. This is shown in Table 1 where additional configurations to complete the series are also included.

Table 1. *Suggested electron configurations (beyond radon) for gaseous atoms of actinide elements)*

$_{89}$Ac	$6d\,7s^2$	$_{97}$Bk	$5f^8\,6d\,7s^2$ (or $5f^9\,7s^2$)
$_{90}$Th	$6d^2\,7s^2$ (or $5f6\,d\,7s^2$)	$_{98}$Cf	$5f^{10}\,7s^2$
$_{91}$Pa	$5f^2\,6d\,7s^2$ (or $5f\,6d^2\,7s^2$)	$_{99}$E	$5f^{11}\,7s^2$
$_{92}$U	$5f^3\,6d\,7s^2$	$_{100}$Fm	$5f^{12}\,7s^2$
$_{93}$Np	$5f^4\,6d\,7s^2$ (or $5f^5\,7s^2$)	$_{101}$Mv	$5f^{13}\,7s^2$
$_{94}$Pu	$5f^6\,7s^2$	102	$5f^{14}\,7s^2$
$_{95}$Am	$5f^7\,7s^2$	103	$5f^{14}\,6d\,7s^2$
$_{96}$Cm	$5f^7\,6d\,7s^2$		

In Sect. II below the evidence pertaining to the relative significance of $5f$ and $6d$ electrons is summarized in some detail. In addition to this principal purpose such a summary is a useful way to present some of the comparative chemistry of the transuranium element group.

II. Evidence relating to the actinide concept.

a) Chemical evidence.

3. Tripositive oxidation state in aqueous solution. The characteristic oxidation number of an actinide series as opposed to a thoride series or a uranide series, for example, is III. The earliest member of the series which clearly shows the III state is uranium, the third member. The III state rapidly becomes stabilized in going up the series and is the predominant one for americium and curium (elements 95 and 96), so much so that it is the only oxidation state which is thermodynamically stable in acidic aqueous solution. Berkelium (element 97) has aqueous oxidation states of III and IV analogous to terbium in the rare-earth series whose sesquioxide Tb_2O_3 can be oxidized to TbO_2. The essential point here is that the half-filled shell configuration, $5f^7$ or $4f^7$, is especially stable; berkelium the eighth member of the actinide series strives to reach it by releasing one $5f$ electron by oxidation; similarly, terbium the eighth member of the lanthanide series achieves the $4f^7$ configuration by conversion to Tb (IV).

The oxidation potentials for the III-IV couple and for the IV-VI couple for the elements of the series show the rapidly increasing stability of the tripositive state in going up the series. These potentials are given in Table 2. CUNNINGHAM[1] has presented a figure showing the free energy of formation of the aqueous ions of the

Table 2. *Some formal oxidation potentials of the actinide elements in acid solution.*

Atomic number	Element	Potential (Volts) III-IV	Potential (Volts) IV-VI
92	uranium	+ 0.6	− 0.33
93	neptunium	− 0.14	− 0.94
94	plutonium	− 0.98	− 1.04
95	americium	− 2.44	− 1.34
96	curium	∼ − 3.00 (estimated)	
97	berkelium	− 1.6	
98−101		(no data)	

[1] B. B. CUNNINGHAM: Paper P/726, "Peaceful Uses of Atomic Energy Proceedings of the Geneva Conference", August 1955, United Nations.

actinide elements. From this figure, reproduced here as Fig. 2, it is quite clear that the tripositive state becomes stabilized with respect to all other aqueous oxidation states as the atomic number increases.

In the case of curium attempts to effect oxidation of tracer amounts to curium (IV) or to curium (VI) have been uniformly unsuccessful. The 162-day Cm^{242} which has had to be used in most experiments is a difficult isotope for the study of oxidation-reduction equilibria because of the reducing effects of the alpha particles, but additional experiments using the 19-year Cm^{244} gave the same negative results. Tetrapositive curium is known in the solid compounds, CmO_2 and CmF_4.

All chemical studies of the elements 97 to 101 have been done on a strictly tracer scale. With the exception of berkelium (element 97) the tracer scale behavior in coprecipitation, solvent extraction, and ion exchange is typical of that for the trivalent actinide elements. No evidence is found for oxidation to a higher state except in the case of berkelium where such oxidation is expected.

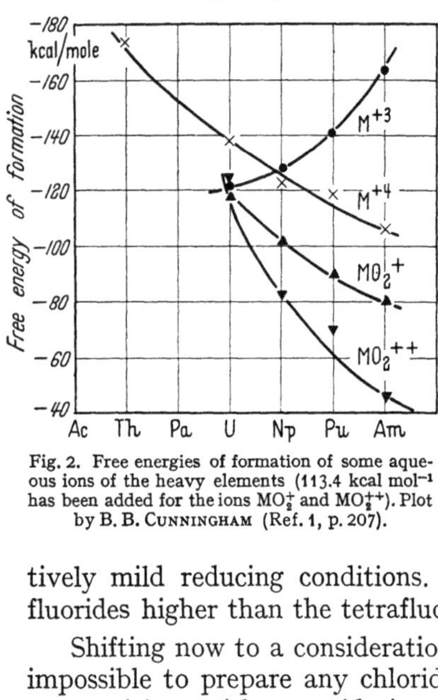

Fig. 2. Free energies of formation of some aqueous ions of the heavy elements (113.4 kcal mol⁻¹ has been added for the ions MO_2^+ and MO_2^{++}). Plot by B. B. Cunningham (Ref. 1, p. 207).

4. Stability of anhydrous halides. The stability of the solid compounds of the various oxidation states of these elements provides notable evidence for the increased stability of the III state in going up the series. This is best illustrated by a consideration of the solid non-oxygenated halides of these elements. Considering first the fluorides, it is noted that no trifluoride appears in thorium or protactinium and that uranium trifluoride can be prepared only under drastic reducing conditions. The trifluorides of neptunium and plutonium can be prepared under relatively mild reducing conditions. For the elements americium and curium no fluorides higher than the tetrafluoride have been prepared.

Shifting now to a consideration of the other halides we note that it has been impossible to prepare any chloride, bromide, or iodide compound of plutonium or americium with an oxidation state higher than III; it has been possible to prepare the chloride and bromide of neptunium in the IV and the III oxidation state in addition to the iodide of oxidation state III, but no compounds of these halogens of oxidation number greater than four are known. In the case of uranium a whole series of chlorides are known with oxidation numbers III, IV, V, and VI; the bromides and iodides are known only for the III and IV state.

Table 3. *Halides of some of the heaviest elements.*

Element	Fluorides				Chlorides				Bromides			Iodides		
$_{89}$Ac				AcF_3				$AcCl_3$			$AcBr_3$			AcI_3
$_{90}$Th			ThF_4				$ThCl_4$	$ThCl_3$		$ThBr_4$	$ThBr_3$		ThI_4	ThI_3
$_{91}$Pa		PaF_5	PaF_4			$PaCl_5$	$PaCl_4$		$PaBr_5$			PaI_5		
$_{92}$U	UF_6	UF_5	UF_4	UF_3	UCl_6	UCl_5	UCl_4	UCl_3		UBr_4			UI_4	UI_3
$_{93}$Np	NpF_6	NpF_5(?)	NpF_4	NpF_3			$NpCl_4$	$NpCl_3$			$NpBr_3$			NpI_3
$_{94}$Pu	PuF_6		PuF_4	PuF_3				$PuCl_3$			$PuBr_3$			PuI_3
$_{95}$Am			AmF_4	AmF_3				$AmCl_3$			$AmBr_3$			AmI_3
$_{96}$Cm			CmF_4	CmF_3										

Table 3 lists all the nonoxygenated halides of the elements from actinium to curium which have been prepared and maintained in the solid state except for the dihalides of thorium.

5. Discussion of evidence on oxidation states. The chemical evidence just discussed indicates that the 5*f* electrons are more readily removed by oxidation than the 4*f* electrons as should be expected on the basis of the predicted lower ionization potentials of 5*f* as compared to the potentials of 4*f* electrons. In a detailed analysis it is necessary to consider the free energy of hydration or of the crystal lattice as well as ionization potentials.

Thus, the tripositive state of thorium cannot exist in aqueous solution. The most stable oxidation state for protactinium is protactinium (V) although there is now rather clear evidence for an easily oxidized protactinium (IV) in aqueous solution[1–5]. Whether protactinium (III) can be formed remains in doubt. BOUISSIÈRES and HAISSINSKY[4] have reported the reduction of protactinium (V) to protactinium (IV) in solution by use of amalgamated zinc, chromous chloride, or other strong reducing agents. FRIED and HINDMAN[5] have dissolved anhydrous $PaCl_4$ in an oxygen-free solution of hydrochloric acid and determined its absorption spectrum. This is characterized by three maxima in the ultraviolet region (2760, 2550, and 2237.5 Å respectively) and by no structure in the visible region. It is of interest that the observed spectrum is remarkably similar to that of cerium (III) (configuration $4f^1$) since protactinium (IV) might be expected to have the configuration $5f^1$ on the basis of the actinide concept. FRIED and HINDMAN estimate the Pa(IV)-Pa(V) couple to be $\sim +0.1$ volts.

In the solid state there is definite evidence for lower oxidation states of thorium and protactinium. Thorium triiodide[6–8], thorium trichloride[9] and thorium tribromide[9] are known as well as the corresponding dihalides[9]. In addition the sesquisulfide[10] and other sulfides of lower oxidation number are known[7,10]. However, thorium triiodide is not isostructural with the heavier actinide triiodides and the lower sulfides are semimetallic in character.

In the case of protactinium, ZACHARIASEN[11,12] and MCCULLOUGH[13] have obtained crystallographic evidence for the dioxide of protactinium with the fluorite structure. ELSON, FRIED, SELLERS and ZACHARIASEN[12,14] have prepared $PaCl_4$ and

---

1 G. BOUISSIÈRES and M. HAISSINSKY: J. Chem. Soc. **1949**, 256 (supplementary issue 2, 54).
2 R. ELSON: Argonne National Laboratory Reports, ANL-4370, Nov. 23, 1949, and ANL-4252, Febr. 17, 1949.
3 M. HAISSINSKY and G. BOUISSIÈRES: Bull. Soc. chim. France **18**, 146 (1951).
4 G. BOUISSIÈRES and M. HAISSINSKY: C. R. Acad. Sci., Paris **226**, 573 (1948). — J. Chem. Soc. **1949**, 253, 554.
5 S. FRIED and J. C. HINDMAN: J. Amer. Chem. Soc. **76**, 4863 (1954).
6 E. HAYEK and TH. REHNER: Experientia, Basel **5**, 114 (1949).
7 J. S. ANDERSON and R. W. M. D'EYE: J. Chem. Soc. **1949**, 244 (supplementary issue 2, 51).
8 E. F. WESTRUM jr.: Private communication, April 1949.
9 E. HAYEK, TH. REHNER and A. FRANCK: Mh. **82**, 575 (1951).
10 E. D. EASTMAN, L. BREWER, L. A. BROMLEY, P. W. GILLES and N. L. LOFGREN: J. Amer. Chem. Soc. **72**, 4019 (1950).
11 W. H. ZACHARIASEN: Report MUC-FWHZ-175, 1946.
12 R. ELSON, S. FRIED, P. SELLERS and W. H. ZACHARIASEN: J. Amer. Chem. Soc. **72**, 5791 (1950).
13 J. D. MCCULLOUGH: Report RL-4. 5. 56, April 1947.
14 R. ELSON, S. FRIED, P. SELLERS: Argonne National Laboratory Report ANL-4545, Sept. 1950, p. 9, see also Chap. 5, N.N.E.S., Vol. 14A, "The Actinide Elements", 1954.

shown it to be isostructural with UCl_4; Zachariasen and Plettinger[1] have identified PaF_4 as isostructural with ThF_4, UF_4, NpF_4, and PuF_4. It seems likely that many other compounds of protactinium (IV) will ultimately be prepared.

It cannot be said whether the electrons in excess of the radon shell are in the $5f$ or the $6d$ configuration for these lower oxidation states of thorium and protactinium or possibly in some combination of these, since this is the region where the binding energies for the two types of electrons come closest to equality. From the behavior of uranium, neptunium, and plutonium it must be deduced that as many as three of the $5f$ electrons are given up fairly readily but with increasing difficulty as the atomic number increases.

In the lanthanide series the tripositive state is dominant throughout the series, but, nevertheless, there are several instances of dipositive oxidation states [samarium (II), europium (II), and ytterbium (II)] from which $4f$ electrons are lost upon oxidation to the corresponding tripositive states. Also the spectroscopic data indicate that in the neutral gaseous atoms there are generally only two electrons (beyond the radon structure) outside the $4f$ shell although the most important oxidation state is certainly III. Cerium has a well-known tetrapositive oxidation state. Praseodymium, the second lanthanide rare-earth, exhibits a tetrapositive oxidation state. Terbium, as we mentioned above, strives to achieve the half-filled configuration $4f^7$ by oxidation to terbium (IV) although this oxidation has been established only for the anhydrous oxide. This oxidation presumably involves the removal of a $4f$ electron. These facts indicate that even in the lanthanide series the inner-shell $4f$ electrons are not bound so tightly that they are quite unavailable for participation in oxidation-reduction reactions. A slight decrease in their binding energy would have resulted in quite different chemical behavior. Or, as Connick[2] has pointed out, if the hydration and lattice free energies of the tripositive ions relative to the ions of other oxidation states were different, the III state might not have its pronounced importance.

The known oxidation states of the lanthanide and actinide elements are summarized in Table 4. The similarities of the two series are apparent as are the differences resulting from the more ready oxidation of the actinide group. The uncertain or unusual states are designated with parentheses; possible states

Table 4. *Oxidation states of lanthanide and actinide elements.*

Lanthanide elements

Atomic number	57	58	59	60	61	62	63	64	65	66	67	68	69	70	71
Element . . .	La	Ce	Pr	Nd	Pm	Sm	Eu	Gd	Tb	Dy	Ho	Er	Tm	Yb	Lu
Oxidation states						2	2							2	
	3	3	3	3	3	3	3	3	3	3	3	3	3	3	3
		4	4						4						

Actinide elements

Atomic number	89	90	91	92	93	94	95	96	97	98	99	100	101	102	103
Element . . .	Ac	Th	Pa	U	Np	Pu	Am	Cm	Bk	Cf	E	Fm	Mv		
Oxidation states	**3**	(3)	(3)	3	3	3	**3**	**3**	**3**	**3**	**3**	**3**	**3**		
		4	4	4	4	**4**	(4)	4	4						
			5	5	**5**	5	5								
				6	6	6	6								

[1] W. H. Zachariasen and H. A. Plettinger: Argonne National Laboratory Report ANL-4515, Sept. 1950, p. 33.

[2] R. E. Connick: J. Chem. Soc. **1949**, 235 (supplementary issue 2, 49).

with oxidation number less than III are omitted where metallic or possible metallic bonding character is involved.

Some of the oxidation states shown in Table 4 are observable only under drastic conditions so that their inclusion here may be misleading. On the other hand to list only those states which are stable in aqueous solution also has shortcomings since the existence of an aqueous ion, besides being limited by the ionization potentials of the electrons, is affected by the specific chemical processes of hydration and complex formation and is arbitrarily confined within the limits of its oxidation and reduction by water. The most stable oxidation states of the actinide elements are printed bold face in Table 4 to show the variation in their stability in going across the series.

6. Crystallographic data—isostructural series. Table 5 shows the crystallographic radii of the tripositive and tetrapositive ions of the actinide elements. The radii of the tripositive lanthanide ions are shown for comparison. As the number of electrons in the *f* shell increases, there is a regular contraction of the ionic radius in both series; in the rare-earth series this is referred to as the "lanthanide contraction". The extensive studies of ZACHARIASEN and his co-workers[1,2] on the crystal structure of the heavy element compounds have shown a large number of isostructural series of compounds. The facts are summarized in Table 6.

The large number of isostructural series and the closely-similar and regularly varying crystal radii indicate a smooth transition in going up the series, a transition most easily explained by the steady filling of an inner shell of electrons.

Table 5. *Crystal radii of lanthanide and actinide element ions.*

Lanthanide series[3]			Actinide series[2]			
Number of 4*f* or 5*f* electrons	Element	Radius Å	Element	Radius Å	Element	Radius Å
0	La^{+3}	1.061	Ac^{+3}	1.11	Th^{+4}	0.99
1	Ce^{+3}	1.034	(Th^{+3})	(1.08)	Pa^{+4}	0.96
2	Pr^{+3}	1.013	(Pa^{+3})	(1.05)	U^{+4}	0.93
3	Nd^{+3}	0.995	U^{+3}	1.03	Np^{+4}	0.92
4	Pm^{+3}	0,979	Np^{+3}	1.01	Pu^{+4}	0.90
5	Sm^{+3}	0.964	Pu^{+3}	1.00	Am^{+4}	0.89
6	Eu^{+3}	0.950	Am^{+3}	0.99		
7	Gd^{+3}	0.938				
8	Tb^{+3}	0.923				
9	Dy^{+3}	0.908				
10	Er^{+3}	0.894				
11	Ho^{+3}	0.881				
12	Tm^{+3}	0.869				
13	Yb^{+3}	0.858				
14	Lu^{+3}	0.848				

7. The metallic state. The metallic state has been prepared for the elements thorium through curium. The properties of this series of metals do not correspond to those of the 5*d* elements, hafnium (element 72) through platinum (element 78). On the contrary, the resemblance of the chemical properties of

[1] W. H. ZACHARIASEN: Phys. Rev. **73**, 1104 (1948). Numerous papers in Acta crystallogr. **1948—1955**. — J. Chem. Phys. **16**, 254 (1948).
[2] W. H. ZACHARIASEN: Chap. 18, N.N.E.S., Vol. 14A "The Actinide Elements", 1954.
[3] D. H. TEMPLETON and C. H. DAUBEN: J. Amer. Chem. Soc. **76**, 5237 (1954).

Table 6. *Unit cell dimensions for isostructural series in the actinide element compounds*[1].

The dioxides, XO_2 cubic symmetry.

Compound	Cell dimension, Å a
ThO_2	5.597 ± 0.001
PaO_2	5.507 ± 0.002
UO_2	5.468 ± 0.001
NpO_2	5.436 ± 0.001
PuO_2	5.397 ± 0.001
AmO_2	5.376 ± 0.001 [2]
CmO_2	5.372 ± 0.003 [2]

The tetrachlorides, XCl_4 tetragonal symmetry.

Compound	Cell dimension, Å	
	a	c
$ThCl_4$	8.490 ± 0.001	7.483 ± 0.001
$PaCl_4$	8.377 ± 0.004	7.482 ± 0.004
UCl_4	8.303 ± 0.001	7.483 ± 0.001
$NpCl_4$	8.27 ± 0.01	7.47 ± 0.01

The trichlorides, XCl_3 hexagonal symmetry.

Compound	Cell dimension, Å	
	a	c
$AcCl_3$	7.64 ± 0.02	4.56 ± 0.02
UCl_3	7.443 ± 0.003	4.321 ± 0.003
$NpCl_3$	7.420 ± 0.010	4.282 ± 0.005
$PuCl_3$	7.395 ± 0.001	4.246 ± 0.001
$AmCl_3$	7.38 ± 0.01	4.25 ± 0.01

The trifluorides, XF_3 hexagonal symmetry.

Compound	Cell dimensions, Å	
	a	c
AcF_3	4.28 ± 0.01	7.55 ± 0.02
UF_3	4.146 ± 0.003	7.348 ± 0.004
NpF_3	4.116 ± 0.001	7.288 ± 0.004
PuF_3	4.095 ± 0.001	7.254 ± 0.001
AmF_3	4.067 ± 0.001	7.225 ± 0.002 [3]
CmF_3	4.04 ± 0.01	7.22 ± 0.02 [4]

The tetrafluorides, XF_4 [5] monoclinic symmetry.

Compound	Cell dimensions, Å			
	a	b	c	β
ThF_4	13.1 ± 0.1	11.0 ± 0.1	8.6 ± 0.1	126° ± 1°
UF_4	12.82 ± 0.06	10.74 ± 0.05	8.41 ± 0.05	126° 10′ ± 30′
NpF_4	12.69 ± 0.06	10.64 ± 0.05	8.33 ± 0.05	126° 10′ ± 30′
PuF_4	12.61 ± 0.06	10.57 ± 0.05	8.28 ± 0.05	126° 10′ ± 30′
AmF_4 [6]	12.49	10.47	8.20	126° ± 1°

The hexapositive compounds, $Na(XO_2)(C_2H_3O_2)_3$ cubic symmetry.

Compound	Cell dimensions Å
Uranium .	10.692 ± 0.001
Neptunium	10.680 ± 0.002
Plutonium	10.664 ± 0.002
Americium	10.6 [7]

The hexafluorides XF_6 orthorhombic symmetry.

Compound	Cell dimensions Å		
	a	b	c
UF_6 [8]	9.900 ± 0.002	8.962 ± 0.002	5.207 ± 0.002
NpF_6 [9]	9.92 ± 0.02	8.99 ± 0.02	5.22 ± 0.02
PuF_6 [10]	isostructural		

The sesquioxides, X_2O_3

hexagonal form.

Compound	Cell dimensions, Å	
	a	c
Ac_2O_3 [11]	4.08 ± 0.01	6.30 ± 0.02
Pu_2O_3 [12]	3.840 ± 0.004	5.957 ± 0.006
Am_2O_3 [13]	3.817 ± 0.005	5.971 ± 0.010

cubic form.

Compound	Cell Dimensions, Å a
Pu_2O_3 [14]	11.03 ± ?
Am_2O_3 [13]	11.03 ± 0.01
Cm_2O_3 [2]	11.00 ± 0.02

The tribromides[15], XBr_3

hexagonal form—UCl_3 type.

Compound	Cell dimensions, Å	
	a	c
$AcBr_3$	8.08 ± 0.04	4.69 ± 0.02
UBr_3	7.942 ± 0.002	4.441 ± 0.002
$\alpha NpBr_3$	7.933 ± 0.005	4.391 ± 0.005

orthorhombic—$PuBr_3$ type.

Compound	Cell dimensions, Å		
	a	b	c
$\beta NpBr_3$	12.67 ± 0.05	4.12 ± 0.03	9.17 ± 0.04
$PuBr_3$	12.64 ± 0.05	4.10 ± 0.03	9.15 ± 0.04
$AmBr_3$	12.6 ± 0.1	4.11 ± 0.04	9.12 ± 0.05

The tri-iodides[15], XI_3 orthorhombic—$PuBr_3$ type.

Compound	Cell dimensions, Å		
	a	b	c
UI_3	14.01 ± 0.06	4.32 ± 0.03	10.01 ± 0.05
NpI_3	14.03 ± 0.06	4.30 ± 0.03	9.95 ± 0.05
PuI_3	14.03 ± 0.06	4.30 ± 0.03	9.92 ± 0.05
AmI_3	14.0 ± 0.1	4.31 ± 0.05	9.9 ± 0.1

The oxychlorides, XOCl tetragonal symmetry

Compound	Cell dimensions, Å	
	a	c
AcOCl[5]	4.25 ± 0.02	7.08 ± 0.03
PuOCl[5]	4.012 ± 0.002	6.782 ± 0.010
AmOCl[13]	4.00 ± 0.01	6.78 ± 0.01

The oxybromides, XOBr.

Compound	Cell dimensions, Å	
	a	c
AcOBr[5]	4.28 ± 0.02	7.41 ± 0.03
PuOBr[5]	4.022 ± 0.004	7.571 ± 0.011

1 Unless other wise indicated data are taken from W. H. ZACHARIASEN: Chap. 18 N.N.E.S. Vol. 14A, "The Actinide Elements" 1954.

2 L. B. ASPREY, F. H. ELLINGER, S. FRIED and W. H. ZACHARIASEN: J. Amer. Chem. Soc. **77**, 1707 (1955).

3 D. H. TEMPLETON and C. H. DAUBEN: J. Amer. Chem. Soc. **75**, 4560 (1953).

4 L. B. ASPREY and F. H. ELLINGER: Atomic Energy Commission Declassified Document, AECD-3627.

5 W. H. ZACHARIASEN: Acta crystallogr. **2**, 388 (1949).

6 L. B. ASPREY: J. Amer. Chem. Soc. **76**, 2019 (1954).

7 L. B. ASPREY, S. E. STEPHANOU and R. A. PENNEMAN: J. Amer. Chem. Soc. **73**, 5715 (1951).

8 J. L. HOARD and J. D. SHOUPE; In: The Chemistry of Uranium, N.N.E.S., Div. VIII, Vol. 5, p. 439. New York: McGraw-Hill Book Co., Inc. 1951.

9 W. H. ZACHARIASEN: U. S. Atomic Energy Commission Report MDDC-1151.

10 J. MALM and B. WEINSTOCK: Paper P/733, Peaceful Uses of Atomic Energy-Proceedings of the Geneva Conference, August 1955, United Nations.

11 W. H. ZACHARIASEN: Acta crystallogr. **2**, 388 (1949).

12 D. H. TEMPLETON and C. H. DAUBEN: University of California Radiation Laboratory Report UCRL-1886, July 14, 1952.

13 D. H. TEMPLETON and C. H. DAUBEN: J. Amer. Chem. Soc. **75**, 4560 (1953).

14 W. H. ZACHARIASEN: Report CK-1530.

15 W. H. ZACHARIASEN: Acta crystallogr. **1**, 265 (1948).

the heavy element metals to those of the rare-earth element metals is quite close. All are highly electropositive and to about the same degree. One notable resemblance to the lanthanide metals is the low density of americium metal (11.7) relative to its preceding neighbors which is precisely the situation with the metallic state of its homologue europium which has a low density with respect to its preceding neighbors. A comparison with tungsten, rhenium, osmium, and iridium shows no such analogy.

8. Ion-exchange behavior. Striking similarities in the chemical behavior of the aqueous tripositive ions of the lanthanide and actinide elements are found in ion-exchange behavior. The very difficult problem of the separation of the rare-earth fission product elements was solved on the wartime Plutonium Project by adsorbing the aqueous ions on top of a column of cation-exchange resin and then selectively eluting the elements with a complexing agent such as ammonium citrate buffered to a suitable p_H[1]. The effect of small differences in complex formation and of small differences in resin adsorbability, both effects presumably influenced by small differences in hydrated ionic radii, are superimposed in this method with the result that neighboring lanthanide elements can be separated from each other.

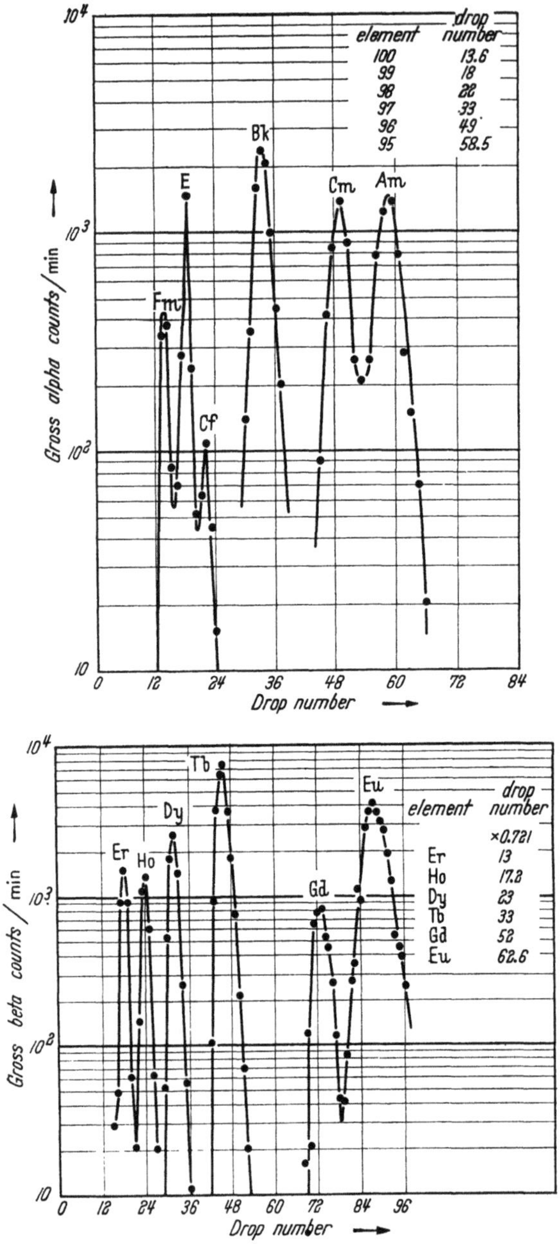

Fig. 3. Elution of trace concentrations of tripositive actinide elements (upper curve) and tripositive lanthanide elements (lower curve) from Dowex-50 cation exchange resin with ammonium lactate eluting solution. Figure reproduced from S. G. Thompson, *et al.*: J. Amer. Chem. Soc. 76, 6229 (1954).

When the chemical similarities of the tripositive lanthanide and actinide elements became apparent, it was only natural that the ion-exchange method should be applied to the latter. Cunningham, Tompkins, and Asprey[2] first used the ion-exchange method to perform a separation of tripositive americium from curium.

When the column elution behavior of the two series is compared, the analogy even to details is striking with some eluting agents. Fig. 3 shows elution curves for lanthanide and actinide element ions. In both cases the mixture of elements was adsorbed on a short column of Dowex-50 cation exchange resin and eluted selectively with lactic acid buffered to a suitable p_H.

A remarkable analogy can be seen in the spacing between the californium - berkelium - curium - americium group and their rare-

[1] Symposium on Ion Exchange: J. Amer. Chem. Soc. **69**, 2769—2881 (1947).
[2] B. B. Cunningham, E. R. Tompkins and L. B. Asprey: Unpublished work 1947.

earth homologues dysprosium-terbium-gadolinium-europium. These spacings reflect the relative changes in ionic radii which presumably determine the relative separations in the ion-exchange adsorption method. It can be concluded, therefore, that the same sequence of changes in the ionic radius is encountered on filling the $5f$ electron shell as occurs on filling the $4f$ shell, therefore, it seems clear that curium represents the midpoint in the actinide transition series of elements in view of its position analogous to gadolinium.

b) Physical evidence.

9. Absorption and fluorescence spectra in aqueous solution and crystals. The absorption spectra of aqueous ions of uranium, neptunium, plutonium, and americium show very sharp absorption bands; similarly, absorption of light in crystal compounds of these elements occurs in very sharp bands. This situation is very reminiscent of the lanthanide series where the absorption is ascribed to electronic transitions involving the $4f$ electrons. It is believed that the upper and lower states in these transitions belong to the same configuration $4f^n$ and differ only in their electronic coupling to different L or S values[1,2]. For an ion with only one $4f$ electron, such as cerium (III) or with thirteen $4f$ electrons, such as ytterbium (III), the number of expected transitions is small. The sharpness of the transition is due to the shielding of the $4f$ electrons by the $5s$ and $5p$ electrons. In the absorption spectra of crystals extra lines due to the STARK effect of the crystal field were observed. The literature in this field is quite extensive, both in experimental studies and theoretical calculations and correlations. Enough is known to be sure of the essential correctness of the interpretation on the basis of $4f$ electrons, but many details are not satisfactorily explained. A brief review is given by YOST, RUSSELL, and GARNER[3] where references to the pre-1946 literature may be found.

It seems highly likely that the sharp absorption spectra of the actinide elements is due to the presence of $5f$ electrons. A number of details may be cited in support of this view. In both series the absorption spectra undergo similar simplification as the middle of the two series is approached, that is, as the elements gadolinium and curium, with their seven $4f$ or $5f$ electrons, are approached. This may be explained qualitatively as being due to the increased stability of the ground state with respect to the first excited state, leading to energy differences which correspond to absorption in the ultraviolet region. Curium (III) exhibits no apparent absorption in aqueous solution at wavelengths above about 3200 Å. Similarly, gadolinium shows no absorption at wavelengths above about 2800 Å.

The homologous ions americium (III) and europium (III) have analogous absorption peaks at 5000 and 4000 Å, respectively, and show striking similarity even in their fine structure. The aqueous-solution absorption spectra of tripositive uranium, neptunium, and plutonium show many sharp absorption peaks in the visible region. Solutions of these ions are highly colored. Sharp absorption bands undoubtedly involving $5f$ electrons have also been found in aqueous solution for the other oxidation states of plutonium and neptunium. The absorption spectra of the transuranium elements will be found later in this article under the discussion of the individual elements. GRUEN[4] has carefully analyzed

---

[1] S. FREED: Phys. Rev. **38**, 2122 (1931).
[2] J. H. VAN VLECK: J. Phys. Chem. **41**, 67 (1937).
[3] D. M. YOST, H. RUSSELL jr. and C. S. GARNER: The Rare-Earth Elements and Their Compounds. New York: John Wiley & Sons, Inc. 1947.
[4] D. M. GRUEN: J. Chem. Phys. **20**, 1818 (1952).

the very similar absorption spectra of NpO_2^+ and PuO_2^{++} and has concluded that the $5f^2$ configuration is involved. In the case of neptunium (VI) which contains only one $5f$ electron no such absorption bands are expected and none are found. Absorption spectra similar to each other are observed for the aqueous ions of the isoelectronic series uranium (IV)-neptunium (V)-plutonium (VI) where the configuration is presumably $5f^2$ and for the isoelectronic series uranium (III) neptunium (IV)-plutonium (V) where the electron configuration is presumably[1] $5f^3$.

JØRGENSEN[2] applied the CONDON-SHORTLEY[3] theory of interaction between a pair of electrons in different coupling schemes to calculate the expected transitions for the $5f^2$ configuration. The calculated values were in good agreement with many of the absorption peaks observed in solutions of uranium (IV). JØRGENSEN concludes that the configuration must be $5f^2$ beyond the emanation shell; $6d^2$ electrons should show broad band absorption and seem to be ruled out clearly. JØRGENSEN[4] also discussed systems with three or more f-electrons and compared the theoretical predictions with absorption spectra of various actinide ions in aqueous solution.

Absorption spectra of crystals are somewhat sharper than solution spectra. The absorption spectrum of solid americium trichloride contains sharp lines of the order of 1 to 5 Å wide even when the measurements are made at room temperature[5]. This sharpness ranks with the finest ever observed, and, in fact, only the spectrum of tripositive europium ion is comparable. The absorption spectrum of the latter originates from a ground state $4f^6\,(^7F_0)$ configuration; it seems likely that the ground state of tripositive americium also contains six $5f$ electrons. The absorption spectra of crystalline uranium tetrachloride, uranium trichloride, neptunium tetrachloride and plutonium trichloride also have sharp lines suggesting rather strongly that transitions take place between upper and lower states in which the $5f^n$ configuration is preserved[5, 6].

It has been pointed out[7] that one difference between the absorption spectra of the actinide as compared to the lanthanide ions is that the relative intensity of absorption is much greater for the principal bands of the former than for the equivalent bands in the latter. In the case of the tripositive ions the intensity of absorption is as much as ten times as great for the transuranium elements. The transuranium element ions of higher charge show even more intense absorption, as much as one hundred times as great as for the tripositive lanthanide ions. These observations are in line with many other lines of evidence which suggest that the $5f$ electrons are much less shielded than $4f$ electrons and hence are much more readily influenced by the external environment.

Fluorescence is well known in a number of rare-earth ions. This fluorescence is almost certainly due to the presence of $4f$ electrons although the complexities in interpretation of the details are very great. Fluorescence has been observed by GRUEN and co-workers[8] in mixed crystals of americium trichloride and lanthanum trichloride. The fluorescence is weakly self-excited at room temperature due to the

[1] R. ROHMER, R. FREYMANN, A. CHEVET and P. HAMON: Bull. Soc. Chem. France **1952**, 603.

[2] C. K. JØRGENSEN: Dan. Mat. fys. Medd. **29**, Nr. 7 (1955).

[3] E. U. CONDON and G. H. SHORTLEY: Theory of Atomic Spectra.: Cambridge Press 1935.

[4] C. K. JØRGENSEN: Dan. Mat. fys. Medd. **29**, Nr. 11 (1955).

[5] S. FREED and F. J. LEITZ jr.: J. Chem. Phys. **17**, 540 (1949).

[6] K. M. SANCIER and S. FREED: Brookhaven National Laboratory Report, BNL-1008, 1951.

[7] R. SJOBLOM and J. C. HINDMAN: J. Amer. Chem. Soc. **73**, 1749 (1951).

[8] D. M. GRUEN, J. G. CONWAY, R. D. MCLAUGHLIN and B. B. CUNNINGHAM: J. Chem. Phys. **24**, 1115 (1956).

radioactivity of Am^{241} and is strongly excited by irradiation with a mercury lamp. Many of the lines correspond exactly with lines found in the absorption spectrum and are interpreted as representing transitions from excited electronic states to the $^7F_0$ level of the ground state multiplet. A comparison of the fluorescence spectrum of americium (III) and its lanthanide homologue, europium (III), shows the difference that transitions from only the two lowest-lying electronic levels above the $^7F_0$ ground state are observed in the americium case whereas transitions from five relatively low-lying excited electronic levels are involved in europium. The wider spacing of the americium levels is in agreement with magnetic susceptibility data to be discussed later.

Fluorescence has also been observed[1] in mixed crystals of plutonium trichloride and lanthanum trichloride. The fluorescence is weakly self-excited and is brought out strongly upon irradiation with a mercury lamp.

10. Magnetic susceptibility data—theoretical considerations. The presence of unpaired electrons in the heaviest elements should and does lead to paramagnetism This magnetic behavior is expected whether the unpaired electrons are in the $5f$ or the $6d$ shell; quantitative measurements of magnetic susceptibilities of compounds of the heaviest elements should in principle lead to the resultant magnetic moments in units of BOHR magnetons and distinguish between possible electron configurations. The considerable amount of experimental data now available, in general, lends rather strong support to the actinide concept, but quantitative agreement of theory and experiment have not been achieved because of complexities in the situation which make prediction of the magnetic moment from the electron configuration more difficult than in the case of the lanthanide rare-earths. As is the case for other lines of evidence, the situation is not conclusive for the early members of the series. Nevertheless for the elements neptunium and plutonium the magnetic evidence provides some of the strongest evidence for the actinide concept.

The usual approximate approach is to calculate a molar susceptibility from the well-known expression:

$$\chi_m = \frac{N B^2}{3kT} g^2 J(J+1) \tag{10.1}$$

where $N =$ AVOGADRO's number,

$g =$ LANDÉ splitting factor $= 1 + \dfrac{S(S+1) + J(J+1) - L(L+1)}{2J(J+1)}$,

$B =$ BOHR magneton $= eh/4\pi mc$,

$\boldsymbol{J} = \boldsymbol{L} + \boldsymbol{S} =$ total angular momentum of atom,

$k =$ BOLTZMANN constant,

$T =$ absolute temperature.

The value of J is taken from the ground state term value calculated according to HUND's rules[2]. RUSSELL-SAUNDERS coupling is assumed and out of the possible terms allowed by the PAULI principle, HUND supposes that the one of highest multiplicity, i.e. greatest value of $2S+1$, lies lowest. In those cases where several values of L are possible with the same multiplicity, the greatest L consistent with this S is chosen. The minimum possible J lies lowest when the $5f$ electrons are less than seven in number; the maximum possible J lies lowest when more than seven f-electrons are present. The predicted ground states for the members of the series are: $5f^1\,(^2F_{5/2})$, $5f^2\,(^3H_4)$, $5f^3\,(^4I_{9/2})$, $5f^4\,(^5I_4)$, $5f^5\,(^6H_{5/2})$, $5f^6\,(^7F_0)$, $5f^7\,(^8S_{7/2})$, $5f^8\,(^7F_6)$, $5f^9\,(^6H_{15/2})$, $5f^{10}\,(^5I_8)$, $5f^{11}\,(^4I_{15/2})$, $5f^{12}\,(^3H_6)$, $5f^{13}\,(^2F_{7/2})$ and $5f^{14}\,(^1S)$.

[1] See footnote 8, p. 216.

[2] F. HUND: Z. Physik **33**, 855 (1925).

The Eq. (10.1) is, strictly speaking, applicable only to gaseous ions and furthermore, it neglects a small diamagnetic term. All experimental measurements for the heavy elements are made not on gaseous ions but on solutions or crystals. The justification for the use of Eq. (10.1) is the one applied previously to the rare-earths, namely, that the inner shell of electrons is shielded from the effects of the solution or crystal environment by the s^2 and p^6 electrons in the next shell. This is nearly completely true for the lanthanides; it must be regarded as an approximation for the heavy elements.

It is possible also to calculate theoretically the magnetic susceptibilities expected for unpaired electrons in the $6d$ shell. In doing this, one usually assumes that the orbital contribution to the magnetic moment is quenched by field effects within the crystal or solution or by HEISENBERG exchange forces. This leads to a "spin-only" value for the magnetic susceptibility given by the expression:

$$\chi_m = \frac{N B^2}{3kT} [4S(S+1)] \tag{10.2}$$

where S is the vector sum of the electron spins of the unpaired electrons.

The calculated molar susceptibilities for $6d^n$ electron configurations, where n is 3 or greater, are much too high to agree with experimental measurements on actinide element ions. The calculated values for one, two, and three unpaired $6d$ electrons are 1245, 3333, and 6250 respectively in c.g.s. units $\times 10^6$. Experimental values for ions with one, two, or three unpaired electrons frequently fall between these values and the ones given by Eq. (10.1). In such cases the decision between $5f^n$ or $6d^n$ configurations is not simple and requires more detailed experiments and analysis with suitable evaluation of all factors which may contribute to quenching of the orbital contribution of f as well as d electrons. Some of these factors will be mentioned in the following review of the experimental data.

11. Magnetic susceptibility data—experimental studies. CALVIN[1] has measured and compared the magnetic susceptibilities of ThS and CeS, analogous compounds which might be expected to possess two magnetic electrons; similarly he measured and compared the magnetic susceptibilities of Th_2S_3 and Ce_2S_3 in which the metallic ion might be expected to possess one magnetic electron. Both cerium compounds are paramagnetic to an amount consistent with the presence of unpaired $4f$ electrons. The thorium compounds are not paramagnetic. This is most reasonably interpreted as indicating the presence of $6d$ rather than $5f$ electrons in thorium. In the sulfides the interatomic distances are so small that the magnetic d electrons on neighboring thorium atoms can pair to form a metallic bond. DAWSON and D'EYE[2] have similarly found Th_2Se_3 to be slightly diamagnetic.

HOWLAND and CALVIN[3] have measured the magnetic susceptibilities of the cations of uranium, neptunium, plutonium, and americium. The results are presented in Fig. 4 together with curium data of CRANE and CUNNINGHAM to be mentioned later[4]. The qualitative similarity of the actinide curve with the lanthanide curve is interpreted as evidence for the $5f^n$ electron configuration for the magnetic electrons. The fact that the experimental effective magnetic moments are generally lower than the theoretical values (given in the figure title) is inter-

[1] M. CALVIN as reported in E. D. EASTMAN *et al:* J. Amer. Chem. Soc. **72**, 4019 (1950); **72**, 2248 (1950).

[2] J. K. DAWSON and R. D'EYE as reported by DAWSON: Nucleonics **10** (9), 39 (1952).

[3] J. J. HOWLAND and M. CALVIN: J. Chem. Phys. **18**, 239 (1950).

[4] W. W. T. CRANE and B. B. CUNNINGHAM: University of California Radiation Laboratory Report UCRL-1220, 1951.

preted as resulting from a STARK effect produced by electric fields of anions and of water dipoles. Partial failure of the RUSSELL-SAUNDERS approximation could also account for some of the discrepancy. CRANE, WALLMANN and CUNNINGHAM[1] made measurements on solid AmF_3 and obtained for the molar susceptibilities, (1040 ± 300) at 295 °K, (1290 ± 300) at 199 °K and (1740 ± 300) at 77 °K, in c.g.s. units $\times 10^6$. GRAF, CUNNINGHAM and co-workers[2] measured the susceptibility of americium metal and obtained a value of $(1000 \pm 250) \times 10^{-6}$ c.g.s. units at 300 °K.

CRANE and CUNNINGHAM[3] have measured the molar magnetic susceptibility of solid CmF_3 in 10 mole percent solid LaF_3 and have found the value to be $(26500 \pm 700) \times 10^{-6}$ c.g.s. units at about 20° C which is a very large value in good agreement with the value expected for the ground state term $^8S_{7/2}$ in curium (III). The similarity of curium (III) in this respect to its lanthanide homologue gadolinium (III) is striking (see Fig. 4).

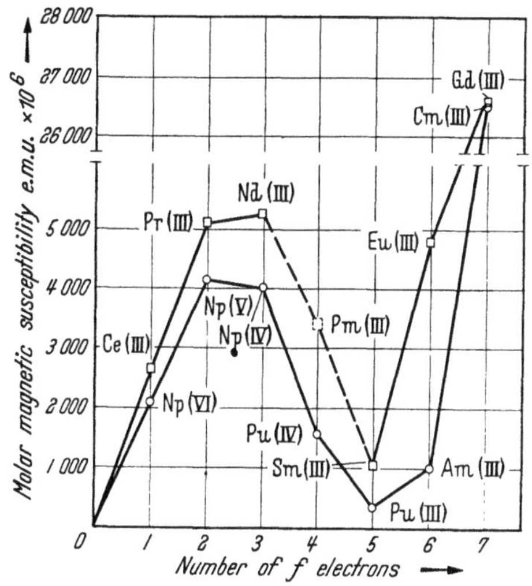

Fig. 4. Molar magnetic susceptibilities of some actinide and lanthanide ions. The values of the susceptibilities given by the simple theory at 20° C are 2730, 5420, 5540, 3040, 300, 0, 27250, multiplied by 10^{-6}. (Data from Ref. 3 and 4, p. 218 and [1].)

Sharp departures from the theoretical susceptibilities occur in both series for the ions with 5 and 6 unpaired electrons. The $^7F_0$ ground state term for europium (III) and for americium (III) should lead to zero magnetic moment; instead, the rather high values listed in Fig. 4 are found. The theoretical value for the $^6H_{5/2}$ ground state term for samarium (III) and plutonium (III) is 300×10^{-6} units. The discrepancy in the case of samarium and europium was resolved completely by VAN VLECK and FRANK[4] who showed that the multiplet splitting in these cases is comparable to kT; Eq. (10.1) above is based on the assumption, not previously stated, that multiplet splitting is large compared to kT, i.e., that only the ground state term contributes to the magnetism. When allowance is made for the contribution of other terms of the multiplet, good agreement is obtained with experiment. This correction is unimportant for the other lanthanide ions. In the case of plutonium (III) and americium (III) the analogous calculation has not been made, but the lower values of the susceptibilities of these ions presumably indicate wider multiplet splitting in the actinide ions. Preliminary interpretation of the fluorescence spectrum of americium bears out this deduction[5].

Another consequence of the VAN VLECK and FRANK theory is that the susceptibility for samarium should pass through a temperature minimum in the neighborhood of 400 °K; such a minimum has been observed experimentally. ELLIOTT

[1] W. W. T. CRANE, J. C. WALLMANN and B. B. CUNNINGHAM: University of California Radiation Laboratory Report, UCRL-846, 1950.

[2] P. GRAF, B. B. CUNNINGHAM *et al:* University of California Radiation Laboratory Report UCRL-3280; J. Amer. Chem. Soc. **78**, 2340 (1956).

[3] See footnote 4, p. 218.

[4] J. H. VAN VLECK and A. FRANK: Phys. Rev. **34**, 1494, 1625 (1929).

[5] D. M. GRUEN, J. G. CONWAY, R. D. MCLAUGHLIN and B. B. CUNNINGHAM: Unpublished results, January 1956.

and LEWIS[1] made measurements on PuF_3 and $Pu_2(C_2O_4)_2 \cdot 9\,H_2O$ from 76 to 300 °K. DAWSON, MANDLEBERG and DAVIES[2] extended measurements on PuF_3 and $PuCl_3$ up to about 600 °K and found a minimum in the susceptibility-temperature curve at about 500 to 555 °K. This is good qualitative evidence that plutonium (III) has the electron configuration $5f^5$.

Susceptibility measurements have been made on a large number of uranium (IV) compounds by many authors[3-18]. The results lie in the range 3200 to 3800×10^{-6} units (at room temperature) which is closer to the $6d^2$ spin-only value of 3333 than to the limiting $5f^2$ value of 5390. The behavior of the susceptibility of $U(SO_4)_2 \cdot 3.26\,H_2O$ and of $U(C_2O_4)_2 \cdot 5\,H_2O$ over a wide range of temperature, however, caused HUTCHISON and ELLIOTT[4] to conclude that uranium (IV) was isoelectronic with praseodymium (III). They also concluded that the crystal fields produce more pronounced perturbing effects on uranium (IV) than on praseodymium (III). On the other hand, the magnetic measurements of DAWSON and LISTER[6] on UO_2 and uranium tetrafluoride diluted with diamagnetic thorium tetrafluoride are interpreted by DAWSON[8] to indicate the configuration $6d^2$ for uranium (IV). Similarly, TRZEBIATOWSKI and SELWOOD[7] feel that their measurements on solid dilute solutions of uranium dioxide in diamagnetic thorium dioxide are best interpreted in terms of $6d$ electrons in uranium (IV) since the value at infinite dilution corresponds to complete quenching of the orbital contribution; similar experiments by DAWSON and LISTER[18] and SLOWINSKI and ELLIOTT[17] lead to the same result. This seems reasonable since uranium is in just the region where the energy difference between the $5f$ and $6d$ levels seems to be within the range of chemical binding energies; however, since the argument for the $6d^2$ configuration is based on the observed quenching of the orbital contribution to the magnetic moment the possibility of this occurring with $5f$ electrons should be studied further.

A number of measurements have been made on plutonium (IV) compounds in which four unpaired electrons may be expected[19-21]. DAWSON has evidence that the paramagnetic electron-exchange interaction contributes materially to

---

[1] N. ELLIOTT and W. B. LEWIS: Abstract 90, reported at the 118th meeting of the American Chemical Society, Chicago, Illinois, Sept. 1950; Report AECD-29100, 1950.
[2] J. K. DAWSON, C. J. MANDLEBERG and D. DAVIES: J. Chem. Soc. **1951**, 2047.
[3] W. SUCKSMITH: Phil. Mag. **14**, 1115 (1935).
[4] C. A. HUTCHINSON jr., and N. ELLIOTT: J. Chem. Phys. **16**, 920 (1948). — Phys. Rev. **73**, 1229 (1948).
[5] R. LAWRENCE: J. Amer. Chem. Soc. **56**, 776 (1934).
[6] J. K. DAWSON and M. W. LISTER: J. Chem. Soc. **1950**, 2181.
[7] W. TRZEBIATOWSKI and P. W. SELWOOD: J. Amer. Chem. Soc. **72**, 4504 (1950).
[8] J. K. DAWSON: J. Chem. Soc. **1951**, 2889; **1952**, 1185.
[9] N. ELLIOTT: Phys. Rev. **76**, 431 (1949).
[10] J. K. DAWSON: J. Chem. Soc. **1951**, 429.
[11] L. SACCONI: Atti Accad. naz. Lincei **6**, 639 (1949).
[12] H. HARALDSEN and R. BAKKEN: Naturwiss. **28**, 127 (1940).
[13] H. BOMMER: Z. anorg. Chem. **247**, 249 (1941).
[14] N. ELLIOTT, L. CORLISS and Y. DELABARRE: Brookhaven National Laboratory Report BNL-39 (AS-3) p. 47, December 1949.
[15] R. STOENNER and N. ELLIOTT: J. Chem. Phys. **19**, 950 (1951).
[16] N. ELLIOTT and W. B. LEWIS: U.S. Atomic Energy Commission Document, AECD-2910, 1950.
[17] E. SLOWINSKI and N. ELLIOTT: Brookhaven National Laboratory Report, BNL-1199.
[18] J. K. DAWSON and M. W. LISTER: J. Chem. Soc. **1952**, 5041.
[19] J. J. HOWLAND and M. CALVIN: J. Chem. Phys. **18**, 239 (1950).
[20] J. K. DAWSON: J. Chem. Soc. **1952**, 1882.
[21] N. ELLIOTT and W. B. LEWIS: Paper 90, Symposium on "The Chemistry of the Actinide Elements", American Chemical Society Meeting, Chicago, Sept. 1950.

a lowering of the susceptibility. He made measurements on plutonium tetrafluoride and plutonium diluted with the isostructural diamagnetic compounds thorium tetrafluoride and thorium dioxide, respectively, and found susceptibilities increasing with dilution, with the extrapolated values at infinite dilution for room temperature (3020×10^{-6}) corresponding essentially to the value predicted (3010×10^{-6}) for the configuration $5f^4$.

The neptunyl ion contains one unpaired electron and the plutonyl ion contains two. Several measurements of the magnetic susceptibility of compounds such as sodium neptunyl acetate[1,2] and sodium plutonyl acetate[3] have been carried out. Detailed theoretical analysis of ELLIOT[4] and of EISENSTEIN and PRYCE[5] has shown that the susceptibility is strongly influenced by the special properties of the linear $[\text{O-Np-O}]^{+2}$ and $[\text{O-Pu-O}]^{+2}$ ions and that the measured low values for the susceptibility are completely consistent with the presence of $5f$ electrons. This theoretical analysis modifies previous interpretations of the data on these ions[6,7].

Predictions of the theory[5] regarding anisotropic susceptibility in single crystals have been confirmed by paramagnetic resonance experiments[8,9]. The presence of a single $5f$ electron in NpO_2^{+2} and of two $5f$ electrons in PuO_2^{+2} may be considered to be definitely established.

GRUEN, MALM and WEINSTOCK[10] measured the magnetic molar susceptibility of crystalline plutonium hexafluoride and found the value 290×10^{-6} at 295 °K and 330×10^{-6} at 81 °K. This small, relatively temperature-independent susceptibility is surprising since other compounds that are isoelectronic with PuF_6 (U^{+4}, NpO_2, and PuO_2^{+2}) have susceptibilities of magnitude 3000 to 4000×10^{-6} at room temperature that are strongly temperature dependent. It seems likely that the two non-bonding electrons in PuF_6 unlike those in U^{+4}, NpO_2^{+1}, and PuO_2^{+2} have paired spins and that the electronic ground state is non-degenerate.

In summary, the interpretation of experimental data has shown that the predictions of the actinide concept are borne out in the following ions: Cm^{+3} ($5f^7$); Am^{+3} ($5f^6$); Pu^{+3} ($5f^5$); Pu^{+4} ($5f^4$); NpO_2^{+1}, PuO^{+2} ($5f^2$); NpO_2^{+2} ($5f^1$) and UO_2^{+2} ($5f^0$). In the case of U^{+4} (two unpaired electrons) and U^{+3} (three unpaired electrons) many workers have interpreted the data in support of $6d$ electrons, but a final decision has not been made. In the case of Th^{+3} (one unpaired electron) and Th^{+2} (two unpaired electrons) the evidence strongly favors a $6d$ assignment. These results are consistent with the chemical evidence in indicating that the $6d$ and $5f$ levels are of equivalent energy in the first few elements of the series, but that the f electron levels definitely lie lower once these elements are passed. The magnetic evidence is perhaps the most definite in showing the presence of f electrons in neptunium and plutonium.

12. Spectroscopic data. Spectroscopic data yield detailed information on the electron configurations of the lowest-lying states for the gaseous atoms for a few

---

[1] D. M. GRUEN: Argonne National Laboratory Report ANL-4469, June 1950.
[2] D. M. GRUEN and C. A. HUTCHISON jr.: J. Chem. Phys. 22. 386 (1954).
[3] J. K. DAWSON: Atomic Energy Research Establishment Report AERE C/R-876, 1952.
[4] R. J. ELLIOTT: Phys. Rev. **89**, 659 (1953).
[5] J. C. EISENSTEIN and M. H. L. PRYCE: Proc. Roy. Soc. Lond., Ser. A **229**, 20 (1955).
[6] J. K. DAWSON: Nucleonics **10** (9), 39 (1952).
[7] J. K. DAWSON and G. R. HALL: Paper, "Peaceful Uses of Atomic Energy-Proceedings of the Geneva Conference", P/440, August 1955, United Nations.
[8] B. BLEANEY, P. M. LLEWELLYN, M. H. L. PRYCE and G. R. HALL: Phil. Mag. **45**, 773, 991, 992 (1954).
[9] C. A. HUTCHISON jr., and W. B. LEWIS: Phys. Rev. **95**, 1096 (1954).
[10] D. M. GRUEN, J. G. MALM and B. WEINSTOCK: J. Chem. Phys. **24**, 905 (1956).

of the actinide elements. Such data lend strong support to the actinide concept. KIESS, HUMPHREYS, and LAUN[1] observed the spectra of uranium under arc and spark excitation and catalogued 9000 lines in the U I spectrum with accurate wavelengths and estimated intensities for the region 2900 to 11000 Å. With the aid of well resolved ZEEMAN patterns the authors determined that the lowest lying term was $^5L_6(5f^3\,6d^1\,7s^2)$. Other important low-lying terms are $^5K_5(5f^3\,6d\,7s^2)$ and $^7M_6(5f^3\,6d^2\,7s)$ with the three $5f$ electrons appearing throughout. SCHUURMANS[2,3] and co-workers confirm this analysis of UI and provide an analysis of the U II spectrum. The lowest-lying term is $^4I_{9/2}(5f^3\,7s^2)$. All important low levels are ascribed to configurations with three $5f$ electrons; namely $(5f^3\,7s^2)$ $(5f^3\,6d\,7s)$ and $(5f^3\,6d^2)$. The important terms are ^4I, ^6L, ^6K, ^6M, and ^6I. Additional work on U II has been carried out by MCNALLY and HARRISON[4,5] and by VAN DEN BOSCH[6]. The information which has been deduced by the analysis of these workers is extensive, but a complete analysis of the exceedingly complex spectra of uranium will require much additional work. The present analysis is very strong evidence for the conception of uranium as the third actinide element analogous to neodymium in the rare-earth series. The term analysis of neodymium spectra[7], however, leads to somewhat different results for the ground state terms. In Nd I the only ground state term identified was $4f^4\,6s^2(^5I)$. In Nd II the low-lying levels and configurations are $4f^4\,6s$ (^6I and ^4I) and $4f^4\,5d$ ($^6L\,^6K$); the ground state term is $^6I_{7/2}$.

In the case of thorium, two groups of investigators[8,9] have interpreted the spectrum of Th (II) due to singly ionized thorium atoms. The lowest-lying energy states belong to $(6d\,7s^2)$, $(6d^2\,7s)$, and $(6d^3)$ electron configurations. A set of prominent low-lying states, however, are attributed to the configurations $(5f\,7s^2)$ and $(5f\,6d\,7s)$.

Additional observations on the gaseous Th^{+2} and Th^{+3} spectra[10–14] show that $6d$ and $5f$ electrons are close in energy. For example, the ground level of Th (III) is $^3F_2(6d^2)$ while the level $^3H_4(5f\,6d)$ lies only 809.9 cm^{-1} above[10].

In the case of the transuranium elements the results from uranium spectra have been used, in lieu of additional information, to make predictions concerning the expected electronic configuration for the neutral gaseous atoms[15]. More recent experimental information has necessitated changes in these configurations which reflect a closer similarity to the corresponding lanthanide elements than had been expected.

---

1 C. C. KIESS, C. J. HUMPHREYS and D. D. LAUN: J. Res. Nat. Bur. Stand. **37**, 57 (1946). — J. Opt. Soc. Amer. **36**, 357 (1946).

2 P. SCHUURMANS: Physica, Haag **11**, 419 (1946).

3 P. SCHUURMANS, J. C. VAN DEN BOSCH, and N. DIJKWEL: Physica, Haag **13**, 117 (1947).

4 J. R. MCNALLY jr., and G. R. HARRISON: Report Y-340, 1949.

5 J. R. MCNALLY jr.: Phys. Rev. **77**, 417 (1950).

6 J. C. VAN DEN BOSCH: Physica, Haag **15**, 503 (1949). — Phys. Rev. **80**, 100 (1950).

7 W. E. ALBERTSON: Astrophys. J. **84**, 26 (1936).

8 J. R. MCNALLY, G. R. HARRISON and H. B. PARK: J. Opt. Soc. Amer. **32**, 334 (1942). — J. R. MCNALLY jr.: J. Opt. Soc. Amer. **35**, 390 (1945).

9 T. L. DE BRUIN, P. SCHUURMANS and P. F. A. KLINKENBERG: Z. Physik **121**, 667 (1943); **122**, 23 (1944).

10 P. F. A. KLINKENBERG: Physica, Haag **16**, 185, 618 (1950).

11 R. J. LANG: Canad J. Res. A **14**, 43 (1946).

12 T. L. DE BRUIN and P. F. A. KLINKENBERG: Proc. Acad. Sci. Amsterd. **43**, 581 (1948).

13 T. L. DE BRUIN, P. F. A. KLINKENBERG and P. SCHUURMANS: Z. Physik **118**, 58 (1941).

14 P. F. A. KLINKENBERG and R. J. LANG: Physica, Haag **15**, 774 (1949).

15 W. F. MEGGERS: Science, Lancaster, Pa. **105**, 514 (1947).

1 hydrogen H 1.0080																	2 helium He 4.003
3 lithium Li 6.940	4 beryllium Be 9.013	5 boron B 10.82											6 carbon C 12.011	7 nitrogen N 14.008	8 oxygen O 16.000	9 fluorine F 19.00	10 neon Ne 20.183
11 sodium Na 22.991	12 magnesium Mg 24.32	13 aluminum Al 26.98											14 silicon Si 28.09	15 phosphorus P 30.975	16 sulfur S 32.066	17 chlorine Cl 35.457	18 argon A 39.944
19 potassium K 39.100	20 calcium Ca 40.08	21 scandium Sc 44.96	22 titanium Ti 47.90	23 vanadium V 50.95	24 chromium Cr 52.01	25 manganese Mn 54.94	26 iron Fe 55.85	27 cobalt Co 58.94	28 nickel Ni 58.69	29 copper Cu 63.54	30 zinc Zn 65.38	31 gallium Ga 69.72	32 germanium Ge 72.60	33 arsenic As 74.91	34 selenium Se 78.96	35 bromine Br 79.916	36 krypton Kr 83.80
37 rubidium Rb 85.48	38 strontium Sr 87.63	39 yttrium Y 88.92	40 zirconium Zr 91.22	41 niobium Nb 92.91	42 molybdenum Mo 95.95	43 technetium Tc 98.91	44 ruthenium Ru 101.1	45 rhodium Rh 102.91	46 palladium Pd 106.7	47 silver Ag 107.880	48 cadmium Cd 112.41	49 indium In 114.76	50 tin Sn 118.70	51 antimony Sb 121.76	52 tellurium Te 127.61	53 iodine I 126.91	54 xenon Xe 131.3
55 cesium Cs 132.91	56 barium Ba 137.36	57–71 La – Lu	72 hafnium Hf 178.6	73 tantalum Ta 180.95	74 tungsten W 183.92	75 rhenium Re 186.31	76 osmium Os 190.2	77 iridium Ir 192.2	78 platinum Pt 195.23	79 gold Au 197.0	80 mercury Hg 200.61	81 thallium Tl 204.39	82 lead Pb 207.21	83 bismuth Bi 209.00	84 polonium Po 210.00	85 astatine At 210	86 radon Rn 222
87 francium Fr 223	88 radium Ra 226.05	89 – 103 Ac–103	104	105	106												

Lanthanides	57 lanthanum La 138.92	58 cerium Ce 140.13	59 praseodymium Pr 140.92	60 neodymium Nd 144.27	61 promethium Pm 146.97	62 samarium Sm 150.43	63 europium Eu 152.0	64 gadolinium Gd 156.9	65 terbium Tb 158.93	66 dysprosium Dy 162.46	67 holmium Ho 164.94	68 erbium Er 167.2	69 thulium Tm 168.94	70 ytterbium Yb 173.04	71 lutetium Lu 174.99
Actinides	89 actinium Ac 227.04	90 thorium Th 232.05	91 protactinium Pa 231.04	92 uranium U 238.07	93 neptunium Np 237.06	94 plutonium Pu 244	95 americium Am 243	96 curium Cm (248)	97 berkelium Bk 249	98 californium Cf 249	99 einsteinium E (254)	100 fermium Fm (257)	101 mendelevium Mv (259)	102	103

Fig. 5. Periodic chart of the elements.

Work on the ZEEMAN effect in plutonium[1] spectra indicated a prominent $^6F(5f^7\,7s)$ state for Pu II. An extended analysis by McNALLY[2] showed that the ground state of Pu II was $^8F_{1/2}(5f^6\,7s^1)$ with $^8F_{3/2}$ lying 2014.7 cm^{-1} above. This ground state is identical to that of Sm II[3], the rare earth homologue of plutonium. One may predict that the ground state of Pu I is $^7F(5f^6\,7s^2)$ by analogy to Sm I.

A preliminary term analysis of the spectrum of americium by FRED and TOMKINS[4] establishes the ground state of Am II as $^9S(5f^7\,7s^1)$ with the splitting between the ^9S and the ^7S multiplet equal to 2498.35 cm^{-1}. This compares to 1669.21 cm^{-1} for the same multiplet splitting in Eu II. Hence the ground state of Am I is probably $^8S(4f^7\,6s^2)$, similar to that of Eu I.

The rare earth homologue of curium is gadolinium. The ground state[3] of Gd I is $^9D(4f^7\,5d^1\,6s^2)$ and of Gd II is $^{10}D(5f^7\,7d^1\,6s^1)$. There are six multiplets lying below 11000 cm^{-1} and the highest two of these are inverted.

$^{10}D(4f^7\,5d^1\,6s^1)$
$^8D(4f^7\,5d^1\,6s^1)$
$^8S(4f^7\,5d^1\,6s^1)$
$^{10}F(4f^7\,5d^1\,6s^1)$
$^8D(4f^7\,5d^1\,6s)$ inverted
$^6D(4f^7\,5d^1\,6s)$ inverted.

The optical spectrum of curium has been observed[5] but no term analysis has been carried out. The isotope shift has been studied[6] using samples of Cm^{242} and Cm^{244}. A shift was observed in 148 lines out of 182 recorded with shifts ranging up to 0.131 cm^{-1}. For some lines the shifts are opposite in sign to those of the majority. These observations constitute qualitative evidence for an electronic configuration in curium similar to gadolinium. The large isotope shift observed for several of the actinide elements is believed to be associated with s^1 and s^2 electrons. Six multiplets near the ground state would explain why such a large percentage of lines show the isotope shift. Inverted multiplets could account. for the shifts in the abnormal direction.

Thus the recent work on plutonium, americium, and curium suggests that these elements are following a pattern similar to samarium, europium, and gadolinium. The succeeding actinide elements in lieu of any experimental information other than chemical information, are shown in Table 1, Sect. 2, with electron configurations analogous to the rare earths.

Fig. 5 is a periodic chart of the elements with the elements above actinium grouped as an actinide series.

B. The individual elements.

I. Neptunium (element 93).

a) History of neptunium.

13. The transuranium elements before the discovery of fission. Not long after the discovery of the neutron FERMI[7] realized that the bombardment of uranium

[1] P. M. GRIFFIN and J. R. McNALLY jr.: J. Opt. Soc. Amer. **45**, 63 (1955).
[2] J. R. McNALLY jr.: J. Opt. Soc. Amer. **45**, 901 (1955).
[3] H. N. RUSSELL: J. Opt. Soc. Amer. **40**, 550 (1950).
[4] M. FRED and F. S. TOMKINS: J. Opt. Soc. Amer. **44**, 824 (1954).
[5] J. G. CONWAY, M. F. MOORE and W. W. T. CRANE: J. Amer. Chem. Soc. **73**, 1308 (1951).
[6] J. G. CONWAY and R. D. McLAUGHLIN: J. Opt. Soc. Amer. **46** (1956).
[7] E. FERMI: Nature, Lond. **133**, 898 (1934).

by neutrons might be expected to lead to the production of one or more radioactive isotopes of uranium which would decay to form atoms of element 93. This prediction was eventually proved to be correct and it is now known that Np^{239} can be produced by the following sequence when uranium is irradiated with slow neutrons.

$$U^{238}(n,\gamma)\,U^{239}; \qquad U^{239} \xrightarrow[\beta^-]{23.5\text{ min}} Np^{239}. \tag{13.1}$$

When fast neutrons are used Np^{237} is produced via the sequence:

$$U^{238}(n,2n)\,U^{237}; \qquad U^{237} \xrightarrow[\beta^-]{6.7\text{ days}} Np^{237}. \tag{13.2}$$

These facts were not determined for many years, however, because between 1934 and the time of the careful radiochemical experiments of HAHN and STRASSMANN[1] published in 1939, it was not realized that nuclear fission occurred when uranium was bombarded with slow neutrons. Many of the fission product activities were incorrectly considered to be transuranium element isotopes. The situation was further confused experimentally by the belief that element 93 should resemble rhenium, that element 94 should resemble osmium, etc. A rather extensive literature developed on the "transuranium" elements 93 to 96 in the period immediately before the discovery of fission. An historical account is given by TURNER[2]. The interesting but confusing experimental results which had become so difficult to understand from the standpoint of known physics or chemistry were quickly and correctly interpreted in terms of fission product activities when the phenomenon of nuclear fission was recognized.

14. The discovery of neptunium—summary of known isotopes. Following the discovery of fission and the realization that most of the activities produced in the bombardment of uranium with slow neutrons were fission products rather than transuranium elements, a new search was begun for element 93. The first certain identification of an isotope of this element was carried out by MCMILLAN and ABELSON[3] at the University of California in 1940; they found the 2.3-day isotope Np^{239} formed as the decay product of the 23-min U^{239} produced from the bombardment of uranium with neutrons; $U^{238}(n,\gamma)\,U^{239} \xrightarrow[23\text{ min.}]{\beta^-} Np^{239}$. MCMILLAN and ABELSON chose the name neptunium (symbol Np) from the planet Neptune which is the first planet beyond Uranus in our solar system. MCMILLAN and ABELSON realized from their chemical experiments that this first transuranium element was not eka-rhenium and that it must be a member of a $5f$ transition series.

When uranium was bombardment with deuterons in the Berkeley 60-inch cyclotron 2.3-day Np^{239} was produced by the (d, n) reaction, and in addition another isotope, 2.0-day Np^{238}, was produced[4]. Much of the initial tracer work on neptunium chemistry was performed with Np^{239} produced by neutron irradiation of uranium or with the Np^{239}-Np^{238} mixture produced by deuteron bombardment of uranium.

The longest-lived isotope of neptunium and the isotope which must be regarded as the most important from the standpoint of chemistry is Np^{237} which has a half-life of 2.2×10^6 years. It was found by WAHL and SEABORG in 1942[5]

---

[1] O. HAHN and F. STRASSMANN: Naturwiss. **27**, 163 (1939).
[2] L. A. TURNER: Rev. Mod. Phys. **12**, 1 (1940).
[3] E. M. MCMILLAN and P. H. ABELSON: Phys. Rev. **57**, 1185 (1940).
[4] G. T. SEABORG, A. C. WAHL and J. W. KENNEDY: Paper 1.4, N.N.E.S., Vol. 14B "The Transuranium Elements", 1949.
[5] A. C. WAHL and G. T. SEABORG: Phys. Rev. **73**, 940 (1948).

Table 7. *Isotopes of Neptunium**.

Isotope and half-life	Type of decay	Energy (Mev)	Method of production	Disintegration** energies (Mev)
Np^{231} 50 min	α	6.28	$U^{238}(d, 9n)$ $U^{233}(d, 4n)$ $U^{235}(d, 6n)$	$Q_\alpha = 6.39$ $Q_{E.C.} = 1.82$(calc.)
Np^{232} 13 min	E.C. γ	 hard	$U^{233}(d, 3n)$	$Q_\alpha = 6.01$ (est.) $Q_{E.C.} = 2.65$ (est.)
Np^{233} 35 min	E.C. $99^+\%$ α $10^{-3}\%$	 5.53	$U^{233}(d, 2n)$ $U^{235}(d, 4n)$ $U^{233}(p, n)$	$Q_{E.C.} = 1.04$ (calc.) $Q_\alpha = 5.39$ (calc.)
Np^{234} 4.4 days	E.C. 99% γ β^+ $5\times10^{-2}\%$ (Ref. 2)	 0.78, 1.57 (Ref. 1) 0.8	$U^{235}(d, 3n)$ $U^{235}(\alpha, p\,4n)$ $U^{235}(p, 2n)$ $U^{233}(d, n)$ $U^{233}(\alpha, p\,2n)$ $Pa^{231}(d, n)$	$Q_{E.C.} = 1.82$ (est.) $Q_\alpha = 5.39$(calc.)
Np^{235} 410 days	E.C. $L/K = 30$ α $3.5\times10^{-3}\%$ γ (Ref. 3)	no γ [0.085, 0.026] metastable $t_{1/2} = 4\times10^{-8}$ sec	$U^{235}(d, 2n)$ $U^{233}(\alpha, p\,n)$	$Q_{E.C.} = 0.17$ (calc.) $Q_\alpha = 5.15$
Np^{236m}*** > 5000 years	β^-		$U^{238}(d, 4n)$	
Np^{236} 22 hours	β^- 57% γ E.C. 43% γ	0.518 (80%), 0.475 (20%) (Ref. 4) 0.0446 (Ref. 5) 0.0452 (Ref. 5)	$U^{235}(d, n)$ $U^{235}(\alpha, p\,2n)$ $U^{238}(d, 4n)$ $Np^{237}(d, p\,2n)$ $U^{233}(\alpha, p)$ $Np^{237}(n, 2n)$	$Q_{\beta^-} = 0.52$ $Q_{E.C.} = 0.91$ (calc.) $Q_\alpha = 5.10$ (calc.)
Np^{237} 2.2×10^6 years	α β stable γ	4.872 (3.1%), 4.816 (3.5%) 4.787 (53%), 4.767 (29%), 4.713 (1.7%), 4.674 (3.3%) 4.644 (6.0%), 4.589 (0.5%) 4.52 (0.02%) (Ref. 6) 0.029^+, 0.0869^+, 0.145, 0.175, 0.200 (Ref. 6) + metastable $t_{1/2} = 3.7\times10^{-8}$ sec 0.033 and 0.076 levels in Np^{237} reached by COULOMB excitation (Ref. 11)	daughter U^{237} $U^{238}(n, 2n)\,U^{237}$	$Q_\alpha = 4.97$ $I = 5/2$
Np^{238} 2.1 days	β^- γ E.C. < 1%	0.27, 1.25 (Ref. 7) 0.044, 0.1022, 0.927, 0.942, 0.986, 1.029 (Ref. 7)	$Np^{237}(n, \gamma)$ $U^{238}(d, 2n)$ $U^{238}(\alpha, p\,3n)$ $Np^{237}(d, p)$	$Q_{\beta^-} = 1.30$ $Q_{E.C.} = 0.12$ (calc.) $Q_\alpha = 4.88$ (calc.)
Np^{239} 2.33 days β^-	β^- γ	0.640, 0.430, 0.330, 0.210 (Ref. 8) 0.3345, 0.3165, 0.2856‡, 0.2777‡, 0.2731, 0.2546, 0.2284‡, 0.2264, 0.2090‡, 0.1818, 0.1061^+, 0.0678, 0.0614^+, 0.0573, 0.0494, 0.0446 (Ref. 8) + metastable $t^{1/2} = 1.93\times10^{-7}$ sec ‡ metastable $t^{1/2} = 1.1\times10^{-9}$ sec	$U^{238}(d, n)$ $U^{238}(\alpha, p\,2n)$ $U^{238}(n, \gamma\beta^-)$	$Q_\beta = 0.72$ $Q_\alpha = 4.56$ (calc.) $I = 1/2$

Table 7. (Continued.)

Isotope and half-life	Type of decay	Energy (Mev)	Method of production	Disintegration** energies (Mev)
Np^{240m} 7.3 min	β^-	0.76 (6%), 1.26 (11%), 1.59 (31%), 2.16 (52%) (Ref. 9)	daughter U^{240}	$Q_{\beta-}=2.16$ $Q_\alpha=4.53$ (est.)
	γ	0.56 (63%), 0.90 (26%), 1.40 (10%) (Ref. 9)		
Np^{240} 63 min	β^-	0.9	$U^{238}(\alpha, p n)$	$Q_{\beta-}=2.05$ $Q_\alpha=4.42$ (est.)
	γ	numerous (Ref. 10)		

* For references to original data not given specifically here, refer to HOLLANDER, PERLMAN and SEABORG, "Table of Isotopes", [Rev. Mod. Phys. **25**, 469 (1953)] and to the compilations of the Nuclear Data Group, National Research Council.

** Disintegration energies Q are revised values from R. A. GLASS, S. G. THOMPSON and G. T. SEABORG: J. Inorg. a. Nucl. Chem. **1**, 3 (1955).

*** Placement of $Np^{236,m}$ with respect to Np^{236} is not known.

[1] R. W. HOFF: Ph. D. Thesis, University of California Radiation Laboratory Report UCRL-2325, Sept. 1953.

[2] R. J. PRESTWOOD *et al:* Phys. Rev. **98**, 1324 (1955).

[3] R. W. HOFF *et al.*: University of California Radiation Laboratory Report UCRL-4517, Dec. 1955.

[4] T. O. PASSELL: Ph. D. Thesis, University of California Radiation Laboratory Report UCRL-2528, 1954.

[5] J. M. HOLLANDER: Private communication 1955.

[6] L. B. MAGNUSSON *et al.*: Abstract A 11 Bulletin of the American Physical Society, Vol. 30, Number 7, Nov. 1955.

[7] J. O. RASMUSSEN *et al.*: Phys. Rev. **99**, 42, 47 (1955).

[8] J. M. HOLLANDER *et al.*: University of California Radiation Laboratory Report UCRL-3222, Sept. 1955. Phys. Rev. **1956**.

[9] J. D. KNIGHT *et al.*: Phys. Rev. **91**, 889 (1953).

[10] R. M. LESSLER: University of California Radiation Laboratory Report UCRL-2647, p. 54, July 1954.

[11] J. O. NEWTON: Nature, Lond. **175**, 1028 (1955).

and is produced by the reaction

$$U^{238}(n, 2n)\,U^{237} \xrightarrow[6.75\text{ days}]{\beta^-} Np^{237}. \tag{14.1}$$

The first isolation of a weighable amount of Np^{237} was carried out by MAGNUSSON and LACHAPELLE[1] at the Metallurgical Laboratory in October 1944. The operation of large nuclear reactors for the production of plutonium has made it possible to isolate many grams of neptunium because it is produced as a by-product by the $(n, 2n)$ reaction. The cross-section for this reaction is approximately one percent of the neutron capture cross-section of U^{238}.

Neptunium-237 has another significance in that it can be regarded as the parent of the $4n+1$ radioactive series missing in nature. The other three series, the thorium series $(4n)$, the uranium-radium series $(4n+2)$, and the actinium series $(4n+3)$ are present in nature because the parent activities Th^{232}, U^{238}, and U^{235}, respectively, are sufficiently long-lived to have persisted through geologic time. The half-life of Np^{237} is about a factor of ten too short for the survival of an appreciable fraction of the neptunium which almost certainly was present at the beginning of the world.

It was predicted[2] that very small amounts of Np^{237} must be present in nature due to its continuous production by the action of naturally occurring fast

[1] L. B. MAGNUSSON and T. J. LACHAPELLE: J. Amer. Chem. Soc. **70**, 3534 (1948); see also Paper 1.7 N.N.E.S., Vol. 14B "The Transuranium Elements", 1949.

[2] G. T. SEABORG: Chem. Engng. News **25**, 358 (1947).

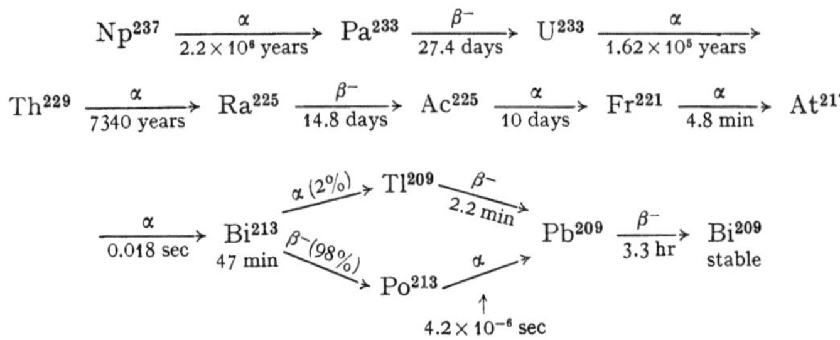

Fig. 6. Decay sequence of the neptunium series[1,2].

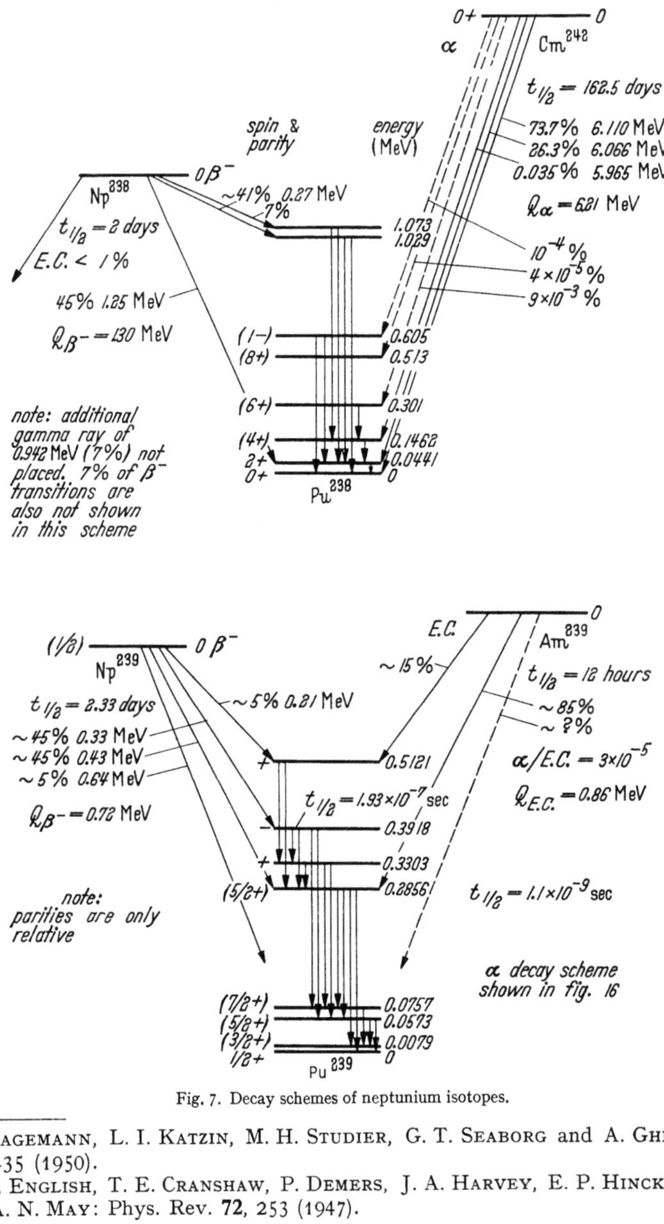

Fig. 7. Decay schemes of neptunium isotopes.

[1] F. Hagemann, L. I. Katzin, M. H. Studier, G. T. Seaborg and A. Ghiorso: Phys. Rev. **79**, 435 (1950).

[2] A. C. English, T. E. Cranshaw, P. Demers, J. A. Harvey, E. P. Hincks, J. V. Jelley and A. N. May: Phys. Rev. **72**, 253 (1947).

neutrons via reaction (14.1). PEPPARD and his co-workers[1] have established the existence of Np^{237} in Belgian Congo pitchblende in concentration corresponding to a ratio of Np^{237} to U^{238} of 1.8×10^{-12}.

The $4n+1$ series is sometimes called the neptunium series by analogy to the uranium and thorium series.

The neptunium series decay sequence is shown in Fig. 6.

A large additional number of neptunium isotopes have been prepared by cyclotron reactions. These isotopes are listed in Table 7. The decay scheme of those isotopes which have received most careful study are given in Fig. 7. Nuclear properties of the transuranium elements as a group will be reviewed in Part C of this report.

b) Chemical properties of neptunium.

15. The oxidation states. Neptunium exhibits interesting and complex chemical properties due to the existence of four oxidation states of roughly

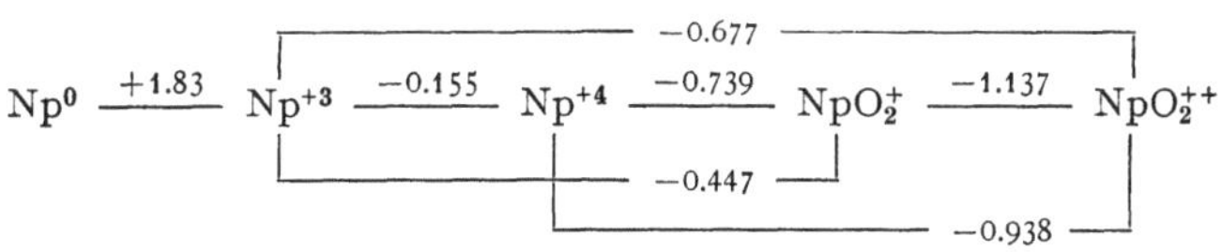

Fig. 8. Formal oxidation potentials of neptunium in perchloric acid[2].

comparable thermodynamic stability. These are the III, IV, V, and VI oxidation states. In aqueous solution the ionic species, neglecting hydration, are Np^{+3}, Np^{+4}, NpO_2^+ and NpO_2^{+2}. The formal oxidation potentials for neptunium in perchloric acid are given in Fig. 8. Only slight complexing of neptunium ion species occurs in dilute hydrochloric acid and nitric acid so that the same potential scheme holds for these acids with but slight changes. In sulfuric acid and other strong complexing media some of the potentials are markedly changed.

The potentials of the couples in basic solutions are quite different because of hydrolysis reactions, but no reliable data are available.

The oxygenated ion NpO_2^{+2} resembles UO_2^{+2} and PuO_2^{+2}. This structure, as well as the NpO_2^{+1} structure, is very stable and persists throughout a wide range of acidity; it is modified only when hydrolysis occurs in solutions of low acidity. This type of ion may be considered unique to the actinide elements and probably results from a special type of hybrid-bond formation involving $5f$ orbitals.

Oxidation-reduction reactions involving interconversion of Np^{+3} and Np^{+4} are rapid because only an electron transfer is involved. A similar situation holds with respect to the NpO_2^+-NpO_2^{+2} couple. Interconversion of members of these two groups such as oxidation of Np^{+4} to NpO_2^+ involves the breaking or making of Np-O bonds so that such reactions can be quite slow. The study of the kinetics of these reactions constitutes an interesting and important part of neptunium chemistry.

Tables 8 and 9 summarize thermodynamic data for the aqueous ions of neptunium.

---

[1] D. F. PEPPARD, G. W. MASON, P. R. GREY and J. F. MECH: J. Amer. Chem. Soc. **74**, 6081 (1952).

[2] D. COHEN and J. C. HINDMAN: J. Amer. Chem. Soc. **74**, 4682 (1952).

Table 8. *Values of ΔF, ΔH, and ΔS at 298 °K iu 1.0 M $HClO_4$ for oxidation reactions involving neptunium Ions*[1].

Reaction	$\Delta\bar{F}$ kcal mole⁻¹	$\Delta\bar{H}$ kcal mole⁻¹	$\Delta\bar{S}$ e. u.
$Np^{+3}_{(aq)} + H^+ \rightleftharpoons Np^{+4}_{(aq)} + \frac{1}{2}H_2$	3.58	−5.7	−31.2
$Np^{+4}_{(aq)} + 2H_2O \rightleftharpoons NpO_2^+(aq) + 3H^+ + \frac{1}{2}H_2$	17.04	35.6	62.3
$NpO_2^+(aq) + H^+ \rightleftharpoons NpO_2^{+2}(aq) + \frac{1}{2}H_2$	26.23	28.1	6.4

Table 9. *Partial molal heats and free energies of formation for neptunium ions in 1.0 M $HClO_4$ at 298 °K*[1].

Reaction	$\bar{H}$ kcal mole⁻¹	$\bar{F}$ kcal mole⁻¹
$Np^\circ + 3H^+ \rightarrow Np^{+3}(aq) + \frac{3}{2}H_2(g)$	−126.9	−126.3
$Np^\circ + 4H^+ \rightarrow Np^{+4}(aq) + 2H_2(g)$	−132.6	−122.7
$Np^\circ + 2H_2O(l) + H^+ \rightarrow NpO_2^+(aq) + \frac{5}{2}H_2(g)$	−97.0	−105.7
$Np^\circ + 2H_2O(l) + 2H^+ \rightarrow NpO_2^{+2}(aq) + 3H_2(g)$	−68.9	−79.5

16. Absorption spectra. The aqueous ions of neptunium are colored; neptunium (III), pale purple; neptunium (IV), yellow-green; neptunium (V) green-blue, and neptunium (VI), pale pink in 1 *M* $HClO_4$ and dark yellow in 1 *M* H_2SO_4.

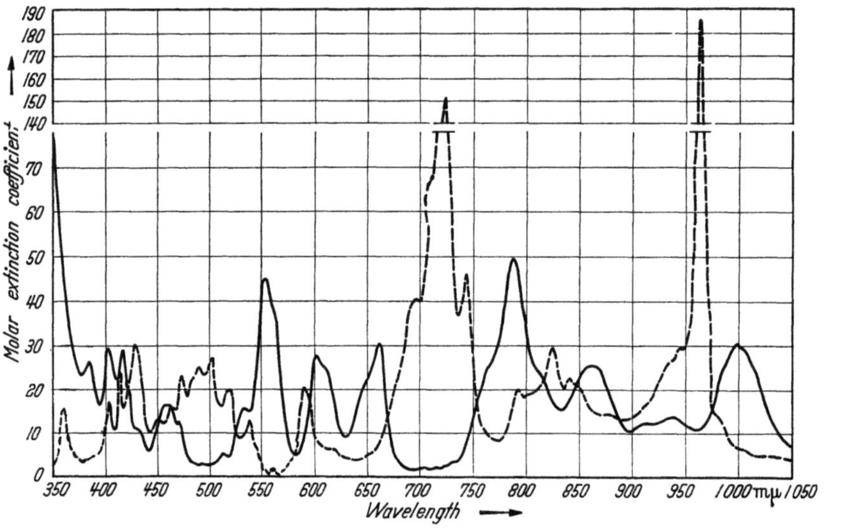

Fig. 9. Absorption spectra of the neptunium (III) and (IV) ions in 1.0 *M* $HClO_4$: ——— Np (III), – – – Np (IV). From Sjoblom and Hindman[2].

These ions have sharp absorption spectra in the visible and near-ultraviolet. These sharp bands are believed to arise from forbidden transitions in the *f*-shell as was mentioned in Sect. 9. Figs. 9 and 10 show the spectra for the neptunium ions in 1.0 molar perchloric acid. The spectra for neptunium (III), neptunium (IV), and neptunium (V) are sufficiently different that spectrophotometric analysis can be used to identify the oxidation state of a neptunium solution or to determine the composition of a mixture of states. Neptunium (VI) does not show any absorption in the visible; this is expected since it has only one unpaired electron.

[1] D. Cohen and J. C. Hindman: J. Amer. Chem. Soc. **74**, 4682 (1952).
[2] R. K. Sjoblom and J. C. Hindman: J. Amer. Chem. Soc. **73**, 1744 (1951).

17. Neptunium (III) in aqueous solution. Neptunium (III) is unstable in aqueous solution in the presence of atmospheric oxygen. However in the presence of an inert atmosphere neptunium can be reduced to neptunium (III) at a mercury pool cathode. Alternatively neptunium can be reduced by hydrogen using a platinum-black catalyst[1]. Neptunium (III) resembles plutonium (III) but very little quantitative information is available on its reactions.

18. Neptunium (IV) in aqueous solution. Neptunium (IV) may be prepared by the reduction of aqueous solutions of neptunium (V) or (VI); in the presence

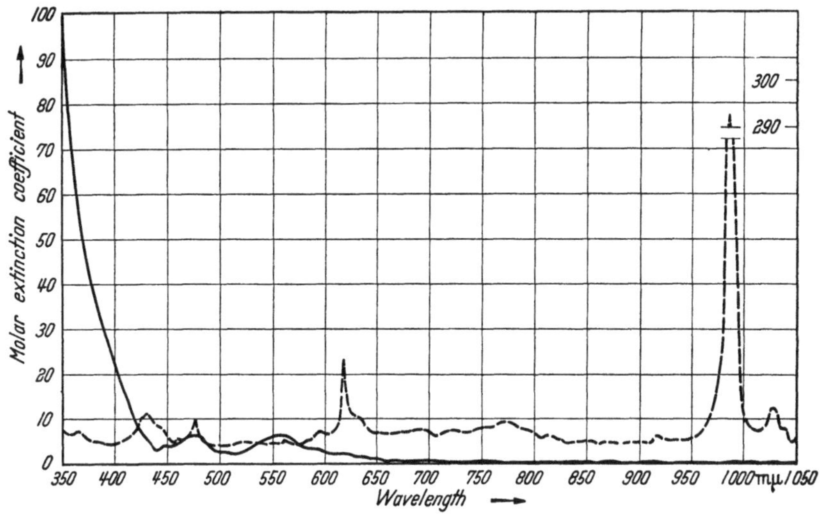

Fig. 10. Absorption spectra of neptunium (V) and (VI) ion in 1.0 *M* $HClO_4$: ——— Np (VI), – – – – Np (V). SJOBLOM and HINDMAN[1].

of oxygen no neptunium (III) is found. The reduction to neptunium (IV) is slow in many instances unless: (1) elevated temperatures are used, (2) a catalyst such as Fe (II) is added, or (3) the (IV) state is stabilized by complexing. Some reagents which are used are potassium iodide and hydrazine in 5 *M* HCl, hydroxylamine or sulfur dioxide in 1 *M* sulfuric acid, and stannous ion in 1 *M* HCl in the presence of iron (II) catalyst.

Neptunium (IV) in 1 *M* nitric acid is unstable with respect to oxidation by nitrate; in 1 *M* HCl or 1 *M* $HClO_4$ it is unstable with respect to oxidation by atmospheric oxygen. In all three solutions, however, the rate of oxidation is slow. In sulfuric acid neptunium (IV) is quite stable because of complex formation. Complexes are also formed with fluoride, phosphate, oxalate, and acetate ions. Complexing of the (IV) state is more pronounced than that of the other oxidation states.

Disproportionation (self-oxidation and reduction) of neptunium (IV) is quite negligible. This is in contrast to the situation in plutonium chemistry.

19. Neptunium (V) in aqueous solution. Neptunium is unique among the actinide elements, except for protactinium, in having a stable pentapositive state in aqueous solution. Uranium (V) is extremely unstable and prepared only with difficulty. Plutonium (V) is an important aqueous oxidation state of that element, but in acidic solution it is unstable toward disproportionation (self-oxidation-reduction to higher and lower oxidation states). Neptunium (V) by contrast is

[1] R. K. SJOBLOM and J. C. HINDMAN: J. Amer. Chem. Soc. **73**, 1744 (1951).

stable in an uncomplexing acidic medium until very high acidities are reached. The disproportionation reaction can be written

$$2\,Np(V) \rightarrow Np(IV) + Np(VI). \quad K = \frac{[Np(IV)][Np(VI)]}{[Np(V)]^2}. \tag{19.1}$$

The value of the equilibrium quotient, K, has been determined by SJOBLOM and HINDMAN[1] for a variety of acid concentrations. These values are given in Table 10.

Table 10. *Value of the disproportionation constant of neptunium (V) as a function of perchloric acid concentration**.

Acid concentration (moles/liter)	$K = \frac{[Np(IV)][Np(VI)]}{[Np(V)]^2}$
1.0	4×10^{-7}
5.34	0.127
7.11	5.78
8.45	169
8.67	200

Neptunium (V) can be prepared by reduction of neptunium (VI) with hydrazine, iodide, oxalate, sulfite, and tin (II). At room temperature the further reduction of neptunium (V) to neptunium (IV) by these reagents is very slow. Neptunium (V) may also be prepared by heating solutions of neptunium (IV) in 1 M HNO_3 or by stoichiometric oxidation of neptunium (IV) with cerium (IV). Neptunium (V) is oxidized to neptunium (VI) by cerium (IV), silver (II), chlorine, bromate ion and other oxidizing agents. It is rapidly reduced by iron (II) in the presence of nitrous acid.

Neptunium (V) forms no important complexes with chloride, nitrate, perchlorate, or sulfate ion. The weak complexing tendency of NpO_2^+ is in strong contrast to the tendency toward complex formation of NpO_2^{+2}, UO_2^{+2}, and PuO_2^{+2}. Some evidence for complex formation between NpO^+ and oxalate ion has been reported[2]. The NpO_2^+ ion does not appear to hydrolyze in solutions of low acidity up to a p_H of 5.7. At higher basicity the hydroxide NpO_2OH precipitates.

20. Neptunium (VI) in aqueous solution. Neptunium (VI) exists in aqueous solution in the form of the ion NpO_2^{+2} and has a very close resemblance in its properties to UO_2^{+2} and PuO_2^{+2}.

In the presence of strong oxidizing agents such as silver (II), bismuthate ion, bromate ion, cerium (IV), etc., neptunium in its lower oxidation states is converted to neptunium (VI).

Very little quantitative data are available on the complexes of NpO_2^{+2}, but these undoubtedly resemble those of UO_2^{+2} and PuO_2^{+2}.

21. Compounds of neptunium. *Neptunium metal* can be prepared by the reduction of neptunium trifluoride or neptunium tetrafluoride with barium metal vapor at 1200° C in a beryllia crucible system[3]. It is a silvery metal melting at $(640 \pm 1)°$ C. It exists in several forms, an orthorhombic form stable from room temperature to $(278 \pm 5)°$ C, density 20.45, and a tetragonal β form of density 19.36 stable from 279° to about 550° C where it is believed to undergo transition to a body-centered cubic γ form[4]. Purified hydrogen reacts with neptunium metal at 50° C to yield a black flaky *hydride* with the composition $NpH_{3.6-3.8}$[3].

Black or purple hexagonal *neptunium trifluoride* has been prepared by the reaction:

$$NpO_2(cryst) + \tfrac{1}{2}H_2(g) + 3\,HF(g) \xrightarrow{500°\,C} NpF_3(cryst) + 2\,H_2O(g).$$

* Data from SJOBLOM and HINDMAN[1].

1 R. K. SJOBLOM and J. C. HINDMAN: J. Amer. Chem. Soc. **73**, 1744 (1951).

2 D. M. GRUEN and J. J. KATZ: J. Amer. Chem. Soc. **75**, 3772 (1953).

3 S. FRIED and N. R. DAVIDSON: J. Amer. Chem. Soc. **70**, 3539 (1948).

4 W. H. ZACHARIASEN: Acta crystallogr. **5**, 660 (1952).

In the presence of a mixture of hydrogen fluoride gas and oxygen gas at 500° C the trifluoride is converted to light green monoclinic *neptunium tetrafluoride*

$$NpF_3(\text{cryst}) + \tfrac{1}{4}O_2(g) + HF(g) \longrightarrow NpF_4(\text{cryst}) + \tfrac{1}{2}H_2O(g).$$

Neptunium pentafluoride has never been isolated but should be quite stable.

Neptunium hexafluoride can be prepared by the reaction of neptunium trifluoride with very pure fluorine gas. The trifluoride is supported on a nickel filament heated electrically to red heat. The volatile hexafluoride is condensed on a liquid nitrogen-cooled surface immediately adjacent to the reaction zone before thermal decomposition can occur.

Reddish brown *neptunium tetrachloride* crystallizing in the tetragonal UCl_4 structure can be prepared by the reaction

$$NpO_2 + CCl_4 \xrightarrow{530^\circ C} NpCl_4 + CO_2.$$

At this temperature neptunium tetrachloride is volatile and sublimes to a cool portion of the apparatus. This volatility is useful in the separation of neptunium and plutonium[1]; when a mixed $NpO_2 - PuO_2$ sample is treated with chlorinating agents a mixture of plutonium trichloride (plutonium does not form a stable tetrachloride) and neptunium tetrachloride is produced. The neptunium tetrachloride is readily sublimed away from the plutonium trichloride at 500 to 530° C.

Neptunium tetrachloride is reduced to *neptunium trichloride* by treatment with hydrogen at 450° C. This compound is white and crystallizes in the hexagonal structure. It is volatile at 750 to 800° C.

Neptunium tribromide is a green compound existing in an α hexagonal form and a β orthorhombic form. It has been prepared by reaction of neptunium dioxide with excess aluminum and aluminum tribromide in a sealed tube at 350 to 400° C.

Neptunium tetrabromide is a reddish-brown compound. It has been prepared by heating neptunium dioxide with excess aluminum tribromide in a quartz capillary at 350° C. It sublimes at 500° C.

Neptunium triiodide has been prepared by reacting excess aluminum triiodide with neptunium dioxide in a sealed tube at 350 to 400° C. *Neptunium tetraiodide* has not been prepared but should be stable.

When neptunium "hydroxide", peroxide, oxalate, or nitrate of any oxidation state are ignited in air to high temperatures *neptunium dioxide* is formed. Other oxides of neptunium are obtained only under special conditions. Neptunium dioxide is a green compound crystallizing in the face-centered cubic form. *Neptunium monoxide* has been prepared as a surface coating on neptunium metal. It is semi-metallic in character. Since uranium has an important oxide of the composition U_3O_8, it is interesting that a similar oxide of neptunium has been prepared. KATZ and GRUEN[2] showed that treatment of various dried hydroxides of neptunium with nitrogen dioxide at 300 to 450° C resulted in the formation of Np_3O_8, a chocolate-brown solid easily soluble in nitric acid. It is isomorphous with U_3O_8. Np_3O_8 is not stable above 500° C. There is no evidence for NpO_3.

Neptunium forms a colorless, flocculent peroxide compound of low solubility; the formula is $NpO_4 \cdot 2\,H_2O$ and it is isomorphous with $UO_4 \cdot 2\,H_2O$.

Neptunium (IV) and (V) are precipitated from alkaline solution as the hydrated "hydroxide" compounds, $Np(OH)_4 \cdot xH_2O$ and $NpO_2OH \cdot xH_2O$. The solubility of the (IV) hydroxide is a few milligrams per liter from strongly alkaline solutions; that of the (V) hydroxide is considerably greater.

---

[1] I. K. SHVETSOV and A. M. VOROBYEV: Paper P/674, "Proceedings of the Geneva Conference-Peaceful Uses of Atomic Energy", August 1955.

[2] J. J. KATZ and D. M. GRUEN: J. Amer. Chem. Soc. **71**, 2106 (1949).

Addition of sodium hydroxide or ammonium hydroxide to aqueous neptunium (VI) solutions results in the precipitation of the *dineptunate compounds*, $(NH_4)_2Np_2O_7 \cdot H_2O$ and $Na_2Np_2O_7 \cdot H_2O$. These are isostructural with the analogous diuranate compounds.

One of the best known solid compounds of hexavalent neptunium is *sodium neptunyl acetate*, $NaNpO_2(C_2H_3O_2)_3$. This pink crystalline compound is isostructural with the analogous compounds of uranium, plutonium, and americium.

Additional solid compounds about which structural information is available include *neptunium sesquisulfide, neptunium oxysulfide, neptunium dioxyfluoride, neptunium oxydichloride, neptunium nitride, trineptunium tetraphosphide, neptunium (IV) oxalate*, and *neptunium disilicide*.

A detailed treatment of the compounds of neptunium with numerous references to the original literature is given by CUNNINGHAM and HINDMAN[1]. Lattice constants for a number of crystalline neptunium compounds are given in Table 6, Sect. 6.

22. Coprecipitation behavior of neptunium. Trace concentrations of neptunium can be isolated from aqueous solution by coprecipitation with certain carrier compounds. This coprecipitation can occur even though the solubility product of the related neptunium compound is not exceeded, but the coprecipitation is most likely to be complete if the carrier compound is similar to a neptunium compound of slight solubility. General types of compounds which are known to "carry" neptunium include the hydroxides, fluorides, phosphates, oxalates, iodates, phenylarsonates, and peroxides. The extent of coprecipitation for most carriers is a function of the oxidation state of neptunium and carriers specific for certain oxidation states can be selected. Radiochemical purification schemes designed to separate neptunium from lighter-element impurities such as the fission products or from neighboring elements invariably depend on the differing behavior of the various oxidation states. These will not be reviewed in detail here, but a few statements will be made about the most important carrier compounds.

Lanthanum fluoride is an excellent carrier for the removal of neptunium (IV) or neptunium (V) from solution. Plutonium in the (III) or (IV) state is also strongly carried. When oxidized to the hexapositive state, neptunium is not carried. In acid solution neptunium (IV) and plutonium (III) can be separated by treatment with bromate ion for 20 min at room temperature followed by precipitation of lanthanum fluoride. The plutonium is not oxidized beyond plutonium (IV) because of a slow rate of oxidation, but the neptunium is oxidized to neptunium (VI) which is not precipitated with lanthanum fluoride.

Zirconium phosphate or zirconium phenyl arsonate are good carriers for neptunium or plutonium in the tetrapositive state but leave these elements in solution when they are in the tripositive or hexapositive state. A good neptunium-plutonium separation is achieved by precipitation of one of these carrier compounds after treatment of the solution with a strong reducing agent such as a mixture of hydrazine and iron (II). In the presence of atmospheric oxygen, neptunium is not reduced beyond neptunium (IV), (carried) while plutonium is converted to plutonium (III) (not carried).

Sodium uranyl acetate is an excellent carrier compound for trace amounts of hexavalent neptunium. The lower oxidation states are not carried.

23. Solvent extraction of neptunium. Neptunium (VI) resembles uranium (VI) in being extractable from aqueous solution into certain organic solvents immiscible

[1] B. B. CUNNINGHAM and J. C. HINDMAN: Chap. 12, N.N.E.S., Vol. 14A "The Actinide Elements".

with water. Ethyl ether has been used extensively in small scale applications, the extractions usually being performed from nitric acid solutions. To achieve appreciable distribution coefficients into the ether phase it is necessary to add high concentrations of neutral nitrate salts such as ammonium nitrate or magnesium nitrate to the aqueous phase. This extraction is rather specific and effects purification of neptunium from all but a few elements in the periodic system. Neptunium (VI) can also be extracted into tributylphosphate, dibutyl carbitol, methyl isobutyl ketone, and other solvents of the ketone or "carbitol" or "cellosolve" type.

Neptunium (V) is poorly extracted by ethyl ether and many other organic solvents. Neptunium (IV) is poorly extracted by ethyl ether or methyl isobutyl ketone. It is moderately well extracted by tributyl phosphate, diluted to a 15 to 30% solution in some other organic solvent such as carbon tetrachloride or kerosene, when the aqueous phase has a high concentration of nitrate salts. Detailed information on the extraction of neptunium has not been published. In general the extractability of the hexapositive actinide element ions into organic solvents increases in the order: $PuO_2^{+2} < NpO_2^{+2} < UO_2^{+2}$. For the tetrapositive ions the order is $Th^{+4} < Np^{+4} < Pu^{+4}$.

The differing behavior of the various oxidation states can be useful in making a separation of two neighboring actinide elements. For example, if an ether solution of hexapositive uranium and neptunium is contacted with a saturated ammonium nitrate solution containing 0.1 *M* HNO_3 and a small amount of iron (II) and urea the neptunium (VI) is rapidly reduced to neptunium (V) and is removed from the ether phase. Uranium remains in the ether.

24. Solvent extraction of the TTA complex of neptunium. A number of inner complex salts of neptunium are known, and the solubility of these salts in organic solvents under suitable conditions of acidity and concentration can be taken advantage of for analytical or purification purposes. The most widely used chelating agent is the beta diketone, α-thenoyltrifluoroacetone, (frequently referred to as TTA) which forms a strong complex with neptunium (IV) readily extractable into benzene or other non-polar solvents. High partition coefficients in favor of the organic phase are obtained in solutions sufficiently acidic (e.g. 0.5 molar H^+) that only a few other ions, notably iron (III), protactinium (V), plutonium (IV), tin (IV), zirconium (IV), cerium (IV), and hafnium (IV), are extracted. Neptunium-plutonium separations are effected by reducing plutonium to plutonium (III) which is not extracted. Iodide ion in the presence of hot concentrated hydrochloric acid is often used for the reduction.

The extraction of the TTA complex of neptunium (V) and (VI) is slight.

25. Ion-exchange behavior of neptunium. The ion-exchange behavior of neptunium has not been reported in detail although ion-exchange techniques are of considerable importance. Neptunium in its three principal solution states forms negatively charged chloride complexes in strong solutions of hydrochloric acid; these complexes are readily removed from solution by exchange with anion resins such as Dowex-1 (a copolymer of styrene and divinyl benzene with quaternary ammonium functional groups). The other actinide elements show similar behavior as is shown in Table 11. The chloride concentration must be quite high for these negatively charged complexes to form; hence adsorption and desorption can be controlled by adjusting the hydrochloric acid concentration. Separation of neighboring actinide elements can be achieved by adjustment of acid concentration or by control of oxidation-reduction conditions. The separation from rare-earth contaminants is very large as these elements do not form anionic complexes in hydrochloric acid.

The exact change in adsorbability with hydrochloric acid concentration depends on the particular exchange resin sample. Rough data for Dowex-1 resin are the following:

Hexapositive ions. Uranium (VI), neptunium (VI), and plutonium (VI) are readily adsorbed from hydrochloric acid $6M$ or greater in concentration. These ions are readily desorbed with 0 to 3 molar hydrochloric acid.

Pentapositive ions. Neptunium (V) is readily adsorbed above 4 molar hydrochloric acid. Protactinium (V) is readily adsorbed above 8 molar hydrochloric acid. Uranium (V) and plutonium (V) are not stable enough to be of radiochemical interest.

Table 11. *Relative adsorption of actinide element ions on anion-exchange resins from strongly acidic hydrochloric acid solutions.*

Hexapositive Ions.	U(VI) > Np(VI) > Pu(VI) All strongly adsorbed
Pentapositive Ions.	Pa(V) and Np(V) Strongly adsorbed
Tetrapositive Ions.	Th(IV) and U(IV) Not adsorbed. Np(IV) < Pu(IV) Strongly adsorbed.
Tripositive Ions. .	Pu(III)—not adsorbed. Americium and higher elements—slight adsorption increasing with atomic number

Tetrapositive ions. Uranium (IV) is adsorbed only from very concentrated hydrochloric acid. Neptunium (IV) is adsorbed above 4 molar hydrochloric acid. Plutonium (IV) is adsorbed strongly above 2.5 molar hydrochloric acid.

Anion exchange methods are attractive because they are rapid and quantitative. Neptunium can be separated from plutonium easily and quantitatively by anion exchange. If both elements are in concentrated hydrochloric acid, iodide ion can be added to reduce neptunium to the (IV) state and plutonium to the (III) state. The solution can then be drawn through a short bed of Dowex-1 resin and washed with concentrated hydrochloric acid. The plutonium passes through quantitatively while the neptunium is adsorbed quantitatively. The neptunium (IV) can then be desorbed with 0.5 molar hydrochloric acid.

Neptunium and plutonium can be separated from a large number of fission product and light element impurities by anion-exchange techniques. The greater number of those elements which form adsorbable negative ions in hydrochloric acid can be removed by procedures based on a detailed knowledge of their adsorption behavior. KRAUS[1] has provided an excellent review of this field.

Uranium, neptunium, and plutonium in all their stable solution oxidation states are readily adsorbed from acid solution on cation-exchange resins such as Amberlite (copolymer of formaldehyde and phenol with carboxylic and phenolic functional groups) and Dowex-50 (copolymer of styrene and divinylbenzene with nuclear sulfonic acid functional groups). Small amounts of these elements can be adsorbed on a small column of resin from rather large volumes of dilute acid solution. This property is very useful for concentration purposes. For adsorption of neptunium it is advisable to adjust the oxidation state to the tetrapositive state since the Np^{+4} ion is very much more strongly bound than is the divalent NpO_2^{+2} ion. The monovalent NpO_2^+ ion is only weakly adsorbed. On long contact with most commercial resins neptunium (VI) is reduced to lower oxidation states.

Neptunium, uranium, and plutonium ions can be desorbed from cation-exchange resin columns by elution with moderately concentrated solutions of hydrochloric acid or nitric acid as would be expected from the mass action influence

[1] K. A. KRAUS: Paper P/837, "Peaceful Uses of Atomic Energy—Prodeedings of the Geneva Conference", August 1955, United Nations.

of the hydrogen ion. However the elution with hydrochloric acid is much speedier, particularly for neptunium (IV), than would be expected from this effect alone because of the strong tendency toward the formation of chloride complexes (see Fig. 30, Sect. 55).

Neptunium can be eluted from cation-exchange resin columns with other eluting agents which form complex ions with neptunium. Data on such elution behavior is scanty.

II. Plutonium (element 94).

a) History of plutonium.

26. Discovery of plutonium—summary of known isotopes. McMillan and Abelson[1] found that Np^{239}, formed by the beta decay of U^{239}, underwent radioactive decay by the emission of beta particles. They were unable to detect the Pu^{239} in their samples because of the long half life of this isotope. The first isotope of element 94 to be identified was Pu^{238} produced in the experiments of Seaborg, McMillan, Wahl, and Kennedy[2] by the bombardment of uranium by 14-Mev deuterons in the University of California cyclotron.

$$U^{238}(d, 2n)\,Np^{238} \xrightarrow[2.1\text{ days}]{\beta^-} Pu^{238}. \tag{26.1}$$

The isotope Pu^{238} was shown to emit alpha particles with a half-life of approximately 50 years (best present value = 90 years). The discoverers proposed the name plutonium (symbol Pu) from Pluto, at that time believed to be the second planet beyond Uranus in our solar system.

In 1941, Kennedy, Seaborg, Segrè, and Wahl[3] isolated and identified Pu^{239} as the decay product of a very intense source of Np^{239}. This isotope Pu^{239} decays by the emission of alpha particles, with the half-life (best value determined later) of 24360 years. These same investigators demonstrated that Pu^{239} underwent nuclear fission with a high cross section when bombarded with slow neutrons[4]. Because of this high cross section (730×10^{-24} cm^2) and because it can be prepared in kilogram quantities in nuclear reactors, it is of very considerable importance in nuclear weapons and in the controlled release of nuclear power in reactors.

In a chain-reacting pile, plutonium is produced via the reactor sequence

$$U^{238}(n, \gamma)\,U^{239}; \quad U^{239} \xrightarrow[23\text{ min}]{\beta^-} Np^{239} \xrightarrow[2.33\text{ days}]{\beta^-} Pu^{239}. \tag{26.2}$$

The first isolation of a weighable amount of plutonium free from all carrier matter and impurities was carried out by Cunningham and Werner[5] in 1942.

Minute amounts of plutonium were found in pitchblende by Seaborg and Perlman[6] and in a series of uranium and thorium ores by Levine and Seaborg[7]. The ratio of plutonium to uranium in pitchblende and monazite ores is approximately 10^{-11} and substantially less in carnotite and fergusonite. All the plutonium

---

[1] E. M. McMillan and P. Abelson: Phys. Rev. **57**, 1185 (1940).

[2] G. T. Seaborg, E. M. McMillan, J. W. Kennedy, and A. C. Wahl: Phys. Rev. **69**, 366 (1946). — G. T. Seaborg, A. C. Wahl and J. W. Kennedy: Phys. Rev. **69**, 367 (1946).

[3] J. W. Kennedy, G. T. Seaborg, E. Segrè, and A. C. Wahl: Phys. Rev. **70**, 555 (1946).

[4] J. W. Kennedy, G. T. Seaborg, E. Segrè and A. C. Wahl: Report A-33, May 29, 1941.

[5] B. B. Cunningham and L. B. Werner: Paper 1.8, N.N.E.S., Vol. 14B, "The Transuranium Elements", 1949.

[6] G. T. Seaborg and M. L. Perlman: Paper 1.3, N.N.E.S., Vol. 14B, "The Transuranium Elements". J. Amer. Chem. Soc. **70**, 1571 (1948).

[7] C. A. Levine and G. T. Seaborg: J. Amer. Chem. Soc. **73**, 3278 (1951).

isolated was Pu^{239} formed from the uranium by the reaction (26.2). The sources of the neutrons include (1) the spontaneous fission of uranium, (2) (α, n) reactions caused by the action of alpha particles from the heavy elements on the nuclei of light elements in the ore, and (3) cosmic rays. In the case of pitchblende ores, the spontaneous fission of uranium and (α, n) reactions contribute the major portion of the neutrons. In thorium ores, neutrons from (α, n) reactions are probably dominant. Peppard and coworkers[1] isolated microgram quantities of

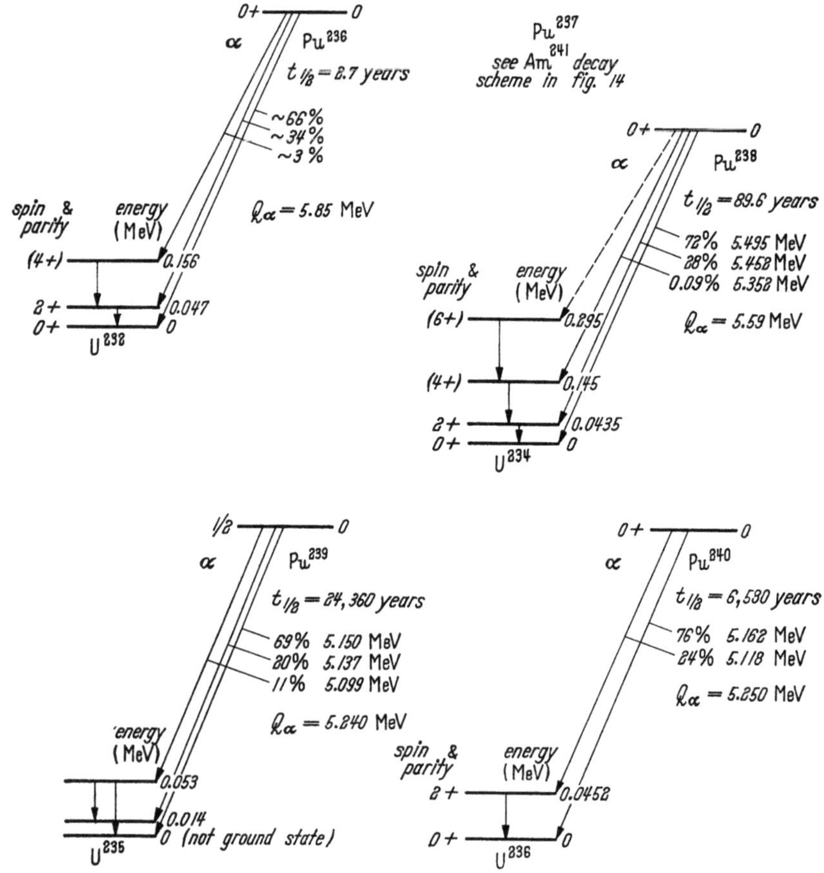

Fig. 11. Decay schemes of plutonium isotopes.

plutonium from Belgian Congo pitchblende concentrates. They examined this plutonium carefully for higher mass plutonium isotopes by neutron irradiation techniques and by mass spectroscopy and proved that it was pure Pu^{239}.

Table 12 lists all known isotopes of plutonium together with their principal decay characteristics. Fig. 11 shows detailed decay schemes for the more carefully studied isotopes. Nuclear data for the transuranium elements as a group are summarized in Sect. C at the end of this article. The lighter isotopes are produced by bombardment of uranium with high energy helium ions while the heavier isotopes are produced by multiple neutron capture reactions in high flux reactors starting with plutonium isotopes of lower mass.

---

[1] D. F. Peppard *et al.*: J. Amer. Chem. Soc. **73**, 2529 (1951).

Table 12. *Isotopes of plutonium**.

Isotope and half life	Type of decay	Energy (Mev)	Method of production	Disintegration energies (Mev)**
Pu^{232} 36 min	E.C. < 98% α > 2%	 6.58	U^{235} (α, 7n) U^{233} (α, 5n)	$Q_{E.C.} = 0.99$ $Q_\alpha = 6.70$ est.
Pu^{233} 20 min (Ref. 11)	α 0.12%	6.3	U^{233} (α, 4n)	$Q_\alpha = 6.49$ est.
Pu^{234} 9.0 hours	E.C. 96% α 4%], No γ < 0.5% (Ref. 1)	 6.19	U^{235} (α, 5n) U^{233} (α, 3n) daughter Cm^{238}	$Q_{E.C.} = 0.48$ calc. $Q_\alpha = 6.30$
Pu^{235} 26 min	E.C. 99⁺% α 0.002%		U^{233} (α, 2n) U^{235} (α, 4n)	$Q_{E.C.} = 1.17$ calc. $Q_\alpha = 5.95$
Pu^{236} 2.7 years α 3.5×10^9 years spon. fission	α β stable γ	5.75 0.045	U^{233} (α, n) U^{235} (α, 3n) U^{238} (α, 6n) Np^{237} (d, 3n) daughter Cm^{240}	$Q_\alpha = 5.85$
Pu^{237} 40 days	E.C. γ X-rays α (Ref. 11)	 0.064 5.36 (79%) 5.65 (2%)	Np^{237} (p, n) U^{235} (α, 2n) Np^{237} (d, 2n)	$Q_{E.C.} = 0.18$ est. $Q_\alpha = 5.72$ est.
Pu^{238} 89.6 years α 4.9×10^{10} years spon. fission	α β stable γ X-ray	5.352 (0.09%) 5.452 (28%) 5.495 (72%) (Ref. 2) 0.0438, 0.099 (Ref. 2) 0.150 (Ref. 2) 0.203 (Ref. 8)	U^{238} (d, 2n) U^{238} (α, 4n) U^{235} (α, n) daughter Np^{238} daughter Cm^{242}	$Q_\alpha = 5.59$
Pu^{239} 24.360 years 5.5×10^{15} years spon. fission	α β stable γ	5.150 (69%) 5.137 (20%) 5.099 (11%) 0.039, 0.0531, 0.100, 0.124, 0.384	U^{238} (α, 3n) Pu^{238} (n, γ) daughter Np^{239}	$Q_\alpha = 5.24$ $I = \frac{1}{2}$
Pu^{240} 6580 years α 1.2×10^{11} years spon. fission	α β stable γ	5.162 (76%) 5.118 (24%) 0.0452	Pu^{239} (n, γ) U^{238} (α, 2n)	$Q_\alpha = 5.25$
Pu^{241} 13.0 years (Ref. 3)	β⁻ α 10^{-3}% γ	0.021 4.91 0.100, 0.145	Pu^{240} (n, γ) U^{238} (α, n)	$Q_\beta = 0.02$ $Q_\alpha = 5.13$ $I = \frac{5}{2}$
Pu^{242} 3.73×10^5 years (Ref. 4) 3.78×10^5 years (Ref. 10) 7.3×10^{10} years spon. fission	α β stable	4.898	Pu^{241} (n, γ) daughter Am^{241m}	 $Q_\alpha = 4.98$
Pu^{243} 4.98 hours	β⁻ γ	0.37 (12%), 0.468 (35%) 0.57 (53%) (Ref. 5) 0.085, 0.12 (Ref. 5)	Pu^{242} (n, γ)	$Q_\beta = 0.56$ $Q_\alpha = 4.71$ calc.

Table 12. (Continued.)

Isotope and half life	Type of decay	Energy (Mev)	Method of production	Disintegration energies (Mev)**
Pu^{244} 7×10^7 years α (Ref. 6) 2.5×10^{10} years spon. fission	α (Ref. 6)		$Pu^{243}(n,\gamma)$ (Ref. 6)	$Q_\alpha=4.57$ est.
Pu^{245} 10.1 hours (Ref. 7, 8, 9)	β^-		$Pu^{244}(n,\gamma)$	$Q_\beta=1.31$ est. $Q_\alpha=4.44$ est.
Pu^{246} 11.2 days (Ref. 7)	β^- (Ref. 7) γ	0.043, 0.103 0.175, 0.220 (Ref. 7)	$Pu^{245}(n,\gamma)$ (Ref. 7)	$Q_\alpha=4.30$ est. $Q_\beta=0.37$

* For references to original data not given specifically here, refer to Hollander, Perlman, and Seaborg, "Table of Isotopes", [Rev. Mod. Phys. **25**, 469 (1953)] and to the compilations of the Nuclear Data Group, National Research Council.

** Revised values taken from Glass, Thompson and Seaborg: J. Inorg. a. Nucl. Chem. **1**, 3 (1955).

[1] R. W. Hoff: Ph. D. Thesis, University of California Radiation Laboratory Report UCRL-2325, Sept. 1953.

[2] F. Asaro and I. Perlman: Phys. Rev. **94**, 381 (1954).

[3] D. R. MacKenzie *et al.*: Phys. Rev. **90**, 327 (1953).

[4] J. P. Butler, M. Lounsbury and J. S. Merritt: Canad. J. Chem. **34**, 253 (1956).

[5] D. W. Engelkemeir *et al.*: Phys. Rev. **90**, 6 (1953).

[6] W. C. Bentley *et al.*: Paper P/809, "Proceedings of the Geneva Conference-Peaceful Uses of Atomic Energy", August 1955, United Nations.

[7] D. W. Engelkemeir *et al.*: J. Inorg. a. Nucl. Chem. **1**, 345 (1955).

[8] C. I. Browne *et al.*: J. Inorg. a. Nucl. Chem. **1**, 254 (1955).

[9] M. H. Studier *et al.*: J. Inorg. a. Nucl. Chem. **1**, 262 (1955).

[10] J. Mech *et al.*: Argonne National Laboratory Report ANL-5348, 1954.

[11] T. D. Thomas *et al.*: In publication Phys. Rev. 1957.

b) The chemical properties of plutonium.

27. Solution oxidation states. In aqueous solution plutonium exhibits the four oxidation states (III), (IV), (V), and (VI) so that its chemical reactions are quite complex. In acidic solution the ion species corresponding to these oxidation states are Pu^{+3}, Pu^{+4}, PuO_2^+, and PuO_2^{+2}. Each ion is hydrated, but no exact information on the hydration is available. The aqueous ions have characteristic colors: Pu^{+3} (blue-lavender), Pu^{+4} (yellow-brown), and PuO_2^{++} (pink orange). Pure solutions of PuO_2^+ have not been prepared and since the ion shows but little absorption of visible light, its appearance in solution is not definitely known; presumably dilute solutions would be colorless and concentrated solutions of PuO_2^+ would have a reddish appearance since the ion absorbs mainly in the shorter and medium wave lengths of the visible spectrum. The colors of the plutonium ions are altered by hydrolysis or complex ion formation.

The formal oxidation potential scheme for plutonium in 1 molar perchloric acid is given in Fig. 12.

The potentials of Fig. 12 are applicable in dilute hydrochloric acid or nitric acid systems with but slight changes because the chloride and nitrate complexes are relatively weak. In stronger complexing media such as sulfate, phosphate, fluoride, or oxalate solutions or in solutions containing organic chelating agents considerable complexing of plutonium occurs. The tendency of the ions to form complexes increases in the order $PuO_2^+ < Pu^{+3} < PuO_2^{+2} < Pu^{+4}$ and the observed potential changes are in the direction expected from this.

Oxidation-reduction interconversions of the Pu^{+3} — Pu^{+4} pair or of the PuO_2^+ — PuO_2^{+2} pair are rapid and readily reversible because only electron transfer is involved. When Pu^{+4} is converted to PuO_2^+ or PuO_2^{+2}, however, chemical bonds must be changed and such reactions may be slow. An interesting part of plutonium chemistry is a study of the reaction kinetics and mechanisms of oxidation-reduction reactions.

$$Pu^0 \xrightarrow{+2.03} Pu^{+3} \xrightarrow{-0.982} Pu^{+4} \xrightarrow{-1.161} PuO_2^+ \xrightarrow{-0.925} PuO_2^{+2}$$

$$Pu^{+4} \xrightarrow{-1.043} PuO_2^{+2}$$

Fig. 12. Formal oxidation potential diagram of plutonium in perchloric acid.

28. Equilibria involving the plutonium aqueous ion species. Plutonium chemistry takes on an additional complexity from the fact that the oxidation potentials connecting the (III), (IV), (V), and (VI) states are so close in value to each other. This means that appreciable concentrations of several oxidation states can exist at equilibrium in the same solution and that pure solutions of intermediate oxidation states undergo disproportionation (self oxidation and reduction reactions).

Disproportionation of plutonium (IV) can be written

$$3\,Pu^{+4} + 2\,H_2O \rightleftharpoons PuO_2^{+2} + 2\,Pu^{+3} + 4\,H^+. \tag{28.1}$$

The reverse reaction is called reproportionation. The equilibrium quotient, K_1, for disproportionation is:

$$K_1 = \frac{[PuO_2^{+2}]\,[Pu^{+3}]^2\,[H^+]^4}{[Pu^{+4}]^3}. \tag{28.2}$$

From the potentials of Fig. 12, K_1 for perchloric acid solutions at room temperature is calculated to be 0.0086. A pure solution of plutonium (IV) would disproportionate in 0.5 molar acid to give a solution consisting of 50.6% plutonium (IV), 33% plutonium (III), and 16.5% plutonium (VI). In 1.0 molar acid the final concentrations would be 72% plutonium (IV), 18.6% plutonium (III), and 9.3% plutonium (VI). The value of K_1 is much greater at elevated temperatures e.g. ~ 3.3 in 0.95 molar hydrochloric acid at 70° C. Disproportionation is greatly suppressed in solutions containing complexing agents because of the greater stabilization of plutonium (IV). The rate of disproportionation is slow at room temperature. For example, 1.93×10^{-3} molar plutonium (IV) in 0.5 molar hydrochloric acid at 25° C does not come to equilibrium with plutonium (III) and plutonium (VI) for about 100 hours. Equilibrium is much more rapid in more concentrated solutions and at elevated temperatures.

Another equilibrium of considerable importance is represented by the equation

$$PuO_2^+ + Pu^{+4} \rightleftharpoons PuO_2^{+2} + Pu^{+3}. \tag{28.3}$$

This equilibrium is unimportant in strongly acidic media because PuO_2^+ is relatively unstable with respect to the other oxidation states and can exist to the extent of less than 1 or 2% at equilibrium. In solutions of lower acidity e.g. < 0.2 molar it is necessary to consider all four plutonium ion species.

The oxidation potentials can be used to calculate the value of the equilibrium quotient:

$$K_2 = \frac{[Pu^{+3}][PuO_2^{+2}]}{[Pu^{+4}][PuO_2^+]} = 9.2. \tag{28.4}$$

Values of this constant have been determined in a number of solutions. This equilibrium shows no acid dependence and since the reaction involves only an electron transfer, equilibrium is rapidly attained. Since the disproportionation reaction previously discussed goes at a slow rate the very striking situation is possible in which four oxidation states of the element are in balance with respect to each other at the same time that three of the oxidation states are out of balance with respect to one another.

Another reaction between plutonium ions which can be written is

$$3\,PuO_2^+ + 4\,H^+ \rightarrow 2\,PuO_2^{+2} + Pu^{+3} + 2\,H_2O. \tag{28.5}$$

This is known as the disproportionation of plutonium (V). In the p_H region just below p_H 1.4 where the concentration of Pu^{+4} is greatly reduced this equilibrium is the principal equilibrium involving the plutonium ions. When the acidity is reduced below p_H 1.4 plutonium (IV) becomes stabilized by hydrolysis and plutonium (V) rapidly becomes less important. In strongly acidic solutions the equilibrium concentration of PuO_2^+ is quite small because of the fourth power acid dependence. Nevertheless reaction (28.5) is still of importance in strongly acid solutions because of the role it plays in the mechanism of many oxidation-reduction reactions.

The equilibrium quotient

$$K_3 = \frac{[PuO_2^{+2}]^2\,[Pu^{+3}]}{[PuO_2^+]^3\,[H^+]^4}$$

is calculated to be 8.9×10^9 in perchloric acid from the potentials of Fig. 12.

The equilibrium quotients for all three reactions discussed in this section are markedly affected by complexing agents such as sulfate, phosphate, oxalate, etc. It is also true that ion concentrations in plutonium solutions may not stabilize at the values predicted by the equilibrium quotients given in this section because of the reducing effect of the alpha radiation of the plutonium. The alpha radiation, particularly in moderately concentrated solutions, causes a slow lowering of the overall oxidation state of the plutonium in solution.

29. Absorption spectra of plutonium ions. Fig. 13 shows the absorption spectra of the aqueous ions of plutonium. The curves for plutonium (III), (IV), and (V) represent measurements taken in 0.5 *M* hydrochloric acid, that for plutonium (VI) in 0.5 *M* nitric acid, and that for colloidal plutonium (IV) in 0.15 *M* hydrochloric acid. Essentially identical spectra are observed in perchloric acid solutions or other media where complexing is slight. The sharp absorption bands are attributed to electronic transitions within the $5f$ shell, preserving the configuration $5f^n$ in upper and lower states as was discussed previously. The distinct peaks characteristic of the individual oxidation states are widely employed in the analysis of plutonium solutions to determine the amount of each oxidation state present.

Table 13. *Wave lengths of absorption bands characteristic of plutonium oxidation states (millimicrons).*

Pu^{+3}	5620	6025	6660	7500	8500	9500
Pu^{+4}	4700	6650	7000	7300	8150	
PuO_2^+	5690	7750				
PuO_2^{+2}	8330	9530	9830			

The principle absorption peaks are listed in Table 13.

30. Hydrolysis of plutonium ions. Plutonium (III) does not undergo appreciable hydrolysis below p_H 7. In strongly alkaline solution it is completely precipitated as $Pu(OH)_3$, the solubility product being approximately 2×10^{-20}. In moist air $Pu(OH)_3$ is oxidized to $Pu(OH)_4$. The formal oxidation potential is +0.95 volts for the reaction

$$Pu(OH)_3 + OH^-_{aq} \rightarrow Pu(OH)_4 + e^- . \qquad (30.1)$$

The green hydroxide of tetrapositive plutonium is noted for its extreme insolubility. The solubility product $K_{SP} = [Pu^{+4}]\,[OH^-]^4$ is estimated to be 7×10^{-56}. The hydroxide begins to form at 0.1 molar hydrogen ion concentration for 0.001 M plutonium (IV).

One rather unusual property of tetrapositive plutonium is a tendency to polymerize into a very stable green-colored colloid in solutions of low acidity ($p_H > 1$). The polymeric or colloidal plutonium (IV) is in the form of cationic aggregates of the hydroxide or hydrous oxide consisting of particles of high molecular weight, probably of the order of tens of millions, with a small overall positive charge. In the p_H range 1 to 3 the polymer is suspended as a dispersed colloid. At higher p_H, particularly above p_H 5, it coagulates readily to a green precipitate. Dissolution of the polymer in acid solution may be quite slow.

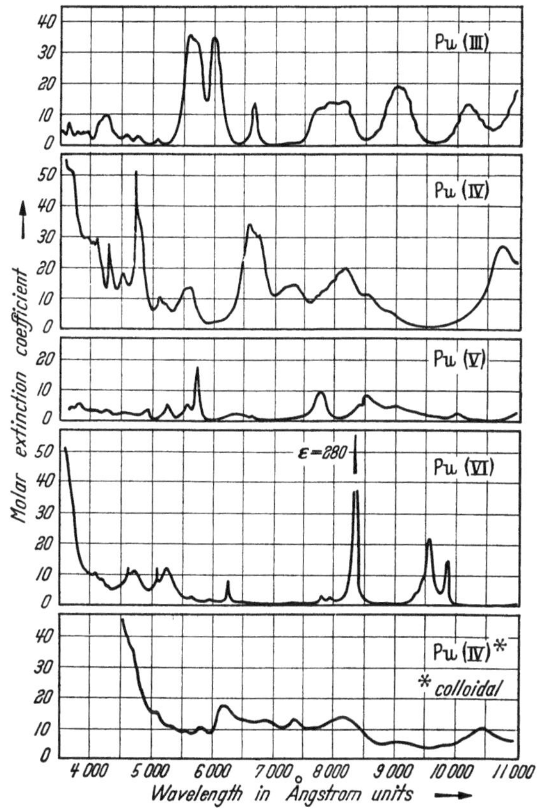

Fig. 13. Summary of absorption spectra of aqueous ions of plutonium in its various oxidation states. Reproduced from R. E. Connick *et al.*: N. N. E. S., Vol. 14 B, "The Transuranium Elements", paper 4.20, p. 588.

There is little evidence for the hydrolysis of PuO_2^+ in basic solutions below p_H 9. The hydroxide $PuO_2(OH)$ is quite soluble.

A solid hydroxide of plutonium (VI) has not been identified but plutonyl ion binds two hydroxide groups near p_H 5.5. $PuO_2(OH)_2$ is the main species in neutral solution. A third hydroxide is added near p_H 9 or 10 to form a plutonate ion which may undergo polymerization. In the presence of such cations as magnesium or calcium at high p_H moderately-insoluble plutonate salts precipitate from solution.

The hydrolytic behavior of the plutonium ions is not known quantitatively so that no reliable oxidation potential scheme can be quoted for alkaline solutions. Connick has proposed the scheme shown in Fig. 14 for 1 molar hydroxide solutions.

$$\underset{\text{solid}}{Pu(OH)_3 \cdot xH_2O} \xrightarrow{+0.95} \underset{\text{solid}}{Pu(OH)_4 \cdot yH_2O} \xrightarrow{-0.76} PuO_2(OH)_{aq} \xrightarrow{-0.26} PuO_2(OH)^-_{3\,aq}$$

$$Pu(OH)_4 \cdot yH_2O \xrightarrow{\sim -0.4} PuO_2(OH)^-_{3\,aq}$$

Fig. 14. Formal potential scheme for plutonium in 1 molar hydroxide[1].

31. Stability of plutonium (V).

It has been noted in previous discussion that neptunium (V) is stable in acid solutions of moderate strength, while plutonium (V)

[1] R. E. Connick: Chap. 8, N.N.E.S. Vol. 14A, "The Actinide Elements", 1952.

because of the disproportionation reaction (28.5) is of little importance above 0.5 molar acidity. This is an important difference in the chemistry of the two elements. The optimum acidity range for the existence of plutonium (V) is p_H 2 to 7. GEVANTMAN and KRAUS[1] noted that a pure 0.0007 molar solution of PuO_2^+ was less than 10% disproportionated in 350 hours at p_H 3.52. Increase of temperature and of plutonium concentration increase the extent of disproportionation. It is doubtful whether PuO_2^+ is thermodynamically stable at any p_H but the uncertainties in the hydrolysis of plutonium (IV) and the free energy of formation of plutonium (IV) polymer make it hard to settle this point with certainty.

32. Oxidation reduction behavior of plutonium. In tracer solutions the mechanism of oxidation or reduction reactions must involve the direct action of the oxidizing or reducing agent on the plutonium ions. Such reactions may be slow when conversion of oxygenated to unoxygenated ions is involved even though the oxidation potential of the solution is favorable to the reaction. In moderately concentrated solutions of plutonium the mechanism of reaction may involve the disproportionation of plutonium (IV) or of plutonium (V). When disproportionation is the rate-determining step, oxidation-reduction reactions can be speeded up by adjustment of conditions to favor disproportionation.

In this section the behavior of plutonium when treated with common reducing and oxidizing agents will be summarized. Most statements refer to behavior at room temperature.

Oxidation of plutonium (III) to plutonium (IV) occurs rapidly with bismuthate ion, bromate ion, iron (III), cerium (IV), dichromate ion, iodate ion, permanganate ion, etc. Oxidation by dilute nitric acid is slow at room temperature; the reaction mechanism involves nitrous acid or the oxides of nitrogen so that the addition of reagents which rapidly destroy nitrous acid will stabilize plutonium (III) in nitric acid. Prevention of the oxidation of plutonium beyond plutonium (IV) is aided by the addition of complexing agents such as sulfate or concentrated nitric acid.

Rapid or relatively rapid reduction of plutonium (IV) to plutonium (III) occurs with ferrous ion, hydroquinone, iodide ion, nitrous acid, hydroxylamine, hydrazine, sulfurous acid, and zinc metal.

Plutonium (IV) is oxidized slowly to higher oxidation states by bromate ion, dichromate ion, permanganate ion, and nitrate ion. Oxidation by bismuthate ion, ceric ion, argentic ion, or by peroxydisulfate ion in the presence of silver ion is rapid.

Reduction of plutonium (VI) to plutonium (V) by iodide ion is very rapid but at room temperature further reduction is slow. Reduction of plutonium (VI) at p_H 1.2 by iodide ion is a good method for the preparation of nearly pure plutonium (V) solutions. Ferrous ion reduces plutonium (VI) rapidly to plutonium (IV) or (III). Ferrous ion may be used in combination with other reducing agents to catalyze reduction below the pentapositive state. Hydrogen peroxide reduces plutonium (VI) to plutonium (V) at a rather rapid rate. Further reduction appears to proceed by the disproportionation mechanism. Hydrogen gas in the presence of platinized platinium will reduce plutonium (VI) to plutonium (V) and ultimately to lower oxidation states. Hydroxylamine, hydrazine, sulfurous

[1] L. H. GEVANTMAN and K. A. KRAUS: Paper 4.16, N.N.E.S., Vol. 14B, "The Transuranium Elements", 1949.

acid, and other reducing agents reduce plutonium (VI) at a moderate but measurable rate to plutonium (V) at room temperature. Further reduction occurs slowly via the disproportionation of plutonium (V).

33. Compounds of plutonium[1]. *Plutonium metal* has astonishing properties[2-5]. It exists in six allotropic forms below its melting point (639° C) at atmospheric pressure, a behavior unique to this metal. Phase-transformation data are given in Table 14. The phase changes were determined by thermal analysis,

Table 14. *Phase data on plutonium*[2].

A. Phase transformation data. (Temperature °C[a]).

Transformation	Dilatometer method	Thermal analysis	Change, ΔV^b
α–β	122 ± 2	122	8.9%
β–γ	206 ± 3	203	2.4
γ–δ	319 ± 5	317	6.7
δ–δ'	451 ± 4	453	−0.4
δ'–ε	476 ± 5	477	−3.0
ε-liquid		639.5 ± 2	

a - on heating only. b - $\Delta V = (V_2 - V_1)/V_1$.

B. Density data.

Phase	Density
α	19.737 (25° C)
β	17.65 (150° C)
γ	17.19 (210° C)
δ	15.92 (320° C)
δ'	15.99 (465° C)
ε	16.48 (500° C)
liquid	16.50 (665° C)

C. Crystal structure data.

Phase	Density
Gamma: face-centered orthorhombic at 210° C a_0=3.1603 A, b_0=5.7624, c_0=10,141	17.19
Delta: face-centered cubic at 320° C a_0=4.6370	15.92
Delta-prime: face-centered tetragonal at 465° C a_0=4.701, c_0=4.489, c/a=0.955	15.99
Epsilon: body-centered cubic at 500° C a_0=3.638	16.48

by dilatometric methods, or by following changes in electrical resistivity or other properties. The transition temperatures are difficult to measure because of the sluggishness of the phase transformation and because of severe hysteresis effects in the heating and cooling curves.

It is possible to observe much sharper transitions if metal of very high purity is studied by special dilatometric techniques. The results of such studies are given in Table 14.

The density of α-plutonium is quite high, 19.737, but this density is seldom achieved because of the difficulty of converting a sample completely to the α-phase. Practical densities fall somewhat below 19.6. The densities of the other phases are markedly lower. The crystal structure of four of the allotropic modifications have been determined (see Table 14C). The electrical resistivity of α-plutonium has the value 145 microohm centimeters, higher than any other "metallic" element at room temperature.

[1] A detailed review of the compounds of plutonium is given by B. B. Cunningham, Chap. 10, N.N.E.S., Vol. 14A, "The Actinide Elements", 1954.

[2] E. R. Jette: J. Chem. Phys. **23**, 365 (1955).

[3] C. S. Smith: Phys. Rev. **94**, 1068 (1954). — Metal Progr. **65**, 81–89 (1954).

[4] W. B. H. Lord: Nature, Lond. **173**, 534 (1954).

[5] J. A. L. Robertson *et al.*: Nature,Lond. **173**. 535 (1954).

Plutonium forms binary intermetallic compounds with aluminum, beryllium, cobalt, iron, manganese, nickel, and silver. These compounds include $PuAl_2$, $PuAl_3$, $PuAl_4$, $PuBe_{13}$, $PuCo_2$, $PuFe_2$, $PuMn_2$, $PuNi_2$, $PuNi_5$, Pu_2Ni_{17}. Some crystallographic data on these compounds has been published by RUNNALLS[1].

The principal plutonium halides and oxyhalides are summarized in Table 15.

Table 15. *Halides and oxyhalides of plutonium*[2,3].

Compound	Color	Crystal structure	Density g/cm³	M.P. (°C)	ΔH formation (kcal per mole)[4]
PuF_6	deep brown gas red or reddish brown solid	orthorhombic	—	50.75	—
PuF_4	light brown	monoclinic	7.0	—	−424
PuF_3	violet	hexagonal tysonite type	9.32	1425	−375
$PuCl_3$	emerald green	hexagonal, UCl_3 type	5.70	760	−230
$PuBr_3$	green	orthorhombic	6.69	681	−198.8
PuI_3	bright green	orthorhombic	6.92	770 est.	−155
PuOF	metallic	tetragonal	9.76	above 1635	—
PuOCl	blue green	tetragonal PbFCl type	8.81	—	−223.6
PuOBr	deep green	tetragonal PbFCl type	9.07	—	−210
PuOI	green	tetragonal PbFCl type	8.46	—	−190

The very insoluble *plutonium trifluoride* is precipitated from acid solutions containing hydrofluoric acid. The crystalline anhydrous compound is readily prepared by treating PuO_2, PuF_4, $PuO_2 \cdot xH_2O$, $Pu(NO_3)_4 \cdot xH_2O$, or $Pu(C_2O_4)_2 \times xH_2O$ with anhydrous hydrogen fluoride-hydrogen mixtures at 500 to 600° C.

Plutonium tetrafluoride is very insoluble and is precipitated in hydrated form with hydrofluoric acid. Anhydrous crystalline PuF_4 is prepared by treatment of plutonium dioxide or oxalate with hydrogen fluoride gas plus oxygen at 550 to 600° C.

Plutonium hexafluoride, the volatile plutonium analogue of uranium hexafluoride, can be made by the reaction of pure gaseous fluorine on plutonium tetrafluoride at 700° C. Successful preparation of the compound depends on the instant condensation of the compound on a cooled surface located close to the reaction zone[5-8]. The melting point of the hexafluoride is 50.75° C; it is by far the most volatile compound of plutonium. It is thermodynamically unstable with respect to dissociation into fluorine and plutonium tetrafluoride and is a

[1] O. J. C. RUNNALLS: Canad. J. Chem. **34**, 133 (1956).

[2] Crystal data from W. H. ZACHARIASEN: Paper 20.6, N.N.E.S., Vol. 14B, "The Transuranium Elements", 1949.

[3] Lattice constants are given in Table 6, Sect. 6.

[4] Heat of formation data from B. B. CUNNINGHAM: Chap. 10, N.N.E.S., Vol. 14A, "The Actinide Elements", 1952.

[5] B. WEINSTOCK and J. G. MALM: Paper P/733, Proceedings of the Geneva Conference. Peaceful Uses of Atomic Energy, August 1955, United Nations. J. Inorg. a. Nucl. Chem. **2**, 380 (1956).

[6] A. E. FLORIN, I. R. TANNENBAUM and J. F. LEMONS: J. Inorg. a. Nucl. Chem. **2**, 368 (1956).

[7] C. J. MANDLEBERG, H. K. RAE, R. HURST, G. LONG, P. DAVIES and K. E. FRANCIS: J. Inorg. a. Nucl. Chem. **2**, 358 (1956).

[8] N. J. HAWKINS, H. C. MATTRAW and W. W. SABOL: Knolls Atomic Power Laboratory Report KAPL—1007, May 1954.

powerful fluorinating agent. The rate of decomposition when stored as a vapor is, however, only about 0.1% per day. When it is stored principally as a solid, the radiation decomposition induced by the intense alpha activity destroys about 1.5% of the hexafluoride per day. Plutonium hexafluoride is isostructural with uranium and neptunium hexafluoride.

A number of *plutonium double fluorides* such as $NaPuF_4$, $NaPuF_5$, $KPuF_5$, NH_4PuF_5, KPu_2F_9, and $CsPu_2F_2$ have been precipitated from aqueous solution and analyzed crystallographically. These double salts are more insoluble than the simple plutonium fluorides.

The compound $PuO_2F_2 \cdot xH_2O$ is very soluble although double salts such as $RbPuO_2F_3$ are somewhat less soluble.

Anhydrous *plutonium trichloride* may be prepared by the action of various strong chlorinating agents on plutonium dioxide at elevated temperatures or by the direct burning of plutonium metal in chlorine at 450° C. The trichloride sublimes at 800° C. The anhydrous compound is hygroscopic. A well defined monohydrate, trihydrate, and hexahydrate of plutonium trichloride are known.

Anhydrous *plutonium tetrachloride* is not stable and has not been prepared, a distinction from neptunium tetrachloride or uranium tetrachloride.

Anhydrous crystalline *plutonium tribromide* may be prepared by the action of anhydrous hydrogen bromide gas on plutonium dioxide or oxalate at elevated temperature or by direct burning of the metal in bromine at 300 to 325° C. At 750 to 800° C the tribromide is appreciably volatile and may be purified by sublimation in a vacuum. The tribromide is very soluble in water. *Plutonium tetrabromide* is apparently not stable at any temperature.

Anhydrous *plutonium triiodide* may be prepared by the action of anhydrous hydrogen iodide gas on plutonium metal at 450° C. Only the oxyiodide is produced by the reaction of hydrogen iodide with the dioxide.

Plutonium oxyfluoride (PuOF), *plutonium oxychloride* (PuOCl), and *plutonium oxyiodide* (PuOI) are well characterized anhydrous compounds.

The principal oxide of plutonium is *plutonium dioxide*, PuO_2. It is obtained by ignition of the "hydroxide", peroxide, oxalate, sulfate or nitrate of plutonium (IV), the oxalate of plutonium (III), or the nitrate of plutonium (VI) in air or oxygen within the temperature range 500 to 1000° C. Plutonium dioxide crystallizes in a well defined face-centered cubic structure. It is a suitable compound for weighing in gravimetric analysis. The calculated density is 11.44. Oxidation to a higher oxide has not been achieved. No oxide similar to the well known uranium oxides, U_3O_8 or UO_3, have been prepared. Plutonium dioxide which has been strongly ignited is extremely difficult to dissolve.

Partially reduced oxides in the composition range Pu_2O_3-Pu_4O_7 have been made by ignition of PuO_2 in vacuo at very high temperatures.

Plutonium monoxide, PuO, has been obtained by the reduction of PuOCl with barium vapor at 1250° C. It may appear on the surface of metal exposed to air. It is semi-metallic in character.

Plutonium (IV) peroxide is formed when hydrogen peroxide is added to acid solutions of plutonium in any oxidation state, particularly in the presence of sulfate ion. The compound is very insoluble and is a convenient choice for the direct precipitation of plutonium. This precipitation effects a high degree of purification from most light elements, cationic impurities and from most heavy elements except zirconium, thorium, and uranium. It is not suitable for gravimetric determination. The peroxide is readily dissolved in strong nitric acid and is catalytically decomposed by ferric ion.

The *hydroxides of plutonium* were discussed in the previous section on hydrolysis.

Other compounds of plutonium (III) or plutonium (IV) with very low to small solubility in acid solution include *plutonium (III) phosphate, plutonium (IV) dimonohydrogen phosphate, plutonium (IV) diphenylarsonate, diplutonium (III) trioxalate,* and *plutonium (IV) dioxalate.*

One of the most important compounds of hexapositive plutonium is *sodium plutonyl acetate,* $NaPuO_2(C_2H_3O_2)_3$. This pink compound is isostructural with the analogous compounds of uranium, neptunium, and americium.

Carbides, nitrides, silicides, and *sulfides.* Plutonium forms several binary compounds which are of interest because of their refractory character. The properties of these compounds are given in Table 16.

Table 16. *Some binary compounds of plutonium.*

Compound	Color and appearance	Crystal structure[1, 2, 3]	Density[4] (g/cm³)	Approx. temp. limit for stability in high vacuum (°C)
PuC . . .	black, semimetallic	cubic, rock salt type	13.99	2000
PuN . . .	brown	cubic, rock salt type	14.22	1500
α-$PuSi_2$ . .	silvery, metallic	tetragonal, $ThSi_2$ type	9.12	1000
β-$PuSi_2$[5]. .	—	hexagonal β-USi_2 type	9.18	—
$PuS_{1.33-1.5}$	black	cubic, Ce_3S_4 type	8.14	1200

The monocarbide is formed by reacting the dioxide in intimate mixture with carbon at about 1600° C, and the mononitride may be obtained by heating the trichloride in a stream of anhydrous ammonia at 900° C; it is prepared more easily, however, by reacting finely divided metal with ammonia at 650° C. Although the lower temperatures are favorable to the production of higher nitrides, none are obtained, in contrast to the uranium-nitrogen system in which compositions up to $UN_{1.75}$ are easily realized.

The disilicide is formed when a slight stoichiometric excess of calcium disilicide is heated with plutonium dioxide in vacuum at about 1550° C. The disilicide is only moderately stable in air and burns slowly to the dioxide when heated to about 700° C.

Plutonium "sesquisulfide" may be prepared by prolonged treatment of the dioxide in a graphite crucible with anhydrous hydrogen sulfide at 1340 to 1400° C, or by reaction of the trichloride with hydrogen sulfide at 900° C.

The most important compounds of plutonium have been mentioned in this section. Additional information on a large number of compounds has been summarized by Cunningham[6].

34. Coprecipitation behavior of plutonium. Concentrations of plutonium too small to exceed the solubility product of any plutonium compound can nevertheless be removed from aqueous solution by coprecipitation with certain carrier compounds. In general, these carrier compounds are related to the plutonium

[1] W. H. Zachariasen: Paper 20.2, N.N.E.S., Vol. 14B, "The Transuranium Elements", 1949.
[2] W. H. Zachariasen: Paper 20.3. Same volume.
[3] W. H. Zachariasen: Paper 20.4. Same volume.
[4] Calculated from x-ray data.
[5] O. J. C. Runnalls and R. R. Boucher: Acta crystallogr. **8**, 592 (1955).
[6] B. B. Cunningham: Chap. 10, N.N.E.S., Vol. 14A, "The Actinide Elements", 1954.

compounds of low solubility. Table 17 gives an indication of the coprecipitation behavior of the three most important oxidation states of plutonium. The data on this topic are very extensive and can only be touched upon briefly in this article. Certain of the more important purification or analytical procedures based upon coprecipitation will be discussed briefly.

The differing behavior of the oxidation states makes possible a very high degree of separation from unwanted impurities by successive precipitation of the carrier compound under different oxidation conditions. Many impurities which coprecipitate with the plutonium can be removed by a "by-product" precipitation of the carrier compounds after adjustment of the oxidation state of the plutonium to a state which does not coprecipitate.

Table 17. *Coprecipitation behavior of trace amounts of plutonium.*

Carrier compound	Pu (III)	Pu (IV)	Pu (VI)
Insoluble hydroxides	*C*	*C*	*C*
Lanthanum fluoride	*C*	*C*	*NC*
Phosphates:			
Zirconium phosphate	*NC*	*C*	*NC*
Bismuth phosphate	*C*	*C*	*NC*
Thorium pyrophosphate	*NC*	*C*	*NC*
Thorium hypophosphate		*C*	*NC*
Oxalates:			
Thorium oxalate	*C*	*C*	*NC*
Uranium (IV) oxalate	*C*	*C*	*NC*
Bismuth oxalate	*C*	*C*	*NC*
Lanthanum oxalate	*C*	*C*	*NC*
Iodates:			
Zirconium iodate		*C*	*NC*
Thorium iodate		*C*	*NC*
Ceric iodate		*C*	*NC*
Sodium uranyl acetate	*NC*	*NC*	*C*
Zirconium phenylarsonate	*NC*	*C*	*NC*
Thorium peroxide		*C*	
Bismuth arsonate		*C*	*NC*

The letter *C* indicates that coprecipitation is nearly quantitative under suitable conditions. The letters *NC* mean that coprecipitation may easily be made less than 1 to 2% under proper conditions.

The *lanthanum fluoride* cycle was used in the very earliest studies of plutonium and is still extensively used in laboratory scale studies.

Neutron-irradiated uranium is dissolved in nitric acid, and lanthanum fluoride is precipitated after reduction of the plutonium with sulfur dioxide. The precipitate "carries" plutonium and some of the fission products, particularly the rare earths. The mechanism of carrying of plutonium (IV) on LaF_3 is the formation of anomalous mixed crystals[1]. The lanthanum is put back into solution and reprecipitated after oxidation of the plutonium to the non-coprecipitated hexapositive state. The rare-earth impurities are removed with this "by-product" lanthanum fluoride precipitate. Neptunium and plutonium can be separated by taking advantage of differences in the rates of oxidation of the two elements to the hexavalent state, neptunium being more quickly oxidized than plutonium. Bromate in at rooom temperature has frequently been used for this selective oxidation of neptunium.

In the *zirconium phosphate cycle* the oxidation state of plutonium is adjusted to plutonium (IV), for example, by treatment with hot concentrated nitric acid and zirconium phosphate is precipitated to carry the plutonium. The plutonium is put back into solution and oxidized to the hexapositive form. Zirconium phosphate is again precipitated to remove phosphate insoluble impurities while plutonium remains in solution.

Neptunium and plutonium can be separated with zirconium phosphate by treatment of the solution with a hydrazine—ferrous ion mixture before precipitation. Plutonium is reduced to the non-carryable (III) state while neptunium is reduced only to the (IV) state which carries quantitatively.

[1] K. Schlyter and L. G. Sillén: Acta chem. scand, Copenh. **4**, 1323 (1950).

Zirconium phenylarsonate is very useful in radiochemical analyses because of its specificity for the removal of tetravalent cations from solution. It is superior to zirconium phosphate in this respect. It can be used to measure the percentage of plutonium present in solution as plutonium (IV) in a mixture of ion species, or it can be used to carry out the separation of neptunium (IV) from plutonium (III), or zirconium (IV) from plutonium (VI) or for similar separations.

These coprecipitation separations are discussed more completely elsewhere[1].

35. The bismuth phosphate process. One of the early industrial processes suggested for the isolation of plutonium is based on bismuth phosphate and lanthanum fluoride as carrier precipitation agents. This process was originally conceived by S. G. THOMPSON, G. T. SEABORG, and their collaborators at the Metallurgical Laboratory of the University of Chicago during World War II[2]. Neutron-irradiated uranium is dissolved in nitric acid and treated with sodium nitrite to fix the plutonium oxidation state at plutonium (IV). Sulfuric acid and nitric acid concentrations are adjusted to prevent precipitation of uranyl phosphate after which bismuth phosphate is precipitated. Plutonium (IV) is coprecipitated quantitatively in spite of the fact that the amount of bismuth left in solution is rather high. Plutonium (III) is carried much less completely than plutonium (IV) in spite of the greater structural similarity of plutonium (III phosphate to bismuth phosphate.

The bismuth phosphate is dissolved in concentrated nitric acid. (The ease of separation of bismuth phosphate precipitates and their relative ease of dissolution were important factors in the choice of bismuth phosphate.) Sodium bismuthate is used to oxidize plutonium (IV) to plutonium (VI). A by-product bismuth phosphate precipitate is then formed to remove phosphate insoluble fission-product impurities leaving oxidized plutonium in solution. Plutonium in the solution is reduced again to plutonium (IV) and coprecipitated with bismuth phosphate. This decontamination cycle is repeated. At this point the carrier is changed to lanthanum fluoride. A lanthanum-fluoride oxidation reduction cycle is gone through for further decontamination but principally for concentration. The final product-bearing lanthanum fluoride precipitate is metathesized to lanthanum hydroxide and dissolved in acid. The plutonium at this point is sufficiently concentrated that final purification can be done without the use of carrier compounds i.e. by direct precipitation of plutonium compounds.

36. Solvent extraction of plutonium. Plutonium resembles the other actinide elements in that a number of organic solvents immiscible with water are capable of extracting it from acidic aqueous solutions. Such extractions are very useful in the separation of plutonium from most of the elements in the periodic system. The differing extraction behavior of the different oxidation states can be a powerful aid in these separations. A list of some of the more carefully investigated solvents given in Table 18 indicates the general type of solvents which are most effective.

Ethyl ether is chiefly used for the extraction of plutonium (VI) from nitric acid solutions containing a high concentration of nitrate salts such as ammonium nitrate, calcium nitrate, magnesium nitrate or aluminum nitrate. The extrac-

---

[1] E. K. HYDE: Chap. 15, N.N.E.S., Vol. 14A, "The Actinide Elements", 1954; E. K. HYDE: Paper P/728, "Proceedings of the Geneva Conference-Peaceful Uses of Atomic Energy", August 1955, United Nations. Volume 7.

[2] S. G. THOMPSON and G. T. SEABORG: "The First Use of Bismuth Phosphate for Separating Plutonium from Uranium and Fission Products", Progress in Nuclear Energy, Ser. 3, Vol. 3, "Process Chemistry", edited by F. R. BRUCE, J. M. FLETCHER, H. H. HYMAN and J. J. KATZ, Pergamon Press Ltd., 1956.

tion is highly specific for hexapositive actinide element ions, except for a very few elements such as the halogens, and is widely used in laboratory research with trace amounts of plutonium.

The other solvents listed in Table 18 exhibit much higher distribution coefficients for plutonium in the tetrapositive and hexapositive oxidation states. For this and a variety of other reasons they are more suitable for large scale or industrial scale use. Tributyl phosphate (TBP) extraction will be discussed to

Table 18. *Organic solvents capable of extracting plutonium from aqueous solution.*

Ethyl-ether	diethyl-cellosolve
Tributyl-phosphate (TBP)	dibutyl-carbitol
Diisopropyl-ketone	dibutoxytetraethylene-glycol ("pentaether")
Methyl-isobutyl-ketone ("hexone")	
Methyl-n-amyl-ketone	triglycol-dichloride

show the effect of a number of variables important in determining the extent of plutonium extraction. TBP is usually diluted to a 15 to 40% solution in inert solvents as kerosene, butyl ether, or carbon tetrachloride to improve the physical characteristics of the solvent. Data quoted here will be for 20% solutions.

The extraction of the actinide element ions from nitrate aqueous systems into TBP increases in the order:

Plutonium ions	Pu(III) ≪ Pu(VI) < Pu(IV)
Tetrapositive ions	Th(IV) < Np(IV) < Pu(IV) High extractability
Hexapositive ions	Pu(VI) < Np(VI) < U(VI) High extractability
Pentapositive ions	Np(V) Limited data indicates very low extractability
Tripositive ions	Pu(III) < Am(III) < Cm(III) < etc. Low extractability.

Some sample data are given in Table 19.

Increasing the nitric acid concentration increases the distribution coefficient, but it is much more effective to increase the concentration of sodium nitrate or other neutral nitrate salts because of the high extractability of the complex $HNO_3 \cdot TBP$ which competes with the plutonium-solvent complex. Furthermore, extraction of impurity ions is much greater in highly acidic systems. Hence extraction may be carried out with concentrated salt systems with just enough nitric acid present to be certain that no hydrolysis of plutonium occurs or else from rather dilute nitric acid solutions without other salting agents. Complexing agents for plutonium such as fluoride, oxalate, sulfate, or phosphate ion will interfere with the extraction.

Table 19. *Distribution ratio,* $E_{\text{aq.}}^{\text{org.}}$, *for plutonium extraction into tributyl phosphate (20% solution in kerosene).*

Aqueous composition	Pu (III)[1]	Pu (IV)[2]	Pu (VI)[1]
1 *M* HNO_3	0.01	1.5	0.7
3 *M* HNO_3	—	6.8	2.26
6 *M* HNO_3	0.01	19.6	2.86
10 *M* HNO_3	—	13	2.27
1 *M* HNO_3, 3 *M* $NaNO_3$	—	19.0	—
1 *M* HNO_3, 5 *M* $NaNO_3$	—	62	—

The fission product elements most difficult to separate by solvent extraction are cerium, zirconium, niobium, iodine, and ruthenium.

Plutonium may be separated conveniently from neighboring elements by solvent extraction. Uranium can be extracted away from plutonium by reduction

[1] J. M. Fletcher: Paper P/413, Proceedings of the Geneva Conference-Peaceful Uses of Atomic Energy, August 1955, United Nations. Volume 7.

[2] H. A. C. McKay: Paper P/441, Proceedings of the Geneva Conference-Peaceful Uses of Atomic Energy, August 1955, United Nations. Volume 7.

of the latter to unextractable plutonium (III); alternatively an organic solvent solution of the two can be contacted with a nitrate salt solution containing a reducing agent capable of reducing plutonium to plutonium (III). A neptunium-plutonium aqueous mixture is easily separated by treatment with a reducing agent to produce a mixture of neptunium (IV) and plutonium (III). The neptunium is then extracted leaving the plutonium behind. Separation of plutonium from americium (III) or higher elements is simple after plutonium is oxidized to a higher oxidation state.

In industrial practice solvent extraction is performed in packed columns or pulsed columns or in a series of mixing-settling chambers in which the aqueous phase and solvent phase pass in counter current flow in a multi-stage process. Such systems incorporate a back-washing scrub section in which impurities with lower distribution coefficients than the desired product are returned to the aqueous phase.

In the industrial process known as the "Redox" process uranium and plutonium are separated from fission products and from each other by extraction with methyl *n*-butyl ketone (hexone). The uranium nitrate dissolver solution is oxidized and contacted with hexone to extract uranium (VI) and plutonium (VI). The hexone layer is scrubbed with an acid deficient aluminum nitrate solution to remove impurities. The hexone is then contacted with an aluminium nitrate solution containing a reducing agent which does not affect the uranium but reduces the plutonium to plutonium (III) and strips it into the aqueous phase. The plutonium product from this cycle is oxidized to prepare it for reextraction and the entire sequence of extraction as oxidized plutonium followed by reduction and stripping as tripositive plutonium is repeated once or twice more.

In the industrial "Purex" process, TBP diluted with a kerosene type solvent is used. The uranium slugs are dissolved in nitric acid, plutonium is fixed as plutonium (IV) and the acid strength is adjusted so that plutonium (IV) and uranium (VI) are coextracted away from the fission products. Impurities are backwashed with a nitric acid solution flowing countercurrent to the solvent stream. In a second part of the process, the solvent is contacted with a nitric acid solution containing a reducing agent. Uranium (VI) is left in the TBP phase while plutonium (III) is removed. Both uranium and plutonium are put through additional purification before complete purification is achieved.

A detailed discussion of such processes is beyond the scope of this article; more information can be found elsewhere[1,2].

37. Solvent extraction of chelate complexes of plutonium. A very sizeable number of organic compounds are known which form compounds with aqueous ions of plutonium which are insoluble or which can be extracted from aqueous solution into non-polar organic solvents. Many of these such as the alpha-hydroxquinolinate, the nitrobenzoate, or the pyrogallate are useful in analysis. A widely used reagent cupferron (the ammonium salt of nitrosophenylhydroxylamine) forms a compound with plutonium (IV) which is extractable into chloroform. Rydberg[3] has made extensive studies on the acetylacetone complexes of plutonium and other heavy elements.

The beta diketone, alpha-thenoyltrifluoroacetone, usually referred to as TTA, forms a chelate compound with plutonium (IV) which is very stable and highly

[1] F. L. Culler: Paper P/822, Proceedings of the Geneva Conference—Peaceful Uses of Atomic Energy, August 1955, United Nations.

[2] Progress in Nuclear Energy, Ser. 3, Vol. 3, "Progress Chemistry", edited by F. R. Bruce, J. M. Fletcher, H. H. Hyman, and J. J. Katz, Pergamon Press Ltd., 1956.

[3] J. Rydberg: Ark. Kem. 9, No. 8, 109 (1955).

extractable into non-polar solvents even from strongly acidic solutions. The equilibrium quotients for complex formation between TTA and various metal ions cover the enormous range from 10^{10} to 10^{-15}. Since plutonium (IV) is among the most strongly complexed ions, the extraction of the TTA-plutonium complex separates plutonium to a very high degree from most impurities including the tripositive actinide or lanthanide elements. From acidic solution the principal contaminating ions are zirconium (IV), hafnium (IV), iron (III), cerium (IV), tin (IV), and niobium (V). Among the heavy elements protactinium (V), thorium (IV), uranium (IV), and neptunium (IV) are extracted. The acid dependence of the extraction is very high (fourth power for tetrapositive ions) and by adjustment of acidity and TTA concentration very sensitive control of the extraction is possible. Hence even those elements with similar equilibrium quotients can be separated. Complexing agents inhibit the extraction of the plutonium (IV).

Plutonium (IV) is readily separated from uranium (VI) as the extraction of the latter under acidic conditions into non-polar solvents is slight. Plutonium is separated from neptunium by TTA extraction under reducing conditions when neptunium (IV) is extracted into the solvent layer leaving plutonium (III) in the aqueous phase.

An industrial process has been studied for the recovery of plutonium and uranium from reactor fuel elements based on the use of TTA extraction[1].

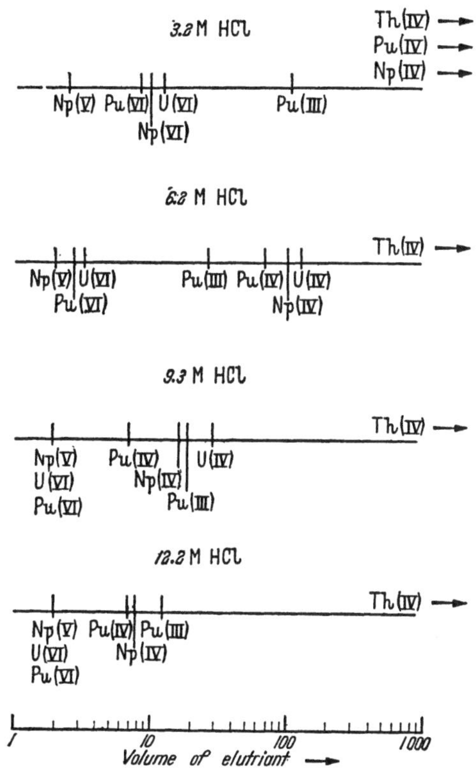

Fig. 15. Summary of elution peak position of thorium, uranium, neptunium, and plutonium in various oxidation states showing the effect of hydrochloric acid concentration on the elution of the elements. Data taken from DIAMOND, STREET, and SEABORG[2].

38. Ion exchange behavior of plutonium. It was mentioned in Sect. 25 under neptunium chemistry that negatively charged complexes of plutonium (VI) and plutonium (IV) are formed in concentrated solutions of hydrochloric acid and that these complexes are readily adsorbed by anion exchange resins. Plutonium (VI) also forms negatively charged nitrate complexes in concentrated solutions of nitric acid which are adsorbed on anion exchange resins. The importance of these properties was discussed in Sect. 25.

Plutonium in all its oxidation states is readily adsorbed on cation exchange resins from dilute acid solution. Highest adsorbability is achieved for the tetrapositive state. This property is of considerable use in concentration of plutonium from large volumes of solution and in the removal of non-cationic impurities. Plutonium may be separated from other cations by selective elution from columns

[1] See footnote 1, p. 252.
[2] R. M. DIAMOND, K. STREET jr. and G. T. SEABORG: J. Amer. Chem. Soc. 76, 1461 (1954).

packed with cation exchange resin. Oxalic acid, citric acid, lactic acid, sulfate ion, ethylenediamine tetraacetic acid, and other complexing agents have been used. Since negatively charged chloride complexes of plutonium are formed in moderately concentrated hydrochloric acid such solutions readily elute plutonium from cation exchange resins. This is shown dramatically in Fig. 15. In this figure data of DIAMOND, STREET, and SEABORG[1] are used. In a series of experiments trace amounts of thorium, uranium, neptunium, and plutonium in various oxidations states were adsorbed on Dowex-50 cation exchange resin and eluted with hydrochloric acid; the elution of the radioactive ions was followed by radiochemical analysis. The relative position of the elution peaks constructed from this series of experiments are presented on Fig. 15 on a logarithmic scale. The separations shown may not be quantitatively reproducible but the schematic representation of the relative elution order is believed to be correct.

The behavior of the tetrapositive ions is particularly striking. Thorium (IV) is a comparatively large highly charged, unoxygenated ion and adheres very strongly to the resin. In fact, the mass action effect of hydrogen ion in any concentration of hydrochloric acid is not sufficient to elute it at a convenient rate. Uranium (IV), neptunium (IV), and plutonium (IV), however, show an increasing tendency toward chloride complex formation and plutonium (IV) in particular can be desorbed rather quickly with concentrated hydrochloric acid.

III. Americium (element 95).

a) History of americium.

39. Discovery of americium—summary of known isotopes. Americium was the fourth transuranium element to be discovered, its first identification at the wartime Metallurgical Laboratory of the University of Chicago by SEABORG, JAMES, MORGAN, and GHIORSO[2,3] taking place late in 1944 and early in 1945. The first isotope to be discovered was Am^{241}, formed by intense neutron irradiation of plutonium in nuclear reactors according to the reactions Pu^{239} (n, γ) Pu^{240} (n, γ) Pu^{241} followed by $Pu^{241} \xrightarrow[13\text{ years}]{\beta^-} Am^{241}$. The isotope Am^{241} decays by the emission of alpha particles with a half life of 470 years. The element was named americium, symbol Am, in honor of the Americas. This name was chosen to point up the strong analogy between element 95 and its homologue, europium, named after Europe, of the lanthanide rare earth series. Americium does not occur in nature.

Americium was first isolated in the form of a pure compound by B. B. CUNNINGHAM[4] at the Metallurgical Laboratory in 1945. Plutonium which is isolated from reactor fuel elements that have been subjected to intensive neutron irradiation contains an appreciable amount of Pu^{241}. Hence plutonium is a convenient source of americium because upon standing Am^{241} grows into the plutonium. Gram quantities of Am^{241} can now be prepared in this way.

A large number of americium isotopes have been prepared by the bombardment of heavy element targets in cyclotrons or by intense neutron irradiation of plutonium or americium samples in reactors. These isotopes and their principal

---

[1] R. M. DIAMOND, K. STREET jr. and G. T. SEABORG: J. Amer. Chem. Soc. **76**, 1461 (1954)

[2] G. T. SEABORG, R. A. JAMES and L. O. MORGAN: Paper 22.1, N.N.E.S., Vol. 14B, "The Transuranium Elements", 1949.

[3] A. GHIORSO, R. A. JAMES, L. O. MORGAN and G. T. SEABORG: Phys. Rev. **78**, 472 (1950).

[4] B. B. CUNNINGHAM: Paper 19.2, N.N.E.S., Vol. 14B, "The Transuranium Elements", 1949.

decay characteristics are listed in Table 20. Decay schemes of the more carefully studied isotopes are given in Fig. 16. The isotope Am^{243} is worth special mention because it is more stable (half life 7930 years) than Am^{241}. This is important because the study of the chemistry of americium using Am^{241} is severely hindered by the remarkable chemical effects caused by the intense alpha radiation of

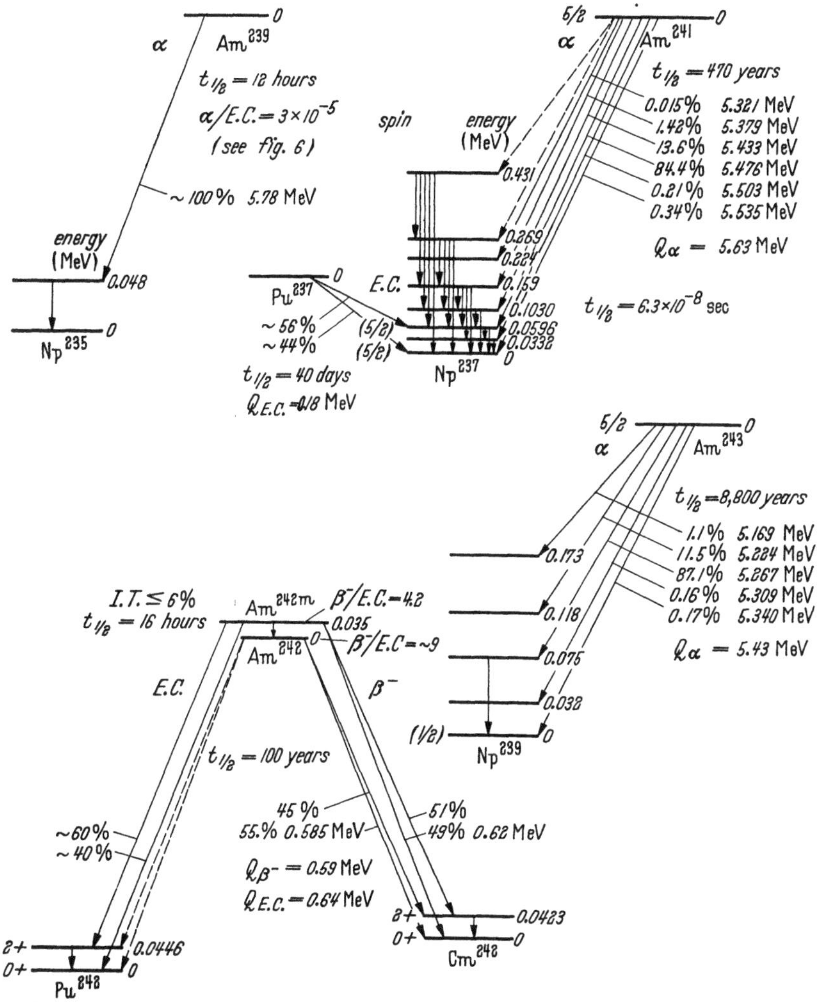

Fig. 16. Decay schemes of americium isotopes.

Am^{241}; the specific activity is 7×10^{13} disintegrations per minute per gram. Most chemical studies up until 1954 were of necessity performed with Am^{241}. Increasing use of Am^{243} is certain as milligram and larger quantities become more generally available. The isotope Am^{243} can be prepared nearly pure isotopically by a sequence of neutron bombardments and chemical separations as follows: The bombardment of Am^{241} with neutrons yields Pu^{242} by the reactions Am^{241} (n, γ) $Am^{242m} \xrightarrow{\text{electron capture}} Pu^{242}$; after chemical separation the Pu^{242} can be transformed to Am^{243} via the reactions $Pu^{242}(n,\gamma)$ $Pu^{243} \xrightarrow[\text{hours}]{\beta^-} Am^{243}$, and the Am^{243} can be chemically separated. An alternate sequence is to prepare Pu^{242} of fairly high

Table 20. *Isotopes of americium**.

Isotope and half-life	Type of decay	Energy (Mev)	Method of production	Disintegration energies (Mev) **
Am^{237} 1.3 hours	E.C. 99^{+}% α 0.005%	 6.01	$Pu^{239}(d, 4n)$ $Pu^{239}(p, 3n)$	$Q_{E.C.} = 1.43$ (calc.) $Q_{\alpha} = 6.11$
Am^{238} 1.86 hours	E.C. γ	 0.98, 0.58, 0.102 (Ref. 1, 2)	$Pu^{239}(d, 3n)$ $Pu^{239}(p, 2n)$	$Q_{E.C.} = 2.25$ (est.) $Q_{\alpha} = 6.02$ (est)
Am^{239} 12 hours	E.C. 99^{+}% α 4.1×10^{-3}% γ (with α) γ (with E.C.)	 5.78 0.048 0.04940, 0.05725, 0.06782, 0.1818, 0.2099, 0.2264, 0.2284, 0.2777, 0.2856 deduced from Np^{239} decay scheme (Ref. 13)	$Np^{237}(\alpha, 2n)$ $Pu^{239}(p, n)$ $Pu^{239}(d, 2n)$ $Pu^{239}(\alpha, p3n)$	$Q_{E.C.} = 0.86$ (calc.) $Q_{\alpha} = 5.93$
Am^{240} 50 hours	E.C. γ	 0.92 (14%), 1.02 (70%), 1.40 (15%) (Ref. 2, 3)	$Np^{237}(\alpha, n)$ $Pu^{239}(d, n)$ $Pu^{239}(\alpha, p2n)$	$Q_{E.C.} = 1.44$ (est.) $Q_{\alpha} = 5.78$ (est.)
Am^{241} 470 years α 461.3 (Ref. 21) 457 (Ref. 22)	α β stable γ	5.535 (0.34%), 5.503 (0.21%) 5.476 (84.2%), 5.433 (13.6%) 5.379 (1.4%), [5.321 (0.015%) 5.241 ? (0.002%) (Ref. 12)] 0.02636, 0.03320^{+}, 0.04346, 0.056, 0.05957^{+}, 0.099, 0.103, 0.113, 0.130, 0.159, 0.165, 0.210, 0.270, 0.328, 0.370, 0.430 (Ref. 4) (See also Ref. 15, 16, 17) $^{+}$ metastable $t_{1/2} = 6.3 \times 10^{-8}$ sec Coulombic excitation of Np^{237} produces levels at 0.332 and 0.076 kev (Ref. 14)	Pu^{239} multiple n capture daugther Pu^{241}	$Q_{\alpha} = 5.63$ $I = \frac{5}{2}$
Am^{242} 16.01 hours	β^{-} 81% E.C. 19% $IT \leq 6$% γ	0.620 0.0422, 0.0445 (Ref. 11)	Pu^{239} multiple n capture $Am^{241}(n, \gamma)$	$Q_{\beta^{-}} = 0.62$ $Q_{E.C.} = 0.67$ $Q_{\alpha} = 5.53$ (calc.)
Am^{242} 100 years	β^{-} 90% E.C. 10% γ	0.585 (Ref. 5) 0.0422, 0.0446 (Ref. 5)	$Am^{241}(n, \gamma)$	$Q_{\beta^{-}} = 0.59$ $Q_{E.C.} = 0.64$ $Q_{\alpha} = 5.50$ (calc.)
Am^{243} 8800 years (Ref. 19) 7740 years (Ref. 20) 7930 years (Ref. 22) 7600 years (Ref. 10)	α β stable γ	5.340 (0.17%), 5.309 (0.16%) 5.267 (87.1%), 5.224 (11.5%) 5.169 (1.1%) (Ref. 6) 0.075	$Am^{242}(n, \gamma)$ Pu^{239} multiple n capture	$Q_{\alpha} = 5.43$ $I = \frac{5}{2}$
Am^{244} 26 min	$\beta^{-} > 94$% E.C. 3.9×10^{-2}% (Ref. 18)	1.5 (Ref. 7)	$Am^{243}(n, \gamma)$	$Q_{\alpha} = 5.34$ (calc.) $Q_{\beta^{-}} = 1.50$
Am^{245} 1.98 hours	β^{-} (Ref. 8) γ	0.905 (Ref. 8) 0.036, 0.120, 0.143, 0.156, 0.232 0.255 (Ref. 8)	Pu^{239} multiple n capture $Pu^{244}(n, \gamma)$ (Ref. 8)	$Q_{\beta^{-}} = 0.90$ $Q_{\alpha} = 5.17$ (est.)
Am^{246} 25 min	β^{-} (Ref. 9) γ	1.2 (Ref. 9) 0.019, 0.103, 0.795, 1,069 (Ref. 9)	Pu^{239} multiple n capture (Ref. 9)	$Q_{\beta} = 2.29$ $Q_{\alpha} = 5.00$ (est.)

* For references to original data not given specifically here, refer to Hollander, Perlman, and Seaborg, "Table of Isotopes" [Rev. Mod. Phys. **25**, 469 (1953)], and to the compilations of the Nuclear Data Group, National Research Council.

** Revised values from Glass, Thompson, and Seaborg: J. Inorg. a. Nucl. Chem. **1**, 3 (1955).

References to table 20.

[1] R. J. CARR: Reported in University of California Radiation Laboratory Report UCRL-2455, Jan. 1954.
[2] R. A. GLASS, R. J. CARR, W. M. GIBSON and J. W. COBBLE, University of California Radiation Laboratory Report UCRL-3599, October 1956.
[3] R. A. GLASS: Ph. D. Thesis, University of California Radiation Laboratory Report UCRL-2560, April 1954.
[4] P. P. DAY: Phys. Rev. **97**, 689 (1955); see also H. JAFFE et al., Phys. Rev. **97**, 142 (1955).
[5] T. O. PASSELL: Ph. D. Thesis, University of California Radiation Laboratory Report UCRL-2528, March 1954.
[6] F. S. STEPHENS *et al.*: Phys. Rev. **98**, 261 A (1955).
[7] A. GHIORSO, *et al.*: Phys. Rev. **94**, 1081 (1954).
[8] C. I. BROWNE, *et al.*: J. Inorg. a. Nucl. Chem. **1**, 254 (1955).
[9] D. ENGELKEMEIR *et al.*: J. Inorg. a. Nucl. Chem. **1**, 345 (1955).
[10] I. PERLMAN *et al.*: University of California Radiation Laboratory Report UCRL-3068, July 1955.
[11] R. W. HOFF *et al.*: Phys. Rev. **100**, 1403 (1955).
[12] L. L. GOLDIN, E. F. TRETYAKOV and G. I. NOVIKOVA: p. 226—249. Proceedings of the Moscow Conference on Atomic Energy, July 1—5, 1955.
[13] J. M. HOLLANDER, W. G. SMITH and J. W. MINELICH: Phys. Rev. **102**, 740 (1956).
[14] J. O. NEWTON: Nature, Lond. **175**, 1028 (1955).
[15] J. M. HOLLANDER, W. G. SMITH and J. O. RASMUSSEN: University of California Radiation Laboratory Report UCRL-3239, 1955. Phys. Rev. **102**, 1372 (1956).
[16] MILSTED, ROSENBLUM and VALADARES: C. R. Acad. Sci., Paris **239**, 259, 700 (1954).
[17] S. A. BARANOV and K. N. SHLYAGIN: Proceedings of Moscow Meeting on Atomic Energy, July 1955, p. 252.
[18] P. R. FIELDS *et al.*: Phys. Rev. **100**, 172 (1955).
[19] H. DIAMOND *et al.*: Phys. Rev. **92**, 1490 (1953).
[20] J. P. HUMMEL: Ph. D. Thesis, University of California 1956. Report UCRL-3456.
[21] G. R. HALL and T. L. MARKIN: J. Inorg. a. Nucl. Chem. **2**, 203 (1956).
[22] P. GRAF, J. WALLMANN and L. GODA: Unpublished results, 1956.

isotopic purity by intense neutron irradiation of Pu^{239}, purify it and return it to the reactor for conversion to Am^{243} via the sequence:

$$Pu^{242}(n,\gamma)\,Pu^{243}; \qquad Pu^{243} \xrightarrow[\text{5.0 hours}]{\beta^-} Am^{243}.$$

b) Chemical properties of americium.

We have noted the decreasing stability of the higher oxidation states in the sequence uranium-neptunium-plutonium. In americium we reach the first element for which the tripositive state characteristic of an actinide series is the most stable solution state. For some time after its discovery the tripositive state was the only solution state for which there was reliable evidence. Later, extensive evidence for other oxidation states was found as described below. For curium and all the higher actinide elements the tripositive state is the only solution state except for berkelium which can be oxidized to berkelium (IV).

40. Americium metal. Americium is a highly electropositive element and the reduction of compounds to the metal can be effected only under anhydrous conditions with powerful reducing agents such as the alkali and alkaline earth metals. For example, WESTRUM and EYRING[1] reduced AmF_3 with barium vapor at 1200° C. The measured density of americium metal is (11.7 ± 0.3) g/cm³ [1]. The structure of americium from x-ray studies[2] is double hexagonal close-packed with $a = (3.642 \pm 0.005)$ Å and $c = (11.76 \pm 0.01)$ Å. The americium radius is 1.82 Å

[1] E. F. WESTRUM jr. and L. EYRING: J. Amer. Chem. Soc. **73**, 3396 (1951).
[2] P. GRAF *et al.*: J. Amer. Chem. Soc. **78**, 2340 (1956).

and the calculated density is (11.87 ± 0.05) g/cm³ in excellent agreement with the experimental value. Americium is the first of the transuranium elements to resemble the lanthanide elements closely in the metallic state. Data on americium metal are summarized in Table 21.

Table 21. *Thermodynamic properties of americium metal.*

Melting point	Softens at 850° C; incomplete melting at 1200° C (Ref. 1)
Heat of reaction with 1.5 *M* HCl	-162.3 ± 2.7 kcal mole^{-1} (Ref. 2)
Heat of formation at 25° C of Am^{+3} (aq.)	-163.2 ± 2.7 kcal mole^{-1} (Ref. 1, 2)
Formal potential for the reaction $Am \rightarrow Am^{+3}$ (aq.) $+ 3e^-$	$+2.32 \pm 0.04$ est. (Ref. 2)
Vapor pressure	$\log p = 7.563 - 13162/T$ over range 1103 to 1453° K (Ref. 3) Assuming ΔC_p of vaporization $= -2$ cal mole^{-1}deg^{-1} $\log p = 11.092 - 13700/T - \log T$
Free energy of vaporization	$\Delta F = 62690 - 50.76\,T - 2.303\,(-2)\,T \log T$ (Ref. 3)
Heat of vaporization at 1273 °K	$\Delta H_{1273} = 60.2$ kcal mole^{-1} (Ref. 3)

[1] E. F. WESTRUM jr. and L. EYRING: J. Amer. Chem. Soc. **73**, 3396 (1951).
[2] B. B. CUNNINGHAM and H. R. LOHR: J. Amer. Chem. Soc. **73**, 2026 (1951).
[3] S. C. CARNIGLIA and B. B. CUNNINGHAM: J. Amer. Chem. Soc. **77**, 1502 (1955).

41. Americium (II). Because of the special stability of the half-filled electron configuration $5f^7$ it might be expected that americium (II) by analogy to samarium (II) would be stabilized enough to permit its observation. Fragmentary evidence[1] for the apparent production of americium (II) has not been confirmed and strenuous efforts to produce it under conditions which readily converted samarium and ytterbium to the dipositive state have all failed[2]. A limit of $> +1.5$ volts has been set on the $Am^{+2} - Am^{+3}$ couple.

Americium metal reacts with a deficiency of oxygen to yield AmO,[3] a compound which is metallic and not ionic in character and similar to NpO and PuO. Similarly a metallic hydride of the formula AmH_2 is known[3].

42. Americium (III). Americium (III) forms a pink-colored solution in dilute acids changing to yellow as the concentration is increased. The absoprtion spectrum of americum (III) is shown in Fig. 17. The absorption peak at 503 millimicrons is so sharp that instrumental defects can cause deviations from the BEER-LAMBERT law when it is used for analysis. The peaks at 503 and 812 millimicrons are somewhat similar to europium (III) but have much greater extinction coefficients which indicates greater influence of the environment on the $5f$ electrons. The sharp absorption line at 503 millimicrons was shown to resolve into four very sharp lines when it was measured with a grating spectrograph[4]. In strong nitric

[1] S. G. THOMPSON, L. O. MORGAN, R. A. JAMES and I. PERLMAN: Paper 19.1, N.N.E.S., Vol. 14B, "The Transuranium Elements", 1949.
[2] R. A. PENNEMAN and L. B. ASPREY: Paper P/838, "Proceedings of the Geneva Conference—Peaceful Uses of Atomic Energy", August 1955, United Nations.
[3] F. H. ELLINGER: Unpublished work cited by R. A. PENNEMAN and L. B. ASPREY, above.
[4] B. J. STOVER, J. G. CONWAY and B. B. CUNNINGHAM: J. Amer. Chem. Soc. **73**, 491 1951).

acid or sulfuric acid marked changes in the extinction coefficients and slight shifts of the major peaks occur because of complexing[1].

The principal compounds of low solubility are the hydroxide, the fluoride, oxalate, and phosphate. The resemblance in structure and solubility of the americium compounds and the rare earth compounds is quite close. The crystal radius of Am^{+3} is nearly the same as that of neodymium. Such compounds as lanthanum trifluoride are commonly used to carry trace concentrations of americium. The separation of tripositive americium from tripositve rare earths can be done by fractional precipitation of certain compounds, for example, the oxalates[2].

The most powerful method for the separation of tripositive americium from the rare earths is the ion exchange elution method. A mixture of americium and rare earth activities can be adsorbed from dilute acid solution on a column of cation resin such as Dowex-50 and then by selective elution with buffered citrate, lactate, hydroxyisobutyrate, ethylenediamine tetraacetic acid or similar agents the elements can be separated. The rare earth element promethium is the most difficult to separate as its elution behavior under most conditions is quite similar to that of americium.

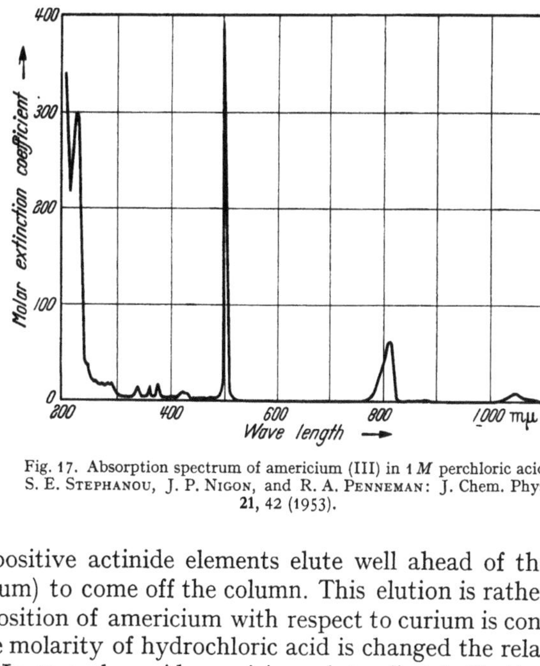

Fig. 17. Absorption spectrum of americium (III) in 1 M perchloric acid. S. E. STEPHANOU, J. P. NIGON, and R. A. PENNEMAN: J. Chem. Phys. 21, 42 (1953).

A more satisfactory separation is achieved if 13 molar hydrochloric acid is used as the eluting agent. Americium and curium and the other tripositive actinide elements elute well ahead of the first rare-earth element (lutetium) to come off the column. This elution is rather interesting when the elution position of americium with respect to curium is considered. It is found that as the molarity of hydrochloric acid is changed the relative order of elution changes. In 13 molar acid americium elutes first indicating a stronger tendency to form chloride complexes[3].

Tripositive americium and curium are poorly extracted by the organic solvents which we have previously discussed for the higher oxidation states of neptunium and plutonium (Table 18, Sect. 36). However, under certain conditions appreciable distribution coefficients between acidic aqueous solutions and undiluted tri-n-butyl phosphate can be achieved. In Fig. 18 the extraction coefficients of tripositive actinide and rare-earth elements are given for extraction into tributylphosphate from $12M$ hydrochloric acid and $15M$ nitric acid. These extraction coefficients are sufficiently different that neighboring elements such as americium and curium can be separated from each other by proper adjustment of conditions[4].

[1] G. N. YAKOVLEV and V. N. KOSYAKOV: Paper P/676, "Proceedings of the Geneva Conference—Peaceful Uses of Atomic Energy", August 1955, United Nations.
[2] J. A. HERMANN: Los Alamos Scientific Laboratory Report LADC-1687, 1954.
[3] R. M. DIAMOND, K. STREET jr. and G. T. SEABORG: J. Amer. Chem. Soc. 76, 1461 (1954).
[4] D. F. PEPPARD, P. R. GRAY and M. M. MARKUS: J. Amer. Chem. Soc. 75, 6063 (1953).

Other methods of purifying americium which depend upon oxidation to a higher oxidation state will be discussed later.

43. Americium (IV). This oxidation state is well known in the solid state as will be discussed in Sect. 46, but there is no evidence for americium (IV) in aqueous solution. This fact is understandable since the oxidation potential of the americium (III)—americium (IV) couple in acid solution is estimated to be −2.44 volts[1].

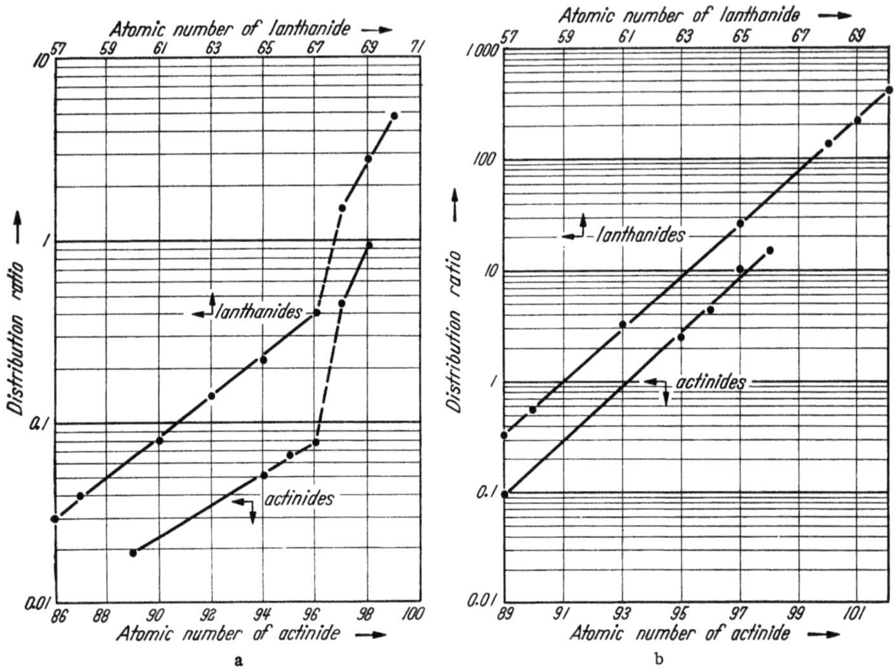

Fig. 18. Distribution coefficients for the extraction of tripositive lanthanide and actinide elements into tributyl phosphate from (A) 12 *N* hydrochloric acid and (B) 15.6 *N* nitric acid. Unpublished data, D. F. PEPPARD, Argonne National Laboratory and P. R. GRAY, University of California.

44. Americium (V). This oxidation state was first prepared by oxidizing americium (III) in potassium carbonate solution using sodium hypochlorite[2]. Americium (V) precipitates as a double carbonate salt of americium and potassium. Alternate oxidizing agents are ozone or peroxydisulfate ion[3,4]. The compound can be dissolved in dilute acid to give a solution of americium (V) with a characteristic absorption spectrum. Slow reduction to americium (III) occurs from the reducing effects of the alpha radiation; in strong acid extensive disproportionation occurs. Tracer amounts of americium can be oxidized in carbonate solution and removed quantitatively from solution by precipitating potassium uranyl carbonate or potassium plutonyl carbonate as a carrier compund[4]. Excellent separation from rare earths and many impurities is achieved.

Americium (V) may also be prepared by reduction of americium (VI) with chloride ion or bromide ion. Solutions of americium (VI) simply on standing are reduced at a rate of approximately 4% per hour because of the reducing effects

---

[1] L. EYRING, H. R. LOHR and B. B. CUNNINGHAM: J. Amer. Chem. Soc. **74**, 1186 (1952).
[2] L. B. WERNER and I. PERLMAN: J. Amer. Chem. Soc. **73**, 495 (1951).
[3] S. E. STEPHANOU, J. P. NIGON and R. A. PENNEMAN: J. Chem. Phys. **21**, 42 (1953).
[4] G. N. YAKOVLEV and D. S. GORBENKO-GERMANOV: Paper P/677, "Proceedings of the Geneva Conference—Peaceful Uses of Atomic Energy", August 1955, United Nations.

of the alpha radiation. No reduction to americium (III) occurs until the hexapositive ion has disappeared completely. GUNN[1] prepared pure solutions of americium (V) by passing americium (VI) in $1\,M$ $HClO_4$ through a short column of Dowex-50 cation exchange resin. The resin instantly reduced the americium to americium (V) which passed through the column while any americium (III) impurity was strongly adsorbed on the resin.

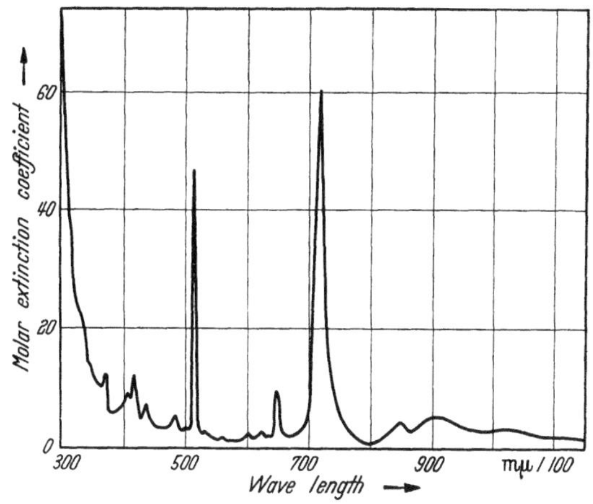

Fig. 19. Absorption spectrum of americium (V) in $1\,M$ $HClO_4$. S. E. STEPHANOU, J. P. NIGON and R. A. PENNEMAN: J. Chem. Phys. 21, 42 (1953).

Several lines of evidence strongly suggest that AmO_2^+ is the ionic species in acidic solution. The similarity in chemistry of americium (V) and PuO_2^+ and NpO_2^+ is large. The solid compounds $RbAmO_2CO_3$ and $NH_4AmO_2CO_3$ have been isolated[2] and identified by x-ray analysis[3]. Infrared spectrum studies favor the AmO_2^+ structure for the ion. The reversibility of the (V)—(VI) couple is evidence for a simple electron transfer mechanism and hexavalent americium is believed to be AmO_2^{++}.

The absorption spectrum of americium (V) has been determined in perchloric acid, hydrochloric acid, and sulfuric acid media by many laboratories. Fig. 19 shows the spectrum in 1 molar perchloric acid where complexing is slight. The peak at 717 millimicrons is used in analysis.

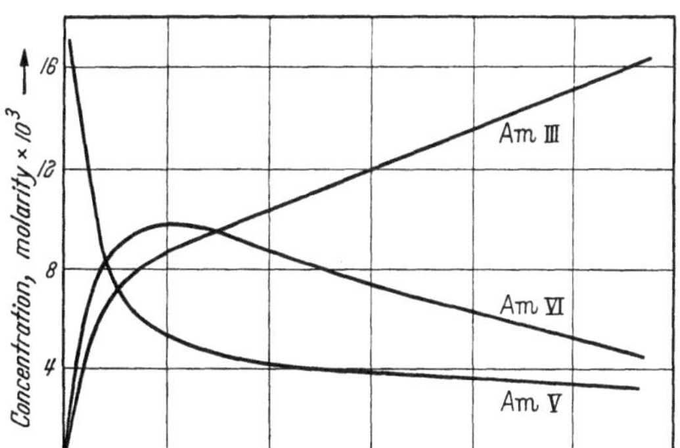

Fig. 20. The disproportionation of americium (V) in $6\,M$ $HClO_4$. From S. E. STEPHANOU, J. P. NIGON and R. A. PENNEMAN: J. Chem. Phys. 21, 42 (1953).

Photospectrometry has been used to follow the disproportionation of americium (V) in perchloric acid[1,4] in hydrochloric acid[5], and in sulfuric acid[6]. Strong acid is required for appreciable disproportionation; this is in sharp constrast to the disproportionation of plutonium (V). The rate curves for disproportionation in $6\,M$ perchloric acid are given in Fig. 20. The disappearance of americium (V) is second order

[1] S. GUNN: Thesis, University of California Radiation Laboratory, declassified report UCRL-2541.

[2] J. P. NIGON *et al.*: J. Phys. Chem. **58**, 403 (1954).

[3] F. H. ELLINGER and W. H. ZACHARIASEN: J. Phys. Chem. **58**, 405 (1951).

[4] S. E. STEPHANOU, L. B. ASPREY and R. A. PENNEMAN: Atomic Energy Commission Unclassified Document AECU-925, 1950.

[5] G. R. HALL and P. D. HERNIMAN: J. Chem. Soc. **1954**, 2214.

[6] G. N. YAKOVLEV and V. N. KOSYAKOV: Paper P/676, "Proceedings of the Geneva Conference—Peaceful Uses of Atomic Energy", August 1955, United Nations.

with respect to the concentration of americium (V) and fourth order with respect to the hydrogen ion concentration so the rate determining step is written:

$$2\,AmO_2^+ + 4\,H^+ \rightleftharpoons AmO_2^{+2} + [Am(IV)] + 2\,H_2O. \qquad (44.1)$$

The americium (IV) is postulated although no spectral evidence for it is seen. It is immediately destroyed by reduction by water, by radiation-produced reducing agents, or by the reaction:

$$Am^{+4} + AmO_2^+ \rightarrow Am^{+3} + AmO_2^{+2}. \qquad (44.2)$$

Combining this last equation the overall equation for disproportionation is

$$3\,AmO_2^+ + 4\,H^+ \rightleftharpoons 2\,AmO_2^{+2} + Am^{+3} + 2\,H_2O. \qquad (44.3)$$

True disproportionation equilibrium is never reached because of self-reduction effects of the alpha particles and eventually all of the americium is converted to americium (III).

The self-reduction of Am^{241}(V) solutions is roughly two percent per hour in perchloric acid[1,2] or sulfuric acid[3]. The rate of self-reduction is appreciably lower in 0.5 molar hydrochloric acid[4]. Marked reduction of this rate is achieved when reagents are added to destroy hydrogen peroxide, one of the breakdown products from water chiefly responsible for the reduction[1].

In pure solutions of the longer-lived isotope, 7930-year Am^{243}, the self-reduction effects interfere to a lesser degree with a study of the chemistry of americium.

45. Americium (VI). This oxidation state was first established by ASPREY, STEPHANOU, and PENNEMAN[5]. Americium (III) in dilute acid solution can be converted to pale-yellow americium (VI) by peroxydisulfate ion or by argentic ion. Oxidation by ceric ion is incomplete; oxidation by ozone and periodic acid is unsuccessful, but anodic oxidation is successful.

ASPREY and PENNEMAN[6] report that quantitative oxidation of americium (III) to americium (VI) with peroxydisulfate ion has been carried out in solutions varying between $10^{-8}M$ and $\sim 0.2\,M$ in americium. The reaction is no more than first order with respect to americium. It is quite essential that the hydrogen ion concentration be kept low to prevent the acid-catalyzed breakdown of peroxydisulfate ion with the ultimate formation of hydrogen peroxide, a reducing agent for americium (VI). A suggested set of conditions are: $H^+ \leq 0.1$ molar; $S_2O_8^= = 0.2$ to 0.5 molar; temperature 90° C; time ~ 15 min. This oxidation can be carried out in perchloric, nitric or sulfuric acid but not in hydrochloric acid.

Americium (V) may be converted to americium (VI) by ozone oxidation in hot perchloric or nitric acid. Cerium (IV) and argentic ion are also effective in this oxidation. Disproportionation of americium (V) in strong acid converts part of it to americium (VI).

Electrolytic oxidation to americium (VI) is incomplete and erratic in a simple cell but GUNN obtained rapid and near quantitative oxidation in a three-compartment cell with the americium placed only in the anode compartment[1]. One molar nitric acid, 1 molar perchloric acid or 6 molar perchloric acid could be used at 0° C.

---

[1] S. GUNN: Thesis, University of California Radiation Laboratory, declassified report UCRL-2541.

[2] See footnote 4, p. 261.

[3] See footnote 6, p. 261.

[4] See footnote 5, p. 261.

[5] L. B. ASPREY, S. E. STEPHANOU and R. A. PENNEMAN: J. Amer. Chem. Soc. **72**, 1425 (1950); **73**, 5715 (1951).

[6] R. A. PENNEMAN and L. B. ASPREY: Report P/838, "Proceedings of the Geneva Conference—Peaceful Uses of Atomic Energy", August 1955, United Nations.

The chemical behavior of americium (VI) is very similar to that of uranium (VI) neptunium (VI), and plutonium (VI) and the ionic species is undoubtedly AmO_2^{+2}. Insoluble fluorides may be precipitated in its presence (preferably with KF) to remove fluoride insoluble impurities. Organic solvents such as ethyl ether and methyl isobutyl ketone will extract americium (VI) from aqueous solution.

The absorption spectrum of AmO_2^{+2} in perchloric acid is given in Fig. 21. The peaks at ~995 and 1667 millimicrons are used in analysis.

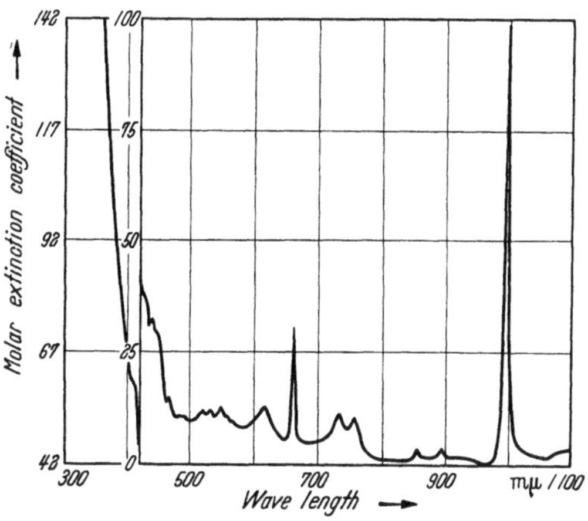

Fig. 21. The absorption spectrum of americium (VI) in 1 *M* $HClO_4$. S. E. STEPHANOU, J. P. NIGON and R. A. PENNEMAN: J. Chem. Phys. 21, 42 (1953).

The formal potential for the $AmO_2^{+}-AmO_2^{+2}$ couple was found to be -1.60 volts by measurements against the cerium (III)—cerium (IV) couple in 1 *M* perchloric acid and against $Hg_2^{+2}-Hg^{+2}$ in 1 *M* perchloric acid using shiny platinum electrodes [1]. A study [2] of the oxidation of americium (III) to americium (VI) by cerium (IV) yielded the value of -1.67 for the half reaction:

$$Am^{+3} + 2H_2O \rightleftharpoons AmO_2^{+2} + 4H^+ + 3e^- . \qquad (45.1)$$

An indirect measurement of GUNN [3] based upon the measurement of the heat of reduction of americium (VI) with ferrous ion and the use of previously known and estimated thermodynamic quantities gave -1.70 volts as the potential for this reaction.

$$Am \xrightarrow{<2.7} Am^{+2} \xrightarrow{>1.5} Am^{+3} \xrightarrow{-2.44} Am^{+4} \xrightarrow{-1.04} AmO_2^{+} \xrightarrow{-1.60} AmO_2^{+2}$$

(Am — Am^{+3}: 2.32; Am^{+3} — AmO_2^{+2}: -1.69; Am^{+3} — AmO_2^{+}: -1.74)

Fig. 22. Formal potential scheme for americium in acid solution.

Fig. 22 is an oxidation potential scheme for americium in acid solution. In this scheme the (III)—(VI) couple is shown as -1.69 volts rather than as either of the two values just cited for internal consistency reasons. The values for the $Am-Am^{+2}$ couple and for the $Am^{+4}-AmO_2^{+}$ couple were obtained simply by difference.

$$Am \xrightarrow{2.71} Am(OH)_3 \xrightarrow{0.4} Am(OH)_4 \xrightarrow{(-0.7)} AmO_2OH \xrightarrow{(-1.1)} AmO_2(OH)_2$$

Fig. 23. Potential scheme for americium in basic solution.

Fig. 23 is an oxidation potential scheme given by LATIMER [4] for basic solution. These are estimated rather than experimental values.

[1] R. A. PENNEMAN and L. B. ASPREY: U.S. Atomic Energy Commission Unclassified Report AECU-936, 1950.

[2] J. P. NIGON: Unpublished work, Los Alamos Scientific Laboratory, reported by PENNEMAN and ASPREY, Paper P/838, footnote.

[3] S. GUNN: Thesis, University of California Radiation Laboratory, declassified report UCRL-2541.

[4] WENDELL M. LATIMER: Oxidation Potentials, second edit., p. 311. New York: Prentice Hall Inc. 1952.

46. The anhydrous compounds of americium. When americium is precipitated as the insoluble hydroxide from aqueous solution and heated in air a black oxide is formed with the formula AmO_2. This may be reduced to orange-red Am_2O_3 through the action of hydrogen at elevated temperatures. The AmO_2 has the fluorite type structure, isostructural with UO_2, NpO_2, and PuO_2. The Am_2O_3 has the cubic sesquioxide structure typical of the rare-earth elements. A second crystal modification of Am_2O_3 is cubic[1]. See Table 6, Sect. 6.

The halogen compounds of americium were first prepared by FRIED[2]. Pink *americium trichloride* has been prepared by treating AmO_2 at 800 to 900° C with carbon tetrachloride and with gaseous hydrogen chloride. Its vapor phase hydrolysis to the oxychloride AmOCl has been studied carefully[3]. The oxychloride is isostructural[1] with PuOCl and PrOCl. *Americium tribromide*, $AmBr_3$, has been prepared by carrying out the following reaction in a sealed tube at 500° C.

$$3\,AmO_2 + 4\,AlBr_3 \rightarrow 3\,AmBr_3 + 2\,Al_2O_3 + \tfrac{3}{2}\,Br_2. \qquad (46.1)$$

Americium triiodide was made by a similar reaction. When americium dioxide is treated with oxygen-hydrogen fluoride mixtures at 600 to 750° C pink *americium trifluoride* is produced whereas neptunium or plutonium dioxides under similar circumstances are converted to the tetrafluoride. ASPREY[4] succeeded in preparing *americium tetrafluoride* by reacting americium dioxide or trifluoride with fluorine gas at one atmosphere pressure and 500° C. Asprey also prepared $KAmF_5$ by fluorine treatment of $KAmO_2CO_3$. All of these halogen compounds have been characterized crystallographically and are isostructural with the analogous plutonium compounds.

The only hexapositive compound of definitely known composition and structure is the lemon-yellow sodium americyl acetate, $NaAmO_2(C_2H_3O_2)_3$, isostructural with similar compounds of uranium, neptunium, and plutonium.

Among pentapositive americium compounds $RbAmO_2CO_3$, $NH_4AmO_2CO_3$, and $KAmO_2F_2$ have been identified by x-ray analysis[5-7].

Lattice constants for a number of crystalline americium compounds are given in Table 6, Sect. 6.

IV. Curium (element 96).

47. Discovery of curium—summary of known isotopes. The first isotope of element 96 to be found was Cm^{242} which was identified in 1944 by SEABORG, JAMES, and GHIORSO[8] among the products of the bombardment of a Pu^{239} target with 32-Mev helium ions. The Cm^{242} was made by the reaction $Pu^{239}(\alpha, n)Cm^{242}$ reaction. The discovery of curium preceded that of americium. The first isolation of curium in the form of a pure compound was made by WERNER and PERLMAN in 1947[9].

The name curium was chosen to emphasize the analogy between the rare-earth element gadolinium which contains seven $4f$ electrons and element 96

[1] D. H. TEMPLETON and C. H. DAUBEN: J. Amer. Chem. Soc. **75**, 4561 (1953).
[2] S. FRIED: J. Amer. Chem. Soc. **73**, 416 (1951).
[3] C. KOCH and B. B. CUNNINGHAM: J. Amer. Chem. Soc. **76**, 1470 (1954).
[4] L. B. ASPREY: J. Amer. Chem. Soc. **76**, 2019 (1954).
[5] J. P. NIGON *et al.*: J. Phys. Chem. **58**, 403 (1954).
[6] F. H. ELLINGER and W. H. ZACHARIASEN: J. Phys. Chem. **58**, 405 (1951).
[7] L. B. ASPREY, F. H. ELLINGER and W. H. ZACHARIASEN: J. Amer. Chem. Soc. **76**, 5235 (1953).
[8] G. T. SEABORG, R. A. JAMES and A. GHIORSO: Paper 22.2, N.N.E.S., Vol. 14B, "The Transuranium Elements", 1949.
[9] L. B. WERNER and I. PERLMAN: Paper 22.5, N.N.E.S., Vol. 14B, "The Transuranium Elements", 1949.

which contains the same number of $5f$ electrons. Gadolinium was named indirectly after J. GADOLIN, the great Finnish investigator of the rare-earth elements; curium was named after PIERRE and MARIE CURIE, famous investigators in the field of natural radioactivity. Curium is of some special interest because it completes the first half of the actinide transition series.

The isotope, Cm^{242}, can also be prepared in a nuclear reactor by the reaction sequence:

$$Am^{241}(n,\gamma)\,Am^{242\,m}; \quad Am^{242\,m} \xrightarrow[16\text{ hours}]{\beta^-} Cm^{242}. \tag{47.1}$$

One can also start with plutonium, thus:

$$Pu^{239}(n,\gamma)\,Pu^{240}(n,\gamma)\,Pu^{241} \xrightarrow[13\text{ years}]{\beta^-} Am^{241}(n,\gamma)\,Am^{242\,m} \xrightarrow[16\text{ hours}]{\beta^-} Cm^{242}. \tag{47.2}$$

Larger amounts can be prepared in these ways for chemical studies than can be made in cyclotrons.

The isotope, Cm^{242}, decays by the emission of alpha particles (the energy of the most prominent group being 6.110 Mev) with a half life of 162.5 days. This is about the half life of polonium which is 138.3 days. In both elements chemical studies are highly hindered by physiological hazards and by chemical effects of the intense alpha radiation. The specific activity of Cm^{242} is about 10^{14} alpha disintegrations per minute per milligram. Water solutions of Cm^{242} are decomposed by this radiation; some of the decomposition products such as hydrogen peroxide interfere with attempted chemical reactions. Precise temperature control of solutions or compounds, as for example, in magnetic susceptibility experiments is difficult because of heating effects. A massive chunk of such curium metal would be heated instantly to incandescence even if it could be prepared; the energy released is 1230 watts per gram. Crystal lattices of compounds are disarranged by recoil effects of the alpha particles.

At the present time all the curium isotopes of mass number 238 to 250 are known. Those lighter than 243 may be prepared by bombardment of plutonium with very high energy helium ions. Those higher in mass number are made by neutron capture or multiple neutron capture reactions in high flux reactors starting with plutonium or curium isotopes of lower mass. The principal decay characteristics of these isotopes are given in Table 22. Decay schemes for those curium isotopes which have been most carefully studied are given in Fig. 24.

The isotope Cm^{244} has a particular importance for chemical studies because it can be prepared with rather high isotopic purity in milligram or greater quantities. The half life is 19.4 years and the specific alpha activity is 43 times less than that of Cm^{242}. It can be made in a high flux reactor via the neutron capture sequence:

$$Pu^{239}(n,\gamma)\,Pu^{240}(n,\gamma)\,Pu^{241}(n,\gamma)\,Pu^{242}(n,\gamma)\,Pu^{243} \xrightarrow[5.0\text{ hr}]{\beta^-} Am^{243}(n,\gamma)\,Am^{244} \xrightarrow[26\text{ min}]{\beta^-} Cm^{244}. \tag{47.3}$$

Chemical purification of the Am^{243} (which may also be prepared as described in Sect. 39) may be desirable before the final transmutation to curium. A sample of 350 milligrams of Pu^{239} irradiated to a total neutron flux of 10^{22} neutrons in the Materials Testing Reactor in Idaho resulted in the production of a minimum of 3 milligrams of curium with the isotopic composition 1.84% Cm^{242}, 95.5% Cm^{244}, 1.27% Cm^{245}, 1.36% Cm^{246}, and 0.016% Cm^{247}[1]. Further neutron irradiation of Cm^{244} produces larger percentages of the higher mass isotopes, Cm^{245} and Cm^{246} of longer half life which are even better for use in chemical investigations.

---

[1] C. M. STEVENS *et al.*: Phys. Rev. **94**, 297 (1954).

Table 22. *Isotopes of curium**.

Isotope and half-life	Type of decay	Energy (Mev)	Method of production	Disintegration energies (Mev) **
Cm^{238} 2.5 hours	E. C. < 90% α > 10%	 6.50	$Pu^{239}(\alpha, 5n)$	$Q_{E.C.} = 1.09$ (est.) $Q_\alpha = 6.63$
Cm^{239} ~3 hours	E. C. ~100%		$Pu^{239}(\alpha, 4n)$	$Q_{E.C.} = 1.74$ (est.) $Q_\alpha = 6.50$ (est.)
Cm^{240} 26.8 days α 1.9×10^6 years spon. fiss.	α	6.25	$Th^{232}(C, 4n)$ $Pu^{239}(\alpha, 3n)$	$Q_{E.C.} = 0.08$ (est.) $Q_\alpha = 6.38$
Cm^{241} 35 days	E. C. 99⁺% α 0.96% (Ref. 1) γ	 0.472 (Ref. 1)	$Pu^{239}(\alpha, 2n)$	$Q_{E.C.} = 0.85$ (est.) $Q_\alpha = 6.30$ (est.)
Cm^{242} 162.5 days α 7.2×10^6 years spon. fiss.	α β stable γ	6.110 (73.7%), 6.066 (26.3%) 5.965 (0.035%) 0.044, 0.100, 0.157 (Ref. 2) 0.210, 0.562, 0.605 (Ref. 5) 0.0441, 0.1019, 0.1577 (Ref. 11)	$Pu^{239}(\alpha, n)$ $Am^{241}(d, n)$ daughter Am^{242} daughter Cf^{246}	$Q_\alpha = 6.21$
Cm^{243} 35 years	α β stable γ	6.003 (1%), 5.985 (6%), 5.777 (79%), 5.732 (12%) 5.672 (2%) (Ref. 9) 0.04464, 0.04940, 0.05725, 0.0614 0.06782, 0.1061, 0.2099, 0,2284, 0.2546, 0.2731, 0.2777, 0.2856, 0.3161, 0.3345 deduced from Np^{239} decay scheme (Ref. 10)	$Cm^{242}(n, \gamma)$ daughter Bk^{243}	$Q_\alpha = 6.19$
Cm^{244} 18.4 years α (Ref. 4) 1.4×10^7 years spon. fiss.	α β stable γ	5.798 (76.7%), 5.755 (23.3%) 5.655 (0.014%) 0.043, 0.100, 0.150 (Ref. 5) 0.0429 (Ref. 11)	$Cm^{243}(n, \gamma)$ daughter Am^{244}	$Q_\alpha = 5.89$
Cm^{245} 2×10^4 years (Ref. 3) 1.4×10^4 years (Ref. 8) 1.2×10^4 years (Ref. 4)	α β stable γ	5.36 (Ref. 3) 0.101 (80%), 0.173 (13%) (Ref. 5)	Pu^{239} multiple n capture daughter Bk^{245} $Cm^{244}(n, \gamma)$	$Q_\alpha = 5.62$
Cm^{246} 4000 years (Ref. 4) 2300 years (Ref. 8) 3×10^7 years spon. fiss.	α β stable	5.36 (Ref. 4)	$Cm^{243}(n, \gamma)$	$Q_\alpha = 5.45$
Cm^{247} (Ref. 6) 4×10^7 years (Ref. 12)	β stable α		Pu^{239} multiple n capture	$Q_\alpha = 5.25$ (est.)
Cm^{248} (Ref. 7) long lived	β stable		Pu^{239} multiple n capture	$Q_\alpha = 5.17$ (est.)
Cm^{249} (Ref. 7) short	β^- (Ref. 7)		Pu^{239} multiple n capture	$Q_\beta = 1.09$ (est.) $Q_\alpha = 5.29$ (est.)

* For references to original data not given specifically here, refer to Hollander, Perlman and Seaborg "Table of Isotopes", [Rev. Mod. Phys. **25**, 469 (1953)] and to the compilations of the Nuclear Data Group, National Research Council.

** Revised values from Glass, Thompson and Seaborg: J. Inorg. a. Nucl. Chem. **1**, 3 (1955).

[1] R. A. Glass, R. J. Carr, W. M. Gibson and J. W. Cobble: University of California Radiation Laboratory Report UCRL-3599, October 1956.

[2] F. Asaro *et al.*: Phys. Rev. **92**, 694 (1953).

[3] E. K. Hulet *et al.*: Phys. Rev. **95**, 1703 (1954).

References of table 22.

[4] A. M. FRIEDMAN *et al.*: Phys. Rev. **95**, 1501 (1954).

[5] I. PERLMAN *et al.*: University of California Radiation Laboratory Report No. UCRL-2932, March 1955 and UCRL-3068, July 1955.

[6] C. M. STEVENS *et al.*: Phys. Rev. **94**, 974 (1954).

[7] W. C. BENTLEY *et al.*: Paper P/809, "Peaceful Uses of Atomic Energy-Proceedings of the Geneva Conference", August 1955, United Nations.

[8] C. I. BROWNE *et al.*: J. Inorg. a. Nucl. Chem. **1**, 254 (1955).

[9] F. ASARO: Private communication 1956.

[10] J. M. HOLLANDER *et al.*: University of California Radiation Laboratory Report No. UCRL-3222, Sept. 1955. Phys. Rev. **102**, 740 (1956).

[11] W. G. SMITH and J. M. HOLLANDER: University of California Radiation Laboratory Report No. UCRL-2974. May 1955, Phys. Rev. **101**, 746 (1956).

[12] H. DIAMOND *et al.*: Phys. Rev. **105**, 679 (1957).

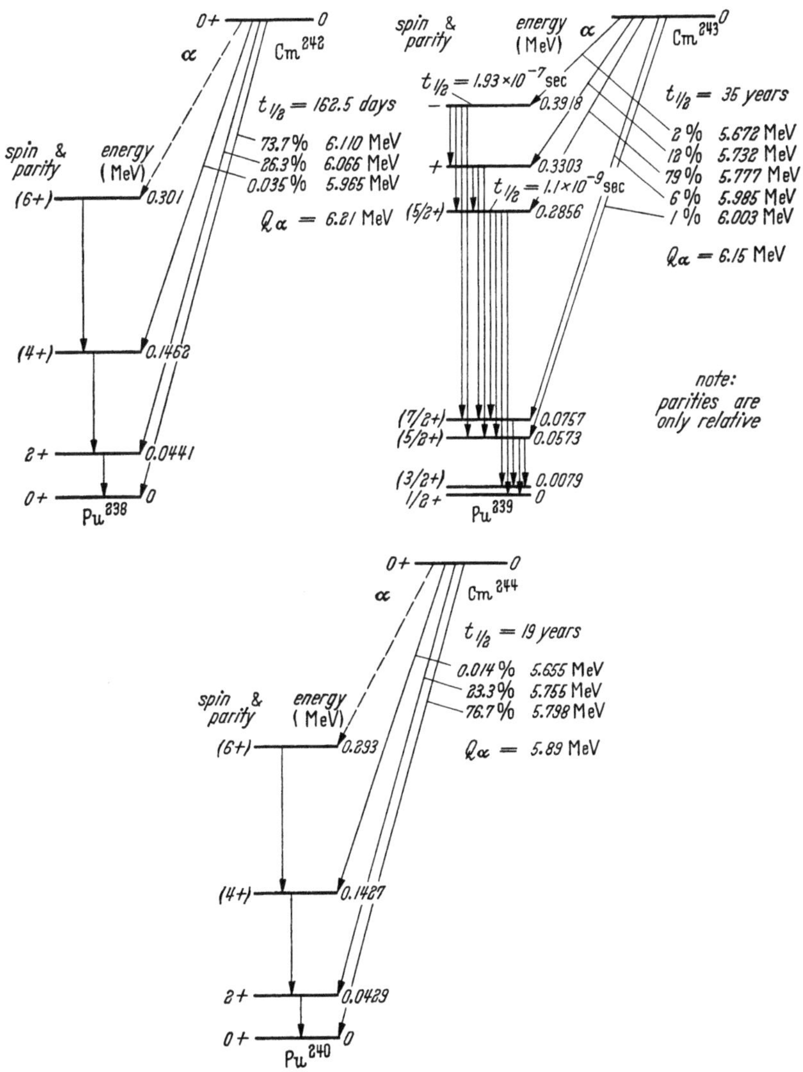

Fig. 24. Decay schemes of curium isotopes.

48. Chemical properties of curium in aqueous solution. Most investigations of curium have been carried out with the tracer technique. In aqueous solution curium is present in the tripositive state, and no reliable evidence for any other solution state has ever been found. A variety of the strongest oxidizing agents such as dichromate, permanganate, peroxydisulfate, argentic ion, and others gave no indication of oxidation under a variety of conditions. Failure to observe oxidation conceivably could be due to slow reaction at tracer concentration so many experiments were repeated at concentration of about 10^{-5} molar with similar results. In solutions containing mixtures of Am^{241} and Cm^{242} it has been possible to oxidize americium to americium (V) in a carbonate solution or to oxidize it to americium (VI) in an acid solution without oxidizing any of the curium. Experiments with Cm^{244} solutions similarly show no evidence for oxidation. It seems unlikely that curium will ever be found in aqueous solution as curium (IV), curium (V), or curium (VI). Any reasonable extrapolation of the increase of the (III)—(IV) or the (III)—(V), or the (III)—(VI) couples with atomic number shows that this must be so (see for example, Fig. 2, Sect. 2).

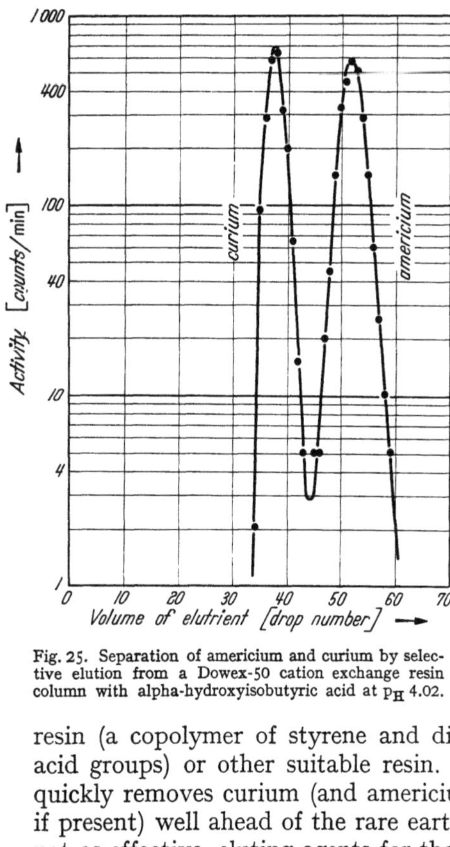

Fig. 25. Separation of americium and curium by selective elution from a Dowex-50 cation exchange resin column with alpha-hydroxyisobutyric acid at p_H 4.02.

Tripositive curium closely resembles the other tripositive actinide and lanthanide ions. It is readily coprecipitated with rare-earth fluorides, oxalates, and hydroxides. In macroscopic concentrations curium trifluoride precipitates directly upon addition of hydrofluoric acid. The chloride, bromide, iodide, perchlorate, nitrate, and sulfate are soluble. Separation from rare earths is moderately difficult and is best carried out by the cation exchange method. The mixture of activities is adsorbed from dilute acid solution on top of a column of Dowex-50 resin (a copolymer of styrene and divinyl benzene containing nuclear sulfonic acid groups) or other suitable resin. Elution with 13 molar hydrochloric acid quickly removes curium (and americium and other tripositive actinide elements if present) well ahead of the rare earths (see Fig. 30, Sect. 55). Alternative, but not as effective, eluting agents for the curium-rare earth separation are buffered solutions of citric acid, lactic acid, or alpha-hydroxyisobutyric acid. Column operations with Cm^{242} are complicated by bubble formation in the solutions due to the alpha activity when amounts greater than 10 micrograms are used. A partial solution is to increase the diameter of the column and to increase the flow rate.

Removal of americium impurity is a problem of frequent occurrence in curium chemistry which may be solved in a variety of ways. For a long time the best method was selective elution from cation-resin columns with buffered citric, lactic, tartaric, or alpha-hydroxyisobutyric acid. See Fig. 25 which shows that complete separation is possible. Americium and curium extract into tributyl phosphate from concentrated hydrochloric acid or nitric acid with appreciable distribution coefficients[1] (see Fig. 18, Sect. 42). These coefficients are sufficiently

[1] D. F. Peppard, P. R. Gray and M. M. Markus: J. Amer. Chem. Soc. **75**, 6063 (1953).

different for the two elements that a good separation is possible. When larger amounts of americium are handled it is convenient to take advantage of the upper oxidation states of americium. Oxidation with peroxydisulfate ion in dilute acid at 70 to 95° C for a few minutes followed by rapid precipitation of a small amount of lanthanum fluoride to carry the curium has been reported by S. E. STEPHANOU and R. A. PENNEMAN[1]. GUNN[2] suggests anodic oxidation at 0° C to americium (VI) at a platinum anode with the active solution in an isolated anode compartment in dilute nitric acid or perchloric acid solution; after oxidation a quick precipitation of lanthanum fluoride removes curium. YAKOVLEV and GORBENKO-GERMANOV[3] suggest oxidation of americium to americium (V) in 50% carbonate solution with ozone, hypochlorite, or peroxydisulfate ion. Americium precipitates from solution as the double carbonate salt. When tracer americium is present the compound potassium uranyl carbonate can be precipitated as a carrier.

The absorption spectrum of curium solutions has no structure in the visible region, and although it absorbs heavily in the ultraviolet it shows no sharp absorption bands. Hence curium solutions cannot be analyzed by the spectrophotometric methods so useful for the other actinide elements. Analysis is done by measuring the alpha particle emission rate (often with the refinement of alpha particle energy determination) or by spectrographic analysis for impurities.

49. Anhydrous compounds of curium. The number of well characterized anhydrous curium compounds is quite small. The metal has been prepared repeatedly on the microgram scale by precipitating curium trifluoride from aqueous solution, drying it, and reducing it in a beryllia crucible in a vacuum system with barium vapor at 1275° C[4]. Conditions are critical for securing a good agglomerate. The metal is silvery and shiny and about as malleable as plutonium metal. Under nitrogen it retains its luster for several hours, then slowly tarnishes. The energy of alpha decay keeps the metal temperature substantially above that of the surroundings. An estimate was made of the density of a spheroidal piece of metal by weighing it and estimating its volume. The unexpectedly low value of approximately 7 may indicate a void.

X-ray studies of curium (Cm^{242}) oxide samples have two limiting difficulties; namely, fogging of the film from the Cm^{242} radiations and destruction of the crystal lattice by alpha recoil effects. Nevertheless some positive measurements have been made. These have been supplemented by measurements made with the longer lived Cm^{244}. A white sesquioxide Cm_2O_3 and a black dioxide CmO_2 have been identified[5]. This was the first clear evidence for a higher oxidation state for curium.

FEAY, CUNNINGHAM, and CONWAY[6] have examined the absorption spectrum of solid CmF_3 and of solid GdF_3 on a 21-foot spectrograph. Marked similarities in the spectra of these homologous element compounds were found. The

[1] S. E. STEPHANOU and R. A. PENNEMAN: J. Amer. Chem. Soc. **74**, 3701 (1952)

[2] S. R. GUNN: Ph. D. Thesis, University of California Radiation Laboratory Declassified Report UCRL-2541.

[3] G. N. YAKOVLEV and D. S. GORBENKO-GERMANOV: Paper P/677, "Proceedings of the Geneva Conference—Peaceful Uses of Atomic Energy", August 1955, United Nations.

[4] J. C. WALLMAN, W. W. T. CRANE and B. B. CUNNINGHAM: J. Amer. Chem. Soc. **73**, 493 (1951).

[5] L. B. ASPREY, F. H. ELLINGER, S. FRIED and W. H. ZACHARIASEN: J. Amer. Chem. Soc. **77**, 1707 (1955).

[6] D. C. FEAY, B. B. CUNNINGHAM and J. CONWAY: Unpublished work; see D. C. FEAY, Ph. D. Thesis, University of California Radiation Laboratory Unclassified Report UCRL-2547, 1954.

prominent lines are compared in Table 23. Below 2850 Å no absorption was found in either compound. These similarities support the belief that the electronic configuration of curium is $5f^7$ analogous to the $4f^7$ structure of gadolinium. On an absolute scale absorption by CmF_3 is much stronger than that by GdF_3.

The magnetic susceptibility of curium trifluoride has been determined at two temperatures using samples of only about 25 micrograms[1]. The molar susceptibility (in c.g.s units $\times 10^6$) is 22500 ± 4000 at 295° K and 58000 ± 6000 at 77° K. The value 22500 is in reasonably good agreement with the theoretical value, 26000, for 295° K, if one assumes an electron configuration $5f^7$ for the ground state configuration.

Table 23. *Prominent lines in the ultraviolet absorption spectra of solid CmF_3 and GdF_3.*

CmF_3 λ, Å	Relative intensity	GdF_3 λ, Å	Relative intensity
2826	2	2834	4
2774	10	2759	10
2680	6	2666	4
2368	4	2334	4

The wavelengths of the most prominent lines in the arc and spark spectra of curium were measured[2] using one microgram samples. The region from 2516 to 5000 Å was covered and over 200 lines were recorded. The isotope shift was observed in curium optical spectra when a mixture of Cm^{242} and Cm^{244} was run; the shift was seen in 148 lines out of the 182 lines recorded[3]. The magnitude of the shift ranged up to 0.131 cm^{-1}.

V. Berkelium (element 97).

50. Discovery of berkelium—summary of known isotopes. Some years elapsed between the identification of americium and curium and the first proof of the synthesis of element 97. This time was needed to overcome the experimental difficulties in working with highly radioactive cyclotron target materials, in improving the necessary ion exchange chemical techniques and in deriving correlations of heavy element nuclear properties to permit prediction of the properties of expected isotopes.

In December 1949 an isotope of the element 97 was synthesized and identified by Thompson, Ghiorso, and Seaborg[4]. The isotope of mass number 243 was made by the (α, 2*n*) reaction in the Berkeley 60-inch cyclotron using milligram amounts of Am^{241} as target material and bombarding with 35-Mev helium ions. Element 97 is a chemical homologue of the rare earth terbium which was named after the town of Ytterby, Sweden; hence the name berkelium (symbol Bk) was proposed in honor of the city of Berkeley.

The isotope, Bk^{243}, decays primarily by orbital electron capture with a half-life of 4.6 hours. Branching decay by alpha particle emission occurs to the extent of about 0.1% and in this decay three groups of alpha particles of energies 6.72, 6.55, and 6.20 Mev are emitted.

A number of other isotopes of berkelium have been made by cyclotron bombardments of americium or curium targets or by intense and prolonged neutron irradiation of plutonium or americium samples. The principal decay characteristics of these isotopes are listed in Table 24 and Fig. 26. The most intense sources

[1] W. W. T. Crane, B. B. Cunningham and J. C. Wallmann: University of California Radiation Laboratory Unclassified Report UCRL-846, 1950.

[2] J. G. Conway, M. F. Moore and W. W. T. Crane: J. Amer. Chem. Soc. **73**, 1308 (1951).

[3] J. G. Conway and R. D. McLaughlin: University of California Radiation Laboratory Report UCRL-3037, 1955. J. Opt. Soc. Amer. **46**, 91 (1956).

[4] S. G. Thompson, A. Ghiorso and G. T. Seaborg: Phys. Rev. **80**, 781 (1950).

Table 24. *Isotopes of berkelium**.

Isotope and half-life	Type of decay	Energy (Mev)	Method of production	Disintegration energies (Mev) **
Bk^{243} 4.5 hours	E. C. 99^{+}% α 0.15% (Ref. 1) γ	 6.72 (30%) 6.55 (53%) 6.20 (17%) 0.740, 0.840, 0.960 following E.C. 0.147, 0.187, 0.540 following α	$Am^{241}(\alpha, 2n)$ $Cm^{242}(d, n)$ $Am^{243}(\alpha, 4n)$	$Q_{E.C.}=1.54$ (calc.) $Q_{\alpha}=6.83$
Bk^{244} 4.4 hours (Ref. 2)	E. C. α 6×10^{-5}% (Ref. 2) γ	 6.66 (100%) (Ref. 2) 0.900, 1.060, 1.160, 1.230, 1.370, 1.500, 1.720 (Ref. 2)	$Am^{241}(\alpha, n)$ $Am^{243}(\alpha, 3n)$	$Q_{E.C.}=2.32$ (est.) $Q_{\alpha}=6.77$
Bk^{245} 5.0 days	E. C. 99^{+}% α 0.1% γ	 5.89 (26%), 6.17 (41%) 6.37 (33%) (Ref. 2, 6) 0.252, 0.380 (Ref. 2) following E. C. 0.165, 0.205, 0.480 following α	$Cm^{244}(d, n)$ $Am^{243}(\alpha, 2n)$	$Q_{E.C.}=0.80$ (calc.) $Q_{\alpha}=6.44$
Bk^{246} 1.9 days (Ref. 2)	E.C. γ	 0.145, 0.800, 0.980, 1.030, 1.130 (Ref. 2)	$Am^{243}(\alpha, n)$	$Q_{E.C.}=1.34$ (est.) $Q_{\alpha}=6.15$ (est.)
Bk^{247} 7×10^{3} years (Ref. 2)	α γ	5.67 (30%), 5.50 (60%) 5.30 (10%) (Ref. 2) 0.085, 0.265, 0.420(?) following α	$Cm^{244}(\alpha, n)$ daughter Cf^{247}	$Q_{\alpha}=5.85$ $Q_{E.C.}=0.04$ (est.)
Bk^{248} 18 hours (Ref. 2) 23 hours (Ref. 7)	β^- (75%) E.C. (25%) $\frac{K}{L+M}\sim1$	0.670	daughter E^{252} $Bk^{247}(n, \gamma)$	$Q_{E.C.}=0.65$ (est.) $Q_{\beta}=0.66$ $Q_{\alpha}=5.52$ (calc.
Bk^{249} 270 days (Ref. 3) 290 days (Ref. 4) 6×10^{8} years spon. fiss.	β^- α 10^{-3}% (Ref. 4) γ	0.100 (Ref. 2) 5.40 (94%), 5.08 (6%) (Ref. 2) 0.320 with α	Pu^{239} multiple n capture	$Q_{\beta}=0.10$ $Q_{\alpha}=5.51$
Bk^{250} 3.13 hours (Ref. 5)	β^- γ	0.9, 1.9 (Ref. 5) 0.9 (Ref. 5)	$Bk^{249}(n, \gamma)$ (Ref. 5)	$Q_{\beta}=1.90$ $Q_{\alpha}=5.73$ (calc.)

* For references to original data not given specifically here, refer to HOLLANDER, PERLMAN and SEABORG, "Table of Isotopes", [Rev. Mod. Phys. **25**, 469 (1953)] and to the compilations of the Nuclear Data Group, National Research Council.

** Revised values from R. A. GLASS, S. G. THOMPSON and G. T. SEABORG: J. Inorg. a. Nucl. Chem. **1**, 3 (1955).

[1] E. K. HULET: Ph. D. Thesis, University of California Radiation Laboratory Report, UCRL-2283, 1953.

[2] A. CHETHAM-STRODE: Ph. D. Thesis, University of California Radiation Laboratory Report, UCRL-3322, 1956.

[3] S. G. THOMPSON *et al.*: Phys. Rev. **93**, 908 (1954).

[4] L. B. MAGNUSSON *et al.*: Phys. Rev. **96**, 1576 (1954).

[5] A. GHIORSO *et al.*: Phys. Rev. **94**, 1081 (1954).

[6] L. B. MAGNUSSON *et al.*: Phys. Rev. **1956**.

[7] E. K. HULET: Phys. Rev. **102**, 182 (1956).

of berkelium have been made in nuclear reactors; samples of berkelium of 10^8 disintegrations per min of Bk^{249} (approximately 0.025 micrograms) have been isolated and considerably larger amounts can be made. All chemical studies to date have been carried out on a tracer scale, but weighable amounts of berkelium can be prepared by multiple neutron capture reactions such as the following:

$$\left.\begin{array}{l} Pu^{239}(n,\gamma)Pu^{240}(n,\gamma)Pu^{241}(n,\gamma)Pu^{242}(n,\gamma)Pu^{243}\xrightarrow[5.0\text{ hour}]{\beta^-}Am^{243}(n,\gamma)Am^{244} \\ \xrightarrow{\beta^-}Cm^{244}(n,\gamma)Cm^{245}(n,\gamma)Cm^{246}(n,\gamma)Cm^{247}(n,\gamma)Cm^{248}(n,\gamma)Cm^{249} \\ \xrightarrow{\beta^-}Bk^{249}. \end{array}\right\} \quad (50.1)$$

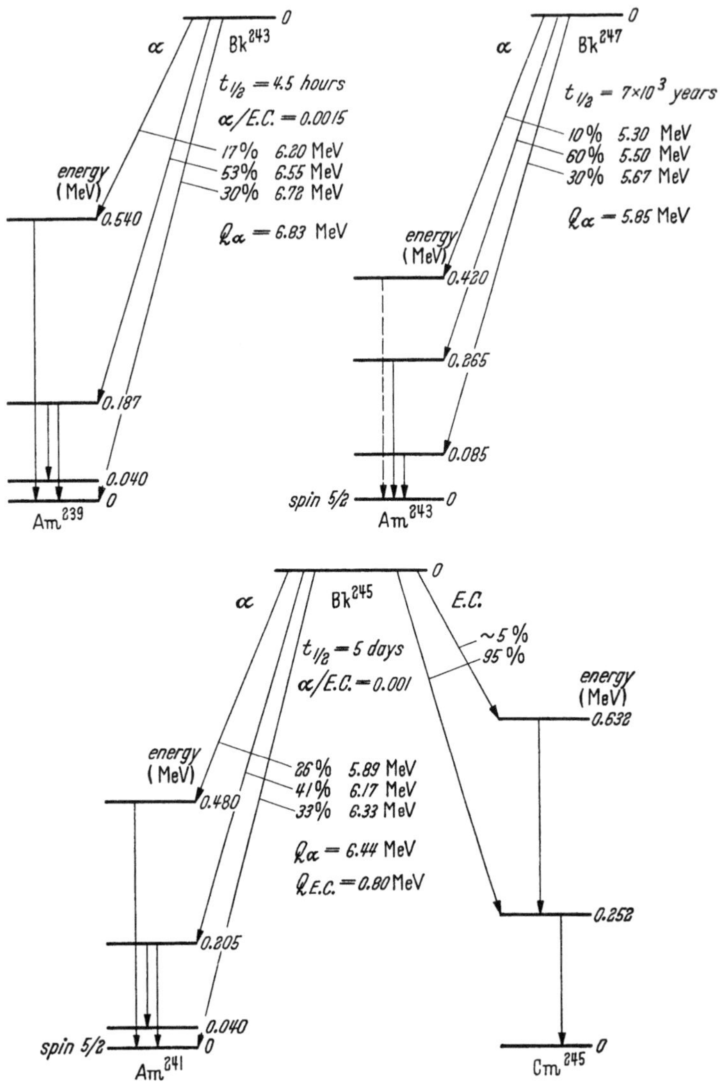

Fig. 26. Decay schemes of berkelium isotopes.

Alternatively one could carry out this sequence in two stages by isolating and chemically purifying the curium fraction, then reinserting it in the reactor for the final conversion to berkelium.

With larger amounts of Bk^{249} or preferably of 7000-year Bk^{247} it will be possible to characterize many chemical compounds and to measure important thermodynamic quantities with accuracy.

51. Chemical properties of berkelium. The chemical properties of the tripositive solution state of berkelium are markedly similar to the neighboring tripositive actinide element ions and to the tripositive rare-earths. Since rare-earth fission products and neighboring actinide elements are always present when berkelium is produced, its separation from these elements is of considerable importance. The ion exchange elution technique is of primary importance. To separate berkelium from Am^{241} cyclotron targets, THOMPSON, GHIORSO, and SEABORG[1] proceeded as follows:

The bombarded Am^{241} was dissolved in nitric acid, precipitated as hydroxide (carrying the berkelium and the rare-earths), separated and redissolved in 0.1 molar nitric acid. Ammonium peroxydisulfate was added and the solution heated to 75° C for $1\frac{1}{2}$ hours. The major portion of the americium was oxidized to fluoride-soluble americium (VI). Removing this bulk of material improves the sharpness and speed of later ion exchange separations. The residual americium (III) was precipitated as the trifluoride to serve as a carrier for berkelium, curium and rare earth fission products. The fluoride precipitate was metathesized to the hydroxide, washed, and dissolved in 0.5 molar perchloric acid. The mixture of radioactivities was absorbed on the top of a 20 cm length by 2 millimeter diameter column of Dowex-50 cation resin. This column was operated at 87° C. The elution was performed by passing 0.25 molar ammonium citrate buffered with citric acid to p_H 3.5 through the column at the rate of one drop every two minutes. Element 97 was found in drops 40 through 45 and for comparison curium was collected in drops 65 to 70. In order to separate element 97 from residual rare-earth contaminants (principally terbium and gadolinium) the berkelium fraction was adsorbed on a second column of Dowex-50 resin of approximately 7 cm length. Elution with 13 molar hydrochloric acid removed the berkelium well ahead of rare earth contaminants.

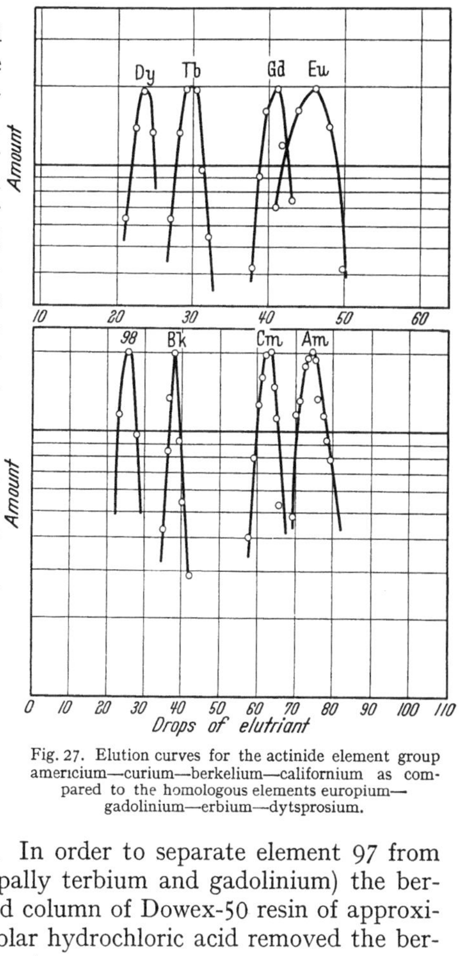

Fig. 27. Elution curves for the actinide element group americium—curium—berkelium—californium as compared to the homologous elements europium—gadolinium—erbium—dytsprosium.

An interesting aspect of the elution behavior of berkelium is the striking similarity in the spacing of the elution peaks in the sequence americium—curium—berkelium—californium as compared to europium—gadolinium—terbium—dysprosium. The spacings are shown in Fig. 27. The gap in the center of the curves is probably a reflection of a break in the change of ionic radius which occurs in both series just after the half-filling of the $4f$ or $5f$ electron shell. The behavior of berkelium in this regard was predicted by BOYD[2].

[1] S. G. THOMPSON, A. GHIORSO and G. T. SEABORG: Phys. Rev. **80**, 781 (1950).
[2] G. E. BOYD: U.S. Atomic Energy Commission Report MDDC 840, April 16, 1947.

Recent studies in ion exchange elution have shown that buffered lactic acid[1,2] and particularly buffered α-hydroxy-isobutyric acid[3,4] are superior to ammonium citrate see Fig. 31, Sect. 55).

The elution of the berkelium ahead of the rare-earths when 13 molar hydrochloric acid is used as eluting agent indicates that berkelium forms stronger chloride complexes than the lanthanide ions. More striking evidence for this is the fact that berkelium can be adsorbed from 13 molar hydrochloric acid onto a column of anion-exchange resin such as Dowex-1. This adsorption is weak and the berkelium can readily be desorbed with 13 molar hydrochloric acid, but there is an appreciable delay after the elution of americium and curium (see Fig. 30, Sect. 55). Stronger adsorption on anion resins is achieved from thiocyanate solution[5].

The coprecipitation behavior of tripositive berkelium is identical with that of other actinide elements and coprecipitation is observed with such typical carrier compounds as lanthanum fluoride or lanthanum hydroxide. The nitrate, chloride, sulfate, bromide, perchlorate, and sulfide salts are apparently soluble. Berkelium can be extracted readily into benzene as a chelate complex of α-thenoyltrifluoroacetone[6] (TTA). At p_H 3.4 the extraction of berkelium and californium is roughly a factor of ten greater than that of americium and curium.

It was anticipated by SEABORG[7] that element 97 as ekaterbium in the actinide transition series would possess a tetrapositive oxidation state. Berkelium (IV) has the half-completed shell configuration $5f^7$ which has a special stability. Such an oxidation state has indeed been found[8] by tracer experiments in which treatment with bromate ion, dichromate ion, or ceric ion caused oxidation of berkelium to a form coprecipitated with zirconium phosphate or zirconium iodate, compounds which are rather specific carriers for tetrapositive ions. The oxidation potential of the berkelium (III)—berkelium (IV) couple is about −1.6 volts[8].

It may be expected that milligram quantities of berkelium in the form of 7000 year Bk^{247} will eventually be available and that the chemistry of the element including characterization of solid compounds and measurement of thermodynamic quantities will be at least as extensive as is presently the case for americium.

VI. Californium (element 98).

52. Discovery of californium—summary of known isotopes. Following within two months the first successful identification of berkelium, the synthesis and identification of an isotope of element 98 was carried out by THOMPSON, STREET, GHIORSO, and SEABORG[9] in February 1950. This first isotope of element 98, tentatively assigned the mass number 244, was produced when microgram amounts

[1] R. A. GLASS: J. Amer. Chem. Soc. **77**, 807 (1955).
[2] L. WISH, E. C. FREILING and L. R. BUNNEY: J. Amer. Chem. Soc. **76**, 3444 (1954).
[3] G. R. CHOPPIN, B. G. HARVEY and S. G. THOMPSON: University of California Radiation Laboratory Report UCRL-3022. J. Inorg. a. Nucl. Chem. **2**, 66 (1956).
[4] G. R. CHOPPIN and R. J. SILVA: University of California Radiation Laboratory Report UCRL-3265, 1956; J. Inorg. a. Nucl. Chem. **3**, 153 (1956).
[5] J. SURLS: Thesis, University of California Radiation Laboratory Report UCRL-3209, 1955.
[6] L. B. MAGNUSSEN and M. L. ANDERSON: J. Amer. Chem. Soc. **76**, 6207 (1954).
[7] G. T. SEABORG: Nucleonics **5** (5), 16 (1949). — Paper 22.1, N.N.E.S., Vol. 14B, "The Transuranium Elements", 1949.
[8] S. G. THOMPSON, B. B. CUNNINGHAM and G. T. SEABORG: J. Amer. Chem. Soc. **72**, 2798 (1950).
[9] S. G. THOMPSON, K. STREET jr., A. GHIORSO and G. T. SEABORG: Phys. Rev. **80**, 790 (1950).

of Cm^{242} (about 10^{11} alpha disintegrations per minute) were bombarded with 35 Mev helium ions in the 60 inch cyclotron of the University of California at Berkeley. The isotope decayed partly by the emission of alpha particles of energy 7.1 Mev and partly by electron capture with a half-life of 45 min. At the time of the first experiments the tentative mass assignment was 244. Later[1] it was established that the correct mass number was 245. Many serious experimental problems had to be overcome in bombarding safely such a highly radioactive target and in swiftly making complete chemical separation of the minute amount of the new element from the chemically very similar heavy elements and the prominent rare-earth fission product activity.

The name californium, symbol Cf, was chosen in honor of the university and the state where the work was done. Californium is the eighth member of the actinide series and is the homologue of dysprosium of the rare-earth series.

A number of californium isotopes ranging in mass from 244 to 254 have been synthesized. Data on californium isotopes are summarized in Table 25. Decay schemes are shown in Fig. 28. A number of these have been made by (α, xn) reactions on curium targets. Another method for their production is the bombardment of targets of lower atomic number with accelerated heavy ions[2,3]. In this manner Cf^{244} and Cf^{246} have been produced by bombardment of uranium targets with carbon ions accelerated to approximately 120 Mev; similarly Cf^{245}, Cf^{246}, Cf^{247}, and Cf^{248} have been produced by the bombardment of uranium targets with nitrogen ions accelerated to 140 Mev. A sample reaction of this type is:

$$U^{238} + C^{12} \longrightarrow Cf^{246} + 4\,{}_0n^1. \tag{52.1}$$

By far the best method for the production of larger amounts of californium is the bombardment of gram or larger amounts of heavy element isotopes such as Pu^{239} in modern high flux reactors. Successive neutron capture reactions proceeding through a sequence such as the following can convert substantial amounts of lighter mass material into high mass californium isotopes. Considerable material is lost through fission in the first few steps, but beyond this point a considerable fraction of the cross-section goes into capture reactions.

$$Pu^{239}(n,\gamma)\,Pu^{240}(n,\gamma)\,Pu^{241}(n,\gamma)\,Pu^{242}(n,\gamma)\,Pu^{243} \xrightarrow[5.0\text{ hour}]{\beta^-} Am^{243}(n,\gamma)$$
$$Am^{244} \xrightarrow[26\text{ min}]{\beta^-} Cm^{244}(n,\gamma)\,Cm^{245}(n,\gamma)\,Cm^{246}(n,\gamma)\,Cm^{247}(n,\gamma)\,Cm^{248}$$
$$(n,\gamma)\,Cm^{249} \xrightarrow[\text{short}]{\beta^-} Bk^{249}(n,\gamma)\,Bk^{250} \xrightarrow[3.13\text{ hour}]{\beta^-} Cf^{250}(n,\gamma)\,Cf^{251}(n,\gamma)$$
$$Cf^{252}(n,\gamma)\,Cf^{253}(n,\gamma)\,Cf^{254}.$$

Experiments with californium fractions isolated from highly irradiated plutonium samples (nvt of the order of 10^{22}) have revealed substantial new information[4,5] on heavier isotopes of the element. The isotopic composition[4] of the californium fraction from such an irradiation was determined mass spectrographically to be 4.3% Cf^{249}, 49% Cf^{250}, 11% Cf^{251}, and 36% Cf^{252}. Data on these isotopes are given in Table 25.

---

[1] A. Chetham-Strode jr., G. R. Choppin and B. G. Harvey: Phys. Rev. **102**, 747 (1956).

[2] A. Ghiorso, S. G. Thompson, K. Street jr. and G. T. Seaborg: Phys. Rev. **81**, 154 (1951).

[3] A. Ghiorso, G. B. Rossi, B. G. Harvey and S. G. Thompson: Phys. Rev. **93**, 257 (1954).

[4] H. Diamond *et al.*: Phys. Rev. **94**, 1083 (1954). — L. B. Magnusson *et al.*: Phys. Rev. **96**, 1576 (1954).

[5] A. Ghiorso, S. G. Thompson, G. R. Choppin and B. G. Harvey: Phys. Rev. **94**, 1081 (1954).

Table 25. *Isotopes of californium**.

Isotope and half-life	Type of decay	Energy (Mev)	Method of production	Disintegration ** energies (Mev)
Cf^{244} 25 min (Ref. 1)	α	7.17 (Ref. 2)	$U^{238}(C, 6n)$ $Cm^{242}(\alpha, 2n)$	$Q_\alpha = 7.29$ $Q_{E.C.} = 0.60$ (est.)
Cf^{245} 44 min (mass reassignment Ref. 1)	E.C. (66%) α (34%)	7.11	$Cm^{242}(\alpha, n)$ $Cm^{244}(\alpha, 3n)$	$Q_{E.C.} = 1.64$ (est.) $Q_\alpha = 7.23$
Cf^{246} 35.7 hours α 2.1×10^3 years spon. fiss.	α γ	6.711 (22%), 6.753 (78%) (Ref. 3) 0.044 (Ref. 3), 0.097 ± 0.003 (1.3×10^{-4}), 0.149 ± 0.005 (3.4×10^{-5}) (Ref. 12)	$Cm^{244}(\alpha, 2n)$ $U^{238}(C, 4n)$ $Cm^{243}(\alpha, n)$ $U^{238}(N, p\,5n)$	$Q_\alpha = 6.86$ $Q_{E.C.} = 0.12$ (est.)
Cf^{247} (Ref. 4) 2.4 hours (Ref. 2)	E.C. γ	0.285, 0.490 (Ref. 2)	$Cm^{244}(\alpha, n)$ (Ref. 2) $U^{238}(N, p\,4n)$ (Ref. 4)	$Q_{E.C.} = 0.77$ (est. $Q_\alpha = 6.62$ (est.
Cf^{248} (Ref. 4) 250 days 7×10^3 years spon. fiss.	α β stable	6.26 (Ref. 5)	$U^{238}(N, p\,3n)$ (Ref. 4) $Cm^{245}(\alpha, n)$ (Ref. 5)	$Q_\alpha = 6.36$
Cf^{249} 400 years α (Ref. 6) 470 years α (Ref. 7) 1.5×10^9 years spon. fiss.	α β stable γ	5.82 (90%) (Ref. 6) 6.190 (3%) 5.91 (6%) (Ref. 11) 0.25, 0.053, 0.265 (?)	daughter Bk^{249}	$Q_\alpha = 6.31$
Cf^{250} 12 years α (Ref. 6) 10 years α (Ref. 7) 1.5×10^4 years spon. fiss.	α β stable γ	5.980 (17%), 6.024 (83%) 0.0428 (0.014%)	daughter Bk^{250} $Bk^{249}(n, \gamma\beta)$	$Q_\alpha = 6.12$
Cf^{251} Long (Ref. 7)	β stable		Pu^{239} multiple n capture	$Q_\alpha = 6.32$ (est.)
Cf^{252} 2.2 years α (Ref. 7) 66 years spon. fiss.	α β stable γ	6.112 (84.5%), 6.069 (15.5%) (Ref. 8) 0.0434, 0.100 (Ref. 11)	Pu^{239} multiple n capture	$Q_\alpha = 6.28$
Cf^{253} 20 days (Ref. 6) 18 days (Ref. 7)	β^-		Pu^{239} multiple n capture	$Q_{\beta^-} = 0.54$ (est.) $Q_\alpha = 6.20$ (est.)
Cf^{254} 85 days (Ref. 9) 60 days (Ref. 10)	spon. fiss. β stable		daughter E^{254}	$Q = 6.03$ (est.)

* For references to original data not given specifically here, refer to Hollander, Perlman and Seaborg, "Table of Isotopes", [Rev. Mod. Phys. **25**, 469 (1953)] and to the compilations of the Nuclear Data Group, National Research Council.

** Revised values from Glass, Thompson and Seaborg: J. Inorg. a. Nucl. Chem. **1**, 3 (1955).

[1] A. Chetham-Strode *et al.*: University of California Radiation Laboratory Report UCRL-3219, Dec. 1955. See also Phys. Rev. **102**, 747 (1956).

[2] A. Chetham-Strode: Ph. D. Thesis, University of California Radiation Laboratory Report UCRL-3322, 1956.

[3] J. P. Hummel *et al.*: Phys. Rev. **98**, 22 (1955).

References of table 25.

[4] A. GHIORSO *et al.*: Phys. Rev. **93**, 257 (1954).
[5] E. K. HULET *et al.*: Phys. Rev. **95**, 1702 (1954).
[6] A. GHIORSO *et al.*: Phys. Rev. **94**, 1081 (1954).
[7] L. B. MAGNUSSON *et al.*: Phys. Rev. **96**, 1576 (1954).
[8] I. PERLMAN *et al.*: Reported in University of California Radiation Laboratory Reports UCRL-2932, March 1955 and UCRL-3068, July 1955. — F. ASARO *et al.*: Pys. Rev. **100**, 137 (1955).
[9] B. G. HARVEY *et al.*: Phys. Rev. **99**, 337 (1955).
[10] W. C. BENTLEY *et al.*: Paper P/809, Peaceful Uses of Atomic Energy—Proceedings of the Geneva Conference, August 1955, United Nations.
[11] F. S. STEPHENS: Ph. D. Thesis, University of California Radiation Laboratory Report UCRL-2970, April 1955.
[12] I. PERLMAN *et al.*: University of California Radiation Laboratory Report UCRL-3068, p. 37, July 1955.

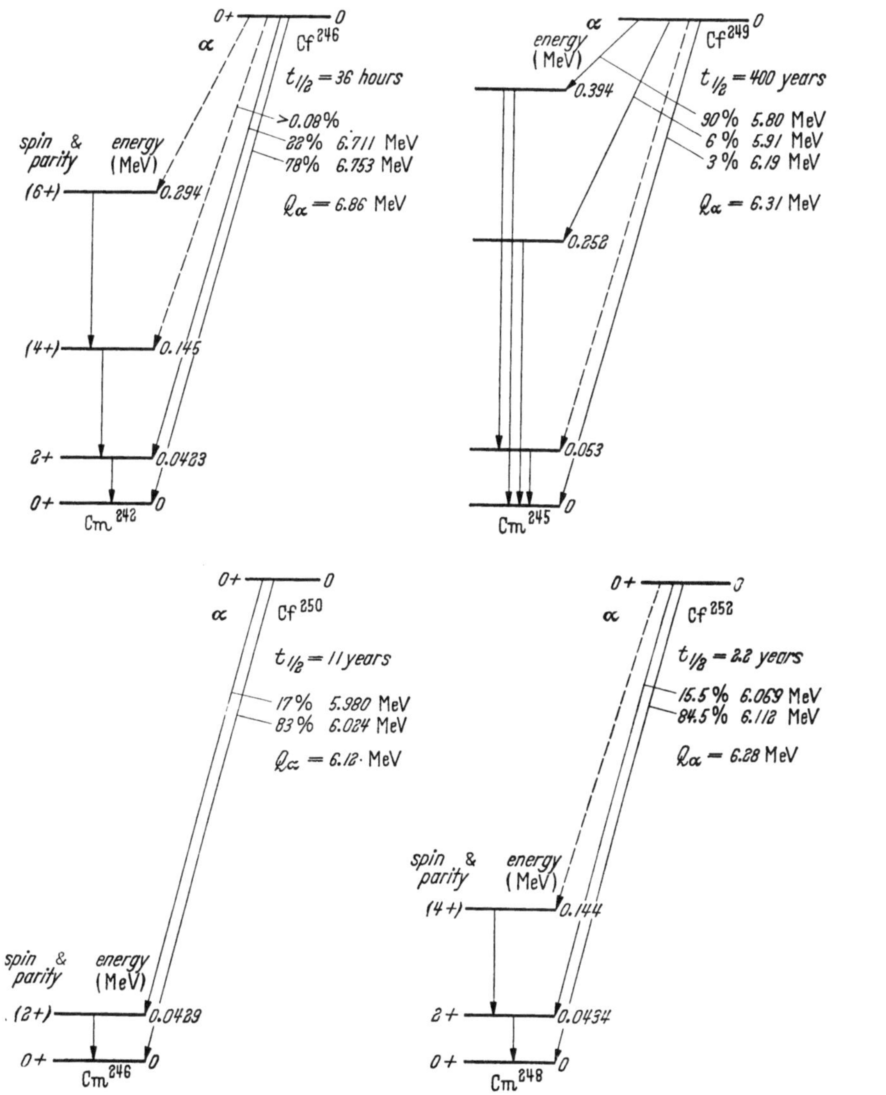

Fig. 28. Decay schemes of californium isotopes.

Eventually it will be possible to prepare the isotopes Cf^{249}, Cf^{250}, Cf^{251}, and Cf^{252} in weighable amounts so that the chemical properties of californium can be investigated in considerably more detail.

The spontaneous fission characteristics of californium are quite important and will be summarized later in this report.

53. Chemical properties of californium[1-3]. The only oxidation state known is the tripositive state. The californium (III)—californium (IV) couple is expected to be much more negative than -1.6 volts which would make californium (IV) unobtainable in aqueous solution. There is a possibility that californium (V) may be found in the future because of the special stability of the $5f^7$ configuration. Determined efforts have been made to oxidize it with such powerful oxidants as peroxydisulfate and bismuthate ions without success. So far as is known the chemistry of californium is nearly identical with that of the other tripositive actinide ions. It can be coprecipitated with such typical carriers as lanthanum fluoride and lanthanum oxalate. The nitrate, sulfate, bromide, chloride, iodide, sulfide and perchlorate appear to be soluble. The principal method of chemical purification is ion exchange. Californium adsorbs on cation exchange resin columns and can be eluted ahead of berkelium with buffered citrate, lactate, alpha-hydroxy-isobutyrate and similar eluting agents (see for example, Fig. 31, Sect. 55). It can be eluted with 13 molar hydrochloric acid from cation exchange columns; it elutes just ahead of berkelium and well ahead of the first rare earth element (see Fig. 30, Sect. 55). Californium forms anionic chloride complexes in 13 molar hydrochloric acid and can be adsorbed on an anion resin column. The californium is readily eluted with more 13 molar hydrochloric acid but with sufficient delay to effect a separation from berkelium, americium, and curium, and, of course, the rare-earth elements which do not adsorb on the column at all (see for example, Fig. 30, Sect. 55).

Our present knowledge of californium chemistry has all been obtained with tracer amounts. Later studies with weighable amounts should provide considerable quantitative information.

VII. Einsteinium (element 99).

54. Discovery of einsteinium—summary of known isotopes. The first identification of an isotope of element 99 and of element 100 came about as the result of careful analysis of heavy element samples from the "Mike" thermonuclear explosion staged by the Los Alamos Scientific Laboratory in November 1952. The uranium in this device was subjected to a very intense instantaneous neutron flux which by multiple neutron capture reactions of a very high order gave rise to very heavy uranium isotopes which rapidly decayed by beta emission into previously uninvestigated heavy isotopes of neptunium, plutonium, americium, curium, berkelium, californium, and of the elements 99 and 100. Three teams of nuclear chemists[4] working at the University of California Radiation Laboratory, the Argonne National Laboratory, and the Los Alamos Scientific Laboratory contributed to the chemical isolation and identification of the new elements in December 1952 and January 1953. The scientists involved were A. Ghiorso, S. G. Thompson, G. H. Higgins, and G. T. Seaborg of the University of Cali-

[1] S. G. Thompson, K. Street jr., A. Ghiorso and G. T. Seaborg: Phys. Rev. **80**, 790 (1950).

[2] K. Street jr., S. G. Thompson and G. T. Seaborg: J. Amer. Chem. Soc. **72**, 4832 (1950).

[3] S. G. Thompson, B. G. Harvey, G. R. Choppin and G. T. Seaborg: J. Amer. Chem. Soc. **76**, 6229 (1954).

[4] Phys. Rev. **99**, 1048 (1955).

Table 26. *Isotopes of einsteinium.*

Isotope and half-life	Type of decay	Energy (Mev)	Method of production
E^{246} min (Ref. 1)	E.C.		$U^{238} + {}_7N^{14}$
E^{247} (?) 7.3 min (Ref. 1)	α	7.35 (Ref. 1)	$U^{238} + {}_7N^{14}$
E^{248} 25 min (Ref. 14)	E.C. >99% α	 6.87 (Ref. 14)	 Cf^{249} $(d, 3n)$ (Ref. 14)
E^{249} 2 hours (Ref. 2)	α E.C. (>99%)	6.76 (Ref. 2)	Bk^{249} $(\alpha, 4n)$ (Ref. 2)
E^{250} (?) 7 hours (Ref. 2)	α, E.C. E.C. (>98%)		Bk^{249} $(\alpha, 3n)$ (Ref. 2)
E^{251} 1.5 days (Ref. 2)	α E.C. >99%	6.48 (Ref. 2)	Bk^{249} $(\alpha, 2n)$ (Ref. 2)
E^{252} ~150 days	α	6.64 (Ref. 2)	Bk^{249} (α, n) (Ref. 2)
E^{253} (Ref. 3) 19.3 days (Ref. 4) 20 days (Ref. 5, 7, 13) 7×10^5 years Spon fiss (Ref. 13)	α	 6.61 (Ref. 4) 6.63 (Ref. 5) 6.633 (91%) 6.592 (8%) 6.543 (1%) 6.491 (0.3%) (Ref. 6)	U^{238} (MNC)* (Ref. 3) Pu^{239} (MNC)* (Ref. 7, 8) β^- daughter Cf^{253} (Ref. 3) E.C. daughter Fm^{253} (Ref. 2)
$E^{254\,m}$ 36 hours (Ref. 5) 37 hours (Ref. 4)	β^- E.C. (0.1%) (Ref. 10, 12) I.T. not ob.	1.1 (Ref. 5) 1.04 (Ref. 13)	E^{253} (n, γ) (Ref. 9, 4)
E^{254} 272 days (Ref. 2) 320 days (Ref. 13)	α	6.44 (Ref. 10)	E^{253} (n, γ) (Ref. 10)
E^{255} 30 days (Ref. 3, 5) 24 days (Ref. 13)	β^-		U^{238} (MNC)* (Ref. 3)
E^{256} short (hours or less)	β^-		E^{255} (n, γ) (Ref. 11)

* MNC means multiple neutron capture.

[1] A. GHIORSO *et al.*: Phys. Rev. **93**, 257 (1954).

[2] B. G. HARVEY, S. G. THOMPSON, G. R. CHOPPIN, A. GHIORSO and A. CHETHAM: Phys. Rev. **104**, 1315 (1956).

[3] A. GHIORSO *et al.*: Phys. Rev. **99**, 1048 (1955).

[4] P. R. FIELDS *et al.*: Phys. Rev. **94**, 209 (1954).

[5] G. R. CHOPPIN *et al.*: Phys. Rev. **94**, 1080 (1954).

[6] J. HUMMEL: University of California Radiation Laboratory, Ph. D. Thesis 1956, UCRL-3456.

[7] M. H. STUDIER *et al.*: Phys. Rev. **93**, 1428 (1954).

[8] S. G. THOMPSON *et al.*: Phys. Rev. **93**, 908 (1954).

[9] B. G. HARVEY *et al.*: Phys. Rev. **93**, 1129 (1954).

[10] B. G. HARVEY *et al.*: Phys. Rev. **99**, 337 (1955).

[11] G. R. CHOPPIN *et al.*: Phys. Rev. **98**, 1519 (1955).

[12] W. C. BENTLEY *et al.*: Paper P/809, "Peaceful Uses of Atomic Energy—Proceedings of the Geneva Conference", August 1955, United Nations.

[13] M. JONES *et al.*: Phys. Rev. **102**, 203 (1956).

[14] A. CHETHAM-STRODE and L. W. HOLM: Phys. Rev. **104**, 1314 (1956).

fornia, M. H. Studier, P. R. Fields, S. M. Fried, H. Diamond, J. F. Mech, G. L. Pyle, J. R. Huizenga, A. Hirsch and W. M. Manning of the Argonne National Laboratory, and C. I. Browne, H. L. Smith, and R. W. Spence of the Los Alamos Laboratory.

The names which have been suggested and accepted for these elements are einsteinium (symbol *E*) after Albert Einstein, for element 99 and fermium (symbol Fm) after Enrico Fermi, for element 100.

Einsteinium and fermium are actinide elements with chemical properties very similar to other actinide elements in the tripositive state. Hence, as in the identification of berkelium and californium, the ion exchange elution technique played a prominent role in the purification and identification. Rare-earth contaminants were removed by adsorbing the contaminated heavy element fractions on Dowex-50 cation resin and eluting with 13 molar hydrochloric acid. The partially purified heavy element fraction was then adsorbed on another column of Dowex-50 resin and eluted carefully with ammonium citrate solution to effect separation of the individual actinide elements. These experiments showed the elution of a 6.6 Mev alpha activity ahead of the element californium and in the expected eka-holmium (element 99) position. Also a new 7.1 Mev alpha activity was found to elute in the eka-erbium (element 100) position. The conclusions from the detailed experiments were the following: An appreciable amount of U^{253} had formed in the initial nuclear explosion. This rapidly decayed by a chain of short-lived beta emissions to Cf^{253} which decayed by beta emission with a half-life of approximately 20 days to E^{253} which decayed with the emission of 6.6 Mev alpha particles with a half-life of about 20 days. Similarly an appreciable amount of U^{255} had formed instantaneously and had decayed via a long beta-emitter chain to E^{255} which decayed by beta emission with a half-life of approximately 30 days to Fm^{255} which decayed by the emission of 7.1 Mev alpha particles with a half-life of approximately 16 hours.

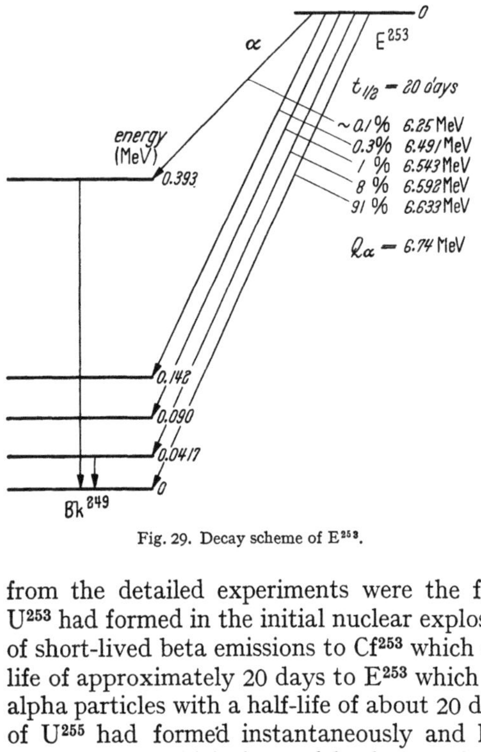

Fig. 29. Decay scheme of E^{253}.

The bombardment of uranium targets[1] with nitrogen ions accelerated to more than 100 Mev energy in a cyclotron has produced small amounts of short-lived activities tentatively assigned to E^{246} and E^{247}.

By neutron irradiation of Pu^{239} in the high flux Materials Testing Reactor in Idaho for extended periods of time (integrated flux between 10^{21} and 10^{22} neutrons) it has been possible to produce californium isotopes containing substantial percentages of higher mass isotopes. (See Sect. 52); by prolonged reirradiation the mass of californium isotopes was increased until the beta emitter Cf^{253} was reached. The daughter isotope, 20 day E^{253} was isolated from these samples[2-4].

[1] A. Ghiorso, G. B. Rossi, B. G. Harvey and S. G. Thompson: Phys. Rev. **93**, 257 (1955).
[2] S. G. Thompson, A. Ghiorso, B. G. Harvey and G. R. Choppin: Phys. Rev. **93**, 908 (1954).
[3] M. H. Studier *et al.*: Phys. Rev. **93**, 1428 (1954).
[4] M. Jones *et al.*: Phys. Rev. **102**, 203 (1956).

Irradiation of E^{253} with neutrons results in the production of the isomer pair[1-3] E^{254m} and E^{254}. E^{254m} decays by beta emission with a half-life of 36 hours into Fm^{254} which decays by spontaneous fission or by the emission of 7.2 Mev alpha particles. E^{254m} also decays by orbital electron capture to Cf^{254} in 0.1% of its disintegrations. E^{254} decays by the emission of 6.44 Mev alpha particles with a half-life of 272 days[3,4]. This is the longest lived form of einsteinium.

Evidence has been found for the isotopes E^{249}, E^{251}, and E^{252} produced in helium ion bombardments of Bk^{249} [4].

Data on the isotopes of einsteinium are summarized in Table 26.

The decay scheme of E^{253} is shown in Fig. 29.

55. Chemical properties of einsteinium. Einsteinium is the tenth member of the actinide $5f$ transition series. To date its chemical properties have been studied only on the tracer scale in aqueous solution. Its behavior in every respect is that expected of a tripositive actinide ion. The chemistry of einsteinium has been described by THOMPSON, HARVEY, CHOPPIN, and SEABORG[5]. Einsteinium will coprecipitate with rare-earth hydroxides or fluorides. It will extract into the solvent, tributyl phosphate, from an aqueous dilute nitric solution highly salted with neutral nitrates[6]. It will form a chelate complex with the beta-diketone thenoyl trifluoroacetone, which is readily extracted into benzene from an aqueous solution of p_H 3.4[7].

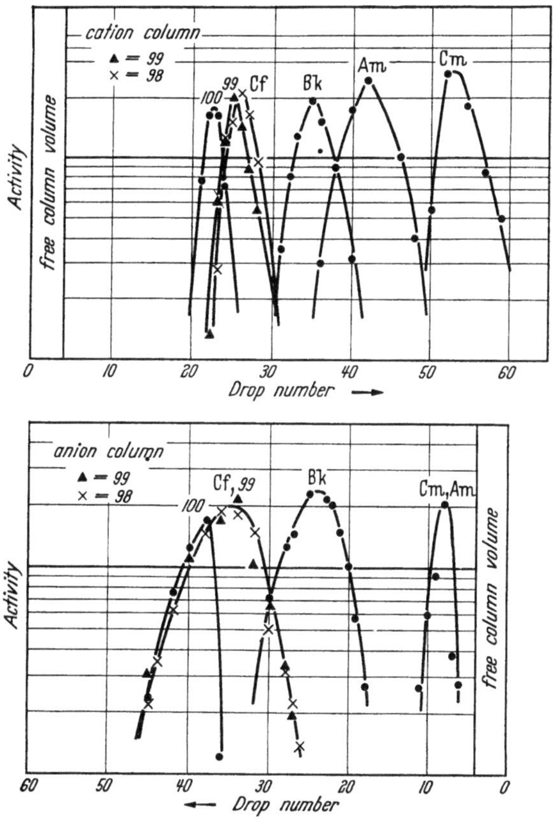

Fig. 30. (Top) Elution of trace amounts of actinide elements from a 5 cm length by 3 mm diameter column of Dowex-50 12% cross-linked collo idal cation exchange resin. Eluting agent—20% alcohol—12.5 molar hydrochloric acid. (Bottom) Elution of trace amounts of actinide elements from 5 cm length by 3 mm diameter column of Dowex-1 anion exchange resin (8% cross-linked) with 13 molar hydrochloric acid. Figure reproduced from S. G. THOMPSON *et al*,; J. Amer. Chem. Soc. 76, 6229 (1954).

The major emphasis in einsteinium chemistry has been concerned with ion exchange elutoin separations of einsteinium from the rare-earths and from the

[1] B. G. HARVEY, S. G. THOMPSON, A. GHIORSO and G. R. CHOPPIN: Phys. Rev. **93**, 1129 (1954).
[2] P. R. FIELDS *et al.*: Phys. Rev. **94**, 209 (1954).
[3] B. G. HARVEY, S. G. THOMPSON, G. R. CHOPPIN and A. GHIORSO: Phys. Rev. **99**, 337 (1955).
[4] B. G. HARVEY, S. G. THOMPSON, G. R. CHOPPIN and A. GHIORSO: Unpublished results, 1955-1956.
[5] S. G. THOMPSON, B. G. HARVEY, G. R. CHOPPIN and G. T. SEABORG: J. Amer. Chem. Soc. **76**, 6229 (1954).
[6] M. H. STUDIER *et al.*: Phys. Rev. **93**, 1428 (1954).
[7] L. B. MAGNUSSEN and M. L. ANDERSON: J. Amer. Chem. Soc. **76**, 6207 (1954).

other actinide elements. Einsteinium is readily absorbed on Dowex-50 cation exchange resin. When 13 molar hydrochloric acid is used as an eluting agent einsteinium is readily desorbed ahead of the rare-earths indicating a stronger tendency to form chloride complexes. Somewhat sharper results are achieved when 20% ethyl alcohol saturated with hydrochloric acid is used as the eluting agent and colloidal Dowex-50 resin, 12% cross-linked is used[1,2]. Fig 30 shows an elution curve using this reagent. The rare earths elute well after curium and are hence very completely separated from einsteinium. The einsteinium elution peak is seen to be cleanly separated from berkelium and lower actinide elements but not from elements 98 or 100.

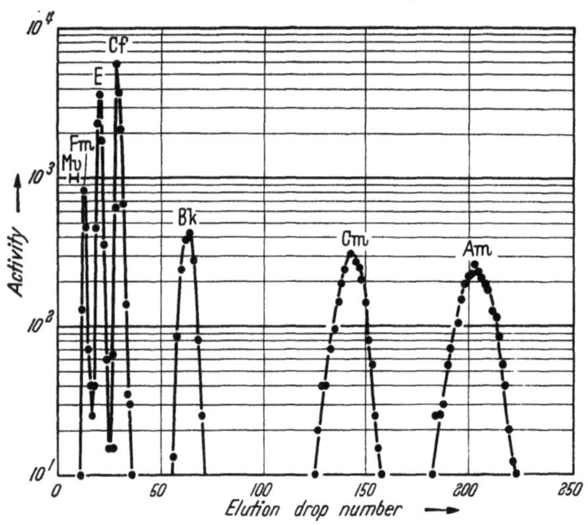

Fig. 31. Elution of actinide elements from a 5 cm × 2 mm column of 12% cross-linked Dowex-50 cation exchange resin by 0.4 molar ammonium alpha-hydroxyisobutyrate. The column was operated at 87° C. Figure reproduced from G. R. Choppin *et al.*: J. Inorg. a. Nucl. Chem. 2, 66 (1956).

After the rare earths are removed by the hydrochloric acid elution separation, the einsteinium is separated from the other actinide elements by adsorbing the mixture of actinide elements on a column of cation exchange resin and eluting with a buffered solution of a suitable organic acid. Some acids which have been studied, in the order of their effectiveness in separating these elements, are citric acid, lactic acid, and α-hydroxyisobutyrate. A sample elution curve with α-hydroxyisobutyrate, the best of these organic acid complexing agents, is shown in Fig. 31[3].

In concentrated solutions of hydrochloric acid einsteinium forms anionic chloride complexes which can be adsorbed on an anion exchange resin such as Dowex-Al. If additional concentrated hydrochloric acid is passed through the anion exchange column the einsteinium is eluted rather quickly from the column, but with a delay sufficient to effect a separation from the actinide elements below californium which elute faster. Fig. 30 shows the extent of the separation which can be achieved. In this separation, the lanthanide elements do not adsorb and appear in the free column volume of elutrient. Instead of hydrochloric acid it is also possible to use very concentrated solutions of lithium chloride. Negatively charged complexes of einsteinium are also found in concentrated solutions of ammonium thiocyanate[4].

Early in 1956 the amounts of einsteinium available for study were in the range of 10^7 alpha disintegrations per minute of 20 day E^{253}. Due to the existence of the long-lived E^{254} isomer weighable amounts of einsteinium will be available eventually for more detailed chemical and physical studies.

[1] See footnote 5, p. 281.

[2] K. Street jr.: Unpublished work.

[3] G. R. Choppin, B. G. Harvey and S. G. Thompson: J. Inorg. a. Nucl. Chem. **2**, 66 (1956). G. R. Choppin and R. Silva: University of California Radiation Laboratory Report UCRL-3265, 1956. J. Inorg. a. Nucl. Chem. **3**, 153 (1956)

[4] J. P. Surls jr.: Unpublished results quoted by S. G. Thompson *et al.* in footnote 5, p. 281; see also J. P. Surls jr., University of California Radiation Laboratory Report UCRL-3209, 1955.

VIII. Fermium (element 100).

56. Discovery of fermium—summary of known isotopes. The first isotope of element 100 to be identified was 100^{255}, a 16 hour activity decaying by the emission of 7.1 Mev alpha particles. It was found in the heavy element fractions from the "Mike" thermonuclear explosion of November 1952 in the same set of experiments which resulted in the discovery of einsteinium, element 99. These experiments are summarized briefly in Sect. 54. It was also recorded in Sect. 54 that the name fermium (symbol Fm) was given to element 100 in honor of ENRICO FERMI.

A second isotope of fermium, Fm^{254}, is produced when californium samples rich in Cf^{252},[1] or einsteinium samples rich in E^{253},[2,3] are subjected to intense neutron irradiation. The reaction sequence is:

$$\begin{array}{l} Cf^{252}(n,\gamma)\,Cf^{253} \\ \quad \beta^- \downarrow 20 \text{ days} \\ \qquad E^{253}(n,\gamma)\,E^{254m} \\ \qquad\quad \beta^- \downarrow 36 \text{ hours} \\ \qquad\qquad Fm^{254} \end{array} \tag{56.1}$$

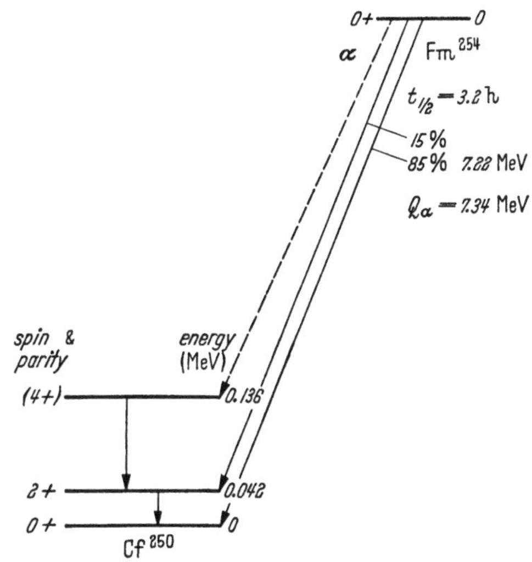

Fig. 32. The decay scheme of Fm^{254}.

Fm^{254} decays by the emission of 7.2 Mev alpha particles with a half-life of 3.2 hours. It also decays by spontaneous fission with a partial half-life of 220 days. The genetic relationship of the 36 hour E^{254m} and Fm^{254} was firmly established by observing the growth of Fm^{254} radiations into a purified einsteinium fraction[2,3].

The isotope Fm^{256} was prepared[4] by irradiating an einsteinium sample containing a small amount of 30 day E^{255} mixed with E^{253}. Spontaneous fission events assigned to Fm^{256} were found in the chemically separated fermium fraction from this bombardment. The observed half-life was 3 to 4 hours.

A Swedish research team[5] bombarded uranium targets with hextuplicately charged O^{16} ions accelerated in a cyclotron to greater than 180 Mev energy. After chemical purification an examination of the fraction corresponding to the element 100 fraction was carried out and some 7.7 Mev alpha activity decaying with a half-life of about ½ hour was found. The tentative mass assignment of this activity was 100^{250}.

Fermium isotopes have been prepared by bombarding targets of californium with helium ions in the Berkeley 60 inch cyclotron[6]. The californium targets

---

[1] B. G. HARVEY, S. G. THOMPSON, A. GHIORSO and G. R. CHOPPIN: Phys. Rev. **93**, 1129 (1954).

[2] P. R. FIELDS *et al.*: Phys. Rev. **94**, 209 (1954).

[3] G. R. CHOPPIN, S. G. THOMPSON, A. GHIORSO and B. G. HARVEY: Phys. Rev. **94**, 1080 (1954).

[4] G. R. CHOPPIN, B. G. HARVEY, S. G. THOMPSON and A. GHIORSO: Phys. Rev. **98**, 1519 (1955).

[5] H. ATTERLING, W. FORSLING, L. W. HOLM, L. MELANDER and B. ASTRÖM: Phys. Rev. **95**, 585 (1954).

[6] B. G. HARVEY and co-workers: University of California Radiation Laboratory, unpublished results, Jan. 1956.

contained a high percentage of the heavy isotopes Cf^{249}, Cf^{250}, Cf^{251}, and Cf^{252}. In this way Fm^{254}, Fm^{253} (not clearly identified), Fm^{252}, and Fm^{250} were observed.

The data on the fermium isotopes are summarized in Table 27. The spontaneous fission half-lives for the even-even isotopes of fermium are of the same order of magnitude as the alpha decay half-lives. This matter will be discussed in a later section of this report.

The decay scheme of Fm^{254} is shown in Fig. 32.

Table 27. *Isotopes of fermium.*

Isotope and half-life	Type of decay	Energy (Mev)	Method of Production
Fm^{250} (?) 30 min (Ref. 1)	α	7.7 (Ref. 1)	$U^{238}(O^{16}, 4n)$ (Ref. 1)
Fm^{250} ~½ hour (Ref. 2)	α	7.43 (Ref. 2)	$Cf^{249-252}(\alpha, xn)$ (Ref. 2)
Fm^{251} (Ref. 3) 6 hours	E.C. >99.5% α	6.90 (Ref. 3)	
Fm^{252} 30 hours (Ref. 2) 22 hours (Ref. 9)	α	7.05 (Ref. 2) 7.04 (Ref. 9)	Pu^{239} (multiple neutron capture) (Ref. 3) $Cf^{249-252}(\alpha, xn)$ (Ref. 2, 9)
Fm^{253} 4.5 days (Ref. 10)	E.C. α	6.85(?) (Ref. 9) 6.94 (Ref. 10)	$Cf^{249-252}(\alpha, xn)$ (Ref. 9, 10)
Fm^{254} 3.2 hours (Ref. 4) 3.3 hours (Ref. 5) 220 days spon. fiss. (Ref. 4, 5) 246 days (Ref. 8)	α spon. fiss.	7.22 7.20 (Ref. 8)	daughter $E^{254\,m}$ Pu^{239} (multiple neutron capture)
Fm^{255} (Ref. 6) 16 hours (Ref. 6) 21.5 hours (Ref. 3, 8)	α	7.1 (Ref. 6)	Pu^{239} (multiple neutron capture) U^{238} (multiple neutron capture) (Ref. 6)
Fm^{256} (Ref. 7) 3 hours (Ref. 3)	spon. fiss.		$E^{255}(n, \gamma)\ E^{256} \xrightarrow{\beta^-}$ (Ref. 7)

[1] H. Atterling *et al.*: Phys. Rev. **95**, 585 (1954).
[2] S. Amiel *et al.*: University of California Radiation Laboratory Report UCRL-3610, 1956.
[3] B. G. Harvey *et al.*: Unpublished results, University of California Radiation Laboratory, 1955.
[4] G. R. Choppin *et al.*: Phys. Rev. **94**, 1080 (1954).
[5] P. R. Fields *et al.*: Phys. Rev. **93**, 1428 (1954).
[6] A. Ghiorso *et al.*: Phys. Rev. **99**, 1048 (1955).
[7] G. R. Choppin *et al.*: Phys. Rev. **98**, 1519 (1955).
[8] M. Jones *et al.*: Phys. Rev. **102**, 203 (1956).
[9] A. M. Friedman *et al.*: Phys. Rev. **102**, 585 (1956).
[10] S. Amiel: University of California Radiation Laboratory Report UCRL-3611, 1956.

57. Chemical properties of fermium. Fermium is the eleventh member of the actinide transition series. The chemical properties have been studied only on a tracer scale. The conclusions from these studies are that fermium exists as the tripositive ion in aqueous solution and that its chemical properties are entirely similar to the tripositive ions of the other actinide elements. The ion exchange elution technique has been used to separate fermium cleanly from rare-earth and actinide element contaminants. These properties of fermium are apparent from the discussion of Sect. 55 and the curves in Fig. 30 and 31 in Sect. 55.

IX. Mendelevium (element 101).

58. Discovery of mendelevium. An isotope of element 101 believed to be 101^{256} was produced and identified by GHIORSO, HARVEY, CHOPPIN, THOMPSON, and SEABORG early in 1955 by bombarding extremely minute targets of E^{253} with 41 Mev helium ions in the Berkeley 60-inch cyclotron[1]. Chemical separations were carried out after bombardment using the ion-exchange elution techniques employed previously in the discovery of elements 97 through 100. In the eka-thulium elution position, i.e., the element 101 elution position, a few spontaneously-fissionable atoms were found. The half-life for the decay of these atoms numbering a combined total of only 17 from the summation of many experiments, was approximately 3.5 hours. The interpretation of the results was that 101^{256} was produced by the reaction:

$$E^{253}(\alpha, n)\,101^{256}. \tag{58.1}$$

The isotope 101^{256} decays by electron capture with a half-life of the order of a half-hour to Fm^{256},

$$101^{256} \xrightarrow[\sim\frac{1}{2}\,\text{hour}]{\text{E.C.}} Fm^{256}.$$

The Fm^{256} then decays by spontaneous fission with a half-life of about 3.5 hours. The radiations of element 101 itself were not observed. These results were later confirmed with substantially greater amounts of E^{253} target material[2] leading to the chemical identification of some hundreds of atoms of the new element.

Special bombardment techniques had to be developed to carry out these experiments effectively, safely, and quickly.

The name mendelevium (symbol Mv) was suggested for element 101 in recognition of the pioneering role of the great Russian chemist, DIMITRI MENDELÉEV in the use of the periodic system of the elements.

It may be expected that several mendelevium isotopes covering the mass number range 251 to 261 will be found with half-lives of the order of a few seconds to an hour[3].

The production of mendelevium by intense neutron irradiation of lighter elements is not promising because it is necessary to build up the mass number of the fermium fraction to 259 before beta decay to an isotope of mendelevium occurs. The sequence would be:

$$\ldots\; Fm^{254}(n,\gamma)\,Fm^{255}(n,\gamma)\,Fm^{256}(n,\gamma)\,Fm^{257}(n,\gamma)\,Fm^{258}(n,\gamma)\,Fm^{259} \downarrow Mv^{259}$$

The difficulty is that Fm^{256} has a half-life for spontaneous fission of only three hours and Fm^{258} probably has a half-life for spontaneous fission of much less than this, so the reaction path to mendelevium is effectively blocked.

59. Chemical properties of mendelevium. Mendelevium is the twelfth member of the actinide transition series. As such it is expected to show only the tripositive oxidation state in solution and is expected to resemble quite closely the chemical properties of the tripositive actinide elements. The ion exchange elution behavior will be of chief importance in radicchemical purification. This behavior may be summarized as follows:

---

[1] A. GHIORSO, B. G. HARVEY, G. R. CHOPPIN, S. G. THOMPSON and G. T. SEABORG: Phys. Rev. **98**, 1518 (1955).

[2] B. G. HARVEY, S. G. THOMPSON, G. R. CHOPPIN and A. GHIORSO: Unpublished work, Jan. 1956.

[3] R. A. GLASS, S. G. THOMPSON and G. T. SEABORG: J. Inorg. a. Nucl. Chem. **1**, 3 (1955).

Mendelevium will adsorb readily on cation exchange resin columns. If buffered citrate, lactate, alpha-hydroxyisobutyrate, or similar agents are used as an eluting agent, mendelevium will elute from the column just ahead of fermium. If saturated hydrochloric acid or 20% ethyl alcohol saturated with hydrochloric acid is used as the eluting agent mendelevium will elute just ahead of fermium and will be more completely separated from the rare earths than any of the tripositive actinide elements of lower atomic number. Mendelevium will adsorb on anion exchange resins from 13 molar hydrochloric acid. If 13 molar hydrochloric acid is used as an eluting agent mendelevium will elute with fermium or slightly after.

C. Systematics of nuclear data in the transuranium element region.

I. Nuclear thermodynamics.

60. Closed decay-energy cycles. The purpose of this section is to summarize in graphical and tabular form the extensive data on disintegration energies for the transuranium element region. For a more complete treatment of this subject and for references to the original literature pertaining to this subject or to basic data, the reader is referred to the extensive articles of Glass, Thompson and Seaborg[1]; Newton[2], Perlman, Ghiorso, and Seaborg[3]; and Huizenga[4]. The data presented in the following figures and tables represent a revision of the data of Glass, et al[1] using newer data available as of January 1956. For these revisions the authors are indebted to Mr. Bruce M. Foreman jr.

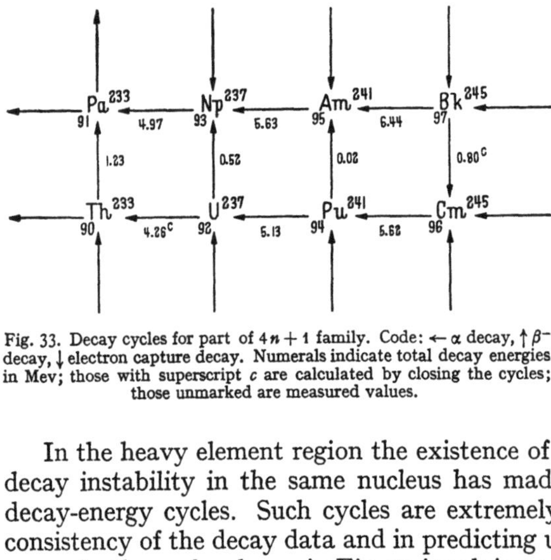

Fig. 33. Decay cycles for part of $4n+1$ family. Code: ← α decay, ↑ β⁻ decay, ↓ electron capture decay. Numerals indicate total decay energies in Mev; those with superscript *c* are calculated by closing the cycles; those unmarked are measured values.

In the heavy element region the existence of alpha decay instability and beta decay instability in the same nucleus has made it possible to construct closed decay-energy cycles. Such cycles are extremely useful in checking the internal consistency of the decay data and in predicting unknown properties. Consider for example, the cycles shown in Fig. 33 involving a few isotopes of the $4n+1$ family of isotopes. In the cycle containing Np^{237}, Am^{241}, Pu^{241} and U^{237} experimental data are available on all four sides of the data cycle and these are mutually consistent, i.e., the summation of decay energies around the complete cycle is zero. In those cases where experimental data or reliable estimates are available for three branches of the cycle, the fourth can then be calculated by difference. A systematic extension of the closed decay-energy cycle method to cover the entire transuranium element region is given in Figs. 34 to 37. Since alpha decay decreases the mass number of a decaying nucleus by four and beta processes leave the mass number unchanged all isotopes in a given figure differ in mass by multiples of four. Hence

[1] R. A. Glass, S. G. Thompson and G. T. Seaborg: J. Inorg. a. Nucl. Chem. **1**, 3 (1955).
[2] J. O. Newton: Progr. Nuclear Phys. **4**, 234—286 (1955).
[3] I. Perlman, A. Ghiorso and G. T. Seaborg: Phys. Rev. **77**, 26 (1950).
[4] J. R. Huizenga: Physica, Haag **21**, 410 (1955).

four figures are required to present the data for the $4n$, $4n+1$, $4n+2$, and $4n+3$ mass types.

In Figs. 34 to 37 approximately one-fourth of the energy values listed represent directly determined values. Approximately one-half of the values were estimated

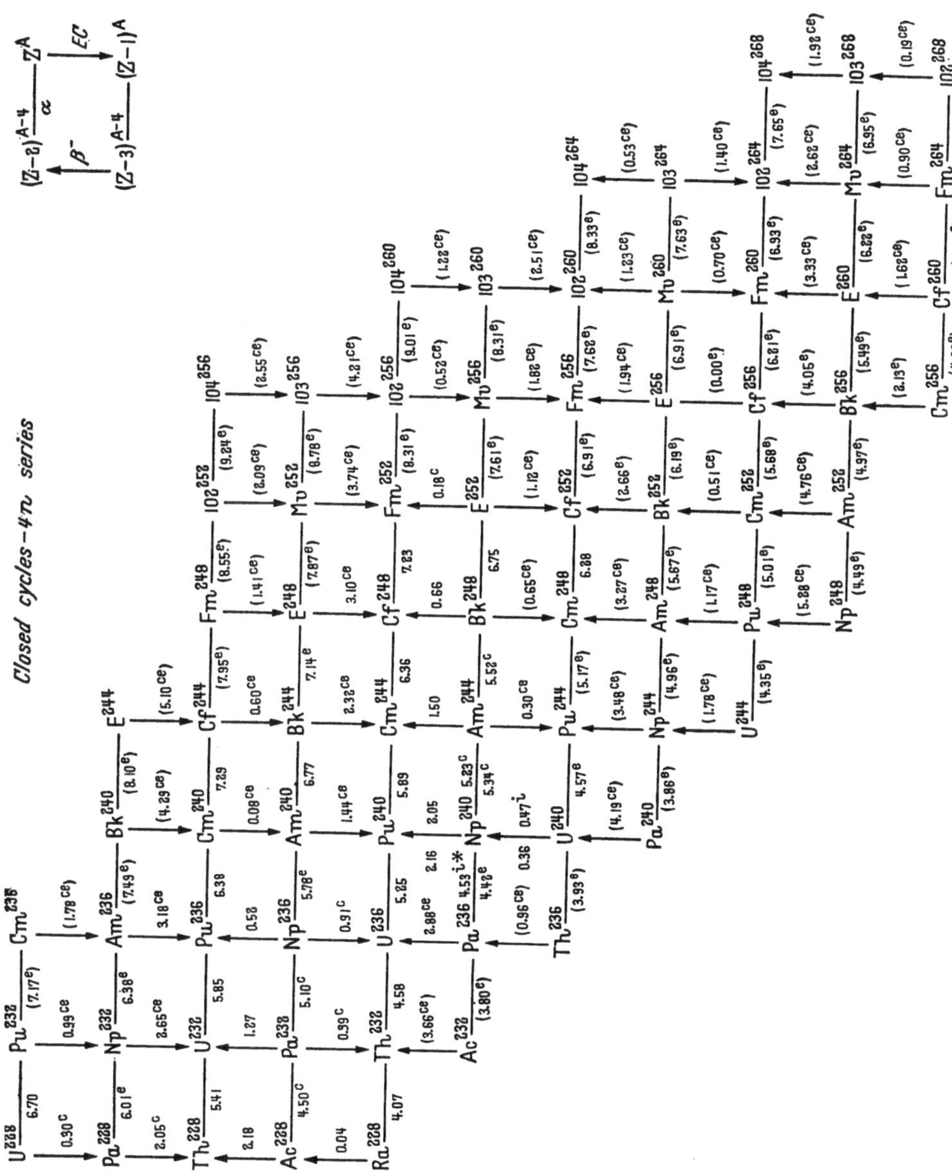

Fig. 34. Closed decay-energy cycles for the 4 *n* family. Code: no superscript, measured energy; *c*, calculated energy; *cn*, calculated with neutron binding energies; *e*, estimated; (), uncertain by more than 0.1 Mev.

from the graphs showing systematic trends in alpha and beta disintegration energies which will be discussed in Sects. 61 and 62. A final group of energies was calculated by closing decay cycles after the experimental and estimated values were entered. Since any given disintegration energy is common to two decay cycles opportunities arise for cross-checking the experimental or estimated values.

In the construction of Figs. 34 to 37 any inconsistencies which arise (usually in the vertical branches) were settled after due consideration of possible sources of error and suitable adjustments were made to give the best possible self-consistent set of decay energies. These energy cycles refer only to the ground-state transitions and provide no information on decay to excited levels in the daughter nucleus

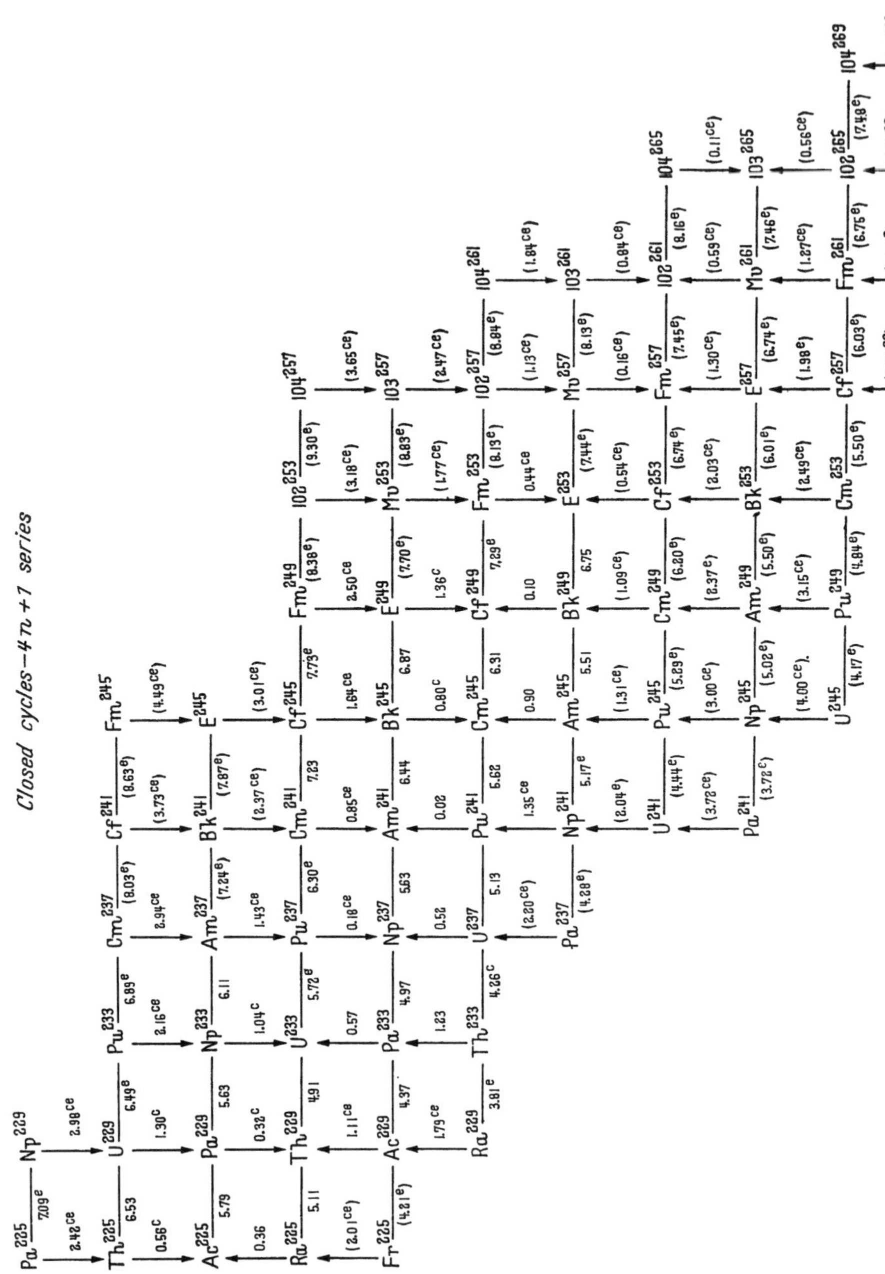

Fig. 35. Closed decay energy cycles for the 4n + 1 family. For code see Fig. 34

The only exception to this is in the case of the isomer pairs of Pa^{234}, Am^{242}, and Np^{240} where data referring to the upper and lower state are included.

61. Alpha decay energy systematics. Fig. 38 is a plot of alpha decay energies versus mass number where the points representing a given element have been connected. The decay energy refers to the ground-state transition and includes the nuclear recoil energy in addition to the energy of the alpha particle. Except for minor irregularities the alpha energies for isotopes of a given element generally decrease in a regular fashion with increasing mass number. Extensions of the

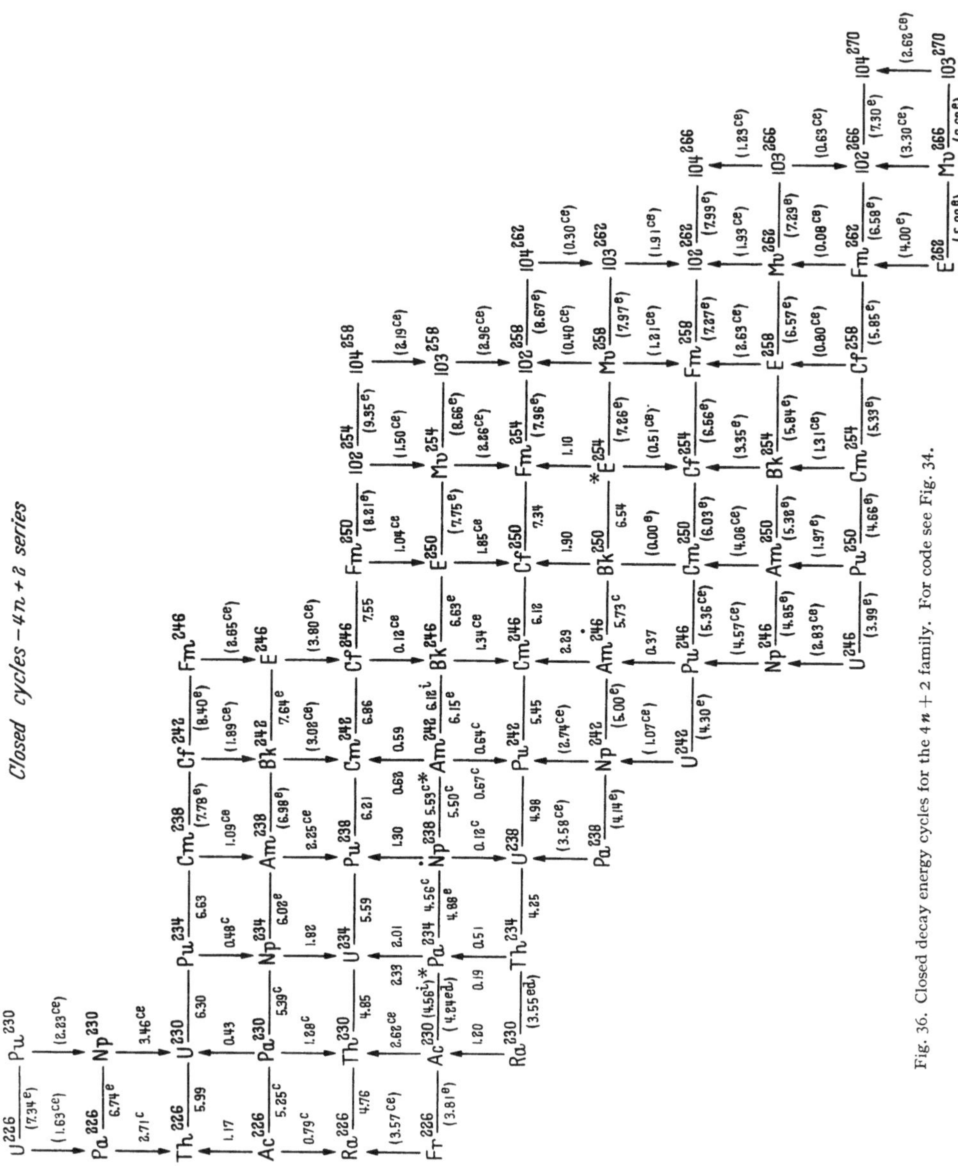

Fig. 36. Closed decay energy cycles for the $4n+2$ family. For code see Fig. 34.

experimental curves make it possible to predict unmeasured or unmeasurable disintegration energies between or beyond the measured ones.

In the curves for the elements californium, einsteinium, and fermium an irregularity occurs which reflects a discontinuity of the energy surface[1] in this mass

[1] "Energy-surface" refers to a surface defined by the energy contents or masses of the nuclides.

region. The shell model of the nucleus proposed by Mayer[1] and Haxel, Jensen and Suess[2] predicts a large discontinuity in nuclear properties for isotopes in the region of 126 neutrons, since according to this theory 126 neutrons represents a completely filled major shell. The alpha disintegration energies for isotopes below

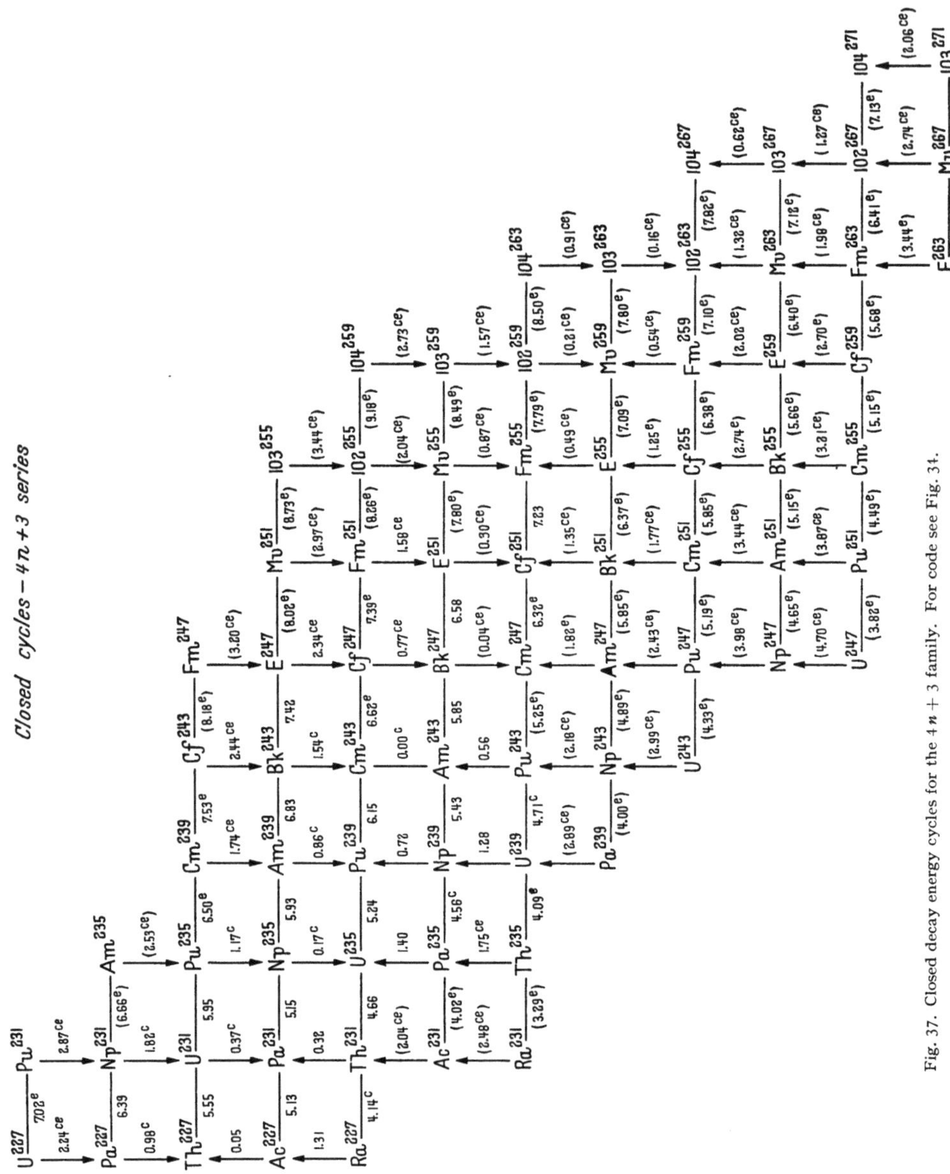

Fig. 37. Closed decay energy cycles for the $4n+3$ family. For code see Fig. 31.

the transuranium element region show a striking discontinuity of 3 to 4 Mev in the mass region which includes isotopes on both sides of the 126 neutron shell. The discontinuity which appears in Fig. 38 is of a much smaller magnitude and has been ascribed to a "subshell" closure at 152 neutrons[3].

[1] M. G. Mayer: Phys. Rev. **75**, 1969 (1949).

[2] O. Haxel, J. H. D. Jensen and H. E. Suess: Z. Physik **128**, 295 (1950).

[3] A. Ghiorso *et al.*: Phys. Rev. **95**, 293 (1954).

In the case of elements beyond element 100 for which no experimental points are available, lines have been placed in Fig. 38 after due consideration of the spacings and slopes of the lines for elements 98, 99 and 100 including an allowance for a probable discontinuity at 152 neutrons. Values taken from these lines are

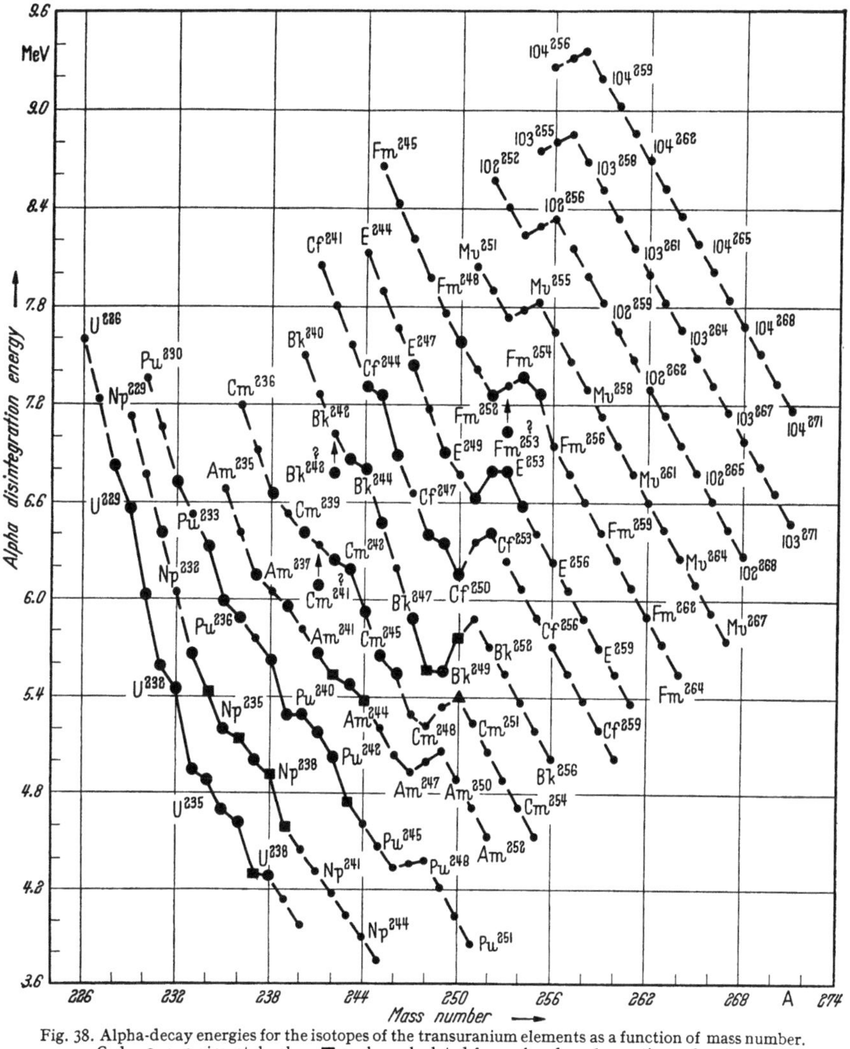

Fig. 38. Alpha-decay energies for the isotopes of the transuranium elements as a function of mass number. Code: ●, experimental value; ■, value calculated from closed cycles, estimated value.

probably subject to more error than values taken from an extension of the curves for lower elements. Data taken from the curves of Fig. 38 were used in the construction of the decay-energy cycles of Fig. 34 to 37.

62. Beta-decay energy systematics. In Fig. 39 the beta-decay energies for the transuranium element isotopes have been correlated with mass number. The upper part of the figure presents the data for even mass types while the lower part refers to the odd mass types.

Similar curves have been given by WAY and WOOD[1] and by SUESS and JENSEN[2]. The figures given here represent revisions of Fig. 8, Reference 1, page 286, carried

[1] K. WAY and M. WOOD: Phys. Rev. **94**, 119 (1954).
[2] H. E. SUESS and J. H. D. JENSEN: Ark. Fysik **3**, 577 (1951).

out by Mr. BRUCE M. FOREMAN, jr. The legend distinguishes those decay energies which are directly measured, those which are estimated, those calculated by closure of suitable decay cycles in Fig. 34 to 37, and those calculated from decay cycles

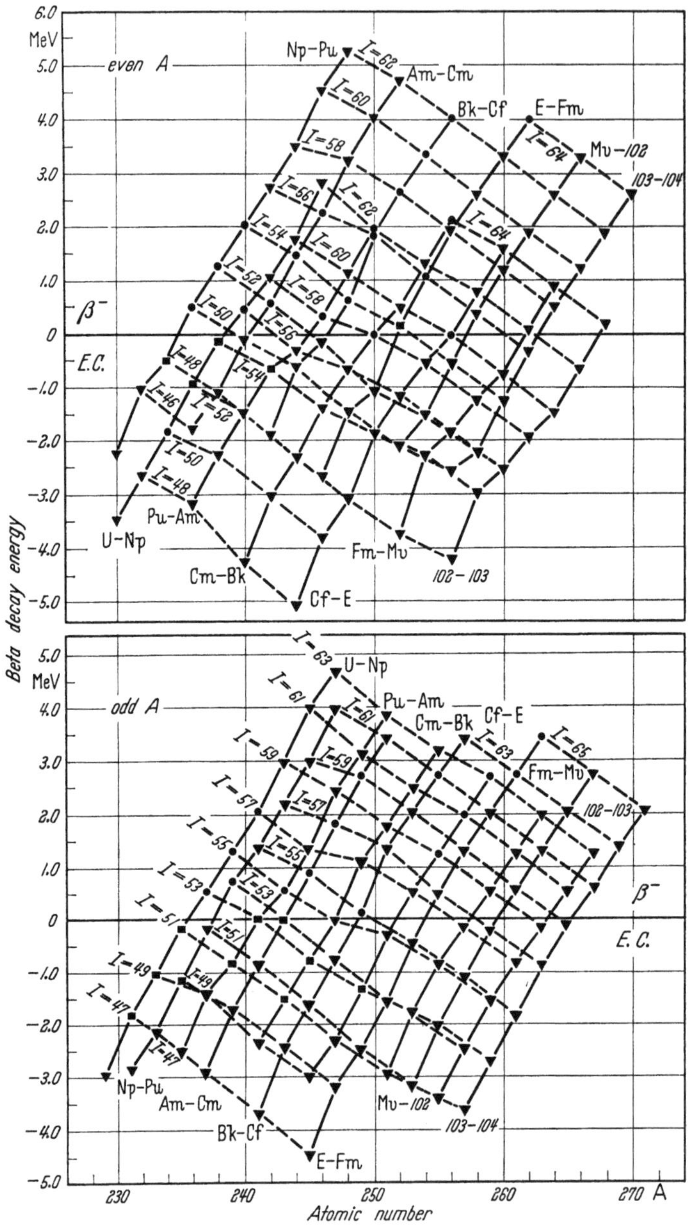

Fig. 39. Beta decay energies for the isotopes of the transuranium elements as a function of mass number. Code: ●, experimental energy; •, estimated energy; ■ energy calculated from closed decay-energy cycles; ▼ calculated from closed cycles which included estimated values. Upper part of curve refers to even mass types. Lower part refers to odd mass types.

involving estimated alpha decay energies taken from Fig. 38. Directly measured decay energies involve a knowledge of the decay scheme as well as of the maximum energies of the beta particle groups. In many instances unfortunately, the decay scheme is not known with certainty.

From Figs. 39 or Figs. 34 to 37, it is possible to determine which transuranium element isotopes are beta stable. This important information is summarized in Table 28.

Table 28. *Beta-stable isotopes of transuranium elements**.

Neptunium	Np^{237}
Plutonium	Pu^{236}, Pu^{238}, Pu^{239}, Pu^{240}, Pu^{242}, Pu^{244}
Americium	Am^{241}, $Am^{243}_{(?)}$
Curium	Cm^{242}, $Cm^{243}_{(?)}$, Cm^{244}, Cm^{245}, Cm^{246}, $Cm^{247}_{(?)}$, Cm^{248}, $Cm^{250}_{(?)}$
Berkelium	None
Californium	Cf^{248}, Cf^{249}, Cf^{250}, Cf^{251}, Cf^{252}, Cf^{254}, $Cf^{256}_{(?)}$
Einsteinium	E^{253}
Fermium	Fm^{252}, Fm^{254}, Fm^{255}, Fm^{256}, (Fm^{257}), (Fm^{258}), ($Fm^{260}_{(?)}$), ($Fm^{262}_{(?)}$)
Mendelevium	(Mv^{259})
102	(258), (260), (261), (262), (263), (264), (266)
103	(265)

* Parentheses enclose unknown isotopes predicted to be beta stable.
A question mark indicates an isotope for which the uncertainties in the data make it impossible to be sure of its beta stability.

63. Masses and binding energies. The energy data for the four mass types presented in Figs. 34 to 37 can be interrelated and used to calculate the exact masses of all the transuranium isotopes provided that at least one mass is determined on an absolute scale and that certain neutron-binding energies are known. Since the most reliable and complete experimental data on absolute masses and on neutron binding energies are available in the region of lead, it is necessary to extend the closed decay-cycle plots to include isotopes in this region.

This has been done by GLASS, THOMPSON and SEABORG[1] who then selected the following "best values" for neutron binding energies to interrelate the four mass types: $Pb^{206-207}$, 6.73 Mev; $Pb^{207-208}$, 7.38 Mev; and $Pb^{208-209}$, 3.87 Mev. These values were selected after considering data from certain nuclear reactions, namely: (a) energy thresholds for reactions such as Pb^{207} (γ, n) Pb^{206}, (b) determination of the capture gamma-ray spectra in a reaction of the type Pb^{207} (n, γ) Pb^{208}, and (c) measurement of the deuteron and proton energies in a reaction like Pb^{208} (d, p) Pb^{209}.

The absolute masses of the nuclides then have been calculated from their energies relative to the absolute mass of Pb^{208} for which the values of 208.0410 ± 0.0015[2] and 208.0409 ± 0.0013[3] have been reported as a result of time-of-flight and magnetic mass spectrographic measurement, respectively. The value adopted for this compilation is 208.04100 (physical scale). In the table which follows this number of significant figures is carried throughout the calculations to be consistent with the *relative* accuracy of many of the energy values.

Additional quantities which enter into these mass calculations are the mass of the helium atom, 4.00387[4]; mass of the neutron 1.00898[4] and energy equivalent of mass, 931.2 Mev[5] per atomic mass unit.

The method and the fundamental data outlined in the past few paragraphs were used to calculate the atomic mass values summarized in Table 29. In this table uncertain values are given in parentheses. Proton and neutron binding energies were calculated for each isotope from mass values using the value 1.00898

[1] R. A. GLASS, S. G. THOMPSON and G. T. SEABORG: J. Inorg. a. Nucl. Chem. **1**, 3 (1955).
[2] P. I. RICHARDS, E. E. HAYS and S. A. GOUDSMIT: Phys. Rev. **85**, 630 (1952).
[3] G. S. STANFORD, H. E. DUCKWORTH, B. C. HOGG and J. S. GEIGER: Phys. Rev. **85**, 1039 (1952).
[4] F. AJZENBERG and T. LAURITSEN: Rev. Mod. Phys. **24**, 321 (1952).
[5] J. W. DUMOND and E. R. COHEN: Rev. Mod. Phys. **25**, 691 (1953).

Table 29. *Isotopic masses (physical scale) and binding energies B_n of last neutron and B_p of last proton (in Mev) for isotopes of transuranium elements.*

	B_n	B_p
92-Uranium		
(226.09962)		(4.24)
227.10158	(6,54)	4.13
228.10195	8.02	5.00
229.10433	6.15	5.07
230.10527	7.48	5.40
231.10795	5.86	5.51
232.10911	7.29	6.14
233.11173	5.92	6.41
234.11343	6.77	6.56
235.11678	5.25	6.48
236.11884	6.44	7.06
237.12211	5.32	7.42
238.12458	6.06	(7.48)
239.12843	4.78	(7.58)
240.13106	5.91	(8.02)
(241.13514)	(4.57)	(7.98)
(242.13789)	(5.79)	(8.73)
(243.14291)	(3.69)	
(244.14549)	(5.96)	
(245.15130)	(2.96)	
(246.15433)	(4.53)	
(247.16076)	(2.38)	
93-Neptunium		
229.10753		2.39
230.10898	7.00	3.24
231.10991	7.50	3.26
232.11195	6.46	3.86
233.11285	7.53	4.10
234.11539	5.99	4.17
235.11696	6.90	4.30
236.11982	5.70	4.75
237.12155	6.75	5.06
238.12471	5.42	5.16
239.12705	6.18	5.28
240.13056	5.10	5.60
241.13295	6.14	5.83
(242.13674)	4.82	(6.08)
(243.13970)	5.61	(5.90)
(244.14358)	4.75	(6.96)
(245.14700)	5.18	(6.18)
(246.15129)	4.36	(7.58)
(247.15571)	4.25	(6.30)
(248.15971)	4.64	(8.56)
94-Plutonium		
(230.11138)		(3.99)
231.11299	(6.86)	3.85
232.11302	8.34	4.69
233.11517	6.36	4.59
234.11591	7.67	4.73
235.11822	6.21	4.95
236.11926	7.39	5.44
237.12175	6.05	5.79
238.12331	6.90	5.94
239.12628	5.60	6.12
240.12836	6.43	6.37
241.13150	5.44	6.71
94-Plutonium (cont.)		
242.13380	6.21	6.78
243.13736	5.05	(7.01)
244.13984	6.05	(7.45)
(245.14378)	(4.70)	(7.40)
246.14638	(5.93)	(8.15)
(247.15143)	(3.66)	(7.45)
(248.15404)	(5.94)	(9.14)
(249.15965)	(3.14)	(7.64)
(250.16249)	(5.71)	
(251.16873)	(2.55)	
95-Americium		
(235.12093)		(2.90)
236.12268	(6.74)	3.43
237.12328	7.80	3.84
238.12573	6.08	3.87
239.12720	6.99	3.96
240.12990	5.85	4.21
241.13147	6.90	4.68
242.13449	5.55	4.79
243.13676	6.25	4.83
244.14017	5.19	4.97
245.14237	6.31	5.23
246.14599	4.99	(5.52)
(247.14882)	(5.72)	(5.31)
(248.15278)	(4.68)	(6.33)
(249.15626)	(5.12)	(5.51)
(250.16037)	(4.53)	(6.90)
(251.16458)	(4.45)	(5.64)
(252.16840)	(4.80)	(7.89)
96-Curium		
(236.12459)		(4.18)
237.12644	(6.64)	4.08
238.12690	7.93	4.21
239.12907	6.34	4.47
240.12999	7.51	4.99
241.13239	6.13	5.27
242.13385	6.99	5.36
243.13676	5.66	5.47
244.13855	6.69	5.91
245.14140	5.71	6.43
246.14353	6.38	6.50
(247.14687)	(5.25)	(6.76)
(248.14927)	(6.13)	(7.17)
(249.15333)	(4.58)	(7.07)
(250.15601)	(5.86)	(7.81)
(251.16088)	(3.83)	(7.11)
(252.16329)	(6.12)	(8.78)
(253.16872)	(3.31)	(7.29)
(254.17137)	(5.89)	
(255.17743)	(2.72)	
(256.17990)	(6.06)	
97-Berkelium		
(240.13460)		(2.44)
(241.13493)	(8.05)	(2.98)
(242.13710)	(6.34)	(3.19)
97-Berkelium (cont.)		
243.13841	(7.14)	3.34
244.14105	5.91	3.59
245.14226	7.23	4.13
246.14497	5.84	4.26
247.14691	6.55	4.43
248.14997	5.52	(4.70)
249.15216	6.32	(4.89)
250.15601	4.77	(5.08)
(251.15898)	(5.60)	(4.82)
(252.16274)	(4.86)	(5.85)
(253.16604)	(5.29)	(5.02)
(254.16996)	(4.71)	(6.42)
(255.17398)	(4.62)	(5.15)
(256.17761)	(4.98)	(7.41)
(257.18220)	(4.09)	(5.44)
98-Californium		
(241.13894)		(3.54)
(242.13913)	(8.18)	(3.67)
243.14103	(6.59)	(3.92)
244.14169	7.75	4.53
245.14402	6.19	4.81
246.14509	7.36	4.94
247.14774	5.90	5.00
248.14926	6.95	5.40
249.15205	5.76	5.64
250.15397	6.57	5.89
(251.15753)	(5.05)	(6.17)
(252.15989)	(6.17)	(6.74)
(253.16386)	(4.66)	(6.54)
(254.16636)	(6.03)	(7.28)
(255.17104)	(4.01)	(6.58)
(256.17326)	(6.29)	(8.25)
(257.17850)	(3.49)	(6.76)
(258.18096)	(6.06)	(8.73)
(259.18683)	(2.90)	
(260.18912)	(6.23)	
99-Einsteinium		
(244.14717)		(1.87)
(245.14726)	(8.28)	(2.40)
(246.14917)	(6.57)	(2.78)
247.15025	(7.36)	2.78
248.15259	6.19	3.07
249.15351	7.50	3.62
250.15596	6.08	3.94
251.15785	6.60	3.97
252.16109	5.35	(4.27)
253.16328	6.32	(4.42)
254.16691	4.98	(4.74)
(255.16969)	(5.77)	(4.48)
(256.17326)	(5.04)	(5.51)
(257.17637)	(5.47)	(4.69)
(258.18010)	(4.88)	(6.08)
(259.18393)	(4.80)	(4.82)
(260.18738)	(5.15)	(7.07)
(261.19179)	(4.26)	(5.10)
(262.19541)	(4.98)	
(263.20050)	(3.63)	

Table 29. (Continued.)

	B_n	B_p
100-Fermium		
(245.15208)		(3.01)
(246.15202)	(8.41)	(3.14)
(247.15369)	(6.81)	(3.38)
(248.15410)	(7.98)	(4.00)
249.15620	(6.41)	4.22
250.15708	7.54	4.26
251.15955	6.06	4.24
252.16090	7.11	4.75
253.16376	5.70	5.10
254.16573	6.52	5.30
(255.16917)	(5.16)	(5.48)
(256.17118)	(6.49)	(6.20)
(257.17498)	(4.83)	(5.99)
(258.17728)	(6.21)	(6.73)
(259.18176)	(4.19)	(6.04)
(260.18381)	(6.46)	(7.70)
(261.18885)	(3.67)	(6.22)
(262.19112)	(6.24)	(8.20)
(263.19680)	(3.07)	(6.29)
(264.19890)	(6.41)	(9.07)
101-Mendelevium		
(251.16274)		(2.31)
(252.16491)	(6.34)	(2.59)
(253.16566)	(7.67)	(3.15)
(254.16815)	(6.03)	(3.48)
(255.17010)	(6.55)	(3.51)
(256.17313)	(5.54)	(3.89)
(257.17515)	(6.49)	(3.89)
(258.17858)	(5.16)	(4.22)
(259.18118)	(5.94)	(3.95)
(260.18456)	(5.22)	(4.98)

	B_n	B_p
101-Mendelevium (cont.)		
(261.18748)	(5.64)	(4.16)
(262.19103)	(5.05)	(5.54)
(263.19468)	(4.97)	(4.27)
(264.19794)	(5.33)	(6.53)
(265.20217)	(4.42)	(4.58)
(266.20560)	(5.16)	
(267.21050)	(3.80)	
102		
(252.16716)		(3.47)
(253.16907)	(6.58)	(3.71)
(254.16977)	(7.71)	(3.75)
(255.17229)	(6.01)	(3.73)
(256.17369)	(7.06)	(4.24)
(257.17636)	(5.88)	(4.58)
(258.17815)	(6.69)	(4.78)
(259.18141)	(5.33)	(4.95)
(260.18324)	(6.66)	(5.67)
(261.18685)	(5.00)	(5.45)
(262.18896)	(6.39)	(6.20)
(263.19326)	(4.36)	(5.51)
(264.19512)	(6.63)	(7.17)
(265.19997)	(3.85)	(5.69)
(266.20206)	(6.41)	(7.68)
(267.20756)	(3.24)	(5.76)
(268.20948)	(6.58)	(8.54)
103		
(255.17599)		(1.79)
(256.17821)	(6.29)	(2.07)
(257.17901)	(7.62)	(2.63)
(258.18133)	(6.20)	(2.95)

	B_n	B_p
103 (cont.)		
(259.18309)	(6.72)	(2.98)
(260.18593)	(5.72)	(3.37)
(261.18775)	(6.67)	(3.38)
(262.19101)	(5.32)	(3.70)
(263.19343)	(6.11)	(3.42)
(264.19662)	(5.39)	(4.45)
(265.19937)	(5.81)	(3.63)
(266.20273)	(5.22)	(5.00)
(267.20620)	(5.14)	(3.73)
(268.20927)	(5.50)	(5.99)
(269.21333)	(4.59)	(4.00)
(270.21658)	(5.32)	
(271.22130)	(3.97)	
104		
(256.18095)		(2.96)
(257.18293)	(6.52)	(3.19)
(258.18368)	(7.66)	(3.23)
(259.18602)	(6.18)	(3.21)
(260.18724)	(7.23)	(3.72)
(261.18973)	(6.05)	(4.05)
(262.19133)	(6.86)	(4.24)
(263.19441)	(5.50)	(4.42)
(264.19606)	(6.83)	(5.14)
(265.19949)	(5.17)	(4.92)
(266.20141)	(6.56)	(5.67)
(267.20553)	(4.53)	(4.98)
(268.20721)	(6.80)	(6.64)
(269.21188)	(4.02)	(5.16)
(270.21377)	(6.59)	(7.16)
(271.21909)	(3.41)	(5.25)

for the mass of the neutron and 1.00814 for the mass of the proton[1]. Values in Table 29 are slightly different from those given previously by GLASS *et al.*[2] reflecting the changes in the decay-energy cycles mentioned previously. HUIZENGA[3] has also provided a complete table of masses for the heavy element region. A general discussion of atomic masses is given by A. H. WAPSTRA in Vol. XXXVIII of this Encyclopedia.

II. Regularities in the rate of alpha decay and in the energy levels observed in alpha decay processes.

64. The rate of alpha decay. A detailed discussion of alpha decay is beyond the scope of this article since this subject is covered by RASMUSSEN and PERLMAN in Vol. XLII of this Encyclopedia. Hence, we shall review briefly only a few regularities pertaining to the transuranium element region.

The original formulation of the one-body theory of alpha decay by GAMOW[4] and CONDON and GURNEY[5] which treats alpha decay as a simple quantum-

[1] See footnote 4, p. 293.
[2] See footnote 1, p. 293.
[3] J. R. HUIZENGA: Physica, Haag **21**, 410 (1955).
[4] G. GAMOW: Z. Physik **51**, 204 (1928).
[5] E. U. CONDON and R. W. GURNEY: Phys. Rev. **33**, 127 (1929). — Nature, Lond. **122**, 439 (1928).

mechanical, barrier-penetration problem accounts rather well in a general way for alpha distintegration rates which vary over the enormous range of 10^{24}. In recent years the large amount of new data which has appeared on artificially prepared isotopes has stimulated many authors to make detailed comparison of the data with various formulations of the GAMOW theory. The principal fact emerging from these studies has been that if attention is restricted to ground-state transitions in even-even nuclei, the simple one body theory accounts with truly remarkable precision for the variation in half-life with atomic number and alpha decay energy[1].

A strictly empirical correlation of the data on even-even alpha emitters is given in Fig. 40. In this figure the logarithm of the alpha half-life (corrected when necessary for branching decay by other decay modes) is plotted versus the alpha disintegration energy. It is seen that the data define a smooth family of curves. The curves in Fig. 40 are theoretical, calculated according to KAPLAN'S[2] formulation of the rigorous equations of PRESTON[3] for the one-body theory. In making such calculations it is necessary to introduce a value for the nuclear radius, r_0. KAPLAN found that the value $r_0 = (1.57 \pm 0.015)\,A^{\frac{1}{3}} \times 10^{-13}$ cm when substituted in the equations for the decay constant λ gave the correct values for the even-even nuclei. PERLMAN and ASARO[4] have shown empirically that for the most precise agreement over the transuranium element range, it is necessary to allow for a steady decrease in r_0 as a function of Z amounting to a few percent in going from plutonium to fermium. It is possible then to account very well for the known alpha decay data on even-even transuranium element isotopes and to predict with considerable confidence the half-lives of unknown even-even isotopes. One other conclusion which follows from the strong dependence of the disintegration constant upon the energy is that the alpha transition leading to the ground state must be the most abundant transition. Experimentally, this is true in every case.

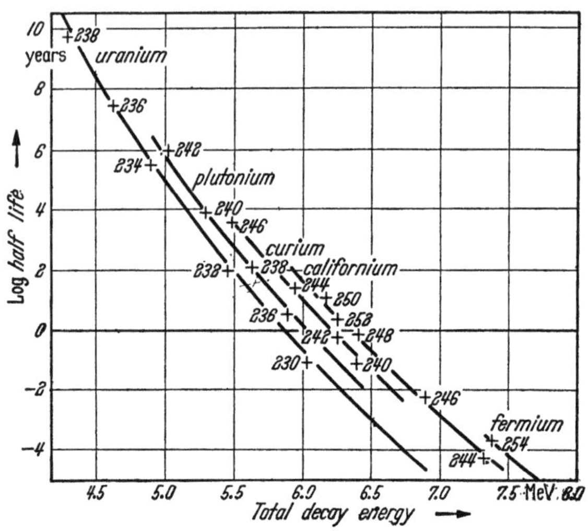

Fig. 40. Half-life of even-even isotopes versus alpha decay energy.

When isotopes with an odd number of neutrons or protons are considered, it is found that the disintegration constants do not vary in the smooth manner predicted by the simple theory. In general, the disintegration constants are considerably longer than would be expected. PERLMAN et al.[1] have introduced the term *departure factor* to give quantitative expression to this deviation. The departure factor (sometimes referred to as hindrance factor) is defined as the ratio of the observed partial half-life (for a specific alpha particle group) to the one calculated from the theory using the nuclear radius parameter obtained from

[1] I. PERLMAN, A. GHIORSO and G. T. SEABORG: Phys. Rev. **77**, 26 (1950).
[2] I. KAPLAN: Phys. Rev. **81**, 962 (1951).
[3] M. A. PRESTON: Phys. Rev. **71**, 865 (1947).
[4] I. PERLMAN and F. ASARO: Ann. Rev. Nucl. Sci. **4**, 157 (1954).

a consideration of the even-even isotopes. Table 30 lists the calculated departure factors for ground-state alpha transitions in transuranium element isotopes containing odd nucleons. The chief facts to be noted are that the departure factor can be large and that the variation in departure factor follows no obvious pattern. Departure factors for transitions to excited levels in the daughter nucleus also vary erratically. It frequently happens that the alpha decay in odd-nucleon types goes more readily to excited levels than to the ground state in spite of the large dependence of alpha decay upon alpha particle energy. Hence, alpha spectra of odd nucleon types are more complex than those of even-even nuclei and in some

Table 30. *Departure factors for ground-state alpha transition of odd-nucleon isotopes*

Alpha emitter	Particle energy	Abundance	Departure factor
Cf^{249}	6.00	0.10	300
Bk^{245}	6.33	0.18	700
Bk^{243}	6.72	0.30	900
Cm^{243}	6.049	<0.5	>26
Am^{243}	5.341	<0.003	$\geqq 1000$
Am^{241}	5.535	0.0034	700
Pu^{241}	5.01	<0.11	>30
Pu^{239}	5.150	0.69	3
Np^{237}	4.86	<0.15	>50

instances the ground state transitions are not observed. Decay schemes of transuranium element alpha emitting isotopes were given previously in Figs. 7, 11, 16, 24, 26, 28, 29 and 32.

It is clear that some parameter or set of parameters not appearing in the simple one-body theory of alpha decay is operating powerfully to retard the emission of alpha particles. The effect of spin change has been considered and found to be incapable of explaining the results, at least in any manner in which spin change has been introduced so far. One suggestion[1] which has not received quantitative formulation has to do with the breakdown of the one-body model in its implication that alpha particles exist as discrete entities within the nucleus and that the probability of alpha-emission depends only upon the frequency factor of this preformed alpha particle in encountering the coulomb barrier and upon the rate of barrier penetration. In the case of odd-proton or odd-neutron isotopes it may be that a considerable delay is involved in assembling the components of the alpha particle. RASMUSSEN[2] has discussed the suggestion in terms of the BOHR-MOTTELSON[3] nuclear model. In the case of the alpha decay of Am^{241} to Np^{237} both parent and daughter nuclei are known to have nuclear spin $^5/_2$. The odd proton which gives rise to the $^5/_2$ spin probably has a different wave function in the two cases; the necessity for a change in the state of this odd nucleon delays the alpha transition rate.

65. Energy level systematics in the decay of even-even alpha emitters. An examination of the decay schemes of the even-even isotopes of plutonium, curium, californium and fermium (see Figs. 11, 24, 28, and 32) reveals a common pattern. The sequence of levels in the daughter nucleus reached by alpha decay have the spin and parity designations 0^+ (ground), 2^+, 4^+, 6^+, 8^+ and the position of these

[1] I. PERLMAN, A. GHIORSO and G. T. SEABORG: Phys. Rev. **77**, 26 (1950).
[2] J. O. RASMUSSEN jr.: Ark. Fysik **7**, Nr. 16 (1953).
[3] A. BOHR and B. R. MOTTELSON: Dan. Mat. fys. Medd. **27**, Nr. 16 (1953). — Beta and Gamma Ray Spectroscopy, Chap. 17. K. SIEGBAHN, Editor. New York: Interscience Publishers 1955. — A. BOHR: Dan. Mat.-Fys. Medd. **26**, Nr. 14 (1952).

levels above ground is approximately 45, 150, 300 and 510 kev respectively. De-excitation of these levels is by a cascade of E 2 gamma transitions with no cross over transitions. The electric quadrupole nature of these transitions have been

Table 31. *Energy in kev of first three excited levels in daughter nucleus of even-even alpha emitters.*

Alpha emitter	Energy of 2^+ level	Energy of 4^+ level	Energy of 6^+ level
Pu^{236}	47	156	—
Pu^{238}	43.5	145	295
Pu^{240}	45	—	—
Cm^{242} *	44	146	301
Cm^{244}	42.9	142.7	293
Cf^{246}	42.3	145	294
Cf^{250}	42.9	—	—
Cf^{252}	43.4	144	—
Fm^{254}	42	136	—

* In the case of Cm^{242} a weak gamma ray of 210 kev is believed to represent the $8^+ \rightarrow 6^+$ transition; the 8^+ level is 513 kev above ground.

characterized by measurement of absolute conversion coefficients, relative L subshell conversion coefficients, and α-γ angular correlations. The regularity of the pattern is evident from Table 31 which lists exact values for the energies of the 2^+, 4^+, and 6^+ states.

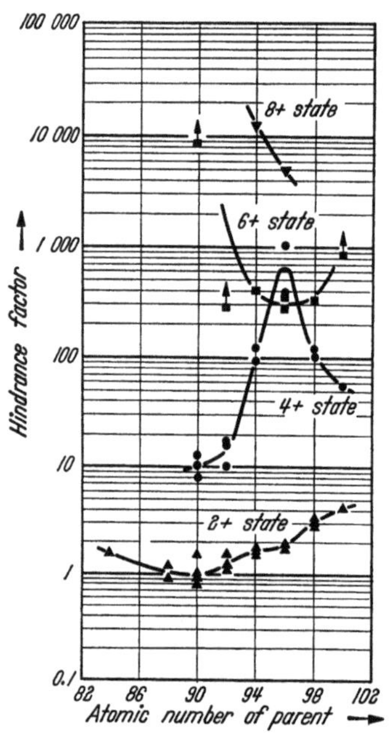

Fig. 41. Departure factors in the alpha decay of even-even isotopes to excited even states in the daughter nucleus. From PERLMAN *et al.* University of California Radiation Laboratory Report UCRL-3068, July 1955, p. 39.

The 2^+ first excited state can be explained by modifications of the shell model of the nucleus or by the collective model of BOHR and MOTTELSON[1]. The collective model is particularly applicable to the transuranium elements as their isotopes appear in a region well removed from a closed shell. The theory predicts a rotational band of levels in which even states (0^+, 2^+, 4^+...) appear for an even-even nucleus. The energy spacing of the levels should be proportional to $\frac{\hbar^2}{2J} I_j(I_j+1)$ where $\frac{\hbar^2}{2J}$ is the rotational energy quantum and I_j is the spin of the j-th even state. This spacing agrees closely with the experimental values given in Table 31. A detailed comparison of the data with the predictions of the collective model makes it appear highly likely that the excited levels observed in alpha decay of even-even nuclei correspond to such rotational bands of levels.

It has been noted that the percentage of transitions leading to these excited levels is less than can be accounted for by the energy dependence of alpha decay; i.e. the departure factor for alpha decay to these levels is greater than unity. This is most readily seen in Fig. 41. Alpha transitions in even-even isotopes to the 2^+ level are slightly but definitely hindered. Transitions to higher even states are hindered by quite large factors.

Odd parity excited states have been observed in the alpha decay of some even-even isotopes in addition to the even states. In the case of Th^{228} decay the well characterized 1^- level appears only 217 kev above the ground state of the Ra^{224}

[1] See footnote 3, p. 297.

daughter[1]. This may be the fundamental state of a rotational band of negative parity with associated 3^- and 5^- levels at 289 and 412 kev, respectively[2]. Similarly in the decay of ionium, Th^{230}, a rotational band of negative parity with levels 1^- (253 kev) 3^- (320 kev) and 5^- (445 kev) is observed in addition to a 2^+, 4^+, 6^+ level sequence at 68, 210 and 416 kev respectively[2].

Such odd parity states have been looked for in the transuranium element region, but only in the case of Cm^{242} has evidence for such a state appeared. In this case the assignment, 1^-, is given to a level at 605 kev which is populated by alpha decay only to the extent 1×10^{-6}. Other high-lying odd parity states probably exist in the transuranium element region, but because of their location above the ground state they are reached in only an extremely small percentage of the alpha transitions and hence are extremely difficult to investigate.

66. Energy level systematics in the decay of odd nucleon alpha-emitters. The alpha spectra of nuclei having an odd number of neutrons or protons are in general more complex than those of even-even nuclei and consequently have not been worked out with the same degree of certainty. Also as a group they show no common generalized pattern such as is the case for the even-even nuclei. The transition to the ground state is frequently highly hindered; in some cases it is not observed.

Nevertheless, some regularities are observable. In particular rotational bands of levels rather similar to those of the even-even nuclei have been identified. The principal difference is that the fundamental level of the rotational band is usually not the ground state in the case of odd-nucleon nuclei. For most odd A nuclei the energies of rotational levels are given by the equation.

$$E_j = \frac{\hbar^2}{2J}\left[I_j(I_j+1) - I_0(I_0+1)\right] \tag{66.1}$$

where J is the moment of the inertia of the surface wave motion, I is the nuclear spin of level j, I_0 the spin of base level.

For odd A nuclei the quantum states representing collective rotational motion based on a given fundamental level I_0 form the spin sequence

$$I = I_0, \quad I_0+1, \quad I_0+2, \quad \ldots$$

all with parity of the fundamental level.

The level system for Np^{237} deduced from the alpha decay of Am^{241}, the beta decay of U^{237} and coulomb excitation experiments on Np^{237} is particularly interesting[3,4]. Consider the level scheme shown in Fig. 42 and the alpha decay scheme of Am^{241} given previously in Fig. 16. 84% of the Am^{241} alpha transitions go to a 59.57 kev level in Np^{237}. This level has nuclear spin and parity ${}^5/_2{}^-$ and is the fundamental level of a rotational band with $\frac{\hbar^2}{2J} = 6.19$ kev. The levels at 103.0 and 158.5 kev with assignments ${}^7/_2{}^-$ and ${}^9/_2{}^-$ respectively are definitely members of this band. A level at $\sim$227 kev reported by BARANOV and SHLYAGIN[5], GOLDIN, *et al.*[6] is probably the ${}^{11}/_2{}^-$ member of this band. It is reached by only 0.015% of the alpha transitions from Am^{241}[6].

[1] I. PERLMAN and F. ASARO: Ann. Rev. Nucl. Sci. **4**, 157 (1954).

[2] F. STEPHENS, F. ASARO and I. PERLMAN: Unpublished results, University of California Radiation Laboratory, 1955.

[3] H. JAFFE, T. O. PASSELL, C. I. BROWNE and I. PERLMAN: Phys. Rev. **97**, 142 (1955).

[4] J. M. HOLLANDER, W. C. SMITH and J. O. RASMUSSEN: Phys. Rev. **102**, 1372 (1956).

[5] S. A. BARANOV and K. N. SHLYAGIN: Proceedings of Moscow meeting on Atomic Energy, July 1955, p. 252.

[6] L. L. GOLDIN, E. F. TRETYAKOV and G. I. NOVIKOVA: Proceedings of Moscow Meeting on Atomic Energy, July 1955, pp. 226–249.

The ground state of Np^{237} has nuclear spin $^5/_2$ but has opposite parity. The levels at 33.20 and at 76.1 kev are believed to be members of a rotational band based on the ground state. NEWTON[1] has produced these levels by coulombic excitation of Np^{237}. The arrows pointing upward denote these excited transitions. The 76.1 kev level has not been observed in the decay of Am^{241} or U^{237}. The spin sequence for this rotational band is $^5/_2{}^+$, $^7/_2{}^+$, and $^9/_2{}^+$ and the quantum of rotation $\frac{\hbar^2}{2J}$ is 4.75 kev. The gamma transitions of Fig. 42 have been well characterized where the transition type is indicated; these data are in agreement with the spin and parity assignments.

The low-lying levels of Np^{237} fall neatly into two rotational bands and indicate the essential correctness of the collective model in this mass region.

The alpha decay of Am^{243} to Np^{239} shows a pattern with considerable similarity to Am^{241} (see Fig. 16). 87% of the alpha transitions go to a 75 kev level in Np^{239} which probably is the fundamental level of a rotational band with spins $^5/_2$, $^7/_2$ and $^9/_2$. The upper levels fall at 118 and 173 kev respectively.

In a number of other cases it seems likely that series of levels will be proved to be associated with collective modes of nuclear motion. It is also probable that other important features of the spectra must bring in the concept of single particle states. Just what these states are and what guides the alpha-decay process to some and not to others are matters for future work.

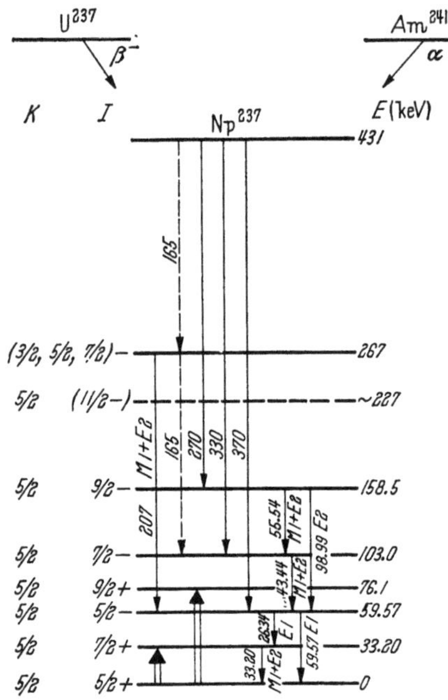

Fig. 42. The energy levels of Np^{237} (from J. M. HOLLANDER *et al.*: Phys. Rev. **102**, 1372 (1956).

III. Nuclear spins.

Direct measurements of nuclear spin for the ground states of isotopes in the transuranium element region have been made in only six cases. These are summarized in Table 32.

IV. Neutron cross-sections and spontaneous fission properties.

67. Neutron fission and capture cross-sections. Since nuclear fission is fully discussed by WHEELER in Vol. XLI, we shall not discuss that important subject here. The values for the slow neutron fission cross-sections and the neutron capture cross-sections of transuranium element isotopes are given in Table 33.

The fission cross-sections in almost every case were determined by fission counting in a double ionization chamber placed in the thermal column of a nuclear reactor. The sample to be measured was placed in one ionization chamber and a monitor sample containing a known amount of an isotope whose fission cross-section was accurately known was placed in the duplicate chamber. With present

[1] J. O. NEWTON: Nature, Lond. **175**, 1028 (1955) and unpublished results, 1955.

Table 32. *Determinations of nuclear spin in transuranium element region.*

Isotope	Spin	Method	Reference
Np^{237}	$^5/_2$	hyperfine structure	F. S. TOMKINS: Phys. Rev. **73**, 1214 (1948).
	$^5/_2$	paramagnetic resonance	B. BLEANEY, P. M. LLEWELLYN, M. H. L. PRYCE and G. R. HALL: Phil. Mag. **45**, 992 (1954).
Np^{239}	$^1/_2$	hyperfine structure	J. G. CONWAY and R. D. MCLAUGHLIN: Phys. Rev. **96**, 541 (1954).
Pu^{239}	$^1/_2$	paramagnetic resonance	B. BLEANEY, P. M. LLEWELLYN, M. H. L. PRYCE and G. R. HALL: Phil. Mag. **45**, 773 (1954).
	$^1/_2$	hyperfine structure	M. VAN DEN BERG and P. F. A. KLINKENBERG: Physica, Haag **20**, 37, 461 (1954).
Pu^{241}	$^5/_2$	paramagnetic resonance	B. BLEANEY, P. M. LLEWELLYN, M. H. L. PRYCE and G. R. HALL: Phil. Mag. **45**, 991 (1954).
Am^{241}	$^5/_2$	hyperfine structure	M. S. FRED and F. S. TOMKINS: Phys. Rev. **89**, 318 (1953).
Am^{243}	$^5/_2$	hyperfine structure	J. G. CONWAY and R. D. MCLAUGHLIN: Phys. Rev. **94**, 498 (1954).

techniques this method is highly sensitive and accurate when samples of suitable isotopic purity are measured. The radiative capture cross-sections quoted in Table 33 were determined by activation methods using samples which were placed close to the reactor core where the neutron spectrum is not well known. In many cases the capture cross-sections were obtained as incidental information from

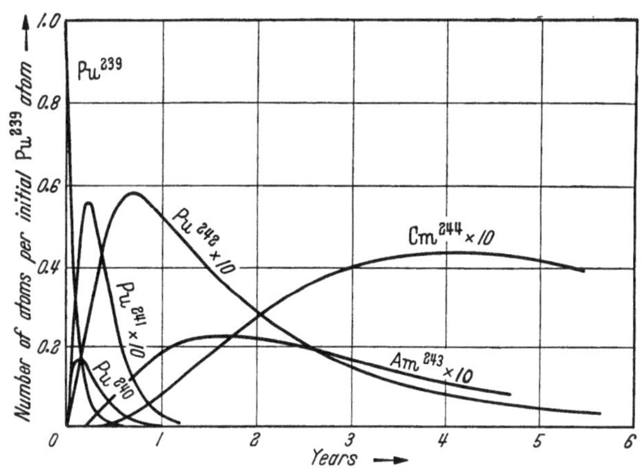

Fig. 43. Production of high mass isotopes by the irradiation of Pu^{239} at a flux of 3×10^{14} neutrons per cm^2 per second The neutron capture and beta decay sequence is:

$$\underset{\substack{\downarrow\\ \sigma_{fiss}=765}}{Pu^{239}}(n,\gamma)\underset{\sigma=385}{Pu^{240}}(n,\gamma)\underset{\substack{\sigma=550\,\downarrow\\ \sigma_{fiss}=1100}}{Pu^{241}}(n,\gamma)\underset{\sigma=350}{Pu^{242}}(n,\gamma)\underset{\sigma=68}{Pu^{243}}\xrightarrow{\beta^-}Am^{243}(n,\gamma)\underset{\sigma=110}{Am^{244}}\xrightarrow{\beta^-}Cm^{244}$$

experiments carried out principally for other purposes. Hence many of these values, particularly for the heaviest isotopes, are subject to revision.

Table 34 lists values of $\bar{\nu}$, the average number of neutrons released in fission; all except four of these values refer to spontaneous fission. Another important quantity is the multiplicity of neutrons from fission defined as the fraction of fissions producing 0, 1, 2, ... neutrons. Table 35 taken from the work of HICKS,

Table 33. *Slow neutron fission cross-sections and pile neutron radiative capture cross-sections for transuranium element isotopes.*

Isotope	σ Fission (Barns)	References	σ Capture (Barns)	References
Np^{234}	900 ± 300	1		
Np^{236} (5000 years)	2800	2		
Np^{237}	0.019 ± 0.003	3	170 ± 20*	4
Np^{238}	1600 ± 100	7		
Np^{239}	< 3	8	80 ± 20	29
Pu^{238}	18 ± 2	9	455 ± 50*	13
			552 ± 7	10
Pu^{239}	750 ± 15	5	315 ± 16*	5
	720	6		
Pu^{240}	4.4	28	530 ± 50	11
Pu^{241}	1091 ± 87	12	350	11
	1100 ± 30	11, 26		
Pu^{242}	< 0.3	11	68.2 ± 1	10
			30 ± 10	14
Pu^{243}			170 ± 90	11
Pu^{244}			1.5 ± 0.3	11
Pu^{245}			260 ± 145	11
Am^{241}	3.0 ± 0.2	15	700 ± 200	5, 17, 18
	3.0**	16		
Am^{242}	6000	17	2000	17
	2500**	16	5500	16
Am^{242m}	3000	19		
	2000**	17		
Am^{243}	< 25**	16	137.5 ± 1.7	10
	< 40	17	115 ± 20	21
			140 ± 50	23
Cm^{242}	< 5**	16	20 ± 10	20
Cm^{243}			250 ± 150	5
Cm^{244}			25 ± 10	21
Cm^{245}	1800 ± 300	11	200 ± 100	21
Cm^{246}			15 ± 10	2
Cm^{247}			180	11
Cm^{248}			4	22
Bk^{249}			1100 ± 300	23
			350	24
Cf^{249}	630*	23	270 ± 100	23
Cf^{250}			~ 1500	24
Cf^{251}			~ 3000	24
Cf^{252}			25	24
			30	23
Cf^{254}			< 2	23
$E^{253} \rightarrow E^{254m}$			240 ± 100	27
			195	25
$E^{253} \rightarrow E^{254}$			7	25
E^{254m}			< 15	23
E^{254}	2000	25	40	25
E^{255}			30	25
Fm^{255}			< 100	25

References of table 33.

* Thermal neutron capture cross section.

** Pile neutron fission cross section.

[1] E. K. Hyde, W. C. Bentley and F. Hagemann: Argonne National Laboratory Report ANL-4152, May 20, 1948.

[2] M. H. Studier, J. E. Gindler and C. M. Stevens: Phys. Rev. **97**, 88 (1955).

[3] A. Ghiorso, D. W. Osborne and L. B. Magnusson: Paper 22.26, N.N.E.S., Vol. 14B, "The Transuranium Elements", 1949.

[4] Unpublished results, Materials Testing Reactor Group, reported by Hughes and Harvey, Ref. 5.

[5] D. J. Hughes and J. A. Harvey: "Neutron Cross-Sections" Brookhaven National Laboratory Report BNL-325, July 1955.

[6] D. J. Hughes and J. A. Harvey: "Heavy Element Cross-Sections" presented at Geneva, August 1955; Addendum to BNL-325.

[7] M. H. Studier, H. H. Hopkins jr., A. Ghiorso and W. C. Bentley: Report CF-3762, 1947.

[8] J. J. Floyd, H. Z. Schofield, J. Halperin and L. B. Borst: Report MON-P-85, March 31, 1946.

[9] G. Reed jr., W. M. Manning and W. C. Bentley: Argonne National Laboratory Report ANL-4112, March 1, 1948.

[10] J. P. Butler, M. Lounsbury and Mrs. J. S. Merritt: Report, Chemical Institute of Canada Symposium on Nuclear and Radiochemistry, Montreal 1955.

[11] W. C. Bentley *et al.*: Paper P/809, "Peaceful Uses of Atomic Energy-Proceedings of the Geneva Conference", August 1955, United Nations.

[12] M. H. Studier, J. R. Huizenga, J. C. Sullivan, W. C. Bentley and J. F. Mech: Argonne National Laboratory Report ANL-4667, May 1951, p. 18. The value of σ fission for Pu^{241} given in this report was recalculated by Jaffey *et al.* (Ref. 26) to the value quoted here.

[13] R. J. Bruehlman, W. C. Bentley and E. K. Hyde: Argonne National Laboratory Report ANL-4215, Oct. 18, 1948.

[14] M. H. Studier *et al.*: Phys. Rev. **93**, 1433 (1954).

[15] A. Ghiorso and B. B. Cunningham: Argonne National Laboratory Report ANL-4045, 1947. Phys. Rev. **82**, 558 (1951).

[16] G. C. Hanna, B. G. Harvey, N. Moss and P. R. Tunnicliffe: Phys. Rev. **81**, 893 (1951).

[17] K. Street jr., A. Ghiorso and S. G. Thompson: Phys. Rev. **85**, 135 (1952).

[18] G. C. Hanna, B. G. Harvey and N. Moss: Phys. Rev. **81**, 486 (1951).

[19] G. Higgins and W. W. T. Crane: Phys. Rev. **94**, 735 (1954).

[20] S. G. Thompson, A. Ghiorso and F. L. Reynolds: Unpublished work cited in Chap. 20, N.N.E.S., Vol. 14A, "The Actinide Elements", 1954.

[21] C. M. Stevens *et al.*: Phys. Rev. **94**, 974 (1954).

[22] B. G. Harvey *et al.*: Unpublished results, University of California Radiation Laboratory, 1955.

[23] B. G. Harvey, H. P. Robinson, S. G. Thompson, A. Ghiorso and G. R. Choppin: Phys. Rev. **95**, 581 (1954).

[24] L. B. Magnusson *et al.*: Phys. Rev. **96**, 1576 (1954).

[25] S. G. Thompson *et al.*: Unpublished results, University of California Radiation Laboratory, 1955.

[26] A. H. Jaffey, M. H. Studier, P. R. Fields and W. C. Bentley: Argonne National Laboratory Report ANL-5397, March 1955.

[27] P. R. Fields *et al.*: Phys. Rev. **94**, 209 (1954).

[28] E. K. Hulet *et al.*: Phys. Rev. **102**, 1621 (1956).

[29] J. Halperin *et al.*: Nucl. Sci. Eng. **1**, 108 (1956).

Ise, and Pyle[1] shows the multiplicity of neutrons for the spontaneous fission of several isotopes. Hammel and Kephart[2] have made similar measurements on Pu^{240} and Diven, Martin, Taschek and Terrell[3] have made measurements on a number of heavy isotopes.

---

[1] D. A. Hicks, J. Ise jr. and R. V. Pyle: University of California Radiation Laboratory Report UCRL-3146 Rev., Nov. 1955. — Phys. Rev. **101**, 1016 (1956).

[2] J. E. Hammel and J. F. Kephart: Phys. Rev. **100**, 190 (1955).

[3] Diven, Martin, Taschek and Terrell: Phys. Rev. **101**, 1012 (1956); see also R. B. Leachman, Paper P/592, "Peaceful Uses of Atomic Energy-Proceedings of the Geneva Conference", August 1955, United Nations. Phys. Rev. **101**, 1005 (1956).

Table 34. *Average number of neutrons released in fission, $\bar{\nu}$.*
sf = spontaneous fission, n = slow neutron fission.

Isotope		Mode of fission	References	Isotope		Mode of fission	References
U^{233}	2.54 ± 0.04	n	1	Pu^{241}	2.91 ± 0.10	n	4
	2.46 ± 0.15	n	10		2.98 ± 0.21	n	10
U^{235}	2.46 ± 0.03	n	1	Pu^{242}	2.26 ± 0.13	sf	6
	2.45 ± 0.15	n	10		2.18 ± 0.09	sf	2
Pu^{236}	1.89 ± 0.20	sf	6	Cm^{242}	3.0 ± 0.3	sf	8
	2.30 ± 0.19	sf	2		2.65 ± 0.09	sf	2
Pu^{238}	2.04 ± 0.10	sf	6		2.33 ± 0.10	sf	6
	2.33 ± 0.08	sf	2	Cm^{244}	2.60 ± 0.12	sf	9
Pu^{239}	2.88 ± 0.04	n	1		2.84 ± 0.09	sf	2
	2.86 ± 0.18	n	10	Cf^{252}	3.82 ± 0.12	sf	2
Pu^{240}	2.257 ± 0.046	sf	5		3.53 ± 0.15	sf	7
	2.02 ± 0.11	sf	6	Fm^{254}	4.05 ± 0.19	sf	3

1 D. J. Hughes and J. A. Harvey: "Neutron Cross-Sections", Brookhaven Nationa Laboratory Report BNL-325, July 1955.
2 D. A. Hicks, J. Ise jr. and R. V. Pyle: University of California Radiation Laboratory Report UCRL-3146, Rev., 1955. Phys. Rev. **101**, 1016 (1956). All values based on 2.257 ± 0.046 for Pu^{240}.
3 G. R. Choppin, B. G. Harvey, D. A. Hicks, J. Ise jr. and R. V. Pyle: Phys. Rev. **107**, 766 (1956).
4 A. H. Jaffey, C. T. Hibdon and R. K. Sjoblom: Argonne National Laboratory Report ANL-5396, 1955. Quoted by W. C. Bentley *et al.*: Paper P/809, "Peaceful Uses of Atomic Energy-Proceedings of the Geneva Conference", August 1955, United Nations.
5 B. C. Diven *et al.*: Phys. Rev. **101**, 1012 (1956).
6 W. W. T. Crane, G. H. Higgins and H. R. Bowman: University of California Radiation Laboratory Report UCRL-4555. Phys. Rev. **101**, 1804 (1956). There is a systematic difference of ~7% between the $\bar{\nu}$ values from this report and those from Ref. 2; this is due to a difference in standardization of neutron counting efficiency.
7 W. W. T. Crane, G. H. Higgins and S. G. Thompson: Phys. Rev. **97**, 242 (1955); erratum Phys. Rev. **97**, 1727 (1955).
8 F. R. Barclay and W. J. Whitehouse: Proc. Phys. Soc. Lond. A **66**, 447 (1953).
9 G. H. Higgins, W. W. T. Crane and S. R. Gunn: Phys. Rev. **99**, 183 (1955).
10 P. A. Egelstaff and J. E. Sanders: Paper P/425, "Peaceful Uses of Atomic Energy-Proceedings of the Geneva Conference", August 1955, United Nations.

In Sects. 39 and 47 we described the preparation of americium and curium by intense neutron irradiation of plutonium. Fig. 43 (p. 301) shows the changing composition of the heavy mass number isotopes as a function of time when small samples of Pu^{239} are irradiated at a flux of 3×10^{14} neutrons per cm^2 per second. The Pu^{239} is rapidly destroyed by fission and by neutron capture reactions. Fig. 43 shows the build up and destruction of the higher mass isotopes of plutonium and the eventual production of substantial amounts of Am^{243} and Cm^{244}.

In the case of a few transuranium element isotopes the value of the cross-sections for fission, for capture, or for inelastic scattering as a function of neutron energy have been determined with some care. The compilations of data on neutron cross-sections prepared by Hughes and Harvey[1] include such data for Np^{237}, Pu^{239}, Pu^{240}, Pu^{241}, and Am^{241}.

1 D. J. Hughes and J. A. Harvey: "Neutron Cross-Sections", Brookhaven National Laboratory Report BNL-325, July 1955. — Addendum to Report BNL-325, "Heavy Element Cross-Sections", presented at Geneva, August 1955.

Table 35*. *Calculated probabilities of emitting ν neutrons per spontaneous fission, P(ν), and the average number of neutrons per spontaneous fission, $\bar{\nu}$, based on $\bar{\nu} = 2.257 \pm 0.046$ for Pu^{240}.*

P(ν)	Pu^{236}	Pu^{238}	Pu^{240}	Pu^{242}	Cm^{242}	Cm^{244}	Cf^{252}
P_0	0.062 ±0.035	0.044 ±0.009	0.041 ±0.009	0.063 ±0.013	0.011 ±0.005	0.001 ±0.004	0.001 ±0.001
P_1	0.156 ±0.090	0.175 ±0.026	0.219 ±0.021	0.192 ±0.034	0.126 ±0.018	0.099 ±0.017	0.021 ±0.007
P_2	0.38 ±0.13	0.384 ±0.026	0.351 ±0.021	0.351 ±0.041	0.323 ±0.018	0.281 ±0.022	0.111 ±0.019
P_3	0.28 ±0.12	0.237 ±0.027	0.241 ±0.020	0.324 ±0.047	0.347 ±0.020	0.365 ±0.018	0.271 ±0.019
P_4	0.096 ±0.086	0.124 ±0.021	0.127 ±0.018	0.033 ±0.026	0.139 ±0.013	0.198 ±0.220	0.326 ±0.018
P_5	0.033 ±0.036	0.036 ±0.009	0.020 ±0.006	0.036 ±0.013	0.050 ±0.009	0.049 ±0.009	0.178 ±0.016
P_6			0.001 ±0.002		0.004 ±0.002	0.007 ±0.002	0.077 ±0.013
P_7					0.001 ±0.001		0.013 ±0.004
P_8							0.003 ±0.001
$\bar{\nu}$	2.30 ±0.19	2.33 ±0.08	2.257 ±0.046	2.18 ±0.09	2.65 ±0.09	2.84 ±0.09	3.82 ±0.12

* D. A. HICKS, J. ISE jr. and R. V. PYLE: Phys. Rev. **101**, 1016 (1956).

English work on the isotopes of uranium and plutonium is reviewed by LYNN and PATTENDEN[1] and by RAFFLE and PRICE[1].

68. Spontaneous fission. The experimental values of spontaneous fission half-lives for heavy element isotopes are collected in Table 36. No adequate theory for the spontaneous fission process has been developed although a number of empirical observations and correlations have been presented.

An inspection of the table reveals a very strong dependence on atomic number, the process rapidly becoming more probable with increasing Z. A second striking fact is that nearly all the spontaneously fissionable nuclei are of the even-even nuclear type. Nuclei containing odd numbers of nucleons decay at a rate many factors of ten slower than do neighboring even-even nuclei. WHITEHOUSE and GALBRAITH[2] and SEABORG[3] observed that the half-lives showed an exponential dependence on Z^2/A. The rough validity of this relationship is shown in Fig. 44 where the logarithm of the partial spontaneous fission half life is plotted against Z^2/A. With the accumulation of more data it became apparent that, although the parameter Z^2/A accounted in this manner broadly for the variation in half-life over a range of Z values, for a given value of Z this parameter did not account

[1] J. E. LYNN and N. J. PATTENDEN: Paper P/423 and J. F. RAFFLE and B. T. PRICE, Paper P/422, "Peaceful Uses of Atomic Energy-Proceedings of the Geneva Conference", August 1955, United Nations.

[2] W. J. WHITEHOUSE and W. GALBRAITH: Nature, Lond. **169**, 494 (1952).

[3] G. T. SEABORG: Phys. Rev. **85**, 157 (1952).

Table 36. *Half-lives for spontaneous fission.*

Isotope	Half-life	Reference
Th^{230}	$\geq 1.5\times 10^{17}$ years	E. SEGRÈ: Phys. Rev. **86**, 21 (1952).
Th^{232}	$> 10^{20}$ years	A. V. PODGURSKAYA *et al.*: Zhur. Eksptl. i Teoret. Fiz. **28**, 503 (1955).
U^{232}	$8\pm 5.5\times 10^{13}$ years	A. H. JAFFEY and A. HIRSCH: Unpublished work 1951.
U^{234}	1.6×10^{16} years	A. GHIORSO *et al.*: Phys. Rev. **87**, 163 (1952).
U^{235}	1.8×10^{17} years	E. SEGRÈ: Phys. Rev. **86**, 21 (1952).
U^{236}	2×10^{16} years	A. H. JAFFEY and A. HIRSCH: Unpublished data 1949.
U^{238}	8×10^{15} years	E. SEGRÈ: Phys. Rev. **86**, 21 (1952).
Pu^{236}	3.5×10^{9} years	A. GHIORSO *et al.*: Phys. Rev. **87**, 163 (1952).
Pu^{238}	4.9×10^{10} years	A. H. JAFFEY and A. HIRSCH: Unpublished data 1947.
Pu^{239}	5.5×10^{15} years	E. SEGRÈ: Phys. Rev. **86**, 21 (1952).
Pu^{240}	1.2×10^{11} years	O. CHAMBERLAIN *et al.*: Phys. Rev. **94**, 156 (1954).
	1.32×10^{11} years	E. M. KINDERMAN: Atomic Energy Commission Declassified Report HW-27660, April 1953.
Pu^{242}	$7.25\pm 0.3\times 10^{10}$ years	Unpublished results, Groups at Argonne National Laboratory and University of California Radiation Laboratory.
Pu^{244}	$2.5\pm 0.8\times 10^{10}$ years	P. R. FIELDS *et al.*: Phys. Rev. **100**, 172 (1955).
Cm^{240}	1.9×10^{6} years	A. GHIORSO *et al.*: Phys. Rev. **87**, 163 (1952).
Cm^{242}	7.2×10^{6} years	A. GHIORSO and H. P. ROBINSON: Unpublished results 1947. — G. C. HANNA *et al.*: Phys. Rev. **81**, 466 (1951).
Cm^{244}	1.4×10^{7} years	A. GHIORSO *et al.*: Phys. Rev. **87**, 163 (1952).
Cm^{246}	3×10^{7} years	Unpublished results, Groups at Argonne National Laboratory and University of California Radiation Laboratory.
Bk^{249}	6×10^{8} years	A. GHIORSO *et al.*: Unpublished results 1955.
	$>2\times 10^{8}$ years	L. B. MAGNUSSON *et al.*: Phys. Rev. **96**, 1576 (1955).
Cf^{246}	$2.1\pm 0.3\times 10^{3}$ years	E. K. HULET *et al.*: Phys. Rev. **89**, 878 (1953).
Cf^{248}	7×10^{3} years	E. K. HULET: Ph. D. Thesis, University of California Unclassified Report UCRL-2283, August 1953.
Cf^{249}	1.5×10^{9} years	A. GHIORSO *et al.*: Unpublished results, 1954.
Cf^{250}	$1.5\pm 0.5\times 10^{4}$ years	A. GHIORSO *et al.*: Phys. Rev. **94**, 1081 (1954). — P. R. FIELDS *et al.*: Nature, Lond. **174**, 265 (1954). — L. B. MAGNUSSON *et al.*: Phys. Rev. **96**, 1576 (1954).
Cf^{252}	66 ± 10 years	L. B. MAGNUSSON *et al.*: Phys. Rev. **96**, 1576 (1954). — A. GHIORSO *et al.*: Phys. Rev. **94**, 1081 (1954).
Cf^{254}	85 ± 15 days	B. G. HARVEY *et al.*: Phys. Rev. **99**, 337 (1955).
	60 ± 12 days	H. DIAMOND *et al.*: Phys. Rev. **94**, 1083 (1954).
E^{253}	3×10^{5} years	P. R. FIELDS *et al.*: Phys. Rev. **94**, 209 (1954). — A. GHIORSO *et al.*: Unpublished results, 1954.
E^{254}	1.5×10^{5} years	A. GHIORSO *et al.*: Unpublished results, 1955.
Fm^{254}	200 days	G. R. CHOPPIN *et al.*: Phys. Rev. **94**, 1080 (1954).
	220 ± 40 days	P. R. FIELDS *et al.*: Phys. Rev. **94**, 209 (1954).
Fm^{255}	>60 years	A. GHIORSO *et al.*: Unpublished results 1955.
Fm^{256}	3 hours	G. R. CHOPPIN *et al.*: Phys. Rev. **98**, 1519 (1955).

for the variation of half-life with the mass number, A. HUIZENGA[1] has pointed out that for a given value of Z the half-life goes through a maximum as A varies. In Fig. 44 such a correlation is shown by drawing lines through the isotopes of each element. Similar graphs have been published by FIELDS and coworkers[2], by GLASS and coworkers[3], and by GHIORSO[4]. Other correlations[5,6] involving the

[1] J. R. HUIZENGA: Phys. Rev. **94**, 158 (1954).

[2] P. R. FIELDS, M. H. STUDIER, L. B. MAGNUSSON and J. R. HUIZENGA: Nature, Lond. **174**, 265 (1954).

[3] R. A. GLASS, S. G. THOMPSON and G. T. SEABORG: J. Inorg. a. Nucl. Chem. **1**, 3 (1955).

[4] A. GHIORSO: Paper P/718, "Proceedings of the Geneva Conference—Peaceful Uses of Atomic Energy", August 1955, United Nations.

[5] A. KRAMISH: Phys. Rev. **88**, 1201 (1952).

[6] M. H. STUDIER and J. R. HUIZENGA: Phys. Rev. **96**, 545 (1954).

ratio of spontaneous fission to alpha half-lives have also been useful for predictive purposes. GHIORSO[1] has noted that the spontaneous fission half lives for even-even isotopes containing more than 152 neutrons are considerably shorter than might be predicted even on the basis of a plot like that in Fig. 44. In this figure, for example, the curve for the isotopes of californium drops off steeply on the high mass side. GHIORSO[1] tentatively relates this effect to a sub-shell closure at 152 neutrons.

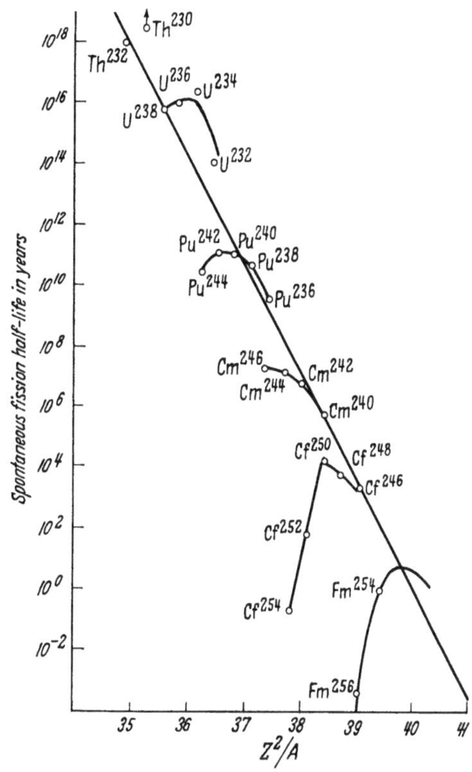

Fig. 44. Spontaneous fission half-life versus Z^2/A.

Spontaneous fission half-lives are competitive with alpha decay half-lives for the higher mass, even-even isotopes of californium and fermium and presumably for higher elements. Cf^{254} decays chiefly by spontaneous fission with a half-life of 85 days; the alpha half-life is expected to be about 100 years. In Fm^{256} the observed mode of decay is spontaneous fission; the observed half-life of 3.5 hours is much shorter than the predicted alpha half-life of about 10 days.

The rapid shortening of spontaneous fission lifetimes makes it unlikely that elements beyond fermium can be made in appreciable quantities by neutron irradiation techniques. According to Fig. 39 it is necessary to reach a mass number of 259 before a beta-emitting fermium isotope is reached. The spontaneous fission rates of Fm^{258} and Fm^{260} will be so large that most of the atoms of these isotopes will be destroyed by this process before they are converted to heavier isotopes by neutron capture. Hence very little build-up of element 101 and higher elements can be expected.

FIELDS *et al.*[2] and BENTLEY *et al.*[3] have discussed the possible usefulness of some of the short-lived spontaneously-fissioning isotopes as sources of neutrons. Cf^{252} is attractive for this purpose because it can be made in appreciable yield by long time irradiation of plutonium (see Sect. 52). This isotope has a neutron emission rate of 3×10^{12} neutrons per second per gram.

Acknowledgements.

The authors wish to extend their thanks to BRUCE M. FOREMAN jr., DONALD STROMINGER, T. DARRAH THOMAS, and Mrs. SUSANNE VANDENBOSCH for valuable assistance in compiling data and to Miss EILEEN CARSON and Miss JEAN SANSOME for assistance in the preparation of the manuscript for publication.

[1] See footnote 4, p. 306.

[2] See footnote 2, p. 306.

[3] W. C. BENTLEY *et al.*: Paper P/809, "Proceedings of the Geneva Conference — Peaceful Uses of Atomic Energy", August 1955, United Nations.

General references.

[1] National Nuclear Energy Series, Plutonium Project Record, Vol. 14A, "The Actinide Elements". G. T. SEABORG and J. J. KATZ, Editors. New York: McGraw Hill Book Co. 1954. — This is a comprehensive summary of the chemical and nuclear properties of the actinide elements including all work published up to 1953—1954. It is the basic reference for the transuranium elements.

[2] National Nuclear Energy Series, Plutonium Project Record, Vol. 14B, "The Transuranium Elements: Research Papers", Edited by G. T. SEABORG, J. J. KATZ and W. M. MANNING. New York: McGraw Hill Book Co. 1949. — This is an extensive collection of original research papers on neptunium, plutonium, americium, and curium covering a large part of the work done in American laboratories during World War II.

[3] SEABORG, GLENN T.: Place in Periodic System and Electronic Structure of the Heaviest Elements. Nucleonics **5** [5], 16 (1949).

[4] Peaceful Uses of Atomic Energy—Proceedings of the International Conference in Geneva, August 1955, United Nations, New York. — The coverage of the physics and chemistry of the transuranium elements is not extensive in these volumes, but the information given here on certain topics such as neutron physics and industrial chemical processing methods goes considerably beyond that previously published.

[5] Progress in Nuclear Energy, Series 3, Vol. 3 "Process Chemistry", Edited by F. R. BRUCE, J. M. FLETCHER, H. H. HYMAN and J. J. KATZ, Pergamon Press, Ltd. 1956. — A collection of papers on the principal methods in use or under consideration for industrial processing of plutonium and uranium.

[6] HUGHES, D. N., and J. A. HARVEY: "Neutron Cross-Sections", Brookhaven National Laboratory Report BNL-325, April 1955; Addendum to Report BNL-325, "Heavy Element Cross-Sections" presented at Geneva Conference August 1955. — An authoritative and extensive collection of tables and graphs on neutron cross-sections for all elements.

[7] NEWTON, J. O.: Nuclear Properties of the Very Heavy Elements. Progr. Nuclear Phys. **4**, 234—286 (1955). — An extensive review of the published literature up to April 1954.

[8] McKAY, H. A. C., and J. MILSTED: The Preparation and Chemistry of the Transuranium Elements. Progr. Nuclear Phys. **4**, 287—329 (1955). — An extensive review of this field.

[9] LATIMER, W. M.: Oxidation Potentials, second edition, Prentice-Hall, Inc. 1952. — Chapter 21 is an extensive compilation of thermodynamic data on the Actinide Elements.

[10] GLASS, R. A., S. G. THOMPSON and G. T. SEABORG: Nuclear Thermodynamics of the Heaviest Elements. J. Inorg. a. Nucl. Chem. **1**, 3 (1955). — A detailed treatment of the material covered in Part C of this article, including many references to original literature and to other authors covering similar material.

[11] HUIZENGA, J. R.: Isotopic Masses $A > 201$. Physica, Haag **21**, 210 (1955). — Masses and binding energies of nuclides with mass number greater than 201. Computed from mass spectrographic data in combination with nuclear disintegration energies.

[12] LISTER, M. W.: The Chemistry of the Transuranic Elements. Chem. Soc. Quart. Rev. **4**, 30 (1950).

[13] PERLMAN, I., and J. O. RASMUSSEN: Alpha radioactivity. In this volume. A detailed review of all experimental data and theoretical treatments of alpha decay.

The Nuclear Photoeffect.

By

GEORGE R. BISHOP and RICHARD WILSON.

With 23 Figures.

1. History and layout. The first photonuclear reaction was demonstrated by CHADWICK and GOLDHABER in 1934 [*1*], who irradiated an ionization chamber filled with deuterium with the 2.62 Mev γ-rays emitted by ThC″. They detected the photoprotons by their ionization in the deuterium gas and measured their energy. By performing a similar irradiation with the γ-rays of 1.8 Mev energy of RaC, and observing almost no ionization due to photoprotons, they were able to deduce that the threshold of the reaction lay between 1.8 and 2.3 Mev. Apart from the intrinsic interest of the new phenomenon of the photodisintegration of a nucleus, they were thus enabled to estimate the mass of the neutron, combining the binding energy with the spectroscopic masses of hydrogen and of deuterium. The measurements indicated that the neutron mass was greater than that of the proton, and in fact that it was unstable energetically against decay by negative electron emission to a proton. Many years later the radioactive decay of the neutron was in fact observed. At the same time CHADWICK and GOLDHABER measured the cross-section for photodisintegration of the deuteron at an energy of 2.62 Mev, and found a value in encouraging agreement with the theoretical one of BETHE and PEIERLS [*2*].

The next photodisintegration of a nucleus to be observed was that of Be^9 by SZILARD and CHALMERS in 1934, using the sensitive method for detecting the photoneutrons which bears their names[1].

Following these pioneering experiments the scope of photonuclear studies was much enlarged by the employment of sources of γ-rays other than those of the naturally occurring radioactive elements. Amongst these new sources were the γ-rays produced in nuclear reactions induced by heavy particles accelerated by conventional high tension sets, and the X-rays produced by electron accelerators, betatrons and synchrotrons. With these machines intense beams of high energy γ-rays became available. One extensively used γ-ray from a nuclear reaction is of 17.6 Mev energy, produced by proton capture on lithium. Electron accelerators, both of the circular and the linear types, have been put in use with maximum energies up to 1 Bev. With the higher energy X-rays it has been possible to produce a wide diversity of photonuclear reactions in all stable nuclear species. These different sources of γ-rays and X-rays set limitations on the type and accuracy of the experiments which can be performed. It is necessary for a proper understanding of the photonuclear experiments to discuss the sources; this is done in Chap. A.

Experiments with low energy γ-rays on deuterium and beryllium have been more and more refined, with the object of measuring accurately the thresholds, the cross-section as a function of energy, and the angular distribution of the

---

[1] L. SZILARD and T. A. CHALMERS: Nature, Lond. **134**, 494 (1934).

products, for comparison with the theoretical predictions. Such measurements are particularly important for the deuteron, because it is the simplest of nuclear systems, and plays a role similar to that of the hydrogen atom in the theory of atomic structures.

Chap. B outlines the theoretical[1] and Chap. C the experimental data on the photodisintegration of the deuteron and show how parameters of the neutron-proton system may be deduced.

The study of photodisintegration of other nuclei which comprises the whole of Chap. D can be roughly separated into two parts; the first part is the study of the mechanism of photon absorption by the nucleus, and the second the study of the subsequent emission of particles or quanta. For heavy elements, and at low energies, these two parts can be clearly separated; the incident photon interacts with the nucleus as a whole, and the excited nucleus then loses energy by evaporating particles according to the well known compound nucleus picture of nuclear reactions. For heavy elements at high energies, and for light elements at all energies, a compound nucleus is not always formed, and the two parts of the photodisintegration process cannot be separated; the "direct" nuclear photoeffect then becomes of importance. At high energies and light nuclei, the nuclear photoeffect can again be simply described by an initial direct interaction, similar to the interaction with the deuteron, followed sometimes by an interaction of an outgoing particle.

The study of the photon absorption process cannot proceed directly, because the photon absorption by atomic electrons is always greater. The nuclear absorption must be derived by adding together the cross-sections for all final states in the reaction, some of which are not always experimentally accessible. The construction of a curve relating the cross section for nuclear absorption of photons to the photon energy can only be made in a few special cases. The curves usually published relating partial cross sections to the photon energy are, at the present stage of knowledge, only of limited interest by themselves.

There is a further difficulty in the study of the absorption process. Most experiments are performed using a continuous distribution of γ-rays from a betatron or synchrotron, and the cross section curve is deduced by a difference method (see Sect. 4). This method is often of inferior accuracy and always of poor resolution. Only in special cases is it possible to study the detailed structure (Sects. 29, 30) of photonuclear processes, but in others it is fortunately possible from γ-ray scattering experiments to show that no detailed structure exists (Sect. 40).

In the study of the decay of the compound nucleus (Sect. 31) there is a similar difficulty. The quantities of interest are the relative yields of different reaction products. Yet again it is often that only a single partial cross section has been studied.

It is here that the studies, initially at Zürich[2], of photonuclear reactions induced by the 17.6 Mev γ-rays from $Li^7(p, \gamma)\,Be^8$ are of particular interest. They have studied the neutrons and protons from a large range of nuclei and were the first to show the inadequacy, for elements near the atomic number of copper, of the compound nucleus picture; for protons, normally inhibited by the potential barrier, were shown to come out with great frequency. For light elements, deuterons and alphas are also frequent.

---

[1] More details on the theory of the deuteron photodisintegration are presented in Chap. VI of HULTHÉN and SUGAWARA's article in Vol. XXXIX of this Encyclopedia.

[2] H. WÄFFLER and O. HIRZEL: Helv. phys. Acta **21**, 200 (1948). — O. HIRZEL and H. WÄFFLER: Helv. phys. Acta **20**, 373 (1947).

For light nuclei, the study of individual levels and processes becomes of interest. The nuclear photoeffect is one of many tools which can be brought to bear upon the energy levels of light nuclei [13]. With good resolution the properties of individual levels may be determined (Sects. 29, 30). For example, WILKINSON[1] has summarized the transition probabilities for many electric dipole γ-ray transitions in light nuclei, some studied by (γ, n) processes, others by (n, γ) or (p, γ). There also the charge independence of nuclear forces can help to simplify the interpretation because of the selection rules in γ-ray processes caused thereby (Sects. 34 to 36). The photodisintegration of light elements has been studied extensively by photographic plates and falls within the scope of the review article by TITTERTON [11].

A. Sources of γ-rays and X-rays.

2. Radioactive sources of γ-rays. Radioactive sources are the most convenient to use, since they are small and require no expensive accelerators. It is also possible to calibrate radioactive sources by measuring a charged particle decay, either directly, or in coincidence with the γ-rays. Thus the absolute β disintegration rate of a radioactive source may now be determined to an accuracy of about 1% by a Geiger counter in a 4π geometry; and γ-ray activities may be similarly determined. The sources of interest, therefore, are those that emit one high energy γ-ray per disintegration, or a known fraction of the disintegrations. Only deuterium, beryllium and lithium-6 are susceptible to disintegration by radioactive sources, but they have been extensively studied.

Table 1. *γ-ray sources.*

Source	Excited nucleus	γ-ray energy
(a) One γ per disintegration		
Na^{24}	Mg^{24}	2.757
RdTh	Pb^{208}	2.618
(b) Other useful sources		
Ga^{72}	Ge^{72}	2.501, 2.507
Mg^{56}	Fe^{56}	2.18
Sb^{124}	Te^{124}	1.7
Ga^{66}	Zn^{66}	2.75, 4.33
Co^{56}	Fe^{56}	3.25, 2.55

3. γ-rays from nuclear reactions. The important γ-ray lines produced by $(p\gamma)$ reactions on the light nuclei are those from F^{19} $(p\alpha\gamma)$, of 6.3 and 7.1 Mev energy;

Table 2.

Reaction	Target	Resonance energy	γ-ray energy Mev	Relative yield	Absolute yield γ-rays per proton (thick target)
$N^{15}(p, \alpha)\,C^{12}$	deposited		4.45 ± 0.04		
$F^{19}(p, \alpha)\,O^{16}$	CaF_2		6.14 ± 0.01	14	2.6×10^{-8}
			6.9	5	
			7.1	2	
$Be^{9}(p, \gamma)\,B^{10}$	metal	998	7.37 ± 0.07		
		1.087	6.8		
$C^{13}(p, \gamma)\,N^{14}$	deposited	554	8.14 ± 0.08		
$B^{11}(p, \gamma)\,C^{12}$	metal		16.7	0.15	
			12.2 ± 0.1	0.2	
$Li^{7}(p, \gamma)\,Be^{8}$	metal		17.6 ± 0.2	1.0	3.7×10^{-7}
			14.8 ± 0.3	0.5	
$T(p, \gamma)\,He^{4}$	deposited		20.4 ± 0.2	0.01	

[1] D. H. WILKINSON: Phil. Mag. 8, 127 (1956).

those from $Li^7(p\gamma)$, of 17.63 and 14.8 Mev; that from the radiative capture of protons on tritium of energy 20 Mev. The absolute yields of some of these reactions have been measured with a precision sufficient at present for photonuclear measurements, and some information is also available on alternative modes of disintegration such as occur in $Li^7(p\gamma)$. The known data are summarised in Table 2.

High tension accelerators of the cascade Cockroft and Walton, or electrostatic van der Graaf types, capable of generating the necessary resonance voltages, usually give maximum currents of the order of 1 milliampere for the former, and 50 microamperes for the latter machine. The available γ-ray intensities are thus quite small, necessitating long irradiations, and, commonly the use of photographic plates.

4. Electron bremsstrahlung. In addition to the discrete sources of γ radiation from excited states of nuclei discussed in Sects. 2 and 3 above there is available for experiment the bremsstrahlung from electron collisions, which is *not* monochromatic and the use of the radiation may therefore be complicated. In order to use the radiation it is necessary to know its characteristics. The details of the calculations do not logically fall into the subject matter of this article, but the conclusions, and their validity, will be summarized.

The most extensive calculations of electron bremsstrahlung have been made using the Born approximation[1]. Thus at high energies they deduce

$$\sigma(E_0, k)\, dk = 4\bar{\sigma}\frac{dk}{k}\left\{\left[1+\left(\frac{E}{E_0}\right)^2-\frac{2}{3}\frac{E}{E_0}\right]\log\frac{183}{Z^{\frac{1}{3}}}+\frac{1}{9}\frac{E_0}{E}\right\} \tag{4.1}$$

where $\bar{\sigma}=Z^2(e^2/mc^2)^2(e^2/\hbar c)$. The cross section for lower energies they display in graphical form. Approximately,

$$k\,\sigma(E_0, k)\, dk = \text{const} \tag{4.2}$$

which is a useful form for qualitative discussion and understanding. The authors used three simplifications which restrict the validity of the calculations for our purpose, but recent work has essentially removed these restrictions which will now be discussed in turn.

1. The screening of the Coulomb field of the nucleus by the atomic electrons enters into the calculations of BETHE and HEITLER in the final stage; their final formulae are correct only if the assumption of a Thomas-Fermi statistical distribution of the atomic electrons is reasonable. The validity of this has been checked by FRAZER[2], WHEELER and PANOFSKY who used more exact wave functions for the atomic electrons. Thus FRAZER showed that for mercury the error is less than $\frac{1}{2}$% at high energies, whereas for atomic hydrogen the error is 3% and for *molecular* hydrogen 5%. Fortunately for photonuclear work only heavy elements are used for sources of γ-radiation so that we can use the uncorrected Bethe Heitler figures.

2. BETHE and HEITLER neglected the pair production in the field of the atomic electrons. This may be included approximately by replacing Z^2 in their formulae by $Z(Z+1)$ which shows that for heavy elements the effect is small. There are small differences from this, which arise because the screening is different,

---

[1] H. A. BETHE and W. HEITLER: Proc. Roy. Soc. Lond., Ser. A **146**, 83 (1934). — H. A. BETHE and J. ASHKIN: In Experimental Nuclear Physics, Vol. 1, Part 2. John Wiley 1953.

[2] P. A. FRAZER: Phys. Rev. **87**, 523 (1952). — J. A. WHEELER and W. E. LAMB: Phys. Rev. **55**, 858 (1939). — D. BERNSTEIN and W. K. H. PANOFSKY: Phys. Rev. **102**, 522 (1956).

the effect of recoil is important, or the effect of identity of the electrons is important. The first of these has been discussed by WHEELER and LAMB and the second and third by VOTRUBA[1], and the effects can now be considered understood. These refinements are small and may be safely ignored for our purposes.

3. The Born approximation is not valid for heavy nuclei, yet it is the heavy nuclei that are used for targets in synchrotrons and betatrons. DAVIES *et al.*[2] have calculated both bremsstrahlung and pair production using Furry wave functions valid to order mc^2/E; their calculations are not, therefore, valid at low energies, but will give the correct result for all Z at high energies. These authors found that the processes of bremsstrahlung and pair production, simply related in Born approximation, are so no longer, so that experimental data on pair production is not of itself a good check on the theory. DAVIES *et al.* suggested that in the limit of complete screening the Born approximation gives the right answer for bremsstrahlung but not for pair production; OLSEN, on the other hand, has shown that if the bremsstrahlung cross section is integrated over all angles, the cross section is lower than the Born approximation in the same way as for the pair production. In the discussion below we have assumed that the bremsstrahlung is experimentally integrated over angles. This has been the case so far. Experimentally both bremsstrahlung and pair production agree with the theory of DAVIES for energies above 50 Mev; at 20 Mev both give a deviation from Born approximation of half the predicted amount[3].

The improved calculation removes the restriction of the Born approximation that the perturbation be small, but retains the assumption, valid at high energies, that the incident and scattered electrons may be represented by plane waves. This is *not* valid at the extreme high energy end of the bremsstrahlung spectrum at any incident energy. DAVIES[4] has further removed this restriction by using spherical waves for the *scattered* electron. DAVIES' refinement is important for it shows that the cross section for producting photons of the maximum energy is *not* zero, as Eq. (4.1) would suggest, but is finite. The theory of electron bremsstrahlung can only be applied simply to experiments with external electron beams where the incident electron current may be measured, and the geometry precisely controlled. For circular accelerators the internal beam is inaccessible, and the thickness of the target is not known with precision.

For a study of photonuclear reactions a monochromatic γ-ray source is desired, but the only available radiation is frequently bremsstrahlung from electrons. Several methods are available for using the bremsstrahlung and deriving information equivalent to that obtained from using monochromatic sources. Firstly the energies of the reaction product may be measured in each photonuclear event and the γ-ray energy inferred therefrom; this method may be applied to the photodisintegration of the deuteron (Sect. 20) and to the photodisintegration of carbon or oxygen into alpha particles (Sect. 33). This method wastes no data.

Secondly, the information to be extracted from an experiment may be restricted. Thus it will be shown (Sect. 22) that the quantity $\int_0^{E_0} \sigma_k\, dk$, where σ_k

[1] V. VOTRUBA: Phys. Rev. **73**, 1468 (1948). The earlier calculations of Borsellino are now known to be in error.

[2] H. DAVIES, H. A. BETHE and L. C. MAXIMON: Phys. Rev. **93**, 788 (1954). — H. OLSEN: Phys. Rev. **99**, 1335 (1955).

[3] K. L. BROWN: Phys. Rev. **103**, 243 (1956). — W. C. BARBER, A. I. BERMAN, K. L. BROWN and W. D. GEORGE: Phys. Rev. **99**, 59 (1955) and references therein.

[4] Private communication. Similar results have been obtained by CONKIE, and by NAGASAKA independently. (All unpublished.)

is the absorption cross section of a nucleus for the γ-ray of energy k, and E_0 is the meson production threshold, is easily interpretable. Moreover, both theory and experiment show that the main contribution to this integrated cross-section comes from a narrow group of energies, so that $\int_0^{E_0} \sigma(k)\, dk$ may be readily calculated from a single measurement of the average σ for a bremsstrahlung spectrum.

Thirdly, a monochromator may be used. The scattered electron from the bremsstrahlung process may be detected by a counter and its energy from its curvature in a magnetic field. If the γ-ray or the secondary particle from a photonuclear reaction are measured in coincidence with this electron, the γ-ray energy may be defined. This procedure has been used at 300 Mev by WEIL[1] and at 22 Mev by GOLDEMBERG[1]. Few experimental results have been obtained by this method, because of the limitations on the counting rate imposed by the coincidence procedure. With the availability of linear electron accelerators, an essentially equivalent, but more direct method becomes possible. The incident electron energy is well defined; if this is used directly to produce a nuclear process, the result is similar to photon produced nuclear processes. The energy of the "equivalent photons" may be defined by measuring the scattered electron energy. This method is discussed further in Sect. 38.

Fourthly a difference method may be employed. The bremsstrahlung formula, as simplified in Eq. (4.2) suggests that if the activity A induced in a nucleus is measured as a function of the incident electron energy, *for the same electron current*, then the function dA/dE_0 evaluated at k, gives directly the cross-section at the photon energy k.

This fourth method has been generally applied but it has several difficulties. The first is the requirement that the electron current be held constant. For a linear accelerator this presents no problem, but for a circular accelerator, where the multiple traversal and the effective target thickness vary with energy, there are difficulties. For the circular accelerators therefore, the data are normalized to a constant electron current by the indirect use of an ionization chamber whose sensitivity can be calculated. This in turn depends critically on the details of the bremsstrahlung spectrum. The details are out of place here, and have been discussed extensively by KATZ and CAMERON[2].

Secondly, Eq. (4.2) is only an approximation and the full equation (4.1), with the Davies correction and the correction for target thickness, must be used. No simple formula can then be deduced, and an iterative procedure must be used to find the final cross-section. This iterative procedure would be of only mathematical interest if $A(E_0)$ were known to high accuracy and free of all uncertainty. This is not the case. In order to make a convergent procedure the "photon difference" method is now usually employed, as discussed by KATZ and CAMERON. The function $A(E_0)$ is arbitrarily smoothed; then the derivative $dA(E_0)/dE_0$ is determined and in turn smoothed and used to determine the further corrections. As a result of this procedure, fine structure is eliminated from the curves, and moreover it becomes difficult to assign errors in any consistent manner to the final cross-section curve[3].

---

[1] J. W. WEIL and B. D. MCDANIEL: Phys. Rev. **92**, 391 (1953). — J. GOLDEMBERG: Phys. Rev. **93**, 1426 (1954).

[2] L. KATZ and A. G. W. CAMERON: Canad. J. Res. **29**, 518 (1951).

[3] RICHARD WILSON: Proc. Phys. Soc. Lond. A **66**, 645 (1953). These two difficulties are discussed later (Sect. 22).

B. Theory of the photodisintegration of the deuteron.

5. Introduction. The original, and most easily understood, method of calculation on photodisintegration processes was by separation of the radiation field into multipoles. This procedure enables the transition probability to be expressed in terms of matrix elements by the usual formula:

$$w = \frac{2\pi}{\hbar} \varrho(k) \, |H_{gf}^l|^2$$

where H_{gf}^l is the matrix element for the transition between the ground state and the final state. Most of this section will be involved only with the calculation of these matrix elements for the deuteron.

We are concerned with integrals of the form

$$H_{gf}^l = e \int \psi_g^* z^l \psi_l \, d\tau$$

for electric multipoles of order l, and

$$H_{gf}^l = \int \psi_g^* \vec{\sigma}_l \psi_l \, d\tau$$

for magnetic multipoles. Here ψ_g and ψ_l are the wave functions for the ground state of the deuteron and the neutron-proton system of angular momentum l. The reason why the expansion into multipoles leads to a simple computation is that there are selection rules on the final state which arise from the orthogonality of the wave functions. FOLDY and BERGER[1] have introduced another computational method which has some merit: the multipole expansion is retained here because of its well known physical meaning.

The calculations presented here are only summaries and are no substitute for the original papers. In particular the original calculations of BETHE and PEIERLS [2] for zero force range have a simplicity and clarity which is unexcelled; and the papers of SCHWINGER [3] have laid the foundation for the inclusion of noncentral forces. The treatment used here follows that of FESHBACH and SCHWINGER [4].

6. SIEGERT's theorem. All the calculations which we shall consider in this section assume that it is valid to use the wave functions for the deuteron and for the final neutron-proton system in conjunction with the appropriate multipole operator.

It is now normally believed that the forces between the neutron and proton are caused, at least in part, by the exchange of mesons. It is not immediately obvious, therefore, that the photodisintegration of the deuteron will be correctly described by calculations which omit all specific reference to mesons. This is made all the more serious by the observation that the mass of the meson is $\frac{1}{8}$ of the nucleon mass and a meson therefore radiates 64 times as readily as a nucleon.

For the electric multipole transitions a good justification can be given for this procedure on general grounds, which justification is termed SIEGERT's theorem[2]. This theorem follows from the requirements of gauge invariance. It is only valid to order v/c in the nucleon velocities, that is up to about 40 Mev.

For magnetic multipole transitions no such theorem applies. The magnetic dipole transition is appreciable, and cannot be correctly calculated except upon

[1] J. M. BERGER: Phys. Rev. **94**, 1698 (1954). — L. L. FOLDY: Phys. Rev. **92**, 178 (1953).

[2] R. G. SACHS and N. AUSTERN: Phys. Rev. **81**, 705 (1951). — J. G. BRENNAN and R. G. SACHS: Phys. Rev. **88**, 824 (1952).

a rigourous meson theory. As discussed in Sect. 9, the effect of a deviation from the use of nucleon wave functions alone, may be included in a phenomenological way. The effect is variously called the effect of "exchange currents" or of an "exchange magnetic moment".

7. Sum rules. It is instructive, for further application to heavy nuclei, to consider here the sum rules as applied to the deuteron (see Sect. 32).

Levinger[1] has considered the experimental data (Chap. C) and applied the detailed corrections to the simple sum rule. He derives the sum rules by three methods; an explicit calculation of the exchange forces leading to a factor $1+0.46x$; by a dispersion theory calculation using theory for the factor $R(k)$; thirdly, using experiment for $R(k)$. These are summarized in Table 3. It is seen that the agreement is good.

Table 3.

	$\int_0^{155\,\text{MeV}} \sigma\, dE$ (Mev × mb)
Experimental	39 ± 2
Ordinary forces	30
Factor $(1+0.46x)$ Plus an *MD* term	38
Dispersion theory: theoretical meson production	38
Dispersion theory: experimental meson production	32

8. Electric dipole. It transpires that at energies from 3 to 50 Mev this is the most important transition, so that it must be calculated accurately.

The ground state of the deuteron is known from other evidence to be predominantly a ^3S state, although there is a 5% admixture of a ^3D state. The important electric dipole transition is the $^3S \to {}^3P$ transition. The first significant feature is that there is, at low energies, only a small interaction in the ^3P state of the neutron-proton system; at one time this was thought to be true at much higher energies (150 Mev) as a result of analyses of neutron-proton scattering. It is now clear, however, from recent polarization and angular distribution measurements in neutron-proton scattering that this is not true. It is not yet clear what interaction to propose, so that none is assumed here, and the wave function ψ_l is taken as that of the ^3P state of the free neutron-proton system. For small energies this wave function involves r^2, so that at low energies the integral has only a small contribution from values of r *inside* the nuclear force range. The only effect of the nuclear force range on H^l_{gf} is then on the normalization of ψ_g. This corresponds to the effective range theory as used in neutron-proton and proton-proton scattering [*4*].

The integral of the radial form is

$$I_l = \int_0^\infty j_l(k\,r)\,\psi_g(r)\,r^3\,d\,r \tag{8.1}$$

and the differential cross section for the electric dipole transitions becomes

$$\sigma_d(\vartheta) = \frac{\pi}{2}\,\frac{e^2}{\hbar\,c}\,\frac{M k\,E_\gamma}{\hbar^2}\,I_1^2 \sin^2\vartheta \tag{8.2}$$

where

$$\hbar^2 k^2 = M(E_\gamma - \varepsilon)\,. \tag{8.3}$$

The ground state function of the deuteron, including the D state may be taken as:

$$\psi_g = B\,\frac{1}{r}\left[u(r) + \frac{1}{\sqrt{8}}\,S_{12}\,w(r)\right]\chi_m. \tag{8.4}$$

[1] J. S. Levinger: Phys. Rev. **97**, 970 (1955).

It is convenient to adopt the normalization

$$(u^2 + w^2) \underset{r\to\infty}{\longrightarrow} e^{-2\gamma r}. \tag{8.5}$$

We now define the triplet effective range ϱ_t

$$\varrho_t = 2 \int\limits_0^\infty \left[e^{-2\gamma r} - (u^2 + w^2)\right] dr. \tag{8.6}$$

Then the ratio of the cross section for finite range to that for zero range is given by:

$$\int (u^2 + w^2)\, dr / \int e^{-2\gamma r}\, dr = 1 - \gamma\, \varrho_t. \tag{8.7}$$

Thus

$$\sigma_e = \frac{8\pi}{3} \left(\frac{e^2}{\hbar c}\right) \left(\frac{1}{\gamma^2}\right) \frac{\varepsilon^{\frac{3}{2}} (E_\gamma - \varepsilon)^{\frac{3}{2}}}{E_\gamma^{\frac{3}{2}}} \frac{1}{1 - \gamma\, \varrho_t}. \tag{8.8}$$

ϱ_t is not quite the effective range (r_t) used in neutron-proton scattering at low energy; the difference is dependent on the potential shape.

At higher energies ($E_\gamma > 10$ Mev) the ^3P wave function is not zero inside the nuclear force range, and the cross section is reduced. This difference is more important for long tailed than for short tailed potentials. HANSSON has made calculations up to 20 Mev for a Yukawa well, which have been extended to 100 Mev for several potentials by SCHIFF and by MARSHALL and GUTH[1].

In order to compare the calculations with experiment it is necessary to known several other experimental numbers. These are enumerated below.

The binding energy of the deuteron $= (2.227 \pm 0.002)$ Mev.

The neutron proton scattering cross section for low energies $= (20.36 \pm 0.10)$ barns.

The neutron proton coherent scattering amplitude $(-3.78 \pm 0.02) \times 10^{-13}$ cm, which leads to an effective range $= (1.71 \pm 0.03) \times 10^{-13}$ cm.

In addition for the magnetic dipole one needs the singlet effective range which may be determined from neutron-proton scattering in the 1—10 Mev region.

The experiments are sufficiently accurate that computations must be made including the effect of potential shape. A good summary of these has been made by SALPETER [4], but it would seem desirable to use exact calculations when possible. For purely central potentials, HULTHÉN and NAGEL have performed these for a Yukawa potential well, and BRUNDELL and ENANDER for an exponential potential well[2]. Unfortunately only SALPETER calculates the magnetic dipole cross section with independent singlet and triplet effective ranges as the data require (Sects. 9 and 16). It should be noted, as it has caused much confusion in the past, even in high places, that the "meson mass" in the papers of HULTHÉN mentioned above is only a constant in the neutron-proton interaction and has no relation to real mesons or to meson field theory.

These and other calculations are performed assuming that the proton has a point charge. Present experiments[3] suggest that this is not true. The effect of the finite size is likely to be important at high energies where it may be confused with effects of non-central forces.

---

[1] L. I. SCHIFF: Phys. Rev. **78**, 733 (1950). — J. F. MARSHALL and E. GUTH: Phys. Rev. **78**, 738 (1950).

[2] L. HULTHÉN and B. C. H. NAGEL: Phys. Rev. **90**, 62 (1953). — I. F. E. HANSSON and L. HULTHÉN: Phys. Rev. **76**, 1163 (1949). — P. O. BRUNDELL and B. ENANDER: K. Tekn. Höqsk. Handl. **1952**, No. 60, 12 pp.

[3] R. HOFSTADTER: Rev. Mod. Phys. **28**, 214 (1956).

9. Magnetic dipole. The relevant matrix element is now

$$H^l_{gf} = \sum_k \mu_k \int \psi_g^* \sigma_k \psi_f \, d\boldsymbol{r} \tag{9.1}$$

where μ_k and σ_k are the magnetic moments and spins of the particles k, and in the case of the deuteron the neutron and proton. Neglecting the ^3D state, the only transition is from the ^3S to the ^1S state.

Then we find

$$H^l_{gf} = \int \psi_g^* \chi_g (\mu_N \sigma_N + \mu_P \sigma_P) \chi_f \psi_f \, d\boldsymbol{r} \tag{9.2}$$

where χ_g and χ_f are spin functions for the ground state and the final state. The operator

$$\mu_n \sigma_n + \mu_p \sigma_p = \tfrac{1}{2} (\sigma_n + \sigma_p)(\mu_n + \mu_p) + \tfrac{1}{2} (\sigma_n - \sigma_p)(\mu_n - \mu_p). \tag{9.3}$$

The total spin $\mathbf{S} = \sigma_n + \sigma_p$, gives zero when it acts between a singlet and a triplet state; moreover it will also give zero between two ^3S substates since there is no radial dependence of the magnetic moment operator, and two ^3S wave functions are orthogonal.

The only important transition probability is the average over all the three separate initial spin substates of the deuteron, because experimentally the deuteron is unpolarized.

As already mentioned, SIEGERT'S theorem is inapplicable to the magnetic dipole operators, and therefore the magnetic dipole calculations are not expected to be completely accurate.

This inaccuracy is due to exchange currents; this may be treated phenomenologically by adding a term to the matrix element. Then

$$|H^l_{gf}|^2 = |\mu_p - \mu_n|^2 \left[\int \psi_g^* \psi_f (1 + f(r))\, d\boldsymbol{r}\right]^2. \tag{9.4}$$

The integral may be rewritten

$$\left[\int u\, u_s (1 + g(r))\, dr\right]^2 \sin^2 \delta_s \tag{9.5}$$

where $u = \psi_f / r$ and $f(r)$, and then $g(r)$, add an exchange current term. The normalization is such that

$$u_s \xrightarrow[r \to \infty]{} \sin(k r + \delta_s)/\sin \delta_s \tag{9.6}$$

where δ_s is the singlet S phase shift. $\int (u^2 + w^2)\, dr$ is normalized to $e^{-2\gamma r}$ as before.

Following FESHBACH and SCHWINGER [4] we may now rearrange this integral in the numerator

$$I = \int_0^\infty u \left[1 + g(r)\right] u_s \, dr \tag{9.7}$$

in terms of the asymptotic values

$$v = e^{-\gamma r}$$

and

$$v_s = \sin(k r + \delta_s)/\sin \delta_s \tag{9.8}$$

We define

$$\bar{\varrho} = 2 \int_0^\infty (v\, v_s - u\, u_s)\, dr - 2 \int_0^\infty u\, u_s\, g(r)\, dr \tag{9.9}$$

in analogy with the triplet effective range. This is useful, for the first term vanishes outside the range of nuclear forces; so also does the second, for $g(r)$ is caused by short range meson effects. $\bar{\varrho}$ is thus insensitive to energy.

By manipulation,

$$\left.\begin{aligned}\sigma_m = \frac{2}{3}\pi\frac{k\gamma}{(\gamma^2+k^2)}\left(\frac{e^2}{\hbar c}\right)(\mu_n-\mu_p)^2\frac{\varepsilon}{Mc^2}(a_s-a_t)^2\times\\ \times\frac{[1-\frac{1}{2}\gamma(a_s\bar{\varrho}-a_t\varrho_t)/(a_s-a_t)]^2}{1-\gamma\varrho_t}\frac{[1+(k^2/2\gamma^2)\gamma(\varrho_s-\bar{\varrho})/(1-\frac{1}{2}\gamma\bar{\varrho}+1/\gamma a)]^2}{1+k^2a_s(a_s+\varrho_s)+\frac{1}{4}k^2a_s\varrho_s}.\end{aligned}\right\}\quad(9.10)$$

a_s and a_t are the singlet and triplet neutron-proton-scattering amplitudes and are given by

$$a_s = \lim_{k\to 0}(\delta_s/k), \qquad (9.11)$$

$$a_t = -\frac{1}{\gamma(1-\frac{1}{2}\gamma r_t)}. \qquad (9.12)$$

Here the slight differences in ϱ_t mentioned earlier are neglected.

The last two factors go to unity for zero force range in agreement with the earlier results. At low energies the second factor always goes to unity.

Thus a measure of σ_m at low energies may be used to measure $\bar{\varrho}$, all other parameters being assumed known. $\bar{\varrho}$ approximately equals $\frac{1}{2}(\varrho_s+\varrho_t)$ in the absence of an exchange moment. Two alternatives have been proposed; to use the value of $\bar{\varrho}$ to determine ϱ_s, assuming a value for the exchange moment, or to assume ϱ_s and calculate the exchange moment. As before, for an exact comparison the shape of the potential must also be considered as has been done by SALPETER. Attempts have been made[1] to assign a value to the exchange moment in a phenomenological way by using data on the magnetic moments of H^3 and He^3; a range dependence of the exchange currents is still required, however. Best estimates give about a 4% contribution, but with a large error. ϱ_s may be found from neutron-proton-scattering at energies in the Mev range, but with a large uncertainty of $\pm 0.3\times 10^{-13}$ cm. At this epoch, therefore, neither of these two procedures has any advantage over the other.

Exchange moment corrections at higher energies remain unexplored and will need an explicit meson theory.

10. Electric quadrupole. These transitions are small at any energy for which the electric dipole effect is adequately understood. It is not necessary, therefore to be able to calculate them as accurately as is necessary to calculate the dipole case. This is fortunate, for it is certain that there is a large interaction of the neutrons and protons in the ^3D state.

If the contribution of the ^3D state of the deuteron is still neglected, we have to consider transition elements of the form

$$\int_0^\infty \psi_g z^2 \psi_2 \, d\tau$$

and the reduced radial integral

$$I_2 = \int_0^\infty u_2 r^2 u_i \, dr. \qquad (10.1)$$

The function u_2 has been taken by MARSHALL and GUTH to be that for a free ^3D state which becomes $u_2 = r j_2(kr)$; SCHIFF made the assumption that the phase shift in the D state can be evaluated in Born approximation, whence

$$u_2 = r\left[\cos\delta_2 j_2(kr) - \sin\delta_2 \eta_2(kr)\right] \qquad (10.2)$$

[1] E.g. R. L. PEASE and H. FESHBACH: Phys. Rev. **81**, 142 (1951). — N. AUSTERN and R. G. SACHS: Phys. Rev. **81**, 710 (1951).

and

$$\sin\delta_2 = k\int_0^\infty V(r)\, j_2^2(k\,r)\, r^2\, d\,r, \tag{10.3}$$

where $V(r)$ is assumed to be a central potential. The difference between these two approaches is too small to be significant compared with other uncertainties.

The chief effect of the electric quadrupole is the interference with the electric dipole term. The electric cross section from dipole and quadrupole terms alone (still neglecting the ^3D state of the deuteron) can be expressed as

$$\sigma_e(\vartheta) = \frac{\sin^2\vartheta}{8\pi}\left[3\sigma_d + 6(5\sigma_d\sigma_q)^{\frac{1}{2}}\cos\delta_1\cos\delta_2\cos\vartheta + 15\sigma_q\cos^2\vartheta\right]. \tag{10.4}$$

Only the first two terms are important.

The angular distribution for electric multipole radiation then becomes of the form:

$$a\sin^2\vartheta\,(1 - 2\beta\cos\vartheta) \tag{10.5}$$

where a and β are constants. β theoretically is approximately equal to v/c where v is the velocity of the nucleons. This is no accident. The term can be considered as a retardation effect in the sum over all electric multipole matrix elements, and such a computation suggests $\beta = v/c$.

11. Non central forces. That noncentral forces are important is well known from the description of the ground state of the deuteron as a $^3S + {}^3D$ admixture. The first important effect of the ^3D state is in the normalization of the ground state wave function and in the effective range theory. This has already been included in Sect. 8. Other effects have however been neglected in Sects. 8 to 10 above.

At high energies, transitions become possible from the ^3D state to the ^3P states by electric dipole. These have been studied theoretically[1]. It is easily seen that the angular distribution is $6 + \sin^2\vartheta$, but the effect is small compared with any of the outstanding discrepancies between theory and experiment.

A more important effect of the noncentral forces is the binding of the particles in the ^3P states. In Sect. 8 this was neglected and justified by using the SERBER force—with *no* interaction in the ^3P states. It is known that above 50 Mev there are considerable interactions in the ^3P states which give rise to polarization in proton-proton scattering.

A correct way to calculate the effect of the noncentral forces would be to use values for the ^3P phase shifts derived from scattering data and perform the calculations of Sect. 8 with the interaction included. It is only recently that these phase shifts have become even approximately known[2]. In particular, FESHBACH finds it necessary to assume a large $^3P_0$ shift to account for the isotropy in proton-proton scattering. This large attractive $^3P_0$ interaction would lead to a large isotropic term in the $^3S \to {}^3P_0$ electric dipole interaction. Various alternatives have been discussed by AUSTERN who calculated the phase shifts used from nuclear models using the Born approximation.

There are, of course, also noncentral force contributions to the photomagnetic effect. Thus there may be contributions from the $^3D_1$ state to the $^1D_2$ state. Although this transition is not large at low energies, it may nevertheless interfere with the ^3S transition. Thus one cross section becomes

$$\sigma_m(\vartheta) = 1 + (5 - 3\cos^2\vartheta)\frac{J_2^2}{4J_0^2} - (3\cos^2\vartheta - 1)\cos(\delta_S - \delta_D)\frac{J_2}{\sqrt{2}\,J_0} \tag{11.1}$$

[1] N. AUSTERN: Phys. Rev. **85**, 283 (1952).

[2] H. FESHBACH and E. LOMON: Phys. Rev. **102**, 891 (1956).

where
$$J_0 = \int_0^\infty u_g\, u_s\, dr\,, \tag{11.2}$$
$$J_2 = \int_0^\infty w_g\, w_s\, dr \tag{11,3}$$
are the overlap integrals of the radial S and D wave functions for the ground and the final states. δ_S and δ_D are the phase shifts for the S and D scattering.

This interference may *reduce* the apparent isotropic component of the photo cross section by adding to the $\sin^2\vartheta$ component. (Unfortunately not enough to resolve the discrepancy in Sect. 18.)

12. Mesonic and exchange effects. The absence of a Siegert theorem for magnetic multipole transitions, leads to a study of the specific meson effects in the magnetic dipole interaction. There is a small effect at zero energy, as can be seen by a phenomenological theory or experiment (see Sect. 18).

At energies above the meson threshold, the following specific mechanism has been suggested[1]. In photomeson production from hydrogen, there is a considerable magnetic dipole term due to a resonance in the $P_{\frac{3}{2}}$ state of the pion and one of the nucleons. Then this suggests the reaction:
$$\gamma + d \rightarrow N^x + N \rightarrow n + p \tag{12.1}$$
where the intermediate state is one excited nucleon and one ordinary nucleon. The same intermediate state would be involved in the reactions
$$\pi^0 + d \rightarrow n + p\,, \tag{12.2}$$
$$\gamma + p \rightarrow p + \pi^0\,, \tag{12.3}$$
$$\pi^0 + p \rightarrow p + \pi^0\,. \tag{12.4}$$

The cross section for reactions 2 and 3 have been measured and that for reaction 4 may be inferred from charge independence.

Then in this approximation
$$\sigma_1 = \sigma_2\,(\sigma_3/\sigma_4) \tag{12.5}$$
with an angular distribution of $2 + 3\sin^2\vartheta$.

That this effect must contribute to the cross section can hardly be denied, but the cross section is only half that found experimentally and less isotropic than found experimentally. The more exact calculations of ZACHARIASEN agrees little better than the qualitative argument above.

13. Related reactions. The reaction $N + P \rightarrow \gamma + D$ is easily seen to be the inverse of the photodisintegration of the deuteron. It has been extensively studied for thermal neutrons.

The cross section is easily deduced from the principle of detailed balancing.

Then
$$\left.\begin{aligned}\sigma_{\text{cap}} &= \frac{3}{2}\left(\frac{E_\gamma}{p}\right)^2 \sigma_{\text{dis}}\\ &= \frac{3}{2}\,\frac{(E+\varepsilon)^2}{M\,\varepsilon}\,\sigma_{\text{dis}}\end{aligned}\right\} \tag{13.1}$$

[1] B. FELD: Nuovo Cim. Suppl. **2**, 145 (1955). — N. AUSTERN: Phys. Rev. **100**, 1522 (1955). — F. ZACHARIASEN: Phys. Rev. **101**, 371 (1956).

where p is the momentum of the nucleons and E_γ/c is the momentum of the incident γ ray. The factor $\frac{3}{2}$ takes into account the spins of the particles and the polarization of the γ-ray. At low energies $\sigma_{\text{dis}} \to 0$ but σ_{cap} remains finite and is measured (Section 15).

The reaction $N+P \to \gamma+N+P$[1] is often called neutron-proton bremsstrahlung. It may be considered as the radiation from the accelerated proton accompanying a scattering. The relation between this and the photodisintegration of the deuteron is seen by considering the possibility of a large interaction in the final neutron-proton state. In the ^3S state this is attractive and for high energy γ-rays leads to a bound deuteron

$$N+P \to \gamma+D. \tag{13.2}$$

Thus the unbound reaction bears the same relation to the bound reaction as in the processes with π mesons:

$$N+P \to \pi^0+D, \tag{13.3}$$

$$N+P \to \pi^0+N+P. \tag{13.4}$$

These radiative scattering reactions have been only qualitatively studied[2]. Their study would yield similar information to the photodisintegration of the deuteron. It is perhaps of interest here to note that for the reaction

$$P+P \to \gamma+P+P \tag{13.5}$$

only electric quadrupole and higher terms will appear, because the electric dipole and part of the magnetic dipole moment vanishes for the two proton system.

14. Polarization experiments. There are two classes of polarization experiments that can be performed. Firstly a polarized γ-ray beam may be used for experimentation. If only electric multipole radiation is important, then no information additional to that obtainable from angular distribution experiments is obtained. However, the complete incoherence of the electric and magnetic multipole effects only arises after averaging over polarization so that experiments of this nature could check the importance of magnetic multipoles.

Secondly the polarisation of outgoing particles may be measured. The forces between the particles in the final state, which has been neglected in the calculations here, will give rise to polarization of the outgoing nucleons even for unpolarized γ-rays. A measurement of the polarization will then give information about the neutron proton scattering phase shifts. A calculation has been performed for beryllium (Section 37) but not yet for deuterium.

C. Photodisintegration of the Deuteron: Experiments.

I. Low energies.

15. The neutron-proton capture cross section. For experiments at energies below 3 Mev the magnetic dipole interaction is the dominant interaction for the photodisintegration of the deuteron. The capture of neutrons by protons at thermal energies likewise proceeds by a magnetic dipole interaction. The measurement of the capture cross section presents a problem because it is smaller than the scattering cross section by a factor of 100. It is, however, easier to measure than the (vanishing) photodisintegration cross section.

---

[1] R. E. CUTKOSKY: Phys. Rev. **103**, 505 (1956). — J. ASHKIN and R. E. MARSHAK: Phys. Rev. **76**, 58, 989 (1949).

[2] RICHARD WILSON: Phys. Rev. **85**, 563 (1952).

There are three types of experiment that have been used to find σ_{cap}. The first measures the mean lifetime of neutrons in a hydrogenous medium with no other capturing elements. This lifetime (t) is related to the capture cross section by the relation

$$4\sigma_{\text{cap}}\, t\, v\, n = 1 \tag{15.1}$$

where v is the velocity of the neutrons and n is the number of hydrogen *atoms* per unit volume in the material. Now we have seen from Sects. 9 and 13 that $\sigma_{\text{cap}}\, v$ is constant for energies below about 50 kev. Thus if fast neutrons are slowed down in water the absorption will become exponential with time after the first few collisions; that is after about 5 μsec. The experiments then require a pulsed neutron source, a large (essentially infinite) tank of water, a neutron detector and a "time sorter". The number of neutrons in the tank is plotted as a function of the time after the neutron pulse and t is determined. In the two experiments of this type that have now been reported the neutrons were produced from a *d-d* neutron source, and the pulsing was accomplished by pulsing the incident deuteron beam.

There are several systematic errors that must be carefully considered and which have been best avoided by MEADS[1]. The water tank was made sufficiently large that corrections for its size could be made experimentally without recourse to theory; the unavoidable entry of the deuteron beam tube into the tank introduced an error for which a correction was made by introducing a similar tube on the opposite side; and the detector was a plastic scintillator which nearly matched the water it displaced both in scattering and absorption. In addition the detector measured the capture γ-rays from a region around it, which further reduced the correction due to the insertion of the measuring counter.

The second type of experiment has been performed by HAMERMESH[2] following the lines of the early experiments of FERMI. A neutron source is placed at the centre of a tank of water sufficiently large that esssentially all the neutrons are slowed down and captured by the hydrogen nuclei. A quantity of sodium iodide is added, sufficiently small that most of the neutrons are captured by the hydrogen yet sufficiently large that the activity induced in the iodine is measurable. After equilibrium has been established, the neutron source is removed, the solution thoroughly stirred, and the iodine activity of an aliquot fraction is measured. Since the oxygen and sodium atoms have small capture cross sections, the iodine activity is given by

$$A_{\text{I}} = K\,\frac{N_{\text{I}}\,\sigma_{\text{I}}}{N_{\text{I}}\,\sigma_{\text{I}} + N_{\text{H}}\,\sigma_{\text{H}}} \tag{15.2}$$

where K would be the activity if *all* the neutrons were captured by the iodine, $N_{\text{I}}, N_{\text{H}}$ are the number of iodine and hydrogen atoms per cm³ of solution and $\sigma_{\text{I}}, \sigma_{\text{H}}$ are the capture cross sections for hydrogen and iodine.

This procedure is now repeated with a small quantity of borax (essentially boron) added. The activity is now given by

$$A_5 = K\,\frac{N_{\text{I}}\,\sigma_{\text{I}}}{N_{\text{I}}\,\sigma_{\text{I}} + N_{\text{H}}\,\sigma_{\text{H}} + N_{\text{B}}\,\sigma_{\text{B}}}\,. \tag{15.3}$$

The ratio

$$R = \frac{A_1}{A_2} = \frac{N_{\text{B}}\,\sigma_{\text{B}} + N_{\text{I}}\,\sigma_{\text{I}} + N_{\text{H}}\,\sigma_{\text{H}}}{N_{\text{I}}\,\sigma_{\text{I}} + N_{\text{H}}\,\sigma_{\text{H}}}\,, \tag{15.4}$$

[1] R. E. MEADS, C. J. ENGLAND, C. H. COLLIE and G. C. WEEKS: Proc. Phys. Soc. Lond. **69**, 469 (1956).

[2] B. HAMERMESH, G. R. RINGO and S. WEXLER: Phys. Rev. **90**, 603 (1953).

or

$$R - 1 = \frac{\sigma_B}{\sigma_H} \frac{N_B}{N_H} \left(1 + \frac{N_I}{N_H} \frac{\sigma_I}{\sigma_H}\right)^{-1}. \tag{15.5}$$

The term in the bracket can be made as small as is compatible with the available intensity, and σ_I does not therefore have to be known well. Thus σ_H/σ_B is determined accurately.

The third type of experiment is the comparison of the effects of specimens of hydrocarbon and boron in a nuclear reactor, using the pile oscillator technique[1]. The difficulty with this technique is again the large scattering cross section of hydrogen. The major inaccuracy of the experiments is the author's estimate of the accuracy of the corrections.

Thus we have now four accurate measurements; two of σ_H directly and two of σ_H/σ_B. The latter are somewhat more accurate, but the value of σ_B is uncertain to within 1%.

Table 4 shows these recent results and the average is taken on the stated errors alone.

Table 4.

Authors	Method	Result (cross section at 2200 m/sec)
Scott	Lifetime with corrections by a theoretical extrapolation	0.323 ± 0.008
von Dardel	Lifetime with corrections by a theoretical extrapolation	0.333 ± 0.003
Meads	Lifetime, all corrections both theoretical and experimental	0.335 ± 0.004
Hamermesh	Ratio to boron using integrating bank	0.329 ± 0.004
Harris	Ratio to boron by pile oscillator	0.332 ± 0.007
	Average	0.3325 ± 0.002

16. Value of the exchange in *n*-*p*-capture. The theory as described in Sects. 12 and 13 can be directly used. From the effective range theory we calculate

$$\bar{\varrho} = (1.87 \pm 0.88) \times 10^{-13} \text{ cm}.$$

The average of the singlet and triplet effective ranges is $(2.05 \pm 0.15) \times 10^{-13}$ cm[2] which is somewhat higher. Thus the fraction of the capture cross section which is caused by the exchange term is (7 ± 5)%. Reversing the argument, and setting the exchange term equal to zero, ϱ_s becomes $(2.04 \pm 0.2) \times 10^{-13}$ cm which is low; by chance these figures from the shape independent approximation here agree almost exactly with that deduced from the Yukawa well from Salpeter's computations including potential shape. For an exponential or a square well, a considerably larger fraction of the exchange moment is found.

Since also there is the dual uncertainty of the shape of the potential well and of the exchange moment, little further progress can be made. It is possible to achieve agreement with a Yukawa potential and a 4% exchange moment[3].

[1] S. P. Harris, C. O. Muelhouse, D. Rose, H. P. Schroeder, G. E. Thomas and S. Wexler: Phys. Rev. **91**, 125 (1953).

[2] Although a large part of the error in ϱ_s is caused by that in ϱ_t the direction is such that the mean has this error.

[3] M. Sugawara: Progr. Theor. Phys. Japan **14**, 535 (1955) gives an exact calculation from experimental data for a Hulthén potential with a repulsive core, and also calculates a 1% exchange moment from a field theory calculation.

17. Total cross section measurements at low energies. Two groups have measured total cross sections at low energies, with similar apparatus. At Oxford radioactive sources were used, and at Cambridge sources of γ-rays from nuclear reactions were employed. In other respects the techniques were similar.

The deuterium was contained in an ionization chamber counter at a pressure (from 5 to 35 atmospheres) sufficiently high that most of the photoprotons were stopped in the gas. The ionization chamber was connected to a low noise pulse amplifier and counter system. The counter was brought close to the source of γ-rays and the disintegrations counted. The cross section is then obtained directly. The following quantities have to be known:

1. The strength, in γ-rays per second, of the source, and the absorption of all containers.
2. The number of deuterium atoms within the sensitive volume of the counter.
3. The distance of the centre of the counter from the source.

Table 5. *Best values for the total cross section for the photodisintegration of the deuteron.*

γ-ray energy (Mev)	Experiment (10^{-28} cm²)	Theory (10^{-28} cm²)
2.504 ± 0.002	11.2 ± 0.4	10.31
2.618 ± 0.002	13.0 ± 0.3	12.51
2.757 ± 0.002	15.0 ± 0.5	15.08
4.45 ± 0.04	24.3 ± 1.7	25.42
6.14 ± 0.01	21.9 ± 1.0	22.11
7.39 ± 0.15	18.4 ± 1.5	19.03
8.14 ± 0.08	18.0 ± 1.3	17.35
12.50 ± 0.21	10.4 ± 1.0	10.62
17.6 ± 0.2	7.7 ± 0.9	6.69

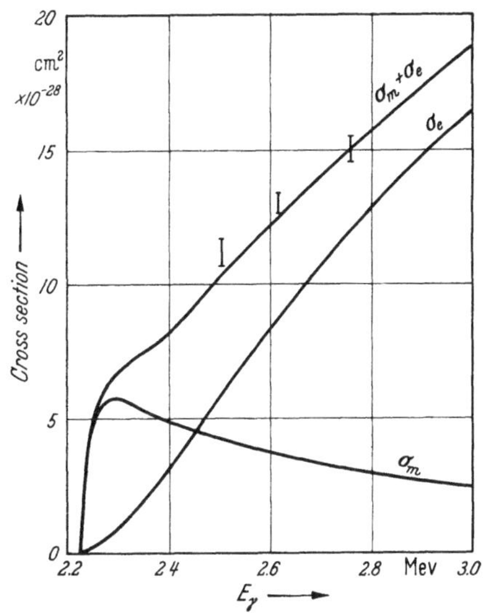

Fig. 1. Cross section for the photodisintegration of the deuteron below 4 Mev; theoretical results are discussed in the text.

The first two quantities are not easily measured and are the main sources of error.

The experimenters at Oxford[1] used sources of 2.504, 2.618 and 2.787 Mev (Sect. 2) and the last two were independently calibrated.

For both these sources, the source strength of the γ-rays was in error by more than 5% in some of the early measurements, for a quoted uncertainty of only 2%. Recent detailed recalibrations have uncovered these early inconsistencies. The measurements were shown to be consistent with different counter fillings and different source preparations to 0.3%.

The experimenters at Cambridge[2] calibrated all their γ-ray sources in terms of the 6.13 Mev γ-ray from $p + F^{19} \rightarrow \alpha + \gamma + O^{16}$. This was done using a thick walled Geiger counter, where the efficiency is expected from theory to vary linearly with the energy. The intensity of the 6.13 Mev γ-ray was measured by the method of α, γ coincidences. The accuracy obtained was not quite so great as in the measurements at Oxford and moreover the values at the different energies are not quite statistically independent, both because of this common error, and the common counter filling. Table 5 and Figs. 1 and 2 show the best

[1] G. Bishop, C. H. Collie, H. Halban, A. Hedgran, K. Siegbahn, S. du Toit and R. Wilson: Phys. Rev. **80**, 211 (1950). — W. R. McMurray and C. H. Collie: Proc. Phys. Soc. Lond. A **68**, 181 (1955). — P. Marin, G. R. Bishop and H. Halban: Proc. Phys. Soc. Lond. A **66**, 608 (1953); A **67**, 1113 (1954).

[2] C. A. Barnes, Carver, G. H. Stafford and D. H. Wilkinson: Phys. Rev. **86**, 359 (1952). — D. H. Wilkinson: Phys. Rev. **86**, 383 (1952).

results, with the theoretical results discussed in *B* plotted also. In computing the magnetic dipole term, the fractional contribution due to exchange effects (Sect. 16) was assumed constant with energy.

Alternatively the results for σ_e can be used to derive a value for the effective range in the *n-p* interaction. Indeed this is a most direct way of obtaining the effective range.

This was first done by Wilkinson for his data. The effective range ϱ_t is deduced directly from the data, but r_t is shown in Table 6.

The average $r_t = (1.74 \pm 0.035) \times 10^{13}$ cm compares favourably with that from liquid mirror experiments[1] [$r_t = (1.71 \pm 0.03) \times 10^{-13}$ cm]. As discussed in Sect. 9 the relation between ϱ_t and r_t is dependent upon the potential shape; the r_t deduced from photo effect measurements would be higher for a square well with an exponential potential intermediate. The Yukawa well clearly gives the best agreement, but this agreement will be destroyed if a repulsive core is assumed or non-central forms included. Thus a square well would make the two determinations of r_t differ by (8 ± 3)% which is a large discrepancy. This illustrates the fact that this comparison is one of the cleanest possible ways of finding the effect of potential shape. The data are not sufficiently concordant that we may regard the Yukawa well as proved. It appears possible that there is still a systematic error in the 2.618 Mev value, and almost certain that a discrepancy exists in the values of σ_m/σ_e as calculated below (Sect. 18).

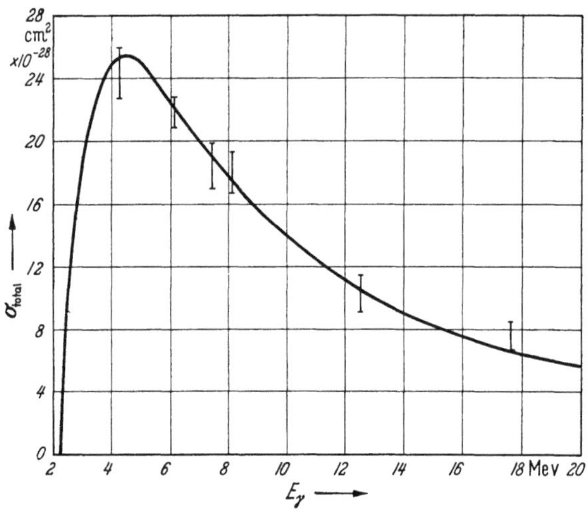

Fig. 2. Cross section for the photodisintegration of the deuteron from 4 to 20 Mev; the theoretical results are those of the low energy theory discussed in the text.

Table 6.

γ-ray energy	r_t
2.618	1.82 ± 0.06
2.757	1.70 ± 0.05
4.45 and above	1.70 ± 0.09
Average	1.74 ± 0.035

18. Angular distribution at low energies. The angular distribution of either the photoprotons or the photoneutrons enables the electric dipole and the magnetic dipole contributions to be separated; for the electric dipole interaction gives the neutrons a $\sin^2 \vartheta$ angular distribution about the incident γ-ray direction, as the noncentral forces are small at these energies. The magnetic dipole gives an isotropic term.

The measurements fall into two groups; first measuring the angular distribution of the neutrons and second that of the protons. In the first measurements, a localized γ ray source of high intensity is placed near a small container of heavy water, and the scattered neutrons are detected by some neutron detector. The neutron detector usually detects slow neutrons with a higher efficiency than fast neutrons, so that double scattering in the heavy water and scattering from the walls of any room are serious sources of error. The only other source

[1] Ringo, Burgy and D. J. Hughes: Phys. Rev. **82**, 344 (1951).

of error is the statistics of the counting, which has always been limited by the available source strength.

The angular distribution of the protons, has been measured by a device which removes the statistics as a limit; the protons are detected in an ionization chamber counter as discussed in Sect. 17. Because of the finite momentum of the incident γ-ray, those photoprotons that travel in the direction of the incident γ-ray have a higher energy than those in the reverse direction. The angular distribution is then directly related to the energy distribution by the formula:

$$E_p = \left(\frac{h\nu - \varepsilon}{2}\right)\left[1 + \frac{h\nu}{c}\sqrt{\frac{M}{h\nu - \varepsilon}}\cos\vartheta + \frac{1}{2}\,\frac{(h\nu)^2}{M c^2 (h\nu - \varepsilon)}\sin^2\vartheta\right] \qquad (18.1)$$

to second order accuracy. The number of events at angle ϑ is given by

$$\frac{d\sigma(\vartheta)}{d\Omega}\, 2\pi \sin\vartheta\, d\vartheta \Big/ \frac{dE}{E} = \sin\vartheta\, d\vartheta$$

to first order, so that the energy distribution follows the angular distribution to first order.

Thus the angular distribution measurement reduces to measuring the energy distribution. The energy distribution was measured on an accurate pulse height analyser. The errors arise from uncertainty in the counter behaviour and from the amplifier noise. The former can lead to spread of pulse heights due to such causes as incomplete collection of the ions. The latter is augmented by pile up of small γ-ray pulses and may be directly measured by measuring the spread of uniform pulse generator pulses injected into the ionization chamber circuit. The accuracy obtainable in this way is still the highest obtainable in such a measurement.

Table 7 gives a summary of the best existing measurements[1]. In this is shown the ratio $3a/2b$ where the angular distribution (in the centre of mass system) is assumed to be of the form $a + b\sin^2\vartheta$. This should then equal σ_m/σ_e.

Table 7. *Best values for the ratio $3a/2b$; because of inconsistencies the error on the 2.757 Mev value is increased.*

γ ray energy Mev	$3a/2b$	
	Experiment	Theory
2.504	0.600 ± 0.02	0.701
2.615	0.360 ± 0.008	0.413
2.757	0.255 ± 0.01	0.260
6.14	0.05 ± 0.05	0.03

The theoretical values are calculated as discussed in Sect. 9 assuming the *np* capture cross section and the theoretical variation of σ_m with energy. It may be seen that there is a large discrepancy. In view of the total cross section measurements, it is not possible to increase σ_e to allow this so that the conclusion has to be drawn that *if* these results are correct, there must be a large $\sin^2\vartheta$ component of the *magnetic* cross section. It is also possible, that the meson exchange contribution decreases with energy. Both these possibilities are unlikely; since the exact calculations predict no such effects.

II. Intermediate energies (15 Mev $< E_\gamma <$ 100 Mev).

19. In the intermediate energy region the only source of gamma rays available is the radiation from betatrons and electron synchrotrons. It is therefore necessary to measure simultaneously the energy and angle of the recoil protons to determine the incident photon energy. In this energy region, this has been done

[1] G. R. Bishop, L. E. Beghian and H. Halban: Phys. Rev. **83**, 1052 (1951). — G. Goldhaber: Phys. Rev. **81**, 930 (1951).

principally by the use of photographic plates[1]. This arises naturally, because the range of the proton is convenient; they may be stopped in an emulsion, yet they can escape the target region. Counter telescopes as discussed in Sect. 20 below are not so simple because of the short range of the protons.

The experimental procedure has in all these cases been similar. In the experiments of Fuller, up to 22 Mev, the plates were placed *inside* a gas target; each track had to be carefully measured for direction and range. In those of Allen, the range of the protons was adequate to enable the plates to be placed at a distance from a gas target with thin windows, and the angle was defined by collimating slits. In the experiments of Whaling, where the energy is higher still a liquid deuterium target is used, and the angle is defined by the size of this target. Thus the higher energy measurements are the simplest to perform.

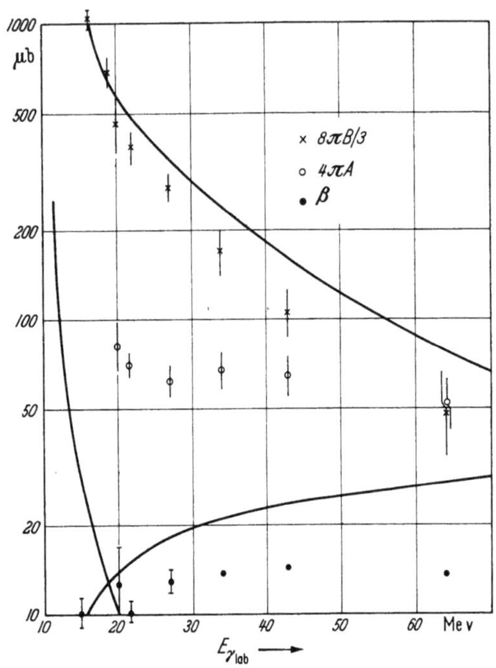

Fig. 3. Plot of the values of β, $4\pi A$ and $8\pi B/3$ from 20 to 100 Mev where the quantities are the best fits to the equation $d\sigma/d\Omega = (A + B\sin^2\vartheta)(1+2\beta\cos\vartheta)$. The theoretical curves are from the low energy theory discussed in Sect. B.

Fuller made no attempt to obtain accurate beam intensity figures, so that his data are useful for angular distributions only. The normalization of Allen and Whalin is uncertain to 5%.

Theoretical considerations in Part B above suggest that it may be instructive to try to fit the differential cross sections at one energy by a formula

$$(d\sigma/d\Omega)_\vartheta = A + B\sin^2\vartheta \times (1+2\beta\cos\vartheta) \tag{19.1}$$

which is given by central forces alone. However the data are better fitted by allowing the retardation term to multiply the term A as well as $B\sin^2\vartheta$. Thus the fits have been to

$$(d\sigma/d\Omega)_\vartheta = (A + B\sin^2\vartheta) \times (1+2\beta\cos\vartheta). \tag{19.2}$$

Thus in Fig. 3 are plotted β, $4\pi A$, and $8\pi B/3$ so obtained together with the predictions for central forces with no P state interaction and no exchange magnetic moment.

The failure of this simple picture is immediately clear. The term $4\pi A$ is nearly constant and much larger than theory; $8\pi B/3$ falls a little *below* that expected for electric dipole on the central force calculations. This suggests that at least part of the electric dipole cross section may contribute to the isotropic term as a result of non-central forces. This cannot at present be disentangled from a magnetic dipole contribution with exchange moments.

The retardation parameter β is less than the theoretical value v/c; however this may come partially from allowing the retardation term to multiply A;

[1] J. Halpern and E. V. Weinstock: Phys. Rev. **91**, 934 (1953). — E. G. Fuller: Phys. Rev. **79**, 303 (1950). — L. Allen jr.: Phys. Rev. **98**, 705 (1955). — E. A. Whalin, B. D. Schriever and A. O. Hanson: Phys. Rev. **101**, 377 (1956).

Fig. 4 shows two fits to the 60 Mev experiments; one is that of ALLEN, and the other is a fit using the theoretical retardation term multiplying only the $\sin^2\vartheta$ part of the cross section. Since none of the theoretical discussions predict a term in $\cos\vartheta$ alone, it is premature to discuss the constants in the empirical fit of ALLEN.

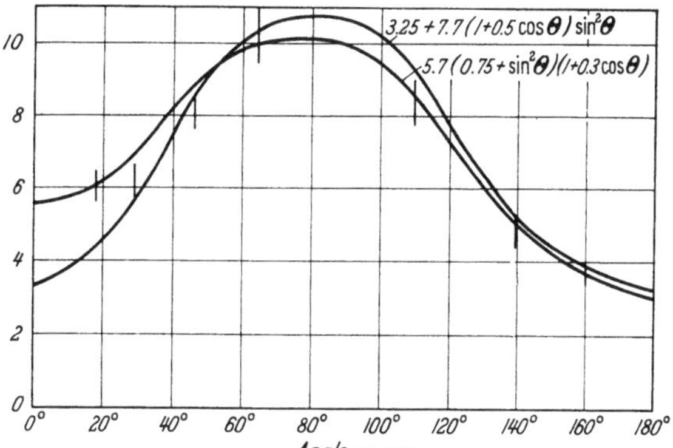

Fig. 4. Alternative fits to the angular distribution data at 70 Mev.

III. High energies (> 100 Mev).

20. At energies greater than 100 Mev meson production becomes possible and it is important to distinguish experimentally mesons from protons, and in addition to distinguish those protons which occur with meson production from the direct photoprocess. The latter is accomplished by studying the kinematics of the reaction.

The direct photodisintegration is a two body process and at a given laboratory angle, the relation between the proton and the photon energy is unique. The production of a pion in the collision involves a third body, and this uniqueness is lost. However, in order to discuss the separation, it is instructive to consider two special cases. The first is that in which the pion is produced on one nucleon in the deuteron with no momentum transfer to the "spectator" nucleon. This is expected to be the dominant process. The kinematics are then the same as in the two body process of the photoproduction of pions from hydrogen. The second is that in which the neutron and pion recoil as an excited nucleon, leaving the proton with the maximum possible energy.

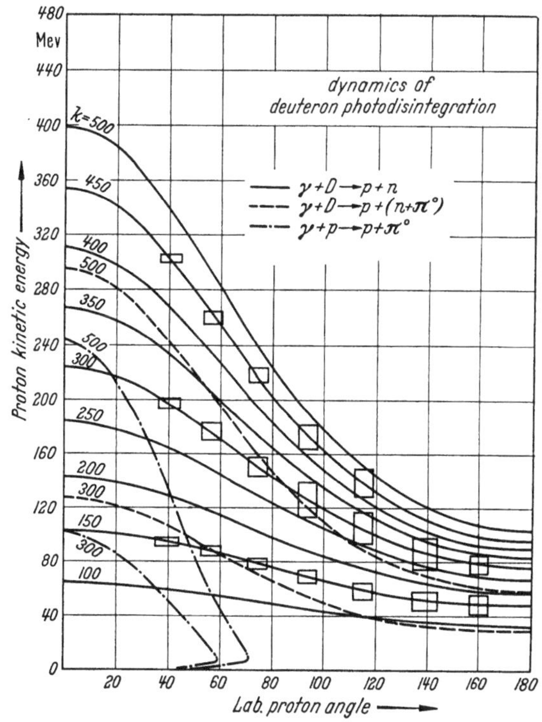

Fig. 5. Separation of protons and mesons for a high energy photoeffect experiment.

These cases are shown in Fig. 5. It is easily seen that for low proton energies and small proton angles, that a confusion is possible. In the experiments that have been performed[1], the maximum energy of the Bremsstrahlung was always altered so that no trouble arose.

The separation of protons from mesons has been performed by a scintillation counter telescope. That of KECK used four scintillation counters. The range

[1] J. C. KECK and A. V. TOLLESTRUP: Phys. Rev. **101**, 360 (1956) and references therein.

was defined by requiring that the particle stopped between the third and fourth counters and produced no pulse in the fourth counter; the specific ionization was defined by measuring the pulse amplitude from the first counter. Thus the energy and the type of particle were defined at each angle.

When the attention is directed to the high energy protons only, the telescope may be separated from the deuterium target by the target walls. KECK was able to use a high pressure deuterium gas target, though liquid deuterium has also been used. The largest correction to the data comes from the use of the proton range to define the energy, because some protons are lost by inelastic collisions before reaching the end of the range. This correction is especially serious for the high proton energies.

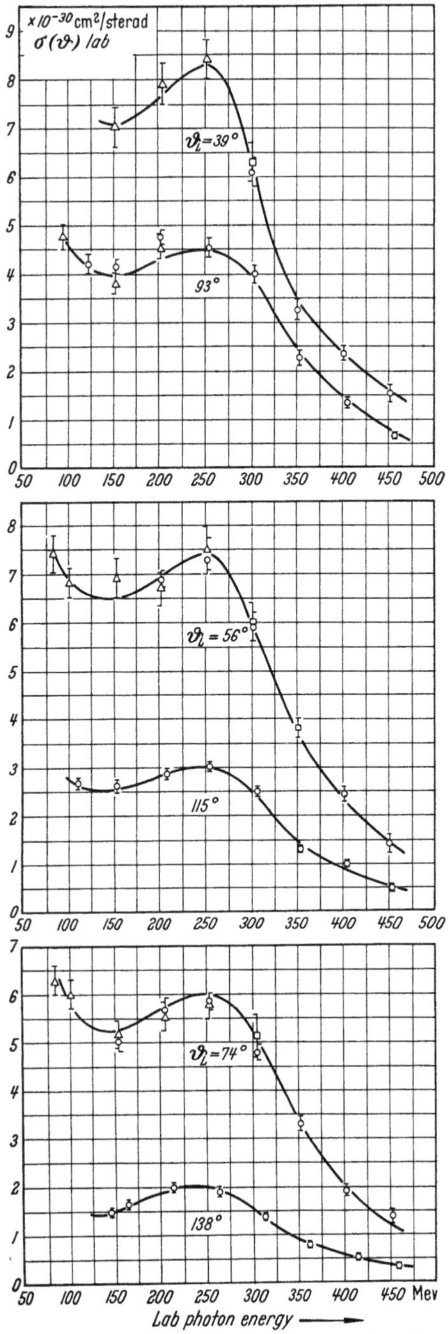

Fig. 6. Angular distribution of photoprotons from disintegration of deuterium by γ-rays from 100 to 450 Mev.

The results of some recent experiments are shown in Fig. 6, where the results of KECK are shown. It is of interest now to compare with theory. The angular distribution expected below 100 Mev no longer appears to fit the data; for there now appears to be a direct $\cos\vartheta$ term. The analysis has therefore proceeded in the simplest possible assumption, of angular distribution (in the centre of mass system)

$$\frac{d\sigma}{d\Omega} = A + B\cos\vartheta + C\cos^2\vartheta. \qquad (20.1)$$

Fig. 7 shows the best values for B/A, C/A and $\sigma_{\text{tot}} = 4\pi(A + C/3)$. A slight peak in σ_t is evident at about $h\nu = 200$ Mev, in qualitative agreement with the effect of meson production and reabsorption (Sect. 12). The difficult feature to understand is the magnitude of the $\cos\vartheta$ term which is given by none of the present theoretical treatments. It must be borne in mind, however, that experiments have not included 0° and 180° and it is conceivable that terms in $\cos^3\vartheta$ and $\cos^5\vartheta$ (allowed by theoretical treatments) could produce the observed forward assymmetry.

It is understood (private communication from Prof. W.K.H. PANOFSKY) that results from the 250 Mev synchrotron at Prof. VEKSLER's laboratory in Moscow are in essential agreement with the American data.

D. Photodisintegration of nuclei beyond deuterium.

I. Threshold measurements.

21. Threshold measurements give information about masses of nuclei. The procedure of threshold measurement has been the same; an electron source of

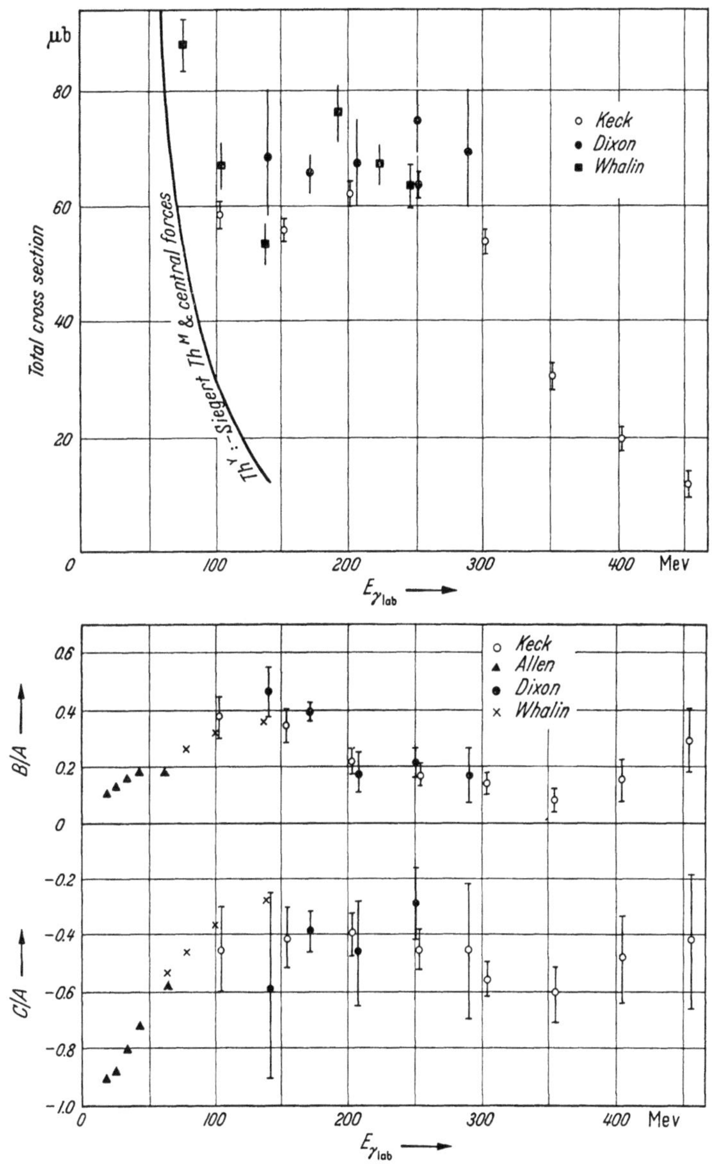

Fig. 7. Best values of B/A; C/A and $\sigma_{tot} = 4\pi(A + C/3)$ where the quantities are best fits to the equation.

variable energy has been used, and the neutron intensity observed as a function of the electron energy. The intensity is then extrapolated to zero, and the intercept on the energy axis gives the threshold energy. It is important to know, from theoretical reasons, that the extrapolation is valid.

The thresholds of deuterium and beryllium are accessible to van de Graaf generators. The theory is here very reliable, and the energy scale is determined by comparison with an electrostatic analyser. The thresholds can therefore be directly used. That for deuterium is, of course, an important constant in the neutron-proton system (Sect. 8). The thresholds are:

$$\text{Beryllium} \quad (1.664 \pm 0.002)\ \text{Mev},$$
$$\text{Deuterium} \quad (2.227 \pm 0.002)\ \text{Mev}$$

being the average of two good determinations[1].

These thresholds agree well with those obtained from mass measurements[2].

For heavier nuclei a betatron is used. The energy is determined from the value of the magnetic field at which the beam is contracted onto the target; this method needs to be calibrated. Deuterium and beryllium provide convenient low energy thresholds for calibration, and thresholds in the light elements C^{12}, N^{14}, O^{16} are usually used as the high energy calibration, for they are known well from mass data. Here the requirement that the theory of the extrapolation be understood becomes very important. The detailed measurements on carbon show that there is relatively little photoneutron production below the energy of the first level above threshold[3]. This can lead to a spurious assignment of the experimental threshold. The experimental threshold data[4] [*6*] are of an accuracy of 100 kev and this particular source of error is smaller; it is of great importance, however, for establishing the energy scale of the betatron in the region where the individual levels are found (see Sect. 29).

II. (γ, p) and (γ, n) cross section.

22. Measurement by activation methods. The most comprehensive data on photonuclear reactions is, not unnaturally, that which it is easiest to determine. This has been the study of the activities produced by bombardment of nuclei by γ-rays. The most common reaction is the (γ, n) process. For example, carbon foil when irradiated by Bremsstrahlung γ-rays becomes radioactive by the reaction $\gamma + C^{12} \rightarrow C^{11} + n$. The resultant C^{11} has a half life of 20 min and the activity is readily counted with a Geiger counter. An activation curve is measured, directly relating the activity induced on a sample to the incident electron energy; this is then converted into a cross section curve by methods described earlier (Sect. 4). The activation data can conveniently be discussed in two separate energy regions. In the region where the cross section increases with energy, a betratron yields the most accurate data, because the electron energy may be controlled to a few kev. Where, however, the cross section *decreases* with energy, the photon difference method involves the subtraction of two large numbers and the beam normalization becomes critical. The best beam normalization procedure is a Faraday cup; for an external linear accelerator beam this is possible, for an internal synchrotron or betatron beam, other, less reliable, methods may be used. The linear accelerator, however, has a large energy spread of about 1 Mev, so that the betatron is far superior for the rising part of the cross section curve.

---

[1] R. C. Mobley and R. A. Laubenstein: Phys. Rev. **80**, 309 (1950). — J. C. Noyes, J. E. van Hoomissen, W. C. Miller and B. Waldman: Phys. Rev. **95**, 396 (1954).

[2] Li, Whaling, W. Fowler and T. Lauritsen: Phys. Rev. **83**, 516 (1951).

[3] B. M. Spicer and A. S. Penfold: Phys. Rev. **100**, 1375 (1955).

[4] R. Scherr, J. Halpern and A. K. Mann: Phys. Rev. **84**, 387 (1951) and references therein.

This division of function is illustrated by the specific case of $C^{12}(\gamma, n)\,C^{11}$. The betatron experiments[1] showed the shape of the cross section curve up to a peak at 22.5 Mev. The accuracy and resolution is now sufficient even to identify a fine structure in these curves (to be discussed in more detail in Sect. 29). However the early experiments were erroneously interpreted to show that the cross section falls to zero beyond the peak. The first suggestion that this might not be so is the data of SAGANE.

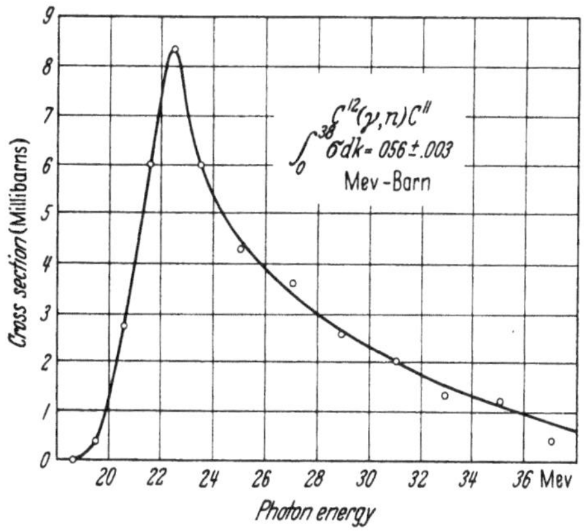

Fig. 8. The cross-section for the reactions $C^{12}(\gamma, n)\,C^{11}$ as a functional of energy upto 36 Mev. The fine structure is smoothed in this experiment.

Linear accelerator experiments, on the other hand, have insufficient resolution for giving any detail of the cross section curve[2]. However, they show very clearly that the cross section does *not* fall to zero above 22.5 Mev. A detailed cross section curve cannot be drawn above 36 Mev, again because of insufficient data. The only possible procedure, therefore, is to *assume* a cross section curve, to calculate an activation curve from this and the bremsstrahlung spectrum, and then to compare this with experiment. Figs. 8 and 9 show how this has been done.

The various assumptions, the predictions from which are illustrated in these curves, are;

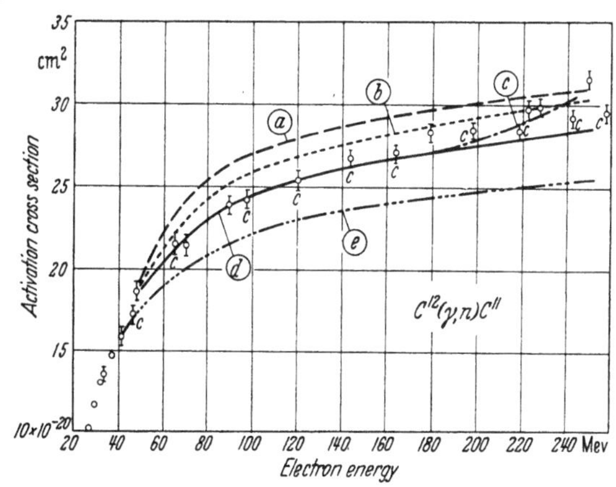

Fig. 9. Activation curve for $C^{12}(\gamma, n)\,C^{11}$ with various theoretically deduced curves for comparison, as discussed in the text.

A. The cross section curve deduced from the data up to 36 Mev, assumed to fall to zero at 36 Mev and stay zero for higher energies.

B. The same as A but with a symmetrical bump added to the cross section curve centered at 50 Mev and of base width 12 Mev; the height of this bump is normalized to give the correct yield at 250 Mev.

C. The contribution due to assumption A, with the addition of a cross section proportional to the inverse cube of the photon energy ($1/k^3$) smoothly joined at 34 Mev.

D. The same as C except that the ($1/k^3$) tail is *not* smoothly joined at 34 Mev, but the cross section is assumed constant from 34 to 44 Mev.

It is seen, therefore that curves C and D bracket the experimental points. Curve A gives yields too low at 250 Mev and curve B is high in the intermediate region. Thus the linear accelerator work shows beyond doubt that such a tail to the activation curve exists.

---

[1] R. SAGANE: Phys. Rev. **84**, 587 (1951). — R. N. H. HASLAM, H. E. JOHNS and R. J. HORSLEY: Phys. Rev. **82**, 270 (1951). — B. M. SPICER and A. S. PENFOLD: Phys. Rev. **100**, 1375 (1955).

[2] W. C. BARBER, W. D. GEORGE and D. REAGAN: Phys. Rev. **98**, 73 (1955).

By contrast, we may consider the reaction $Cu^{63}(\gamma, n)\,Cu^{62}$ which has also been studied with a linear accelerator[1]. This reaction has no observable tail beyond 30 Mev. This difference is believed to be a general difference between light and heavy nuclei.

Studies of (γ, n) activation cross sections have been extensive; some (γ, p) (γ, pn) (γ, d) $(\gamma, 2n)$ activation cross section curves have been studied but not so many. This is for two reasons. Some of the betatrons used have had a maximum energy of 25 Mev which excluded a detailed study of these more complex reactions. Also, the cross sections for $(\gamma, 2n)$ reactions are frequently lower than for (γ, n) reactions, and the final nucleus has an inconveniently short life. Even more difficult to study is the inelastic γ-ray scattering. A lower limit may be given to this in certain specific cases, where the nucleus has an isomeric state; the excitation of this excited state is studied in the same way as the (γ, n) activation curve is studied. This gives a figure for those inelastic γ-ray scatterings that lead to the isomeric state; by calculation an estimate may then be made of the total inelastic γ-ray scattering. It can be 10% of the (γ, n) cross section[2]. For comparison with theory, however (cf. Part III of this chapter), it is useful to obtain a curve of the total absorption cross section and not just the partial cross sections. For heavy nuclei particularly, enough is known to add these together, for an accuracy of 20%.

23. Measurement by neutron or proton detection. The neutrons from a (γ, n) reaction may be conveniently detected in a neutron counter. The neutrons are expected to be low energy "evaporation neutrons", and for these neutrons a BF_3 counter embedded in paraffin wax (the so-called long counter) has a sensitivity independent of energy[3]. Under these conditions, a cross section curve can be obtained from a plot of neutron counts versus electron energy. The curve, suitably normalized, should be (and in general is) identical with the cross section curves from activation experiments, until an energy is reached which allows $(\gamma, 2n)$ reactions to take place. Abover the $(\gamma, 2n)$ threshold, the cross sections for neutron counting will be given by:

$$\sigma = \sigma(\gamma, n) + \sigma(\gamma, pn) + 2\sigma(\gamma, 2n) + 2\sigma(\gamma, 2np) + 3\sigma(\gamma, 3n) \quad \text{etc.}$$

Since these curves include many partial cross sections, they more nearly equal the true γ-ray absorption cross sections[4]. (γ, p) reactions may also be studied by measuring the protons directly. In one sense this is more difficult, in another, more comprehensive. It is hard to measure the protons from a photonuclear reaction without simultaneously measuring the energy and angle of emission; no equivalent of the long counter exists. These energy and angle distributions give information which will be discussed later (Parts IV and V of this chapter respectively).

24. General features and table of giant resonance parameters. The cross section curves all show a sharp peak followed by a fall off on the high energy side. When first discovered, there was a conjecture that the fall off of the (γ, n) cross section above the peak was due to the competition of $(\gamma, 2n)$ and (γ, pn) processes, and that the total absorption had no peak, and no resonance behaviour. It is now known that the $(\gamma, 2n)$ and (γ, pn) cross sections, while their existence broadens

---

[1] A. I. Berman and K. L. Brown: Phys. Rev. **96**, 83 (1954).

[2] J. L. Burkhardt, E. J. Winhold and T. H. Dupree: Phys. Rev. **100**, 199 (1955). — C. S. del Rio and V. L. Telegdi: Phys. Rev. **90**, 439 (1953).

[3] A. O. Hanson and J. L. McKibben: Rev. Sci. Instrum. **72**, 673 (1947).

[4] The most complete data are: L. W. Jones and K. M. Terwilliger: Phys. Rev. **91**, 699 (1953).

the peak, are not large enough to alter the energy for the maximum cross section and only affect the width slightly. This resonance character has suggested the theoretical interpretations, and the large cross section has suggested the dipole character of the resonance.

The data for C^{12} are perhaps worth discussing in detail to illustrate the problem involved in interpreting the data. The possible processes are listed in Table 8, together with the parameters measured as discussed in the text. It is clear that the (γ, n) and (γ, p) dominate the situation below 27 Mev; each have a peak at about 22 Mev so that the total photon absorption must have a peak at this energy also. This is clear for carbon.

Table 8. *Nuclear photoeffect on carbon 12.*

	Threshold Mev	Peak cross section millibarns	Energy at peak Mev	$\int \sigma\, dE$ Mev-barns
(γ, n)	18.73	8.3	22.5	56 (to 38 Mev)
				80 (to 250 Mev)
(γ, p)	15.9	34 ± 8	21.5	63 (to 30 Mev)
$(\gamma, 2n)$	32.0	not measured		
(γ, pn)	27.4	not measured		
$(\gamma, 3\alpha)$		0.1	18 and 25	
$(\gamma, n) + (\gamma, pn) + 2(\gamma, 2n) + \cdots$			21.5	
(γ, γ')		not measured		

For slightly heavier nuclei, aluminum for example, the peak for (γ, n) is at 19.5 Mev which is also the threshold for (γ, pn). It is suggestive, therefore, that the fall off after the peak might be due to competition from (γ, pn) or $(\gamma, 2n)$ and not to a genuine peak in the photon absorption. The peak in the neutron production cross section $\left((\gamma, n) + \gamma(pn) + 2(\gamma\, 2n) + \cdots\right)$ at the same energy disproves this. Thus the peak of the giant resonance would seem to be a peak in the photon absorption. This is in contrast to the measurements of photon scattering (Sect. 13) which suggest a higher energy for the photon absorption.

This difference is one of the many unanswered questions in the field.

Bearing these difficulties of interpretation in mind, we now present Figs. 10 to 12, some summary graphs of the presently available data[1].

The graph of Fig. 11 shows that for heavy elements the integrated cross section up to 25 Mev exhausts the dipole sum for 50% exchange forces. For light elements, on the other hand, the high energy tail is important.

The data of different laboratories on absolute cross sections have been difficult to reconcile; because of the greater ease of monitoring, the absolute data from linear accelerators is to be preferred.

The graph of Fig. 10 shows that the peak energy of the giant resonance is a smooth function of A. This is in contrast to the threshold energy for neutron emission. This slow variation with A is strikingly brought out by a comparison of the photon absorption in C^{12} and of C^{13}[2] [*12*], shown in Fig. 13. The threshold for $C^{12}(\gamma, n)C^{11}$ is 18.7 Mev; that for $C^{13}(\gamma, n)C^{12}$ is 4.9 Mev. Yet the peaks of the giant resonance occur at similar energies.

[1] R. Montallbetti, L. Katz and J. Goldenberg: Phys. Rev. **91**, 659 (1953). — R. Nathans and J. Halpern: Phys. Rev. **93**, 437 (1954). — L. W. Jones and K. M. Terwilliger: Phys. Rev. **91**, 699 (1953). — J. Goldenberg and J. L. Lopez: Phys. Rev. **99**, 1054 (1955) and references [7] to [*10*].

[2] B. C. Cook: Phys. Rev. (in press).

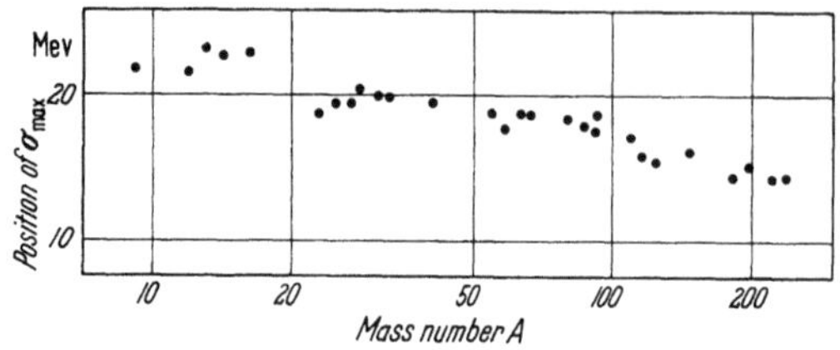

Fig. 10. The peak energy of the giant resonance plotted versus $A^{\frac{1}{3}}$ as measured by (γ, n) activation or by neutron measurements. This may be compared with Fig. 22.

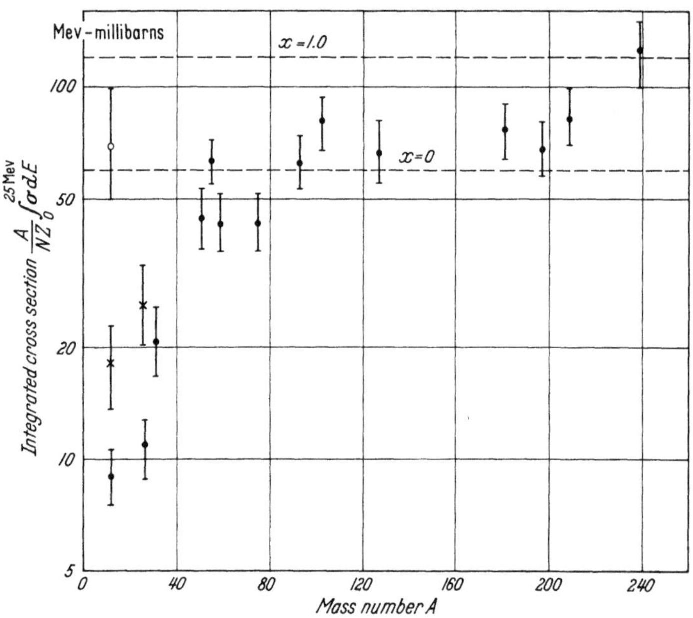

Fig. 11. The integrated cross section for (γ, n) plotted as a function of A. The points are from neutron measurements and include the integration up to 25 Mev only; the crosses include integration up to 200 Mev, and the circles include an estimate of (γ, p) and other processes.

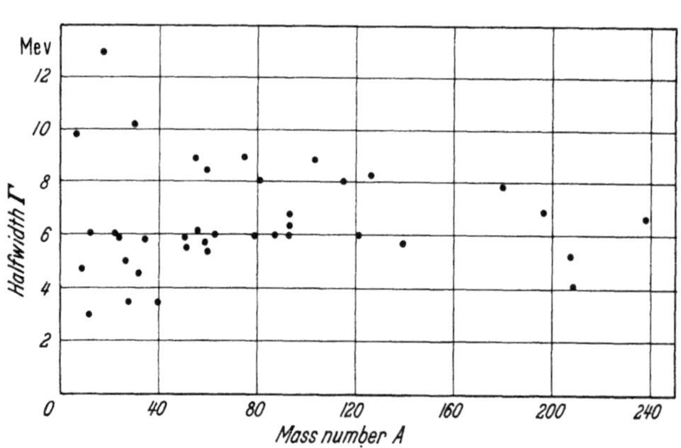

Fig. 12. The width of the giant resonance plotted as a function of A, as determined by (γ, n) activation or neutron measurements.

The data on C^{13} also bring out another feature of the photonuclear effect in light nuclei. A resonance in photon absorption is found at (13.2 ± 1) Mev as well as the giant resonance at 25 Mev. This resonance for photon absorption has previously been noticed for O^{16}, N^{14} and F^{19} and, being smaller than the giant resonance is called a pygmy resonance[1]. This is often attributed to quadrupole, instead of dipole, absorption.

The graph of Fig. 12 shows that the width of the giant resonance shows a smooth dependence on A within the large experimental errors with the exception of magic nuclei such as C^{12}, which show an anomously narrow resonance. Figs. 11 and 13 show that the change in width between C^{12} and C^{13} are compensated by a change in the peak cross section so that the integrated cross section does not change appreciably.

These three features, the large integrated cross section varying smoothly with A, the variation of E_{max} with A, and the narrow widths for magic nuclei, are those which any theory of the giant resonance must explain.

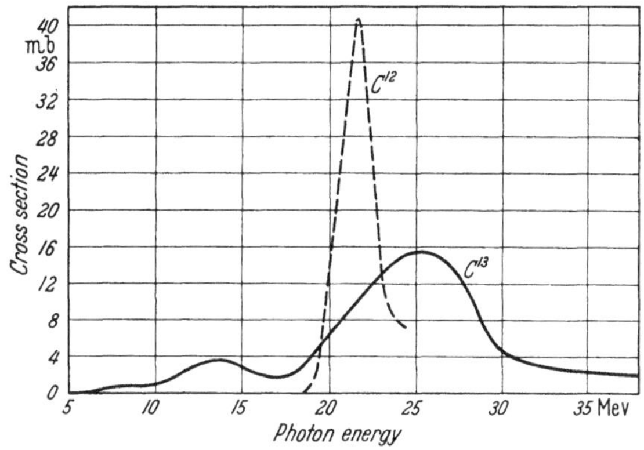

Fig. 13. The total photon absorption in C^{12} and C^{13} as a function of energy.

III. Theories of giant resonance.

25. Sum rules. Experimentally it is easier to measure $\int_0^{k_1} \sigma(k)\, dk$ than to measure $\sigma(k)$ at any one energy. This has already been discussed in Sects. 4 and 22. However, if $\sigma(k)$ is the absorption cross section for *all* nuclear processes, then this integral finds an interpretation. This is the f-sum rule[2].

If the forces between nucleons are all "ordinary" forces, that is, if they can be described by a potential with no especial exchange properties, then the following may be readily derived, independent of the positions of the nucleons.

$$\int_0^\infty \sigma(k)\, dk = \frac{\pi e^2 \hbar}{M} \frac{NZ}{A}. \tag{25.1}$$

As was shown by Siegert and later discussed in detail by Levinger and Bethe, this formula must be modified when exchange forces between the nucleons are assumed. This unfortunately reduces the generality of the result, and calculations must be performed upon a specific model. Levinger has calculated the result expected upon an independent particle model of the nucleus, and finds a multiplying factor $(1+0.8x)$, with an error of perhaps ± 0.2, where x is the fraction of exchange force in the neutron-proton interaction. An inclusion of correlation between nucleons would cause this result to be increased somewhat.

[1] Horsley, Haslam and Johns: Canad. J. Res. **32**, 238 (1954).

[2] J. Levinger and H. A. Bethe: Phys. Rev. **78**, 115 (1950). — J. S. Levinger: Phys. Rev. **91**, 122 (1955). — J. S. Levinger: Phys. Rev. **101**, 733 (1956). — R. G. Sachs and N. Austern: Phys. Rev. **81**, 705 (1951). — J. S. Levinger and D. C. Kent: Phys. Rev. **95**, 418 (1950). — Y. K. Kholchlov: Dokl. Akad. Nauk USSR. **97**, 239 (1954).

It has been suggested that the real mass M of the nucleon can be reduced to take account of internucleon correlations.

This has been suggested on the basis of ideas of Brueckner[1], that a velocity dependent nucleon-nucleus interaction can be represented by reducing the nucleon mass. This velocity dependent interaction is intimately connected with the internucleon correlations, and if the mass M is reduced, the correlations must not be otherwise introduced or they will be included twice.

A different approach has been that of Gell-Mann[2]. The sum rule may be derived from a dispersion relation; this then enables the exchange forces to be included by integrating over meson production cross sections. Then

$$\int_0^\mu \sigma(k)\, dk = \frac{NZ}{A}\,\frac{e^2\hbar}{M} + \frac{1}{\pi}\int_\mu^\infty dk\,[1 - R(k)]\,[Z\,\sigma_p(k) + N\,\sigma_n(k)] \tag{25.2}$$

where $\sigma_p(k)$ and $\sigma_n(k)$ are the cross sections for meson productions for individual nucleons, and the factor under the integral on the right hand side gives the difference between the meson production of free protons and neutrons and the nucleons in a nucleus.

These two approaches appear to give the same results to within about 20%, however it is not yet clear to what extent this may be fortuitous or due to the same assumptions. The sum rule therefore can only be expected to give an answer correct to 30 to 40%; but this is still the best theoretical prediction available.

Levinger and Bethe have gone further than evaluating the sum rule. They also evaluate the harmonic mean energy

$$k_H = \frac{\int \sigma(k)\, dk}{\int \sigma(k)\, \frac{dk}{k}} \tag{25.3}$$

and the mean energy

$$\bar{k} = \frac{\int k\,\sigma(k)\, dk}{\int \sigma(k)\, dk}\,. \tag{25.4}$$

These two results now depend very critically upon the *correlation* between the positions of the nucleons. Experimentally we know, that since the plot of $\sigma(k)$ versus k is sharply peaked, then k_H and $\bar{k}$ must be close together. Levinger and Bethe evaluated these mean energies, assuming no correlations, with gross disagreement with experiment. Even on an independent particle model, however, correlations exist because of the Pauli principle acting between unlike nucleons. These have the effect of raising both the mean energies. It is important to note that only the correlations in the ground state of a nucleus matter here.

Table 9 taken from Levinger shows the results of typical calculations for the harmonic mean and mean energies for tantalum. The first column assumes a Fermi gas nucleus with radius $r_0 A^{\frac{1}{3}}$, the second and third include antisymmetry correlations, the fourth further includes exchange effects. Dynamic correlations, such as are needed to explain the high energy photoeffect (Sect. 34), are excluded but will raise the energies.

It is clear that the nuclear correlations are all important for detailed calculations of the photoeffect, but that their inclusion gives reasonable fits to experi-

[1] K. A. Brueckner: Phys. Rev. **100**, 36 (1955).

[2] M. Gell-Mann, M. L. Goldberger and G. W. E. Thirring: Phys. Rev. **95**, 1612 (1954).

ment. It is not clear, however, that the calculations of the mean and harmonic mean energies have any advantage over the complete calculation of the cross section curve.

Table 9. *Mean energies for photon absorption by tantalum.*

	Fermi gas without antisymmetrization	Fermi gas with antisymmetrization		Exchange included	Experiment
	$r_0=1.5$	$r_0=1.5$	$r_0=1.2$	$r_0=1.2$	
$\bar{k}_H$ (Mev)	1.4	6.0	8.7	12.2	15
k (Mev)	16	6.5	10.4	30	20

26. The collective model. The sum rules indicate that the considerations of the correlation between nucleons affects, very critically, both the integrated cross section and the mean energies for photon absorption. Some correlation between the nucleons is seen to be necessary to account for the experiments.

The first model suggested to explain the larged peaked dipole absorption cross section is that of GOLDHABER[1]. This is a classical model and assumes that all the protons are moving (collectively) in the opposite direction from the neutrons. Various assumptions may now be made about the forces between the neutrons giving a variety of models.

Firstly, the restoring force acting upon the protons may be assumed proportional to the displacement. This gives a constant energy.

Secondly, the neutrons and protons can be treated as compressible fluids constrained within the nucleus. This gives $E_{\max}$ proportional to $A^{-\frac{1}{3}}$.

Thirdly, the neutrons and protons may be considered incompressible but interpenetrating. Then the energy is proportional to $A^{-\frac{1}{6}}$. As can be seen (Sects. 24 and 39) there are discrepancies and it is not clear experimentally whether the $A^{-\frac{1}{3}}$ variation or the $A^{-\frac{1}{6}}$ variation is most plausible. The constant in the curve can be made, within the limited accuracy, to give good agreement with experiment.

The integrated cross section gives exactly the sum rule result for ordinary forces, as indeed it should, for the model includes all the dipole vibrations and makes no mention of exchange forces[2]. Further resonances are expected due to quadrupole etc. excitations.

The width of the resonance is interpreted as a frictional effect between the nucleons. A value for the width has been calculated by WILDERMUTH[3] from the time for a neutron in the neutron fluid to be scattered by a proton in the proton fluid.

This hydrodynamical model has been further studied by STEINWEDEL, JENSEN and DANOS[3] who calculate the value of the constants.

It appears that this model automatically involves compound nucleus formation. While this may be a good approximation for heavy nuclei, for light nuclei an appreciable amount of direct photoeffect occurs. Moreover the complete internucleon correlation involved seems inconsistent with the applicability of the shell model to ground states of nuclei.

The success of the Goldhaber-Teller theory has led to several attempts to derive similar results by a correct quantum mechanical description of the nucleus.

---

[1] M. GOLDHABER and E. TELLER: Phys. Rev. **74**, 1046 (1948).

[2] J. H. D. JENSEN and P. JENSON: Z. Naturforsch. **5**a, 343 (1950).

[3] H. STEINWEDEL and J. H. D. JENSEN: Z. Naturforsch. **5**a, 343 (1950). — M. DANOS and H. STEINWEDEL: Z. Naturforsch. **6**a, 217 (1951). — K. OKAMOTO: Progr. Theor. Phys. **15**, 75 (1956). — K. WILDERMUTH: Z. Naturforsch. **10**a, 447 (1955).

It seems that any such attempt must take into account the nuclear correlations. One such attempt is that of FERENTZ[1] who also find $E_{max} = W_0 A^{-\frac{1}{3}}$ and an integrated cross section "exhausting" the dipole sum.

This relation for E_{max} had also been derived by MIGDAL[2] from sum rule considerations similar to those of JENSEN.

27. WILKINSON's model. WILKINSON[3] has gone further than LEVINGER and has used the independent particle model to give a complete description of the nuclear photoeffect in the light nuclei. Since the giant dipole resonance is large he begins by searching for those dipole transitions which have a width larger than the Weisskopf single particle width. These would give the main contribution to the nuclear photoeffect. They may be found by considering the enhancement factors calculated by LANE and RADICATI[4], for a transition in which a nucleon falls from a level $l+1+\frac{1}{2}$ or $l+1-\frac{1}{2}$ to a closed shell with $l\pm\frac{1}{2}$. The associated radiative width Γ_γ is larger than the single particle width $\Gamma_{\text{S.P.}}$ by the factors $\frac{3}{2}(l+2)$ and $\frac{3}{2}l(2l+3)/(2l+1)$, respectively. These factors do not arise from specific dynamic correlations giving a collective motion as in the Goldhaber-Teller theory, but are a consequence of the Pauli principle acting in closed shells, as in LEVINGER's calculation of the sum rules. For most nuclei two neutron closed shells and two proton closed shells contribute to the transition probability, at energies that for a harmonic oscillator potential are degenerate, and even for a square well do not spread more than the width of a giant resonance. The integrated cross section is given by

$$\int_0^\infty \sigma\, dE = 4\pi^2 e^2 \frac{E_n}{\hbar c} R^2 \sum \frac{l+1}{2l+3} F_{jl} D_l \tag{27.1}$$

where the sum is over the closed shells. R is the nuclear radius and $D_l R^2$ is the square of the overlap integral,

$$D_l^{\frac{1}{2}} R = \int_0^\infty \psi_i\, r\, \psi_f\, d\tau. \tag{27.2}$$

F_{jl} is the enhancement factor for the j-shell of the l-orbit. If this expression is evaluated, by substituting the observed values of E_{max}, it describes reasonably well the trend of integrated cross section with A. The magnitude is correct for the light elements and too small by a factor of 2 for the heavy elements. The failure at large A may be accounted for either by experimental or theoretical uncertainties. Thus the experimental results contain $(\gamma, 2n)$ and other processes in the measured cross sections, the contributions of $M1$ and $E2$ transitions, and also of $E1$ transitions not due to the giant dipole resonance but in the same energy range. The above theoretical expression also excludes wavefunctions with radial nodes, which are more important for heavy elements.

Some idea of the resonance energy E_m is gained by considering the energy differences between successive states in an infinitely deep well with the nuclear radius $R = r_0 A^{\frac{1}{3}}$. This gives $E_m = 36$ Mev for $A = 6$, falling to $E_m = 10$ Mev for $A = 240$. If one now allows for the energy level shift due to the finite nuclear binding energy this estimate falls from $E_m = 20$ Mev to $E_m = 9$ Mev. These values are too low by about 30% in the heavy elements. This has already been discussed in the section on the sum rules. WILKINSON has thus supplemented the sum rules by relating the absorption to shell model states.

[1] M. FERENTZ, M. GELL-MANN and D. PINES: Phys. Rev. **92**, 836 (1953).
[2] A. MIGDAL: J. exper. theor. Phys. USSR. **15**, 81 (1945).
[3] D. H. WILKINSON: Proceedings of the Asterdam Conference 1956.
[4] A. M. LANE, and L. A. RADICATI: Proc. Phys. Soc. Lond. A **67**, 167 (1954).

In order to obtain agreement with the experimental values for E_{max}, WILKINSON suggests using an effective mass $M/2$ as BRUECKNER uses in other fields. As discussed earlier in Sect. 25 this is now introducing a correlation between nucleons in addition to that implied by the exclusion principle, and is coming closer to the collective model.

WILKINSON then shows how sometimes a direct photoprocess results and sometimes a compound nucleus is formed. The analysis of low-energy neutron scattering by nuclei with the cloudy crystal ball model, leads to the conclusion that the mean free path of a nucleon in a nucleus may be as much as $\lambda = 2\times 10^{-12}$ cm. A nucleon in a free orbit in the ground state of the target nucleus absorbs a γ-ray quantum and is elevated to a higher shell model orbit. The excited nucleon remains in its orbit until collision with other nucleons causes a redistribution of its energy to the whole nucleus. This is governed by the value of λ. If the nucleons are emitted before this redistribution takes place a direct photoprocess occurs; if after, compound nucleus formation results. The ratio of direct photoprocess to compound nucleus formation can then be calculated to be given by

$$C = \frac{2k P\cdot\frac{\hbar^2}{m R}}{2W} \tag{27.3}$$

where the numerator is the product of nucleon wave number, penetrability P, and the single particle reduced width $\hbar^2/mR$, and the denominator is the imaginary part of the cloudy crystal ball potential. The absorption to form a compound nucleus leads to a broadening, and for heavy nuclei at least, coalescence, of the individual levels involved. The broadening is given by $\hbar$ times the frequency of collision v/λ, where v is the average nucleon velocity in the nucleus, and amounts to about 2 Mev, which is less than the average width of the giant dipole resonances. The total width Γ_t is determined mainly by the following factors: 1. The damping width of about 2 Mev. 2. The spread in absorption energy of the shells. 3. The contributions from valence nucleons, which straggle widely in energy. 4. During absorption in one shell there may be recoupling of valence nucleons. It is interesting to note experimentally that for closed shell nuclei the values of Γ_t are less than average (Sect. 24, Fig. 12). Thus Γ_t is in the average about 6 Mev, but O^{16} has $\Gamma_t = 3$ Mev, for Si^{28} $\Gamma_t = 3.5$ Mev, and for Ca^{40}, $\Gamma_t = 3.2$ Mev. For these nuclei the independent particle model is expected to work particularly well since they show rather widely spaced levels near the ground state, indicating a rather purer ground state wavefunction, and also of course they have no valence nucleons.

The systematics of E_m also demonstrate another source of support for the independent particle model. For absorption only within closed shells and for an infinitely deep well we should expect a break in E_m at each shell filling. The sharp break is smoothed out into a bend, owing to the finite well depth, and absorption in partly filled shells, but such a bend is found in the filling of the $1d_{\frac{5}{2}}$ shell ($A=28$).

The angular distribution of the direct photonucleons can be calculated for the Wilkinson model, and gives for ejection from the l-orbit

$$W(\vartheta) = 1 + \frac{1}{2}\left(1 + \frac{2}{l}\right)\sin^2\vartheta. \tag{27.4}$$

A previous attempt by COURANT to calculate direct photoprocesses yields a smaller value for C, the ratio of direct to compound nucleus effects, and a more isotropic angular distribution. COURANT'S theory was partially successful in accounting for the relatively large number of high-energy photoprotons emitted compared with photoneutrons. The ratios predicted by the statistical model differed from those observed by factors of as much as one thousand. COURANT'S calculations account for the discrepancy to within a factor of twenty throughout the periodic table, but he omitted the enhancement factors referred to above, and considered transitions from l to $l \pm 1$ only, for the excited nucleons. WILKINSON has calculated the ratio G of the partial cross sections given by his theory to those obtained by COURANT, in which case the agreement with experiment becomes much improved (Table 10).

The Wilkinson theory is thus seen to be quite successful in predicting several general features of the nuclear photoeffect. Separate evidence exists confirming the existence of independent particle states at the high excitation energies corresponding to the giant dipole resonance, though the integrated cross section, at least, depends mainly upon the applicability of the independent particle model to the ground states of nuclei, as the sum rule calculations (Sect. 25) show. The plausibility of these states depends strongly on the value of λ, which is found to be large from quite different experimental data. In this case the single particle excitations must also exist, and we see from the above that they account for a large fraction of the dipole sum rule. Since the giant resonances also nearly exhaust the dipole sum, it seems that the only way to avoid a gross violation of the sum rules is to associate the single particle excitations with the giant resonance.

Table 10.

Orbital-momentum	1	2	3	4	5
Mass number A .	16	40	88	140	250
Ratio G	65	52	36	31	23

28. Synthesis of theories. The "collective" theory of Sect. 26 gives quite a good representation of a fair number of experimental data; E_m, Γ, and $\int \sigma dE$. These are also given by the independent particle model of WILKINSON. The two approaches seem radically different. Yet in the different field of nuclear spectroscopy, the independent particle model and the collective model of AAGE BOHR have had to be reconciled. It seems worth while therefore to attempt such a synthesis here and this has been done by LEVINGER[1]. The agreement between the two models, and between the independent particle model and experiment, is only obtained by including internucleon correlations (Sect. 25). This may arise in the independent particle model by the degeneracy of the single particle levels, which for a harmonic oscillator potential is exact.

The agreement in $\int \sigma dE$ is clear; this quantity is not sensitive to the model (Sect. 25).

A qualitative explanation of the agreement in Γ comes from the fact that Γ is due to the same effect in each, nucleon-nucleon interactions, as modified by the exclusion principle. Although in the collective model more nucleons are assumed to move, they move more slowly and their mean free path is longer.

The predictions of the two models that are hard to reconcile are those relating to the formation or otherwise of a compound nucleus. By these tests then, the theories must be compared, as will be done in Sect. 31.

BRINK [12] has suggested a detailed reason why the independent particle model and the collective model give the same result. He shows that the independent particle model wave function can be considered, for a harmonic oscillator

[1] J. S. LEVINGER: Phys. Rev. (in press).

potential, as a product of wave functions corresponding to different particle motions. When the total wave function is operated upon by the dipole operator, only the part corresponding to the collective motion of all protons against all neutrons, changes. This suggests that the independent particle model wave function already includes collective motion. Any attempt to attempt a synthesis by allowing a linear combination of the independent particle model and collective wave functions would then be wrong.

IV. Fine structure and level widths.

29. Fine structure. During work on the energy calibration of betatrons (Sect. 21), which demanded high accuracy in establishing the activities produced by (γ, n) reactions in C^{12} and O^{16} at energies just above threshold, a series of well-defined and reproducible discontinuities in the slope of the activation curve were established[1] (Figs. 14 and 15). The breaks have been found both by the activation method, and by measuring the photoneutrons directly[2]. They have been found for other elements as well, occurring at different energies for different elements, which suggests strongly that they are really a property of the nuclear photoeffect and not of the operation of a betatron. The interpretation of this fine structure is that in these nuclei absorption of photons takes place, at least in part, into well defined nuclear levels of the target nucleus. Thus these reactions are the inverse of the radiative capture of light particles by light nuclei, well known at lower nuclear excitation energies.

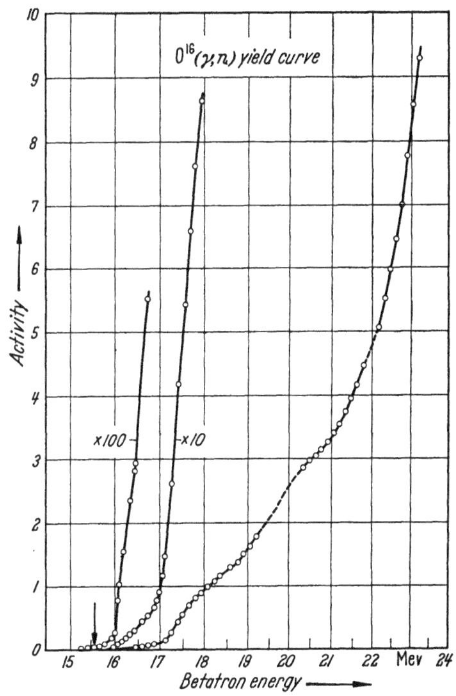

Fig. 14. Activation curve for $O^{16}(\gamma, n)\,O^{15}$ showing breaks in the activation curve.

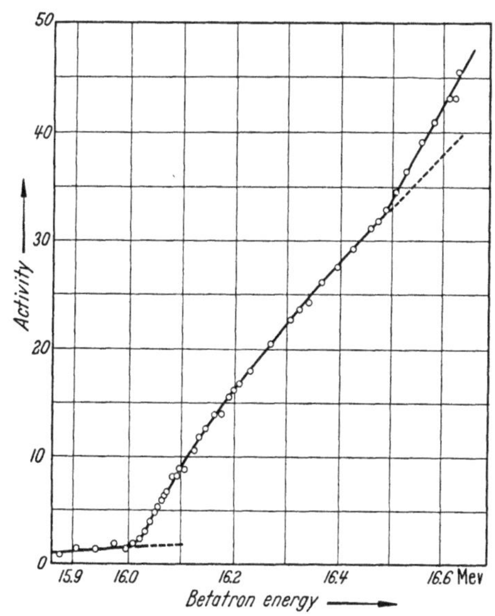

Fig. 15. As Fig. 14, showing detail in the region of the breaks.

[1] L. Katz, R. N. H. Haslam, A. G. W. Cameron and R. Montalbetti: Phys. Rev. **95**, 464 (1954).

[2] J. Goldemberg and L. Katz: Phys. Rev. **95**, 471 (1954). — A. S. Penfold, and B. M. Spicer: Phys. Rev. **100**, 1377 (1955). — R. Rasile and C. Schuhl: C. R. Acad. Sci., Paris **240**, 2399, 2512 (1955); **241**, 387 (1955).

The cross section for the photonuclear reaction may be written

$$\sigma_{\gamma,n} = \sigma_c(\gamma)\, G(n) \tag{29.1}$$

where $\sigma_c(\gamma)$ is the cross-section for photon absorption, and $G(n)$ is the probability that a neutron is emitted, in competition with other processes. The number of photoneutrons is given by,

$$n(E_0) = \int_0^{E_0} P\,\sigma(\gamma, n)\, dE = \int_0^{E_0} P\,\sigma_c(\gamma)\, G(n)\, dE \tag{29.2}$$

for a betatron energy of E_0, where $P(E, E_0)$ is the number of photons of energy E incident on the sample. The fine structure might be attributed to discontinuities in P, $\sigma_c(\gamma)$ or $G(n)$. Discontinuities in P might arise from photon absorption in the irradiated samples, but at these energies the effective electronic absorption energies do not change rapidly with photon energy. Discontinuities in $G(n)$ could only arise by the competition of other deexcitation processes with neutron emission, but the agreement concerning the position and magnitude of the breaks, between activation and direct photoneutron detection experiments, make this unlikely. Thus the experiments demonstrate fine structure in $\sigma_c(\gamma)$. These results are a crowning success for the efforts made to control betatron operation with an energy stability of ± 5 kev.

Further evidence for the finestructure, and its interpretation, is found from a study of the spectra of the photonucleons. For (γ, p) in oxygen and carbon this has been observed in nuclear emulsions to give groups of photoprotons with energies corresponding to the excitation of various levels in the residual nucleus[1].

Confirmation that sharp levels at these excitation energies can exist in C^{12} comes from a study of the reaction $B^{11}(p, n)\, C^{11}$ [2], which gives energy levels in the compound nucleus C^{12} agreeing with the positions of some of the fine structure discontinuities. Similar fine structure has been observed in the reaction $C^{12}(\gamma, 3\alpha)$ and $O^{16}(\gamma, 4\alpha)$ [3] (see Fig. 17).

If the explanation of the breaks as being due to resonance absorption by nuclear levels is correct then the formation and the disintegration of such states should be governed by the Breit-Wigner formula. A detailed derivation of this formula where one particle is a photon has been given by PEASLEE[4], using a perturbation treatment for the initial photon absorption and boundary fitting for the emitted particle. An explicit expression is obtained for the radiation width Γ which depends on the many particle structure of the compound nucleus state, and the dependence is sufficiently complicated to suggest that one particle estimates of the width would not apply. PEASLEE suggests that the flat shoulder which occurs below the giant dipole resonance in the excitation curves may represent incoherent electric dipole absorption rather than magneticdipole or electric quadrupole excitation, which was predicted on general grounds by BLATT and WEISSKOPF [5]. However the sharp resonances account for the whole of the (γ, p) and (γ, n) cross sections in this energy region without requiring a continuous component. For purposes of comparison we can calculate the radiation widths corresponding to the observed integrated cross section in the following manner.

---

[1] S. A. JOHANSSON and B. FORKMAN: Phys. Rev. **99**, 1031 (1955). — W. E. STEPHENS, A. K. MANN, B. J. PATTON and E. J. WINHOLD: Phys. Rev. **98**, 839 (1955). — W. E. STEPHENS and A. K. MAN: Phys. Rev. **98**, 241 (A) (1955).

[2] BLASER, BOEHM, MARMIER and SCHERRER: Helv. phys. Acta **24**, 30 (1951).

[3] F. K. GOWARD and J. J. WILKINS: A. E. R. E. Report A/M 127 Harwell.

[4] D. C. PEASLEE: Phys. Rev. **88**, 812 (1952).

The Breit-Wigner formula may be written

$$\sigma(\gamma, n) = \pi \lambda_\gamma^2 \frac{S_0 \Gamma_\gamma \Gamma_n}{(E - E_r)^2 + \frac{1}{4}\Gamma^2} \tag{29.3}$$

where S_0 is the statistical factor, $S_0 = \frac{1}{2}(2J+1)/(2I+1)$, for a system taken from a state of initial spin I to a final state of spin J by means of a γ-ray quantum with two states of polarisation. If this is integrated over energy E to give the integrated cross section σ_i, we have,

$$\sigma_i = \int \sigma \, dE = 2\pi^2 \lambda_\gamma^2 S_0 \Gamma_\gamma \Gamma_n/\Gamma. \tag{29.4}$$

The ratio Γ_n/Γ can be calculated from statistical theory[1] to be about $\frac{1}{3}$, and this is supported by the relative yields of photoprotons to photoneutrons. From this formula, combined with the experimental σ_i, we can estimate Γ_γ. The result is given in Tables 11 and 12, for oxygen and carbon respectively, supposing dipole and quadrupole γ-ray absorption (which alters the statistical factor). For comparison the values of Γ_γ calculated from the WEISSKOPF one particle relations are also given.

It is clear that the experimental widths are considerably smaller than the theoretical ones for electric dipole transitions. Now it is known that the experimental values for Γ_γ for other electric dipole transitions in the light nuclei show a remarkable constancy about a value close to the WEISSKOPF[2]. One these

Table 11.

Quantum energy Mev	$\sigma_i \times 10^3$ Mev-barns	Γ^{dipole} ev	Γ^{quad} ev	Γ^{E1} ev	Γ^{M1} ev	Γ^{E2} ev
15.98	0.078	5.3	3.2	2840	85	4.4
16.44	0.045	3.3	2.0	3100	93	5.1
16.67	0.071	5.2	3.1	3240	97	5.45
16.85	0.142	10.6	6.4	3340	100	5.75
16.92	0.222	16.8	10.1	3380	102	5.9
17.04	0.575	44.2	26.5	3450	104	6.1
17.44	0.236	19.2	11.5	3630	112	6.8
17.54	0.185	15.2	9 1	3760	114	7.05
17.69	0.185	15.5	9.3	3850	116	7.35
17.88	0.077	6.5	3.9	3980	120	7.8
18.48	0.117	10.6	6.4	4400	132	9.1
18.91	0.511	48.5	29.2	4700	142	10.3
19.3	1.5	144.0	86.4	5050	150	11.5
20.7	2.1	240.0	144.0	6150	186	16.3
21.9	7.5	960.0	576.0	7320	218	21.2

Table 12.

Quantum energy Mev	$\sigma_i \times 10^3$ Mev-barns	Γ^{dipole} ev	Γ^{quad} ev	Γ^{E1} ev	Γ^{M1} ev	Γ^{E2} ev
19.3	0.23	22.6	13.6	4150	151	7.65
19.8	0.66	68.0	41.2	4500	162	8.8
20.1	1.20	129.0	77.5	4700	170	9.75
20.5	1.40	156.0	93.5	4950	181	10.4
20.7	2.00	227.0	136.5	5130	186	11.2
21.1	5.20	610.0	368.0	5420	198	12.0
21.6	5.80	715.0	430.0	5820	212	13.7
22.4	5.50	727.0	438.0	6500	236	16.2
22.8	33.00	4550.0	2740.0	6860	248	17.7

[1] H. MORINAGA: Phys. Rev. **97**, 1185 (1955).
[2] D. H. WILKINSON: Phil. Mag. **44**, 450 (1953).

facts would therefore suggest that the transitions are not electric dipole but rather magnetic dipole or electric quadrupole. Distinction between these two modes of excitation cannot be made on the basis of radiation width alone; measurement of the angular distribution of the photonucleons would assist in the separation and would also lead to an assignment of the spins and parities of the nuclear states.

Since at least one unhindered electric dipole transition is known at lower excitation energies for O^{16}, it is necessary to explain why electric dipole transitions do not occur in the photonuclear reaction just above threshold. It is significant that a strong peak in the reaction $O^{16}(\gamma, \alpha)C^{12}$, is observed at 17.5 Mev, and fine structure has been observed in this reaction[1]. The interpretation this suggests is that these levels have isotopic spin $T=0$, in which case the electric dipole transitions would be inhibited by the isotopic spin selection rule. In fact there does seem to be a trend for the observed radiation widths (Tables 11 and 12), to become more closely comparable to the theoretical widths for $E\,1$ transitions as the energy of excitation increases. This would suggest a stronger mixing of $T=0$ and $T=1$ isotopic spin states as the individual resonance levels approach in energy the giant dipole resonance.

The expected angular distributions of photonucleons are easily calculated, for a given set of spins and parities, by the methods of BLATT and BIEDENHARN. The spins and parities of compound states formed in O^{16} or C^{12} by $E\,1$, $M\,1$, $E\,2$, absorption, are 1^-, 1^+, 2^+, respectively. The ground states and lower excited states that can energetically be formed by the (γ, n) and (γ, p) reactions on these target nuclei have spins and parities $^1/_2{}^-$ and $^3/_2{}^-$ for the residual nuclei. Thus for $E\,1$ absorption the most probable orbital momentum for the outgoing nucleon is zero, and for $M\,1$ or $E\,2$ absorption equal to unity. Then for the $E\,1$ case the angular distribution would be isotropic; for the other cases there is the complication that the total angular momentum of the emitted particles can have two values $l \pm \frac{1}{2}$. The ratio of the probabilities of occurence α^2 of the two values of total angular momentum can be calculated if a coupling scheme for the angular momenta is assumed. In this connection it should be rememberred that L-S coupling is perhaps favoured over j-j coupling in the light nuclei.

It is of interest to determine the width of these resonances. The direct method described above is not sensitive; but indications are that widths are in the range 20 to 50 kev. Indirect methods, may be used. γ-ray scattering (Sect. 39) yields a figure for $\int \sigma^2 dE$; so also does a self absorption process[2]. By a comparison with $\int \sigma dE$ an indication is given that widths are larger than the limits suggested by the direct methods. For heavy nuclei there seem to be no separable resonances.

The break up of the giant resonance into individual levels is not contrary to WILKINSON's theory, as might at first appear. The single particle states which he assumes split into separate states when the interaction with the core is taken into account. The several states then collectively display the aspects of a single particle state.

30. Level widths. The widths of the resonances are clearly of great interest. Below the thresholds for particle emission, the widths of nuclear levels are entirely the radiative widths (Γ_γ). Above the thresholds, there will be competition from particle emission, and the levels broaden until eventually they merge into

[1] MILLER and A. G. W. CAMERON: Canad. J. Phys. **31**, 723 (1953). — H. STOLL: Helv. phys. Acta **27**, 395 (1954).

[2] R. WILSON: Phys. Rev. (in press).

a continuum. The widths of the levels just above particle thresholds have been extensively studied by (p, γ) processes [13].

The direct methods for obtaining the widths are the (p, γ) process when applicable, which has a resolution of about 1 kev; the measurement of protons in the (γ, p) process, which up to now has only had a resolution of 200 kev, and the use of the photon difference method to study the (γ, n) process which has a poor resolution also.

In particular cases, it has been possible to use the variable γ-ray energy from Li^7 (p, γ) Be^8 to study the photodisintegration over a limited energy range near 17 Mev[1].

A third type of measurement is of self absorption; for example if an absorption coefficient of carbon is measured for photons using C^{12} (γ, n) C^{11} as a detector the resultant cross section is proportional to $\int \sigma^2 dE / \int \sigma\, dE$, where the integral is carried out over the incident γ-ray spectrum[2]. By a comparison with $\int \sigma dE$ for the level concerned, the width Γ can be found. The same procedure can be applied where the detector is not energy selective but follows an elastic scattering, and has been used extensively to study the levels below, and just above, the particle thresholds (Sect. 39).

The results of all these measurements are sketchy, and not in complete agreement with each other. The general picture is as expected. The 15 Mev level in C^{12}, for example, is less than 1 kev broad; yet the widths of the levels in the giant resonance in C^{12} are shown by self absorption to be greater than 50 kev. In the nuclei Cu^{63} and Zn^{64}, studies by $Li^7 (p, \gamma)$ Be^8, structure was observed, but the levels were not even separable at 17 Mev.

This is also confirmed by data on elastic scattering combined with the dispersion relations, which also yield values of $\int \sigma^2 dE$.

V. Compound nucleus versus direct photoeffect.

31. The discussion of Parts II, III, and IV of this chapter (Sects. 22 to 30 with the exception of WILKINSON's theory) is principally concerned with the absorption of photons. It is of course of interest to measure the subsequent emission of particles or quanta. There are, as mentioned summarily in the introduction, two extreme cases. In the first case, the energy of the incident photon is distributed throughout all the nucleons of the target nucleus before any particles are emitted. This is the compound nucleus formation as at first envisaged by BOHR. At the other extreme, the usual event is the emission of charged particles essentially instantaneously. This latter extreme occurs for high energy photon absorption and is discussed further in part VII (Sects. 34 to 36) below.

The collective absorption model of GOLDHABER and TELLER (Sect. 26) necessarily implies a compound nucleus formation.

If the photon absorption always led to compound nucleus formation, then the energy distribution and relative numbers of emission of photons and neutrons is calculable by statistical theory[3]. The energy distribution of the neutrons will be Maxwellian with the mean energy appropriate to the nuclear excitation energy; sometimes two neutrons in cascade will be emitted, the second from

[1] D. ST. P. BUNBURY: Proc. Phys. Soc. Lond. A **67**, 1106 (1954). — J. G. CAMPBELL: Austral. J. Phys. **8**, 449 (1955).

[2] R. N. H. HASLAM, R. J. HORSLEY, H. E. JOHNS and L. B. ROBINSON: Canad. J. Res. **31**, 636 (1953). — RICHARD WILSON: Phys. Rev. (in press). — The author would like to take this opportunity of apologizing to Dr. HASLAM and associates for not having previously observed their paper.

[3] K. J. LECOUTEUR: Proc. Phys. Soc. Lond. A **63**, 259 (1950).

a lower nuclear excitation. Photon emission will be inhibited by the Coulomb barrier, so that photoprotons will be less extractable from heavy nuclei than from light nuclei. Thus the ratio of (γ, p) processes to (γ, n) processes is expected to fall from near unity for light elements to about 10^{-5} for heavy elements. Small deviations about a smooth curve will, of course, occur depending upon the binding energies of the nuclei formed by subtracting a proton or a neutron.

Evidence that this picture cannot always be correct, is the work on photoprotons from heavy elements. These come about 100 times too frequently to be accounted for by a compound nucleus effect; they must be due to some direct interaction even at low γ-ray energies. This does not discredit the compound nucleus formation as the principal process; only a small addition of a direct effect is necessary to explain the results. Thus we may regard a study of the photoprotons from heavy elements as a selection of those processes for which only the direct effect is operative. It is therefore of interest to study these to decide whether Wilkinson's theory (Sect. 27) or for example Courant's theory is correct. Wilkinson's theory suggests that even the direct protons go through a well defined single particle state at the giant resonance, whereas Courant suggests no such final state interaction.

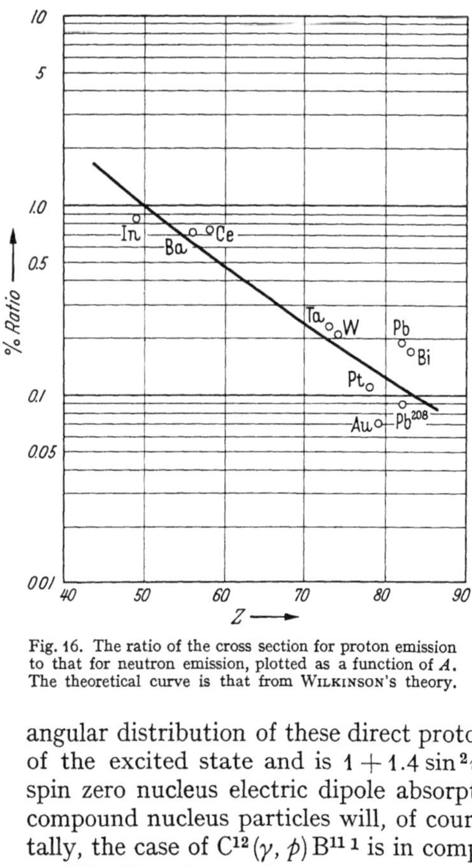

Fig. 16. The ratio of the cross section for proton emission to that for neutron emission, plotted as a function of A. The theoretical curve is that from Wilkinson's theory.

The comparison with theory has taken three forms: (i) the ratio of proton emission to total absorption is shown in Fig. 16 and illustrates that this is well given by theory; (ii) the angular distribution of these direct protons should be given by the characteristics of the excited state and is $1 + 1.4 \sin^2\vartheta$, for a spin 1 state, Eq. (27.4). For a spin zero nucleus electric dipole absorption always leads to a spin 1 state. The compound nucleus particles will, of course, be emitted isotropically. Experimentally, the case of $C^{12}(\gamma, p)B^{11}$ [1] is in complete agreement with this.

(iii) The direct photoprotons and photoneutrons must display the giant resonance, which is a function of the absorption of the photon; the apparent resonance will be shifted to a slightly higher energy for proton emission because the effect of the Coulomb barrier becomes rapidly less for high energies [2]. If the Goldhaber-Teller model is held to explain the giant resonance, then an *additional* effect must be invoked to account for a direct photoeffect. It is hard to see why this additional effect should display the same giant resonance.

(iv) The energy distribution of the photoprotons is peaked at about 11 Mev, compared with 6.5 Mev for evaporation protons affected by the Coulomb barrier [3], which in itself shows that a direct photoeffect is taking place.

---

[1] J. Halpern and A. K. Mann: Phys. Rev. **83**, 370 (1951). — A. K. Mann, W. E. Stephens and D. H. Wilkinson: Phys. Rev. **97**, 1184 (1955).

[2] F. Ferrero, A. O. Hanson, R. Malvano and C. Tribuno: Amsterdam Conference 1956.

[3] M. E. Toms and W. E. Stephens: Phys. Rev. **98**, 626 (1955) and references therein.

That compound nucleus formation is the main process for elements of copper and heavier, is suggested by activation data. Thus one may compare the reactions:

$$\begin{array}{lll} Zn^{64}(\gamma, n)\,Zn^{63}; & Cu^{63}(p, n)\,Zn^{63}; & Ni^{60}(\alpha, n)\,Zn^{63}; \\ Zn^{64}(\gamma, 2n)\,Zn^{62}; & Cu^{63}(p, 2n)\,Zn^{63}; & Ni^{60}(\alpha, 2n)\,Zn^{62}; \\ Zn^{64}(\gamma, p\,n)\,Cu^{62}; & Cu^{63}(p, p\,n)\,Cu^{62}; & Ni^{60}(\alpha, p\,n)\,Cu^{62}. \end{array}$$

These all proceed through the same compound nucleus Zn^{64}, if formed. Thus at any given excitation, the ratios Zn^{63}; Zn^{62}; Cu^{62} should be the same. This holds for the charged particle reactions[1], but to date only the integrated cross sections have been measured for the photon induced reactions[2]. Because of the different shapes of the cross section curves, only the Zn^{62} and Cu^{62} can be exactly compared (unless precise data on $Cu^{63}(\gamma, n)Cu^{62}$ and $Cu^{63}(\gamma, 2n)Cu^{61}$ can be used to assist a comparison of the (γ, n) and $(\gamma, 2n)$ processes in zinc).

This gives complete accord with the compound nucleus picture.

For light elements it is probable that compound nucleus formation is *not* the dominant interaction.

VI. Effects of charge independence.

32. Isotopic spin selection rules. The assumption of charge independence in nuclear forces has led to some definite predictions about the role of electric dipole transitions in photonuclear reactions. This has followed from the formalism of isotopic spin. There results a simplification in the photonuclear processes for light nuclei which are otherwise so complex. The processes here considered are only an extension to higher γ-ray energies, of γ-transitions found in (p, γ) reactions. The treatment here follows that of MacDonald[3].

The nonrelativistic Schrödinger Hamiltonian for the interaction of a system of nucleons with an electromagnetic field can be written

$$H = -\sum_{i=1}^{A}\left\{\frac{e}{Mc}\boldsymbol{P}_i\cdot\boldsymbol{A}(r_i)\left(\frac{1}{2}-t_{\xi i}\right)+\left[\mu_P\left(\frac{1}{2}-t_{\xi i}\right)+\mu_N\left(\frac{1}{2}+t_{\xi i}\right)\right]\sigma_i\cdot\left[\nabla\times\boldsymbol{A}(r_i)\right]\right\}. \tag{32.1}$$

where the isotopic spin operator t_ξ has the eigenvalues $+\frac{1}{2}$ for a neutron and $-\frac{1}{2}$ for a proton. This interaction can be separated into the sum of a scalar H_0 in isotopic spin space and the ξ-component of a vector H_1.

$$H_0 = -\sum_{i=1}^{A}\left\{\frac{e}{2Mc}\boldsymbol{P}_i\cdot\boldsymbol{A}(r_i)+\frac{1}{2}(\mu_P+\mu_N)\,\sigma_i\cdot\left[\nabla\times\boldsymbol{A}(r_i)\right]\right\}, \tag{32.2}$$

$$H_1 = \sum_{i=1}^{A}\left\{\frac{e}{Mc}\boldsymbol{P}_i\cdot\boldsymbol{A}(r_i)+(\mu_P-\mu_N)\,\sigma_i\cdot\left[\nabla\times\boldsymbol{A}(r_i)\right]\right\}t_{\xi i}. \tag{32.3}$$

The probability of electric dipole transitions induced by H_0 is zero, since we can take the centre of mass of the nucleus as the origin and find $\sum_{k=1}^{A} r_k = 0$. Then only H_1 can produce transitions.

Since H_1 is the ξ-component of a vector in isotopic spin space, we can see that there must be a change of one unit (vectorially) in the isotopic spin. This

[1] S. Ghoshal: Phys. Rev. **80**, 939 (1950).
[2] K. Strauch: Phys. Rev. **81**, 973 (1951).
[3] D. C. Peaslee and V. L. Telegdi: Phys. Rev. **92**, 126 (1953). — M. Gellman and V. L. Telegdi: Phys. Rev. **91**, 169 (1953). — W. M. MacDonald: Phys. Rev. **98**, 60 (1955); **100**, 51 (1955).

leads at once to the selection rules

$$T = 0, \ \pm 1, \qquad T \neq 0,$$
$$T = \ \pm 1, \qquad T = 0.$$

Higher order terms in the electric dipole transition elements can remove the absolute ban on $T=0$ to $T=0$ transitions; these effects and the effects of the neutron-proton mass difference have been taken into account by MACDONALD and shown to be small, enough that any deviation from the selection rule can be attributed to impurity of isotopic spin states.

The Coulomb forces, being charge dependent, can of course mix isotopic spin states, and cause a violation of the selection rule. If the impurity of the states is small, the electric dipole is still however prohibited by a large factor.

In addition to these isotopic spin selection rules, the concept of pure isotopic spin states can be used to predict the ratio of (γ, n) to (γ, p) cross sections. For a $T=0$ nucleus, the electric dipole absorption at the giant resonance will lead to $T=1$ states. Then the ratio of (γ, n) to (γ, p) cross sections should be unity[1]. It transpires that in this region this is the best test of the purity of the isotopic spin states.

One of the purposes of an experimental study is the measurement of the purity of these isotopic spin states.

33. Special cases. In order to illustrate the effect of this selection rule upon photodisintegration processes, the following reactions will be discussed in detail

$$\gamma + \mathrm{Li}^6 \rightarrow d + \alpha, \tag{33.1}$$

$$\gamma + \mathrm{C}^{12} \rightarrow \mathrm{Be}^8 + \alpha \rightarrow 3\alpha, \tag{33.2}$$

$$\gamma + \mathrm{O}^{16} \rightarrow \mathrm{C}^{12} + \alpha \rightarrow 4\alpha. \tag{33.3}$$

In each of these cases, the nucleus on the left hand side has $T=0$, and the final states have also $T=0$ so that the electric dipole transition is forbidden. For further studies see [*11*].

The reaction (33.1) has a low threshold (1.6 Mev) and has been studied with radioactive sources. The cross section at 2.76 Mev is very small ($<10^{-30}$ cm^2) illustrating the operation of the selection rule[2].

The reaction (33.3) has been studied in photographic emulsions using radiation from betatrons and synchrotrons. Studies were first made by TELEGDI[3] using the 17 Mev nuclear γ-ray from protons on lithium. In a photographic emulsion the reaction shows as a star of three α-particles. The α-particles will, in general, be entirely contained within the emulsion and it is therefore possible to measure their energy by a range determination, and the angles. Thus it is possible to verify that it is indeed a three body process and not a four body process by a consideration of momentum balance. In addition, by adding together the α-particle energies it is possible to measure the γ-ray energy responsible for the reaction. A measurement with mixed monochromatic 17 and 14 Mev rays is usually performed first, and by measuring the particle energies it is possible to

---

[1] F. C. BARKER and A. K. MANN: In press. — D. H. WILKINSON: Phil. Mag. **47**, 379 (1956). — H. MORINAGA: Phys. Rev. **97**, 444, 1185 (1955).

[2] E. W. TITTERTON and T. A. BRINKLEY: Aus J. Phys. **7**, 350 (1954). — H. B. GLENN: Phys. Rev. **88**, 418 (1952). — A. GAMBA and V. WATAGHIN: Nuovo Cim. **10**, 174 (1953). — P. JENSEN and K. GIS: Z. Naturforsch. **8**a, 137 (1953).

[3] H. HANNI, V. L. TELEGDI and W. ZÜNTI: Helv. phys. Acta **21**, 203 (1948). — V. L. TELEGDI: Phys. Rev. **84**, 600 (1951).

determine the experimental energy resolution which is about 10%. Moreover the absolute cross sections at 17 and 14 Mev are determined.

Data[1] taken with betatrons and synchrotrons can be interpreted when the resolution is known. The cross section for the reaction is found to rise to a peak at about 18.5 and 29 Mev and fall to zero at 32 Mev (Fig. 17). This is in contrast to the (γ, n) reaction in C^{12}, discussed in Sect. 25, which has a peak at 22.5 Mev and does not fall to zero at 60 Mev.

It is assumed that the reaction takes place via the intermediate state

$$\gamma + C^{12} \rightarrow Be^8 + \alpha.$$

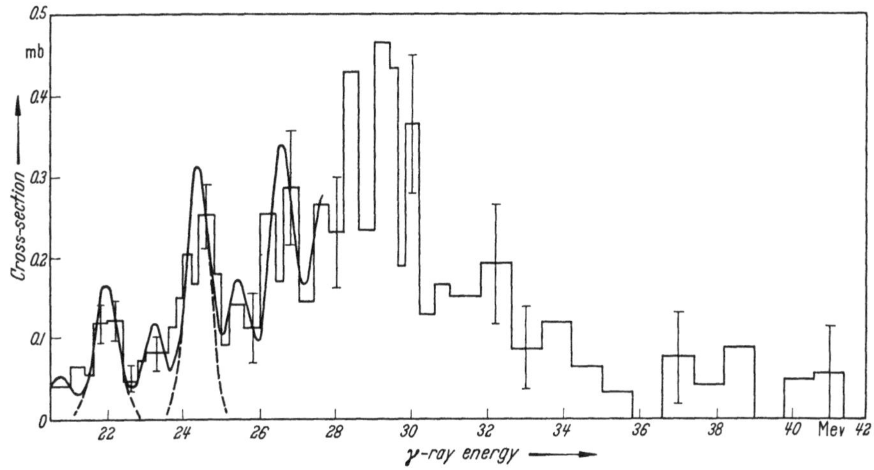

Fig. 17. The cross section for $C^{12}(\gamma, 3\alpha)$ as a function of energy.

From kinematic arguments it is found that

$$E^* = \tfrac{1}{2}(E_B + E_C) - E_B E_C \cos\varphi$$

where E^* is the excitation energy of Be^8 plus the binding energy of the ground state. Of course it is not known how to label the α-particles A, B and C so that three values of E^* can be obtained for each photodisintegration, only one of which can be correct. A plot of *all* values of E^* so determined can, however, yield peaks at certain values of E^*.

By this means it has been shown that, below 20 Mev, most of the disintegrations take place via the 3 Mev level of Be^8. Above 20 Mev, 86% of the disintegrations proceed via levels in Be^8 at 16.8 and 17.6 Mev.

Once the correct E^* for a given disintegration has been decided, it is possible to study the angular distributions and correlations of the α-particles, and this shows that electric dipole γ-ray absorption is responsible, and that the spin and parity of the 16.8 Mev state are 2^+.

These facts find a ready explanation in the isotopic spin selection rules.

Below 20 Mev, only $T=0$ states in Be^8 are energetically accessible to photodisintegration. Since an α-particle and C^{12} nuclei have $T=0$, the electromagnetic transition must be $T=0$ to $T=0$ and therefore cannot be electric dipole. The peak in the cross section for photodisintegration at about 18 Mev is that caused

[1] F. K. Goward and J. J. Wilkins: Proc. Roy. Soc. Lond., Ser. A **217**, 357 (1953); Ser. A **228**, 376 (1955). — J. H. Carver, H. J. Hay and E. W. Titterton: Phil. Mag. **46**, 841 (1955).

by electric quadrupole absorption. Since the carbon nucleus has spin zero, the reaction must therefore proceed via an excited state of C^{12} of spin 2, and then to the ground state (0^+) or the first excited state (2^+). The first excited state, though energetically unfavoured is spin favoured. Above 20 Mev other states of Be^8 become available which are $T=1$ states. The reaction now proceeds, by the electric dipole transition, so that a second peak in the cross section appears. All the transitions now, however, proceed via the $T=1$ states. The 16.8 and 17.1 Mev levels in Be^8 are therefore classed as $T=1$ states.

It should now be noted that the isotopic spin selection rule forbids $T=1$ states to decay into two $T=0$ particles. Only a small admixture of $T=0$ in the wavefunctions for their states are necessary to explain the breakdown of this rule, since α-particle lifetimes are much shorter than γ-ray lifetimes. The isotopic spin selection rule is therefore *not* as rigid as the angular momentum conservation rule which forbids the decay into two α-particles of the well known 1^+ 17.6 Mev level in the same nucleus. Indeed the ratio of (γ, n) to (γ, p) cross sections in C^{12} at 23 Mev is not unity; Barker and Mann have shown that this demands an admixture of 1 to 10% of $T=0$ states into the wavefunction of the $T=1$ states. Wilkinson shows that a similar situation exists for O^{16}.

It is noteworthy that the electric dipole reaction $\gamma + C^{12} \rightarrow 3\alpha$ has a peak at about 29 Mev and falls to zero by 36 Mev, whereas the reaction $\gamma + C^{12} \rightarrow C^{11} + n$ has a *lower* peak at 23 Mev but does not fall to zero even at 60 Mev. The higher energy for the peak in the first case might be due to the high threshold for disintegrations into the $T=1$ states; a (γ, n) process might be produced readily by a direct photoeffect or by a "Levinger effect" (Sect. 34) with subsequent reabsorption of the proton. A (γ, α) process would appear to need an intermediate compound nucleus formation. This may explain the absence of the high energy tail in the (γ, α) reaction.

It is not known at what energy the total photon absorption has a maximum (see a discussion of the discrepancies in Sect. 27).

VII. High energy photoprocesses.

34. Theory of Levinger[1]. At high γ-ray energies the wavelength becomes smaller than the internucleon distance in a nucleus. The γ-ray will therefore prefer to interact with only one nucleon at a time. If it interacts with a proton, the only process which can conserve energy and momentum simultaneously is elastic scattering, which has a low cross section. Occasionally, however, two nucleons will chance to be within a wavelength, and then a disintegration process can occur with a probability so large that the elastic scattering process can be ignored. At high enough energies, the probability of three nucleons being within a wavelength is small, and if we assume, as is reasonable, that a three-body interaction is not enhanced over a two-body interaction, this can be ignored.

Two cases may be considered. A proton may be close to another proton; then the electric dipole moment of the system vanishes and part of the magnetic dipole moment also (Sect. 13) so that only electric quadrupole disintegration can occur. Secondly the proton may be close to a neutron, and the pair of nucleons will than behave in many ways like a deuteron. This has given rise to the name of the "Pseudodeuteron model".

This model gives clear general predictions. The pseudodeuteron gives the main cross sections, and the calculations may be made to follow those of the photo-

[1] J. S. Levinger: Phys. Rev. **84**, 43 (1951). — Y. K. Khokhlov: Dokl. Akad. Nauk USSR. **23**, 241 (1952).

disintegration of the deuteron; indeed, the cross section may be related to the cross section for the deuteron. Firstly the qualitative features; for every γ-ray absorption, a neutron and a proton will be emitted in opposite directions in the centre of mass system of the pseudodeuteron. One of these may be subsequently reabsorbed, particularly in heavy nuclei. An angular correlation between the proton and the neutron will be caused by the momentum distribution of the pseudodeuteron inside the nucleus. This distribution may be considered as the vectorsum of the distribution of the two nucleons separately; the momentum distribution of a single nucleon in a nucleus has been determined by several methods with essentially concordant results[1]. The momentum distribution also affects the energies of the outgoing proton and neutron which would otherwise each be half the γ-ray energy.

For γ-ray energies sufficiently high that the effect of the internal momentum of the nucleons is small, to each photoproton energy corresponds a γ-ray energy. Then if an experiment is performed with a bremsstrahlung distribution of X-rays, the distribution of photoproton energies will correspond to the cross section for the process, and will cease abruptly at half the maximum X-ray energy. Such an abrupt break is, in fact, observed, although the momentum distribution has a sufficient effect that there are *some* photoprotons of more than half the maximum X-ray energy.

The detailed predictions are more difficult to make. DEDRICK[2] has made some detailed numerical calculations, assuming that the γ-ray will interact with the neutron and proton by electric dipole and quadrupole only and neglecting noncentral forces; these calculations follow those of SCHIFF on the deuteron exactly (see Sect. 8). A figure has to be assumed for the nuclear volume, to calculate the probability of the neutron and proton being close enough together.

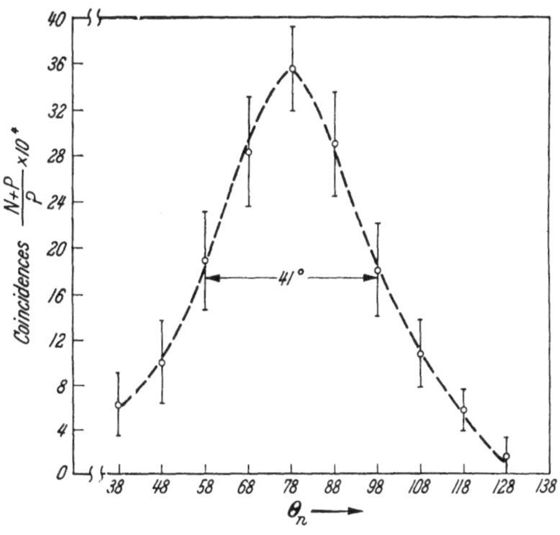

Fig. 18. The angular correlation between photoneutrons and photoprotons for disintegration of carbon by 300 Mev rays. The theoretical curve assumes that each nucleon has a Gaussian momentum distribution with a 1/e value of 17 Mev/c.

35. Experiments at high energies. The calculations appear to be well borne out by experiment. For example ODIAN[3] and also BARTON, have found that neutrons and protons are indeed produced in opposite directions. It is convenient to specify a particular γ-ray energy by measuring the energy of the proton. When this is done, the angular correlation between the neutrons and protons from carbon is shown in Fig. 18. On the same figure, the similar curve for a deuterium target is given to show the instrumental resolution. The theoretical curve is calculated assuming a single nucleon has a gaussian momentum distribution with a 1/e width of 17 Mev, as given by other data; the agreement is good. That quadrupole absorption is small is shown by a search for two coincident protons. The number is less than 4% of the coincident neutron and proton and

[1] K. A. BRUECKNER, R. J. EDEN and N. C. FRANCIS: Phys. Rev. **98**, 1445 (1955).
[2] K. DEDRICK: Phys. Rev. **100**, 58 (1955).
[3] A. C. ODIAN, P. C. STEIN, A. WATTENBERG, B. FELD and A. WEINSTEIN: Phys. Rev. **102**, 837 (1956).

is consistent with them being produced by secondary (n, p) events in the target nucleus. The energy distribution of the photoprotons has been measured by many authors[1]. At the higher energies only measurements using the complete bremsstrahlung spectrum have been made, and have produced the break in the photoproton spectrum as predicted. More detailed measurements have been made at 100 Mev using the photon difference method to separate out the effect of a monochromatic γ-ray energy. The energy spectrum of photoprotons from carbon at three angles is shown in Fig. 19 and compared with Dedrick's predictions. In the two comparison curves the nucleus was assumed to be uniform with a radius equal to 1.2 or $1.4 A^{\frac{1}{3}} \times 10^{-13}$ cm; subsequent reabsorption of the particles has been included.

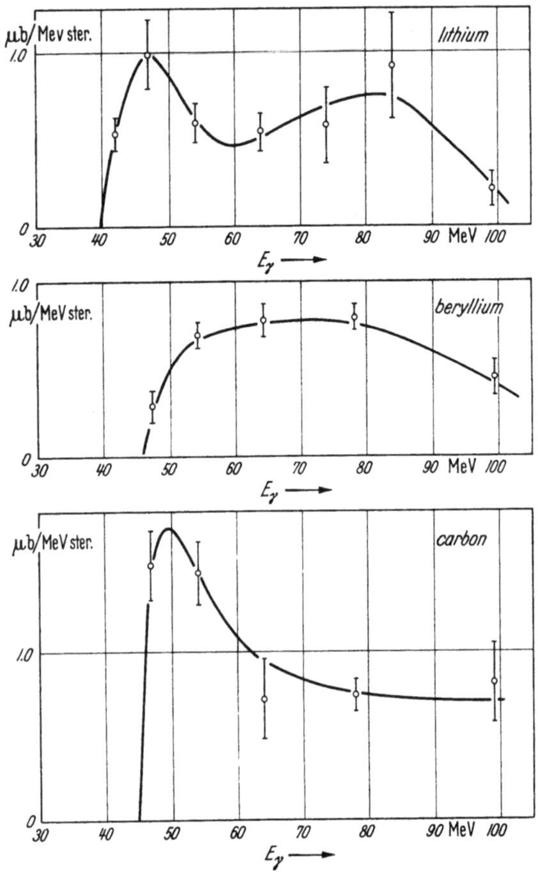

Fig. 19. The cross section for production of photoprotons of 37 Mev at 90° by γ-rays on carbon as a function of γ-ray energy.

Thus it is seen that the high energy photonuclear effect is moderately well understood. The experiments of Odian *et al.*, can therefore be interpreted directly to give a nucleon momentum distribution. The data of Odian and Barton are not in complete agreement; in general, however, the conclusion is the same as reached by other experiments.

36. Anomalies. There remain some anomalies, however. The process $C^{12}(\gamma, n)$ has been discussed earlier (Sect. 25) where it was shown that the data show that the partial cross section for this process alone at 80 Mev is 10^{-28} cm² which is a large fraction of the total absorption on Levinger's model, of about 6×10^{-28} cm². If the absorption takes place entirely by the Levinger model, C^{11} can only be formed by a subsequent reabsorption of the proton, which at this energy would lead to a highly excited state of C^{11}. It might be expected that a second neutron would be evaporated off. Yet Lawson[2] has shown that the cross section for $C^{12}(\gamma, 2n)\,C^{10}$ is small. Moreover a more complicated direct interaction seems implausible at these energies for the cross section for $C^{12}(\gamma, 3\alpha)$ becomes small above 30 Mev (Sect. 33).

Whitehead *et al.*, have measured excitation functions for the production of protons of a specific energy; these are then converted into cross sections as a

[1] At 300 Mev, B. T. Feld, R. D. Dodbole, A. Odian, F. Scherb and A. Wattenberg: Phys. Rev. **94**, 1000 (1954). 30 to 100 Mev, C. Whitehead, McMurray, M. I. Aitken, N. Middlemas and C. H. Collie: Thesis Oxford 1956.

[2] J. L. Lawson and M. L. Perlman: Phys. Rev. **74**, 1190 (1948).

function of γ-ray energy. The graphs for production of 37 Mev protons from lithium, beryllium and carbon at 90° are shown in Fig. 19. The prediction of DEDRICK is a fairly constant cross section. The cross section at low energies is much higher than expected. The dip in the cross section at about 60 Mev is particularly puzzling and for lithium at forward angles even becomes zero.

These effects are presumably due to a change of mechanism from the Levinger model at high energies to the direct photoproton emission discussed in Sects. 22 to 31. No detailed explanation is yet forthcoming.

VIII. Special problems.

37. The photodisintegration of beryllium. The photodisintegration of Be^9 has a threshold at 1.8 Mev; after deuterium it was therefore, the first nucleus to be studied. Firstly it is of importance to note that at low energies the reaction proceeds by an initial two particle process, with a subsequent decay of the Be^8 produced.

$$\gamma + Be^9 \rightarrow Be^8 + n \qquad (37.1)$$

and not

$$\gamma + Be^9 \rightarrow \alpha + \alpha + n. \qquad (37.2)$$

The threshold for the second reaction is lower by 100 kev but no disintegrations are found.

The total cross section rises sharply above threshold to 10^{-26} cm², and then falls to 5×10^{-27} cm² at 2.5 Mev with a subsequent rise (Fig. 20). The angular distribution of the neutrons is isotropic just above threshold, becoming $\sin^2\vartheta$ at 2.8 Mev.

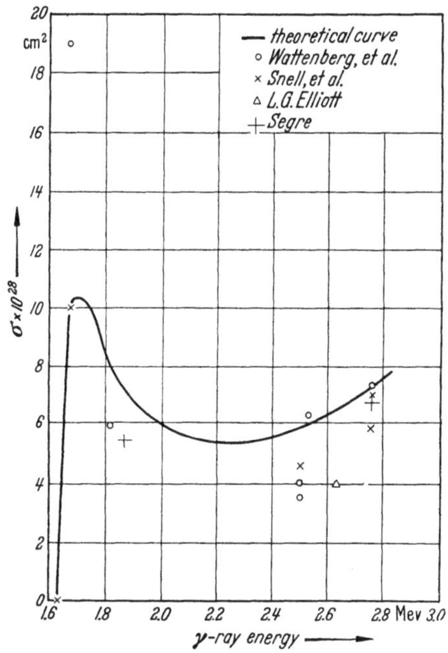

Fig. 20. The cross section for photodisintegration of beryllium versus energy; the theoretical curve is that of MULLIN and GUTH.

The independent particle model of the nucleus considers the motion of one nucleon in a field due to the rest of the nucleons; it appears that a favourable case for applying this model to photodisintegration problems should be Be^9. The low threshold for liberating a neutron suggests that this should be a good approximation up to the threshold for disintegration of the particle core. Using this approximation, MULLIN and GUTH[1] have discussed the available low energy data. The fit to the data is good. Experimentally however the approximation fails above 5 Mev when breakup of the core becomes possible.

This model enables one to calculate the polarization of the outgoing neutron[2]. It transpires that a large polarization is predicted. No experimental study of the polarization of the outgoing particles has yet been undertaken, though it is clear that a wealth of information lies hidden here.

Above 5 Mev the (γ, n) reaction on beryllium exhibits the same giant resonance as other elements; this is no longer the simple two-body direct interaction discussed here.

[1] E. GUTH and J. MULLIN: Phys. Rev. **76**, 234, 682 (1949).

[2] W. CZYZ and J. SAWICKI: Bull. Acad. Pol. Sci. **4**, 141 (1956). — Nuovo Cim. **3**, No 5, 864 (1956).

38. Disintegration by electrons. The cross section for electrodisintegration of nuclei may be related to that for photodisintegration by a method due to WEIZSÄCKER and WILLIAMS[1], known as the method of virtual quanta. They showed that within the limits of their approximation, the beam of electrons may be regarded as a beam of photons. Even for more exact treatments this concept of virtual quanta is useful. In Born approximation the spectrum of virtual quanta differs according to the multipolarity of the transition being induced. In this respect the phenomenon is the inverse of internal conversion. The spectrum of virtual quanta becomes:

$$\left[1+\left(\frac{E}{E_0}\right)^2\right]\log\left[\frac{2E_0E}{\mu k}\right]-A \tag{38.1}$$

with E_0 the energy of the incoming electrons, μ their rest energy and $k=(E_0-E)$. A is the contribution from the scalar potential and is $-2E/E_0$ for electric dipole, zero for magnetic dipole and $8E^2/3k^2$ for electric quadrupole. This may be compared with the bremsstrahlung formula (4.1) above. No correction for screening due to the atomic electrons is normally necessary, because the electrons have a small impact parameter for these inelastic collisions. A correction may be necessary for screening by the nuclear size.

This formula has been checked by BROWN[2] experimentally who found it valid up to an atomic number of 29. Brown measured the *ratio* of electro disintegration to photodisintegration by passing an electron beam through a stack of foils. The first foil measures only electrodisintegration and the last foil measures electrodisintegration plus disintegration by the photons produced in the intervening radiator. The *ratio* was easily and accurately measured, and should equal the ratio of the formula (38.1) to (4.1).

Experiments[3] have been carried out, for example, for Be^9, which (Sect. 37) is believed to disintegrate by an electric dipole process. The result is in good agreement with theory. For nuclei of atomic number 29 (Cu^{63}), agreement is moderate for electric dipole disintegration. For Ta^{181}, however, agreement is bad. This may be due to neglect of the monopole transition which can occur by electrons and not at all by photons. Alternatively it could be disagreement with the theory.

Disintegration by electrons, however, is most important when the scattered electron energy is measured. The energy of the incident virtual quantum is then defined. In addition the angular distribution of the scattered electron may be measured which will yield still more information. A start has been made with experiments of this nature using the 600 Mev linear electron accelerator at STANFORD[4]. In this work 200 Mev electrons were used to study the first excited states of carbon. The *results* confirm the level assignments previously known and do not fall within the scope of this article. The *method* is, however, general, and would seem to be a very important tool to unravel the mysteries of the giant dipole of the nuclear photoeffect.

39. Nuclear elastic scattering of photons[5]. The nuclear absorption of photons must be accompanied by an elastic scattering. Theoretically the scattering ampli-

---

[1] E. J. WILLIAMS: Kgl. danske Vid. Selsk. **13**, No. 4 (1935).

[2] K. L. BROWN and RICHARD WILSON: Phys. Rev. **93**, 443 (1954).

[3] W. C. BARBER: Private communication (*14*]. — K. L. BROWN and RICHARD WILSON: Phys. Rev. **93**, 443 (1954).

[4] FREGEAU and R. HOFSTADTER: Phys. Rev. **99**, 1503 (1955). — R. HELM: Thesis, Stanford 1955.

[5] M. GELLMANN, GOLDBERGER and THIRRING: Phys. Rev. **95**, 1612 (1954). — E. G. FULLER and E. HAYWARD: Phys. Rev. **98**, 1537 (1955); **101**, 692 (1956).

tudes in the forward direction may be derived from the absorption cross sections. Of course, the cross section for photon scattering in the forward direction is predominantly due to scattering by various electronic processes. Fortunately, however, the relations also apply to the hypothetical process where both the electronic scattering and absorption are considered absent. Then the imaginary part of the forward scattering amplitude is given by the total cross section, and the real part is given by the scattering amplitude at zero energy plus integrals over the absorption cross section at a band of energies.

The two theorems used can then be written

$$\operatorname{Im} E(k,0) = \frac{k}{4\pi}\left[\sigma_{\text{tot}}(k)\right], \tag{39.1}$$

$$\operatorname{Re} E(k,0) - \operatorname{Re} f(0,0) = \frac{k^2}{2\pi^2} P \int \frac{\sigma_{\text{tot}}(k')\, dk'}{k'^2 - k^2} \tag{39.2}$$

where P is the Cauchy principal value of the integral.

The nuclear elastic scattering may be experimentally distinguished from the electronic scattering by measuring at large angles. At energies below about 20 Mev the nuclear form factor is of the order of unity for all angles; equivalently it can be said that only the dipole term in the multipole expansion is of importance. The angular distribution of the photons may then be assumed to be and for many cases it has been experimentally found to be that of an electric dipole and the dispersion relations may be used.

Since γ-ray scattering contributes little to the total cross section we may then write:

$$\sigma_{\text{scatt}}(k) = \frac{k^2}{6\pi}\left[\sigma_a^2(k) + \left(\frac{2k}{\pi} P \int \frac{\sigma_a(k')\, dk'}{k'^2 - k^2}\right)^2\right] + \frac{8\pi}{3} D^2 + \frac{8\pi}{3} D k^2 P \int \frac{\sigma_a(k')\, dk'}{k'^2 - k^2} \tag{39.3}$$

where D is the Thompson (zero energy) scattering amplitude. The measurements of FULLER and HAYWARD use a betatron and a scintillation γ-ray detector. The detector is placed in the direct beam and the bias is set so that only the highest energy γ-rays from the betatron are detected. The detector is then placed to detect only scattered photons. The ratio of the counts in the two cases, together with geometrical factors, gives the scattering cross section. Fig. 21 illustrates the results.

The experiments are useful in three ways in elucidating features about the giant dipole resonance.

Firstly, the cross section for scattering will largely follow the cross section for photon absorption. For light elements, where the measured (γ, n) cross sections do not give all the cross section, the peak of the photon scattering curve may give a clearer indication of the position of the peak in the photon absorption than any other experimental datum, though discrepancies exist (Sect. 27).

Secondly, the two sides of Eq. (39.3) may be compared. If the cross section at the giant dipole is indeed a smooth curve, and not composed of individual resonances, then it is valid to replace the right hand side of Eq. (39.3) by the appropriate average value. Since in the important term σ_a appears squared, the existence of individual resonances would increase the right hand side. The scattering cross section as measured is inserted into the left hand side; if any inelastic scattering were present, a *smaller* figure should be inserted. FULLER and HAYWARD find complete agreement for copper and agreement at the lower energy tail of the giant dipole for aluminium. Thus the assumptions are simultaneously verified. For aluminium the scattering is higher than predicted above the giant dipole.

Fuller and Hayward suggest that there may be some γ-ray absorption in this region which has so far been missed by activation experiments. Indeed they take the peak of the γ-ray scattering curve to be the peak of the giant resonance and plot this in Fig. 22. This is to be compared with Fig. 10, obtained

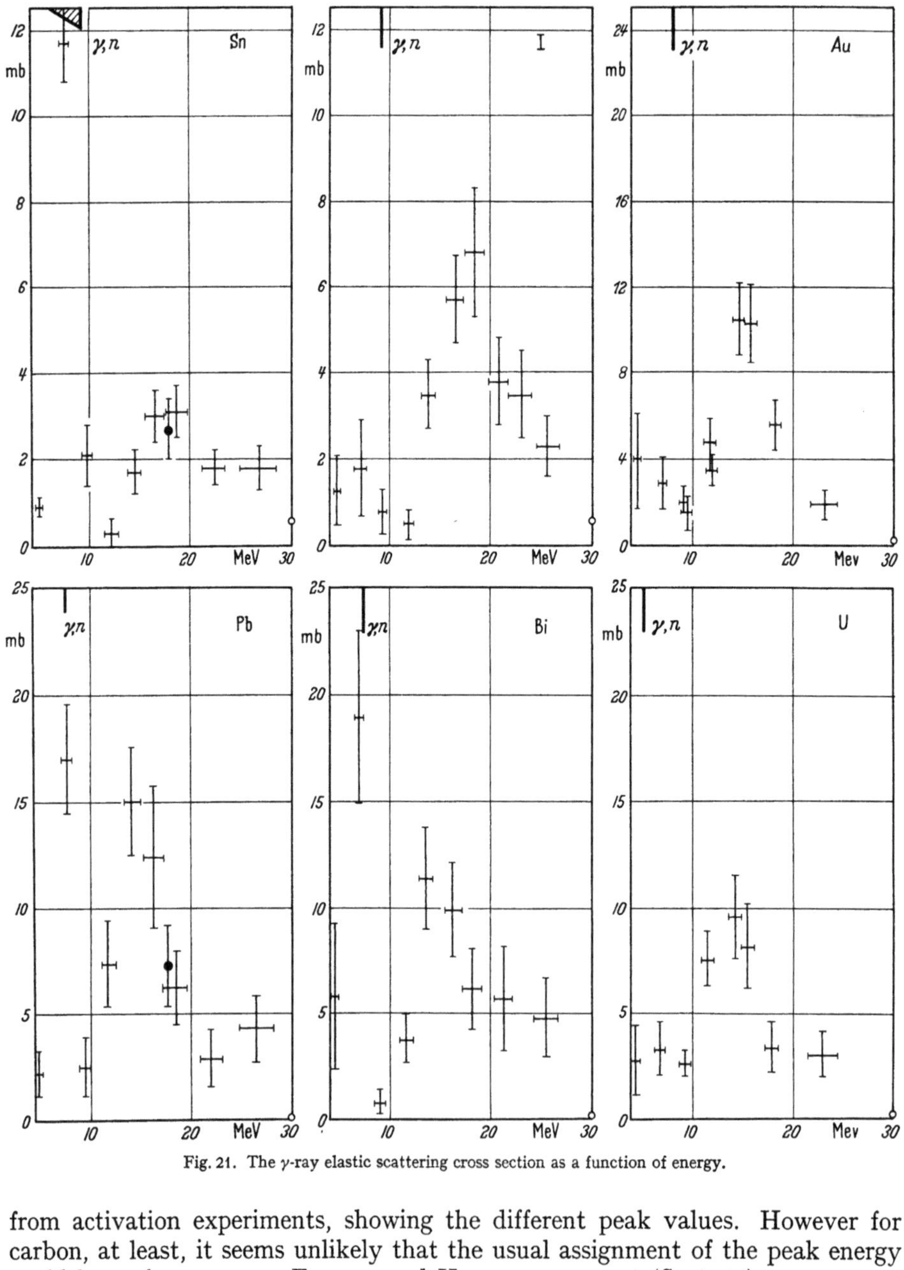

Fig. 21. The γ-ray elastic scattering cross section as a function of energy.

from activation experiments, showing the different peak values. However for carbon, at least, it seems unlikely that the usual assignment of the peak energy could be as far wrong as Fuller and Hayward suggest (Sect. 24).

The elastic photon scattering experiments give a further check that the photon cross sections for heavy nuclei are not composed of individual levels. An absorption coefficient has been measured for those photons which are elastically scattered.

Then, if the scattering cross section consists of resonances, the absorption coefficient will be appreciably higher than that normally measured. A small effect has been found for energies about 5 Mev, but none in the region of the giant dipole. Unfortunately data are not available at the time of writing for the elements (carbon and oxygen) for which individual level structure (Sect. 29) has been observed.

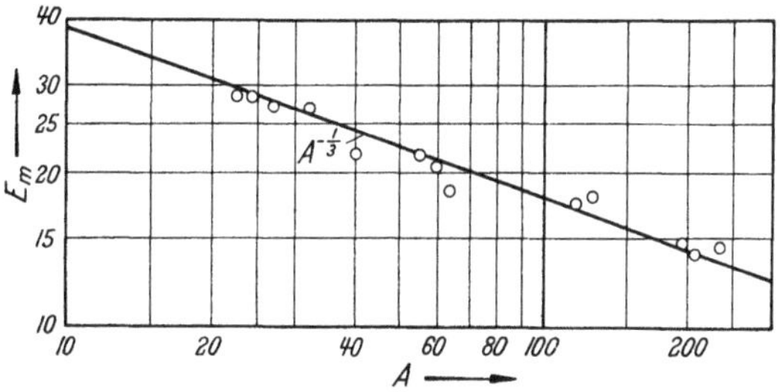

Fig. 22. The peak in the photon scattering cross section as a function of A; this may be compared with Fig. 10.

40. Photofission. The first experiments on photofission involved the use of the monoenergetic γ-rays from nuclear reactions[1]. These gave cross-sections of a few millibarns in U^{238} and Th^{232} for γ-ray energies of 6.3 and 17.5 Mev. Later X-rays of higher energies were used to measure the cross-section as a function of energy, the photofission thresholds, the distribution in mass of the fission fragments, and the angular distribution of the fragments[2]. Measurements of the cross-section demand the use of a differential ionisation chamber to avoid the effects of the sudden pulse of ionisation produced by the X-ray beam, which would otherwise swamp the ionisation produced by the fission fragments. The cross-section has a maximum at low energies occurring at about 17 Mev[3] and after a small fall rises with increasing photon energy.

The threshold energy for photofission of U^{238} is found to be 5.3 Mev, which is lower than the energy predicted from the liquid-drop model of 7 Mev[4], but indications from other data are that the threshold energy is considerably lower than these theoretical estimates. The photofission threshold energy can be calculated from the threshold energy E_n for fission by thermal neutrons, if the binding energy E_{bn} of the last neutron is known

$$E_{ph} = E_{bn} + E_n\left(\frac{A-1}{A}\right). \tag{40.1}$$

Seaborg[5] has proposed that the variation of E_{ph} with Z^2/A is best represented for the heavy elements by the equation

$$E_{ph} \approx \left(18.6 - 0.36\,\frac{Z^2}{A}\right) \text{ Mev.} \tag{40.2}$$

Table 13 of photofission thresholds is given, with the theoretical values calculated on the liquid drop model[6].

The distribution in mass of the fission fragments is one of the most interesting data to come from photofission experiments. The experiments are done by the chemical separation of the fragments, and assessment of their relative probability of formation by counting their radioactivity. The usual assymmetric double humped curve is found for the mass distribution; the outstanding feature of these studies is the decrease in the peak-to-trough yield ratio of this yield-

[1] R. O. Haxby, W. E. Shapp, W. E. Stephens and W. H. Wells: Phys. Rev. **59**, 57 (1941).
[2] G. C. Baldwin and G. S. Klaiber: Phys. Rev. **71**, 3 (1947).
[3] R. B. Duffield and J. R. Huizenga: Phys. Rev. **89**, 1042 (1953). — J. Gindler and R. B. Duffield: Phys. Rev. **94**, 759 (1954).
[4] S. M. Frankel and N. Metropolis: Phys. Rev. **72**, 924 (1947).
[5] G. T. Seaborg: Phys. Rev. **88**, 1429 (1952).
[6] H. W. Koch, J. McElhinney and E. L. Gasteiger: Phys. Rev. **77**, 329 (1950).

mass curve, as the energy of the bombarding radiation is increased. Thus the curve Fig. 23 obtained for the fission yields of U^{238} at quantum energies of a little above the threshold shows a peak-to-trough ratio quite comparable to that obtained by thermal neutron fission of U^{235}. Care must be taken when making these measurements to avoid fission caused by the photoneutron, produced with a larger cross-section than the photofission[1].

Table 13.

Nucleus	Photofission threshold Mev	Theoretical threshold Mev; Frankel and Metropolis
Pu^{239}	5.31 ± 0.27	4.9
U^{235}	5.08 ± 0.15	7.0
U^{238}	5.31 ± 0.25	6.1
U^{233}	5.18 ± 0.27	5.7
Th^{232}	5.40 ± 0.22	

At higher energies the peak-to-trough ratio decreases from values of 300 at 7 Mev to 4 at 300 Mev, bombarding energy[2]. This dependence of the yield-mass curve on the energy available for fission is also found for fission caused by fast neutrons and α-particles[3]. It is explained by assuming that at higher energies symmetric fission of the excited nucleus becomes more and more probable compared with the asymmetric fission typical of the process at excitation energies only slightly above the fission threshold. The discussion of the various theories proposed to explain this competition is not properly the subject of this present article, but is should be noted that symmetrical fission was the prediction of the original theories of the fission process, based on the fact that for the mass formula adopted in these calculations the total energy release is largest for symmetrical fission[4]. In order to explain asymmetrical fission several modifications of the theory were introduced, which include shell structure effects and the variation of barrier penetration with the mass of the fragments[5]. Such theories have been summarised and their inadequacies discussed in terms of the collective model, blending the individual particle and the liquid drop models of the nucleus. More recently a successful statistical model calculation has been made, based on a mass curve corrected for deviations which are attributed to the effects of nuclear shells. This gives a calculated mass distribution curve of fission products of U^{235} induced by thermal neutrons which fits very well the experimental data[6].

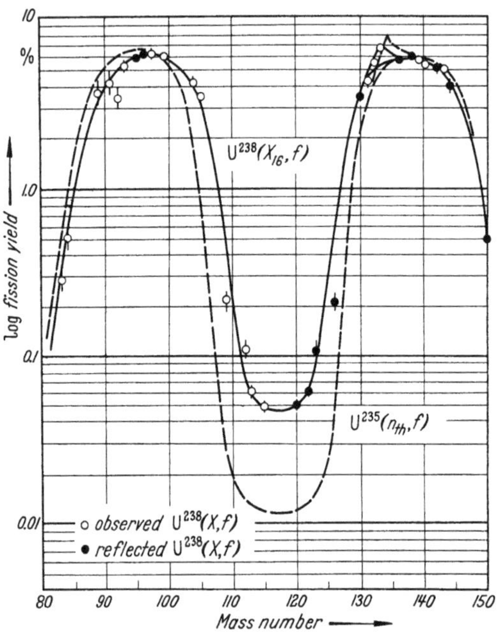

Fig. 23. Mass distribution of the products of photofission of U^{238} by photons of maximum energy 16 Mev. Curve for thermal fission of U^{235} for comparison.

Fine structure, which is present in the thermal neutron fission of U^{235} at fragment masses 100 and 134 may also be present in the photofission of U^{238}

[1] H. G. Richter and C. D. Coryell: Phys. Rev. **95**, 1550 (1954).
[2] R. A. Schmitt and N. Sugarman: Phys. Rev. **95**, 1260 (1954).
[3] R. W. Spence and G. P. Ford: Ann. Rev. Nucl. Sci. **2**, 399 (1953).
[4] N. Bohr and J. A. Wheeler: Phys. Rev. **56**, 426, 1065 (1939).
[5] R. D. Hill: Phys. Rev. **98**, 1272 (1955).
[6] P. Fong: Phys. Rev. **102**, 434 (1956).

at the same masses[1]. This may be the result of preferential formation of a 50 or 82 neutron configuration in the fission act, or a neutron "boil-off" effect after the fission.

Anisotropy in the angular distribution of photofission fragments from thorium has been observed by counting the total β-activity of the fragments caught at various angles with respect to the γ-ray beam[2]. The angular distribution of the unseparated fragments was found to be compatible with the form $a+b\sin^2\vartheta$ with respect to the incident γ-ray beam. The ratio b/a was 0.41 ± 0.05. A second experiment, designed to measure the anisotropy as a function of the mass ratio of the fragments, showed that the angular distribution became more pronounced as the mass ratio increased. In this experiment an average was taken between the directions making 0 and 180° with the incident beam; thus no exact information was obtained regarding the relative amounts of odd and even powers of $\cos\vartheta$ entering into the photofission process. This would be most interesting in aiding interpretation of fission[3].

The observed angular anisotropy conflicts with the liquid drop model of the fission process. Here it is assumed that from the moment the energy of excitation is absorbed to the moment that this energy is concentrated on the mode of deformation leading to fission, a complicated redistribution of the energy occurs. Thus one would expect no correlation between the direction of incidence of the photon and the direction of emergence of the fragments. A schematic picture of the process, based on the collective model has been discussed by HILL and WHEELER[4].

General bibliography.

Historical.

[1] CHADWICK, J., and M. GOLDHABER: Proc. Roy. Soc. Lond., Ser. A **151**, 479 (1935). — The discovery of the process and its use for measurement of the neutron mass.

[2] BETHE, H. A., and R. A. PEIERLS: Proc. Roy. Soc. Lond., Ser. A **148**, 146 (1935). — First quantum theory of the deuteron with a calculation of the photoeffect. This form the basis for all more accurate discussions.

Theory of the deuteron.

[3] RARITA, W., J. SCHWINGER and H. A. NYE: Phys. Rev. **59**, 209 (1941). — RARITA, W., and J. SCHWINGER: Phys. Rev. **59**, 436, 556 (1941). — Outlines the method of inclusion of non central forces in a systematic way.

[4] SCHWINGER, J., and H. FESHBACH: Phys. Rev. **84**, 194 (1951). — BETHE, H. A., and C. LONGGUIRE: Phys. Rev. **77**, 647 (1950). — SALPETER, E. E.: Phys. Rev. **82**, 60 (1951). — These give the theory of the effective range as applied to the photonuclear effect. In addition they discuss the small effects of potential shape.

[5] J. M. BLATT and V. F. WEISSKOPF: Theoretical Nuclear Physics. New York: Wiley 1952.

Photodisintegration of nuclei beyond deuterium.

[6] PATTER, D. M. VAN, and W. WHALING: Rev. Mod. Phys. **26**, 402 (1954). — Nuclear disintegration energies.

[7] STRAUCH, K.: Ann. Rev. Nucl. Sci. **2**, 105 (1953).

[8] CHASTEL, R.: J. Phys. Radium **14**, 50, 56 (1953).

[9] ELAINE TOMS, M.: Naval Research Lab. Report Bibliography of photonuclear reactions, 1955 and supplement No. 1, 1956.

[10] SHAW, G. L., and E. C. GREGG: Case Institute of Technology. Report 1955.

[7] to [10] give excellent tables of the characteristics of the giant resonance.

[11] TITTERTON, E. N.: Progr. Nucl. Phys. **4**, 1 (1955). — Discusses the photodisintegration of light nuclei, particularly using the photographic plate technique.

[12] Proceedings of the Photonuclear Conference at Chicago, November, 1956.

[13] AJZENBERG, F., and T. LAURITSEN: Rev. Mod. Phys. **24**, 321 (1952). — Energy Levels of Light Nuclei.

---

[1] L. E. GLENDENIN, E. P. STEINBERG, M. G. INGHRAM and D. C. HESS: Phys. Rev. **84**, 860 (1955).

[2] E. J. WINHOLD, DEMOS and I. HALPERN: Phys. Rev. **87**, 1139 (1952).

[3] A. W. FAIRHALL, I. HALPERN and E. J. WINHOLD: Phys. Rev. **94**, 733 (1954).

[4] R. D. HILL and J. A. WHEELER: Phys. Rev. **89**, 1102 (1953).

Angular Correlations.

By

S. Devons and L. J. B. Goldfarb.

With 51 Figures.

A. Introduction.

1. Subject. The study of the angular distribution and correlation of nuclear radiations is applied to two broad classes of phenomena. The first of these, to which the term "angular correlation" is normally applied, concerns the decay, by the emission of successive radiations[1], of radioactive nuclei, and of nuclei which have been produced in unstable excited states. These radiations may be detected at particular directions with respect to the decaying nuclei and the relative probability of detection of the cascade is then a function, *inter alia*, of these directions. In addition to this directional correlation, which refers to directions of emission only, other angle variables may be involved, e.g. if the efficiency of the detectors of radiation depends also on the state of polarisation of the radiations, and in such cases the appropriate correlation functions are more complex functions of several angles. Furthermore, if the original system of decaying nuclei is not isotropic, in a statistical sense, or if the cascade of radiation occurs in the presence of some anisotropic disturbance, such as an extra-nuclear electromagnetic field, then the correlation function will involve angles relating the directions of emission and of polarisation to those characterising the original or external anisotropies, as well as angles characteristic of the relative directions of emission and polarisation themselves.

The second application is to the study of nuclear reactions. The use of a differential reaction or scattering cross-section, $\sigma(\vartheta)\,\delta\Omega$, as a measure of the absolute probability of a reaction with emission (or scattering) in a solid angle $\delta\Omega$ at an angle ϑ to the incident parallel beam, represents the simplest form of angular correlation between the directions of absorption of one radiation and emission of a successive one. Such correlations are conventionally referred to, insofar as they involve relative rather than absolute values of $\sigma(\vartheta)$, as angular distributions. In this case, as in the case successive emissions, more complex angular correlations may arise if the incident beam of particles, or the target nuclei, are polarised, or if the radiation detectors are polarisation-sensitive. In this connection it should be borne in mind that *any* of the products of the reaction may be detected by detectors which are sensitive to polarisation. Again, as in the first group of phenomena, external fields may be present which involve specific directions in space, and reference to these directions will be required

[1] We frequently use the term "radiation" to denote generally γ-radiation, β-ν particles, internal conversion electrons, nucleons, α-particles, etc. In many cases, when an unstable nucleus breaks up into two parts, we can in principle treat either part as the "radiation", but we normally consider this to be the lighter fragment, in accordance with the general experimental practice of detecting this particle.

as well as relative angular coordinates specifying the relative directions of absorption, emission, and polarisation.

When a nuclear reaction is followed by the emission of several radiations in cascade, then the most general angular-correlation properties involve the relative directions of the absorbed radiation (i.e. the incident beam), and the successive radiations, the angles specifying any initial or detected polarisation, together with the angles relevant to any external disturbing fields, if these are present. Such general processes, involving what have been referred to as correlations and distributions, would normally result in considerable complication if studied in explicit detail. Fortunately, the interpretation of experimental work very rarely requires the complication which results from this degree of generality; indeed, the object of experiment seems usually to be achieved more readily by studying simpler processes. However, from a purely formal point of view, the relative angular correlation of a succession of either emissions or absorptions, under the general conditions described, forms the subject of investigation.

In addition to the study of successive radiations, as in decay of the unstable nucleus in nuclear reactions, there are also nuclear processes in which several radiations may be emitted simultaneously (e.g. β-decay, electron-positron emission, etc.). Formally, of course, such multiple emissions can be treated as a succession of emissions and/or absorptions, but as will be seen, there is a definite distinction between the treatment of such multiple-radiation and of the cascade-radiations of an unstable nucleus, apart from the specific aspect of simultaneity. In many respects the process of multiple emission is more closely related to nuclear reactions and scattering than to successive emissions, since in the former case the time-sequence is less clear-cut than in the latter. For example, if the reaction is described in terms of an incident beam of exactly defined energy, then both absorption and emission processes are continuous, and the treatment of the two processes as consecutive is no longer strictly correct.

From the viewpoint of angular correlations, the distinction, which is not sharp, between processes which can be treated as successive emissions or absorptions and other processes is based on factors other than the strict time-sequence, namely, the nature of the nuclear states involved. The extent to which both types of process will be treated will be determined by the aim and scope of the present article, which we now discuss explicitly.

2. Scope. It is evident from the above remarks that the study of angular correlations in nuclear processes is not so much a distinct part of nuclear physics as a particular aspect of a large variety nuclear processes, and that a complete account of the angular-correlation aspects of all such processes would take us into practically every branch of nuclear physics. In the present article, therefore, the aim will be to describe and illustrate the basic principles, methods, and aims, underlying these aspects of nuclear phenomena insofar as they form a coherent background to a large diversity of particular topics.

One of the main aims of the study of angular correlations is the determination of the symmetry properties of nuclear states, particularly excited states (total angular momentum or "spin"[1] and parity), and all three classes of problems mentioned above share this aim. In addition angular correlation studies enable properties of the interaction between nuclei and radiation (nucleons, electromagnetic, electron-neutrino, etc.) to be studied. These interactions are related not only to the basic symmetry properties but also to more detailed properties

[1] For brevity we use the term "spin" without qualification to mean total angular momentum when applied to a nuclear state.

of the nuclear states such as coupling between nucleons, approximate j-j or l-s descriptions (where relevant), isotopic spin (of light nuclei), and in general to the detailed structure of nuclear states.

The basic quantum-mechanical principles involved in the interpretation of angular correlation experiments, and in the design of such experiments, concern the close relationship between the angular momenta and parities of the nuclear states and the emitted and absorbed radiations and their symmetry properties under spatial rotation and inversion of space and time. These symmetry properties do not, in themselves, provide information about the absolute probability of emission or absorption, but this is not the primary concern in the study of angular correlations. The full power of the general quantum-mechanical principles in providing an interpretation of the phenomena is most fully exemplified in the case of processes involving nuclear states and radiation fields or wave-functions which are all characterised by a single value of total angular momentum and well-defined parity, which in most cases implies symmetry properties determined by a unique set of quantum numbers. In the simplest situations we start with an assembly of nuclei in a particular state with definite spin and parity which are randomly oriented, i.e. statistically isotropic. These nuclei emit (or absorb) radiation which is detected (or incident) in a particular direction. The resultant nuclei are assumed to be again in a state of definite spin and parity, but in this state the nuclei are no longer randomly oriented, since the state has been "prepared" anisotropically, i.e. by emission (or absorption) in a particular direction. Consequently further radiation from the nuclei in this new state will not be, even statistically, isotropic, but will be correlated with the direction of the first radiation. If, furthermore, both first and second radiations are characterised by a unique angular momentum (and, if they have intrinsic spin, also by a unique combination of orbital and intrinsic angular momentum), then the quantum-mechanical laws for addition of angular momenta determine completely, from the spins of the nuclear states and the angular momentum of the first radiation, the anisotropy resulting from its emission; and from the spins and angular momentum of the second radiation the anisotropy of the latter, i.e. the angular correlation between the two radiations.

The processes which are dealt with in this article can be characterised by the extent to which the general quantum-mechanical principles require supplementation by further, more detailed, interpretation. The need for such additions arises when the nuclear states or radiations comprise components with different symmetry properties and some reference of the relative contributions of the different components must enter into the interpretation. Such additional information may be provided by the study of the angular correlations themselves or may be obtained from other information about nuclear properties. If a succession of emissions and absorptions at each stage involves only nuclear states with unique values of spin and parity, then the conservation of angular momentum will restrict the range of possible angular momenta for the radiations. Moreover, for the paritcle and radiation energies appropriate to the study of the properties of individual nuclear states, the de Broglie wavelength, $\lambda\!\!\!^{-}$, characterising the motion of the radiation in the centre-of-mass frame of reference is sufficiently large, in comparison with the size of the nuclei or range of interaction of nuclear particles ($\lambda\!\!\!^{-} \gg R$), that the radiations of higher angular momenta, L, play a rôle of rapidly diminishing importance. Provided then that a process can be described as the succession of emission or absorption of single radiations, leaving a nucleus with well-defined spin and parity, the interpretation of the angular distributions will only be of moderate complexity and the general principles need little supplementation. For such processes also the determination of the spins and parities of all the nuclear states involved is naturally an appropriate aim.

When all the nuclear states involved in a process can no longer be characterised by unique spin values and well-defined parity, the interpretation, at least within the framework indicated above, becomes, usually, much more complicated. If, however, the states without precise values of spin and parity can still be described as a combination of a very small number of superposed states each with definite spin and parity, as for example in resonance reactions involving only a small amount of overlapping of resonance levels, then the interpretation of angular correlations in terms of the symmetry properties of the successive nuclear states is still, although complicated, the appropriate one. There are many processes, referred to in Sect. 1, for which the description in terms of successive absorption or emission of individual radiations, with nuclear states that have, at each stage, even approximate symmetry properties, is no longer a suitable one. For example, in processes involving the absorption of radiation of high energy, the density of levels of the nucleus formed after absorption may by sufficiently large that the description in terms of a superposition of several levels of simple symmetry is neither practicable nor advantageous. In addition, for high-energy radiation ,the value of $\lambda\!\!\!^{-}$ may be of the same order or smaller than R so that the rôle of high angular-momentum components is not reduced, i.e. the relation $L\lambda\!\!\!^{-} \gtrsim R$ may be satisfied for a substantial range of L values. Under these circumstances, the relatively simple treatment based on general quantum-mechanical principles is no longer possible. Other simplifying features may arise, such as the use of semiclassical approximations, or the simplification that results from statistical averaging over many nuclear levels, or, on account of the generally reduced interaction at high energies, the applicability of approximate methods such as the Born approximation; but these are more concerned with the general mechanics of nuclear reactions in which the angular correlations form an incidental rather than a predominant part. Such processes will only be dealt with here in order to illustrate their relation to those involving nuclear states and radiations with essentially simpler symmetry properties. They are dealt with fully in the article on nuclear reactions by BREIT *et al.* in Volume XLI of this Encyclopedia.

There are processes, other than high-energy ones, in which it is not possible to consider that the emission or absorption of each individual radiation results in a nuclear state of well-defined symmetry. The simultaneous emission of two radiations already mentioned, non-resonant or "potential" scattering, Coulomb excitation, and other types of inelastic scattering, stripping, and similar processes usually referred to as "direct processes", are examples of such processes. In these processes the nuclear states before and after the emission of two radiations, or the absorption of one and the emission of another, do normally have well-defined spin and parity, so that the two radiations together are characterised by a transfer of a unique, or a small range of, angular momentum, and definite parity. The interpretation of angular correlations in these processes is more complicated than in the case where the radiations can be treated singly, since it is now necessary to treat also the detailed relationship between two radiations (e.g. the combined β-particle-neutrino field) or to provide some detailed, and possibly approximate, theory of the direct process (e.g. the mechanics of stripping or inelastic scattering). These detailed theories do not form part of the main theme of the subject matter of this article but are dealt with fully in other places (for example the articles on β-decay in Vol. XLI and nuclear reactions in Vol. XL, but since the study of angular correlations in these cases often shares, with the simpler processes, the common aim of investigation of the symmetry properties of nuclear states, and since moreover such processes are often part of a sequence involving the simpler type of process (e.g. β-decay or stripping followed by γ-emission), a discussion of the relevant aspects of such processes is included in the present article.

From the experimental point of view, the methods used in angular-correlation studies can be broadly divided into those involving absorption and emission (reactions and scattering), and those involving emission only (cascade emission), a division which, of course, does not coincide precisely with the theoretical division made above. Absorption implies a collimated beam of incident radiation, usually charged, which nowadays is almost invariably provided directly or indirectly by artificially accelerated particles. Both absorption-emission and cascade-emission processes utilise similar basic methods of detection which are not numerous. There is a much larger variety in the type and method of preparation of the source of radiation, i.e. in the preparation of the nuclear states. Here we can distinguish between processes in which the nuclei involved can be treated as effectively isolated (except of course for their interaction with the radiations being studied), and those processes in which the interaction of the nuclei with extra-nuclear fields play a significant rôle. These fields may arise from electromagnetic interaction of the nuclear moments with atomic electrons, macroscopic fields, or local fields produced by the medium in which the nuclei are embedded. The modification of the angular correlations by such fields provides methods for studying the stationary-state electromagnetic properties of excited nuclei, an important and interesting aspect of angular-distribution studies which has been developed recently. Such studies also provide a means of exploring the local fields of atomic origin, such as those at nuclear sites in crystal lattices, but the discussion of such topics lies the scope of the present article.

A particular class of angular-correlation problems arises when the nuclei studied have been partially aligned or polarised by thermodynamic methods. Although the theoretical interpretation of such phenomena has much in common with the processes we have previously mentioned, we shall not deal with this topic in this article, since the subject of nuclear alignment generally is dealt with in a separate article in this volume.

We have stated that processes involving single radiations and nuclear states of well-defined symmetry properties form a basis for our discussion of angular correlation in terms of general quantum-mechanical principles. It is clear that these same principles apply to systems other than nuclei which emit successive radiations and which can be described by states of appropriate symmetry properties. Up to the present the application of these methods, both experimental and theoretical, has been almost exclusively to the study of nuclei, but some application has been made to other fields, e.g. the decay with the emission of several radiations of hyperons and heavy mesons, or angular correlations between the directions of emission of such decay products and directions associated with collision processes in which these particles are produced.

3. Historical. The study of angular correlations in nuclear processes, in the sense in which we shall be mainly concerned, is a comparatively recent development; the greater part of the development of the subject has occurred in the past decade. In the broader sense, however, the threads of development can be traced back to the earliest phases of the study of atomic nuclei.

RUTHERFORD's classical investigations, in 1911, demonstrated strikingly that the angular distribution of α-particles scattered by heavy nuclei could be interpreted in terms of an effectively fixed nucleus and an interaction between this nucleus and the α-particle with a simple Coulomb repulsive force even for distances of approach as small as 10^{-12} cm. Subsequent investigations, in the period up to 1930, involving similar techniques, revealed in a qualitative way the departure from the simple Coulomb law of force when the distance of approach became still smaller, which classical-mechanical arguments indicated was possible for light nuclei. With the development of quantum-mechanics, the interpretation of scattering distributions in terms of the form of interaction was put on a quantitative basis, but the material available for such analysis was of limited scope. Later investigations of nuclear scattering were concerned particularly with "elementary" nuclear particles, e.g. the demonstration of the essential isotropy of the neutron-proton scattering, and the nuclear component of proton-proton scattering (for the energies of a few Mev or less which were used) showed the nuclear forces to be essentially short-ranged. These early scattering experiments have set the pattern for numerous investigations. Generally speaking, in such measurements one is not dealing, except in special circumstances, with a system of well-defined angular momenta—the incident beam comprises all components of momentum—so that the theoretical interpretation of such measurements lies rather outside the scope of the general principles referred to in the previous section.

In the early scattering experiments only limited precision and resolution were possible using α-particles from natural radioactive sources and the detecting instruments then current. Consequently only broad trends were observed rather than the detailed behaviour of particular nuclei or nuclear levels. Improved methods of detection, the development of the dispersion theory of nuclear resonances, and its application to the resonance scattering of charged, spinless particles and spinless nuclei, stimulated a few attempts at the investigation of individual nuclear levels by α-particle scattering; but the main developments in this field have been made by using highly canalised and monoenergetic beams of artificially accelerated particles, more refined methods of detection, and theoretical analysis applicable to nuclei and particles with spin.

Early studies of nuclear reactions were also limited by considerations of intensity and resolution, and in addition attention was focussed on dynamical features involving energy and linear momentum rather than differential cross-sections, particularly as no theoretical framework for the interpretation of these latter existed. The rapid development of machines for producing intense, canalised beams removed the experimental limitations, and the parallel theoretical developments indicated at least one basis, that of the compound-nucleus theory, for the interpretation of angular distributions. The first extensive treatment, in which the relevant quantum-mechanical features were displayed, was given by MYERS in 1938[1]. Later investigations emphasised the rôle of discrete resonances in simplifying the interpretation by reducing the number of factors introduced *ad hoc*, and in permitting the exploitation of the general quantum-mechanical principles[2]. A particularly fruitful application of the study of angular distribution has been to the study of nucleon-capture reactions, since these reactions are very frequently characterised by discrete resonances. In this application the limitations

[1] R. D. MYERS: Phys. Rev. **54**, 361 (1938).

[2] E. GERGUOY: Phys. Rev. **58**, 503 (1940). — C. L. CRITCHFIELD and E. TELLER: Phys. Rev. **60**, 10 (1941). — G. BREIT and B. T. DARLING: Phys. Rev. **71**, 402 (1947). — E. EISNER and R. SACHS: Phys. Rev. **72**, 680 (1947).

of early experiments[1] arose from the lack of energy-selectivity of the γ-detecting technique used, a limitation which has been largely removed by the development of scintillation-photomultiplier detectors of γ-radiation. The further developments of the theory, in particular the analysis of nuclear reactions by WIGNER and EISENBUD [*43*], and the analytical procedures of RACAH [*3*] and FANO [*8*] have resulted in a powerful and compact theoretical framework in which to interpret angular distributions. This and the parallel development of experimental technique, especially the increased detection efficiency and energy resolution of both particles and γ-radiations, have brought the study and interpretation of angular distributions from resonance reactions to a high degree of perfection.

The interpretation of the angular distributions of direct, as opposed to compound nucleus and resonance-type reactions, is a more recent development. The first comprehensive theory, based on the mechanism proposed by OPPENHEIMER and PHILLIPS[2] for low-energy, and SERBER[3] for high-energy processes, but taking into account the detailed properties of individual nuclei, was given by BUTLER[4] to explain the characteristic results observed in deuteron (d, p) and (d, n) reactions[5]. The first "stripping" experiments used the photographic emulsion technique, which by allowing simultaneous detection at all angles with moderate energy resolution helped to overcome the limited incident beam intensities available. Improvements in acceleration machines have made possible, and the application of the stripping technique to reactions producing a complex spectrum of product particles has made desirable, the use of more refined detection techniques, in particular magnetic-spectrometer analysis of the reaction products. The original theory of stripping has been refined and extended by many investigators, and the theory of the complementary processes ("pick-up") in which a nucleon is captured from, rather than by, the bombarded nucleus has been developed along similar lines. Altogether, the study of the differential cross-section in stripping reactions has been extensively exploited and provides a most powerful tool for the investigation of nuclear levels, particularly bound levels, whose properties are revealed only rather indirectly from a study of compound-nucleus type reactions.

The study of the angular correlations of successive nuclear radiations arose out of the application of the coincidence technique to the elucidation of β-γ cascades of radioactive nuclei. The possibility of utilising this correlation in the interpretation of the radiation processes was first mentioned by DUNWORTH (1940)[6] and the pioneer theoretical study was made by HAMILTON [*20*] (1940). HAMILTON used the language of perturbation theory to study γ-γ correlation processes for a restricted range of multipole emissions. Although this treatment was of limited applicability and did not bring out the full generality of the underlying principles, and in particular the close connection between the theory of angular correlations and of differential cross-sections, it did contain many of the salient features of the later theoretical work. One important aspect, to which attention was drawn, is the effect of extra-nuclear surroundings on the directional properties of emitted radiation. In the study of nuclear reactions this effect is generally negligible since the lifetime of the nuclear states involved is usually too short for

[1] S. DEVONS and M. G. N. HINE: Proc. Roy. Soc. Lond., Ser. A **199**, 56 (1949).
[2] J. R. OPPENHEIMER and M. PHILLIPS: Phys. Rev. **48**, 500 (1935).
[3] R. SERBER: Phys. Rev. **72**, 1008 (1947).
[4] S. T. BUTLER: Proc. Roy. Soc. Lond., Ser. A **208**, 559 (1951).
[5] A. C. HELMHOLZ, E. M. MACMILLAN and D. C. JEWELL: Phys. Rev. **72**, 1003 (1947) (High-energy stripping). — H. B. BURROWS, W. M. GIBSON and J. ROTBLAT: Phys. Rev. **80**, 1095 (1950) (Low-energy stripping).
[6] J. V. DUNWORTH: Rev. Sci. Instrum. **11**, 167 (1940).

the interaction of the nucleus with external fields to be appreciable. Detailed calculations of this interaction, particularly that between the nuclear magnetic moment and the electronic moment (characterised by the hyperfine structure of atomic spectra), were first made by GOERTZEL [*21*] (1946).

The generalisation of the theory of cascade radiations and the expression of the results in a form which could be readily applied to experimental results has been the work of numerous investigators (GARDNER [*23*], FIERZ [*24*], 1949; FALKOFF and UHLENBECK [*25*], 1950; LLOYD [*28*], 1950; SPIERS [*26*], 1950; RACAH [*27*], 1951; FANO [*8*], 1952; COESTER and JAUCH [*29*], 1953). In addition the theory has been extended to cover the treatment of multiple correlations of several radiations in cascade, and of cascades which include β-γ emission and emission of internally-converted electrons, as well as γ-radiation and nuclear particles (cf. BIEDENHARN and ROSE [*9*], 1953).

The incentive for this vigorous theoretical development has come from the remarkably rapid improvement in the technique of studying angular correlations. The probability of detection in "coincidence" of two radiations of a cascade, emitted randomly in time, depends on the product of the efficiencies of the two detectors. As long as relatively inefficient detectors, such as Geiger counters, were used the measurement of angular correlations was a tedious process. The use of scintillation counters has changed this situation dramatically. Not only has the measurement of an angular correlation been made less time-consuming, but the further possibility of using the energy resolution of the scintillation counter to select particular γ-transitions and other features to determine γ-ray polarisations (DEUTSCH, 1950)[1] has greatly widened the scope of these investigations. The present-day sophisticated experimental techniques involving short resolving-time (10^{-8} sec or less) coincidence counting, coupled with efficient scintillation counters and automatic pulse-size recording, make angular-correlation studies a powerful tool of wide scope for the investigation of the symmetry properties of low-lying nuclear states and of the radiations they emit. The theoretical interpretation of the results is greatly facilitated, indeed it is made trivially easy in many cases, by the compilation of many useful tables of algebraic functions of angular-momentum quantum numbers suitable for use with current theoretical formulae, a development which was foreshadowed by the early tabulations of CONDON and SHORTLEY (1935) [*2*] and the development of the appropriate quantum-mechanical algebra by RACAH (1942), [*3*], both in the investigation of atomic spectra.

The most recent development is the detailed study of the effects on angular correlations of extra-nuclear fields, both of atomic and molecular origin and externally applied macroscopic fields. A knowledge of the former type of disturbance has proved necessary for a reliable interpretation of correlation experiments whenever the lifetimes of any of the nuclear states (other than the initial one) involved in a cascade is sufficiently long. The theory of these interactions, insofar as the limited knowledge of the fields acting on a radiating nucleus in a solid or liquid permits, has been investigated particularly by ALDER (1952) [*60*], ABRAGAM and POUND (1953) [*61*], and COESTER (1954) [*63*]. For the purpose of determining the nuclear symmetry properties, it is desirable to eliminate the effects of external fields so that the formula developed for field-free nuclei are applicable. Many experimental techniques have been devised to this end. Alternatively, if the extra-nuclear fields are known, or can be estimated, and their effects on the angular correlations studied, then information about extra-nuclear

[1] E. L. BRADY and M. DEUTSCH: Phys. Rev. **78**, 558 (1950). — F. METZGER and M. DEUTSCH: Phys. Rev. **78**, 551 (1950).

coupling (e.g. magnetic dipole coupling with external magnetic fields or electric quadrupole coupling with the electric field gradient) can be obtained. In this way measurements have been made, particularly by the Zürich group of investigators[1], of magnetic moments of some short-lived (10^{-7} to 10^{-8} sec) nuclear states, and some estimate of nuclear quadrupole moments. The angular correlations observed in such cases are dependent on the time-interval between detection of successive radiations, provided this interval is not much longer than the natural lifetimes for emission of the radiations. Indeed, in the most refined measurements "coincidences" are recorded with a specific time-interval between successive radiations, the ideal being an uncertainty in time (i.e. resolving time) which is small compared with the lifetime for emission from the state whose properties are being studied. In a similar fashion, the effect of an external magnetic field on the angular distribution of γ-radiation produced by inelastic proton scattering has recently been studied, and the magnetic moments of excited states of light nuclei measured.

In summary the present situation is as follows: for processes involving effectively "isolated" nuclei in which all states have well-defined symmetry, a complete comprehensive theory exists which is readily applicable to experiment. Likewise the experimental methods for investigating such processes have reached a high degree of perfection. For processes involving states without well-defined symmetry, the interpretation may be complicated by the necessary introduction of *ad hoc* parameters; or in the case of "direct" processes the interpretation, albeit approximate, in terms of a particular simple mechanism, is sufficiently successful to provide a powerful means of investigation of nuclear states. For nuclei not isolated from their surroundings, theory is limited by incomplete knowledge of extra-nuclear fields, and experiments, which provide information about the electric and magnetic moments of excited states of nuclei, are limited to particularly favourable cases.

B. Mathematical preliminaries.

The theory of angular correlations is mainly concerned with the representation of quantum-mechanical systems in terms of eigenstates of angular-momentum operators, and with the transformations connecting different representations of this type. The theory of such transformations was first developed in connection with the theory of complex spectra [*2*], [*3*], where it has been found convenient to introduce explicit functions to describe the different schemes of representation. Unfortunately, the notation used for these functions is far from a universal one, and it will prove useful to define some of them in this preliminary section, and to introduce the notation that will be used throughout the article[2].

4. Clebsch-Gordan coefficients. Consider a system composed of two non-interacting parts, each of which is characterised by its total angular momentum and a component in some specified direction. We use Latin letters, $a, b, c, \ldots$, to denote total angular-momentum quantum numbers, and the corresponding Greek letters, $\alpha, \beta, \gamma, \ldots$, to denote their components.

We can represent the state of the combined system by the product wave-function $\psi(a\alpha r_a)\,\psi(b\beta r_b)$, where r_a, r_b, represent the eigenvalues of other operators

[1] H. AEPPLI, H. ALBERS-SCHÖNBERG, H. FRAUENFELDER and P. SCHERRER: Helv. phys. Acta **25**, 339 (1952).

[2] The reader is referred to a report CERN-55-26 by A. R. EDMONDS [*17*] "Angular Momentum in Quantum Mechanics", for a more detailed discussion of these definitions, together with a display of the numerous different notations used by different authors. A list of the notations used by us is given at the end of the article.

which commute with $a, \alpha; b, \beta$, respectively. Such product wavefunctions are degenerate with respect to α, β. We can construct from them a wavefunction $\psi(a\,b\,c\,\gamma\,r)$ corresponding to a state with well-defined total angular momentum $(c\,\gamma)$ with r equivalent to r_a, r_b, viz.,

$$\psi(a\,b\,c\,\gamma\,r) = \sum_{\alpha\beta} (a\,\alpha, b\,\beta | a\,b\,c\,\gamma)\, \psi(a\,\alpha\,r_a)\, \psi(b\,\beta\,r_b)\,, \tag{4.1}$$

or, using the Dirac notation,

$$|a\,b\,c\,\gamma\,r\rangle = \sum_{\alpha\beta} |a\,b\,\alpha\,\beta\,r_a r_b\rangle \langle a\,\alpha, b\,\beta | a\,b\,c\,\gamma\rangle\,. \tag{4.1a}$$

The coefficient, $\langle a\,\alpha, b\,\beta | a\,b\,c\,\gamma\rangle$, is called a vector-coupling or Clebsch-Gordan coefficient [1], and to avoid some repetition we shall use the notation $(a\alpha, b\beta | c\gamma)$. These coefficients are real and vanish unless the values of the quantum numbers are consistent with the conservation of angular momentum, i.e. unless $\alpha+\beta=\gamma$, and $|a-b| \leq c \leq a+b$, in integral steps[1].

The transformation (4.1) is unitary so that

$$\left.\begin{aligned} &\sum_{\alpha\beta} (a\,\alpha, b\,\beta | c\,\gamma)\, (a\,\alpha, b\,\beta | c'\,\gamma') = \delta_{c c'}\, \delta_{\gamma\gamma'}\,, \\ \text{and}\qquad &\sum_{c\gamma} (a\,\alpha, b\,\beta | c\,\gamma)\, (a\,\alpha', b\,\beta' | c\,\gamma) = \delta_{\alpha\alpha'}\, \delta_{\beta\beta'}\,. \end{aligned}\right\} \tag{4.2}$$

Thus we can also write

$$\psi(a\,\alpha\,r_a)\, \psi(b\,\beta\,r_b) = \sum_{c\gamma} \psi(a\,b\,c\,\gamma\,r)\, (a\,\alpha, b\,\beta | c\,\gamma)\,. \tag{4.3}$$

We consider now that the two separate parts of the system interact to form a state denoted by $\psi(c'\,\gamma'\,r')$. The separate angular momenta, $\boldsymbol{a}, \boldsymbol{b}$, may not be constants of motion in the new state. The probability of an initial state, denoted by (4.1), transforming by virtue of some mutual interaction, $\mathcal{O}$, into this new state will depend on matrix elements of the type $\left(\psi^*(c'\gamma' r'), \mathcal{O}\,\psi(a b c \gamma r)\right)$. These matrix elements will vanish unless $c=c', \gamma=\gamma'$, since the operator $\mathcal{O}$ is rotationally invariant[2].

Thus we see from (4.1), (4.3), that the vector-coupling coefficients determine the components of the original states $\psi(a\alpha r_a) \cdot \psi(b\beta r_b)$, which have the symmetry properties appropriate to the production of the state $\psi(c\,\gamma\,r')$. Thus the *relative* contributions of the different initial states, with component parts having different spatial symmetry denoted by the various values of α and β, to the final state $\psi(c\,\gamma\,r')$, are determined entirely by the vector-coupling coefficients which we can regard as having a purely geometrical significance. The absolute contributions are determined by the matrix elements of $\mathcal{O}$, and these represent the physical factors.

We could of course consider the reverse process in which a system denoted by $\psi(c\,\gamma\,r')$ breaks up into two non-interacting parts $\psi(a\,\alpha\,r_a) \cdot \psi(a\,\beta\,r_b)$, and the same considerations apply.

---

[1] This relation between a, b, c, is often referred to as the "triangle condition", and written $\Delta(abc)$. The condition implies that the lengths a, b, c, form a real triangle, including a straight line.

[2] The matrix element will also vanish unless the two wavefunctions $\psi(c'\gamma' r')$ and $\psi(c\,\gamma\,r)$ have the same parity, i.e. the same transformation properties with respect to reflection in the origin. This condition acts, of course, as a selection rule in restricting the combinations of the type (4.1) which result in non-vanishing matrix elements of any interaction operator of the type $\mathcal{O}$. It does not affect the relative contributions of different components, α, β, to the non-vanishing matrix elements.

We can also use the vector-coupling coefficients to transform a wavefunction, which represents two independent degrees of freedom, each characterised by a pair of angular-momentum quantum numbers, such as the orbital $(l\,\lambda)$ and spin $(s\,\sigma)$ motion of a particle, into wavefunctions characterised by total angular-momentum quantum numbers, viz.

$$\psi(l\,\lambda)\,\psi(s\,\sigma) = \sum_{L\Lambda} (l\,\lambda, s\,\sigma|L\Lambda)\,\psi(l\,s\,L\Lambda)\,. \tag{4.4}$$

In view of the rôle played by the vector-coupling coefficients in the combination of eigenstates of angular momentum, and the close quantum-mechanical relationship between angular-momentum quantum numbers and spatial symmetry properties, the importance of the vector-coupling coefficients (and the more elaborate coefficients described below) can hardly be over-emphasised.

The following are some relations between vector-coupling coefficients:

$$\left.\begin{aligned}(a\,\alpha, b\,\beta|c\,\gamma) &= (-)^{a+b-c}\,(b\,\beta, a\,\alpha|c\,\gamma)\,,\\ &= (-)^{a+b-c}\,(a-\alpha, b-\beta|c-\gamma)\,,\\ &= (-)^{a-\alpha}\,\frac{\hat{c}}{\hat{b}}\,(a\,\alpha, c-\gamma|b-\beta)\,,\\ &= (-)^{b+\beta}\,\frac{\hat{c}}{\hat{a}}\,(c-\gamma, b\,\beta|a-\alpha)\,,\end{aligned}\right\} \tag{4.5}$$

where we use the notation $\hat{a} \equiv \sqrt{2a+1}$, $\hat{b} \equiv \sqrt{2b+1}$, etc. Also

$$(a\,\alpha, 0\,0|c\,\gamma) = \delta_{a\,c}\,\delta_{\alpha\gamma}, \tag{4.6}$$

which, using (4.5) and (4.6), yields

$$(a\,\alpha, b\,\beta|0\,0) = \frac{(-)^{a-\alpha}}{\hat{a}}\,\delta_{a\,b}\,\delta_{\alpha-\beta}\,. \tag{4.7}$$

5. Racah coefficients. If we consider a system composed of three parts, characterised by $a\,\alpha$, $b\,\beta$ and $c\,\gamma$, we can regard the state characterised by $d\,\delta$ as having been obtained by a linear superposition of states corresponding to $a\,\alpha$ and $b\,\beta$, to yield a resultant, characterised by $e\,\varepsilon$, which is then combined with a state with $c\,\gamma$ to yield, finally, the required state with $d\delta$. This is expressed by

$$|a\,b(e)\,c\,d\,\delta\rangle = \sum_{\varepsilon\gamma\alpha\beta} |a\,\alpha\rangle\,|b\,\beta\rangle\,|c\,\gamma\rangle\,(a\,\alpha, b\,\beta|e\,\varepsilon)\,(e\,\varepsilon, c\,\gamma|d\,\delta)\,. \tag{5.1}$$

(We now omit the remaining quantum numbers r in our expressions since they are not pertinent to the present discussion.)

Alternatively, we can obtain a state with $d\,\delta$ by combining a state with $a\,\alpha$ with the linear superposition of states $b\,\beta$ and $c\,\gamma$, the latter combination yielding $f\,\varphi$, so that

$$|a\,b\,c(f)\,d\,\delta\rangle = \sum_{\alpha\beta\gamma\varphi} |a\,\alpha\rangle\,|b\,\beta\rangle\,|c\,\gamma\rangle\,(b\,\beta, c\,\gamma|f\,\varphi)\,(a\,\alpha, f\,\varphi|d\,\delta)\,. \tag{5.2}$$

There is, again, a unitary transformation connecting these two representations of the same system, given by

$$\langle a\,b(e)\,c\,d|a\,b\,c(f)\,d\rangle = \hat{e}\,\hat{f}\,W(a\,b\,d\,c; e\,f)\,, \tag{5.3}$$

where the W functions, which were first introduced by RACAH [3], are called Racah coefficients. Expressed in terms of the Clebsch-Gordan coefficients,

$$\left.\begin{aligned} &W(a\,b\,c\,d;e\,f)\,\delta_{cc'}\,\delta_{\gamma\gamma'} \\ &\quad=\frac{1}{\hat{e}\,\hat{f}}\sum_{\alpha\beta\delta\varepsilon\varphi}(a\,\alpha,b\,\beta|e\,\varepsilon)\,(e\,\varepsilon,d\,\delta|c\,\gamma)\cdot(b\,\beta,d\,\delta|f\,\varphi)\,(a\,\alpha,f\,\varphi|c'\,\gamma')\end{aligned}\right\}\tag{5.4}$$

or

$$\left.\begin{aligned} &(a\,\alpha,b\,\beta|e\,\alpha+\beta)\,(e\,\alpha+\beta,d\,\delta|c\,\alpha+\beta+\delta) \\ &\quad=\sum_f \hat{e}\,\hat{f}\,(b\,\beta,d\,\delta|f\,\beta+\delta)\cdot(a\,\alpha,f\,\beta+\delta|c\,\alpha+\beta+\delta)\,W(a\,b\,c\,d;e\,f).\end{aligned}\right\}\tag{5.5}$$

We refer to the transformation (5.3) as one between two different "coupling" schemes of angular momenta.

In the development of the theory, we shall need to calculate the sum of products of Clebsch-Gordan coefficients over all possible magnetic quantum numbers. In the early treatments of the subject, this procedure was found to be so formidable that the results were not presented in closed form. However, Eq. (5.5) and equivalent ones involving Racah coefficients by successive application allow the summations to reduce to very simple results. The triangular conditions for $W(abcd,ef)$, viz. $\Delta(abc)$, $\Delta(adc)$, $\Delta(bdf)$, and $\Delta(afc)$, are very useful in providing selection rules. In their consideration of angular distributions for scattering and reactions, BIEDENHARN, BLATT, and ROSE [6], [7], [9] introduced a related function

$$Z(a\,b\,c\,d;e\,f)=i^{f-a+c}\,\hat{a}\,\hat{b}\,\hat{c}\,\hat{d}\,(a\,0,c\,0|f\,0)\,W(a\,b\,c\,d;e\,f).\tag{5.6}$$

6. The 9-j symbol. It should be pointed out that WIGNER has defined a 3-j symbol which differs from the Clebsch-Gordan coefficient by a phase factor, but which exhibits a high degree of symmetry. Similarly, there was introduced the 6-j symbol, which depends on 6 angular momenta and differs by a phase factor from the Racah coefficient [17].

If we now consider the connection between two schemes of coupling four angular momenta (e.g. from $\boldsymbol{L}\cdot\boldsymbol{S}$ coupling to $\boldsymbol{J}\cdot\boldsymbol{J}$ coupling), this is effected by a unitary transformation

$$\langle a\,b\,(c),a'\,b'\,(c')\,d|a\,a'\,(e),b\,b'\,(f)\,d\rangle=\hat{c}\,\hat{c}'\,\hat{e}\,\hat{f}\begin{Bmatrix}a & a' & e\\ b & b' & f\\ c & c' & d\end{Bmatrix}.\tag{6.1}$$

The symbol in parentheses on the right-hand side depends on nine angular momenta and is called the Wigner 9-j symbol. It is also referred to as the Fano function and is related to the χ-function of HOPE [12] [χ is equal to the entire expression on the right-hand side of (6.1)] and to the S function of SCHWINGER[1] (S differs by a phase factor from the 9-j symbol). The 9-j symbol is highly symmetrical, being invariant to an interchange of rows and columns and to even permutations of rows and/or columns. A single change of two rows or columns multiplies the symbol by $(-)^{\sigma}$, where $\sigma=a+b+c+a'+b'+c'+e+f+d$. Expressed in terms of Clebsch-Gordan coefficients and Racah coefficients, (6.1)

[1] J. SCHWINGER: On Angular Momentum, Report N.Y.O. 3071. White Plains, New York: Nuclear Development Associates Inc. 1952.

becomes

$$\left.\begin{aligned}&\hat{e}\hat{f}\hat{c}\hat{c}'\begin{Bmatrix} a & a' & e\\ b & b' & f\\ c & c' & d\end{Bmatrix}\delta_{d d'}\\ &\quad=\sum_{\substack{\alpha\beta\alpha'\beta'\\ \gamma\gamma'\varepsilon\varphi}}(a\,\alpha, b\,\beta|c\,\gamma)\,(a'\,\alpha', b'\,\beta'|c'\,\gamma')\cdot(-)^{a'-\alpha'}(a\,\alpha, a'-\alpha'|e\,\varepsilon)\times\\ &\qquad\times(-)^{b'-\beta'}(b\,\beta, b'-\beta'|f\,\varphi)\,(-)^{c'-\gamma'}(c\,\gamma, c'-\gamma'|d\,\delta)\cdot(e\,\varepsilon, f\,\varphi|d'\,\delta)\end{aligned}\right\}\tag{6.2}$$

$$=\hat{e}\hat{f}\hat{c}\hat{c}'\sum_k(2k+1)\,W(a\,a'\,d\,f; c\,k)\,W(b\,f\,c'\,a'; b'\,k)\,W(a\,b\,d\,c'; c\,k). \tag{6.3}$$

Other useful relations are given in the Appendix[1].

7. Rotation matrices. We shall also need the transformation properties of eigenfunctions of angular momentum under the rotation of the coordinate axes of representation. Such eigenfunctions, associated with angular momenta $a\alpha$, are transformed into new eigenfunctions with the same value a, thus:

$$\psi_a^{\alpha'}(\vartheta'\,\varphi')=\sum_\alpha\psi_a^\alpha(\vartheta\,\varphi)\,\mathscr{D}^a_{\alpha\alpha'}(\vartheta_1\vartheta_2\vartheta_3); \tag{7.1}$$

or, in Dirac notation,

$$|a\,\alpha'\rangle=\sum_\alpha|a\,\alpha\rangle\langle\alpha|\,\mathscr{D}^a(\mathscr{R})\,|\alpha'\rangle, \tag{7.1 a}$$

where $\mathscr{D}^a_{\alpha\alpha'}$ are the elements of a unitary matrix in the α-representation, usually called the rotation matrix, carrying the coordinate system, yielding quantum numbers, $a\alpha$, into a new coordinate system with quantum numbers, $a\alpha'$, by means of the rotation $\mathscr{R}=(\vartheta_1\vartheta_2\vartheta_3)$, and (ϑ,φ), (ϑ',φ') are the coordinates of a point with respect to each coordinate system.

The angles $\vartheta_1,\vartheta_2,\vartheta_3$, are Euler angles, which are convenient parameters, describing an arbitrary displacement about the origin. This is effected by means of a rotation, ϑ_1, about the z-axis of the initial coordinate system, followed, successively, by a rotation, ϑ_2, about the y-axis of the coordinate system resulting from the first rotation and a rotation, ϑ_3, about the new z-axis[2].

Since $\mathscr{D}^a(\mathscr{R})$ is unitary,

$$\sum_\alpha|\mathscr{D}^a_{\alpha\alpha'}|^2=1, \tag{7.2}$$

and

$$\int\mathscr{D}^{a\,*}_{\alpha\beta}(\mathscr{R})\,\mathscr{D}^{a'}_{\alpha'\beta'}(\mathscr{R})\sin\vartheta_2\,d\vartheta_1\,d\vartheta_2\,d\vartheta_3=\frac{8\pi^2}{\hat{a}^2}\,\delta_{a\,a'}\,\delta_{\alpha\alpha'}\,\delta_{\beta\beta'}. \tag{7.3}$$

It follows that

$$\psi_a^\alpha(\vartheta,\varphi)=\sum_{\alpha'}\psi_a^{\alpha'}(\vartheta',\varphi')\,\mathscr{D}^{a\,*}_{\alpha\alpha'}(\vartheta_1\vartheta_2\vartheta_3). \tag{7.4}$$

In the terminology of group theory, the eigenfunctions ψ_a^α transform according to the $2a+1$-dimensional irreducible representation, $\mathscr{D}^a$, of the spatial (three-dimensional) rotation group, the transformation being depicted by (7.1). If a set of $2L+1$ functions transform under rotations according to $\mathscr{D}^L$ these functions constitute the components of an irreducible tensor of rank L. One such set of tensors is represented by the spherical harmonics $Y_l^\lambda(\vartheta\,\varphi)$, here defined according

[1] Cf. also Appendix 2 and 3 of the article by J. P. ELLIOTT and A. M. LANE in Vol. XXXIX of this Encyclopedia.

[2] We use a right-handed coordinate system and assume all rotations to be positive rotations. This differs from the conventions of WIGNER [1], while the order of the Euler angles differs from that used by ROSE [18].

to the convention of CONDON and SHORTLEY [2], $Y_l^\lambda = (-)^\lambda Y_l^{-\lambda *}$. Using the group-theoretical properties of $\mathcal{D}^a_{\alpha\alpha'}$, we have

$$\sum_{\alpha''} \mathcal{D}^a_{\alpha\alpha''}(\mathcal{R}_1)\, \mathcal{D}^a_{\alpha''\alpha'}(\mathcal{R}_2) = \mathcal{D}^a_{\alpha\alpha'}(\mathcal{R}_1 \mathcal{R}_2), \tag{7.5}$$

and

$$\mathcal{D}^a_{\alpha\alpha'}(\mathcal{R}) = \mathcal{D}^{a*}_{\alpha'\alpha}(\mathcal{R}^{-1}), \tag{7.6}$$

where $\mathcal{R}_1 \mathcal{R}_2$ represents the product of two successive rotations $\mathcal{R}_1$ and $\mathcal{R}_2$, and $\mathcal{R}^{-1}$ represents the inverse rotation to $\mathcal{R}$, so that $\mathcal{R}^{-1} = (-\vartheta_3 - \vartheta_2 - \vartheta_1)$ if $\mathcal{R} = (\vartheta_1 \vartheta_2 \vartheta_3)$. Also

$$\mathcal{D}^{a*}_{\alpha\alpha'}(\mathcal{R}) = (-)^{\alpha-\alpha'}\, \mathcal{D}^a_{-\alpha-\alpha'}(\mathcal{R}). \tag{7.7}$$

Applying (7.1) to the inverse of (7.7),

$$\sum_{\alpha'\beta'} \mathcal{D}^a_{\alpha'\alpha}\, \mathcal{D}^b_{\beta'\beta}\, \psi_a^{\alpha'} \psi_b^{\beta'} = \sum_c (a\,\alpha, b\,\beta | c\,\gamma) \sum_{\gamma'} \mathcal{D}^c_{\gamma\gamma'}\, \psi_c^{\gamma'},$$

leading to a relation for the product of two rotation matrix elements,

$$\mathcal{D}^a_{\alpha'\alpha}\, \mathcal{D}^b_{\beta'\beta} = \sum_{c\gamma'\gamma} (a\,\alpha', b\,\beta' | c\,\gamma')\, (a\,\alpha, b\,\beta | c\,\gamma)\, \mathcal{D}^c_{\gamma'\gamma}. \tag{7.8}$$

The rotation matrix reduces simply for α and/or $\alpha' = 0$ to

$$\left.\begin{aligned}
\mathcal{D}^a_{\alpha 0}(\vartheta_1 \vartheta_2 \vartheta_3) &= \frac{\sqrt{4\pi}}{\hat{a}}\, Y_a^{\alpha *}(\vartheta_2 \vartheta_1), \\
\mathcal{D}^a_{\alpha 0}(0, \vartheta_2 \vartheta_3) &= \sqrt{\frac{(a-|\alpha|)!}{(a+|\alpha|)!}}\, (-)^{\alpha\,\delta_{\alpha,|\alpha|}}\, P_a^{|\alpha|}(\cos\vartheta_2), \\
\mathcal{D}^a_{00}(\vartheta_1 \vartheta_2 \vartheta_3) &= P_a(\cos\vartheta_2),
\end{aligned}\right\} \tag{7.9}$$

where $P_a^\alpha(\cos\vartheta_2)$ and $P_a(\cos\vartheta_2)$ represent the associated functions of the first kind of Legendre and the Legendre polynomials, respectively, and are related by

$$P_a^\alpha(x) = (1 - x^2)^{\alpha/2} \frac{d^\alpha}{dx^\alpha}\left[P_a(x)\right], \tag{7.10}$$

where $x = \cos\vartheta_2$ and $\alpha \geqq 0$.

8. Density matrices. In the study of angular correlations, we deal not with individual systems which are fully defined quantum-mechanically, but with an assembly of systems, about which we have only limited information which is not sufficient to provide a complete quantum-mechanical description but rather a statistically significant one. For such systems an appropriate formalism is that of density matrices.

We consider an assembly of N similar systems, the state of one such system, the n-th, being described by the wavefunction ψ_n. $u(q)$ represents a complete ortho-normal set of wavefunctions, so that we can write

$$\psi_n = \sum_q a_n(q)\, u(q). \tag{8.1}$$

The density matrix form is defined as[1],

$$\left.\begin{aligned}
\varrho_{qq'} &= \frac{1}{N} \sum_{n=1}^{N} a_n^*(q')\, a_n(q), \\
&= \overline{\overline{a_n^*(q')\, a_n(q)}},
\end{aligned}\right\} \tag{8.2}$$

[1] R. C. TOLMAN: Principles of Statistical Mechanics. Oxford 1938.

where the double bar indicates averaging over the whole assembly[1]. Equivalently, (DIRAC[2]), if we denote by P_n the proportion of the systems of the assembly in the pure state ψ_n, etc., then we can write

$$\varrho_{qq'} = \frac{1}{\sum\limits_n P_n} \sum_{n=1}^{r} P_n a_n^*(q')\, a_n(q);$$

or, normalising so that

$$\left.\begin{aligned} &\sum_{n=1}^{r} P_n = 1, \\ &\varrho_{qq'} = \sum_{n=1}^{r} P_n \{a_n^*(q')\, a_n(q)\}. \end{aligned}\right\} \tag{8.3}$$

Here r denotes the number of *different* states in the assembly. $\varrho_{qq'}$ is clearly a Hermitian matrix, i.e. $\varrho_{qq'}^* = \varrho_{q'q}$. We notice also that the occurrence of the real probability factors, P_n, outside the { } in (8.3), which contains only terms with the same n, means that $\varrho_{qq'}$ is independent of the relative phases of the wave-functions ψ_n in the states labelled by different n, since these phases cancel in the products $a_n^*(q')\, a_n(q)$. Thus, physically, the description in terms of the density matrix (8.3) corresponds to the incoherence of the *different* states, ψ_n, of the assembly.

If we use the special representation in which the basic states coincide with the actual states in the assembly, so that

$$\psi_n = \sum_q u(q)\, \delta_{q q_n},$$

then, corresponding to (8.3), we have

$$\varrho_{qq'} = \sum_{n=1}^{r} P_n\, \delta_{q q_n} \delta_{q' q_n}, \tag{8.4}$$

i.e. $\varrho_{qq'}$ is a diagonal matrix with elements $P_1, P_2, \ldots P_n, \ldots P_r$. For the particular case of all members of the assembly in the same pure state, ψ_{q_n}, then the matrix reduces to a single diagonal term

$$\varrho_{qq'} = \delta_{q q_n} \delta_{q' q_n}. \tag{8.5}$$

The density matrix provides all the significant information about the physical properties of the assembly. The average value of any observable, represented by F, can readily be shown to be given by:

$$\overline{\overline{F}} = \mathrm{Tr}(F_\varrho) = \mathrm{Tr}(F_\varrho).$$

In particular, the probability that any member of the assembly is in a given state u_q is given by

$$\sum_{n=1}^{r} P_n a_n^*(q)\, a_n(q) = \varrho_{qq}. \tag{8.6}$$

We notice also from (8.4) that for *any* assembly $\mathrm{Tr}(\varrho) = 1$, and that for an assembly of identical states ("pure" state) also $\mathrm{Tr}(\varrho^2) = 1$. For a mixed assembly $\mathrm{Tr}(\varrho^2) \leq 1$. A pure state corresponds to complete specification, i.e. maximum information, whereas minimum information is available when the systems are

---

[1] We shall also frequently use the Dirac notation for density matrices: $\langle q | \varrho | q' \rangle$.

[2] P. A. M. DIRAC: Principles of Quantum Mechanics, p. 139. Oxford 1935.

uniformly spread over all the possible states, and then, in the representation (8.4),

$$\varrho_{q q'} = \frac{\delta_{q q'}}{r} \tag{8.7}$$

and

$$\mathrm{Tr}\,(\varrho^2) = \frac{1}{r}. \tag{8.8}$$

The extent to which the assembly is not random can be measured (FANO [8]) by the quantity:

$$\mathscr{P} = \left[\mathrm{Tr}\,(\varrho^2) - \frac{1}{r}\right] \frac{r}{r-1} \tag{8.9}$$

which has a maximum of unity for a "pure" state and a minimum of zero for uniform distribution over all available states.

For systems whose wavefunctions can be factorised, the density matrix can likewise be compounded from separate factors.

In a manner quite analogous to the use of the density matrix, an efficiency matrix can be employed to describe the properties of a detection equipment whose response is not sufficient to ensure that the system studied is in a pure quantum-mechanical state. The efficiency matrix, like the density matrix, refers to an assembly of such pure states.

More detailed properties of efficiency and density matrices will be given later in relation to particular applications.

C. General theory of directional correlation for processes involving states of well-defined symmetry.

I. Formalism for the basic correlation process involving two successive radiations.

We shall deal in this Chap. C, with one of the main classes of problems in angular correlation theory, and one for which the general theory requires the minimum of supplementation. This class comprises the directional-correlation of two radiations which can be considered to be emitted, or absorbed, in succession, and in the process the nuclear states involved at each stage have well-defined spin and parity. We shall also assume, in this chapter, that extra-nuclear perturbations are negligible, and that radiations are unpolarised, or alternatively the detectors of the radiations are not sensitive to the state of polarisation of the radiations. This is the problem originally dealt with by HAMILTON [20], who treated the special case of successive emission of γ-radiation. We shall not restrict ourselves here to a particular type of radiation and shall consider both emission and absorption. It should be pointed out that the restriction to states of well-defined symmetry is not made on the grounds of formal simplicity, since very simple generalisations, involving only slight formal changes, are required in order to treat the more general problem. From a physical point of view, however, the more general processes include a much wider range of phenomena in which, also, the significance of the general theory varies greatly.

Three theoretical procedures will be presented in order of increasing sophistication. The first is typical of those early treatments which made use of simple quantum-mechanical properties of angular momentum without explicit treatment of the dynamical processes. States of the system are represented directly by wavefunctions appropriate to pure quantum-mechanical states, and the averaging over all the unobserved or unspecified variables is performed explicitly. Such a treatment is, physically, quite transparent. This is followed by a more generalised treatment, which was introduced by RACAH [27], and is similar to the contemperaneous treatments of LLOYD [28] and ALDER [60]. RACAH's starting point, which is an expansion of the Hamiltonian in terms of a product of terms depending separately on the nucleus and the states of the radiation, was previously used by FALKOFF and UHLENBECK [25]. The essential point of his treatment is the reduction of the problem to the determination of "radiation parameters", which completely characterise the radiation whatever its type. The final treatment (FANO [8], COESTER and JAUCH [29]), deals with the problem in a more elegant fashion, using the concepts of statistical and efficiency tensors. It is distinguished by a direct attack on the problem, by means of (i) density matrices which serve to describe the assemblies of states which are associated with the nuclei and radiations, and (ii) statistical tensors, constructed from the density matrices, which are used to represent the states of the system or its parts, and which have convenient rotational transformation properties. It is this procedure which will be used in subsequent sections of this article. The first three sections are devoted to the two-stage cascade decay problem using the above procedures. Following this, the formalism is extended to include the related problem of absorption of radiation followed by emission, and the emission or absorption in succession of more than two radiations.

9. The wavefunction formalism. Our designation of states, which will be used throughout, is a follows: successive nuclear states in a cascade, or in other transformations, are characterised by their angular momentum quantum numbers, $a\,\alpha$, $b\,\beta$, $c\,\gamma$, ..., respectively. The radiation accompanying the transitions will similarly be designated by $L_1 \Lambda_1$, $L_2 \Lambda_2$, $L_3 \Lambda_3$, ..., etc. If the radiation properties include intrinsic spin, the states corresponding to $L\Lambda$ may further be characterised by the quantum numbers $l\lambda$ and $s\sigma$, which represent the orbital momentum and spin respectively[1]. For compactness we shall frequently denote the set of quantum numbers L, l, s, by the single letter $\mathscr{L}$.

Our procedure follows that of SPIERS [5], [26] and GARDNER [23]. Each radiation is described by wavefunctions of the type

$$\Phi^{\Lambda}_{Lls}(r\,\Omega) \equiv \sum_{\sigma} \Phi^{\Lambda}_{Lls}(r\,\Omega\,\sigma)\,\chi^{\sigma}_{s} = \sum_{\sigma} a_{\sigma}\,f_{l}(r)\,(l\Lambda-\sigma, s\,\sigma\,|\,L\Lambda)\,Y^{\Lambda-\sigma}_{l}(\Omega)\,\chi^{\sigma}_{s}. \tag{9.1}$$

The wavefunction $\Phi^{\Lambda}_{Lls}(r\,\Omega)$ is obtained by a linear superposition of spin states corresponding to $s\,\sigma$ and orbital angular momentum states, $l\lambda\,(\lambda=\Lambda-\sigma)$. a_{σ} designates the polarisation, and $f_{l}(r)$ is the radial function for the outgoing radiation into the direction $\Omega \equiv (\vartheta, \varphi)$.

The first stage of decay is depicted by

$$\boldsymbol{a} = \boldsymbol{b} + \boldsymbol{L}_1. \tag{9.2}$$

We introduce the wavefunction ψ^{α}_{a} which describes the system before decay. After decay, the *overall* system is still characterised by angular-momentum quantum numbers, a, α; but it is now composed of two parts, the radiation with states represented by $\Phi^{\Lambda}_{L_1 l_1 s_1}(r_1\,\Omega_1)$, and the residual nucleus represented by ψ^{β}_{b} These states are coupled together as in (4.1),

$$\psi^{\alpha}_{a} = \sum_{L_1 l_1} B_{b L_1 l_1} \sum_{\beta \Lambda_1} (a\,\alpha\,|\,b\,\beta, L_1\,\Lambda_1)\,\psi^{\beta}_{b}\,\Phi^{\Lambda_1}_{L_1 l_1 s_1}(r_1\,\Omega). \tag{9.3}$$

[1] For γ-radiation the separation of spin and orbital motion is not possible. In this case we may still use the letter l, but to denote the parity of the radiation only, i.e. to distinguish between electric and magnetic multipoles of the same order L.

The quantity $B_{bL_1l_1}$ is related to the probability amplitude for the process of decay. Similarly, we can describe the system of nucleus of angular momentum c plus the second radiation represented by L_2, in terms of ψ_b^β, so that the wavefunction for the whole system is

$$\left.\begin{aligned}\psi_a^\alpha = \sum_{l_1 l_2 L_1 L_2} B_{bL_1l_1} B_{cL_2l_2} \sum_{\beta\gamma\Lambda_1\Lambda_2} (a\,\alpha|\,b\,\beta,\,L_1\,\Lambda_1)\,(b\,\beta|\,c\,\gamma,\,L_2\,\Lambda_2)\times \\ \times\,\Phi_{L_1l_1s_1}^{\Lambda_1}(r_1\,\Omega_1)\,\Phi_{L_2l_2s_2}^{\Lambda_2}(r_2\,\Omega_2)\,\psi_c^\gamma(\xi)\,.\end{aligned}\right\}\tag{9.4}$$

Introducing

$$\mathscr{F}_{\alpha\beta}(r_1\,\Omega_1\,\sigma_1) \equiv \sum_{L_1 l_1} B_{bL_1l_1}(a\,\alpha|\,b\,\beta,\,L_1\,\Lambda_1)\,\Phi_{L_1l_1s_1}^{\Lambda_1}(r_1\,\Omega_1\,\sigma_1)\tag{9.5}$$

with a similar expression for $\mathscr{F}_{\beta\gamma}$, we can rewrite the expression for the wavefunction describing the system after emission of the two radiations as

$$\psi_a^\alpha(\text{final}) = \sum_{\beta\gamma\Lambda_1\Lambda_2\sigma_1\sigma_2} \mathscr{F}_{\alpha\beta}(r_1\,\Omega_1\,\sigma_1)\,\mathscr{F}_{\beta\gamma}(r_2\,\Omega_2\,\sigma_2)\,\psi_c^\gamma(\xi)\,\chi_{s_1}^{\sigma_1}\,\chi_{s_2}^{\sigma_2}\,.\tag{9.6}$$

The probability that the two radiations are emitted into the directions defined by Ω_1 and Ω_2 is obtained by summing or integrating over all the unobserved coordinates, ξ, σ_1 and σ_2, and since no particular substate, α, of the initial system is singled out in an experiment, by averaging over α. The correlation function for the two radiations is therefore

$$W(\Omega_1\,\Omega_2) = \frac{1}{\hat{a}^2} \sum_{\alpha\sigma_1\sigma_2\gamma} \left|\sum_\beta \mathscr{F}_{\alpha\beta}(r_1\,\Omega_1\,\sigma_1)\,\mathscr{F}_{\beta\gamma}(r_2\,\Omega_2\,\sigma_2)\right|^2,\tag{9.7}$$

where $\hat{a}^2 = (2a+1)$. The summation over Λ_1 and Λ_2 is unnecessary since $\beta+\Lambda_1=\alpha$, and $\gamma+\Lambda_2=\beta$, according to the properties of Clebsch-Gordan coefficients. Also the orthogonality of the nuclear wavefunctions and the spin function allows the summation over γ, σ_1 and σ_2, to be taken outside the square-modulus.

The angular correlation is now expressed in terms of functions whose angular dependence is given in (9.1). Since the summation of β is inside the square-modulus in (9.7), the correlation function contains cross-terms of the type:

$$\mathscr{F}_{\alpha\beta}(r_1\,\Omega_1\,\sigma_1)\,\mathscr{F}_{\beta\gamma}(r_2\,\Omega_2\,\sigma_2)\,\mathscr{F}_{\alpha\beta'}^*(r_1\,\Omega_1\,\sigma_1)\,\mathscr{F}_{\beta'\gamma}^*(r_2\,\Omega_2\,\sigma_2)\tag{9.7a}$$

as well as terms of the type:

$$|\mathscr{F}_{\alpha\beta}(r_1\,\Omega_1\,\sigma_1)|^2\,|\mathscr{F}_{\beta\gamma}(r_2\,\Omega_2\,\sigma_2)|^2.\tag{9.7b}$$

That is to say the correlation function cannot be factorised into terms which indicate separately the contribution from the substates β, β', etc. This interference between the (coherent) contributions from the different substates $\beta, \beta', \ldots$, can be eliminated by choosing as the axis of quantisation the direction of emission of one of two radiations, let us say the first. Thus, referring to (9.1) and noting that $Y_l^{\Lambda-\sigma}(0)=0$, unless $\Lambda-\sigma=0$, we see that $\beta=\alpha-\Lambda_1=\alpha-\sigma_1$. For a given value of σ_1 there is only one non-vanishing value of β. The angular correlation depends only on the *relative* directions of the two radiations and is given by[1]

$$W(0,\Omega_2) = \frac{1}{\hat{a}^2} \sum_{\alpha\beta\gamma} \left\{\left(\sum_{\sigma_1} |\mathscr{F}_{\alpha\beta}(r_1\,0\,\sigma_1)|^2\right)\left(\sum_{\sigma_2} |\mathscr{F}_{\beta\gamma}(r_2\,\Omega_2\,\sigma_2)|^2\right)\right\},\tag{9.8}$$

[1] Although β and the rest of the variables are not independent, it is convenient to retain, formally, all the summations in (9.8) and (9.9).

or
$$W(\vartheta) = \sum_{\alpha\beta\gamma} P_{\alpha\beta}(0)\, P_{\beta\gamma}(\vartheta), \tag{9.9}$$

where $P_{\alpha\beta}(0)$ is the relative probability of emission of the first radiation, with α, β referred to an axis of quantisation in the direction of emission, and $P_{\beta\gamma}(\vartheta)$ is the relative probability of emission of the second radiation per solid angle, $d\Omega$, at an angle, ϑ, with respect to the first direction. Since we are interested in the angular distribution, we ignore all multiplicative factors in (9.9). The result (9.9) is, of course, independent of which radiation is chosen to define the axis of quantisation, but depends only on the angle between the two radiations.

The absence of interference terms of the type (9.7a) from the directional-correlation function, when the axis of quantisation is taken along the direction of one (the first) of the radiations, has a straightforward physical interpretation. Since the initial nucleus is randomly oriented and the polarisation of the radiation is not observed, the "spin" states $(a\alpha)$, $(s_1\sigma_1)$, are incoherent. Also, by the choice of axis, $\Lambda_1 = 0$ and $\beta = \alpha - \sigma_1$, so that the intermediate states corresponding to different values of β are also incoherent. Alternatively we can regard the observation of the direction only, Ω_1, of the first radiation as resulting in an assembly which has symmetry with respect to rotation about this direction Ω_1. An assembly composed of independent (incoherent) states with different β will be rotationally invariant. A rotation will change the factor $e^{i\beta\varphi}$ by a phase only and therefore leave invariant quadratic products with the same β, but not those with different values of β.

The above treatment can apply to γ-radiation by use of the electromagnetic vector potential, which serves as a wavefunction for a spin-one particle[1]. Both electric and magnetic multipoles are described by (9.1); electric multipoles correspond to parity $(-)^L$ and $l = L \pm 1$, while magnetic multipoles correspond to parity $(-)^{L+1}$ and $l = L$, Eq. (9.9) corresponds to the results obtained by HAMILTON [*20*], FALKOFF and UHLENBECK [*25*], and others.

In the early stages of development of the theory, solutions of (9.9) were usually presented in the form
$$W(\vartheta) = \sum_k a_{2k} \cos^{2k}\vartheta, \tag{9.10}$$

where k is an integer with $k \leq L_1, L_2$, or b. The evaluation of a_{2k} required laborious summation over magnetic quantum numbers of products of Clebsch-Gordan coefficients, and explicit formulae were calculated for various values of L.

A substantial simplification of the calculations was made by GARDNER [*23*], who utilised the transformations involving Racah coefficients to present results in a closed form, valid for all L. We will illustrate the procedure for the simple case of spinless particles (e.g. alpha-particle emission). In this case, $\Phi^{\Lambda}_{Lls} \sim Y^{\lambda}_l(\Omega)$ and (9.7) becomes
$$W(\vartheta) \sim \sum_{\alpha\gamma\lambda_2} |(a\,\alpha|b\,\alpha, l_1\,0)\,(b\,\alpha|c\,\gamma, l_2\,\lambda_2)\, Y^{\lambda_2}_{l_2}(\vartheta)|^2, \tag{9.11}$$

where the sum over λ_2 is equivalent to summation over β, since $\beta = \gamma + \lambda_2$.

By (7.8) to (7.10), we find
$$|Y^{\lambda_2}_{l_2}(\vartheta)|^2 = \sum_k \frac{(-)^{\lambda_2}\hat{l}_2^2}{4\pi} (l_2\,0, l_2\,0|k\,0)\,(l_2 - \lambda_2, l_2\,\lambda_2|k\,0)\, P_k(\cos\vartheta), \tag{9.12}$$

so that
$$W(\vartheta) \sim \sum_k A_k\, P_k(\cos\vartheta) \tag{9.13}$$

[1] W. HEITLER: Proc. Cambridge Phil. Soc. **37**, 112 (1936).

where

$$A_k = (l_2 0, l_2 0 | k 0) \sum_{\alpha \gamma \lambda_2} (-)^{\lambda} (a\alpha | b\alpha, l_1 0)^2 (b\alpha | c\gamma, l_2 \lambda_2)^2 (l_2 - \lambda_2 l_2 \lambda_2 | k 0). \qquad (9.14)$$

The sum over γ, λ_2 of the product of the last three Clebsch-Gordan coefficients is proportional to a Racah coefficient, using (A. *11*). Thus,

$$\sum_{\gamma \Lambda_2} (b\alpha | c\gamma, l_2 \lambda_2)^2 (l_2 - \lambda_2, l_2 \lambda_2 | k 0) = (-)^{c-b} \hat{b} \hat{k} (b\alpha, k 0 | b\alpha) W(b b l_2 l_2; k c). \quad (9.15)$$

A similar summation over α of the remaining set of products of three Clebsch-Gordan coefficients gives

$$A_k = \hat{b}^2 W(b b l_1 l_1; k a) W(b b l_2 l_2; k c) (l_1 0, l_1 0 | k 0) (l_2 0, l_2 0 | k 0). \qquad (9.16)$$

Results similar to (9.16) are obtained for other types of radiation, such as γ-rays or particles with spin. The complexity involved in the summation over magnetic quantum numbers now resides completely in the Racah coefficients, for which algebraic tables are available. For large values of the parameters in W, the results become sufficiently elaborate to require numerical tabulation. The complete absence of any dependence of the final result on magnetic quantum numbers is natural for observations of a process which has no overall preferred direction. If the detectors are sensitive to polarisation in some particular direction, the wavefunction for the radiation has a more complicated angular dependence than given in (9.1), so that there will be interference between the transitions, with Eq. (9.9) no longer valid. The simplification introduced in the choice of quantisation axis is, in fact, a restrictive one for the more general type of problem. This led to a more formal procedure introduced independently by RACAH, ALDER, and LLOYD. Instead of choosing the axis of quantisation along the propagation direction of one of the radiations, the radiations are treated on a more symmetrical basis. The eigenfunctions for the radiations, which are expressed in terms of an arbitrary coordinate system, are transformed into eigenfunctions quantised along their specific propagation directions by means of the rotation matrix. This leads to an angular correlation function which is expressed in terms of the representation, $\mathscr{D}^L$, and which reduces to $P_l(\cos\vartheta)$ for the special case of direction-direction correlations. The method is characterised by the use of certain quantities to describe the radiation, which are called parameters of radiation. The main points will be sketched below.

10. The parameters of radiation. Again we consider the problem of a two-stage cascade decay. HAMILTON, using perturbation theory, showed that, for emission of two gamma rays, the angular correlations is of the form

$$W(\Omega_1 \Omega_2) = \mathscr{S} \sum_{\alpha\gamma} \Big| \sum_{b\beta} \langle c\gamma | H_2 | b\beta \rangle \langle b\beta | H_1 | a\alpha \rangle \Big|^2, \qquad (10.1)$$

which, for sharp b, reduces to

$$\left. \begin{aligned} \mathscr{S} \sum_{\alpha\beta\beta'\gamma} &\langle b\beta \Omega_1 \sigma_1 | H_1 | a\alpha \rangle \langle b\beta' \Omega_1 \sigma_1 | H_1 | a\alpha \rangle^* \times \\ &\times \langle c\gamma \Omega_2 \sigma_2 | H_2 | b\beta \rangle \langle c\gamma \Omega_2 \sigma_2 | H_2 | b\beta' \rangle^*, \end{aligned} \right\} \qquad (10.2)$$

where the summation $\mathscr{S}$ is over all the non-observed properties of the radiation. Multiplicative constants are ignored in the formula for W. H_1 and H_2 represent the interaction Hamiltonians for emission of radiation, and are Hermitian. This formula can be extended to all types of radiation, and to absorption as well as emission. For particles we cannot justifiably use perturbation procedures, but can achieve a similar result by replacing the interaction Hamiltonian by a suitable interaction operator $\mathcal{O}$, and the probability for transition from state A (nucleus + radiation) to state B during a time δt varies as $\delta t |\langle A | \mathcal{O} | B \rangle|^2$. $\mathcal{O}$ differs from H in that its matrix is not necessarily Hermitian.

The matrix elements in (10.1) can be arranged so that W is the product of two matrices,

$$W(\Omega_1 \Omega_2) = \sum_{\beta\beta'} E^{(1)}(\beta\beta')\, E^{(2)}(\beta'\beta), \tag{10.3}$$

where

$$\left.\begin{aligned} E^{(1)}(\beta\beta') &= \mathcal{S}_1 \sum_{\alpha} \langle b\beta\,\Omega_1\sigma_1 | H_1 | a\alpha\rangle \langle b\beta'\,\Omega_1\sigma_1 | H_1 | a\alpha\rangle^* \\ \text{and} \qquad E^{(2)}(\beta'\beta) &= \mathcal{S}_2 \sum_{\gamma} \langle c\gamma\,\Omega_2\sigma_2 | H_2 | b\beta\rangle \langle c\gamma\,\Omega_2\sigma_2 | H_2 | b\beta'\rangle^*. \end{aligned}\right\} \tag{10.4}$$

In this manner, as in (9.9), the angular correlation is represented by the products of factors referring separately to each transition. The method consists of reducing each matrix element to the product of two terms, one depending on the nucleus, and the other on the properties of the radiation, so that, for example,

$$\langle b\beta\,\Omega_1\sigma_1 | H_1 | a\alpha\rangle = \sum_{L_1 l_1 \Lambda_1} \langle \Omega_1\sigma_1 | L_1 l_1 \Lambda_1\rangle \langle b\beta\, L_1 l_1 \Lambda_1 | H_1 | a\alpha\rangle, \tag{10.5}$$

where l_1 is restricted to values consistent with conservation of parity. Each term in (10.5) is, in turn, subject to a change in representation. Thus, the first term, which represents an eigenfunction of the radiation referred to an arbitrary coordinate axis, is expressed in terms of eigenfunctions quantised along the direction of the radiation by means of the rotation matrix (7.1),

$$\langle \Omega_1\sigma_1 | L_1 l_1 \Lambda_1\rangle = \sum_{\mu} \langle 0\,\sigma_1 | L_1 l_1 \mu_1\rangle\, \mathcal{D}^{L_1 *}_{\Lambda_1 \mu_1}(\mathcal{R}_1), \tag{10.6}$$

where $\mathcal{R}_1 = (\vartheta_1\,\vartheta_2\,\vartheta_3)$ denotes the rotation carrying the z-axis into the direction of the first radiation. The second term in (10.5) is transformed by a change of representation to

$$\langle b\beta\, L_1 l_1 \Lambda_1 | H_1 | a\alpha\rangle = \sum_{k\varkappa} (b\beta, L_1\Lambda_1 | k\varkappa)\, \langle b\, L_1 l_1 (k\varkappa) | H_1 | a\alpha\rangle. \tag{10.7}$$

Since H_1 is invariant under rotations, the matrix element is diagonal in the angular-momentum representation, and is independent of magnetic quantum number; hence (10.7) is simplified to

$$\langle b\, L_1 l_1 (k\varkappa) | H_1 | a\alpha\rangle = \delta_{ak}\,\delta_{\alpha\varkappa} \langle b \| \mathcal{L}_1 \| a\rangle. \tag{10.8}$$

This reduction is an expression of the Wigner-Eckart theorem, and $\langle b \| \mathcal{L}_1 \| a\rangle$ is the reduced matrix element, as defined, for example by FRAUENFELDER [*14*a]. So the matrix element in (10.5) becomes

$$\langle b\beta\,\Omega_1\sigma_1 | H_1 | a\alpha\rangle = \sum_{L_1 l_1 \Lambda_1 \mu_1} \langle b \| \mathcal{L}_1 \| a\rangle (b\beta, L_1\Lambda_1 | a\alpha) \langle 0\,\sigma_1 | L_1 l_1 \mu_1\rangle\, \mathcal{D}^{L_1 *}_{\Lambda_1 \mu_1}(\mathcal{R}_1). \tag{10.9}$$

With the use of some Racah algebra, the first transition is described by

$$\left.\begin{aligned} E^{(1)}(\beta\beta') = \sum_{L_1 L_1' l_1 l_1' k_1 \varkappa_1 \nu_1} & c_{k_1\varkappa_1}(\mathcal{L}_1' \mathcal{L}_1)\, (b\beta, b-\beta' | k_1\nu_1)\, (-)^{a-L_1-\beta'} \times \\ & \times W(b\,b\,L_1 L_1'; k_1 a)\, \mathcal{D}^{k_1}_{\nu_1\varkappa_1}(\mathcal{R}_1) \langle b \| \mathcal{L}_1 \| a\rangle \langle b \| \mathcal{L}_1' \| a\rangle^*, \end{aligned}\right\} \tag{10.10}$$

where $c_{k_1\varkappa_1}(\mathcal{L}_1 \mathcal{L}_1')$ is the radiation parameter, introduced by RACAH, which is characteristic of the emitted radiation and of the specification of the means of detection of the radiation. It is expressed as

$$c_{k_1\varkappa_1}(\mathscr{L}_1\mathscr{L}_1') = \mathscr{S}\sum_{\mu\mu'}(-)^{L_1'-\mu'}(L_1\mu, L_1'-\mu'\,|\,k_1\varkappa_1)\cdot\langle 0\sigma|\,L_1 l_1\mu\rangle^*\langle 0\sigma|\,L_1' l_1'\mu'\rangle. \quad (10.11)$$

The symbol, $\mathscr{S}$, represents a summation over the spin-states of the radiation and depends on the characteristics of the detector apparatus. In our subsequent treatment, (Sect. 11), this is readily handled by the efficiency matrix $\langle\sigma|\,\varepsilon\,|\sigma'\rangle$, whose elements are determined for special cases. The summation, $\mathscr{S}$, is replaced by $\sum_{\sigma\sigma'}\langle\sigma|\,\varepsilon\,|\sigma'\rangle$ which multiplies the terms in (10.11). If the interaction can be described by a Hamiltonian perturbation, the reduced matrix elements are real as a consequence of the invariance of the interaction under time inversion (LLOYD [28], COESTER [45], BIEDENHARN and ROSE [9]). In general, these reduced matrix elements are neither real nor Hermitian, but satisfy[1]:

$$\hat{b}\,\langle a\,\|\,\mathscr{L}_1\,\|\,b\rangle = (-)^{a+L_1-b}\,\hat{a}\,\langle b\,\|\,\mathscr{L}_1\,\|\,a\rangle^*. \quad (10.12)$$

Similarly, we find

$$\left.\begin{aligned} E^{(2)}(\beta'\beta) = \sum_{L_2L_2'l_2l_2'k_2\varkappa_2\nu_2} &\hat{b}^2\,c^*_{k_2\varkappa_2}(\mathscr{L}_2\mathscr{L}_2')\,(b\,\beta, b-\beta'\,|\,k_2\nu_2)\,(-)^{c+\beta'-L_2-k_2-2b}\times\\ &\times W(b\,b\,L_2\,L_2';\,k_2c)\,\mathscr{D}^{k*}_{\nu_2\varkappa_2}(\mathscr{R}_2)\,\langle c\,\|\,\mathscr{L}_2\,\|\,b\rangle\langle c\,\|\,\mathscr{L}_2'\,\|\,b\rangle^*.\end{aligned}\right\} \quad (10.13)$$

The general correlation is obtained by means of (10.3). Observing that

$$\sum_{\beta\beta'}(b\,\beta, b-\beta'\,|\,k_1\nu_1)\,(b\,\beta, b-\beta'\,|\,k_2\nu_2) = \delta_{k_1k_2}\,\delta_{\nu_1\nu_2} \quad (10.14)$$

and

$$\sum_{\nu}\mathscr{D}^k_{\nu\varkappa_1}(\mathscr{R}_1)\,\mathscr{D}^{k*}_{\nu\varkappa_2}(\mathscr{R}_2) = \sum_{\nu}\mathscr{D}^k_{\nu\varkappa_1}(\mathscr{R}_1)\,\mathscr{D}^k_{\varkappa_2\nu}(\mathscr{R}_2^{-1}) = \mathscr{D}^k_{\varkappa_2\varkappa_1}(\mathscr{R}_2^{-1}\,\mathscr{R}_1), \quad (10.15)$$

where $\mathscr{R}_2^{-1}\mathscr{R}_1$ is the rotation carrying the coordinate system for the second radiation into the coordinate system of the first radiation, we find

$$\left.\begin{aligned} W(\Omega_1\Omega_2) = (-)^{a+c-2b}\hat{b}^2 \sum_{\substack{L_1L_1'L_2L_2'\\ l_1l_1'l_2l'\\ k\varkappa_1\varkappa_2}} &(-)^{k-L_1-L_2}\,W(b\,b\,L_1L_1';\,k\,a)\,W(b\,b\,L_2L_2';\,k\,c)\times\\ &\times\langle c\,\|\,\mathscr{L}_2\,\|\,b\rangle\langle c\,\|\,\mathscr{L}_2'\,\|\,b\rangle^*\langle b\,\|\,\mathscr{L}_1\,\|\,a\rangle\langle b\,\|\,\mathscr{L}_1'\,\|\,a\rangle^*\times\\ &\times c_{k\varkappa_1}(\mathscr{L}_1'\mathscr{L}_1)\,c^*_{k\varkappa_2}(\mathscr{L}_2\mathscr{L}_2')\,\mathscr{D}^k_{\varkappa_2\varkappa_1}(\vartheta_1\vartheta_2\vartheta_3).\end{aligned}\right\} \quad (10.16)$$

The angular dependence is contained in the factor $\mathscr{D}^k_{\varkappa_1\varkappa_2}(\mathscr{R}_2^{-1}\mathscr{R}_1)$ which is, of course independent of the original, arbitrary, axis with respect to which $\mathscr{R}_1$ and $\mathscr{R}_2$ are defined. This equation is completely general, which is evident by the complexity of the summation. The variables include the possibility of mixtures of radiations, e.g. magnetic-dipole and electric-quadrupole emission, in each transition. If the measurement of radiation is insensitive to polarisation or if it responds to circular polarisation only, $c_{k\varkappa}$, which is an eigenfunction of the

---

[1] This corresponds to the definition of reduced matrix elements used by FRAUENFELDER [14a], except that these matrix elements are now generalised to be possibly complex if the interaction is not a Hamiltonian. In the notation of [9], the matrix element (10.7) is represented by $\langle j_1 m_1|\,T(LM\pi)\,|\,jm\rangle$, where $T(LM\pi)$ is a Hermitian nuclear tensor operator chosen so that the interaction remains invariant under time inversion (Sect. 14), i.e. so that $T^+(LM\pi) = (-)^{L-M}\,T(L-M\pi)$. Other authors, RACAH [3], FANO [8], etc., express the Wigner-Eckart theorem (10.7) and define Hermitian tensor operators differently, so that their angular-correlation expressions appear to differ by phase factors. However, this is accounted for by the different definitions used for the reduced matrix elements. In what follows, we will asume the generalised operators to be Hermitian, so that (10.12) is still valid.

operator $L_z = -i\hbar \frac{\partial}{\partial \vartheta_3}$ with eigenvalue, $\varkappa$, vanishes unless $\varkappa = 0$. This is because ϑ_3 refers to a rotation about the direction of motion, and only this latter direction is observed, so that the result is independent of ϑ_3. If measurements of both radiations are insensitive to polarisation or if they respond to circular polarisations, $\mathscr{D}^k_{\varkappa_2 \varkappa_1}$ becomes $P_k(\cos \vartheta_2)$.

Thus it is seen that the essential problem in the calculation of an angular correlation is the determination of the parameters of radiation, $c_{k\varkappa}$. These parameters are dependent on the nature of the radiation, its direction of motion, and the measurement of its polarisation. BIEDENHARN and ROSE have given expressions for these parameters for the cases of gamma radiation, α-particle emission, and also for conversion electrons and β-particles, which will be referred to later (Chap. F, G).

Comparison with the results using the wavefunction procedure [compare (9.13) with (10.16)] shows that the former treatment deals with $\varkappa = 0$, i.e. with cases where only the direction of motion is observed. Thus, only measurements relative to the direction of motion are important, which indicates the reason for the simplicity introduced in quantising along the direction of motion The more general case of $\varkappa \neq 0$ can include detection of linearly polarised radiation along some direction, and thus involve other angles.

Although (10.16) is sufficiently general to deal with most problems of angular correlation, the method suffers from the tedious summations over magnetic quantum numbers. This cannot be eliminated, although a list of all possible parameters of radiation and of the different formulae for more complicated cases than a simple cascade could be prepared and used for any particular problem. Nevertheless, it would be advantageous to have a treatment which gives greater physical or geometrical significance to the various parameters occurring in the formalism. It would also be more logical in such a procedure to avoid introducing pure states and subsequently averaging over all the states introduced, consistent with the experiment, but from the outset, to incorporate the limitations of the measurement. This is done by means of the formalism, using density matrices, and, associated with them, statistical tensors and efficiency tensors—a procedure arrived at independently by FANO [*8*] and COESTER and JAUCH [*29*], [*11*].

11. Statistical and efficiency tensors. In this third treatment we shall restrict ourselves, at the beginning, to the case of a two-stage process, so as to compare the procedure with the two others described in Sects. 9 and 10. It should be pointed out, however, that an attractive feature of this treatment is the elegant and natural extensions that can be applied to the more complicated processes, e.g. triple cascades.

Starting with the introduction of density matrices and efficiency matrices, which represent the initial spin states, the experimental arrangement for the production and detection of the radiations, and if necessary the residual nuclear states, statistical tensors and efficiency tensors are constructed. The angular correlation is then expressed in terms of these tensors. They also serve to represent the various substates, and have a particular mathematical convenience since their components are transformed under a rotation of coordinate axes in the same manner as spherical harmonics of a particular degree. In order to form the tensors describing the resultant state, it is necessary to change the angular-momentum coupling scheme which describes the initial state of nucleus plus radiation. This change of coupling is effected by means of a transformation, as in (6.1) and (6.2). The complete summation over magnetic quantum numbers is carried out automatically, thus eliminating the considerable practical difficulties which beset the early calculations, and which, to a certain extent, still persist in the procedure described in the previous section.

First we will describe the properties of the density matrix as it relates to our particular problem. The initial state of our system before interaction is characterised by the total angular momenta, $(a\,\alpha)$, $(a\,\alpha')$, ..., of the whole system, and by other quantum numbers denoted by $A_0\,A_0'$, ... After the interaction the whole system, nucleus plus radiation, still has the same total angular momenta, and can be characterised by the quantum numbers $(a\,\alpha)$, $(a\,\alpha')$, ..., and $A_x\,A_x'$. The corresponding density matrices are written

$$\varrho_0\colon \langle a\,\alpha\,A_0|\varrho|\,a\,\alpha'\,A_0'\rangle, \tag{11.1}$$

$$\varrho_x\colon \langle a\,\alpha\,A_x|\varrho|\,a\,\alpha'\,A_x'\rangle. \tag{11.2}$$

The relation between the density matrices before and after interaction is given in terms of the interaction operator (Sect. 10) by the equation, COESTER [29], [45]:

$$\left.\begin{aligned}\langle a\,\alpha\,A_x|\varrho|\,a\,\alpha'\,A_x'\rangle &= \text{const}\sum_{A_0A_0'}\langle a\,A_x\,|\mathcal{O}|\,a\,A_0\rangle\times\\ &\times\langle a\,\alpha\,A_0|\varrho|\,a\,\alpha'\,A_0'\rangle\,\langle a\,A_0'\,|\mathcal{O}^+|\,a\,A_x'\rangle,\end{aligned}\right\} \tag{11.3}$$

in which $\mathcal{O}$ is independent of α, α'. Since we are dealing with processes in which all the nuclear states are well-defined, we can factorise the $\mathcal{O}$ matrix elements in the following manner:

$$\langle a\,A_x\,|\mathcal{O}\,|\,a\,A_0\rangle = \langle a\,A_x\,|\mathcal{O}_2|\,a\,A_i\rangle\,\langle a\,A_i\,|\mathcal{O}_1|\,a\,A_0\rangle, \tag{11.4}$$

where A_i represents the well-defined intermediate state. If the first process is an absorption, we can use the S-matrix formalism with the same essential features as (11.3). In Eq. (11.3), $\mathcal{O}$ is given by $R = S - 1$, whose matrix elements are independent of components of angular momentum. For cascade emission we can regard the transformation of the density matrix as representing a time-development:

$$\left.\begin{aligned}\langle a\,\alpha\,A_t|\varrho|\,a\,\alpha'\,A_t'\rangle &= \text{const}\sum_{A_0A_0'}\langle a\,\alpha\,A_0|\varrho|\,a\,\alpha'\,A_0'\rangle\times\\ &\times\langle a\,A_t|\,\mathcal{O}\,|\,a\,A_0\rangle\,\langle a\,A_t'|\,\mathcal{O}\,|\,a\,A_0'\rangle^*.\end{aligned}\right\} \tag{11.5}$$

The factorised matrix elements that appear in (11.4) are equivalent to the reduced matrix elements in (10.8). If we apply (11.3) to the case of cascade radiation, $a\to b\to c$, where we can use perturbation formalism, we can write:

$$\left.\begin{aligned}\langle a\,\alpha\,A_t|\varrho|\,a\,\alpha'\,A_t'\rangle &= \text{const}\sum_{A_0A_0'}\langle a\,\alpha\,A_0|\varrho|\,a\,\alpha'\,A_0'\rangle\times\\ &\times\langle c\,L_2\,l_2\,L_1\,a\,\|H\|\,b\,L_1\,l_1\,a\rangle\,\langle c\,L_2'\,l_2'\,L_1'\,a\,\|H\|\,b\,L_1'\,l_1'\,a\rangle^*\times\\ &\times\langle b\,L_1\,l_1\,a\,\|H\|\,a\rangle\,\langle b\,L_1'\,l_1'\,a\,\|H\|\,a\rangle^*.\end{aligned}\right\} \tag{11.6}$$

In the above expression, the additional quantum numbers, A_t, A_t', are seen to involve quantum numbers $b, c, L_1, L_2, L_1'L_2', l_1, l_2, l_1', l_2'$. The reduced matrix elements refer to separate transitions between well-defined states, $b\to c$ and $a\to b$, so that we denote them by the abbreviated notation used previously. In addition, if the cascade transition starts from a nucleus which is in a well-defined state and randomly oriented, we obtain from (8.7) the density matrix of the original state:

$$\langle a\,\alpha\,A_0|\varrho|\,a\,\alpha'\,A_0'\rangle = \frac{1}{\hat{a}^2}\,\delta_{\alpha\alpha'}\,\delta_{A_0A_0'}. \tag{11.7}$$

Hence we write (11.6) in the form

$$\left.\begin{aligned}\langle a\,\alpha\,A_t\,|\,\varrho\,|\,a\,\alpha'\,A_t'\rangle = \frac{\text{const}}{\hat{a}^2}\,\delta_{\alpha\alpha'}\langle c\,\|\,\mathscr{L}_2\,\|\,b\rangle\,\times\\ \times\langle c\,\|\,\mathscr{L}_2'\,\|\,b\rangle^*\langle b\,\|\,\mathscr{L}_1\,\|\,a\rangle\langle b\,\|\,\mathscr{L}_1'\,\|\,a\rangle^*.\end{aligned}\right\}\tag{11.8}$$

The notation $\langle b\|\mathscr{L}_1\|a\rangle$ etc. will be used quite generally whenever the interaction operator can be factorised, e.g. this reduced matrix element will be used to represent the transformation $\boldsymbol{a}\to\boldsymbol{b}+\boldsymbol{L}_1(l_1)$, and is not restricted to the meaning that the interaction is a Hamiltonian operator characterised by $\mathscr{L}_1$.

In an experimental arrangement we are concerned with the detection of one or more of the products of the nuclear transformations. This is described by the use of an efficiency matrix, ε, (COESTER and JAUCH [*29*], FANO [*8*]) as a certain matrix, diagonal in a q-representation, with elements ε_q, so that the probability of detection of different particles by a counter arrangement is

$$W=\sum_q \varepsilon_q\langle q\,|\,\varrho\,|\,q\rangle=\operatorname{Tr}(\varepsilon\,\varrho)\,. \tag{11.9}$$

For example, if our system consists of a nucleus plus radiation, then the probability of detection of the radiation in a certain direction with no measurements as regards its polarisation can be written as

$$W(\Omega)=\sum_{\beta\sigma}\langle b\,\beta\,\sigma\,\Omega\,|\,\varrho\,|\,b\,\beta\,\sigma\,\Omega\rangle=\operatorname{Tr}(\varepsilon\,\varrho)\,, \tag{11.10}$$

where ε is a matrix having elements such as

$$\langle b\,\beta\,\mathscr{L}\,\Lambda\,|\,\varepsilon\,|\,b\,\beta'\,\mathscr{L}'\,\Lambda'\rangle=\delta_{\beta\beta'}\sum_{\sigma\sigma'}\langle\mathscr{L}\,\Lambda\,|\,\Omega\,\sigma\rangle\langle\Omega\,\sigma'\,|\,\mathscr{L}'\,\Lambda'\rangle\,\delta_{\sigma\sigma'}\,. \tag{11.11}$$

In the general case of a polarisation-sensitive detector, the efficiency matrix is not necessarily diagonal in the variables σ describing polarisation (cf. Sect. 29).

The efficiency matrix for the products of a nuclear transformation is equal to the product of the efficiency matrices for each of the product fragments.

It is convenient to introduce a set of statistical tensors in the following manner. Consideration of $\langle a\alpha\,|\,\varrho\,|\,a'\alpha'\rangle$ shows that it is composed of products of expansion coefficients of eigenstates corresponding to quantum numbers $a\alpha$ and $a'\alpha'$. The matrix therefore transforms under rotation as the Kronecker product $\boldsymbol{a}\times\boldsymbol{a}'$, which is reducible into a sum of terms of rank k, where k varies in steps of one from $a+a'$ to $|a-a'|$, in a similar manner to the combination of vectors $\psi_a^\alpha\,\psi_b^\beta$ as in (4.1).

The statistical tensor $\varrho_{k_a\varkappa_a}(aA,\,a'A')$ is defined as follows:

$$\langle a\,\alpha\,A\,|\,\varrho\,|\,a'\,\alpha'\,A'\rangle=\sum_{k_a\varkappa_a}(-)^{a'-\alpha'}\,(a\,\alpha,\,a'-\alpha'\,|\,k_a\,\varkappa_a)\;\varrho_{k_a\varkappa_a}(a\,A,\,a'\,A')\,, \tag{11.12}$$

with $\boldsymbol{k}_a=\boldsymbol{a}-\boldsymbol{a}'$. Because of the unitary property (4.2) of the Clebsch-Gordan coefficients, we can invert (11.12) thus:

$$\varrho_{k_a\varkappa_a}(a\,A,\,a'\,A')=\sum_{\alpha\alpha'}(-)^{a'-\alpha'}\,(a\,\alpha,\,a'-\alpha'\,|\,k_a\varkappa_a)\langle a\,\alpha\,A\,|\,\varrho\,|\,a'\,\alpha'\,A'\rangle. \tag{11.13}$$

The particular choice of phase ensures that $\varrho_{k_a\varkappa_a}$ is a Hermitian tensor[1] satisfying

$$\varrho^*_{k_a\varkappa_a}(a\,A,\,a'\,A')=(-)^{a'-a+\varkappa_a}\,\varrho_{k_a-\varkappa_a}(a'\,A',\,a\,A)\,. \tag{11.14}$$

[1] A. R. EDMONDS: Ref. [*17*], p. 56.

The statistical tensor $\varrho^*_{k_a \varkappa_a}$ is an irreducible tensor of rank k_a, which transforms according to the $2k_a+1$-dimensional irreducible representation $\mathcal{D}^{k_a}$ of the rotation group, and satisfies equations of the type (7.1). From its definition (11.12), it is clear that a knowledge of the statistical tensors is entirely equivalent to a knowledge of the density matrix.

The statistical tensor corresponding to (11.7) is obtained by using (4.7) and noting that

$$\langle a\,\alpha|\varrho|a\,\alpha'\rangle = \frac{(-)^{a+\alpha'}}{\hat{a}}\,(a\,\alpha, a-\alpha'|0\,0)\,, \tag{11.15}$$

so that

$$\varrho_{k_a \varkappa_a}(a\,A_0, a\,A_0') = \frac{\delta_{k_a 0}\,\delta_{\varkappa_a 0}\,\delta_{A_0 A_0'}}{\hat{a}}\,. \tag{11.16}$$

Combining (11.8) and (11.16) and omitting the irrelevant constant, we have

$$\varrho_{k_a \varkappa_a}(a\,A, a'\,A') = \frac{\delta_{k_a 0}\,\delta_{\varkappa_a 0}\,\delta_{a\,a'}}{\hat{a}} \langle c\|\mathcal{L}_2\|b\rangle\langle c\|\mathcal{L}_2'\|b\rangle^*\langle b\|\mathcal{L}_1\|a\rangle\langle b\|\mathcal{L}_1'\|a\rangle^*. \tag{11.17}$$

In a similar fashion, the efficiency tensor is introduced with

$$\varepsilon_{k_a \varkappa_a}(a\,A, a'\,A') = \sum_{\alpha\alpha'} (-)^{a'-\alpha'}(a\,\alpha, a'-\alpha'|k_a \varkappa_a)\langle a\,\alpha\,A|\varepsilon|a'\,\alpha'\,A'\rangle \tag{11.18}$$

and

$$\langle a\,\alpha\,A|\varepsilon|a'\,\alpha'\,A'\rangle = \sum_{k_a \varkappa_a} (-)^{a'-\alpha'}(a\,\alpha, a'-\alpha'|k_a \varkappa_a)\,\varepsilon_{k_a \varkappa_a}(a\,A, a'\,A')\,. \tag{11.19}$$

Hence, (11.9) can be written, either in terms of the density and efficiency matrices, as

$$W = \sum_{a a' \alpha \alpha' A A'} \langle a\,\alpha\,A|\varepsilon|a'\,\alpha'\,A'\rangle\langle a'\,\alpha'\,A'|\varrho|a\,\alpha\,A\rangle; \tag{11.20}$$

or, in terms of statistical tensors and efficiency tensors, as

$$W = \sum_{a a' A A' k_a \varkappa_a} \varrho_{k_a \varkappa_a}(a\,A, a'\,A')\,\varepsilon^*_{k_a \varkappa_a}(a\,A, a'\,A')\,. \tag{11.21}$$

The quantum numbers, A, A' usually serve to represent the various possible angular momentum states which characterise the system after the decay has taken place. The tensors are introduced because of the convenient relations describing the transformations from one representation to another. The transformations arise in the representation of tensors such as $\varepsilon_{k_a \varkappa_a}$ in terms of the tensors of the decay products. Thus, noting that the efficiency matrix of the system can be represented as the product of the individual matrices, if we consider the first decay with $\boldsymbol{a} = \boldsymbol{b} + \boldsymbol{L}_1$ and transform our representation by means of (4.1), we have

$$\left.\begin{aligned}\langle a\,\alpha\,A|\varepsilon|a'\,\alpha'\,A'\rangle = \sum_{\beta\beta'\Lambda\Lambda'} &\langle b\,\beta\,B|\varepsilon|b'\beta'\,B'\rangle\langle \mathcal{L}_1 \Lambda_1|\varepsilon|\mathcal{L}_1' \Lambda_1'\rangle\times\\ &\times(a\,\alpha|b\,\beta, L_1 \Lambda_1)(b'\beta', L_1'\Lambda_1'|a'\alpha'),\end{aligned}\right\} \tag{11.22}$$

where B, B', represent the remaining quantum numbers, aside from $b\,\beta, b'\beta'$, describing the system after decay. Transforming (11.22) in terms of the corresponding efficiency tensors by means of (11.19), we find that $\varepsilon_{k_a \varkappa_a}(a\,A, a'\,A')$ can be expressed in terms of $\varepsilon_{k_b \varkappa_b}(b\,B, b'\,B')\,\varepsilon_{k_1 \varkappa_1}(\mathcal{L}_1\,\mathcal{L}_1')$ multiplied by the product of five Clebsch-Gordan coefficients. On comparison with (6.1), this becomes

$$\left.\begin{aligned} \varepsilon_{k_a \varkappa_a}(a\,A,\,a'\,A') = \sum_{k_1 \varkappa_1 k_b \varkappa_b} \varepsilon_{k_b \varkappa_b}(b\,B,\,b'\,B')\,\varepsilon_{k_1 \varkappa_1}(\mathscr{L}_1\,\mathscr{L}_1') \times \\ \times\,(k_a \varkappa_a \,|\, k_b \varkappa_b,\,k_1 \varkappa_1)\,\hat{a}\,\hat{a}'\,\hat{k}_b \hat{k}_1 \begin{Bmatrix} b & L_1 & a \\ b' & L_1' & a' \\ k_b & k_1 & k_a \end{Bmatrix}. \end{aligned}\right\} \tag{11.23}$$

The physical significance of this transformation is made clear if we note that

$$\left.\begin{aligned} \boldsymbol{a} &= \boldsymbol{b} + \boldsymbol{L}_1, & \boldsymbol{a} - \boldsymbol{a}' &= \boldsymbol{k}_a, \\ \boldsymbol{a}' &= \boldsymbol{b}' + \boldsymbol{L}_1', & \boldsymbol{b} - \boldsymbol{b}' &= \boldsymbol{k}_b, \end{aligned}\right\} \quad \boldsymbol{L}_1 - \boldsymbol{L}_1' = k_1, \tag{11.24}$$

and

$$\langle b\,b'(k_b),\,L_1 L_1'(k_1),\,k_a \,|\, b\,L_1(a),\,b'\,L_1'(a'),\,k_a\rangle = \hat{a}\,\hat{a}'\,\hat{k}_b \hat{k}_1 \begin{Bmatrix} b & L_1 & a \\ b' & L_1' & a' \\ k_b & k_1 & k_a \end{Bmatrix}, \tag{11.25}$$

so that the efficiency tensor can be expressed in terms of the product of the efficiency tensor of the decay fragments, provided that we change the coupling scheme of the latter from

$$\langle b\,L_1(a),\,b'\,L_1'(a'),\,k_a| \rangle \quad \text{to} \quad \langle b\,b'(k_b),\,L_1 L_1'(k_1),\,k_a| \rangle.$$

In a similar manner, $\varepsilon_{k_b \varkappa_b}(b\,B,\,b'\,B')$ can be expressed in terms of $\varepsilon_{k_c \varkappa_c}(c\,C,\,c'\,C')$ and $\varepsilon_{k_2 \varkappa_2}(\mathscr{L}_2\,\mathscr{L}_2')$, corresponding to the decay $\boldsymbol{b} = \boldsymbol{c} + \boldsymbol{L}_2$.

The efficiency tensor for each radiation is found by referring it to represenations corresponding to the observed properties of the radiation. Thus,

$$\varepsilon_{k \varkappa}(\mathscr{L}_1\,\mathscr{L}_1') = \sum_{\Lambda \Lambda'} (-)^{L'-\Lambda'}\,(L\Lambda,\,L'-\Lambda' \,|\, k\varkappa)\,\langle L\,l\,\Lambda \,|\, \varepsilon \,|\, L'\,l'\,\Lambda'\rangle \tag{11.26}$$

$$\left.\begin{aligned} = \sum_{\Lambda \Lambda \sigma \sigma'} (-)^{L'-\Lambda'}\,(L\Lambda,\,L'-\Lambda' \,|\, k\varkappa)\,\langle L\,l\,\Lambda \,|\, \Omega\,\sigma\rangle \times \\ \times\,\langle \sigma \,|\, \varepsilon \,|\, \sigma'\rangle\,\langle \Omega\,\sigma' \,|\, L'\,l'\,\Lambda'\rangle, \end{aligned}\right\} \tag{11.27}$$

where $\langle \sigma | \varepsilon | \sigma' \rangle$ depends on the polarisation sensitivity of the counter arrangement and is equal to $\delta_{\sigma\sigma'}$ if the counter is insensitive to polarisation. The expansion coefficients are Hermitian, so that

$$\langle L\,l\,\Lambda \,|\, \Omega\,\sigma\rangle = \langle \Omega\,\sigma \,|\, L\,l\,\Lambda\rangle^*. \tag{11.28}$$

These have been encountered in the procedure described in Sect. 10. As in this latter procedure, the radiation is represented in terms of a coordinate system whose axis of quantisation coincides with the direction of emission by means of the rotation matrix $\mathscr{D}$ (10.6). The procedure is thus symmetrical with respect to both radiations, in contrast to the early procedure. By means of (10.6), (7.7), and (7.8), we find

$$\left.\begin{aligned} \varepsilon_{k\varkappa}(\mathscr{L}\,\mathscr{L}') = \sum_{\sigma \sigma' \mu \mu' \varkappa'} (-)^{L'-\mu'}\,(L\mu,\,L'-\mu' \,|\, k\varkappa')\,\langle 0\,\sigma \,|\, L\,l\,\mu\rangle^* \times \\ \times\,\langle 0\,\sigma' \,|\, L'\,l'\,\mu'\rangle\,\langle \sigma \,|\, \varepsilon \,|\, \sigma'\rangle\,\mathscr{D}^k_{\varkappa\varkappa'}(\mathscr{R}). \end{aligned}\right\} \tag{11.29}$$

If we compare the above expression with the radiation parameter $c_{k\varkappa}(L L')$ as defined in (10.11), and consider the summation $\mathscr{S} f(\sigma\sigma')$ to be equivalent to $\sum_{\sigma\sigma'} \langle \sigma | \varepsilon | \sigma' \rangle f(\sigma\,\sigma')$ we find

$$\varepsilon_{k\varkappa}(\mathscr{L}\,\mathscr{L}') = \sum_{\varkappa'} c_{k\varkappa'}(\mathscr{L}\,\mathscr{L}')\,\mathscr{D}^k_{\varkappa\varkappa'}(\mathscr{R}). \tag{11.30}$$

Both $\varepsilon_{k_1\varkappa_1}(\mathscr{L}_1\mathscr{L}_1')$ and $\varepsilon_{k_2\varkappa_2}(\mathscr{L}_2\mathscr{L}_2')$ are expressed as in (11.30), reducing the problem to a determination of the radiation parameters.

The nucleus is left in a sharp eigenstate with quantum number, c, and no further measurement is made on the final state, so that the efficiency matrix is

$$\langle c\gamma C|\varepsilon|c'\gamma' C'\rangle = \delta_{cc'}\delta_{CC'}\delta_{\gamma\gamma'}, \tag{11.31}$$

and the corresponding efficiency tensor is

$$\varepsilon_{k_c\varkappa_c}(cC, c'C') = \hat{c}\,\delta_{k_c0}\delta_{\varkappa_c0}\delta_{CC'}\delta_{cc'}. \tag{11.32}$$

Assembling (11.17), (11.21), (11.23) and (11.32), the angular correlation is

$$\left.\begin{aligned} W \sim \frac{1}{\hat{a}} \sum_{\substack{k_a k_b k_c \varkappa_a \varkappa_b \varkappa_c \\ k_1 k_2 \varkappa_1 \varkappa_2 \\ L_1 L_1' L_2 L_2' l_1 l_1' l_2 l_2'}} &\langle c\|\mathscr{L}_2\|b\rangle\langle c\|\mathscr{L}_2'\|b\rangle^*\langle b\|\mathscr{L}_1\|a\rangle\langle b\|\mathscr{L}_1'\|a\rangle^* \times \\ &\times \delta_{k_a0}\delta_{\varkappa_a0}\varepsilon^*_{k_1\varkappa_1}(\mathscr{L}_1\mathscr{L}_1')\,\varepsilon_{k_2\varkappa_2}(\mathscr{L}_2\mathscr{L}_2')\,\hat{a}^2\hat{k}_1\hat{k}_b \begin{Bmatrix} b & L_1 & a \\ b & L_1' & a \\ k_b & k_1 & k_a \end{Bmatrix} \hat{b}^2\hat{k}_2\hat{k}_c \times \\ &\times \begin{Bmatrix} c & L_2 & b \\ c & L_2' & b \\ k_c & k_2 & k_b \end{Bmatrix} (k_b\varkappa_b, k_1\varkappa_1|k_a\varkappa_a)(k_c\varkappa_c, k_2\varkappa_2|k_b\varkappa_b)\,\hat{c}\,\delta_{k_c0}\delta_{\varkappa_c0}, \end{aligned}\right\} \tag{11.33}$$

which, using the properties of the Clebsch-Gordan coefficients and 9-j symbols (see the appendix), reduces to

$$\left.\begin{aligned} W \sim \hat{b}^2(-)^{a+c-2b} \sum_{\substack{L_1L_1'L_2L_2' \\ l_1l_1'l_2l_2' \\ k\varkappa}} &(-)^{k-L_1-L_2}\,W(b\,b\,L_1L_1';k\,a)\,W(b\,b\,L_2L_2';k\,c)\times \\ &\times \varepsilon_{k\varkappa}(\mathscr{L}_1'\mathscr{L}_1)\,\varepsilon^*_{k\varkappa}(\mathscr{L}_2\mathscr{L}_2')\langle c\|\mathscr{L}_2\|b\rangle\langle c\|\mathscr{L}_2'\|b\rangle^*\langle b\|\mathscr{L}_1\|a\rangle\langle b\|\mathscr{L}_1'\|a\rangle^*. \end{aligned}\right\} \tag{11.34}$$

Expressing $\varepsilon_{k\varkappa}(\mathscr{L}\mathscr{L}')$ in terms of the radiation parameters, the angular correlation $W(\Omega_1\Omega_2)$ is found to agree exactly with the result previously obtained in (10.16),

$$\left.\begin{aligned} W(\Omega_1\Omega_2) \sim \hat{b}^2(-)^{a+c-2b} \sum_{\substack{L_1L_1'L_2L_2'l_1l_1' \\ l_2l_2'k\varkappa_1\varkappa_2}} &(-)^{k-L_1-L_2}W(b\,b\,L_1L_1';k\,a)\times \\ &\times W(b\,b\,L_2L_2';k\,c)\,c_{k\varkappa_1}(\mathscr{L}_1'\mathscr{L}_1)\,c^*_{k\varkappa_2}(\mathscr{L}_2\mathscr{L}_2')\times \\ &\times \mathscr{D}^k_{\varkappa_2\varkappa_1}(\vartheta_1\vartheta_2\vartheta_3)\langle c\|\mathscr{L}_2\|b\rangle\langle c\|\mathscr{L}_2'\|b\rangle^*\langle b\|\mathscr{L}_1\|a\rangle\langle b\|\mathscr{L}_2'\|a\rangle^*. \end{aligned}\right\} \tag{11.35}$$

We can interpret the above result in a direct fashion. The physical characteristics of the radiating system show up entirely in the reduced matrix elements, which act essentially as weighting factors selecting the contributions from the various angular momenta of the radiation. The influence of the measurement apparatus shows up in the determination of the radiation parameters, $c_{k\varkappa}(\mathscr{L}\mathscr{L}')$, which are simply related to the efficiency tensors of the radiation. The angular dependence manifests itself in the rotation matrix $\mathscr{D}^k_{\varkappa_2\varkappa_1}(\mathscr{R})$, where $\mathscr{R}$ corresponds to the rotation from one propagation direction to the other. The remaining expressions, the Racah coefficients, arise as coefficients in the change of coupling schemes, necessary to describe the system of nucleus plus radiation before and after the emission of the radiation, and they involve purely geometrical transformations.

Although the procedures of Racah *et al.* are characterised by the introduction of the radiation parameters $c_{k\varkappa}(\mathscr{L}\mathscr{L}')$ or equivalent quantities, the significance

$c_{k\varkappa}$ becomes much clearer in the procedure of Sect. 11 a the expansion coefficient of the efficiency tensor in terms of the rotation matrices (11.30). It should be pointed out that RACAH does give $c_{k\varkappa}$ an equivalent meaning to the above one, although it appears in a less transparent fashion.

12. Evaluation of correlation for absorption and emission. In Sect. 11 we have considered, primarily, the case of the successive emission of two radiations from an unstable nucleus. The methods used there apply as well to absorption of radiations. Let us consider a reaction which takes place through the absorption of radiation having various partial waves with angular momenta $\mathscr{L}_1$, and the subsequent emission of a radiation characterised by $\mathscr{L}_2$. The total system here is clearly not one of well-defined total angular momentum, since the incident plane wave contains components having all values of angular momentum. However, if we assume that the reaction proceeds through an isolated resonance level, which is a well-defined state with definite total angular momentum, then we can regard the initial system as comprising the target nucleus, a, together with those components of the incident radiation, $\mathscr{L}_1$, which together form a state of definite parity and total angular momentum equal to the angular momentum, b, of the resonance level. We assume the target nuclei are randomly oriented. After the absorption of the radiation we have a compound nucleus with spin b. The initial state is represented by the product of two density matrices, or their corresponding statistical tensors, whith are transformed to a representation wherein the angular-momentum coupling scheme is altered from $\langle a a'(k_a), L_1 L_1'(k_1), k_b|\rangle$ to $\langle a L_1(b), a L_1'(b), k_b|\rangle$ similar to (11.25). The statistical tensor for the whole system after the interaction is found by means of (11.3), (11.4), so that

$$\left.\begin{aligned}\varrho_{k_b\varkappa_b}(b\,B, b\,B') = \sum_{\substack{k_1 k_a \varkappa_1 \varkappa_a\\ a\,a' L_1 L_1' l_1 l_1'}} \varrho_{k_a\varkappa_a}(a\,a')\,\varrho_{k_1\varkappa_1}(\mathscr{L}_1\mathscr{L}_1')\,\times\\ \times (k_b\varkappa_b|\,k_a\varkappa_a, k_1\varkappa_1)\cdot \hat{b}^2\hat{k}_1\hat{k}_a\begin{Bmatrix} a & L_1 & b\\ a' & L_1' & b\\ k_a & k_1 & k_b\end{Bmatrix}\times\\ \times\langle a\,A_t|\,\mathscr{O}|\,a\,A_0\rangle\langle a'\,A_t'|\,\mathscr{O}|\,a'\,A_0'\rangle^*.\end{aligned}\right\}\qquad(12.1)$$

The angular correlation is expressed, in a manner similar to (11.21), as

$$W = \sum_{b\,B\,B' k_b\varkappa_b} \varrho_{k_b\varkappa_b}(b\,B, b\,B')\,\varepsilon^*_{k_b\varkappa_b}(b\,B, b\,B'),\qquad(12.2)$$

where the individual statistical tensors are

$$\varrho_{k_a\varkappa_a}(a\,a') = \frac{1}{\hat{a}}\,\delta_{a'a}\,\delta_{k_a 0}\,\delta_{\varkappa_a 0},\qquad(12.3)$$

and

$$\left.\begin{aligned}\varrho_{k\varkappa}(\mathscr{L}\mathscr{L}') = \sum_{\sigma\sigma'\mu\mu'\varkappa'} (-)^{L'-\mu'}(L\,\mu, L'-\mu'|\,k\,\varkappa')\,\langle 0\,\sigma|\,L\,l\,\mu\rangle^*\langle 0\,\sigma'|\,L'\,l'\,\mu'\rangle\times\\ \times\langle\sigma|\,\varrho\,|\sigma'\rangle\,\mathscr{D}^k_{\varkappa\varkappa'}(\mathscr{R}).\end{aligned}\right\}\qquad(12.4)$$

$\langle\sigma_1|\,\varrho\,|\sigma_1'\rangle$ describes the state of polarisation of the incident beam (and is $\frac{1}{\hat{s}^2}\,\delta_{\sigma_1\sigma_1'}$ if the beam is unpolarised). Eq. (12.4) can be written, similarly to (11.30), as:

$$\varrho_{k_1\varkappa_1}(\mathscr{L}_1\mathscr{L}_1') = \sum_{\varkappa_1'} c_{k_1\varkappa_1'}(\mathscr{L}_1\mathscr{L}_1')\,\mathscr{D}^{k_1}_{\varkappa_1\varkappa_1'}(\mathscr{R}_1).\qquad(12.5)$$

As long as we are dealing with unpolarised beams and detectors insensitive to polarisation, both $\langle\sigma|\,\varrho\,|\sigma'\rangle$ and $\langle\sigma|\,\varepsilon\,|\sigma'\rangle$ are simply multiples of δ-functions, so that Eqs. (11.30) and (12.5) are only formally different. We obtain the resultant angular correlation

$$\left.\begin{aligned} W(\Omega_1\Omega_2) \sim \frac{\hat{b}^4}{\hat{a}^2}(-)^{a+c-2b}\sum(-)^{-L_1-L_2} W(b\,b\,L_1 L_1'; k\,a)\, W(b\,b\,L_2 L_2'; k\,c)\times \\ \times c_{k\varkappa_1}(\mathscr{L}_1\mathscr{L}_1')\, c^*_{k\varkappa_2}(\mathscr{L}_2\mathscr{L}_2')\, \mathscr{D}^k_{\varkappa_2\varkappa_1}(\vartheta_1\vartheta_2\vartheta_3)\times \\ \times\langle c\|\mathscr{L}_2\|b\rangle\langle c\|\mathscr{L}_2'\|b\rangle^*\langle a\|\mathscr{L}_1\|b\rangle\langle a\|\mathscr{L}_1'\|b\rangle^*, \end{aligned}\right\} \quad (12.6)$$

where the summation is over $L_1 L_1' L_2 L_2' l_1 l_1' l_2 l_2' k \varkappa_1 \varkappa_2$. This angular correlation differs from that obtained in (10.16) for the case of emission by (i) the substitution of $c_{k_1\varkappa_1'}(\mathscr{L}_1'\mathscr{L}_1)$ by $c_{k_1\varkappa_1'}(\mathscr{L}_1\mathscr{L}_1')$, and (ii) the multiplication by $\frac{\hat{b}^2}{\hat{a}^2}(-)^{k_1}$ and (iii) the appearance of matrix elements $\langle a\|\mathscr{L}_1\|b\rangle$, etc., which are absorption instead of emission matrix elements. For unpolarised radiation, Eq. (12.6) is clearly symmetric on interchanging initial and final states.

II. The parameters of radiation for some specific radiations.

In order to show what the angular correlation reduces to in its simplest forms, we shall find the values of the parameters of radiation $c_{k\varkappa}$ for two types of radiation, (1) heavy particle radiation with or without spin, and (2) gamma radiation. (In both cases we deal here with polarisation-insensitive detectors.) These will then be applied to the formula (10.16), to obtain explicit correlation formulae for several double cascade decay processes.

13. Heavy particles. The radiation is assumed to be in a plane-wave state, travelling in a direction characterised by $\Omega=(\vartheta,\varphi)$, and in a spin state designated by σ. If the coordinate z-axis coincides with the direction of propagation, the plane wave can be expanded into a sum of spherical waves by the Rayleigh plane-wave expansion,

$$\Psi(0\,\sigma) = e^{i\mathfrak{k}z}\chi_s^\sigma = \sqrt{4\pi}\sum_l j_l(\mathfrak{k}r)\,\hat{l}\,(i^l Y_l^0)\,\chi_s^\sigma, \quad (13.1)$$

where $\mathfrak{k}$ denotes the wave number and $j_l(\mathfrak{k}r)$ the spherical Bessel function. The phase i^l is retained with Y_l^0 to ensure that the equations satisfy invariance with respect to time reversal[1]. Combining the spin and angular-momentum functions vectorially, the wavefunction $\Psi(0\sigma)$ becomes

$$\Psi(0\,\sigma) \equiv \langle 0\,\sigma|\rangle = \text{const}\sum_{L l \Lambda}\hat{l}\,(l\,0, s\,\sigma|L\,\Lambda)\langle L\,\Lambda\,l|\rangle, \quad (13.2)$$

where

$$\langle L\,\Lambda\,l|\rangle \sim \sum_\lambda (l\,\lambda, s\,\Lambda-\lambda|L\,\Lambda)\, i^l Y_l^\lambda \chi_s^{\Lambda-\lambda} j_l(\mathfrak{k}r). \quad (13.3)$$

With respect to an arbitrary frame, which transforms itself into the frame fixed to the propagation by means of a rotation,

$$\langle\Omega\,\sigma|\rangle = \text{const}\sum_{L\Lambda l\Lambda'}\hat{l}\,(l0, s\sigma|L\Lambda)(L\Lambda'\,l|\rangle\,\mathscr{D}^L_{\Lambda'\Lambda}(\mathscr{R}), \quad (13.4)$$

$$\equiv \sum_{L\Lambda' l}\langle\Omega\,\sigma|L\Lambda'\,l\rangle\langle L'\Lambda'\,l|\rangle, \quad (13.5)$$

so that

$$\langle\Omega\,\sigma|L\Lambda\,l\rangle = \text{const}\sum_{\Lambda'}\hat{l}\,(l\,0, s\,\sigma|L\Lambda')\,\mathscr{D}^{L\,*}_{\Lambda\Lambda'}(\mathscr{R}). \quad (13.6)$$

---

[1] See also Eq. (14.7) and footnote to Eq. (24.2).

Also

$$\langle \Omega\,\sigma | L\,\Lambda\,l\rangle = \sum_{\Lambda'} \langle 0\,\sigma | L\,\Lambda'\,l\rangle\, \mathscr{D}^{L\,*}_{\Lambda\Lambda'}(\mathscr{R}). \tag{13.7}$$

Thus

$$\langle 0\,\sigma | L\,\Lambda\,l\rangle = \frac{\hat{l}}{\sqrt{4\pi}}\,(l\,0, s\,\sigma | L\Lambda), \tag{13.8}$$

where the constant $1/\sqrt{4\pi}$ is chosen so that $\langle \Omega\,\sigma | L\Lambda\,l\rangle$ is normalised to unity for the case $\sigma=0$.

Restricting ourselves to the case where the polarisation is not observed, so that

$$\langle \sigma | \varepsilon | \sigma'\rangle = \frac{\delta_{\sigma\sigma'}}{\hat{s}^2}, \tag{13.9}$$

the radiation parameters $c_{k0}(\mathscr{L}\mathscr{L}')$, (10.11), for heavy particles with spin are

$$\left.\begin{aligned} c_{k0}(\mathscr{L}\mathscr{L}') &= \frac{\hat{l}\,\hat{l}'}{4\pi\,\hat{s}^2}\sum_{\sigma}(-)^{L'-\sigma}(L\,\sigma, L'-\sigma | k\,0)\,(l\,0, s\,\sigma | L\,\sigma)\,(l'\,0, s\,\sigma | L'\,\sigma)\\ &= \frac{\hat{l}\,\hat{l}'\,\hat{L}\,\hat{L}'}{4\pi\,\hat{s}^2}\,(l\,0, l'\,0 | k\,0)\,(-)^{L+k-s}\,W(l\,l'\,L\,L'; k\,s). \end{aligned}\right\} \tag{13.10}$$

This expression corresponds to absorbed radiation; for emitted radiation it must be multiplied by $\hat{s}^2$.

If the radiation consists of particles with spin, $s=\frac{1}{2}$, e.g. neutrons or protons, the radiation parameters are

$$c_{k0}(\mathscr{L}\mathscr{L}') = \frac{\hat{L}\,\hat{L}'}{8\pi}\,(-)^{L-\frac{1}{2}+k}\left(L\,\tfrac{1}{2}, L'-\tfrac{1}{2}\,\middle|\,k\,0\right). \tag{13.11}$$

The above corresponds to absorption; for emission this must be multiplied by 2.

If we are concerned with alpha-particles ($s=0$),

$$c_{k0}(\mathscr{L}\mathscr{L}') = \frac{\hat{l}\,\hat{l}'}{4\pi}\,(-)^{l'}\,(l\,0, l'\,0 | k\,0), \tag{13.12}$$

following (6.17).

For charged particles there are phase shifts (ξ_l) in the spherical waves introduced by the (screened) Coulomb field of the nucleus. These are frequently included in the radiation parameters (13.11), (13.12), (e.g. BIEDENHARN and ROSE [9]). We have used for our base states the spherical waves for uncharged radiation, and taken into account the effects of the Coulomb field in the reduced matrix elements, so that our radiation parameters are independent of energy and nuclear charge.

14. Gamma radiation. Electromagnetic radiation is treated by considering the vector potential of the photons as a wavefunction for a particle of spin one. The multipole expansion of the field leads to two types of multipoles, electric and magnetic, which arise for a given value of L, but which have different parities. Thus magnetic multipoles correspond to $l=L$ only, with parity, $p=(-)^{L+1}$. Electric multipoles satisfy $p=(-)^L$ with $l=L\pm1$. In addition, we must have $\sigma=\pm1$ (but not 0), which describes the condition that longitudinal photons do not exist. The states can be described by quantum numbers L, Λ and p, and the expansion coefficients become

$$\left.\begin{aligned}\langle 0\,\sigma|L\Lambda\,p\rangle &= \sum_{l=L-1}^{L+1}\langle 0\,\sigma|L\Lambda\,l\rangle\,\langle l\,|\,p\rangle,\\ &= \sum_{l=L-1}^{L+1}\frac{\hat{l}}{\sqrt{4\pi}}\,(l\,0,1\,\sigma|\,L\Lambda)\,\langle l\,|\,p\rangle,\end{aligned}\right\}\tag{14.1}$$

where $\langle l|p\rangle$ is determined for each type of radiation and satisfies

$$\sum_{l=L-1}^{L+1}|\langle l\,|\,p\rangle\,|^2=1.\tag{14.2}$$

Since $l=L$ for magnetic multipoles, use of Table 1 shows that $\sigma=0$ is prohibited, and that

$$\left.\begin{aligned}&\langle 0\,\sigma\,|\,L\Lambda\,p\rangle_{\text{magnetic}}\\ &\quad= e^{i\varphi_m}\,\delta_{\sigma\Lambda}\frac{\hat{L}}{\sqrt{8\pi}}\,\sigma,\end{aligned}\right\}\tag{14.3}$$

where the phase, φ_m, is real. Using $\sigma=\pm 1$ and the restriction for electric multipoles, we have, from Table 1 and (14.1),

$$\frac{\langle L-1\,|\,p\rangle}{\langle L+1\,|\,p\rangle}=\sqrt{\frac{L+1}{L}}.\tag{14.4}$$

Table 1. *Clebsch-Gordan coefficients for* $\langle l\,0,1\,\sigma|L\,\sigma\rangle$.

l	$\sigma=0$	$\sigma=\pm 1$
$L-1$	$\sqrt{\frac{L}{2L-1}}$	$\sqrt{\frac{L-1}{2\,(2L-1)}}$
L	0	$-\sigma\sqrt{\frac{1}{2}}$
$L+1$	$-\sqrt{\frac{L+1}{2L+3}}$	$\sqrt{\frac{L+2}{2(\,2L+3)}}$

With (14.2) we find

$$\langle 0\,\sigma|L\Lambda\,p\rangle_{\text{electric}}=e^{i\varphi_e}\,\delta_{\sigma\Lambda}\,\frac{\hat{L}}{\sqrt{8\pi}}.\tag{14.5}$$

The relative phase of φ_m to φ_e must yet be determined. This is done by reference to GOERTZEL's [21] expression for the vector potential (suitably normalised in terms of its multipoles,

$$\langle 0\,\sigma|\rangle=\pi\sum_{L}i^L\,\hat{L}\,[A_{L\Lambda}^{\text{magnetic}}+i\,\sigma\,A_{L\Lambda}^{\text{electric}}].\tag{14.6}$$

If the interaction is invariant with respect to time inversion, the wavefunction $A_{L\Lambda}$ must be defined so that

$$K_t A_{L\Lambda}=(-)^{L-\Lambda}A_{L-\Lambda}\tag{14.7}$$

for both electric and magnetic multipoles, where the time-reversal operator, K_t in this case, is the charge-conjugate operator (WIGNER and EISENBUD [43]). Using ROSE's expression[1] for $A_{L\Lambda}$, we can define new quantities, $\bar{A}_{L\Lambda}$, which satisfy (14.7):

$$\bar{A}_{L\Lambda}^{\text{magnetic}}\equiv i^L\,A_{L\Lambda}^{\text{magnetic}},\tag{14.8}$$

$$\bar{A}_{L\Lambda}^{\text{electric}}\equiv i^{L\pm 1}\,A_{L\Lambda}^{\text{electric}},\tag{14.9}$$

so that the proper vector potential is

$$\langle 0\,\sigma|\rangle=\pi\sum_{L}\hat{L}\,[\bar{A}_{L\Lambda}^{\text{magnetic}}+\sigma\,\bar{A}_{L\Lambda}^{\text{electric}}].\tag{14.10}$$

[1] M. E. ROSE: Ref. [18], p. 30.

The arbitrariness is not quite removed by the condition of invariance with respect to time inversion. We thus find:

$$\langle 0\,\sigma | L\,\Lambda\,p\rangle_{\text{magnetic}} = \pm\,\sigma\,\delta_{\sigma\Lambda}\frac{\hat{L}}{\sqrt{8\pi}}\,; \quad p = (-)^{L}, \tag{14.11a}$$

$$\langle 0\,\sigma | L\,\Lambda\,p\rangle_{\text{electric}} = \delta_{\sigma\Lambda}\frac{\hat{L}}{\sqrt{8\pi}}\,; \quad p = (-)^{L+1}. \tag{14.11b}$$

For both pure electric or both pure magnetic multipole gamma radiation,

$$c_{k\,0}(L\,L) = \frac{\hat{L}^2(-)^{L-1}}{8\pi}(L\,1, L-1\,|\,k\,0)\,. \tag{14.12}$$

If L and L' are unequal, they may refer (*a*) to like multipoles, in which case

$$c_{k\,0}(L\,L') = \frac{\hat{L}\,\hat{L}'(-)^{L'-1}}{8\pi}(L\,1, L'-1\,|\,k\,0)\,, \tag{14.13}$$

or (*b*) to unlike multipoles, in which case

$$c_{k\,0}(L\,L') = \pm\frac{\hat{L}\,\hat{L}'(-)^{L'-1}}{8\pi}(L\,1, L'-1\,|\,k\,0)\,. \tag{14.14}$$

(14.12) to (14.14) refer to absorption of gamma rays; for emission, they must be multiplied by 2.

III. Directional correlation for unpolarised radiations.

15. Directional correlation functions for two unpolarised cascade radiations. As is to be expected, unpolarised radiation is described by parameters of radiation $c_{k\varkappa}$ which vanish if $\varkappa \neq 0$. The general angular correlation formula (10.16) in this case reduces to

$$\left.\begin{aligned} W(\Omega_1\Omega_2) \sim (-)^{a+c-2b}\,\hat{b}^2 \sum_{\substack{L_1 L_1' L_2 L_2' \\ l_1 l_1' l_2 l_2' k}} (-)^{k-L_1-L_2}\,W(b\,b\,L_1\,L_1';k\,a)\,\times \\ \times\,W(b\,b\,L_2 L_2';k\,c)\cdot\langle c\|\mathcal{L}_2\|b\rangle\,\langle c\|\mathcal{L}_2'\|b\rangle^*\times \\ \times\,\langle b\|\mathcal{L}_1\|a\rangle\,\langle b\|\mathcal{L}_1'\|a\rangle^*\cdot c_{k\,0}(\mathcal{L}_1'\,\mathcal{L}_1)\,c_{k\,0}(\mathcal{L}_2\mathcal{L}_2')\,P_k(\cos\vartheta)\,, \end{aligned}\right\} \tag{15.1}$$

where $P_k(\cos\vartheta)$ represent the ordinary Legendre polynomials.

If we make use of the Hermitian character of the efficiency and statistical tensors, (11.14) say, it follows that

$$c_{k\varkappa}(\mathcal{L}\,\mathcal{L}') = (-)^{L'-L+\varkappa}\,c^*_{k-\varkappa}(\mathcal{L}'\,\mathcal{L})\,. \tag{15.2}$$

If, further, we express $\langle c\|\mathcal{L}_2\|b\rangle$ in terms of $\langle b\|\mathcal{L}_2\|c\rangle$ by means[1] of (10.12), it then follows that (15.1) may be rewritten using a compact notation [cf. (11)]

$$\left.\begin{aligned} W(\Omega_1\Omega_2) &= \sum_k A_k[1,2]\,P_k(\cos\vartheta)\,, \\ &= \sum_k A_k(a,b\,b)\,A_k(b\,b,c)\,P_k(\cos\vartheta)\,. \end{aligned}\right\} \tag{15.3}$$

This latter notation draws attention to the fact that each of the terms $A_k[1,2]$ may be split into two factors, one of which depends entirely on the first transition

[1] Provided that the process can be described by a Hermitian tensor operator [*17*], this is valid and $A_k(b\,b,c)$, as written in (15.4.), is correct. Otherwise, $A_k(b\,b,c)$ must be rewritten in terms of $\langle c\|\mathcal{L}_2\|b\rangle$, etc., by reapplication of (10.12).

and the other on the second. These factors are given by:

$$\left.\begin{aligned} A_k(a,bb) &= \sum_{\mathscr{L}_1\mathscr{L}_1'} \langle b\|\mathscr{L}_1\|a\rangle\langle b\|\mathscr{L}_1'\|a\rangle^* c_{k0}^*(\mathscr{L}_1\mathscr{L}_1')\times \\ &\qquad\times\hat{b}(-)^{a-b+k-L_1'}\,W(b\,b\,L_1L_1';ka), \\ A_k(bb,c) &= \sum_{\mathscr{L}_2\mathscr{L}_2'} \langle b\|\mathscr{L}_2\|c\rangle^*\langle b\|\mathscr{L}_2'\|c\rangle\, c_{k0}^*(\mathscr{L}_2\mathscr{L}_2')\times \\ &\qquad\times\hat{b}(-)^{c-b+k-L_2'}\,W(b\,b\,L_2L_2';kc). \end{aligned}\right\} \tag{15.4}$$

The grouping of the quantum numbers in A_k determines the choice of the factorised matrix elements in each of the above expressions. The factors A_k are normalised so that the angular correlation is expressed in standard form, i.e. with $A_0(a, bb) = A_0(bb, c) = 1$ [1]. To ensure this, the matrix element is measured in units such that if the radiation is confined to a single L_1,

$$|\langle b\|\mathscr{L}_1\|a\rangle|^2 = 4\pi. \tag{15.5}$$

We are dealing at present with an overall system of well-defined symmetry, and in particular well-defined parity. This is the parity of the initial nuclear state, a, in the case of cascade emission, and of the compound nuclear state, b, in the case of resonance reactions. In addition the parities of all the nuclear states involved are assumed to be well-defined, so that the radiations also have well-defined parities. Hence the efficiency tensor describing one of the radiations (11.27) is invariant under the parity transformation if the measurements of the spin do not correspond to a value of $\langle\sigma|\varepsilon|\sigma'\rangle$ which changes under this transformation. This of course is *a fortiori* the case if no spin measurements are made. From the properties of the rotation matrix (Sect. 7) we see then that under these circumstances only even values of k can occur in (11.30), and therefore the correlation function, (15.1), involves only even values of k. This is also apparent from the examination of the radiation parameters c_{k0} in particular cases (Sect. 14) [2].

A further generalisation can be made with respect to the complexity of the angular correlation. A consideration of the rank of the efficiency tensor $\varepsilon_{k\varkappa}(\mathscr{L}\mathscr{L}')$ from (15.1) shows that k cannot exceed $2L$ or $2L'$. Since the angular correlation can be expressed in terms of products $\varepsilon_{k_1\varkappa_1}(\mathscr{L}_1\mathscr{L}_1')\ \varepsilon_{k_b\varkappa_b}(bb')$, or $\varepsilon_{k_1\varkappa_1}(\mathscr{L}_1\mathscr{L}_1')$ $\varepsilon_{k_2\varkappa_2}(\mathscr{L}_2\mathscr{L}_2')$, and only those values of k_1, k_2, and k_2, are inserted into (11.33) that are equal to each other, it follows that k is less than or equal to the minimum of $(2L_1, 2L_2, 2b)$. This result can also be derived formally by making use of the triangular conditions for $W(abcd; ef)$ referred to in the mathematical preliminaries. The general correlation formula (10.16) contains $W(bbL_1L_1'; ka)$ and $W(bbL_2L_2'; kc)$. We must therefore have $\Delta(bbk)$, $\Delta(L_1L_1'k)$, and $\Delta(L_2L_2'k)$, so that $k \leq$ smallest of the integers $(2L, 2L_2, 2b)$. For the case of radiation of particles with spin, we can obtain the further condition $\Delta(ll'k)$ from Eq. (13.10), so that in addition $k \leq$ smallest of integers $(2l_1, 2l_2)$.

[1] The symmetry of (15.4) and the standard form in which the angular correlation is expressed are ensured by the choice of phases made in (10.12). The repetition of b in $A_k(a, bb)$ etc. emphasises that the intermediate state is sharp, and allows for a generalisation of A_k for the case of ill-defined intermediate states (cf. Chap. E).

[2] It may be pointed out that the requirement that the overall system has well-defined parity, or that *all* the nuclear states are of well-defined parity, is not strictly necessary for this result. It is sufficient, for the above argument, for either one of the radiations to have well-defined parity. In most practical cases this condition is equivalent to the more restrictive one.

Since k is even, half integral values of L, say $L = L' + \frac{1}{2}$, require that L be replaced by L' in our condition on the complexity of k. It is clear from this result that if the angular momenta of any of the radiations involved or of the indermediate state equal either 0 or $\frac{1}{2}$, the resultant angular distribution is isotropic.

The theorem that the angular distribution, for unpolarised radiations, can be expressed as an even power series of $\cos\vartheta$ involving no higher power than $\cos^{2n}\vartheta$, where n is restricted to the minimum of l_1, l_2, and b, follows from (15.1). An equivalent theorem was derived by YANG [22] for nuclear reactions from group-theoretical arguments concerning the invariance properties of the system under spatial rotations and under inversion, and the derivation is in principle equivalent to the one given here. YANG also extended the theorem to the cases of γ-γ correlation and γ-e^- correlation.

Reference to the initial form of the angular-correlation function in (10.1) shows that it is invariant with respect to interchange of final and initial nuclear states and of both radiations if the interaction is described by a Hamiltonian. This can also be verified by a consideration of the final formula (10.16). If we use (15.2) and express $\langle a\|\mathcal{L}_1\|b\rangle$ in terms of $\langle b\|\mathcal{L}_1\|a\rangle^*$ by (10.12), and if we further use the properties of the rotation matrix (7.7), (7.8), we find $W(\Omega_1\Omega_2)$ and $W(\Omega_2\Omega_1)$ agree, except for a factor which arises because of the lack of symmetry in averaging over initial and summing over final states of the system. If the angular correlations are expressed in standard form, they will agree exactly. This argument also applies to arbitrary interactions, provided that (10.12) is satisfied, and no regard is given to polarisation effects. In this case, we make use of the fact that the coefficient of the matrix elements is symmetrical with respect to $\mathcal{L}$ and $\mathcal{L}'$, so that it is only the real part of the matrix elements that occurs in the angular correlation. The symmetry of the angular correlation is then evident. A similar result applies to reactions.

If each of the radiations is pure, the angular correlation (15.1), except for constant factors, is denoted by

$$\left.\begin{aligned} W_{\mathcal{L}_1\mathcal{L}_1,\,\mathcal{L}_2\mathcal{L}_2}(\vartheta) &= \sum_k (-)^{-L_1-L_2}\, c_{k0}(\mathcal{L}_1\,\mathcal{L}_1)\, c_{k0}^*(\mathcal{L}_2\,\mathcal{L}_2)\times \\ &\quad\times W(b\,b\,L_1L_1;\,k\,a)\cdot W(b\,b\,L_2L_2;\,k\,c)\,P_k(\cos\vartheta)\,. \end{aligned}\right\}\qquad(15.6)$$

If one of the radiations is a superposition of radiations of different L, the values of L are either all even or all odd for particles, but for γ-radiation both even and odd L-values are possible provided all components have the same parity. [The most common type of γ-radiation mixture is $E(L+1)+M(L)$; in particular the mixture of electric quadrupole, $E2$, and magnetic dipole, $M1$.]

For example, if the first radiation is pure L_1l_1, and the second mixed L_2l_2, $L_2'l_2'$, then we can express the directional-correlation function as:

$$\left.\begin{aligned} W(\vartheta) &= |\langle c\|\mathcal{L}_2\|b\rangle|^2\, W_{\mathcal{L}_1\mathcal{L}_1,\,\mathcal{L}_2\mathcal{L}_2}(\vartheta) + |\langle c\|\mathcal{L}_2'\|b\rangle|^2\, W_{\mathcal{L}_1\mathcal{L}_1,\,\mathcal{L}_2'\mathcal{L}_2'}(\vartheta) + \\ &\quad + 2\,\mathrm{Re}\left[\langle c\|\mathcal{L}_2\|b\rangle\,\langle c\|\mathcal{L}_2'\|b\rangle^*\right] W_{\mathcal{L}_1\mathcal{L}_1,\,\mathcal{L}_2\mathcal{L}_2'}(\vartheta)\,. \end{aligned}\right\}\qquad(15.7)$$

The first two terms are the correlation functions, $W_{\mathcal{L}_1\mathcal{L}_1,\,\mathcal{L}_2\mathcal{L}_2}(\vartheta)$ and $W_{\mathcal{L}_1\mathcal{L}_1,\,\mathcal{L}_2'\mathcal{L}_2'}(\vartheta)$ for the pure radiations as given by (15.6) weighted by the appropriate reduced matrix elements. The third term represents the "interference" between the two components L_2, L_2', and is given by:

$$\left.\begin{aligned} W_{\mathcal{L}_1\mathcal{L}_1,\,\mathcal{L}_2\mathcal{L}_2'}(\vartheta) &= \sum_k W(b\,b\,L_1L_1;\,k\,a)\,W(b\,b\,L_2L_2';\,k\,c)\times \\ &\quad\times\left[(-)^{L_2}\,c_{k0}^*(\mathcal{L}_2\,\mathcal{L}_2')\,(-)^{L_1}\,c_{k0}(\mathcal{L}_1\,\mathcal{L}_1)\right] P_k(\cos\vartheta)\,. \end{aligned}\right\}\qquad(15.8)$$

The values of k are even, ranging from 0 to the smallest of the integers $(2L_1, 2l_1, 2L_2, 2l_2, 2b)$ and $(2L_1, 2l_1, 2L_2', 2l_2', 2b)$ in $W_{\mathscr{L}_1\mathscr{L}_1,\mathscr{L}_2\mathscr{L}_2}(\vartheta)$ and $W_{\mathscr{L}_1\mathscr{L}_1,\mathscr{L}_2'\mathscr{L}_2'}(\vartheta)$ respectively, and from 0 to the smallest of $(l_2+l_2', L_2+L_2', 2l_1, 2L_1, 2b)$ in $W_{\mathscr{L}_1\mathscr{L}_1,\mathscr{L}_2\mathscr{L}_2'}(\vartheta)$. There is no contribution to this last term for $k=0$, since $W(b\,b\,L_2\,L_2'; 0\,c)$ vanishes if $L_2 \neq L_2'$.

BIEDENHARN and ROSE [9] introduced a useful method for presenting correlation factors of radiations other than gamma radiation in terms of those derived for gamma radiation. Denoting $c_{k0}(LL')$ by $c_{k0}(LL';\gamma)$ for the emission of a gamma ray, they defined particle parameters, $b_k(\mathscr{L}\mathscr{L}'; x)$ for particle x, by:

$$b_k(\mathscr{L}\mathscr{L}'; x) \equiv \frac{c_{k0}(\mathscr{L}\mathscr{L}'; x)}{c_{k0}(\mathscr{L}\mathscr{L}';\gamma)}, \tag{15.9}$$

so that, for example, of x and y represent two different types of radiation which are pure,

$$W_{xy}(\vartheta) = \sum_k b_k(\mathscr{L}_1\mathscr{L}_1; x)\, b_k(\mathscr{L}_2\mathscr{L}_2; y)\cdot A_k[1,2]\, P_k(\cos\vartheta). \tag{15.10}$$

The selection of the γ radiation as the standard is purely a matter of convenience.

16. Directional correlation for three successive radiations. The angular correlation of three successive γ-rays was first found by BIEDENHARN, ARFKEN, and ROSE[1], and the theory was also applied to the case where an emission process was replaced by an absorption process. The complexity and highly tedious nature of the summation over magnetic quantum numbers, even on making use of the parameters of radiation, allowed only very simple cases to be considered. The method of statistical tensors and efficiency tensors enables the procedures used for the double cascade process to be extended to the problem of triple cascades and the more general multiple cascade, and later work by COX and TOLHOEK[2], COESTER and JAUCH [29], and SATCHLER [30], avoid this complication by the use of this tensor formalism.

We shall consider first the successive emission of three radiations, whose polarisations are not detected. The procedure is already indicated in the work following (11.23). Thus the efficiency tensor for the system $\varepsilon_{k_a\varkappa_a}(aA, a'A')$ is expressed in terms of products $\varepsilon_{k_1\varkappa_1}(\mathscr{L}_1\mathscr{L}_1')\,\varepsilon_{k_2\varkappa_2}(\mathscr{L}_2\mathscr{L}_2')\,\varepsilon_{k_3\varkappa_3}(\mathscr{L}_3\mathscr{L}_3')\,\varepsilon_{k_d\varkappa_d}(dd)$ with $c_{k\varkappa'}(\mathscr{L}\mathscr{L}')=0$, unless $\varkappa'=0$, and

$$\varepsilon_{k_d\varkappa_d}(dd) = \hat{d}\,\delta_{k_d 0}\,\delta_{\varkappa_d 0}. \tag{16.1}$$

We find

$$\left.\begin{aligned} W(\Omega_1\Omega_2\Omega_3) &\sim \frac{1}{\hat{a}}\sum R_{da}\,\delta_{k_a0}\,\delta_{\varkappa_a0}\,\mathscr{D}^{k_1*}_{\varkappa_1 0}(\mathscr{R}_1)\,\mathscr{D}^{k_2*}_{\varkappa_2 0}(\mathscr{R}_2)\,\mathscr{D}^{k_3*}_{\varkappa_3 0}(\mathscr{R}_3)\times\\ &\times c^*_{k_1 0}(\mathscr{L}_1\mathscr{L}_1')\,c^*_{k_2 0}(\mathscr{L}_2\mathscr{L}_2')\,c^*_{k_3 0}(\mathscr{L}_3\mathscr{L}_3')\,(\hat{a}^2\hat{k}_1\hat{k}_b)\,(\hat{b}^2\hat{k}_2\hat{k}_c)\,(\hat{c}^2\hat{k}_3\hat{k}_d)\times\\ &\times\begin{Bmatrix} b & L_1 & a\\ b & L_1' & a\\ k_b & k_1 & k_a\end{Bmatrix}\begin{Bmatrix} c & L_2 & b\\ c & L_2' & b\\ k_c & k_2 & k_b\end{Bmatrix}\begin{Bmatrix} d & L_3 & c\\ d & L_3' & c\\ k_d & k_3 & k_c\end{Bmatrix}(k_b\varkappa_b, k_1\varkappa_1|k_a\varkappa_a)\times\\ &\times (k_c\varkappa_c, k_2\varkappa_2|k_b\varkappa_b)\,(k_d\varkappa_d, k_3\varkappa_3|k_c\varkappa_c)\,\hat{d}\,\delta_{k_d0}\,\delta_{\varkappa_d0}, \end{aligned}\right\} \tag{16.2}$$

where the summation is over:

$$k_a k_b k_c k_d k_1 k_2 k_3 L_1 L_2 L_3 l_1 l_2 l_3 L_1' L_2' L_3' l_1' l_2' l_3' \varkappa_a \varkappa_b \varkappa_c \varkappa_d \varkappa_1 \varkappa_2 \varkappa_3$$

[1] L. C. BIEDENHARN, G. B. ARFKEN and M. E. ROSE: Phys. Rev. **83**, 586 (1951).

[2] J. A. M. COX and H. A. TOLHOEK: Physica, Haag **19**, 1178 (1953).

and

$$R_{da} \equiv \langle d \| \mathscr{L}_3 \| c \rangle \langle c \| \mathscr{L}_2 \| b \rangle \langle b \| \mathscr{L}_1 \| a \rangle \cdot \langle d \| \mathscr{L}'_3 \| c \rangle^* \langle c \| \mathscr{L}'_2 \| b \rangle^* \langle b \| \mathscr{L}'_1 \| a \rangle^* . \quad (16.3)$$

This reduces to

$$\left.\begin{aligned} W(\Omega_1 \Omega_2 \Omega_3) \sim (-)^{a-b-c+d} \hat{b}^2 \hat{c}^2 & \sum_{\substack{L_1 L_2 L_3 L'_1 L'_2 L'_3 \\ l_1 l_2 l_3 l'_1 l'_2 l'_3 \\ k_1 k_2 k_3}} \frac{(-)^{k-L'_1-L_3}}{\hat{k}_3^2} \times \\ & \times W(b\, b\, L_1 L'_1; k_1 a)\, W(c\, c\, L_3 L'_3; k_3 d) \times \\ & \times \begin{Bmatrix} c & L_2 & b \\ c & L'_2 & b \\ k_3 & k_2 & k_1 \end{Bmatrix} c^*_{k_1 0}(\mathscr{L}_1 \mathscr{L}'_1)\, c^*_{k_2 0}(\mathscr{L}_2 \mathscr{L}'_2)\, c^*_{k_3 0}(\mathscr{L}_3 \mathscr{L}'_3) \times \\ & \times \sum_{\varkappa_1 \varkappa_2 \varkappa_3} Y_{k_1}^{\varkappa_1 *}(\Omega_1)\, Y_{k_2}^{\varkappa_2 *}(\Omega_2)\, Y_{k_3}^{\varkappa_3}(\Omega_3)\, (k_1 \varkappa_1, k_2 \varkappa_2 | k_3 - \varkappa_3)\, R_{da} . \end{aligned}\right\} \quad (16.4)$$

In terms of the compact notation introduced in (15.3),

$$W(\Omega_1 \Omega_2 \Omega_3) = \sum_{k_1 k_2 k_3} A_{k_1}(a, b\, b)\, A_{k_3}(c\, c, d) \cdot R_{k_1 k_2 k_3}(b\, b\, c\, c) \cdot S_{k_1 k_2 k_3}(\Omega_1 \Omega_2 \Omega_3), \quad (16.5)$$

where the angular dependent term $S_{k_1 k_2 k_3}$ is

$$S_{k_1 k_2 k_3}(\Omega_1 \Omega_2 \Omega_3) = (4\pi)^{\frac{3}{2}} \sum_{\varkappa_1 \varkappa_2 \varkappa_3} Y_{k_1}^{\varkappa_1 *}(\Omega_1)\, Y_{k_2}^{\varkappa_2 *}(\Omega_2)\, Y_{k_3}^{\varkappa_3}(\Omega_3)\, (k_1 \varkappa_1, k_2 \varkappa_2 | k_3 \varkappa_3), \quad (16.6)$$

the intermediate radiation is described by

$$R_{k_1 k_2 k_3}(b\, b\, c\, c) = \sum_{L_2 L'_2 l_2 l'_2} \langle c \| \mathscr{L}_2 \| b \rangle \langle c \| \mathscr{L}'_2 \| b \rangle^* c_{k_2 0}(\mathscr{L}_2 \mathscr{L}'_2) \frac{\hat{b}\, \hat{c}}{\hat{k}} (-)^{k_3} \begin{Bmatrix} c & c & k_3 \\ L_2 & L'_2 & k_2 \\ b & b & k_1 \end{Bmatrix} \quad (16.7)$$

and the first and third radiations are described by $A_{k_1}(a, b\, b)$, $A_{k_3}(c\, c, d)$, respectively, as defined in (15.4).

If the intermediate radiation is not observed, and a correlation is made between the first and third radiation, this correlation function is found by averaging (16.5) over all possible values of Ω_2, i.e. by integrating $S_{k_1 k_2 k_3}(\Omega_1 \Omega_2 \Omega_3)$ over Ω_2. Using the orthonormal properties of the spherical harmonics, we find that $k_2 = \varkappa_2 = 0$, so that

$$\int S_{k_1 k_2 k_3}(\Omega_1 \Omega_2 \Omega_3)\, d\Omega_2 = 4\pi\, \hat{k}^2 \cdot P_k(\cos\vartheta)\, \delta_{k_1 k_3}\, \delta_{k_2 0}, \quad (16.8)$$

where ϑ is the angle between the first and third radiations. Also for this case,

$$R_{k 0 k}(b\, b\, c\, c) = \sum_{L_2 l_2} |\langle c \| \mathscr{L}_2 \| b \rangle|^2 \frac{\hat{b}\, \hat{c}}{\hat{k}^2\, \hat{L}_2} (-)^{L_2 - b - c} c^*_{00}(\mathscr{L}_2 \mathscr{L}_2)\, W(b\, b\, c\, c; k L_2), \quad (16.9)$$

so that the angular correlation takes on the form

$$W(\vartheta) = \sum_k A_k[1, 3]\, P_k(\cos\vartheta) \quad (16.10)$$

where

$$A_k[1, 3] = A_k(a, b\, b)\, \bar{A}_k(b\, c)\, A_k(c\, c, d), \quad (16.11)$$

with

$$\bar{A}_k(b\, c) = \frac{1}{4\pi} \sum_{L_2 l_2} \hat{b}\, \hat{c}\, (-)^{b+c-L_2} W(b\, b\, c\, c; k L_2)\, |\langle c \| \mathscr{L}_2 \| b \rangle|^2 . \quad (16.12)$$

The constants are chosen so that $W(\vartheta)$ is of normal form, i.e. $A_0[1, 3] = 1$ according to (15.5). Notice that when the intermediate radiation is not observed, there are no interference terms of the type $\langle c \| \mathcal{L}_2 \| b\rangle \langle c \| \mathcal{L}_2' \| b\rangle^*$. The different $\mathcal{L}_2$-components occur incoherently.

In addition, k is even, having a maximum value

$$k \leq \text{minimum}\,(2L_1, 2l_1, 2L_3, 2l_3, 2b, 2c). \tag{16.13}$$

A special simplification of the general angular correlation can be made by observing a pair of radiations, labelled by $n(p)$ and $n(q)$, which are either parallel or antiparallel, and correlating these with the third, labelled by $n(r)$, making an angle ϑ_r with respect to either of the others. (Since k is even, there is a front-back symmetry so that both parallel and antiparallel radiation present the same correlation with respect to the r-radiation.) If the direction of propagation of the p-radiation determines the polar axis, then $\varkappa_{n(p)} = \varkappa_{n(q)} = 0$, and the angular dependence becomes

$$S_{k_1 k_2 k_3}(\vartheta_r) = \hat{k}_1 \hat{k}_2 \hat{k}_3 \,(k_1 0, k_2 0 | k_3 0)\, P_{k_{n(r)}}(\cos\vartheta_r) \tag{16.14}$$

where $n(p)$ and $n(q)$ are any two of 1, 2, and 3, and $n(r)$ is the remaining label.

For the case of a triple correlation where the first radiation is absorbed instead of emitted, there is no change in the result, except for the substitution of $c^*_{k_1 \varkappa_1}(\mathcal{L}_1 \mathcal{L}_1')$ by $c^*_{k_1 \varkappa_1}(\mathcal{L}_1' \mathcal{L}_1)$ and the multiplication of $A_{k_1}(a, bb)$ by $\frac{\hat{b}^2}{\hat{a}^2}(-)^{k_1}$.

The angular dependence is contained in the term $S_{k_1 k_2 k_3}(\Omega_1 \Omega_2 \Omega_3)$, which, for a polar axis defined by the direction of the captured particle, is

$$S_{k_1 k_2 k}(\Omega_2 \Omega_3) = 4\pi \hat{k}_1 \sum_{k} Y^{\varkappa *}_{k_3}(\vartheta_2 0)\, Y^{\varkappa}_{k_3}(\vartheta_3 \varphi_3)\,(k_1 0, k_2 \varkappa | k_3 \varkappa). \tag{16.15}$$

17. Directional correlations for many radiations (SATCHLER [30]). Expression (16.5) can be generalise to describe the directional correlation of any three radiations from an arbitrary cascade where the intermediate radiations are not observed.

To specify a general cascade, we will modify our notation as follows: inserting the values of the angular momenta of the radiations, emitted or absorbed at each transition, between the initial and resultant nuclear angular-momentum states, the cascade is represented as

$$a, \mathcal{L}_1, b, \mathcal{L}_2, c, \ldots, q, \mathcal{L}_{n(q)}, r, \ldots, v, \mathcal{L}_{n(v)}, w. \tag{17.1}$$

Using the direction of L_1 as the polar axis, the angular correlation between radiations $\mathcal{L}_1$, $\mathcal{L}_{n(q)}$ and $\mathcal{L}_{n(v)}$ is

$$\left.\begin{aligned} W(\vartheta_{n(q)}, \Omega_{n(v)}) \sim \sum_{k_1 k_2 k_3} A_{k_1}(a, bb)\, A_{k_3}(vv, w)\, R_{k_1 k_2 k_3}(qqrr)\, S_{k_1 k_2 k_3}(\vartheta_{n(q)} \Omega_{n(v)}) \times \\ \times U_{k_1}(L_2 b c)\, U_{k_1}(L_3 c d) \ldots U_{k_1}(L_{n(p)} p q)\, U_{k_3}(L_{n(r)} r s) \ldots U_{k_3}(L_{n(v)} u v), \end{aligned}\right\} \tag{17.2}$$

where

$$\left.\begin{aligned} R_{k_1 k_2 k_3}(qqrr) = \sum_{\mathcal{L}_{n(q)} \mathcal{L}'_{n(q)}} \langle r \| \mathcal{L}_{n(q)} \| q\rangle \langle r \| \mathcal{L}'_{n(q)} \| q\rangle^* c^*_{k_2 0}(\mathcal{L}_{n(q)} \mathcal{L}'_{n(q)}) \times \\ \times \frac{\hat{q}\,\hat{r}}{\hat{k}_3} (-)^{k_3} \begin{Bmatrix} q & q & k_1 \\ L_{n(q)} & L'_{n(q)} & k_2 \\ r & r & k_3 \end{Bmatrix}. \end{aligned}\right\} \tag{17.3}$$

We employ the notation used by SATCHLER, although the terms as defined here do not coincide exactly with his. A_k is defined as in (15.4)

$$\left.\begin{aligned}&A_{k_1}(a, b\,b)\\&\quad=\sum_{\mathscr{L}_1\mathscr{L}_1'}\langle b\|\mathscr{L}_1\|a\rangle\langle b\|\mathscr{L}_1'\|a\rangle^* c^*_{k_1 0}(\mathscr{L}_1\mathscr{L}_1')\,\hat{b}\,(-)^{a-b-k_1-L_1'}\,W(b\,b\,L_1L_1';k\,a)\,.\end{aligned}\right\}\quad(17.4)$$

The only influence of the unobserved radiation on the angular correlation is contained in the expression $U_{k_1}(L_1\,b\,c)$ etc., normalised so that $U_0(L_1\,b\,c)=1$.

$$U_{k_1}(L_2\,b\,c)=\hat{b}\,\hat{c}\,(-)^{b+c-L_1}\,W(b\,b\,c\,c;k_1L_2)\,.\qquad(17.5)$$

These terms can be verified by referring to (16.2). Thus the unobserved radiation, $\mathscr{L}_2$, contributes to $W(\vartheta_{n(q)},\Omega_{n(v)})$, a multiplicative term, varying as

$$\left.\begin{aligned}&\langle b\|\mathscr{L}_2\|c\rangle\langle b\|\mathscr{L}_2'\|c\rangle^*\,\mathscr{D}^{k_2 *}_{k_2 0}(\Omega_2)\,c^*_{k_2 0}(\mathscr{L}_2\mathscr{L}_2')\times\\&\qquad\times\hat{b}^2\,\hat{k}_2\,\hat{k}_c\begin{Bmatrix}c & L_2 & b\\ c & L_2' & b\\ k_c & k_2 & k_1\end{Bmatrix}\cdot(k_2\varkappa_2,k_c\varkappa_c|k_1\varkappa_1)\,,\end{aligned}\right\}\quad(17.6)$$

which is summed over $k_2\varkappa_2$ and integrated over all angles of propagation of L_2. This results in $k_c=k_1$, $\mathscr{L}_2=\mathscr{L}_2'$, $k_2=\varkappa_2=0$, and a reduction to (17.5), where the coefficient of $(-)^{b+c-L_1}\,W(b\,b\,c\,c;k_1L_2)$ is adjusted so that $U_0(L_2bc)=1$. If the radiations are mixed, they contribute incoherently to U_k, and are weighted according to the ratio of the nuclear matrix elements. In addition, the terms are independent of the nature of the radiations, since $c^*_{00}(\mathscr{L}\mathscr{L}')$ is the same for all kinds of radiation.

From parity considerations, k_1, k_2, and k_3, are all even. The complexity of the correlation is determined by the Clebsch-Gordan coefficients and the triangular conditions for the Racah coefficients. We have:

$$\left.\begin{aligned}&\Delta(k_1k_2k_3)\,,\\&\Delta(k_1l_1l_1')\,\Delta(k_1L_1L_1')\,\Delta(k_1b\,b)\,\Delta(k_1c\,c)\ldots\Delta(k_1q\,q)\,,\\&\Delta(k_2L_{n(q)}L_{n(q)}')\,\Delta(k_2l_{n(q)},l_{n(q)}')\,,\\&\Delta(k_3L_{n(v)}L_{n(v)})\,\Delta(k_3l_{n(v)}l_{n(v)}')\,\Delta(k_3r\,r)\,\Delta(k_3s\,s)\ldots\Delta(k_3v\,v)\,,\end{aligned}\right\}\quad(17.7)$$

so that the unobserved radiations have no effect on the complexity of the correlation.

An interesting theorem concerning emission of radiation in a multiple-cascade has been given by WENESER and HAMILTON[1]. It is restricted to "basic" transitions of any character, which are transitions between nuclear states whose spins form a monotonic sequence, and for which the angular momentum of the radiation is the lowest possible. Thus $a=b+L_1$, $b=c+L_2,\ldots$, or $a=b-L_1$, $b=c-L_2,\ldots$, etc. The theorem states that the directional correlation between any two radiations of the cascade is independent of the number of transitions intervening between them, of the character of the intervening radiation and their order of occurrence, and of the spins of the intermediate states.

Thus, if L_1 and $L_{n(q)}$ are observed,

$$W(\vartheta)=\sum_k A_k(a,b\,b)\,U_k(L_2b\,c)\,U_k(L_3c\,d)\ldots U_k(L_{n_p}p\,q)\,A_k(q\,q,r)\,P_k(\cos\vartheta)\,,\qquad(17.8)$$

[1] J. WENESER and D. R. HAMILTON: Phys. Rev. 92, 321 (1953).

where U_k is defined in (17.5). These products of Racah functions, which are contained in the above equation, can be found to reduce to a quite simple form,

$$W(f f L_1 L_1; k a)\, W(f f L_{n(q)} L_{n(q)}; k r), \tag{17.9}$$

where $a = f + L_1$ and $f = L_{n(q)} + r$; proving the above theorem. RABOY and KROHN[1] have shown that this theorem holds, free of the requirement that the first radiation be basic and that a be part of the monotonic sequence. Expressed more generally, their result is that the angular correlations between one radiation L_1 and each of two succeeding radiations $L_{n(q)} L_{n(q)}$ are similar, provided that:

(i) $L_{n(q)} = L_{n(v)}$ and both transitions are basic,

(ii) the unobserved intermediate radiations between $L_{n(q)}$ and $L_{n(v)}$ are also basic,

(iii) the spins for the nuclear states involved in these intermediate radiations form a monotonic sequence.

The sequence can also be reversed in time, i.e. $L_{n(v)}$ and $L_{n(q)}$ can precede L_1.

This result can be extended to correlations between three radiations or any multiple correlation. For example, the triple correlation between L_1, $L_{n(q)}$, $L_{n(v)}$, is independent of the unobserved radiations between $L_{n(q)}$ and $L_{n(v)}$ provided the above three conditions are satisfied.

D. Particular processes involving states of well-defined symmetry.

I. Experimental investigation of directional correlation of cascade emissions

18. Experimental methods for γ-cascades. One of the most widely studied types of angular correlation is that between the directions of γ-radiations emitted successively from a radioactive (statistically isotropic) source. Such radiations almost invariably occur between states of well-defined total angular momentum and parity, so that the theory of Chapt. C is directly applicable, provided the effects of extra-nuclear fields can be ignored or eliminated. In addition, a γ-radiation always carries away some angular momentum, so that correlation studies involving them always provide some information about the spins of the nuclear states. (This in contrast, for example, to β-γ correlations where the correlation function is isotropic if the β-transition is allowed (cf. Sect. 49) and to α-γ correlations where the α-particle may be emitted with zero orbital angular momentum, (cf. Sect. 29). Experimentally, the particular feature which enhances the study of γ-correlations is the small interaction, and especially scattering, suffered by the γ-radiation when it traverses any material which may be between the source and detector. In the case of very low energy γ-radiation (of the order or less than 0.1 Mev), this interaction does introduce difficulties, and in addition strong internal conversion reduces the intensity of the detected radiation.

The experimental methods will be considered under three headings: (i) the detecting equipment, (ii) the source, (iii) interpretation of experimental results. Many of these considerations apply more generally than to the γ-γ cascade.

(i) The basic equipment for studying γ-γ correlations comprises two γ-detectors, one normally fixed and the other movable about a fixed axis, which passes through the centre of the source (taken as the origin of the coordinate system). The positions of the counters are represented by $\boldsymbol{r}_1$ and $\boldsymbol{r}_2$ ($|\boldsymbol{r}_1|$ and $|\boldsymbol{r}_2|$ constant

[1] S. RABOY and V. E. KROHN: Phys. Rev. 98, 24 (1955).

(see Fig. 1). The plane containing $\boldsymbol{r}_1$ and $\boldsymbol{r}_2$ is fixed and passes through the origin. Normally each detector has cylindrical symmetry about the appropriate direction $\boldsymbol{r}$ †.

Related pairs of γ-rays are detected by observing the coincident signals from the two detectors. For γ-cascades in which the time interval ($\gtrsim 10^{-9}$ sec) between successive emissions of the two radiations is so short that extra-nuclear effects can be neglected, the resolving time of the coincidence circuit is made as short as practicable, since even then it is usually greater than the time interval between the two nuclear events. The usual considerations that apply to coincidence technique[1] apply in the case of angular correlation studies with the special feature that the requirements of high angular resolution and precision conflict with high ratio of real to random coincidence counts and high statistical accuracy, which emphasises the need for the shortest resolving times that are technically possible, and consistent with the physical time delays in the γ-cascade, together with maximum efficiency of the γ-detectors. In addition, if other γ-radiation is present as a background, and contributes to the random but not the desired coincidence count, some energy selectivity of the detectors is also desirable.

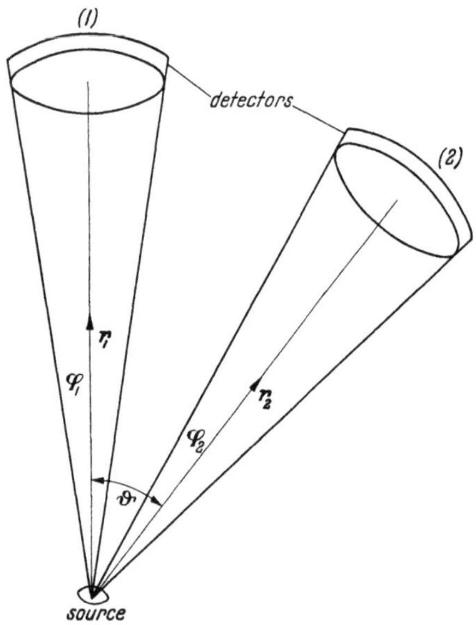

Fig. 1. Geometry of γ-γ correlation measurement.

Crystal scintillation counters, normally activated sodium iodide crystals of thickness appropriate to the γ-radiation energy, coupled to fast photomultiplier tubes, are almost invariably used as γ-detectors. In some arrangements each detector responds to both γ-rays of the cascade, in which case energy resolution is only possible if the two radiations have similar energies. If a simple system is used in which the coincidence circuit registers when a pulse of appropriate magnitude is recorded by each detector, then the resolving time of the coincidence circuit is limited by the speed of operation of the energy discriminating system (commonly 10^{-7} sec or greater). To overcome this limitation, a more complex arrangement is often used involving two coincidence systems, one fast ($\sim 10^{-8}$ to 10^{-9} sec), which determines the time correlation of the two radiations without refined energy discrimination, and another slower system ($\sim 10^{-6}$ sec), which enables pulses of appropriate amplitude to be selected (Fig. 2a).

A modified arrangement, Fig. 2b, can be used in which the whole spectrum of pulses from one counter can be observed, thereby permitting the angular correlations of one γ-ray with all other γ-rays which occur in (fast) coincidence with it to be studied simultaneously.

The energy discrimination is normally used so as to select only those pulses which correspond to the absorption of the full energy of the γ-radiation in the crystal detector (the so-called "photoelectric peak"), which not only eliminates

† We refer here to cylindrical symmetry about the axis of the detector, for the efficiency of detection. Polarisation-sensitive detectors are considered later (Sect. 33).

[1] A general account of this experimental technique is given by A. C. G. Mitchell (Chap. 8) and R. E. Bell (Chap. 18) in: Ref. [*14*].

γ-radiation not part of the γ-cascade being studied, but also usually eliminates the disturbance due to scattering from one detector into the other, an effect which is very sensitive to the angular separation of the detectors. This form of disturbance can also be reduced by suitable lead-shielding of the detectors[1].

(ii) For sources of high specific activity, the source strength is limited by the rate of random coincidence. For sources of low specific activity the maximum source strength is determined by consideration of scattering and absorption in the source material (or its support), or by limitation of size compatible with good geometry. In order to eliminate the effects of scattering, the correlation can

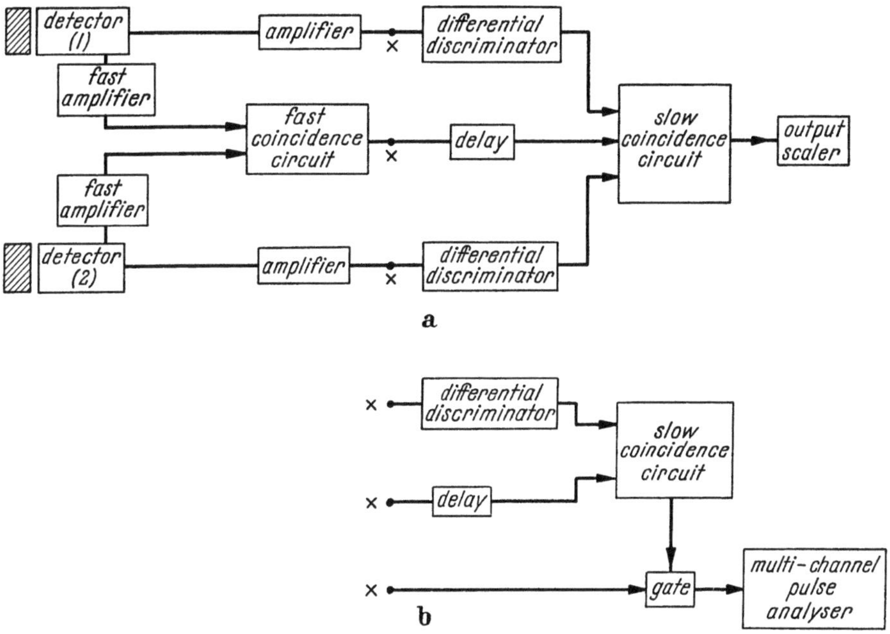

Fig. 2a and b. (a) Fast-slow coincidence arrangement. (b) Modification for multi-channel energy discrimination.

be studied for several source thicknesses, and the results extrapolated to zero thickness. Effects due to finite source size and possible displacement from the origin can similarly be examined by making measurements with several sources. In addition to size, the physico-chemical composition of the source material may have to be chosen so as to eliminate the effects of extra-nuclear fields. Sources in liquid form show more nearly undisturbed correlation functions than solids (cf. Chap. H), and the disturbance decreases as the viscosity of the liquid is reduced. Measurement of some feature of the angular correlation, e.g. the ratio of the coincidence rate at $\pi/2$ and π (normally referred to as the "anisotropy"), with variation of solvent, dilution, or temperature, usually enables the conditions for the elimination or reduction of the effect of extra-nuclear fields to be ascertained, or alternatively the observed anisotropy can be extrapolated to give the value for zero disturbance.

Sometimes the lifetime (or limits to it) of the intermediate nuclear state is known from independent evidence (e.g. direct measurement, Coulomb excitation) to be of the same order or greater than the minimum resolving time. Significant measurements of the correlation can then be made with different resolving times

[1] E.g. J. J. KRAUSHAAR and M. GOLDHABER: Phys. Rev. 89, 1081 (1953).

or the coincidences can be made with a delay (into the channel detecting the first radiation), which is greater than the resolving time, but less than or of the same order as the decay time of the intermediate state. In this way the correlation can be determined as a function of the time-interval between the emission of the radiations, as opposed to the time-integrated correlation function which is normally measured. A correlation function which is not isotropic and does not change with the introduction of such delay normally implies absence of any extra-nuclear disturbance. In cases of complex γ-cascades the same intermediate level may be involved in the study of several γ-γ correlations. If an unperturbed correlation is observed for any one γ-γ pair involving such a level, then the perturbation can be ignored for other correlations[1].

(iii) After elimination of extra-nuclear disturbance and correction for random coincidences, the observed variation of coincidence rate with angle between the detector axes $(\boldsymbol{r}_1, \boldsymbol{r}_2)$ does not give the correlation function directly but must still be corrected for distortion on account of the finite size of the source and detectors. The latter of these is usually the more important, and has been discussed by many authors [*14a*], [*71*]. For detectors with axial symmetry (as in Fig. 1), the effect of this finite solid angle is to reduce the anisotropy without changing the form of the correlation function. If the ideal correlation function is expressed in the form [Eq. (15.3)]:

$$W(\vartheta) = \sum_k A_k[1, 2]\, P_k(\cos\vartheta), \tag{18.1}$$

then the observed correlation function $W'(\vartheta)$ can be expressed in the form (first given by FRANKEL [*70*]):

$$W'(\vartheta) = \text{const}\left\{1 + \sum_{k>0} A_k g_k P_k(\cos\vartheta)\right\},$$

where

$$g_k = g_k^{(1,1)} g_k^{(2,2)} \leq 1 \tag{18.2}$$

and

$$g_k^{(1,1)} = \frac{\int\limits_0^{\varphi_1} P_k(\cos\vartheta)\, \varepsilon_{11}(\varphi)\, d(\cos\varphi)}{\int\limits_0^{\varphi_1} \varepsilon_{11}(\varphi)\, d(\cos\varphi)}, \tag{18.3}$$

and a similar expression holds for $g_k^{(2,2)}$.

φ_1 is the half-angle of detector (1) (see Fig. 1), and $\varepsilon_{11}(\varphi)$ the efficiency of this detector for radiation (1) incident at an angle φ to the central vector $\boldsymbol{r}_1$. In (18.3) it is assumed that each detector is sensitive to one radiation only. If each detector is sensitive to *both* radiations with efficiences $\varepsilon_{11}(\varphi)$ and $\varepsilon_{12}(\varphi)$ for detector (1) and radiations 1 and 2 respectively, and similarly $\varepsilon_{21}(\varphi)$ and $\varepsilon_{22}(\varphi)$, then with the obvious modifications of (18.2), (18.3), we have for g_k in (18.2):

$$g_k = \tfrac{1}{2}\left[g_k^{(1,1)} g_k^{(2,2)} + g_k^{(1,2)} g_k^{(2,1)}\right]. \tag{18.4}$$

For detectors of arbitrary shape, the distortions introduced may lead to new terms in (18.1) as well as to changes in the coefficients of the P_k's. In most practical cases the detectors posses such symmetry that no higher P_k terms are introduced, but the distortion effect does not generally take the simple form (18.3).

[1] E.g. D. E. ALBURGER and A. W. SUNYAR: Phys. Rev. **99**, 695 (1955).

The quantities g_k can either be estimated from the geometry of the counter and the absorption properties of the counter material[1], or measured using a highly canalised beam of γ-radiation of appropriate energy[1,2], or in the special case of γ-radiation with energy near to that of annihilation radiation (0.5 Mev) the collinear properties of the two annihilation quanta can be used[3].

The effect of finite-source size has been evaluated explicitly only for extension along the rotation axis (i.e. the direction $(\boldsymbol{r}_1 \times \boldsymbol{r}_2)$], where the effect on the correlation function is similar to that expressed in (18.2). In general, displacement of the source from the origin will introduce spurious terms into the correlation function in addition to modifying the coefficients. Some, but not complete, elimination of the distortion can be effected by using for $W'(\vartheta)$ the ratio of the coincidence rate to the single counting rate in the movable counter as a function of the (apparent) value of ϑ.

Scattering, if it is cylindrically symmetrical about the direction of propagation and the same for all positions of the movable detector, gives rise to attenuation factors similar to the g_k's in Eq. (18.2).

The experimental results, corrected for effects other than the geometrical factors, are of a set of values, $W'(\vartheta_i)$, of the coincidence rate for the angles ϑ_i. The counting rates are normally sufficiently large for a least-squares analysis to be appropriate, and this is normally applied to find the best fit to the results by a function of this form:

$$W'(\vartheta) = \sum_{k=0}^{\text{maximum}} Q_k P_k(\cos\vartheta), \tag{18.5}$$

where $k_{\max}$ represents the highest term statistically significant, but in any case is limited by the number of observations $W'(\vartheta_i)$. The determination of $k_{\max}$, i.e. the highest coefficient Q_k which is statistically significant, is often of considerable importance, since the presence of a particular k-value, apart from the exact value of A_k, may be decisive in choosing between alternative interpretations (cf. Sect. 19, 20). Comparing Eqs. (18.5) and (18.1), we get the result

$$A_k = \frac{1}{g_k}\,\frac{Q_k}{Q_0}. \tag{18.6}$$

This procedure is only strictly correct if $A_k \ll 1$ for all k, since the Q_k's obtained by fitting Eq. (18.5) are not independent[4]. It is also possible to arrange the sequence of measurements so as to simplify the task of fitting the results, by least squares, to an expression of type (18.5)[5].

The maximum useful angular aperture of the γ-detectors, which is reached when the increased statistical accuracy obtained from larger detectors becomes offset by the decrease in correlation due to diminishing values of g_k, obviously depends on the highest term, $A_k P_k(\cos\vartheta)$, for which the correlation is being examined. For ideal cylindrically symmetrical detectors with $\varepsilon(\varphi)$ constant, the optimum values of φ are easily calculated (e.g. for $k=2$, $\varphi_{\text{opt}}=42°$; and for $k=4$, $\varphi_{\text{opt}}=27°$); but detectors as large as this are not practically useful for precise measurements, since their use would involve undue reliance on uniformity of the detector and on the accuracy of estimation of the attenuation factors g_k.

---

[1] J. J. Lawson and H. Frauenfelder: Phys. Rev. **91**, 639 (1953).

[2] J. J. Kraushaar, J. W. Mihelich and A. W. Sunyar: Phys. Rev. **95**, 456 (1954). — E. D. Klema and F. K. McGowan: Phys. Rev. **92**, 1469 (1953).

[3] E. L. Church and J. J. Kraushaar: Phys. Rev. **88**, 419 (1952).

[4] E. Breitenberger: Proc. Phys. Soc. Lond. A **69**, 489 (1956).

[5] P. C. Price: Phil. Mag. **45**, 237 (1954).

In many cases the overall performance of the equipment and procedures can be tested by using a γ-cascade for which the correlation is well known experimentally and theoretically, e.g. the cascade in Ni^{60} which is of the type $4\to 2\to 0$, and the theoretical expected correlation

$$W(\vartheta) = 1 + 0.1020\, P_2(\cos\vartheta) + 0.0091\, P_4(\cos\vartheta)$$

has been checked to a high degree of precision[1].

19. γ-γ directional correlation: Application of theory. The theoretical formula for γ-γ correlation using detectors which are not sensitive to polarisation is obtained by using the appropriate radiation parameters $c_{k0}(LL)$ and $c_{k0}(LL')$, given by (14.12) to (14.14), in the expressions (15.3) for the angular correlation, i.e.

$$W(\vartheta) = \sum_k A_k(a, b\,b)\, A_k(b\,b, c)\, P_k(\cos\vartheta), \tag{19.1}$$

where the coefficients A_k are, neglecting irrelevant constant factors,

$$\left.\begin{aligned} A_k(a, b\,b) &\sim \sum_{L_1 L_1'} (-)^{a-b+k-1}\, \hat{L}_1 \hat{L}_1' \langle b \| L_1 \| a\rangle \langle b \| L_1' \| a\rangle (-)^{0,1} \times \\ &\quad \times (L_1 1, L_1' - 1 \mid k0)\, W(b\,b\,L_1 L_1'; k\,a), \\ A_k(b\,b, c) &\sim \sum_{L_2 L_2'} (-)^{c-b+k-1}\, \hat{L}_2 \hat{L}_2' \langle b \| L_2 \| c\rangle \langle b \| L_2' \| c\rangle (-)^{0,1} \times \\ &\quad \times (L_2 1, L_2' - 1 \mid k0)\, W(b\,b\,L_2 L_2'; k\,c). \end{aligned}\right\} \tag{19.2}$$ [2]

In each of these expressions *one* of the reduced matrix elements can be normalised arbitrarily since we are only interested in relative values of $W(\vartheta)$.

In the simplest cases each transition is a pure multipole, i.e. only a single value of $L_1 (= L_1')$, and $L_2 (= L_2')$, occurs in (19.1), and the correlation function is independent of the reduced matrix elements.

$$\left.\begin{aligned} W(\vartheta) &\sim \sum_k \hat{L}_1^2 \hat{L}_2^2 (L_1 1, L_1 - 1 \mid k0)\, (L_2 1, L_2 - 1 \mid k0)\, (-)^{a+c-2b} \times \\ &\quad \times W(b\,b\,L_1 L_1; k\,a)\, W(b\,b\,L_2 L_2; k\,c)\, P_k(\cos\vartheta), \end{aligned}\right\} \tag{19.3}$$

or in the form (19.1) we can equate $A_k(a, bb)$ to the function $F_k(L_1 L_1 a b)$ (BIEDENHARN and ROSE [9], FALKOFF and UHLENBECK [25]) defined by:

$$\left.\begin{aligned} A_k(a, b\,b) &= F_k(L\,L\,a\,b), \\ F_k(L_1 L_1\, a\, b) &= (-)^{a-b-1}\, \hat{b}\, \hat{L}_1^2 (L_1 1, L_1 - 1 \mid k0)\, W(b\,b\,L_1 L_1; k\,a), \end{aligned}\right\} \tag{19.4}$$

and similar expressions for $A_k(bb, c)$ and $F_k(L_2 L_2 c b)$. The functions F_k have been tabulated [9] for many integral values of L and integral and half-integral values of a and b.

For γ-transitions of mixed multipolarity, it has not so far been found necessary to consider more than two multipole components, normally differing by one unit of angular momentum $(L, L+1)$. The mixing ratio, δ, is defined as the ratio of reduced matrix elements.

$$\delta_{(1)} \equiv \frac{\langle b \| L+1 \| a\rangle}{\langle b \| L \| a\rangle}, \quad \text{or} \quad \delta_{(2)} = \frac{\langle b \| L+1 \| c\rangle}{\langle b \| L \| c\rangle} \tag{19.5}$$

for the first and second radiations respectively.

---

[1] R. M. STEFFEN: Ref. [15], p. 298.

[2] The $\pm$-sign in the expressions for $c_{k0}(LL')$ applies only to the cases $L \neq L'$. The ambiguity arises from the two possible conventions for the choice of electromagnetic potentials representing the electric multipoles (14.9).

For a γ-γ cascade with one pure (L_1) and one mixed (L_2, L_2+1) radiation, we can write for the coefficients of the correlation function (19.1),

$$\left.\begin{aligned} &A_k(a, b\,b) = F_k(L_1 L_1\, a\, b)\,, \\ &A_k(b\,b, c) = F_k(L_2 L_2\, c\, b) + \delta^2 F_k(\overline{L_2+1}\;\overline{L_2+1}\, c\, b) \pm 2\delta\, F_k(L_2 \overline{L_2+1}\, c\, b)\,. \end{aligned}\right\} \qquad (19.6)$$

The functions $F_k(L_2\overline{L_2+1}\,cb)$ are of the form:

$$F_k(L_2\overline{L_2+1}\,c\,b) \equiv (-)^{c-b-1}\,\hat{b}\,\hat{L}_2\,\widehat{\overline{L_2+1}}\;G_k(L_2\,\overline{L_2+1}\,c\,b) \qquad (19.7)$$

with

$$G_k(L_2\overline{L_2+1}\,c\,b) \equiv (L_2 1, \overline{L_2+1} - 1 \mid k0)\, W(b\,b\,L_2\overline{L_2+1}\,; k\,c)\,.$$

Numerical values for the G_k's have also been tabulated [9] (see also Ferentz and Rosenzweig, reference to mathematical Tables [9]).

The extension to the case of both transitions mixed is obvious; both sets of terms $A_k(a, bb)$ and $A_k(bb, c)$ take the form of the latter in (19.6), and there will be a different mixing ratio, $\delta_{(1)}$ and $\delta_{(2)}$, for each transition.

The finite number of terms in the expression for $W(\Omega_1\Omega_2)$, i.e. the numerical value of $k_{\max}$, depends as discussed in Sect. 15 on the values of L_1, L_2 and b.

In the case of γ-transitions following radioactive decay, when only levels of moderate energy (up to a few Mev) are normally excited, the second transition if it is sufficiently rapid for extra-nuclear coupling to be ignored, will be of low multipolarity, in practice either dipole or quadrupole[1]. For these cases we can therefore write the correlation function as:

$$\begin{aligned} W(\vartheta) = A_0(a, bb)\, A_0(bb, c) &+ A_2(a, bb)\, A_2(bb, c)\, P_2(\cos\vartheta) \\ &+ A_4(a, bb)\, A_4(bb, c)\, P_4(\cos\vartheta)\,, \end{aligned}$$

or

$$W(\vartheta) = A_0[1, 2] + A_2[1, 2]\, P_2(\cos\vartheta) + A_4[1, 2]\, P_4(\cos\vartheta) \qquad (19.8)$$

with the A_4's different from zero only if both transitions contain some components of quadrupole or higher multipolarity.

If both transitions are pure, then $A_0[1, 2] = 1$, with the normalisation indicated above, and $F_1(L L' a\, b) = \delta_{LL'}$. The coefficients $A_2[1, 2]$ and $A_4[1, 2]$ are determined directly by assigning integral or half-integral values to the spins of the nuclear states, and the multipole order, i.e. a, b, c, and L_1, L_2, respectively. The parities of the states or of the radiations do not enter explicitly.

For mixed transitions, on the other hand, the correlation function involves the non-discrete variable, δ. In the case of one pure (L_1) and one mixed (L_2, L_2+1) transition, the angular correlation given by (19.1), (19.6), (19.7), can also be written, as in (15.7), thus:

$$W(\vartheta) = W_{L_1 L_1, L_2 L_2}(\vartheta) + \delta^2 W_{L_1 L_1, L_2+1\, L_2+1}(\vartheta) + 2\delta\, W_{L_1 L_1, L_2 L_2+1}(\vartheta)\,, \qquad (19.9)$$

where the first terms are the normalised correlation functions for the pure transitions L_1, L_2, and L_1, L_2+1, and the last is the interference term[2].

---

[1] Correlations involving two γ-rays emitted with a large time interval ($\gg 10^{6}$ sec) do not lead to useful results for two reasons: firstly, the correlation will be greatly distorted or destroyed by extra-nuclear fields; secondly, the resolving time of the coincidence apparatus will, of necessity, be large, so that a workable ratio of real to random coincidences will only be possible with very weak sources ($N\tau \lesssim 1$), and the experimental results will consequently be of low accuracy.

[2] We have chosen the positive sign for δ in the interference term. Since the matrix element varies as $A_{L\Lambda}$, the choice of a particular sign in Eq. (14.9) determines the sign of δ in (19.5), and the sign of $c_{k0}(LL')$ in (14.14). If the other sign were chosen, the product of δ and $c_{k0}(LL')$ would be unaltered.

Expressing $W(\vartheta)$ in the form with the coefficient of the constant term $A_0[1, 2]$ equal to unity, we get

$$\left.\begin{aligned} W(\vartheta) &= 1 + A_2[1, 2]\, P_2(\cos\vartheta) + A_4[1, 2]\, P_4(\cos\vartheta) \\ &= 1 + \frac{1}{1+\delta^2} \sum_{k=2,4} [F_k(L_1 L_1 a b)\, F_k(L_2 L_2 c b) + \\ &\quad + \delta^2 F_k(L_1 L_1 a b)\, F_k(\overline{L_2+1}\;\overline{L_2+1}, c b) + \\ &\quad + 2\delta F_k(L_1 L_1 a b)\, F_k(L_2\, \overline{L_2+1}\, c b)]\, P_k(\cos\vartheta)\,. \end{aligned}\right\} \tag{19.10}$$

From the experimental angular correlation of the type indicated in (19.10), one in general obtains two quantities A_2, A_4, and each of these may be compared with the theoretically predicted quantities $A_2[1, 2]$ and $A_4[1, 2]$ in (19.10). With a particular assignment of values to a, b, c, and the L's the comparison for each of the coefficients A_2, A_4, yields a quadratic in δ and therefore two possible values of δ.

If neither of the coefficients A_2, A_4, is zero, then from both comparisons a unique value of δ may be deduced if the experimental precision is sufficient. For a correlation in which the $P_4(\cos\vartheta)$ term is absent (e.g., E 2 and M 1 mixed preceded or followed by M 1 pure), then the ambiguity in the value of δ may be removed by reference to other data, for example the value of the internal-conversion coefficients for the mixed γ-transition. The same applies if only a single measurement of the correlation is made, e.g. that of the anisotropy $W(\pi)/W(\pi/2)$, whether the A_4 term is zero or not.

For correlations involving two mixed transitions, and therefore two ratios of reduced matrix elements $\delta_{(1)}$ and $\delta_{(2)}$, the unambiguous assignment of the discrete parameters and the determination of the δ's from the experimental values of the two coefficients of the terms $P_2(\cos\vartheta)$ and $P_4(\cos\vartheta)$ may not be possible, and considerable knowledge about the transitions additional to the measurement of the correlation may be required.

In general, higher precision is required in measuring angular correlations involving mixed transitions than in those with only pure transitions, since not only will the best set of parameters, spins, multipolarities, and δ's be more, difficult to ascertain reliably, but the value of δ is not quantised (in contrast to the spins etc.), and the precision with which it is determined depends on the refinement of the measurement.

The measurement of angular correlation is normally supplement by other information about the nuclear states and γ-transitions, as well as the internal-conversion coefficients already mentioned. For example, the choice between possible γ-multipolarities may be helped by knowledge of the absolute probabilities (lifetimes) of the transitions, or their relative values if more than one transition occurs from a single level. The spins of the ground states of the nuclei involved are often known from direct experimental measurement, and the possible spins and parities of excited states may be limited by the values of the absolute probabilities of β-transitions to these states. Internal-conversion coefficients for transitions of the same L (and energy) are different for different parities (i.e. electric or magnetic multipoles) and can thus supplement directional-correlation measurements as well as help in their interpretation. Direct information about the parities can also be obtained from polarisation-correlation measurements, as discussed later (Sect. 31).

Correlation between two γ-radiations which are not directly consecutive can, of course, be studied experimentally in exactly the same manner as those involved in a direct cascade, provided always that the particular radiations can be resolved energetically by the γ-detectors. The correlation functions are given for the case of one intermediate, unobserved, γ-transition by Eqs. (16.10) to (16.12), with the appropriate values of the radiation parameters. Generally when a complex γ-cascade is being studied, the angular correlations can be measured for several combinations of the set of transitions, including both cascades involving pairs of transitions in direct sequence, and others with intervening radiation. In this way not only can the spins of the levels

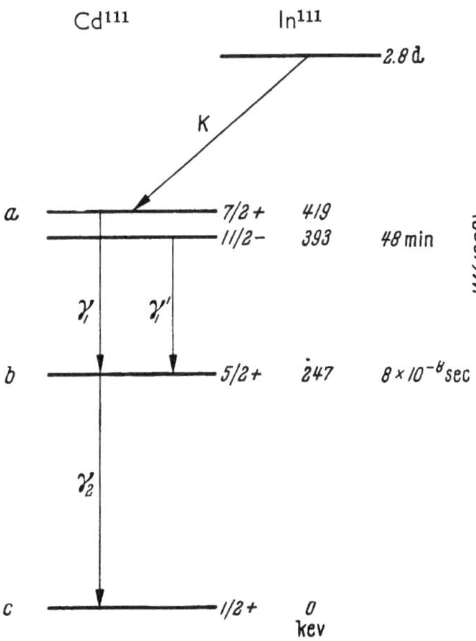

a

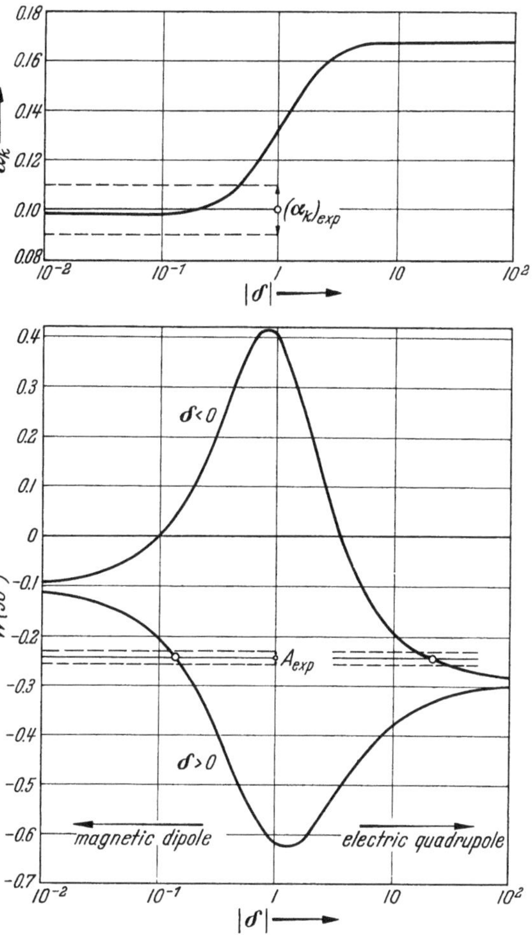

b

Fig. 3a and b. (a) Level-scheme for Cd^{111}. The cascade discussed in the text is labelled γ_1, γ_2. (b) Calculated and experimental values of the anisotropy, A, and K-conversion coefficient, α_K, as a function of the $E2/M1$ mixing ratio, $|\delta|$ (cf. H. FRAUENFELDER [*14a*)], p. 558).

and transition multipolarities be determined unambiguously, but also the time sequence of the γ-radiations can sometimes be ascertained. Thus for a triple cascade L_1, L_2, L_3, between four levels, the assignment of spins a, b, c, d, on the basis of the correlations between the radiations L_1, L_2, and L_2, L_3, respectively can be made in two ways, but this ambiguity is removed by measuring the correlation L_1, L_3 (BIEDENHARN and ROSE [*9*], p. 772).

The simultaneous observation of three (or more) γ-transitions in a γ-cascade, although it might, in principle, be useful in elucidation of decay schemes, is, from an experimental point of view, not very practicable. In a typical case, the introduction of the third detector, with, for example, a solid angle of 1% and an efficiency of $\frac{1}{4}$, would reduce the coincidence counting rate by a factor 400, which would make the accumulation of statistically useful information tediously slow. In addition, the correlation in such a case would be, in general, a complex function of three angles, although particular, simple, correlations, such as, for example, that given by Eq. (16.15), might be useful in special circumstances.

The above discussion will now be illustrated by some specific examples.

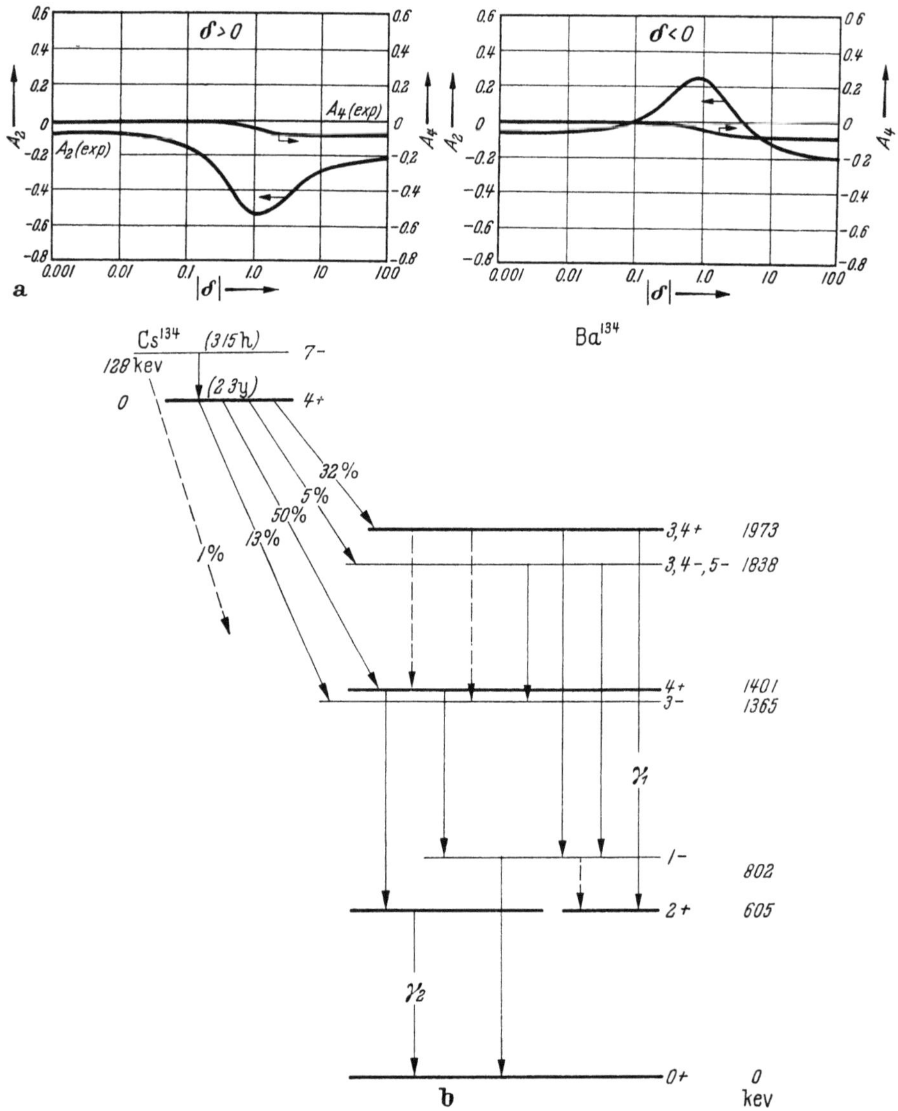

Fig. 4a and b. (a) Calculated and experimental coefficients of $P_2(\cos\vartheta)$ and $P_4(\cos\vartheta)$ in the γ-γ correlation shown as a function of the $E2/M1$ mixing ratio [cf. E.D. KLEMA: Phys. Rev. **100**, 68 (1955)]. (b) Level-scheme for Ba^{134}. The cascade discussed in the text is labelled γ_1, γ_2 [cf. G. L. KEISTER, E. B. LEE and F. H. SCHMIDT: Phys. Rev. **97**, 451 (1955)].

20. γ-γ correlation: Illustrations. The first example is the γ-γ correlation for the radiations of Cd^{111}, following K capture in In^{111}. The observed correlation is consistent with the spin values shown in Fig. 3a. The calculated anisotropy, $\frac{W(\pi)}{W\left(\frac{\pi}{2}\right)} - 1$, as a function of δ for the first transition [using Eq. (19.10)], is shown for both signs of δ in Fig. 3b[1], together with the experimental value[2]. The theoretical K-conversion coefficient, on the assumption of an $M1$-$E2$ mixture, depends on the magnitude, but not sign, of δ, and the comparison of experimental and theoretical values of α_k (Fig. 3b) clearly settles the correct value of δ. This comparison brings out the important feature that the angular correlation, which contains the interference term δ, is a far more sensitive method of determining small admixtures of one multipole with another than is measurement of internal conversion, which involves δ^2 only.

The second illustration is a measurement by KLEMA[3] of the γ-cascade in Ba^{134} following the β-decay of Cs^{134} (Fig. 4), Cs^{134}, with known spin (4^+) decays to a level of Ba^{134} with excitation energy 1.973 Mev, which in turn radiates two γ-radiations in cascade via a level (2^+) at 0.605 Mev (Fig. 4). The γ-γ angular correlation (1.368 and 0.605 Mev) was consistent with the commonly occurring scheme $4^+ \to 2^+ \to 0$, with both transitions quadrupole, but the assignment 4 (either $\pm$ parity) to the upper level was inconsistent with the known values of the internal-conversion coefficients. The scheme $2 \to 2 \to 0$ with the first transition mixed, $E2$, $M1$, could be fitted to the measured correlation with δ, the $E2{:}M1$ ratio, negative. However, an assignment 2^+ to the level at 1.973 Mev was inconsistent with the known characteristics of the preceding β-decay. The assignment 3^+ to this level was inconsistent with internal-conversion data. However, a mixture $M2-E1$ for the first transition, from a 3^- level, gave a consistent interpretation with δ; the $M2{:}E1$ reduced matrix element ratio $= -0.22$.

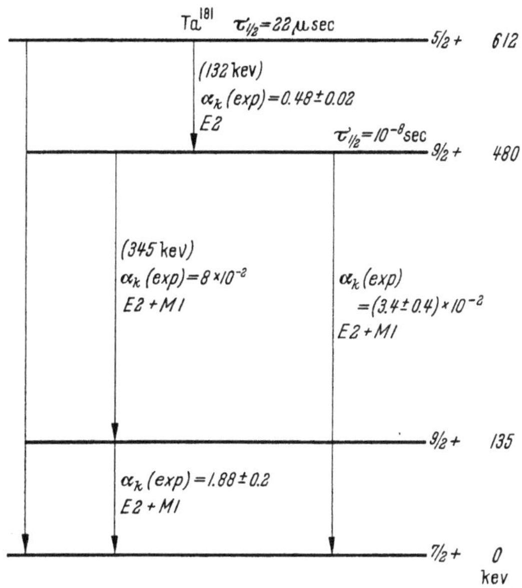

Fig. 5. Level-scheme for Ta^{181} [cf. F. K. MCGOWAN: Phys. Rev. 93, 472 (1954)].

This unusual result, a mixture of $E1$ and $M2$ of comparable amplitudes, is less surprising when it is recalled that for medium and heavy nuclei, low-energy $E1$ transitions normally have reduced matrix elements much smaller than those predicted from the simple, independent-particle picture, even when the nuclei are of the odd-proton type.

It should be noticed that in cases of mixed, just as for pure, transitions, the angular correlation, alone, provides no information about parity, i.e. mixtures $E2-M1$ and $M2-E1$ give rise to identical correlations. Considerations of likely values of reduced matrix elements are not always reliable, but as the above examples show, information about the internal conversion is, if available and sufficiently precise, decisive in removing this ambiguity, as well as resolving possible ambiguity in the choice of δ.

---

[1] H. AEPPLI, H. FRAUENFELDER and M. WALTER: Helv. phys. Acta **24**, 335 (1951).
[2] C. L. MCGINNIS: Phys. Rev. **81**, 734 (1951).
[3] E. D. KLEMA: Phys. Rev. **100**, 66 (1955).

In the third illustration of the angular-correlation technique the situation is more involved, but nevertheless a fully consistent interpretation has been achieved[1].

In the β-decay of Hf^{181}, the low levels of the product nucleus Ta^{181} are excited, and there is information about partial β-decay lifetimes and γ-internal conversion coefficients, together with some knowledge of the lifetimes of γ-transitions and guidance from the shell model. The relevant decay scheme is shown in Fig. 5.

Table 2.

Cascade	Coefficients: Experimental values	
	A_1	A_2
132, 480 kev	-0.280 ± 0.004	-0.058 ± 0.003
132, 345	0.110 ± 0.012	0.029 ± 0.014
345, 135	0.202 ± 0.009	0.053 ± 0.014

Transition	Ratio of $E2/M1$ (derived values)
132 (612→480)	Pure $E2$
480 (480→0)	-1.25
345 (480→135)	1.00 ± 0.2
	0.95 ± 0.1
135 (135→0)	0.50 ± 0.05

The special problems that arose in this investigation were firstly, the half-life of the level at 480 kev was known to be 10^{-8} sec, so that some effects of extra-nuclear fields were expected for transitions from this level; secondly, the close similarity in the energy of the transitions $(612 \to 480) = 132$ kev and $(135 \to 0) = 135$ kev, which made energy resolution for these two transitions impractical; and lastly, the occurrence of cascades with both transitions mixed.

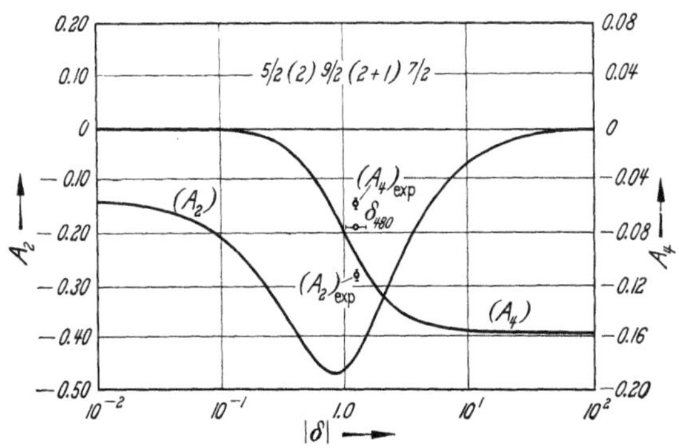

Fig. 6. Calculated and observed angular-correlation coefficients for 132 to 480 kev cascade. δ_{480} represents the value deduced from the experimental K-conversion coefficient of the 480 kev-transition. This result indicates that the intermediate state is perturbed [cf. F. K. McGowan: Phys. Rev. 93, 475 (1954)].

The correlations were studied using a fast-slow coincidence unit and with pulse-height analysis in each detector channel (Fig. 2, Sect. 18). Firstly, measurements of the anisotropy of the correlation of the 132 kev radiation with the 480 kev radiation was examined for effects of source material using relatively long time resolution $(18 \times 10^{-8}$ sec) i.e. measuring the time-integrated correlations. Maximum anisotropy was found with a source of Hf dissolved in HF, and for these sources the correlation was further measured with shorter resolving time, $7 \cdot 10^{-9}$ sec, and with delays of 6×10^{-9} and 10^{-8} sec (in the 132 kev detector channel). No significant difference in the anisotropy was observed for the three different timing arrangements, suggesting that extranuclear effects were small or absent.

Measurements of the correlation between 132 and 345 kev radiations were made using a delay (in the 132 kev detector) of 10^{-8} sec, which with $\tau \sim 7 \cdot 10^{-9}$ sec was sufficient to eliminate contributions from the 345 to 135 kev cascade; (the 135 kev level is short-lived). A correction was also applied to allow for some contribution to correlation from pulses (continuous Compton distribution) originating from the 480 kev transition, entering the 345 kev radiation detector. Since both 480 and 345 kev radiations have the same time-characteristics, this correc-

[1] F. K. McGowan: Phys. Rev. 93, 471 (1954).

tion could be made reliably. To obtain the correlation between the 345 and 135 kev transition, measurements were made with different delay-times (both positive and negative) in the 345 kev detector. With these measurements, supplemented by a calibration of the time-resolution using positron-annihilation radiation, the contributions of the two cascades 132 to 345 kev and 345 to 135 kev could be separated. This latter correlation was found to be independent of source conditions, which confirmed that the source-sensitive 132 to 345 kev correlation had been successfully eliminated.

All three correlations were of the form $A(1+\alpha_2 P_2+\alpha_4 P_4)$, and from the six coefficients, α, a spin scheme consistent with these values and other data was reduced. The results, which involved counting some 7×10^6 coincidences at an average rate of a few counts per second, are shown in Table 2, and Fig. 6, 7.

Consistent values for the mixing ratios and conversion coefficient of the 480 kev transition indicated that even with sources exhibiting the maximum anisotropy, some perturbation remained. Nevertheless, the measurements, with other data, were sufficient to establish the spin scheme completely and to give moderately precise values of the mixing ratios.

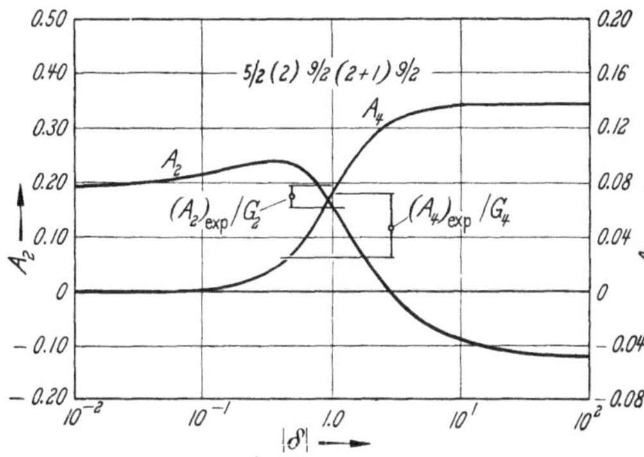

Fig. 7. Angular-correlation coefficients for the 132 to 345 kev-cascade. The experimental values of A_2 and A_4 are modified by attenuation coefficients G_2, G_4 chosen to give a consistent interpretation of the results [cf. F. K. McGowan: Phys. Rev. 93, 475 (1954)].

21. α-γ cascades in radioactive nuclei: Experimental methods. Directional correlation between α-particles emitted in radioactive decay and γ-radiation in the residual nucleus provides information about the spin and parity changes in the α-process (the states of two different nuclei are involved now), and the symmetry properties of the low-lying levels of the heavy product nucleus. Since the probability of α-emission is very sensitive to energy (generally much more so than β-emission) but relatively less sensitive to spin change, only levels of the product nucleus with energies less than a few hundred kev are normally excited, often with low intensity. Furthermore, the subsequent γ-transitions are, normally, strongly internally converted. The overall result is often very low intensity of γ-radiation (as compared with the rate of α-emission forming the main radioactive decay transition). In order to avoid excessive α-particle scattering, which would destroy the correlation, only very thin sources ($\sim 1\ \mathrm{mg/cm^2}$ or less) can be employed. The need for a detecting system which has the largest solid-angle for collection and maximum efficiency, consistent with the required angular resolution, is, then, quite evident.

An additional feature of correlations following α-decay is the large recoil energy possessed by the nucleus following the α-emission, of the order 10^5 ev, as compared with the few ev for a typical β- or γ-transition. This recoil motion produces a violent disturbance to the local atomic surroundings, in the course of which the recoiling nucleus may be subject to strong electromagnetic fields which can perturb the correlation[1]. In addition, since we are concerned only with low-

[1] Most of the recoil motion will be dissipated in a very short time ($\sim 10^{-13}$ sec), too short in fact for any likely fields to cause appreciable nuclear reorientation. The recoil motion will be virtually exhausted by the time γ-emission occurs, but the position of the recoil nucleus then may well be one where it is subject to strong perturbing fields.

energy states, the time interval between α- and γ-emission is likely to be significant from the point of view of extra-nuclear perturbations[1]. Variation of the physical properties of the source is more difficult in this case in view of the other conditions imposed on it to avoid scattering of the α-particles.

Many measurements of α-γ correlation have been made using techniques similar to those employed for γ-γ correlation[2] with obvious modification of source and detector for α-particles, and with the space between source and this detector evacuated. Separation of the α-particle groups is not usually practicable owing to their small energy difference, but pulse-size analysis of the output of γ-detector

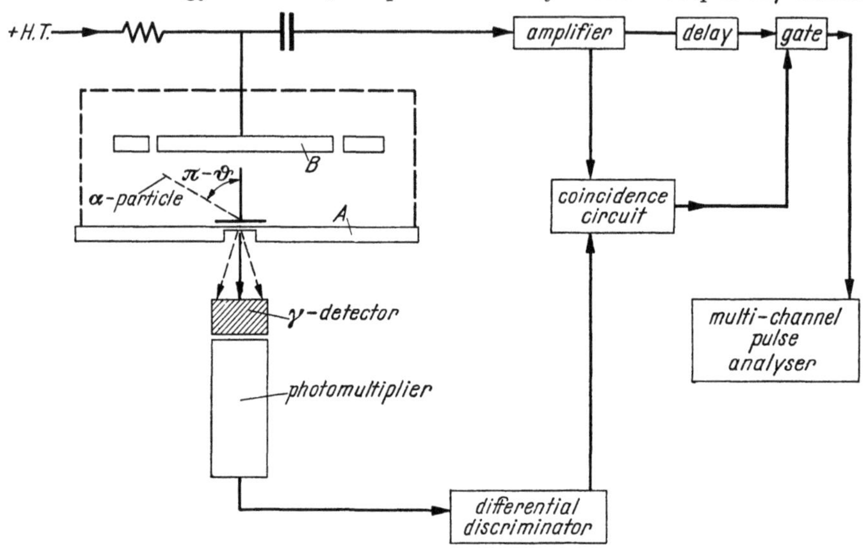

Fig. 8. Schematic diagram showing principle of α-γ correlation measurements.

enables the correct α-γ cascade to be selected. Accurate measurements of cascades involving low-intensity α-particle groups are very tedious, particularly as the source strength may be drastically limited either by the need to avoid undue scattering, or to keep the number of random coincidences, which are augmented by the presence of other intense α-particle groups, at a reasonable level.

A great improvement in the conventional technique has been achieved (VALLADAS *et al.*)[3] by an arrangement in which nearly all the α-particles emitted in coincidence with a particular γ-ray are simultaneously recorded together with the relative angle between α-particle and γ-ray. The method is illustrated diagrammatically in Fig. 8.

The radioactive source is placed on the inner side of one plate of a parallel plate ionised chamber (argon and CO_2 filled), AB, and the α-particle pulses detected in channel (1), using only the fast component of the pulse, i.e. that due to electron collection. This pulse is proportional, for a particular α-particle group, to $\left(1 - \frac{R\cos(\pi-\vartheta)}{d}\right)$, where R is the distance from the source to the centroid of

[1] It should be noticed in this respect that the presence of strong internal conversion reduces the lifetime of radiating states, thus enhancing the significance of α-γ, as well as α-conversion electron measurements.

[2] E.g. G. M. TEMMER and J. M. WYKOFF: Phys. Rev. **92**, 913 (1953). — F. STEPHENS jr., F. A. SARO and I. PERLMAN: Phys. Rev. **96**, 1568 (1954). — J. W. WEALE: Proc. Phys. Soc. Lond **68**, 35 (1955).

[3] G. VALLADAS, J. TEILLAC, P. FALK-VAIRANT and P. BENOIST: J. Phys. Radium **16**, 125 (1955).

the ionisation charges it produces, d is the separation of the plates of the chamber, and $(\pi-\vartheta)$ the angle between the axis of the chamber and the α-particle track, and with a symmetrical γ-detector placed in the axis, ϑ is then the ordinary correlation angle. By recording the α-particle pulses of different sizes in coincidence with the appropriate γ-detector pulse in channel 2, the full angular-correlation function is studied in a single set of measurements.

The system used possessed full cylindrical symmetry so that correction for finite source size (assuming the source a circular disc on the axis) and finite detector size could be made in a fashion similar to that for finite detectors of cylindrical symmetry (cf. Sect. 18). An additional correction for finite angular resolution in the α-detector, which in this case arose from the finite range of pulse-amplitudes recorded in each channel of the pulse analyser, was also made.

Since in this type of experiment measurements can be made at several (~10) angles simultaneously, and the full cone of solid angle $2\pi\sin\vartheta\,\delta\vartheta$ is used, which represents a further gain of ~20 in a typical experiment, the overall gain in efficiency is a factor of the order 200. This advantage is partially offset by the longer resolving times necessary with ionisation-chamber detectors (10^{-6} sec in the above experiment), as compared with scintillation counters, but for sources of low specific activity, when the correlation-function may take several hundred hours to measure, the increased efficiency of the arrangement is invaluable.

22. α-γ cascades in radioactive nuclei: Interpretation and illustrations. Since the α-particle is spinless, a transition involving a particular L-value is also associated with a definite parity change: even L, even parity, or no parity difference between initial and final nuclei; odd L, odd parity, or parity change (in contrast to the situation in the case of γ-emission). α-γ correlations provide information, therefore, about the parities as well as the spins of the nuclear states involved. Moreover, for states with well-defined symmetry, with which we are concerned here, the conservation of parity limits the L-values that are possible in a particular transition to all even or all odd values. If either of the nuclear states has zero spin, then only a single L-value will be possible, and this only if the spin change is appropriate (even spin change, L even, and conversely). If either state has spin $\frac{1}{2}$, again only one L-component is possible, the actual L-values depending on whether there is, or is not, a parity difference between initial and final nuclear states. In all other cases, several component L's are, in principle, possible, and each component involves a different reduced matrix element.

For α-particles with energies much less than the nuclear barrier height, with which we are concerned here, the relative values of these matrix elements decrease with increasing L-values, and although this decrease may not be very rapid, the accuracy and conditions of measurement are not likely to justify a more elaborate interpretation than one involving two L-components (differing by two units). In this case, the α-γ correlation, for pure γ-transition, will involve a real parameter δ representing the ratio of the real reduced matrix-elements for the components $L, L+2$, and in addition a further phase factor, $\cos\xi$, for the relative phase difference, $\xi_{L+2}-\xi_L$, between the two components due to phase shifts, ξ_L, ξ_{L+2}, introduced by the potential barrier[1]. The reduced matrix elements can be

[1] This phase difference may be calculated from known nuclear dimensions. In the limiting case of α-particle energy ≪ barrier height, the phase shift is independent of nuclear radius

$$\xi_{L+2}-\xi_L=\operatorname{arc\,tan}\frac{\eta}{L+2}+\operatorname{arc\,tan}\frac{\eta}{L+1}-\pi,$$

where

$$\eta=\frac{Z^2e^2}{\hbar}\sqrt{\frac{2M_\alpha}{E_\alpha}},$$

[J. Seed and A. P. French: Phys. Rev. **88**, 1007 (1952).]

factorised into a real part and a phase factor:

$$\langle b \| L \| a \rangle = \langle b \| L \| a \rangle' \, e^{i \xi_L}.$$

For the emission of an α-particle with unique angular momentum $L_1 (a \to b)$, followed by a pure γ-transition $L_2 (b \to c)$, the angular correlation (15.6) with the radiation parameters (13.12) becomes:

$$W(\vartheta) \sim \sum_k (L_1 0, L_1 0 | k 0) \, W(b b L_1 L_1; k a) \, F_k (L_2 L_2 c b) \, P_k (\cos \vartheta), \qquad (22.1)$$

with $F(L_2 L_2 c b)$ describing the γ-transition defined as in (19.4). For a mixed $L_1, \overline{L_1 + 2}$ α-particle emission, we have

$$\left.\begin{aligned} W(\vartheta) \sim \{ & \hat{L}_1^2 (L_1 0, L_1 0 | k 0) \, W(b b L_1 L_1; k a) + \\ & + \delta^2 \, \widehat{\overline{L_1 + 2}}^2 \, (\overline{L_1 + 2}\, 0, \overline{L_1 + 2}\, 0 | k 0) \, W(b b \overline{L_1 + 2} \, \overline{L_1 + 2}; k a) + \\ & + 2 \delta \cos \xi \, \hat{L}_1 \widehat{\overline{L_1 + 2}} (\overline{L_1 + 2}\, 0, L_1 0 | k 0) \, W(b b \overline{L_1 + 2} \, L_1; k a) \} \times \\ & \times F_k (L_2 L_2 c b) \, P_k (\cos \vartheta), \end{aligned}\right\} \qquad (22.2)$$

with

$$\delta = \frac{\langle b \| L_1 + 2 \| a \rangle'}{\langle b \| L_1 \| a \rangle'}.$$

The phase-difference enters only as $\cos \xi$ in the interference factor. For a pure α-transition L_1 followed by mixed γ-multipole $L_2, L_2 + 1$, the correlation is given by:

$$W(\vartheta) \sim \sum_k (L_1 0, L_1 0 | k 0) \, W(b b L_1 L_1; k a) \, A_k (b b, c) \, P_k (\cos \vartheta), \qquad (22.3)$$

where $A_k (b b c)$ is given explicitly in (19.6).

The extension to more complicated cases with mixed α- and γ-transitions is straightforward (15.1).

We can also express results of the type (22.2) in terms of the parameters b_k defined by Eq. (15.9). For α-particles we have:

$$b_k (L L'; \alpha) = \frac{-(L\, 0, L' 0 | k\, 0)}{(L\, 1, L' - 1 | k\, 0)} = \frac{-2 \sqrt{L L' (L + 1)(L' + 1)}}{k (k + 1) - L (L + 1) - L' (L' + 1)}, \qquad (22.4)$$

In place of (22.2) we can write:

$$\left.\begin{aligned} W(\vartheta) \sim \sum_k b_k (L\, L + 2; \alpha) \, [& F_k (L_1 L_1 a\, b) + \delta^2 F_k (\overline{L_1 + 2} \, \overline{L_1 + 2} \, a\, b) + \\ & + 2 \delta \cos \xi \, F_k (L_1 \overline{L_1 + 2} \, a\, b)] \cdot F_k (L_2 L_2 c\, b) \, P_k (\cos \vartheta). \end{aligned}\right\} \qquad (22.5)$$

The effect of extra-nuclear fields (which for the reasons given above is normally appreciable for α-radioactive cascades) such as arise in a statistically isotropic source material, e.g. a randomly oriented assemblage of micro-crystals, is to modify the coefficient A_k by factors ("attenuation factors") each less than unity (cf. Sects. 50, 51). As in the case of γ-γ correlation, the interpretation of the correlation is normally supplemented by γ-internal-conversion data and other independent information.

As an example of an α-γ correlation which has been studied by both the conventional experimental technique and the method of VALLADAS *et al.* described above, we consider the α-decay of Ionium, or Th^{230} to Ra^{226} (Fig. 9). Both nuclei are of the even-even type and have spin zero, and the conversion coefficient of the 67.8 kev transition indicates that it is $E2$; the cascade is therefore completely defined as

$$0^+ \xrightarrow{L=2} 2^+ \xrightarrow{L=2} 0^+$$

and application of (6.24) leads immediately to the correlation function[1]:

$$W(\vartheta) = 1 + \tfrac{5}{7} P_2(\cos\vartheta) - \tfrac{12}{7} P_4(\cos\vartheta).$$

The measured correlation $(\alpha_1 - \gamma_1)$, corrected for finite resolution, was found to be $(1 + 0.339\,P_2 - 0.969\,P_4)$, and is shown in Fig. 9b. It is consistent with the assumed spin sequence and the assumption of some extra-nuclear attenuation. The ratio of the attenuation coefficients is consistent with the theoretical value and their order of magnitude indicates electric-quadrupole perturbation of reasonable magnitude.

The high collection efficiency of the method of VALLADAS *et al.* makes it possible to measure cascades of quite low intensity, such as the $\alpha_2 - \gamma_2$ cascade (Fig. 9c).

In this cascade the spins of the intermediate state are not definitely established by the imprecise value of the conversion coefficient of γ_2; but the relatively weak competing 210 kev transition indicates a multipolarity for γ_2 not higher than quadrupole. The corrected experimental results, $W(\vartheta) = 1 + 0.376\,P_2 - 0.224\,P_4$, are consistent with a perturbed correlation of either of two types:

$$0^+ \xrightarrow{L=2} 2^+ \xrightarrow{E\,2 \text{ and } M\,1} 2^+,$$

or

$$0^+ \xrightarrow{L=4} 4^+ \xrightarrow{E\,2} 2^+.$$

The latter alternative leads to reasonable values of the attenuation coefficients, and this spin value is also consistent with the interpretation of the level as one of rotation on the collective model.

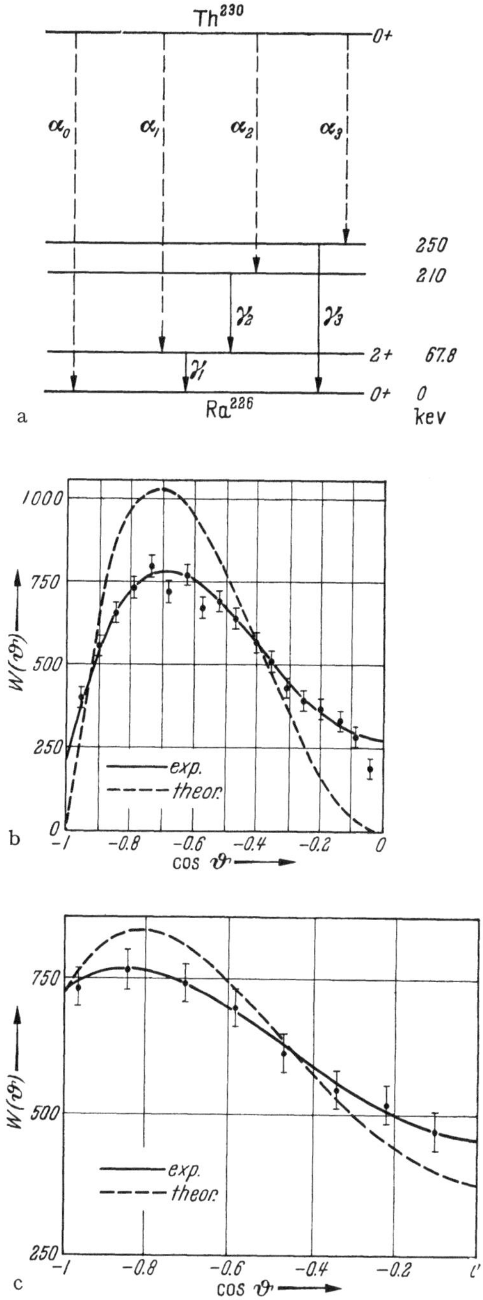

Fig. 9a–c. (a) Level-scheme for α-decay of Th^{230} and level-scheme of Ra^{226} [cf. G. VALLADAS *et al.*: J. Phys. Radium **16**, 130 (1952)]. (b) Calculated and observed angular correlation for the cascade $\alpha_1 - \gamma_1$. (c) Calculated and observed angular correlation for the cascade $\alpha_2 - \gamma_2$.

[1] In this specially simple case the correlation can be derived readily from the more elementary treatment. $W(\vartheta) \sim |\varphi^{\Lambda}_{L l s}(r\Omega\sigma)|^2$ with $L = 2$, $s = 1$, $\Lambda = 0$, $p = (-)^2 = 1$, σ appropriate to an $E\,2$ γ-transition.

II. Resonance reactions.

23. Directional correlations for reactions involving a single resonance level[1]. We have already discussed in Sect. 12 the close similarity between the theory of angular correlations for cascade processes and reactions proceeding through a single isolated resonance level. In both cases we can consider that at each stage in the process a single radiation is emitted (or absorbed), leaving (or producing) a nucleus in a state of well-defined symmetry. There are, however, differences both of a theoretical and of a practical nature which we shall discuss before dealing with reaction processes in detail.

The energy of the complete system in a reaction can, in principle, be sufficiently well-defined that both the processes of absorption and emission must be considered as continuous processes. On the other hand, we can imagine the reaction induced by "switching on" the absorption process (i.e. the incident beam of particles) for a very short time, $\tau\left(\tau \ll \frac{\hbar}{\Gamma} (=\tau_b)\right.$, where Γ is the total width and τ_b the corresponding lifetime of the compound resonance state$\left.\right)$, in which case the time-sequence of absorption and emission will be a real one (as in a cascade), but the energy of the system will be spread over a range greater than the resonance-level width. If the resonance level is truly isolated, then the symmetry properties of the intermediate state will be identical whether or not the energy is sharply defined or spread over the whole resonance, or indeed in any intermediate situation, but this does not mean that the angular correlation will necessarily be independent of energy if only one resonance is excited. The *form* of the angular correlation function *will* be energy-independent, in the above sense, but the reduced matrix elements, and the phase shifts for charged particles, are energy-dependent quantities (and different matrix elements have different energy dependence), so that if the reaction involves, either in the emission process or absorption process, mixed transitions, and consequently the ratio of reduced matrix elements enters into the correlation function, this will be energy-dependent and possibly significantly so. In the majority of reactions, however, where the correlation has been studied, and where conditions approximate to those of a single resonance level, the width of the level is only a small fraction of the energy of either incident or emergent particles, so that the variation of reduced matrix elements (apart, of course, from the dominant resonance factor which is present in all the matrix-elements) is small, and the change with energy of their *relative* values is commonly ignored.

Since resonance reactions, by definition, involve an unbound intermediate state, and with excitation usually much higher than the type of energy involved in a radioactive decay, the lifetime of this intermediate state is normally far too short ($\ll 10^{-12}$ sec) for the effect of extra-nuclear fields on the correlation to be appreciable. In the γ-cascades following such a reaction, the lower levels of the product nucleus may be excited, so that in considering either the correlations between members of such cascades, or the angular distribution of any radiation not proceeding directly from the resonance level, it may be necessary to examine the possible effects of extra-nuclear field parturbation.

There is a difference in the customary treatment of particle (or radiation), intrinsic spin in cascade and reaction processes, which derives not from any difference in principle but from the manner in which the theory of angular correlations developed separately for cascades and reactions. For radioactive cascades

[1] More detailed discussions of reaction processes are given in Chap. E and in the article by G. BREIT, Vol. XLI of this Encyclopedia.

the radiation is normally described in terms of components, L, of total angular momentum. For γ-radiation, although it is possible to treat it formally as having intrinsic spin 1, it is not possible to separate orbital and spin angular momentum, and for α-particles, with zero intrinsic spin, orbital and total angular momentum are identical. In most reactions, on the other hand, the incident particle (e.g. $n, p, d, \ldots$) has intrinsic spin, and this can be, and usually is, treated separately from the orbital motion, in what is known as the "channel-spin" formalism [*43*], [*6*]. The wavefunctions (or the density matrices etc.) of the overall compound state are formed by a coupling scheme in which the spins of both the absorbing nucleus and absorbed particle are first coupled together, to give a particular channel spin, and this channel spin is coupled with the angular momentum of the relative orbital motion to give the total angular momentum of the system. This coupling system has the advantage that, for unpolarised beams and target nuclei, the contributions from different channel spins add incoherently, or in terms of the general treatment of Sect. 11 the statistical tensors describing the spins take the particularly simple form: δ-functions. Reactions can be treated just as completely in terms of total and orbital angular momenta, L and l, where the coupling scheme involves first coupling l the orbital, and s the intrinsic spin of the radiation, to give L, and then coupling L and the spin of the absorbing nucleus. This scheme has the advantage that it gives simpler results where strong spin-orbit coupling, as in the shell model, is in evidence. These are not, of course, the only coupling schemes which can be employed. For example, if the nucleus can be described by a wavefunction with Russell-Saunders coupling (or a superposition of such), then the orbital momenta of the target nucleus and the relative motion can be coupled together to give a total orbital angular momentum $\mathfrak{L}$, and likewise the spins are coupled to a total $\mathfrak{S}$, and then $\mathfrak{L}$ and $\mathfrak{S}$ are coupled to give the total angular momentum. This coupling scheme is useful in dealing with very light nuclei. From the general standpoint of angular correlations the three angular momenta involved in a single emission or absorption process: the nucleus and of the "radiation" and the angular momentum of the relative motion, can be coupled together in any sequence, but the general expressions for the correlation may become much simpler, in a particular case, according as one or another coupling scheme is used.

For particles with spin there are wider possibilities of mixed transitions than occur for spinless particles or γ-radiation. Not only may different L-components be involved in a transition, but a particular L-component may be formed from more than one l-component. In the channel-spin formalism, we can have mixing of different channel-spins for the same l-value (incoherent) or mixing of different l-values for the same channel-spin (coherent). It should also be noticed that for a reaction the properties of both the absorbed and emitted "radiation" can be described formally in terms of the properties (S-matrix etc.) of a single system, that corresponding to the resonance level of the compound nucleus. This is in contrast to cascade processes where the properties of two (or more) nuclear states are involved[1].

In one special case, when the reaction process proceeds by the absorption of the (unpolarised) particle S-wave component only ($l=0$), or the components corresponding to $L=0$ or $\frac{1}{2}$, then the angular distribution is isotropic, of course, and correlations of succeeding cascades can be treated as from a statistically isotropic source.

[1] Cf. L. C. BIEDENHARN and M. E. ROSE [*9*], Sect. II, D.

24. Coupling schemes in resonance reactions. We now consider in more detail the application of the general theory to the resonance reaction with unpolarised incident beam and polarisation-insensitive detectors. Consider the absorption of a particle, characterised by $\boldsymbol{L}_1 = \boldsymbol{l}_1 + \boldsymbol{s}_1$, by a target with spin $\boldsymbol{a}$, with the subsequent emission of a particle with $\boldsymbol{L}_2 = \boldsymbol{l}_2 + \boldsymbol{s}_2$. It is assumed that the reaction proceeds through an isolated resonance level with spin $\boldsymbol{b}$. The angular distribution, using this coupling scheme, is provided directly by the general formalism by inserting the radiation parameters appropriate to the absorbed and emitted radiations (13.10) in (15.11).

The angular distribution has the form

$$\left.\begin{aligned} W(\vartheta) = \frac{\text{Constant}}{\hat{s}_1^2 \hat{a}^2} \sum_k \sum A_k(\mathscr{L}_1 \mathscr{L}_1'; a\,b\,b)\, A_k(\mathscr{L}_2 \mathscr{L}_2; c\,b\,b) \times \\ \times \langle c\,\mathscr{L}_2 b \,|\, \mathcal{O} \,|\, a\,\mathscr{L}_1 b\rangle \langle c\,\mathscr{L}_2' b \,|\, \mathcal{O} \,|\, c\,\mathscr{L}_1' b\rangle^* P_k(\cos\vartheta). \end{aligned}\right\} \tag{24.1}$$

The unlabelled summation is over $l_1 l_1' L_1 L_1' l_2 l_2' L_2 L_2'$.

For the case of absorption followed by emission we can use a reaction matrix similar to the conventional R matrix, except that in (24.1) we have a coupling scheme different from the normal one for channel spin. The coupling is indicated by the notation, although we have omitted to indicate s_1 and s_2, since these are constants for all components of the reaction.

The coefficients A_k which refer to each stage of the process are similar to the coefficients $A_k(a, bb)$, etc. of (15.4), except that here the interaction matrix is expressed separately. Introducing the function Z' related to Z, as defined in (5.6), by[1]

$$Z'(a\,b\,c\,d; e\,f)\, i^{f-a+c} = Z(a\,b\,c\,d; e\,f), \tag{24.2}$$

we can express the coefficients in (24.1) thus:

$$\left.\begin{aligned} A_k(\mathscr{L}_1 \mathscr{L}_1'; a\,b\,b) &= (-)^{a-b-s_1} \hat{b}^2 Z'(l_1 L_1 l_1' L_1'; a\,b)\, W(b\,b\,L_1 L_1'; k\,a), \\ A_k(\mathscr{L}_2 \mathscr{L}_2'; c\,b\,b) &= (-)^{c-b-s_2} \hat{b}^2 Z'(l_2 L_2 l_2' L_2'; c\,b)\, W(b\,b\,L_2 L_2'; k\,c). \end{aligned}\right\} \tag{24.3}$$

Expression (8.1) can also be used to give the absolute cross-section for elastic resonance scattering if the constant is put equal to $\frac{1}{4}\lambdabar^2$ ($\lambdabar$ is the reduced wavelength in the centre-of-mass coordinate system). The total cross-section becomes:

$$\sigma = \pi \lambdabar^2 \sum_{lL} \frac{\hat{b}^2}{\hat{a}^2 \hat{s}_1^2} |\langle a\,\mathscr{L} b \,|\, \mathcal{O} \,|\, a\,\mathscr{L} b\rangle|^2. \tag{24.4}$$

The appearance of b in Eq. (24.1) indicates that there is a unique value for the total angular momentum of the whole system, i.e. $\boldsymbol{a} + \boldsymbol{L}_1$, $\boldsymbol{c} + \boldsymbol{L}_2$, etc $= \boldsymbol{b}$. This would normally correspond to a well-defined intermediate state, that of the compound nucleus, and then the reaction matrix can be factorised:

$$\langle c\,\mathscr{L}_2 b \,|\, \mathcal{O} \,|\, a\,\mathscr{L}_1 b\rangle = \langle c \| \mathscr{L}_2 \| b \rangle \langle a \| \mathscr{L}_1 \| b \rangle \cdot f(l_2)\, f(l_1), \tag{24.5}$$

[1] It will be noticed that the functions $Z'(abcd; ef)$ appear in the general formula in contrast to the function $Z(abcd; ef)$ which appear in BLATT and BIEDENHARN's [6] treatment of reactions and scattering. As pointed out by R. HUBY, Proc. Phys. Soc. Lond. **67**, 1103 (1954), these authors used expressions for the scattered wave corresponding to those introduced at an earlier stage by WIGNER and EISENBUD [43], but which fail to satisfy the conditions for invariance under time reversal (14.7). As a consequence, the results of BLATT and BIEDENHARN are correct only if Z is replaced by Z' wherever that quantity occurs in the cross-section.

each factor referring to one stage of the reaction, and in order to be able to treat the reduced matrix-elements as real we have also separated out the phase factors, $f(l)$, introduced by Coulomb fields.

It should be pointed out that although (24.4) gives an expression for elastic scattering through an isolated resonance level, in any actual case (with the possible exception of resonance scattering of γ-radiation) it is not possible to ignore other, coherent, contributions to the scattering such as that due to Coulomb or nuclear potentials, and other, distant, levels. These other contributions will not, in general, involve states of the complete system with the same symmetry properties as those due to the resonance level, and the energy-dependence of the reduced matrix elements will be quite different. Consequently, the angular distribution of the elastic scattering will in general be quite energy-sensitive, particularly so near resonances (see Sects. 38, 39).

Expression (24.3) simplifies[1] for the important case $s=\frac{1}{2}$, since

$$\hat{l}\,\hat{l}'\,W(l\,L, l'\,L'; \tfrac{1}{2}k)\,(l\,0, l'\,0|k\,0) = (L\,\tfrac{1}{2}, L'-\tfrac{1}{2}|k\,0), \tag{24.6}$$

so that

$$Z'(l\,L\,l'\,L'; \tfrac{1}{2}k) = \hat{L}\,\hat{L}'\,(L\,\tfrac{1}{2}, L'-\tfrac{1}{2}|k\,0). \tag{24.7}$$

Alternatively we can use the particular parameters of radiation for spin-$\frac{1}{2}$ particles (13.11), and arrive at a cross-section similar to (24.1), (24.3), in which Z' is replaced by the right-hand side of (24.7) thus:

$$A_k(\mathcal{L}\mathcal{L}'; a\,b\,b) = (-)^{a-b-\frac{1}{2}}\,\hat{b}^2\,\hat{L}\,\hat{L}'\,(L\,\tfrac{1}{2}, L'-\tfrac{1}{2}|k\,0)\cdot W(b\,b\,L\,L'; k\,a). \tag{24.8}$$

If γ-radiation is absorbed, and subsequently emitted:

$$W(\vartheta) = \frac{\lambda\!\!\!^{-2}\,\hat{b}^2}{4a^2}\sum_k A_k(a, b\,b)\,A_k(b\,b, c)\,P_k(\cos\vartheta), \tag{24.9}$$

where the functions $A_k(a, b\,b)$ and $A_k(b\,b, c)$ are as defined by Eq. (15.2).

Channel-spin formalism. This formalism, corresponding to the coupling scheme $\mathbf{S}_1=\boldsymbol{a}+\mathbf{s}_1$, $\mathbf{S}_2=\boldsymbol{a}+\mathbf{s}_2$, is that most commonly used to describe nuclear reactions. $\mathbf{S}_1$, $\mathbf{S}_2$, are the "channel spins" of the absorption and emission processes respectively; $\mathbf{s}_1$, $\mathbf{s}_2$, are the intrinsic spins of absorbed and emitted particles. For an isolated resonance level, b,

$$\mathbf{S}_1+\boldsymbol{l}_1=\boldsymbol{b}=\mathbf{S}_2+\boldsymbol{l}_2 \tag{24.10}$$

(we cannot use this representation for γ-radiation, since, as already mentioned, spin and orbital angular momentum cannot be separated).

The angular correlation for absorption, followed by emission, is expressed by (12.2) with the following expressions for $\varrho_{k_b\varkappa_b}(b\,B, b\,B')$, and $\varepsilon_{k_b\varkappa_b}(b\,B, b\,B')$,

$$\left.\begin{aligned}
\varrho_{k_b\varkappa_b}(b\,B, b\,B') = {} & \sum \varrho_{k_a\varkappa_a}(a\,a)\,\varrho_{k_{s_1}\varkappa_{s_1}}(s_1\,s_1)\,\varrho_{k_{l_1}\varkappa_{l_1}}(l_1\,l_1')\,\times \\
& \times \sum_{k_{S_1}k_{S_1'}} \hat{S}_1\,\hat{S}_1'\,\hat{k}_a\,\hat{k}_{s_1}\begin{Bmatrix} a & s_1 & S_1 \\ a & s_1 & S_1' \\ k_a & k_{s_1} & k_{S_1}\end{Bmatrix}\hat{b}^2\,\hat{k}_{S_1}\,\hat{k}_{l_1}\begin{Bmatrix} S_1 & l_1 & b \\ S_1' & l_1' & b \\ k_{S_1} & k_{l_1} & k_b\end{Bmatrix}\times \\
& \times (k_a\varkappa_a, k_{s_1}\varkappa_{s_1}|k_{S_1}\varkappa_{S_1})\,(k_{S_1}\varkappa_{S_1}, k_{l_1}\varkappa_{l_1}|k_b\varkappa_b)\,\times \\
& \times \langle S_2 l_2 b|R|S_1 l_1 b\rangle\langle S_2' l_2' b|R|S_1' l_1' b\rangle^*,
\end{aligned}\right\} \tag{24.11}$$

[1] G. R. Satchler: Proc. Phys. Soc. Lond. A **67**, 471 (1954); A **68**, 1037 (1955).

with the first summation over $k_a\, k_{s_1}\, k_{l_1}\, \varkappa_a\, \varkappa_{s_1}\, \varkappa_{l_1}\, l_1\, l_1'\, S_1\, S_1'$, and

$$\left.\begin{aligned} \varrho_{k_a \varkappa_a}(a\,a) &= \frac{1}{\hat{a}}\,\delta_{k_a 0}\,\delta_{\varkappa_a 0}, \\ \varrho_{k_{s_1} \varkappa_{s_1}}(s_1\,s_1) &= \frac{1}{\hat{s}_1}\,\delta_{k_{s_1} 0}\,\delta_{\varkappa_{s_1} 0}, \\ \varrho_{k_{l_1} \varkappa_{l_1}}(l_1\,l_1') &= \frac{\hat{l}_1\,\hat{l}_1'(-)^{l_1'}}{4\pi}\,(l_1 0, l_1' 0\,|\,k_{l_1} 0)\,\mathcal{D}^{k_{l_1}}_{\varkappa_{l_1} 0}(\mathcal{R}_1). \end{aligned}\right\} \tag{24.12}$$

Also:

$$\left.\begin{aligned} \varepsilon_{k_b \varkappa_b}(b\,B, b\,B') = \sum \varepsilon_{k_c \varkappa_c}(c\,c)\,\varepsilon_{k_{s_2} \varkappa_{s_2}}(s_2\,s_2)\,\varepsilon_{k_{l_2} \varkappa_{l_2}}(l_2\,l_2') \times \\ \times \sum_{k_{S_1} \varkappa_{S_2}} \hat{b}^2\,\hat{k}_{s_2}\,\hat{k}_{l_2} \begin{Bmatrix} S_2 & l_2 & b \\ S_2' & l_2' & b \\ k_{S_2} & k_{l_2} & k_b \end{Bmatrix} \hat{S}_2\,\hat{S}_2'\,\hat{k}_{S_2}\,\hat{k}_{l_2} \begin{Bmatrix} c & s_2 & S_2 \\ c & s_2 & S_2' \\ k_c & k_{s_2} & k_{S_2} \end{Bmatrix} \times \\ \times\,(k_b \varkappa_b\,|\,k_{S_2} \varkappa_{S_2}, k_{l_2} \varkappa_{l_2})\,(k_{S_2} \varkappa_{S_2}\,|\,k_c \varkappa_c, k_{s_2} \varkappa_{s_2}), \end{aligned}\right\} \tag{24.13}$$

with the first summation over $k_c\, k_{s_2}\, k_{l_2}\, \varkappa_c\, \varkappa_{s_2}\, \varkappa_{l_2}$ and

$$\left.\begin{aligned} \varepsilon_{k_c \varkappa_c}(c\,c) &= \hat{c}\,\delta_{k_c 0}\,\delta_{\varkappa_c 0}, \\ \varepsilon_{k_{s_2} \varkappa_{s_2}}(s_2\,s_2) &= \hat{s}_2\,\delta_{k_{s_2} 0}\,\delta_{\varkappa_{s_2} 0}, \\ \varepsilon_{k_{l_2} \varkappa_{l_2}}(l_2\,l_2') &= \frac{\hat{l}_2\,\hat{l}_2'(-)^{l_2'}}{4\pi}\,(l_2 0, l_2' 0\,|\,k_{l_2} 0)\,\mathcal{D}^{k_{l_2}}_{\varkappa_{l_2} 0}(\mathcal{R}_2). \end{aligned}\right\} \tag{24.14}$$

After some algebra, this leads to the result:

$$\left.\begin{aligned} W(\vartheta) = \frac{\lambda^2}{4\hat{a}^2\hat{s}_1^2} \sum_{k l_1 l_1' l_2 l_2'} \sum_{S_1 S_2} A_k(l_1 b\, l_1' b;\, S_1)\, A_k(l_2 b\, l_2' b;\, S_2) \times \\ \times \langle S_2 l_2 b\,|R|\,S_1 l_1 b\rangle \langle S_2 l_2' b\,|R|\,S_1 l_1' b\rangle^* P_k(\cos\vartheta), \end{aligned}\right\} \tag{24.15}$$ [1]

where

$$A_k(l_1 b\, l_1' b;\, S_1) = (-)^{b+S_1} Z'(l_1 b\, l_1' b;\, S_1 k), \tag{24.16}$$

and a similar expression holds for $A_k(l_2 b\, l_2' b;\, S_2)$, which corresponds to the results of WIGNER and EISENBUD [43], and BLATT and BIEDENHARN [6]. Following WIGNER and EISENBUD, we can define the real quantities $g_{xSl} = \pm\sqrt{\Gamma_{xSl}}$, where Γ_{xSl} is the partial width of the resonance level, and x is an additional label to indicate the manner of break-up of the compound nucleus, and the reaction matrix can be expressed:

$$\langle l_2 S_2 x_2 b\,|R|\,l_1 S_1 x_1 b\rangle = e^{i[\xi(x_1 l_1) - \zeta(x_1) + \xi(x_2 l_2) - \zeta(x_2)]}\, \frac{i\, g_{x_1 S_1 l_1}\, g_{x_2 S_2 l_2}}{E_0 - E + i\,\frac{\Gamma}{2}}, \tag{24.17}$$

where Γ is the total level width, $\left(\Gamma = \sum_{xSl} \Gamma_{xSl}\right)$, E_0 is the resonance energy, ξ is the phase shift for nuclear and Coulomb potential, $\zeta(x)$ is the phase-factor, independent of l_1, determined by the screening radius of the Coulomb field, which does not appear in the final expression.

---

[1] The normalisation in (24.15) is taken so as to agree with that used in (24.4) for the case $s_1 = s_2 = 0$.

Using (24.17), (24.15), we get

$$\left.\begin{aligned} W(\vartheta)\left(\equiv\sigma(\vartheta)\right) &= \frac{\lambda\!\!\!^{-2}}{4\hat{a}^2\hat{s}_1^2}\sum_k\sum\frac{(-)^{S_1-S_2}}{(E-E_0)^2+\frac{\Gamma^2}{4}}Z'(l_1 b l_1' b;\, S_1 k)\times \\ &\times Z'(l_2 b l_2' b;\, S_2 k)\cdot g_{x_1 S_1 l_1} g_{x_2 S_2 l_2} g_{x_1 S_1 l_1'} g_{x_2 S_2 l_2'}\times \\ &\times\cos\left[\xi(x_1 l_1)+\xi(x_2 l_2)-\xi(x_1 l_1')-\xi(x_2 l_2')\right]\cdot P_k(\cos\vartheta), \end{aligned}\right\}\quad (24.18)$$

with the unlabelled summation over $l_1 l_1' S_1 l_2 l_2' S_2$. We notice that in (24.15) and (24.18) the summation over S_1, S_2 corresponds to the simple addition of the separate angular correlations, corresponding to the fact that the different channel spins are incoherent.

In the special case of a single channel spin and single value of l for both stages of the reaction, this result simplifies to

$$W(\vartheta)\sim\sum_k Z'(l_1 b l_1 b;\, S_1 k)\, Z'(l_2 b l_2 b;\, S_2 k)\, P_k(\cos\vartheta).$$

Expressions (24.1) and (24.15) yield, of course, equivalent results and are related by a simple transformation of representation.

SATCHLER [*16*] has shown how to derive the relative contributions of the angular-momentum components in one representation (say the L-component in the L, l formalism) to the component represented by a specific value of angular momentum (say a specific channel in the channel-spin representation) in the other representation.

The matrix element in (24.15) can be written as

$$\left.\begin{aligned} &\langle c\, s_2(S_2)\, l_2 b\,|R|\, a\, s_1(S_1)\, l_1 b\rangle \\ &\quad=\sum_{L_1 L_2}\langle S_2|L_2\rangle\langle a\, l_2 s_2(L_2)\, b\,|R|\, a\, l_1 s_1(L_1)\, b\rangle\langle L_1|S_1\rangle, \end{aligned}\right\}\quad (24.19)$$

where $\langle L_1|S_1\rangle=\langle S_1|L_1\rangle$ is the unitary transformation connecting two representations of the system with angular momentum b.

$$\langle L_1|S_1\rangle=\langle a\, l_1 s_1(L_1)\, b\,|\, a\, s_1(S_1)\, l_1 b\rangle=(-)^{s_1+L_1-l_1}\hat{S}_1\hat{L}_1 W(a\, s_1 b\, l_1;\, S_1 L_1). \quad (24.20)$$

In accordance with the assumption of a single resonance level, R can be factorised:

$$\langle c\, s_2(S_2)\, l_2 b\,|R|\, a\, s_1(S_1)\, l\, b\rangle=\langle c\,\|S_2 l_2\|\, b\rangle\langle b\,\|S_1 l_1\|\, a\rangle, \quad (24.21)$$

where we use for the factors of R a notation similar to that used for reduced matrix elements (cf. Sect. 11), so that, for example,

$$\langle b\,\|S_1 l_1\|\, a\rangle=\sum_{L_1}\langle a\,\|\mathcal{L}_1\|\, b\rangle(-)^{l_1+s_1-L_1}\hat{S}_1\hat{L}_1 W(a\, s_1 b\, l_1;\, S_1 L_1), \quad (24.22)$$

which gives the relative contributions of different L_1 components to the incident channel of spin S_1.

In case where a "Russell-Saunders" (or "$\mathfrak{L}$-$\mathfrak{S}$") coupling scheme is appropriate, it is of interest to find the relative contribution of different L_1 to a process in which absorption (or emission) leads from a nuclear state designated $\boldsymbol{a}=\boldsymbol{a}^{(s)}+\boldsymbol{a}^{(l)}$ to another $\boldsymbol{b}=\boldsymbol{b}^{(s)}+\boldsymbol{b}^{(l)}$, where $\boldsymbol{a}^{(s)}$ and $\boldsymbol{a}^{(l)}$ denote the separate contributions of orbital and intrinsic spin angular momenta to the total spin, $\boldsymbol{a}$, and similarly $\boldsymbol{b}^{(s)}$, $\boldsymbol{b}^{(l)}$, and $\boldsymbol{a}^{(l)}+\boldsymbol{l}_1=\boldsymbol{b}^{(l)}$; $\boldsymbol{a}^{(s)}=\boldsymbol{s}_1=\boldsymbol{b}^{(s)}$. This is given by the following relation, (6.1):

$$\left.\begin{aligned} &\langle c\, L_2 b\,|R|\, a^{(l)} l_1(b^{(l)}),\, a^{(s)} s\,(b^{(s)}),\, b\rangle \\ &\quad=\sum_{L_1}\langle c\, L_2 b\,|R|\, a^{(l)} a^{(s)}(a),\, l_1 s_1(L_1)\, b\rangle\langle a\, L_1|\, b^{(l)} b^{(s)}\rangle, \end{aligned}\right\}\quad (24.23)$$

where

$$\langle b^{(l)} b^{(s)} \,|\, a L_1 \rangle = \langle a L_1 \,|\, b^{(l)} b^{(s)} \rangle = \hat{a}\,\hat{L}_1\,\hat{b}^{(l)}\,\hat{b}^{(s)} \begin{Bmatrix} a^{(l)} & a^{(s)} & a \\ l & s & L_1 \\ b^{(l)} & b^{(s)} & b \end{Bmatrix}. \tag{24.24}$$

Expressions such as (24.23) may be useful for reactions involving light nuclei where states may be characterised, approximately, by a single value of intrinsic spin and orbital momentum (e.g. $\boldsymbol{a}^{(s)}$, $\boldsymbol{a}^{(l)}$), or more generally where the description of a state involves expansion in terms of several eigenfunctions of $\boldsymbol{a}^{(s)}$ and $\boldsymbol{a}^{(l)}$.

25. Resonance reactions (absorption) followed by successive emissions. The considerations of Sect. 24 can be extended to deal with reactions followed by cascade emission by application of the general results given in Sects. 16 and 17. The most frequently studied correlations of this type are (a) the angular distribution of radiation *not* directly emitted from the resonance level, and (b) angular correlations between the radiation emitted directly from the resonance level and that immediately following it. In the first of these, only one angle is involved in the distribution, which involves an average over all directions of emission and polarisations of the intermediate, unobserved, radiations. In the second case, the directions of all three radiations (one absorption and two emissions) are known, and we have, from the theoretical standpoint, a triple correlation.

The angular distribution of radiations emitted after an intermediate unobserved radiation is given by (16.10) modified so that the initial process corresponds to absorption. (It should be noticed that complexity of the distribution formula is still limited by the smallest of the quantities L_1, b, c, L_3, but it is not limited by the value of L_2 for the unobserved radiation.)

$$W(\vartheta) = \sum_k A'_k(a, b\,b)\,\bar{A}_k(b\,c)\,A_k(c\,c, d)\,P_k(\cos\vartheta), \tag{25.1}$$

with $A'_k(a, b\,b)$ differing from $A_k(a, b\,b)$ in that $c_{k\varkappa}(\mathscr{L}_1\,\mathscr{L}'_1)$ is replaced by $c_{k\varkappa}(\mathscr{L}'_1\,\mathscr{L}_1)$ and with $\langle b \| \mathscr{L}_1 \| a \rangle$ replaced by $\langle a \| \mathscr{L}_1 \| b \rangle$, etc.

The interpretation of the angular distribution, involving as it does the spin values of four levels and three multipolarities, often cannot be made unambiguous without some supplementary information. However, since angular distributions can be measured relatively easily with high precision, the results they provide are of comparable value to the less exact, and more elaborate, measurements of the angular correlation of successive radiations following the resonance absorption. For these latter measurements it is usually convenient and sufficient to measure the correlation for some fairly simple angular relation between the incident beam and one (not necessarily the first, although frequently so) of the cascade radiations, e.g. with one radiation detected either in the same direction, or in one normal to the incident beam. In the first of these arrangements, the correlation reduces to a function of one angle, ϑ, only, irrespective of the plane in which ϑ is measured.

In the second arrangement, the angular correlation depends on the plane in which it is measured. This is usually done either in the plane normal to the incident beam or in the plane containing the incident beam (and the two radiations), or in both. In these cases the expression (16.15) for the angular correlations yields results as follows:

(a) Radiation (2) following absorption detected in the same direction as the incident particle (1). Succeeding radiation (3) detected at angle ϑ to this:

$$\left.\begin{aligned} W(\vartheta_{3,2}) = W(\vartheta_{3,1}) &= \sum_{k_1 k_2 k_3} A'_{k_1}(a, b\,b)\,A_{k_3}(c\,c, d)\,\times \\ &\times R_{k_1 k_2 k_3}(b\,b\,c\,c)\,\hat{k}_1\,\hat{k}_2\,\hat{k}_3\,(k_1 0, k_2 0 \,|\, k_3 0)\,P_k(\cos\vartheta_{3,2}), \end{aligned}\right\} \tag{25.2}$$

(b) Radiation (2) following absorption detected at $\pi/2$ to incident radiation (1). Succeeding radiation (3) observed at ϑ_{13} with respect to (1). Radiation (1) to (3), co-planar:

$$\left.\begin{aligned} W(\vartheta_{1,3}) \equiv W\left(\frac{\pi}{2} - \vartheta_{2,3}\right) &= \sum_{k_1 k_2 k_3} A'_{k_1}(a, b\,b)\, A_{k_3}(c\,c, d) \times \\ &\times R_{k_1 k_2 k_3}(b\,b\,c\,c)\, T_{k_1 k_3; k_2} \sum_{k \geq 0} B_{k_1 k_3; k_2 \varkappa}\, P^{\varkappa}_{k_3}(\cos\vartheta_{1,3}), \end{aligned}\right\} \tag{25.3}$$

where

$$T_{k_1 k_3; k_2} = \frac{\hat{k}_1 \hat{k}_2 \hat{k}_3 (-)^{k_2/2}}{2^{k_2}}, \tag{25.4}$$

$$B_{k_1 k_3; k_2 \varkappa} = \frac{\sqrt{(k_2 - \varkappa)!\,(k_2 + \varkappa)!}}{\left(\frac{k_2 + \varkappa}{2}\right)!\left(\frac{k_2 - \varkappa}{2}\right)!} (-)^{k_2/2} (k_1\, 0, k_2\, \varkappa \mid k_3\, \varkappa)\, (2 - \delta_{\varkappa 0}), \tag{25.5}$$

and $k_1 k_2 k_3$ and $\varkappa$ are even,

(c) Radiation (2) following absorption detected at $\pi/2$ to incident radiation (1). Succeeding radiation (3) observed at ϑ_{23} with respect to (2). Plane containing (2), (3), perpendicular to (1):

$$\left.\begin{aligned} W(\vartheta_{2,3}) &= \sum_{k_1 k_2 k_3} A'_{k_1}(a, b\,b)\, A_{k_3}(c\,c, d)\, R_{k_1 k_2 k_3}(b\,b\,c\,c) \times \\ &\times T_{k_2 k_3; k_1} \sum_{\varkappa \geq 0} B_{k_2 k_3; k_1 \varkappa} (-)^{\varkappa/2} P^{\varkappa}_{k_3}(\cos\vartheta_{2,3}), \end{aligned}\right\} \tag{25.6}$$

where $k_1 k_2 k_3$ and $\varkappa$ are even:

In the next two sections we consider the experimental investigations of processes discussed in Sects. 23 to 25, with illustrations of particular reactions.

26. Resonance reactions involving γ-radiation. The conditions for the occurrence of isolated resonances are that the partial, and hence the total, level widths, Γ_{xSl} and Γ, should be small in comparison with the level spacing, D, for the energies with which the compound nucleus is formed. Such conditions occur with light nuclei (large D), particularly when the compound nucleus can only dissociate either into charged particles, for which processes the partial widths are seriously reduced by the Coulomb barrier, or by emission of γ-radiation, for which the weak electromagnetic interaction ensures a small partial width. Thus radiative capture of protons and α-particles[1], (p, γ) and (α, γ), of low energy provide common examples of isolated levels[2]. The inverse processes, (γ, p), (γ, α), satisfy the conditions equally at the appropriate energies, but the difficulties of providing homogeneous γ-radiations of variable energy makes the study of such reactions at discrete resonances difficult experimentally. It should also be noticed that the two reactions do not permit the same processes to be studied when excited states of the initial or product nucleus are involved. Neutron capture, (n, γ), reactions are less easily studied, again because of the relative difficulty of providing canalised beams of monoenergetic reactions, and more fundamentally because, in the absence of a potential barrier, reaction widths comparable with level-spacings are found except for very light nuclei or very low energies[3].

[1] Capture of other incident particles (other than slow neutrons) is not generally observed, since emission of energetic heavy particles is a much more likely process.

[2] With increasing energy not only do the partial proton and α-particle widths increase rapidly, but the emission of neutrons, with large partial width, becomes a possibility.

[3] For "slow-neutron" reactions, (n, γ), (n, α), etc., whether involving isolated resonances or not, there is no correlation with the incident beam. However, the compound nucleus, although statistically isotropic, is not necessarily in a state of well-defined total angular-momentum; but the different spin states are incoherent, so that cascades following slow neutron absorption can always be reduced to the simpler problem of cascades from a normal isotropic source.

Absorption and re-emission of γ-radiation, with either the same or different energy (resonance fluorescence) via a bound level (i.e. one which cannot dissociate into heavy particles), satisfies the conditions for an isolated resonance very well, but the experimental difficulties are in general formidable. Over the range of energies covered by the resonance level (typically a few volts or less), the resonance cross-section is likely to be much larger than other, non-resonant, coherent, contributions to the cross-sections; but in practice the range of energies in the incident radiation is likely to be much greater than the resonance width, and this other radiation produces a background of incoherent scattering. So far such processes have been studied experimentally only in a few cases, and these have been for transitions with known spin changes and multipolarities. The interest has been the determination of the absolute reduced matrix elements which are not obtained from correlation studies.

Measurements of the angular distribution of γ-radiations in (p, γ) reactions have been made using magnetically analysed proton beams of well-defined energy (van de Graaff or Cockroft-Walton type accelerators) up to a few Mev. Many resonances have been studied in nuclei ranging from deuterium to calcium. Sodium iodide scintillation crystals are almost invariably used as detectors, and since the γ-spectrum is often complex, some form of pulse-analysis is usually necessary. If the output of the detector is recorded in a multi-channel analyser, then, by studying the variation of the observed pulse-spectrum with position of the detector, the angular distribution of several γ-radiations can be studied simultaneously. The correction of the observations for finite solid angle of detector is simpler in this than in the cascade-correlation case. The collimation of the proton beam is normally sufficiently good for it to be regarded as ideal, and the size of source is likewise normally much smaller than that of the detector. Consequently the factors g_k entering the observed correlation (18.2) take the simpler form of a single term such as $g_k^{(1,1)}$, (18.3). Similar considerations affect the maximum useful aperture. For straightforward angular-distribution measurements, the problem of random counts does not arise, but the available intensity may be limited by physical conditions at the source. Since also the source strength may fluctuate irregularly due to variation in beam intensity, or energy, and changes in conditions at the source, it is usually necessary to monitor the source by measurement of the γ-intensity with a detector at a fixed position. The difference between laboratory and centre-of-mass coordinates, due to the finite proton momentum, is often appreciable, and the results must be corrected accordingly.

Unlike the analysis of correlations in radioactive cascades, the interpretation of p-γ correlations is not frequently helped by knowledge of the K, L-conversion coefficients of the radiations. These coefficients are very small for energetic radiations and light nuclei, and in most cases have not been measured. However, there are compensating features. Some knowledge of multipolarity may be available from the study of internal electron-pair creation (Sect. 48), and information about the spin and parity of the resonance level may be obtained more directly from measurements of the elastic scattering (Sects. 39, 41). From the study of the level structure of neighbouring light nuclei, isotopic spin may be assigned to some levels, and this in turn may be of help in assigning γ-transition multipolarities[1]. One further feature of resonance capture processes, in the cases where there are many levels differing by small energy steps, but which can still be treated as approximately "isolated", lies in the possibility of determining the

[1] As in the case of γ-cascades, polarisation measurements can also be used to establish parity changes (cf. Sect. 33).

sequence of γ-transitions by observing the change in energy of the components of the γ-spectrum in passing from one resonance to the next. In transitions which take place directly from the resonance level, the γ-transition energy changes by the energy difference between resonances, but for other γ-transitions there is no such change.

The general expression for the angular distribution of γ-radiation, $b\to c$, (L_2, L_2') following proton capture, $a\to b$, $(\mathscr{L}_1, \mathscr{L}_1')$ with initial nucleus spin a, resonance level spin b, and final state c, is given in the $\mathscr{L}$-representation, by the appropriate modification of (24.1),

$$\left.\begin{aligned} W(\vartheta) \sim \sum_k \sum A_k(\mathscr{L}_1 \mathscr{L}_1'; a b b)\, A_k(L_2 L_2'; b b c) \times \\ \times \langle c L_2 b | R | a \mathscr{L}_1 b\rangle \langle c L_2' b | R | a \mathscr{L}_1' b\rangle^* P_k(\cos\vartheta), \end{aligned}\right\} \tag{26.1}$$

where $A_k(\mathscr{L}_1 \mathscr{L}_1'; a b b)$ is defined as in (24.8) and $A_k(L_2 L_2'; b b c)$ is the same as $A_k(b b, c)$ in (19.2), but omitting the reduced matrix-elements, i.e.

$$\sum_{L_2 L_2'} A_k(L_2 L_2'; b b c) \langle c \| L_2 \| b\rangle \langle c \| L_2' \| b\rangle = A_k(b b, c).$$

The unlabelled summation is over $l_1 l_1'$, $L_1 L_1'$, $L_2 L_2'$, with $L_1 = l_1 \pm \frac{1}{2}$, $L_1' = l_1' \pm \frac{1}{2}$, and successive values of l_1, l_1', differ by two units (conservation of parity).

In the special case when the target nucleus has zero spin, $a=0$, then $b=L$, and also, since the proton spin is $\frac{1}{2}$, only one l_1-value is possible. Hence the summation is only over $L_2 L_2'$. It is usually necessary to consider only the mixing of two possible γ-components, L_2 and L_2+1, as in the case of cascade-correlation processes, so that (26.1) simplifies to

$$\left.\begin{aligned} W(\vartheta) = \sum_k A_k(\mathscr{L}_1 \mathscr{L}_1'; 0 b b)\, [F_k(L_2 L_2 c b) + \delta^2 F_k(\overline{L_2+1}\,\overline{L_2+1}\, c b) + \\ + 2\delta F_k(L_2 \overline{L_2+1}\, c\, b)]\, P_k(\cos\vartheta), \end{aligned}\right\} \tag{26.2}$$

where the F_k's are as in (19.6), (19.7); and $L_1 = b$.

For an isolated resonance, the R matrix can be factored so that δ is the ratio of the γ-radiation reduced widths,

$$\delta = \frac{\langle b \| L_2 + 1 \| c\rangle}{\langle b \| L_2 \| c\rangle}.$$

For pure multipole γ-radiation, and target nucleus with zero-spin, we have the particularly simple result:

$$W(\vartheta) \sim \sum_k (b \tfrac{1}{2}, b - \tfrac{1}{2} | k 0)\, (L_2 1, L_2 - 1 | k 0)\, W(b b L_2 L_2; k c)\, P_k(\cos\vartheta) \tag{26.3}$$

with k (even) $< 2b$.

The correlation (26.1) can also be expressed in mixed-representation, using the channel-spin to describe the proton absorption process and the $\mathscr{L}$-representation, of course, for the γ-emission:

$$W(\vartheta) = \sum_{S_1 l_1 l_1'} A_k(l_1 b l_1' b; S_1)\, A_k(b b, c) \cdot \langle b \| S_1 l_1 \| a\rangle \langle b \| S_1 l_1' \| a\rangle^* P_k(\cos\vartheta), \tag{26.4}$$

where $A_k(l_1 b l_1' b; S_1)$ is as in (24.16), and $\langle b \| S_1 l_1 \| a\rangle$, $\langle b \| S_1 l_1' \| a\rangle$ are the factors of the R matrix which refer to the absorption process. (As previously, all summations except over $L_2 L_2'$ disappear if the target spin is zero, $a=0$.)

Study of the angular distribution of γ-radiation succeeding the direct capture γ-ray is, apart from the more realistic possibility of extra-nuclear effects, a similar problem experimentally to the study of direct capture radiation. For γ-γ correlations following capture, most of the experimental considerations are similar to those applicable to other γ-γ correlation studies (Sects. 19, 20), with the additional feature that the correlation now depends on the relative orientation of the plane containing the target and the two γ-detectors, and the proton beam. The angular distributions (25.1) and the correlation functions (25.2), (25.3), (25.6), can be readily applied to the process $(p_1, \gamma_2, \gamma_3)$:

(a) Angular distribution of γ_3 (with respect to p_1); γ_2 unobserved,

$$\left.\begin{aligned} W(\vartheta) = \sum_{k S l_1 l_1'} A_k(l_1 b l_1' b; S)\, \bar{A}_k(b c)\, A_k(c c, d) \times \\ \times \langle b \| S l_1 \| a\rangle \langle b \| S l_1' \| a\rangle^* P_k(\cos\vartheta), \end{aligned}\right\} \tag{26.5}$$

(b) Angular correlation (γ_2, γ_3); γ_2 parallel to p_1,

$$\left.\begin{aligned} W(\vartheta) = \sum_{\substack{k_1 k_2 k_3 \\ S l_1 l_1'}} A_{k_1}(l_1 b l_1' b; S)\, A_{k_3}(c c, d)\, R_{k_1 k_2 k_3}(b b c c) \times \\ \times \hat{k}_1 \hat{k}_2 \hat{k}_3 (k_1 0, k_2 0 | k_3 0) \langle b \| S l_1 \| a\rangle \langle b \| S l_1' \| a\rangle^* P_k(\cos\vartheta), \end{aligned}\right\} \tag{26.6}$$

(c) Angular correlation (γ_2, γ_3); γ_2 at 90° to p_1 and γ_3 at ϑ to p_1; $p_1 \gamma_2 \gamma_3$ coplanar,

$$\left.\begin{aligned} W(\vartheta) = \sum_{\substack{k_1 k_2 k_3 \\ S l_1 l_1'}} A_{k_1}(l_1 b l_1' b; S)\, A_{k_3}(c c, d)\, R_{k_1 k_2 k_3}(b b c c)\, T_{k_1 k_2; k_3} \times \\ \times \sum_{\varkappa \geq 0 \text{ and even}} B_{k_1 k_3; k_2 \varkappa} \langle b \| S l_1 \| a\rangle \langle b \| S l_1' \| a\rangle^* P_{k_3}^{\varkappa}(\cos\vartheta), \end{aligned}\right\} \tag{26.7}$$

(d) Angular correlation (γ_2, γ_3); γ_2 at 90° to p_1, and γ_3 at ϑ to p_1; p_1 is perpendicular to plane $(\gamma_2 \gamma_3)$,

$$\left.\begin{aligned} W(\vartheta) = \sum_{\substack{k_1 k_2 k_3 \\ S l_1 l_1'}} A_{k_1}(l_1 b l_1' b; S)\, A_{k_3}(c c, d)\, R_{k_1 k_2 k_3}(b b c c)\, T_{k_2 k_3; k_1} \times \\ \times \sum_{\varkappa \geq 0, \text{ and even}} (-)^{\varkappa/2} B_{k_2 k_3; k_1 \varkappa} P_{k_3}^{\varkappa}(\cos\vartheta) \langle b \| S l_1 \| a\rangle \langle b \| S l_1' \| a\rangle^*. \end{aligned}\right\} \tag{26.8}$$

27. Examples. One of the earliest γ-transitions to be examined from the point of view of angular-correlations[1] involves the reaction Li^7 (p, γ) Be^8, in which γ-transitions, of energy 17.6 and approximately 14.6 Mev, occur from the resonance level to the ground and first excited states of Be^8. Apart from a small anisotropy which is attributable to interference with a neighbouring level, both components of the γ-radiation at resonance are isotropic, which is obviously consistent with $l=0$, $L=\frac{1}{2}$, for the incident proton. This result[2] is also consistent with $l=1$ for the proton if the contributions of the two possible channel spins, $\frac{3}{2}+\frac{1}{2}=\mathbf{2}$, and $\frac{3}{2}+\frac{1}{2}=\mathbf{1}$, are assumed to be in the ratio 5:1. This assumption is also consistent with the analysis of the observed elastic scattering of protons by Li^7 at this resonance[3]. Isotropy of the γ-radiation would also follow from the assumption that only the $l=1$, $L=\frac{1}{2}$, component of the incident proton beam were involved in the resonance, and this also leads, of course, to

[1] S. DEVONS and M. G. N. HINE: Proc. Roy. Soc. Lond., Ser. A **199**, 56 (1949).
[2] S. DEVONS and G. R. LINDSEY: Proc. Phys. Soc. Lond. A **63**, 1202 (1950). — M. B. STEARNS and B. D. MCDANIEL: Phys. Rev. **82**, 450 (1951).
[3] W. D. WARTERS, W. A. FOWLER and C. C. LAURITSEN: Phys. Rev. **91**, 917 (1953).

the same channel-spin mixture. A further possibility is the assumption of pure Russell-Saunders coupling. With the usual spectroscopic notation, Li^7 is designated $2P_{\frac{1}{2}}$ ($a^{(l)}=1$, $a^{(s)}=\frac{1}{2}$); and the resonance level $3S_1$ ($b^{(l)}=0$, $b^{(s)}=1$), so that $\boldsymbol{a}^{(l)}+\boldsymbol{L}_1(=1)=\boldsymbol{0}$, and $\boldsymbol{a}^{(s)}+\boldsymbol{S}_1(=\frac{1}{2})=\boldsymbol{1}$. This interpretation also predicts isotropy of the γ-radiation and the same contributions from the two spin channels, (CHRISTY[1]). CHRISTY has examined several other similar processes in light nuclei [$H^3(p,\gamma)He^4$; $B^{11}(p,\gamma)C^{12}$; etc.], and found that the correlations can be interpreted on the basis of Russell-Saunders ("$\mathfrak{L}$-$\mathfrak{S}$") coupling rather better than on the assumption of a single L-component ("j-j" coupling). For heavier nuclei, beyond the region where the first P shell is filling, one would not expect this trend to continue, and for most of the actual range of nuclei in which the (p,γ) process has been studied a more complicated coupling is likely.

The present-day techniques used in studying and analysing capture γ-radiation are well illustrated in a recent study[2] of the reaction $Mg^{24}(p,\gamma)Al^{25}$. Some special features of this reaction are: (a) the low binding energy of the proton in Al^{25}, which means that comparatively low excited states of Al^{25} are produced as resonance levels. (b) the zero spin of the bombarded nucleus, Mg^{24}. Consequently, there is only a single channel spin involved, and since parity is conserved only a single value of l or L contributes to a particular resonance, and the theoretical analysis is thereby simplified. (c) the product nucleus is radioactive and the resulting positron-annihilation radiation is a source of disturbance. For proton energies below about 2 Mev, proton capture is (apart from elastic scattering) the most significant process. Above this energy, inelastic scattering of protons leaving Mg^{24} in an excited state of 1.37 Mev becomes the dominant process.

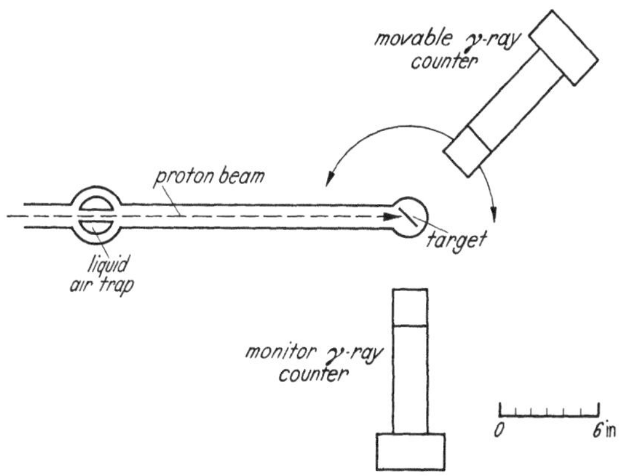

Fig. 10. Schematic arrangement for angular correlations and distributions in (p,γ) reactions [cf. G. A. BARTHOLOMEW *et al.*: Canad. J. Phys. **33**, 443 (1955)].

A beam of protons, mono-energetic to ± 1 kev, produced by an electrostatic generator, bombarded a thin (~ 0.1 mg/cm^2) target of isotopically separated Mg^{24} supported on a tantalum foil. The γ-radiation was detected by two large sodium iodide crystals (5″ diameter and 4″ long) coupled to photomultipliers, one fixed (at 270° to the proton beam), and the other movable in range $\cos\vartheta = 1.0$ to -0.8 (see Fig. 10). The fixed γ-detectors could be used either to monitor the radiation or in a coincidence arrangement with the movable crystal, in which case a gated spectrometer arrangement was employed (cf. Fig. 2b). The response of the γ-detectors was examined using radioactive sources or reactions producing simple γ-spectra, so that the complicated spectra observed in the actual measurements could be resolved into their separate components. The correction for finite solid angles of the detectors was determined by making measurements (using a suitable resonance) with different geometrical arrangements.

[1] R. F. CHRISTY: Phys. Rev. **89**, 839 (1953).

[2] A. E. LITHERLAND, E. B. PAUL, G. A. BARTHOLOMEW and H. E. GOVE: Phys. Rev. **102**, 208 (1956); also L. L. GREEN, J. J. SINGH and J. C. WILLMOTT: Phil. Mag. **46**, 982 (1955).

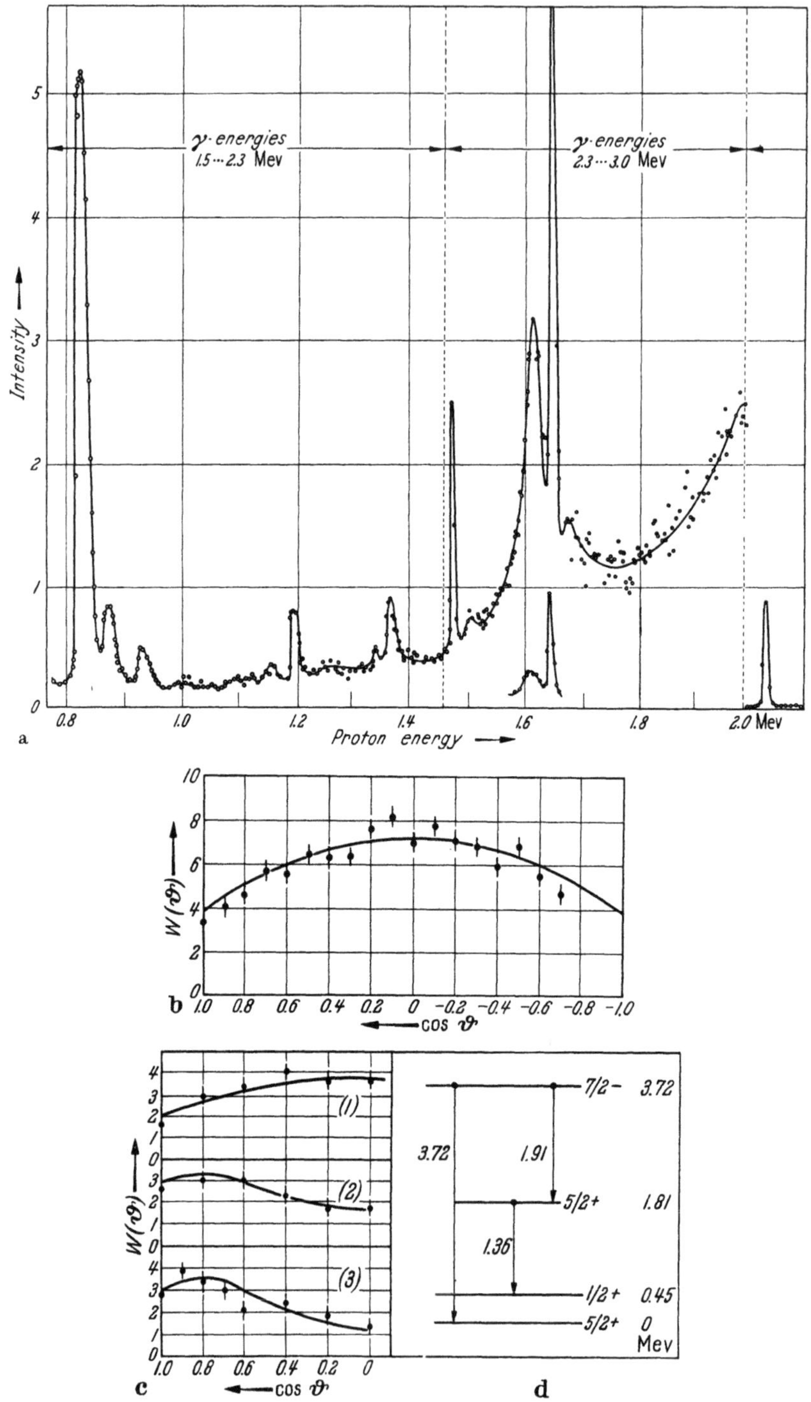

Fig. 11a–d. γ-radiations from $Mg^{24}(p, \gamma)\,Al^{25}$ [cf. A. E. Litherland *et al.*: Phys. Rev. **102**, 211, 213 (1956)]. (a) Yield of γ-rays from the reactions $Mg^{24}(p, \gamma)\,Al^{25}$ and $Mg^{24}(p, p')\,Mg^{24}$. (b) Angular distribution of 3.172 Mev γ-ray (with respect to proton beam) at 1.49 Mev resonance. (c) (1) Angular distribution of 1.91 Mev γ-ray at same resonance. (2) Angular distribution of 1.36 Mev γ-ray at same resonance. (3) Coincidence-correlation of 1.91 Mev (counter fixed at 270° with respect to beam) and 1.36 Mev γ-rays (counter in range $0 \leq \vartheta \leq 90°$). (d) Energy-level diagram showing γ-rays studied.

Angular distributions of the principal γ-radiations were measured at all resonances, coincidence measurements with fixed detectors were made at some resonances to establish the decay scheme, and some triple correlations were measured with one detector fixed at 270° and the other movable between 0° and 90°, i.e. with all three directions, beam and both γ-radiations, coplanar. The angular distributions were of the type discussed above: (a) direct γ-radiation from the resonance level [Eq. (26.2)], (b) indirect γ-radiation, intermediate radiation unobserved [Eq. (26.5)], (c) correlations between direct γ-radiation and that succeeding it [Eq. (26.7)]. (In all cases $a=0$, $S_1=\frac{1}{2}$, and the summations over $l_1 l_1'$ and $L_1 L_1'$ reduce to a single term.)

The proton energy range up to 2 Mev covers the range of levels of Al^{25} from 2.3 to 4.2 Mev, and these and previous measurements enable the spins and parities of five resonances in this level, together with those of four lower states, to be assigned unambiguously. A part of the excitation curve, and a typical set of measurements at one level, are shown in Fig. 11.

It should also be pointed out that in studying the radiation from capture processes, the measurement of the cross-sections provides information about the absolute value of the reduced matrix elements, either of the initial capture process or the subsequent radiation from the resonance level, or both[1], [cf. also Eq. (24.15)]. The information about the spins, parities, and γ-multipolarities, coupled with that of the absolute values of the reduced matrix-elements, provides a very powerful means of examining the properties of nuclear levels. In the present example the information was compared with the predictions of the "collective model" of the nucleus[2].

The study of (α, γ) processes[3] presents problems similar to those discussed above. The correlation functions (apart, of course, from questions of absolute probability) are, in the case of isolated resonances, identical with those given for the (α, γ) cascade, (Sect. 21). Study of the (α, γ) capture process is, however, not widely applicable, since for α-particle energies sufficient for reasonable cross-sections to be obtained, the competing, and more probable, processes (α, n), (α, p), occur, except for very light nuclei ($Z \lesssim 10$). A similar limitation occurs, of course, for (p, γ) processes, but the more gradual dependence of barrier penetrability on energy, and the possibly different threshold for the (p, n) reaction, results in a much wider range of useful application.

The inverse processes, i.e. the "photo-nuclear" reactions (γ, p), (γ, α), etc., have been extensively studied[4], but mostly with radiation of high energy ($\gtrsim 10$ Mev) produced by electron bremsstrahlung and not therefore precisely mono-energetic. Even when the disintegrations resulting from a limited range of energies are isolated, the conditions are not usually such that the theory for a single resonance level is applicable. For example, the common observation of angular distributions with terms $P_k(\cos\vartheta)$ having odd k indicates that one is not dealing with a system of well-defined parity, and that *a fortiori* the process does not take place through an isolated resonance level of well-defined symmetry.

28. Resonance reactions with nucleonic particles. Apart from elastic scattering, which requires special consideration, the reactions in this class, which have been chiefly studied and for which the conditions of isolated resonances are satisfied, are (p, α), (α, p), (α, n) and inelastic scattering of protons and α-particles,

[1] S. Devons: Excited States of Nuclei, Chap. III, § 9. Cambridge 1949.
[2] A. Bohr and B. R. Mottelson: Ref. [*14*], Chap. XVII.
[3] E.g. P. C. Price: Proc. Phys. Soc. Lond. A **68**, 553 (1955).
[4] E.g. B. Spicer: Phys. Rev. **100**, 791 (1955). — S. A. E. Johansson: Phys. Rev. **97**, 439 (1955). — P. Stöll: Helv. phys. Acta **27**, 395 (1954).

all with low-energy particles (up to a few Mev), and light nuclei[1]. In addition, when the product nucleus is left in an excited state, the angular distribution of the subsequent γ-radiation can be studied, and also its correlation with the preceding particle. In the former case the particle emitted from the resonance

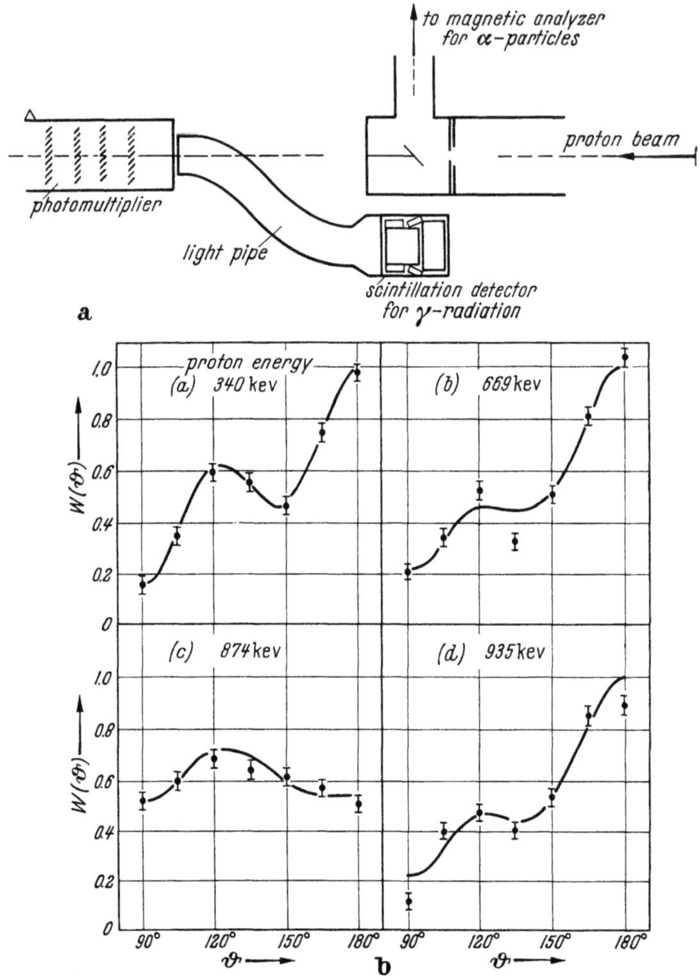

Fig. 12a and b. (a) Experimental arrangement of a $(p, \alpha\gamma)$ correlation. (b) Experimental results for α-γ correlation in $F^{19}(p, \alpha\gamma)\,O^{16}$-measurements made in the plane perpendicular to proton beam for four proton resonances at the proton energies indicated [cf. J. SEED and A. P. FRENCH: Phys. Rev. **89**, 1007 (1952)].

is not observed, and the general experimental arrangements are the same as for the measurement of (p, γ, γ) processes. When the emitted particle is detected, either in an angular distribution or correlation measurent, the experimental arrangements are more complicated. Since for the criterion of discrete resonances to be satisfied the emitted particles have, essentially, low energy, their detection in the presence of a background of scattered particles (and, possibly, particle groups corresponding to other reaction processes) often requires the use of spectro-

[1] E.g. E. S. SHIRE, J. R. WORMALD, G. LINDSAY-JONES, A. LUNDEN and A. G. STANLEY: Phil. Mag. **44**, 1197 (1954). $B^{10}(\alpha, p)$, (α, n), $B^{11}(\alpha, p)$. — R. W. PETERSON, W. A. FOWLER and C. C. LAURITSEN: Phys. Rev. **96**, 1250 (1954). $F^{19}(p, \alpha)$, (p, p'). — P. H. STELSON: Phys. Rev. **96**, 1584 (1954). $Na^{23}(p, \alpha)$.

graphic resolution with consequent reduction in solid angle for detection. Under such circumstances correlation measurements between a resolved particle group and subsequent γ-radiation can be made with only moderate precision, although possibly sufficient to lead to unambiguous interpretation. Simple angular distribution measurements, on the other hand, can be made with much greater precision.

Extensive studies have been made of the reactions of the type (p, α, γ) [1] which serve to illustrate the measurements and their analysis. Angular distribution of the α-particle, γ-radiation, and (α, γ) correlations, have been measured.

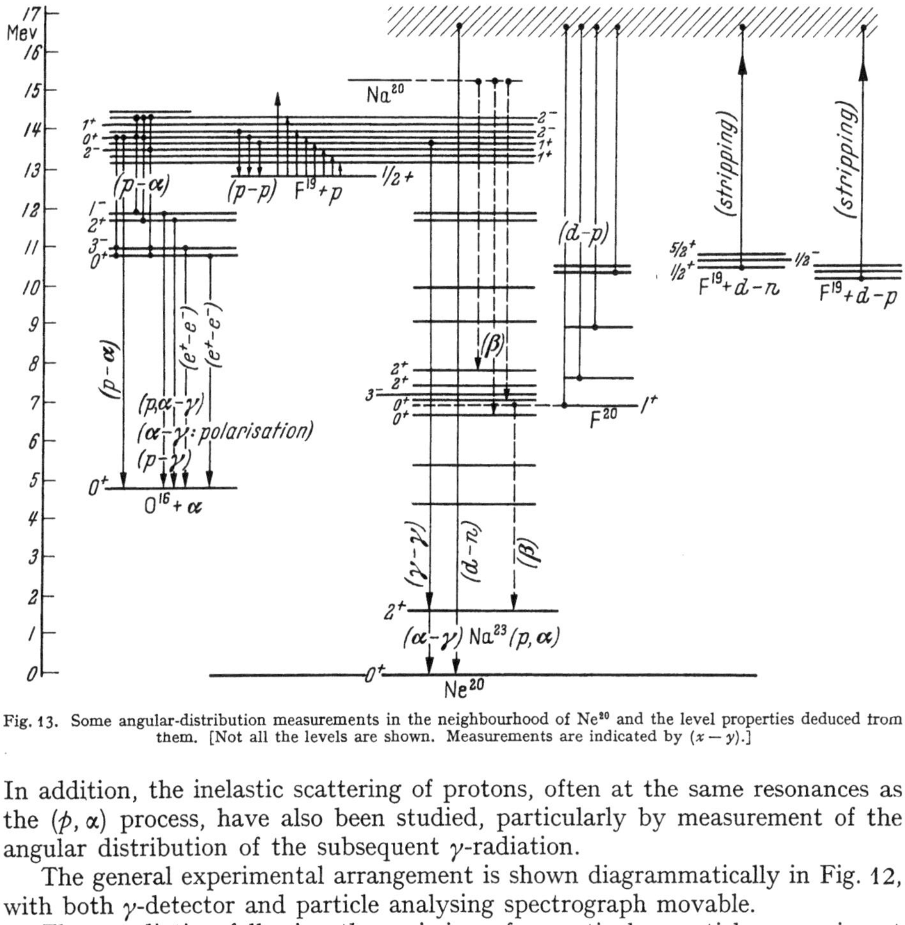

Fig. 13. Some angular-distribution measurements in the neighbourhood of Ne^{20} and the level properties deduced from them. [Not all the levels are shown. Measurements are indicated by $(x - y)$.]

In addition, the inelastic scattering of protons, often at the same resonances as the (p, α) process, have also been studied, particularly by measurement of the angular distribution of the subsequent γ-radiation.

The general experimental arrangement is shown diagrammatically in Fig. 12, with both γ-detector and particle analysing spectrograph movable.

The γ-radiation following the emission of a particular particle group is not generally very complex, so that simple pulse-analysis of the γ-detector output is the most that is usually necessary.

For the α-particle distribution, we can apply Eq. (24.1) with the simplification that $L_2 = l_2$, $L_2' = l_2'$, and $L_1 = l_1 \pm \frac{1}{2}$, etc. Alternatively, we can use the channel-spin formalism (24.15), and in this case $S_1 = a \pm \frac{1}{2}$, and $S_2 = c$ (zero spin of α-particle), so that the summation over S_2 disappears. Conservation of parity restricts the summations to differences of two units in successive values of l_1, l_1', and l_2, l_2'.

[1] E.g. A. A. Kraus: Phys. Rev. **94**, 975 (1954). — J. Seed and A. P. French: Phys. Rev. **88**, 1007 (1952).

The angular distribution of γ-radiation following α-emission (α-particle not observed) is given by an expression of the type (26.5) with the appropriate modification to $\bar{A}_k(bc)$ [cf. Eq. (16.12)]. The same expression, again with the appropriate $\bar{A}_k(bc)$, gives the angular distribution of γ-radiation following proton inelastic scattering, assuming that the condition of an isolated resonance is applicable.

For triple correlations in reactions (p, α, γ), (p, p', γ), etc., we can use the general result (16.5) with results similar to those shown in (26.6), (26.7), (26.8).

As an illustration we may consider the angular correlations of the processes resulting from the proton bombardment of F^{19} (which shows many "discrete" resonances for proton energies up to about 2 Mev). These have been extensively studied[1] and have led to information about the resonance levels (of Ne^{20}), the spins and parities of the lowest few states of O^{16}, F^{19}, and Ne^{20}. (p, α), (p-(α unobserved)γ) and (p-(proton unobserved)γ) angular distributions, (α, γ) correlations, (γ, γ) correlation following capture, have all been measured for some of the resonance levels. In addition, the interpretation of the measurements is consistent with other information such as elastic scattering[2], γ-polarisation measurements[3], and internal paircreation[4]. Fig. 13 shows the main symmetry properties of some of the nuclear states which have been established by correlation measurements.

III. Angular correlations with polarised radiations.

29. General theory. In the preceding account we have limited our discussions to correlations involving only the directions of propagation of the absorbed or emitted radiations, and we have dealt only with unpolarised incident beams, detectors which are insensitive to the state of polarisation of the radiations (or the product nuclei), and initial nuclei which are randomly oriented. In a more general treatment of angular correlations we must consider the possibility that the incident particles (if any) are polarised, the target nucleus or the initial, decaying, nucleus is partially polarised, and that the products of the reaction or the cascade decay are detected with polarisation-sensitive detectors. The term "polarised" is used here in the most general sense. A collection of nuclei with spin a, b, etc. is called polarised, partially or completely, when it is characterised by a distribution amongst the substates, $(a\,\alpha)$, $(b\beta)$, etc., which differs from the normal distribution of equal population and random phases. Similarly, a beam of particles or radiation may be considered as an assembly of systems with a population distribution in different, completely specified, quantum-mechanical states defining, *inter alia*, spin orientation or polarisation $(s\sigma)$. Again, if the population amongst these states differs from one of equal population and random phases, the beam is called polarised. In the special case of all systems being in the same pure, quantum-mechanical state, then, except for particles with zero spin, such an assembly will always be polarised in the sense described. Of course, all the above conditions are not normally met with simultaneously in a single experiment, but, formally at least, the general theory based on the use of density matrices and statistical tensors can be adapted to meet this general

---

[1] Cf. F. AJZENBERG and T. LAURITSEN: Rev. Mod. Phys. **27**, 77 (1953), p. 150.

[2] T. S. WEBB, F. B. HAGEDORN, W. A. FOWLER and C. C. LAURITSEN: Phys. Rev. **99**, 138 (1955). — E. U. BARANGER: Phys. Rev. **99**, 145 (1955).

[3] L. W. FAGG and S. S. HANNA: Phys. Rev. **92**, 372 (1953).

[4] S. DEVONS, G. GOLDRING and G. R. LINDSEY: Proc. Phys. Soc. Lond. A **67**, 413 (1954).

situation. Instead of describing the particular spin states by the simple δ-function density matrices, more elaborate functions appropriate to the mode of preparation or detection are required.

We should note also the symmetry between radiation and nuclear polarisation, which is well brought out in this treatment. Thus, in a nuclear reaction, we can treat in a similar fashion both the polarisation of the incident beam and of the target nucleus, although quite different experimental considerations are involved. Likewise, the polarisation of the product nuclei and radiation can be treated similarly. Different significance would be attached here to the polarisation of the nucleus and of the radiation. Measurement of the former would normally, but not necessarily, imply direction-correlation measurements of radiations emitted by this nucleus, i.e. a correlation involving successive radiations. Polarisation-sensitive detection of the radiation can also be made by a further directional correlation (for example, an additional nuclear scattering or reaction measurement of the radiation). but more direct methods which do not involve further *nuclear* processes are possible, particularly for γ-radiation. For cascade reactions where the initial nucleus is randomly oriented or polarised, either the first radiation or the associated product nucleus could be detected by polarisation-sensitive detectors. In the first case we could use detectors which do not involve further nuclear processes. In the latter, polarisation measurement would normally mean directional measurements of further radiations emitted by this nucleus: that is to say, a complete description of the polarisation of the intermediate state in a double cascade includes the directional correlation. We could then regard all directional-correlation measurements as a special case of polarisation measurement of the intermediate state, in the sense that *full* information about the polarisation of this state (in terms, say, of the observations made on the preceding radiations) gives the directional properties of the subsequent radiation. It should be noticed, however, that a particular experimental arrangement for measuring polarisation may not give a *full* description of the state of polarisation (statistical tensors or density matrix) of the radiation or nucleus in question. Likewise, part of the complete system, e.g. the γ-radiation from an unpolarised decaying nucleus, may not reveal any polarisation, whereas the remaining part, the nucleus, with the radiation, does so. We may mention, in particular, polarisation-effects associated with systems having axial symmetry, such as the radiation incident or detected in a particular direction, and the associated recoil nucleus. The appropriate description of the system from the point of view of the polarisation either of radiation or nucleus would be an assembly with non-random distribution amongst the substates, but with the *phases* of the different substates random when the axis of symmetry is chosen as axis of quantisation (i.e. a diagonal matrix). Such systems obviously exhibit no "transverse" polarisation effects (cf. Sect. 32).

We shall limit our discussion to a few of the many possible types of polarisation-correlation measurements. In a nuclear reaction the incident beam could be partially polarised (target unpolarised), and measurements made of the resultant angular distribution. A similar measurement would be the angular distribution using a polarisation-sensitive detector of the products of a disintegration using an unpolarised beam and target. In a correlation we might measure the angular correlation from an unpolarised source using one, or both, detectors sensitive to polarisation; or the correlation could be measured at one, or more, fixed angles as a function of the characteristics of one or both detectors ("polarisation-direction" and "polarisation-polarisation" correlations). The directional properties of radiation emitted from a polarised source are treated fully elsewhere in this volume[1], so that here we shall only indicate briefly the relation of this subject to the present topic.

If we are dealing with a target having polarised nuclei, or a polarised incident beam in a reaction, then the information concerning its state of polarisation is contained in the appropriate density matrix or the corresponding statistical tensor: $\varrho_{k_a \varkappa_a}(a\, a)$ for the target nuclei with spin a, and $\varrho_{k\varkappa}(\mathscr{L}\,\mathscr{L}';\, s)$ for a beam of particles with spin s. Likewise, the detection of radiation with a polarisation-sensitive detector is represented by an efficiency tensor, $\varepsilon_{k\varkappa}(\mathscr{L}\,\mathscr{L}';\, s)$, which includes the efficiency-tensor representing the specific polarisation properties: $\varepsilon_{k_s \varkappa_s}(s\, s)$. In both cases of polarised radiations, the information regarding polarisation is also contained in the corresponding radiation parameters, which we now denote by $c_{k\varkappa}(\mathscr{L}\,\mathscr{L}';\, s)$ in order to emphasise the fact that polarisation effects are involved [cf. Eqs. (11.30), (12.5)].

In constructing the density-matrices for polarised radiation, we may be able to represent the radiation as an incoherent superposition of pure states ψ_r each present to the extent of a fraction p_r of the whole assembly, and in this case the density matrix takes the simple diagonal form $\langle r|\varrho|r'\rangle = p_r\, \delta_{r r'}$. By a change in the representation, $\psi_{r'} \to T\psi_r$, ϱ will, in general, transform to the non-diagonal matrix $T^{-1}\varrho\, T$. Similarly, if the detector responds with an efficiency, ε_r, to systems in the pure state, ψ_r, and the response to an assembly of such systems is simply the sum of the individual responses (incoherent in phase), then the efficiency matrix has the diagonal form $\langle r|\varepsilon|r'\rangle = \varepsilon_r \delta_{rr'}$, and can be transformed in the same way as the density matrix. The nature of the polarisation-sensitive detector may be such that its response cannot conveniently be characterised in this simple way, and correspondingly the efficiency-matrix cannot be represented in diagonal form. The non-diagonal matrix elements can be obtained in the general case by considering the response of the detector to pure states resulting from the coherent superposition of base states. For example, if the efficiency of detection in the state represented by $\psi_n = \sum_r a_r(n)\, \psi_r$, is ε_n, then the non-diagonal efficiency matrix is: $\langle r\,|\varepsilon|\, r'\rangle = \sum_n \{a_r(n)\, a_{r'}^*(n)\}\, \varepsilon_n$.

In most cases where we are dealing with the polarisation of radiation, we are concerned with radiation propagated in a particular direction (incident beam or detected particles). The spin-matrices and tensors are expressed in terms of parameters characterising the analysers or detectors, and these characteristics are frequently associated with some directional properties, i.e. angles measured with respect to the direction of propagation of the radiation and some other axis. In special cases the analyser or detector may possess an axis of symmetry, and then the statistical (or efficiency) tensors determine the intensity or radiation transmitted (or detected) for various orientations of the axis of the analyser (detector), the direction of propagation being kept fixed. We can refer to these intensities, suitably normalised, as the transmission intensity (or detection efficiency) of the analyser (detector).

In the usual manner, tensors are formed corresponding to the density and efficiency matrices. For example, if the matrices are defined with reference to the spin states $(s\sigma)$ alone, the corresponding tensor, $\tau_{k_s \varkappa_s}(s)$ is:

$$\sum_{\sigma\sigma'} (-)^{s-\sigma'}\, (s\sigma,\, s-\sigma'\,|\,k_s \varkappa_s)\, \langle s\sigma|\,\tau\,|\, s\sigma'\rangle\,. \tag{29.1}$$

[1] See the article on Oriented Nuclei by R. J. BLIN-STOYLE and M. A. GRACE, in this volume.

In addition, we have the inverse relation to (29.1):

$$\langle s\sigma|\tau|s\sigma'\rangle = \sum_{k_s\varkappa_s} (-)^{s-\sigma'}(s\sigma, s-\sigma'|k_s\varkappa_s)\,\tau_{k_s\varkappa_s}(s). \tag{29.2}$$

The symbol, τ, refers to either density or efficiency matrices and tensors.

With the specification of the matrices (or tensors) in some convenient representation, the radiation parameters are obtained by means of Eq. (12.4). [Although gamma radiation can be interpreted as a particular type of radiation with spin $s=1$, we defer discussion of gamma-ray polarisation till Sect. 31.] Expressed in terms of spin matrices, the radiation parameters for particles of spin s are:

$$\left.\begin{aligned} c_{k\varkappa}(\mathscr{L}\mathscr{L}'; s) = \sum_{\sigma\sigma'\mu\mu'} &(-)^{L'-\mu'}(L\mu, L'-\mu'|k\varkappa)\times \\ &\times\langle 0\sigma|L\mu l\rangle^*\langle 0\sigma'|L'\mu' l'\rangle\langle s\sigma|\tau|s\sigma'\rangle, \end{aligned}\right\} \tag{29.3}$$

where the eigenfunctions are all quantised along the direction of propagation with $\langle 0\sigma|L\mu l\rangle$ given by (13.8). It may be more convenient to express the parameters in terms of the spin tensors, in which case, using (29.2) and the expression for the 9-j function, we find

$$c_{k\varkappa}(\mathscr{L}\mathscr{L}'; s) = \frac{\hat{L}\hat{L}'\hat{l}\hat{l}'}{4\pi}(-)^{l'}\sum_{k_l k_s}(l0, l'0|k_l 0)(k_l 0, k_s\varkappa|k\varkappa)\,\hat{k}_l\hat{k}_s\begin{Bmatrix} l & l' & k_l \\ s & s & k_s \\ L & L' & k\end{Bmatrix}\tau_{k_s\varkappa}(s). \tag{29.4}$$

We can arrive at (29.4) in an equivalent fashion by expanding the tensor $\varepsilon_{k\varkappa}(\mathscr{L}\mathscr{L}')$ (or the statistical tensor if required) in terms of the efficiency tensors describing the orbital motion and the spin, $\varepsilon_{k_l\varkappa_l}(ll')$ and $\varepsilon_{k_s k_s}(ss)$, in the manner of (11.23). Using the fact that the spin axes are defined by the directions of motion, we find, using (7.8),

$$\left.\begin{aligned} \varepsilon_{k\varkappa}(\mathscr{L}\mathscr{L}') = \sum_{\varkappa'} c_{k\varkappa'}\mathscr{D}^k_{\varkappa\varkappa'} = \sum_{k_l k_s\varkappa'} &c_{k_l 0}(ll')\,c_{k_s\varkappa'}(ss)\,\hat{k}_l\hat{k}_s\hat{L}\hat{L}'\times \\ &\times\begin{Bmatrix} l & l' & k_l \\ s & s & k_s \\ L & L' & k\end{Bmatrix}(k_l 0, k_s\varkappa'|k\varkappa')\,\mathscr{D}^k_{\varkappa\varkappa'}, \end{aligned}\right\} \tag{29.5}$$

where $c_{k_l 0}(ll')$ is given by (13.12) and refers to spinless particles. Since this applies equally to statistical tensors, $c_{k_s\varkappa}(ss)$ can be written as $\tau_{k_s\varkappa}(s)$, and the identification of (29.5) with (29.4) is complete.

Expression (29.4) reduces to that for unpolarised particles, $c_{k0}(\mathscr{L}\mathscr{L}')$, if we let $k_s=0$, in which case $k_l=k$ and $\varkappa=0$. It is also closely related, in form, to the functions:

$$G_k\begin{Bmatrix} s & l & L \\ k_s & & k \\ s & l' & L'\end{Bmatrix}$$

introduced by SIMON [84]. Although k_s can assume all values from 0 to $2s$, it is clear from the properties of the Clebsch-Gordan coefficient $(l0, l'0|k_l 0)$ that k_l must be even, if the radiation has well-defined parity.

The general angular correlation or angular distribution formulae are obtained by inserting (29.4) in (10.16)[1]. The resulting expressions are very cumbersome for the general case.

[1] R. J. BLIN-STOYLE: Proc. Phys. Soc. Lond. A **64**, 700 (1951). — A. SIMON and T. A. WELTON: Phys. Rev. **90**, 1036 (1953). — A. SIMON: Phys. Rev. **92**, 1050 (1953).

We will consider that the polarisation of the second radiation is detected in a two-cascade process. These results can be immediately extended to reactions by interchanging $\mathscr{L}$ and $\mathscr{L}'$ in the radiation parameters of the first stage. We will point out the appropriate modifications (if any) to be made for polarisation in the first stage or for any other possible experimental case. The direction-polarisation correlation function is of the form:

$$\left.\begin{aligned} W(\Omega_1\,\Omega_2) \sim \sum (-)^{k-L_1-L_2}\, W(b\,b\,L_1\,L_1';\,k\,a)\; W(b\,b\,L_2\,L_2';\,k\,c)\,\times \\ \times\,(l_2\,0,\,l_2'\,0\,|\,k_l\,0)\,\hat{k}_l\,\hat{k}_s \begin{Bmatrix} l_2 & l_2' & k_l \\ s_2 & s_2 & k_s \\ L_2 & L_2' & k \end{Bmatrix} \hat{L}_2\,\hat{L}_2'\,\hat{l}_2\,\hat{l}_2'\,(-)^{l_2'}\times \\ \times\,\langle c\,\mathscr{L}_2(b)\,\mathscr{L}_1\,a\,|\,\mathcal{O}\,|\,a\rangle\,\langle c\,\mathscr{L}_2'(b)\,\mathscr{L}_1'\,a\,|\,\mathcal{O}\,|\,a\rangle^*\times \\ \times \sum_{\varkappa_1\varkappa_2} (k_l\,0,\,k_s\,\varkappa_2\,|\,k\,\varkappa_2)\; c_{k\,\varkappa_1}(\mathscr{L}_1'\,\mathscr{L}_1)\; \tau^*_{k_s\varkappa_2}(s_2)\; \mathscr{D}^k_{\varkappa_2\varkappa_1}(\mathscr{R}_2^{-1}\,\mathscr{R}_1)\,, \end{aligned}\right\} \tag{29.6}$$

with $\tau(s_2) \equiv \varepsilon(s_2)$. The summation is over $k\,k_l\,k_s\,L_1\,l_1\,L_1'\,l_1'\,L_2\,l_2\,L_2'\,l_2'$. If the first radiation is unpolarised or its polarisation is not measured, $\varkappa_1 = 0$, and $\mathscr{D}^k_{\varkappa_2\varkappa_1}(\mathscr{R}_2^{-1}\mathscr{R}_1)$ becomes $\frac{\sqrt{4\pi}}{\hat{k}}\,Y_k^{\varkappa_2^*}(\vartheta_2\,\vartheta_1)$, where the Euler angles corresponding to $(\mathscr{R}_2^{-1}\mathscr{R}_1)$ are $(\vartheta_1\,\vartheta_2\,0)$. The directions of propagation of the two radiations are designated by unit vectors $\mathfrak{k}_1$ and $\mathfrak{k}_2$, so that ϑ_2 represents the angle between $\mathfrak{k}_1$ and $\mathfrak{k}_2$, ϑ_1 the angle between the y-axis used to describe radiation 2, and $\mathfrak{k}_2\times\mathfrak{k}_1$ the normal to the plane containing the two radiations. If a polarisation measurement is made on the first radiation but not on the second, (29.6) can be used, provided that we interchange some of the suffices. For incoming radiation, $c_{k\varkappa}(\mathscr{L}_1'\,\mathscr{L}_1)$ is replaced by $c_{k\varkappa}(\mathscr{L}_1'\,\mathscr{L}_1')$ and $\tau(s_1) \equiv \varrho(s_1)$.

As already mentioned, we can equally as well study the polarisation of the nuclei produced in a reaction or in cascade decay by interchanging the rôles played by the spin of radiation 2 and the spin of the nucleus, i.e. by interchanging c and s_2 and replacing L_2 by J_2. In this manner, we couple the nuclear spin to the relative orbital angular momentum (which is possible, since our considerations are confined to the centre of mass system), so that, instead of the coupling scheme, $\boldsymbol{b} = \boldsymbol{c} + \boldsymbol{L}_2$, with $\boldsymbol{L}_2 = \boldsymbol{l}_2 + \boldsymbol{s}_2$, we now have $\boldsymbol{b} = \boldsymbol{s}_2 + \boldsymbol{J}_2$, with $\boldsymbol{J}_2 = \boldsymbol{l}_2 + \boldsymbol{c}$. The angular correlation (distribution) for the emission (absorption) of radiation 1 followed by the emission of radiation 2 and a residual nucleus, where the direction, (Ω_2), and polarisation of the nucleus is measured, is given by the appropriate modification of (29.6), viz.,

$$\left.\begin{aligned} W(\Omega_1\,\Omega_2) = \sum (-)^{k-L_1-J_2}\, W(b\,b\,L_1\,L_1';\,k\,a)\; W(b\,b\,J_2\,J_2';\,k\,s_2)\,\times \\ \times\,(l_2\,0,\,l_2'\,0\,|\,k_l\,0)\cdot \hat{k}_l\,\hat{k}_c \begin{Bmatrix} l_2 & l_2' & k_l \\ c & c & k_c \\ J_2 & J_2' & k \end{Bmatrix} \hat{J}_2\,\hat{J}_2'\,\hat{l}_2\,\hat{l}_2'\times \\ \times\,(-)^{l_2'}\langle s_2\,J_2(b)\,\mathscr{L}_1\,a\,|\,\mathcal{O}\,|\,a\rangle\,\langle s_2\,J_2'(b)\,\mathscr{L}_1'\,a\,|\,\mathcal{O}\,|\,a\rangle^*\times \\ \times \sum_{\varkappa_1\varkappa_2} (k_l\,0,\,k_c\,\varkappa\,|\,k\,\varkappa_2)\; c_{k\,\varkappa_1}(\mathscr{L}_1'\,\mathscr{L}_1)\cdot \tau^*_{k\varkappa_2}(c)\; \mathscr{D}_{\varkappa_2\varkappa_1}(\mathscr{R}_2^{-1}\,\mathscr{R}_1)\,, \end{aligned}\right\} \tag{29.7}$$

with L_1 and L_1' interchanged if radiation 1 is ingoing. The summation is over $k\,k_c\,k_l\,L_1\,l_1\,L_1'\,l_1'\,J_2\,l_2,\,J_2'\,l_2'$. The tensor, $\tau_{k\varkappa_2}(c)$ is completely specified by the manner of measurement of the polarisation state of the nucleus.

Again, we can consider polarisation effects in the first stage of the above process and thereby study nuclear reactions involving target nuclei which are aligned.

Although (29.6) provides a general formula for the treatment of radiation of arbitrary spin, it is useful to specialise our considerations to the cases of spin-$\frac{1}{2}$ particles and γ-rays, which are the most common types of radiation encountered. Rather than insert the values of s into the formula, we shall find the radiation parameters, *ab initio*, and insert these into the general correlation formula, (10.16). The spin-tensors will be evaluated by use of the explicit eigenfunctions for these radiations. These special cases will be followed by a general discussion concerning radiation of arbitrary spin, s.

30. Polarisation of spin-$\frac{1}{2}$ particles. For the sake of definiteness we shall discuss in the first place a beam of particles prepared in a state of partial polarisation. For spin-$\frac{1}{2}$ particles there are only two base states required, so that we can decompose the partially polarised beam into two parts, a fraction P which is completely polarised (all particles in the same pure state—not necessarily the base state), and a fraction $(1-P)$ completely unpolarised (equal population but random phases in both base states). This is fully equivalent to the description of the beam as having a fraction $\frac{1}{2}(1+P)$ of the particles in one pure state and the incoherent fraction $\frac{1}{2}(1-P)$ in the orthogonal state. When only two states are involved, it is somewhat more convenient to refer to the fractions polarised and unpolarised.

In an arbitrary representation, the density matrix takes the form

$$(1-P)\begin{pmatrix}\frac{1}{2} & 0\\ 0 & \frac{1}{2}\end{pmatrix}+P\begin{pmatrix}\tau(\ \frac{1}{2}\ \frac{1}{2}) & \tau(\ \frac{1}{2}\ -\frac{1}{2})\\ \tau(-\frac{1}{2}\ \frac{1}{2}) & \tau(-\frac{1}{2}\ -\frac{1}{2})\end{pmatrix}. \tag{30.1}$$

The elements of the second matrix are completely determined by the polarisation properties of the fraction P (or equivalently by the "direction" of polarisation), and its trace is unity; $\tau(\frac{1}{2}\,\frac{1}{2})+\tau(-\frac{1}{2}\,-\frac{1}{2})=1$. Referred to some particular axis of quantisation, specified by ϑ, φ, the density matrix becomes

$$(1-P)\begin{pmatrix}\frac{1}{2} & 0\\ 0 & \frac{1}{2}\end{pmatrix}+P\begin{pmatrix}1 & 0\\ 0 & 0\end{pmatrix}, \tag{30.2}$$

so that the matrix elements $\tau(\ldots)$ in (30.1) are functions of the angles ϑ, φ. The actual matrix elements describing the polarisation in the general representation can be determined by writing for the state corresponding to complete polarisation in the direction (ϑ,φ)[1]:

$$\cos\frac{\vartheta}{2}\left|\frac{1}{2}\right\rangle+\sin\frac{\vartheta}{2}\,e^{i\varphi}\left|-\frac{1}{2}\right\rangle, \tag{30.3}$$

where the eigenfunctions $|\frac{1}{2}\rangle, |-\frac{1}{2}\rangle$ correspond to $\sigma=\pm\frac{1}{2}$. The density matrix, $\langle\sigma|\varrho|\sigma'\rangle$, or, if we wish to refer either to the density or efficiency matrices, the "spin-matrix", $\langle\sigma|\tau|\sigma'\rangle$, is then given by

$$\begin{pmatrix}\frac{1-P}{2}+P\cos^2\frac{\vartheta}{2} & \frac{P}{2}\sin\vartheta\,e^{-i\varphi}\\ \frac{P}{2}\sin\vartheta\,e^{i\varphi} & \frac{1-P}{2}+P\sin^2\frac{\vartheta}{2}\end{pmatrix}, \tag{30.4}$$

and the corresponding spin-tensors are:

$$\tau_{00}=\frac{1}{\sqrt{2}},\qquad \tau_{10}=\frac{P}{\sqrt{2}}\cos\vartheta,\qquad \tau_{1\pm1}=\mp\frac{P}{2}\sin\vartheta\,e^{\mp i\varphi}. \tag{30.5}$$

We introduce the polarisation-vector, $\boldsymbol{P}$, whose components in any direction are equal to the expectation value, $\overline{\mathbf{s}}\div\frac{1}{2}$ in that particular direction. Using the

[1] P. A. M. Dirac: The Principles of Quantum Mechanics, 3rd Edit., p. 151. Oxford 1947.

eigenfunctions for the partially polarised beam represented above, which is an incoherent superposition of states corresponding to (30.3) with states which are random, we find:

$$\left.\begin{aligned} \bar{s}_x &\equiv \frac{P_x}{2} = \frac{1}{2} P \sin\vartheta \cos\varphi\,, \\ \bar{s}_y &\equiv \frac{P_y}{2} = \frac{1}{2} P \sin\vartheta \sin\varphi\,, \\ \bar{s}_z &\equiv \frac{P_z}{2} = \frac{1}{2} P \cos\vartheta\,. \end{aligned}\right\} \tag{30.6}$$

The spin-matrix $\langle\sigma|\tau|\sigma'\rangle$ can then be rewritten in terms of the polarisation-vector $\boldsymbol{P}$ as

$$\frac{1}{2}\begin{pmatrix} 1+P_z & P_x - i P_y \\ P_x + i P_y & 1 - P_z \end{pmatrix} = \frac{1}{2}(1 + \boldsymbol{P}\cdot\boldsymbol{\sigma})\,, \tag{30.7}$$

where $\boldsymbol{\sigma}$ denotes the usual spin-$\frac{1}{2}$ matrix operator.

Using a representation in which the polarisation-vector lies along the z-axis, this becomes

$$\frac{1}{2}\begin{pmatrix} 1+P & 0 \\ 0 & 1-P \end{pmatrix}. \tag{30.8}$$

For an *incident* beam in a prepared state of polarisation, the vector, $\boldsymbol{P}$, specifies this polarisation, or alternatively we can regard $\boldsymbol{P}$ as specifying the physical properties of the apparatus which prepares the actual beam from an unpolarised one. We notice that P is equal to or less than unity[1]. $P=1$ corresponds to a completely polarised beam (pure state) and $P=0$ to a completely unpolarised one. The direction of $\boldsymbol{P}$ relates to the directional characteristics of the state of the completely polarised part. For an *outgoing* detected beam, which may have any polarisation, the same matrix (30.7) or (30.8) can be used to describe the polarisation-properties of the detector. In this case, P equals the difference between the sensitivities of the detector to a certain completely polarised beam (i.e. pure state) which gives maximum response, $\frac{1}{2}(1+P)$, and the response to a completely polarised beam representing the orthogonal state, which will be the minimum response, $\frac{1}{2}(1-P)$. (The sum of these responses is here normalised to unity.) P in this case has its maximum possible value, unity if there is some completely polarised beam (or equivalently, some "orientation" of the polarisation-sensitive detector), for which the response is zero; and of course the response to polarisation in the orthogonal state will be unity. For a general detector partially sensitive to polarisation, the minimum response to some completely polarised beam will be $\frac{1}{2}(1-P)$ and the maximum $\frac{1}{2}(1+P)$. P may therefore be regarded as the "polarisation-efficiency" of the detector. The direction of $\boldsymbol{P}$ characterises the directional properties (with respect to polarisation) of the detector. In the present case, it is the axis of quantisation of the pure state which produces the maximum response in the detector.

The radiation parameters are determined from (29.3) and (30.7) and are, for outgoing particles[2],

$$c_{k0}\left(\mathcal{L}\mathcal{L}';\frac{1}{2}\right) = \frac{\hat{L}\hat{L}'}{4\pi}\left(L\,\frac{1}{2}, L' - \frac{1}{2}\middle|k\,0\right)(-)^{L+k-\frac{1}{2}}, \tag{30.9a}$$

$$c_{k\pm1}\left(\mathcal{L}\mathcal{L}';\frac{1}{2}\right) = \frac{\hat{L}\hat{L}'}{8\pi}\left(L\,\frac{1}{2}, L'\,\frac{1}{2}\middle|k\,1\right)(-)^{L-L'+l'-1}\cdot(P_x \mp i P_y). \tag{30.9b}$$

[1] We denote $|\boldsymbol{P}|$ by P.

[2] The radiation parameters displayed in (30.9) refer only to even values of k, which are the only ones used in the chapter. More generally, for odd values of $l+l'+k$, $c_{k0}(\mathcal{L}\mathcal{L}';\frac{1}{2})$ is a function of P_z.

Inserting the above quantities into the general correlation formula (10.16), we find the form for the angular correlation between two radiations in a cascade, where the polarisation of the second radiation is detected, is identical to that for the unpolarised case, except that $P_k(\cos\vartheta)$ is replaced by (SATCHLER [87]):

$$P_k(\cos\vartheta) + i\, f_k(L\,L', \tfrac{1}{2})\,(-)^{l'}\, P_k^1(\cos\vartheta)\, \boldsymbol{P}\cdot\boldsymbol{n}, \tag{30.10}$$

where $\boldsymbol{n}$ lies along $\mathfrak{k}_1 \times \mathfrak{k}_2$, and

$$f_k\left(L\,L', \frac{1}{2}\right) = \frac{(-)^{L'+\frac{1}{2}}}{\sqrt{k(k+1)}}\;\frac{(L\,\tfrac{1}{2}, L'\,\tfrac{1}{2}\,|\,k\,1)}{(L\,\tfrac{1}{2}, L'-\tfrac{1}{2}\,|\,k\,0)}, \tag{30.11}$$

$$= \frac{(-)^{L'+\frac{1}{2}}}{2k(k+1)}\left[(2L'+1) + (-)^{k+L+L'}(2L+1)\right], \tag{30.12}$$

$$= (-)^{2L-k} f_k(L'\,L, \tfrac{1}{2}). \tag{30.13}$$

No assumption has been made concerning the nature of the first radiation in the above formulae. If polarisation measurements are made of the first radiation, which consists of spin-$\frac{1}{2}$ particles, with no such measurements made on the second radiation, the latter not being necessarily restricted to spin-$\frac{1}{2}$ particles, $P_k(\cos\vartheta)$ is again to be replaced by (30.10), although $\boldsymbol{n}$ must also be replaced by $-\boldsymbol{n}$. These results also apply to reactions where either the incoming or outgoing radiation is composed of spin-$\frac{1}{2}$ particles whose polarisation is being studied. In contrast to cascade correlations, the angular distribution is modified according to (30.10), with $\boldsymbol{n} = \mathfrak{k}_1 \times \mathfrak{k}_2$ in *both* cases.

Since $2L$ is odd, $f_k(LL', \frac{1}{2}) = -f_k(L'L, \frac{1}{2})$ if the radiation has well-defined parity. An interchange of L with L' does not affect any of the coefficients of (30.10) appearing in the general correlation or distribution, except for the interaction matrices. As a result, the part of $W(\Omega_1\,\Omega_2)$ which is independent of polarisation depends on the real part of $\mathcal{O}\,\mathcal{O}'^*$, where $\mathcal{O}$ and $\mathcal{O}'$ represent the interaction matrix elements as a function of the angular momentum variables of the radiation, when unprimed and when primed, while the other part of $W(\Omega_1\,\Omega_2)$ depends on the imaginary part of $\mathcal{O}\,\mathcal{O}'^*$. Thus the product of $\mathcal{O}\,\mathcal{O}'^* P_k(\cos\vartheta)$ which occurs for unpolarised radiation, is replaced by

$$\mathrm{Re}\,[\mathcal{O}\,\mathcal{O}'^*]\, P_k(\cos\vartheta) + (-)^{l'+1}\,\mathrm{Im}\,[\mathcal{O}\,\mathcal{O}'^*]\,\boldsymbol{P}\cdot\boldsymbol{n}\cdot f_k(L\,L', \tfrac{1}{2})\, P_k^1(\cos\vartheta), \tag{30.14}$$

where Re and Im represent the real and imaginary parts of complex quantities. If the matrix elements are real, as they would be if perturbation theory or the Born approximation is used, polarisation effects vanish. Also, if the radiation is pure, i.e. $L = L'$, it follows from (30.13) that $W(\Omega_1\,\Omega_2)$ is independent of the polarisation.

An important feature of (30.14) is that for a partially polarised beam only the component of $\boldsymbol{P}$ normal to the plane of the two radiations appears in the polarisation-sensitive part of the correlation; or, for a polarisation-sensitive detector characterised by $\boldsymbol{P}$, it is only the component of $\boldsymbol{P}$ normal to the plane of the two radiations which gives the polarisation dependent part of the angular distribution (WOLFENSTEIN [80]).

A further discussion concerning polarisation of spin-$\frac{1}{2}$ particles is given in Sect. 32, where we consider reactions in the channel-spin representation.

31. Polarisation for γ-radiation. The wavefunction for photons is taken to be the vector potential, in terms of which the plane-wave states are represented as in Sect. 14, with the spin states given proper specification. Since there are only transverse waves, the radiation can involve only two independent states. Thus, a partially polarised beam of γ-rays can be described, like spin-$\frac{1}{2}$ particles, as the incoherent superposition of two completely polarised beams. The usual description is in terms of the eigenstates of $\sigma = \pm 1$, with the wavefunction written as:

$$|\boldsymbol{A}\rangle = [\alpha_+|+1\rangle + \alpha_-|-1\rangle]\, e^{itz}, \tag{31.1}$$

where $|+1\rangle$ corresponds to a state of left-handed circular polarisation with $\sigma = +1$. The vector potential can also be expanded in terms of unit vectors along the x- and y-axes:

$$|\boldsymbol{A}\rangle = [\alpha_x|\boldsymbol{e}_x\rangle + \alpha_y|\boldsymbol{e}_y\rangle]\, e^{itz}. \tag{31.2}$$

The base states $|\pm 1\rangle$, which behave like components of a spin-1 tensor, can be expressed in terms of $|\boldsymbol{e}_x\rangle$, $|\boldsymbol{e}_y\rangle$, as:

$$|\pm 1\rangle = \mp \frac{1}{\sqrt{2}}[|\boldsymbol{e}_x\rangle \pm i|\boldsymbol{e}_y\rangle], \quad |0\rangle = |\boldsymbol{e}_z\rangle, \tag{31.3}$$

so that

$$\alpha_\pm = \mp \frac{1}{\sqrt{2}}(\alpha_x \pm i\,\alpha_y). \tag{31.4}$$

To complete the description of the polarisation, we introduce a third set of base states, with axes 1 and 2 inclined at $\pi/4$ to the x-y axes, so that

$$|\boldsymbol{A}\rangle = [\alpha_1|\boldsymbol{e}_1\rangle + \alpha_2|\boldsymbol{e}_2\rangle]\, e^{itz} \tag{31.5}$$

and

$$\alpha_\pm = \mp \tfrac{1}{2}[(\alpha_1 - \alpha_2) \pm i(\alpha_1 + \alpha_2)]. \tag{31.6}$$

All the wavefunctions are seen to involve four real terms, as does the spin-$\frac{1}{2}$ matrix[1]. It is convenient to introduce the STOKES' parameters (FANO [8]) to describe the polarisation of γ-rays. These are given by

$$I = 1 = |\alpha_x|^2 + |\alpha_y|^2 = |\alpha_-|^2 + |\alpha_+|^2 = |\alpha_1|^2 + |\alpha_2|^2, \tag{31.7}$$

$$P_1 = |\alpha_x|^2 - |\alpha_y|^2, \tag{31.8}$$

$$P_2 = |\alpha_1|^2 - |\alpha_2|^2 \tag{31.9}$$

and

$$P_3 = |\alpha_+|^2 - |\alpha_-|^2. \tag{31.10}$$

For a polarisation-sensitive detector the STOKES' parameters represent the sensitivity of the detector for different types of polarised γ-radiation. For partially polarised beams, the parameters represent the composition in terms of these same states of polarisation. I represents the overall intensity (or sensitivity) and P_3, for example, represents the excess of intensity (or response) for left-handed over right-handed circular polarisation.

The matrix τ, has elements [referred to the base set used in (31.1)]:

$$\begin{pmatrix} |\alpha_+|^2 & 0 & \alpha_+\alpha_-^* \\ 0 & 0 & 0 \\ \alpha_+^*\alpha_- & 0 & |\alpha_-|^2 \end{pmatrix}, \tag{31.11}$$

which become

$$\tfrac{1}{2}\begin{pmatrix} 1+P_3 & 0 & -(P_1 + i\,P_2) \\ 0 & 0 & 0 \\ -(P_1 - i\,P_2) & 0 & 1 - P_3 \end{pmatrix}. \tag{31.12}$$

[1] In dealing with spin-$\frac{1}{2}$ particles, one parameter was removed by the normalisation; the other three were P_x, P_y and P_z or P, ϑ and φ.

The corresponding tensors are obtained using (29.1):

$$\tau_{00}=\frac{1}{\sqrt{3}};\quad \tau_{10}=\frac{P_3}{\sqrt{2}};\quad \tau_{1\pm1}=0;\quad \tau_{2\pm2}=-\frac{1}{2}(P_1\pm i P_2). \tag{31.13}$$

The radiation parameters for outgoing waves are found from (29.3):

$$\left.\begin{aligned} c_{k0}(L L') &= \frac{\hat{L}\hat{L}'}{4\pi}(-)^{L'-1}(L1, L'-1\,|\,k0),\\ c_{k\pm1}(L L') &= 0,\\ c_{k\pm2}(L L') &= (\pm)_{L'}\frac{\hat{L}\hat{L}'}{4\pi}(-)^{L'}(L1, L'1\,|\,k2)\frac{P_1\pm i P_2}{2}, \end{aligned}\right\} \tag{31.14}$$

where $(\pm)_{L'}$ is $(-)^{L'}$ multiplied by the parity of the radiation associated with L'. It will be noticed[1] that the radiation parameters, (31.14), do not involve P_3. In addition, if τ refers to an efficiency tensor and the polarisation-orientation of the detector can be altered, then clearly P_2 can be obtained from P_1, since both refer to measurement of essentially the same quantity, but with different reference axes (reorientation of the detector is equivalent to a change in the state of polarisation to which the detector responds), so that, in such cases, we can ignore P_2 and characterise the radiation parameter for the detected radiation as a function of the parameter P_1 only (and its orientation). P_1 is now a measure of the polarisation-efficiency and the orientation of the detector.

Let ϑ_1 be the angle between the x-axis used in determining P_1, (31.8), and the plane containing both radiations. We find that if one of the emitted radiations consists of γ-rays whose polarisation is detected, the correlation function differs from the "unpolarised" correlation in that $P_k(\cos\vartheta)$ is replaced by

$$P_k(\cos\vartheta)+(\pm)_{L'}f_k(L L')\,P_k^2(\cos\vartheta)\cdot P_1\cdot\cos 2\vartheta_1, \tag{31.15}$$

where

$$f_k(L L')=f_k(L' L)=-\sqrt{\frac{(k-2)!}{(k+2)!}}\,\frac{(L1, L'1\,|\,k2)}{(L1, L'-1\,|\,k0)}, \tag{31.16}$$

$$=\frac{(L-L')(L+L'+1)}{(k+2)(k+1)k(k-1)},\quad \text{if}\quad k+L+L' \text{ is odd}, \tag{31.17}$$

$$=\frac{-2L(L+1)}{[k(k+1)-2L(L+1)](k+2)(k-1)},\quad \text{if}\quad L=L'. \tag{31.18}$$

Identical results hold for reactions and for correlations, and for first and second stages of a correlation. Furthermore, the linear-polarisation term in $W(\Omega_1\Omega_2)$ only involves the real part of the reduced matrix element, in contrast to the situation (Sect. 30) with spin-$\frac{1}{2}$ particles. The quantity $(\pm)_{L'}$, which occurs in (31.15), depends explicitly on the parity of the radiation, and is $+1$ if the radiation consists of electric multipoles and is -1 if the radiation is magnetic, so that correlations with polarisation-sensitive detectors provide a means of determining the parities of nuclear states.

32. Polarisation for spin-s particles. In general, a particle of spin s will have spin tensors whose rank ranges from zero to a maximum value, $2s$, and the corresponding density matrix will have $(2s+1)^2$ independent components, including normalisation. For spin-$\frac{1}{2}$ particles, the polarisation is described by the total intensity and the three components of $\boldsymbol{P}$. The STOKES' parameters, which are four in number, fully describe the polarisation of γ-rays, in spite of the fact

[1] We are dealing here with radiations of well-defined parity and for these, this statement is true only for even values of k. See footnote 2, p. 440.

that spin-1 particles would normally involve nine parameters. This is because of the transverse character of the electromagnetic waves, which results in $\tau_{1\pm1} = \tau_{2\pm1}$ being identically zero. Also, τ_{20} and τ_{00} are not independent of each other.

Although it is possible to give some simple physical interpretation to the tensors for the radiations considered in Sects. 30 and 31, the physical picture becomes more complex for spins equal to or greater than unity. Thus, the specification of polarisation of deuterons, for example with $s=1$, requires knowledge of the second-rank tensor, τ_{20}, which usually does not vanish (LAKIN [86]). Nevertheless, certain interesting results can be obtained which involve the polarisation associated with the rank-one spin tensors for particles of arbitrary spin-s.

The spin tensors, $\tau_{k\varkappa}$, which are defined in terms of the corresponding matrices by

$$\tau_{k\varkappa} = \sum_{\sigma\sigma'} (-)^{s-\sigma'} (s\sigma, s-\sigma' \mid k\varkappa) \langle s\sigma \mid \tau \mid s\sigma' \rangle \tag{32.1}$$

can also be defined as the mean value of a tensor operator $T_{k\varkappa}$ (SIMON and WELTON [84]):

$$\tau_{k\varkappa} = \overline{T}_{k\varkappa}, \tag{32.2}$$

where

$$\overline{T}_{k\varkappa} = \sum_{\sigma\sigma'} \langle s\sigma \mid \tau \mid s\sigma' \rangle \langle s\sigma' \mid T_{k\varkappa} \mid s\sigma \rangle . \tag{32.3}$$

Therefore, the tensor operator, $T_{k\varkappa}$ can be written as

$$T_{k\varkappa} = \sum_{\sigma\sigma'} (-)^{s-\sigma'} (s\sigma, s-\sigma' \mid k\varkappa) \mid s\sigma \rangle \langle s\sigma' \mid . \tag{32.4}$$

Substitution of the values of the Clebsch-Gordan coefficients in the above expression leads to simple expressions for these operators for $k=1$. In particular, we find:

$$T_{10} = \frac{s_z}{\hat{s}} \sqrt{\frac{3}{s(s+1)}} \tag{32.5}$$

and

$$T_{1\pm1} = \mp \frac{1}{\hat{s}} \sqrt{\frac{3}{2s(s+1)}} (s_x \mp i s_y), \tag{32.6}$$

where s_x, s_y and s_z are the spin operators.

We define the "vector" polarisation, $\boldsymbol{P}$, for a particle of spin s as that vector having components satisfying

$$P_x = \frac{\overline{s}_x}{s}, \quad P_y = \frac{\overline{s}_y}{s}, \quad P_z = \frac{\overline{s}_z}{s}. \tag{32.7}$$

It then follows from (32.2) that the spin tensors of rank 1 can be expressed in terms of the vector polarisation, $\boldsymbol{P}$ by

$$\tau_0 = \frac{1}{\hat{s}} \sqrt{\frac{3s}{s+1}} P_z; \quad \tau_{1\pm1} = \mp \frac{1}{\hat{s}} \sqrt{\frac{3s}{2(s+1)}} (P_x \mp i P_y). \tag{32.8}$$

It is clear that the normalisation of the tensor is arbitrary so long as the tensors that are used in the expresssions for the angular correlation are all normalised in the same manner. For the case $s=\frac{1}{2}$, the expressions (32.8) agree with those given in (30.5), but differ from corresponding tensors used by SIMON and WELTON[1]. Their tensors, $T'_{k\varkappa}$ are defined so that they are irreducible[2], from which

[1] A. SIMON and T. A. WELTON: Phys. Rev. **93**, 1435 (1954).

[2] U. FANO [8], Appendix C.

fact $T'_{k\varkappa}$ can be made to satisfy certain commutation relations[1] with the angular-momentum operators, J_x, J_y, J_z.

$$\left.\begin{aligned}[J_x \pm i\, J_y, T'_{k\varkappa}] &= \sqrt{(k \mp \varkappa)(k \pm \varkappa + 1)}\; T'_{k\varkappa\pm 1}\\ [J_z, T'_{k\varkappa}] &= \varkappa\, T'_{k\varkappa}.\end{aligned}\right\} \tag{32.9}$$

We list the tensor operators for ranks 0, 1 and 2 used by SIMON and WELTON:

$$\left.\begin{aligned}
T'_{00} &= 1; \quad T'_{10} = \frac{s_z}{\sqrt{s(s+1)}};\\
T'_{1\pm1} &= \mp \frac{1}{\sqrt{2s(s+1)}}\,(s_x \pm i\, s_y);\\
T'_{20} &= \frac{1}{2s(s+1)}\,[3 s_z^2 - s(s+1)];\\
T'_{2\pm1} &= \mp \frac{\sqrt{6}}{2s(s+1)}\,[(s_x \pm i\, s_y)\, s_z + s_z\,(s_x \pm i\, s_y)];\\
T'_{2\pm2} &= \frac{\sqrt{6}}{2s(s+1)}\,(s_x \pm i\, s_y)(s_x \pm i\, s_y).
\end{aligned}\right\} \tag{32.10}$$

We now make some observations concerning the angular correlations and distributions where one radiation is unpolarised and the other radiation involves "vector"-polarisation, i.e. spin tensors of rank one. Referring to (29.6), we note that k_l must equal k, since the Clebsch-Gordan coefficient $(k_l\, 0, 1\, \varkappa_2 |\, k\, \varkappa_2)$ vanishes unless $k = k_l$ or $k_l \pm 1$, and both k and k_l are even. Also, $(k\; 0, 1\, \varkappa_2 |\, k\; \varkappa_2)$ vanishes if $\varkappa_2 = 0$, allowing us to conclude (i) that the angular distribution in a reaction is independent of the degree of "vector" polarisation of the incident beam of particles, if this polarisation is in the direction of the incident radiation, and (ii) if the detectors are such that they are polarisation-sensitive only to "vector" polarisation in the direction of motion, then the resultant correlation is the same as for polarisation-insensitive detectors. (Similar conclusions hold for polarisations described by spin-tensors of any odd rank, i.e. odd k_s of the type $\tau_{k_s} 0$).

Combining (32.8) and (29.6), we find, after some algebra, that the angular distribution, if either the incoming or outgoing beams involves "vector" polarisation, differs from that obtained without regard to polarisation by the replacement of $P_k(\cos\vartheta)$ by

$$P_k(\cos\vartheta) + i\, \boldsymbol{n} \cdot \boldsymbol{P}\, (-)^{l'}\, f_k(\mathscr{L}\mathscr{L}', s)\, P_k^1(\cos\vartheta), \tag{32.11}$$

where $\boldsymbol{n} = \mathfrak{k}_1 \times \mathfrak{k}_2$, is a unit vector normal to the plane of the reaction and

$$\left.\begin{aligned}
f_k(\mathscr{L}\mathscr{L}', s) &= (-)^{k+1} f_k(\mathscr{L}'\mathscr{L}, s)\\
&= \frac{3(-)^{k+s-L}\,\hat{k}\,\hat{s}}{W(l\,l'\,L\,L'; k\,s)} \sqrt{\frac{s}{k(k+1)(s+1)}} \begin{Bmatrix} l & l' & k\\ s & s & 1\\ L & L' & k \end{Bmatrix},
\end{aligned}\right\} \tag{32.12}$$

which, by use of (A.26) reduces to

$$f_k(\mathscr{L}\mathscr{L}', s) = \frac{3(-)^{l'}\,[L(L+1) - L'(L'+1) - l(l+1) + l'(l'+1)]}{2k(k+1)(s+1)}. \tag{32.13}$$

For correlations the same result holds, except that if it is the first radiation whose polarisation is measured, $\boldsymbol{n}$ is replaced by $-\boldsymbol{n}$. If we substitute $s = \frac{1}{2}$ in

[1] G. RACAH: Phys. Rev. **62**, 438 (1942).

the above expression and make use of the relation, valid for the case $s=\frac{1}{2}$,

$$\left.\begin{aligned}&L(L+1)-L'(L'+1)+l'(l'+1)-l(l+1)\\&\qquad=\frac{(-)^{l'+L'+\frac{1}{2}}}{2}\left[2L'+1+(-)^{k+L+L'}(2L+1)\right],\end{aligned}\right\}\tag{32.14}$$

we arrive at the expression $f_k(LL',\frac{1}{2})$ given in (30.12).

The conclusions concerning the dependence of the polarisation part of $W(\Omega_1\Omega_2)$ on the imaginary part of the interaction matrix element follow as for the case of spin-$\frac{1}{2}$ particles. In addition, there are no polarisation effects if the radiation is pure, i.e. $L=L'$. Finally, it is only that component of the "vector" polarisation that is perpendicular to $\mathfrak{k}_1$ and $\mathfrak{k}_2$ that appears in the angular distribution; or, if the detector is sensitive to "vector" polarisation when this is in the direction $\boldsymbol{P}$, it is only the response in the direction normal to $\mathfrak{k}_1$ and $\mathfrak{k}_2$ that affects the correlation.

The above analysis also applies to polarised nuclei, if the formalism is modified as in (29.7). Thus angular distributions involving "vector" polarisation of nuclei include terms such as

$$P_k(\cos\vartheta)+i\,\boldsymbol{n}\cdot\boldsymbol{P}\,(-)^{l'}f_k(l\,J\,l'\,J',a)\,P_k^1(\cos\vartheta)\tag{32.15}$$

where $\boldsymbol{J}$ represents the vector sum of $\boldsymbol{l}_1$ and $\boldsymbol{a}$ or $\boldsymbol{l}_2$ and $\boldsymbol{b}$. The expression for $f_k(lJl'J',a)$ is given by (32.13) if the symbols are appropriately permuted. We can similarly treat reactions and correlations involving outgoing nuclei that are "vector"-polarised.

A similar analysis concerning polarisation of spin-s (and spin-$\frac{1}{2}$) particles can be made in the channel-spin representation. The characteristic feature here is that the spin of the radiation is coupled to that of the nucleus (target or recoil). If we seek information concerning the polarisation of the outgoing radiation, or if we are interested in the effects of polarisation of the ingoing radiation, it is necessary to decouple this polarisation spin from the channel spin, and the result is a more complicated expression for the angular distribution. In the L-representation, the spin is already decoupled, so that the modifications arising from polarisation [e.g. Eq. (32.15)] are not as complex; on the other hand, this simplicity is gained at the expense of a less compact form for the unpolarised distribution.

As is to be expected, the polarisation modifications are similar in form whether we are dealing with one or the other stage of a cascade or a reaction. The channel-spin representation is normally used to deal with reactions; therefore we will restrict ourselves to these and first consider reactions involving an incoming polarised beam of particles.

The angular distribution, given in (12.2), requires that we calculate the product $\varrho_{k_b\varkappa_b}(bB,bB')\,\varepsilon^*_{k_b\varkappa_b}(bB,bB')$. The expression for the efficiency tensor is unaltered from that given in (24.14), but $\varrho_{k_b\varkappa_b}(bB,bB')$ must include the effect of polarisation of the first radiation. This is found by coupling the relative orbital angular momentum to the channel spin whose statistical tensor, $\varrho_{k_S\varkappa_S}(S_1,S_1')$, is given by

$$\left.\begin{aligned}\varrho_{k_S\varkappa_S}(S_1S_1')&=\frac{\hat{S}_1\hat{S}_1'\hat{k}_S}{\hat{a}}\,\varrho_{k_s\varkappa_s}(s_1s_1)\begin{Bmatrix}a & s_1 & S_1\\ a & s_1 & S_1'\\ 0 & k_s & k_S\end{Bmatrix}\\&=(-)^{a+k_s-s_1-S_1'}\,\varrho_{k_s\varkappa_s}(s_1s_1)\,\frac{\hat{S}_1\hat{S}_1'}{\hat{a}^2}\,W(s_1s_1S_1S_1';k_sa),\end{aligned}\right\}\tag{32.16}$$

by virtue of the lack of information concerning the spin state of the target nucleus. The resultant expression for $\varrho_{k_b \varkappa_b}(b B, b B')$ becomes

$$\left.\begin{aligned}\varrho_{k_b \varkappa_b}(b B, b B') = {} & \frac{1}{4\pi \hat{a}^2} \sum_{\substack{k_s \varkappa_s k_l \varkappa_l \\ l_1 l'_1 S_1 S'_1}} \varrho_{k_s \varkappa_s}(s_1 s_1)\, \hat{b}^2 \hat{k}_s \hat{k}_l \hat{S}_1 \hat{S}'_1 \hat{l}_1 \hat{l}'_1 \times \\ & \times (-)^{a+k_s-s_1-S'_1+l'_1} (l_1 0, l'_1 0 \mid k_l 0)\,(k_s \varkappa_s, k_l \varkappa_l \mid k_b \varkappa_b) \times \\ & \times \begin{Bmatrix} S_1 & l_1 & b \\ S'_1 & l'_1 & b \\ k_s & k_l & k_b \end{Bmatrix} W(s_1 s_1 S_1 S'_1; k_s a) \cdot \mathcal{D}^{k_l}_{\varkappa_l 0}(\mathcal{R}_1) \langle S_2 l_2 b \mid R \mid S_1 l_1 b\rangle \times \\ & \times \langle S_2 l'_2 b \mid R \mid S'_1 l'_1 b\rangle^* . \end{aligned}\right\} \tag{32.17}$$

In evaluating the angular distribution by means of (12.2), we first perform the summation over $\varkappa_l$, $\varkappa_s$, and $\varkappa_b$, i.e., we must calculate:

$$\sum_{\varkappa_l \varkappa_s \varkappa_b} (k_s \varkappa_s, k_l \varkappa_l \mid k_b \varkappa_b)\, \mathcal{D}^{k_l}_{\varkappa_l 0}(\mathcal{R}_1)\, \mathcal{D}^{k_b}_{\varkappa_b 0}(\mathcal{R}_2^{-1})\, \varrho_{\varkappa_s k_s}(s_1 s_1) \tag{32.18}$$

which, because of (12.5), becomes

$$\sum_{\varkappa_l \varkappa_s \varkappa_b \varkappa'_s} \big(k_s \varkappa_s, k_l \varkappa_l (k_b \varkappa_b)\, \tau_{k_s \varkappa'_s}(s_1 s_1)\, \mathcal{D}^{k_l}_{\varkappa_l 0}(\mathcal{R}_1)\, \mathcal{D}^{k_b}_{\varkappa_b 0}(\mathcal{R}_2^{-1})\, \mathcal{D}^{k_b}_{\varkappa_s \varkappa'_s}(\mathcal{R}_1)\big) . \tag{32.19}$$

Using some of the properties of the rotation matrices (Sect. 7), this finally reduces to

$$\sum_{\varkappa_s} \tau_{k_s \varkappa_s}(s_1 s_1)\,(k_s \varkappa_s, k_l 0 \mid k_b \varkappa_s)\, \mathcal{D}^{k_b}_{0 \varkappa_s}(\mathcal{R}_1 \mathcal{R}_2^{-1}) . \tag{32.20}$$

The angular distribution can be expressed as the sum of terms, each containing two factors. Thus,

$$W(\Omega_1 \Omega_2) = \sum_k [W_k(\Omega_1 \Omega_2)]_0\, \mathcal{P}_k(\Omega_1 \Omega_2) \tag{32.21}$$

where $[W_k(\Omega_1 \Omega_2)]_0$ is the k-th component of the unpolarised distribution and $\mathcal{P}_k(\Omega_1 \Omega_2)$ represents the polarisation factor. After some algebra, the latter is found to be

$$\left.\begin{aligned}\mathcal{P}_k = {} & \sum_{k_s \varkappa_l \varkappa_s S'_1} \frac{\hat{s}_1 \tau_{k_s \varkappa_s}(s_1 s_1)\,(k_s \varkappa_s, k_l 0 \mid k \varkappa_s)\, \mathcal{D}^k_{0 \varkappa_s}(\vartheta_1 \vartheta_2 \vartheta_3)}{\tau_{00}(s_1 s_1)\, P_k(\cos\vartheta)} \times \\ & \times \frac{(-)^{a-s_1+k_s-S_1-S'_1+b+l'_1} W(s_1 s_1 S_1 S'_1; k_s a)\, \hat{k}_s \hat{k}_l \hat{S}_1 \hat{S}'_1 \begin{Bmatrix} S_1 & l_1 & b \\ S'_1 & l'_1 & b \\ k_s & k_l & k \end{Bmatrix}}{W(b b l_1 l'_1; k S_1)} . \end{aligned}\right\} \tag{32.22}$$

For the case $s = \frac{1}{2}$, there is considerable simplification since $k_s = 1$ and $k = k_l$. The spin tensors are defined in (30.5). We find

$$\left.\begin{aligned}\mathcal{P}_k = {} & 1 - i\, \boldsymbol{P} \cdot \boldsymbol{n}\, \frac{P^1_k(\cos\vartheta)}{P_k(\cos\vartheta)} \sqrt{\frac{6}{k(k+1)}}\, \frac{(-)^{a+\frac{1}{2}+l'_1+b-S_1} \hat{k} \hat{S}_1}{W(b b l_1 l'_1; 1 S_1)} \times \\ & \times \sum_{S'_1} \hat{S}'_1 (-)^{-S'_1} W(\tfrac{1}{2} \tfrac{1}{2} S_1 S'_1; 1 a) \begin{Bmatrix} S_1 & l_1 & b \\ S'_1 & l'_1 & b \\ 1 & k & k \end{Bmatrix} , \end{aligned}\right\} \tag{32.23}$$

where $\boldsymbol{n} = \mathfrak{k}_1 \times \mathfrak{k}_2$.

If the incoming beam is unpolarised and measurement is made of the polarisation of the outgoing spin-$\frac{1}{2}$ particles, the expression for $\mathcal{P}_k$ in (32.23) is found by substituting c for a and the suffix 2 for 1, and by leaving $\boldsymbol{n}$ unaltered.

SIMON and WELTON [*84*] derived certain conditions for the observation of polarisation (or for the effects produced by polarisation) by using the conventional reaction theory (BLATT and BIEDENHARN [*6*]) and performing the appropriate geometrical summation. Our results can be compared with theirs provided that we drop the assumption that the intermediate state is well-defined. If this is done, the expression for $\mathcal{P}_k$ (32.22) is unchanged, except for the replacement of

$$\frac{(-)^b \begin{Bmatrix} S_1 & l_1 & b \\ S_1' & S_1' & b \\ k_s & k_l & k \end{Bmatrix}}{W(b\,b\,l_1\,l_1';\,k\,S_1)}$$

by

$$\frac{(-)^{b'} \begin{Bmatrix} S_1 & l_1 & b \\ S_1' & l_1' & b' \\ k_s & k_l & k \end{Bmatrix}}{W(b\,b'\,l_1\,l_1';\,k\,S_1)}\,.$$

Polarisation which is related to *odd* values of k_s will not be detected (or will produce no effect on the correlations) if any of the following conditions appertain:

(a) there is only *one* value of either l_1 or l_2, and this equals zero,

(b) the intermediate state has spin 0 or $\frac{1}{2}$ (both these conditions result in isotropic correlation functions),

(c) for a specific process, defined as $(\varkappa_1, \varkappa_2)$, there is only a single set of values of the parameters $l_1 S_1, l_2 S_2$, in the reaction matrix $\langle l_2 S_2 \varkappa_2 | R | l_1 S_1 \varkappa_1 \rangle$,

(d) all the reaction matrix-elements have the same phase. The polarisation depends on the factors: $[R R'^* - R' R^*]$, where R' differs from R in that it is a function of angular momenta written with primed symbols. Since the elements of the R-matrix are real, if either perturbation theory or the Born approximation is used, there is neither polarisation nor an effect due to polarisation in such cases.

There are additional conditions applying to polarisation which are related to *any* value of k_s. Polarisation effects are absent if the system can be represented in channel-spin formalism and if:

(e) only the channel spin, $S_2=0$, is present in the final state, or

(f) there is no spin-orbit coupling.

Further considerations have been given [*84*] to reactions involving incoming polarised beams where it is found that the outgoing beams show no polarisation if $k_{s_1}+k_{s_2}$ is odd, and either

(a) the Born approximation or perturbation theory is used, or

(b) $\varkappa_{s_1}+\varkappa_{s_2}=0$.

The complexity rules are similar to those obtained for unpolarised angular correlations (Sect. 15). If there is a largest effective value of l_1, l_2, or b,

$$k \leq \text{minimum}\,(2l', 2l+k_s, 2b+k_s) \quad \text{if} \quad k_s \text{ is even}$$

$$k \leq \text{minimum}\,(2l', 2l+k_s-1, 2b+k_s-1) \quad \text{if} \quad k_s \text{ is odd,}$$

and k is in all cases even if the radiation has well-defined parity.

33. Polarisation: Experimental study of γ-radiation. The application of polarisation measurements to the study of processes involving states and radiations of well-defined symmetry is limited by two sorts of considerations—one theoretical and the other experimental. From the theoretical viewpoint, we have seen that there are definite conditions to be satisfied if polarisation effects are to be present at all. For example, with spin-$\frac{1}{2}$ particles which are commonly involved in reactions, there occur only the zero-rank tensor (which gives the ordinary unpolarised correlation) and the first-rank tensor, for which the special restrictions associated with odd-rank tensors are applicable. From the experimental viewpoint, the measurement of polarisation properties of detected radiation is a far more complicated matter than measurement of direction of propagation. Up to the present at least, there has been overwhelming emphasis in the study of cascade-radiations and reactions, on extracting the maximum information from the easier and more precise directional-correlation measurements, and only in a relatively small number of cases have polarisation measurements been made. In a similar way, the preparation of a canalised beam is normally a much simpler, or at least commoner, procedure than the preparation of a polarised beam or target, or a source of partially aligned radioactive nuclei.

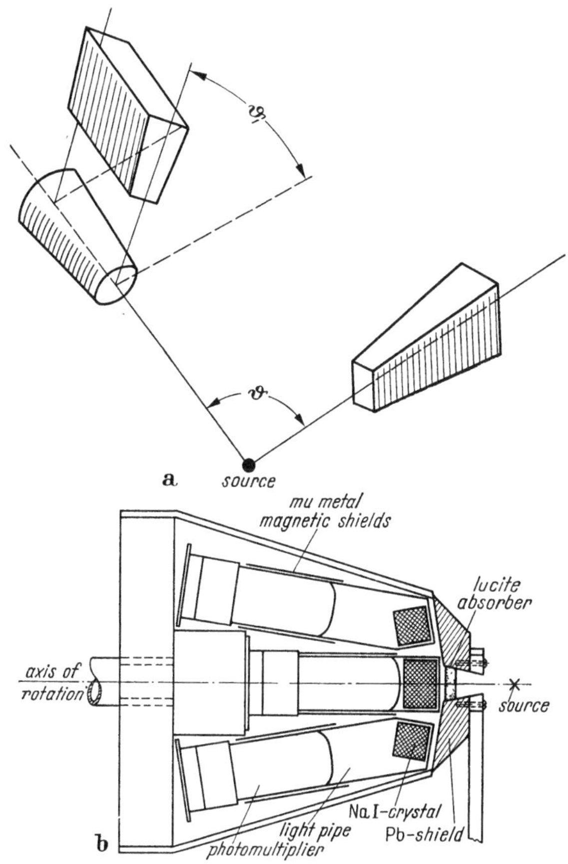

Fig. 14a and b. (a) Schematic arrangement for γ-polarisation measurements [cf. F. METZGER and M. DEUTSCH: Phys. Rev. 78, 551 (1950)]. (b) Arrangement for crystals and photomultipliers in a three-crystal γ-polarimeter [cf. J. J. KRAUSHAAR and M. GOLDHABER: Phys. Rev. 89, 1081 (1953)].

Gamma radiation in cascade emissions and reactions is, however, quite generally susceptible to polarisation measurements, if the directional correlation shows some anisotropy. Moreover, since the parity of the radiation is not explicitly revealed by directional correlation alone, there is a particular motive for polarisation measurements in this case in order to determine the radiation parity. Most of the measurements that have been made are of γ-cascades, or of γ-radiation following reactions with well-defined resonance levels, where the conditions of well-defined symmetry apply. We shall consider in this section only such measurements in which polarisation-sensitive detectors are used. Apart from the general verification of the theoretical possibilities[1], the main purpose—to distinguish between two possible parity values—can be achieved with limited accuracy in the measurements.

For γ-radiation of low energy (of the order 2 Mev or less), polarisation-sensitive detectors have been used which depend on the correlation between the directions

[1] F. METZGER and M. DEUTSCH: Phys. Rev. 78, 551 (1950).

of propagation of primary and Compton-scattered γ-rays and the polarisation vectors ($\mathcal{E}$ vector). There is a maximum probability for the scattered radiation to be in a direction perpendicular to the polarisation. A typical detector assembly[1] and the schematic arrangements in a correlation experiment are shown in Fig. 14. The γ-radiation is detected in the central scintillation-crystal through the excitation produced by the scattered Compton electron. The secondary scattered γ-radiation is detected in either one of the pair of side-crystals. The average angle between primary and secondary γ-radiation is rather less than $\pi/2$. Coincidences are recorded between counts in the central and either side detector. In the case of angular distributions, the polarisation can be obtained directly from this response of the detector. In cascade correlation, this detector response must be recorded, of course, in coincidence with the other radiation of the cascade, and the triple coincidence rate can be studied as a function of the normal correlation angle and also as a function of the orientation of the polarisation detector about the direction of the detected radiation as axis.

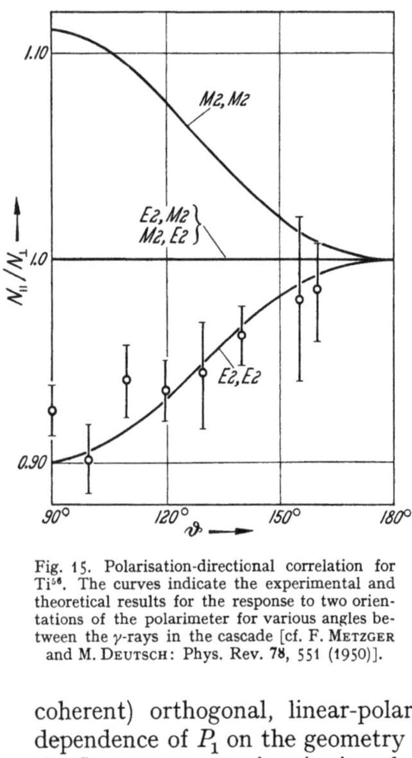

Fig. 15. Polarisation-directional correlation for Ti^{56}. The curves indicate the experimental and theoretical results for the response to two orientations of the polarimeter for various angles between the γ-rays in the cascade [cf. F. METZGER and M. DEUTSCH: Phys. Rev. 78, 551 (1950)].

Polarisation detectors of this type can be described in terms of a quantity P_1 and an angle ϑ_1 which measures the orientation of the polarimeter with respect to the plane containing both radiations of the correlation. The value of P_1 will depend on the reference zero, used in defining the orientation. This zero is chosen so that the maximum response is obtained with $\vartheta_1 = 0$ or $\pi/2$; also P_1 will then have its maximum value. The angle 0 is the conventional choice; Eq. (31.15) then gives the angular correlation with an arbitrary orientation of the detector[2]. In any particular position the detector response can be regarded as the sum of its responses to two (incoherent) orthogonal, linear-polarised radiations. A detailed treatment of the dependence of P_1 on the geometry of the detector, on the basis of the theory of the Compton-scattering, is given by METZGER and DEUTSCH.

Results obtained in one of such polarisation measurements are shown in Fig. 15. These clearly indicate how the ambiguity in parity assignment can readily be removed.

Instead of using calculations involving the detailed geometry of the detector, more experimental methods can be used to measure the polarisation efficiency of detectors of this type. For example, the polarisation of scattered γ-radiation can be calculated on the basis of the Klein-Nishina formula, the radiation scattered from material of varying thickness can be measured with the polarisation detector, and the results can be extrapolated to zero thickness[1]. Alternatively, a well-established γ-cascade, for which the polarisation-correlation as a function

[1] J. J. KRAUSHAAR and M. GOLDHABER: Phys. Rev. 89, 1081 (1953).

[2] We could of course regard P_1 as a variable quantity, changing as the orientation of the detector is altered. Provided the variation of P_1 with orientation is known, measurement of the response for two orientations is sufficient.

of P_1 can be predicted reliably, can be used to calibrate a particular detector, provided the appropriate γ-energies can be chosen. For γ-radiation of energy near 0.5 Mev, the polarisation properties of the two γ-radiations produced in electron-positron annihilation can be used[1].

As a particular example which is applicable to most experimental conditions, we consider a cascade in which one of the radiations is of mixed multipolarity, and this one is detected by a polarisation-sensitive detector.

The correlation function in this case is given according to (31.15) by:

$$\left.\begin{aligned} W(\Omega_1\Omega_2) = 1 + \sum_k A_k(a, b\,b)\,\{&A_k(b\,b, c)\,P_k(\cos\vartheta) + [F_k(L_2 L_2 c\,b)\times\\ &\times(\pm)_{L_2} f_k(L_2 L_2) + \delta^2 F_k(\overline{L_2+1}\;\overline{L_2+1}\,c\,b)\times\\ &\times(\pm)_{L_2+1} f_k(\overline{L_2+1}\;\overline{L_2+1}) + 2\delta\,F_k(L_2\overline{L_2+1}\,c\,b)\times\\ &\times(\pm)_{L_2+1} f_k(L_2\overline{L_2+1})]\,P_1\cdot P_k^2(\cos\vartheta)\cos 2\vartheta_1\}, \end{aligned}\right\} \tag{33.1}$$

where the function $A_k(a, bb)$ refers to the (first) radiation detected without polarisation, $A_k(bb, c)$ to the second mixed transition, and these functions and $F_2(L_2L_2cb)$ etc. are defined as in (19.2), (19.4), (19.7). $f_k(L_2L_2)$ etc. are defined in (31.16). $A_k(a, bb)$ might refer to any type of unpolarised radiation.

We can write (33.1) in the form

$$W(\Omega_1\Omega_2) = C(\vartheta) + D(\vartheta)\,P_1\cos 2\vartheta_1. \tag{33.2}$$

In a typical polarisation measurement, the value of W is measured at the fixed angle $\vartheta = \pi/2$, with $\vartheta_1 = 0$ and $\pi/2$. Denoting these values by $W(\pi/2, 0)$, $W(\pi/2, \pi/2)$, and their ratio by r, we have

$$r = \frac{W\left(\frac{\pi}{2}, 0\right)}{W\left(\frac{\pi}{2}, \frac{\pi}{2}\right)} = \frac{C\left(\frac{\pi}{2}\right) + P_1 D\left(\frac{\pi}{2}\right)}{C\left(\frac{\pi}{2}\right) - P_1 D\left(\frac{\pi}{2}\right)} \quad \text{or} \quad r = \frac{1 + x + x P_1}{1 + x - x P_1} \tag{33.3}$$

where $x = \dfrac{D(\pi/2)}{C(\pi/2) - D(\pi/2)}$ is a measure of the ratio of polarised to unpolarised radiation in the (second) transition (L_2, L_2+1). Its sign will depend on the parity of this radiation. The measured effect, xP_1 may then be considered as the product of the radiation-polarisation x, and the polarisation efficiency, P_1. Similar polarisation measurements have also been made for γ-radiation following α-emission[2] and β-decay[3] (see Sects. 22, 46).

For γ-energies much greater than the electron rest-mass, the direction of the Compton-scattered radiation is most probably in the forward direction, close to the direction of the primary radiation, so that polarimeters of the type illustrated in Fig. 14 become difficult to construct with high polarisation-efficiency. Above about 3 Mev, the fact that the photo-disintegration of the deuteron occurs predominantly through an electric-dipole transition can be utilised[4]. Neglecting other multipole contributions to the photo-disintegration, the correlation in the centre-of-mass frame of reference between the direction of the proton and neutron produced and the polarisation ($\mathscr{E}$-vector) of radiation is of the form $\cos^2\Theta$.

---

[1] B. L. Robinson and L. Madansky: Phys. Rev. **88**, 1065 (1952).
[2] A. P. French and J. O. Newton: Phys. Rev. **85**, 1041 (1952).
[3] D. Hamilton, A. Lemonick and F. M. Pipkin: Phys. Rev. **92**, 1191 (1953).
[4] D. H. Wilkinson: Phil. Mag. **43**, 659 (1952). — L. W. Fagg and S. S. Hanna: Phys. Rev. **92**, 372 (1953).

We consider, for example, the photo-nucleons produced by the γ-radiation from a reaction and observed in a plane normal to the direction of propagation. The γ-radiation may be considered to be "detected" if a pair of photo-nucleons are produced and the direction of their motion makes an angle ϑ_1 with the reaction plane. Then the angular correlation with such a "detector" is given by (31.15) with $P_1 = 1$, since a pure linearly polarised beam would give zero response if the photo-nucleons are detected at an angle $\pi/2$ with the polarisation plane. Alternatively, if measurements are made at a fixed reaction angle ϑ and the distribution of photo-nucleons as a function of their directions of motion ϑ_1 in the plane normal to the radiation is studied, then (31.15), with ϑ fixed, gives this distribution function. Such a distribution can also be interpreted in terms of an incoherent superposition of two γ-radiations components, one polarised in the reaction plane, and the other normal to it. The intensities of these components are provided by (31.15) with $P_1 = 1$ and $\vartheta_1 = 0$ and $\pi/2$ respectively; the resultant distribution of photo-nucleon directions is obtained by using these intensities together with the $\cos^2 \Theta$ distribution in the photo-disintegration.

Experimentally, if no coincident-timing is involved, the distribution of the photo-nucleons can be measured by using as a detector a photographic emulsion containing D_2O and examining the photo-proton tracks. Measurements of this sort have been made in connection with the angular distribution of γ-radiation resulting from proton capture in light nuclei[1]. Polarisation measurements are particularly useful in the case of high-energy radiation in light nuclei, where one alternative method of assigning parity—by measurement of internal conversion—is not practicable.

Polarisation-sensitive detectors can also be based on the correlation between the polarisation and the directions of positron and electron produced in pair-creation[2], and for low energies on the correlation with direction-emission of photo-electrons[3].

E. Processes involving states not having well-defined symmetry.

I. Reactions and inelastic scattering.

34. Characteristics of Compound system. Nuclear reactions, for the most part, do not proceed through single isolated resonance levels. Consequently, it is not generally possible to represent the state which results from the absorption of the incident radiation as being one of well-defined spin and parity. The incident beam contains components of all angular momentum, many of which may be involved in the reaction, so that the overall system is represented by an assembly of states, each of which may in turn be a superposition of several, and possibly very many, states of well-defined symmetry. Previously, also, a state has been regarded as "sharp" if the spin and parity were well-defined, but we must now consider the possibility that even when this is so the intermediate state cannot be considered "sharp" in the same sense as an isolated resonance level[4]. In principle the same situation can arise for cascade emission if the state resulting after the first emission is not an isolated one. However, since emission leads to

[1] E.g. I. S. HUGHES and D. SINCLAIR: Proc. Phys. Soc. Lond. A **69**, 125 (1956).

[2] M. M. MAY: Phys. Rev. **84**, 265 (1951).

[3] F. L. HEREFORD and J. P. KEUPER: Phys. Rev. **90**, 1043 (1953).

[4] For example, the absorption of slow neutrons ($l = 0$; $L = s = \frac{1}{2}$) by a nucleus of spin zero and parity p results only in states ($\frac{1}{2}\, p$), but the compound nucleus formed may be in a region of several overlapping levels of the type ($\frac{1}{2}\, p$). This example is of course trivial from the point of view of angular correlations.

nuclei of low excitation energy, where the level width is usually very much smaller than level spacing, the states involved in cascade emission can almost invariably be treated as sharp, and in particular as having well-defined symmetry. We shall, therefore, in the present discussion apply our remarks to reactions.

The basic theory developed for reactions involving well-defined (Sects. 11, 12) states can be extended readily, with few formal changes, to include the present more general situation, and the resultant angular distribution formulae are of similar construction; but since the qualitative interpretation may be substantially different, we shall first discuss some differences on a physical basis. Firstly, we may consider the degree of complexity of the compound system formed by the absorption of the first particle. It may be possible to describe this system in terms of the superposition of only a few states which separately have the character of resonance levels. In this case the general theory used previously, when modified suitably, would be the appropriate one, and the complexity of such an analysis might not be excessive. For example, if we can represent the state of the compound system as arising from the superposition of only two resonance levels, then there will be two sets of parameters entering the theoretical expression, each set being similar to the set required in the theory of an isolated resonance, and in addition a complex ratio factor, representing the relative contributions of the two resonance levels in the description of the compound system. We should notice that since some of the parameters corresponding to each resonance level, and expecially the complex ratio-factor, will in general be energy-dependent, the correlation functions will be energy dependent and often sensitively so. Such marked energy sensitivity may occur even if both separate resonance levels have the same spin and parity, in which case the whole system can be considered as having well-defined symmetry. This feature is in marked contrast to the correlation for an isolated resonance reaction. Another marked difference occurs if the compound system can be represented in terms of states with opposite parity. Directional-correlation functions are no longer restricted to the type $\sum_k A_k P_k(\cos\vartheta)$, $k=$ even, since the absence of odd k terms is a consequence of parity conservation. The presence of odd k terms, i.e. the lack of symmetry about $\vartheta=\pi/2$, in the correlation function is frequently used as a criterion in checking whether more than a single resonance level is effective in a reaction.

If the description of the compound system requires the superposition of more than two or three states, then the theory of the type we have discussed ceases to be practically useful for the interpretation of experiments. In the extreme situation of the superposition of a very large number of states, certain general features of the correlation may be significant. Thus, if the many states can be considered as levels of a compound nucleus each characterised by a spin and parity, but with a phase factor that is equally likely to be positive or negative, the terms in the distribution arising from the superposition will, on the average, cancel. The resulting superposition will be approximately the same as the simple addition of many angular distributions, each of which is characteristic of an isolated resonance level. Such a distribution will have approximate symmetry about $\vartheta=\pi/2$, and this *approximate* symmetry may be regarded as the indication that a reaction proceeds through the formation of a "compound nucleus". It is also unlikely that the angular distribution in such cases is very markedly anisotropic or has appreciable contributions from terms $A_k P_k(\cos\vartheta)$ with large values of k. Although the superposition may involve contributions from many levels, there will be only a limited range of values of the total angular momentum of the system, and many of the contributions to the overall correlation function are likely to result in the largest terms having small values of k (including zero).

It is always possible, in principle, to formulate a theory of the angular distributions in any nuclear reaction on the basis of contribution from expressions each corresponding to a particular value of the total angular momentum, b, of the system[1], even though the actual situation does not correspond to the physical formation of a superposition of such resonance states. For each value of b there would also be a range of possible values of (L, l), for both ingoing and outgoing radiation, limited by conservation of angular momentum and parity. In addition to a large measure of coherence amongst the terms corresponding to different values of b, L, and l, there would be numerous interference terms in the distribution function, and the theory might involve a very large number of "physical" parameters, real and complex, apart from the quantum numbers describing the symmetry properties. Clearly the main burden of such a "theory" would be in assigning values to the "physical" parameters, and this, of course, would imply a detailed physical theory or model for the reaction. However, even if such a theory could be formulated in terms of the type of spherical-wave expansion indicated (quite different representations may well be more suitable), the rôle of the general quantum-mechanical principle in such a theory would be quite secondary as compared with that of the particular physical theory.

Many nuclear reactions are characterised by angular distributions which show contributions from a large range of L and l, say up to 10 or 20, and are markedly anisotropic and asymmetric (with respect to $\vartheta = 90°$), but are not very sensitive to energy. All these features indicate a mechanism other than many overlapping levels of a compound nucleus. They are usually designated as "direct" processes, for which the interpretation in terms of a well-defined intermediate state or superposition of such is inappropriate. If the correlation for such processes could be interpreted in terms of contributions characterised by (b, L, l), then one would expect some regular variation (of both amplitude and phase) for these contributions with increasing values of angular momenta, characteristic of the particular reaction. (This is in marked contrast to what is expected from a large number of levels of the compound nucleus.) However, as already remarked, the theory of such direct processes is normally best treated not as a generalisation of the case of well-defined angular-momentum states, but rather in a manner suited to the *particular* nature of the process. In the present section we shall limit our discussion to reactions of the compound-nucleus type.

35. Theoretical correlations for several overlapping levels of the compound nucleus. The angular distribution, with unpolarised beams and detectors, can be represented by Eq. (12.2) with the summation extended to include different states, B, B', and different angular momenta, b, b', of the intermediate state:

$$W(\vartheta) = \sum_{b b' B B' k_b \varkappa_b} \varrho_{k_b \varkappa_b}(b B, b' B')\, \varepsilon^*_{k_b \varkappa_b}(b B, b' B'). \tag{35.1}$$

The statistical tensor $\varrho_{k_b \varkappa_b}$ and the efficiency tensor $\varepsilon_{k_b \varkappa_b}$ can be expressed either in the $(\mathcal{L})$ representation or the channel-spin (l, S) representation. The evaluation of the expressions for $W(\vartheta)$ follows the same general lines as the treatment, in Sect. 24, of the case of an isolated resonance with unpolarised incident beams and polarisation-insensitive detectors.

$\mathcal{L}$-representation. The reaction is represented by $(a A \mathcal{L}_1 \mathcal{L}_1') \to (b B\, b' B') \to (c C \mathcal{L}_2 \mathcal{L}_2')$. The efficiency and density tensors are expressed in the usual

[1] We denote by b both the total angular momentum b and the parity of a state.

manner in terms of the radiation parameters, (Sect. 13),

$$\varepsilon_{k0}(LL') \quad \text{or} \quad \varrho_{k0}(LL') = \sum_{\varkappa} c_{k\varkappa}(\mathscr{L}\mathscr{L}')\,\mathscr{D}^k_{\varkappa 0}(\mathscr{R}). \tag{35.2}$$

In place of (12.6) we have the result:

$$\left.\begin{aligned} W(\vartheta) = \frac{\lambda^2}{4\hat{a}^2}(-)^{a+c} \sum_{bb'} \hat{b}^2 \hat{b}'^2 (-)^{2b'} \sum_{\substack{kl_1l_2l_1'l_2' \\ L_1L_2L_1'L_2'}} (-)^{-L_1-L_2} c_{k0}(\mathscr{L}_1\mathscr{L}_1')\, c^*_{k0}(\mathscr{L}_2\mathscr{L}_2')\times \\ \times W(bb'L_1L_1';ka)\,(W(bb'L_2L_2';kc)\times \\ \times q(bb';L_1L_1',L_2L_2',l_1l_1',l_2l_2')\,P_k(\cos\vartheta)\times \\ \times \operatorname{Re}[\langle c\mathscr{L}_2 b|\mathscr{O}|a\mathscr{L}_1 b\rangle\langle c\mathscr{L}_2' b'|\mathscr{O}|a\mathscr{L}_1' b'\rangle^*]. \end{aligned}\right\} \tag{35.3}$$

The summations occurring in (35.3) can be contracted, for the purpose of practical computation, so that individual terms are not duplicated. This is ensured by the quantity $q(\ldots)$, defined as:

$$q(bb';L_1L_1',L_2L_2',l_1l_1',l_2l_2') = \begin{cases}0\\2\end{cases},$$

when $b \lessgtr b'$ or when $b = b'$ and any of the following further conditions hold:

$$\begin{aligned} &L_1 \lessgtr L_1', \\ &L_1 = L_1' \quad \text{and} \quad L_2 \lessgtr L_2', \\ &L_1 = L_1', \quad L_2 = L_2' \quad \text{and} \quad l_1 \lessgtr l_1', \\ &L_1 = L_1', \quad L_2 = L_2', \quad l_1 = l_1' \quad \text{and} \quad l_2 \lessgtr l_2'. \end{aligned}$$

Also

$$q(bb;L_1L_1,L_2L_2,l_1l_1,l_2l_2) = 1.$$

The summations over $\mathscr{L}$ are further restricted by the triangle conditions such as $\Delta(l\,s\,L)$, $\Delta(a\,L_1\,b)$, $\Delta(c\,L_2\,b)$. Notice also the summation over b, b' includes both b and parity, since the latter is not well-defined.

The reduced matrix elements of the interaction operators do not necessarily represent individual levels of a compound nucleus, since even if description in terms of such intermediate states is valid, several such states may be included in the same value of b. If each b corresponds to a distinct level, then the reduced matrix elements may be factorised [cf. Eq. (24,5)]:

$$\langle c\mathscr{L}_1 b|\mathscr{O}|a\mathscr{L}_1 b\rangle = \langle c\|\mathscr{L}_2\|b\rangle\langle a\|\mathscr{L}_1\|b\rangle f(l_1)\,f(l_2). \tag{35.4}$$

The factors $\langle c\|\mathscr{L}_2\|b\rangle$, $\langle b\|\mathscr{L}_1\|a\rangle$, cannot, however, be treated as real, even when the phase factors due to Coulomb fields, $f(l)$, have been separated out, since in the expansion of (35.3) there will occur terms of the type

$$\langle c\|\mathscr{L}_2\|b\rangle\langle a\|\mathscr{L}_1\|b\rangle\langle c\|\mathscr{L}_2'\|b'\rangle^*\langle a\|\mathscr{L}_1'\|b'\rangle^* \tag{35.5}$$

which refer to two levels, b, b', so that the relative phase of the respective matrix elements is significant. For a single level, b, only products of the type

$$\langle c\|\mathscr{L}_2\|b\rangle\langle a\|\mathscr{L}_1\|b\rangle\langle c\|\mathscr{L}_2'\|b\rangle^*\langle a\|\mathscr{L}_1'\|b\rangle^* \tag{35.6}$$

occur and all the significant phase differences have been removed in factors $f(l)$[1].

[1] Cf. references [6], [45].

Eq. (35.3) can also be written in the form:

$$\left.\begin{aligned} W(\vartheta) = \frac{\lambda\!\!\!^{-2}}{4\hat{a}^2\hat{s}_1^2} \sum_{b b' k} \sum_{\substack{L_1 L_1' L_2 L_2' \\ l_1 l_1' l_2 l_2'}} & A_k(l_1 L_1 l_1' L_1', a b b')\, A_k(l_2 L_2 l_2' L_2'; c b b') \times \\ & \times q(b b'; L_1 L_1', L_2 L_2', l_1 l_1', l_2 l_2') \times \\ & \times \mathrm{Re}\left[\langle c \mathcal{L}_2 b | \mathcal{O} | a \mathcal{L}_1 b\rangle \langle c \mathcal{L}_2' b' | \mathcal{O} | a \mathcal{L}_1' b'\rangle^*\right] P_k(\cos\vartheta), \end{aligned}\right\} \tag{35.7}$$

where $A_k(l_1 L_1 l_1' L_1'; a b b')$ has the following values:

(i) For particles of spin s_1:

$$(-)^{a-b'-s_1}\, \hat{b}\, \hat{b}'\, Z'(l_1 L_1 l_1' L_1'; s_1 k)\, W(b b' L_1 L_1'; k a). \tag{35.8a}$$

(ii) For particles of spin $\frac{1}{2}$:

$$(-)^{a-b'-\frac{1}{2}}\, \hat{b}\, \hat{b}'\, \hat{L}_1 \hat{L}_1' (L_1 \tfrac{1}{2}, L_1' -\tfrac{1}{2} | k 0)\, W(b b' L_1 L_1'; k a). \tag{35.8b}$$

(iii) For γ-radiation:

$$(-)^{a-b'+L_1'-L_1+k-1}\, \hat{b}\, \hat{b}'\, \hat{L}_1 \hat{L}_1' (L_1 1, L_1' -1 | k 0)\, W(b b' L_1 L_1'; k a). \tag{35.8c}$$

In this case $A_k(l_1 L_1 l_1' L_1'; a b b')$ is replaced by $A_k(L_1 L_1'; a b b')$. Similar expressions hold for $A_k(l_2 L_2 l_2' L_2', c b b')$. We notice that although each factor $A_k(\ldots)$ in (35.7) involves only one radiation, we cannot now write the coefficient of $P_k(\cos\vartheta)$ in the correlation as a simple product of the two factors, each referring to one stage of the process.

In some cases where it is advantageous to express the angular distribution in terms of the reduced matrix elements of the interaction operators, we may use alternative expressions:

$$W(\vartheta) \sim \sum_{k b b'} A_k'(a, b b')\, A_k(b b', c)\, P_k(\cos\vartheta), \tag{35.8d}$$

where

$$\left.\begin{aligned} A_k'(a, b b') = \sum_{\mathcal{L}_1 \mathcal{L}_1'} & \langle a \| \mathcal{L}_1 \| b\rangle' \langle a \| \mathcal{L}_1' \| b'\rangle'^*\, f(l_1)\, f^*(l_1') \times \\ & \times c_{k0}^*(\mathcal{L}_1' \mathcal{L}_1)\, \hat{b}' (-)^{a-b'-L_1'+k}\, W(b b' L_1 L_1'; k a), \end{aligned}\right\} \tag{35.8e}$$

and[1]

$$\left.\begin{aligned} A_k(b b', c) = \sum_{\mathcal{L}_2 \mathcal{L}_2'} & \langle b \| \mathcal{L}_2 \| c\rangle'^* \langle b' \| \mathcal{L}_2' \| c\rangle'\, f(l_2)\, f^*(l_2') \times \\ & \times c_{k0}^*(\mathcal{L}_2 \mathcal{L}_2')\, \hat{b}\, (-)^{c-b-L_2'+k}\, W(b b' L_2 L_2'; k c). \end{aligned}\right\} \tag{35.8f}$$

Channel-spin representation. The reaction is now represented, as in Sect. 24, as $(a_1 s_1 l_1 l_1' S_1 S_1') \to (b B b' B') \to (c s_2 l_2 l_2' S_2 S_2')$, and we can write in place of (24.15):

$$\left.\begin{aligned} W(\vartheta) = \frac{\lambda\!\!\!^{-2}}{4\hat{a}^2\hat{s}_1^2} \sum_{S_1 S_2} \sum_{k l_1 l_1' l_2 l_2' b b'} & A_k(l_1 b l_1' b'; S_1)\, A_k(l_2 b l_2' b'; S_2)\, q(b b'; l_1 l_1', l_2 l_2') \times \\ & \times \mathrm{Re}\left[\langle l_2 S_2 b | R | l_1 S_1 b\rangle \langle l_2' S_2 b' | R | l_1' S_1 b'\rangle^*\right] P_k(\cos\vartheta). \end{aligned}\right\} \tag{35.9}$$

$$q(b b'; l_1 l_1', l_2 l_2') = \left.\begin{matrix} 0 \\ 2 \end{matrix}\right\} \text{ if } b \lesseqgtr b', \text{ or, if } b = b' \text{ and any of the following hold:}$$

$$l_1 \lesseqgtr l_1', \quad l_1 = l_1' \text{ and } l_2 \lesseqgtr l_2',$$

$$= 1, \quad \text{if } b = b', \quad l_1 = l_1', \quad l_2 = l_2',$$

and

$$A_k(l_1 b l_1' b'; S_1) = (-)^{b'+S_1} Z'(l_1 b l_1' b'; S_1 k). \tag{35.10}$$

[1] See footnote before (15.3).

This result is equivalent to that of BLATT and BIEDENHARN [6]. In the channel-spin formalism the expression for $W(\vartheta)$, when no polarisation effects are present, is rather simpler than the corresponding expression in the $\mathscr{L}$-formalism. This is because the summation over S_1 and S_2 for random spins is simply equivalent to the addition of the separate angular distributions, whereas in the summation over $\mathscr{L}$ the different incoherent spins are involved together with the coherent l-components. This simplicity of the S-representation would be absent if polarisation effects were included.

The interaction operator in (35.9) is the usual R matrix. As in the $\mathscr{L}$-representation, the matrix elements do not necessarily represent individual states of the compound nucleus. If we represent the individual states of the same (b) by the suffix (n), then, remembering that each matrix element of R correspond to definite l_1 and l_2, we can write, e.g.

$$\langle l_2 S_2 b | R | l_1 S_1 b \rangle = \sum_n \langle c \| l_2 S_2 \| b_n \rangle' \langle b_n \| l_1 S_1 \| a \rangle' \Phi(b_n) f(l_1) f(l_2), \quad (35.11)$$

where the reduced matrix elements are real and the complex part is separated out in the phase factors $\Phi(b_n)$, which refer to the states of different n and which are independent of l_1, l_2, etc.

For the case in which each b corresponds to a single distinct level, we can write more simply:

$$\langle l_2 S_2 b | R | l_1 S_1 b \rangle = \langle c \| l_2 S_2 \| b \rangle' \langle b \| l_1 S_1 \| a \rangle' \Phi(b) f(l_1) f(l_2), \quad (35.12)$$

where the reduced matrix elements are again real.

The separation of the R-matrix elements into a real part and a complex phase-factor, $\Phi(b)$, is normally used when a reaction is analysed in terms of the superposition of resonance levels. A similar separation could also have been made in the $\mathscr{L}$-representation, for example in Eq. (35.4).

A more general discussion of the reaction matrix, and in particular the number of real parameters required to specify the matrix-elements completely, is given by BLATT and BIEDENHARN [6] (cf. also G. BREIT, Vol. XLI of this Encyclopedia).

Polarised radiations. The $\mathscr{L}$-representation is particularly suited to the description of effects involving polarised radiations, since the specification of these can be included in the expressions for the radiation parameters. On the other hand, in the channel-spin representation, the polarisation specification must be separated from the tensor corresponding to channel-spin.

If the parity of the compound nucleus is not well-defined, the distribution normally involves odd as well as even k-values[1]. In any case (k either even or odd), the polarisation-radiation parameters may be calculated as in Sects. 30 and 31 and the modifications to the correlation for γ-radiation and spin-$\frac{1}{2}$ particles as in (31.15) still remain valid in this more general case. The same holds true for "vector"-polarisation of spin-s particles, even though k may be odd, since the only contribution arises from the $k=k_l$ term.

From (30.14) we see that the polarisation effects for spin-$\frac{1}{2}$ arise from the imaginary part of $(\mathcal{O}\mathcal{O}'^*)$. For an isolated resonance, as already mentioned, the quantity is often real, but if $\mathcal{O}$, $\mathcal{O}'$ refer to two different levels, $\mathcal{O}\mathcal{O}'^*$ will, in general, be complex. Hence polarisation effects will normally occur where there is interference between two or more levels of the compound nucleus.

The angular distribution for an ingoing beam with vector-polarisation, $\boldsymbol{P}$, and polarisation-insensitive detectors (or unpolarised incident beam and detector

[1] We may notice that, in principle, the parity of the intermediate state in a correlation may be well-defined and the radiations not so (cf. Sect. 15, footnote 2, p. 395).

with polarisation efficiency $\boldsymbol{P}$) is of the form

$$W_{1,2}(\vartheta) = A_{1,2}(\vartheta) + \boldsymbol{n}\cdot\boldsymbol{P}\, B_{1,2}(\vartheta), \tag{35.13}$$

where $\boldsymbol{n}=\mathfrak{k}_1\times\mathfrak{k}_2$ and $A_{1,2}(\vartheta)$ and $B_{1,2}(\vartheta)$ involve the real and imaginary parts of $\mathcal{O}\,\mathcal{O}'^*$,

$$\left.\begin{aligned} A_{1,2}(\vartheta) &\sim \sum A_k(l_1 L\, l_1' L_1'; a\, b\, b')\, A_k(l_2 L_2 l_2' L_2'; c\, b\, b')\,\times \\ &\quad\times q(\ )\,\mathrm{Re}\,\mathcal{O}\,\mathcal{O}'^*\, P_k(\cos\vartheta), \end{aligned}\right\} \tag{35.14}$$

and

$$\left.\begin{aligned} B_{1,2}(\vartheta) &\sim \sum A_k(l_1 L_1 l_1' L_1'; a\, b\, b')\, A_k(l_2 L_2 l_2' L_2'; c\, b\, b')\,\times \\ &\quad\times q(\ )\,(-)^{l'+1} f_k(L\,L', \tfrac{1}{2})\,\mathrm{Im}\,\mathcal{O}\,\mathcal{O}'^*\, P_k^1(\cos\vartheta). \end{aligned}\right\} \tag{35.15}$$

If $\boldsymbol{P}$ is in the direction of $\boldsymbol{n}$, then $W_{1,2}=A_{1,2}(\vartheta)+|\boldsymbol{P}|\,B_{1,2}(\vartheta)$. Examination of the above expressions shows that for the reverse reaction, i.e. absorption of spin-$\frac{1}{2}$ particles represented by $\mathcal{L}_2$ and the emission of particles $\mathcal{L}_1$, the angular distribution is

$$\left.\begin{aligned} W_{2,1}(\vartheta) &= A_{2,1}(\vartheta) + (\mathfrak{k}_2\times\mathfrak{k}_1)\cdot\boldsymbol{P}\, B_{2,1}(\vartheta) \\ &= A_{1,2}(\vartheta) - (\mathfrak{k}_2\times\mathfrak{k}_1)\cdot\boldsymbol{P}\, B_{1,2}(\vartheta) \\ &= W_{1,2}(\vartheta). \end{aligned}\right\} \tag{35.16}$$

This result can also be reached from (35.13) by an argument based on the time-reversal properties of the process.

Application to cascade-radiations. The equations given so far in this section apply to reactions (i.e. ingoing followed by outgoing radiation). For two outgoing radiations, we can use the same expressions if we make the following changes:

For the radiation parameter $c_{k\varkappa}(\mathcal{L}_1\mathcal{L}_1')$ we make the substitution

$$c_{k\varkappa}(\mathcal{L}_1\mathcal{L}_1') \to c_{k\varkappa}(\mathcal{L}_1'\,\mathcal{L}_1)\,\frac{\hat{a}^2}{\hat{b}\,\hat{b}'}\,(-)^{k+b-b'}. \tag{35.17}$$

Triple correlations. The above discussion can be extended in a straightforward manner to deal with triple cascades in which one (or more) of the intermediate stages involves overlapping levels; for example, a reaction followed by further γ-emission.

The correlation function $W(\Omega_1\Omega_2\Omega_3)$ for three consecutive emissions, with both intermediate states possibly not sharp, is given by

$$W(\Omega_1\Omega_2\Omega_3) = \sum_{k_1k_2k_3}\sum_{b\,b'c\,c'} A_{k_1}(a, b\,b')\, A_{k_3}(c\,c', d)\, R_{k_1k_2k_3}(b\,b'\,c\,c')\, S_{k_1k_2k_3}(\Omega_1\Omega_2\Omega_3), \tag{35.18}$$

with $S_{k_1k_2k_3}(\Omega_1\Omega_2\Omega_3)$ as previously defined, Sect. 16, and

$$R_{k_1k_2k_3}(b\,b'\,c\,c') = \sum_{\mathcal{L}_2\mathcal{L}_2'} \langle c\,\|\mathcal{L}_2\|\, b\rangle \langle c'\,\|\mathcal{L}_2'\|\, b'\rangle^*\, c_{k_20}^*(\mathcal{L}_2\mathcal{L}_2')\,\frac{\hat{b}'}{\hat{k}_3\,\hat{c}}\,(-)^{k_3}\begin{Bmatrix} c & c' & k_3 \\ L_2 & L_2' & k_2 \\ b & b' & k_1 \end{Bmatrix}. \tag{35.19}$$

$A_{k_1}(a, b\,b')$, $A_{k_3}(c\,c', d)$ are given by:

$$\left.\begin{aligned} A_{k_1}(a, b\,b') &= \sum_{\mathcal{L}_1\mathcal{L}_1'} \langle b\,\|\mathcal{L}_1\|\, a\rangle' \langle b\,\|\mathcal{L}_1'\|\, a\rangle^{*\prime} f(l_1)\, f^*(l_1')\, c_{k_10}^*(\mathcal{L}_1\mathcal{L}_1')\,\times \\ &\quad\times \hat{b}\,(-)^{a-b-L_1'+k_1}\, W(b\,b'\,L_1 L_1'; k_1\, a), \end{aligned}\right\} \tag{35.20a}$$

$$\left.\begin{aligned} A_{k_3}(c\,c', d) &= \sum_{\mathcal{L}_3\mathcal{L}_3'} \langle c\,\|\mathcal{L}_3\|\, d\rangle^{*\prime} \langle c'\,\|\mathcal{L}_3'\|\, d\rangle' f(l_3)\, f^*(l_3')\, c_{k_30}^*(\mathcal{L}_3\mathcal{L}_3')\,\times \\ &\quad\times \hat{c}\,(-)^{d-c-L_3'+k_3}\, W(c\,c'\,L_3 L_3'; k_3\, d) \end{aligned}\right\} \tag{35.20b}$$

[cf. Eq. (35.8f)].

If the intermediate radiation ($\mathscr{L}_2$) is not observed, the correlation function for the first and third radiations is

$$W(\vartheta) = \sum_k A_k[1,3]\, P_k(\cos\vartheta), \tag{35.21}$$

where

$$A_k[1,3] = \sum_{b\,b'\,c\,c'} A_k(a, b\,b')\, \bar{A}_k(b\,b'\,c\,c')\, A_k(c\,c', d), \tag{35.22}$$

with

$$\bar{A}_k(b\,b'\,c\,c') = \frac{1}{4\pi\,\hat{c}^2} \sum_{L_2} \langle c \| \mathscr{L}_2 \| b\rangle \langle c' \| \mathscr{L}_2 \| b'\rangle^* \cdot W(b\,b'\,c\,c';\, k\,l_2)\, \hat{b}'\, \hat{c}\, (-)^{b'+c-L_2}. \tag{35.23}$$

If the above formalism is to be used for a double-cascade process, it should be borne in mind that the quantity $\bar{A}_k(b\,b'\,c\,c')$ in (35.23) reduces to $(-)^k/\hat{b}^2$, so that for double-cascade emission

$$W(\vartheta) = \sum_{k\,b\,b'\,c\,c'} A_k(a, b\,b')\, A_k(c\,c', d)\, \frac{(-)^k}{\hat{b}^2}\, \delta_{b\,c}\, \delta_{b'\,c'}\, P_k(\cos\vartheta). \tag{35.24}$$

If the first radiation is absorbed instead of emitted, then $A_k(a, b\,b')$ is replaced by $A'_k(a, b\,b')$, which is

$$A'_k(a, b\,b') = \sum_{\mathscr{L}_1 \mathscr{L}'_1} (-)^{a-b'-L'_1+k}\, \hat{b}' \langle a \| \mathscr{L}_1 \| b\rangle \langle a \| \mathscr{L}'_1 \| b'\rangle^*\, c^*_{k0}(\mathscr{L}'_1 \mathscr{L}_1) \cdot W(b\,b'\,L_1\,L'_1;\, k\,a) \tag{35.25}$$

[cf. Eq. (35.8e)]. In addition, the quantity $R_{k_1 k_2 k_3}$ is to be replaced by

$$R_{k_1 k_2 k_3} = \sum_{L_2 L'_2} \langle c \| \mathscr{L}_2 \| b\rangle \langle c' \| \mathscr{L}'_2 \| b'\rangle^*\, c^*_{k_2 0}(\mathscr{L}_2 \mathscr{L}'_2)\, \frac{\hat{b}^2\, \hat{b}'}{\hat{c}\, \hat{k}_3}\, (-)^{k_3-k_1} \begin{Bmatrix} c & c' & k_3 \\ L_2 & L'_2 & k_2 \\ b & b' & k_1 \end{Bmatrix}. \tag{35.26}$$

36. Interference between two overlapping levels. We shall discuss as an illustration of the theory of the previous section, and its application to the analysis of the unpolarised directional-correlation, reactions in which a compound nucleus is formed and can be represented by the superposition of two states. In this case the complication is limited, and it is possible to relate the measurements to the general theory, particularly if the angular distribution is studied at different energies so as to vary the contributions from the two levels.

The expression (35.9), applied to the case of two levels, labelled b, b', contains contributions of two sorts, one with factors of the type:

$$\sum_b \langle l_2\, S_2\, b | R | l_1\, S_1\, b\rangle \langle l'_2\, S_2\, b | R | l'_1\, S_1\, b\rangle^*, \tag{36.1}$$

and the other with factors

$$\sum_{b \neq b'} \langle l_2\, S_2\, b | R | l_1\, S_1\, b\rangle \langle l'_2\, S_2\, b' | R | l'_1\, S_1\, b'\rangle^*. \tag{36.2}$$

The first type of factor is the same as those which occur for an isolated resonance level. Using the results of WIGNER and EISENBUD [*43*], cf. Eq. (24.17), the contribution from a particular set of values $l_1 l'_1 l_2 l'_2$ can be expressed in the form [cf. (24.18)]:

$$\frac{g^{(b)}_{x_1 S_1 l_1}\, g^{(b)}_{x_2 S_2 l_2}\, g^{(b)}_{x_1 S_1 l'_1}\, g^{(b)}_{x_2 S_2 l'_2}}{[E_0(b) - E]^2 + \frac{1}{4}\Gamma^2(b)} \cos\left[\xi(x_1 l_1) + \xi(x_2 l_2) - \xi(x_1 l'_1) - \xi(x_2 l'_2)\right], \tag{36.3}$$

in which, as previously, the x's represent other properties of the particular channels, $E_0(b)$ and $\Gamma(b)$ characterise the levels b, and the g's are real.

In (36.2) we have the interference terms of the type:

$$\left.\begin{aligned}\sum\{[\langle l_2 S_2 b|R|l_1 S_1 b\rangle\langle l_2' S_2 b'|R|l_1' S_1 b'\rangle^* + \text{c.c.}] + \\ + [\langle l_2' S_2 b|R|l_1' S_1 b\rangle\langle l_2 S_2 b'|R|l_1 S_1 b'\rangle^* + \text{c.c.}]\},\end{aligned}\right\} \tag{36.4}$$

where c. c. signifies the complex conjugate of the expression which precedes it. Expressing the matrix elements of R, (24.17), in the form:

$$\langle l_2 S_2 b|R|l_1 S_1 b\rangle = f(x_1 l_1)\, f(x_2 l_2)\, \Phi(b,E)\, g^{(b)}_{x_1 S_1 l_1}\, g^{(b)}_{x_2 S_2 l_2}, \text{ etc.} \tag{36.5}$$

where the f's are phase factors, Φ is complex, and the g's are real[1], (36.4) becomes

$$\left.\begin{aligned}\sum\{[f(x_1 l_1)\, f(x_2 l_2)\, f^*(x_1 l_1')\, f^*(x_2 l_2')\, \Phi(b,E)\, \Phi^*(b' E) + \text{c.c.}] \times \\ \times\, g^{(b)}_{x_1 S_1 l_1} g^{(b)}_{x_2 S_2 l_2} g^{(b')}_{x_1 S_1 l_1'} g^{(b')}_{x_2 S_2 l_2'} + \\ + [f^*(x_1 l_1)\, f^*(x_2 l_2)\, f(x_1 l_1')\, f(x_2 l_2')\, \Phi^*(b' E)\, \Phi(b E) + \text{c.c.}] \times \\ \times\, g^{(b')}_{x_1 S_1 l_1} g^{(b')}_{x_2 S_2 l_2} g^{(b)}_{x_1 S_1 l_1'} g^{(b)}_{x_2 S_2 l_2}\}.\end{aligned}\right\} \tag{36.6}$$

The factors in the [...] can be written as

$$2\cos[\xi(x_1 l_1) + \xi(x_2 l_2) - \xi(x_1 l_1') - \xi(x_2 l_2') + \varphi(b) - \varphi(b')]\, a(E\, b\, b') \tag{36.7}$$

and

$$2\cos[\xi(x_1 l_1') + \xi(x_2 l_2') - \xi(x_1 l_1) - \xi(x_2 l_2) + \varphi(b) - \varphi(b')]\, a(E\, b\, b'),$$

where $\varphi(b)$, $\varphi(b')$, are the phase-factors for resonances b, b', and $a(E\ b\ b')$ is the real positive factor

$$[\{[E_0(b) - E]^2 + \tfrac{1}{4}\Gamma^2(b)\}\{[E_0(b') - E]^2 + \tfrac{1}{4}\Gamma^2(b')\}]^{-\frac{1}{2}}. \tag{36.8}$$

We see then that in addition to the phase differences relating different l, l' components (and the corresponding amplitude terms, the factors $g^{(b)}$, $g^{(b')}$), the additional phase angle $(\varphi(b) - \varphi(b'))$ enters into the correlation function. Near either resonance this phase angle changes rapidly with energy from approximately $-\pi/2$ for $\frac{(E_0 - E)}{\Gamma} \gg 1$ to $+\pi/2$ for $\frac{(E - E_0)}{\Gamma} \gg 1$.

From (35.9), (35.10), we see also that since the levels corresponding to b, b', may have opposite parities, and therefore even and odd values of l_1, l_1' and l_2, l_2' may occur, the interference terms can have odd as well as even k-values; but the terms with odd k can only be of the interference type. When such terms are present, they give rise to asymmetry about the $\vartheta = \pi/2$ plane, and this asymmetry is a pronounced function of energy.

For example, if only one pair of values l_1, l_2 is associated with b and one l_1', l_2' with b', then only the first terms in the first half of (36.4) are present. The complete angular distribution can be written:

$$\left.\begin{aligned} W(\vartheta) = \frac{\lambda\!\!\!^{-2}}{4\hat{a}^2\hat{s}_1^2}\sum_{S_1 S_2}(-)^{S_1 - S_2}\sum_k\Bigg\{&\frac{Z'(l_1 b\, l_1 b;\, S_1 k)\, Z'(l_2 b\, l_2 b;\, S_2 k)\, [g^{(b)}_{x_1 S_1 l_1} g^{(b)}_{x_2 S_2 l_2}]^2}{[E_0(b) - E]^2 + \frac{1}{4}\Gamma^2(b)} \\ &+ \frac{Z'(l_1' b'\, l_1' b';\, S_1 k)\, Z'(l_2' b'\, l_2' b';\, S_2 k)\, [g^{(b')}_{x_1 S_1 l_1'} g^{(b')}_{x_2 S_2 l'}]^2}{[E_0(b') - E]^2 + \frac{1}{4}\Gamma^2(b')} \\ &+ 2Z'(l_1 b\, l_1' b';\, S_1 k)\, Z'(l_2 b\, l_2' b';\, S_2 k)\, g^{(b)}_{x_1 S_1 l_1} g^{(b)}_{x_2 S_2 l_2} g^{(b')}_{x_1 S_1 l_1'} g^{(b')}_{x_2 S_2 l_2'} \times \\ &\times \frac{\cos[\xi(x_1 l_1') + \xi(x_2 l_2') - \xi(x_1 l_1) - \xi(x_2 l_2) + \varphi(b) - \varphi(b')]}{\{[E_0(b) - E]^2 + \frac{1}{4}\Gamma^2(b)\}^{\frac{1}{2}}\{[E_0(b') - E]^2 + \frac{1}{4}\Gamma^2(b')\}^{\frac{1}{2}}}\Bigg\}\, P_k(\cos\vartheta). \end{aligned}\right\} \tag{36.9}$$

[1] The expression (24.17) with real g's is only strictly applicable in the case of a single isolated resonance, and is not therefore, applicable to the interference terms involving significant contributions from two levels. The simplified treatment here assumes that the complexity of the g's is absorbed in the complex factors $\Phi(bE)$, which include the normal resonance phase shifts, $\arctan\dfrac{2[E - E_0(b)]}{\Gamma(b)}$.

The first terms are due to the separate levels, b, b', and the last term is the interference term (b, b'). Eq. (36.10) is applicable to a reaction with channel spins zero or $\frac{1}{2}$.

The summations in expressions such as (35.7) and (35.9) are limited by the usual conditions for the non-vanishing of the W- and Z-coefficients, and the conservation of parity. The complexity of the angular distribution is similarly limited (cf. Sect. 15). In practice there are the additional features of the Coulomb and centrifugal barriers which frequently reduce the contributions from high orbital angular momenta and so simplify the interpretation.

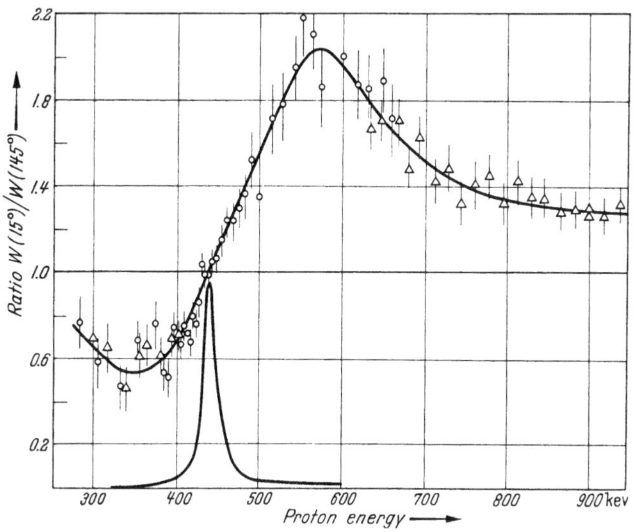

Fig. 16. Variation of asymmetry with proton energy in the angular distribution of γ-radiation from the reaction $Li^7(p, \gamma) Be^8$ [cf. S. DEVONS and M. G. N. HINE: Proc. Roy. Soc. Lond., Ser. A **199**, 56 (1949)].

Experimentally, the study of reactions with interfering levels of the compound nucleus has similar features to the study of simpler resonance reactions (Sects. 26 to 28), with the additional important factor that now the study of the variation of the distribution with energy plays a most important rôle. The level density for reactions in light nuclei (with moderate bombarding energy) is often sufficiently small for the interpretation, in terms of pairs of interfering levels, to be possible. Typical measurements of this sort have been made for (p, γ), (p, α), and (p, n) reactions[1,2,3]. Some results of the measurements for the reaction $Li^7(p, \gamma)$) are shown in Fig. 16, which shows the variation of the asymmetry, and hence of the interference term, over a wide range of energies. The results can be interpreted in terms of the superposition of a sharp level (at 0.44 Mev) of even parity interfering with a weak background produced by a distant level (or levels) of opposite parity. The rapid change in asymmetry near the sharp resonance is due to the phase change, $\Phi(b)$, in passing through the resonance. Due to the low amplitude of the interfering level the phase change can be followed over an energy range which is very wide compared with the level width.

---

[1] S. DEVONS and M. G. N. HINE: Proc. Roy. Soc. Lond., Ser. A **199**, 56 (1949). $Li^7(p, \gamma)$. H. E. GOVE and E. B. PAUL: Phys. Rev. **91**, 463 (1953). $B^{11}(p, \gamma)$.

[2] A. V. COHEN: Phil. Mag. **44**, 583 (1953). $O^{18}(p, \alpha)$. — D. M. THOMSON *et al.*: Proc. Phys. Soc. Lond. **65**, 745 (1952). $B^{11}(p, \alpha)$. — A. V. COHEN and A. P. FRENCH: Phil. Mag. **44**, 1259 (1953). $N^{15}(p, \alpha)$.

[3] R. K. ADAIR: Phys. Rev. **96**, 709 (1954). $Li^7(p, n)$.

In the analysis of reactions involving γ-radiation, we cannot, of course, use channel-spin formalism throughout. Either the $\mathscr{L}$-representation can be used, or, if one only stage in the process is γ-radiation, the interfering levels of the compound nucleus can be studied in the mixed representation.

37. Complex reactions. Apart from the detailed analysis of resonance reactions in terms of one or two overlapping levels, many experimental results are assessed on the basis of the qualitative nature of the angular distribution. The general differences between reactions proceeding through the formation of a compound nucleus and through direct processes have already been discussed. The interference terms of the type (36.4) depend, as can be seen from (36.10), on the relative phases of the interfering levels, and, where many levels overlap, some of these terms will give rise to positive and others to negative coefficients for the odd k-terms which produce asymmetry with respect to the $\vartheta=\pi/2$ plane. The non-interference terms will be simply additive and give rise to a large, average, isotropic component, together with some terms with even k, which may arise from both interference and non-interference terms. The sort of angular distributions to be expected under these conditions have been discussed by WOLFENSTEIN[1]. The interference terms are neglected, and the degree of anisotropy is estimated for the direct terms, using a statistical nuclear model for nuclear level densities. It is pointed out that when the residual nucleus has a small spin value, then the orbital momentum of the final system is closely related to the total angular momentum. The average of this latter component along the axis of the incident beam is zero, and hence that of the outgoing beam will be small, corresponding to angular distributions which have maximum values lying on the axis of the incident beam. The effect is naturally most pronounced when the maximum values of the outgoing orbital momenta are large (e.g., when energetic neutrons are produced).

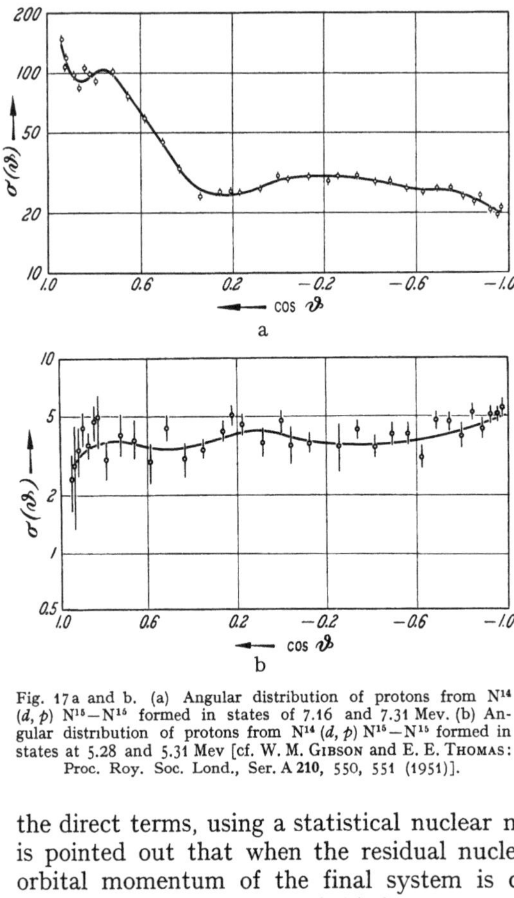

Fig. 17a and b. (a) Angular distribution of protons from N^{14} (d, p) $N^{15}-N^{16}$ formed in states of 7.16 and 7.31 Mev. (b) Angular distribution of protons from N^{14} (d, p) $N^{16}-N^{15}$ formed in states at 5.28 and 5.31 Mev [cf. W. M. GIBSON and E. E. THOMAS: Proc. Roy. Soc. Lond., Ser. A **210**, 550, 551 (1951)].

These considerations are frequently applied to reactions of the (d, p) and (d, n) type where both the direct "stripping" mechanisms and compound-nucleus formation are possible, and also occasionally to the converse process, "pickup" reactions of the type (n, d), (n, α), (d, H^3), (d, He^3). At low energies, particularly, the simple "stripping" and "pickup" processes may be seriously modified by large Coulomb-field effects, and the expected asymmetry may not

[1] L. WOLFENSTEIN: Phys. Rev. **82**, 690 (1951).

be very pronounced. In addition, the components with small orbital momentum may dominate in the formation of the compound nucleus, also on account of barrier factors. Nevertheless, even with the same incident particles and different outgoing groups of particles all having energies of the same order of magnitude, the angular distributions often appear to be of two types, one with marked asymmetry and the other approximately symmetrical in the $\pi/2$ plane, indicating that the direct stripping mechanism depends on the detailed characteristics of the nuclear states as well as more general factors[1]. One should mention, in this connection, that the absolute cross-section for stripping processes decreases rapidly with increasing orbital angular momentum, whereas the same trend is not necessarily present in compound-nucleus formation. The differences between "direct" and "compound-nucleus" processes with overlapping levels are illustrated in Fig. 17, 18.

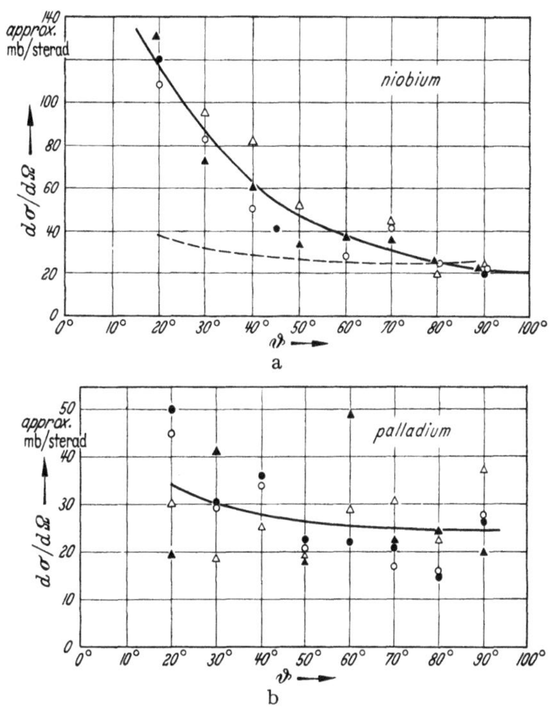

Fig. 18a and b. (a) Angular distribution of tritons from Nb^{93} (p, t) Nb^{91}. The dashed line reproduces the results shown in (b). (b) Angular distribution of tritons from the (p, t) reaction in palladium (natural isotopic abundance) [cf. B. L. Cohen and T. H. Handley: Phys. Rev. 93, 516 (1954)].

In the above discussion of nuclear reactions we have considered as the chief source of complexity, the nature of the intermediate, compound state. If we try to apply analysis of the same type to reactions in which two complex nuclear fragements are produced and where the individual states of the *residual* nuclei are not resolved, such as fission, then, in addition to the complexity of the compound nucleus we have to consider the complexity of the product nuclei. Consequently, we can only expect to explain, on the basis of the statistical model with conservation of angular momentum, very general trends. Even if it could be assumed that only a single level of the intermediate compound nucleus were predominant, the above type of analysis would still not lead to very precise conclusions. However, quite pronounced and systematic angular correlations have been observed in fission processes, and these have been successfully interpreted, at least semi-quantitatively, by A. Bohr[2], using arguments based on the collective model.

For example, photo-fission[3] of even-even nuclei (Th^{232}, U^{238}) shows an angular distribution of the type: $W(\vartheta) = 1 + A \sin^2 \vartheta$, with A largest near the fission

[1] E.g. B. L. Cohen and J. H. Handley: Phys. Rev. **93**, 514 (1954). (p, t) reaction. — N. T. S. Evans and W. C. Parkinson: Proc. Phys. Soc. Lond. A **67**, 684 (1954). $B^{10}(d, p)$. — W. M. Gibson and E. E. Thomas: Proc. Roy. Soc. Lond., Ser. A **210**, 543 (1951). (d, p) and (d, α) reactions.

[2] A. Bohr: Proceedings of the International Conference on the Peaceful Uses of Atomic Energy, Vol. 2, p. 151. New York: United Nations 1956.

[3] E. J. Winhold, P. T. Demos and I. Halpern: Phys. Rev. **87**, 1139 (1952); see also Ref. 2.

threshold and tending to zero at high γ-energies. This result is interpreted[1] in terms of electric-dipole ($E1$) absorption with the production of an intermediate "rotational" (1^-) state. The effective level spacing in the highly-deformed compound nucleus ("fission channel") is assumed to be quite large, since most of the excitation energy is absorbed as potential energy in the surface deformation. The resulting relatively simple wavefunction for the deformed nucleus leads to an angular distribution of the observed type, if it is assumed that the second stage of the fission process is sufficiently fast that little exchange of angular momentum occurs between the relative motion of the fission fragments and their internal degrees of freedom.

The level structure of the compound nucleus is less simple for odd-A nuclei and also, of course, for even-even nuclei at higher excitation energies, and the angular distributions are consequently less markedly anisotropic.

Fission induced by the absoption of fast neutrons and protons shows asymmetry of the type: $W(\vartheta) = 1 + A\cos^2\vartheta$, for example[2,3] with 14 Mev neutrons and $Th^{233,235,238}$ and Np^{237} where $0.2 < A < 0.5$, and with thorium and 23 Mev protons where $0.1 < A < 0.3$. This general trend is to be expected since, as in reactions previously discussed on a "statistical" basis, conservation of angular momentum results in outgoing particle wave-functions with small components of angular momentum along the direction of the ingoing beam and this leads to maxima in the distribution at $\vartheta = 0$ and π. For these large nuclei and nucleon energies, quite large angular-momentum quantum numbers are involved so that semi-classical approximations may also be used[1].

II. Elastic scattering.

38. Elastic scattering of uncharged particles. In dealing with the elastic scattering of particles by nuclei, we have a system in which many values of total angular momentum are in general involved, since in addition to any resonance processes there is also the "direct" scattering process which arises from the average nuclear potential, together with the Coulomb potential if the particles are charged. The representation of the interaction between ingoing (or outgoing) particles and the nucleus as a simple static potential (apart from the resonance effects) is of course only an approximation to the real state of affairs, but nevertheless over a limited range of energies it is a reasonable one. For moderate energies and small values of the orbital angular momentum, it is usual to regard the potential scattering as arising from a spherically symmetrical central force (e.g. "hard-sphere" or "clouded-crystal ball" scattering), i.e. spin-orbit coupling is neglected. In the usual channel-spin formalism, this potential scattering is assumed to be independent of channel-spin and the channel-spin is considered as a constant of motion.

We shall first illustrate the application of the formalism to scattering by dealing with the simple case of potential scattering. The reaction matrix is diagonal in l and S and the matrix elements are independent of S, i.e. we can write

$$\langle l_1 S_1 b | R | l_2 S_2 b \rangle = (e^{2i\zeta_{l_1}} - 1)\, \delta_{l_1 l_2}\, \delta_{S_1 S_2}, \tag{38.1}$$

where the ζ_l's are the usual "phase-shifts".

[1] See footnote 2, p. 463.

[2] W. C. DICKINSON and J. E. BROLLEY jr.: Phys. Rev. **90**, 308 (1953); **94**, 640 (1954).

[3] B. L. COHEN, W. H. JONES, G. H. McCORMICK and B. L. FERRELL: Phys. Rev. **94**, 625 (1954). — B. L. COHEN, B. L. FERRELL-BRYAN, D. J. COOBE and M. K. HOLLINGS: Phys. Rev. **98**, 685 (1955).

The general formula for the distribution (35.9) reduces to:

$$W(\vartheta) = \frac{\lambda^2}{4\hat{s}_1^2\hat{a}^2} \sum_{\substack{l l' b b' \\ S k}} \hat{l}^2 \hat{l}'^2 (l0, l'0 | k0)^2 \hat{b}^2 \hat{b}^2 W^2(l b l' b'; S k) \times \\ \times \operatorname{Re}\{(e^{2i\zeta_l} - 1)(e^{-2i\zeta_{l'}} - 1)\} P_k(\cos\vartheta). \qquad (38.2)$$

From the properties of the Racah coefficient:

$$\sum_{b b' S} \hat{b}^2 \hat{b}'^2 W^2(l b l' b'; S k) = \sum_{S b'} \frac{\hat{b}'^2}{\hat{l}'^2} = \sum_S \hat{S}^2, \qquad (38.3)$$

(since $\boldsymbol{b}' = \boldsymbol{l}' + \boldsymbol{S}$ and $\sum_{b'} \hat{b}'^2 = \hat{l}'^2 \cdot \hat{S}^2$).

Also $\boldsymbol{S} = \boldsymbol{a} + \boldsymbol{s}$, so that

$$\sum_{S b b'} \hat{b}^2 \hat{b}'^2 W^2(l b l' b'; S k) = \hat{a}^2 \hat{s}_1^2. \qquad (38.4)$$

Using (38.3), (38.4), in (38.2), we find

$$W(\vartheta) = \lambda^2 \sum_{l l' k} \hat{l}^2 \hat{l}'^2 (l0, l'0 | k0)^2 q(l l') \sin\zeta_l \sin\zeta_{l'} \cos(\zeta_l - \zeta_{l'}) P_k(\cos\vartheta), \quad (38.5)$$

where $q(ll') = {2 \atop 0}\}$ if $l \lessgtr l'$, and $q(ll) = 1$.

Although all the components of angular momentum do strictly participate in the elastic scattering, in practice the contributions decrease rapidly (i.e. ζ_l tend to zero) with increasing angular momentum, l, in the case of uncharged particles of moderate energy. The general problem of analysing the observed scattering in terms of phase-shifts, or elements of the R-matrix, and relating these to a particular form of interaction lies outside the scope of the present article. We are concerned here with the frequent use of elastic-scattering measurements as a means of establishing the properties of individual resonance levels of a compound nucleus, and we shall therefore discuss the problem of scattering where a single resonance level is present, as well as the potential scattering. In order to analyse the scattering in this context, it is not necessary to evaluate the potential scattering explicitly, but rather to determine the phase-shifts introduced by the potential. These can usually be obtained by study of the scattering in the neighbourhood of the resonance, at energies where the resonance scattering is much reduced, but the potential scattering is not greatly changed.

Detailed calculations, using channel-spin formalism, of the scattering angular-distribution for unpolarised particles have been given by BLATT and BIEDENHARN [6], and we give their results. Expressing the distribution in the form:

$$W(\vartheta) = \frac{\lambda^2}{\hat{a}^2 \hat{s}_1^2} \sum_k A_k P_k(\cos\vartheta), \qquad (38.6)$$

the coefficients A_k can be split into three terms

$$A_k = \sum_{S_1 S_2} [R_k(S_1 S_2) + \hat{a}^2 \hat{s}_1^2 \sum_{l l'} U_k(l l' \zeta_l \zeta_{l'}) + I_k(S_1 S_2)], \qquad (38.7)$$

where R_k denotes the resonance, U_k the potential contributions, and I_k the interference between resonance and potential scattering. R_k is given by (24.18), $x_1 = x_2$, and U_k, which is independent of S, is given in terms of the usual phase-shifts, ζ_l and $\zeta_{l'}$, as in (38.5) [apart from the factor $\lambda^2 P_k(\cos\vartheta)$, which is already in (39.6)]. The interference term is given by:

$$\left.\begin{aligned} I_k(S_1 S_2) = &-\hat{b}^2 \sum_{l=l_{\min}}^{b+S_{\max}} \sum_{l'=|l-k|}^{l+k} \hat{l}'^2 (l0, l'0 \mid k0)^2 \times \\ &\times \frac{\Gamma_{xl}}{\sqrt{(E-E_0)^2+\frac{1}{4}\Gamma^2}} \sin\zeta_{l'} \sin(\varphi + 2\zeta_l + \zeta_{l'}), \end{aligned}\right\} \qquad (38.8)$$

where φ is defined as in (36.8) and $\Gamma_{xSl} = [g_{xSl}]^2$ is the total level-width for emission into channel x with orbital momentum l, i.e.

$$\Gamma_{xl} = \sum_S \Gamma_{xSl}. \qquad (38.9)$$

Expression (38.6) refers to the total elastic scattering, with or without change in channel spin, and corresponds therefore to the normally measured scattering with unpolarised beams.

The summation over $S_1 S_2$ (for R) is for all possible channel spins, i.e. $|a - s_1| \leq S_1 \leq (a + s_1)$. The summation of k is from 0 to ∞, except for the resonance term where the usual limits are applicable. $l_{\min}$ is the smallest value of l that satisfies the triangle relation $\Delta(lSb)$, with S chosen to give the smallest l.

It should be emphasised that this treatment of elastic scattering is only approximate. We known that for nucleons of greater energy (> about 50 Mev the "potential" scattering produces strong polarisation effects. However, at these high energies we do not encounter discrete resonance phenomena, so that the above treatment is in any case inapplicable. The approximate treatment of potential scattering as channel-spin independent is probably reasonable for energies less than 30 Mev, but even at low energies polarisation due to scattering may be detected when resonance processes are present.

39. Elastic scattering of charged particles. Expression (38.6) can be applied to the scattering of charged particles if the phase-shifts appropriate to the combined Coulomb and nuclear potentials are used in place of the nuclear-potential phase-shifts. However, the Coulomb potential, even when allowance is made for atomic screening, is long-ranged, so that there are appreciable contributions from large l-values which decrease only slowly with increasing l. In practice the atomic screening can be ignored since it affects only the scattering at very small angles (which would otherwise tend to become infinite), and this is not usually measured. BLATT and BIEDENHARN [6] give results on the assumption that the nuclear potential is that corresponding to an impenetrable sphere, an approximation which is probably better for charged-particle scattering than uncharged with moderate energies. Coulomb wavefunctions are used in order to avoid the slow convergence when ordinary spherical waves are used. We give here only the results of the calculations for the angular distribution without polarisation. The distribution function is the sum of six parts:

1. The resonance scattering:

$$\frac{\lambda\!\!\!^{-2}}{\hat{a}^2 \hat{s}_1^2} \sum_k R_k(S, S)\, P_k(\cos\vartheta).$$

2. The "Rutherford" scattering:

$$Z^2 \operatorname{cosec}^4 \frac{\vartheta}{2}, \quad \text{with} \quad Z = \frac{z z' e^2}{2 M v^2}.$$

3. 4. Two terms representing the correction to the Rutherford scattering due to finite nuclear size (hard sphere):

$$\lambda\!\!\!^{-2}\sum_k\sum_{l=0}^{\infty}\sum_{l'}\hat{l}^2\hat{l}'^2(l0,l'0\,|\,k0)^2\sin\xi_l\sin\xi_{l'}\cos[(2\psi_l+\xi_l)-(2\psi_{l'}+\xi_{l'})]\,P_k(\cos\vartheta)-$$

$$-2\lambda\!\!\!^{-}Z\sum_{l=0}^{\infty}\hat{l}^2\sin\xi_l\cos\left[2\eta\log\sin\frac{\vartheta}{2}+2\psi_l+\xi_l\right]\operatorname{cosec}^2\frac{\vartheta}{2}\,P_l(\cos\vartheta),$$

where

$$e^{2i\xi_l}=\frac{G_l(R)-iF_l(R)}{G_l(R)+iF_l(R)};\quad e^{2i\psi_l}=\frac{(l+i\eta)(l-1+i\eta)\dots(1+i\eta)}{(l-i\eta)(e-1-i\eta)\dots(1-i\eta)},\quad \eta=\frac{zz'}{\hbar v},$$

and $F_l(R)$, $G_l(R)$ are regular and irregular Coulomb wavefunctions evaluated at the nuclear radius R.

5. The interference between the resonance and Rutherford scattering:

$$\frac{\lambda\!\!\!^{-}Z\hat{b}^2}{\hat{a}^2\hat{s}_1^2}\sum_{l_{\min}}^{b+s_1+a}\frac{\Gamma_{xl}}{[(E-E_0)^2+\Gamma^2/4]^{\frac{1}{2}}}\operatorname{cosec}^2\frac{\vartheta}{2}\times$$

$$\times\sin\left[2\eta\log\sin\frac{\vartheta}{2}+2\psi_l+2\xi_l+\varphi\right]P_l(\cos\vartheta).$$

6. The correction to 5 due to finite nuclear size:

$$-\frac{\lambda\!\!\!^{-2}\hat{b}^2}{\hat{a}^2\hat{s}_1^2}\sum_{k=0}^{\infty}\sum_{l_{\min}}^{b+s_1+a}\sum_{l'}\hat{l}'^2(l0,l'0\,|\,k0)^2\sin\xi_{l'}\times$$

$$\times\frac{\Gamma_{xl}}{\sqrt{(E-E_0)^2+\Gamma^2/4}}\sin(\varphi+2\psi_l+2\xi_l-2\psi_{l'}-\xi_{l'})\,P_k(\cos\vartheta).$$

Terms 2, 3 and 4, are relatively slowly varying with energy, since they do not involve the characteristics of the resonance level.

The result simplifies considerably for zero-spin nuclei or particles, but all six contributions from all terms are present.

40. Scattering of neutrons: Experimental. Since the width Γ_{xl} for neutrons is not reduced by any Coulomb-barrier, it is only with neutrons of low energy and nuclei of small level density that discrete resonance phenomena can be studied. Otherwise, many measurements of the scattering of neutrons have been made to examine the average interaction between neutron and nucleus[1]. Large level spacings occur for light nuclei, and especially for nuclei of the even-even type and closed shell nuclei (e.g. He^4, C^{12}, O^{16}), and many such nuclei have been studied using neutrons of energies up to a few Mev obtained from (p, n) and (d, d) reactions.

A particular feature of scattering by these exceptionally stable nuclei is the absence of competing reactions or inelastic scattering for the neutron energies of interest. This permits a convenient method of studying the scattering by the nuclei by observing the ionisation produced by the recoil-nuclei when the scattering material is in the form of a gas in a cloud chamber or proportional counter[2]. In the latter case the size of the ionisation pulse due to the recoil nucleus is a direct measure of the scattering angle. Some measurements of the scattering of neutrons by O^{16} are illustrated in Fig. 19. In this case three levels in O^{17}, at energies 4.56, 5.08, and 5.39 Mev were examined using neutron energies from

[1] H. Feshbach, C. E. Porter and V. F. Weisskopf: Phys. Rev. **96**, 448 (1954). This paper contains many references to earlier theoretical and experimental work.

[2] E.g. R. K. Adair: Phys. Rev. **92**, 1491 (1953). O^{16}. — J. L. Fowler and C. H. Johnson: Phys. Rev. **98**, 728 (1955). N^{14}.

0.4 to 1.4 Mev. Only the S-wave ($l=0$) phase-shift due to potential scattering was significant for neutrons of these low energies.

In the scattering of neutrons of 2.4 to 3.6 Mev by C^{12}, an interesting result was observed by MEIER *et al.*[1]. S-wave scattering was observed to be in agreement with hard-sphere scattering, but the p-wave phase shifts indicated two interfering resonances, both of the type $^3/_2{}^+$, but differing in width by a factor 20.

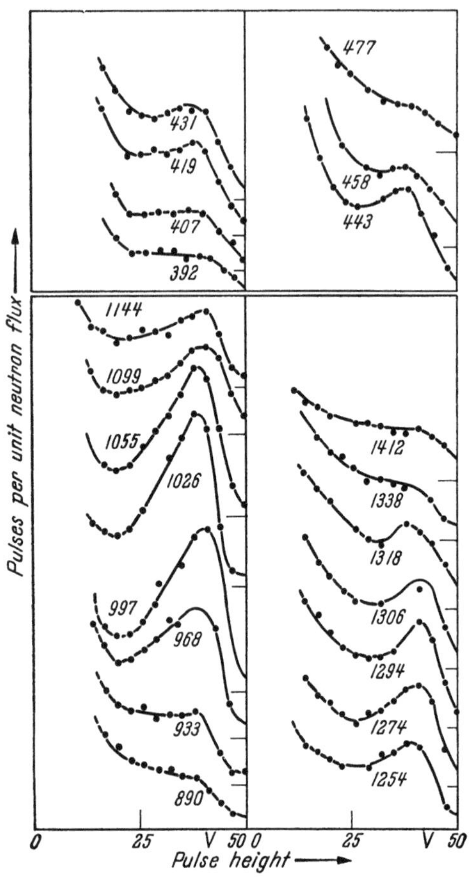

Fig. 19. Recoil-ion energy distribution resulting from bombardment of CO_2 with mono-energetic neutrons with energies from 392 to 1412 kev. The maximum energy oxygen recoil at each neutron energy produced a pulse of ~ 45 v [cf. R. K. ADAIR: Phys. Rev. **92**, 1492, Fig. 1 (1952)].

It should be pointed out that from the measured phase-shifts, as a function of energy, the scattering matrix elements and hence the polarisation as a function of energy and angle can also be deduced. In this connection it should also be noticed that using neutrons produced in a particular direction from a reaction, as for example from the (d, d) reaction in the above measurements, some significant polarisation may be present in the neutron beam used for the scattering experiments, in which case the scattering will be modified and the angular distribution will depend not only on the scattering angle ϑ but also on the azimuthal angle. Such effects were examined by MEIER *et al.* and are also discussed below (Sect. 42).

In general, scattering measurements with neutrons are not capable of high precision. Corrections must be made for finite angular resolution, multiple scattering, and variation of detector sensitivity, etc. Nevertheless, the absence of the complications due to Coulomb fields is a substantial compensation.

41. Charged-particle scattering: Experimental. The scattering of protons and α-particles, and in one or two cases deuterons, has been extensively used in the study of the discrete resonance levels of light nuclei. A high degree of energy homogeneity and canalisation of the incident beam is readily attainable with present-day techniques, and to utilise these fully the scattering must be observed from "thin" layers of the scattering material. There are several methods of achieving these conditions of high resolution which are usually necessary if the resonance scattering is not be swamped by the Rutherford scattering. Gas targets[2] are commonly used, but of course there are limitations in the choice of suitable material, and also in the absence of unwanted nuclei. Scattering foils can be used, and if there are other nuclei present the unwanted scattering

[1] R. W. MEIER, P. SCHERRER and C. TRUMPY: Helv. phys. Acta **27**, 577 (1954).

[2] For recent examples of this technique see: A. GALONSKY, R. A. DOUGLAS, W. HAEBERLI, M. T. MCELLISTRAM and H. T. RICHARDS: Phys. Rev. **98**, 586, 590 (1955). — T. LAURITSEN, T. HUUS and S. G. NILSSON: Phys. Rev. **92**, 1501 (1953).

can be separated by spectrometric analysis of the scattered particles. In this way, also, scattering from the surface layers of a thick scatterer, or from a thin deposit on a thick support, can be isolated[1]. A typical experimental arrangement for use with a gas target is shown in Fig. 20[2], and the result of a typical measurement, the scattering of α-particles of energies 2 to 5 Mev, by O^{16} in Fig. 21.

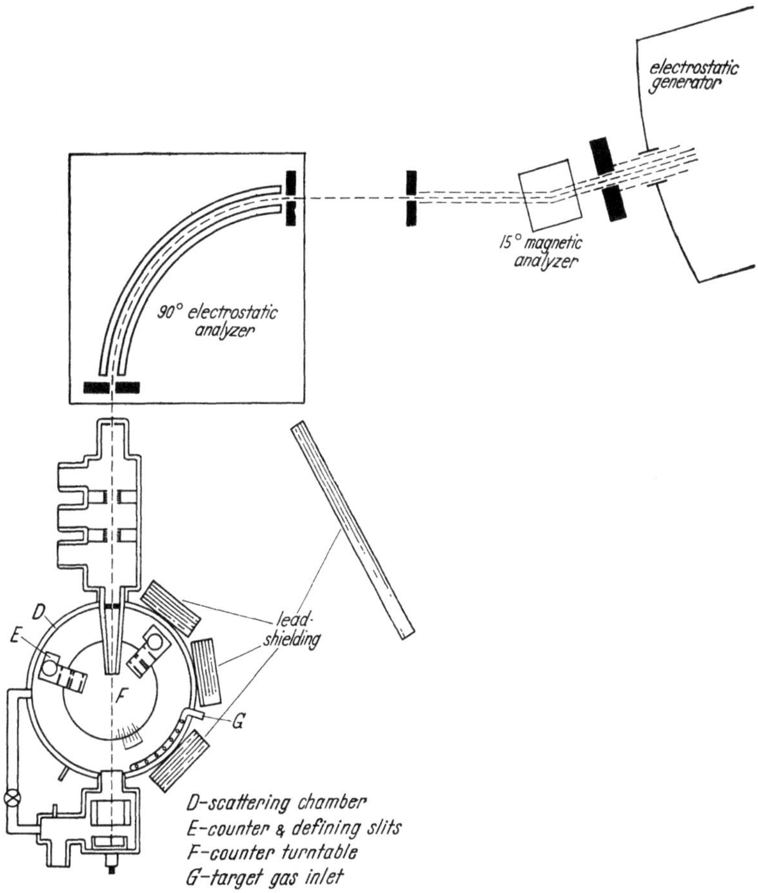

Fig. 20. Schematic diagram showing arrangement for measuring elastic scattering of charged particles using gas targets [cf. J. R. CAMERON: Phys. Rev. 90, 839 (1953)].

The interpretation of the scattering of α-particles and spinless nuclei is relatively simple. For each isolated resonance there can be only one l-value ($l = b$), and then only for levels of appropriate parity. Moreover, with spinless, light nuclei, the conditions are such that at the energies in question no competing reactions (including inelastic scattering) are usually possible, so that the resonance scattering is represented by simple phase-shifts, without absorption, and a single phase-shift $[\xi_l + \varphi(b)]$ is sufficient for each resonance. Using the same assumptions and notation as in Sect. 39, i.e. assuming "hard-sphere" scattering plus scattering by narrow resonance levels, the scattering cross-section can in this

[1] For example, W. D. WARTERS, W. A. FOWLER and C. C. LAURITSEN: Phys. Rev. **91**, 917 (1953). — A. A. KRAUS: Phys. Rev. **93**, 1308 (1954). — G. DEARNALEY: Phil. Mag. **45**, 1213 (1954).

[2] J. R. CAMERON: Phys. Rev. **90**, 839 (1953).

case be written more simply as [1,2]:

$$\left.\begin{aligned} W(\vartheta) = \lambda\!\!\!^{-2} \Big| \frac{\eta}{2} \operatorname{cosec}^2 \frac{\vartheta}{2} \exp\cdot\left(i\eta \log \operatorname{cosec}^2 \frac{\vartheta}{2}\right) - \\ - \sum_l (2l+1)\, P_l(\cos\vartheta)\cdot\cos\left(\xi_l + \varphi(b)\right) \exp\left[i\left(\xi_l + \varphi(b)\right)\right] \exp 2i\psi_l \Big|^2 . \end{aligned}\right\} \quad (41.1)$$

The radius of the "hard-sphere" scattering is chosen so as to give the best fit with the experimental data at energies away from the resonances, and in this

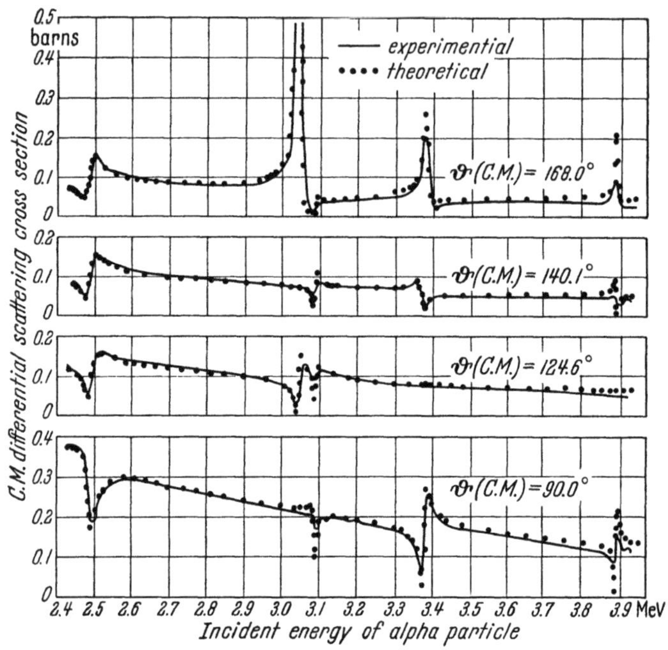

Fig. 21. Elastic scattering of α-particles by oxygen [cf. J. R. Cameron: Phys. Rev. 90, 843 (1953)].

way the potential phase-shifts are determined. The resonance phase-shifts $\varphi(b)$ can be determined by graphical methods [2].

The analysis can be helped by choosing a set of scattering angles for which each of the terms $P_1(\cos\vartheta)$, $P_2(\cos\vartheta)$ ..., in turn vanishes. For example in Fig. 21 the pair of close levels at α-particle energies of 3.045 and 3.090 Mev can be studied separately this way. In the absence of competing processes, the parameters describing the resonance (α-particle width, or total width), and resonance energy can be estimated even though the level width is too small for direct determination. The general agreement of theory and experiment is shown in the figure.

Proton scattering by spinless nuclei can be analysed in a similar manner, provided no competing process occurs, since in this case also each level can only be produced by a single l-component. With protons and nuclei of non-zero spin, the situation is usually more complicated, since apart from the possibility of more than one l-value, competing reactions are usually possible (p, n; p, α; etc.). If the width of an isolated resonance is sufficiently large so as to be directly measurable, then in this case the full characteristics of the resonance, as far as

[1] R. A. Laubenstein and M. J. W. Laubenstein: Phys. Rev. **84**, 18 (1951).

[2] H. L. Jackson and A. I. Galonsky: Phys. Rev. **84**, 401 (1951); **89**, 370 (1953).

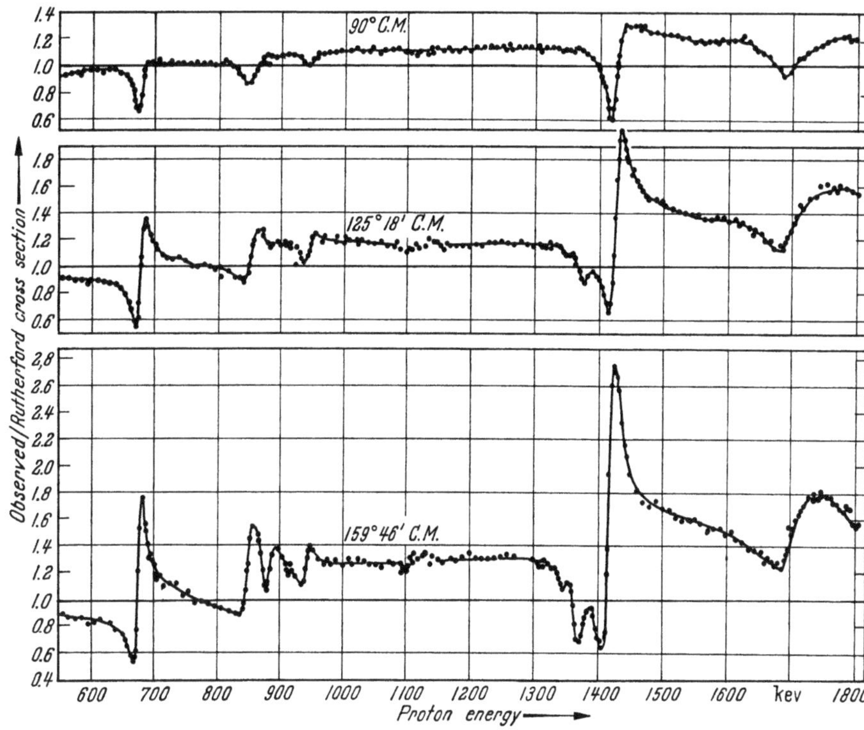

Fig. 22. Elastic scattering of protons by F^{19} [cf. T. S. WEBB *et al.*: Phys. Rev. 99, 138 (1955)].

scattering is concerned, can be determined. Fig. 22 illustrates some measurements[1] of the level spectrum of Ne^{20} by means of the scattering of protons by F^{19}.

For charged particles of much higher energy, of the order 20 Mev or more, the scattering distributions are no longer markedly dependent on the detailed characteristics of the particular nucleus, but depend rather on the average nuclear interaction and the interference between this and the Coulomb scattering. The interpretation in terms of phase-shifts (Sect. 39) is still of course possible, but many of these are now appreciable over a wide energy range and not just one or two near the energies associated with a particular resonance.

A typical variation of $W(\vartheta)$ for 22 Mev α-particles is shown in Fig. 23[2].

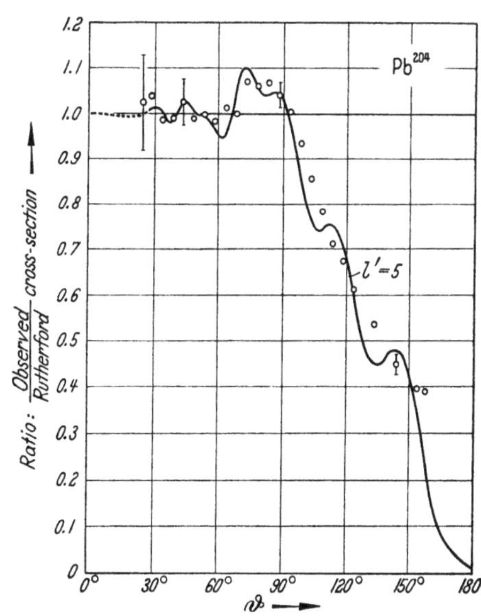

Fig. 23. Elastic scattering of 22 Mev α-particles by Pb. Theoretical curve is based upon a nuclear model with finite surface thickness [cf. N. S. WAHL *et al.*: Phys. Rev. 97, 728 (1955)].

42. Polarisation in reactions and scattering. For both elastic scattering and

[1] T. S. WEBB, F. B. HAGEWORN, W. A. FOWLER and C. C. LAURITSEN: Phys. Rev. **99**, 138 (1955).

[2] N. S. WALL *et al.*: Phys. Rev. **97**, 726 (1955).

reactions involving overlapping levels, we have seen, in the previous sections, that generally some polarisation occurs due to the interference between different levels or between these and the potential scattering. It has also been pointed out (Sect. 30) that the change in the angular correlation due to introduction of polarisation can be treated by analogous methods for the cases of a beam prepared in a given state of polarisation, and the case of a detector which has corresponding polarisation-sensitive properties. Any reaction or scattering process which produces polarisation may, therefore, also be employed as the basis of a polarisation-sensitive detector.

We consider the double process indicated in Fig. 24, and restrict our considerations to spin-$\frac{1}{2}$ particles. We consider the process to be defined by the plane of the paper and represent the first and second reaction or scattering processes by (1) and (2) respectively. The incident beam, which is unpolarised, is designated by $\mathfrak{k}_1$ and the outgoing beam of process (1) by $\mathfrak{k}_2$. A and B represent alternative positions of a polarisation-insensitive detector.

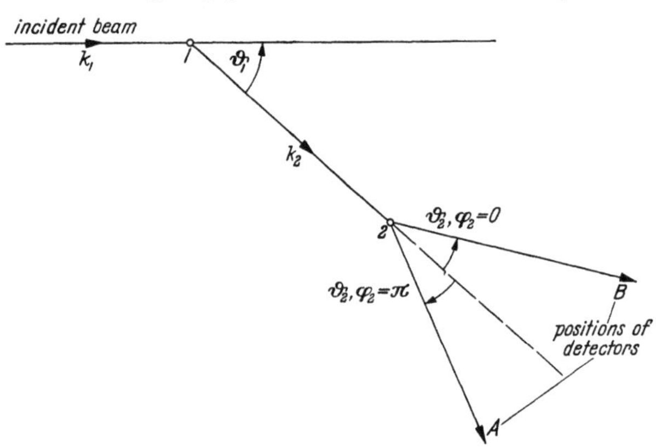

Fig. 24. Schematic arrangement for polarisation measurements in a double-scattering experiment.

Suppose that the particles $\mathfrak{k}_2$ were in a pure state with spin along the direction $\boldsymbol{n}=\mathfrak{k}_1\times\mathfrak{k}_2$. Then, applying the relationship (30.10) or (30.14) to the reaction (2), we can obtain correlation functions, for polarisation-insensitive detectors, of the form

$$W_{\substack{A\\B}}(\vartheta_2) \sim C_2(\vartheta_2) \pm D_2(\vartheta_2), \tag{42.1}$$

where the two signs refer to the two alternative detector positions, A and B, and C_2 and D_2, characterise the process (2).

If the particles $\mathfrak{k}_2$ were a combination of a fraction P polarised, and $(1-P)$ unpolarised, then the correlation would be (Sect. 30):

$$W(\vartheta_2) \sim C_2(\vartheta_2) \pm P D_2(\vartheta_2), \tag{42.2}$$

so that the ratio of response of the counter in positions A, B, provides a measure of the polarisation P, if the ratio $\frac{D_2(\vartheta_2)}{C_2(\vartheta_2)}$ is known.

In a similar way we can consider the detection of the particles $\mathfrak{k}_2$ in a reaction (1) by means of an ideal (unit polarisation-efficiency) polarisation-sensitive detector, with its axis directed along $\boldsymbol{n}$. For the angular distribution in this process (1), we can write similarly

$$W(\vartheta_1 \pm 1) \sim C_1(\vartheta_1) \pm D_1(\vartheta_1), \tag{42.3}$$

where the $\pm$ sign now refers to parallel and anti-parallel orientations of the ideal detector. We can therefore represent the radiation $\mathfrak{k}_2$ resulting from the process (1) at ϑ_1 as comprising amounts $C_1(\vartheta_1)+D_1(\vartheta_1)$ in the direction $\boldsymbol{n}$ and $C_1(\vartheta_1)-D_1(\vartheta_1)$ in the direction $-\boldsymbol{n}$, or, equivalently, as a fraction $D_1(\vartheta_1)/C_1(\vartheta_1)$ which is polarised, and $[1-D_1(\vartheta_1)/C_1(\vartheta_1)]$ which is unpolarised. We can therefore write in place of (42.2) the overall correlation in terms of the properties of processes (1) and (2):

$$W_{\substack{A\\B}}(\vartheta_1\,\vartheta_2) \sim 1 \pm \frac{D_1(\vartheta_1)}{C_1(\vartheta_1)}\,\frac{D_2(\vartheta_2)}{C_2(\vartheta_2)}\,. \tag{42.4}$$

The ratios $\left(\frac{D_1}{C_1}\cdot\frac{D_2}{C_2}\right)$ can be interpreted as the fractional polarisation produced by process (1) and (2) respectively at the corresponding angles[1].

Double correlations and measurements of this type have been used frequently to measure polarisation effects, particularly in elementary scattering processes. We consider here only a typical application to the measurement of the properties of a nuclear resonance level.

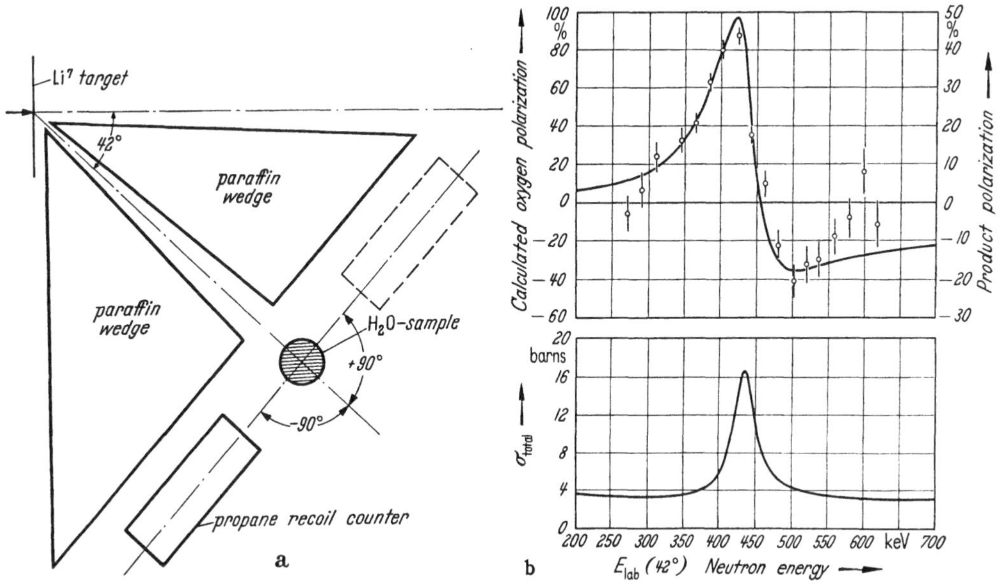

Fig. 25a and b. (a) Experimental arrangement for measuring polarisation arising from the Li^7 (p, n) Be^7 reaction. (b) The lower curve shows the total cross-section for neutrons elastically scattered by O^{16}. The upper curve using the left ordinate shows the theoretical polarisation for neutrons scattered through 90° by O^{16} and is compared with the measured values of $\frac{D_1(42°)}{C_1(42°)}\cdot\frac{D_2(90°)}{C_2(90°)}$ which are indicated by use of the right ordinate [cf. H. B. Willard *et al.*: Phys. Rev. 95, 1360 (1954)].

The experimental arrangements and results used in an experiment involving (1) a neutron-producing reaction, and (2) the elastic scattering of these neutrons, are shown in Fig. 25[2].

The first reaction was Li^7 (p, n) Be^7, which has a resonance for a proton energy of about 2.2 Mev, and polarisation is produced by interference with broad levels at higher or lower energies. The second process was the elastic scattering by O^{16}. There is a sharp resonance for neutron energies of 435 kev, which is due to the $l=1$, $L=\frac{3}{2}$ component ($P_{\frac{3}{2}}$ level in O^{17}). There is interference between this level and the predominantly S-wave "potential" scattering. The lower curve in Fig. 25b shows the variation in total cross-section for the scattering of O^{16}, and the upper curve the value of both, calculated using the phase-shifts obtained from the measurements and the analysis of the elastic scattering (cf. Sect. 38). Comparison of this latter with the measured quantity $\frac{D_1(42°)}{C_1(42°)}\cdot\frac{D_2(\pi/2)}{C_2(\pi/2)}$, also shown in Fig. 25, enables the value

[1] J. V. Lepore: Phys. Rev. 79, 137 (1950).
[2] H. B. Willard, J. K. Blair and J. D. Kington: Phys. Rev. 95, 1359 (1954).

of D_1/C_1 to be deduced. In this case the polarisation from the reaction Li^7 (p, n) Be^7 appears approximately constant at about $\frac{1}{2}$ over the energy range of about 0.25 Mev.

Similar measurements have been made with neutrons produced in other reactions, e.g. $H^2(\alpha, n)$ He^3 reaction, and the scattering in C^{12} has also been employed as the second polarisation-detecting process[1]. The polarisation of the neutron produced in the $H^2(\alpha, n)$ He^3 reaction can be compared with the proton polarisation—also measured[2] by a second scattering process—produced in the mirror reaction $H^2(\alpha, p)$ H^3.

F. Internal conversion electrons.

43. Theory. The characteristics of the nuclear levels and γ-radiations involved in a cascade or reaction may, in principle, be examined equally by correlation studies of the internal-conversion electrons as well as the γ-radiations. There are obvious advantages of such a procedure for heavy elements or γ-transitions of low energy where the probability of internal conversion may be far greater than that of direct γ-radiation. Moreover, the directional correlation of the conversion electrons depends on the parity of the transitions in contrast to the situation for γ-radiation. In general the extent of the correlation with conversion electrons is comparable with that for γ-radiation. There are, however, complications in the practical application of this method. Firstly, electrons, and particularly low-energy electrons with which one is most likely to be concerned, are readily scattered in the source and in neighbouring material, so that the accurate observation of the true correlation imposes severe conditions on the preparation of the source (or target). Secondly, the correlation is dependent on the atomic shell in which conversion occurs, so that the particular conversion electron-group must be isolated. This is again easiest in the case of heavy elements and low-energy transitions, for which the fractional difference between the energies of K, L, etc. conversion electrons is greatest. If the conversion electrons arise in the first stage of a cascade process, then there is the added complication that the atomic state that prevails when the second radiation occurs is undergoing rearrangement consequent on the removal of the K-electron, so that unknown perturbations of the correlation may be present.

From the theoretical point of view the interpretation of internal conversion involves not only nuclear and radiation states but also the atomic states with which the virtual radiation emitted by the nucleus interacts. The correlation depends not only on the symmetry properties of the nuclear levels and radiation, but also the nuclear charge Z and electron energy. The electron-radiation interaction is described in terms of the matrix elements for the transition of the electrons from the original bound state to the plane-wave state in which they are detected by the action of a multipole electromagnetic field. In dealing with the states of the electron, it is necessary, for the general case, to use relativistic wavefunctions to describe its motion. General approximate treatments of the process have been given[3]; a complete relativistic treatment and numerical results was first given by ROSE, BIEDENHARN, and ARFKEN[4], whose theory we outline

[1] R. W. MEIER, P. SCHERRER and G. TRUMPY: Helv. phys. Acta **27**, 577 (1954). — BUDDE and P. HUBER: Helv. phys. Acta **28**, 49 (1955).

[2] G. R. BISHOP, G. PRESTON, J. M. WESTHEAD and H. H. HALBAN: Nature, Lond. **170**, 113 (1952).

[3] J. W. GARDNER: Proc. Phys. Soc. Lond. A **62**, 763 (1949); A **64**, 218, 1136 (1951)

[4] M. E. ROSE, L. C. BIEDENHARN and G. B. ARFKEN: Phys. Rev. **85**, 5 (1952); also [9], p 752.

below. The detailed calculations are limited to K-conversion, and we shall also restrict our discussion to this. It might be mentioned that for many transitions in the heavy elements, the γ-energy is insufficient for K-conversion, and in others this energy is so small as to make correlation measurements with K-conversion electrons extremely difficult.

We may regard the detection of the conversion-electron as a particular, indirect, form of γ-detection, and from this point of view the problem is to find the tensor-parameters of the γ-radiation for this type of "detector". These tensor parameters may then be used in the general correlation formula (10.16) to give the correlation of the conversion-electrons with any other process (including another conversion-electron). For the sake of definiteness we refer here to a cascade-correlation in which the first radiation is a conversion-electron.

This first stage, radiation and internal conversion, is described by matrix elements of the form[1]:

$$\sum_{\mathscr{L}_1 \Lambda_1} \langle \Omega\sigma | Q | \mathscr{L}_1 \Lambda_1 L_e = \tfrac{1}{2} \rangle \langle \mathscr{L}_1 \Lambda_1 b | H | a \rangle. \tag{43.1}$$

The first factors represent the electronic transition from an initial state, $L_e = \frac{1}{2}$, to a final state, $|\Omega\sigma\rangle$, by the absorption of a virtual photon with quantum numbers $\mathscr{L}_1$ (angular momentum and parity). The second factors describe the emission of this virtual photon in the nuclear transition from state a to state b. The statistical tensor for the whole system, nucleus plus electron, differs from that corresponding to double-cascade emission (11.17) by additional multiplicative factors, $\langle \| Q \| \rangle \langle \| Q \| \rangle^*$ and $\varrho_{k_e \varkappa_e}$, where the latter is equal to $\frac{1}{\sqrt{2}} \delta_{k_e 0} \delta_{\varkappa_e 0}$, since the K-electrons are not observed in their initial state.

The efficiency tensor for the whole system is constructed in a manner similar to (11.23), except that $\varepsilon_{k_1 \varkappa_1}(L_1 L_1')$ which represents the efficiency tensor for the observation of the virtual radiation is further coupled to $\varepsilon_{k_e \varkappa_e}(\frac{1}{2}\,\frac{1}{2})$, by the relation:

$$\varepsilon_{k_1 \varkappa_1}(L_1 L_1')\, \varepsilon_{k_e \varkappa_e}(\tfrac{1}{2}\,\tfrac{1}{2}) = \sum_{k_f \varkappa_f} \varepsilon_{k_f \varkappa_f}(L_f L_f')\,(k_1 \varkappa_1, k_e \varkappa_e | k_f \varkappa_f) \cdot \hat{k}_1 \hat{k}_e \hat{L}_f \hat{L}_f' \begin{Bmatrix} L_1 & \frac{1}{2} & L_f \\ L_1' & \frac{1}{2} & L_f' \\ k_1 & k_e & k_f \end{Bmatrix}. \tag{43.2}$$

Since the evaluation of the angular correlation requires a summation over k_e and $\varkappa_e$, we can replace these quantities by zero in (43.2) and obtain for the angular correlation an expression such as (11.34) except that $\varepsilon_{k\varkappa}(L_1 L_1')$ is replaced by

$$\tfrac{1}{2}\, \varepsilon_{k\varkappa}(L_f L_f')\, \hat{L}_f \hat{L}_f'\, W(L_1 L_1' L_f L_f'; k\, \tfrac{1}{2})\, (-)^{k + \frac{1}{2} - L_1' - L_f} \tag{43.3}$$

and there occurs the product of the reduced matrix elements for the electronic transition. The suffix f refers to the state of the outgoing electron.

The evaluation of the efficiency tensor $\varepsilon_{k_1 \varkappa_1}(L_f L_f')$ is made in a manner similar to that used for other radiations which are detected as plane waves, except that here, Dirac plane-wave spinors are used. If no measurement of polarisation is involved, and Coulomb fields are neglected, then we find [9, Eq. (8.7)]:

$$\left.\begin{aligned} \varepsilon_{k\varkappa}(L_f L_f') &= i^{l_f(K') - l_f(K)} (-)^{k_f + L_f' - \frac{1}{2}} \hat{l}_f \hat{l}_f' \sqrt{|K K'|}\, S(K)\, S(K') \times \\ &\times (l_f 0, l_f' 0 | k_f 0)\, W(l_f l_f' L_f L_f'; k\, \tfrac{1}{2})\, \mathscr{D}^k_{\varkappa 0}(\mathscr{R}_1). \end{aligned}\right\} \tag{43.4}$$

We discard the suffix 1 which refers to the first radiation, since our results apply equally to either the first or the second radiation. We use the quantum number K, whose sign, $S(K)$, and magnitude, $|K|$, determine L_f, l_f, and the parity, thus:

$$L_f = |K| - \tfrac{1}{2}; \quad l_f = |K| + \tfrac{1}{2}[S(K) - 1]; \quad \text{parity} = (-)^{l(K)+1}, \tag{43.5}$$

[1] N. Tralli and G. Goertzel: Phys. Rev. **83**, 399 (1951).

and, of course, $L_f = l_f \pm \frac{1}{2}$. The initial state of the electron is described by $K = -1$ (S-state).

The angular correlation for conversion-electrons, and pure multipoles $L = L'$, can be obtained then from the correlation for radiation by substituting, for the radiation parameters $c_{k0}(LL)$, the quantity $c_{k0}(\mathcal{L}\mathcal{L})$,

$$\left.\begin{aligned} c_{k0}(\mathcal{L}\mathcal{L}) = \sum_{KK'} & \hat{l}_f \hat{l}'_f \,|K K'|\, (-)^L\, i^{l_f(K') - l_f(K)}\, e^{i(\xi_K - \xi_{K'})} \times \\ & \times (l_f 0, l'_f 0 \,|\, k 0)\, S(K)\, S(K')\, W(l_f l'_f L_f L'_f; k\tfrac{1}{2}) \times \\ & \times W(L L L_f L_f; k\tfrac{1}{2}) \cdot \langle K \| \mathcal{L} \| -1 \rangle \cdot \langle K' \| \mathcal{L} \| -1 \rangle^* . \end{aligned}\right\} \tag{43.6}$$

Here, $\xi_K, \xi_{K'}$, are the phase-shifts in the electron wavefunctions introduced by the (screened) Coulomb field which was neglected in (43.4)[1]. The reduced matrix elements $\langle K \| \mathcal{L} \| -1 \rangle$ etc. describe the electron transition from state -1 to state K, by the radiation $\mathcal{L}$. [These correspond to the quantities $Q(KL\sigma)$ defined by ROSE *et al.*[2].] Since we are dealing with states of well-defined parity, k assumes even values only. Also, for pure multipoles, $K = L$ or $-L-1$, and $l(K) = L$ for both values of K. $c_{k0}(\mathcal{L}\mathcal{L})$ is obtained therefore by evaluating the Racah-coefficients with $K = L$ or $-L-1$:

$$c_{k0}(\mathcal{L}\mathcal{L}) = c_{k0}(L L; \gamma)\, b_k(\mathcal{L}\mathcal{L})\, c(\mathcal{L}). \tag{43.7}$$

$c(\mathcal{L})$ is the usual internal conversion-coefficient for L-multipole conversion-electrons, i.e. for either EL or ML multipole electrons, depending on the parity of $\mathcal{L}$. The factors $b_k(\mathcal{L}\mathcal{L})$ are given by:

$$b_k(\mathcal{L}\mathcal{L}; \text{electric}) = 1 + \frac{k(k+1)}{2L(L+1) - k(k+1)}\, \frac{L}{2L+1}\, \frac{|L+1+T_e|^2}{L(L+1) + |T_e|^2}, \tag{43.8}$$

$$b_k(\mathcal{L}\mathcal{L}; \text{magnetic}) = 1 + \frac{k(k+1)}{2L(L+1) - k(k+1)}\, \frac{L(L+1)}{2L+1}\, \frac{|1 - T_m|^2}{L+1+L|T_m|^2} \tag{43.9}$$

where

$$T_e \equiv L \frac{e^{i\delta_L} \langle L \| \mathcal{L} \| -1 \rangle}{e^{i\delta_{-L-1}} \langle -L-1 \| \mathcal{L} \| -1 \rangle}; \qquad T_m \equiv -\frac{L+1}{L}\, \frac{e^{i\delta_{L+1}} \langle L+1 \| \mathcal{L} \| -1 \rangle}{e^{i\delta_{-L}} \langle -L \| \mathcal{L} \| -1 \rangle}$$

and

$$\delta_K = \xi_K + \frac{\pi}{2}\left[\frac{1 + S(K)}{2} - |K|\right]$$

$b_k(\mathcal{L}\mathcal{L})$ is normalised in (43.8), (43.9), so that $b_0 = 1$. It is convenient to tabulate the values of $b_k(\mathcal{L}\mathcal{L})$ in terms of $b_2(\mathcal{L}\mathcal{L})$. This is done by means of the relationship

$$b_k(\mathcal{L}\mathcal{L}) = 1 + \frac{k(k+1)\,[L(L+1) - 3]}{3\,[2L(L+1) - k(k+1)]}\,[b_2(\mathcal{L}\mathcal{L}) - 1]. \tag{43.10}$$

$b_2(\mathcal{L}\mathcal{L})$ is a function of Z (atomic number), the electron energy, and $\mathcal{L}$.[3] The results for mixed transitions may be illustrated by the case of the commonly

---

[1] Although our convention is to introduce the dependence on the Coulomb field in the reduced matrix elements (Sect. 13), we here conform with the procedures of BIEDENHARN and ROSE and include them in the expressions for the radiation parameters. This is done in order that the reader may use the tabulated values of the particle parameters $b_k(LL)$, (43.8), (43.9), which are defined in terms of the Coulomb phase-shifts.

[2] M. E. ROSE, G. H. GOERTZEL, B. I. SPINRAD, J. HARR and P. STRONG: Phys. Rev. **83**, 79 (1951). — M. E. ROSE, L. C. BIEDENHARN and G. B. ARFKEN: Phys. Rev. **85**, 5 (1952).

[3] Tabulated values of b_2 for 12 values of Z from 10 to 96, for $\mathcal{L}$ from 1 to 5 both electric and magnetic, and several values of electron energy from $0.3\,mc^2$ to $10\,mc^2$ are given by BIEDENHARN and ROSE [9].

occurring $ML + E(L+1)$. In this case we have $K(\text{magn}) = -L, L+1$; $K(\text{elec}) = L+1, -L-2$, and we find:

$$\left.\begin{aligned} b_k(L_{\text{magn}}, L+1_{\text{elec}}) = \sqrt{\frac{L(L+2)}{(2L+1)(2L+3)}} \times \\ \times \frac{\operatorname{Re}\left[e^{i(\eta_e-\eta_m)}\left(1+\frac{L+1}{L T_m}\right)^*\left(1-\frac{L+1}{T_e}\right)\right]}{\sqrt{\left(1+\frac{L+1}{L|T_m|^2}\right)\left(1+\frac{(L+2)(L+1)}{|T_e|^2}\right)}}. \end{aligned}\right\} \tag{43.11}$$

η_e, η_m, represent the phases of the reduced matrix elements for internal conversion, for electric and magnetic multipole radiation, respectively:

$$e^{i\eta_m} = \frac{\langle L+1\|L\|-1\rangle}{|\langle L+1\|L\|-1\rangle|}, \qquad e^{i\eta_e} = \frac{\langle L+1\|L+1\|-1\rangle}{|\langle L+1)\|L+1\|-1\rangle|}.$$

The correlation function for such a mixed transition followed by a pure multipole $\mathscr{L}_2$ can be written as

$$\left.\begin{aligned} W(\vartheta) = \sum_k \big[c(\mathscr{L}_1)\, b_k(\mathscr{L}_1\mathscr{L}_1)\, F_k(L_1 L_1 a b) + \delta^2 c(\mathscr{L}_1+1) \times \\ \times b_k(\overline{\mathscr{L}_1+1}\;\overline{\mathscr{L}_1+1})\, F_k(\overline{L_1+1}\;\overline{L_1+1}\, a b) + \\ + 2\delta\sqrt{c(\mathscr{L}_1)\, c(\mathscr{L}_1+1)}\, b(\mathscr{L}_1\overline{\mathscr{L}_1+1})\, F_k(L_1\overline{L_1+1}\, a b)\big] \times \\ \times F_k(L_2 L_2 c b)\, P_k(\cos\vartheta) \end{aligned}\right\} \tag{43.12}$$

where $\mathscr{L}_1$ and $\mathscr{L}_1+1$ refer to magnetic and electric-type radiations respectively. δ is the usual mixing ratio, as defined for γ-emission in Eq. (19.5). It refers to the reduced matrix elements for emission of γ-radiation only, not for the transitions as a whole. We note that the factor $b(\mathscr{L}_1\overline{\mathscr{L}_1+1})$ in the interference term is independent of k (43.11). As in the case of mixed γ-radiations, the presence of the interference term implies that the correlation function is more sensitive to mixtures than the overall internal-conversion coefficient which involves only $c(\mathscr{L}_1)$, $c(\mathscr{L}_1+1)$ and δ^2. In addition, the correlation functions, either of γ-radiation or conversion-electron, offer the possibility of measuring the sign of δ.

At high energies ($\gg mc^2$) the values of $b_k(\mathscr{L}\mathscr{L})$ approach unity asymptotically, i.e. the correlation functions are the same for electrons and γ-radiation, both of which, for pure transitions, depend only on the initial and final spins and L. For mixtures of all electric and all magnetic multipoles, this asymptotic behaviour still obtains, but for mixed electric and magnetic transitions, the sign of the interference term is opposite to that for γ-radiation.

In the low-energy, non-relativistic limit (energy $\ll mc^2$), spin-orbit coupling in the electron motion disappears, and pure electric-transitions are Z-independent, except for the coefficient $c(\mathscr{L})$. We then have

$$b_k(\mathscr{L}\mathscr{L})_{\text{elec}} = \frac{2L(L+1)}{2L(L+1) - k(k+1)}. \tag{43.13}$$

For magnetic transitions the spin-orbit interaction of the electron can never be ignored; both the coefficients $b_k(\mathscr{L}\mathscr{L})_{\text{magn}}$ and the mixed coefficients are dependent on Z and energy, even in the non-relativistic limit.

44. Internal conversion: Experimental. For all transitions other than dipole, b_2 is greater than unity and positive, although for higher values of k, b_k may be negative and less than unity. In the region of transition energies of interest, say less than 1 Mev, $b_2(E1)$ is usually negative and numerically of the order unity; $b_2(M1)$ usually positive and much smaller than unity. For comparable information, then, the necessary accuracy of measurement is of the same order for both γ-radiation and conversion-electron correlations. The two distinctive properties of conversion-electron measurement are (a) the stringent requirements for the conditions at and near the source to avoid scattering, (b) the need to separate K-conversion electrons from those in other atomic shells. The first of these is most important for low-energy and the second for higher-energy transitions.

The scattering is kept to a minimum by using sources of high specific activity mounted on thin foils. Sources of thickness ranging from a few hundred $\mu g/cm^2$ to less than one $\mu g/cm^2$ for low-energy electrons, have been used[1] with supporting foils in the region $100-10$ $\mu g/cm^2$.

The need for thin sources may present special difficulties in the case of transitions which are sufficiently slow to be effected by extra-nuclear interaction. For these the material of the source may also have to satisfy the requirements for minimum perturbation of the correlation, and the two simultaneous requirements are not easily met. Liquid sources, in general, show less extra-nuclear perturbation than solid materials, and thin, evaporated (~ 100 mg/cm^2), liquid sources of carrier-free Zn^{111} in gallium have been used successfully in electron-correlation measurements[2]. Measurements were made with this source (γ-cascade in Cd^{111}, 0.173 Mev, 0.247 Mev) for all four types of correlation, γ-γ, γ-e, e-γ, and e-e, and the absence of extra-nuclear perturbation was verified.

Scattering of electrons between source and electron-detectors is usually avoided by placing both in vacuum or displacing the intervening air by hydrogen gas[3,4]. No complete treatment of the scattering problem has been given, but some estimates have been made[5].

Several techniques have been used for resolving the particular electron group corresponding to conversion in the K-shell. These include deflection-type β-spectrometers[6], scintillation counters using thin organic phosphors[7], and cloud chambers[8]. In many cases only partial resolution of the K and L electrons is achieved, but by selecting the appropriate part of the experimental spectrum the K-conversion component can be isolated.

The effects of extra-nuclear perturbation, arising from the excitation of the atomic electrons by internal conversion, are illustrated in some recent measurements of cascades in Ta^{181} and Hg^{197} (see Fig. 26).

In the case of 164 kev to 133 kev cascade in Hg^{197}, angular correlations have been measured for K, L and M electrons followed by γ-radiation with sources of various physico-chemical composition[9]. The correlation was found to be

---

[1] E.g. E. BREITENBERGER: Proc. Phys. Soc. Lond. A **69**, 453(1956).

[2] F. GIMMI, E. HEER and P. SCHERRER: Z. Physik **138**, 394 (1954). — J. Phys. Radium **16**, 600 (1955).

[3] E.g. H. FRAUENFELDER, M. WALTER and W. ZÜNTI: Phys. Rev. **77**, 557 (1950).

[4] F. K. McGOWAN: Phys. Rev. **92**, 523 (1953).

[5] M. WALTER, O. HUBER and W. ZÜNTI: Helv. Phys. Acta **23**, 697 (1950); also S. FRANKEL: Phys. Rev. **83**, 673 (1951).

[6] D. T. STEVENSON and M. DEUTSCH: Phys. Rev. **83**, 2202 (1951). — K. SIEGBAHN: Ark. Fysik **2**, Nr. 10 (1952). — N. GOLDBERG and S. FRANKEL: Phys. Rev. **93**, 1425 (1954). — F. GIMMI, E. HEER and P. SCHERRER: Z. Physik **138**, 394 (1954).

[7] E. BREITENBERGER: Proc. Phys. Soc. Lond. A **69**, 453, 566 (1956). — F. K. McGOWAN: Phys. Rev. **92**, 523 (1953).

[8] R. R. ROY and M. L. GOES: C. R. Acad. Sci., Paris **238**, 581 (1954) (correlation with preceding α-emission).

[9] H. H. COBURN and S. FRANKEL: Phys. Rev. **99**, 671 (1955); see also E. BREITENBERGER: Proc. Phys. Soc. Lond. A **69**, 566 (1956).

insensitive to environment, and not appreciably dependent on the time delay between the detection of the two radiations (cf. Sect. 50). Since the actual correlation was much less than the theoretical, unperturbed, value, it was deduced that perturbations of short duration arise from the disturbance to the electronic surroundings. The interaction between the intermediate nucleus and the disturbed electrons is probably magnetic.

Expressing the correlation function as $W(\vartheta) = 1 + G_2 A_2 b_2 P_2(\cos\vartheta) + G_4 A_4 b_4 P_4(\cos\vartheta)$, where G_2, G_4, are the attenuation factors due to extra-nuclear effects (cf. Sect. 51), and b_2, b_4, are defined as in Eqs. (43.9) etc., GIMMI *et al.*[1] found the values $G_2 A_2 b_2 = 0.102 \pm 0.005$ and $G_4 A_4 b_4 = 0 \pm 0.007$. Combined with the information about the K-conversion and K/L ratio for the 133 kev transition, which indicates that this transition is predominantly $E\,2$, this result led to the conclusion that the cascade was: ${}^{13}/_2 \xrightarrow{(M4)} {}^{5}/_2 \xrightarrow{(E2)} {}^{1}/_2$ and that G_2 was about $^1/_2$ for the K-electron-γ-correlation.

Similar conclusions were obtained from a comparison of the γ-γ, K-electron-γ, and γ-K-electron angular correlations in the 132 kev to 480 kev cascade in Ta^{181} (Fig. 26b)[3]. Here the γ-γ correlation indicates very small extra-nuclear perturbation effects, whereas the K-electron-γ-correlation again indicates a value of G_2 about $\frac{1}{2}$. The difference between the two correlations clearly indicates that the attenuation of the angular correlation is due, in this case also, to the disturbance to the atomic electrons caused by the internal-conversion process.

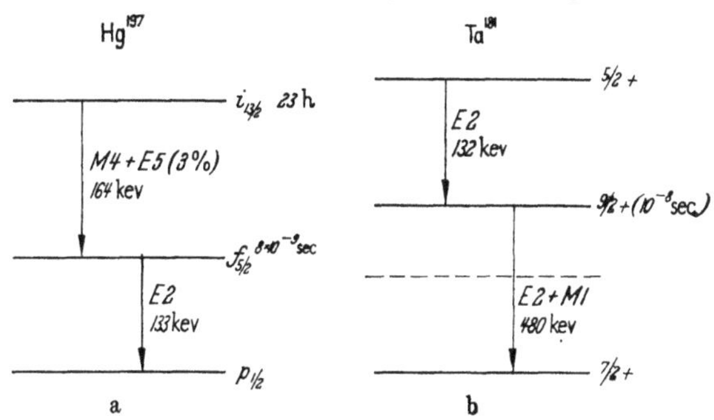

Fig. 26a and b. (a) γ-transitions in Hg^{197}. (b) γ-transitions in Ta^{181}

G. Processes involving simultaneous emission or absorption of two radiations.

In this Chapter we shall be concerned with processes in which the transitions occurring between well-defined nuclear states involve more than one single radiation, for example two emitted or one absorbed and one emitted radiation. The angular-momentum change in the process is shared between the two radiations, so that each separate radiation is no longer associated with a unique angular momentum, or even, necessarily, with a small limited range of angular momenta.

For such processes two sorts of angular correlations may be studied. Firstly, there is the correlation between the two simultaneous radiations themselves, and secondly, the correlation between some feature of this process considered as a simple entity, and some preceding or succeeding radiation occurring between well-defined nuclear states. Such studies are useful in connection both with the determination of the spins and parities of the states involved, and also with the elucidation of the detailed mechanism responsible for the process involving two radiations. In respect of the latter type of investigation, we are concerned with the particular details of many different types of nuclear process rather

[1] F. GIMMI, E. HEER and P. SCHERRER: Helv. phys. Acta 28, 470 (1955).

than with the general features of angular correlation. Hence we shall discuss such processes only briefly, our main purpose being to indicate the rôle they play in the investigation of the symmetry properties of nuclear states.

I. Nuclear and electromagnetic interactions.

45. Stripping reactions. The characteristic features of deuteron reactions can be attributed to the loose binding of the neutron and proton which form the deuteron, and to the associated large spatial extent of the deuteron. In contrast with the collision between a nucleus and a single nucleon, a deuteron collision may involve predominantly only one of its constituents, the probability of the other constituent reaching the nuclear surface being negligible. The reaction is not characterised, therefore, by an even approximately well-defined quantum-mechanical state of the deuteron plus nucleus as a whole. On the contrary, the angular momentum of the "non-interacting" nucleon, and thus of the whole system, can take a wide range of values extending far beyond the values characteristic of compound-nucleus-type reactions of comparable energy. When analysed in terms of states having a particular value of angular momentum for the whole system, deuteron reactions are then quite complicated and have little in common, from the viewpoint of angular correlations, with cascade emissions and resonance reactions.

The original experimental observation[1] of angular distributions involving high-order Legendre polynomials clearly demonstrated this implication. However, the fact that in stripping reactions only one nucleon interacts appreciably with the target nucleus leads to a different type of theoretical simplification in which the *linear* momenta of the particles play an important rôle. With deuterons of moderate energy, where discrete levels of the final nucleus are involved, this simplification leads to direct relationships between the angular distributions and the *angular* momenta transferred in the stripping process. Thus stripping reactions, like the reactions and angular correlations already described, provide a powerful tool in the study of the symmetry properties of nuclear levels. Moreover, the relationships between the *linear* momenta in the stripping process lead to the result that the theory of the angular correlation between the particles produced in stripping and any subsequent radiation is very closely related to the theory of resonance reactions proceeding through states of well-defined angular momentum.

We shall discuss here stripping reactions mainly from the point of view of determination of nuclear symmetry-properties; but we shall, first consider the mechanisms and approximations for a wide range of deuteron energies, since this will provide some insight into the interpretation of stripping at those energies which are most appropriate to our purpose[2].

We may distinguish roughly three ranges of deuteron energy for which the stripping reaction has qualitatively different features. For deuteron energies ($E_D =$ a few Mev or less), which are small compared with the electrostatic potential at the nuclear surface, $V(R)$, we have the possibility that the neutron of the incident deuteron may penetrate to the nuclear surface, although the proton or the deuteron as a whole has only a small probability of doing so. As a

[1] A. C. HELMHOLTZ, E. M. MCMILLAN and D. C. SAVILL: Phys. Rev. **72**, 1003 (1947). — H. B. BURROWS, W. M. GIBSON and J. ROTBLAT: Phys. Rev. **80**, 1095 (1950).

[2] A comprehensive theoretical treatment of the stripping reaction is given by G. BREIT in Vol. XLI. A more detailed account of particular reactions is contained in Sect. 52 of W. E. BURCHAM's article in Vol. XL. See also [*13*].

result, the cross-section for the (d, p) reaction is enhanced, as compared with what would be expected on the basis of compound-nucleus formation ("Oppenheimer-Philips" process). At the relevant energies, both the ingoing deuteron and outgoing proton wavefunctions may be strongly affected by Coulomb forces. The simple relationships between linear momenta, referred to abovc, are obscured by this feature, as well as by the added complication due to possible compound-nucleus formation, so that simple angular-correlation features are not prominent.

At the other extreme of high deuteron-energy, of the order 100 Mev or more, where the collision energy is much greater than the deuteron binding energy and Coulomb forces can be neglected, the collision time is much smaller than the characteristic period of the internal deuteron motion. For a large nucleus SERBER [90] shows that this condition is given by $E_D > \frac{2\varepsilon_D R}{a}$, where ε_D and a represent the binding energy and average n-p separation of the deuteron, and R the nuclear radius. Either component of the deuteron may interact with the nucleus leaving the other component with, approximately, the same velocity that it possessed just before the collision occurred. The momentum of the "stripped" nucleon is then compounded of that due to translation of the deuteron as a whole, $\frac{1}{2}(2M_D E_D)^{\frac{1}{2}}$ and the internal momentum, with an average component normal to the translational motion of the order $(M\varepsilon_D)^{\frac{1}{2}}$, ($M$ is the nucleon mass). On this basis SERBER was able to explain the observed angular spread of stripped nucleons, the mean angle with respect to the deuteron beam being approximately $\vartheta_D = (\varepsilon_D/E_D)^{\frac{1}{2}}$. ϑ_D is about 6° for $E_D = 180$ Mev. The angular distribution was examined more quantitatively by considering the probability $\Pi(\mathfrak{K}_p)$ that the proton [we consider the (d, p) reaction for the sake of definiteness] has the momentum $\mathfrak{K}_p$ due to the internal motion of the deuteron. This is given by

$$\Pi(\mathfrak{K}_p) = \frac{1}{(2\pi)^3} \left| \int e^{-i\mathfrak{K}_p \cdot \boldsymbol{r}} \varphi_D(\boldsymbol{r})\, d\boldsymbol{r} \right|^2, \tag{45.1}$$

where $\varphi_D(\boldsymbol{r})$ is the internal deuteron wavefunction, $\boldsymbol{r} = \boldsymbol{r}_n - \boldsymbol{r}_p$, and $\mathfrak{K}_p$ is associated with the momentum of the stripped-off proton by

$$\mathfrak{k}_p = \mathfrak{K}_p + \tfrac{1}{2}\mathfrak{k}_D. \tag{45.2}$$

A simple S-wavefunction can be adopted for φ_D and the spins of the nucleons are neglected. If it is further assumed that the probability of simultaneous interaction of the emitted protons with the nucleus absorbing the neutron is negligible at these energies (i.e. an assumed transparent nucleus), then (45.1), weighted by $\mathfrak{k}_p^2 d\mathfrak{k}_p$ and integrated over $\mathfrak{k}_p$ (which is approximately constant and equal to $\frac{1}{2}\mathfrak{k}_D$), gives the approximate angular distribution as:

$$\sigma(\vartheta) \sim \frac{\vartheta_D}{2\pi(\vartheta^2 + \vartheta_D^2)^{\frac{3}{2}}}, \tag{45.3}$$

which is highly peaked in the forward direction. This angular distribution is not sensitive to the assumed interaction between target nucleus and proton. Serber calculates the distribution for the case of completely absorbing nucleus by limiting the integration (45.1) to the region for which only the neutron is inside the target nucleus and the proton outside. The resulting angular distribution is only slightly different from (45.3). The effect of Coulomb forces, which result in a small difference between (d, p) and (d, n) distributions, is taken into account by classical considerations of the deflections suffered by ingoing deuteron and outgoing protons.

A feature of very high-energy stripping is, as indicated above, the absence of any detailed description of the process for which one of the nucleons is absorbed. This absorption is observed to take place equally for all nucleons which cross the nuclear surface irrespective of their exact energy or angular momentum. When we turn to the third energy region, for which the deuteron (and proton) energies are sufficiently large that Coulomb forces can be considered as secondary effects, but still the energies are not so large that the discrete energy-level structure of the residual nucleus can be ignored, the process of nuclear capture must be treated more exactly. If the stripping reaction leaves a product nucleus in a well-defined state (either bound or virtual), then the consequences of conservation of angular momentum and energy must be included in the considerations. The qualitative effect of these factors can be seen from a simple classical argument [15]. We again treat the (d, p) reaction for definiteness, neglect Coulomb forces, and assume, as before, that only the neutron interacts directly with the target nucleus. The effect on the proton is indirect through the neutron-proton interaction which determines the deuteron wavefunction. The reaction is denoted by $a+d \to b+p$, where a, b, denote target and product nuclei (or their spins) respectively. b is assumed to be formed by the absorption of neutrons with the transfer of definite angular momentum, l_n.

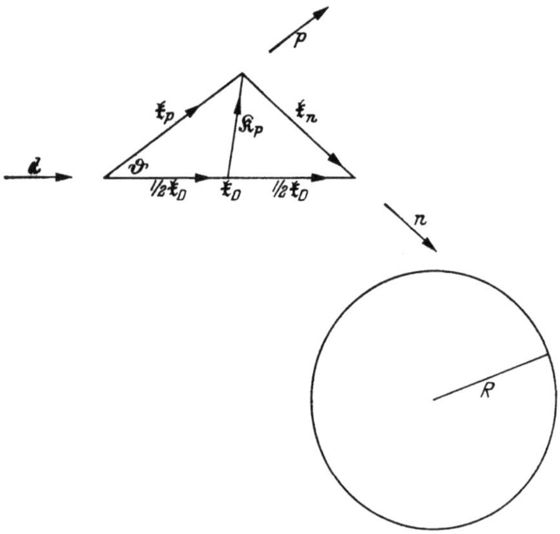

Fig. 27. Momentum relations in deuteron-stripping reactions (cf. R. HUBY [13], Fig. 3).

If the proton is observed with (linear) momentum $\mathfrak{k}_p$, then, as in (45.2), there is a corresponding contribution $\mathfrak{K}_p$ from the internal motion of the deuteron: $\mathfrak{K}_p = \mathfrak{k}_p - \frac{1}{2}\mathfrak{k}_D$. If ϑ is the angle between $\mathfrak{k}_p$ and $\mathfrak{k}_D$, then[1] (Fig. 27):

$$\mathfrak{K}_p = \sqrt{\left|\mathfrak{k}_p - \frac{\mathfrak{k}_D}{2}\right|^2 + 2\,\mathfrak{k}_p\,\mathfrak{k}_D \sin^2\frac{\vartheta}{2}}\,. \tag{45.4}$$

For this value of $\mathfrak{k}_p$, the neutron momentum, $\mathfrak{k}_n$, is determined by

$$\mathfrak{k}_n = \mathfrak{k}_D - \mathfrak{k}_p, \tag{45.5}$$

$$\mathfrak{k}_n = \sqrt{|\mathfrak{k}_D - \mathfrak{k}_p|^2 + 4\,\mathfrak{k}_p\,\mathfrak{k}_D \sin^2\frac{\vartheta}{2}}\,. \tag{45.6}$$

Now $\mathfrak{k}_n$ is restricted by the requirements of angular momentum to values satisfying

$$\mathfrak{k}_n R \geq \hbar\, l_n, \tag{45.7}$$

if it is assumed that absorption of the neutron takes place only for impact parameters less than R. Conditions (45.7) and (45.6) specify a possible range of values of ϑ. For each value of ϑ there will be, according to (45.4), a corresponding value of $\mathfrak{K}_p$; but the relative probability of different $\mathfrak{K}_p$ values is determined by the

[1] We neglect, for the present, the recoil motion of the target nucleus.

deuteron wavefunction, as in (45.1), so that for each value of l_n there will be corresponding range of values of ϑ, the direction of emission of the proton. The final differential cross-section may be considered determined by three factors: (i) the factor $\Pi(\mathfrak{K}_p)$ which determines the relative probability of the internal momentum possessing the value $\mathfrak{K}_p$, (ii) a factor dependent on l_n and $\mathfrak{k}_n$ (and therefore $\mathfrak{K}_p$), expressing the probability that the neutron with momentum $\mathfrak{k}_n$ appears at the nuclear surface R with the appropriate angular momentum l_n, and (iii) a factor involving the familiar reduced matrix element, statistical factors, and Clebsch-Gordan coefficients which describe the probability that the neutron wave with its associated values of l_n and $\mathfrak{k}_n$, is absorbed by the target nucleus a to form the product nucleus b. It should be noticed that in considering the process of absorption of the neutron, the normal relation between momentum and energy is not applicable since this absorption is not of a physically free neutron. The energy of the proton is determined by the conservation of energy for the overall process.

The first quantum-mechanical treatment of stripping at intermediate energies was given by BUTLER [*91*]. Subsequently many variations of the theory have been given, all of which contain the same basic feature (the interaction of one nucleon only in the transfer of momentum), but differ in the assumptions and approximations made in other respects [*92*] to [*94*]. The wavefunction for the system for $r_n > R$ is taken to be

$$\Phi \sim e^{i\mathfrak{k}_D \cdot (\boldsymbol{r}_n + \boldsymbol{r}_p)} \varphi_D(\boldsymbol{r}_n - \boldsymbol{r}_p)\, \chi_D\, \psi_{a\alpha} \tag{45.8}$$

where χ_D is the spin wavefunction for the deuteron and $\psi_{a\alpha}$ the target-nucleus wavefunction. For $r_n < R$, the wavefunction is compounded of terms each of which is the product of the proton wavefunction corresponding to a plane-wave and that of the compound-state formed by the absorption of the neutron by the nucleus. The two wavefunctions must be joined smoothly at the boundary $r_n = R$. The incident wave can be analysed into Fourier components corresponding to different outgoing values of $\mathfrak{k}_p$, with the following result:

$$\Phi \sim \int d\mathfrak{k}_p\, e^{i\mathfrak{k}_p \cdot \boldsymbol{r}_p}\, G(\mathfrak{K})\, e^{i\mathfrak{k}_n \cdot \boldsymbol{r}_n}\, \chi_D\, \psi_{a\alpha} \tag{45.9}$$

where $G(\mathfrak{K})$, like $\Pi(\mathfrak{K})$ introduced in (45.1), determines the probability that there should be the appropriate momentum contributions from the internal deuteron motion. From (45.9) we can see that the probability amplitude for the production of a proton with a particular $\mathfrak{k}_p$, described by $e^{i\mathfrak{k}_p \cdot \boldsymbol{r}_p}$, contains, *inter alia*, a factor which is the probability that a neutron plane-wave incident on the nucleus $(a\,\alpha)$, together described by the wavefunction $e^{i\mathfrak{k}_n \cdot \boldsymbol{r}_n}\psi_{a\alpha}$, should result in a product compound nucleus $(b\,\beta)$[1]. This is just the result to be expected from the momentum relationships displayed in Fig. 27.

The interaction between the neutron and the target nucleus has been considered in varying degrees of approximation. The simplest treatments use a Born-approximation (e.g. [*92*]), although in BUTLER's original analysis [*91*] a more exact wave-fitting procedure, involving the wavefunctions of the product nucleus b, was adopted. The resultant stripping cross-section has the form:

$$\frac{d\sigma(\mathfrak{k}_p)}{d\Omega} \sim G(\mathfrak{K}_p) \sum_{l_n} \Gamma_{l_n} |\mathcal{J}_{l_n}(\mathfrak{k}_n R)|^2, \tag{45.10}$$

where Γ_{l_n} is the "reduced width" for the emission of a neutron with angular momentum l_n, by the product nucleus in state b and the function $\mathcal{J}$ depends on

[1] The deuteron spin wavefunction is omitted since its inclusion does not materially affect these considerations (cf. DALITZ [*94*]).

the particular approximation used. [In the Born approximation, $\mathcal{J}_{l_n} = j_{l_n}(\mathfrak{k}_n R)$, where j_{l_n} is a spherical Bessel function.] $\mathfrak{K}_p$, $\mathfrak{k}_p$, and $\mathfrak{k}_n$, and the direction, ϑ, of the proton, are related by Eqs. (45.2), (45.4), and (45.6).

As indicated in (45.10), the contribution to the differential cross-sections of different l_n-components of the captured neutron wave are added incoherently, so that the results of stripping theory can be presented in terms of the separate distributions for each l_n-component. Typical results, obtained by BUTLER [91], are illustrated in Fig. 28.

The main feature of the stripping cross-sections is the appearance of pronounced maxima in the cross-sections at angles which generally increase with increasing l_n. For $l_n = 0$, the principle maximum is at $\vartheta = 0$, and for $l_n > 0$ at finite angles except for protons produced with very small energy. In a particular reaction several values of l_n may be possible without violation of conservation of angular momentum and parity. The relative peak values depend, apart from the values of Γ_{l_n}, on the maxima of the functions $\mathcal{J}_{l_n}$ [Eq. (45.10)], and these decrease fairly rapidly with increasing l_n. The stripping cross-section characteristically emphasises, then, the contribution from the lowest l_n-values. In fact the subsidiary maxima in the theoretical stripping curves are not usually observed in practice. The approximate nature of the theory, and particularly the neglect of other mechanisms for the d-p reaction, make it reasonable that only the dominant features of the theory, namely the principle peak in the cross-section, should correspond with what is observed.

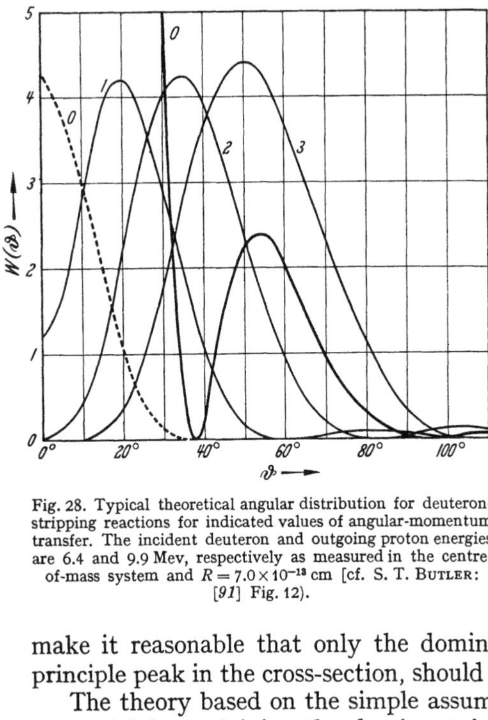

Fig. 28. Typical theoretical angular distribution for deuteron-stripping reactions for indicated values of angular-momentum transfer. The incident deuteron and outgoing proton energies are 6.4 and 9.9 Mev, respectively as measured in the centre-of-mass system and $R = 7.0 \times 10^{-13}$ cm [cf. S. T. BUTLER: [91] Fig. 12).

The theory based on the simple assumption indicated above is conspicuously successful in explaining the dominant features of the stripping process. Many attempts [94] have been made to introduce refinements in the theory, but these are at the expense of the intrinsic simplicity and require detailed assessment of many secondary physical factors. Whilst it is possible to extend the range of applicability of the theory in this way, its main success lies in the simple interpretation of the stripping process under conditions when the secondary effects are negligible or small.

The most obvious correction, which might permit the application of the theory to lower energies, is the allowance for Coulomb forces by replacing the plane waves with appropriate Coulomb wavefunctions. The wavefunction for the complex deuteron in the Coulomb field has been expressed by several different approximations (cf. GRANT [94]). Explicit formulae and numerical evaluations for these effects have been obtained. Corrections of comparable magnitude are introduced if the nuclear interaction, previously ignored, between the outgoing proton and the residual nucleus, is also taken into account. The combination of this nuclear scattering and Coulomb effects can result in major alteration of the form of the cross-section, particularly if the deuteron energy is only of the same order of magnitude as the Coulomb potential barrier. At higher energies these effects may modify the angular distribution, but not sufficiently to make the assignment of l_n ambiguous.

The simultaneous occurrence of compound-nucleus and stripping processes in the (d, p) and (d, n) reactions has been considered in detail by GRANT [94] in an approximation which includes both Coulomb effects and the nuclear scattering of the emitted particles. The overall differential cross-section is resolved into a sum of terms each corresponding (as in the simple stripping case) to a particular value[1] of l_n (and $L_n = l_n \pm \frac{1}{2}$). There is no interference between these different contributions[2], although for particular l_p, L_p values there are terms due to the interference between compound-nucleus and stripping amplitudes as well as their separate contributions.

In the previous discussion we have ignored the recoil motion of the target and product nuclei. To allow for these, $\mathfrak{k}_D$ must be replaced by $\mathfrak{k}'_D$ which defines the wave-number for the relative motion of deuteron and target nucleus before the collision, and $\mathfrak{k}_p$ by $\mathfrak{k}'_p$, which is the corresponding quantity for the relative motion of proton and product nucleus after the collision. $\mathfrak{K}_p$ is again given, now in terms of the reduced momenta, by an expression like (45.2). However, $\mathfrak{k}_n$ in (45.5) is replaced by $\mathfrak{k}'_n$ defined by:

$$\mathfrak{k}'_n = \mathfrak{k}'_D - \frac{A}{A+1}\,\mathfrak{k}'_p \tag{45.11}$$

where A is the mass number of the target nucleus, and in the expansion (45.9) $\boldsymbol{r}_p$ now represents the proton position with respect to the final nucleus as origin. The angular distributions (45.10) are with respect to the centre-of-mass coordinate frame of reference.

Cascade radiation. From (45.9) we see that the captured neutrons behave like plane waves, $e^{i\mathfrak{k}'_n \cdot \mathfrak{r}_n}$, incident along the directions $\mathfrak{k}'_n$. If therefore the proton is observed along a particular direction $\mathfrak{k}_p$ (and the deuteron direction $\mathfrak{k}_D$ is also known, of course) simultaneously with the observation of a subsequent radiation, x, from the product nucleus, b, then the angular correlation between the proton and deuteron on the one hand and the radiation, x, on the other, can be expressed in terms of the angular distribution in an n-x reaction. Here n represents a spin-$\frac{1}{2}$ particle (neutron or proton) and its effective incoming direction is $\mathfrak{k}'_n$, which is readily seen to be the direction of recoil of the nucleus b (in the laboratory frame of reference). With $\mathfrak{k}'_n$ as axis of reference, the angular distribution of x, in coincidence with protons in the direction $\mathfrak{k}'_p$, is then given by an expression of the form (24.1), where the different l_n-components are weighted, as in (45.10), by the amplitudes $\mathcal{J}_{l_n}(\mathfrak{k}'_n R)$, i.e.,

$$\left.\begin{aligned} W(\vartheta) \sim \sum_k \sum A_k(\mathcal{L}_n \mathcal{L}'_n; a\, b\, b)\, A_k(\mathcal{L}_x \mathcal{L}'_x; c\, b\, b) \langle a \| \mathcal{L}_n \| b\rangle' \times \\ \times \langle a \| \mathcal{L}'_n \| b\rangle'^* f(l_n)\, f^*(l'_n)\, \mathcal{J}_{l_n}(\mathfrak{k}'_n R)\, \mathcal{J}_{l'_n}(\mathfrak{k}'_n R) \langle c \| \mathcal{L}_x \| b\rangle' \times \\ \times \langle c \| \mathcal{L}'_x \| b\rangle'^* f(l_x)\, f^*(l'_x)\, P_k(\cos\vartheta). \end{aligned}\right\} \tag{45.12}$$

The unlabelled summation extends over $l_n\, l'_n, L_n\, L'_n, l_x\, l'_x, L_x\, L'_x$. We use the formula for an isolated resonance level with unique angular momentum, b, since, in the stripping reactions of interest here, the captured nucleon forms, with the target nucleus, either a bound state or a discrete virtual state. The residual

[1] Or of the corresponding l_p and L_p in the case of the (d, n) reaction.

[2] In general, of course, there would be interferences between contributions corresponding to different orbital momenta of incoming deuteron or outgoing proton if the (d, p) reaction were analysed in the usual manner adopted for compound-nucleus reactions (cf. Sect. 34).

state after both the stripping process and the subsequent radiation, x (usually γ-radiation), is denoted by c, and the process x is characterised by $\mathcal{L}_x$, $\mathcal{L}'_x$, and the reduced matrix-elements $\langle c \| \mathcal{L}_x \| b\rangle'$. $\mathcal{L}_n \mathcal{L}'_n$ refer to the angular momenta of the nucleon (neutron) captured in the stripping process. The reduced widths in (45.10) are related to the matrix elements $\langle a \| \mathcal{L}_n \| b\rangle'$ by:

$$\Gamma_{l_n} = \sum_{L_n} |\langle a \| \mathcal{L}_n \| b\rangle' f(l_n)|^2. \tag{45.13}$$

If the nucleon (proton) produced in the stripping process is observed at the appropriate angles where the major contribution to (45.10) is from a single value of l_n, then the summation over $\mathcal{L}_n$, $\mathcal{L}'_n$ in (45.10) reduces to the simpler summation over L_n, L'_n. If more than one l_n value contributes, then there is interference between the different l_n-amplitudes. A further simplification occurs if also only one L_n value is important, as might be expected in the strong spin-orbit coupling approximations. The correlation can then be written:

$$\left.\begin{aligned} W(\vartheta) \sim & \sum_k A_k(L_n L_n; a b b)\, \Gamma_{l_n} |\mathcal{J}_{l_n}(\mathfrak{k}'_n R)|^2 \times \\ & \times \sum_{L_x L'_x} A_k(L_x L'_x; c b b) \langle c \| L_x \| b\rangle' \langle c \| L'_x \| b\rangle'^* f(l_x) f^*(l'_x)\, P_k(\cos\vartheta). \end{aligned}\right\} \tag{45.14}$$

Explicit expressions for the factors A_k, in particular for the case of spin-$\frac{1}{2}$ particles, have already been discussed in Sect. 24 [Eqs. (24.3), (24.7)]. The emission of γ-radiation following nucleon capture, which is the most useful application of (45.12), (45.14), has been treated in Sect. 26.

Polarisation. On the basis of the simple theory outlined above no polarisation of the particles produced (protons) in stripping is to be expected. This is the case even when the simple S-wavefunction for the deuteron is replaced by a more exact wavefunction which contains S and D components, (DALITZ [*94*]). This result is a consequence of the physical assumption that the proton spin is coupled only to the neutron-spin (in the deuteron wavefunction), and since there is no prepared orientation of the neutron spin in its absorption by the target nucleus, there will be no preferred orientation of the proton spin. When, however, secondary effects are taken into consideration, some degree of polarisation in the protons may be expected.

NEWNS[1] uses semi-classical arguments to consider the polarisation that may arise from secondary interaction of the protons produced in stripping with the product nucleus. Thus he finds that a proton, produced in the direction $\mathfrak{k}_p$ and which escapes interaction with the product nucleus, will be associated, in the deuteron, with a neutron whose component, λ_1, of orbital angular momentum, l_1, is predominantly positive when measured with respect to the axis $\mathfrak{k}_p \times \mathfrak{k}_D$. Bearing in mind that the proton and neutron spins in the deuteron are parallel, it follows that if the neutron absorption is predominantly with the coupling $L_1 = l_1 + \frac{1}{2}$, where the spin and orbital angular momentum are parallel, then the protons will be polarised in the positive direction $\mathfrak{k}_p \times \mathfrak{k}_D$. Conversely, if the coupling $L_1 = l_1 - \frac{1}{2}$ predominates, then the polarisation will be in the opposite direction. Maximum polarisation is to be expected when $\mathfrak{k}_p$ is at right angles to $\mathfrak{k}_D$, assuming of course that the protons observed at this angle are produced in a stripping mechanism. A knowledge of L_1 can help in the determination of the spin b since the nuclear spin $\boldsymbol{a}$ combines vectorially with $\boldsymbol{L}$ to form $\boldsymbol{b}$. Polarisation is also possible, of course, if there is compound-nucleus formation, and may also arise if the effects of Coulomb forces, neglected in the simple treatment, are included (GRANT [*97*])[2].

[1] H. C. NEWNS: Proc. Phys. Soc. Lond. A **66**, 477 (1953).
[2] Also J. YOCCOZ: Proc. Phys. Soc. Lond. A **67**, 813 (1954).

Other processes. The reactions which are the inverse of stripping (the so-called "pick-up" reactions[1]) can be expected to show features similar to stripping. Thus, according to the principle of detailed balance, we can express the cross-section for, say, a (p, d) reaction, $b+p\to a+d$, in terms of the corresponding (d, p) reaction, $a+d\to b+p$, by

$$\frac{\sigma_{(p,d)}(\vartheta)}{\sigma_{(d,p)}(\vartheta)}=\left(\frac{\hat{a}\,\hat{s}_D}{\hat{b}\,\hat{s}_p}\right)^2\frac{\mathfrak{k}_D^2}{\mathfrak{k}_p^2}, \tag{45.15}$$

where a, b, s_p, s_d, represent the spins of the nuclear states and proton and deuteron. This relation applies only, of course, when a, b are the identical states of the two nuclei, a, b, in the two cases. One can picture the "pick-up" process as an interaction between the bombarding particle (proton) and a single nucleon (neutron) of the target nucleus resulting in the formation of a deuteron. Just as in the case of stripping, the cross-section involves the reduced width for emission of the single nucleon from the nucleus b. In practice different nuclear states will be involved in a particular (d, p) reaction and the inverse (p, d) reaction, since in the first case a will be in the ground state and excited states of b will be produced, and in the second case b is in the ground state and excited states of a are produced.

Processes similar to deuteron stripping, and its inverse, can be expected for the collision of any complex particle with a nucleus when only one nucleon at a time interacts with the target nucleus. In particular, the processes (He^3, d) and (H^3, d) and the inverse reactions show stripping features at the appropriate energies[2]. The angular distribution for stripping of mass-3 particles is similar to that for deuteron stripping, the main modification being the replacement of the function $G(\mathfrak{k}_p)$ by a function which determines the probability that the nucleon captured from the mass-3 particle shall have the appropriate momentum. The function replacing $G(\mathfrak{k}_p)$ cannot be predicted theoretically so simply as in the deuteron case, since the wavefunction is no longer a single two-particle one.

46. Stripping: Experimental. Many measurements have been made, particularly of (d, p) and (d, n) cross-sections, which can be interpreted in terms of the stripping mechanism[3]. The results give the l-values of the captured nucleon and hence a direct relation between the parities of the target nucleus and the states of the product nucleus, and also limits for the angular momenta of the final states[4]. In addition, when absolute measurements are made, some estimate is provided of the reduced widths for nuclear emission from the *product nuclear states*, and thus of the wavefunctions of these states[5]. The results are least ambiguous when the deuteron energies employed are sufficient for Coulomb effects to be small, and for reactions at these energies the peaks in the stripping cross-sections are located at relatively small forward angles (at least for smal l-values where the contribution from stripping is significant). For the lightest nuclei, the level spacing in the product nucleus is large, so that only a few particle groups are produced and relatively simple methods can be used to resolve these. For most nuclei, however, many particle groups with close-spaced energies are

---

[1] S. T. BUTLER and E. E. SALPETER. — Phys. Rev. **88**, 133 (1953). — H. C. NEWNS: Proc. Phys. Soc. Lond. A **65**, 916 (1952).

[2] See, for example, A. D. HOLMGREN *et al.*: Phys. Rev. **95**, 1544 (1954). — S. H. LEVINE, R. S. BENDER and J. N. MCGRUER: Phys. Rev. **97**, 1249 (1955).

[3] For a review of the early experimental work see R. HUBY (1953) [*95*].

[4] Only when $l_n=0$ and the target has zero spin, $a=0$, do stripping results alone lead to an unambiguous assignment for the spin of the product nucleus, b.

[5] See, for example, T. AUERBACH and J. B. FRENCH: Phys. Rev. **98**, 1276 (1955). — J. B. FRENCH and B. J. RAZ: Phys. Rev. **98**, 1523 (1955).

produced, and if charged (protons), high-resolution spectrometry is required. An example of such an arrangement is shown in Fig. 29. In this example ad-

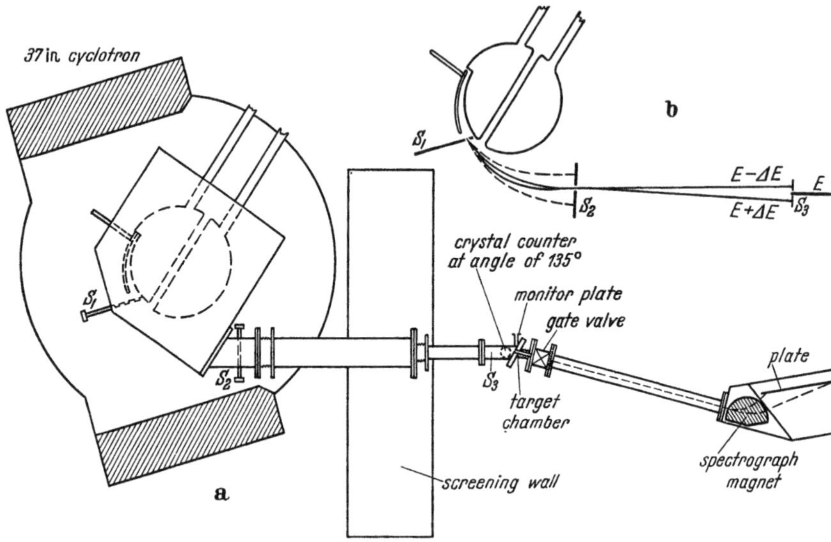

Fig. 29a and b. Experimental arrangement for stripping reactions with analysis of charged particles [cf. T. S. GREEN and R. MIDDLETON: Proc. Phys. Soc. Lond. A 69, 16 (1956)].

vantage has been taken of the fact that the relevant information is obtainable from studying the cross-section at small angles to simplify the target and spectrometer arrangement.

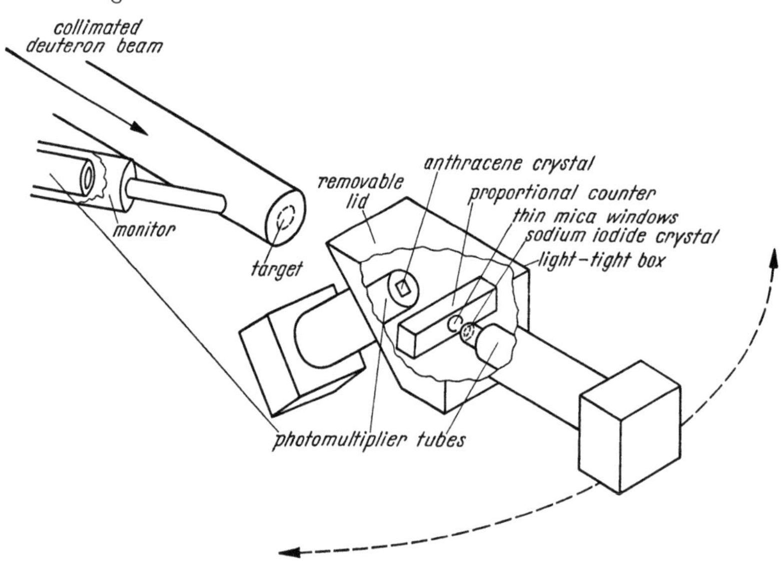

Fig. 30. Neutron-spectrum analyser used for study of (d, n) reactions [cf. J. M. CALVERT *et al.*: Proc. Phys. Soc. Lond. A 68, 1018 (1955)].

Measurements of (d, n) stripping cross-sections are generally of lower resolution and absolute precision. Neutron groups are separated by analysing the recoil protons produced in hydrogenous material in one of a variety of ways. An apparatus used in a recent investigation[1] is shown in Fig. 30. Here the recoil protons are produced in an anthracene crystal which also acts as a scintillation detector. The recoil protons also pass through a proportional counter with their entrance and exit windows and finally into a sodium-iodide crystal scintillation detector. Coincident pulses in all three detectors define the direction of the recoil proton along the direction of the incident neutrons, and from the pulse sizes in the anthracene and sodium-iodide detectors the neutron energy can be inferred. The energy resolution obtained was about 6%. Fig. 31 shows the results obtained for the process F^{19} (d, n) Ne^{20}, and are typical.

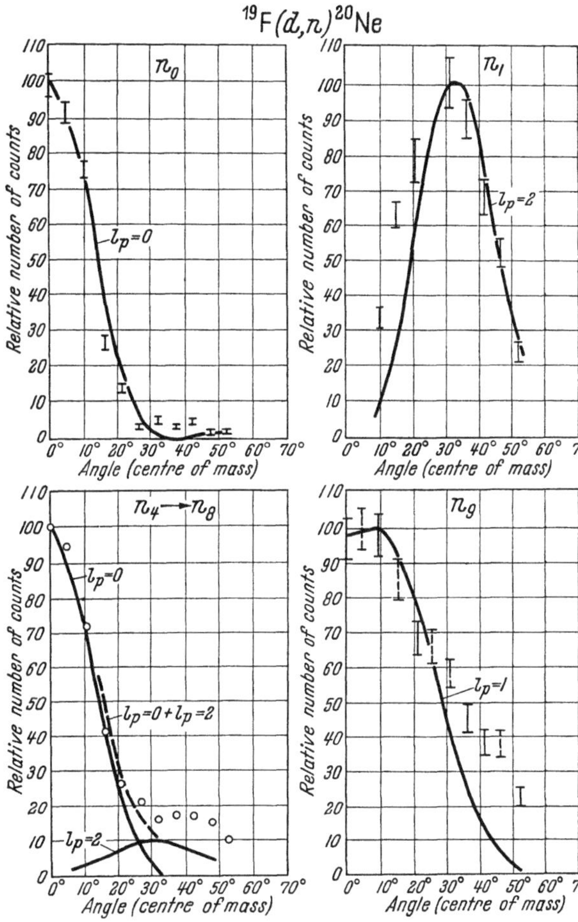

Fig. 31. Angular distributions for the stripping reactions F^{19} (d, n) Ne^{20} [cf. J. M. Calvert *et al.*: Proc. Phys. Soc. Lond. A 67, 1023 (1955)].

The ambiguities in the assignment of spins to the states of the product nucleus can be removed, in principle, by measurement of the polarisation of the particles produced or of their correlation with subsequent radiations, but only very few such measurements have been made. The angular correlation between the protons and γ-radiation in some $(d, p\,\gamma)$ reactions have been studied in order to test the predictions of the simple stripping theory[2,3]. For example, in the reaction Si^{28} $(d, p\,\gamma)$ Si^{29}, in which Si^{29} is produced in the first excited state, $L_n = 2$ is well established from the observed stripping cross-section[4], and the symmetry of this state is well established as $3/2^+$, so that $L_n = \frac{3}{2}$ is also known. With a reasonable assumption as to the multipolarity of the γ-radiation, the p-γ correlation can be predicted [cf. Eq. (45.14)], and a large anisotropy is indicated. The correlation has been measured, using the arrangement shown in Fig. 32, both with the γ-radiation detected in the plane of the reaction, $(\mathfrak{k}_d, \mathfrak{k}_p)$, and in the azimuthal plane, i.e. in the plane perpendicular to the recoil axis, $\mathfrak{k}_n$. The correlation in the reaction plane showed

[1] J. M. Calvert, A. A. Jaffe and E. E. Maslin: Proc. Phys. Soc. Lond. A **68**, 1017 (1956).
[2] R. M. Williamson and S. A. Cox: Phys. Rev. **99**, 631, 644 (1955).
[3] K. W. Allen *et al.*: Proc. Phys. Soc. Lond. A **69**, 705 (1956).
[4] J. R. Holt and T. N. Marsham: Proc. Phys. Soc. Lond. A **66**, 467 (1953).

symmetry about the neutron axis, as is expected theoretically. Some anisotropy (~20%) was observed in the azimuthal plane in contradiction to the predictions of the simple theory.

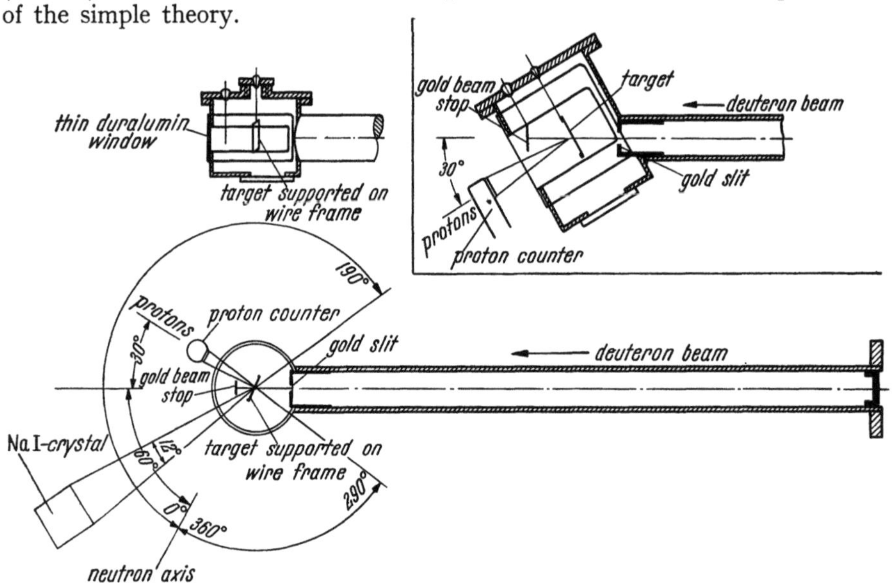

Fig. 32. Arrangement for studying angular correlations in a $(d, p\gamma)$ process [cf. K. W. ALLEN *et al.*: Proc. Phys. Soc. Lond. A 69, 705 (1956)].

The study of stripping reactions can provide information about the properties of the bombarding particle as well as those of the target nucleus. Thus, by a comparison of the two stripping reactions C^{12} (d, p) C^{13} and the pick-up reaction C^{13} (d, H^3) C^{12}, in both of which the same pair of nuclear states is involved, it is possible[1] to estimate the probability of finding the H^3 nucleus in the configuration of a deuteron ground state plus a second neutron. The theoretical and experimental cross-sections for the two processes are shown in Fig. 33.

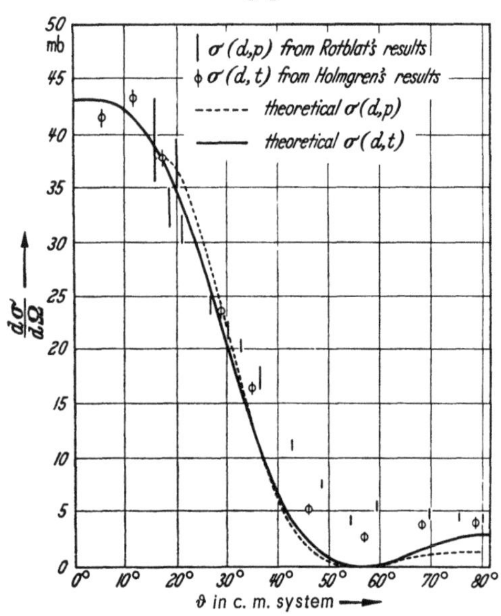

Fig. 33. Comparison of (d, p) and (d, t) angular distributions [cf. A. WERNER: Nuclear Phys. 1, 9 (1956)].

47. Direct interactions. In this section we consider some other reaction processes which, like stripping, do not involve either a single well-defined state of a compound nucleus or a "statistical" superposition of such states, but which, nevertheless, if they represent a valid approximation to the true state of affairs, result in characteristically simple angular correlations, both for the reaction itself and for subsequent radiations. Two such reaction mechanisms will be discussed. The first, which emphasises the shell-model features of the nucleus, deals with the direct collision between the impinging nucleon and a single

[1] A. WERNER: Nuclear Phys. 1, 9 (1956).

nucleon of the target nucleus, without involving the internal motions of the remainder of the target nucleus. In the second process, which emphasises the "collective" aspects of nuclear dynamics, the interaction postulated is between the incoming nucleon and the surface deformation oscillations of the target nucleus, resulting in an inelastic scattering.

α) Direct nucleon-nucleon interaction. This mechanism was originally proposed[1] to account for the large (n, p) cross-sections observed with heavy elements[2] and which were difficult to reconcile with the statistical compound-nucleus theory. It is expected to be applicable to the collisions of nucleus of energy ~ 10 Mev and upwards, which result in inelastic collisions with or without charge exchange. The assumed mechanism closely parallels deuteron stripping, and the predicted results are also similar. Just as in the case of stripping one of the colliding particles is treated as a loose association of two parts; here it is the target nucleus which is so treated, with only the least-bound nucleon (the nuclear "tail") interacting with the incoming nucleon and the remainder of the nucleus passive during the collision. As a result of the interaction [the (n, p) reaction is considered for definiteness] the incident neutron is captured and the originally bound proton is ejected. (Notice that if the process were the exact parallel, in reverse, of stripping, the neutron would be captured by the proton rather than by the target nucleus core.)

We have to deal in this process with two states of attachment of a single nucleon, the original proton state and the final neutron state, as compared with only one state, that of the captured neutron, in deuteron stripping, so that the angular-momentum relations in this (n, p) process are less clear-cut, and the interpretation in terms of the single-particle properties of particular nuclear states is less readily made. The transfer of angular momentum, L, to the target nucleus can be characterised by a set of values between $(l_p + l_n)$ and $(l_p - l_n)$ where l_p, l_n are respectively the orbital momenta of the least-bound proton [single-particle wavefunction: $\psi_1(\boldsymbol{r}_p)$] in the original nucleus and of the captured neutron: $\psi_2(\boldsymbol{r}_n)$. The orbital momenta l_p, l_n are assumed to be well-defined. The interaction is restricted to the region outside the nuclear surface proper, $r \gtrsim R$, since for closer interactions the assumption that only the one constituent nucleon is involved would not be valid.

In a first approximation the distortion of the incoming neutron waves (wave vector $\mathfrak{k}_n$) and outgoing proton wave, (wave vector $\mathfrak{k}_p$) by the nuclear potential is ignored. The interaction can then be described by the matrix element (SATCHLER [96]).

$$\langle l_n \lambda_n \mathfrak{k}_p | \mathcal{O} | l_p \lambda_p \mathfrak{k}_p \rangle = \iint\limits_{r_n, r_p \gtrsim R} d\boldsymbol{r}_n \, d\boldsymbol{r}_p \, \psi_{l_n}^{\lambda_n *}(\boldsymbol{r}_n) \, e^{-i \mathfrak{k}_p \cdot \boldsymbol{r}_p} V_{np} \, \psi_{l_p}^{\lambda_p}(\boldsymbol{r}_p) \, e^{i \mathfrak{k}_n \cdot \boldsymbol{r}_n}. \quad (47.1)$$

Using a zero-range approximation for the nucleon-nucleon interaction,

$$V_{np} = V_0 \, \delta(\boldsymbol{r}_n - \boldsymbol{r}_p)$$

gives immediately:

$$\langle l_n \lambda_n \mathfrak{k}_p | \mathcal{O} | l_p \lambda_p \mathfrak{k}_n \rangle = V_0 \int\limits_{r \gtrsim R} d\boldsymbol{r} \, e^{i \mathfrak{k} \cdot \boldsymbol{r}} f_{l_n}^*(r) \, f_{l_p}(r) \, Y_{l_n}^{\lambda_n *}(\vartheta, \varphi) \, Y_{l_p}^{\lambda_p}(\vartheta, \varphi) \quad (47.2)$$

where $\mathfrak{k} = \mathfrak{k}_n - \mathfrak{k}_p$, and f_{l_n}, f_{l_p} represent the radial parts of the single particle wavefunctions.

[1] N. AUSTERN, S. T. BUTLER and H. MCMANUS: Phys. Rev. **92**, 350 (1953).

[2] E. B. PAUL and R. L. CLARKE: Canad. J. Phys. **31**, 267 (1953). See also: Statistical Aspects of the Nucleus. Brookhaven Conference Report, Brookhaven, New York 1955.

The "plane-wave" function $e^{i\mathfrak{k}\cdot\mathbf{r}}$ can be expanded in terms of spherical waves, with the direction $\mathfrak{k}$ as axis. The integration over angle variables then gives:

$$\left.\begin{aligned}&\langle l_n \lambda_n \mathfrak{k}_p | \mathcal{O} | l_p \lambda_p \mathfrak{k}_n \rangle \\ &\sim V_0 \hat{l}_p \hat{l}_n \sum_L \{ i^L (l_p 0, l_n 0 | L 0) (-)^{\lambda_p} (l_p \lambda_p, l_n \lambda_n | L 0) \int_{r \geq R} j_L(\mathfrak{k}_r) f_{l_n}^*(r) f_{l_p}(r) r^2 dr \}\end{aligned}\right\} \quad (47.3)$$

with $\mathfrak{k} = \sqrt{\mathfrak{k}_p^2 + \mathfrak{k}_n^2 - 2\mathfrak{k}_p \mathfrak{k}_n \cos\vartheta}$ and the range of the summation over L is limited by the condition $\Delta(L l_n l_p)$. Parity conservation limits L to all odd or all even values. Since f_{l_n}, f_{l_p} decrease rapidly with r for $r > R$, and $j_L(\mathfrak{k}r)$ is slowly varying with $\mathfrak{k}r$ for $r \approx R$, a further reasonable approximation can be made by replacing $j_L(\mathfrak{k}r)$ by its value on the nuclear surface, $j_L(\mathfrak{k}R)$. With this simplification, and on averaging over initial states $|\lambda_p\rangle$ and summing over final states (47.3) gives for the angular dependence of the (n, p) cross-section:

$$\sigma(n, p) \sim \sum_L (l_n 0, l_p 0 | L 0)^2 |j_L(\mathfrak{k}R)|^2. \quad (47.4)$$

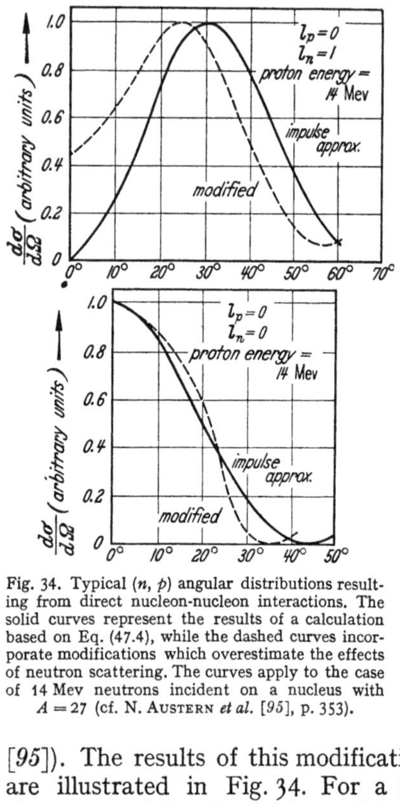

Fig. 34. Typical (n, p) angular distributions resulting from direct nucleon-nucleon interactions. The solid curves represent the results of a calculation based on Eq. (47.4), while the dashed curves incorporate modifications which overestimate the effects of neutron scattering. The curves apply to the case of 14 Mev neutrons incident on a nucleus with $A = 27$ (cf. N. AUSTERN *et al.* [95], p. 353).

The similarity to the angular distribution in simple stripping theory [cf. Eq. (45.10)] is obious. Each term in (47.4) corresponding to a particular L value has the characteristic Bessel-function dependence, and contributions corresponding to different L values add incoherently. The first maximum in the differential crosssection is determined by the lowest L value.

Some estimate of the effect of neglecting the interaction between incident neutrons and the target nucleus core, can be obtained by modifying the model so that all incident neutron partial waves with $l_n \leq \mathfrak{k}_n R$ are absorbed and others are unaffected (AUSTERN, BUTLER and MCMANUS [95]). The results of this modification as well as those of the unmodified theory are illustrated in Fig. 34. For a nuclear coupling scheme more general than that assumed for the single-particle, the initial and final nuclear states are characterised by parities and total angular momenta $\boldsymbol{a}$, $\boldsymbol{b}$. The restriction on the L values is relaxed:

$$|\boldsymbol{a} + \boldsymbol{b} + \mathbf{1}|_{\text{maximum}} \geq L \geq |\boldsymbol{a} + \boldsymbol{b} + \mathbf{1}|_{\text{minimum}} \quad (47.5)$$

but the restriction of L to all even or all odd values (according as the states a, b, have the same or opposite parities) still remains, in accordance with the interpretation of L as the transfer of orbital momentum in the process.

The form of (47.2) indicates that the direction $\mathfrak{k}$, which is the direction of recoil motion, is an axis of symmetry for the process, i.e. one can consider the transfer of angular momentum L as having zero component along the direction $\mathfrak{k}$. Hence, if the contribution corresponding to a particular $\mathfrak{k}_p$ and L is isolated, and $\mathfrak{k}$ is defined, (e.g. by detecting the emitted proton in the direction $\mathfrak{k}_p$ corresponding to a maximum in the cross-section) any subsequent radiation should show axial symmetry about the direction $\mathfrak{k}$. If this subsequent radiation also has well-defined parity, as would be the case for the normal γ-transitions from the excited states, b, of the residual nucleus, then there will also be back-to-front symmetry for this radiation.

In the above discussion we have considered only the transfer of orbital angular momentum, but in addition there may be a change of spin angular momentum, as indicated in (47.5), depending on the nature of the nucleon interaction. The angular distribution with respect to $\mathfrak{k}$ for the subsequent radiation, in coincidence with protons in a particular direction, will then have the same form as the capture along the direction $\mathfrak{k}$, of an incoherent mixture of particles of spin 1 and spin 0. For the general case of several l_n, l_p, and L values simultaneously present, the resulting expressions for the correlation will be quite complicated, but the general feature of symmetry with respect to the $\mathfrak{k}$ axis remains.

A number of experimental investigations[1] of the differential cross-section of the (n, p) (p, n) and inelastic proton scattering have been made with nucleons of energy 10 to 30 Mev, and the results show the general features of forward maxima which the theory predicts (e.g. Fig. 34). Detailed confirmation of the rather crude theory is hardly to be expected. No detailed measurements have been made of the correlation of the inelastic scattered particles with subsequent radiation.

β) Surface interactions. We consider the target nucleus to be described in accordance with the description of the collective model developed by BOHR and and MOTTELSON[2].

The nuclear surface, $\varrho(\vartheta, \varphi)$ is defined by the coordinates $\alpha_{b\beta}$:

$$\varrho(\vartheta, \varphi) = R\left[1 + \sum_{b\beta} \alpha_{b\beta} Y_b^\beta(\vartheta, \varphi)\right] \tag{47.6}$$

and the Hamiltonian of the target nucleus is also a function of these coordinates. The total Hamiltonian for the system (incident neutron + target nucleus) is $H_0 + V(r)$ where $V(r)$ represents the interaction between neutron + target, and r the nucleon position. If the target nucleus is represented by a uniform potential well, then

$$V(r, \alpha_{b\beta}) \begin{cases} = -V_0, & r < \varrho(\vartheta, \varphi), \\ = 0, & r > \varrho(\vartheta, \varphi). \end{cases} \tag{47.7}$$

V depends on both neutron and surface-deformation coordinates, and this interaction represents the possibility of inelastic neutron scattering resulting in surface excitation. For small deformations the interaction can be approximated by a δ-function at the nuclear surface:

$$\left.\begin{aligned} V(r, \alpha_{b\beta}) &\approx V_0(r) + V_0 R\,\delta(r - R) \sum_{b\beta} \alpha_{b\beta} Y_b^\beta(\vartheta, \varphi), \\ &= V_0(r) + V', \end{aligned}\right\} \tag{47.8}$$

with

$$V_0(r) = -V_0; \quad r < R$$
$$V_0(r) = \quad 0; \quad r > R.$$

[1] E.g. B. L. COHEN: Phys. Rev. **98**, 49 (1955) [(p, n) for 23 Mev protons and target nuclei with $Z = 12$ to 92]. — J. BENVENISTE, R. G. FINKE and E. A. MARTINELLI: Phys. Rev. **101**, 655 (1956) [(p, p') for levels up to 22 Mev in Be^9]. — R. M. EISBERG: Phys. Rev. **103**, 645 (1956) [(p, p') for 10 Mev protons and A^{40}].

[2] A. BOHR and B. R. MOTTELSON: Dan. Mat.-Fys. Medd. **27**, No. 16 (1953).

The first term in (47.8) is the normal "optical" complex potential[1] which is related to the normal elastic scattering and absorption of neutrons on undeformed neutrons. The second term, V', is the interaction responsible for the direct inelastic scattering.

The most significant surface oscillations are the quadrupole ones described by $\alpha_{2\beta}$, and for these the quantity $\sum_{2\beta} Y_2^\beta(\vartheta,\varphi)$ is proportional to the intrinsic nuclear quadruple moment[2], Q. In the simplest treatment (BRINK, HAYAKAWA and YOSHIDA [97]), the term V' is treated as a first order perturbation. V_0 is not so treated—that is, the nuclear wavefunction used in evaluating the matrix elements of V' are distorted waves appropriate to the potential V_0.

If the Born approximation is used, as was done in considering direct reactions in (a) above, then the interaction, V_0 is ignored and plane waves are used in evaluating the matrix elements of V'. For quadrupole deformation, the differential cross-section has the same angular dependence as in the case of direct interaction [Eq. (47.4)]:

$$\sigma \sim (V_0 R Q)^2 |j_2(\mathfrak{k} R)|^2 \tag{47.9}$$

where, as before, $\mathfrak{k} = |\mathfrak{k}_{\text{initial}} - \mathfrak{k}_{\text{final}}|$ represents the recoil momentum. This angular dependence is characteristic of the Born-approximation for scattering due to a surface interaction ($r = R$) and with the transfer of two units of angular momentum.

When the distorting effect of the potential V_0 is taken into account, the differential cross-section can be expressed in terms of the radial wavefunctions for this square well potential with ingoing and outgoing waves described by wave numbers $\mathfrak{k}_i$ and $\mathfrak{k}_f$ respectively. Denoting these wavefunctions by $u_{l_0}(\mathfrak{k}_i r)$ and $v_l(\mathfrak{k}_f r)$ respectively, the angular dependence of the cross-section is given by (BRINK, HAYAKAWA and YOSHIDA [97]):

$$\sigma(\vartheta) \sim \sum_\beta \Big| \sum_{l l_0} \hat{l}\, \hat{l}_0\, u_{l_0}(\mathfrak{k}_i r)\, v_l(\mathfrak{k}_f r)\, (2-\beta, l\beta\,|\,l_0 0)\,(l0, l_0 0\,|\,20)\, Y_l^\beta(\vartheta,\varphi) \Big|^2. \tag{47.10}$$

The expansion of (47.10) and the summation over magnetic quantum numbers gives:

$$\left.\begin{aligned} \sigma(\vartheta) \sim \sum_{L l l_0 l_0' l'} & u_{l_0}(\mathfrak{k}_i r)\, v_l(\mathfrak{k}_f r)\, u_{l_0'}^*(\mathfrak{k}_i r)\, v_{l'}^*(\mathfrak{k}_f r)\, (\hat{l}_0 \hat{l}_0' \hat{l} \hat{l}')^2 \times \\ & \times (l_0 0, l0\,|\,20)\,(l_0' 0, l' 0\,|\,20)\,(l_0 0, l_0' 0\,|\,20)\,(l0, l'0\,|\,L0)\, P_L(\cos\vartheta). \end{aligned}\right\} \tag{47.11}$$

The contributions l, l_0 which are of importance are the same as those which are significant in the optical model used for neutron scattering. Since on the basis of this model $l = 0, 2$ give large contributions when that of $l = 1$ is small, and conversely, the interference terms in (47.11) which result in asymmetry will be small. The magnitude of this surface scattering has been estimated by BRINK and YOSHIDA and HAYAKAWA for neutrons of a few Mev energy, and with Q estimated from the values of the static-quadrupole moments or electric-quadrupole transition moments of neighbouring nuclei. Their results indicate that surface interaction is of comparable importance to inelastic scattering by compound-nucleus formation.

The validity of the above treatment, which regards the surface interaction as a small perturbation compared with the optical potential V_0, has been examined in detail by YOSHIDA [98] who gives a more exact treatment. The assumption that the distorting potential can be equated to the optical potential used to

---

[1] H. FESHBACH, C. E. PORTER and V. F. WEISSKOPF: Phys. Rev. 96, 448 (1954).

[2] See Ref. 2, p. 493.

explain elastic scattering, is only justified if the surface interactions are small. Strong surface interactions will affect the elastic scattering in second and higher orders and therefore the optical (i.e. the distorting) potential must be adjusted so as to fit the observed scattering in the presence of strong surfaces interactions. Excited states of the target nucleus which are coupled by weak interactions are dealt with by including their effects, in an approximate manner, in the distorting potential.

For the specially simple case where the surface interaction is represented by (47.8) the strongly coupled states are treated by non-perturbation methods. YOSHIDA also considers the modification of the statistical (compound-nucleus) inelastic scattering which results from excluding those reaction channels which are associated with strongly coupled states. The theoretical results are found to be quite sensitive to the choice of the distorting potential. The main difference between the distorting potential and the effective optical potential (i.e. the complex potential which would produce the observed scattering in the absence of surface excitation is the smaller value of the imaginary part of the former. This is to be expected since the effective potential includes the absorption which is dealt with explicitly by the surface interaction.

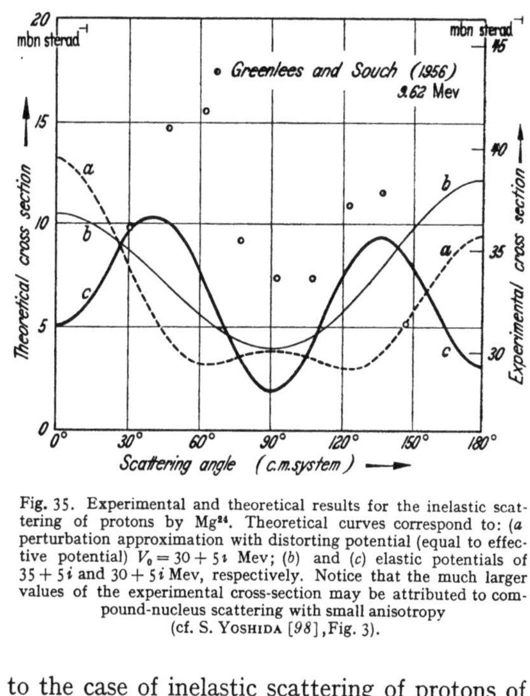

Fig. 35. Experimental and theoretical results for the inelastic scattering of protons by Mg^{24}. Theoretical curves correspond to: (*a* perturbation approximation with distorting potential (equal to effective potential) $V_0 = 30 + 5i$ Mev; (*b*) and (*c*) elastic potentials of $35 + 5i$ and $30 + 5i$ Mev, respectively. Notice that the much larger values of the experimental cross-section may be attributed to compound-nucleus scattering with small anisotropy (cf. S. YOSHIDA [98], Fig. 3).

The theory has been applied to the case of inelastic scattering of protons of about 10 Mev by Mg^{24}, resulting in the excitation of the latter from the ground 0^+ state, to the first excited 2^+ state, and estimating Q from the value of the quadrupole moment of Al^{27} and Na^{23}. The effective elastic potential is calculated from the experimental results[1, 2] for the elastic scattering of protons by Mg^{24}. The theoretical predictions together with the experimental results[2] are shown in Fig. 35. The agreement between theory and experiment appears qualitatively reasonable. However, the observed asymmetry is rather larger than is theoretically predicted, and moreover, the cross-section and angular distribution appear to change with small changes in proton energy far more rapidly than would be expected on the basis of the model used.

In some other cases[3] the observed inelastic scattering cross-section does not exhibit marked asymmetry. YOSHIDA has been able to explain these results in terms of a δ-function surface interaction with conventional spin-orbit dependence, but not with a consistent value of the distorting potential.

[1] G. E. FISHER: Phys. Rev. **94**, 704 (1954).
[2] R. W. GREENLEES and A. E. SOUCH: Proc. Phys. Soc. Lond. A **69**, 686 (1956).
[3] $C^{12}(p, p')$. W. E. BURCHAM, W. M. GIBSON, A. H. HOSSAIN and J. ROTBLAT: Phys. Rev. **92**, 1266 (1954). — $Ne^{20}(p, p')$. R. G. FREEMANTLE, D. J. ROWSE and J. ROTBLAT: Phys. Rev. **96**, 1268 (1954).

The sensitivity of the theoretical results to the exact value of the distorting potential suggests that the use of a square-well potential and δ-function surface interaction are likely to be poor approximations, particularly as the use of a "rounded-off" square-well potential gives better agreement with the observed elastic scattering. However, it is difficult to see how a refinement of the theory of this sort could explain such qualitative features as the large asymmetry and its rapid variation with energy in some cases.

48. Internal pair-creation. Inelastic scattering of charged particles. The processes of internal electron-pair creation and inelastic charged-particle scattering, by virtue of an electromagnetic interaction, have many features in common with each other and with internal conversion and the β-process. These two latter processes are treated in analogous ways, but this treatment is not so readily applicable to pair-creation and inelastic scattering for the following reasons: Both pair-creation and inelastic scattering involve the transition of a free particle from an initial to final state in the continuum, accompanied by a nuclear transition between well-defined states. In the case of pair-creation the free particle is initially an electron in a negative energy state, and finally in a positive energy state (plus a "hole" in the negative energy state). The nuclear transition is a de-excitation normally associated with a γ-transition. For inelastic scattering, both the initial final states are of positive energy (e.g. non-relativistic motion for "Coulomb-excitation" and relativistic plane waves for high-energy electron scattering), and the nuclear transition is the transition from the ground to some excited state. In both types of processes a well-defined amount of angular momentum is transmitted to the nucleus, but since both initial and final states of the particle are free-states, this is possible without violation of angular-momentum conservation for an indefinite range of initial and final angular momenta of this particle. Moreover, since the interaction is long-range (electromagnetic), many angular-momentum components will in fact be involved. This is in contrast with ordinary internal-conversion (K, L etc.) where, although we again have an electron transition by virtue of an electromagnetic interaction and the final electron state is free, the initial state is a bound one with definite angular momentum (e.g. zero for K-conversion, cf. Sect. 43), so that there are only limited possibilities for the angular momentum of the final free state. In the β-process on the other hand both initial and final particle states (neutrino and electron) are free, but here the interaction is short-range (δ-function, cf. Sect. 49) and this feature limits the range of angular momenta involved.

Because of these features pair-creation and inelastic scattering processes are often dealt with by more direct (and frequently approximate) methods than the general procedures described previously.

α) Internal pair-creation. The study of internal pair-creation associated with a transition between two nuclear states is most useful for high transition energies ($\gg 2mc^2$) and light nuclei. For such transitions the ordinary atomic internal-conversion is smallest, whereas the probability of pair-creation increases with energy, is practically independent of Z, and is largest for the lowest multipolarities, although much less sensitive to multipolarity than is atomic internal-conversion. For very large transition energies the total internal pair-creation coefficient (i.e. the ratio of the number of electrons pairs to γ-transitions) tends to become independent of multipolarity, but the angular correlation between electron and positron remains multipole-sensitive and becomes independent of transition energy. This correlation depends also on the parity of the transition, i.e. it is different for electric and magnetic multipoles. The absolute magnitude of the internal conversion coefficient is rather small (in the range 0.001 to 0.01 for typical transitions), so that the study of the correlation in internal pair-creation is in principle a useful one for high-energy transitions. Its application on the other hand is rather limited and the more straightforward measurement of the angular distribution and correlation of the associated γ-radiation is usually more easily employed.

Since the most significant application is to high-energy transitions in light nuclei, where the effect of nuclear charge on the electron wavefunctions is small, we shall consider only the approximate theory (Born approximation) for such cases [*101*]. The interaction, H, responsible for the transition of the positron from the initial negative energy and momentum state, $-E_+, -\boldsymbol{p}_+, \sigma_+$, to the final state, $E_-, \boldsymbol{p}_-, \sigma_-$, is assumed to be the same electromagnetic field as is responsible for γ-radiation, and is expanded in the usual multipole fashion. This assumption involves neglect of the interaction between the electron-positron field and the electromagnetic field *inside* the nucleus. The main contribution to the matrix element, $\langle a\,\alpha\,\boldsymbol{p}_+\sigma_+|H|b\,\beta\,\boldsymbol{p}_-\sigma_-\rangle$, comes from the region $\mathfrak{k}r \lesssim 1$ (where $\mathfrak{k} = E_+ + E_-$ in units $\hbar = c = m = 1$). For the transitions of interest $\mathfrak{k}R \ll 1$ (R is nuclear radius) so that contributions from the region $r < R$ are neglected. Only if these contributions from the internal nuclear fields are neglected can the pair-creation process be analysed in terms of simple multipole fields ($V_{L\Lambda}$, $A_{L\Lambda}$), each giving an independent contribution, i.e.

$$H = \sum_{L\Lambda} [V_{L\Lambda} + \boldsymbol{\alpha}\cdot(\boldsymbol{A}_{L\Lambda}(\text{elec}) + \boldsymbol{A}_{L\Lambda}(\text{magn}))],$$

where

$$\left.\begin{aligned} \boldsymbol{A}_{L\Lambda}(\text{elec}) &= \left[\frac{2\pi\mathfrak{k}}{L(L+1)}\right]^{\frac{1}{2}} \chi_{L-1}(\mathfrak{k}r)\left[r\,\text{grad} + \frac{L\,\boldsymbol{r}}{r}\right] Y_L^\Lambda(\vartheta,\varphi), \\ V_{L\Lambda} &= -i\left[\frac{2\pi L\,\mathfrak{k}}{L+1}\right]^{\frac{1}{2}} \chi_L(\mathfrak{k}r)\,Y_L^\Lambda(\vartheta,\varphi) \end{aligned}\right\} \tag{48.1}$$

for electric radiation, and

$$\boldsymbol{A}_{L\Lambda}(\text{magn}) = -i\left[\frac{2\pi\mathfrak{k}}{L(L+1)}\right]^{\frac{1}{2}} \chi_L(\mathfrak{k}r)\,(\boldsymbol{r}\times\text{grad})\,Y_L^\Lambda(\vartheta,\varphi);\quad V_{L\Lambda} = 0,$$

for magnetic radiation. The χ_L are spherical Hankel functions and the fields are normalised to unit quantum per second. $L\Lambda$ refers to the angular momenta changes in the transition (ROSE [*101*]). As in the case of γ-radiation it is normally assumed that a single, or at most two multipole contributions are significant. The probability, W, of emission of electron and positron in the solid angles $d\Omega_-$ and $d\Omega_+$ and energies E_- and E_+ to $E_+ + dE_+$, respectively, normalised per unit γ-quantum emitted, is then

$$\left.\begin{aligned} W(E_+E, \Omega_+\Omega_-)\,d\Omega_+\,d\Omega_-\,dE_p &= \frac{e^2\mathfrak{k}}{32\pi^3}\,p_+\,p_-\,E_+\,E_-\times \\ &\times \sum |\langle \boldsymbol{p}_+\sigma_+|\sum(V_{L\Lambda} + \boldsymbol{\alpha}\cdot\boldsymbol{A}_{L\Lambda})|\boldsymbol{p}_-\sigma_-\rangle|^2\,d\Omega_+\,d\Omega_-\,dE_+ \end{aligned}\right\} \tag{48.2}$$

where $|\boldsymbol{p}_+\sigma_+\rangle$, $|\boldsymbol{p}_-\sigma_-\rangle$. in the present approximation, are plane-wave spinors and the first summation is over the unobserved spins and the second over the multipolarities of the radiation, if more than one is significant.

In order to obtain the positron-electron angular correlation, (48.2) must firstly be averaged over all values of Λ corresponding to all the transitions $|a\,\alpha\rangle \to |b\,\beta\rangle$, and the expression then integrated over all values of Ω_-, for a particular Ω_+ with the angle, Θ, between Ω_+ and Ω_- kept constant. This correlation will only be significant, i.e. independent of the actual directions Ω_-, Ω_+, in the case where the initial nuclear states $|a\alpha\rangle$ are random and equally populated (i.e. isotropic). In this case the summation over Λ can be replaced by an integration over all space directions of the Ω_+, or some other characteristic, e.g. the direction $\boldsymbol{q} = \boldsymbol{p}_+ + \boldsymbol{p}_-$, of the electron-positron pair.

The result is given by ROSE for pure multipoles:

Electric multipoles

$$\left.\begin{aligned} W_L(\Theta\, E_+\, E_-) \sim \frac{p_+ p_-}{q^2}\,\frac{q^{2L}}{(\mathfrak{k}^2 - q^2)^2}\Big\{&\Big[(2L+1)\Big(E_+ E_- + 1 - \frac{p_+ p_-}{3}\cos\Theta\Big)\Big] + \\ &+ \Big[L\Big(\frac{q^2}{\mathfrak{k}^2} - 2\Big)(E_+ E_- - 1 + p_+ p_- \cos\Theta)\Big] + \frac{L-1}{3}\, p_+ p_- \times \\ &\times \Big[\frac{3}{q^2}(p_- + p_+ \cos\Theta)(p_+ + p_- \cos\Theta) - \cos\Theta\Big]\Big\}. \end{aligned}\right\} \quad (48.3\,\mathrm{a})$$

Magnetic multipoles

$$W_L(\Theta E_+ E_-) \sim \frac{p_+ p_- q^{2L}}{(\mathfrak{k}^2 - q^2)^2}\Big[1 + E_+ E_- - \frac{p_+ p_-}{q^2}(p_- + p_+ \cos\Theta)(p_+ + p_- \cos\Theta)\Big] \quad (48.3\,\mathrm{b})$$

where

$$q = [p_+^2 + p_-^2 + 2 p_+ p_- \cos\Theta]^{\frac{1}{2}}.$$

The most pronounced dependence of $W(\Theta)$ on Θ is produced by the first factors in (48.3): $\big(q^{2L}/(\mathfrak{k}^2 - q^2)^2\big)$. For small angles, $\Theta < \Theta_0$ defined by $\cos\Theta_0 = 1 - \frac{1}{\mathfrak{k}}$, $W_L(\Theta)$ is independent of L and depends on $\mathfrak{k}$. For large angles $\Theta \gg \Theta_0$, the reverse is true, $W(\Theta)$ depending primarily on L (and parity) and only slightly on energy. For the determination of multipolarity, then, measurements must be made at large angles, where the absolute pair-production rate is small. The large probability of pair production at high energies with small Θ corresponds to production in the radiation zone of the multipole field, in which case the processes is naturally independent of the multipole nature. The general features of $W_L(\Theta)$ are shown in Fig. 36.

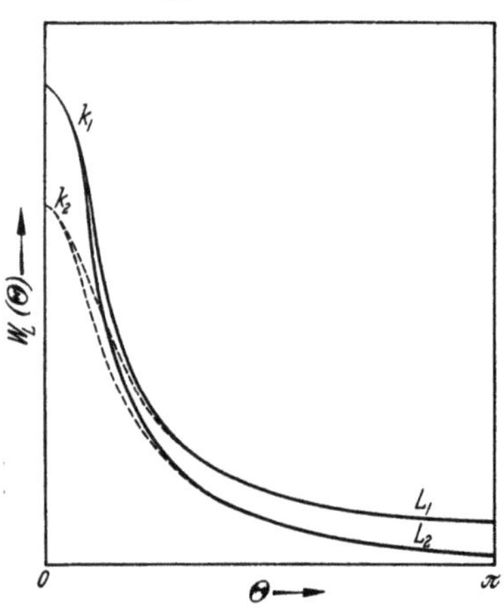

Fig. 36. Dependence of $W_L(\Theta)$ and k and L.

When the nuclear states $|a\alpha\rangle$ are not random and equally populated, then (48.2) must be summed accordingly. In a practical case the excited nucleus may be produced in a nuclear reaction, and if the direction of the incident beam is chosen as axis of quantisation, then the different states $|a\alpha\rangle$ will be incoherent and the summation in (48.2) involves a simple weighting (cf. Sect. 9).

The problem of anisotropic emission of pairs has been examined by GOLDRING[1]. He characterises the anisotropy by integrating the function $W(E_+ E_- \Omega_+ \Omega_-)$ over δ, the azimuthal angle of $\boldsymbol{p}_+$ about $\boldsymbol{q}$. In this way the relative probability of creation of a pair with a particular direction of $\boldsymbol{q}$ is determined, and it is established that the independence of the probability with respect to the direction $\boldsymbol{q}$ is a necessary and sufficient condition for the process to be isotropic in the full sense. In the limit of high transition energies,

[1] G. GOLDRING: Thesis 1953. Imperial College of Science Proc. Phys. Soc. Lond. A **66**, 341 (1953).

the angular dependence of the probability on $\boldsymbol{q}$ is the same for pair emission and γ-radiation.

In the general case the anisotropy of the γ-radiation can provide the necessary information for estimating the modification to the pair-correlation function. The simplest correlation functions are obtained when both electron and positron are observed in the plane perpendicular to the symmetry axis (i.e. axis of quantisation). For mixed ML, ML' or EL, EL', in this plane, the interference term has the form of a pure $\frac{1}{2}(L+L')$ emission. Interference of the type ML', EL vanishes, in this plane, if the correlation is integrated symmetrically over positron and electron energies. For isotropic emission, all interference terms vanish.

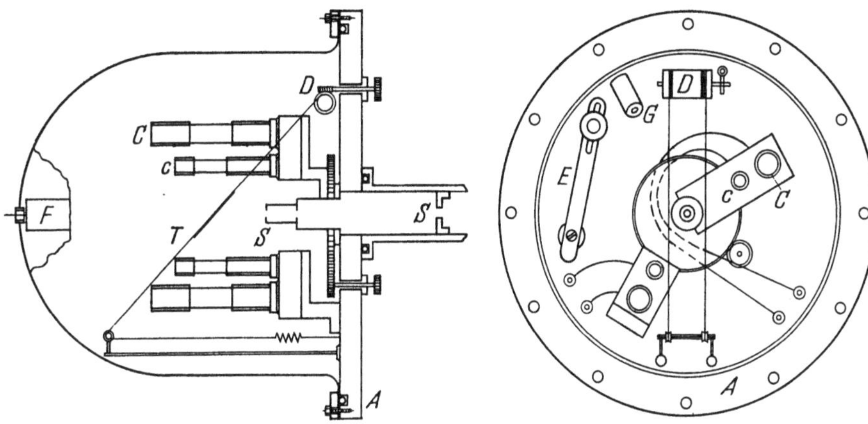

Fig. 37. Apparatus for measuring angular correlations of electron-proton pairs from light nuclei [cf. S. Devons *et al.* Proc. Phys. Soc. Lond. A **67**, 138 (1954)]. *A* Vacuum Vessel, *Cc* Counters, *G* Monitor Counter, *T* Target, *S* Beam-Defining Slits.

The above discussion refers to the condition in which the static nuclear charges can be neglected. Detailed calculations have also been made for some heavy nuclei and low energies (Jaeger and Hulme [*100*]) and in the Schrödinger approximation[1]. In the low-energy region, however, the total probability of pair-creation is sensitive to multipolarity and is more easily measured than the angular correlation. The validity of the Born approximation has been examined in some detail by Horton and Phipps[2].

The main difficulties in the experimental investigation of the electron-positron angular correlation in internal pair-creation arise from the small value of the pair-conversion coefficient, and the need to detect the small number of electrons against a large γ-background. Spurious electron-positron pairs created in material near to the true sources (target) must be kept to a minimum. Scattering of the electrons and positrons is a most serious factor, even when measurements are made at large angles where the correlation function is only slowly varying with angle. This is due to the presence of a large intensity of pairs produced with small angular separation (Fig. 36), so that even a very small amount of scattering of these particles into larger angles can seriously distort the measurements.

A few measurements of angular correlations have been made using cloud-chamber, magnetic spectrometer[3], and counter telescope techniques[4]. Only

[1] M. E. Rose and G. E. Uhlenbeck: Phys. Rev. **44**, 948 (1935). — G. K. Horton: Proc. Phys. Soc. Lond. A **60**, 457 (1948).

[2] G. K. Horton and E. Phipps: Phys. Rev. **96**, 1066 (1954).

[3] G. Harris: Proc. Phys. Soc. Lond. A **67**, 153 (1954). — K. Siegbahn: Manne Siegbahn, p. 199. Uppsala 1951.

[4] S. Devons and G. Goldring: Proc. Phys. Soc. Lond. A **67**, 413 (1954).

in the last of these have the experimental conditions satisfied the stringent requirements with respect to scattering and at the same time results of sufficient statistical accuracy been obtained to permit quantitative deductions. These measurements have been made of some high-energy γ-transitions following reactions in light nuclei [e.g. (p, γ)]. In order to reduce the scattering to a minimum, the whole equipment, target and counter-detectors, was contained in an evacuated vessel (Fig. 37).

Maximum intensity was obtained by measuring the angular correlation effectively integrated over all energies, i.e. $\int_1^{\mathfrak{k}-1} W(\Theta, E_+ E_-)\, dE_+$. Practically, the limits of this integration, $E_{\min}$ and $\mathfrak{k} - E_{\min}$, are set by the minimum energy, $E_{\min}$ for which a particle can traverse the thin-walled counter "telescopes". (Since these low-energy electrons tend to be isotropic, their elimination enhances the correlation.) Correlation measurements were made with the plane containing source and detectors perpendicular to the incoming beam, in order to reduce the complications arising from over—all anisotropy discussed above. The results for the 17.5 Mev γ-transition following proton capture in Li^7 are shown in Fig. 38. Within the limited accuracy there is agreement between theory and experiment with an assignment for the multipole transition ($M1$) which has been established in other, independent ways[1].

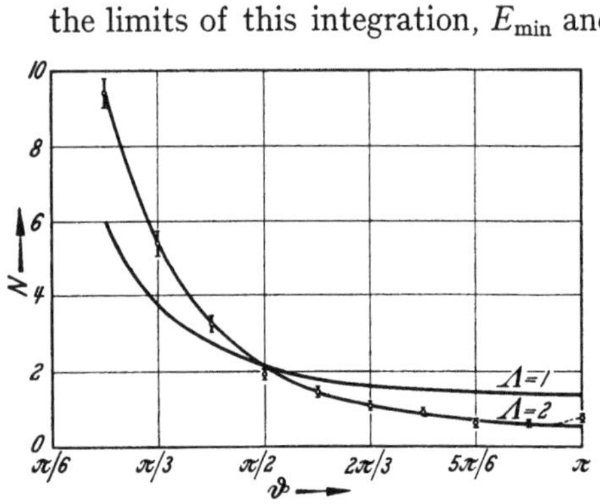

Fig. 38. Angular correlation of pairs from Li^7 (p, γ). Curve labelled $\Lambda = 1$ is the theoretical curve for $E1$ and $\Lambda = 2$ that for $M1$. The theoretical curve for $E2$ is nearly the same as that for $M1$. Both curves (1) and (2) are normalised to coincide at $\pi/2$ [cf. S. DEVONS and G. GOLDRING: Proc. Phys. Soc. Lond. A **67**, 414 (1954)].

In one special case, pure monopole transitions, pair-creation represents one of the most probable processes since the corresponding γ-emission is not allowed. Providing competing γ-transitions to other nuclear levels are also absent, then the properties of pair-creation can be studied with some precision. Such favourable conditions are present for the transition from the first excited state of O^{16}; $[0^+ \text{(6 Mev)} \rightarrow 0^+ \text{(ground)}]$. The energy-integrated electron-positron angular correlation has been measured[2] in this case and found to agree well with theoretical predictions. The field responsible for pair-emission in this case is that inside the nuclear surface, since the field outside is the simple Coulomb time-independent one. The theory in this special case has been examined in detail and higher-order corrections to the single theory are estimated (DALITZ [*102*]) to be only of the order 1%. This does not, however, mean that the approximations used in dealing with γ-radiation transitions are equally valid.

β) Inelastic scattering of charged particles[3]. We refer here to processes in which the whole of the interaction responsible for the inelastic collision is electromagnetic in nature, and the essentially short-range nuclear interaction is ab-

[1] An order-of-magnitude estimate indicates that the difference of the radiation field from the ideal multipole form (48.1) inside the nucleus might lead to departure from the results (48.3) which were appreciable at large angles ($\vartheta \gtrsim \pi/2$). Such discrepancies between theory and experiment as have been observed cannot be conclusively attributed to theoretical rather than experimental shortcomings.

[2] S. DEVONS, G. GOLDRING and G. R. LINDSEY: Proc. Phys. Soc. Lond. A **67**, 134 (1954).

[3] A full discussion of Coulomb excitation is given in the review article by K. ALDER, A. BOHR, T. HUUS, B. MOTTELSON and A. WINTHER: Rev. Mod. Phys. **28**, 432 (1956).

sent. Such processes are to be contrasted with the inelastic scattering processes involving, for example, compound-nucleus formation or nuclear-surface interactions which have been discussed previously, and in which the electromagnetic forces play an essentially minor rôle. For an inelastic scattering process to be interpreted entirely in terms of electromagnetic interaction it is necessary for the effect of nuclear forces to be negligible, and this occurs in two broad classes of phenomena:

(i) The mutual repulsion between the interacting particles is sufficiently large, and their relative energy sufficiently small, to ensure that they do not approach closely enough for nuclear forces to be involved—and yet the probability of an inelastic collision is appreciable. This is the case with heavy charged particles (protons, α-particles or heavier nuclei) of moderate energies (a few Mev per nucleon) in collision with medium and heavy nuclei, for which also γ-transitions of low energy ($\ll 1$ Mev) are possible. Inelastic collisions of this sort are normally (but inappropriately) referred to as Coulomb excitation processes.

(ii) The interaction between electrons (and μ-mesons) and nuclei over a wide range of energy (up to $\sim$ Gev) is primarily electromagnetic in nature. The inelastic scattering of electrons by nuclei is very closely the reverse of the pair-creation process, except that for the scattering case both initial and final electron states are positive energy ones. The study of this nuclear inelastic electron scattering, in the presence of strong elastic scattering and radiation background, is best achieved by using high-energy electrons ($\gtrsim 100$ Mev). For electrons of these energies the approximation used in dealing with pair-creation, namely that the interaction can be expanded in terms of ideal point multipoles is not valid, since the condition $\mathfrak{k} R \ll 1$ is not well satisfied. In fact the full interpretation of high-energy electron scattering involves a more detailed picture of the nuclear states than is required for the description of low-energy scattering and "Coulomb" excitation. The discussion of high-energy electron phenomena is rather beyond the range of the present article, particularly since both experiment and theory in this field are in the early stages of rapid development[1].

In this section we shall continue with a brief discussion of the low-energy inelastic scattering.

This process has been the subject of theoretical investigation for some time[2], although the earlier calculations were mainly concerned with electric-dipole transitions. More recently the investigations have been extended to included higher multipole transitions. In fact the most frequently observed inelastic collisions of this sort involve the excitation of low-energy $E\,2$ transitions, especially those with large transition multipole moments which are associated with rotations and vibrations of highly deformed nuclei having large intrinsic quadrupole moments[3]. Low-energy $E\,1$ transitions with large transition moments are rare.

The inelastic collision is normally detected, and the cross-section for the process measured, by observing the γ-radiation resulting from the de-excitation of the target nucleus. The variation of the total cross-section with energy depends on the multipolarity of the transition and the reduced matrix elements. (It is for the determination of these latter that the process of Coulomb excitation is most important.) The angular distribution of the γ-radiation with respect

[1] An excellent review of the subject has been written recently by R. HOFSTADTER: Rev. Mod. Phys. **28**, 214 (1956).

[2] L. LANDAU: Phys. Z. Sowjet. **1**, 88 (1932). — V. WEISSKOPF: Phys. Rev. **53**, 1018 (1938). — C. MULLIN and E. GUTH: Phys. Rev. **82**, 141 (1951). — R. HUBY and H. C. NEWNS: Proc. Phys. Soc. Lond. A **64**, 619 (1951). — K. A. TER-MARTIROSYAN: J. exp. theor. Phys. USSR **22**, 284 (1952).

[3] A. BOHR and B. R. MOTTELSON: Dan. Mat.-Fys. Medd. **27**, No. 16 (1953).

to incident particles, involves the nuclear spin as well as the transition multipolarity. In addition the differential cross-section for inelastic scattering itself, and the angular correlations between the inelastic scattered particles and the subsequent γ-radiation can also be studied.

The original calculation of ALDER and WINTHER [*104*] of the angular correlation between the incoming particle and the γ-ray follows the semi-classical treatment of TER-MARTIROSYAN [*103*], where the trajectory of the particle is considered classically as a Kepler hyperbolic orbit. The condition for the validity of such a treatment is that

$$\eta = \frac{Z_1 Z_2 e^2}{\hbar v} \gg 1, \tag{48.4}$$

where Z_1 and Z_2 refer to the charges of the incident particle and the target nucleus, and v is the initial velocity of the particle. In addition, the nuclear interaction can be ignored since the low energy implicit in (48.4) limits the penetration through the Coulomb barrier. Time-dependent perturbation theory is used, except that for the perturbation, represented by

$$V = \sum_{p=1}^{Z_2} \frac{Z_1 e^2}{|r_p - r(t)|}, \tag{48.5}$$

$r(t)$ represents the classical trajectory of the particle, the coordinates of the protons in the nucleus being denoted by r_p. This approximation is appropriate to electric multipoles. From a knowledge of the population of the excited levels of the nucleus, the angular correlation involving subsequent γ-radiation is found to be

$$W(\vartheta) = \sum_{k=0,2,4} A_k[1, 2]\, a_k\, P_k(\cos\vartheta) \tag{48.6}$$

where A_k represents the usual angular correlation coefficients for γ-γ transitions (19.1) and a_k denotes the specific Coulomb-excitation parameters. These parameters, expressed as a function of $\xi = \frac{\Delta E}{2E}\eta$, where E is the kinetic energy of the particle in the centre-of-mass system of the particle and target, and ΔE represents the energy lost by the particle, were evaluated by ALDER and WINTHER [*104*] by performing a numerical integration over the trajectory of the particle. Subsequent experiments disagreed with this angular correlation, but, as discovered by BREIT, EBEL and RUSSELL [*104*], the discrepancy is on account of a numerical error in the formula for a_2.

If no coincidence measurement is made which involves detection of the particle following excitation, there is a close relationship between this process and processes such as internal conversion and β-decay followed by γ-ray emission[1]. The first "radiation" corresponds to a virtual photon which is absorbed by the nucleus and observation is made of this photon through its virtual emission by the incident, inelastically scattered particle. In the absence of information concerning the direction of this particle after sattering, the efficiency tensor for this particle, $\varepsilon_{\bar{k}\bar{\varkappa}}(\bar{l}\,\bar{l}')$ is represented simply by the following:

$$\varepsilon_{\bar{k}\bar{\varkappa}}(\bar{l}\,\bar{l}') = \hat{\bar{l}}\, \delta_{\bar{l}\bar{l}'}\, \delta_{\bar{k}0}\, \delta_{\bar{\varkappa}0}. \tag{48.7}$$

As in internal conversion, the Hamiltonian for the process is factorised into a matrix element involving the nuclear transition and a Coulomb matrix element

[1] L. C. BIEDENHARN, J. L. MCHALE and R. M. THALER: Phys. Rev. **100**, 376 (1955).

describing the electromagnetic process. The angular correlation is

$$W = \sum_{k_b \varkappa_b b B B' \bar{l} \bar{l}'} \varrho_{k_b \varkappa_b}(b\,B, b\,B')\, \varrho_{\bar{k}\bar{\varkappa}}(\bar{l}\,\bar{l}')\, \varepsilon^*_{k_b \varkappa_b}(b\,B, b\,B')\, \varepsilon^*_{\bar{k}\bar{\varkappa}}(\bar{l}\,\bar{l}'). \tag{48.8}$$

The tensor $\varrho_{k_b \varkappa_b}$ is expressed in terms of $\varrho_{k_1 \varkappa_1}(L_1 L_1')$ and $\varrho_{k_a \varkappa_a}(a\,a')$ as in (12.1). The statistical tensor $\varrho_{k_1 \varkappa_1}(L_1\,L_1')$ is then coupled with $\varrho_{\bar{k}\bar{\varkappa}}(\bar{l}\,\bar{l}')$. Thus,

$$\left.\begin{aligned} \varrho_{k_1 \varkappa_1}(L_1 L_1')\, \varrho_{\bar{k}\bar{\varkappa}}(\bar{l}\,\bar{l}') &= \sum_{k_l \varkappa_l} \varrho_{k_l \varkappa_l}(l\,l')\,(k_l \varkappa_l \mid k_1 \varkappa_1, \bar{k}\,\bar{\varkappa})\, \hat{l}\,\hat{l}'\,\hat{k}_1 \hat{\bar{k}} \times \\ &\times \begin{Bmatrix} L_1 & \bar{l} & l \\ L_1' & \bar{l}' & l' \\ k_1 & \bar{k} & k_l \end{Bmatrix} \langle \bar{l} \,\|\, Q(L_1) \,\|\, l \rangle \langle \bar{l}' \,\|\, Q(L_1') \,\|\, l' \rangle^*\, e^{i(\xi_l - \xi_{l'})} \end{aligned}\right\} \tag{48.9}$$

where $\langle \| Q(L) \| \rangle$ represent reduced Coulomb multipole matrix elements. We assume no spin-interaction of the incident particle, so that $\varrho_{k_l \varkappa_l}(l\,l')$ is given as

$$\left.\begin{aligned} \varrho_{k_l \varkappa_l}(l\,l') &= c_{k_l 0}(l\,l')\, \mathscr{D}^{k_l}_{\varkappa_l 0}(\mathscr{R}_1) \\ &= \frac{\hat{l}\,\hat{l}'}{4\pi}\,(-)^{l'}\,(l\,0, l'\,0 \mid k_l 0)\, \mathscr{D}^{k_l}_{\varkappa_l 0}(\mathscr{R}_1). \end{aligned}\right\} \tag{48.10}$$

From the above equations, we finally obtain the result that the angular correlation for Coulomb excitation differs from that corresponding to the emission of two radiations by the replacement of the first radiation parameter by

$$\left.\begin{aligned} \sum_{l l' \bar{l}} (\hat{l}\,\hat{l}')^2\,(-)^{l'}\, e^{i(\xi_l - \xi_l')}\,(l\,0, l'\,0 \mid k\,0)\, W(l\,l'\,L_1 L_1'; \bar{l}\,k) \times \\ \times \langle \bar{l} \,\|\, Q(L_1) \,\|\, l \rangle \langle \bar{l} \,\|\, Q(L_1') \,\|\, l' \rangle^*. \end{aligned}\right\} \tag{48.11}$$

The reduced Coulomb multipole matrix elements have been evaluated by BIEDENHARN *et al.*[1] for arbitrary types of multipoles, and in particular, for electric quadrupole excitation.

The results can be given in terms of Coulomb-excitation parameters, $a_k(\mathscr{L})$ which represent the ratio of the radiation parameter (48.11) to that for a γ-ray of the same multipole character (14.14). In particular, for electric-quadrupole excitation,

$$a_k(2e) = \frac{\sum_{l \bar{l} l'} \hat{l}^2\, \hat{l}'^2\, e^{i(\xi_l - \xi_{l'})}\,(-)^{l'} \langle \bar{l} \| Q(L) \| l \rangle \langle \bar{l} \| Q(L') \| l' \rangle^*\, W(l\,\bar{l}\,k\,2; 2\,l')\,(l0, l'0/k0)}{-\,5\,(2\,1, 2-1 \mid k\,0) \sum_{l \bar{l}} \hat{l}^2\, |\langle \bar{l} \| Q(L) \| l \rangle|^2}. \tag{48.12}$$

The numerator in the expression (48.12) is usually defined as $b_k(\mathscr{L})$ and $b_0(\mathscr{L})$ is referred to as the excitation function. Explicit values of $b_k(\mathscr{L})$ for electric-quadrupole excitation are found to be

$$\left.\begin{aligned} b_0 &= \sum_l \left[\frac{l(l+1)(2l+1)}{(2l-1)(2l+3)}\, I^2(l,l) + \frac{3(l+1)(l+2)}{2(2l+3)}\, I^2(l+2,l) + \frac{3l(l+1)}{2(2l-1)}\, I^2(l-2,l)\right], \\ b_2 &= \sum_l \left[\frac{3(l+1)(l+2)(l+3)}{(2l+3)^2}\, I^2(l+2,l) + \frac{3l(l-1)(l-2)}{(2l-1)^2}\, I^2(l-2,l) - \right. \\ &\quad - \frac{l(l+1)(2l-3)(2l+1)(2l+5)}{(2l+3)^2(2l-1)^2}\, I^2(l,l) - 6\cos(\xi_l - \xi_{l+2})\, \frac{l(l+1)(l+2)}{(2l+3)^2} \times \\ &\quad \left. \times\, I(l+2,l)\, I(l,l) - 6\cos(\xi_l - \xi_{l-2})\, \frac{l(l-1)(l+1)}{(2l-1)^2}\, I(l-2,l)\, I(l,l)\right], \end{aligned}\right\} \tag{48.13}$$

---

[1] Cf. Appendix I of Ref. [2], p. 130. The reduced matrix elements differ from those in (48.11) by factors $i^{l-\bar{l}}$, since the basic functions used are Y_l^λ instead of $i^l Y_l^\lambda$. The latter are necessary if the matrix elements are to be invariant with respect to time inversion. BIEDENHARN *et al.* obtain the correct final results by introducing the factor $i^{l-l'}$ in their expression for $\varrho_{k_l \varkappa_l}(l l')$.

with a more complicated expression for b_4. In the above formulae, $I(l, l')$ are given by

$$I(l, l') = \int_0^\infty \frac{dr}{r^3} F_l(\eta_{\text{init}}, \mathfrak{k}_{\text{init}} r)\, F_{l'}(\eta_{\text{fin}}, \mathfrak{k}_{\text{fin}} r), \qquad (48.14)$$

F_l represents the radial Coulomb wave functions, and the suffices *init* and *fin* refer to initial and final values, respectively. The summation over l in the expression for b_k introduces difficulties, since $I(l, l')$ does not converge rapidly for large values of l. By use of recursion relations and transformation of the integrals into more useful forms[1], numerical values of b_k have been found for various values of $\varrho = \mathfrak{k}_{\text{init}}/\mathfrak{k}_{\text{fin}}$ and η. BIEDENHARN and collaborators [*106*], have, in addition, obtained numerical results for electric-dipole excitation and for magnetic-dipole excitation.

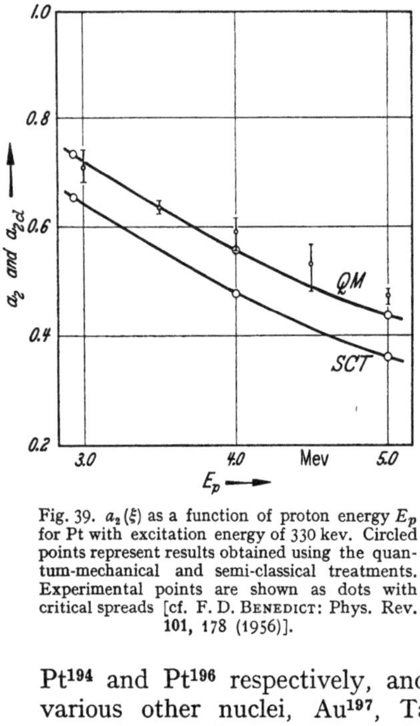

Fig. 39. $a_2(\xi)$ as a function of proton energy E_p for Pt with excitation energy of 330 kev. Circled points represent results obtained using the quantum-mechanical and semi-classical treatments. Experimental points are shown as dots with critical spreads [cf. F. D. BENEDICT: Phys. Rev. **101**, 178 (1956)].

A detailed analysis of the semi-classical treatment by BREIT and collaborators[2] shows that for a certain interval of energy and assuming no energy loss, i.e. $\xi = 0$, the integrals involved in the semi-classical treatment agree quite closely with the integrals that occur in quantum-mechanical calculations. In fact, if the semi-classical treatment is modified by using the W.K.B. approximation for the evaluation of the Coulomb wave functions, the evaluations of the total cross-section for Coulomb excitation differ by less than a few percent from the result of a quantum-mechanical calculation.

Verification of the theoretical predictions for the angular correlation is provided by the experiments of STELSON and McGOWAN[3] who measured 330 and 358 kev γ-rays from Pt^{194} and Pt^{196} respectively, and later observed the angular distribution from various other nuclei, Au^{197}, Ta^{181}, $Ag^{107,109}$, $Pd^{106,108,110}$ and Rh^{103}. BENEDICT[4] has used the semi-classical approximation to calculate $I(l, l')$ for the excitation of the 330 kev γ-ray level of Pt^{194}. By inserting these values into the above equations or into the equations obtained by the treatment of Ter-Martirosyan, he found good agreement between the quantum-mechanical predictions and experiment for the values of a_2 for various values of the energy. On the other hand, the semi-classical treatment introduces marked deviations. This is illustrated in Fig. 39. The coefficient of $P_4(\cos\vartheta)$ is not considered, since its magnitude is not very great over the energy range that is of interest.

The angular correlation between inelastically scattered particle and subsequent γ-radiation has been examined theoretically by BREIT and coworkers [*107*],

[1] L. C. BIEDENHARN, J. L. McHALE and R. M. THALER: Phys. Rev. **100**, 376 (1955). — K. ALDER and A. WINTHER: Dan. Mat.-Fys. Medd. **29**, Nos. 18, 19 (1955).

[2] G. BREIT and P. B. DAITCH: Phys. Rev. **96**, 1447 (1956). — F. D. BENEDICT, P. B. DAITCH and G. BREIT: Phys. Rev. **101**, 171 (1956). — R. L. GLUCKSTERN, J. P. LAZARUS and G. BREIT: Phys. Rev. **101**, 175 (1956).

[3] P. H. STELSON and F. K. McGOWAN: Phys. Rev. **98**, 249 (1955); **99**, 127 (1955).

[4] F. D. BENEDICT: Phys. Rev. **101**, 178 (1956).

particularly in relation to an interesting possibility of observing, in Coulomb excitation, the interaction between the static quadrupole moment of the excited nucleus and the electric field produced by the inelastically scattered particle. Such an interaction could produce a re-orientation of the target nucleus, after its excitation, which would in turn modify the angular distribution and correlation properties predicted by the ordinary theory which neglects this higher-order effect. The conditions which are likely to be most favourable for detecting and measuring this orientation are discussed by BREIT *et al.* [*107*].

II. β-Decay.

49. β-Decay. Angular-correlation studies involving β-decay, like those of the other processes discussed in this section, fall into two categories. First there is the β-decay process itself, in which, associated with a nuclear transition, there is emitted both an electron and neutrino, or equivalently the process may be described as the absorption of an anti-neutrino and the emission of an electron[1]. From a theoretical viewpoint one can examine the directional (and in principle polarisation) correlation between the absorbed anti-neutrino and the emitted electron; from the experimental viewpoint the direction of recoil of the transforming nucleus is determined, and from this and the β-particle momentum the anti-neutrino direction is inferred from the conservation of momentum. The purpose of such experimental studies, which present formidable problems even in favourable cases, is the study of the nature of the interaction responsible for β-decay itself, rather than the investigation of the symmetry of the nuclear states involved, for which other methods are more readily available.

The second type of process is one in which the β-decay is part of a cascade, most commonly the β-γ sequence, and the angular correlation studied is that between the emitted electron and the subsequent γ-radiation. The neutrino (or recoil nucleus) is undetected, that is, we deal with the β-γ correlation averaged for all directions of the absorbed anti-neutrino. Experimentally, the study of such correlations is similar to that of other cascades, particularly those involving conversion-electrons, although the continuous nature of the electron energy spectrum introduces important differences. Such limited studies as have been made of cascades of this sort have been concerned with the determination of both the properties of the β-decay interaction and the symmetry properties of the nuclear states involved. Owing to the much greater complexity of the β-decay interaction than that responsible, for example, for γ-radiation, unequivocal conclusions are not easily derived from the study of cascades involving β-decay.

We shall first discuss briefly the results obtained from the study of the correlations in the β-decay process itself, since these have a direct and important bearing on the interpretation of cascade processes[2].

---

[1] We refer for definiteness to the case of β-emission. Positron emission is formally equivalent to the absorption of an electron and the emission of a neutrino. In the subsequent formulae, the change to positron emission is accompanied by the change $Z \to -Z$ and $C_T \to -C_T$ [(H. A. TOLHOEK and S. R. DE GROOT: Physica, Haag **16**, 456 (1953)].

[2] Throughout this article we have assumed that parity is conserved in all interactions. It has recently been pointed out by T. D. LEE and C. N. YANG [Phys. Rev. **104**, 254 (1956)] that the previously existent experimental evidence did not exclude the possibility that parity is not conserved in the "weak" interactions (such as those involved in β-decay or in meson-decay processes). Among the experiments proposed to test this possibility are measurement of: (1) the angular correlation of β-particles, emitted from oriented cobalt nuclei and (2) the angular distribution of electrons in the π-μ-e decay. Since the article was written, such experiments have been made [C. S. WU, E. AMBLER, R. W. HAYWARD, D. D. HOPPES and R. P. HUDSON, Phys. Rev. **105**, 1413 (1957) (β-decay); R. L. GARWIN, L. M. LEDERMAN and

(*a*) *β-ν correlation*. The interaction Hamiltonian in β-decay can be compounded, as is well known, from five independent, relativistically invariant terms, each bilinear in nuclear and lepton wavefunctions, and which are characterised by the Lorentz-transformation properties of the factors from which they are formed. Thus

$$H = C_S H_S + C_V H_V + C_T H_T + C_A H_A + C_P H_P, \tag{49.1}$$

where S, V, T, A, P, refer to scalar, vector, tensor, axial-vector, and pseudo-scalar interactions, and the coefficients C are real constants. In evaluating this interaction it is usual to use non-relativistic expressions for the nuclear matrix elements (expanding, if necessary, in powers of nucleon velocity). The lepton term is expanded in powers of $(E r/\hbar c)^2$ (E is the energy of the transition and the leading non-vanishing term is retained). In general, the β-decay is not necessarily represented by all the terms in (49.1). In addition, the symmetry properties of the nuclear states involved in particular β-transitions may result in a further simplification. For allowed transitions, i.e. those for which the first term, of zero power in $\left(\frac{E r}{\hbar c}\right)^2$, in the expansion is non-vanishing, the nuclear selection rules are:

$a - b = 0$, no parity change: C_S and C_V,

$a - b = 0, \pm 1$, no parity change (except $a = b = 0$): C_T and C_A.

The pseudoscalar coupling results in allowed transitions only when there is a parity change.

For an allowed transition involving only one term of (49.1), the shape of the β-spectrum is independent of the type of the interaction; even if more than one type of interaction is involved, the spectrum is only slightly different, if at all. Apart then from demonstrating that the pseudoscalar term is zero or small, evidence from β-spectra and decay-lifetimes of allowed transitions does not enable the nature of the interaction to be determined. The shapes of β-spectra of forbidden transitions are dependent on the nature of the interaction, but the unambiguous analysis is not possible owing to the many unknown terms which can contribute in most cases.

The β-ν correlation, on the other hand, is dependent on the interaction even for allowed transitions, where the theoretical analysis is straightforward. Assuming C_P is negligible, the electron-anti-neutrino correlation is given by (DE GROOT and TOLHOEK [*52*], ROSE [*14*]):

$$W_\pm(\vartheta, E) = f(E)\left[1 \mp \frac{B}{E} + A\,\frac{v}{c}\cos\vartheta\right], \tag{49.2}$$

M. WEINRICH, Phys. Rev. **105**, 1415 (1957)] (decay of μ-meson)], and the result do in fact, show that parity is not conserved in these processes.

Consequently many statements made in this chapter need revision in order to take into account this violation of one of the fundamental principles of quantum mechanics used in angular-correlation theory. In particular, the most general interaction Hamiltonian for β-decay includes besides the five terms in (49.1) five additional terms which owe their origin to lack of parity conservation. Nevertheless, the expression for the β-ν and β-ν correlations, which are presented in this section, remain valid except for modifications of the coupling constants. For a detailed discussion, see the paper by LEE and YANG and the article on β-decay by L. NORDHEIM and W. GREULING to appear in Vol. XLI of this Encyclopedia. For a summary of the developments in the field of β-decay prior to the recent investigation, see e.g. the review by E. J. KONOPINSKI and L. M. LANGER: Ann. Rev. Nucl. Sci. **2**, 261 (1953), or [*14*]: M. E. ROSE: Theory of Allowed β-Decay, Chap. IX; C. S. WU: Experiments on the Shape of β-Spectra, Chap. XI.

where v and E refer to the velocity and total energy of the electron, the upper and lower signs refer to positron and electron-emission, respectively, and

$$A = \frac{\frac{1}{3}(C_T^2 - C_A^2) + (C_V^2 - C_S^2)\,\varrho}{C_T^2 + C_A^2 + (C_V^2 + C_S^2)\,\varrho}, \tag{49.3}$$

$$B = 2\sqrt{1 - \left(\frac{Z e}{\hbar c}\right)^2}\,\frac{\varrho\, C_S C_V + C_T C_A}{\varrho(C_S^2 + C_V^2) + C_T^2 + C_A^2}, \tag{49.4}$$

$$\varrho = \frac{|\langle|\mathbf{1}|\rangle|^2}{|\langle|\sigma|\rangle|^2}. \tag{49.5}$$

The expression for $f(E)$ includes the effect of the Coulomb field and is given, for example, by ROSE [*14*].

The nuclear matrix elements in (49.5) indicate the form of the interaction in the non-relativistic approximation; the operator **1** refers to Fermi interactions obeying the selection rule $a-b=0$, and σ to the Gamow-Teller interactions for which $a-b=0, \pm 1$. The term $\mp B/E$ in (49.2) also occurs in the theoretical expression for the shape of allowed β-spectra (the so-called "Fierz interference" term), but the experimentally observed β-spectra indicate[1] that $B=0$. This implies that the interactions cannot involve both S and V or both T and A. The angular correlation (49.2) is particularly simple for pure interactions: A takes the values -1, 1, $\frac{1}{3}$ and $-\frac{1}{3}$ for S, V, T and A respectively.

The choice between S and V, and between T and A, and the relative strengths of the two types of interaction can therefore be made by examining the β-ν correlation for two or more transitions in which the nuclear selection rules ensure different contributions from Fermi and Gamow-Teller interactions in the two transitions.

In practice, the correlation is made between the electron and the nuclear recoil. The transformation between the correlations of the two types has been examined in detail by KOFOED-HANSEN [*54*]. Electron recoil-nucleus correlations in the allowed β-decay $He^6 \rightarrow Li^6$, involving only a Gamow-Teller interaction, have been measured by RUSTAD and RUBY[2]. Experimentally conditions here are also favourable to precise measurements, since the large energy of the β-transition and the light nucleus result in large recoil energies. A schematic diagram of the apparatus is shown in Fig. 40. For two different electron energy ranges the values of A obtained were 0.36 ± 0.11 and 0.31 ± 0.14, as compared with the theoretical value of $+\frac{1}{3}$ for T and $-\frac{1}{3}$ for A, showing that the Gamow-Teller interaction is predominantly T.

Similar experiments by MAXON, ALLEN, and JENTSCHKE[3] using the $Ne^{19} \rightarrow F^{19}$ β-transition, for which both Fermi and Gamow-Teller interactions are possible, indicate that the Fermi contribution is predominantly scalar. To establish the ratio C_S/C_T it is necessary to assess the ratio of matrix elements, ϱ, on the basis of a nuclear model.

The β-ν correlation in the decay of the neutron has also been examined recently by measurement of momenta at a fixed angle[4]. From these correlation measurements, together with spectrum and lifetime of particular β-processes[5], one can

---

[1] H. M. MAHMOUD and E. J. KONOPINSKI: Phys. Rev. **88**, 1266 (1952). — O. KOFOED-HANSEN and A. WINTHER: Phys. Rev. **89**, 526 (1953). — J. DAVIDSON and D. C. PEASLEE: Phys. Rev. **92**, 1584 (1953).

[2] B. M. RUSTAD and S. L. RUBY: Phys. Rev. **97**, 991 (1955).

[3] D. R. MAXON, J. S. ALLEN and W. K. JENTSCHKE: Phys. Rev. **97**, 109 (1955).

[4] S. A. ROBSON: Phys. Rev. **100**, 933 (1955).

[5] In particular the lifetime for neutron decay and the β-spectrum of C^{14}, cf. J. B. GERHART: Phys. Rev. **95**, 288 (1954); C. S. WU [*14*], Chap. XI.

conclude that the β-interaction is, at least predominantly, S and T with possibly a small P-component, and the ratio C_S/C_T is about unity.

(b) β-γ correlations. The study of β-γ angular correlations, like that of γ-γ cascades, can be utilised to study the nature of the β (or γ transitions) and the spins and parities of the nuclear states involved. There are, however, many extra features associated with β-transitions that render cascades involving them much more complicated, less equivocal, and less generally applicable than γ-γ cascades.

Firstly, all allowed transitions lead to isotropy. This feature can be used to check the allowed nature of a β-transition, but this fact can usually be deduced more readily from the shape of the β-spectrum or the β-decay lifetime ("ft" values). In general, forbidden β-transitions result in some anisotropy in β-γ correlations, but the extent of the anisotropy is not necessarily related in a simple fashion with the departure of the spectrum from the allowed shape. Since the energy-distribution of the electrons is continuous, the correlation can be studied as a function of the electron energy (the neutrino or recoil nucleus is not, in practice, observed). The interpretation of the correlation must, of course, be consistent with the observed uncorrelated β-spectrum and lifetime.

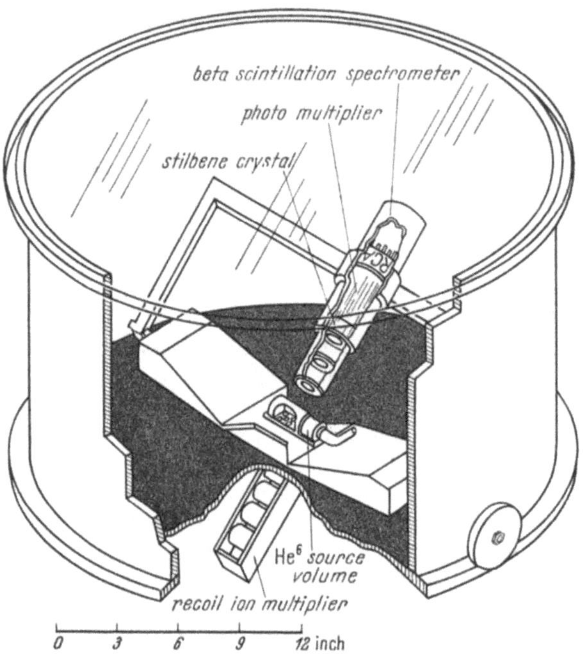

Fig. 40. Schematic sketch of the apparatus used to determine the electron-recoil correlation in the decay of He^6. Both source and spectrometer are located in the vacuum of the bell-jar [cf. B. M. Rustad and S. L. Ruby: Phys. Rev. 97, 994 (1955)].

The major complication arises from the fact that the β-interaction is not simple. We have seen that there are, in general, contributions from at least two types of interaction, scalar and tensor. Moreover, each interaction separately can introduce terms which have different symmetry properties with respect to ordinary three-dimensional spatial rotations. For a particular order-of-forbiddenness the matrix-elements of the interaction generally include (i) terms corresponding to different total angular momentum carried away by the electron-neutrino pair, and (ii) different contributions either from the same or from different basic operators (S and T) but corresponding to the same total angular momentum. In the quadratic products that appear in the expressions for both the β-spectrum and the angular correlation, interference terms (mixed products of matrix-elements) also occur. Interference between terms of the type (i) affects the angular correlation, similarly to multipole mixtures in γ-transitions, but does not affect the β-spectra. Interference involving type (ii) terms affects both the β-spectrum shape and the angular correlation.

The kinetic energy of the electron in β-decay is usually of the same order as, or smaller then (often much smaller than), its potential energy at the nuclear surface. Consequently the nuclear radius and charge, R, Z, and the electron energy, E, are important parameters in the correlation function, quite apart from the usual presence of Z and E in the Fermi function. Approximations which depend on the magnitude of these and other factors have usually to be made and the magnitude of the contributions from the individual interaction terms estimated for each particular case studied.

The first calculations of the angular correlation in β-γ cascades were made by FALKOFF and UHLENBECK [*51*]. Nuclear charge was neglected and the various matrix-element terms were considered independently. This work was extended by YAMADA and MORITA[1] so as to include the effects of nuclear charge, but neglecting finite nuclear size, and mixing of terms with a pure interaction. SPIERS and BLIN-STOYLE[2], using a formalism involving the total angular momentum, L, of the electron-neutrino field similar to that described in Sect. 9, extended the considerations to an arbitrary degree of forbiddenness, where previously calculations were made only up to second-order forbidden processes. BIEDENHARN and ROSE [*9*], using a similar L-representation, treated the problem using statistical tensors in a manner which is formally similar to that used for the internal-conversion process.

In the case of internal-conversion, the electron is regarded as transformed from a bound state to an outgoing plane wave by virtue of an electromagnetic interaction (the absorption of a virtual photon). Likewise in β-decay an incoming anti-neutrino is transformed into an outgoing electron by virtue of a point interaction between the electron-neutrino and nuclear fields. The matrix-elements describing the interaction may be separated into the product of two covariant elements which refer respectively to the nuclear and lepton (electron-neutrino) wavefunctions, which is contracted into a scalar and evaluated at a point. Thus the matrix element is of the form:

$$\sum_i C_i \sum_\mu \langle \psi_b | \mathcal{O}_{i\mu} | \psi_a \rangle \cdot \langle \psi_e | \mathcal{O}_{i\mu} | \psi_\nu \rangle \tag{49.6}$$

where ψ_b, ψ_a, ψ_e and ψ_ν denote, respectively, the final and initial nuclear wavefunctions, and the electron and neutrino wavefunctions, $\sum\limits_\mu$ denotes contraction, and $\sum\limits_i$ is over the appropriate basic interactions. The lepton matrix element is evaluated at values of the lepton position vector, $\boldsymbol{r}_l = \boldsymbol{r}_n$, where $\boldsymbol{r}_n$ is a nucleon position.

Biedenharn and Rose transform the matrix element into the L-representation by the following procedure. For those components of the interaction which transform in 3-dimensional space as scalars, they expand the δ-function:

$$\delta(\boldsymbol{r}_n - \boldsymbol{r}_l) = \frac{1}{r^2}\,\delta(r_n - r_l) \sum_{L\Lambda} Y_L^{\Lambda +}(\boldsymbol{r}_n)\, Y_L^\Lambda(\boldsymbol{r}_l), \tag{49.7}$$

whilst for those which transform as vectors, they use the expression for the unit dyadic δ-function:

$$\boldsymbol{I}\,\delta(\boldsymbol{r}_n - \boldsymbol{r}_l) = \frac{1}{r^2}\,\delta(r_n - r_l) \sum_{JL\Lambda} \boldsymbol{\Phi}_{JL}^\Lambda(\boldsymbol{r}_n)\, \boldsymbol{\Phi}_{JL}^\Lambda(\boldsymbol{r}_l), \tag{49.8}$$

where $J = L$ or $L \pm 1$ and $\boldsymbol{\Phi}_{JL}^\Lambda$ are vector operators with special properties. The Pauli adjoint symbol $^+$, is defined by: $\psi^+ = i\,\psi^*\gamma_4$.

---

[1] M. YAMADA and M. MORITA: Progr. Theor. Phys. **8**, 431 (1952); **10**, 641 (1953). — M. MORITA: Progr. Theor. Phys. **10**, 363 (1953). — Y. KATO and M. MORITA: Progr. Theor. Phys. **13**, 276 (1955).

[2] J. A. SPIERS and R. J. BLIN-STOYLE: Proc. Phys. Soc. Lond. A **65**, 809 (1952).

The operators for the lepton fields with different 3-dimensional rotational and reflection symmetries are listed in Ref. [9], Table 5. The values of L and parity are sufficient to specify the interaction operator only in the case of scalar and pseudoscalar interactions. In the case of other interactions an additional index n is used to specify different terms with the same L and parity values (same $\mathscr{L}$).

The lepton wavefunctions are represented by spinors with quantum numbers K and K_ν referring to electron and neutrino respectively (cf. Sect. 43 on internal conversion). The reduced matrix elements for the leptons, $\langle K \| \mathscr{L} n \| K_\nu \rangle$, are tabulated in Ref. [9], Table 6, in terms of radial wavefunctions evaluated at the nuclear radius and for a point-charge nucleus[1]. As compared with the initial electron state in the case of internal conversion, the initial anti-neutrino state is not represented simply by the quantum number, $K_\nu = -1$. The statistical tensor is however still represented by δ-functions:

$$\varrho_{k_\nu \varkappa_\nu}(K_\nu K'_\nu) = \frac{1}{\hat{L}_\nu} \delta_{k_\nu 0} \delta_{\varkappa_\nu 0} \delta_{K_\nu K'_\nu}. \tag{49.9}$$

As in internal conversion, the observation is made of the outgoing electron, and it is the efficiency tensor for the virtual radiation that appears in the general angular-correlation formula. By coupling the neutrino states with the virtual radiation, we find as in (43.3) an expression in the angular correlation varying as:

$$\frac{\sqrt{K K'}}{\hat{L}_\nu^2} (-)^{L_\nu + k - L - L'} W(L L' L_1 L'_1; k L_\nu) \, \varepsilon_{k\varkappa}(K K'). \tag{49.10}$$

In (49.10), L, L' characterise the β-transition, and just as in gamma-ray emission refer to the multipole expansion that is used, while L_1, L'_1 refer to the total angular momentum of the emitted electron. $\varepsilon_{k\varkappa}(K K')$ refers to the efficiency tensor for observation of an electron in an asymptotic plane-wave state and is associated with angular momentum defined by K, K', where $L_1 = |K| - \frac{1}{2}$ etc. This tensor, which provides the relationship between the geometry of the observations and the physical process, is given by (43.4):

$$\left.\begin{aligned} \varepsilon_{k\varkappa}(K K') = (-)^{k + L'_1 - \frac{1}{2}} i^{l_1(K') - l_1(K)} \hat{l}_1 \hat{l}'_1 \sqrt{|K| \, |K'|} \, S(K) \, S(K') \times \\ \times (l_1 0, l'_1 0 | k 0) \, W(l_1 l'_1 L_1 L'_1; k \tfrac{1}{2}) \, \mathscr{D}^k_{\varkappa 0}(\mathscr{R}_1). \end{aligned}\right\} \tag{49.11}$$

l_1 and l'_1 are abbreviations for $l(K)$ and $l(K')$. There is a triangle condition $\Delta(l_1 l_\nu L)$, where l_ν is similarly an abbreviation for $l(K_\nu)$. In the evaluation of the angular correlation, summation is made over all possible quantum numbers, K, K', and K_ν. The summation over K_ν is incoherent corresponding to lack of observation of the neutrino (49.9). Two sets of reduced matrix elements occur: (1) those characterising the electronic transition and denoted by the products $\langle K \| \mathscr{L} n \| K_\nu \rangle \langle K' \| \mathscr{L}' n' \| K_\nu \rangle^*$, and (2) the reduced nuclear matrix elements which are expanded according to the order of forbiddenness, with significance given only to the lowest non-vanishing terms. The reduced nuclear matrix-elements $\langle b \| \mathscr{L} n \| a \rangle$ are real and are defined as usual by

$$\langle \psi_{b\beta} | \mathscr{O} | \psi_{a\alpha} \rangle = (b\beta, L\Lambda | a\alpha) \langle b \| \mathscr{L} n \| a \rangle. \tag{49.12}$$

[1] HAUSER [56] points out that the two approximations—a point charge nucleus and evaluation of the matrix elements at $r = R$ lead to quite small errors of opposite sign, and that the overall approximation is not likely to be significant in relation to present experimental accuracy.

The general result for the correlation between β-radiation and a succeeding radiation characterised by L_2, L_2', is given by the usual formula (15.3):

$$W(\vartheta) = \sum_k A_k(a, b b)\, A_k(b b, c)\, P_k(\cos\vartheta) \tag{49.13}$$

where $A_k(a, bb)$ describes the β-process and is given as in (15.4) by:

$$\left.\begin{aligned} A_k(a, b b) = \sum_{n n'} \sum_{L L'} \langle b \| \mathcal{L} n \| a\rangle \langle b \| \mathcal{L}' n' \| a\rangle\, c^*_{k 0}(\mathcal{L}\mathcal{L}') \times \\ \times\, \hat{b}\, (-)^{a-b+k-L'}\, W(b b L L'; k a) \end{aligned}\right\} \tag{49.14}$$

and

$$\left.\begin{aligned} c_{k 0}(\mathcal{L}\mathcal{L}') = \sum_{K K' K_\nu} (-)^{L_\nu - L - \frac{1}{2}}\, i^{l_1(K') - l_1(K)} \cdot \frac{\hat{l}_1 \hat{l}_1'}{\hat{L}_\nu^2}\, |K K'|\, S(K)\, S(K') \times \\ \times (l_1 0, l_1' 0 \mid k 0)\, W(l_1 l_1' L_1 L_1'; k \tfrac{1}{2})\, W(L L L_1 L_1'; k L_\nu) \times \\ \times \langle K \| \mathcal{L} n \| K_\nu\rangle \langle K' \| \mathcal{L}' n' \| K_\nu\rangle^* . \end{aligned}\right\} \tag{49.15}$$

The evaluation of $W(\vartheta)$ involves explicit, approximate calculations of the matrix elements, for example on the basis of S and T interaction of approximate equal strength. In the special cases when no interference is present, "particle parameters" can be defined, as for other radiations, by comparison with γ-emission. Thus the "particle" parameters $b_k(L L', n n'; \beta)$ are defined by Biedenharn and Rose [9]:

$$b_k(L L', n n'; \beta) = \frac{c_{k 0}(\mathcal{L} n, \mathcal{L}' n')}{c_{k 0}(L L'; \gamma)} \tag{49.16}$$

and are tabulated for pure interactions and first and second forbidden transitions The effect of the finite nuclear size on the lepton wavefunctions in the calculation of matrix elements is neglected.

When the β-γ correlation is measured with electrons of a particular energy selected from the β-spectrum, the correlation will generally decrease with decrease in electron energy. The total angular momentum transferred by the electron and anti-neutrino may be a constant, but as the electron energy approaches zero its wavefunction will approach spherical symmetry, and the β-γ correlation will vanish. In the limit of zero energy this can be seen from (49.10), since $K = \pm 1$ for these electrons and $W(L L' \frac{1}{2} \frac{1}{2}; k L_\nu)$ vanishes for even k unless $k = 0$. If the β-spectrum is of exactly allowed shape, i.e. the energy correction factor is independent of electron momentum, then also $K = \pm 1$ and there is no correlation. The extent of the departure from the allowed shape is not, however, necessarily a direct indication of the degree of departure from isotropy. In the first place small interaction-terms may appear linearly ("interference") in the correlation expression and only quadratically, and therefore negligible except in a higher approximation, in the β-spectrum. In addition, "accidental" cancellation due to interference between S and T terms can lead to a reduction in what would otherwise be a dominant interaction term giving an allowed shape and isotropy.

Hauser [56] has made detailed calculations of the correlation functions for first and second forbidden transitions assuming the presence of S, T and P interactions, and allowing for finite nuclear size. Since in most practical cases the electron kinetic energy is small compared to its potential energy on the nuclear surface, the terms in the correlation are classified according to the power of $\left(\frac{Z e^2}{E R}\right)$. In the limit $\frac{Z e^2}{E R} \to \infty$, the spectrum is the allowed shape and the correlation zero. However, as indicated above, for large finite values the effect on the spectrum can be negligible but that on the angular correlation quite significant. HAUSER also calculates the effect on the correlation in *allowed* transitions (satisfying Gamow-Teller selection rules only, i.e. only tensor interaction) of a small pseudoscalar interaction. His results are consistent with small or zero correlation[1] in Na^{24} and Na^{22} and the upper limits set to C_P from measurements of the spectrum shape[2] of He^6 and B^{12}.

The experimental investigation of β-γ correlation presents technical problems which are similar to those met in the case of internal conversion-electrons. There is an additional disturbance which arises from the inner-bremsstrahlung which

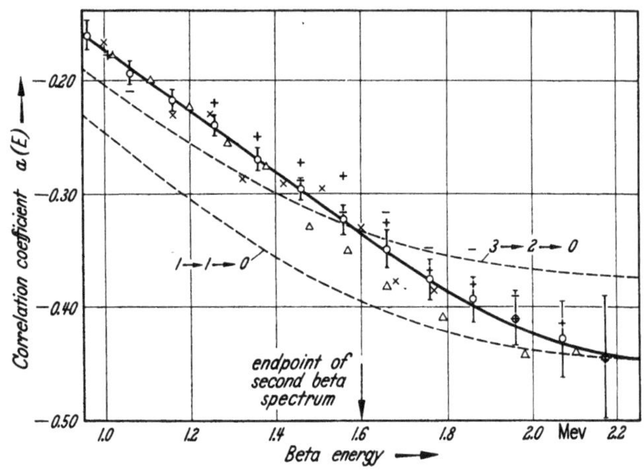

Fig. 41. Comparison of the theoretical and experimental values of the β-γ correlation coefficient $a(E)$ as a function of the β-energy E. $a(E) = \frac{w(\pi, E) - w(\pi/2, E)}{w(\pi/2, E)}$ [cf. E. K. DARBY: Canad. J. Phys. **29**, 576 (1951)].

accompanies β-emission and which is strongly correlated in angle with respect to it[3]. Many β-γ cascades have been investigated in which the β-decay is known to be allowed, or the spectrum of allowed shape, and the correlation is as expected, zero. A few nuclei ranging from K^{42} to Tm^{170} show measurable anisotropies. Detailed interpretation of the correlation provides information about the β-interaction in cases when the analysis of the spectrum shape is ambiguous. For example, the β-spectrum may depart from the allowed shape, but may be either second forbidden or non-unique first forbidden. The differential angular correlation is, however, quite different in the two cases (e.g. in the latter case only $k=0, 2$, may appear, but in the former $k=0, 2, 4$, may be present). MORITA and YAMADA established the decay of Sb^{124} in this way to be first forbidden and the β-γ cascade $3^- \xrightarrow{\beta} 2^+ \xrightarrow{E2} 0^+$ (cf. [55]). The experimental correlation is shown in Fig. 41.

Measurements have also been made[4] of the β-γ polarisation correlation for several forbidden β-transitions. The γ-polarisation was similar to that used in γ-γ polarisation measurements (Sect. 33). The main feature of such experiments is that they permit the parity changes in the transitions to be examined.

[1] D. T. STEVENSON and M. DEUTSCH: Phys. Rev. **83**, 1202 (1954).
[2] J. FUJITA and M. YAMADA: Progr. Theor. Phys. **10**, 518 (1953).
[3] T. B. NOVEY: Phys. Rev. **89**, 672 (1953). — P. BOLGIANO, L. MADANSKY and F. RASETTI: Phys. Rev. **89**, 679 (1953).
[4] D. R. HAMILTON, A. LEMONICK and F. M. PIPKIN: Phys. Rev. **92**, 1191 (1953).

H. Extra-nuclear fields.

50. Qualitative discussion. In the preceding part of this article we have treated all nuclei as ideally isolated, apart, of course, from the interaction with the absorbed or emitted radiations. We referred, in Sect. 18, to the possible distortions of the correlation which may result from the departure from this ideal condition. In practice the sources and target materials used are subject to a variety of forces due to extra-nuclear environment, and some perturbation of the correlation is invariably present, although often to a quite negligible extent. In the present section we shall consider these disturbances in more detail. There are three aspects to such phenomena:

(i) One may wish to find the ideal, unperturbed correlation, in which case either it is necessary to establish the absence of perturbations, or to assess their magnitude, or, for example to vary the conditions of measurements and extrapolate to zero perturbation.

(ii) The nature and magnitude of the perturbation provides information about the interaction (electromagnetic) of the intermediate nuclear states with their environment. If the unperturbed correlation is known, by measurement or calculation, and the symmetry properties of nuclei and radiation can be deduced, then, provided also the nature of the perturbing influence can be assessed the static electromagnetic properties of these nuclear states can be inferred. Thus, whilst the unperturbed correlation of γ-radiation, for example, can provide information about the dynamic behaviour (relative transition multipole-moments), the influence of extra-nuclear fields also yields information about the static magnetic dipole and electric quadrupole moments, and possibly higher moments.

(iii) The influence of extra-nuclear fields may also be useful as a technique for exploring these fields, rather than nuclear properties: for example, the field gradients in crystalline sources or the fluctuating fields in liquids. We shall deal here mainly with the first two aspects of perturbed correlations, and only incidentally with the last aspect.

We consider first the circumstances in which the interaction between the nucleus and its environment may produce measurable or disturbing effects. The only interactions which have, so far, proved significant are those between the nuclear magnetic moment, μ, and extra-nuclear magnetic fields, and between the electric quadrupole moment and external electric-field gradients[1].

We notice that since anisotropic correlations only occur if the intermediate, perturbed, states have spin greater than $\frac{1}{2}$, for such nuclear states both magnetic dipole and electric quadrupole moments are in general possible. Thus, in contrast for example to hyperfine structure and nuclear resonance studies, we cannot, in angular-correlation effects, completely separate electric and magnetic perturbations. The external field may be either constant or varying in time on a time scale determined by the lifetime of the intermediate nuclear state. In many cases, for example an externally applied magnetic field or the electric field gradient in a crystalline material, it may be treated classically; in other cases, for example the interaction between the nuclear and electronic magnetic moments of an isolated atom, the overall system of nucleus and externally interacting agency must be treated as a complete quantum-mechanical system. For the sake of definiteness, we refer, in the following discussion, to a cascade process involving two emitted γ-radiations, although most of the remarks will apply to correlation processes in general.

[1] No measurable effects due to higher moments have been, found, although some attempts to examine these have been made, e.g. E. HEER and R. RÜETSCHI: Helv. phys. Acta **28**, 525 (1955).

In the case of an unperturbed correlation, the time that elapses between the emission of the first and second radiation is not significant. If it were possible to separate out successive radiations with a specified time-interval, then the correlation would be independent of this interval. For perturbed correlations, on the other hand, we may regard the perturbation as essentially switched on after the emission of the first radiation, so that the magnitude of the disturbance depends on the time spent by the nucleus in the intermediate state, before emission of the second radiation; and the angular correlation $W(\vartheta, t)$, will depend therefore on the time-interval between the emission of the radiations. In practice, measurements are made with a finite resolving time in the coincidence circuit, and there may be in addition a time-delay in the channel detecting the first radiation. If the experimental resolving time, τ, is much longer than the lifetime of the intermediate state, $\tau_b\left(=\frac{1}{\lambda_b}\right)$, and no delay is introduced, then the measurement is of the total, "integrated", correlation function, $\frac{1}{\tau_b} W(\vartheta, t)\,\mathrm{e}^{-\lambda_b t}\,dt$. With the resolving time less than, or comparable with τ_b, and with, possibly, some delay, T, then the correlation function will be given by

$$W(\vartheta, T, \tau) = \frac{1}{\tau_b}\int\limits_{T-\tau}^{T+\tau} W(\vartheta, t)\,\mathrm{e}^{-\lambda_b t}\,dt. \tag{50.1}$$

In practice, the possibility of measuring the differential correlation, i.e. corresponding to a specific time-interval, depends on the value of τ_b in relation to the experimental technique used.

α) Static fields. (a) Magnetic. One of the simplest cases is the perturbation introduced by a static magnetic field, $\mathcal{H}$. First, we may notice that if the direction of this field is along the direction of emission of the first radiation, and this is also chosen as axis of quantisation, then in the intermediate states, $|b\beta\rangle$, β is a constant of motion, i.e. no transitions between the different substates are induced by the magnetic field. In addition, by reasons of symmetry, the phases of these substates are arbitrary (cf. Sect. 9) so that there can be no change in the angular correlation. Indeed, the imposition of a large magnetic field in this direction, by introducing a large energy difference between the substates corresponding to different values of β, can inhibit transitions between these states that might otherwise arise from other causes, and so restore the correlation to its unperturbed value. From the general symmetry of the correlation, the same result follows if $\mathcal{H}$ is directed along the direction of either radiation. If $\mathcal{H}$ is not directed along one of the radiations, (and the quantisation axis still is), its presence will cause the amplitudes of the originally incoherent substates, $|b\beta\rangle$ to change with time, and the correlation will be modified. Alternatively, we may choose the quantisation axis along the direction $\mathcal{H}$ which does not coincide with either of the radiations. In this case, the amplitudes of the substates $|b\beta\rangle$ will be partially coherent, and their relative phases time-dependent. In this general case the correlation is no longer dependent simply on the angle between the two radiations, but depends on the relative orientation of these and the magnetic field.

Classically we may regard the nuclei in states $|b\beta\rangle$ as precessing round the direction $\mathcal{H}$ with the frequency ω, equal $\frac{\mu\mathcal{H}}{b\hbar}$ or $g\mathcal{H}$, where μ is the magnetic moment, $b\hbar$ is the angular momentum, and g is the gyromagnetic ratio of the states $|b\rangle$. This frequency is independent of β (i.e. the magnetic splitting of the levels is equidistant). With a field of 10^4 gauss and $\frac{\mu}{b\hbar}$ of the same order as for a nucleon, the precession frequency is of the order 10^8 radians/sec. The limiting perturbation that can be detected depends on the sensitivity of the experimental arrangement, but one might regard such a perturbation as detectable if the lifetime of the intermediate state is sufficiently long that $\tau_b\,\omega \gtrsim 0.1$. Somewhat larger magnetic fields may arise from the nuclear-electron interaction in free atoms; but besides the difficulty of achieving such conditions in practice, the fields cannot be treated as static, classical ones. The magnetic fields inside normal solid materials are generally much smaller.

(b) Electric. Nuclear states with spin greater than or equal to 1 in general have electric quadrupole moments which interact with electric field gradients, thereby perturbing the angular correlation. For fields with axial symmetry, the field gradient is characterised by a single quantity (and its direction), $\frac{\partial \mathcal{E}_z}{\partial z}$, and the nuclear quadrupole moment by a scalar Q[1]. With the axis of symmetry as the quantisation axis, β is a constant of motion, and the precession frequency is given by

$$\omega_Q = \frac{E_Q(\beta) - E_Q(\beta')}{\hbar}$$

with

$$E_Q(\beta) = \frac{3\beta^2 - b(b+1)}{4b(2b-1)}\, Q\, \frac{\partial \mathcal{E}_z}{\partial z}. \tag{50.2}$$

$E_Q(\beta)$ is degenerate with respect to the sign of β so that precession occurs in both directions, in contrast to the magnetic case. In addition, the precession frequency is no longer independent of β, [$E_Q(\beta)$ is not equi-spaced], but can be one of several multiples of a fundamental frequency obtained from (50.2) by putting $\beta = 1$, $\beta' = 0$ (for b integral) or $\beta = \frac{3}{2}$, $\beta' = \frac{1}{2}$ (b half-integral). As in the magnetic case, the angular correlation is unperturbed if the symmetry axis is along the direction of either radiation, and if the quadrupole interaction is large enough, perturbations from other sources can, in principle, be suppressed.

The static electric-field gradients which are present at the nuclear sites in solid materials give rise to precession frequencies of the order $2\cdot 10^{10}$ radians/sec or less, so that using the same criterion as above, one might expect to detect effects on correlations if the lifetime of the intermediate state is of the order 10^{-11} sec or greater. If the radiating nuclei are embedded in a single crystal and they all occupy similar well-defined lattice positions, then $\frac{\partial \mathcal{E}_z}{\partial z}$ will be the same, and in the same direction, for all nuclei of the source. In this case the correlation function will depend on the relative directions of the symmetry axis and the two radiations, and not simply on the angle ϑ between the radiations. For a micro-crystalline source, on the other hand, which is macroscopically isotropic, $\frac{\partial \mathcal{E}_z}{\partial z}$ is random in direction, and the resultant angular correlation is obtained by averaging over all directions of the symmetry axis with respect to the directions of the radiations, and it is independent of the orientation of the source.

For both types of static field the variation of the correlation with time interval is essentially periodic so that the correlation is never completely destroyed.

[1] More generally the electric field gradient and the quadrupole moment are tensors. In this special case the axes can be chosen so that the interaction is proportional to the product of two scalars, $Q\,\frac{\partial \mathcal{E}_z}{\partial z}$.

This is true even for a micro-crystalline "isotropic" source, since one may regard some of the micro-crystals as aligned along the direction of one of the radiations, and, for these, no change in the correlation occurs. Measurement of the correlation, with a time interval $2\pi/\omega$ (ω is the "fundamental" precession frequency) and ideally small resolving time, reproduces the original, unperturbed correlation. The time-integrated correlation function, on the other hand, approaches an asymptotic value, which usually shows a diminished degree of correlation, which is referred to as the "hard-core" value. In one case of special symmetry static fields can lead to complete suppression of the correlation: if the symmetry axis of the perturbing field $\boldsymbol{n}$ is perpendicular to both radiations, and the directions $\pm\boldsymbol{n}$ cannot be distinguished (for example, if each detector is equally sensitive to both radiations).

Time-dependent fields. When the fields acting on the nucleus in the intermediate state are time-dependent, then transitions between the different substates $|b\beta\rangle$, $|b\beta'\rangle$, etc. occur which eventually lead to a normal random population, that is, the correlation is completely destroyed. A time-scale for this process can be characterised by a relaxation time, or its reciprocal λ. The criterion for a large perturbation is then, $\lambda\tau_b$ of the order or greater than unity. In practice a change in the angular correlation might be detected for values of $\lambda\tau_b$ as small as 0.05. The relaxation constant, λ, is proportional to a correlation time, τ_{corr}, which is a measure of the rapidity of the fluctuations of $\left(\frac{\partial \mathcal{E}_z}{\partial z}\right)$, and to the quantity $Q^2 \overline{\overline{\left|\frac{\partial \mathcal{E}_z}{\partial z}\right|^2}}$. If τ_{corr} is very small (so that $\lambda\tau_b \ll 1$), then the correlation is unperturbed; we may regard the perturbing field as fluctuating too rapidly to cause appreciable precession. As τ_{corr} increases the angular-correlation is reduced[1]. This trend continues as long as $\tau_{\text{corr}} \ll \tau_b$ and $\tau_{\text{corr}} \ll \tau_\omega$, the characteristic precession time given by a relation of the form (50.2) with some average value of $\frac{\partial \mathcal{E}_z}{\partial z}$. When, however, τ_{corr} becomes greater than τ_w and τ_b, the situation goes over into the static case, and in limit of large τ_{corr} and $\omega\tau_b \gg 1$, the correlation is reduced to the ordinary, static, "hard-core" value (see Fig. 42). If $\tau_b \gg \tau_{\text{corr}}$, and τ_{corr} is of the *same order (or greater than)* τ_w, then the angular correlation is destroyed.

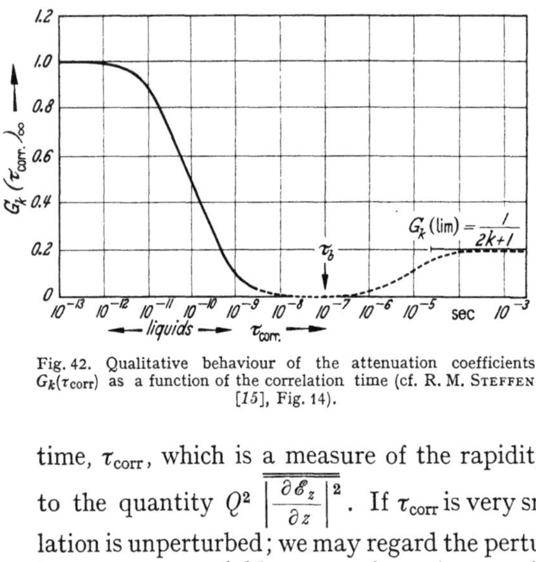

Fig. 42. Qualitative behaviour of the attenuation coefficients $G_k(\tau_{\text{corr}})$ as a function of the correlation time (cf. R. M. STEFFEN [*15*], Fig. 14).

Significant time-dependent perturbations occur in liquid sources, the correlation times being of the order 10^{-11} sec, that is of a rather smaller magnitude than the precession times. The correlation times increase, and the angular correlation diminishes, as the viscosity of the medium increases. Extrapolation to zero viscosity therefore gives the true, unperturbed, angular correlation.

If the radiating nucleus occupies a normal lattice site in a solid, then there will still be some time-dependence of the perturbing field, related to the normal

[1] The relaxation constant, λ, is multiplied by a numerical constant which depends on so that the coefficients of the different terms $P_k(\cos\vartheta)$ in the angular correlation do not k, decrease at the same rate [cf. Eq. (51.53)]. all

spin-lattice relaxation time. However, the relaxation times are usually sufficiently long for the static approximation to apply. The disturbance suffered by the nucleus or its surroundings due to the emission of the first radiation or the preceding process (particularly in the case of γ-cascade following K-capture, and α-γ cascades) may lead to time-dependent perturbations which can appreciably affect the correlation. In the case of α-emission the recoil nucleus will be displaced violently in the lattice; most of the recoil motion will be dissipated in a time of the order 10^{-12} sec, but the nuclear surroundings at the time of emission of the second radiation of the cascade may be far from the normal ones. The hole in the K-shell resulting from K-capture will rapidly (in a time less than 10^{-9} sec) move to the outer atomic layers by Auger processes, but such processes cease to be possible when the outermost layer is reached. The time taken for this hole in the outer layer to be filled is of a different order for a conductor and an insulator. For a metal, the duration of the disturbance may be limited to 10^{-12} sec, so that the perturbation of the correlation is small. For an insulator it is much larger, $\sim 10^{-8}$ sec, and as a result the correlation is greatly affected.

β) Non-classical fields. The perturbation which may arise from the interaction between the nuclear moment and that of a free atom cannot be treated in the same way as that due to a "classical" field. If we assume that the electronic state of the atom is constant, then the nuclear spin $\boldsymbol{b}$ will be coupled to the atomic spin $\boldsymbol{J}$ to give a total $\boldsymbol{F}$, $(J+b \geqq F \geqq |J-b|)$, and β is therefore not a constant of motion. Classically $\boldsymbol{b}$ precesses around $\boldsymbol{F}$, but due to random orientation of $\boldsymbol{J}$, the direction of $\boldsymbol{J}$ will not be unique. The situation is nevertheless similar to the static field case in that the perturbation does not destroy the correlation completely, even, if τ_b is very large. The precession frequencies are directly related by the hyperfine splitting, and are typically of the order 10^9 radians/sec.

Coupling between the nuclear and atomic spin can occur, for example, in the case of paramagnetic solids $(J=\frac{1}{2})$, but, here, the perturbation will eventually destroy the correlation on account of the coupling of the *atomic* spin with the lattice. The time scale for this effect is determined, *inter alia*, by the paramagnetic spin-relaxation time, which is temperature-dependent. This type of perturbation, which arises from magnetic coupling, can be suppressed by a sufficiently large external magnetic field in the direction of one of the cascade radiations.

The paramagnetic perturbations, as for perturbation in solids generally, may be macroscopically anisotropic (e.g. as in a single crystal), so that the angular-correlation function may not be a simple function of the angle between the radiations alone.

51. Theory of perturbed correlations. The first detailed calculations (GOERTZEL [*21*]) dealt with the case of nuclear-electron spin interaction, i.e. the atomic hyperfine-structure interaction. This theory was extended to include the effect of applied static magnetic fields by GOERTZEL, and by ALDER [*30*], and BIEDENHARN and ROSE [*9*], using the formalism of RACAH [*28*]. A more recent treatment by ABRAGAM and POUND [*61*] includes a great simplification of the earlier procedures and generalisations to cover time-dependent interactions. The use of density matrices and the tensor formalism is advantageous here as in the case of unperturbed correlations, as has been shown by COESTER [*63*], but because of their relative simplicity we shall first outline some of the calculations of ABRAGAM and POUND.

The starting point is the result, derived, by HAMILTON [*20*] using the perturbation formalism, for the undisturbed angular correlation of two

γ-radiations[1]:

$$\left.\begin{aligned} W(\Omega_1\Omega_2) = \mathscr{S}\sum_{\alpha\beta\beta'\gamma} \langle c\gamma\Omega_2\sigma_2|H_2|b\beta\rangle\langle b\beta\Omega_1\sigma_1|H_1|a\alpha\rangle \times \\ \times\langle a\alpha|H_1|b\beta'\Omega_1\sigma_1\rangle\langle b\beta|H_2|c\gamma\Omega_2\sigma_2\rangle. \end{aligned}\right\} \tag{51.1}$$

H_1, H_2 represent the interaction between nuclei and radiation only, and $\mathscr{S}$ is as usual the summation over the unobserved properties of the radiations. If, in addition to the interactions H_1, H_2, there is a Hamiltonian, H', which describes the interaction of the nucleus (in the intermediate state) with its environment, and which is assumed to act from the time, $t=0$, at which the first radiation of the cascade is emitted, until a later time, t, when the second radiation occurs, then during this time interval the change in the states $|b\beta\rangle$ can be represented by a time development operator $U(t\,0)$ which satisfies:

$$i\hbar\frac{\partial}{\partial t}U(t\,0) = H'\,U(t\,0)\ , \tag{51.2}$$

$$U(0\,0) = 1. \tag{51.3}$$

The solution of (51.2), for a time-independent interaction, H', is simply $\exp\frac{-iE't}{\hbar}$ where E' is an eigenvalue of the perturbing Hamiltonian. For a more general interaction we can write symbolically:

$$U(t\,0) = \exp\left[-\frac{i}{\hbar}\int_0^t H'(t)\,dt\right]. \tag{51.4}$$

If we assume that the Hamiltonian, H', acts on the nucleus for a time, t, at which time the second radiation is emitted, the angular correlation takes on the form:

$$\left.\begin{aligned} W(\Omega_1\Omega_2) = \mathscr{S}\sum_{\alpha BB'\gamma}\langle c\gamma|H_2U|bB\rangle\langle bB|H_1|a\alpha\rangle \times \\ \times\langle a\alpha|H_1|bB'\rangle\langle bB'|U^\dagger H_2|c\gamma\rangle, \end{aligned}\right\} \tag{51.5}$$

where we have, for brevity, omitted writing out the quantum numbers which refer to the radiations, and we denote by B, B', characteristics other than b specifying the intermediate state.

Since the nucleus in the intermediate state, $|b\beta\rangle$, decays with lifetime τ_b, the correlation function (51.5) must be weighted by the factor $\frac{e^{-t/\tau_b}dt}{\tau_b}$, which represents the probability that the second radiation will be emitted during the interval $t, t+dt$. If the correlation is measured for all pairs of radiations with time interval between t_1 and t_2, it will be given by:

$$\left.\begin{aligned} W(\Omega_1\Omega_2) = \mathscr{S}\sum_{\alpha BB'\gamma}\frac{1}{\tau_b}\int_{t_1}^{t_2} dt\,e^{-t/\tau_b}\langle c\gamma|H_2U|bB\rangle\langle bB|H_1|a\alpha\rangle \times \\ \times\langle a\alpha|H_1|bB'\rangle\langle bB'|U^\dagger H_2|c\gamma\rangle. \end{aligned}\right\} \tag{51.6}$$

If the interaction H' is static, we can choose $|bB\rangle$ to be eigenfunctions of H'. (For example, in the special case of a field with axial symmetry, the eigenfunctions would be denoted by $|b\beta\rangle$ provided the axis of quantisation for β is along the symmetry axis of the field.) These eigenfunctions of H' are also eigenfunc-

[1] The discussion can be generalised to include any type of emitted radiation provided the Hamiltonian can be replaced by a more general operator $\mathcal{O}$ with similar physical significance.

tions of $U(t\,0)$ with eigenvalues $\exp\left[\frac{-i}{\hbar} E_{bB}\,t\right]$, where E_{bB} denote the eigenvalues of H'.

If we extend the limits of integration in (51.6) from 0 to ∞ (i.e. the total time-integrated angular correlation), we obtain the result (first derived by GOERTZEL):

$$W(\Omega_1 \Omega_2) = \mathscr{S} \sum_{\alpha B B' \gamma} \frac{\langle c\gamma | H_2 | b\,B\rangle \langle b\,B | H_1 | a\alpha\rangle \langle a\alpha | H_1 | b\,B'\rangle \langle b\,B' | H_2 | c\gamma\rangle}{1 - \frac{i}{\hbar} \tau_b (E_B - E_{B'})}. \tag{51.7}$$

For an axially symmetrical static field, B in (51.7) can be replaced by β, the component of b along the direction of the symmetry axis. For a static field, generally the states $|b\,B\rangle$ can be represented as linear combinations of states $|b\,\beta\rangle$. In the case of non-classical fields we can still express the states as linear combinations of states $|b\beta\rangle$, but appropriate designation of the atomic state is necessary, so that the nucleus together with environment is represented as a complete quantum-mechanical system. In the particular case of hyperfine coupling, where the interaction proportional to $\boldsymbol{b} \cdot \boldsymbol{J}$, $|b\,B\rangle$ in (51.7) is replaced by $|F\,\Phi\rangle$, which can be expanded as:

$$|F\,\Phi\rangle = \sum_{\beta\zeta} |b\,\beta\rangle\, |J\zeta\rangle\, (F\,\Phi | b\,\beta, J\zeta) \tag{51.8}$$

and $E_B - E_{B'}$ is replaced by $E_{F\Phi} - E_{F'\Phi'}$. Here, $J\zeta, F\,\Phi$ refer to the angular momenta of the atomic system and of the complete system (consisting of atom and nucleus) respectively, and $E_{F\Phi}$ etc. refer to the energies of the complete system.

The summation in (51.7), which refers only to static fields and time-integrated correlations, has been effected by ALDER [30] using Racah techniques (cf. Sect. 10), who obtains, for example, the angular correlation for an axially symmetric field in the form:

$$W(\Omega_1 \Omega_2) = \sum_{k_1 k_2 \varkappa} \mathrm{I}(k_1)\, \mathrm{II}(k_2)\, \mathrm{III}(k_1 k_2, \varkappa)\, Y_{k_1}^{\varkappa}(\Omega_1)\, Y_{k_2}^{\varkappa *}(\Omega_2) \tag{51.9}$$

where

$$\mathrm{I}(k_1) \equiv \frac{\sqrt{4\pi}}{\hat{b}} A_{k_1}(a, b\,b)\,, \tag{51.10}$$

$$\mathrm{II}(k_2) \equiv \frac{\sqrt{4\pi}}{\hat{b}} A_{k_2}(b\,b, c) \tag{51.11}$$

and

$$\mathrm{III}(k_1 k_2, \varkappa) \equiv \sum_{\beta\beta'} \frac{(b\beta', k_1\varkappa | b\beta)\,(b\beta', k_2\varkappa | b\beta)}{1 - \frac{i}{\hbar}\tau_b (E_\beta - E_{\beta'})}. \tag{51.12}$$

ALDER's treatment has been generalised by ABRAGAM and POUND so as to include more general fields, and in particular time-dependent fields, and correlations with finite time intervals. Expression (51.7) is no longer applicable. From (51.5) they obtain the more general result for the hypothetical correlation with ideal time characteristics in the coincidence apparatus:

$$W(\Omega_1 \Omega_2, t) = \sum_{k_1 k_2 \varkappa_1 \varkappa_2} \mathrm{I}(k_1)\, \mathrm{II}(k_2)\, \mathrm{III}(k_1 k_2 \varkappa_1 \varkappa_2 t)\, Y_{k_1}^{\varkappa_1}(\Omega_1)\, Y_{k_2}^{\varkappa_2 *}(\Omega_2)\,. \tag{51.13}$$

Here, $\mathrm{I}(k_2)$, $\mathrm{II}(k_2)$ are as above, and III is as follows:

(i) Classical fields.

$$\left.\begin{aligned} \mathrm{III}(k_1 k_2 \varkappa_1 \varkappa_2 t) = \sum_{\beta\beta'\beta''\beta'''} & (b\beta', k_1\varkappa_1 | b\beta)\,(b\,\beta''', k_2\varkappa_2 | b\beta'') \times \\ & \times \langle\beta | U^\dagger | \beta''\rangle \langle\beta''' | U | \beta'\rangle; \end{aligned}\right\} \tag{51.14}$$

(ii) Nuclear-atomic interaction.

$$\left.\begin{aligned}\mathrm{III}\,(k_1\,k_2\,\varkappa_1\,\varkappa_2\,t) = \sum_{\beta\beta'\beta''\beta'''\zeta\zeta'} &(b\beta',k_1\varkappa_1|b\beta)\,(b\beta''',k_2\varkappa_2|b\beta'')\times\\ &\times\langle\beta\zeta|U^{\dagger}|\beta''\zeta'\rangle\langle\beta'''\zeta'|U|\beta'\zeta\rangle.\end{aligned}\right\}\tag{51.15}$$

In obtaining this expression J, the atomic angular momentum is assumed to be constant and both b and J are implicit in the matrix elements of U.

The time-integrated angular correlation is obtained directly from (51.13) by replacing $\mathrm{III}\,(k_1\,k_2\,\varkappa_1\,\varkappa_2\,t)$ by $\mathrm{III}\,(k_1\,k_2\,\varkappa_1\,\varkappa_2)$, defined as:

$$\mathrm{III}\,(k_1\,k_2\,\varkappa_1\,\varkappa_2) = \int\limits_0^{\infty}\mathrm{III}\,(k_1\,k_2\,\varkappa_1\,\varkappa_2\,t)\,\mathrm{e}^{-t/\tau_b}\,\frac{dt}{\tau_b}.\tag{51.16}$$

From their definitions it is clear that the factors III(...) involve only the properties of the intermediate state and the nuclear environment, and are independent of the properties of the radiation. In the absence of perturbations $\beta''=\beta$, $\beta'''=\beta'$, and $\zeta'=\zeta$, so that

$$\mathrm{III}\,(k_1\,k_2\,\varkappa_1\,\varkappa_2\,t) = \frac{\hat{b}^2}{\hat{k}_1^2}\,\delta_{k_1k_2}\,\delta_{\varkappa_1\varkappa_2}\tag{51.17}$$

and the time-integrated angular correlation obtained from (51.15) and (51.12) reduces to that for an unperturbed correlation.

We now consider the formulation of the theory using the general method of efficiency and density matrices and the corresponding tensors. From Sect. 11, the general result for the angular correlation (COESTER [*48*]) requires the evaluation of the statistical and efficiency tensors in the expression,

$$W(\Omega_1\Omega_2) = \sum_{k_a\varkappa_a A_t A'_t}\varrho_{k_a\varkappa_a}(a\,\alpha\,A_t,\,a\,\alpha'\,A'_t)\,\varepsilon^{*}_{k_a\varkappa_a}(a\,\alpha\,A_t,\,a\,\alpha'\,A'_t).\tag{51.18}$$

The time development of the density matrix, taking into account the interaction with the surroundings of the intermediate state, has been examined by COESTER whose results we quote. In order to conform with the designation of time used by COESTER and also in Sect. 11, we assume that at times, $t=0$, the system consists solely of nuclei with angular momenta $\boldsymbol{a}$, so that the density matrix at time, $t=0$, is given by

$$\langle a\,\alpha\,A_0|\varrho|\,a\,\alpha'\,A'_0\rangle = \frac{1}{\hat{a}^2}\,\delta_{\alpha\alpha'}\,\delta_{A_0A'_0}.\tag{51.19}$$

This definition of $t=0$ differs from our choice in (51.3), where it is assumed that the first radiation is *emitted*, at time $t=0$. If the system is described at some later time, $t=T$, during which only a single radiation has been emitted, then the density matrix is related to the initial density matrix by:

$$\left.\begin{aligned}\langle a\,\alpha\,A_T|\varrho|\,a\,\alpha'\,A'_T\rangle \sim \frac{1}{N}\sum_{i\alpha\alpha'A_0A'_0}&\int\limits_0^T dt'\,\mathrm{e}^{-\left(\frac{T-t'}{\tau_b}+\frac{t'}{\tau_a}\right)}\times\\ \times\langle a\,\alpha\,A_0|\varrho^i|\,a\,\alpha'\,A'_0\rangle\,U^i(T\,t')\,&\langle b\,\|\mathscr{L}_1\|\,a\rangle\langle b\,\|\mathscr{L}'_1\|\,a\rangle\,U^i(T\,t')^{\dagger}.\end{aligned}\right\}\tag{51.20}$$

Each of the N nuclei in the source is described by a matrix ϱ^i and

$$\varrho = \frac{1}{N}\sum_i\varrho^i.\tag{51.21}$$

For the cascade emission of two radiations, the density matrix is given by:

$$\left.\begin{aligned}\langle a\,\alpha\, A_T\,|\varrho|\, a\,\alpha'\, A_T'\rangle \sim \frac{1}{N}\sum_i \int_0^T dt' \int_0^{t'} dt''\, \mathrm{e}^{-\left(\frac{t'-t''}{\tau_b}+\frac{t''}{\tau_a}\right)} \times \\ \times \langle c\,\|\mathcal{L}_2\|\, b\rangle\, U^i(t'\,t'')\,\langle b\,\|\mathcal{L}_1\|\,a\rangle\,\langle b\,\|\mathcal{L}_1'\|\,a\rangle\, U^i(t'\,t'')^\dagger \langle c\,\|\mathcal{L}_2''\|\,b\rangle.\end{aligned}\right\} \quad (51.22)$$

We proceed as in the case of an unperturbed correlation and insert the tensor corresponding to (51.22) into (51.18), and expand the efficiency tensor for the system in terms of the tensor of the first radiation and the residual nucleus. The angular correlation is expressed in terms of the operators $U^i(t'\,t'')$ which operate on the eigenfunctions used in construction of the efficiency matrix for the nucleus in the state b. Thus we have

$$\left.\begin{aligned}U^i(t'\,t'')\, U^i(t'\,t'')^\dagger \langle b\,\beta\,\zeta\, B\,|\,\varepsilon\,|\, b\,\beta'\,\zeta\, B'\rangle^* = \sum_{\beta''\beta'''\zeta'} \langle b\,\beta''\,\zeta' B\,|\,\varepsilon\,|\, b\,\beta'''\,\zeta' B'\rangle^* \times \\ \times \langle \beta'''\,\zeta'\,|\,U^i(t'\,t'')\,|\,\beta'\,\zeta\rangle^* \langle \beta''\,\zeta'\,|\,U^i(t'\,t'')\,|\,\beta\,\zeta\rangle.\end{aligned}\right\} \quad (51.23)$$

B represents the quantum numbers additional to $b\,\beta\,\zeta$ required to specify the states, and ζ the quantum-mechanical state (if relevant) of the interacting environment. The efficiency matrix in (51.23) is used to construct the associated efficiency tensor which, again as in the unperturbed case, is expanded in terms of the tensors for the second radiation and the residual nucleus.

Using these tensors from (51.22) and (51.23), and carrying out the summations, we finally obtain the result for the angular correlation at some time, T,

$$\left.\begin{aligned}W_T(\Omega_1\Omega_2) = \sum_{k_1 k_2 \varkappa_1 \varkappa_2} A_{k_1}(a, b\,b)\, A_{k_2}(b\,b, c)\,\langle k_2\,\varkappa_2\,|\,G_T|\, k_1\,\varkappa_1\rangle^* \times \\ \times \frac{4\pi}{\hat{k}_1\,\hat{k}_2}\, Y_{k_1}^{\varkappa_1 *}(\Omega_1)\, Y_{k_2}^{\varkappa_2}(\Omega_2)\end{aligned}\right\} \quad (51.24)$$

where $A_{k_1}(a, b\,b)$, $A_{k_2}(b\,b, c)$ are, as previously, defined by (15.4), and

$$\langle k_2\,\varkappa_2\,|\,G_T|\, k_1\,\varkappa_1\rangle \sim \frac{1}{N}\sum_{i\zeta\zeta'} \int_0^T dt' \int_0^{t'} \frac{dt''}{\tau_a}\, \mathrm{e}^{-\left(\frac{t'-t''}{\tau_b}+\frac{t''}{\tau_a}\right)} \langle k_2\,\varkappa_2\,\zeta'\,\zeta'|\,G^i(t'\,t'')|\,k_1\,\varkappa_1\,\zeta\,\zeta\rangle \quad (51.25)$$

with

$$\left.\begin{aligned}\langle k_2\,\varkappa_2\,\zeta'\,\zeta'\,|\,G^i(t'\,t'')\,|\,k_1\,\varkappa_1\,\zeta\,\zeta\rangle = \sum_{\beta\,\beta'\beta''\beta'''} \langle \beta'''\,\zeta'\,|\,U^i(t'\,t'')\,|\,\beta'\,\zeta\rangle^* \times \\ \times \langle \beta''\,\zeta'\,|\,U^i(t'\,t'')\,|\,\beta\,\zeta\rangle\,(-)^{2b-\beta'-\beta'''}(b\,\beta, b-\beta'\,|\,k_1\,\varkappa_1)\,(b\,\beta'', b-\beta'''\,|\,k_2\,\varkappa_2)\,.\end{aligned}\right\} \quad (51.26)$$

In the foregoing analysis, no statements have been made concerning the characteristics or means of detection of the radiation. Thus, the correlation also applies to processes involving the observation of polarisation effects. In addition, k_1 and k_2 are, of course, restricted to even values if the radiation states are of well-defined parity. In the absence of any perturbation, and assuming that the polarisation is not observed,

$$\langle k_2\,\varkappa_2\,\zeta'\,\zeta'\,|\,G^i(t'\,t'')\,|\,k_1\,\varkappa_1\,\zeta\,\zeta\rangle = \delta_{\zeta\zeta'}\,\delta_{\varkappa_1\varkappa_2}\,\delta_{k_1 k_2} \quad (51.27)$$

so that (51.24) reduces to the unperturbed correlation:

$$W(\vartheta) = \sum_k A_k(a, b\,b)\, A_k(b\,b, c)\, P_k(\cos\vartheta)\,. \quad (51.28)$$

The expression (51.25) is simplified for static fields since $U^i(t' t'')$ can be written as $\exp\left(\frac{-iE'(t'-t'')}{\hbar}\right)$, i.e. U^i depends only on the difference of the times of the two emissions and does not involve t', t'' explicitly. We can then express $G(t' t'')$ so that it has only a functional dependence on $t = t' - t''$, where now t assumes the meaning attached to it in the earlier treatment, for example in (51.2). $G(t' t'')$ will therefore now be written as $G(t)$. The matrix of $G(t)$ is closely related to $\mathrm{III}(k_1 k_2 \varkappa_1 \varkappa_2 t)$: from (51.15) and (51.26) we see that

$$\sum_{\zeta\zeta'} \langle k_2 \varkappa_2 \zeta' \zeta' \,|\, G^i(t) \,|\, k_1 \varkappa_1 \zeta \zeta \rangle^* = \frac{\hat{k}_1 \hat{k}_2}{\hat{b}^2} \,\mathrm{III}\,(k_1 k_2 \varkappa_1 \varkappa_2 t). \tag{51.29}$$

We now consider some illustrative examples of the general theory.

(a) Micro-crystalline solid. For such a macroscopically isotropic source we obtain the perturbed angular correlation for a double cascade by averaging over all directions of the perturbing field, which is equivalent to a fixed direction of field, and averaging over all directions Ω_1, Ω_2, but keeping $\mathcal{R}_1 \mathcal{R}_2^{-1}$ fixed. This averaging leads to a simple correlation which depends only on the angle between the two radiations. We have

$$\left.\begin{aligned} \frac{4\pi}{\hat{k}_1 \hat{k}_2} \overline{Y_{k_1}^{\varkappa_1 *}(\Omega_1)\, Y_{k_2}^{\varkappa_2}(\Omega_2)} &= \int d\Omega_1\, \mathcal{D}_{\varkappa_1 0}^{k_1 *}(\mathcal{R}_1)\, \mathcal{D}_{\varkappa_2 0}^{k_2}(\mathcal{R}_2) \\ &= \sum_{\varkappa} \int d\Omega_1\, \mathcal{D}_{\varkappa_1 0}^{k_1 *}(\mathcal{R}_1)\, \mathcal{D}_{\varkappa 0}^{k_2}(\mathcal{R}_2 \mathcal{R}_1^{-1})\, \mathcal{D}_{\varkappa_2 \varkappa}^{k_2}(\mathcal{R}_1) \\ &= P_k(\cos\vartheta)\, \delta_{\varkappa_1 \varkappa_2}\, \delta_{k_1 k_2}. \end{aligned}\right\} \tag{51.30}$$

The perturbing field is treated classically so that $\zeta' = \zeta$ in (51.25). The perturbed angular correlation is given by using (51.30) in (51.24), so that

$$W(\Omega_1 \Omega_2) = \sum_k A_k[1, 2]\, G_k\, P_k(\cos\vartheta), \tag{51.31}$$

where

$$G_k = \frac{1}{\hat{k}^2} \sum_{\varkappa} \langle k\varkappa \,|\, G_{00} \,|\, k\varkappa \rangle^*,$$

when time-integrated in accordance with (51.16), is termed the attenuation factor, and $A_k[1, 2]$ denotes the coefficients of the unperturbed angular correlation. These attenuation factors have been evaluated explicitly, for various types of crystalline field, by ABRAGAM and POUND. There are two broad classes of fields:

(i) fields, having axial symmetry,

$$G_k\,(\text{axial}) = \frac{1}{\hat{k}^2} \sum_{\varkappa\beta\beta'} \frac{(b\beta, b-\beta' \,|\, k\varkappa)^2}{1 + \frac{\tau_b^2}{\hbar^2}(E_\beta - E_{\beta'})^2}\, q(\beta\beta'), \tag{51.32}$$

where $q(\beta\beta') = \left.\begin{matrix}0\\2\end{matrix}\right\}$ according to whether $\beta \lessgtr \beta'$, and $q(\beta\beta) = 1$;

(ii) fields, without axial symmetry,

$$\left.\begin{aligned} G_k\,(\text{non-axial}) = \frac{1}{\hat{k}^2} \sum_{\beta\beta'\beta''\beta'''\varkappa BB'} & (-)^{2b-\beta'-\beta'''} (b\beta, b-\beta' \,|\, k\varkappa)\, (b\beta'', b-\beta''' \,|\, k\varkappa) \times \\ \times & \frac{\mathrm{Re}\left[\langle B|\beta\rangle \langle B'|\beta'''\rangle \langle \beta''|B\rangle \langle \beta'|B'\rangle\right]}{1 + \frac{\tau_b^2}{\hbar^2}(E_B - E_{B'})^2}\, q(BB'; \beta\beta'''), \end{aligned}\right\} \tag{51.33}$$

where $|B\rangle = \sum_{\beta} |\beta\rangle\langle\beta|B\rangle$, and $q(BB'; \beta\beta') = \left.\begin{matrix}0\\2\end{matrix}\right\}$ according to whether $B \lessgtr B'$ and if $B = B'$ as to whether $\beta \lessgtr \beta'$; $= 1$ if $B = B'$ and $\beta = \beta'$. The above expressions illustrate a general feature of static interactions with a source that is macroscopically isotropic, namely, that the angular correlation is never completely destroyed even if the perturbation is strong or the lifetime τ_b, very long, i.e. for example $\frac{\tau_b(E_\beta - E_{\beta'})}{\hbar} \gg 1$. Even in the limit $\frac{\tau_b(E_\beta - E_{\beta'})}{\hbar} \to \infty$, for $\beta \neq \beta'$, G_k in (51.32) and (51.33) will not vanish owing to the contribution from terms with $\beta = \beta'$. There may also be contributions arising from degeneracy of the states, i.e. $E_\beta = E_{\beta'}$ when $\beta \neq \beta'$. These depend on the symmetry properties of the perturbing field, for example the electric quadrupole interaction in different crystalline structures ([*61*], Sect. IV). The contributions from the terms $\beta = \beta'$ arise in all cases irrespective of degeneracy and yield lower limits to the values of the attenuation coefficients in the limit of strong perturbations. These lower limits are the "hard core" values of the attenuation coefficients, and are given by

$$G_k(\lim) = \frac{1}{\hat{k}^2} \sum_\beta (b\beta, b-\beta \mid k0)^2 = \frac{1}{2k+1}. \tag{51.34}$$

The "hard core" values refer to the time-integrated correlations only. For finite delay times other (and possibly smaller) values of the attenuation coefficients are found.

(b) Interaction with free atom. The matrix U is diagonal in the representation $F\Phi$. Its matrix elements in (51.26) can be evaluated either by transforming the eigenfunctions $|b\beta\rangle \cdot |J\zeta\rangle$ in terms of $|F\Phi\rangle$ by means of Clebsch-Gordan coefficients and summing over magnetic quantum numbers (Alder [*60*]), and the angular correlation is obtained by (51.24), (51.25) and (51.26). Otherwise we can derive the angular correlation by expanding $\varepsilon_{k_a \varkappa_a}$ in terms of $\varepsilon_{k_b \varkappa_b}\, \varepsilon_{k_1 \varkappa_1}$, and using an expansion valid if $\boldsymbol{F} = \boldsymbol{b} + \boldsymbol{J}$ and no measurement is made of the electronic states ($\varepsilon_{k_J \varkappa_J} = \hat{J}\, \delta_{k_J 0}\, \delta_{\varkappa_J 0}$). Then

$$\left.\begin{aligned} \varepsilon_{k_F \varkappa_F}(FF') &= \varepsilon_{k_b \varkappa_b} \hat{J}\hat{F}\hat{F}'\hat{k}_F \begin{Bmatrix} b & J & F \\ b & J & F' \\ k_b & 0 & k_F \end{Bmatrix} (k_b \varkappa_b, 00 \mid k_F \varkappa_F) \\ &= \varepsilon_{k_b \varkappa_b} \hat{F}\hat{F}' (-)^{J+k_F-F-b}\, W(bbFF'; k_F J). \end{aligned}\right\} \tag{51.35}$$

The angular correlation is given by (51.24) with the following expressions for the matrix elements of G_∞:

$$\left.\begin{aligned} \langle k_2 \varkappa_2 | G_\infty | k_1 \varkappa_1 \rangle = \frac{1}{\hat{J}^2} \sum_{FF'} \hat{F}^2 \hat{F}'^2 W(bbFF'; k_1 J)\, W(bbFF'; k_2 J) \times \\ \times \sum_{\Phi\Phi'} \frac{(F\Phi, F'-\Phi' \mid k_1 \varkappa_1)(F\Phi, F'-\Phi' \mid k_2 \varkappa_2)}{1 - \frac{i\tau_b}{\hbar}(E_{F\Phi} - E_{F'\Phi'})}. \end{aligned}\right\} \tag{51.36}$$

For an isolated atom, with equi-distant hyperfine splitting, this expression simplifies to:

$$\langle k_2 \varkappa_2 | G_\infty | k_1 \varkappa_1 \rangle = \frac{1}{\hat{J}^2} \sum_{FF'} \frac{\hat{F}^2 \hat{F}'^2 W^2(bbFF'; k_1 J)}{1 + \frac{\tau_b^2}{\hbar^2}(E_F - E_{F'})^2}\, q(FF')\, \delta_{k_1 k_2}\, \delta_{\varkappa_1 \varkappa_2}, \tag{51.37}$$

where $q(FF') = \left.\begin{matrix}0\\2\end{matrix}\right\}$ where $F \lessgtr F'$ and $q(FF) = 1$. Then, for isolated atoms (with random orientation) the time-integrated correlation is of the form (51.31) with the attenuation factor given by:

$$G_k = \frac{1}{\hat{J}^2} \sum_{FF'} \hat{F}^2 \hat{F}'^2 \frac{W^2(bbFF'; kJ)}{1 + \frac{\tau_b^2}{\hbar^2}(E_F - E_{F'})^2}\, q(FF') \tag{51.38}$$

with $G_0 = 1$, $G_k = 1$ for $(E_F - E_{F'})\,\tau_b = 0$. This expression again shows the feature of a non-vanishing "hard-core" value, which arises from terms in the summation with $F' = F$.

(c) Applied magnetic field. In the two previous illustrations there was no preferred direction in the source and hence the correlation functions were functions of a single angle ϑ, as in the unperturbed case. In the present example we consider a uniform magnetic field, $\mathcal{H}$, applied in a direction perpendicular to the plane containing the two radiations. No new angles are involved, since the two radiations already specify this direction, but the sense of this direction is now significant, whereas without the field it is not so (the correlation is symmetrical with respect to the first and second radiations). The correlation is obtained from (51.24) with the following value for $\langle | G(t) | \rangle$:

$$\langle k_2 \varkappa_2 \zeta' \zeta' \,|\, G(t) \,|\, k_1 \varkappa_1 \zeta \zeta \rangle = \mathrm{e}^{i\varkappa\omega t}\, \delta_{k_1 k_2}\, \delta_{\varkappa_1 \varkappa_2}\, \delta_{\zeta \zeta'}\,; \quad \omega = \frac{\mu \mathcal{H}}{b\hbar} \tag{51.39}$$

so that

$$W(\vartheta, t) = \sum_{\varkappa} \mathrm{e}^{i\varkappa\omega t}\, W(\vartheta, 0)\,. \tag{51.40}$$

Now, $W(\vartheta, 0)$ can also be written as:

$$\left.\begin{aligned} W(\vartheta, 0) &= \sum_{\varkappa} B_\varkappa(0)\, \mathrm{e}^{i\varkappa\vartheta} \\ \text{or} \quad W(\vartheta, 0) &= \sum_{\varkappa \geq 0} B_\varkappa(0)\, (2 - \delta_{\varkappa 0}) \cos \varkappa\vartheta \end{aligned}\right\} \tag{51.41}$$

with

$$B_\varkappa(0) = \sum_k A_k(a, b\,b)\, A_k(b\,b, c)\, \frac{(k+\varkappa)!\,(k-\varkappa)!}{2^{2k}\left[\left(\frac{k+\varkappa}{2}\right)!\right]^2 \left[\left(\frac{k-\varkappa}{2}\right)!\right]^2} \tag{51.42}$$

so that

$$\left.\begin{aligned} W(\vartheta, t) &= \sum_{\varkappa} B_\varkappa(0)\, \mathrm{e}^{i\varkappa(\vartheta - \omega t)} \\ \text{or} \quad W(\vartheta, t) &= \sum_{\varkappa \geq 0} B_\varkappa(0)\, (2 - \delta_{\varkappa 0}) \cos[\varkappa(\vartheta - \omega t)]\,. \end{aligned}\right\} \tag{51.43}$$

For the time-integrated correlation function we get:

$$\left.\begin{aligned} W(\vartheta) &= \sum_{\varkappa} \int_0^\infty B_\varkappa(0)\, \frac{dt}{\tau_b}\, \mathrm{e}^{\frac{-t}{\tau_b} + i\varkappa(\vartheta - \omega t)} \\ &= \sum_{\varkappa \geq 0} \frac{B_\varkappa(0)}{\sqrt{1 + \varkappa^2 \omega^2 \tau_b^2}} \cos[\varkappa(\vartheta - \Delta\vartheta)]\, (2 - \vartheta_{\varkappa 0})\,. \end{aligned}\right\} \tag{51.44}$$

Hence the magnetic field results in an attenuation factor of $(1 + \varkappa^2 \omega^2 \tau_b^2)^{-\frac{1}{2}}$ together with a precession of the correlation function through an angle

$$\Delta\vartheta = \frac{1}{\varkappa} \arctan(\varkappa\omega\tau_b)\,. \tag{51.45}$$

If $\omega\tau_b \ll 1$, then this angle is approximately equal to the classical value, i.e. $\Delta\vartheta \sim \omega\tau_b$.

(d) Liquid sources. The rapidly fluctuating electric fields in a liquid interact with the nuclear quadrupole moment in the intermediate state and modify the angular correlation. Since there is no macroscopically preferred direction in a liquid, there is only a simple correlation involving the angle between the radiations (for a double cascade). The absence of any special direction in the environment enables the direction of emission of the first radiation to be chosen conveniently

as the axis of quantisation. This results in $\varkappa_1=0$ and $\varkappa_2=0$ [for example in (51.13)]. The time-differential angular correlation is then:

$$W(\Omega_1\Omega_2,t)=\sum_{k_1k_2}A_{k_1}(a,bb)\,A_{k_2}(bb,c)\,\frac{4\pi}{\hat{k}_1\hat{k}_2}\,Y_{k_1}^0(0)\,Y_{k_2}^0(\Omega_2)\,\langle k_2\,0\,\zeta'\zeta'\,|\,G(t)\,|\,k_1\,0\,\zeta\zeta\rangle, \tag{51.46}$$

where

$$\left.\begin{aligned}&\langle k_2\,0\,\zeta'\zeta'\,|\,G(t)\,|\,k_1\,0\,\zeta\zeta\rangle\\&\quad=\sum_{\beta\beta'}|\langle\beta'\,|\,U(t)\,|\,\beta\rangle|^2\,(-)^{2b-\beta-\beta'}\,(b\beta,b-\beta\,|\,k_1 0)\,(b\beta',b-\beta'\,|\,k_2 0)\,\delta_{\zeta\zeta'}.\end{aligned}\right\} \tag{51.47}$$

The quantity $|\langle\beta'\,|\,U\,|\,\beta\rangle|^2$, denoted by $W_{\beta\beta'}(t)$ by ABRAGAM and POUND, measures the probability that the intermediate nucleus, in the state $|\beta\rangle$ at $t=0$, makes a transition to the state $|\beta'\rangle$ at time t. This probability has to be evaluated and averaged over all nuclear sites in the liquid source.

From first-order perturbation theory,

$$W_{\beta\beta'}(t)=\frac{1}{\hbar^2}\left|\int_0^t H'_{\beta\beta'}(t')\,e^{-i\omega_{\beta\beta'}t'}\,dt'\right|^2. \tag{51.48}$$

Here $H'_{\beta\beta'}(t)$ represents the fluctuating quadrupole interaction of the nucleus with the surroundings and $\omega_{\beta\beta'}$ is the angular frequency corresponding to the energy difference between the substates β,β', in the absence of the fluctuating interaction. The averaged quantity $\overline{\overline{W_{\beta\beta'}(t)}}$ can be expressed in terms of the "correlation function" $\overline{\overline{H'_{\beta\beta'}(t')\,H'_{\beta\beta'}(t'-\tau)}}=G_{\beta\beta'}(\tau)$, which, as a function of τ, is related to the frequency spectrum of the fluctuations, thus,

$$\overline{\overline{W_{\beta\beta'}(t)}}=\frac{1}{\hbar^2}\left|\int_0^t e^{-i\omega_{\beta\beta'}\tau}\,G_{\beta\beta'}(\tau)\cdot(t-\tau)\,d\tau+\int_{-t}^0 e^{-i\omega_{\beta\beta'}\tau}\,G_{\beta\beta'}(\tau)\,(t+\tau)\,d\tau\right|^2. \tag{51.49}$$

The "correlation time" for the fluctuations may be defined so that for $\tau=\tau_{\mathrm{corr}}$, $G(\mathrm{i})$ is appreciably less than $G(0)$: for example, it is assumed that

$$G_{\beta\beta'}(\tau)=G_{\beta\beta'}(0)\,e^{-t/\tau_{\mathrm{corr}}}. \tag{51.50}$$

The correlation time, τ_{corr}, is usually sufficiently small (for non-viscous liquids) for the conditions: (i) $\tau_{\mathrm{corr}}\ll\omega_{\beta\beta'}^{-1}$, and (ii) the transition probability for a time interval equal to τ_{corr} is much smaller than unity, i.e.

$$\left(\frac{\tau_{\mathrm{corr}}}{\hbar}\right)^2\overline{\overline{|H'_{\beta\beta}|^2}}\ll 1,$$

to be valid. With these assumptions, ABRAGAM and POUND have evaluated $\overline{\overline{W_{\beta\beta'}(t)}}$ and obtain the result:

$$\overline{\overline{W_{\beta\beta'}(t)}}=\sum_k(-)^{2b-\beta-\beta'}\,(b\beta,b-\beta\,|\,k0)\,(b\beta',b-\beta'\,|\,k0)\,e^{-\lambda_k t}, \tag{51.51}$$

where

$$\lambda_k=\frac{3}{80}\left(\frac{Q}{\hbar}\right)^2\overline{\overline{\left(\frac{\partial^2 V}{\partial z'^2}\right)^2}}\,\tau_{\mathrm{corr}}\,\frac{k(k+1)\,[4b(b+1)-k(k+1)-1]}{[b(2b-1)]^2}. \tag{51.52}$$

Here z' is measured along the direction of the largest field gradient in a coordinate system moving with the fluctuating field.

The time-differential angular correlation is given by (51.47), (51.51) and (51.52). It is:

$$W(\vartheta,t)=\sum_k A_k[1,2]\,G_k(t)\,P_k(\cos\vartheta) \tag{51.53}$$

where $G_k(t)=e^{-\lambda_k t}$ are the time-differential attenuation coefficients.

The time-integrated angular-correlation is obtained from the weighted integral of $G_k(t)$:

$$\bar{G}_k = \int_0^\infty e^{-t/\tau_b} G_k(t) \frac{dt}{\tau} = \frac{1}{1+\lambda_k \tau_b} \tag{51.54}$$

and

$$W(\vartheta) = \sum_k A_k[1,2]\, \bar{G}_k\, P_k(\cos\vartheta). \tag{51.55}$$

For large values of $\lambda_k \tau_b$, $\bar{G}_k$ tends to zero (except for $k=0$) so that, as is to be expected, the correlation is destroyed.

52. Experimental results. The experimental methods of investigating the interaction with extra-nuclear fields through their influence on angular correlations, can be divided broadly into two groups. In the first of these the extra-nuclear environment, such as the chemical composition, physical properties, or crystalline states, is changed; in the second the time relationships between the successive radiations are controlled (total and "delayed" coincidences). The first method is usually simpler if the aim is restricted to the determination of the true, unperturbed, correlation; both types of variation may be employed if the actual interaction is studied. We may notice also that one can produce externally applied magnetic fields of sufficient magnitude ($\sim 10^4$ gauss) to affect measurably the angular correlation if the lifetime of the intermediate state is not too short ($\lesssim 10^{-9}$ sec), but electric-field gradients of sufficient intensity are only available on the atomic scale.

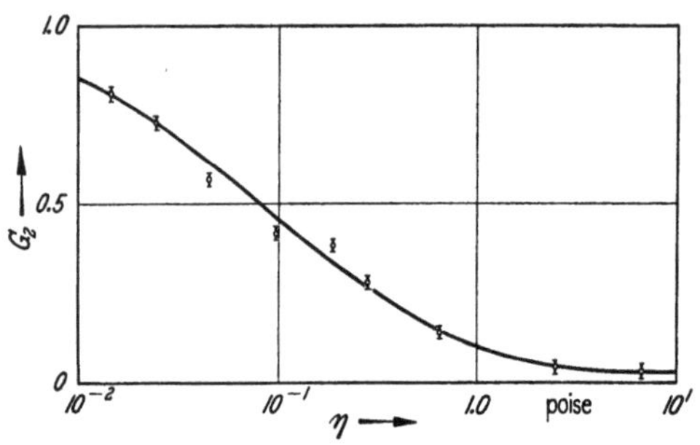

Fig. 43. Variation of the attenuation coefficient, G_2, with viscosity of the source embedded in water in water-glycerine mixtures for the γ-γ cascade of Cd^{111} [cf. H. ALBERS-SCHÖNBERG *et al.*: Helv. phys. Acta **27**, 656 (1954)].

Determination of the true correlation [*60*], [*14a*], [*15*]. The usual cause of perturbation is the electric-quadrupole interaction[1], either static (micro-crystal) or time-dependent (liquid). The electric-field gradient is eliminated if the source is an ideal cubic metal crystal and the radiating nuclei occupy regular lattice sites. In practice one must take into account possible lattice imperfections as well as the dislocations due to preceding radiations, so that unperturbed correlations with solid sources are not easily attained. True angular correlations are also obtainable, in principle, with non-cubic crystals, provided the symmetry axis of the crystal is directed along the direction of one of the radiation. The use of single-crystal sources is, however, confined in practice to the detailed study of perturbations.

Liquid sources, for which the fluctuation "correlation times" are very short, e.g. liquid metals or dilute solutions, are very frequently used to determine the ideal angular-correlation. In particular the variation of the anisotropy with viscosity and the extrapolation to zero viscosity provides a useful method for sources in solute form. Fig. 43 shows the variation of attenuation coefficient G_2

[1] Detailed investigations of quadrupole-couplings in the cascade In^{111} (K-capture) $\rightarrow Cd^{111}$, $\gamma-\gamma$, have been made by H. ALBERS-SCHÖNBERG, E. HEER and P. SCHERRER: Helv. Phys. Acta **27**, 637 (1954) and R. M. STEFFEN: Phys. Rev. **103**, 116 (1956).

for the γ_1, γ_2 cascade in Cd^{111} obtained with a source of $InCl_3$ in water-glycerine solution of various viscosities.

The true correlation is obviously given by the time-differential angular correlation with a time interval between the successive radiations sufficiently small to ensure that no appreciable perturbation occurs. Practically, the anisotropy can be studied using finite resolving times and delay intervals. The theoretically predicted curves for static and time-dependent interactions in the same Cd^{111} cascade are shown in Fig. 44, (a) for the ideal case, and (b) for a more realistic situation with the resolving time equal to the lifetime of the intermediate state and both equal to time for "precession" through an angle $\pi/2$ (i.e. $\pi/2\omega$). The experimental curves, Fig. 45, obtained with rather inexact timing conditions show qualitative agreement and indicate how the ideal correlation can be inferred[1].

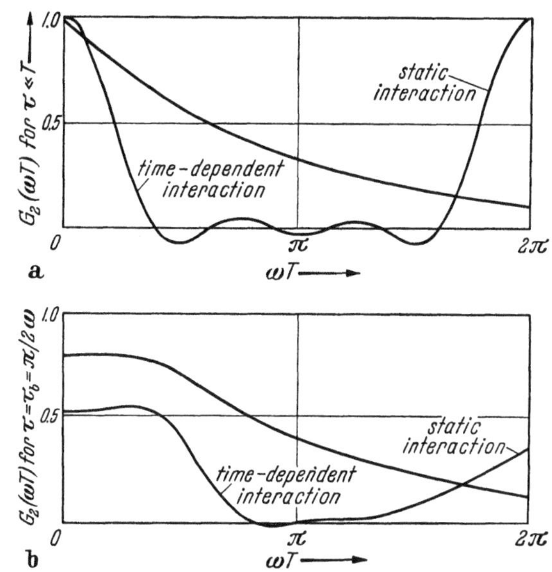

Fig. 44a and b. Theoretical curves for the attenuation coefficient for the γ-γ cascade of Cd^{111} as a function of the delay of the delay time for (a) an ideal case and (b) a case where the resolving time equals $\pi/2\omega$ [cf. H. ALBERS-SCHÖNBERG *et al.*: Helv. phys. Acta **27**, Fig. 6 (1954)].

The true correlation may, as previously mentioned, be restored by the application of a strong magnetic field in the direction of one of the radiations. It is necessary that the energy separation of the substates due to the magnetic interaction be large compared with the other interactions tending to perturb the angular correlation. This method has not, so far, been used decisively in experimental work.

Some indication of the extent of the attenuations observed experimentally in the In^{111} γ-γ cascades is shown in Table 3 ([*14a*], p. 590).

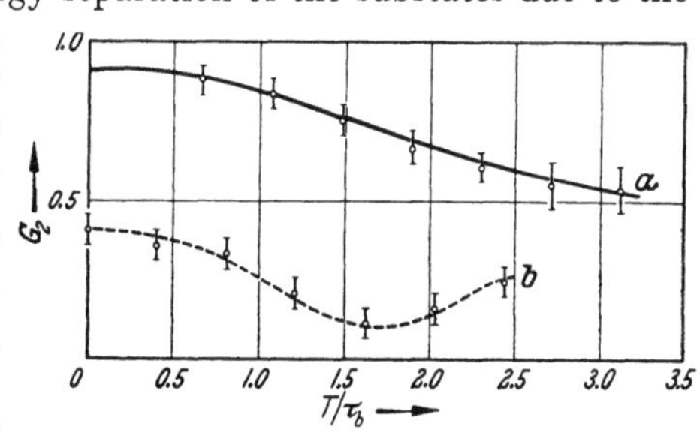

Fig. 45. Experimental variation of the attenuation coefficient as a function of the delay time for (*a*) a non-static quadrupole-interaction ($InCl_3$ in a mixture of glycerine and water) and (*b*) a static quadrupole interaction (powdered source of metallic indium) [cf. H. ALBERS-SCHÖNBERG *et al.*: Helv. phys. Acta **27**, Fig. 7 (1954)].

It is interesting to notice that the attenuation factors for the metallic sources are much the same for the cascade $\gamma_1 - \gamma_2$ which follows K-capture and for the cascade $\gamma_1' - \gamma_2$ which follows an isomeric transition. Both cascades involve the same intermediate state, so that the perturbation cannot be attributed to the effects of the transition preceding the cascade which are quite different in the two cases. For the crystalline sources, however, the difference in attenuation is quite pronounced, and the fact that for the γ_1-γ_2 cascade the attenuation is far below the "hard-core" value indicates the presence of "time-dependent" effects due to the preceding K-capture.

[1] See also [*64*], p. 590.

Table 3. *Experimental attenuation coefficients G_2 for the γ-γ cascades in* Cd^{111}.
The cascade $\gamma_1-\gamma_2$ follows K-capture in In^{111}, the cascade $\gamma_1'-\gamma_2$ starts from the 48-minute isomeric level in Cd^{111}. The G_2 values are based on an assumed unperturbed anisotropy of -0.245, which may be too large so that the G_2 values may be too small. The "hard-core" value of G_2 is 0.2 ([*64*], p. 594).

Source		Attenuation factor G_2 cascade $\gamma_1-\gamma_2$	Attenuation factor G_2 cascade $\gamma_1'-\gamma_2$
Molten metal	Cd	0.86 ± 0.07	0.97 ± 0.08
	In	0.86 ± 0.05	—
Solid metal	Cd	0.32 ± 0.04	0.40 ± 0.07
	In	0.35 ± 0.02	—
Aq. solution	$CdCl_2$	—	0.88 ± 0.05
	$InCl_3$	0.90 ± 0.02	—
Solid	$CdCl_2$	—	0.19 ± 0.06
	$InCl_3$	0.04 ± 0.02	—

Measurement of nuclear-magnetic moments[1] [*62*], [*14a*], [*15*]. Provided the extra-nuclear interactions in the source material are not large, and the lifetime of the intermediate state, b, is sufficient, the magnetic moment of this state may be determined by observing the effect of an applied magnetic field on the angular correlation.

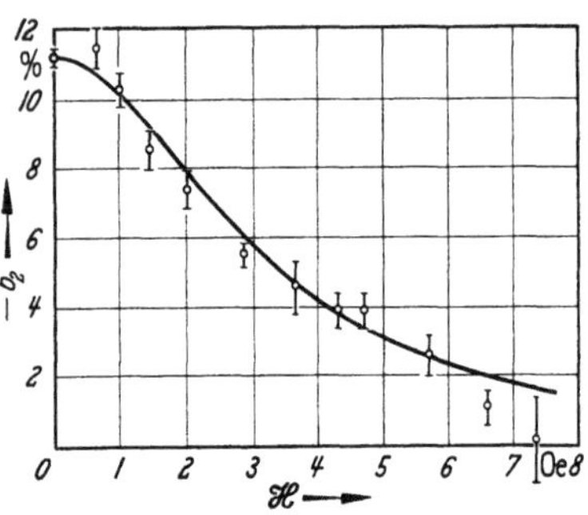

Fig. 46. Coefficient of $\cos 2\vartheta$ in the angular correlation for the γ-γ cascade of Cd^{111} as a function of the applied transverse magnetic field [cf. H. AEPPLI *et al.*: Helv. phys. Acta **25**, 365 (1952)].

If also the detectors (in the case of a cascade correlation) can distinguish between the radiations, then a magnetic field perpendicular to both radiations results in an attenuation together with a rotation of the angular correlation as given by Eqs. (51.44) and (51.45). For the ideal time-differential correlation, with other perturbations completely absent, only the rotation is present. If the two detectors are equally sensitive to both radiations of the cascade, then only an attenuation occurs. In the earlier experiments the attenuation of the correlation in the In^{111} cascade was measured as a function of the applied magnetic field and the approximate value of the magnetic moment deduced (cf. Fig. 46).

More accurate values of the moment can be obtained by observing the rotation of the correlation as a function of the field using delayed coincidences and short resolving times[1]. For example, the magnetic field, $\mathcal{H}$, is applied alternatively in the positive and negative z-directions and measurements made in each case of the number of delayed coincidences at a fixed angle $\left(\vartheta = \frac{3\pi}{4} \text{ or } -\frac{3\pi}{4}\right)$. The ratio, R, the rate of coincidences in the fixed direction is thus determined. The variation of R with $\mathcal{H}$ is shown in Fig. 47. This ratio is unity for $\mathcal{H}=0$, and again for $\mathcal{H}$ such that $\omega\tau = \pi/2$. (The angles are chosen so as to give the maximum rate of variation of R with magnetic field.) In the relatively favourable case of In^{111}, the magnetic moment of the intermediate state has been measured to an accuracy of some 5%.

[1] H. AEPPLI, H. ALBERS-SCHÖNBERG, H. FRAUENFELDER and P. SCHERRER: Helv. phys. Acta **25**, 339 (1952). — V. KROHN, T. B. NOVEY and S. RABOY: Phys. Rev. **98**, 1187 (1955). — R. M. STEFFEN and W. ZOBEL: Phys. Rev. **103**, 126 (1956).

[2] R. STEFFEN [*15*]. — V. E. KROHN and S. RABOY: Phys. Rev. **97**, 1017 (1955).

Measurements of a similar type for the angular distribution in a reaction (the inelastic scattering of protons by F^{19} leaving this nucleus in the $^5/_2{}^+$ level

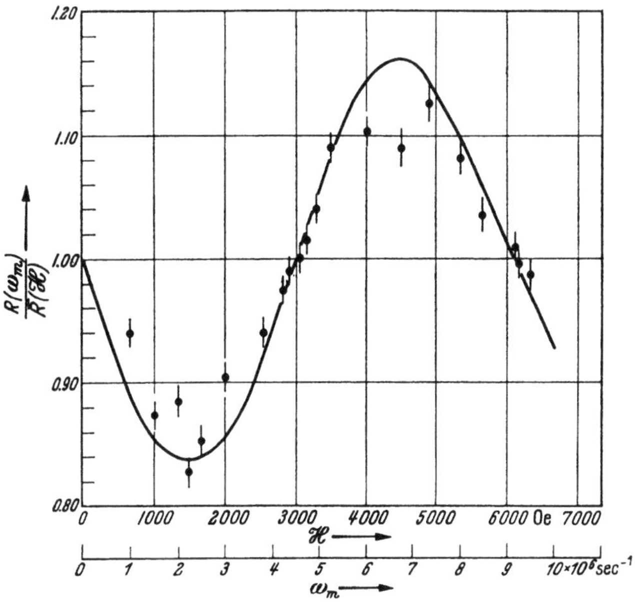

Fig. 47. Variation of R, the ratio of the rates of coincidence at fixed angles $\vartheta = \pm 3\pi/4$ for alternative signs of either $\mathcal{H}$, the applied magnetic field or the Larmor frequency—Comparison of theory with experiment. [Cf. R. M. STEFFEN and W. KOBEL: Phys. Rev. 103, 132 (1956)].

at 0.197 Mev) have also been made[1]. The angular distribution is of the form:

$$W(\vartheta) = \int_0^\infty \left\{1 + A_2 e^{-\lambda_2 t} P_2[\cos(\vartheta - \omega t)]\right\} e^{-\lambda_b t} dt$$

where λ_2 characterises the attenuation in the source due to quadrupole perturbations and $\lambda_b = 1/\tau_b$. The ratio $W(\vartheta)/W(-\vartheta)$, $\vartheta = 30°$, was measured for a liquid target and with $\mathcal{H}$ applied perpendicular to the reaction plane. This ratio passes through a maximum when

$$\omega \approx \frac{\lambda_b}{2}\left(1 + \frac{\lambda_2}{\lambda_b}\right)$$

(cf. Fig. 48). The contribution from λ_2, which is probably small, can be estimated by varying the viscosity of the target material, and an estimate of the magnetic moment obtained.

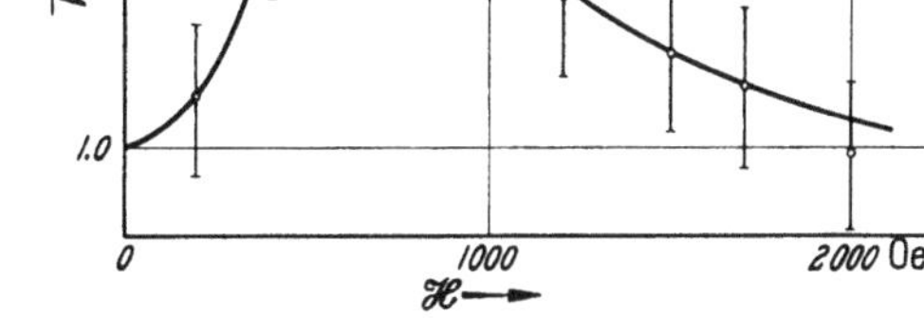

Fig. 48. Variation of the ratio $\frac{W(30°)}{W(-30°)}$ as a function of the magnetic field which is perpendicular to the reaction plane for the reaction $F^{19}(p, p')\,F^{19}$ (cf. P. LEHMANN *et al.*: C. R. Acad. Sci., Paris 241, 700 (1955)].

Electric-quadrupole interactions [*61*], [*14a*]. A comparison of the static-interaction in a micro-crystal line source and the true correlation can provide a measure of

[1] P. LEHMANN, A. LEVEQUE and M. FIEHRER: C. R. Acad. Sci., Paris **241**, 700 (1955); see also P. B. TREACY: Nature, Lond. **176**, 923 (1955), and W. R. PHILLIPS and G. A. JONES: Phil. Mag. **6**, 576 (1956).

the quadrupole-interaction $Q\frac{\partial\mathscr{E}}{\partial z}$, and if $\frac{\partial\mathscr{E}}{\partial z}$ can be estimated, of Q. In order to interpret the result in terms of a static interaction [cf. Eqs. (51.32) and (51.33)], the source conditions must be such that the effects of preceding radiations, giving rise to time-dependent interactions, are not important, for example as in a metallic source.

More direct investigation of the quadrupole interaction has been made by using single crystals of Indium containing radioactive In^{111} and measuring the anisotropy as a function of the orientation of the crystal axis with respect to the directions of the radiations[2]. The agreement of the experimental results with theory confirmed that the interaction was essentially static quadrupole, and that therefore, *inter alia*, the disturbances due to the radiations did not seriously disturb the lattice position of the In^{111} nuclei. Measurements were also made of the combined effect of a magnetic field, along the direction of the symmetry axis, together with the quadrupole-coupling from the crystal itself. Since the magnetic precession is unidirectional and the electric quadrupole occurs in both directions, the anisotropy, as a function of magnetic field, first increases, reaches a maximum for $\omega_{\text{magn}} = \frac{1}{2}\omega_Q$ and then decreases. Such measurements can in principle determine the magnitude of both magnetic and electric interactions.

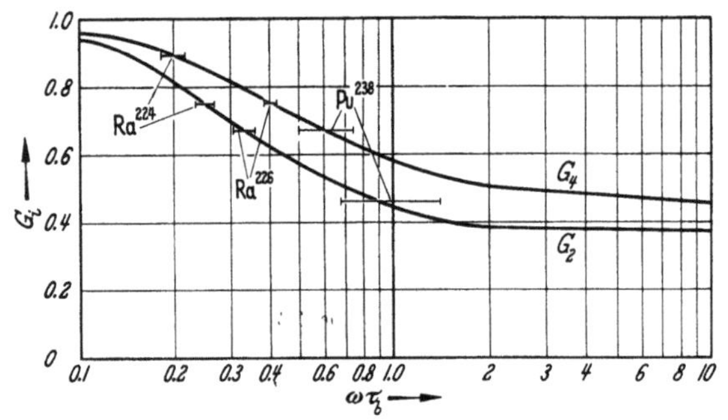

Fig. 49. Theoretical attenuation coefficients G_2, G_4 as a function of $\omega\tau_b$ for the α-γ directional correlation in Ra^{224}, Ra^{226}, Pu^{238} cascades assuming pure quadrupole interaction. The experimental values of ω are inferred from the measured values of correlations and lifetime [cf. H. FRAUENFELDER [*14a*], p. 587).

A characteristic differences between electric and magnetic perturbations is the theoretical result that the attenuation factors G_k decrease monotonically with k for magnetic but not for electric quadrupole interactions. This is illustrated in Fig. 49, which shows the measured attenuation coefficients[2] for α-γ cascades in Ra^{224} and Ra^{226}, and Pu^{238}. The theoretical curves are on the assumption of pure quadrupole interaction.

When several levels of the same nucleus are excited by α-emission and it is, reasonable to assume a similar environment, $\partial\mathscr{E}_z/\partial z$ for all the intermediate states, then the ratios $Q_b\tau_b/Q_{b'}\tau_{b'}$ for the different levels, b, b', can be deduced from the attenuation of the corresponding α-γ angular correlations[3]. These quantities are of particular interest in connection with the collective "rotational" levels of heavy nuclei.

I. Application of theory to mesons, heavy mesons and hyperons.

53. Heavy mesons and hyperons. As a result of the successful application of angular correlation theory in determining symmetry properties of nuclei, attempts have recently been made to apply the formalism to the interpretation

[1] H. ALBERS-SCHÖNBERG, E. HEER, T. B. NOVEY and P. SCHERRER: Helv. phys. Acta **27**, 547 (1954).

[2] J. C. D. MILTON and J. S. FRASER: Phys. Rev. **95**, 628 (1954).

[3] G. VALLADAS, J. TEILLAC, P. FALK-VAIRANT and P. BENOIST: J. de Phys. **16**, 125 (1955).

of phenomena involving hyperons and heavy mesons so as to determine their spins and parities. The achievements are generally limited by the scanty experimental data that are presently available.

The simplest enquiry is to test whether lower limits can be set to the spins of these new particles where these are greater than 0 or $\frac{1}{2}$. In some of the reactions in which these new particles participate this is a direct consequence of any observed anisotropy in the angular distribution. However, in all the cases so far considered, the events are till too few in number for any definite conclusion to be reached by using angular correlation measurements as a basis.

We shall first discuss three decay schemes that illustrate direct applications of the angular correlation theory. They are: (1) the cascade decay of Ξ^- and Σ^0-particles, (2) the decay of the τ^+-meson, and (3) the decay involving associated production of hyperons and heavy mesons.

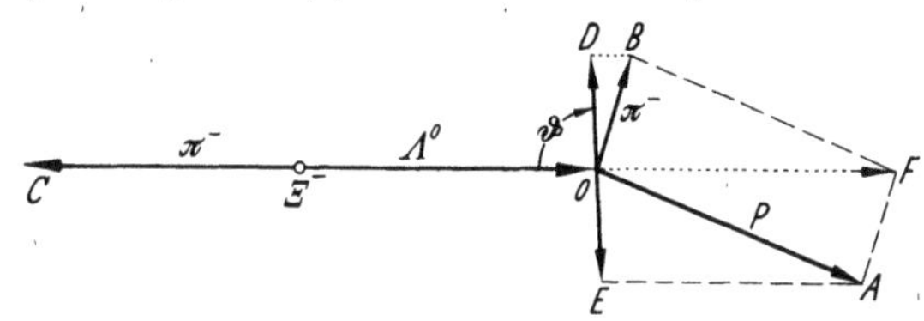

Fig. 50. Momentum diagram for the description of the cascade, $\Xi^- \to \pi^- + \Lambda^0$; $\Lambda^0 \to \pi^- + p$, in terms of the rest system of the Λ^0-particle.

Further fruitful application of the theory can be made concerning meson reactions currently studied. Here only a few relative or total angular momenta play a significant rôle since the meson energies are generally low. By a comparison of the experimental cross-section with the predicted angular distribution it should be possible to establish certain relations concerning the relative strengths of the various modes or channels through which the reaction proceeds. Although the usual treatment is through conventional phase-shift analyses, similar results are obtained using the methods of Chap. E. Thus, instead of considering each reaction in a specific manner, a very general formula is used which involves arbitrary angular momenta and which applies to all types of radiation. Some of the ambiguity in the interpretation of the angular correlations may be removed by the study of polarisation effects and these are, therefore, also considered.

We will refer to three reactions: (1) the scattering of π-mesons by nucleons, e.g. $p+\pi^+ \to p+\pi^+$, (2) photo-meson production, e.g. $\gamma+p \to n+\pi^+$, and (3) the absorption of π-mesons by deuterium or its inverse process, i.e., $\pi^- + d \rightleftharpoons p+p$.

α) Cascade decay of Ξ^- and Σ^0-particles. GATTO[1] proposed the study of the angular correlation of emitted particles in the cascade decay of the Ξ^- particle and the Σ^0-particle, i.e.,

$$\Xi^- \to \Lambda^0 + \pi^-, \quad \Lambda^0 \to p + \pi^-, \tag{53.1}$$

$$\Sigma^0 \to \Lambda^0 + \gamma, \quad \Lambda^0 \to p + \pi^- \tag{53.2}$$

in order to gain some information concerning the spins and parities of these hyperons. Both processes demand essentially the same treatment in determining the angular correlation. We will first consider the decay of the Ξ^--particle.

To conform with our notation, we designate the spin and parity[2] of the Ξ^--particle and the Λ^0-particle by a, p_a and b, p_b, respectively, while L_1 and L_2 denote the relative orbital angular momenta of the π-mesons for each stage of the decay, as measured in the rest system of the Ξ^--particle and the Λ^0-particle.

[1] R. GATTO: Nuovo Cim. **2**, 841 (1955); **3**, 665 (1956); **4**, 196 (1956).

[2] By parity, we mean the intrinsic parity relative to that of a proton. In the following analysis and throughout the rest of this chapter, conservation of parity is assumed (see Sect. 49).

The correlation angle, ϑ, is the angle between the two π-mesons, as measured in the Λ^0-particle rest system, and is determined by measurement of the momenta of the various emitted particles together with a simple kinematical transformation to the hyperon rest system. This is illustrated[1] in Fig. 50.

Conservation of angular momentum and parity requires that

$$\boldsymbol{a} = \boldsymbol{b} + \boldsymbol{L}_1, \qquad \boldsymbol{b} = \boldsymbol{L}_2 + \tfrac{1}{2}, \tag{53.3}$$

$$p_a = p_b(-)^{L_1+1}, \qquad p_b = (-)^{L_2+1}. \tag{53.4}$$

The last relation follows from the pseudoscalar property of the π-meson. Assuming that both the Λ^0- and the Ξ^--particles have well-defined states, it is clear from (53.3) and (53.4) that L_2 is unique, and that L_1 is restricted to either all even or all odd integral values, the number of such values being the minimum of $\frac{2a+1}{2}$ and $\frac{2b+1}{2}$. Thus if $b = \frac{3}{2}$, $a = \frac{5}{2}$, say, L_1 assumes either two even or two odd integral values.

The angular correlation is obtained from (10.16) by inserting into the formula the radiation parameters for spinless particles (13.12), which results in

$$\left.\begin{aligned} W(\vartheta) &\sim |\langle \tfrac{1}{2} \| L_2 \| b\rangle|^2 \sum_{L_1 L_1' k} \hat{L}_2^2 \hat{L}_1 \hat{L}_1' (L_1 0, L_1' 0 | k 0)(L_2 0, L_2 0 | k 0) \times \\ &\quad \times W(b b L_1 L_1'; k a)\, W(b b L_2 L_2; k \tfrac{1}{2}) \langle b \| L_1 \| a\rangle \langle b \| L_1' \| a\rangle^* P_k(\cos\vartheta) \sim \\ &\sim \sum_{L_1 L_1' k} \hat{L}_1 \hat{L}_1' (L_1 0, L_1' 0 | k 0)(b \tfrac{1}{2}, b - \tfrac{1}{2} | k 0)\, W(b b L_1 L_1'; k a) \times \\ &\quad \times \langle b \| L_1 \| a\rangle \langle b \| L_1' \| a\rangle^* P_k(\cos\vartheta). \end{aligned}\right\} \tag{53.5}$$

Since L_1 takes on two values which differ by 2, if the parities of the Ξ^- and Λ^0-particles are well-defined, the angular correlation involves the mixing parameter, $\delta = \frac{\langle b \| L_1 + 2 \| a\rangle}{\langle b \| L_1 \| a\rangle}$, which is real if the interaction is assumed to have properties similar to that of the Hamiltonian. k is even and has a maximum value determined by the maximum of $2b$, $2L_1$, $2L_2$.

If the hyperon spin, b, is $\frac{1}{2}$, there is clearly no correlation. GATTO has calculated some of the angular correlations that obtain for $b = \frac{3}{2}$, where various assumptions are made concerning the parities of the particles and the value of a. Eq. (53.5) can be used for higher values of b or a. In particular, if $b = \frac{3}{2}$ and $a = \frac{1}{2}$, both L_1 and L_2 are either 1 or 2 according to whether p_a and p_b are $+1$ or -1 respectively. In this case, the angular distribution is simply

$$\left.\begin{aligned} W(\vartheta) &\sim \sum_k (\tfrac{3}{2} \tfrac{1}{2}, \tfrac{3}{2} - \tfrac{1}{2} | k 0)^2 P_k(\cos\vartheta) \sim \\ &\sim 1 + P_k(\cos\vartheta), \end{aligned}\right\} \tag{53.6}$$

which is independent of the parity of either hyperon. If $b = \frac{3}{2}$ and $a \geq \frac{3}{2}$, two expressions for the angular correlation result according to whether $p_a = \pm p_b$, in consequence of which L_1 is odd or even respectively. Nevertheless, it is impossible to find either p_a or p_b.

Up to the present, very few identifications have been made of the Ξ^--particle (the number is less than a dozen in all)[2], so that (53.5) is hardly useful. With the accumulation of a sufficient number of events it should be possible to determine the spin and parities of the hyperons; however, by then these spins will probably have been deduced from other considerations.

[1] R. GATTO: Nuovo Cim. **2**, 841 (1955).

[2] Sixth Annual Rochester High-Energy Physics Conference, 1956.

Proof for the existence of the Σ^0-particle rests primarily in the lack of energy and momentum balance in some cases in the reaction $\pi^- + p \to \theta^0 + \Lambda^0$. Instead, the reaction proceeds as $\pi^- + p \to \theta^0 + \Sigma^0$, where the Σ^0-particle decays into a Λ^0-particle and a γ-ray. As yet, no γ-rays have been observed. Assuming that the Σ^0-cascade proceeds as in (2), the angular correlation between the γ-ray and π-meson is

$$\left.\begin{aligned} W(\vartheta) \sim \sum_{k L_1 L_1'} \hat{L}_1 \hat{L}_1' (L_1 1, L_1' -1 \mid k 0)(b \tfrac{1}{2}, b - \tfrac{1}{2} \mid k 0)\, W(b b L_1 L_1'; k a) \times \\ \times \langle b \| L_1 \| a \rangle \langle b \| L_1' \| a \rangle P_k(\cos\vartheta) \end{aligned}\right\} \quad (53.7)$$

where the reduced matrix elements are real. Instead of the parity relation (53.4), we have $p_a = p_b p_1$, where $p_1 = (-)^{L_1}$ if L_1 corresponds to an electric type multipole, and $p_1 = (-)^{L_1+1}$ if the multipole is magnetic type. For given values of p_a and p_b, both types of radiation contribute, and more than two values of L_1 can be associated with a set of values of a and b, although the character of the radiation, typified by L_1, is determined by the parity conservation rule.

Again, in principle experiment can decide upon the parity of the Σ^0-particle relative to that of the Λ^0-particle but cannot fix the intrinsic parity of either particle. In particular, if $a = \frac{3}{2}$, $b = \frac{3}{2}$ and $L_1 \to M\,1, E\,2$, or $M\,3, \ldots$, then $p_a = p_b$, while if $L_1 \to E\,1, M\,2, E\,3, \ldots$, then $p_a = -p_b$. GATTO gives explicit values for the angular correlation with $\frac{1}{2} \leq a \leq \frac{7}{2}$ and $\frac{3}{2} \leq b \leq \frac{7}{2}$ in terms of ξ which is related to the mixing ratio for the two lowest order multipoles that are allowed for the transition.

β) The decay of the τ^+-meson. One of the most widely studied of the "new" particles is the τ^+-meson, which decays into three charged π-mesons, i.e. $\tau^+ \to \pi^- + 2\pi^+$. An alternative decay, $\tau^+ \to \pi^+ + 2\pi^0$, is possible; however, this event is much more difficult to observe. With the advent of the Bevatron and Cosmotron, the few τ^+-mesons that were found in the cosmic radiation have been greatly supplemented in number so that there now exists a fairly precise value of the τ^+-meson mass. Because the secondaries are charged, it is usually possible to find their momenta, and, since the τ^+-meson decays at rest (a fact indicated by the coplanarity of the particles), one can also check that these momenta combine vectorially to form a triangle. It is thus clear that from a knowledge of the magnitudes of the momenta (or the energies) of the secondaries, the angles between any of the secondaries can be found. In consequence there exists a close relationship between energy correlations and correlations involving the angles.

The decay can be depicted as a double cascade decay of the τ^+-meson of spin a and parity p_a, with the emission of a π^--meson of angular momentum L_1 in the τ^+-meson rest system followed by the emission of two π^+-mesons, whose relative angular momentum is L_2, as measured in their own centre-of-mass system. The angular correlation between the π^--meson and one of the π^+-mesons referred to the centre-of-mass of the like mesons is given by

$$\left.\begin{aligned} W(\vartheta) &\sim \sum_{b b' L_1 L_1' L_2 L_2' k} \hat{b}\hat{b}' (-)^{b+b'} \hat{L}_1 \hat{L}_1' \hat{L}_2 \hat{L}_2' (L_1 0, L_1' 0 \mid k 0)(L_2 0, L_2' 0 \mid k 0) \times \\ &\quad \times W(b b' L_1 L_1'; k a)\, W(b b' L_2 L_2'; k c)\, \mathrm{Re}\{\langle L_2 L_1 a | \mathcal{O} | a \rangle \langle L_2' L_1' a | \mathcal{O} | a \rangle^*\} \times \\ &\quad \times q(b b'; L_1 L_1', L_2 L_2')\, P_k(\cos\vartheta) \\ &\sim \sum_{L_1 L_1' L_2 L_2' k} \hat{L}_1 \hat{L}_1' \hat{L}_2 \hat{L}_2' (L_1 0, L_1' 0 \mid k 0)(L_2 0, L_2' 0 \mid k 0)\, W(L_1 L_1' L_2 L_2'; k a) \times \\ &\quad \times q(L_1 L_1'; L_2 L_2')\, \mathrm{Re}\{\langle L_2 L_1 a | \mathcal{O} | a \rangle \langle L_2' L_1' a | \mathcal{O} | a \rangle^*\}\, P_k(\cos\vartheta). \end{aligned}\right\} \quad (53.8)$$

This corresponds to the result given by DALITZ[1]. By reason of the identical nature of the two π^+-mesons and of their Bose-statistics, L_2 must be even. Since the π-mesons are pseudoscalar, the parity, p_a, is given by[2]

$$p_a = p_b(-)^{L_1+1} = (-)^{L_1+1}. \tag{53.9}$$

In attempts to render (53.8) into a more useful form, several assumptions have been made. DALITZ and FABRI[3] assumed that the matrix elements are related to the penetrabilities through the centrifugal barrier, which vary roughly as $\left(\frac{p_1}{mc}\right)^{L_1}$, $\left(\frac{p_2}{mc}\right)^{L_2}$, where p_1 and p_2 are the momenta of the unlike and like mesons as measured in the centre-of-mass system of the like mesons, and m is the π-meson mass. Since the momenta are small as compared to mc, the most important contribution to the angular correlation comes from those values of L_1 and L_2 such that L_1+L_2 corresponds to the lowest possible value. The Coulomb-interaction between the mesons is neglected, which implies ignoring the particles of very low energy.

Denoting the properties of the τ^+-meson by $[a, p_a]$, we list in Table 4 the pairs of L_1 and L_2 which are of lowest order of various values of $[a, p_a]$, together with the angular dependence of (53.8). The division is made according to whether the process of decay into two π-mesons is allowed or not. For this to occur, it is necessary that $p_a = (-)^a$. The angle, ϑ, is very easily found in the non-relativistic approximation. If the three momenta as measured in the laboratory system are drawn emanating from the point of decay, then ϑ is the angle between (1) the line joining the extremities of the momentum vectors of the like mesons, and (2) the line at 180° to the momentum vector of the unlike meson. The experimental data[4] are not very conclusive, although they most favour isotropy, which corresponds to [0 —] or [1 +]. On the other hand, if relativistic effects are not ignored, these corrections can alter some distributions by as much as 30%, as pointed out by EISENBERG *et al.*[5]. Their more careful investigation of the validity of the various assumptions shows that if the interaction is only dependent on the lowest value of L_1+L_2, only [0 —] is consistent with the experimental data. If there is a dependence on L_1 and L_2 separately, other classifications are likely, *viz.* [3 +] and [1 +].

Table 4. *Lowest values of L_1 and L_2 consistent with $[a, p_a]$ and the resultant angular correlation.*

Two π-meson decay forbidden.				Two π-meson decay allowed.			
$[a, p_a]$	L_1	L_2	$W(\vartheta)$	$[a, p_a]$	L_1	L_2	$W(\vartheta)$
0 —	0	0	1	1 —	2	2	$p_1^4 p_2^2 \sin^2\vartheta \cos^2\vartheta$
1 +	1	0	p_1^2	2 +	1	2	$p_1^2 p_2^4 \sin^2\vartheta$
2 —	0	2	$p_1^4 + p_2^4 +$	3 —	2	2	$p_1^4 p_2^4 \sin^2\vartheta (5+3\cos^2\vartheta)$
	2	0	$+ p_1^2 p_2^2 (3\cos^2\vartheta - 1)$				

A slightly different approach uses the Dalitz triangle plot (cf. footnotes 1 and 3), which is a two-dimensional plot in which events are plotted as points in a circle which inscribes an equilateral triangle. (If the kinematics are treated relativistically the circle is replaced by a curve of the third degree (cf. footnote 3,

[1] R. H. DALITZ: Phil. Mag. **44**, 1068 (1953). — Phys. Rev. **94**, 1046 (1954).
[2] Conservation of parity is assumed.
[3] E. FABRI: Nuovo Cim. **11**, 479 (1954).
[4] K. GOTTSTEIN: Sixth Annual Rochester High Energy Physics Conference, 1956
[5] Y. EISENBERG, E. LOMON and S. ROSENDORFF: Nuovo Cim. **4**, 610 (1956).

p. 534). The points are plotted such that their distances from the sides of the triangle vary with the kinetic energies of the mesons. By studying the density of events inside the circle, it is possible, to estimate $[a, p_a]$. The recent data[1], which are mainly derived from events observed in photographic emulsions exposed to the K^+-meson beam from the Bevatron and Cosmotron, show that $[a, p_a]$ is most probably $[0\ -]$ or less probably $[2\ -]$, although $[4\ -]$ and $[3\ +]$ are not ruled out definitely.

Thus the evidence indicates that only the assignments in the first column of Table 4 are likely. The exclusion of any of the assignments in the second column implies the result that the two π-meson decay cannot occur, by conservation of parity, and that the θ-meson which decays into two π-mesons is essentially a different particle[2].

These procedures have been modified by FABRI and COSTA and TAFFARA[3]. The latter suggest measurement of the angular distributions between the two like mesons, by measuring the density of events in three segments of the circle, corresponding to ϑ between 0 and $\frac{\pi}{3}$, $\frac{\pi}{3}$ and $\frac{2\pi}{3}$, and $\frac{2\pi}{3}$ and 2π; while Fabri classifies the unlike meson according to whether its energy is greater than, less than, or intermediate to the energy of the π^+ mesons and finds the relative probabilities as a function of $[a, p_a]$.

Attempts have been made to correlate the plane of decay with the direction of emission of the τ^+-mesons which are emitted either from cosmic-ray stars[4] or which travel along the K^+-meson beams[5]. The assumption is made that, if the τ^+-meson has a non-zero spin, upon emission it will probably have some preferred polarisation which remains unaltered in spite of some depolarisation before the meson comes to rest in the emulsion.

The statistical tensor for the τ^+-meson, $\varrho_{k_a \varkappa_a}$, is no longer represented by a δ-function, since there exists now a preferred direction in space along the direction of motion of the K^+-beam. Instead of the angular correlation varying as

$$\sum_{\varkappa} \mathscr{D}^k_{\varkappa 0}(\mathscr{R}_1)\, \mathscr{D}^{k*}_{\varkappa 0}(\mathscr{R}_2) = P_k(\cos\vartheta)$$

where ϑ is the angle between the radiations L_1 and L_2 as measured in the decay plane—a result derived assuming lack of knowledge concerning the initial state, given knowledge concerning the direction of motion, the correlation now varies as

$$\sum_{\varkappa_1 \varkappa_2} Y^{\varkappa_1}_{k_1}(\vartheta_1, \varphi)\, Y^{\varkappa_2}_{k_2}(\vartheta_2, \varphi)\, (k_1 \varkappa_1, k_2 \varkappa_2 \mid k_a \varkappa_a)$$

where ϑ_1 and ϑ_2 are measured in the decay plane with $\vartheta = \vartheta_1 - \vartheta_2$ and where φ is the angle between this plane and the direction of the τ^+-meson. If the distribution is isotropic in φ, either the spin is zero or the τ^+-meson is unpolarised, while if the distribution exhibits some φ-dependence, the spin is greater than

---

[1] J. OREAR, G. HARRIS and S. TAYLOR: Phys. Rev. **102**, 1076 (1956). — R. HADDOCK: Phys. Rev. **100**, 1803 (1955). — B. T. FELD, A. C. ODIAN, D. M. RITSON and A. WATTENBERG: Phys. Rev. **100**, 1539 (1955).

[2] The agreement obtained in measurements of lifetime and mass for the τ- and θ-mesons, in spite of their opposite intrinsic parities led to the supposition that these particles represent two different decay modes of the same particle. This served as the motivation for the speculations of LEE and YANG (see the footnote in Sect. 49) concerning the lack of conservation of parity in weak interactions.

[3] G. COSTA and L. TAFFARA: Nuovo Cim. **3**, 169 (1955).

[4] M. TEUCHER, W. THIRRING and H. WINZELER: Nuovo Cim. **1**, 733 (1955).

[5] J. OREAR, G. HARRIS and S. TAYLOR: Phys. Rev. **102**, 1076 (1956).

zero and there is some preferred state of polarisation. The latter situation provides definite proof that the spin is non-zero. On the other hand, if no φ-dependence is observed, the information is quite inconclusive, for even if the τ^+-mesons are polarised upon emission the polarisation may be too weak to be observed, or the depolarisation effect may be very large.

The analysis of OREAR *et al.* (cf. footnote 5, p. 535) of their recent data shows that the distribution is uniform. As a consequence, no new information concerning the τ-meson properties is provided.

Collecting the various conclusions, the $[0\,-]$ case is favoured, although no definite value yet exists for $[a, p_a]$. It is only a matter of time before an ample supply of information will be made available from the high-energy accelerating machines, and an unambiguous conclusion will be formed.

γ) Associated production of hyperons and heavy-mesons. The operation of the high-energy machines has permitted a detailed study of the production of hyperons and heavy mesons. The first of such studies were made by FOWLER *et al.*[1], who caused the 1.37 Gev π^--meson beam to pass from the Brookhaven Cosmotron through a hydrogen diffusion chamber and observed reactions of the type:

$$\pi^- + p \to \Lambda^0 + \theta^0; \quad \Lambda^\circ \to p + \pi^-; \quad \theta^0 \to \pi^- + \pi^+, \tag{53.10}$$

$$\pi^- + p \to \Sigma^0 + \theta^0; \quad \Sigma^0 \to \Lambda^0 + \gamma, \tag{53.11}$$

$$\pi^- + p \to \Sigma^- + \theta^0; \quad \Sigma^- \to n + \pi^-. \tag{53.12}$$

Reactions such as (53.10) involve the associated production of Λ^0 and θ^0-particles, manifesting themselves in a pair of V-shaped events. Study of the angular correlation between the production plane and that containing the secondaries of the hyperon provides another means of investigating the spin of the hyperon.

Again, we can treat this reaction in a fashion similar to double-cascade decay. The original system is the p-π^- system which emits a θ-meson with relative angular momentum L_1 (as defined with respect to the p-π^- centre of mass system), with a residual hyperon. The latter particle subsequently decays into its secondaries which have a relative angular momentum of L_2 between them. In this case, the initial system is not at rest, so that the momentum vectors, corresponding to the emitted particles, are no longer coplanar.

Assuming that the two radiations corresponding to angular momenta L_1 and L_2 make an angle ϑ with respect to each other as measured in the rest frame of the hyperon, the angular correlation is of the form $\sum_k A_k P_k(\cos\vartheta)$, where

$$\left.\begin{aligned} P_k(\cos\vartheta) &= \sum_{\varkappa} \mathcal{D}^k_{\varkappa 0}(\mathcal{R}_1)\, \mathcal{D}^{k*}_{\varkappa 0}(\mathcal{R}_2) \\ &= 2\sum_{\varkappa\ge 0} \frac{(k-\varkappa)!}{(k+\varkappa)!}\, P_k^{\varkappa}(\cos\vartheta_1)\, P_k^{\varkappa}(\cos\vartheta_2)\cos\varkappa\varphi, \end{aligned}\right\} \tag{53.13}$$

where φ is the angle between the two decay planes, and ϑ_1, ϑ_2 are the angles that the two momenta make with respect to an arbitrary direction of the decaying hyperon, as measured in the rest system of the hyperon (Fig. 51).

To obtain the angular correlation, an integration must be made over all the variables with the exception of φ. Thus the angular correlation varies as $\sum_{\varkappa} b_{\varkappa}\cos\varkappa\varphi$ where $\varkappa$ is even and positive, having a maximum value of k which, in turn, does not exceed $2b$, where b is the spin of the hyperon. If the angular correlation does

---

[1] W. B. FOWLER, R. P. SHUTT, A. M. THORNDIKE and W. L. WHITTEMORE: Phys. Rev. **91**, 1287 (1953); **98**, 121 (1955).

show a dependence on φ, this is clear evidence that the hyperon spin is greater than $\frac{1}{2}$. Observation of the largest value of $\varkappa$ sets a lower bound to the spin of the hyperon. TREIMAN *et al.*[1] have also estimated the limits of the coefficient, $b_\varkappa$, which, in some cases, can provide a lower bound to the spin larger than that obtained by determining $\varkappa_{max}$. This analysis can also extend to the θ-mesons by reversing the rôles played by the hyperon and the θ-meson.

Collisions between π^--mesons and free hydrogen nuclei are best for this purpose, since, apart from any other interaction, if a collision is made with a complex nucleus the Fermi momentum of the target nucleon will modify any correlation that may exist. This means that collisions in the walls of the chamber cannot be relied upon too well. Because of the difficulty of identifying the events, only about a dozen events have been found by BALLAM *et al.*[2], FOWLER *et al.* (cf. footnote 1, p. 536), and WALKER and SHEPHERD[3]. In all cases, the dihedral angle is found to be less than $\pi/4$, in contrast to an expected uniform distribution from 0 to $\pi/2$, for a spin $\frac{1}{2}$ particle, so that from this evidence alone the spin of the hyperon is $\frac{3}{2}$ or greater. On the other hand, the distribution with dihedral angle for the θ-meson production is fairly uniform. It is therefore consistent to consider the spin of the θ-meson as 0, although higher values are not ruled out. It must be borne in mind that the statistics are still too poor for unambiguous conclusions.

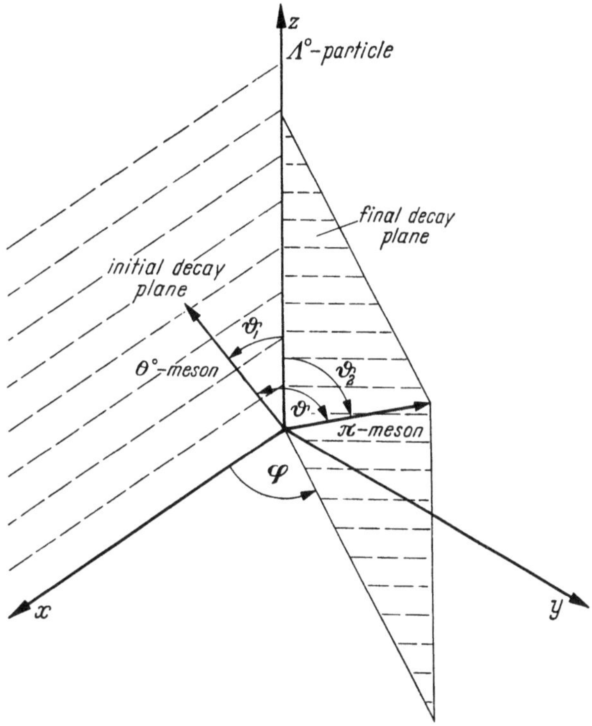

Fig. 51. Momentum diagram for the process of associated production of Λ^0 and θ^0 particles, $\pi^- + p \to \Lambda^0 + \theta^0$; $\Lambda^0 \to p + \pi^-$; $\theta^0 \to \pi^- + \pi^+$, in terms of the rest system of the Λ^0-particle.

This evidence is in contrast to that obtained by cosmic-ray workers[4] who do not observe any correlation.

Some preliminary reports have been given by STEINBERGER[5] which contradict the results obtained by the Brookhaven cloud-chamber workers. Using a propane bubble chamber, whose geometry is more convenient than that of the cloud-chamber, and identifying only interactions with the hydrogen in the chamber,

[1] S. B. TREIMAN, G. T. REYNOLDS and A. L. HODSON: Phys. Rev. **97**, 244 (1955). — S. B. TREIMAN and H. W. WILD jr.: Phys. Rev. **100**, 879 (1955).

[2] J. BALLAM, A. L. HODSON, W. MARTIN, R. R. RAU, G. T. REYNOLDS and S. B. TREIMAN: Phys. Rev. **97**, 245 (1955).

[3] W. D. WALKER and W. D. SHEPHERD: Phys. Rev. **101**, 1810 (1956).

[4] B. V. SREEKANTAN, A. PEVSNER and G. SANDRI: Phys. Rev. **99**, 642 (1955). — S. D. JAMES and R. A. SALMERON: Phil. Mag. **46**, 571 (1955).

[5] J. STEINBERGER: Sixth Annual Rochester High-Energy Conference, 1956.

he identified 19 Λ^0-particles, 3 Σ^0-particles, and 15 are not certain. Combining his results with that of FOWLER *et al.* (cf. footnote 1, p. 536), he found no correlation for either the hyperons or the θ^0-mesons. It was also stated that the bubble chamber technique is particularly suited to this experiment and that in a short time a copious supply of such events will be available for interactions.

An alternate mode of study concerns the correlation of the direction of one of the decay secondaries associated with the hyperon with the direction of the incident π^--meson beam[1]. This is equivalent to a triple correlation, where the second radiation (θ^0-meson) is considered as unobserved. The complexity of the correlation is independent of the nature of the unobserved correlation, and can give information concerning the hyperon spin.

54. Meson reactions. In this section we will consider the following reactions which represent three of the fundamental processes by which π-mesons interact with nucleons:

$$\pi + n \to \pi' + n', \tag{54.1a}$$

$$\gamma + n \to \pi + n', \tag{54.1b}$$

$$n + n' \rightleftharpoons \pi + d. \tag{54.1c}$$

Here, n and n' refer to nucleons and the π-mesons are neutral or charged as the case may be. Reaction (a) represents the scattering of mesons by nucleons and includes the possibility of charge exchange, while reaction (b) designates photomeson production and (c) refers to nucleon-nucleon production of mesons and deuterons and the inverse process, the absorption of π-mesons by deuterium.

The experimental evidence, to date, shows that the angular distribution of the π-mesons is consistent with

$$W \sim A + B\cos\vartheta + C\cos^2\vartheta, \tag{54.2}$$

although, in some cases, B is zero or can be neglected. The complexity of (54.2) indicates that it is only necessary to consider, for most of the energies at which experiments have been performed, mesons which interact with relative orbital angular momenta less than or equal to unity. This limitation restricts the reactions to a few channels and it is feasible to calculate the relative amount of interaction in each of them. A great simplification in the analysis is provided by the assumption of charge independence, although experiments by no means prove this assumption.

The solutions in all three cases are not unique. There exist two sets of solutions that cannot be distinguished by means of angular distribution measurements[2]. They are referred to as (1) the Fermi solution where the dominant interaction is through the intermediate state corresponding to $b = \frac{3}{2}$, $p = +1$, and isobaric spin $T = \frac{3}{2}$, and (2) the Yang solution, where the main interaction is via $b = \frac{1}{2}$, $p = +1$, $T = \frac{3}{2}$. In principle, it should be possible to distinguish between them by taking polarisation effects into account; unfortunately, there are severe experimental limitations to this procedure.

Rather than dwell on the results with their implications for meson theory, we will indicate how angular distributions can be calculated by direct use of the general formulae of Chap. E. Considerations similar to these have been made by FELD. The initial treatments of FERMI and others are simple in nature since

---

[1] L. WOLFENSTEIN: Phys. Rev. **74**, 786 (1954). — R. K. ADAIR: Phys. Rev. **100**, 1540 (1955).

[2] Unless specifically stated, we assume that an angular-distribution measurement is one in which no detection is made of the polarisation of the particles.

only a few angular momentum states are involved, and perhaps give greater insight into the problems. On the other hand, the availability of general formulae applicable to all types of involved processes and types of radiation is of great value in itself. If these formulae are applied to these elementary reactions, they reduce immediately to products of a few Clebsch-Gordan coefficients and Racah coefficients which are readily evaluated.

α) Scattering of π-mesons by nucleons. The experimental data concerning the scattering of positive and negative π-mesons by hydrogen show that, for meson energies up to about 250 Mev, the angular distribution is consistent with (54.2), and that it is very unlikely that meson D-phase shifts play an important rôle. The limitation of the meson angular momenta to values 0 and 1 restricts the number of available channels to three. This can be seen as follows: from parity considerations, l_1 and l_2 are either both even or both odd. Furthermore, they cannot differ by two or more. Since conservation of angular momenta requires that $\boldsymbol{b}=\boldsymbol{a}+\boldsymbol{l}_1=\boldsymbol{c}+\boldsymbol{l}_2$, and a and c are both $\frac{1}{2}$, it is clear that $l_1=l_2=l$ and $l_1'=l_2'=l'$. Corresponding to $l=0$, the intermediate state is characterised by $b=\frac{1}{2}$, $p=-1$, since the π-meson is pseudoscalar with negative intrinsic parity. When $l=1$, two intermediate states are allowed corresponding to $b=\frac{3}{2}$ and $b=\frac{1}{2}$ with $p=+1$ in both cases. The amplitudes for the reaction, $\langle\frac{1}{2}lb\,|R|\,\frac{1}{2}lb\rangle$ are denoted by x, y and z for the three cases and are identified in Table 5.

Table 5. *Reaction amplitudes for π^--p scattering.*

l	b	R
0	$\frac{1}{2}$	x
1	$\frac{1}{2}$	y
1	$\frac{3}{2}$	z

The angular distribution of the π-mesons is given by (35.7) with $A_k(\ldots)$ given by (35.8a), where $s=0$ and $a=c=\frac{1}{2}$. Thus,

$$A_k(l\,l\,l'\,l';\tfrac{1}{2}\,b\,b') = (-)^{\frac{1}{2}-b'}\,\hat{b}\,\hat{b}'\,(b\,\tfrac{1}{2},b'-\tfrac{1}{2}\,|\,k\,0) \tag{54.3}$$

and the angular distribution, $W(\vartheta)$, is given by

$$W(\vartheta) \sim \sum_{b\,b'\,l\,l'\,k} \hat{b}^2\,\hat{b}'^2\,(b\,\tfrac{1}{2},b'-\tfrac{1}{2}\,|\,k\,0)^2\,q(b\,b';l\,l')\,P_k(\cos\vartheta)\,\mathrm{Re}\,(R\,R'^*) \tag{54.4}$$

$$\left.\begin{aligned}&\sim |x|^2+|y-z|^2+2\cos\vartheta\,\mathrm{Re}\,[(2z+y)\,x^*]+\\&\qquad\qquad+3\cos^2\vartheta\,[|z|^2+2\,\mathrm{Re}\,z\,y^*],\end{aligned}\right\} \tag{54.5}$$

which is in agreement with the results of ANDERSON *et al.*[1].

Since the angular distribution is measured with respect to the centre-of-mass system of the nucleon and meson, an identical distribution is to be expected for the protons. This is borne out by application of (35.8b) to (35.7) with $a=0$ and $s=\frac{1}{2}$. We find

$$A_k(l\,L\,l'\,L';0\,b\,b') = (-)^{b-\frac{1}{2}}\,\hat{b}\,\hat{b}'\,(b\,\tfrac{1}{2},b'-\tfrac{1}{2}\,|\,k\,0) \tag{54.6}$$

and $W(\vartheta)$ remains as in (54.4).

The result is usually written in terms of scattering phase-shifts, ξ, which are related to the reaction matrices, R, by $R=\mathrm{e}^{2i\xi}-1$. It holds for scattering of π^- as well as π^+-mesons, and also for charge exchange scattering. In the latter two cases, the reaction matrices are linear combinations of matrices corresponding to the two isobaric spin states $T=\frac{1}{2}$ and $\frac{3}{2}$. The Fermi and Yang solutions are two independent solutions which agree with the observed angular distributions. In the former description, almost all the scattering is associated with $b=\frac{3}{2}$ and $T=\frac{3}{2}$, while in the Yang solution, the $b=\frac{1}{2}$, $T=\frac{3}{2}$ state predominates, although the $b=\frac{3}{2}$, $T=\frac{3}{2}$ state is also important.

[1] H. L. ANDERSON, E. FERMI, R. MARTIN and D. E. NAGLE: Phys. Rev. **91**, 155 (1953).

To distinguish between these solutions, one can calculate the distribution to be expected for the recoil protons using polarisation-sensitive detectors[1]. From (30.10) we note that the distribution is modified, if polarisation effects are considered, by replacing $P_k(\cos\vartheta)$ which occurs in the formulae by

$$P_k(\cos\vartheta) + (-)^{l'} f_k(b b', \tfrac{1}{2})\, \boldsymbol{P}\cdot\boldsymbol{n}\, P_k^1(\cos\vartheta) \tag{54.7}$$

where $f_k(bb', \frac{1}{2})$, which is given by (30.11) to (30.13), reduces to simple values. The result is that for detectors registering only for spins parallel or antiparallel to $\boldsymbol{n}$, the normal to the scattering plane, $W(\vartheta)$ is replaced by

$$W_\pm(\vartheta) = W_0 \mp 2\sin\vartheta\, \mathrm{Im}\,[(z-y)\,x^*] \mp 6\sin\vartheta\cos\vartheta\, \mathrm{Im}\, z\, y^* \tag{54.8}$$

where the sign is $+$ or $-$ according as the proton is parallel or antiparallel to $\boldsymbol{n}$, and W_0 is the unpolarised angular distribution. The result takes on a simpler form if we introduce l, m and n defined by

$$l = 2ix, \qquad m = 2i(2z+y), \qquad n = 2i(z-y) \tag{54.9}$$

in which case the angular distribution becomes

$$W_\pm(\vartheta) \sim |l + m\cos\vartheta \pm i n \sin\vartheta|^2. \tag{54.10}$$

Fermi calculated the left-right asymmetry, represented by the parameter, $P = \frac{W_+ - W_-}{W_+ + W_-}$. For an energy of the π-meson equal to 120 Mev, this is found to be 0.66, which is quite appreciable. In addition, the Fermi and Yang solutions give polarisations which differ in sign and which show a different variation with angle, so that a search for polarisation effects would seem to be rewarding.

(b) Photo-meson production. Both positively charged and neutral π-mesons are produced in γ-p collisions according to

$$\gamma + p \to \pi^+ + n, \tag{54.11a}$$

$$\gamma + p \to \pi^0 + p. \tag{54.11b}$$

Experiments with γ-ray energies varying from the threshold for the reactions up to $\sim$450 Mev indicate that the angular distribution is satisfied by (54.2), although for the case of π^0-meson production B is so small that it may be neglected for certain energies. The usual assumption made is that the meson angular momenta are restricted to $l_2 = 0$ and 1. Corresponding to $l_2 = 0$, the intermediate state is characterised by $b = \frac{1}{2}$ and $p = -1$, which implies that the absorbed γ-ray is of the type $E1$. The reaction matrix, R, is denoted by x in this case. For $l_2 = 1$, parity conservation requires that $p = +1$. If $b = {}^1\!/_2{}^+$, then the γ-ray is $M1$-type, and if $b = {}^3\!/_2{}^+$, both $M1$ and $E2$ γ-radiations are possible. The reaction amplitudes for these three are y, z, and w respectively. These amplitudes are all identified in Table 6.

Table 6. *Reaction amplitudes for photo-meson production.*

l_2	$[bp]$	L_1	R
0	${}^1\!/_2{}^-$	$E1$	x
1	${}^1\!/_2{}^+$	$M1$	y
1	${}^3\!/_2{}^+$	$M1$	z
1	${}^3\!/_2{}^+$	$E2$	w

[1] E. FERMI: Phys. Rev. **91**, 947 (1953). — S. HAYAKAWA, M. KAWAGUCHI and S. MINAMI: Progr. Theor. Phys. **12**, 355 (1954).

The angular distribution for the π-meson is given by (35.7) with $a=c=\frac{1}{2}$ and $A_k(\ldots)$ is represented by:

$$\left.\begin{aligned} A_k(L_1 L_1'; \tfrac{1}{2} b b') = (-)^{L_1'-L_1-b'+k-\frac{1}{2}} \hat{b}\hat{b}' \hat{L}_1 \hat{L}_1' \times \\ \times (L_1 1, L_1' {-1} \mid k0)\, W(b b' L_1 L_1'; k \tfrac{1}{2}), \end{aligned}\right\} \quad (54.12)$$

$$A_k(l_2 l_2 l_2' l_2'; \tfrac{1}{2} b b') = (-)^{\frac{1}{2}-b'} \hat{b} b' (b \tfrac{1}{2}, b' - \tfrac{1}{2} \mid k0). \quad (54.13)$$

This yields for the angular distribution:

$$\left.\begin{aligned} W(\vartheta) = \sum_{b b' L_1 L_1' l_2 l_2' k} (-)^{k+L_1'-L_1-2b'} \hat{b}^2 \hat{b}'^2 \hat{L}_1 \hat{L}_1' (L_1 1, L_1' {-1} | k0) W(bb' L_1 L_1'; k \tfrac{1}{2}) \times \\ \times (b \tfrac{1}{2}, b' - \tfrac{1}{2} \mid k0)\, q(b b'; L_1 L_1', l_2 l_2')\, P_k(\cos\vartheta)\, \mathrm{Re}(R R'^*). \end{aligned}\right\} \quad (54.14)$$

With the evaluation of the necessary Racah coefficients and Clebsch-Gordan coefficients, the angular distribution becomes

$$\left.\begin{aligned} W(\vartheta) \sim |x|^2 + |y|^2 + \tfrac{5}{2}|z|^2 + \tfrac{3}{2}|w|^2 + \mathrm{Re}\left[y^*(z - \sqrt{3} w) + \sqrt{3} z^* w\right] + \\ + 2\,\mathrm{Re}\left[x^*(y - z) + \sqrt{3} z^* w\right] \cos\vartheta + \\ + \left\{-\tfrac{3}{2}|z|^2 + \tfrac{3}{2}|w|^2 - 3\,\mathrm{Re}\left[y^*(z - \sqrt{3} w)\right] + \sqrt{3}\,\mathrm{Re}\, z^* w\right\} \cos^2\vartheta \end{aligned}\right\} \quad (54.15)$$

in agreement with the result of GELL-MANN and WATSON[1], who used a different approach. The use of angular-momentum arguments similar to those used here was first proposed by FELD[2]. His result differs from (54.15) and refers to the distribution of the recoil nucleon instead of the meson. Although identical angular distributions should occur, the reaction amplitudes refer to two different types of radiation, and, if they differ, their coefficients in the angular distribution should also differ. In contrast to (54.13), if the radiation corresponds to the nucleon, then $c=0$ and from (35.8b) we have

$$A_k(l_2 L_2 l_2' L_2'; 0 b b') = (-)^{k+b-\frac{1}{2}} \hat{b}\hat{b}' (b \tfrac{1}{2}, b' - \tfrac{1}{2} \mid k0)\, \delta_{b L_2} \delta_{b' L_2'} \quad (54.16)$$

so that

$$A_k(l_2 L_2 l_2' L_2'; 0 b b') = (-)^{b-b'-k-1} A_k(l_2 l_2 l_2' l_2'; \tfrac{1}{2} b b'). \quad (54.17)$$

Applying (54.17) to (54.14), the angular distribution of the recoil nucleons is then given by

$$\left.\begin{aligned} W(\vartheta) \sim |x'|^2 + |y'|^2 + \tfrac{5}{2}|z'|^2 + \tfrac{3}{2}|w'|^2 + \mathrm{Re}\left[y'^*(-z' + \sqrt{3} w') + \sqrt{3} z'^* w'\right] + \\ + 2\,\mathrm{Re}\left[x'^*(-y' - z' + \sqrt{3} w')\right] \cos\vartheta + \\ + \left\{-\tfrac{3}{2}|z'|^2 + \tfrac{3}{2}|w'|^2 - 3\,\mathrm{Re}\left[y'^*(-z' + \sqrt{3} w') + \sqrt{3} z'^* w'\right\} \cos^2\vartheta. \end{aligned}\right\} \quad (54.18)$$

The primes are affixed to the reaction matrices in order to distinguish them from those appearing in (54.15). To check that the angular distributions in the two cases do, in fact, agree, we must compare $\langle s_2 c\, l_2(L_2)\, b \mid R \mid \tfrac{1}{2} L_1 b\rangle$ with $\langle s_2 l_2(L_2)\, c\, b \mid R \mid \tfrac{1}{2} L_1 b\rangle$. Their relation is given by

$$\langle s_2 l_2(L_2)\, c\, b \mid R \mid \tfrac{1}{2} L_1 b\rangle = \sum_{\overline{L}_2} \langle s_2 c\, l_2(\overline{L}_2)\, b \mid R \mid \tfrac{1}{2} L_1 b\rangle \langle L_2 \mid \overline{L}_2\rangle \quad (54.19)$$

where

$$\left.\begin{aligned} \langle L_2 \mid \overline{L}_2\rangle &= (-)^{c+l_2-L_2} \hat{L}_2 \hat{L}_2' W(s_2 l_2 b c; L_2 \overline{L}_2) \\ &= (-)^{c+l_2-b} \delta_{b\overline{L}_2}, \quad \text{if } s_2 = 0. \end{aligned}\right\} \quad (54.20)$$

[1] M. GELL-MANN and K. M. WATSON: Ann. Rev. Nucl. Sci. **4** (1954).
[2] B. T. FELD: Phys. Rev. **89**, 330 (1953). — Nuovo Cim. **12**, 425 (1954).

It follows from (54.20) that (54.18) can be rewritten in terms of the reaction amplitudes used in (54.15) by multiplying the coefficients by $(-)^{2c+l_2+l_2'-b-b'}=(-)^{b-b'-k-1}$. But this is the negative of the factor by which the $A_k(\ldots)$'s have to be multiplied so that the two expressions for $W(\ldots)$ are identical. Expression (54.18) agrees with that of FELD, except for a change in sign for terms involving z' linearly.

If observation is made of the recoil proton by a polarisation-sensitive detector the angular distribution is obtained by modifying $P_k(\cos\vartheta)$ as in (54.7). Corresponding to measurements in the $+\boldsymbol{n}$ and $-\boldsymbol{n}$ direction, the angular distributions are:

$$W_{\pm}(\vartheta)=W_0(\vartheta)\pm\operatorname{Im}\left[(z-\sqrt{3}\,w-2y)\,x^*\sin\vartheta+3\sin\vartheta\cos\vartheta\,(\sqrt{3}\,w-z)\,y^*\right]. \quad (54.21)$$

FELD has calculated this for γ-ray energies from 200 to 450 Mev. He finds, for charged mesons, values of $P(\vartheta)$ for both Fermi and Yang solutions which are quite large and differ in sign and angular distribution. There is a limitation in the technique of polarisation measurement. On the other hand, for neutron-meson production, the Fermi solution gives no polarisation, in contrast to the appreciable effects for the Yang solution. Here the experimental difficulties are great, both as far as energy and angle of emission are concerned.

γ) The reaction $n+n'\rightleftharpoons d+\pi$. The third set of reactions concerns the associated production of π-mesons and deuterons in collisions of two nucleons and its inverse, the absorption of π-mesons by deuterium. The assumption of charge independence for these reactions has been treated at length (ROSENFELD[1]) and it is found to be consistent with experiment. Indeed, the reactions

$$n+p\rightleftharpoons d+\pi^- \quad (54.22)$$

have been studied experimentally at the same neutron energy[2] (400 ± 20 Mev) and identical angular distributions

$$W(\vartheta)\sim 0.2+\cos^2\vartheta \quad (54.23)$$

were obtained. The latter distribution is fairly similar to those found for the production of π^+-mesons by p-p collisions. From threshold to proton energies of ~450 Mev, this is given by

$$W(\vartheta)\sim a+\cos^2\vartheta \quad (54.24)$$

where a varies from 0.2 to 0.4.

In the following discussion we restrict ourselves to the production of π^+-mesons by p-p collisions. It is convenient to consider this process in the channel-spin representation. Assuming that the mesons have sufficiently low energies, we will only consider meson angular momenta, $l_2=0$ and 1. The intermediate state giving rise to mesons in the s-state correspond to $b=1$, $p=-1$. From the Pauli exclusion principle, such a state can only arise from protons in the $^3P_1$ state. Also, corresponding to $l_2=1$, the intermediate state is $0+$ and arises from protons in the $^1S_0$ state, while the only other intermediate state, characterised by $2+$, is associated with protons in the $^1D_2$ state. The reaction matrices $\langle c\,S_2 b|R|a\,S_1 b\rangle$ are denoted by x, y and z and correspond to the $^1S_0$, $^3P_1$, and $^1D_2$ states of the two-nucleon system.

---

[1] A. H. ROSENFELD: Phys. Rev. **96**, 139 (1954).

[2] R. DURBIN, H. LOAR and J. STEINBERGER: Phys. Rev. **84**, 581 (1951). — R. H. HILDEBRAND: Phys. Rev. **89**, 1090 (1953).

The angular distribution is given by (35.9), where $A_k(\ldots)$ is given as in (35.10). The values of l and S corresponding to the different channels are displayed in Table 7.

With the evaluation of the Clebsch-Gordon coefficients and the Racah coefficients, the angular distribution is found to be

$$\left.\begin{aligned} W(\vartheta) \sim |x|^2 + 3\,|y|^2 + \tfrac{5}{2}\,|z|^2 + \sqrt{10}\,\mathrm{Re}\, x\, z^* + \\ + 3\cos^2\vartheta\,\left[\tfrac{5}{2}\,|z|^2 - \sqrt{10}\,\mathrm{Re}\, x\, z^*\right], \end{aligned}\right\} \tag{54.25}$$

which is in agreement with the results of MARSHAK and MESSIAH[1] and FELD[2] except for some probable misprints in both of their final formulae. An identical distribution is obtained for the inverse process corresponding to π^+-meson absorption in deuterium.

Table 7. *Parameters for the reaction* $p + p \rightleftharpoons \pi^0 + d$.

S_1	l_1	p-p state	b	S_2	l_2	R
0	0	$^1S_0$	0	1	1	x
1	1	$^3P_1$	1	1	0	y
0	2	$^1D_2$	2	1	1	z

Corresponding to the polarisation of one of the nucleons, the angular distribution must be multiplied by the polarisation factor $\mathscr{P}_k$ which is given in (32.23), except for minor changes since the intermediate state is not well-defined. This introduces an extra contribution to the angular distribution, given by

$$\left.\begin{aligned} 6 \sum \boldsymbol{P}\cdot\boldsymbol{n}\,\sin\vartheta\,(-)^{1+l_1'-b'+S_2-S_1'}\,\hat{b}^2\,\hat{b}'^2\,\hat{l}_1\,\hat{l}_1'\,\hat{l}_2\,\hat{l}_2'\,\hat{S}_1\,\hat{S}_1'\,(l_1 0, l_1' 0\,|\,k 0)\,\times \\ \times\,(l_2 0, l_2' 0\,|\,k 0)\; W(l_2\, b\, l_2'\, b'; S_2 1)\; W(\tfrac{1}{2}\tfrac{1}{2} S_1 S_1'; |\, 1\tfrac{1}{2}) \begin{Bmatrix} S_1 & l_1 & b \\ S_1' & l_1' & b' \\ 1 & 1 & 1 \end{Bmatrix} \times \\ \times\, q(b\, b'; l_1 l_1', l_2 l_2')\;\mathrm{Im}\,(R_b R_{b'}^*) \end{aligned}\right\} \tag{54.26}$$

where the summation is over $b\, b'\, l_1\, l_1'\, l_2\, l_2'\, S_1\, S_1'\, S_2$. The different coefficients in (54.26) are quite simply evaluated, by making use of Table 7. Corresponding to whether the polarisation is parallel or antiparallel to $\boldsymbol{n} = \mathfrak{k}_1 \times \mathfrak{k}_2$, the angular distribution is found to be

$$W(\vartheta) = W_0(\vartheta) \pm \sqrt{3}\,\sin\vartheta\,\mathrm{Im}\,[(\sqrt{2}\,x^* + \sqrt{5}\,z^*)\,y] \tag{54.27}$$

in agreement with the results of FELD[2], MARSHAK and MESSIAH[1], WOLFENSTEIN[3], and MANDL and REGGE[4].

FELD has calculated the left-right asymmetries and found rather large results. The geometry in this case is more favourable for measurement of polarisation effects than in the preceding two cases. He finds that the polarisation for Fermi and Yang solutions are nearly equal in magnitude, but opposite in sign, so that to distinguish between them the experimental procedure must be capable of measuring the sign of the polarisation.

Appendix.

Further special properties of the Clebsch-Gordan coefficients, Racah coefficients, and Wigner 9-*j* symbols.

In the calculations of explicit cases of angular correlation, there frequently occur the *Clebsch-Gordan coefficients*, $(L0, L'0|k0)$, and $(L1, L'-1|k0)$. Their

[1] R. E. MARSHAK and A. M. L. MESSIAH: Nuovo Cim. **11**, 337 (1954).
[2] B. T. FELD: Nuovo Cim. **12**, 425 (1954).
[3] L. WOLFENSTEIN: Phys. Rev. **98**, 766 (1955).
[4] F. MANDL and T. REGGE: Phys. Rev. **99**, 1478 (1955).

analytical expressions are:

$$(L\,0, L'\,0\,|\,k\,0) = (-)^{k+n}\hat{k}\,\frac{\Delta(L\,L'\,k)\,n!}{(n-L)!\,(n-L')!\,(n-k)!}\,, \tag{A.1}$$

where $2n \equiv L + L' + k =$ even integer, and

$$\Delta(L\,L'\,k) = \sqrt{\frac{(L+L'-k)!\,(L'+k-L)!\,(k+L-L')!}{(L+L'+k+1)!}}\,, \tag{A.2}$$

$$\left.\begin{aligned}\frac{(L\,1, L'-1\,|\,k\,0)}{(L\,0, L'\,0\,|\,k\,0)} &= \frac{k(k+1) - L(L+1) - L'(L'+1)}{2\sqrt{L\,L'(L+1)(L'+1)}}\\ &\qquad \text{if } L + L' - k = \text{even integer,}\end{aligned}\right\} \tag{A.3}$$

$$\left.\begin{aligned}\frac{(L\,1, L'-1\,|\,k\,0)}{(L\,0, L'+1\,0\,|\,k\,0)} &= -\frac{1}{2}\sqrt{\frac{(L+L'+k+2)(L'+k+1-L)(L'+L+1-k)(k+L-L')}{L\,L'(L+1)(L'+1)}}\\ &\qquad \text{if } L + L' - k = \text{odd integer.}\end{aligned}\right\} \tag{A.4}$$

We shall also need relations:

$$\frac{(L\,1, L'\,1\,|\,k\,2)}{(L\,1, L'-1\,|\,k\,0)} = (L'-L)(L+L'+1)\sqrt{\frac{(k-2)!}{(k+2)!}} \quad \text{if } L + L' + k \text{ is odd}, \tag{A.5}$$

$$= \frac{2k(k+1)\,L(L+1)}{k(k+1) - 2L(L+1)}\sqrt{\frac{(k-2)!}{(k+2)!}} \quad \text{if } k \text{ is even} \tag{A.6}$$

and

$$\frac{(L\tfrac{1}{2}, L'\tfrac{1}{2}\,|\,k\,1)}{(L\tfrac{1}{2}, L'-\tfrac{1}{2}\,|\,k\,0)} = \frac{(2L'+1) + (-)^{k+L+L'}(2L+1)}{2\sqrt{k(k+1)}}\,. \tag{A.7}$$

The *Racah coefficient* $W(a\,b\,c\,d; e\,f)$ satisfies

$$\langle a\,b(e)\,c\,d\,|\,a\,b\,c(f)\,d\rangle = \hat{e}\hat{f}\,W(a\,b\,d\,c; e\,f)\,, \tag{A.8}$$

$$\langle a\,b\,c(e)\,d\,|\,a\,c(f)\,b\,d\rangle = (-)^{b+c-e}\hat{e}\hat{f}\,W(a\,c\,d\,b; f\,e)\,, \tag{A.9}$$

$$\left.\begin{aligned}&\sum_{\alpha\beta\delta\varepsilon\varphi}(a\,\alpha, b\,\beta\,|\,e\,\varepsilon)(e\,\varepsilon, d\,\delta\,|\,c\,\gamma)(b\,\beta, d\,\delta\,|\,f\,\varphi)(a\,\alpha, f\,\varphi\,|\,c'\,\gamma')\\ &\qquad = \hat{e}\hat{f}\,W(a\,b\,c\,d; e\,f)\,\delta_{cc'}\,\delta_{\gamma\gamma'}\,,\end{aligned}\right\} \tag{A.10}$$

$$\left.\begin{aligned}&\sum_{\beta\delta\varepsilon}(a\,\alpha, b\,\beta\,|\,e\,\varepsilon)(e\,\varepsilon, d\,\delta\,|\,c\,\gamma)(b\,\beta, d\,\delta\,|\,f\,\varphi)\\ &\qquad = \hat{e}\hat{f}\,W(a\,b\,c\,d; e\,f)(a\,\alpha, f\,\varphi\,|\,c\,\gamma)\,,\end{aligned}\right\} \tag{A.11}$$

$$\left.\begin{aligned}&(a\,\alpha, b\,\beta\,|\,e\,\alpha+\beta)(e\,\alpha+\beta, d\,\delta\,|\,c\,\alpha+\beta+\delta)\\ &\quad = \sum_f \hat{e}\hat{f}(b\,\beta, d\,\delta\,|\,f\,\beta+\delta)(a\,\alpha, f\,\beta+\delta\,|\,c\,\alpha+\beta+\delta)\,W(a\,b\,c\,d; e\,f)\,,\end{aligned}\right\} \tag{A.12}$$

$$\left.\begin{aligned}W(a\,b\,c\,d; e\,f) &= W(b\,a\,d\,c; e\,f) = W(c\,d\,a\,b; e\,f)\\ &= W(d\,c\,b\,a; e\,f) = (-)^{b+c-e-f}\,W(a\,e\,f\,d; b\,c)\,,\end{aligned}\right\} \tag{A.13}$$

$$\left.\begin{aligned}&= W(a\,c\,b\,d; f\,e) = W(c\,a\,d\,b; f\,e)\\ &= W(d\,b\,a\,c; f\,e) = (-)^{a+d-e-f}\,W(b\,e\,f\,c; a\,d)\,,\end{aligned}\right\} \tag{A.14}$$

$$W(a\,b\,c\,d; 0\,f) = (-)^{b+c-f}\,\frac{\delta_{ab}\,\delta_{cd}}{\hat{a}\,\hat{c}}\,, \tag{A.15}$$

$$W(a\,b\,c\,0; e\,f) = \frac{\delta_{bf}\,\delta_{ce}}{\hat{e}\,\hat{f}}\,, \tag{A.16}$$

$$(b\,0, d\,0\,|\,f\,0)\,\hat{b}\,\hat{d}\,W(a\,b\,c\,d; \tfrac{1}{2}\,f) = (a\,\tfrac{1}{2}, c-\tfrac{1}{2}\,|\,f\,0)\,, \tag{A.17}$$

$$\sum_e \hat{e}^2\,W(a\,b\,c\,d; e\,f)\,W(a\,b\,c\,d; e\,f') = \delta_{ff'}\,\frac{1}{\hat{f}^2}\,. \tag{A.18}$$

The *Wigner 9-j symbol* $\begin{Bmatrix} a & a' & e \\ b & b' & f \\ c & c' & d \end{Bmatrix}$ satisfies

$$\left.\begin{aligned} \hat{e}\hat{f}\hat{c}\hat{c}' \begin{Bmatrix} a & a' & e \\ b & b' & f \\ c & c' & d \end{Bmatrix} \delta_{dd'} = \sum_{\alpha\beta\alpha'\beta'\gamma\gamma'\varepsilon\varphi} & (a\alpha, b\beta \mid c\gamma)(a'\alpha', b'\beta' \mid c'\gamma') \times \\ & \times (-)^{a'-\alpha'}(a\alpha, a'-\alpha' \mid e\varepsilon)(-)^{b'-\beta'}(b\beta, b'-\beta' \mid f\varphi) \times \\ & \times (-)^{c'-\gamma'}(c\gamma, c'-\gamma' \mid d\delta)(e\varepsilon, f\varphi \mid d'\delta), \end{aligned}\right\} \tag{A.19}$$

$$\begin{Bmatrix} a & a' & e \\ b & b' & f \\ c & c' & d \end{Bmatrix} = \sum_k \hat{k}^2 W(a a' d f; e k)\, W(b f c' a'; b' k)\, W(a b d c'; c k), \tag{A.20}$$

$$\left.\begin{aligned} \begin{Bmatrix} a & a' & e \\ b & b' & f \\ c & c' & d \end{Bmatrix} &= \begin{Bmatrix} a & b & c \\ a' & b' & c' \\ e & f & d \end{Bmatrix} = \begin{Bmatrix} a' & e & a \\ b' & f & b \\ c' & d & c \end{Bmatrix} = \begin{Bmatrix} c & c' & d \\ a & a' & e \\ b & b' & f \end{Bmatrix} \\ &= (-)^{\sigma} \begin{Bmatrix} a' & a & e \\ b' & b & f \\ c' & c & d \end{Bmatrix} = (-)^{\sigma} \begin{Bmatrix} a' & b' & c' \\ a & b & c \\ e & f & d \end{Bmatrix}, \end{aligned}\right\} \tag{A.21}$$

where $\sigma = a + b + c + a' + b + c' + e + f + d$. Some orthogonality relations:

$$\sum_{cc'} \hat{e}\hat{f}\hat{c}\hat{c}' \begin{Bmatrix} a & a' & e \\ b & b' & f \\ c & c' & d \end{Bmatrix} \hat{e}'\hat{f}'\hat{c}\hat{c}' \begin{Bmatrix} a & a' & e' \\ b & b' & f' \\ c & c' & d \end{Bmatrix} = \delta_{ee'}\,\delta_{ff'}, \tag{A.22}$$

$$\sum_{ef} \hat{e}\hat{f}\hat{c}\hat{c}' \begin{Bmatrix} a & a' & e \\ b & b' & f \\ c & c' & d \end{Bmatrix} \hat{e}\hat{f}\hat{\bar{c}}\hat{\bar{c}}' \begin{Bmatrix} a & a' & e \\ b & b' & f \\ \bar{c} & \bar{c}' & d \end{Bmatrix} = \delta_{c\bar{c}}\,\delta_{\bar{c}'c'}. \tag{A.23}$$

In particular,

$$\begin{Bmatrix} a & a' & e \\ b & b' & f \\ c & c' & 0 \end{Bmatrix} = \delta_{cc'}\,\delta_{ef'} \frac{(-)^{e+c-b'-a}}{\hat{e}\,\hat{c}} W(b b' a a'; e c), \tag{A.24}$$

$$\begin{Bmatrix} a & a' & 0 \\ b & b' & 0 \\ c & c' & 0 \end{Bmatrix} = \frac{\delta_{aa'}\,\delta_{bb'}\,\delta_{cc'}}{\hat{a}\,\hat{b}\,\hat{c}}, \tag{A.25}$$

$$\begin{Bmatrix} l & l' & k \\ s & s & 1 \\ L & L' & k \end{Bmatrix} = \frac{L(L+1) - l(l+1) - L'(L'+1) + l(l+1)}{2\sqrt{s(s+1)\,k(k+1)}} \begin{Bmatrix} l & l' & k \\ s & s & 0 \\ L & L' & k \end{Bmatrix}. \tag{A.26}$$

Any 9-j symbol is invariant with respect to interchange of rows and columns and to even permutations of rows or columns. An odd permutation multiplies the 9-j symbol by $(-)^{\sigma}$.

Index of notation.

This index of notation is intended to serve the reader as a guide to the more important symbols used in the article. The need for a unified notation, which is reasonably compact, is made the more necessary because of the wide scope of our treatment. As a result, in some cases, the choice of symbols differs from the conventional one. The sections accompanying the definitions correspond to the part of the text where the symbol is first to be found.

$\boldsymbol{a}, \boldsymbol{b}, \boldsymbol{c}, \ldots$	Total angular-momentum vectors, usually referring to successive nuclear states in units of $\hbar$	Sect. 9
$a, b, c, \ldots$	Total angular-momentum quantum numbers	Sect. 4
$\hat{a}, \hat{b}, \hat{c}, \ldots$	$= \sqrt{2a+1}, \sqrt{2b+1}, \sqrt{2c+1}, \ldots$	Sect. 4
$A, B, C, \ldots$	Quantum numbers, other than $a\alpha$, $b\beta$, $c\gamma$, ... characterising the system	Sect. 11
A, A'	Quantum numbers, other than $a\alpha$, $a\alpha'$ characterising the system	Sect. 11
A_0, A_0'	initially;	
A_x, A_x'	after interaction;	
A_t, A_t'	after cascade emission;	
A_i, A_i'	in well-defined intermediate state.	
a_k	Coulomb-excitation parameters	Sect. 48
a_{2k}	Coefficient of $\cos^{2k}\vartheta$ in angular correlation $W(\vartheta) = \sum_k a_{2k} \cos^{2k}\vartheta$	Sect. 9
A_k	General coefficient of $P_k(\cos\vartheta)$ in the angular correlation $W(\vartheta) = \sum_k A_k P_k(\cos\vartheta)$	Sect. 9
C_S, C_V, C_T, C_A, C_P	Coupling constants for the scalar, vector, tensor axial-vector and pseudoscalar fields arising in β-decay	Sect. 49
E	Kinetic energy of particle in Coulomb excitation	Sect. 48
E'	Eigenvalue of time-dependent interaction Hamiltonian	Sect. 51
E_D	Kinetic energy of the deuteron in stripping processes	Sect. 45
E_0	Resonance energy	Sect. 24
$-E_+, E_-$	Initial and final energies of positron in nuclear-pair formation	Sect. 48
F	Quantum number representing resultant of nuclear spin and atomic spin	Sect. 51
$\overline{\overline{F}}$	Average value of observable F over assembly	Sect. 8
g_k	Solid-angle distortion factor	Sect. 18
G_k	Attenuation factor owing to external fields	Sect. 51
$G_k(\text{lim})$	Hard-core value of attenuation factor	Sect. 51
H'	Time-dependent interaction Hamiltonian	Sect. 51
H	Interaction Hamiltonian	Sect. 10
I	One of Stokes' parameters	Sect. 31
J	Quantum number describing angular momentum of atomic system surrounding nucleus	Sect. 51
J_i	Quantum number arising in the coupling of the spin of the target to the relative orbital angular momentum involving i-th radiation	Sect. 29
K	Quantum number determining parity-parity $= (-)^{l(K)+1}$	Sect. 43
K	Parity quantum number for neutrino $(K_\nu = K(l_\nu))$	Sect. 49
K_t	Time-reversal operator	Sect. 14
$l(K)$	$= \lvert K \rvert + \frac{1}{2}[S(K) - 1]$	Sect. 43
l_i	Quantum number corresponding to orbital angular momentum	Sect. 4
$\boldsymbol{l}_n$	Orbital angular momentum of incoming neutron in (d, p) reactions	Sect. 45
L_i	Total angular-momentum quantum number corresponding to i-th radiation	Sect. 4
L_f	Total angular momentum quantum number of emitted conversion-electron $(= \lvert K \rvert - \frac{1}{2})$	Sect. 43
$\boldsymbol{n}$	Unit vector perpendicular to the plane of two radiations $(= \mathfrak{k}_1 \times \mathfrak{k}_2)$	Sect. 30
n	Quantum number specifying interaction terms in β-decay corresponding to the same values of L and parity	Sect. 49
p	Parity quantum number	Sect. 14
$\boldsymbol{P}$	Polarisation vector for spin-$\frac{1}{2}$ and spin-s particle	Sects. 30, 32
P	Polarisation efficiency of detector of spin-$\frac{1}{2}$ particle	Sect. 30
P_1, P_2, P_3	Stokes' parameters	Sect. 31
q	$= \lvert \boldsymbol{p}_+ + \boldsymbol{p}_- \rvert$	Sect. 48
Q	Nuclear quadrupole moment	Sect. 50
Q_k	Coefficient of $P_k(\cos\vartheta)$ in the observed angular correlation	Sect. 18
R	Reaction matrix $(= S - 1)$	Sects. 11, 32
R	Nuclear radius	Sect. 45
$\overline{\boldsymbol{s}}$	Expectation value of $\boldsymbol{s}$	Sect. 30
s_i	Spin quantum number for the i-th radiation	Sect. 4
S_i	Channel-spin quantum number resulting from the coupling of target spin to the particle spin	Sect. 24

S	Scattering matrix	Sects. 11, 32
$S(K)$	Sign of K	Sect. 43
x	Label denoting particles involved in build-up and disintegration of compound nucleus	Sect. 24
$W(\vartheta)$	Correlation function for two radiations in terms of the relative angle, ϑ, between the radiations	Sect. 9
$W'(\vartheta)$	Observed angular correlation	Sect. 18
$\alpha, \beta, \gamma, \ldots$	Components corresponding to $a, b, c, \ldots$	Sect. 4
$\alpha_{b\beta}$	Coordinates of nuclear surface according to collective-model description	Sect. 47
α_K	K-conversion coefficient	Sect. 20
Γ	Total level width $\left(=\sum_{xSl} \Gamma_{xSl}\right)$	Sect. 24
Γ_{l_n}	Reduced width for emission of neutron with angular momentum, l_n, in units of $\hbar$ in (d, p) reactions	Sect. 45
$\delta_{(i)}$	Mixing ratio of i-th stage	Sect. 18
δ_{ij}	Kronecker δ-symbol	Sect. 4
ε	Efficiency matrix	Sect. 11
$-\varepsilon_D$	Binding energy of the deuteron	Sect. 45
ζ	Component corresponding to J	Sect. 51
ζ_l	l-th-phase shift for potential scattering	Sect. 38
η	$=\dfrac{Z_1 Z_2 e^2}{\hbar v}$	Sect. 48
Θ	Angle between electron and positron in positron-electron angular correlation	Sect. 48
$\vartheta_1, \vartheta_2, \vartheta_3$	Euler angles describing a displacement about the origin	Sect. 7
λ_b	Decay constant for intermediate state $\left(=\dfrac{1}{\tau_b}\right)$	Sect. 50
λ_i	Component corresponding to l_i	Sect. 4
$\lambda\!\!\!^{-}$	Reduced wavelength in the centre-of-mass coordinate system	Sects. 24, 35
Λ_i	Component corresponding to L_i	Sect. 4
ϱ	Density matrix	Sect. 8
$\varrho(\vartheta, \varphi)$	Radius of nuclear surface according to collective-model description	Sect. 47
$\varrho_{qq'}$	Density matrix element in the q-representation $(=\langle q \vert \varrho \vert q' \rangle)$	Sect. 8
σ_i	Component corresponding to s_i	Sect. 4
τ	Symbol representing either density or efficiency matrix	Sect. 29
τ	Experimental resolving time	Sect. 50
τ_b	Lifetime of intermediate state	Sect. 50
φ_i	Half-angle of i-th detector	Sect. 18
Φ	Component corresponding to F	Sect. 50
χ_s^σ	Eigenfunction for spin s and component σ	Sect. 13
ξ	Parameter in Coulomb excitation $\left(=\dfrac{\Delta E}{2E}\,\eta\right)$	Sect. 48
ξ_l	Phase-shift for l-th spherical wave introduced by (screened) Coulomb field of nucleus	Sect. 13
ω	Precessional frequency of nucleus in electromagnetic fields	Sect. 50
Ω_i	Direction of i-th radiation $(=\vartheta_i \varphi_i)$	Sect. 9
b	Symbol corresponding to intermediate state of angular momentum b and a certain parity	Sect. 33
$\mathcal{E}$	Electric-field intensity	Sect. 50
$\mathcal{H}$	Magnetic-field intensity	Sect. 50
$\mathfrak{k}$	$=E_+ + E_-$	Sect. 48
$\mathfrak{k}_i$	Propagation vector of i-th radiation	Sect. 30
$\mathfrak{k}_D, \mathfrak{k}_n, \mathfrak{k}_p$	Momenta of incoming deuteron, neutron and stripped-off proton, respectively, in (d, p) reactions assuming an infinitely heavy target	Sect. 45
$\mathfrak{k}'_D, \mathfrak{k}'_n$	Momenta of incoming deuteron and neutron in (d, p) reactions (with respect to the centre-of-mass of the deuteron and target nucleus)	Sect. 45
$\mathfrak{k}'_p$	Momentum of stripped-off proton in (d, p) reactions (with respect to the centre-of-mass system of the proton and product nucleus)	Sect. 45
$\mathfrak{K}'_p$	Momentum of proton due to internal motion of deuteron	Sect. 45
$\mathfrak{L}$	Sum of orbital angular momenta of target nucleus and the relative motion	Sect. 23
$\mathcal{L}_i$	Compact representation of different quantum numbers associated with the i-th radiation, denoting L_i, l_i, s_i and parity	Sect. 9

$\mathcal{O}$	General interaction operator	Sects. 4, 30
$\mathcal{O}$	General interaction matrix element	Sect. 30
$\mathcal{O}'$	General interaction matrix element with angular-momentum variables primed	Sect. 30
$\mathcal{P}$	Assesses degree of randomness $\left(=\left(\operatorname{tr}\varrho^2-\frac{1}{r}\right)\frac{r}{r-1}\right)$, where r denotes the different states in assembly)	Sect. 8
$\mathcal{R}$	$=\vartheta_1\vartheta_2\vartheta_3$	Sect. 7
$\mathcal{R}_i$	Describes rotation carrying z-axis into direction of i-th radiation	Sect. 10
$\mathcal{S}$	Summation over all non-observed properties of nucleus and radiation, e.g. spin states and characteristics of detector-apparatus	Sect. 10
$\mathfrak{S}$	Sum of spins of target nucleus and of particle	Sect. 23
$A_k[1, 2]$	Coefficient of $P_k(\cos\vartheta)$ in angular correlation of two emitted radiations	Sect. 15
$A_k(a, bb')$	Coefficient of $P_k(\cos\vartheta)$ in angular correlation of two emitted radiations corresponding to the first stage of the process	Sects. 15, 35
$A_k(bb', c)$	Coefficient of $P_k(\cos\vartheta)$ in angular correlation of two emitted radiations corresponding to the second stage of the process	Sects. 15, 35
$A_k'(a, bb')$	Coefficient of $P_k(\cos\vartheta)$ in angular distribution corresponding to the absorption of radiation	Sects. 25, 35
$A_k[1, 3]$	Coefficient of $P_k(\cos\vartheta)$ in the triple correlation where the intermediate radiation is not observed	Sect. 16
$\bar{A}_k(bb')$	Factor in expressing for triple correlation corresponding to unobserved, intermediate radiation $(=\bar{A}_k(bc)$, where b and c refer to well-defined nuclear states)	Sects. 16, 35
$A_k(l_1 L_1 l_1' L_1'; abb')$	Coefficient of $\mathcal{O}\mathcal{O}' P_k(\cos\vartheta)$ in the angular distribution corresponding to the absorption and emission processes using the $\mathcal{L}$-representation	Sects. 24, 35
$A_k(l_2 L_2 l_2' L_2'; cbb')$	Coefficient of $RR'P_k(\cos\vartheta)$ in angular distribution for reactions using the channel-spin representation	Sects. 24, 35
A_{LA}	Vector potential for L-th multipole of the electro-magnetic field	Sects. 14, 48
$b_k(\mathcal{L}\mathcal{L}'; x)$	Particle-correlation factor for x-radiation $\left(=\frac{c_{k0}(\mathcal{L}\mathcal{L}'; x)}{c_{k0}(LL')}\right)$	Sect. 15
$b_k\left(\mathcal{L}\mathcal{L}'; \begin{matrix}\text{elec}\\ \text{magn}\end{matrix}\right)$	Conversion-electron correlation factor $\left(=\frac{c_{k0}(\mathcal{L}\mathcal{L}')}{c_{k0}(LL')\,c(\mathcal{L})}\right)$	Sect. 43
$b_k(\mathcal{L}\mathcal{L}'; nn'; \beta)$	Particle-correlation factor for β-particles	Sect. 49
$c(\mathcal{L})$	Internal-conversion coefficient	Sect. 43
$c_{k\varkappa}(\mathcal{L}_i\mathcal{L}_i')$	Radiation parameter for i-th radiation (arbitrary radiation)	Sect. 10
$c_{k\varkappa}(LL')$	Radiation parameter for i-th radiation (γ-rays)	Sect. 14
$c_{k\varkappa}(\mathcal{L}_i\mathcal{L}_i'; s)$	Radiation parameter for particle spin-s	Sect. 29
ΔE	Energy lost by particles in Coulomb excitation	Sect. 48
$\mathcal{D}^a$	Rotation matrix element in the a-representation	Sect. 7
elec	Electric	Sect. 14
$\varepsilon_{ij}(\varphi)$	Efficiency of i-th detector to j-th radiation	Sect. 18
$\varepsilon_{k_a\varkappa_a}(aA, a'A')$	Efficiency tensor for system characterised by (aa')	Sect. 11
$\varepsilon_{k_i\varkappa_i}(\mathcal{L}_i\mathcal{L}_i')$	Efficiency tensor for i-th radiation	Sect. 11
$\varepsilon_{k_b\varkappa_b}(bB, b'B')$	Efficiency tensor of nucleus following emission or absorption of 1-st radiation	Sect. 11
$\varepsilon_{k_e\varkappa_e}(\frac{1}{2}\frac{1}{2})$	Efficiency tensor for conversion-electron in initial state	Sect. 43
$E1, E2, \ldots$	Electric dipole, quadrupole, ... radiation	
$f(l_i)$	Phase factor of i-th radiation introduced by Coulomb fields	Sect. 24
$F_k(LL'ab)$, $G_k(LL'ab)$	Tabulated functions for γ-ray emission	Sect. 18
$\Phi(b_n)$	Phase factor corresponding to resonance part of reaction matrix element	Sect. 35
$\varphi(b)$	Phase angle corresponding to $\Phi(b)$ i.e. $\Phi(b)=\sqrt{\lvert\Phi(b)\rvert}\exp i\varphi(b)$	Sect. 36
g_{xSl}	$=\pm\sqrt{\Gamma_{xSl}}$	Sect. 24
Γ_{xSl}	Partial width of resonance level	Sect. 24
Im	Imaginary part	Sect. 30
$I_k(S_1S_2)$	Component of A_k in elastic scattering corresponding to interference between resonance and potential scattering	Sect. 38
$j_l(kr)$	Spherical Bessel function	Sect. 13

$\langle \sigma | \varepsilon | \sigma' \rangle$ Efficiency matrix element in σ-representation . . . . Sect. 11
$\langle b\beta | H | a\alpha \rangle$ Interaction Hamiltonian matrix element in transition between states $|a\alpha\rangle$, $|b\beta\rangle$ . . . . Sect. 10
$\langle b\beta | \mathcal{O} | a\alpha \rangle$ General interaction matrix element between states $|a\alpha\rangle$, $|b\beta\rangle$ . . . Sect. 10
$\langle b \| \mathcal{L}_1 \| a \rangle$ Reduced matrix element for first stage of process involving the emission of radiation . . . . Sect. 10
$\langle b \| L \| a \rangle'$ Real part of reduced matrix element . . . . Sect. 22
$\langle b \| S_1 l_1 \| a \rangle$ Factorised matrix element in channel-spin representation . . . . . Sect. 24
$\langle S_2 l_2 b | R | S_1 l_1 b \rangle$ Matrix element in channel-spin representation . . . . Sect. 24
$\langle k_2 \varkappa_2 | G_T | k_1 \varkappa_1 \rangle$ Factor in angular correlation function at time T, describing effect of external field . . . . Sect. 51
$\langle K \| \mathcal{L} \| -1 \rangle$ Reduced matrix element for electronic transition from $K = -1$ to K by means of radiation of multipole $\mathcal{L}$ . . . . Sect. 43
$\langle K \| \mathcal{L} n \| K_\nu \rangle$ Reduced matrix element for transition involving leptons in β-decay Sect. 49

Bibliography.

(* denotes extensive bibliography.)

General references, survey articles and reviews.

[1] WIGNER, E. P.: Gruppentheorie. Braunschweig 1931.

[2] CONDON, E. U., and G. H. SHORTLEY: Theory of Atomic Spectra. Cambridge 1935.

[3] RACAH, G.: Theory of Complex Spectra. Phys. Rev. **62**, 438; **63**, 367 (1942) (transformation theory used for the analysis of complex spectra).

[4] SPIERS, J. A.: Directional Effects in Radioactivity. Nat. Res. Council Canada, Publ. No. 1925, 90 pp., 1949 (extensive discussion of properties of angular-momentum states with application to radioactive emissions).

*[5] DEUTSCH, M.: Angular Correlations in Nuclear Reactions. Rep. Progr. Phys. (Physical Society London) **14**, 196—226 (1950) (early survey of experiments involving angular correlations, including an account of elementary theory and a description of experimental technique).

*[6] BLATT, J. M., and L. C. BIEDENHARN: The Angular Distribution of Scattering and Reaction Cross-Sections. Rev. Mod. Phys. **24**, 258—272 (1952) (theory of reactions with unpolarised particles. Racah algebra and S-matrix formalism is used).

[7] BIEDENHARN, L. C., J. M. BLATT and M. E. ROSE: Some Properties of Racah and Associated Coefficients. Rev. Mod. Phys. **24**, 249—257 (1952).

[8] FANO, U.: Statistical Matrix Techniques and Their Applications to the Direction Correlation of Radiations. Nat. Bur. Standards Report NBS 1214. Phys. Rev. **90**, 577 (1952, 1953) (detailed formal development of the statistical-tensor and efficiency-tensor methods).

*[9] BIEDENHARN, L. C., and M. E. ROSE: Theory of Angular Correlation of Nuclear Radiations. Rev. Mod. Phys. **25**, 729—777 (1953) (comprehensive exposition of theory of correlations in cascade radiations, involving α, β and γ radiations and conversion-electrons—using Racah formalism mainly [27]).

*[10] FRAUENFELDER, H.: Angular Correlation of Nuclear Radiation. Ann. Rev. Nucl. Sci. **2**, 129—162 (1953) Stanford (early survey of experiments and dealing mainly with cascade correlations).

[11] COESTER, F.: Gamma-Gamma Angular Correlations. Argonne Nat. Lab. Report ANL 5316, 29 pp., 1954 (concise presentation of statistical-tensor and efficiency-tensor method).

[12] JAHN, H. A., and J. HOPE: Symmetry Properties of the Wigner 9-j Symbol. Phys. Rev. **93**, 318 (1954).

*[13] HUBY, R.: Stripping Reactions. Progr. Nucl. Phys. **3**, 177 (1953) London (survey of theoretical and experimental investigations, including results of deuteron-stripping reactions and some allied processes).

*[14] SIEGBAHN, K., editor: Beta- and Gamma-Ray Spectroscopy. Amsterdam 1955.

[14a] FRAUENFELDER, L.: Angular Correlations, Chap. XIX, pp. 531—599 (general survey of cascade correlations including the effects of extra-nuclear perturbations—mainly experimental).

[15] STEFFEN, R. M.: Extra-nuclear Effects on Angular Correlations of Nuclear Radiations. Adv. Physics (Phil. Mag.) **4**, No. 15, 293—362 (1955) (elementary theory and comprehensive survey of experiments).

*[16] SATCHLER, G. R.: Some Aspects of Nuclear Structure being Some Applications of the Group-Theoretic Properties of Angular Momentum Eigenstates. Thesis Oxford University 1955 (theory of angular correlations using statistical tensors).

*[17] EDMONDS, A. R.: Angular Momentum in Quantum Mechanics. CERN 55—26, Copenhagen 1955 (detailed presentation of definitions and mathematical relationships involving functions of angular-momentum variables).

[18] ROSE, M. E.: Multipole Fields, 99 pp. New York 1955 (description of the electromagnetic field with particular emphasis given to internal conversion and emission of radiation using the transformation properties of the multipole fields).

[19] FEENBERG, E., and G. E. PAKE: Notes on the Quantum Theory of Angular Momenta, 54 pp. Cambridge, Mass. 1953 (contains a discussion of angular-momentum transformations together with applications to nuclear moments).

General theory of correlations in cascade processes.

[20] HAMILTON, D. R.: On Directional Correlation of Successive Quanta. Phys. Rev. **58**, 122 (1940) (first treatment of γ-γ correlations, using time-dependent perturbation theory).

[21] GOERTZEL, G.: Angular Correlation of γ-Rays. Phys. Rev. **70**, 897 (1946) (extension of the treatment in [20], including the first investigation into the effects of external fields on the angular correlation).

[22] YANG, C. N.: On the Angular Distribution in Nuclear Reactions and Coincidence Measurements. Phys. Rev. **74**, 764 (1948) (some general theorems concerning angular distributions, using group-theoretical methods).

[23] GARDNER, J. W.: Directional Correlation between Successive Internal-Conversion Electrons. Proc. Phys. Soc. Lond. A **62**, 763 (1949).

[24] FIERZ, M.: Zur Theorie der Multipolstrahlung. Helv. phys. Acta **22**, 489 (1949).

[25] FALKOFF, D. L., and G. E. UHLENBECK: On the Directional Correlation of Successive Nuclear Radiations. Phys. Rev. **79**, 323 (1950) (extension of early theory to arbitrary radiations).

[26] SPIERS, J. A.: On the Directional Correlation of Successive Nuclear Radiations. Letter in Phys. Rev. **80**, 491 (1950) (generalisation of treatment of [25], using only the properties of angular momentum—a summary of the theory developed in [4]).

[27] RACAH, G.: Directional Correlation of Successive Nuclear Radiations. Phys. Rev. **84**, 910 (1951) (formalism of the tensor parameters of radiation).

[28] LLOYD, S. P.: The Angular Correlation of Two Successive Nuclear Radiations. Phys. Rev. **85**, 904 (1952) (general formulation by means of group-theoretical methods, summarising the author's previous work).

[29] COESTER, F., and J. M. JAUCH: Theory of Angular Correlations. Helv. phys. Acta **26**, 3 (1953) (formalism of statistical and efficiency tensors).

[30] SATCHLER, G. R.: The Angular Correlation of Three Nuclear Radiations. Phys. Rev. **94**, 1304 (1954).

Theory of angular distributions in nuclear reactions.

[40] MYERS, R. D.: The Angular Distribution of Resonance Disintegration Products. Phys. Rev. **54**, 361 (1938) (early use of symmetry arguments with regard to nuclear reactions involving isolated resonance levels).

[41] BREIT, G., and B. T. DARLING: Note on Calculation of Angular Distributions in Resonance Reactions. Phys. Rev. **71**, 402 (1947) (two general theorems concerning properties of angular-momentum transformation coefficients).

[42] EISNER, E., and R. SACHS: Note on Angular Distributions in Nuclear Reactions. Phys. Rev. **72**, 680 (1947) (general theorem with regard to complexity of angular distributions).

[43] WIGNER, E. P., and L. EISENBUD: Higher Angular Momenta and Long-Range Interaction in Resonance Reactions. Phys. Rev. **72**, 29 (1947) (formal theory of reactions and scattering in terms of the compound nucleus).

[44] BLATT, J. M., and V. F. WEISSKOPFF: Theoretical Nuclear Physics, Chap. VIII, IX and X, p. 311—564. New York 1952 (nuclear reactions).

[45] COESTER, F.: The Symmetry of the S-Matrix. Phys. Rev. **89**, 619 (1953).

β-decay processes.

[50] KONOPINSKI, E. J., and G. E. UHLENBECK: On the Fermi Theory of β-Radioactivity. Phys. Rev. **60**, 308 (1941) (early presentation of the theory).

[51] FALKOFF, D. L., and G. E. UHLENBECK: On the Beta-Gamma Angular Correlation. Phys. Rev. 79, 334 (1950).
[52] GROOT, S. R. DE, and H. A. TOLHOEK: On the Theory of Beta-Radioactivity. Physica, Haag 16, 456 (1950) (one of the early treatments of the beta-neutrino angular correlation).
[53] SPIERS, J. A., and R. J. BLIN-STOYLE: A Formulation of Beta-decay Theory for Forbidden Transitions of Arbitrary Order. Proc. Phys. Soc. Lond. A 65, 801, 809 (1952) (formulation using a representation characterising the electron-plus-neutrino angular momenta).
[54] KOFOED-HANSEN, O.: Theoretical Angular Correlations in Allowed Beta Transitions. Dan. Mat.-Fys. Medd. 28 No. 9, 19 pp. (1954) (transformation of β-neutrino angular correlation into functions of the recoil particle).
[14b] ROSE, M. E.: The Theory of Allowed β-Decay, Chap. IX, pp. 273—292. 1955.
[14c] KONOPINSKI, E.: The Theory of Forbidden β-Decay, Chap. X, pp. 292—314. 1955.
[14d] KOFOED-HANSEN, O.: Neutrino Recoil Experiments, Chap. XI, pp. 357—373. 1955.
[55] MORITA, M.: Explicit Formulae of Beta-Gamma Directional Correlation. Progr. Theor. Phys. 14, 27 (1955) (generalisation of earlier work of [51]).
[56] HAUSER, I.: Directional Correlation of β-Rays, 97 pp. 1956. Thesis State University of Iowa

External fields.

[60] ALDER, K.: Contribution to the Theory of Directional Correlation. Helv. phys. Acta 25, 235 (1952) (early treatment of perturbations due to external fields, using Racah algebra).
[61] ABRAGAM, A., and R. V. POUND: Influence of Electric and Magnetic Fields on Angular Correlations. Phys. Rev. 92, 943 (1953) (contains generalisations of earlier treatments in [21] and [60] and deals with specific examples).
[62] ALDER, K., H. ALBERS-SCHÖNBERG, E. HEER and T. B. NOVEY: The Measurement of Nuclear Moments of Excited States by Angular Correlation Methods. Helv. phys. Acta 26, 761 (1953) (particular discussion is given of experimental arrangements).
[63] COESTER, F.: Influence of Extranuclear Fields on Angular Correlations. Phys. Rev. 93, 1304 (1954) (external fields using density matrices).

Experimental procedures in angular correlations.

[70] FRANKEL, S.: The Effect of Scattering on Angular Correlation Measurements. Phys. Rev. 83, 673 (1951) (corrections for finite solid angles of detectors and multiple scattering in source).
[71] ROSE, M. E.: The Analysis of Angular Correlation and Angular Distribution Data. Phys. Rev. 91, 610 (1953) (corrections for finite solid angles of detectors and discussion of counting rates and errors).
*[72] BREITENBERG, E.: Remarks on the Least-Square Reductions of Angular Correlations Data. Proc. Phys. Soc. Lond. A 69, 489 (1956).
[14e] MITCHELL, A. C. G.: The Coincidence Method, Chap. VII, pp. 201—223. 1955.
[14f] BELL, R. E.: Measurement by Delayed Coincidences. Comparison Methods, Chap. XVIII, pp. 494—520. 1955.

Polarisation.

[80] WOLFENSTEIN, L.: Theory of Proposed Reactions Involving Polarised Protons. Phys. Rev. 75, 1664 (1949).
[81] FANO, U.: Remarks on the Classical and Quantum-Mechanical Treatment of Partial Polarization. J. Opt. Soc. Amer. 39, 859 (1949) (on the use of STOKES' parameters in the treatment of γ-radiation).
[82] BLIN-STOYLE, R. J.: Polarised Nuclear Reactions. Proc. Phys. Soc. Lond. A 64, 700 (1951).
[83] DALITZ, R. H.: On Polarised Particle Beams. Proc. Phys. Soc. Lond. A 65, 175 (1952) (use of statistical tensors in the designation of polarisation).
[84] SIMON, A., and T. A. WELTON: Production of Polarized Particles in Nuclear Reactions. Phys. Rev. 90, 1036 (1953).
[85] SIMON, A.: Theory of Polarized Particles and γ-Rays in Nuclear Reactions. Phys. Rev. 92, 1050 (1953) ([84], extended to include polarised initial beams).
[86] LAKIN, W.: Spin Polarization of the Deuteron. Phys. Rev. 98, 139 (1955).
[87] SATCHLER, G. R.: Polarisation in Nuclear Reactions. Proc. Phys. Soc. Lond. A 68, 1041 (1955) (results for polarisation in the L-representation).

[88] TOLHOEK, H. A.: Electron Polarization, Theory and Experiment. Rev. Mod. Phys. **28**, 277 (1956) (description of polarisation using density matrices, being a summary of previous work by author—represents a survey of theory and experiment involving electron polarisation).

Direct interactions.

[90] SERBER, R.: The Production of High-Energy Neutrons by Stripping. Phys. Rev. **72**, 1008 (1947).

[91] BUTLER, S. T.: Angular Distribution from (d, p) and (d, n) Nuclear Reactions. Proc. Roy. Soc. Lond., Ser. A **208**, 559 (1951).

[92] BHATIA, A. B., K. HUANG, R. HUBY and H. C. NEWNS: Angular Distribution in (d, p) and (d, n) Reactions. Phil. Mag. **43**, 485 (1952) (Born-approximation calculations).

[93] SATCHLER, G. R., and J. A. SPIERS: Angular Distribution of γ-Radiation following a Deuteron-Stripping Reaction. Proc. Phys. Soc. Lond. A **65**, 980 (1952).

[94] HUBY, R.: Theory of the Deuteron-Stripping Reaction. Proc. Roy. Soc. Lond., Ser. A **215**, 385 (1952).

DALITZ, R. H.: Some Features of the Deuteron-Stripping Process. Proc. Phys. Soc. Lond. A **66**, 28 (1953).

GRANT, I. P.: Theory of (d, p) and (d, n) Reactions. Proc. Phys. Soc. Lond. A **67**, 981; A **68**, 244 (1954, 1955).

HOROWITZ, J., and A. M. L. MESSIAH: Sur les Reactions (d, p) et (d, n). J. Phys. Radium **14**, 695 (1955).

TOBOCMAN, W., and M. H. KALOS: Numerical Calculation of (d, p) Angular Distributions. Phys. Rev. **97**, 132 (1955)
(considerations of Coulomb-effects and other corrections).

[95] AUSTERN, N., S. T. BUTLER and H. MCMANUS: Angular Distributions from (n, p) Nuclear Reactions. Phys. Rev. **92**, 350 (1953) (treatment in terms of direct nucleon-nucleon interaction).

[96] SATCHLER, G. R.: Gamma-Radiation following Surface Scattering of Nucleons. Proc. Phys. Soc. Lond. A **68**, 1037 (1955).

[97] BRINK, D. M.: Inelastic Scattering of Neutrons by Deformed Nuclei. Proc. Phys. Soc. Lond. A **68**, 994 (1955).

HAYAKAWA, S., and S. YOSHIDA: Inelastic Scattering of Neutrons by Rotational Excitation. Progr. Theor. Phys. **14**, 1 (1955).

[98] YOSHIDA, S.: The Inelastic Scattering of Nucleons by the Surface Interaction. Proc. Phys. Soc. Lond. A **69**, 668 (1956).

Pair formation and inelastic scattering.

[100] JAEGER, J. C., and H. R. HULME: The Internal Conversion of γ-Rays with the Production of Electrons and Positrons. Proc. Roy. Soc. Lond., Ser. A **148**, 708 (1935) (calculation of electric dipole and quadrupole coefficients using Dirac hydrogen-like wave functions for $Z = 84$).

[101] ROSE, M. E.: Internal Pair Formation. Phys. Rev. **76**, 678 (1949) (Born-approximation calculation for radiation of arbitrary multipole type).

[102] DALITZ, R. H.: On Radiative Corrections to the Angular Correlation in Internal Pair Creaton. Proc. Roy. Soc. Lond., Ser. A **206**, 521 (1951) (detailed calculation for monopole transition).

[14g] WILSON, R.: Internal Pair Formation, Chap. XX, p. 636—649 1955 (survey of theory and experiment).

[103] TER-MARTIROSYAN, K. A.: The Excitation of Nuclei by the Coulomb Field of Charged Particles. J. exp. theor. Phys. USSR. **22**, 284 (1952) (in Russian) (semi-classical treatment for dipole excitation).

BOHR, A., and B. R. MOTTELSON: Collective and Individual Particle Aspects of Nuclear Structure, Appendix VI. Dans. Mat.-Fys. Medd. **27**, No. 16 (1953) (contains a summary of the above calculation).

[104] ALDER, K., and A. WINTHER: The Theory of Coulomb Excitation of Nuclei. Letters in Phys. Rev. **91**, 1578; **96**, 237 (1953, 1954) (application of [103] to correlation with γ-rays following Coulomb excitation).

*[105] BREIT, G., M. E. EBEL and J. E. RUSSELL: Gamma-Ray Angular Distribution in Coulomb Excitation. Phys. Rev. **101**, 1504 (1956) (quantum mechanical calculation for quadrupole excitation).

*[106] BIEDENHARN, L. C., J. L. MCHALE, R. M. THALER and M. GOLDSTEIN: Quantum Calculation of Coulomb Excitation. Phys. Rev. **100**, 376; **101**, 662; **102**, 1567 (1955, 1956).

[107] BREIT, G., R. L. GLUCKSTERN and J. E. RUSSELL: Reorientation Effect in Coulomb Excitation. Phys. Rev. **103**, 727 (1956).
*[108] HOFSTADTER, R.: Electron Scattering and Nuclear Structure. Rev. Mod. Phys. **28**, 214 (1956).

References to mathematical tables.

The following list is a representative sample of the numerical tabulation of the mathematical functions used in the article and is not intended to serve as an exhaustive index of work in this field.

[1] CONDON, E. U., and G. H. SHORTLEY: Theory of Atomic Spectra, Tables 1^3, 2^3, 3^3, 4^3. Cambridge (1935) [algebraic tables of $(a\alpha, b\beta|c\gamma)$ for $b=\frac{1}{2}, 1, \frac{3}{2}, 2$).
[2] BIEDENHARN, L. C., J. M. BLATT and M. E. ROSE: Some Properties of the Racah and Associated Coefficients. Rev. Mod. Phys. **24**, 249 (1952) [algebraic tables of $W(lLl'L'; s\,k)$ for $s=\frac{1}{2}, 1, \frac{3}{2}, 2$].
[3] ALDER, K.: Contribution to the Theory of Directional Correlation. Helv. phys. Acta **25**, 253 (1952) (general formulae); pp. 254—257 (algebraic table of Racah coefficients).
[4] YAMADA, M., and M. MORITA: On the β-Ray Angular Correlation. Prog. Theor. Phys. **8**, 444 (1952) [algebraic table of $(a\alpha, 3\beta|c\gamma)$].
[5] BIEDENHARN, L. C.: Tables of the Racah Coefficients. Oak Ridge Nat. Laboratory, ORNL 1098, 1952 [algebraic tables of $W(lLl'L'; sk)$ for $s=\frac{1}{2}, 1, \frac{3}{2}, 2$, and numerical tabels for $s=0$ to 3 in half-integral steps and $k=0, 1, 2, \ldots, 8$).
[6] BIEDENHARN, L. C.: Revised Z-Tables of the Racah Coefficients. ORNL 1501, 1953 (numerical tables of $Z(lLl'L'; sk)$ for same values of parameters as in [5]).
[7] SIMON, A., J. H. VAN DER SLUIS and L. C. BIEDENHARN: Tables of the Racah Coefficients. ORNL 1679, 1952.
[8] SIMON, A.: Numerical Table of the Clebsch-Gordan Coefficients. ORNL 1718, 1952.
[9] FERENTZ, M., and N. ROSENZWEIG: Table of F-Coefficients. Argonne Nat. Laboratory, ANL 5234, 1953 [numerical tables of $F(LL'j_1j)$ for integral and half-integral values of j_1 and j from 0 to 12).
[10] SHARP, W. T., J. M. KENNEDY, B. J. SEARS and M. G. HOYLE: Tables of Coefficients for Angular Distribution Analysis. Atomic Energy of Canada, Limited, AECL 97, 1953 [includes numerical tables of $Z(lLl'L'; sk)$, $Z_1(LjL'j; Ik)$ (related to the F-coefficients), $W(lLl'L'; sk)$ $W(jj_1jj_1; Lk)$, $\begin{Bmatrix} a & b & c \\ a' & b & c \\ g & h & k \end{Bmatrix}$, $\begin{Bmatrix} a & b & c \\ a & e & f \\ k & k & 1 \end{Bmatrix}$, $(l0, l'0|k0)$, and $(Ll, L'-1|k0)$].
[11] KENNEDY, K. M., B. J. SEARS and W. T. SHARP: Tables of X-Coefficients, AECL 106, 1954 (numerical tables of the 9-j symbol).
[12] ARIMA, A., H. HORIE and Y. TANABE: Generalised Racah Coefficient and its Applications. Progr. Theor. Phys. **11**, 143 (1954) (definitions and useful formulae involving the 9-j symbol).
[13] SATO, M.: Tables of $W(abcd; ef)$ for $e=3, \frac{7}{2}, 4$ and $\frac{9}{2}$. Progr. Theor. Phys. **13**, 405 (1955).
[14] KENNEDY, L. M., and M. J. CLIFF: Transformation Coefficients Between LS and j-j coupling. AECL 224, 1955 (numerical tables of $\hat{L}\hat{S}\hat{j}_1\hat{j}_2 \begin{Bmatrix} l_1 & \frac{1}{2} & j_1 \\ l_2 & \frac{1}{2} & j_2 \\ L & S & J \end{Bmatrix}$ for l_1 and l_2 not exceeding 5).

Oriented Nuclei.

By

R. J. BLIN-STOYLE and M. A. GRACE.

With 24 Figures.

A. Oriented nuclear spin systems.

1. Introduction. In the past few years there have been considerable advances in the techniques used for the investigation of nuclear properties in the low energy field. The result of this work has been to accumulate a great wealth of experimental data on nuclear energy levels and on the interactions between nuclei. This has led to the development of nuclear models which have proved to be highly successful in interpreting much of the experimental material.

Information such as the spins and parities of nuclear states and their magnetic and electric multipole moments, together with the relative transition probabilities for different modes of decay are of particular significance in testing and improving these models. One of the methods developed recently for studying nuclei depends on the orientation of the spin axes of nuclei or nucleons with respect to some axis fixed in space. Because of the association between the angular momentum properties of a nuclear system and directional effects in the absorption or emission of radiation by such a system, this nuclear ordering will be accompanied by anisotropic effects in the interaction of the nuclei with radiation, whether particle or electromagnetic. By investigating these directional effects information may be gained on the angular momentum and parity changes taking place during a nuclear disintegration.

As early as 1934 a method had been proposed by GORTER[1] (and independently in 1935 by KURTI and SIMON[2]) by which nuclei might be oriented at a low temperature by the direct interaction between the nuclear magnetic moment and a magnetic field. In 1937 the importance of such a system of nuclei as a means of investigating nuclear disintegrations was pointed out by HALBAN[3], who suggested studying the interaction between slow neutrons and oriented nuclei. In the following years a number of further methods of producing nuclear orientation were proposed which depended on the coupling between the nucleus and its atomic electrons and in this way avoided the use of the large magnetic fields required by the direct interaction method. It was not, however, until the recent investigations by microwave methods had revealed the detailed form of the strong fields existing within crystalline solids, and the magnetic cooling technique had made very low temperatures readily accessible that it became possible to realize such an oriented system. This was first achieved, following the method proposed by BLEANEY[4] in 1951, by using very low temperatures to populate differentially the hyperfine structure (hfs) states in a paramagnetic crystal which exhibited

[1] C. J. GORTER: Phys. Z. **35**, 923 (1934).
[2] N. KURTI and F. E. SIMON: Proc. Roy. Soc. Lond., Ser. A **149**, 152 (1935).
[3] H. HALBAN: Nature, Lond. **140**, 425 (1937).
[4] B. BLEANEY: Proc. Phys. Soc. Lond. A **64**, 315 (1951). — Phil. Mag. **42**, 441 (1951).

magnetic anisotropy. Since that time a number of other techniques, namely, microwave and optical methods, have been suggested for obtaining nuclear orientation although so far it has only been those employing very low temperatures which have been applied to the study of nuclear disintegrations[1].

In the field of nuclear reactions and scattering processes spatial orientation may be introduced into the system whenever the direction of the beam of incident particles is defined. It is therefore to be expected that in many cases there will be anisotropic effects associated with the emitted or scattered particles. By observing these effects it is then possible to deduce information about the mechanism of the nuclear reaction or scattering process initiated by the bombarding particles. Unfortunately, it is generally not possible to deduce all the information necessary to a complete understanding of the nuclear process. However, the ambiguities and uncertainties which arise in the interpretation of angular distributions occurring in such processes can be reduced and even eliminated in some cases, by orienting the initial components of the process and possibly using polarization sensitive detectors in the final stage.

Of particular interest are those experiments in which the bombarding particles are oriented. In the last few years it has, in fact, been found possible to produce beams of oriented nucleons using scattering processes, nuclear reactions and electromagnetic phenomena. Such beams have then been used for the investigation and elucidation of various nuclear reactions and scattering processes. Further, in one or two cases oriented beams of neutrons have been used in conjunction with target nuclei oriented by the low temperature methods already mentioned.

One additional remark needs to be made. From our definition of orientation it is clear that nuclear orientation will be obtained whenever a nuclear system is coupled in some way to a fixed spatial direction. The degree of nuclear ordering may differ insignificantly from that of a random distribution but nevertheless must be finite. In this article it is proposed to confine attention to those systems in which the population difference which is obtainable is sufficiently large to enable anisotropic phenomena associated with nuclear processes to be observed. This, in general, requires population differences of the order of 1% or greater.

In Part A a general discussion of nuclear orientation is given. Parts B and C of this article are concerned with the production and properties of oriented nuclei, while the later Part D deals with beams of oriented particles and their properties.

2. Definition. Consider a system of identical nuclei each of spin I in field free space. (The term "spin" is used here to denote the total angular momentum of the nucleus and no distinction is drawn between contributions to it from the orbital motion of the nucleons and their intrinsic spins.) In the absence of external fields and of interaction between the nuclei these states will all have the same energy, the $2I+1$ nuclear magnetic substates of the ground-state being degenerate. On the application of a field this degeneracy may be partially or totally lifted and in the case of a magnetic field applied in an arbitrary direction, defining the z-axis of the system, the degeneracy will be lifted completely. Thus the $2I+1$ magnetic states, distinguished by their magnetic quantum numbers M which can hold values from $-I$ to $+I$ in unit steps, may be interpreted as representing different orientations of the spin vector I with respect to the z-axis.

In a statistical ensemble of many such nuclei, with spins oriented at random, the probability $W(M)$ of finding a nucleus in a substate M is $1/(2I+1)$ and is, of

[1] From the view point of low temperature physics this field has been treated by D. de Klerk, Vol. XV, p. 203—209, this Encyclopedia.

course, independent of M. An assembly in which $W(M)$ varies with M is said to be oriented with respect to the z-axis along which M is defined.

Such ordering is restricted in the sense that it is referred to a well defined axis in space. In general, it is possible to have more complicated orderings which cannot be described solely in terms of the probabilities $W(M)$ of having a component of angular momentum M along a given axis.

3. General formulation. The formulation of this problem in the general case has been discussed by various authors[1,2]. Consider an ensemble of identical free particles each with spin I. Such an ensemble can be characterised by the contributing pure states Ψ and their relative abundance $\omega(\Psi)$, where $\sum_{\Psi}\omega(\Psi)=1$. The expectation value of any operator O for this ensemble is then given by

$$\langle O\rangle_{\text{ensemble}}=\sum_{\Psi}\omega(\Psi)\,(\Psi,O\,\Psi) \tag{3.1}$$

where $(\Psi,O\,\Psi)$ indicates the expectation value of O for the pure state Ψ.

We can now expand the states Ψ in terms of a basic set of states ψ_I^M ($M=I$, $I-1\cdots-I$) according to a relation of the form

$$\Psi=\sum_{M}a_M(\Psi)\,\psi_I^M. \tag{3.2}$$

Substituting for Ψ in (3.1) then gives

$$\langle O\rangle_{\text{ensemble}}=\sum_{MM'}\Big(\sum_{\Psi}\omega(\Psi)\,a_M(\Psi)\,a_{M'}^*(\Psi)\Big)\,(\psi_I^{M'},O\,\psi_I^M)=\sum_{MM'}\varrho_{MM'}\,O_{M'M}$$

where

$$\varrho_{MM'}=\sum_{\Psi}\omega(\Psi)\,a_M(\Psi)\,a_{M'}^*(\Psi)\quad\text{and}\quad O_{M'M}=(\psi_I^{M'},O\,\psi_I^M).$$

This can be written formally as $\mathrm{Tr}(\varrho O)$ where ϱ is referred to as the *density matrix*[1] for the ensemble and will clearly be a $(2I+1)$ by $(2I+1)$ matrix in spin space. Also, if the state vectors Ψ are normalized to unity, it follows that $\mathrm{Tr}(\varrho)=1$. The expectation value of the operator O for the ensemble is now expressed as the trace of the product of two matrices ϱ and O, of which the former has all the information about the population distribution among the basic states ψ_I^M and the latter is the matrix of O with respect to these states.

Alternatively, the state of the ensemble may be specified in terms of the expectation values of a complete set of operators having specific geometrical properties. In particular, it is useful to consider tensor operators $T_q^{(k)}$ ($q\leq 2I$; $q\geq k\geq -q$) which transform as a result of rotations of the coordinate axes ("space rotations") like components of irreducible tensors, i.e., according to the same law as the spherical harmonics $Y_q^{(k)}(\vartheta,\varphi)$. It should be noted that because of the restrictions on k and q there are $(2I+1)^2$ elements of the density matrix. These tensors are defined so as to agree with the definitions given by RACAH[3] and are so normalised that

$$T_q^{(0)}=P_q\big(I_z/[I(I+1)]^{\frac12}\big)$$

where $P_q(x)$ is the usual Legendre polynomial in the variable x, and $\boldsymbol{I}$ is the spin operator for a particle of the ensemble.

---

[1] U. FANO: National Bureau of Standards, Report 1214, 1951.
[2] R. M. DALITZ: Proc. Phys. Soc. Lond. A **65**, 175 (1952).
[3] G. RACAH: Phys. Rev. **62**, 442 (1942).

Some of the simpler tensor operators are listed below in terms of components I_x, I_y, I_z of the spin operator $\boldsymbol{I}$

$$T_0^{(0)} = 1\,,$$

$$T_1^{(0)} = \frac{I_z}{\sqrt{I(I+1)}}\,, \qquad T_1^{(\pm 1)} = \mp \frac{I_x \pm i I_y}{\sqrt{2I(I+1)}}\,,$$

$$T_2^{(0)} = \frac{3I_z^2 - I(I+1)}{2I(I+1)}\,, \qquad T_2^{(\pm 1)} = \mp \sqrt{\frac{3}{8}}\, \frac{(I_x \pm i I_y) I_z + I_z (I_x \pm i I_y)}{I(I+1)}\,,$$

$$T_2^{(\pm 2)} = \sqrt{\frac{3}{8}}\, \frac{(I_x \pm i I_y)(I_x \pm i I_y)}{I(I+1)}\,.$$

The above two ways of describing the ensemble are completely analogous to the two alternative descriptions of an electric charge distribution, namely by specifying either the charge density $\varrho(\boldsymbol{r})$ or the complete set of electric moments (monopole, dipole, quadrupole, etc.) of the distribution.

Since the tensor operators and the elements of the density matrix form complete sets, it should be possible to expand the density matrix in terms of the above tensor moments. It has been shown by SIMON[1] that the appropriate expansion is

$$\varrho_{MM'} = \sum_q (-)^{I-M'} \frac{(2I)!\sqrt{2q+1}}{[(2I-q)!\,(2I+q+1)!]} \cdot \frac{C(IIq; M-M')}{P_q(\sqrt{I(I+1)})} \langle T_q^{(M-M')} \rangle_{\text{ensemble}}$$

where $\langle T_q^{(M-M')} \rangle_{\text{ensemble}}$ is the expectation value of $T_q^{(M-M')}$ for the ensemble. We now consider the form taken by these expectation values when the spins are oriented symmetrically with respect to a given axis.

4. Polarization and alignment. In the particular case in which the spins of the ensemble are oriented symmetrically with respect to a fixed axis in space, the ordering can be described completely by the tensors $T_q^{(0)}$ since by symmetry the expectation values of $T_q^{(k)}$ $(k \neq 0)$ for the ensemble will vanish. It is then convenient to introduce the concepts of "polarization" and "alignment". The ensemble is said to be polarized about the fixed axis if the expectation value of $T_1^{(0)}$ is non-vanishing and the polarization is then defined as

$$P_1 = \sqrt{\frac{I+1}{I}} \langle T_1^{(0)} \rangle_{\text{ensemble}}.$$

In terms of the population factors $W(M)$ this can be written

$$P_1 = I^{-1} \sum_M W(M) \cdot M\,. \tag{4.1}$$

P_1 clearly represents the ratio of the average spin along the axis of polarization to the actual nuclear spin. It is so normalised that it always lies between $+1$ and -1 and is zero for random orientation. In addition, a unit vector $\boldsymbol{n}$ along the axis of polarization can be introduced and a polarization vector $\boldsymbol{P}_1$ thereby defined according to the relation

$$\boldsymbol{P}_1 = \boldsymbol{n}\, P_1\,. \tag{4.2}$$

Because of the restriction $q \leq 2I$ it follows that for spin $\frac{1}{2}$ particles, the ensemble is completely characterised by the polarization vector P_1.

Another important type of ordering occurs when $W(M)$ depends quadratically on M. In this case P_1 vanishes but the mean value of M^2 may differ from

[1] A. SIMON: Phys. Rev. **92**, 1050 (1953) (Appendix A).

that of a system with random orientation. The system is then said to be aligned. A convenient means of defining this alignment has been suggested by STEENBERG[1] and can be written in terms of $T_2^{(0)}$

$$P_2 = \frac{2(I+1)}{2I-1} \langle T_2^{(0)} \rangle_{\text{ensemble}}$$

or in terms of $W(M)$

$$P_2 = \frac{3\Sigma M^2 W(M) - I(I+1)}{I(2I-1)}. \tag{4.3}$$

Again the expression has been normalized in such a way that P_2 lies between $+1$ and -1 and is zero for random orientation.

B. Production and properties of oriented nuclei.

5. Introduction. In experiments involving oriented nuclei there are two distinct parts. There is firstly the problem of attaining the oriented system: this is mainly an atomic problem since the methods available for orienting nuclei all depend on the nuclei being in a suitable electronic environment. The second part of the experiment is concerned with the nuclear physics of such a system. Discussion of the methods for obtaining orientation will be limited to those details which are necessary to an understanding of the nuclear aspects of the experiments. Some aspects of these methods have been discussed in earlier review articles [*1*] to [*3*].

6. Production of oriented nuclei. Two groups of methods have been proposed by which nuclear orientation can be obtained. On the one hand there are the methods which depend on the separation in energy of the nuclear magnetic states and then ensure that these are unequally populated by reducing the temperature of the system so far that the lower lying states become preferentially populated: these will be known as the very low temperature methods. On the other hand, there are the optical and microwave methods which depend on the atomic absorption and emission of radiation; in these preferential nuclear magnetic state populations may be obtained by emission of radiation through different channels from those by which it was absorbed. The following Sect. 7 will be devoted to discussion of the former methods which are the only ones which have been utilized so far. The latter group of methods will be discussed in the subsequent Sect. 8.

7. Very low temperature methods. The application of magnetic or inhomogeneous electric fields to the magnetic or electric multipole moment of the nucleus causes each substate M to possess a different energy $E(M)$. For a system of nuclei in thermal equilibrium the population of these states will then be governed by a Boltzmann function $a_M = A\,e^{-E(M)/kT}$. At temperatures where $\beta = [E(M) - E(M+1)]/kT$ is small, the populations are closely the same and the nuclear orientations are random. However, when β is of the order of unity these populations will be unequal and there is consequently some degree of nuclear orientation.

Apart from the application of external fields there are a number of ways in which use is made of the ionic and crystalline fields present in certain solids. For their description it is convenient to use the spin Hamiltonian of ABRAGAM and PRYCE[2]. Through the action of the internal crystalline fields and external magnetic fields the normally degenerate levels of a free ion become separated in

[1] N. R. STEENBERG: Thesis, Oxford 1953.

[2] A. ABRAGAM and M. H. L. PRYCE: Proc. Roy. Soc. Lond., Ser. A **205**, 135 (1951).

energy. ABRAGAM and PRYCE have shown that the energy levels of such an ion in a paramagnetic crystal possessing axial symmetry about the z-axis, in the presence of both crystalline fields and an external magnetic field may be represented by a Hamiltonian of the following form:

$$\left.\begin{aligned}\mathscr{H} = \beta_B\,[g_{\parallel} H_z S_z + g_{\perp}(H_x S_x + H_y S_y)] + D\,[S_z^2 - \tfrac{1}{3} S(S+1)] + \\ + A\,S_z I_z + B(S_x I_x + S_y I_y) + Q\,[I_z^2 - \tfrac{1}{3} I(I+1)] - (\mu/I)\,\boldsymbol{H}\cdot\boldsymbol{I}\end{aligned}\right\}\quad(7.1)$$

β_B = Bohr magneton.

The first term represents the splitting of the electronic levels by the external field and is formulated in terms of an effective electron spin S defined by setting the multiplicity of the electronic levels equal to $2S+1$. $g_{\parallel}$ and $g_{\perp}$ are the ionic g-factors parallel and perpendicular to the z-axis. The term in D represents the splitting of the electronic levels by the crystalline electric field. I is the nuclear spin and the A and B terms represent the hfs splitting due to the interaction between the nuclear magnetic moment and the unfilled electron shells which give rise to the magnetic properties of a paramagnetic ion. The Q term is the nuclear electric quadrupole splitting in the crystalline electric field gradient,

$$Q = 3e\frac{\partial^2 V}{\partial z^2} \times \frac{1}{4I(2I-1)} \times \text{(nuclear electric quadrupole moment)}$$

where V is the electric field potential at the nucleus. The last term corresponds to the direct coupling between the external magnetic field and the nuclear magnetic dipole moment μ.

Such a spin Hamiltonian has been used successfully to interpret many paramagnetic resonance experiments (see, for instance, the review articles of BLEANEY and STEVENS[1] and BOWERS and OWEN[2]) in which microwave frequencies are used to induce transitions among the lower levels of paramagnetic ions in magnetically dilute single crystals.

Four methods have been proposed for orienting nuclei (see, for instance, SIMON, ROSE and JAUCH[3]) and for their description it is convenient to suppose that in (7.1) some of the terms vanish.

(a) Direct interaction of the nuclear magnetic dipole moment[4,5,6], with a large external magnetic field causes the nuclear magnetic states to separate by an amount $\Delta E = \mu H/I$, in the absence of contribution by hfs, and is characterized by the $(\mu/I)\,\boldsymbol{H}\cdot\boldsymbol{I}$ term only. The population of individual states will then be proportional to

$$\left.\begin{aligned} a_M &= \mathrm{e}^{M\beta}\Big/\sum_M \mathrm{e}^{M\beta} \\ \text{where}\qquad \beta &= \mu H/\mathrm{k}TI. \end{aligned}\right\}\quad(7.2)$$

Since nuclear states of $+M$ and $-M$ are different in energy this mechanism leads to nuclear polarization, i.e. to non-vanishing values of P_1. To achieve appreciable orientation β must be of the order of unity which implies, for reasonable values of μ and I, very large magnetic fields and very low temperatures. For instance, for $I=2$ and $\mu=1$ nuclear magneton a value of H/T of 5×10^7 gauss/° K would be required to make β unity: thus for $H=50000$ gauss a temper-

[1] B. BLEANEY and K. W. H. STEVENS: Rep. Progr. Physics **16**, 108 (1953).
[2] K. D. BOWERS and J. OWEN: Rep. Progr. Physics **18**, 304 (1955).
[3] A. SIMON, M. E. ROSE and J. M. JAUCH: Phys. Rev. **84**, 1155 (1951).
[4] See footnote 1, p. 555.
[5] See footnote 2, p. 555.
[6] F. E. SIMON and C. R. CONG: Sur le magnétisme, **3**, 1 Strasbourg 1939.

ature of 0.001 °K would be required. Such experiments are clearly very difficult and so far only one has been reported in which nuclear orientation produced in this way has been detected by nuclear physics as distinct from nuclear resonance methods.

In this experiment which was reported by DABBS, ROBERTS and BERNSTEIN[1] stable In^{115} was polarized using a field of 11000 gauss at a temperature T of 0.043 °K and a polarization P_1 of 0.02 was obtained. This polarization was detected by observing the change in transmission of the material for polarized slow neutrons.

This appears to be the most important method of orienting nuclei since it depends only on the existence of a finite nuclear moment and does not require any additional ionic or solid state properties.

(b) A method which avoids the need for such intense magnetic fields has been proposed independently by GORTER[2] and ROSE[3]. Their suggestion is based on the fact that the unfilled electron shell produces a magnetic field at the nucleus of a paramagnetic ion which is known from data on hyperfine structure to be of the order of 10^5 gauss. At sufficiently low temperatures each nucleus will orient itself with respect to the electronic field in an energetically most favourable direction. Nuclear orientation will follow automatically if the electronic magnetic moments are oriented, and a small external field of the order of a few hundred gauss suffices for this at the very low temperatures at which the nuclear orientation will become ordered. This method is described by the A, B and g terms in Eq. (7.1) and leads to nuclear polarization. The cooling is produced either by a separate paramagnetic salt (GORTER) or by the substance to be investigated acting as its own cooling agent (ROSE). AMBLER *et al.*[4] and BERNSTEIN *et al.*[5] have given accounts of experiments in which this method has been shown to work. It should also be pointed out that in ROSE's method, where the ions to be oriented act as their own cooling agent in the adiabatic demagnetization process, the requirement of cooling and also the maintenance of a residual magnetic field are to some extent conflicting, since the final temperature reached is proportional to the magnetic field acting on the ions.

Two other methods have been put forward by POUND[6] and by BLEANEY[7] in which the orientation of the nucleus results directly or indirectly from the action of local electric fields. The direction of these fields is fixed with respect to the crystallographic axes, and to obtain spatial ordering a single crystal (or several with their axes parallel) must be used. Another significant difference from the foregoing methods is that nuclear alignment results without polarization.

(c) The method suggested by POUND makes use of the hfs due to the nuclear electric quadrupole moment. The degeneracy associated with the orientation of the nucleus is resolved through the interaction of the quadrupole moment with the gradient of the electric field. In general such gradients must be at least of the order of 10^{15} e.s.u. to give an energy separation of the required magnitude, and they can be produced only by an asymmetrical distribution of the electron cloud immediately round the nucleus. Such a distribution can be set up in homopolar

[1] J. W. T. DABBS, L. D. ROBERTS and S. BERNSTEIN: Phys. Rev. **98**, 1512 (1955).

[2] C. J. GORTER: Physica, Haag **14**, 504 (1948).

[3] M. E. ROSE: Phys. Rev. **75**, 213 (1949).

[4] E. AMBLER, M. A. GRACE, H. HALBAN, N. KURTI, H. DURAND, C. JOHNSON and H. R. LEMMER: Phil. Mag. **44**, 216 (1953).

[5] S. BERNSTEIN, L. D. ROBERTS, C. P. STANFORD, J. W. T. DABBS and T. E. STEPHENSON: Phys. Rev. **94**, 1243 (1954).

[6] R. V. POUND: Phys. Rev. **76**, 1410 (1949).

[7] See footnote 4, p. 555.

bonds, and if these bonds have a parallel orientation for all similar nuclei, as may be the case in a suitable organic compound, nuclear orientation will set in at sufficiently low temperatures. A preliminary account of an experiment in which this method of nuclear orientation has been used to align U^{233} has been reported by DABBS *et al.*[1].

(d) The method due to BLEANEY like method (c) is based on the fact that the different orientations of the nuclear moments relative to a crystal axis may have different energies. In this method this energy difference is due to the interaction between the inhomogeneous crystalline electric field and the electronic magnetic moments through the intermediary of the electronic orbits. The same mechanism is responsible for the anisotropy of the magnetic susceptibility. The strong electric field set up by the neighbouring ions or molecules in the crystal in the absence of external magnetic fields defines an axis of quantization. The interaction is so strong that already at comparatively high temperatures (10 to 100° K) only the lowest electronic states will be occupied. If the electric fields responsible for this effect have axial symmetry, alignment of the electronic magnetic moments relative to the axis of the single crystal will take place. This will lead at sufficiently low temperatures (about 0.01° K) to nuclear alignment through hfs coupling. They are most conveniently reached by adiabatic demagnetization of the paramagnetic single crystal. In this method H is zero and only the D, A and B terms contribute to the spin Hamiltonian. The earliest successful experiments malking use of this effect were reported by the groups at the Universities of Oxford[2] and Leiden[3].

Although in the above classification certain terms in the spin Hamiltonian are assumed to vanish, and the examples quoted approximate to these ideal conditions, in most cases their contribution will be significant and thus the degree of nuclear orientation will deviate from the limiting case where the interaction is simple. The distinction between the different orientation mechanisms is then no longer sharply defined.

Such are the processes which have been suggested as possible methods of obtaining oriented nuclei using very low temperatures. Some of the experimental conditions which distinguish these different mechanisms are summarized in Table 1.

Table 1. *Summary of low temperature methods of nuclear orientation.*

	Method	Temperature °K	Applied field	Specimen
(a)	External field (Polarization)	0.01	10^5-10^6 gauss	Non magnetic external cooling
(b)	Magnetic hfs (Polarization)	0.01	10^2 gauss	Paramagnetic crystals whose ions exhibit hfs
(c)	Electric hfs (Alignment)	0.01	none	Single crystal, ions with E. Q moment, suitable internal electric fields
(d)	Magnetic hfs (Alignment)	0.01	none	Single crystal, suitable internal fields: ions exhibiting hfs

[1] J. W. T. DABBS, L. D. ROBERTS and G. W. PARKER: Bull. Amer. Phys. Soc., Ser. II **4**, 207 (1956).

[2] J. M. DANIELS, M. A. GRACE and F. N. H. ROBINSON: Nature, Lond. **168**, 780 (1951).

[3] C. J. GORTER, O. J. POPPEMA, M. J. STEENLAND and J. A. BEUN: Physica, Haag **17**, 1050 (1951).

8. Optical and microwave methods. *α) The optical pump.* KASTLER[1] has shown that when an atomic system absorbs polarized resonance radiation atomic orientation may arise following the subsequent emission of radiation. Nuclear orientation accompanies this if the atomic ground-state exhibits hfs and this has been verified experimentally by BROSSEL, CAGNAG and KASTLER[2] using nuclear resonance methods. Consider, for instance, the absorption of circularly polarized $(\sigma+)$ resonance radiation by atomic sodium in its ground state $^2S_{\frac{1}{2}}$. Then, neglecting the effect of nuclear spin, the D_1 line will only induce transitions from the $m=-\frac{1}{2}$ state of the ground state to the $m=+\frac{1}{2}$ state of the first excited state $^2P_{\frac{1}{2}}$ (see Fig. 1).

Atoms in this state now have two possible modes of decay, either by emitting $\sigma+$ radiation to the $m=-\frac{1}{2}$ state or by emitting π radiation to the $m=+\frac{1}{2}$ state of the ground-state. Thus, if interaction between the atoms and other material can be avoided, atoms will accumulate in the $m=+\frac{1}{2}$ state. Similar considerations show that the same will be true of the D_2 line and transitions to the other member of the P doublet. Now the ground-state of atomic sodium is paramagnetic and exhibits hfs. By extending these arguments it can be shown that through the hfs coupling the nuclear magnetic states will become differentially populated and therefore nuclear orientation will be obtained.

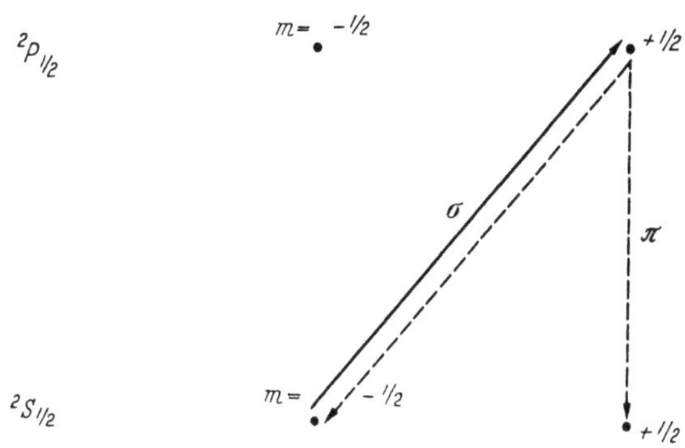

Fig. 1. Diagram illustrating the absorption of circularly polarized resonance radiation (σ^+) by atomic sodium in its ground state $^2S_{\frac{1}{2}}$.

This method has not so far been applied to the study of nuclear disintegration phenomena but estimates of the degree of orientation obtainable suggest that it is likely to yield detectable effects.

β) Microwave methods. The possibility of obtaining nuclear orientation in a metal through relaxation mechanisms in an electron spin resonance from the conduction electrons was pointed out by OVERHAUSER[3]. If a magnetic field is applied to a metal, a population difference between the two possible spin states of the conduction electrons will be set up. On the application of a radiofrequency field normal to the direction of this steady field electron spin flips will be induced when the frequency ω is related to the steady field H_0 by

$$\hbar\omega = g\,\beta_B H_0.$$

The net absorption of power at resonance by the electrons will then be proportional to the excess of electrons in one spin orientation over the other. For large R.F. power the resonance will tend to saturation corresponding to equalization of the two electron spin state populations. However, at the same time electrons from the energetically higher spin state will flip back to the energetically more favourable orientation by one of a number of possible relaxation mechanisms. One of these, the hfs coupling of the form $\boldsymbol{I}\cdot\boldsymbol{S}$ between the con-

[1] A. KASTLER: Proc. Phys. Soc. Lond. A **67**, 853 (1954).
[2] J. BROSSEL, B. CAGNAC and A. KASTLER: J. de Phys. Radium **15**, 6 (1954).
[3] A. W. OVERHAUSER: Phys. Rev. **92**, 411 (1953).

duction electrons and the nuclei will cause the nuclei to flip over in the opposite sense to the electrons since the total angular momentum must be conserved in the process. Thus nuclei will tend to accumulate in one preferred orientation, i.e. a nuclear polarization will be set up. It turns out that the factor governing the preponderance of nuclei in this orientation state is of the form

$$\exp\left(g_N \beta_N + g_E \beta_B S\right) H_0/\mathrm{k}T$$

where S describes the degree of saturation of the electron spin resonance. That is to say, the nucleus appears to have its nuclear moment enhanced and reaches that of the electron at saturation. Thus for $H_0 = 10^4$ gauss, $T = 2^\circ$ K, $I = \frac{3}{2}$ and $S = 0.8$ a value for P_1 of 0.42 is expected.

The essential feature of this method of nuclear orientation, apart from having a mechanism whereby energy can be conserved and at resonance transitions in one direction can be inhibited, appears to be that the predominant method of electron spin relaxation must be through a coupling with the nucleus which conserves the total spin of the system.

It has been pointed out by BLOCH[1] and by KORRINGA[2] that this nuclear orientation mechanism is not confined to metals but should for instance operate in ionic compounds which exhibit paramagnetic resonance. At saturation of the paramagnetic resonance relaxation through coupling with the nuclear spins of the form $\boldsymbol{I} \cdot \boldsymbol{S}$ will leave the nuclear spins oriented if this represents the predominant coupling between the nuclear spin system and the lattice.

Although a number of successful experiments have been performed which demonstrate this effect namely, the enhancement of nuclear resonance (see, for instance, CARVER and SLICHTER[3], BELJERS, VAN DER KINT and VAN WIERENGEN[4]), the method has not so far been applied to obtaining high degrees of nuclear polarization.

The principal difficulties in such a method seem to lie in getting sufficient power into the sample to obtain saturation of the resonance and at the same time maintaining the low temperature required to achieve reasonable degrees of nuclear orientation.

9. Orientation parameters. Having enumerated the different mechanisms whereby a nuclear orientation may be realized, it is now necessary to set up a series of parameters which will describe completely the state of orientation of the system.

For a system of nuclei whose spin vectors I possess cylindrical symmetry with respect to a certain fixed axis in the system (e.g. the direction of an externally applied magnetic field or an axis of crystalline symmetry), the degree of nuclear orientations is uniquely described by the values of the population functions $W(M)$ for the different magnetic states.

As described in section 4 the orientation in the case of cylindrical symmetry is completely described by the expectation values for the ensemble of the tensors $T_q^{(0)}$. These expectations can be expressed in terms of the $W(M)$ by the relation

$$\langle T_q^{(0)} \rangle = \sum_M W(M) \cdot P_q\left(M/\sqrt{I(I+1)}\right)$$

where $P_q(x)$ is a Legendre polynomial. Apart from a normalizing factor, $P_q\left(M/\sqrt{I(I+1)}\right)$ is identical with $C(I q I; M 0)$. Thus, following GRAY and

[1] F. BLOCH: Phys. Rev. **93**, 944 (1954).
[2] J. KORRINGA: Phys. Rev. **94**, 1388 (1954).
[3] T. R. CARVER and C. P. SLICHTER: Phys. Rev. **92**, 212 (1953).
[4] H. G. BELJERS, L. VAN DER KINT and J. S. VAN WIERINGEN: Phys. Rev. **95**, 1683 (1954).

SATCHLER[1], we define orientation parameters $B_q(I)$ by the relation

$$B_q(I) = \sum_M (2q+1)^{\frac{1}{2}} C(I q I; M 0)\, W(M)$$

where $q \leq 2I$. $\left(\text{For } q > 2I,\ B_q(I) = 0\right)$. These B_q differ only in normalization from $\langle T_q^{(0)} \rangle$.

We have preferred to adhere to the $B_q(I)$ as defined in this way because of their direct application to the angular distribution through the tabulated functions F_q of BIEDENHARN and ROSE[2]. These are identical with the functions derived and discussed by STEENBERG[3], namely

$$\sum_M (2q+1)^{\frac{1}{2}} \Pi_q(M I)\, N(I q I)\, W(M)$$

and differ from the spin moment functions $f_q(I)$ defined by TOLHOEK and COX[4] only in normalization, thus

$$B_q(I) = \binom{2q}{q} I^q \left[(2q+1)(2I+1)(2I-q)!/(2I+q-1)\right]^{\frac{1}{2}} f_q(I).$$

In general it is necessary to determine the energy levels of the ionic system by solving the secular equation for the perturbed system and then determine the contributions by the nuclear states to the different ionic states. This evaluation is straightforward where the energy matrix is diagonal but becomes laborious where, for example, finite values of B contribute off diagonal matrix elements and admix the nuclear states. The Boltzmann factor $e^{-E(M)/kT}$ then determines the population distribution between the different nuclear states. By expanding the Boltzmann factor in powers of $E(M)/kT$ one can find closed forms for B_q in the case where $E(M)/kT \ll 1$. For these the first non-vanishing terms are of order $\left(E(M)/kT\right)^q$ and therefore where $E(M)/kT \ll 1$ the B_q will decrease rapidly as q increases. In the angular distribution of radiation it turns out that only the even values of q will contribute to the anisotropy. Since $B_0 = 1$, the first contributing B_q will be B_2 and B_4 will in general be small compared with B_2 when $E(M)/kT \ll 1$. A convenient method for deriving expressions for the B_q has been employed by SIMON, ROSE and JAUCH[5] to whom the reader is referred for details of the method of calculation. The expressions for B_1 and B_2 are:

$$B_1 = 2\sqrt{3} \sum_M M\, W(M),$$

$$B_2 = \sum_M \frac{\left[M^2 - \frac{1}{3} I(I+1)\right] W(M)}{\sqrt{\frac{1}{5} I(I+1)(2I-1)(2I+3)}}$$

and in the case where $E(M)/kT$ is small the closed forms of B_2 are given below for particular forms of the interaction energy $\mathcal{H}$.

(1) For orientation by the direct interaction method where $\mathcal{H}$ is given by

$$\mathcal{H} = g_N \beta_N H_z I_z,$$

$$B_1 = \sqrt{\frac{3}{I(I+1)}} \left[\frac{1}{2} \operatorname{Cot} \frac{\beta}{2} - \frac{2I+1}{2} \operatorname{Cot} \frac{2I+1}{2} \beta\right]$$

which for $\beta \ll 1$ becomes $\beta \sqrt{\frac{I(I+1)}{3}}$ and gives $P_1 = \frac{1}{3} \beta (I+1)$,

$$B_2 = \frac{\beta^2}{6} \sqrt{\frac{1}{5} I(I+1)(2I-1)(2I+3)} \quad \text{for} \quad \beta \ll 1$$

[1] T. P. GRAY and G. R. SATCHLER: Proc. Phys. Soc. Lond. A **68**, 349 (1955).
[2] L. C. BIEDENHARN and M. E. OSSE: Rev. Mod. Phys. **25**, 729 (1953).
[3] N. R. STEENBERG: Proc. Phys. Soc. Lond. A **65**, 791 (1952).
[4] H. A. TOLHOEK and J. A. M. COX: Physica, Haag **19**, 101 (1953).
[5] See footnote 3, p. 560.

where $\beta = g_N \beta_N H_z/\mathrm{k}T$. It will be noticed that the degree of polarization P_1 for low degrees of orientation is linear in $1/T$ in contrast to the behaviour of the alignment parameter B_2 which is quadratic in $1/T$. This behaviour is important since it shows that for low degrees of orientation polarization phenomena will be, in general, greater than alignment phenomena.

(b) For orientation by the method of magnetic hfs (polarization) in the case where $S=\frac{1}{2}$ and

$$\mathscr{H} = g_{\parallel} \beta_B H S_z + A S_z I_z + B(S_x I_x + S_y I_y)$$

we have

$$B_1 = -\frac{S}{3}\sqrt{\frac{I(I+1)}{3}}\,\beta(1+2x)\frac{M_e}{M_\infty} \quad \text{for} \quad \beta(=A/\mathrm{k}T) \ll 1$$

and $g_{\parallel}\beta_B H \gg A$. M_e is the electronic magnetization, M_∞ is the saturation magnetization and $x = B/A$. Under this condition

$$P_1 = -S(I+1)\frac{\beta}{3}\frac{M_e}{M_\infty}(1+2x).$$

The negative sign enters through the fact that the positive values of A imply that the states of antiparallel spin lie lowest.

Similarly

$$B_2 = \frac{2}{3}\beta^2\sqrt{\frac{1}{5}I(I+1)(2I-1)(2I+3)}\left[1 - \frac{2x^2 \operatorname{Tan}(\vartheta/2)}{\vartheta}\right]$$

for $\beta \ll 1$ but no restriction on the value of $\vartheta = g_{\parallel}\beta_B H/\mathrm{k}T$.

(c) In the method of alignment employing the nuclear quadrupole interaction

$$\mathscr{H} = Q\left[I_z^2 - \tfrac{1}{3}I(I+1)\right],$$

and we have

$$B_2 = -\frac{1}{3}\frac{Q}{\mathrm{k}T}\sqrt{\frac{1}{5}I(I+1)(2I-1)(2I+3)} \quad \text{for} \quad \frac{Q}{\mathrm{k}T} \ll 1.$$

(d) Where orientation is obtained by the magnetic hfs (alignment) method for

$$\mathscr{H} = D\left(S_z^2 - \tfrac{1}{3}S(S+1)\right) + A I_z \cdot S_z + B(S_x I_x + S_y I_y),$$

$$B_2 = \frac{2}{9}\beta^2 S(S+1)\sqrt{\frac{1}{5}I(I+1)(2I-1)(2I+3)}\left[1 - x^2 - \frac{(2+x^2)(2S-1)(2S+3)D}{30\,\mathrm{k}T}\right]$$

where in addition to $\beta(=A/2\mathrm{k}T)$, $D/\mathrm{k}T$ is also $\ll 1$.

(e) While for

$$\mathscr{H} = A S_z I_z + B(S_x I_x + S_y I_y),$$

$$B_2 = (1-x^2)\frac{2\beta^2}{9}\sqrt{\frac{1}{5}I(I+1)(2I-1)(2I+3)}\,S(S+1),$$

for $\beta \ll 1$ where $x = B/A : \beta = A/2\mathrm{k}T$.

Closed forms for the B_q have only been developed for small values of β. The tabulations of B_2, B_4 shown in Appendix I (p. 608) have been prepared by GRAY and SATCHLER for the case of orientation through the A term alone, and are valid for all values of β for $S=\frac{1}{2}$. These can be corrected to take account of small contributions through the B terms. Also shown are tabulations for B_2, B_4 for B finite and A zero again for $S=\frac{1}{2}$. Again corrections can be made for small contributions by the A terms.

Similarly the functions $f_q(I)$ are displayed graphically by Poppema[1] for values of the spin I up to 5 and for q up to 4, and for the case where orientation arises through the A term alone.

10. The effect of interionic forces. The influence of interionic forces of the magnetic spin-spin and electron exchange type have not been included in the Hamiltonian for the ionic ground state. These additional terms have been considered by Steenberg[2] in the approximation where $A/\mathrm{k}T$ is small. For large external fields ($H \to \infty$) the influence of such forces is zero: the degrees of orientation (P_1 and P_2) are unaltered by their presence and depend only on the terms in the Hamiltonian [Eq. (7.1)]. For H small, these interactions contribute through a term quadratic in H and therefore for $H=0$ leave B_1 and B_2 unaltered. Furthermore, for B_2 the interaction term is also zero if $x(=B/A)$ is zero for any value of H. Steenberg concludes that only under special limited conditions can these interactions seriously disturb the values of the parameter B_2 deduced from the spin Hamiltonian neglecting these interactions.

11. Properties and detection of oriented nuclei. A number of methods are available for the detection of oriented nuclei. Spiers[3] has pointed out that if the nuclei are radioactive the angular distribution of the radiation, whether particle or electromagnetic, will depend both on the degree of orientation of the parent nuclei and on the angular momenta involved in the transition. Thus the detection of an anisotropic distribution of radiation demonstrates the existence of oriented nuclei. Associated with this angular distribution will be a polarization of the radiation. So long as no observation of the particle spin is made, the angular distribution will only depend on the average values of the even powers of M, i.e. on $M^{2r} W(M)$ where r is integral. The spatial distribution of the radiation, therefore, depends on the even spin moments P_2, P_4 etc., and it is thus a property specific to aligned nuclei and not automatically associated with polarized nuclei.

On the other hand one can study the interaction of the oriented nuclei with beams of particles or γ-rays. Thus the absorption and scattering of polarized slow neutrons by polarized nuclei investigated by the Oak Ridge group[4,5] and the experiments to detect circularly polarized γ-rays by the Leiden group[6] have demonstrated the presence of polarized nuclei: in these experiments non-vanishing of the odd spin moments is essential to the observation of these effects.

For completeness it should be mentioned that the method of nuclear resonance might also be applied. Radiofrequency radiation in resonance will change the population of the magnetic sublevels and this process could in principle be detected either by a change in the angular distribution of nuclear radiation or by standard radiofrequency methods: this has been discussed by Bloembergen and Temmer[7].

In the following sections an outline is given of the theoretical derivation of the angular distributions of radiation from aligned nuclei and then this is followed by an account of properties specific to polarized nuclei.

---

[1] O. J. Poppema: Thesis Groningen 1954.

[2] N. R. Steenberg: Phys. Rev. **93**, 678 (1954).

[3] J. A. Spiers: Nature, Lond. **161**, 807 (1948). — Directional effects in radioactivity. N.R.C.C. publication 1949.

[4] See footnote 5, p. 561.

[5] L. D. Roberts, S. Bernstein, J. W. T. Dabbs and C. P. Stanford: Phys. Rev. **95** 105 (1954).

[6] J. C. Wheatley, W. J. Huiskamp, A. N. Diddens, M. J. Steenland and H. A. Tolhoek: Physica, Haag **21**, 841 (1955).

[7] N. Bloembergen and G. M. Temmer: Phys. Rev. **89**, 883 (1953).

12. Angular distribution and polarization of radiation from aligned nuclei. It is convenient to distinguish between those properties which are specific to polarized nuclei and those which are associated with oriented nuclei in general whether polarized or aligned. The properties of aligned nuclei will be described first.

A theoretical investigation of the angular distribution of radiation from oriented radioactive nuclei has been made by SPIERS[1] for the case of a single emission and later extended by STEENBERG[2], by TOLHOEK, COX and DE GROOT[3], and by GRAY and SATCHLER[4] to cases where several radiations are emitted in cascade. An earlier review of these has been given by TOLHOEK and DE GROOT [1].

For single emission the angular distribution of particles of total angular momentum I, with z-component m is simply

$$I_m(\vartheta) = \int |\Phi_I^m(\vartheta, \varphi, \mu)|^2 \, d\mu \tag{12.1}$$

where Φ_I^m is the angular part of the particle wave-function, ϑ, φ its angular co-ordinates and the integration has been taken over the spin co-ordinates of the particle μ. This corresponds to observation of the particle distribution with spin insensitive detectors. Now for a particle of total angular momentum I with $I_z = m$

$$\Phi_I^m(\vartheta, \varphi, \mu) = \sum_{\sigma=-s}^{+s} \sum_{L=|I-s|}^{I+s} b_L \, C(L\,S\,I; m-\sigma\,\sigma) \, Y_L^{m-\sigma}(\vartheta, \varphi) \, \chi_s^\sigma(\mu)$$

$Y_L^{m-\sigma}(\vartheta\varphi)$ is a spherical harmonic; $\chi_s^\sigma(\mu)$ is a spin wave function corresponding to a spin S and z-component σ (see for instance SPIERS[1]). C is a vector addition coefficient: b_L may be complex and is related to the probability of the orbital angular momentum L occurring. The next step in the development is to perform the summation for the particular type of particle which is being considered.

α) γ-radiation. For γ-radiation of multipolarity L, $I_m(\vartheta)$ reduces to

$$I_m^L(\vartheta) = \sum_k a_k(m\,L) \, P_k(\cos\vartheta) \tag{12.3}$$

where

$$a_k(m\,L) = (2k+1)\left[1 - \frac{k(k+1)}{2L(L+1)}\right] C(k\,L\,L; 0\,0) \, C(k\,L\,L; 0\,m) \tag{12.4}$$

and k can take values from 0 to $2L$: only even k values will contribute since $C(k\,L\,L\colon 0\,0)$ vanishes for k odd. $P_k(\cos\vartheta)$ is the Legendre polynomial of order k.

It is now possible to determine the angular distribution of radiation $I_M^L(\vartheta)$ from a state of angular momentum I_0 and z-component M in a transition to a state of angular momentum I_1. For if L is the angular momentum of the radiation

$$I_M^L(\vartheta) = \sum_m |C(I_1\,L\,I_0; M-m\,m)|^2 \, I_m^L(\vartheta) \tag{12.5}$$

where for γ-radiation $I_m(\vartheta)$ is given above.

---

[1] See footnote 3, p. 567.

[2] N. R. STEENBERG: Phys. Rev. **84**, 1051 (1951). — Proc. Phys. Soc. Lond. A **65**, 791 (1952); A **66**, 391, 399 (1953). — Canad. J. Phys. **31**, 204 (1953); Thesis, Oxford 1953.

[3] H. A. TOLHOEK and J. A. M. COX: Physica, Haag **18**, 357, 359 (1952). — H. A. TOLHOEK: Physica, Haag **18**, 1257 (1952). — J. A. M. COX: Physica, Haag **18**, 1262 (1952). — H. A. TOLHOEK and S. R. DE GROOT: Physica, Haag **17**. 81 (1951). — H. A. TOLHOEK and J. A. M. COX: Physica, Haag **19**, 101 (1953). — CHR. D. HARTOGH, H. A. TOLHOEK and S. R. DE GROOT: Physica, Haag **20**, 1310 (1954). — J. A. M. COX S. R. DE GROOT and CHR. D. HARTOGH: Physica, Haag **19**, 1119, 1123 (1953).

[4] See footnote 1, p. 565.

Thus for single emission, if the magnetic substates (M) each have a population $W(M)$ the total angular distribution is

$$I(\vartheta) = \sum_M W(M) \sum_m |C(I_1 L I_0; M - m)|^2 I_m^L(\vartheta). \tag{12.6}$$

Such an angular distribution reduces to a series of even Legendre polynomials and can be written in the form

$$I(\vartheta) = \sum_{k=0}^{\lambda} B_k F_k P_k(\cos\vartheta) \tag{12.7}$$

where λ is $2L$ or $2I_0$ whichever is smaller. B_k is the orientation parameter (see Sect. 11) and is a measure of the population distribution among the substates of I_0. F_k depends on the angular momentum properties of the transition $I_0 \xrightarrow{L} I_1$ and is defined for γ-radiation by

$$F_k = (-1)^{I_1 - I_0 - 1} \sqrt{2I_0 + 1}\,(2L+1)\, C(L L k; 1 - 1)\, W(I_0 I_0 L L; k I_1) \tag{12.8}$$

where W is a Racah coefficient. These F_k functions are identical with the angular correlation functions which have been tabulated by BIEDENHARN and ROSE[1] and also by FERENTZ and ROSENZWEIG[2]: they have also been discussed by STEENBERG although his nomenclature is slightly different. We have preferred the nomenclature of BIEDENHARN and ROSE because of the completeness of the tabulation of the F_k values and because of its easy extension to the theoretical treatment of mixed transitions and the study of polarized radiations.

We can extend this treatment to cover the situation where the radiation which is observed occurs in a cascade of transitions and is preceded by one or more radiative transitions. Consider a series of transitions of the form

$$I_0 \xrightarrow{L_1} I_1 \longrightarrow \cdots I_{n-1} \xrightarrow{L_n} I_n \longrightarrow I_{n+1} \cdots \text{etc}$$

in which the radiation L_n is observed. The angular distribution of L_n can then be expressed as

$$I(\vartheta) = \sum_k B_k U_k F_k P_k(\cos\vartheta) \tag{12.9}$$

where U_k is a function of the angular momenta of all transitions preceding that observed. U_k is given by

$$U_k = U_k^{(1)} U_k^{(2)} \ldots U_k^{(n-1)} \tag{12.10}$$

where each $U_k^{(m)}$ refers to the particular unobserved transition $I_{m-1} \xrightarrow{L_m} I_m$ and is given by

$$\left.\begin{aligned} U_k^{(m)} = {} & \sqrt{(2I_{m-1}+1)(2I_m+1)}\,\times \\ & \times (-1)^{I_{m-1} - I_m - L_m} W(I_{m-1} I_{m-1} I_m I_m; k L_m). \end{aligned}\right\} \tag{12.11}$$

B_k is the orientation parameter for the initial state I_0 of the cascade and F_k refers to the observed transition L_n and is given by (12.8) with $I_0 I_1$ and L replaced by I_{n-1}, I_n and L_n respectively.

We here assume that the time between successive emission processes is so short that the populations of intermediate nuclear states are governed only by the initial equilibrium distribution and the character of the preceding emission processes in the cascade. When this time is not short disturbance to the populations of the states may occur through such processes as the interaction of the nuclear spin with the thermal vibrations of the lattice; recoil of the nucleus from its lattice position through regions of different fields; changes in magnetic or

[1] See footnote 3, p. 565.
[2] M. FERENTZ and N. ROSENZWEIG: ANL—5324.

electric field gradient at the nucleus following a β process due to the atomic rearrangement or to ionisation or excitation of the atom. Maximal information will only be gained from a nuclear orientation experiment when the absence of disturbance has been established or its magnitude determined.

Similarly for the angular distribution of γ-radiation following a β-transition

$$I_0 \xrightarrow{i_\beta} I_1 \xrightarrow{L} I_2$$

we have

$$I(\vartheta) = \sum_k B_k(I_0, T) \cdot U_k(i_\beta I_0 I_1) \cdot F_k(L I_2 I_1) \cdot P_k(\cos\vartheta). \tag{12.12}$$

More complex cascades can be considered by simple extension of the above expression.

Where the γ-transition is mixed, Eq. (12.12) must be modified by replacing $F_k(LI_2I_1)$ by $F_k'(LI_2I_1)$. In terms of the functions tabulated by FERENTZ and ROSENZWEIG[1], $F_k'(LI_2I_1)$ is given by

$$F_k'(LI_2I_1) = [F_k(LI_2I_1) + \delta^2 F_k(L+1\, I_2I_1) + 2\delta F_k(L, L+1\, I_2I_1)]/[1+\delta^2] \tag{12.13}$$

for a 2^L pole transition in which is admixed an amplitude δ of $L+1$ pole radiation. Clearly this expression takes account of the coherence relationship between the admixed components and approximates in the limits of $\delta = 0$ and ∞ to the expressions applicable to the pure transitions.

It should be noted that where unobserved mixed transitions precede that which is being studied, account must also be taken of the admixture. In this case, however, the cross terms due to the interference of the admixed components vanish and so one has only to replace $U_k(L)$ by

$$[U_k(L) + \delta^2 U_k(L')]/[1+\delta^2]. \tag{12.14}$$

For the case where the angular distribution of the plane polarization of the γ-radiation is being studied we must use the expression

$$I(\vartheta\varphi) = \sum_k B_k U_k [F_k'(LI_2I_1)\cdot P_k(\cos\vartheta) + F_k''(LI_2I_1)\cos 2\varphi \cdot P_k^{(2)}(\cos\vartheta)]/[1+\delta^2] \tag{12.15}$$

where $F'(LI_2I_1)$ is given in (12.13) and

$$\left.\begin{aligned} F_k''(LI_2I_1) = (-1)^{\sigma L}\left[\frac{(k-2)!}{(k+2)!}\right]\Bigg[&\frac{2k(k+1)L(L+1)}{k(k+1)-2L(L+1)} F_k(LI_2I_1) - \\ -\delta^2\frac{2k(k+1)(L+1)(L+2)}{k(k+1)-2(L+1)(L+2)} F_k(L+1\,I_2I_1) &- 4\delta(L+1)\cdot F_k(L, L+1\,I_2I_1)\Bigg]\end{aligned}\right\} \tag{12.16}$$

for the polarization of a mixed 2^L-, 2^{L+1}-pole transition in which δ is defined as before. σ depends on the electric or magnetic character of the 2^L transition and is 0 for a magnetic transition and 1 for an electric transition: $P_k^2(\cos\vartheta)$ is the associated Legendre polynomial. φ describes the inclination of the polarization vector and is the angle between the polarization vector and the plane containing the z-direction and the direction of propagation of the quantum: it should be noted that this is different from the convention adopted by BIEDENHARN and ROSE, the present angle being complementary to theirs.

Much of the experimental work on oriented nuclei has been devoted to the study of the angular distribution and polarization of γ-radiation from radioactive nuclei and is described in detail in a later section (p. 575).

[1] See footnote 2, p. 569.

β) α-particles. SPIERS[1] has considered the directional effects associated with α-decay from aligned nuclei. For a transition $I_0 \xrightarrow{L} I_1$ in which only one value L of the orbital angular momentum contributes

$$W(\vartheta) = \sum_k 2L(L+1)\, B_k(I_0 T)\, F_k(L I_1 I_0)\, P_k(\cos\vartheta)/[2L(L+1) - k(k+1)]$$

where as usual k is even and cannot exceed $2L$ or $2I_0$. It is seen that for outgoing S-wave particles no angular distribution different from isotropic is to be expected. In general where mixed waves are concerned the ratios of the admixed wave amplitudes and their phase differences are required. The method should prove valuable in assessing the contributions due to different outgoing waves. A brief account of an experiment in which an anisotropic distribution of α-particles has been obtained has been given by members of the Oak Ridge National Laboratory[2,3]. They found for the α-particles from U^{233} and Np^{237} an anisotropy of about 6% at 1.1° K.

γ) β-rays. If β-radioactive nuclei are oriented it might be expected that the β-particles would show an angular distribution different from isotropy. TOLHOEK[4] has considered the theory of this process and concludes that for allowed transitions no deviation from isotropy is to be expected[5]. In forbidden β-transitions deviations from spherical symmetry will only occur for those β-transitions which show deviations from the linear Kurie plot, i.e., for the same β-transitions for which the β-γ correlation function differs from isotropy when the β-ray is followed by a γ-ray. For the unique first forbidden group with $\Delta I = \pm 2$ the directional distribution for β rays in a transition $I_0 \to I_1$, is given by

$$W(\vartheta) = 1 - B_2(I_0 T) \cdot \mu \sqrt{\frac{(2I_0+3)(I_0+1)}{5I_0(2I_0-1)}}\, P_2(\cos\vartheta) \qquad (12.17)$$

for $I_1 = I_0 - 2$, and by

$$W(\vartheta) = 1 - B_2(I_0 T) \cdot \mu \sqrt{\frac{I_0(2I_0-1)}{5(I_0+1)(2I_0+3)}}\, P_2(\cos\vartheta) \qquad (12.18)$$

for $I_1 = I_0 + 2$, where $\mu = (E^2-1)/[E^2 - 1 + (E_0 - E)^2]$: E, E_0 are the total β-ray energy and the maximum energy of the β spectrum. It is seen that the maximum effect is observed for the highest energy electrons and for a decreasing spin sequence.

δ) Fission fragments. It has been suggested that the slow neutron fission of oriented nuclei might give rise to an anisotropic distribution of the fission fragments. The form of this angular distribution will depend on the relative amounts of outgoing angular momenta of the fragments and this in turn will be related to purely nuclear properties such as the nature of the distortion produced in the disintegrating nucleous. It is possible that experiments of this sort could lead to interesting information on nuclear deformations.

13. Conclusions. The existence of these properties associated with nuclear orientation leads to two distinct types of experiment. On the one hand, one may

---

[1] See footnote 3, p. 567.
[2] See footnote 1, p. 562.
[3] L. D. ROBERTS, J. W. T. DABBS, G. W. PARKER and R. D. ELLISON: Bull. Amer. Phys. Soc., Ser. II **4**, 207 (1956).
[4] H. A. TOLHOEK: Physica, Haag **18**, 1257 (1952).
[5] This result is based on the assumption that the β-decay interaction is invariant under space reflection. LEE and YANG [Phys. Rev. **104**, 254 (1956)] have recently proposed that this assumption may not be valid and that under these circumstances a backward-forward asymmetry may arise. An experiment by WU, AMBLER, HAYWARD, HOPPES and HUDSON (Phys. Rev. to be published) on Co^{60} shows that such an asymmetry is present.

use these properties to investigate different orientation mechanisms: the nucleus in this case is used as a probe to study atomic properties. On the other hand, one may employ an established nuclear orientation mechanism and investigate purely nuclear properties.

Consider, for example, the angular distribution of radiation from oriented radioactive nuclei. This can be written in the form

$$W(\vartheta) = 1 + B_2(T)\, A_2 P_2(\cos\vartheta) + B_4(T)\, A_4 P_4(\cos\vartheta) + \cdots.$$

From measurements of this angular distribution the coefficients $(B_k A_k)$ can be determined. If a nucleus is chosen whose decay scheme and spin assignments have been established, then the A_k are known and the B_k can be evaluated as functions of temperature. Since apart from temperature the only other factor on which B_k depends is the interaction between the nucleus and its electronic environment, this is a method for studying this interaction. Experiments in which nuclear orientation has been used to study the hfs in a cobalt metal crystal and to investigate neutron capture damage in an ionic crystal have been reported by the Oxford group[1].

Conversely, if a material is selected (e.g. a crystalline solid) in which the interactions are completely understood then the parameters B_k are known as functions of T and the A_k can be determined. From the A_k information can be gained on the angular momenta involved in the radiative transition.

From the theoretical treatment of the previous sections the following general conclusions can be drawn.

No angular distribution different from isotropic can be expected from states preceded in a cascade by a state of angular momentum 0 or $\frac{1}{2}$.

Particle emission is isotropic if L, the orbital angular momentum of the particles emitted, is zero. If L is non-zero, the angular distribution will characterize the angular momenta of the nuclear levels and the initial substate populations.

Under conditions where the radiation is emitted from a completely oriented state the angular distribution of the particles will depend only on L and the angular momenta $I_{\text{initial}}\ I_{\text{final}}$ of the levels: thus a single measurement of the angular distribution at these temperatures may give I_{initial}, I_{final} and L. Where nuclear sublevels other than the lowest are occupied, the angular distribution will yield information on their relative populations. Thus when these are equally spaced due, for instance, to magnetic hfs coupling the Boltzmann factor, $\beta = \mu H/\mathrm{k}TI$ may be found: if H and T are known, a value for μ the nuclear magnetic moment of the initial state may be inferred.

Similar conclusions may be drawn from observations of γ-radiation from oriented nuclei whether or not they are preceded by a previous emission process. In cases where previous emission has occurred, the angular distribution will involve the angular momenta of the previous processes. Thus in the observation of γ-radiation following a β process it may be possible to determine the angular momentum carried away in the β process. This gives information on the relative contributions of Fermi and Gamow-Teller type interactions in the β-decay process. Such information is not easily accessible for allowed β transitions by more conventional methods.

Associated with this anisotropic distribution of radiation there will be a partial or total transverse polarization of the γ-radiation. The plane of the electric vector will depend on the direction of observation, the multipolarity of the

[1] M. A. GRACE, C. E. JOHNSON, N. KURTI, R. G. SCURLOCK and R. T. TAYLOR: Conference de Physique des basses temperatures, Paris, 2–8 Sept. 1955 (Paper Nos. 150, 159).

transition and the electric or magnetic character of the transition. It is thus possible to determine not only the angular momenta involved in a transition (from the angular distribution of the radiation) but also the parity change.

If the angular distribution and the plane of polarization of the γ-radiation are measured for the same value of B_k then in cases where the transition is mixed both the phase and the mixing ratio may be inferred.

14. Properties specific to nuclear polarization. So far the discussion has only related to the properties of systems with finite values for B_k when k is even, i.e. aligned systems. There are a number of processes apart from nuclear resonance when finite values for B_k when k is odd are required in order that detectable effects should be observed. This corresponds to nuclear polarization. These processes depend on the vector properties of the radiation spin such as:

α) Circularly polarized γ-rays.

β) Electron spin polarization in β decay.

γ) Processes involving the interaction of polarized particles with polarized nuclei.

α) Circularly polarized γ-radiation. STEENBERG[1], and TOLHOEK and COX[2] have drawn attention to the fact that the γ-radiation from polarized radioactive nuclei will contain circularly polarized components. The degree of circular polarization can be deduced from the general expression for the intensity of radiation emitted at an angle ϑ and is proportional to

$$\sum_{k \text{ odd}} B_k F_k P_k(\cos\vartheta).$$

As was pointed out earlier B_1 varies as $1/T$ and therefore circular polarization effects should be apparent under lower degrees of orientation than are required for angular distribution effects. It will be seen also that, since $P_1(\cos\vartheta)$ is greatest for $\vartheta = 0$ or π, and zero for $\vartheta = \pi/2$, these effects will be observed most strongly for radiation emitted along the polarization axis. Moreover, since B_1 depends for its sign on the sign of the hfs interaction (i.e. on the sign of A), a measurement of the sense of the circular polarization gives the sign of A and therefore the sign of the nuclear moment.

A successful experiment demonstrating this effect has been described by WHEATLEY *et al.*[3] (see Sect. 16).

β) Electron spin polarization in β-decay. TOLHOEK and DE GROOT[4] have developed expressions for the polarization of the β-particles from polarized radioactive nuclei, for the case of allowed β-transitions and a general β-interaction. These show that in general a spin polarization of the electron will be found. In the particular case of observing the electron polarization in a direction of emission perpendicular to the axis of polarization and for pure axial-vector or tensor interactions these expressions for the polarization reduce to

$$P_1 = \frac{B_1}{E}\sqrt{\frac{(I_0+1)(2I_0+1)}{3I_0}} \quad \text{for} \quad I_0 \xrightarrow{\beta} I_0 - 1,$$

$$P_1 = \frac{B_1}{E}\sqrt{\frac{(2I_0+1)}{3I_0(I_0+1)}} \quad \text{for} \quad I_0 \xrightarrow{\beta} I_0,$$

$$P_1 = \frac{B_1}{E}\sqrt{\frac{I_0(2I_0+1)}{3(I_0+1)}} \quad \text{for} \quad I_0 \xrightarrow{\beta} I_0 + 1,$$

[1] N. R. STEENBERG: Proc. Phys. Soc. Lond. A **66**, 391 (1953).

[2] H. A. TOLHOEK and J. A. M. COX: Physica, Haag **19**, 101 (1953).

[3] See footnote 6, p. 567.

[4] H. A. TOLHOEK and S. R. DE GROOT: Physica, Haag **17**, 81 (1951).

where E is the total energy (rest energy plus kinetic energy of the emitted particles in units of $m_e c^2$).

The electron polarization in this direction is transverse to the direction of emission and is greatest for non-relativistic electrons where $E \sim 1$. In the case of electrons emitted along the axis of polarization the spin polarization is longitudinal. For pure scalar and polar vector interactions no spin polarization will be found.

Electron spin polarization has been produced and detected by a double scattering process using electron beams from high voltage accelerators. Although only single scattering would have to be achieved, the application of this method to polarized nuclei presents formidable difficulties, notably in the prodigious β-ray sources required and in the electron detection within the low temperature cryostat.

γ) *Interactions involving polarized particles* will be discussed in Part D.

15. Effect of extranuclear fields. The influence of extranuclear fields in angular correlation experiments between successive nuclear radiations has been studied both theoretically (ABRAGAM and POUND[1] and others) and experimentally (see FRAUENFELDER[2] for a review of these topics) and many of these effects are now understood. The interaction between the nuclear spin and its environment through magnetic or electrostatic coupling causes the assumption of negligible external influences implicit in the simple theory of angular correlation to break down and for this reason the experimentally observed effect deviates from the theory. In order to explain quantitatively the observed results it is necessary to possess detailed knowledge of the nuclear spin coupling with these external fields for the intermediate nuclear state which becomes perturbed. Only in certain cases is this knowledge sufficient to enable the calculation to be carried through.

It might be expected that similar effects could vitiate the interpretation of nuclear orientation results in cases where the radiation which is observed is one of the lower members of a cascade and might even arise from the orienting field itself.

STEENBERG[3] has considered the influence of a static perturbation on a γ-ray emitting state excited by a previous transition, in the approximation where only the lowest order terms in the expansion in powers of $A/\mathsf{k}T$ are assumed to contribute, i.e. $A/\mathsf{k}T$ is small. Expressions are derived for the parameters $W(M)$ in the case where both the orienting and perturbing interactions are of the form

$$\mathscr{H} = g\beta_B H S_z + A S_z I_z + B(S_x I_x + S_y I_y)$$

and $S = \frac{1}{2}$. It is concluded that so long as $g\beta H \gg A'$ where A' is the value of A for the intermediate state, no perturbation effects will be observed; for reasonable values of the constants this requires $H \sim 10^3$ gauss. Also if B' is zero there will be no disturbance for any value of H. Both these conclusions are valid for all initial degrees of orientation i.e. for all values of $A/\mathsf{k}T$. Where these conditions do not hold, terms of the order $A/\mathsf{k}T$ will be introduced into the expression for $\langle I_z^2 \rangle$ and STEENBERG concludes that nuclear orientation may arise through aniso-

[1] A. ABRAGAM and R. V. POUND: Phys. Rev. **92**, 943 (1953).

[2] H. FRAUENFELDER: Beta and gamma spectroscopy, Chap. 19 (I), Ed. K. SIEGBAHN. North Holland Publishing Co. 1955.

[3] N. R. STEENBERG: Phys. Rev. **95**, 982 (1954).

tropic hfs in the intermediate state even though no orientation arises in the initial state. Such orientation arises through static anisotropic hfs and is not dependent on any relaxation parameters.

Where the values of B/A and B'/A' are the same no first order terms in A/kT occur. Such a situation might be expected to arise immediately after a β^--transition before the electron configuration has had time to pick up another electron and settle down to its new configuration. The terms quadratic in A/kT either enhance or decrease the orientation depending on whether the spin sequence in the preceding transition is increasing $(I_0 \xrightarrow{\beta} I_0 + i_\beta)$ or decreasing $(I_0 \xrightarrow{\beta} I_0 - i_\beta)$.

STEENBERG has also considered the case where $A' = B'$, i.e. the intermediate state has isotropic hfs. The change in the degree of orientation depends on the magnitude of $A'\tau/2\hbar$ where τ is the lifetime of the intermediate state. If this parameter is of the order of 1, then disturbing effects may be observed. Taking a reasonable value for A' ($\sim 10^{-18}$ erg) it is found that for lifetimes longer than about 10^{-10} sec detectable disturbance of the intermediate state may be expected. It is also concluded that such reorientation can only alter the anisotropy of the emitted radiation by about 40%: it cannot destroy it. This is analogous to the "hard core" effect in angular correlation measurements.

Disturbance due to non-static perturbations arising from effects due to recoil, autoionization, K-capture and readjustment of the atomic electrons have not been considered but might well be expected to have serious effects, as in angular correlation measurements where complete destruction has been observed. Evidence for such disturbance has been reported by BISHOP *et al.*[1].

C. Experimental results with oriented nuclei.

Much of the experimental work so far reported has been devoted to the study of γ-radiation from radioactive sources. The earliest successful experiments which were reported by groups at Oxford[2-4] and Leiden[2,5,6] were carried out so as to demonstrate that nuclear orientation could be obtained and both groups worked with Co^{60} included in a paramagnetic crystal (Tutton salt) in which alignment was obtained by the method of magnetic hfs (alignment). In later work more detailed study of γ-ray emitting isotopes of cobalt and other paramagnetic elements of the iron and rare earth groups has been made using the alignment and polarization methods of magnetic hfs. At the same time the polarization of nuclei through the magnetic hfs (polarization) method has been studied by the Oak Ridge group using beams of polarized slow neutrons.

16. Angular distribution and polarization of gamma radiation. α) *Alignment of cobalt 60.* In the experiment described by BLEANEY *et al.*[4] nuclear alignment of radioactive cobalt 60 was obtained by the method of magnetic hfs (alignment). The principle of the method is to choose a crystal which possesses suitable internal

---

[1] G. R. BISHOP, M. A. GRACE, C. E. JOHNSON, A. C. KNIPPER, H. R. LEMMER, J. PEREZ Y JORBA and R. G. SCURLOCK: Phil. Mag. **46**, 951 (1955).

[2] See footnote 2, 3, p. 562.

[3] B. BLEANEY, J. M. DANIELS, M. A. GRACE, H. HALBAN, N. KURTI and F. N. H. ROBINSON: Phys. Rev. **85**, 688 (1952).

[4] B. BLEANEY, J. M. DANIELS, M. A. GRACE, H. HALBAN, N. KURTI, F. N. H. ROBINSON and F. E. SIMON: Proc. Roy. Soc. Lond., Ser. A **221**, 170 (1954).

[5] C. J. GORTER, H. A. TOLHOEK, O. J. POPPEMA, M. J. STEENLAND and J. A. BEUN: Physica, Haag **18**, 135 (1952). — O. J. POPPEMA, J. A. BEUN, M. J. STEENLAND and C. J. GORTER: Physica, Haag **18**, 1235 (1952).

[6] O. J. POPPEMA, M. J. STEENLAND, J. A. BEUN and C. J. GORTER: Physica, Haag **21**, 233 (1955).

fields so that at sufficiently low temperatures the cobalt nuclei become oriented with respect to the crystal axes. Moreover, it is convenient to select a crystal in which other paramagnetic ions can be included whose energy levels are suitable for the attainment of very low temperatures by the magnetic cooling method.

The crystal selected for this experiment was copper rubidium sulphate ($CuSO_4 \cdot Rb_2SO_4 \cdot 6H_2O$) which is a member of the isomorphous series of double sulphates known as Tutton salts. Such a salt possesses suitable internal fields and has the added advantage that either metallic ion can be replaced by other ions of similar ionic dimensions and valency: thus in this case the copper ion may be replaced by any divalent ion of the iron transition group or by zinc. The temperature reached on demagnetization depends on the hfs in the magnetic ions and on the interactions between them. These interactions

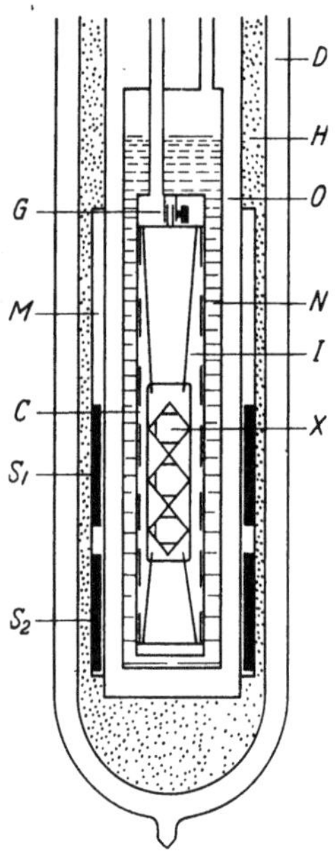

Fig. 2.

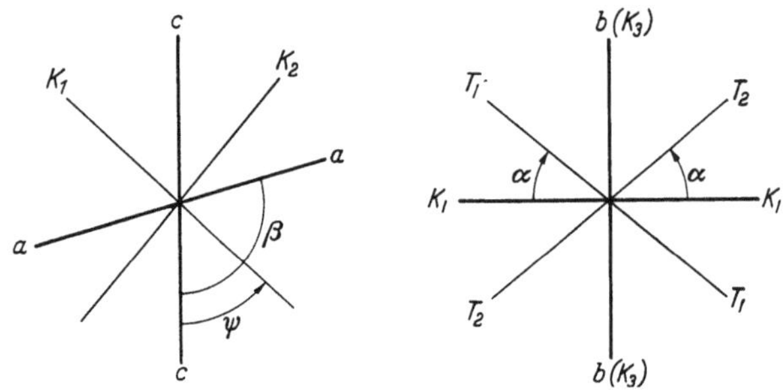

Fig. 3.

Fig. 2. Section through the apparatus. *D*, Pyrex Dewar; *H*, liquid hydrogen; *O*, outer vacuum jacket; *N*, liquid helium; *I*, inner vacuum jacket; *X*, crystals of Tutton salt, supported from cradle *C*, located by grub screw *G*. *M* is a mutual inductance with two secondaries S_1 and S_2.

Fig. 3. Diagram of crystal axes in the Tutton salts.

may be reduced (and hence lower temperatures may be obtained) by magnetic dilution. In the present case approximately 90% of the copper ions were replaced by zinc in order to obtain such magnetic dilution. The crystal was grown from aqueous solution containing cobalt 60 and a small quantity of the isomorphous cobalt replaced some of the copper ions.

Paramagnetic resonance experiments have shown that the ionic spin (S) is effectively $\frac{1}{2}$ and therefore no D terms appear in the spin Hamiltonian which can be represented by

$$\mathcal{H} = A\, S_z\, I_z + B(S_x\, I_x + S_y\, I_y).$$

Also A is very much greater than B and therefore the nuclear magnetic states are approximately equally spaced with a separation $A/2$ and with the state $I_z = \pm I$ lying lowest. At temperatures such that $\mathrm{k}T$ is of the order of $A/2$ considerable alignment should be obtained. For Co^{60} it was apparent from the experiment that $A/2\mathrm{k}$ must be about 0.01° K.

The experimental procedure was therefore to mount the crystal (or several in parallel crystallographic orientation) in the low temperature cryostat and to cool them to about 0.9° K by keeping them in thermal contact with liquid helium boiling under reduced pressure. The section through the low temperature apparatus in Fig. 2 shows the essential features of the equipment including the mutual inductance coils used for determining the magnetic temperature of the crystals. The crystals were then isothermally magnetized in a field of about 20000 gauss, thermally isolated and finally adiabatically demagnetized: the final temperature reached by the crystals was about 0.01° K. The relationship between the temperature (T^*) measured magnetically and the absolute temperature (T) was determined from auxiliary experiments. Counters were then placed round the crystals and cryostat in directions fixed with respect to the axis of alignment and the γ-ray counting rate observed as the crystals slowly returned to 0.9° K where the γ-ray angular distribution was known to be isotropic. The directions of the maxima and minima in the γ-ray angular distribution are along and perpendicular to the axis of alignment. In these crystals there are in each unit cell two positions for the divalent metallic ions whose symmetry (tetragonal) axes do not normally coincide. For cobalt these axes are inclined at an angle of about 74°: in consequence half the cobalt ions become aligned with respect to one of these (T_1) and the remainder with respect to the other (T_2) (see Fig. 3). The most favourable directions for observing the γ-radiation now become the internal bisector K_1 of these two directions and the direction K_2 normal to the $T_1 T_2$ plane. It was along these two directions K_1, K_2 that the counters were mounted.

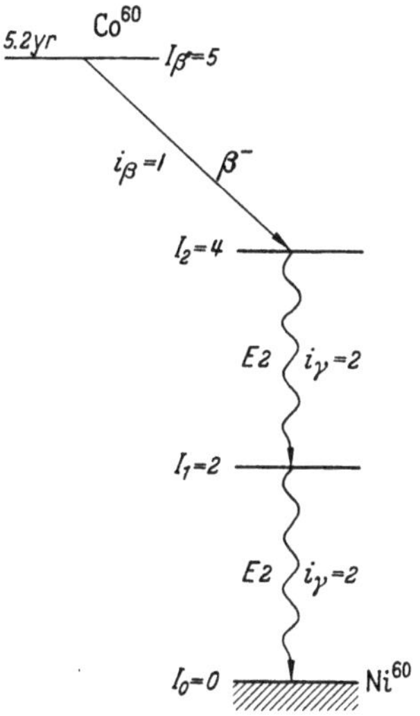

Fig. 4.

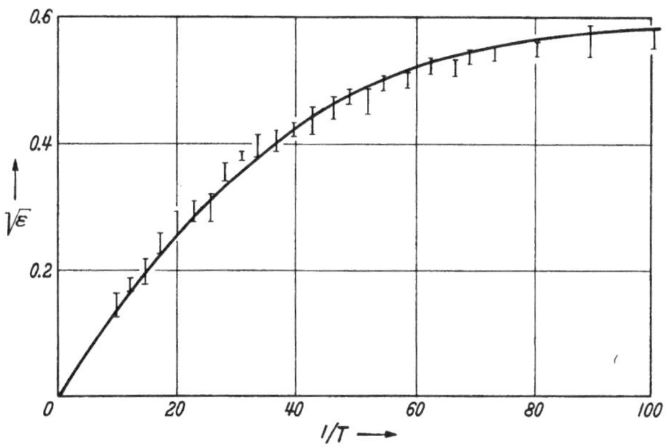

Fig. 5.

Fig. 4. Decay scheme of Co^{60}.

Fig. 5. The dependence of the anisotropy of the γ-radiation from Co^{60} on $1/T$. The solid line represents the theoretical curve for a nuclear magnetic moment for Co^{60} of 3.5 nuclear magnetons.

The well-known decay scheme of Co^{60} (Fig. 4) indicates that the two γ-rays are emitted in cascade following the β-transition: very weak β- and γ-transitions have been omitted in this figure. From angular correlation experiments it is known that these are both pure electric quadrupole and are emitted in the spin sequence 4-2-0. It therefore follows that each of these γ-rays is emitted in a state of angular momentum which is the minimum compatible with the level spins ("stretched configuration"). As a consequence, for a given degree of orientation of the initial state both γ-rays will have the same angular distribution. For this reason it was unnecessary to attempt energy resolution between the two.

In Fig. 5 is shown the variation of $\sqrt{\varepsilon}$ with $1/T$ where ε is the anisotropy of the γ-radiation defined by

$$\varepsilon = \frac{W(\pi/2) - W(0)}{W(\pi/2)}. \tag{16.1}$$

$W(\vartheta)$ is the γ-ray intensity measured in the $K_1 K_2$ plane at an angle ϑ with respect to the K_1 axis. The function of ε used for plotting, namely $\sqrt{\varepsilon}$, was chosen because of the expected quadratic dependence of ε on $1/T$ in the region of small ε. Because of the two possible positions for the cobalt ions in each unit cell the maximum value of $\sqrt{\varepsilon}$ corresponding to complete saturation of the nuclear spins turns out to be $\cos^2 37°$ i.e. 0.64. Values of $\sqrt{\varepsilon}$ up to 0.55 were obtained and their variation with temperature was in satisfactory agreement with theory as is evidenced by the solid curve which was calculated for a value of I_0 of 5 and a nuclear moment of 3.5 n. m. Sufficient accuracy was not available to distinguish between this spin assignment and other less likely choices. The value of the nuclear moment enters through the population function parameters B_k which depend on temperature through the term $\beta = A/2kT$ where A is the hfs parameter for Co^{60}. $A(Co^{60})/A(Co^{59})$ is just the ratio of the nuclear g-factors $g(Co^{60})/g(Co^{59})$. From the fitting of the theoretical curve to the experimental results a value for βT i.e. of $A(Co^{60})$ is found. Thus, using the known values for $g(Co^{59})$ and $A(Co^{59})$ determined by paramagnetic resonance in a similar crystal, a value of $g(Co^{60})$ and hence $\mu(Co^{60})$ was obtained. Errors and uncertainties arising from γ-ray scattering in the cryostat walls, the difference between magnetic and absolute temperatures and failure to fit the experimental points exactly by a theoretical curve cause the authors to attribute a possible error of $\pm 12\%$ to the value for the nuclear moment of 3.5 n. m. Corrections were applied to take account of the small but known value of B in the spin Hamiltonian which causes the levels both to deviate from complete equality of level spacing and to be mixed with neighbouring nuclear magnetic states. It is interesting to note that a recent value of $\mu = (3.800 \pm 0.007)$ n. m. determined by paramagnetic resonance[1] is in good accord with these results and gives confidence that their interpretation is essentially correct. This experiment has been described in some detail since it forms a pattern for many similar experiments in which nuclear alignment in a paramagnetic ion is produced through interaction with the crystalline field.

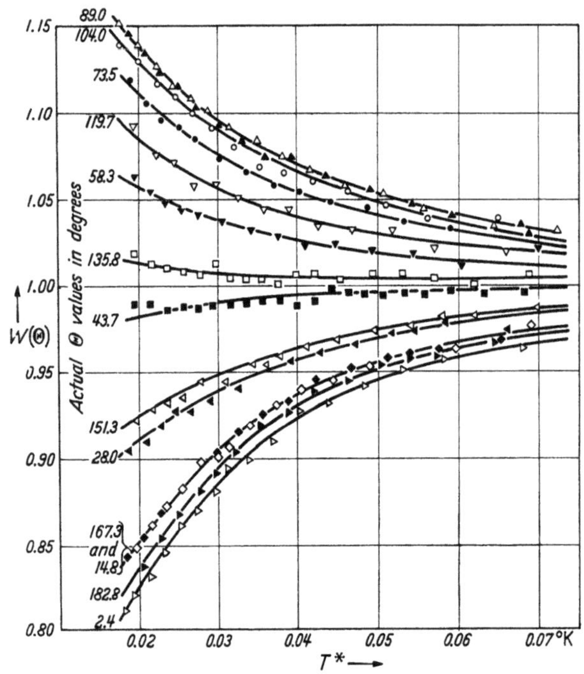

Fig. 6. Normalized intensity $W(\Theta)$ of γ-radiation from Co^{60} as a function of T^*, for different values of Θ. Open points refer to one counter and solid points to another. Again in this crystal there are two axes of alignment. The angle Θ in this figure is different from that used in the text; it is the angle between the direction of emission of the γ-ray and the internal bisector of the angle between these two axes of alignment (K_1 axis). Thus it is equal to $\cos^{-1}(\cos\vartheta/\cos 33°)$.

A rather similar series of experiments was reported by the Leiden group at about the same time. Single crystals of $CoSO_4 \cdot (NH_4)_2SO_4 \cdot 6H_2O$ were used. Two types of experiment were performed[2]. In one the sample was demagnetized

[1] W. Dobrowolski, R. V. Jones and C. D. Jeffries: Phys. Rev. **101**, 1001 (1956).

[2] See footnote 6, p. 575.

from a known value of H/T using stable cobalt as cooling agent and the Boltzmann factor β of the cobalt 59 calculated from the known amount of entropy removed: (the sample was screened from heat leakage by embedding it in chrome alum). β for the cobalt 60 was deduced from ε for the γ radiation. Comparison of the two values of β gave an approximate value of μ for cobalt 60 of 3.9 nuclear magnetons. In the second case the sample in the absence of chrome alum was demagnetized using copper as cooling agent and the dependence of $W(\vartheta)$ on the magnetic temperature T^* found for a number of values of ϑ (see Fig. 6). Comparison of this with the theoretical dependence of $W(\vartheta)$ on β in the region ($\sim 0.07°$ K) where the T/T^* correction is small (1.5%) gave a value for μ (Co^{60}) of (4.3 ± 0.2) nuclear magnetons.

Experiments by BISHOP *et al.*[1] using the same type of crystal containing Co^{60} have confirmed that the γ-radiation is plane polarized and in such a sense that it can be inferred that the quadrupole transitions are both electric. The highest degree of plane polarization is to be expected for γ-radiation emitted in directions perpendicular to the axis of orientation. In the case of this biaxial crystal the most favourable direction is that of the external bisector (K_3) of these two axes. Their method was to scatter the γ-radiation emitted in this direction in the organic scintillator of a scintillation counter (see Fig. 7). The intensity of the γ-radiation scattered in the plane normal to the direction of propagation of the incident γ-ray was then observed in directions parallel and perpendicular to the axis of alignment. Reliance is here placed on the azimuthal dependence of the Compton scattering cross-section with the direction of the electric vector of the primary radiation. This same method has been described in detail by DEUTSCH[2] in connection with polarization-angular correlation experiments. The degree of polarization is related to the intensities $N_\parallel$ and $N_\perp$ by

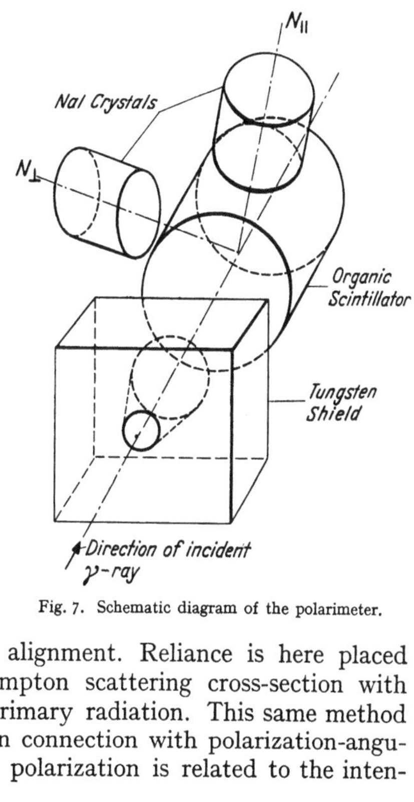

Fig. 7. Schematic diagram of the polarimeter.

$$P = \frac{\left(1 - \frac{N_\perp}{N_\parallel}\right)(R+1)}{\left(1 + \frac{N_\perp}{N_\parallel}\right)(R-1)} \tag{16.2}$$

where P is defined by

$$P = \frac{I_\perp - I_\parallel}{I_\perp + I_\parallel},$$

$I_\perp$, $I_\parallel$ being the intensities of the radiation emitted with the electric vector perpendicular to (σ radiation) or parallel to (π radiation) the alignment axis: R is the analysing ratio of the polarimeter and depends on the geometry and the γ-ray energy. In Fig. 8 is shown the dependence of $N_\parallel/N_\perp$ on $1/T$. That this ratio is greater than 1 shows that the radiation is preferentially of σ type.

[1] G. R. BISHOP, J. M. DANIELS, G. GOLDSCHMIDT, H. HALBAN, N. KURTI and F. N. H. ROBINSON: Phys. Rev. **88**, 1432 (1952).

[2] F. METZGER and M. DEUTSCH: Phys. Rev. **78**, 551 (1950).

Combining this with the knowledge that the transitions are quadrupole leads to the conclusion that the radiations are electric multipoles.

β) Polarization of cobalt 60. The earliest experiment in which nuclear polarization has been produced was reported by Ambler *et al.*[1]. In this case nuclear polarization was obtained by the method of magnetic hfs (polarization) (method *b* of Sect. 7). Two important difficulties arose in the design of this experiment. Firstly, as has been pointed out in Sect. 7, nuclear alignment can arise through the A, B terms alone in zero field. Therefore, in order to demonstrate that nuclear "polarization" has arisen through the application of the magnetic field it is necessary either to use a detection method which is sensitive to nuclear "polarization" as distinct from "alignment" (e. g. circular polarization of γ-rays or neutron polarization experiments); or to arrange that anisotropic γ-ray effects are not present in the absence of an external field. One way of achieving the latter situation would be to employ a powdered specimen in which case alignment through the crystalline field for the specimen as a whole will give zero effect. Another way would be to choose a crystal in which the magnetic ions show isotropic hfs (i. e. $A = B$) and for which $S = \frac{1}{2}$: in this case there will be no orientation in zero field. Ambler *et al.* chose the latter method and used a single crystal of cerium magnesium nitrate, $Ce_2(NO_3)_6 \cdot Mg_3(NO_3)_6 \cdot 24\,H_2O$. Some of the divalent magnesium ions were replaced by Co^{60}. Paramagnetic resonance experiments had shown that for two of the three possible positions available for cobalt ions in each unit cell A and B are approximately equal. Since for these ions $S = \frac{1}{2}$, in the absence of a polarizing field, the amount of nuclear orientation should be very small.

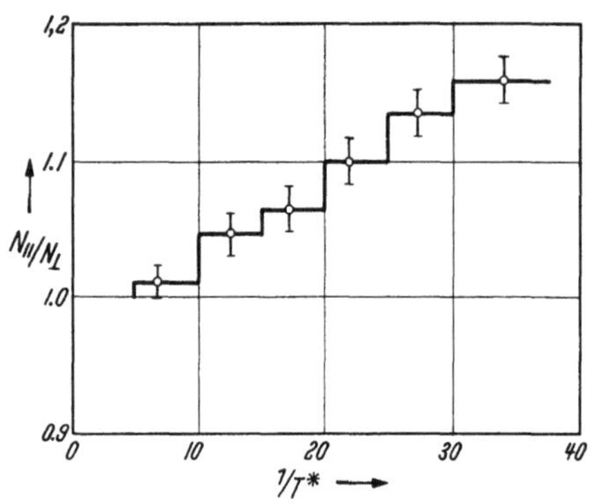

Fig. 8. Polarization of γ-rays from Co^{60}: the dependence of the ratio of the counting rates $N_{\parallel}/N_{\perp}$ on $1/T^*$.

The second difficulty lies in cooling the crystal. If the cooling agent is included in the crystalline material in which the nuclear polarization is to be produced, then in general the application of the polarizing magnetic field will raise the temperature of the system and the hfs level population will be approximately equalized: thus no anisotropic effects will be observed. This difficulty can be avoided if the cooling agent is separated from the sample by a thermal link (see, for instance, Roberts *et al.*[2]). The cooling agent may then be magnetically isolated from the specimen and the low temperature may be maintained in spite of the presence of the magnetic field. Another way of avoiding this difficulty makes use of the highly anisotropic susceptibility of certain paramagnetic crystals: this was the method selected by Ambler *et al.* In cerium magnesium nitrate the g factor for the cerium ions is highly anisotropic: $g_{\perp}$ is large and $g_{\parallel}$ very small. In their experiment cerium was the principal magnetic constituent of the crystals and was therefore the principal cooling agent. The crystals may be cooled by demagnetizing along an axis of easy magnetization ($g_{\perp}$) and the cobalt ions may then be polarized by applying the polarizing field along the symmetry axis ($g_{\parallel}$). Because $g_{\parallel}$ is small for the cerium ions negligible rise in temperature will take place and yet the cobalt nuclei will become polarized.

[1] See footnote 4, p. 561.
[2] See footnote 5, p. 567.

On application of a polarizing field of 500 gauss values of ε up to 0.5 were obtained whereas in zero field ε did not exceed 0.15. Part of this effect in zero field at least could be attributed to those cobalt ions (one third of the total) for which A is greater than B. It was concluded therefore that the application of the magnetic field produced nuclear orientation and, because of the vector character of the interaction, this must have resulted in nuclear polarization although no experiment which distinguished between alignment and polarization was carried out (see, however, Sect. 16δ).

γ) *Angular correlation of successive γ-rays from oriented Co^{60}.* The theory of this process has been investigated by COX and TOLHOEK[1] who have shown that the angular correlation depends not only on the characteristics of the individual radiations and the spin sequence, but also on the degree of orientation of the parent state and the directions of emission of the two radiations in relation to the axis of orientation. The correlation parameters depend in more detail on the orientation parameters than do those of the angular distribution of the two radiations taken separately and therefore this process is in principle a source of information unobtainable from straightforward angular distribution measurements. However, rather high degrees of orientation are required before these benefits become apparent and in the only case hitherto investigated, that of the γ-rays from Co^{60} in cerium magnesium nitrate studied by JASTRAM, SAPP and DAUNT[2], no additional information was obtained. Their experiment consisted of observing the dependence of the angular correlation on the anisotropy of the two γ-rays. This was found to be linear in the orientation region which they studied and in agreement with theoretical prediction, thus affording further support for the theory of these angular distributions. The conclusions from these experiments are that such angular correlation measurements are feasible at low temperatures although so far it has not been possible to obtain the high degree of orientation and accuracy in the angular measurements necessary to yield new information on nuclear or crystalline structure.

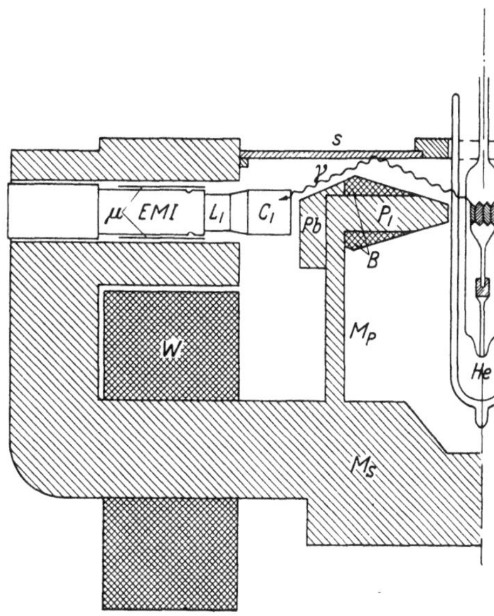

Fig. 9. Schematic diagram of the apparatus for producing and measuring circularly polarized γ-rays. W and B are the magnetizing coils for, respectively, the saturating magnet M_s and the polarizing magnet M_p. The γ-ray detectors consist of NaI (Tl) crystal C, light pipe L, and photomultiplier EMI. S is the scattering Armco-iron.

δ) *Circularly polarized γ-radiation from polarized Co^{60}.* An experiment demonstrating the emission of circularly polarized γ-radiation has been described by WHEATLEY *et al.*[3] In their experiment they utilized the strong magnetic anisotropy of cerium magnesium nitrate (AMBLER *et al.*[4]) to enable them to cool the

[1] J. A. M. COX and H. A. TOLHOEK: Physica, Haag **18**, 359 (1952). — J. A. M. COX: Physica, Haag **18**, 1262 (1952).
[2] P. S. JASTRAM, R. C. SAPP and J. G. DAUNT: Phys. Rev. **101**, 1381 (1956).
[3] See footnote 6, p. 567.
[4] See footnote 4, p. 561.

crystal by demagnetizing along the $g_{\perp}$ direction for cerium and then applying the polarizing magnetic field along the $g_{\parallel}$ direction which caused negligible rise in temperature of the crystal. In this way Co^{60} in the crystal was polarized with a field of about 280 gauss. The γ-radiation emitted along or close to the direction of the polarizing field was then scattered in a sheet of iron which was magnetically polarized in the direction parallel or anti-parallel to the direction of nuclear polarization. The scattered γ-radiation was then detected in a scintillation counter. The experimental arrangement is indicated in Fig. 9 which shows one of the scintillation counters (L_1 C_1) and the scattering plate (S) which is polarized through the iron yoke and the coil (W). P_1 is one of the poles of the magnet producing the field for polarizing the radioactive crystal which also acts as a screen against direct radiation. Now the differential Compton scattering cross-section is given by

$$\frac{d\sigma}{d\omega} = \left(\frac{d\sigma}{d\omega}\right)_{\text{unpolarized}} + \left(\frac{d\sigma}{d\omega}\right)_{\text{plane polarized}} + \left(\frac{d\sigma}{d\omega}\right)_{\text{circularly polarized}}.$$

The contribution to the cross-section which is sensitive to circular polarization depends on the proportion of electrons in the iron which are polarized by the magnetic field: for iron this could be as high as 0.08 and in this experiment reached a value of 0.06. This cross-section also depends on the relative orientations of the spin vector of the electrons and the circular polarization vector of the γ-radiation. Thus reversal of either the iron polarizing field or the crystal polarizing field alters the sign of this term. For forward scattered radiation, however, the first term is much greater than the others and therefore only a small change in scattered intensity is to be expected. An effect of about 3% was observed on reversal of the relative directions of the two polarizing fields and its change with temperature agreed satisfactorily with the behaviour to be expected from angular distribution measurements of the γ-radiation. They were able to conclude that circularly polarized γ-radiation from the polarized nuclei had been detected and that the sense of this polarization was in agreement with a positive sign for the nuclear moment. This sign is in accord with predictions of the shell model.

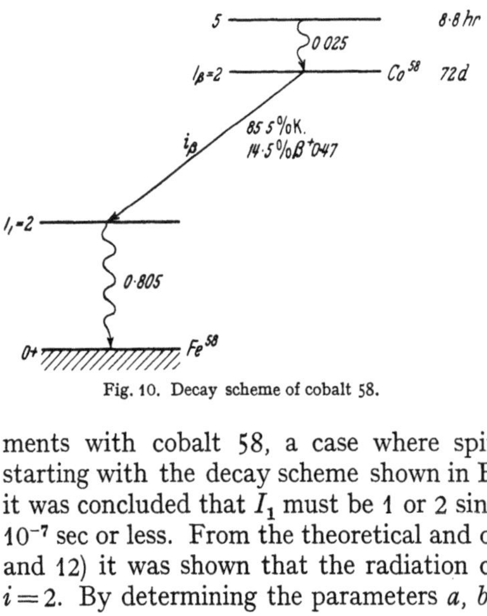

Fig. 10. Decay scheme of cobalt 58.

17. Summary of radioactive nuclei studied by nuclear orientation. α) *Cobalt 58.* Using their earlier method DANIELS *et al.*[1] have carried out nuclear alignment experiments with cobalt 58, a case where spin assignments were uncertain. Thus, starting with the decay scheme shown in Fig. 10 since $I_2 = 0$ (even-even nucleus), it was concluded that I_1 must be 1 or 2 since the lifetime of the excited state was 10^{-7} sec or less. From the theoretical and observed angular distributions (Figs. 11 and 12) it was shown that the radiation could not be dipole and therefore $I_1 = i = 2$. By determining the parameters a, b, of the angular distribution

$$W(\vartheta) = 1 - a\cos^2\vartheta - b\cos^4\vartheta$$

[1] J. M. DANIELS, M. A. GRACE, H. HALBAN, N. KURTI and F. N. H. ROBINSON: Phil. Mag. **43**, 1297 (1952).

it was shown from the dependence of b on temperature that the most satisfactory agreement was obtained for $I_0 = 2$ and $i_\beta = 1$. If no parity change occurred in the β-transition (this was suggested by the comparative half-life of the transition) it was concluded that an axial-vector or tensor interaction was predominant. A value for μ of (3.5 ± 0.3) nuclear magnetons was found.

Measurements of the plane of polarization of this γ-radiation by BISHOP *et al.*[1] showed that this transition is electric in character and therefore the excited state of iron 58 has even parity.

An essentially similar experiment has been reported by WHEATLEY *et al.*[2] who studied cobalt 58 and cobalt 60 in the same crystal.

β) Cobalt 56. This isotope decays by K-capture and β^+ emission with the excitation of a complex γ-ray spectrum which is shown in the decay scheme of Fig. 13. This was investigated by GALLAHER *et al.*[3] and, as in other studies on cobalt isotopes, it was included in a Tutton salt of similar constitution to that used by BLEANEY *et al.*[4] (see Sect. 16 α). It was cooled by adiabatic demagnetization of the Cu ions. The angular distribution of the γ-radiation was deduced from measurements of the pulse height distributions in a pair of scintillation counters mounted along the K_1 and K_2 axes. The maximum observed values of ε are given in table 2:

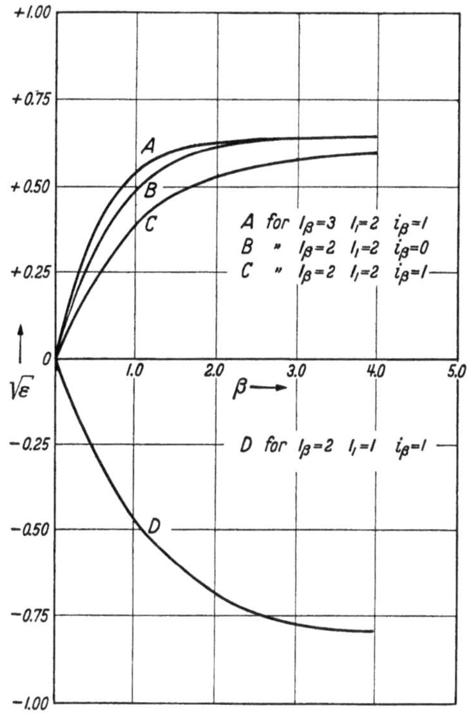

Fig. 11. Variation of anisotropy with $\beta\,(=A/2kT)$ for cobalt 58 (theory).

Table 2. *The γ-ray anisotropies and multipole assignments for Co[56].*

γ-radiation (MeV)	ε	Spin of parent state	Multipolarity
0.845	$+0.10 \pm 0.01$	2^+	$E\,2$
1.24	$+0.09 \pm 0.02$	4^+	$E\,2$
1.75	-0.07 ± 0.02	$3^\pm$, $5^\pm$	
2.30	-0.05	3^+, 4^+	
2.60	-0.15 ± 0.06	3^+, 3^-	
3.25	$+0.12 \pm 0.02$	4^+	$E\,2$

From the temperature dependence of ε for the 0.845 Mev γ-ray in the region where $1/T^*$ is small (and therefore presumably closely equal to $1/T$), a value for

[1] See footnote 1, p. 579.
[2] J. C. WHEATLEY, D. F. GRIFFING and R. D. HILL: Phys. Rev. **99**, 334 (1955).
[3] L. J. GALLAHER, CH. WHITTLE, J. A. BEUN, A. N. DIDDENS, C. J. GORTER and M. J. STEENLAND: Physica, Haag **21**, 117 (1955). — O. J. POPPEMA, J. G. SIEKMAN and R. VAN WAGENINGEN: Physica, Haag **21**, 223 (1955).
[4] See footnote 4, p. 575.

μ of (2.8 ± 0.9) n. m. was obtained assuming that the spin I_0 is 4. In deducing this value it was necessary to make certain assumptions about the absence of

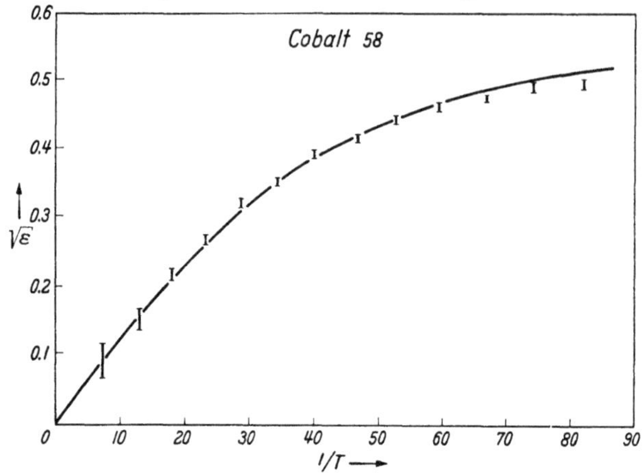

Fig. 12. Variation of anisotropy with temperature showing experimental points and theoretical curve for $I_\beta = 2$, $I_1 = 2$, $i_\beta = 1$ fitted at $1/T = 52$.

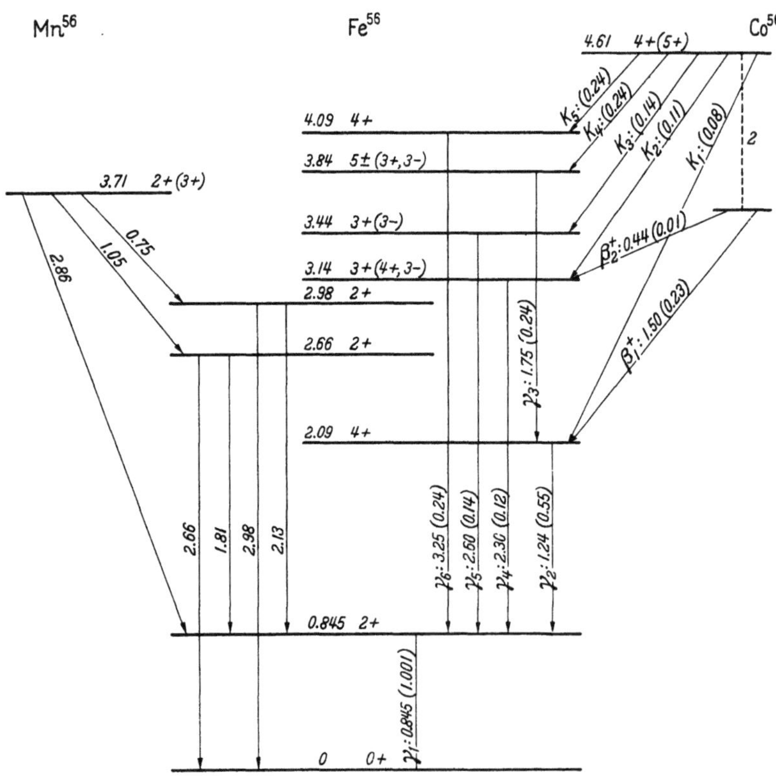

Fig. 13. Level scheme of Fe^{56}. The numbers in the figure indicate: the energies (in MeV) and proposed spins and parities (in decreasing order of probability) for the levels; the energies (in MeV) for the β- and γ-transitions in the decay of Co^{56} and Mn^{56}; the relative intensities for all transitions in the decay of Co^{56} only.

influence of mixed γ-ray transitions leading to the excitation of the 0.845 MeV state and absence of mixtures in the β transitions which were assumed to involve minimum spin changes. This value of nuclear moment is somewhat smaller than that determined by JONES *et al.*[1] of 3.855 n. m. (see also BAKER *et al.*[2]) using paramagnetic resonance methods. These authors confirm that the value of I_0 is 4 as has been assumed above.

By combining the measurements of the anisotropies with other observations on the decay scheme it was possible to restrict the choices of spins of the excited states. Their conclusions on the level spins and the multipole character of the γ-radiations are shown in table 2.

γ) Cobalt 57. The general features of the decay scheme of Co^{57} are shown in Fig. 14. This isotope was studied in the copper Tutton salt by BISHOP *et al.*[3] who showed that although the anisotropy of the γ-radiation, due almost entirely to γ_2, did not exceed 0.004 there appeared to be a significant polarization of about 10% which varied with temperature in the expected way for $M1$ radiation. From the known value for the nuclear moment of Co^{57} large values both for ε ($\sim$0.5) and for the polarization are to be expected if the γ-rays are pure and the parent γ-ray emitting state is unperturbed. These authors conclude that the γ-ray was mixed with $\delta(E2/M1) = +0.19 \pm 0.02$ and also that the parent state was perturbed with an attentuation factor λ of 0.11 where λ is defined by the relation

$$W(\vartheta) = 1 + \lambda B_2 U_2 F_2 P_2(\cos\vartheta).$$

This perturbation seems to occur within a lifetime of the order of 5×10^{-9} sec.

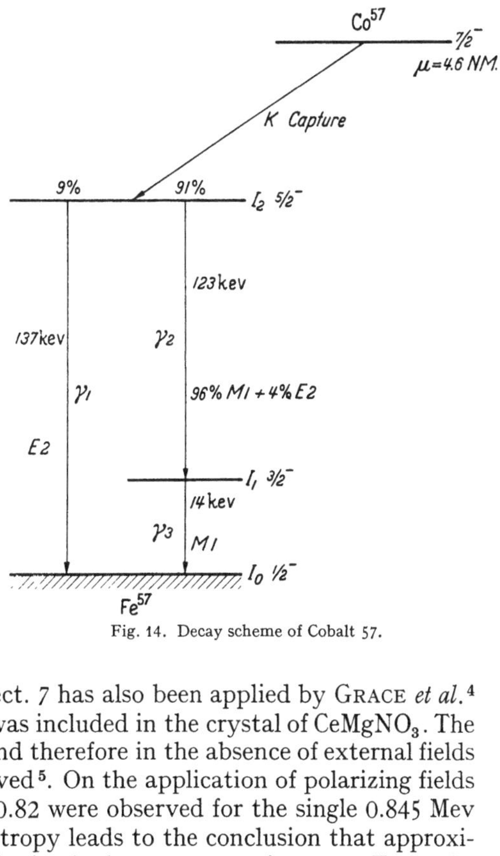

Fig. 14. Decay scheme of Cobalt 57.

δ) Manganese 54. Method (b) of Sect. 7 has also been applied by GRACE *et al.*[4] to the polarization of Mn^{54}. The Mn^{54} was included in the crystal of $CeMgNO_3$. The hfs of Mn in this crystal is isotropic and therefore in the absence of external fields no anisotropic effects should be observed[5]. On the application of polarizing fields of about 1000 gauss values of ε up to 0.82 were observed for the single 0.845 Mev γ-ray (see Fig. 15). Such a large anisotropy leads to the conclusion that approximately 90% of the Mn^{54} nuclei must be in the lowest magnetic state. From the sign of the anisotropy and from polarization measurements[6] on the γ-rays it was concluded that the transition must be $E2$. Moreover, since the residual nucleus is of the even-even type, its ground state must be 0^+ and therefore the 0.845 Mev state 2^+.

[1] R. V. JONES, W. DOBROWOLSKI and C. D. JEFFRIES: Phys. Rev. **102**, 738 (1956).
[2] J. M. BAKER, B. BLEANEY, P. M. LLEWELLYN and P. F. D. SHAW: Proc. Phys. Soc. Lond. A **69**, 353 (1956).
[3] See footnote 1, p. 575.
[4] M. A. GRACE, C. E. JOHNSON, N. KURTI, H. R. LEMMER and F. N. H. ROBINSON: Phil. Mag. **45**, 1192 (1954).
[5] In fact, in the absence of external magnetic fields a small anisotropy of the γ-radiation arises from residual interactions.
[6] G. R. BISHOP, J. M. DANIELS, H. DURAND, C. E. JOHNSON and J. PEREZ Y JORBA: Phil. Mag. **45**, 1192 (1954).

From the large anisotropy it is clear that very little loss of orientation occurs in the K-capture transition and it is therefore concluded that the angular momentum change in this transition is minimal, i.e. the transition is stretched; if $I_0=2$ the transition is of the form $i_\beta=0$ and is therefore of the allowed Fermi type and if $I_0=3$ the transition involves $i_\beta=1$ and is of the allowed Gamow-Teller type. In the former case the admixture of tensor or axial-vector interactions in the β-transition is negligibly small ($<10\%$).

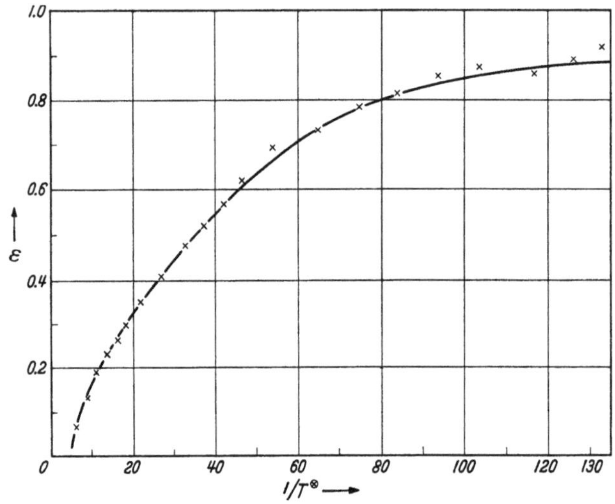

Fig. 15. Variation of anisotropy ε with $1/T^*$ for Mn^{54} in an external field of 1000 gauss. A correction for the finite aperture of the counters has been made.

ε) Cerium 141. Two sets of measurements on Ce^{141} have been reported, the earliest by AMBLER, HUDSON and TEMMER[1,2] using cerium magnesium nitrate and the second set by CACHO *et al.*[3] using neodymium ethyl sulphate. In the first case alignment in zero field is obtained by method (d) of Sect. 7. In this crystal B is greater than A and the nuclei become aligned in a plane perpendicular to the symmetry axis of the crystal, i.e. the nuclear state with $I_z=\pm\frac{1}{2}$ lies lowest

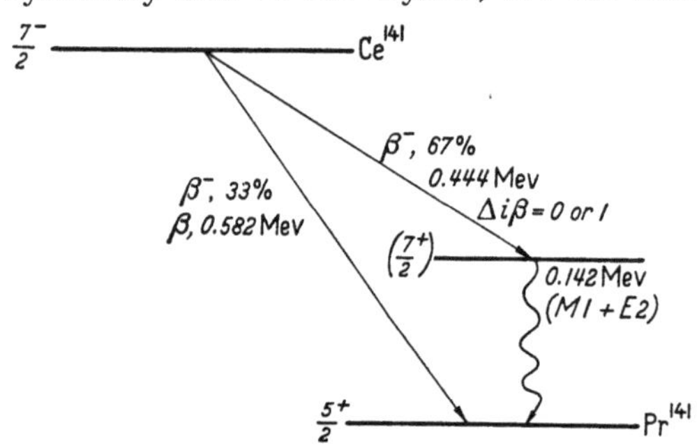

Fig. 16. Decay scheme of Ce^{141}.

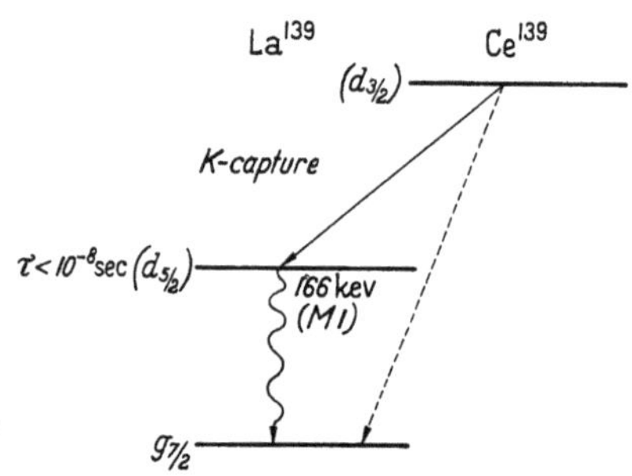

Fig. 17. Decay scheme of Ce^{139}.

and becomes preferentially populated. The decay scheme of Ce^{141} is shown in Fig. 16. The maximum value of ε was found to be about $+0.12$. This led the authors to deduce the spin assignment shown in the figure and that the radiation was principally $M1$. It was also shown that the nuclear moment is (0.16 ± 0.06) n.m.

In the experiments of CACHO *et al.*[3] a crystal of cerium ethyl sulphate was used and "axial" alignment was obtained, that is to say, the state with $I_z=\pm I$ lies lowest and at low temperatures becomes preferentially populated. Anisotropies down to -0.18 were obtained. Measurements on the plane of polarization of the

[1] E. AMBLER, R. P. HUDSON and G. M. TEMMER: Phys. Rev. **97**, 1212 (1955).
[2] E. AMBLER, R. P. HUDSON and G. M. TEMMER: Phys. Rev. **101**, 196 (1956).
[3] C. F. M. CACHO, M. A. GRACE, C. E. JOHNSON, A. C. KNIPPER, R. G. SCURLOCK and R. T. TAYLOR: Phil. Mag. **46**, 1237 (1955).

γ-radiation confirmed the magnetic character of the dipole transition and also indicated a small admixture of $E2$ radiation $\delta(E2/M1) = +0.08 \pm 0.02$. Values of the nuclear moment of (0.75 ± 0.2) n.m. or (0.66 ± 0.16) n.m. were inferred according as i_β is 1 or 0 respectively. The origin of the discrepancy between the value for the nuclear moment determined using the double nitrate salt and that using the ethyl sulphate is unknown but presumably arises either through perturbation of the excited residual nucleus during its finite lifetime $(1.9 \times 10^{-9}$ sec) in the cerium magnesium nitrate crystal or through the presence of additional interactions not included in the spin Hamiltonian. Since almost any interaction tends to reduce anisotropic effects, this implies that the nuclear moments which have been reported must represent lower limits to the true moment of Ce^{141}.

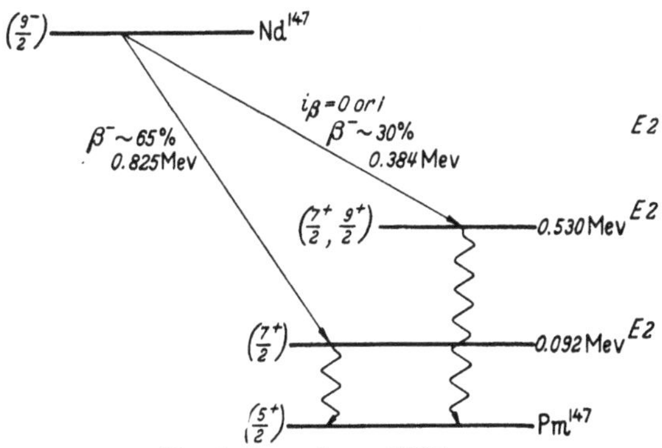

Fig. 18. Decay scheme of Nd^{147}.

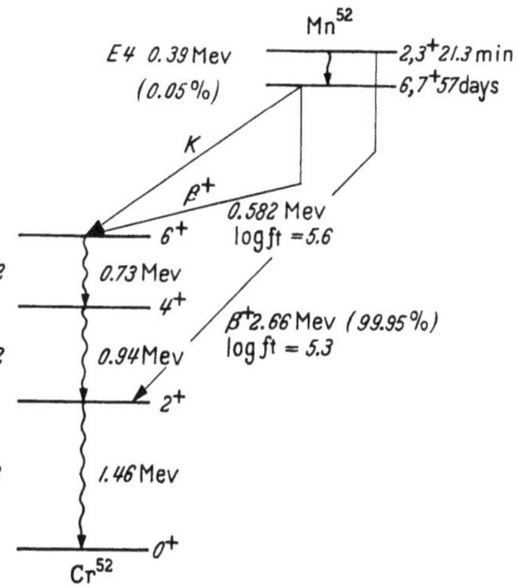

Fig. 19. Decay scheme of Mn^{52}.

ζ) Cerium 139. Ce^{139} has been studied by AMBLER *et al.*[1] also in cerium magnesium nitrate. Their measurements of the γ-ray anisotropy (maximum about 0.10) lead to the spin assignments of Fig. 17 and to the conclusion that an admixture of $E2$ radiation is present in the predominantly $M1$ transition with a value of $\delta(E2/M1)$ of about $+0.2$. This value was inferred from an assumed value for the nuclear moment of Ce^{139} of 0.75 nuclear magnetons.

η) Neodymium 147. AMBLER *et al.*[1,2] using Nd^{147} in cerium magnesium nitrate have studied the angular distributions of the 92 and 530 keV γ-rays excited in Pm^{147} by the decay of this nucleus (see Fig. 18). As in the case of the cerium isotopes alignment in zero field by method (d) of Sect. 7 is obtained with B greater than A which causes planar alignment. No detectable anisotropy of the 92 keV γ-ray was found while values of ε down to -0.39 were found for the 530 keV γ-ray. These authors conclude that the lack of anisotropy of the 92 keV γ-ray originates in an $M1/E2$ mixture in which $\delta(E2/M1) = \pm 0.17$. That of the 530 keV γ-ray leads to the assignment of spins shown in Fig. 18 and to the identification of the γ-ray as $E2$. The value of the nuclear moment of Nd^{147} from this data is (0.22 ± 0.05) nuclear magnetons.

ϑ) Manganese 52. HUISKAMP *et al.*[3] have polarized 5.7 day Mn^{52} in cerium magnesium nitrate using magnetic fields up to 1000 gauss. Their technique was similar to that employed in the polarization of Mn^{54} by GRACE *et al.*[4] described

[1] See footnote 2, p. 586.
[2] See footnote 1, p. 586.
[3] W. J. HUISKAMP, M. J. STEENLAND, A. R. MIEDEMA, H. A. TOLHOEK and C. J. GORTER: Physica, Haag **22**, 587 (1956).
[4] See footnote 4, p. 585.

in Sect. 17δ. The decay scheme of this isotope (Fig. 19) shows that there are three γ-rays in cascade. The anisotropies of these observed at the lowest temperature with a magnetic field of 1000 gauss all lay within the limits 0.42 ± 0.05. From this it was concluded that the γ-ray transitions were all "stretched" and quadrupole and hence spins 0, 2, 4, 6 were assigned to the levels. From other evidence on the decay scheme even parity is assigned to the levels which lead to the identification of the γ-rays as electric quadrupole.

D. Polarized nuclear reactions.

18. Introduction. In the following sections we shall be concerned with nuclear reactions and scattering processes in which one or more components are oriented. The treatment is confined to low energy processes, high energy processes being more suitably discussed in conjunction with nucleon-nucleon scattering phenomena. In order to introduce the nomenclature a short discussion is given first of the theory of nuclear reactions and scattering processes in which no account is taken of the possibility of producing or detecting oriented nuclear spins. After this, consideration is given to nuclear processes in which the components may be polarized (as distinct from aligned or any other higher degree of orientation). Following the theoretical discussion an account is given of the experimental work which has been carried out to date on low energy nuclear processes in which polarization is produced or measured.

19. Unpolarized components. Consider a nuclear reaction[1] denoted symbolically by

$$a + \mathrm{X} \rightarrow \mathrm{Y} + b$$

in which particle a collides with nucleus X and particle b emerges at an angle ϑ (in the centre of mass system) to the direction of the incident beam and Y is the residual nucleus.

The system before collision is described by three quantities, the channel index α which defines the type of incoming particle (neutron, proton, etc.) as well as the state of the target nucleus, the channel spin S and the orbital angular momentum l. The channel spin S is the total spin angular momentum in the entrance channel and is formed by the vector addition of the intrinsic spin i of the incoming particle and the spin I of the target nucleus. It can take values from $I+i, \ldots |I-i|$. The state of the system after the reaction will be denoted by primed quantities.

The properties of such a reaction are most conveniently described in terms of the elements $S^{J\pi}_{\alpha' S' l'; \alpha S l}$[2,3] of the scattering matrix which are related to the probability that a colliding system having a total angular momentum J and parity π should go from an incident state described by the quantities $\alpha S l$ to the final state $\alpha' S' l'$. The "good" quantum numbers for this reaction are J and π. (For pure elastic scattering S is related to the phase shift by the relation $S = e^{2i\eta}$.)

The total cross section for the reaction between well defined initial and final states α and α' can then be expressed in terms of the $S^{J\pi}_{\alpha' S' l'; \alpha S l}$ as follows[2]

$$\sigma_0(\alpha'\alpha) = \frac{4\pi \lambda\!\!\!^{-2}_\alpha}{(2I+1)(2i+1)} \sum_{\pi=\pm 1} \sum_J \sum_S \sum_l \sum_{S'} \sum_{l'} |\delta_{\alpha'\alpha}\delta_{S'S}\delta_{l'l} - S^{J\pi}_{\alpha' S' l'; \alpha S l}|^2 \quad (19.1)$$

[1] The term nuclear reaction is taken to include the special case of elastic scattering.
[2] J. M. BLATT and L. C. BIEDENHARN: Rev. Mod. Phys. **24**, 258 (1952).
[3] J. M. BLATT and V. F. WEISSKOPF: Theoretical Nuclear Physics. New York: John Wiley & Sons 1952.

where λ_α is the de Broglie wavelength for the incident particles and δ is the Kronecker delta. Clearly, if many channels and angular momenta contribute to the reaction, then little information is available about individual elements of the scattering matrix.

On the other hand, if it is possible to measure the angular distribution of the reaction products, then more information may be obtained since the theoretical expression for the angular distribution involves the S matrix elements in different combinations from those in σ_0. Thus, theoretically[1]

$$d\sigma_0(\alpha'\alpha) = \frac{\lambda_\alpha^2}{(2I+1)(2i+1)} \sum_{L=0}^{\infty} B_L(\alpha'\alpha)\, P_L(\cos\vartheta)\, d\Omega \tag{19.2}$$

where

$$B_L(\alpha'\alpha) = \sum_{S'S} \frac{(-1)^{S'-S}}{4} \sum_{J_1 J_2 l_1 l_2 l_1' l_2' \pi_1 \pi_2} Z(l_1 J_1 l_2 J_2;\, SL) \cdot Z(l_1' J_1 l_2' J_2;\, S'L) \times$$
$$\times \operatorname{Re}\left[(\delta_{\alpha'\alpha}\delta_{S'S}\delta_{l_1' l_1} - S^{J_1\pi_1}_{\alpha' S' l_1';\alpha S l_1})^* (\delta_{\alpha'\alpha}\delta_{S'S}\delta_{l_2' l_2} - S^{J_2\pi_2}_{\alpha' S' l_2';\alpha S l_2})\right].$$

Here Re stands for real part and the Z coefficients are defined by

$$Z(abcd;\, ef) = i^{f-a+c}[(2a+1)(2b+1)(2c+1)(2d+1)]^{\frac{1}{2}} W(abcd;\, ef)\cdot C(acf;\, 00)$$

where $W(abcd;\, ef)$ is a Racah coefficient[2] and $C(acf;\, 00)$ is a Clebsch-Gordon coefficient. $P_L(\cos\vartheta)$ is a normalized Legendre function.

As an example one might consider a nuclear reaction (not elastic scattering) in which only two of the S matrix elements S_1 and S_2 are non-vanishing. In this case from a measurement of σ_0 one would obtain the magnitude of an expression of the form $a|S_1|^2 + b|S_2|^2$ whilst from a measurement on the angular distribution it may be possible to deduce the value of combinations of $S_1^* S_2 + S_2^* S_1$, $|S_1|^2$ and $|S_2|^2$ and hence to obtain more detailed information about S_1 and S_2 although it may still not be possible to determine them exactly.

The expression for the total and differential cross sections given so far have been deduced on the assumption that the initial components of the reaction have their spins oriented at random and that the detectors of the final products are insensitive to the different spin orientations. However, if these restrictions are removed then even more information may be obtained about the S matrix elements. As discussed in Sect. 3, to specify completely the spin state of a particle of spin i requires a knowledge of all irreducible spin tensor moments

$$T_q^{(k)} \; (|k| \leq q \leq 2i).$$

However, in view of the complexity of the results for the most general form of particle orientation and the fact that all experimental work to date is concerned with beams of particles of spin $\frac{1}{2}$, we shall confine ourselves primarily to discussing only the problem of the polarization (i.e. $q=1$) of the components of a nuclear reaction.

20. Production of polarized particles in nuclear reactions. The general expression for the polarization of the final particles in a nuclear reaction initiated by unpolarized particles has been deduced by various authors[3-5]. Here we adhere to the formulation of SIMON and WELTON[5] and use the notation of BLATT and BIEDENHARN[1].

---

[1] See footnote 2, p. 588.
[2] L. C. BIEDENHARN, J. M. BLATT and M. E. ROSE: Rev. Mod. Phys. **24**, 249 (1952).
[3] R. J. BLIN-STOYLE: Proc. Phys. Soc. Lond. A **64**, 700 (1951).
[4] L. WOLFENSTEIN: Phys. Rev. **75**, 1664 (1949).
[5] A. SIMON and T. A. WELTON: Phys. Rev. **90**, 1036 (1953).

The asymptotic form of the outgoing wave for the reaction already discussed can be written[1],

$$\psi_{\alpha' S'} = i\,\lambda_\alpha \left(\frac{v_\alpha}{v_{\alpha'}}\right)^{\frac{1}{2}} \frac{\exp(i\,k_{\alpha'}, r_{\alpha'})}{r_{\alpha'}}\, \varphi'_\alpha \sum_{m'_S=-S'}^{S'} q_{\alpha' S' m'_S;\alpha S m_S}(\vartheta, \varphi)\, \chi_{S' m'_S} \tag{20.1}$$

where v_α and k_α are the relative velocity and wave number, respectively in the entrance channel α and similarly for the outgoing channel; $\varphi_{\alpha'}$ is the product of internal wave functions of the nucleus Y and particle b corresponding to the specification α'; $\chi_{S' m^s}$ is the final channel spin wave function and

$$\left.\begin{aligned} q_{\alpha' S' m'_S;\alpha S m_S}(\vartheta,\varphi) &= (\pi)^{\frac{1}{2}} \sum_{J l l' \pi} i^{(l-l')} (2l+1)^{\frac{1}{2}}\, C(l\,S\,J;0\,m_S) \times \\ &\times C(l'\,S'\,J; m_S - m_{S'}\, m_{S'})\,(\delta_{\alpha'\alpha}\,\delta_{S'S}\,\delta_{l'l} - S^{J\pi}_{\alpha' S' l';\alpha S l})\, Y_{l'}^{m_S - m'_S}(\vartheta,\varphi)\,. \end{aligned}\right\} \tag{20.2}$$

The *differential polarization* of the outgoing particle (spin i') in the outgoing channel α' can be defined to be

$$d\boldsymbol{P}_{1\alpha'\alpha} = \langle \sum (\psi_{\alpha' S'} |\, \boldsymbol{i}' \,| \psi_{\alpha' S''})\, r_{\alpha'}^2\, v_\alpha^{-1}\, v_{\alpha'} \rangle_{\text{av}}\, d\Omega \tag{20.3}$$

where the sum is over all final channel spins and the final channel spin directions and an average is taken over the initial states. Here $\boldsymbol{i}'$ is the spin operator for the outgoing particle b whose polarization is to be measured.

The polarization is then defined by

$$P_{1\alpha'\alpha} = \frac{1}{i'}\,\frac{d\,|\boldsymbol{P}_{1\alpha'\alpha}|}{d\sigma_{\alpha'\alpha}}$$

and the polarization vector $\boldsymbol{P}_1$ is given by

$$\boldsymbol{P}_1 = \boldsymbol{p}\cdot P_{1\alpha'\alpha},$$

where

$$d\boldsymbol{P}_{1\alpha'\alpha} = \boldsymbol{p}\cdot d\,|\boldsymbol{P}_{1\alpha'\alpha}|\,.$$

For calculation purposes it is more convenient to evaluate the expectation values of those linear combinations of the components of the spin operator i which transform as spherical harmonics. In particular, for the polarization, the quantities

$$T_1^{(\pm 1)} = \pm \frac{i'_x \pm i'_y}{\sqrt{2i'(i'+1)}}\,, \qquad T_1^{(0)} = \frac{i'_z}{\sqrt{i'(i'+1)}}$$

are used.

By using Racah techniques, the expression for $d\boldsymbol{P}_{1\alpha'\alpha}$ can be written in terms of the unit vector $\boldsymbol{n} = (\boldsymbol{k}_\alpha \times \boldsymbol{k}_{\alpha'})/|(\boldsymbol{k}_\alpha \times \boldsymbol{k}_{\alpha'})|$ as follows:

$$d\boldsymbol{P}_{1\alpha'\alpha} = \boldsymbol{n}\, d\,|P_{1\alpha'\alpha}|\,,$$

where

$$\left.\begin{aligned} d\,|P_{1\alpha'\alpha}| &= \frac{1}{4}\lambda_\alpha^2 \frac{\sqrt{2i'(i'+1)(2i'+1)}}{(2I+1)(2i+1)} \sum i^{l_2 - l_1 + l'_1 - l'_2} \times \\ &\times \mathrm{Re}\,[i\,(\delta_{\alpha'\alpha}\,\delta_{l' l_1}\,\delta_{S'_1 S} - S^{J_1\pi_1}_{\alpha' l'_1 S';\alpha l_1 S})^* (\delta_{\alpha'\alpha}\,\delta_{l'_2 l_2}\,\delta_{S'_2 S} - S^{J_2\pi_2}_{\alpha' l'_2 S'_2;\alpha l_2 S})] \times \\ &\times (-1)^{I'-i'-S+l'_1+J_1-S'_1} [(2l_1+1)(2l_2+1)(2l'_1+1)(2l'_2+1)(2S'_1+1)(2S'_2+1)]^{\frac{1}{2}} \times \\ &\times (2J_1+1)(2J_2+1)\cdot C(l_1 l_2 L;0\,0)\cdot C(l'_1 l'_2 L;0\,0)\cdot W(i'\,S'_1 i'\,S'_2; I'\,1)\times \\ &\times W(l_1 J_1 l_2 J_2; S\,L)\cdot X(J_1 l'_1 S'_1; J_2 l'_2 S'_2; L\,L\,1)\cdot P_L^{(1)}(\vartheta)\, d\Omega \end{aligned}\right\} \tag{20.4}$$

where the sum is over $J_1\,J_2\,\pi_1\,\pi_2\,l_1\,l_2\,l'_1\,l'_2\,S'_1\,S'_2\,S$ and L.

---

[1] See footnote 5, p. 589.

Here the X coefficient[1] is defined by

$$X(a\,b\,c;d\,e\,f;g\,h\,i) = (-)^S \sum_z \left[(2z+1)\, W(b\,c\,d\,g;z\,a)\, W(d\,b\,f\,h;z\,e)\cdot W(g\,c\,h\,f;z\,i)\right]$$

where $S = a + b + \cdots + i$. $P_L^{(1)}(\vartheta)$ is a normalized associated Legendre function.

21. Properties of the polarization. Inspection of the form of $|d\boldsymbol{P}_{1\alpha'\alpha}|$ given in Eq. (20.4) enables various properties of the polarization and selection rules to be deduced. These have been pointed out by various authors[2-4] and conveniently gathered together by Simon and Welton[4]. They are listed below:

(a) The polarization is always normal to the plane of the reaction.

(b) If the initial or the final states involve only S waves, there can be no polarization of the outgoing particles.

(c) If only levels of the compound nucleus having $J=0$ or $\frac{1}{2}$ are effective, there will be no polarization unless interference arises between $J=\frac{1}{2}$ states of opposite parity.

(d) If only channel spin 0 is effective for the final channel spin, the polarization vanishes.

(e) Polarization results from the interference of different subchannels (i.e. partial waves or final channel spins) contributing to the reaction and therefore if there is only a single non-zero element of the scattering matrix, the polarization will vanish.

(f) If there is no spin-orbit coupling, the polarization is zero.

(g) If the largest effective values of the incident orbital angular momentum, final orbital angular momentum and total angular momentum are l, l' and J respectively, then L in Eq. (20.4) must be less than or equal to $2l$, $2l'$ and $2J$. In addition, L must be even if the interfering states have the same parity.

(h) Since the elements of the S matrix resulting from the use of perturbation theory or Born approximations with a real potential are all real, it follows from the form of (20.4) that in these approximations $d|\boldsymbol{P}_{1\alpha'\alpha}|$ is zero.

Inspection of (20.4) shows that the S-matrix elements are now involved in different combinations from the way in which they occur in the unpolarized angular distribution, so that by using polarization sensitive detectors more information may be obtained about these elements.

22. Higher spin moments. So far only the problem of the *polarization* of the outgoing particle i' has been considered. If $i'=\frac{1}{2}$, then this gives complete information about the spin orientation of the particle. However, for $i' > \frac{1}{2}$ higher spin moments exist and then it is necessary to calculate the expectation value of the general irreducible tensor operator of rank q, $T_q^{(k)}$ discussed in Sect. 3 which is formed out of the spin operators i'.

Simon and Welton[4] define the *differential tensor moment* to be

$$d\,T_{q\alpha'\alpha}^{(k)} = \left\langle \sum (\psi_{\alpha' s'}\,|\,T_q^{(k)}\,|\,\psi_{\alpha' s''})\, r_{\alpha'}^2\, v_\alpha^{-1}\, v_{\alpha'}\right\rangle_{\text{av}} d\Omega \tag{22.1}$$

where the sum is taken over all final channel spins and final channel spin directions, and an average is taken over the initial states. They are then able to deduce an extremely complicated, but general, expression for this quantity similar to (20.4).

---

[1] U. Fano: National Bureau of Standards Report 1214, p. 48.

[2] See footnote 3, p. 589,

[3] See footnote 4, p. 589.

[4] See footnote 5. p. 589.

23. Initial particles oriented. SIMON and WELTON[1] have investigated the more complex case of a nuclear reaction in which the spins of the initial and final components are oriented in an arbitrary fashion, the orientations being specified in terms of the spin tensor moments $T_q^{(k)}$. The most important case, however, is that in which only incident particles of spin $\frac{1}{2}$ are polarized the an investigation is made of the effect of such polarization on the angular distribution of the reaction products. This effect is, in fact, used as a means of measuring the polarization of the incident particles. The form of this effect can be deduced at once by applying time reversal arguments to the results of Sect. 20 and using the symmetry properties of the S matrix[2,3]. It then follows that the angular distribution of the products of the reaction in which the incident particles of spin half are polarized (polarization vector $\boldsymbol{p}_1$) has the form

$$d\sigma'_{\alpha'\alpha} = d\sigma_{\alpha'\alpha}[1 + \boldsymbol{n}\cdot\boldsymbol{p}_1 P_{1\alpha\alpha'}(\vartheta)] \tag{23.1}$$

where $P_{1\alpha\alpha'}(\vartheta)$ is the function given in Sect. 20 but for the time reversed reaction $\alpha'\to\alpha$ (i.e. with all primed and unprimed quantities interchanged). $\boldsymbol{n}$ is a unit vector normal to the plane containing the directions of motion of the initial and final particles and the polarization dependent term in general gives rise to azimuthal asymmetry of the reaction products.

In order that there shall be an effect of this type it is clear that the selection rules set out in Sect. 21, which must be satisfied in order that $P_{1\alpha'\alpha}(\vartheta)$ shall be non-vanishing, must also apply in this case. In particular, polarization along the axis of incidence produces no change in the angular distribution. It can be seen from (23.1) that the effect of polarizing the incident particles is to give an azimuthal variation to the angular distribution. Thus if P_1 is perpendicular to the direction of incidence

$$d\sigma'_{\alpha'\alpha} = d\sigma_{\alpha'\alpha}[1 + p_1 \cos\varphi\, P_{1\alpha\alpha'}(\vartheta)]$$

and the polarization can be detected by measuring the ratio R of the intensities of the final particles at $\varphi = 0$ and π, i.e.

$$R = \frac{1 + p_1 P_{1\alpha\alpha'}}{1 - p_1 P_{1\alpha\alpha'}}. \tag{23.2}$$

Such an azimuthal variation always implies polarization of the initial components and may again be used to determine more information about the S-matrix elements.

24. Bombarding particles and target nuclei polarized. This case has been treated by BLIN-STOYLE[4] and SIMON[5] in the most general fashion. Experimentally, however, the only situation of interest to date is that in which thermal (S wave) polarized neutrons are captured by a polarized target nucleus and a measurement made of the capture cross section. This case has been treated separately by ROSE[6]. He points out that in the scattering or absorption of polarized S wave neutrons by polarized nuclei there should be a dependence of the cross section for the process on the relative orientation of neutrons and nuclear spins. For scattering, the cross section σ can be shown to be related to the cross section σ_0

---

[1] See footnote 5, p. 589.
[2] R. J. BLIN-STOYLE: Proc. Phys. Soc. Lond. A **65**, 452 (1952).
[3] F. COESTER: Phys. Rev. **84**, 1259 (1951).
[4] See footnote 3, p. 589.
[5] A. SIMON: Phys. Rev. **92**, 1050 (1953).
[6] M. E. ROSE: Phys. Rev. **75**, 213 (1949).

for unpolarized particles by an expression of the form

$$\sigma = \sigma_0 (1 + \varrho) \tag{24.1}$$

where

$$\varrho = 2I\, P_1(n)\, P_1(N) \left[\frac{b}{a} \cdot \frac{1 - b/2a}{1 + I(I+1)\, b^2/a^2}\right] \tag{24.2}$$

where $P_1(n)$ and $P_1(N)$ are the neutron and nuclear polarizations and a, b are the scattering amplitudes defined by

$$a = \frac{4\pi}{k} \frac{I\eta' + (I+1)\,\eta''}{2I+1}, \qquad b = \frac{4\pi}{k} \frac{\eta'' - \eta'}{2I+1}.$$

Here k is the neutron wave number and η', η'' are the scattering phase shifts corresponding to the compound states $J = I - \frac{1}{2}$, $I + \frac{1}{2}$.

Where contributions from more than one state are concerned, resort must be made to the more general expression given above using the appropriate phase shifts. In general, although the signs of η cannot be determined, their relative signs can.

For scattering or capture by a single resonant state,

$$\sigma = \sigma_0 \left[1 - P_1(n)\, P_1(N)\right] \qquad \text{if} \qquad J = I - \tfrac{1}{2}, \tag{24.3}$$

$$\sigma = \sigma_0 \left[1 + P_1(n)\, P_1(N) \frac{I}{I+1}\right] \text{ if } J = I + \frac{1}{2}. \tag{24.4}$$

Under these circumstances the angular momentum J of this state can be determined by observing the *direction* in which σ changes when the neutron polarization is switched on.

25. Gamma rays in nuclear reactions. The S-matrix formalism, used so far, may be extended to include the possibility of gamma rays in nuclear reactions In this case the vector potential plays the role of the "wave function" for the particle[1]. It is then possible to investigate, in particular, the case of a nuclear reaction which results in the emission of radiation. General expressions for the angular distribution and polarization of outgoing gamma rays resulting from a nuclear reaction initiated by components which may or may not be polarized have been deduced by SIMON[2]. He obtains two interesting selection rules:

(a) Circularly polarized radiation cannot result from a nuclear reaction initiated by unpolarized particles.

(b) Plane polarized radiation cannot be the product of a nuclear reaction initiated by S-wave particles of spin $\frac{1}{2}$.

It has also been pointed out that where the target nuclei are oriented the angular distribution of γ-radiation following the capture of unpolarized S-wave particles will in general deviate from isotropy so long as the capturing nucleus has a spin greater than $\frac{1}{2}$. The compound states thus formed will be oriented along the original axis of orientation and any radiation emitted by them will have an angular distribution about this axis which is characteristic of the change of angular momentum in the reaction. It is a feature of this type of experiment that it leads to anisotropies which for an S-interaction could not be observed without nuclear orientation. For γ-ray emission such anisotropies will be weak unless there is either a strong ground state transition, or a means of selecting the first γ-ray of the cascade by which the compound nucleus decays to the ground state.

[1] See footnote 3, p. 567.
[2] See footnote 5, p. 592.

E. The production and detection of polarized particles.

Methods for the production and detection of polarized particles cannot conveniently be separated since any process that can be used for the production of polarized particles can generally be used for their detection.

In order that polarized particles should be produced or detected there must be some sort of coupling of the particle spin with a fixed spatial direction. There are three main ways of defining such a direction:

(i) By an external magnetic field.

(ii) By an oriented nuclear or atomic system.

(iii) By the orbital angular momentum of the particle (spin-orbit coupling).

In the following sections an account is given of the various experiments that have been carried out in this connection to date, together with a discussion of their theoretical treatment.

26. External magnetic field. Experiments of the Stern-Gerlach type in which particles are passed through an external inhomogeneous magnetic field are well known. In such experiments particles with different spin orientations suffer different deflections so that in principle particles with a given spin orientation (i.e. polarized particles) can be selected. Such a method seems hardly practicable for producing intense beams of polarized nucleons. The method has been used, however, by SHERWOOD *et al.*[1] for detecting the polarization of neutrons scattered by a magnetic material (see Sect. 27).

27. Magnetic scattering of thermal neutrons. The first successful experiments resulting in a beam of polarized nucleons were due to a suggestion by BLOCH[2] and involve the scattering of slow neutrons by magnetized ferromagnetic materials. In this case the polarization depends on the interaction between the magnetic moment of the neutron and the oriented atomic magnetic moments and is therefore of the type (ii) above. This method is discussed in considerable detail elsewhere in this Encyclopedia[3] and here only the fundamental ideas of the process are outlined.

The effect depends on the interference between the purely nuclear scattering and that due to the atomic electrons. The total scattering amplitude for the process can be written as $a \pm f_m(\vartheta, \varphi)$ where a is the scattering length for the particular nucleus under consideration and is independent of the scattering angles ϑ and φ if it is assumed that only S waves are effective. $f_m(\vartheta, \varphi)$ is proportional to the atomic form factor and is the amplitude representing the magnetic scattering, and the alternative signs refer to the two possible neutron spin orientations with respect to the direction of magnetization.

Typical values for a and the mean value of f_m for thermal neutrons scattered by magnetized iron are $a \approx 8 \times 10^{-13}$ cm and $f_m \approx 1.3 \times 10^{-13}$ cm so that the total scattering cross section can be written:

$$\sigma = \int |a \pm f_m(\vartheta, \varphi)|^2 d\Omega \approx Q_0 \pm P \tag{27.1}$$

where

$$Q_0 = 4\pi a^2 \text{ and } p = 2a \int f_m(\vartheta, \varphi) \sin\vartheta d\vartheta d\varphi.$$

If the beam passes through a thickness l of the ferromagnetic material the degree of polarization P_1 for the transmitted beam is

$$P_1 = \operatorname{Tan}(N l p) \tag{27.2}$$

where N is the number of scattering atoms per unit volume.

[1] J. E. SHERWOOD, T. E. STEPHENSON and S. BERNSTEIN: Phys. Rev. **96**, 1546 (1954).

[2] F. BLOCH: Phys. Rev. **50**, 259 (1936).

[3] G. R. RINGO in Vol. XXXII, p. 623 seq.

A difficulty of this method is that small departures from saturation of the magnetization lead to a considerable reduction in the magnitude of the polarization due to precession of the neutron spin caused by fluctuations in the magnetic field as the neutron passes from one domain to another.

Only thermal neutrons can be appreciably polarized in this way since $f_m(\vartheta, \varphi)$ decreases rapidly with increase in neutron energy.

Recently it has become possible to produce even more highly polarized beams by reflecting neutrons from magnetized sheets of ferromagnetic materials.

28. Spin-orbit coupling methods. *α) Polarization effects in Coulomb scattering.* The earliest method suggested for the polarization of elementary particles was to make use of the spin-orbit coupling occurring in the Coulomb scattering of electrons[1,2]. SCHWINGER[3] has proposed a similar method for the polarization of neutrons employing the spin-orbit interaction arising from the motion of the neutron magnetic moment in the nuclear Coulomb field. For an initially unpolarized beam of neutrons, he shows that the scattered particles have a polarization given by

$$\boldsymbol{P}_1(\vartheta) = \boldsymbol{n} \cdot \frac{2\,\mathrm{Re}\,[i f_0(\vartheta)]\,\gamma \cot\frac{\vartheta}{2}}{|f_0(\vartheta)|^2 + \gamma^2 \cot^2\frac{\vartheta}{2}} = \boldsymbol{n} \cdot P_1(\vartheta) \tag{28.1}$$

where for a nucleus of charge Z, $\gamma = \frac{1}{2}\,\mu_n(\hbar/Mc)(Ze^2/\hbar c)$ where μ_n = neutron magnetic moment. $f_0(\vartheta)$ is the amplitude of the wave scattered by specifically nuclear forces and $\boldsymbol{n}$ is a unit vector perpendicular to the plane of the scattering.

As in Sect. 22, it follows that for incident neutrons with polarization $\boldsymbol{p}_1$ the angular distribution of the scattered neutrons is

$$d\sigma' = \left\{|f_0(\vartheta)|^2 + \gamma^2 \cot^2\frac{\vartheta}{2}\right\}\{1 + \boldsymbol{p}_1 \cdot \boldsymbol{n}\, P_1(\vartheta)\} \tag{28.2}$$

Schwinger points out that in order for the electric field to be fully effective in producing spin-dependent scattering, the major portion of this scattering should take place outside the nucleus ($r \gg R$) but well within the screening radius of the atomic electrons ($r \ll a$). In order to satisfy this condition he estimates that the useful range of scattering is restricted to

$$\frac{1}{ka} \ll 2\sin\frac{\vartheta}{2} \ll \frac{1}{kR}\,.$$

For 1 MeV neutrons scattered by lead practically complete polarization is to be expected for a scattering angle $\vartheta = 1.5°$. At larger angles he estimates that $P_1 = 0.8$ for $\vartheta = 3°$; $P_1 = 0.47$ for $\vartheta = 6°$; $P_1 = 0.32$ for $\vartheta = 9°$.

Similar calculations by SAMPLE[4] give a value for R [see Eq. (23.2)] of 14.5 for completely polarized 3.1 MeV neutrons at a scattering angle of 0.5° and this decreases to 1.6 at 5°. This corresponds to values of P_1 in the inverse process of 0.87 for $\vartheta = 0.5°$ and 0.23 for $\vartheta = 5°$.

Experimentally this effect has been used by LONGLEY *et al.*[5] to detect the polarization of neutrons emitted in the $D-D$ reaction (Sect. 29α). They scattered the polarized neutrons in lead through angles $\vartheta = 3$ to 4° and measured the ratio

---

[1] N. F. MOTT: Proc. Roy. Soc. Lond., Ser. A **124**, 425 (1929).
[2] N. F. MOTT: Proc. Roy. Soc. Lond., Ser. A **135**, 429 (1932).
[3] J. SCHWINGER: Phys. Rev. **73**, 407 (1948).
[4] J. T. SAMPLE: Canad. J. Phys. **34**, 36 (1956).
[5] H. J. LONGLEY, R. N. LITTLE and J. M. SLYE: Phys. Rev. **86**, 419 (1952).

$R = \frac{1+p_1 P_1(\vartheta)}{1-p_1 P_1(\vartheta)}$. Although a significant polarization ($p_1 = 0.4 \pm 0.2$) was obtained, a large background precluded the accurate measurement of this quantity.

β) Nucleons scattered by nuclei of spin zero. (Individual resonance effects). The possibility of producing polarized nucleons from a scattering process by a resonance level that is split by spin-orbit coupling was originally suggested by SCHWINGER[1] (see also LEPORE[2]). The general formula for the polarization of particles scattered in this way given in Eq. (20.4) applies, and to estimate the magnitude of this effect it is only necessary to replace the scattering matrix elements by the appropriate resonance forms. As an illustrative example we consider the particular case of low energy protons (say 1—3 MeV) scattered by helium nuclei.

Phase shift analysis of the differential scattering cross section of unpolarized protons by helium nuclei indicates that the scattering can be interpreted in terms of non-resonant S wave scattering and $p_{\frac{3}{2}}$ and $p_{\frac{1}{2}}$ resonance scattering. Thus the only non-vanishing scattering matrix elements are those corresponsing to the $s_{\frac{1}{2}}$, $p_{\frac{1}{2}}$ and $p_{\frac{3}{2}}$ states.

The appropriate scattering matrix elements are then

$$S(s_{\frac{1}{2}}) = \exp(2 i \eta_0),$$

$$S(p_{\frac{1}{2}}) = 1 + \frac{i\Gamma}{E_{\frac{1}{2}} - E - \frac{1}{2} i \Gamma},$$

$$S(p_{\frac{3}{2}}) = 1 + \frac{i\Gamma}{E_{\frac{3}{2}} - E - \frac{1}{2} i \Gamma}$$

where η_0 is the S phase shift, $E_{\frac{1}{2}}$ and $E_{\frac{3}{2}}$ are the relevant resonance energies and Γ their widths which are assumed to be the same.

Substituting these expressions into the general formula, one obtains[2] for the polarization of scattered neutrons with energy E

$$\boldsymbol{P}_1(E, \vartheta) = \frac{A B^* + B A^*}{A A^* + B B^*} \cdot \boldsymbol{n} = P_1(E, \vartheta)\, \boldsymbol{n}$$

where

$$A(\vartheta) = \frac{1}{k}\left\{\exp(i \eta_0) \sin \eta_0 + \left(\frac{\Gamma}{E_{\frac{3}{2}} - E - \frac{1}{2} i \Gamma} + \frac{1}{2} \frac{\Gamma}{E_{\frac{1}{2}} - E - \frac{1}{2} i \Gamma}\right) \cos \vartheta\right\},$$

$$B(\vartheta) = \frac{\sin \vartheta}{2 i k}\left\{\frac{\Gamma}{E_{\frac{3}{2}} - E - \frac{1}{2} i \Gamma} - \frac{\Gamma}{E_{\frac{1}{2}} - E - \frac{1}{2} i \Gamma}\right\},$$

and $\boldsymbol{n}$ is a unit vector normal to the scattering plane.

An experimental investigation of the above effect has been carried out by HEUSINKVELD and FREIER[3]. The phase shift analysis of CRITCHFIELD and DODDER[4] of the elastic scattering of protons by He in the energy range 0.95 to 3.58 MeV does not establish whether the $P_{\frac{1}{2}} - P_{\frac{3}{2}}$ doublet in Li^5 is normal or inverted and two sets of phase shifts are obtained. Using these two sets the polarization to be expected for the two cases is shown in Figs. 20 and 21. It can be seen that polarizations up to 100% are predicted in both cases for certain energies and scattering angles. In order to measure the polarization HEUSINKVELD and FREIER[3] performed a double scattering experiment in which the first scattering polarized the protons and the second acted as an anlayser, a search being

[1] J. SCHWINGER: Phys. Rev. **69**, 681 (1946).
[2] J. V. LEPORE: Phys. Rev. **79**, 137 (1950).
[3] M. HEUSINKVELD and G. FREIER: Phys. Rev. **85**, 80 (1952).
[4] C. L. CRITCHFIELD and D. C. DODDER: Phys. Rev. **76**, 602 (1949).

made for an azimuthal variation in the angular distribution. If the first scattering is through an angle ϑ_1 at energy E_1, then the resulting polarization vector can be written $P_1(E_1\vartheta_1)\,\boldsymbol{n}_1$ where $\boldsymbol{n}_1$ is a unit vector normal to the plane of the first scattering. Denoting the second scattering by the suffix 2, it then follows as in section 23 that the variation in azimuthal intensity after the second scattering is given by

$$d\sigma' = d\sigma_1(E_1\vartheta_1)\cdot d\sigma_2(E_2\vartheta_2)\times$$
$$\times[1 + P_1(E_1\vartheta_1)\cdot P_1(E_2\vartheta_2)\,\boldsymbol{n}_1\cdot\boldsymbol{n}_2]$$

where $d\sigma(E,\vartheta)$ is the differential cross section for an unpolarized scattering.

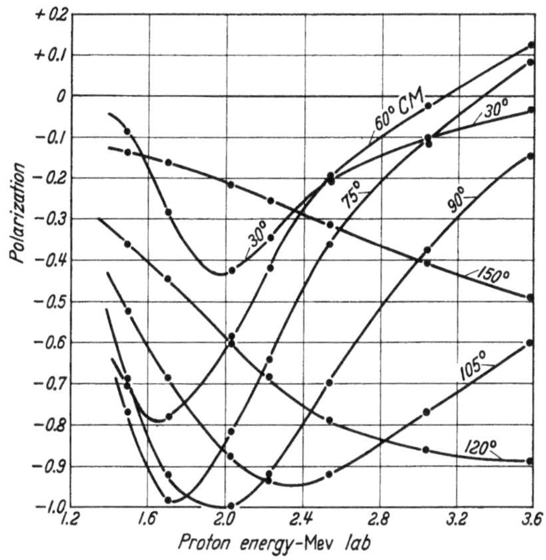

Fig. 20. Expected polarization of protons scattered from He^4 at various center-of-mass angles as a function of incident proton energy. All curves are for the case of the inverted $P_{\frac{3}{2}}-P_{\frac{1}{2}}$ energy levels in Li^{5*}.

Both scatterings took place in the same plane and through 90° in the centre of mass system (since to a first approximation this gives the greatest polarization effect). The ratio in intensity between the forward double scattering and backward double scattering directions (see Fig. 22) is then given by

$$R = \frac{1 + P_1(1)\,P_1(2)}{1 - P_1(1)\,P_1(2)}$$

where $\boldsymbol{n}_1\cdot\boldsymbol{n}_2$ has been given the values ± 1.

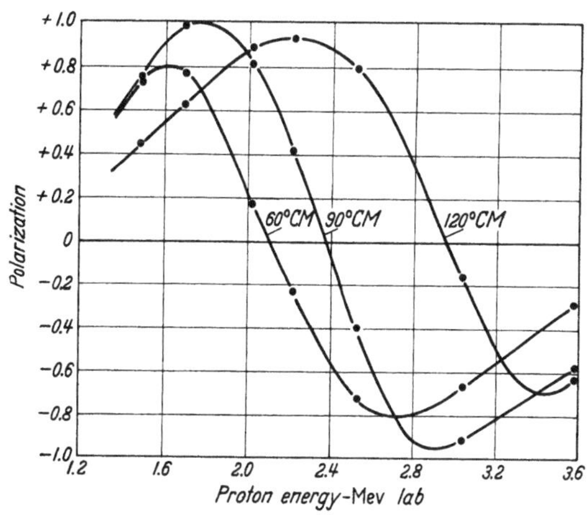

Fig. 21. Expected polarization of protons scattered from He^4 at various center-of-mass angles as a function of incident proton energy. All curves are for the case of the normal $P_{\frac{1}{2}}-P_{\frac{3}{2}}$ energy levels in Li^{5*}.

The proton beam from an electrostatic generator passed into a scattering chamber which provided for the double scattering. The collimating system between the two large gas volumes defined a beam of protons scattered at 90° in the first interaction, while selection of the protons scattered at 90° in the second interaction was made through analysis of the proton tracks in nuclear emulsion plates used for measuring the yield. One plate was placed in a position to detect those protons scattered in a backward direction with respect to the incident proton beam, and another was placed in a position to detect those scattered in the forward direction. The planes of the two scattering interactions were held nearly coplanar by the collimating system between the first and second scattering volumes, so that the expressions $\boldsymbol{n}_1\cdot\boldsymbol{n}_2$ took approximately the values ± 1.

The results obtained by counting proton tracks in the forward and backward plates for two incident proton energies are given in Table 3 together with the theoretical predictions for Inverted and Normal doublets. The lack of exact agreement between theoretical and experimental results can be attributed to the broad geometry used in the experiment. The theoretical values are also subject to the experimental errors in the data from which the original phase shifts were calculated. With this additional correction the ratio, as obtained experimentally, is in statistical agreement with the ratio as obtained from the phase shift data for the inverted doublet and consequently shows that the $P_{\frac{3}{2}} - P_{\frac{1}{2}}$ doublet in Li^5 is inverted.

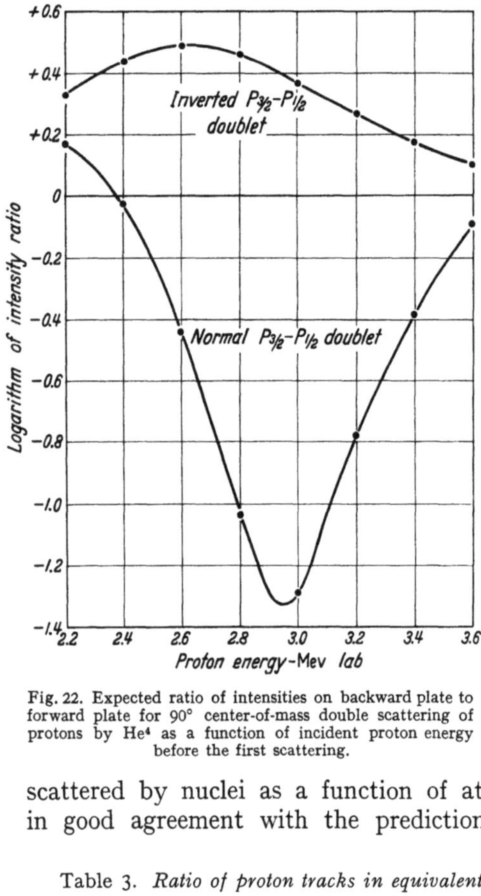

Fig. 22. Expected ratio of intensities on backward plate to forward plate for 90° center-of-mass double scattering of protons by He^4 as a function of incident proton energy before the first scattering.

Similar effects to the above in which polarization is produced by spin-orbit coupling manifested in resonance phenomena of individual levels in the compound nucleus might also be expected for other spin zero nuclei nuclei (e.g. C^{12}, O^{16} and Pb^{208}) in which levels of the compound nucleus are well separated. Indeed Huber and Baumgartner[1] have used scattering by C^{12} and O^{16} (see also Ricamo[2]) to measure the polarization of neutrons produced in the D—D reaction (see Sect. 29 where further details are given).

γ) Neutron scattering at low energies by nuclei. (Average over many resonances.) Feshbach, Porter and Weisskopf[3] have shown that the results of measurements of the total neutron cross-sections of nuclei and angular distributions of neutrons scattered by nuclei as a function of atomic number and neutron energy are in good agreement with the predictions of a simple nuclear model. In this

Table 3. *Ratio of proton tracks in equivalent strips in the forward and backward plates.*

	No. of tracks on backward plate	No. of tracks on forward plate	Ratio— backward to forward	Theoretical ratio for ideal ray	
				Inverted doublet	Normal doublet
3.25 MeV	364	191	1.9	2.6	1/20.2
3.50 MeV	61	33	1.85	1.9	1/6.6

model a square well with a range $R \approx 1.45 A^{\frac{1}{3}} \times 10^{-13}$ cm and depth $V_0(1+i\xi)$ is used where $V_0 \approx +40$ MeV and ξ, which represents the effect of absorption of the neutron to form a compound state, is ≈ 0.03. Such a model appears reasonable in view of the success of the independent particle model of the nucleus, since the small absorption coefficient results in a long mean free path of the nucleons in nuclear matter. It should be stressed here, however, that such a

[1] P. Huber and E. Baumgartner: Helv. phys. Acta **26**, 420 (1953).
[2] R. Ricamo: Helv. phys. Acta **26**, 423 (1953).
[3] H. Feshbach, C. E. Porter and V. F. Weisskopf: Phys. Rev. **96**, 448 (1954).

description of nuclear reactions and scattering is only valid when an average of results is taken over a neutron energy spread much greater than the energy separation between levels of the compound nucleus and would not, for example, give an adequate description of the scattering of neutron by He^4 where there is considerable separation (≈ 3 MeV) between the $P_{\frac{3}{2}}$ and $P_{\frac{1}{2}}$ levels.

Now on the individual particle model there is also introduced a strong spin orbit force of the form $f(r)\,\boldsymbol{L}\cdot\boldsymbol{S}$ and it would appear to be logical to include such a force in the FESHBACH, PORTER and WEISSKOPF description. This force would modify the theoretical expressions for the angular distributions and total cross section; and in particular it would be expected to produce polarization of the scattered neutrons.

The simplest assumption to make is that $f(r)$ is constant (A say). In this case, $A\,\boldsymbol{L}\cdot\boldsymbol{S}$ has eigenvalues $\frac{1}{2}A\,l$ for $j=l+\frac{1}{2}$ and $-\frac{1}{2}A\,(l+1)$ for $j=l-\frac{1}{2}$; that is, the nucleons in each partial wave "l" see a different potential for each j,

$$V(+) = V_0(1+i\xi) + \tfrac{1}{2}A\,l \quad \text{for } r<R; \quad =0 \text{ for } r\geq R \text{ if } j=l+\tfrac{1}{2},$$
$$V(-) = V_0(1+i\xi) - \tfrac{1}{2}A\,(l+1) \quad \text{for } r<R; \quad =0 \text{ for } r\geq R \text{ if } j=l-\tfrac{1}{2}.$$

For elastic scattering the S-matrix elements reduce to the partial wave reflection factors $\eta(j\,l)$ for these potentials[1] (see also SATCHLER[2]):

$$S^{J_1}_{J_2 j_2 l_2;\,J_0 j_1 l_1} = \eta\,(j_1\,l_1)\,\delta_{j_2 j_1}\cdot\delta_{l_2 l_1}\cdot\delta_{J_2 J_0}$$

where J_0, J_1 and J_2 are the spins of initial, compound and final nuclei and the suffixes 1, 2 refer to initial and final states.

In the case of low energy neutrons (~ 400 keV) it is probably sufficient to consider only S and P waves. The calculations are then straightforward and have been performed by (for instance) ADAIR[3] and OKAZAKI[4]. Adair's results for the polarization $P_1\,(\vartheta)$ at $\vartheta=90°$ and for $A=-1$ MeV and -2 MeV are given in Fig. 23 where $P_1\,(90°)$ is plotted as a function of the nuclear radius.

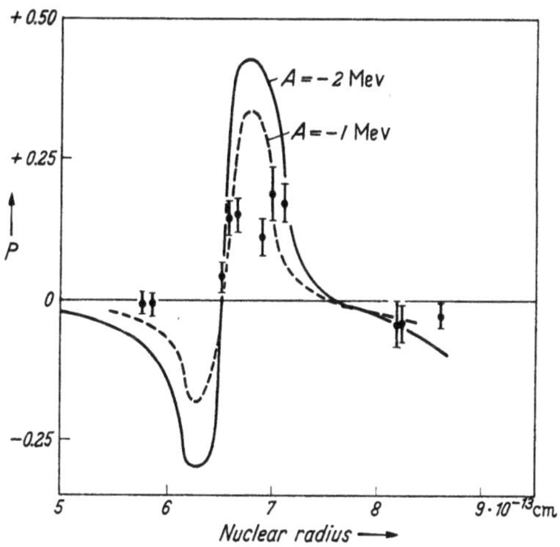

Fig. 23. The points show the measured polarizations of 400-keV neutrons produced by scattering from heavy nuclei as a function of nuclear radius, R. The solid curve shows values obtained when a potential term equal to -2 MeV ($\boldsymbol{L}\cdot\boldsymbol{S}$) is added to the potential used by FESHBACH, PORTER, and WEISSKOPF. The dashed curve represents the results of calculations for a spin-orbit term of -1 MeV ($\boldsymbol{L}\cdot\boldsymbol{S}$).

Adair has carried out an experimental investigation of this effect using polarized neutrons produced from the $Li^7\,(p,n)\,Be^7$ reaction (see Sect. 27) and measured the polarization of the neutrons by scattering from O^{16} in the manner described in Sect. 27 β. In this case it is sufficient to consider the non-resonant S wave scattering and the $P_{\frac{3}{2}}$ resonant scattering from a level in O^{17} at about 435 keV neutron energy. Neutrons emitted at a laboratory angle of 50° with respect to the proton beam in the $Li^7\,(p\,n)\,Be^7$ reaction were scattered by liquid oxygen and those scattered at 90° in the centre of mass system by the oxygen

[1] See footnote 3, p. 598.
[2] G. R. SATCHLER: Proc. Phys. Soc. Lond. A **68**, 1041 (1955).
[3] R. K. ADAIR, S. E. DARDEN and R. E. FIELDS: Phys. Rev. **96**, 503 (1954).
[4] A. OKAZAKI: Phys. Rev. **99**, 55 (1955).

were counted by a hydrogen recoil proportional counter. The proton bombarding energy was 2.262 MeV. The reaction and the scattering both took place in the same plane, the counter occupying both possible positions for this arrangement so that a measurement of the left-right asymmetry in the intensity of scattered neutrons could be obtained.

For the oxygen scattering the S wave phase shift η_0 was estimated from the non-resonant scattering cross section σ using the relation $\sigma = \frac{4\pi}{k^2}\sin^2\eta_0$, its sign being determined from slow neutron interference measurements. The S-matrix element for the $P_{\frac{3}{2}}$ state has the form given in Sect. 28 β for scattering by He^4 and is expressed in terms of Γ and $E_{\frac{3}{2}}$ for the $P_{\frac{3}{2}}$ level. Thus from the left right asymmetry and in terms of the above quantities an estimate could be made of the polarization of the neutron beam from the $Li^7(p\,n)Be^7$ reaction. Adair obtained $P_1 = -0.53 \pm 0.06$.

This beam of neutrons was then scattered through 90° by a selection of nuclei and a measurement was made of the left-right asymmetry under the same conditions as for the scattering by O^{16}. The left-right asymmetry is again given by an expression of the form (23.2), i.e.

$$R = \frac{1 + P_1(1)\,P_1(2)}{1 - P_1(1)\,P_1(2)}$$

where $P_1(1) = 0.53 \pm 0.06$ and $P_1(2)$ is the polarization which would be produced by the nucleus under consideration when scattering neutrons through 90°. Knowing R and $P_1(1)$, $P_1(2)$ can be deduced. The values obtained by ADAIR are plotted in Fig. 23 where it can be seen that there is qualitative agreement between theory and experiment. No doubt better agreement could be obtained by using (a) a more realistic smoothed out potential well, and (b) a more realistic spin-orbit interaction of the Thomas form, i.e. $\frac{1}{r}\cdot\frac{dV}{dr}\,\boldsymbol{L}\cdot\boldsymbol{S}$ which varies with the size of the nucleus[1]. Calculations to check these ideas, however, have not so far been reported.

Similar experiments to those of ADAIR *et al.*[2] have been carried out by OKAZAKI[3], who obtained comparable results.

29. Polarization in reactions of transmutation. Polarization of neutrons and protons produced in nuclear reactions has been predicted in several cases and verified in one or two of these. In the following an account, both theoretical and experimental, is given of such reactions.

α) The D+D reaction. The D+D reaction can take place in two alternative ways resulting in the emission of either neutrons or protons; thus

$$_1D^2 + {}_1D^2 \begin{cases} \nearrow {}_2He^3 + {}_0n^1 \\ \searrow {}_1H^3 + {}_1H^1 \end{cases}$$

Theoretical analysis of the reaction by BEIDUK, PRUETT and KONOPINSKI[4] shows that in order to explain the observed angular distribution of the outgoing neutrons and protons up to deuteron energies of about 0.15 MeV it is necessary to introduce a strong spin-orbit coupling and P wave effect. As indicated in the selection rules given in Sect. 20, these are necessary conditions for the production of polarized particles and it might therefore be expected that the outgoing neutrons and protons are polarized.

[1] R. J. BLIN-STOYLE: Phil. Mag. **46**, 973 (1955).
[2] See footnote 3, p. 599.
[3] See footnote 4, p. 599.
[4] F. M. BEIDUK, J. R. PRUETT and E. J. KONOPINSKI: Phys. Rev. **77**, 622 (1950).

An analysis of the reaction for deuteron bombarding energies below 300 keV where only S and P waves are expected to give appreciable contributions has been made by various authors[1-3]. The results are expressed in terms of energy dependent S matrix elements corresponding to the different transitions allowed by parity and angular momentum considerations, the energy dependence being introduced through Coulomb penetrability factors. It is found that the polarization vector of the nucleons emitted at angle ϑ in the centre of mass system has the form

$$\boldsymbol{P}_1(E,\vartheta)=\boldsymbol{n}\,C\,\frac{A(E)\cdot\sin\vartheta\cdot\cos\vartheta}{1+A(E)\cos^2\vartheta} \tag{29.1}$$

where C is a function of *energy independent* parts of the S matrix elements and $A(E)$ is the *energy dependent* coefficient of $\cos^2\vartheta$ in the expression for the angular distribution of the emitted particles. $\boldsymbol{n}$ is a unit vector normal to the plane of the reaction defined by $\boldsymbol{n}=\boldsymbol{k}_i\times\boldsymbol{k}_0/|\boldsymbol{k}_i\times\boldsymbol{k}_0|$. Experimentally, for both neutrons and protons, $A(E)$ is found to be positive varying from approximately 0 at $E=0$ to ≈ 1.25 at $E=300$ keV. Within this energy range then it follows that maximum polarization is obtained for values of ϑ between 45° to 60° and 135° to 150°.

Various experiments have been performed which confirm that both the protons and the neutrons are polarized.

The neutron polarization has been detected by LONGLEY *et al.*[4] using Coulomb scattering as already described in Sect. 28 α and found to be (40 ± 20) %, the neutrons being detected by a liquid scintillator coupled to a photomultiplier. Unfortunately, no indication was given of the energy of the deuteron beam.

To detect the polarization of protons produced in the reaction, BISHOP *et al.*[5] scattered the protons by He^4 and used the effect described in Sect. 27β. A beam of 300 keV deuterons was collimated on to a heavy ice target, thus producing 10^8 protons per second. Protons escaping at an angle of 120° to the deuteron beam passed through an aluminium window into a scattering chamber containing helium at atmospheric pressure. Protons which were scattered through an angular range (75° to 120° centre of mass) were then detected by a proportional counter and the whole arrangement of scattering volume and counter could be rotated around the axis of the scattering chamber. At the moment of being scattered the protons had an energy of (2 ± 0.2) MeV. The proton counting-rate was measured forward and backward in the plane of incident deuteron and emitted proton beams and in the two positions normal to this plane. The rates observed were

Forward	1.80 ± 0.07
Backward	2.97 ± 0.11
Normal positions	2.30 ± 0.08, 2.30 ± 0.09

According to Eq. (23.1) the azimuthal distribution is of the form $d\sigma'(\vartheta,\varphi)=d\sigma[1+P_1 B\cos\varphi]$ where P_1 is the proton polarization, φ is the angle between the normals to the scattering planes. From the curves given in Fig. 20 for an inverted He^5 doublet, for an energy (2.0 ± 0.2) MeV and from the geometry of the experiment, $B=0.87$. The value of $P_1 B$ obtained from the above data is 0.25 ± 0.05 and hence the polarization of the D+D protons at 300 keV deuteron bom-

[1] See footnote 2, p. 592.
[2] M. CINI: Nuovo Cim., Ser. IX **8**, 1007 (1951).
[3] M. FIERZ: Helv. Phys. Acta **25**, 629 (1952).
[4] See footnote 5, p. 595.
[5] G. R. BISHOP, G. PRESTON, J. M. WESTHEAD and H. HALBAN: Nature, Lond. **170**, 113 (1952).

barding energy is 0.30 ± 0.06. In the above expression (29.1) for the polarization, this implies that $C \approx +0.75$ and that the polarization is in the direction of $\boldsymbol{n}$.

HUBER and BAUMGARTNER[1] measured the polarization of the neutrons produced in the D+D reaction by investigating the azimuthal distribution resulting from the scattering of the neutrons by carbon. Neutrons emitted between 48° to 54° in the centre of mass system were scattered by the carbon and detected with a Hornyak detector together with a photomultiplier. Measurements of the intensity of neutrons scattered through an angle of 90° by the carbon were made at the azimuthal angles $\varphi = 90°$ and 270°. Precision experiments by HUBER and BALDINGER[2] on the scattering of neutrons by C^{12} had enabled phase shifts to be deduced and hence the polarization effects in this scattering could be calculated as in Sect. 27β. From the observed azimuthal variation in intensity of the scattered D+D neutrons it was deduced that neutrons produced at an angle of 50° in the centre of mass system by 600 keV deuterons are polarized with $P_1 = 0.2 \pm 0.05$.

RICAMO[3] performed a very similar experiment and deduced that for neutrons produced at 45° in the D+D reaction with an energy of 3.32 MeV the polarization is $\gtrsim 0.25$.

More recently experiments have been performed by MEIER *et al.*[4] using scattering of the neutrons by carbon as a means of estimating the polarization. Preliminary measurements on the angular distribution of D+D neutrons scattered by carbon with nine different energies (2.4 to 3.65 MeV) were interpreted in terms of various phase shifts, namely, $s_{\frac{1}{2}}$, $p_{\frac{1}{2}}$, $p_{\frac{3}{2}}$, $d_{\frac{3}{2}}$, $d_{\frac{5}{2}}$. The phase shifts were then used as in the experiments already described, to deduce the polarization effect in the scattering. At a deuteron energy of 600 keV the following expression was obtained for the polarization vector of neutrons emitted at an angle ϑ in the centre of mass system from the D—D reaction

$$\boldsymbol{P}_1 (E_D = 600\,\text{keV}, \vartheta) = \boldsymbol{n} \frac{(0.165 \pm 0.017) \sin 2\vartheta}{1 + 1.15 \cos^2\vartheta + 0.75 \cos^4\vartheta} . \tag{29.2}$$

(The $\cos^4\vartheta$ term occurs in the denominator since at $E_D = 600$ keV, D waves are also effective.) P_1 has a maximum value (0.108 ± 0.012) for $\vartheta = 122°$ which is in agreement with HUBER and BAUMGARTNER's[5] result if the phase shifts deduced in the present experiment are used to analyse that experiment. It is also to be noted that the direction of the polarization is the same as that of the protons produced at 300 keV as measured by BISHOP *et al.*[6].

β) Li^7 (p, n) Be^7 reaction. ADAIR[7] has made an analysis of the experimental data concerning the Li^7 (p, n) Be^7 reaction. His investigation indicates that the resonance at 2.25 MeV proton energy is due to the influence of a spin 3, isotopic spin 1, even parity level of Be^8, 19.2 MeV above the ground state and that the background reaction intensity is predominantly the result of a spin 1 state of odd parity and a spin 2 state of odd parity and isotopic spin 0. Li^7 having odd parity, the spin 3 state is formed from incident P waves combined with a quintet spin state[8], whilst the odd parity states result from incident S waves combined

---

[1] See footnote 1, p. 598.
[2] P. HUBER and E. BALDINGER: Helv. phys. Acta **25**, 442 (1952).
[3] See footnote 2, p. 598.
[4] R. W. MEIER, P. SCHERRER and G. TRUMPY: Helv. phys. Acta **27**, 577 (1954).
[5] See footnote 1, p. 598.
[6] See footnote 5, p. 601.
[7] R. K. ADAIR: Phys. Rev. **96**, 709 (1954).
[8] The spins of Li^7 and proton are $\frac{3}{2}$ and $\frac{1}{2}$ respectively so that quintet and triplet incident channel spin states ($S = 2$ and 1) can be formed.

with the quintet and triplet states. The S-matrix element for the P wave transition was chosen to have the usual resonance form whilst the energy dependence of the S wave S-matrix elements entered through COULOMB and angular momentum penetrability factors, the parameters involved being chosen to accord with experimental results for the angular distribution and total cross section.

Using these expressions for the S-matrix elements, ADAIR then calculated the magnitude of the neutron polarization in this reaction. Fig. 24 shows this polarization for neutrons emitted at angles of 60°, 90° and 120° in the centre of mass system as a function of proton energy. The measurement of the polarization at an angle of 66° in the centre of mass system at a proton energy of 2.23 MeV has already been described (see Sect. 28). It gave the result $P_1 = -0.53 \pm 0.06$ compared with the value -0.45 obtained theoretically. Considering the approximations made in the calculations, the agreement seems satisfactory. Similar agreement has also been obtained by OKAZAKI[1].

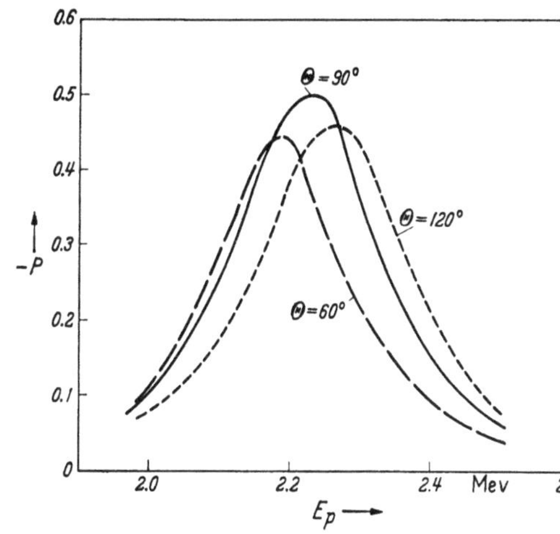

Fig. 24. Values of the polarization P of neutrons produced by the Li (p, n) reaction.

WILLARD *et al.*[2] have also measured the above polarization by the same method for neutrons emitted at 42° in the laboratory system. It was found that P_1 had a nearly constant value of 0.50 ± 0.04 for neutrons with energies from 300 to 350 keV in the laboratory system. This is larger than the values of ADAIR[3] or OKAZAKI[1].

In connection with this reaction WILLARD *et al.*[4] have pointed out that neutrons elastically scattered by Li^7 have a strong P wave $(J = 3^+)$ resonance at 254 keV incident energy and that the interference with the S wave potential scattering background should give rise to fairly large polarization. They used neutrons of 280 keV from the $Li^7(p, n)Be^7$ reaction at a proton energy of 2.2 MeV, emitted at an angle of 90° in the centre of mass system. These were scattered from a Li^7 sample through angles of $\pm 55°$ and $\pm 90°$ in the centre of mass system. The averaged values for the product polarizations obtained from the measured asymmetry ratios, corrected for multiple scattering were

$$P_1^{(1)}(90°)\, P_1^{(2)}(55°) = -0.20 \pm 0.04 \quad \text{and} \quad P_1^{(1)}(90°)\, P_1^{(2)}(90°) = -0.27 \pm 0.05$$

respectively. Assuming that the S-wave potential scattering channel spins are statistically mixed, the calculated polarization values for elastic neutron scattering by Li^7 through angles of 55° and 90° at $E_{n,\mathrm{lab}} = 280$ keV are

$$P_1^{(2)}(55°) = -0.34 \quad \text{and} \quad P_1^{(2)}(90°) = -0.37.$$

This gives an average value for $P_1(90°) = 0.66 \pm 0.09$. If the S-wave potential is all channel spin 2, the average value becomes $P^{(2)}(90°) = 0.59 \pm 0.08$.

---

[1] See footnote 4, p. 599.
[2] H. B. WILLARD, J. K. BAIR and J. D. KINGTON: Phys. Rev. **95**, 1359 (1954).
[3] See footnote 7, p. 602.
[4] H. B. WILLARD, J. K. BAIR, H. O. COHN and J. D. KINGTON: Bull. Amer. Phys. Soc., Ser. II **1**, 54 (1956).

γ) Stripping experiments. NEWNS[1] (see also HOROWITZ and MESSIAH[2] and YOCCOZ[3]) suggested that the particles emitted from (d, p) and (d, n) reactions should be polarized in the plane perpendicular to the reactions. In a (d, p) reaction, for example, in the simplest formulation of stripping theory (BUTLER[4], BHATIA *et al.*[5]) it is assumed that the proton does not interact with the nucleus. In this case the protons are unpolarized. If, however, the proton interacts, then polarization may result. NEWNS[1] describes such an interaction in terms of the impenetrability of the nucleus to the proton so that for a given angle of proton emission some are absorbed by the nucleus thus giving a predominance of proton spins in a particular direction. He formulates these ideas quantitatively using the Born approximation and approximating the deuteron wave function $\psi(\boldsymbol{r})$ by a Dirac delta function $\delta(\boldsymbol{r})$ where $\boldsymbol{r}$ is the neutron-proton separation. This corresponds to assuming that the size of the deuteron is very small compared with that of the nucleus.

The result obtained for the maximum value of the polarization in a (d, p) reaction if it is assumed that only one value of j and l (total and orbital angular momenta of the neutron entering the nucleus respectively) contribute to the scattering is

$$\boldsymbol{P} = \boldsymbol{n} \frac{\pm \sum\limits_{m \geq 0} m \frac{(l-m)!}{(l+m)!} \left(\frac{2^m \sqrt{\pi}}{\Gamma(\frac{1}{2}(l+m)+1)\,\Gamma(\frac{1}{2}(1-l-m))} \right)^2}{3\left(j+\frac{1}{2}\right) \sum\limits_{m \geq 0} \frac{(l-m)!}{(l+m)!} \left(\frac{2^m \sqrt{\pi}}{\Gamma(\frac{1}{2}(l+m)+1)\,\Gamma(\frac{1}{2}(1-l-m))} \right)^2} \tag{29.3}$$

where the negative sign is taken if $j = l + \frac{1}{2}$ and the positive sign if $j = l - \frac{1}{2}$. (Here $\boldsymbol{n} = \boldsymbol{k}_d \times \boldsymbol{k}_p / |\boldsymbol{k}_d \times \boldsymbol{k}_p|$.) For particular values of j and l we obtain the following polarizations. It is only possible to obtain an upper value to the polarization since a nuclear matrix element is also involved whose exact magnitude

	$l=0$	1	2	3
$j=l+\frac{1}{2}$	0	−16.67%	−13.33%	−18.75%
$j=l-\frac{1}{2}$	0	+33.33%	+20.00%	+25.00%

cannot be estimated. If these results are accepted it is sufficient to obtain the sign of the polarization to distinguish between $j = l + \frac{1}{2}$ and $j = l - \frac{1}{2}$. Identical results will apply for (d, n) reactions if no account is taken of Coulomb effects.

CHESTON[6] has also calculated the polarization to be expected in a (d, p) reaction assuming that the proton scattering by the final nucleus can be described by a potential of the form used by ADAIR[7] (for example) in Sect. 28 γ to describe the neutron-nucleus interaction, i.e.

$$V = V_0(1 + i\,\xi) + A\,\boldsymbol{L}\cdot\boldsymbol{S}.$$

As a particular example he calculates the polarization to be expected for the ground state proton group in the C^{12} (d, p) C^{13} reaction for a deuteron bombarding energy of 3.29 MeV. For this reaction the angular distribution measured by

[1] H. C. NEWNS: Proc. Phys. Soc. Lond. A **66**, 477 (1953).
[2] J. HOROWITZ and A. M. L. MESSIAH: J. de Phys. Radium **14**, 731 (1954).
[3] J. YOCCOZ: Proc. Phys. Soc. Lond. A **67**, 813 (1954).
[4] S. T. BUTLER: Proc. Roy. Soc. Lond., Ser. A **208**, 559 (1951).
[5] A. B. BHATIA, K. HUANG, R. HUBY and H. C. NEWNS: Phil. Mag. **43**, 485 (1952).
[6] W. B. CHESTON: Phys. Rev. **96**, 1590 (1954).
[7] See footnote 3, p. 599.

HOLMGREN[1] is consistent with the known spin assignment ($^1/_2{}^-$) for the ground state of C^{13}.

Using

$$V_0 = -19\,\text{MeV} \quad \text{for} \quad r \leq R,$$
$$V_0 = 0 \qquad\qquad\qquad r > R,$$
$$\xi = 0.05,$$
$$A = -2\,\text{MeV}$$

where R is the nuclear radius, the following values for P_1 (where $\boldsymbol{P}_1 = \boldsymbol{n} P_1$ with $\boldsymbol{n}$ as defined above) were obtained for emission angle ϑ:

ϑ	P_1 (%)
0	0
10	−22
20	−14
30	−17
40	−26

The captured neutron on this scheme would be in a $j = l - \frac{1}{2}$ state and according to NEWNS' formulation P_1 should be positive.

Recently experiments on the proton polarization from various (d, p) reactions have been reported by JURELAND and JENTSCHKE[2]. The targets were bombarded by 11.9 MeV deuterons and the polarization of the liberated protons was measured by scattering them off He^4 near the $p_{\frac{3}{2}}$ resonance of Li^5. Assuming the most probable spin assignments, the sign of the polarization was found to be opposite to that predicted by NEWNS. For the ground state proton group from $C^{12}(d, p)\,C^{13}$, the sign and order of magnitude of the polarization agreed with that predicted by CHESTON[3] for 3.29 MeV deuterons. The experimental results are given in Table 4.

Table 4. *Polarization of protons from (d, p) reactions.*

Reaction	Levels of C^{13} or Si^{29}	Proton angle (center of mass)	Polarization
$C^{12}(d, p)\,C^{13}$	Ground state	$15.5 \pm 0.5°$	-0.186 ± 0.032
$C^{12}(d, p)\,C^{13}$	Ground state	$18.5 \pm 0.5°$	-0.151 ± 0.040
$C^{12}(d, p)\,C^{13}$	Ground state	$22.5 \pm 0.5°$	-0.116 ± 0.038
$C^{12}(d, p)\,C^{13}$	Ground state	$35.5 \pm 0.5°$	-0.051 ± 0.057
$C^{12}(d, p)\,C^{13}$	Ground state	$68.5 \pm 0.5°$	-0.49 ± 0.13
$Si^{28}(d, p)\,Si^{29}$	4.934 Mev	$14.5 \pm 0.5°$	$+0.089 \pm 0.035$
$Si^{28}(d, p)\,Si^{29}$	6.380 Mev	$14.5 \pm 0.5°$	-0.067 ± 0.043

The polarization of protons from the stripping of 4 MeV deuterons in carbon at 30° has been measured by HILLMAN[4] by scattering in helium. A value of $P_1 = -58 \pm 13\%$ was obtained which is larger than, but of the same sign as, the predictions of Cheston.

δ) Production of polarized protons using polarized thermal neutrons. WOLFENSTEIN[5] has suggested that it may be possible to produce polarized protons by

[1] H. HOLMGREN: Ph. D. Thesis, University of Minnesota 1954.
[2] A. C. JURELAND and W. JENTSCHKE: Bull. Amer. Phys. Soc., Ser. II 193 (1956).
[3] See footnote 6, p. 604.
[4] P. HILLMAN: Phys. Rev. **104**, 176 (1956).
[5] L. WOLFENSTEIN: Phys. Rev. **75**, 1664 (1949).

using polarized thermal neutrons (resulting from magnetic scattering) as projectiles in an (n, p) reaction.

Grant (private communication) has shown that in any (n, p) reaction induced by S neutrons with polarization P_1 along the z-axis, the polarization of the protons lies in the plane containing the z-axis and the direction of motion of the outgoing protons and has components $P_\perp$ and $P_\parallel$ respectively perpendicular and parallel to this direction. If the proton is travelling at an angle ϑ to the z-axis, then

$$P_\perp = P_1 A \sin\vartheta; \qquad P_\parallel = P_1 B \cos\vartheta$$

where A and B are dependent on the reaction under consideration. In particular, if the outgoing protons are emitted in an S state $A = B$ and their polarization is parallel to that of the incident neutrons.

Thus by appropriate choice of ϑ a beam of polarized protons is obtained, the polarization being in any desired direction.

No experiments of this type have so far been performed. It has been suggested by WOLFENSTEIN[1], however, that the (n, p) reactions of N^{14} and He^3 may be satisfactory sources of polarized protons. He estimates that with fully polarized incident thermal neutrons, the polarization of the outgoing protons will be of the order 20 to 60%.

ε) Interaction of polarized particles with polarized nuclei. The earliest experiment on the interaction of nucleons with polarized nuclei was reported by GORTER *et al.*[2]. In their experiment they looked for a change in the transmission of unpolarized slow neutrons through a sample of gadolinium sulphate when the absorbing gadolinium nuclei were polarized and unpolarized.

Theoretically it can be shown that

$$\frac{J_x - J_x'}{J_x'} = \mathrm{Cos}\,(n\,\tau\,x)$$

where J_x and J_x' are the intensities with polarized and unpolarized nuclei respectively. x is the thickness of the absorber, n the number of nuclei per cm^3 and

$$\tau = \frac{I\,P_1}{2I+1}\,(\sigma_{I+\frac{1}{2}} - \sigma_{I-\frac{1}{2}}).$$

Here P_1 is the polarization of the absorber, I the nuclear spin and $\sigma_{I\pm\frac{1}{2}}$ are the absorption cross sections for the two possible states of the compound nucleus. No significant effect was observed and this was probably due to a failure to obtain nuclear polarization.

Manganese 55. The observation of the variation in the neutron activation cross-section of Mn^{55} with the relative orientation of the neutron and nuclear spins has been described in some detail by BERNSTEIN *et al.*[3]. Their experiment consisted in producing a beam of polarized neutrons by transmission of pile neutrons through a block of magnetized iron and then absorbing these in manganese ammonium sulphate. The manganese was cooled to 0.2° by demagnetizing from 16400 gauss to 2350 gauss and was then polarized by method (b) of Sect. 7 in the residual field of 2350 gauss. Under these conditions M_e/M_∞ is about 0.85 and P_1 about 0.166. The former figure was found from susceptibility measurements and the latter calculated from the known parameters of the

[1] See footnote 4, p. 589.

[2] C. J. GORTER, D. DE KLERK, O. J. POPPEMA and HL. DE VRIES: Physica, Haag **15**, 679 (1949).

[3] See footnote 5, p. 561.

spin Hamiltonian. It was then found that on reversal of the relative spin orientations of neutrons and nuclei that the activation of the manganese changed by 0.0338 the greater activation occurring when the spins are antiparallel.

If the cross-section dependence arises from a single resonance then for the state where the spin of the resonance level I is given by

$$I = I_{\mathrm{Mn}^{55}} + \frac{1}{2}, \qquad \sigma = \sigma_1 \left[1 + P_1(n) \cdot P_1(N) \frac{I}{I+1}\right]$$

and for

$$I = I_{\mathrm{Mn}^{55}} - \frac{1}{2}, \qquad \sigma = \sigma_2 \left[1 - P_1(n) \cdot P_1(N)\right]$$

where σ_1, σ_2 are the cross sections in the absence of neutron polarization for the formation of these two states and $P_1(n)$ and $P_1(N)$ are the polarizations of the neutron beam and nuclei respectively. In the present experiment $P_1(n)$ was 0.32.

In the interpretation of the results it was known that two resonances at least contributed to the effects whose compound state spins I and Breit-Wigner resonance parameters had been measured. However, using these and making suitable corrections for depolarizing effects and the spectral distribution of the pile neutrons it was to be expected that enhanced activity would arise from parallel spin orientations in contrast to the observed result. In subsequent measurements, however, a third resonance level was found and uncertainty introduced into the values of the spins of the states. It is not possible, therefore, to conclude that the polarization results are in accord with theoretical predictions.

Samarium 149. A similar experiment demonstrating the spin dependence of neutron cross-sections was reported subsequently by the Oak Ridge[1] group who studied the polarization of a beam of neutrons when transmitted by a sample of polarized Sm^{149}.

Samarium ethyl sulphate was cooled by embedding one end of a copper rod in a pill of the powdered material, the other end of the copper rod being in a pill of iron ammonium alum which was cooled by adiabatic demagnetization. The samarium was polarized by applying a field of 11000 gauss which produced a polarization $P_1(N)$ of about 0.12 deduced from the results of the experiment. Such a polarization was in accord with the known hfs of Sm^{149} and the final temperature of the specimen. Sm^{149} exhibits a neutron resonance at 0.094 ev and therefore it is to be expected that an unpolarized pile neutron beam transmitted through the sample will become polarized when the sample is polarized. The polarization of the neutrons was detected by reflecting the beam from a crystal of magnetically polarized magnetite whose angle of reflection was set to detect 0.07 ev neutrons. From earlier experiments on the efficiency (P) of the magnetic analyzer ($\sim 90\%$) the nuclear polarization $P_1(N)$ produced in the sample may be calculated from the counting rates due to neutrons reflected from the magnetic crystal (a) when the magnetic fields on the polarizer and analyzer are parallel (C_1) and (b) when in addition depolarizing iron foils are placed between polarizer and analyzer (C_2). In fact

$$1 + P_1(N) \cdot P_1(n) = C_1/C_2.$$

The value of C_1/C_2 was 0.882 ± 0.007 which, when corrected for the effect of neutrons reflected in higher orders, reduces to 0.787 ± 0.024. Associated with this effect should be an increase in transmission by the Sm^{149} sample when the nuclei are polarized: such an effect of about 3% was observed.

[1] See footnote 5, p. 567.

The authors were able to conclude from their results that the Sm was more transparent when neutron and nuclear spins were antiparallel than vice versa and therefore concluded that the 0.094 ev resonance corresponded to a spin of $J=I+\frac{1}{2}$ in agreement with the results of neutron scattering and transmission experiments.

Appendix.

The orientation parameters B_1, B_2 and B_4 have been evaluated for different values of the nuclear spin for special cases of the interaction

$$\mathcal{H}=A\,S_z I_z+B\,(S_x I_x+S_y I_y).$$

In the following Table A1 B_k is tabulated as a function of β for the case where $B=0$, $\beta=A\,S_z/kT$. In the case where B is finite but $x\,(=B/A)$ is small $B_2(x)=(1-x^2)\,B_2(0)$ for small β, i.e. $\beta\sim 0.1$.

In Table A2 B_2 and B_4 are shown for $I=\frac{7}{2}$ and 5 when $A=0$ and $S=\frac{1}{2}$. Again for small β and for $x>1$

$$B_2(x)=(1-1/x^2)\,B_2\,(x=\infty).$$

Table A 1.

	$I=1$				$I=2$		
β	$-B_1$	B_2	$B_4\equiv 0$	β	$-B_1$	B_2	B_4
0.1	0.081	0.002		0.1	0.141	0.008	0.000
0.2	0.162	0.009		0.2	0.278	0.033	0.000
0.3	0.241	0.021		0.3	0.4085	0.071	0.000
0.4	0.318	0.037		0.4	0.530	0.1215	0.001
0.5	0.392	0.055		0.5	0.640	0.181	0.003
0.6	0.4625	0.078		0.6	0.739	0.2435	0.005
0.7	0.529	0.103		0.7	0.827	0.313	0.009
0.8	0.592	0.130		0.8	0.903	0.381	0.014
0.9	0.650	0.158		0.9	0.969	0.448	0.020
1.0	0.7045	0.188		1.0	1.027	0.513	0.027
1.5	0.914	0.335		1.5	1.2115	0.776	0.075
2.0	1.042	0.458		2.0	1.304	0.944	0.129
2.5		0.547		2.5		1.044	0.174
3.0		0.607		3.0		1.104	0.207
∞	1.225	0.707		∞	1.414	1.195	0.267

	$I=\frac{3}{2}$				$I=\frac{5}{2}$		
β	$-B_1$	B_2	$B_4\equiv 0$	β	$-B_1$	B_2	B_4
0.1	0.115	0.005		0.1	0.170	0.012	0.000
0.2	0.221	0.020		0.2	0.334	0.048	0.000
0.3	0.327	0.043		0.3	0.488	0.104	0.001
0.4	0.428	0.075		0.4	0.624	0.175	0.003
0.5	0.523	0.113		0.5	0.745	0.255	0.007
0.6	0.611	0.1565		0.6	0.850	0.340	0.014
0.7	0.691	0.203		0.7	0.940	0.426	0.023
0.8	0.764	0.252		0.8	1.015	0.509	0.036
0.9	0.829	0.302		0.9	1.079	0.588	0.048
1.0	0.888	0.352		1.0	1.132	0.662	0.064
1.5	1.0935	0.575		1.5	1.296	0.963	0.160
2.0	1.203	0.734		2.0	1.372	1.105	0.254
2.5		0.837		2.5		1.199	0.3265
3.0		0.901		3.0		1.2545	0.3765
∞	1.342	1.000		∞	1.464	1.336	0.463

Table 1A. (Continued.)

$I=3$				$I=4$			
β	$-B_1$	B_2	B_4	β	$-B_1$	B_2	B_4
0.1	0.198	0.017	0.000	0.1	0.254	0.029	0.000
0.2	0.389	0.066	0.0005	0.2	0.490	0.1085	0.0015
0.3	0.559	0.141	0.002	0.3	0.693	0.224	0.007
0.4	0.710	0.233	0.007	0.4	0.860	0.357	0.019
0.5	0.838	0.333	0.015	0.5	0.991	0.493	0.039
0.6	0.945	0.436	0.027	0.6	1.094	0.622	0.068
0.7	1.033	0.536	0.043	0.7	1.1735	0.739	0.1035
0.8	1.111	0.630	0.063	0.8	1.236	0.844	0.145
0.9	1.164	0.717	0.087	0.9	1.286	0.935	0.190
1.0	1.212	0.795	0.116	1.0	1.324	1.031	0.237
1.5	1.356	1.076	0.257	1.5	1.438	1.280	0.4655
2.0	1.422	1.232	0.384	2.0	1.4885	1.4165	0.637
2.5		1.319	0.4765	2.5		1.4915	0.751
3.0		1.369	0.538	3.0		1.534	0.824
∞	1.500	1.443	0.640	∞	1.549	1.5955	0.939

$I=\frac{7}{2}$				$I=\frac{9}{2}$			
β	$-B_1$	B_2	B_4	β	$-B_1$	B_2	B_4
0.1	0.227	0.023	0.000	0.1	0.282	0.036	0.001
0.2	0.440	0.086	0.001	0.2	0.538	0.133	0.002
0.3	0.628	0.181	0.004	0.3	0.752	0.269	0.011
0.4	0.788	0.294	0.012	0.4	0.921	0.421	0.032
0.5	0.920	0.4135	0.025	0.5	1.0505	0.570	0.057
0.6	1.026	0.531	0.045	0.6	1.148	0.708	0.095
0.7	1.110	0.641	0.070	0.7	1.222	0.829	0.142
0.8	1.177	0.742	0.101	0.8	1.279	0.935	0.185
0.9	1.231	0.832	0.135	0.9	1.324	1.026	0.249
1.0	1.275	0.912	0.171	1.0	1.359	1.104	0.306
1.5	1.402	1.187	0.360	1.5	1.460	1.358	0.567
2.0	1.459	1.333	0.513	2.0	1.505	1.486	0.7555
2.5		1.414	0.619	2.5		1.555	0.873
3.0		1.460	0.687	3.0		1.594	0.9475
∞	1.528	1.5275	0.798	∞	1.567	1.651	1.064

$I=5$				$I=5$			
β	$-B_1$	B_2	B_4	β	$-B_1$	B_2	B_4
0.1	0.400	0.042	0.000	0.9	1.365	1.106	0.311
0.2	0.586	0.158	0.004	1.0	1.397	1.182	0.387
0.3	0.810	0.315	0.015	1.5	1.490	1.424	0.665
0.4	0.981	0.484	0.0395	2.0	1.532	1.5445	0.861
0.5	1.108	0.644	0.077	2.5		1.609	0.984
0.6	1.202	0.788	0.127	3.0		1.646	1.060
0.7	1.271	0.919	0.1845	∞	1.581	1.698	1.177
0.8	1.324	1.017	0.2475				

Table A 2. $A = 0,\ \beta = \frac{B}{2\mathrm{k}T}$

$I = \frac{7}{2}$			$I = 5$		
β	$-B_2$	B_4	β	$-B_2$	B_4
0.1	0.022	0.000	0.1	0.022	0.000
0.2	0.080	0.001	0.2	0.133	0.001
0.4	0.228	0.011	0.4	0.339	0.029
0.6	0.361	0.029	0.6	0.479	0.071
0.8	0.458	0.057	0.8	0.574	0.1245
1.0	0.529	0.092	1.0	0.645	0.180
1.5	0.653	0.1805	1.5	0.755	0.309
2.0	0.735	0.2625	2.0	0.835	0.415
2.5	0.793	0.336	2.5	0.881	0.496
3.0	0.837	0.4005	3.0	0.914	0.563
∞	1.091	1.026	∞	1.076	0.981

Review articles.

[1] GROOT, S. R. DE and H. A. TOLHOEK: β and γ ray spectroscopy (Ed. K. SIEGBAHN), Chap. 19 (III), p. 613. Amsterdam: North Holland Publishing Co. 1955.

[2] BLIN-STOYLE, R. J., M. A. GRACE and H. HALBAHN: β and γ ray spectroscopy (Ed. K. SIEGBAHN), Chap. 19 (II), p. 600. Amsterdam: North Holland Publishing Co. 1955.

[3] BLIN-STOYLE, R. J., M. A. GRACE and H. HALBAN: Progress in Nuclear Physics III, p. 63 (Ed. O. FRISCH). London: Butterworth 1950.

Sachverzeichnis.

(Deutsch-Englisch.)

Bei gleicher Schreibweise in beiden Sprachen sind die Stichwörter nur *einmal* aufgeführt.

Subject Index.

(English-German.)

Where English and German spelling of a word is identical the German version is omitted.

MIX
Papier aus verantwortungsvollen Quellen
Paper from responsible sources
FSC® C105338

If you have any concerns about our products,
you can contact us on
ProductSafety@springernature.com

In case Publisher is established outside the EU,
the EU authorized representative is:
Springer Nature Customer Service Center GmbH
Europaplatz 3, 69115 Heidelberg, Germany

Printed by Libri Plureos GmbH
in Hamburg, Germany